25 *Years*
Previous Solved Papers

GATE 2025

Chemical Engineering

- Fully solved with explanations
- Analysis of previous papers
- Topicwise presentation

MADE EASY PUBLICATIONS

www.madeeasy.in

MADE EASY Publications Pvt. Ltd.
Corporate Office: 44-A/4, Kalu Sarai (Near Hauz Khas Metro Station), New Delhi-110016 | **Ph. :** 9021300500
Email : infomep@madeeasy.in | **Web :** www.madeeasypublications.org

GATE - 2025

Chemical Engineering

Topicwise Previous GATE Solved Papers (2000-2024)

Edition

1st Edition : 2021
2nd Edition : 2022
3rd Edition : 2023

4th Edition : 2024

Preface

B. Singh (Ex. IES)

Over the period of time the GATE examination has become more challenging due to increasing number of candidates. Though every candidate has ability to succeed but competitive environment, in-depth knowledge, quality guidance and good source of study is required to achieve high level goals.

The first edition of **GATE 2025 Solved Papers : Chemical Engineering** has been divided into topicwise sections. At the beginning of each subject, analysis of previous papers are given to improve the understanding of subject.

I have true desire to serve student community by way of providing good source of study and quality guidance. I hope this book will be proved an important tool to succeed in GATE examination. Any suggestions from the readers for the improvement of this book are most welcome.

B. Singh (Ex. IES)
Chairman and Managing Director
MADE EASY Group

GATE-2025

Chemical Engineering

CONTENTS

Process Calculations

UNIT I

CONTENTS

Process Calculations

Syllabus

Steady and unsteady state mass and energy balances including multiphase, multi-component, reacting and non-reacting systems. Use of tie components; recycle, bypass and purge calculations; Gibb's phase rule and degree of freedom analysis.

Analysis of Previous GATE Papers

Exam Year	1 Mark Ques.	2 Marks Ques.	5 Marks Ques.	Total Marks
2000	1	–	–	**1**
2001	1	1	–	**3**
2002	1	3	–	**7**
2003	–	4	–	**8**
2004	–	4	–	**8**
2005	–	2	–	**4**
2006	–	2	–	**4**
2007	–	4	–	**8**
2008	–	7	–	**14**
2009	1	1	–	**3**
2010	–	2	–	**4**
2011	–	2	–	**4**
2012	–	2	–	**4**
2013	–	3	–	**6**
2014	–	3	–	**6**
2015	–	1	–	**2**
2016	1	1	–	**3**
2017	–	2	–	**4**
2018	2	–	–	**2**
2019	–	–	–	**–**
2020	–	2	–	**4**
2021	1	1	–	**3**
2022	–	1	–	**2**
2023	1	1	–	**3**
2024	2	2	–	**6**

1 Introduction and Basic Concepts

1.1 The molar composition of a gas is 10% H_2, 10% O_2, 30% CO_2 and balance H_2O. If 50% H_2O condenses, the final mole percent of H_2 in the gas on a dry basis will be
(a) 10% (b) 5%
(c) 18.18% (d) 20% **[2000 : 1 M]**

1.2 An aqueous solution of 2.45% by weight H_2SO_4 has a specific gravity of 1.011. The composition expressed in normality is
(a) 0.2500 (b) 0.2528
(c) 0.5000 (d) 0.5055 **[2003 : 2 M]**

1.3 A saturated solution at 30°C contains 5 moles of solute (M.W. = 50 kg/kmol) per kg of solvent (M.W. = 20 kg/kmol). The solubility at 100°C is 10 moles of the solute per kg of the solvent. If 10 kg of the original solution is heated to 100°C, then the weight of the additional solute that can be dissolved in it, is
(a) 0.25 kg (b) 1 kg
(c) 2 kg (d) 3.34 kg
[2010 : 2 M]

1.4 A homogeneous azeotropic distillation process separates an azeotropic *AB* binary feed using a heavy entrainer, *E*, as shown in the figure. The loss of *E* in the two product streams is negligible so that *E* circulates around the process in a closed-circuit. For a distillation column with fully specified feed(s), given operating pressure, a single distillate stream and a single bottoms stream, the steady-state degrees of freedom equals 2. For the process in the figure with a fully specified *AB* feed stream and given column operating pressures, the steady-state degrees of freedom equals

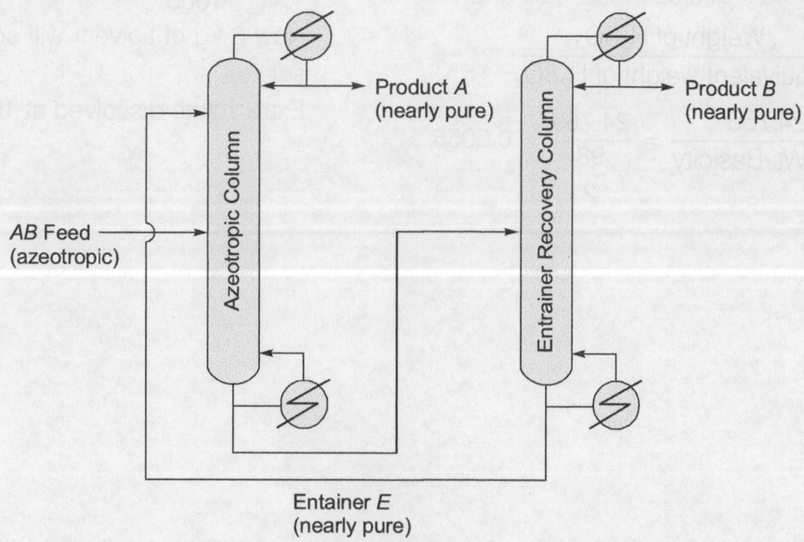

(a) 3 (b) 4
(c) 5 (d) 6 **[2024 : 1 M]**

Answers Introduction and Basic Concepts

1.1 (d) **1.2** (d) **1.3** (c) **1.4** (c)

Explanations Introduction and Basic Concepts

1.1 (d)

Assuming 100 mol of wet gas, 50% water condenses.

Component	Wet basis (mol)	Dry basis (mol)
H_2	10	10
O_2	10	10
CO_2	30	30
H_2O	(100 – 50) = 50	
Total	100	

$$\text{Mol \% of } H_2 \text{ on dry basis} = \frac{10}{50} \times 100 = 20\%$$

1.2 (d)

Aqueous solution of 2.45% by weight H_2SO_4 has specific gravity of 1.011.

So, density of solution is 1.011×10^3 gram/L

2.45% by weight H_2SO_4

So, in 1 litre solution weight of H_2SO_4

$$= 1011 \times 0.0245 = 24.769 \text{ gram}$$

$$\text{Normality} = \frac{\text{Weight of } H_2SO_4}{\text{Equivalent weight of } H_2SO_4}$$

$$= \frac{24.769}{\text{Mol. Wt./Basicity}} = \frac{24.769}{\frac{98}{2}} = 0.5055$$

1.3 (c)

Saturated solution at 30°C contains 5 moles of solute (M.W. 50 kg/kmol) per kg of solvent (M.W. 20 kg/kmol).

Solubility at 100°C is 10 moles of solute per kg of solvent.

$$\text{At } 30° \text{ solubility} = \frac{5 \text{ mole of solute}}{\text{kg of solvent}}$$

$$= 5 \times \frac{50}{1000} = 0.25 \text{ kg solute/kg solvent}$$

$$\text{or } \frac{0.25 \text{ kg solute}}{(1+0.25) \text{ kg solution}} = 0.2 \text{ kg solvent/kg sol.}$$

10 kg solution means 10 × 0.2 = 2 kg solute and 8 kg solvent.

$$\text{Now at } 100°\text{C solubility} = \frac{10 \text{ mole solute}}{\text{kg solvent}}$$

$$= 10 \times \frac{50}{1000} = 0.5 \text{ kg of solute/kg solvent}$$

Now 8 kg of solvent will contain 8 × 0.5 = 4 kg solute.

Extra solute dissolved at 100°C = 4 – 2 = 2 kg.

■■■■

2 Material Balance Calculations

2.1 The reaction $A \rightarrow 2B + C$ takes place in a catalytic reactor (see diagram below). The reactor effluent is sent to a separator. The overall conversion of A is 95%. The product stream from the separator consists of B, C and 0.5% of A entering the separator, while the recycle stream consists of the remainder of the unreacted A and 1% of B entering the separator. Calculate the
(a) single pass conversion of A in the reactor.
(b) molar ratio of recycle to feed.

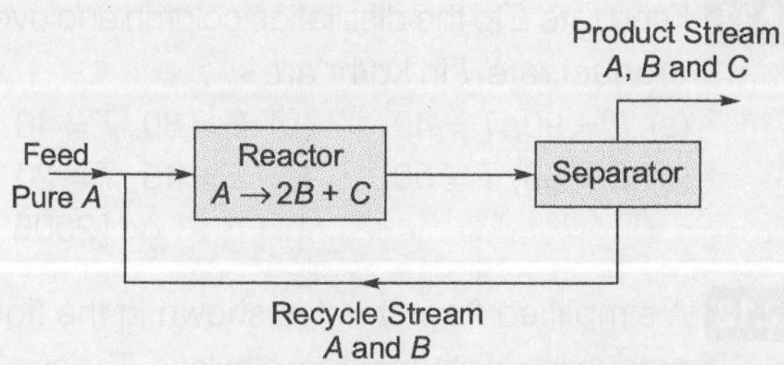

[2000 : 5 M]

2.2 A butane isomerization process produces 70 kmol/h of pure isobutane. A purge stream, removed continuously, contains 85% n-butane and 15% impurity (mole %). The feed stream is n-butane containing 1% impurity (mole %). The flow rate of the purge stream will be:
(a) 3 kmol/h (b) 4 kmol/h
(c) 5 kmol/h (d) 6 kmol/h

[2001 : 2 M]

2.3 Fresh orange juice contains 12% (by weight) solids and the rest water. 90% of the fresh juice is sent to an evaporator to remove water and subsequently mixed with the remaining 10 % of fresh juice. The resultant product contains 40% solids. The kg of water removed from 1 kg fresh juice is
(a) 0.4 (b) 0.5
(c) 0.6 (d) 0.7 [2002 : 2 M]

2.4 1 kg of a saturated aqueous solution of a highly soluble component A at 60°C is cooled to 25°C. The solubility limits of A are (0.6 kg A)/(kg water) at 60°C and (0.2 kg A)/(kg water) at 25°C. The amount, in kgs, of the crystals formed is
(a) 0.4 (b) 0.25
(c) 0.2 (d) 0.175

[2002 : 2 M]

2.5 In the hydrodealkylation of toluene to benzene, the following reactions occur

$$C_7H_8 + H_2 \rightarrow C_6H_6 + CH_4$$
$$2C_6H_6 \Leftrightarrow C_{12}H_{10} + H_2$$

Toluene and hydrogen are fed to a reactor in a molar ratio 1 : 5. 80% of the toluene gets converted and the selectivity of benzene (defined as moles of benzene formed per moles of toluene converted) is 90%. The fractional conversion of hydrogen is
(a) 0.16 (b) 0.144
(c) 0.152 (d) 0.136

[2002 : 2 M]

2.6 $Na_2SO_4.10\ H_2O$ crystals are formed by cooling 100 kg of 30% by weight aqueous solution of Na_2SO_4. The final concentration of the solute in the solution is 10%. The weight of crystals is
(a) 20 (b) 32.2
(c) 45.35 (d) 58.65

[2003 : 2 M]

2.7 80 kg of Na_2SO_4 (molecular weight = 142) is present in 330 kg of an aqueous solution. The solution is cooled such that 80 kg of $Na_2SO_4.10H_2O$ crystals separate out. The weight fraction of Na_2SO_4 in the remaining solution is
(a) 0.00 (b) 0.18
(c) 0.24 (d) 1.00

[2004 : 2 M]

2.8 A metal recovery unit (MRU) of intake capacity 5000 kg/hr treats a liquid product from a plant and recovers 90% of the metal in the pure form. The unrecovered metal and associated liquid are sent to a disposal unit along with the untreated product from the plant (see figure below). Find the flow rate (m_6) and the weight fraction of the metal (w_6). The liquid product flow rate is 7500 kg/hr of composition 0.1 (wt. fraction). Assume steady state.

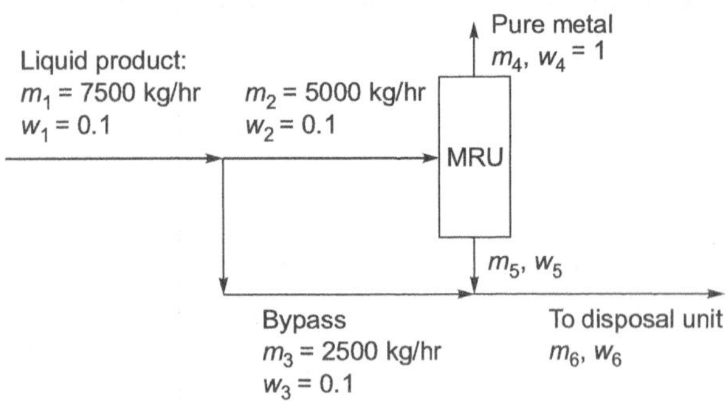

(a) $m_6 = 7500$ kg/hr, $w_6 = 0.0$
(b) $m_6 = 7050$ kg/hr, $w_6 = 0.04255$
(c) $m_6 = 4500$ kg/hr, $w_6 = 0.1712$
(d) $m_6 = 5600$ kg/hr, $w_6 = 0.0314$

[2005 : 2 M]

2.9 A feed stream (S_1) at 100 kg/hr and containing only A mixes with recycle stream S_5 before entering the reactor (see figure below), where the reaction $A \rightarrow B$ takes place. The operation is at steady state. The stream S_3 leaving the reactor is separated, without either phase or composition change, into two streams S_4 and S_5. If the mass fraction of B in S4 is 0.95 and total flow rate of S_5 is 10 kg/hr, then the ratio of flow rates of streams (S_3/S_5), and the flow rate of A in S_3 are, respectively,

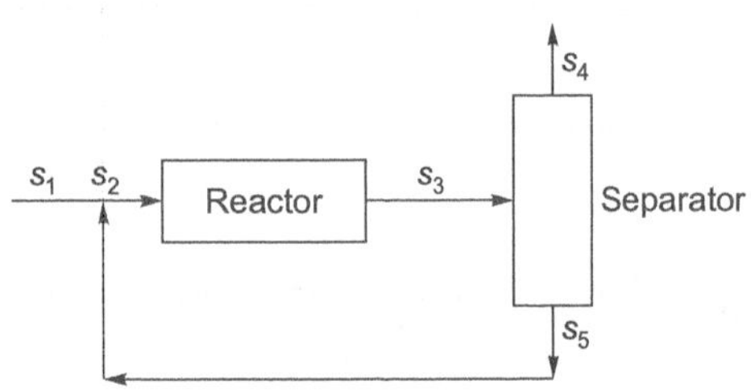

(a) 11 and 110 kg/hr
(b) 24 and 240 kg/hr
(c) 11 and 5.5 kg/hr
(d) 70 and 330 kg/hr

[2005 : 2 M]

Statement for Linked Answer Question (2.10 & 2.11)

Solvent C is used to extract solute B selectively from 100 kg/hr feed mixture $A + B$ in a steady state continuous process shown below. The solubility of C in the raffinate and the solubility of A in the extract are negligible. The extract is distilled to recover B in the bottom product. The overhead product is recycled to the extractor. The loss of solvent in the bottoms is compensated by make up solvent S_d. The total flow rate of the solvent stream S going to the extractor is 50 kg/hr. The mass fractions (X_i's) of some selected streams are indicated in the figure below.

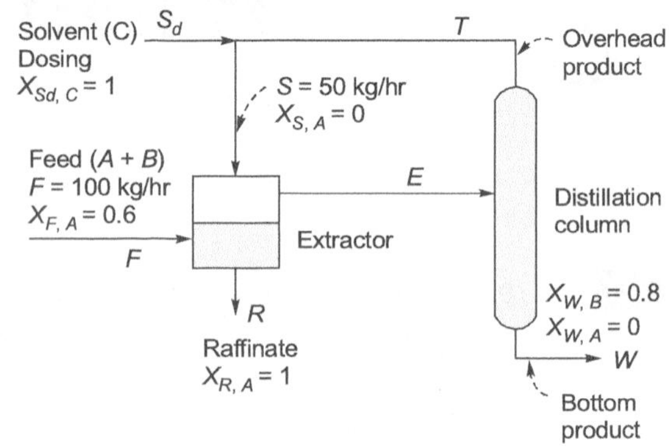

2.10 Distillation bottoms flow rate W and solvent dosing rate S_d in kg/hr are

(a) $W = 50$, $S_d = 50$
(b) $W = 100$, $S_d = 20$
(c) $W = 10$, $S_d = 50$
(d) $W = 50$, $S_d = 10$

[2006 : 2 M]

2.11 Feed rate E to the distillation column and overhead product rate T in kg/hr are

(a) $E = 90$, $T = 40$
(b) $E = 80$, $T = 40$
(c) $E = 90$, $T = 50$
(d) $E = 45$, $T = 20$

[2006 : 2 M]

2.12 A simplified flow sheet is shown in the figure for production of ethanol from ethylene. The conversion of ethylene in the reactor is 30% and the scrubber following the reactor completely separates ethylene (as top stream) and ethanol and water as bottoms. The last (distillation) column gives an ethanol-water azeotrope (90 mol% ethanol) as the final product and water as waste. The recycle to purge ratio is 34.

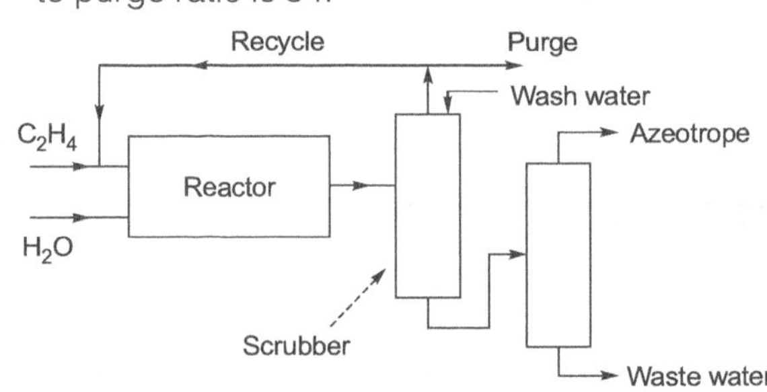

The reaction is : $C_2H_4\,(g) + H_2O\,(g) \rightarrow C_2H_5OH\,(g)$

For an azeotrope product rate of 500 mols/hr, the recycle gas flowrate in mols/hr is

(a) 30
(b) 420
(c) 1020
(d) 1500

[2007 : 2 M]

2.13 For the same process, if fresh H_2O feed to the reactor is 600 mol/hr and wash water for scrubbing is 20% of the condensables coming out of the reactor, the water flowrate in mols/hr from the distillation column as bottom is

(a) 170 (b) 220
(c) 270 (d) 480

[2007 : 2 M]

2.14 A 35% Na_2SO_4 solution in water, initially at 50°C, is fed to a crystallizer at 20°C. The product stream contains hydrated crystals $Na_2SO_4.10H_2O$ in equilibrium with a 20 wt% Na_2SO_4 solution. The molecular weights of Na_2SO_4 and $Na_2SO_4.10H_2O$ are 142 and 322, respectively. The feed rate of the 35% solution required to produce 500 kg/hr of hydrated crystals is

(a) 403 kg/hr (b) 603 kg/hr
(c) 803 kg/hr (d) 1103 kg/hr

[2008 : 2 M]

2.15 600 kg/hr of saturated steam at 1 bar (enthalpy 2675.4 kJ/kg) is mixed adiabatically with superheated steam at 450°C and 1 bar (enthalpy 3382.4 kJ/kg). The product is superheated steam at 350°C and 1 bar (enthalpy 3175.6 kJ/kg). The flow rate of the product is

(a) 711 kg/hr (b) 1111 kg/hr
(c) 1451 kg/hr (d) 2051 kg/hr

[2008 : 2 M]

2.16 Carbon black is produced by decomposition of methane:

$$CH_4\,(g) \rightarrow C\,(s) + 2H_2\,(g)$$

The single pass conversion of methane is 60%. If fresh feed is pure methane and 25% of the methane exiting the reactor is recycled, then the molar ratio of fresh feed stream to recycle stream is

(a) 0.9 (b) 9
(c) 10 (d) 90

[2008 : 2 M]

Common Data for Questions (2.17 to 2.18):

Methane and steam are fed to a reactor in molar ratio 1 : 2. The following reactions take place,

$$CH_4\,(g) + 2H_2O\,(g) \rightarrow CO_2\,(g) + 4H_2\,(g)$$
$$CH_4\,(g) + H_2O\,(g) \rightarrow CO\,(g) + 3H_2\,(g)$$

where CO_2 is the desired product, CO is the undesired product and H_2 is a byproduct. The exit stream has the following composition:

Species	CH_4	H_2O	CO_2	H_2	CO
Mole %	4.35	10.88	15.21	67.39	2.17

2.17 The selectivity for desired product relative to undesired product is

(a) 2.3 (b) 3.5
(c) 7 (d) 8

[2008 : 2 M]

2.18 The fractional yield of CO_2 is ____ (where fractional yield is defined as the ratio of moles of the desired product formed to the moles that would have been formed if there were no side reactions and the limiting reactant had reacted completely)

(a) 0.7 (b) 0.88
(c) 1 (d) 3.5

[2008 : 2 M]

2.19 The fractional conversion of methane is

(a) 0.4 (b) 0.5
(c) 0.7 (d) 0.8

[2008 : 2 M]

2.20 A dehumidifier (shown below) is used to completely remove water vapor from air.

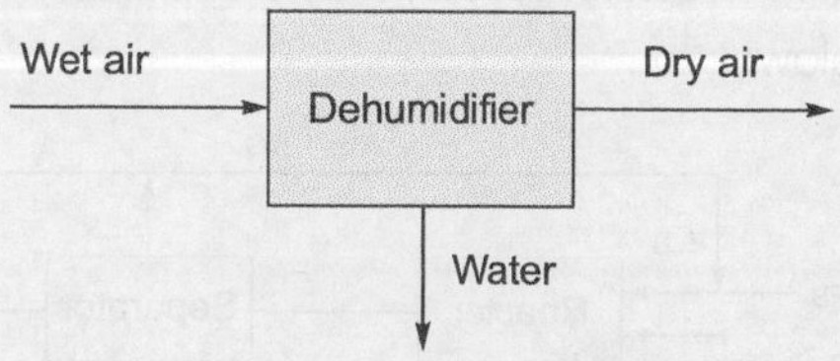

Which one of the following statements is true?

(a) Water is the only tie component.
(b) Air is the only tie component.
(c) Both water and air are tie components.
(d) There are no tie components.

[2009 : 1 M]

2.21 Pure water (stream *W*) is to be obtained from a feed containing 5 wt % salt using a desalination unit as shown below.

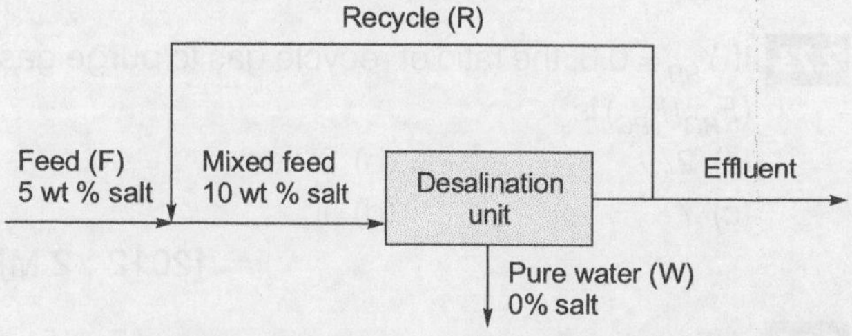

If the overall recovery of pure water (through steam W) is 0.75 kg/kg feed, then the recycle ratio (R/F) is

(a) 0.25 (b) 0.5
(c) 0.75 (d) 1.0

[2009 : 2 M]

2.22 Ammonia is synthesized at 200 bar and 773 K by the reaction $N_2 + 3H_2 \rightleftharpoons 2NH_3$. The yield of ammonia is 0.45 mol/mol of fresh feed. Flow sheet for the process (along with available compositions) is shown below.

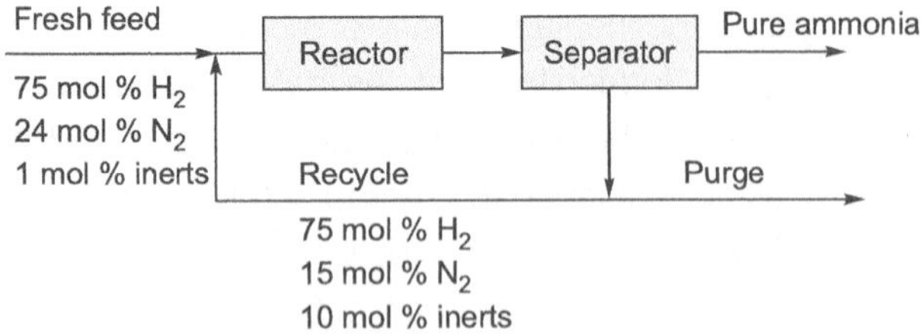

The single pass conversion for H_2 in the reactor is 20%. The amount of H_2 lost in the purge as a percentage of H_2 in fresh feed is

(a) 10 (b) 20
(c) 45 (d) 55 [2011 : 2 M]

Common Data for Questions 2.23 and 2.24:

The reaction $A_{(liq)} + B_{(gas)} \rightarrow C_{(liq)} + D_{(gas)}$, is carried out in a reactor followed by a separator as shown below:

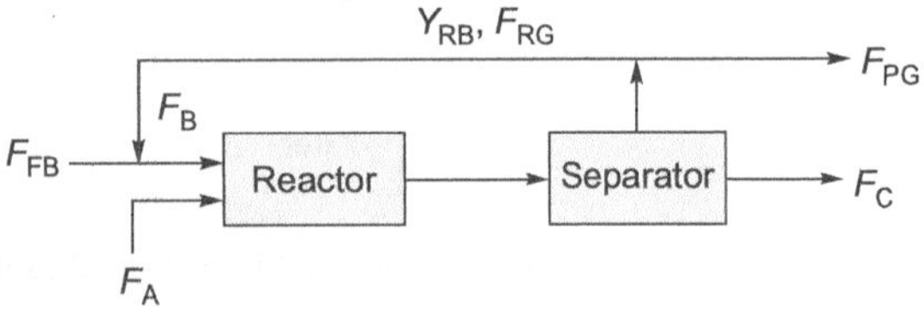

Notation:

Molar flow rate of fresh B is F_{FB}
Molar flow rate of A is F_A
Molar flow rate of recycle gas is F_{RG}
Mole fraction of B in recycle gas is Y_{RB}
Molar flow rate of purge gas is F_{PG}
Molar flow rate of C is F_C
Here, $F_{FB} = 2$ mol/s; $F_A = 1$ mol/s, $F_B/F_A = 5$ and A completely converted.

2.23 If $Y_{RB} = 0.3$, the ratio of recycle gas to purge gas (F_{RG}/F_{PG}) is

(a) 2 (b) 5
(c) 7 (d) 10

[2012 : 2 M]

2.24 If the ratio of recycle gas to purge gas (F_{RG}/F_{PG}) is 4 then Y_{RB} is

(a) $\frac{3}{8}$ (b) $\frac{2}{5}$

(c) $\frac{1}{2}$ (d) $\frac{3}{4}$

[2012 : 2 M]

Common Data for Questions 2.25 and 2.26:

A reverse osmosis unit treats feed water (F) containing fluoride and its output consists of a permeate stream (P) and a reject stream (R). Let C_F, C_P, and C_R denote the fluoride concentrations in the feed, permeate, and reject streams, respectively. Under steady state conditions, the volumetric flow rate of the reject is 60 % of the volumetric flow rate of the inlet stream, and $C_F = 2$ mg/L and $C_P = 0.1$ mg/L.

2.25 The value of C_R in mg/L, up to one digit after the decimal point, is ______.

[2013 : 2 M]

2.26 A fraction f of the feed is bypassed and mixed with the permeate to obtain treated water having a fluoride concentration of 1 mg/L. Here also the flow rate of the reject stream is 60% of the flow rate entering the reverse osmosis unit (after the bypass). The value of f, up to 2 digits after the decimal point, is ______. [2013 : 2 M]

2.27 Two elemental gases (A and B) are reacting to form a liquid (C) in a steady state process as per the reaction $A + B \rightarrow C$. The single-pass conversion of the reaction is only 20% and hence recycle is used. The product is separated completely in pure form. The fresh feed has 49 mol% of A and B each along with 2 mol% impurities. The maximum allowable impurities in the recycle stream is 20 mol%. The amount of purge stream (in moles) per 100 moles of the fresh feed is ______. [2014 : 2 M]

2.28 The schematic diagram of a steady state process is shown below. The fresh feed (F) to the reactor consists of 96 mol% reactant A and 4 mol% inert-I. The stoichiometry of the reaction is A→C. A part of the reactor effluent is recycled. The molar flow rate of the recycle stream is 0.3 F. The product stream P contains 50 mol% C. The percentage conversion of A in the reactor based on A entering the reactor at point 1 in the figure (up to one decimal place) is ______.

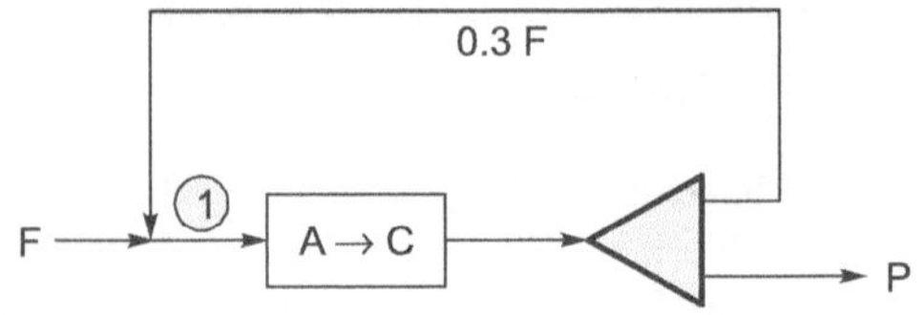

[2015 : 2 M]

2.29 A liquid mixture of ethanol and water is flowing as inlet stream P into a stream splitter. It is split into two streams, Q and R, as shown in the figure below:

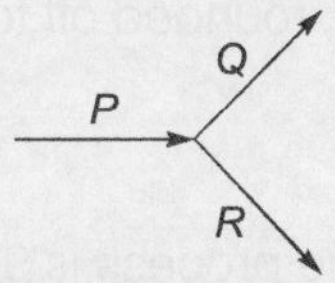

The flowrate of P, containing 30 mass% of ethanol, is 100 kg/h. What is the least number of additional specification(s) required to determine the mass flowrates and compositions (mass%) of the two exit streams?

(a) 0 (b) 1
(c) 2 (d) 3 [2016 : 1 M]

2.30 All aqueous salt-solution enters a crystallizer operating at steady state at 25°C. The feed temperature is 90°C and the salt concentration in the feed is 40 weight %. The salt crystallizes as a pentahydrate. The crystals and the mother liquor leave the crystallizer. The molecular weight of the anhydrous salt is 135. The solubility of the salt at 25°C is 20 weight %.

The feed flowrate required for a production rate of 100 kg/s of the hydrated salt, rounded to the nearest integer, is ________ kg/s.

[2017 : 2 M]

2.31 A feed stream containing pure species L flows into a reactor, where L is partly converted to M as shown in the figure.

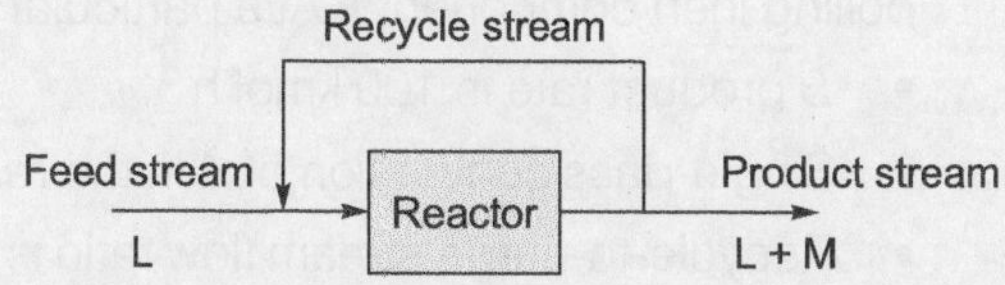

The mass flow rate of the recycle stream is 20% of that of the product stream. The overall conversion of L (based on mass units) in the process is 30%. Assuming steady state operation, the one-pass conversion of L (based on mass units) through the reactor is

(a) 34.2% (b) 30%
(c) 26.3% (d) 23.8%

[2020 : 2 M]

2.32 Consider the process flowsheet in the figure. An irreversible liquid-phase reaction $A \rightarrow B$ (reaction rate $-r_A = 164\, x_A$ kmol m^{-3} h^{-1}) occurs in a 1 m^3 continuous stirred tank reactor (CSTR), where x_A is the mole fraction of A. A small amount of inert, I, is added to the reactor. The reactor effluent is separated in a perfect splitter to recover pure B product down the bottoms and a B-free distillate. A fraction of the distillate is purged and the rest is recycled back to the reactor. At a particular steady state, the product rate is 100 kmol h^{-1}, the recycle rate is 200 kmol h^{-1} and the purge rate is 10 kmol h^{-1}. Given the above information, the inert feed rate into the process is __________ kmol h^{-1} (rounded off to two decimal places).

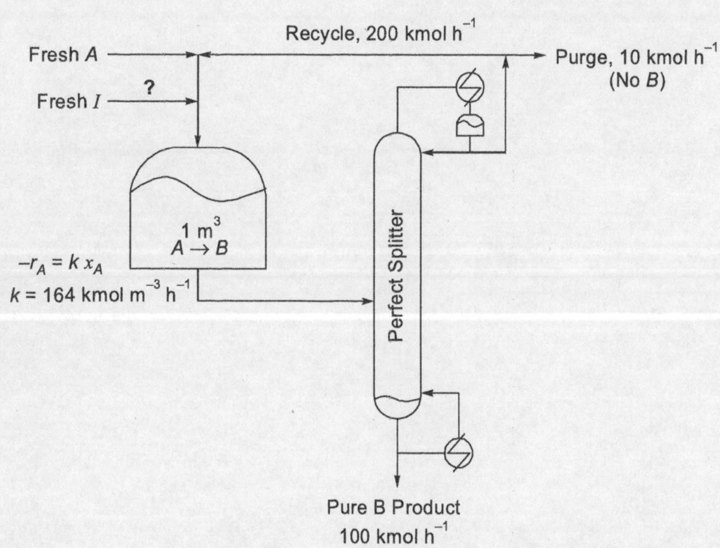

[2022 : 2 M]

2.33 Orsat analysis showing the composition (in mol %, on a dry basis) of a stack gas is given in the table below. The humidity measurement reveals that the mole fraction of H_2O in the stack gas is 0.07. The mole fraction of N_2 calculated on a wet basis is _____ (rounded off to two decimal places).

Species	N_2	CO_2	CO	O_2
mol %	65	15	10	10

[2023 : 2 M]

2.34 Consider the reaction $N_2(g) + 3H_2(g) \rightleftharpoons 2NH_3(g)$ in a continuous flow reactor under steady-state conditions. The component flow rates at the reactor inlet are $F^0_{N_2} = 100 \text{ mol s}^{-1}$, $F^0_{H_2} = 300 \text{ mol s}^{-1}$, $F^0_{inert} = 1 \text{ mol s}^{-1}$. If the fractional conversion of H_2 is 0.60, the outlet flow rate of N_2, in mol s^{-1}, rounded off to the nearest integer, is ________.

[2024 : 1 M]

2.35 Consider the process in the figure for manufacturing *B*. The feed to the process is 90 mol% *A* and a close-boiling inert component *I*. At a particular steady-state:

- *B* product rate is 100 kmol h^{-1}.
- Single-pass conversion of *A* in the reactor is 50%.
- Recycle-to-purge stream flow ratio is 10.

The flow rate of *A* in the purge stream in kmol h^{-1}, rounded off to 1 decimal place, is ________.

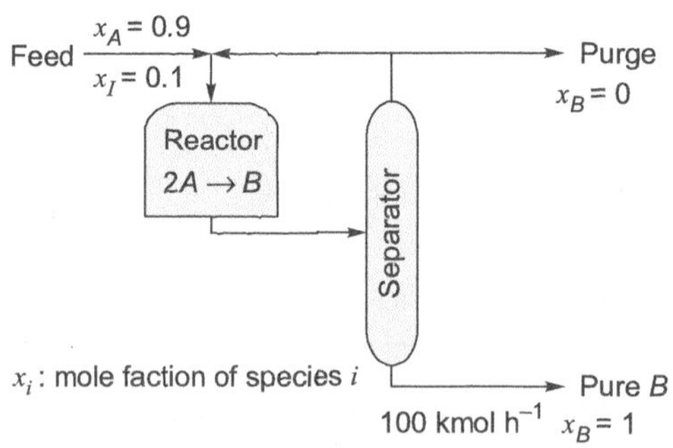

[2024 : 2 M]

■■■■

Answers Material Balance Calculations

2.1	(8.675%, 9.97)	**2.2**	(c)	**2.3**	(d)	**2.4**	(c)	**2.5**	(c)	**2.6**	(d)
2.7	(b)	**2.8**	(b)	**2.9**	(c)	**2.10**	(d)	**2.11**	(a)	**2.12**	(c)
2.13	(d)	**2.14**	(c)	**2.15**	(d)	**2.16**	(b)	**2.17**	(c)	**2.18**	(a)
2.19	(c)	**2.20**	(b)	**2.21**	(b)	**2.22**	(a)	**2.23**	(b)	**2.24**	(a)
2.25	(3.27)	**2.26**	(0.264)	**2.27**	(10)	**2.28**	(45.33)	**2.29**	(b)	**2.30**	(200)
2.31	(c)	**2.32**	(0.99)	**2.33**	(0.61)	**2.34**	(40)	**2.35**	(18.2)		

Explanations Material Balance Calculations

2.1 (8.675%, 9.97)

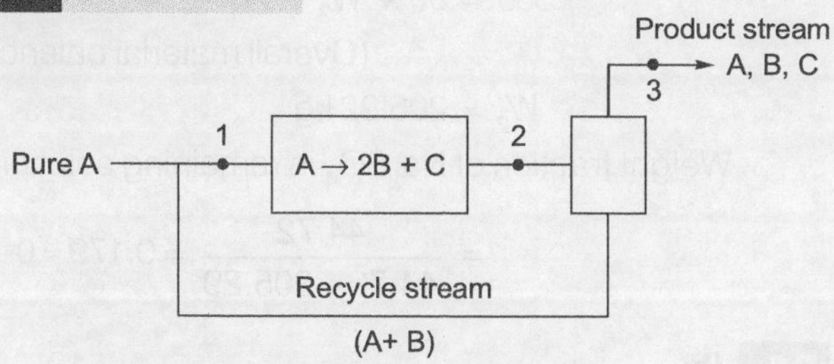

Basis 100 moles of A in pure feed

Overall conversion of A = 95%

So moles of A at point 3 $(n_{A_3}) = 100 \times 0.05$

$= 5$ moles

Product stream from separator contains 0.5% of A entering the separator

$$n_{A_3} = 0.005 n_{A_2}$$

$$n_{A_2} = \frac{5}{0.005} = 1000 \text{ moles}$$

A in recycle stream $= n_{A_2} - n_{A_3}$

$= 1000 - 5 = 995$ moles

At point 1 $n_{A_1} = 100 + 995 = 1095$ moles

(a) So conversion per pass $= \dfrac{1095 - 1000}{1095} \times 100$

$= 8.675\%$

Total moles of A converted in the reactor

$= n_{A_2} - n_{A_1}$

$= 1095 - 1000 = 95$ moles

Moles of B formed in reactor = Moles of B entering the separator $n_{B_2} = 95 \times 2 = 190$ moles

Recycle stream contains 1% of B entering the separator $n_{B_4} = 0.01 n_{B_2} = 1.9$ moles

Total moles in recycle = 995 + 1.9 = 996.9 moles

(b) Molar ratio of recycle to feed

$$= \frac{996.9}{100} = 9.969 \approx 9.97$$

2.2 (c)

Butane isomerization process produces 70 kmol/hr of pure butane as shown in figure below.

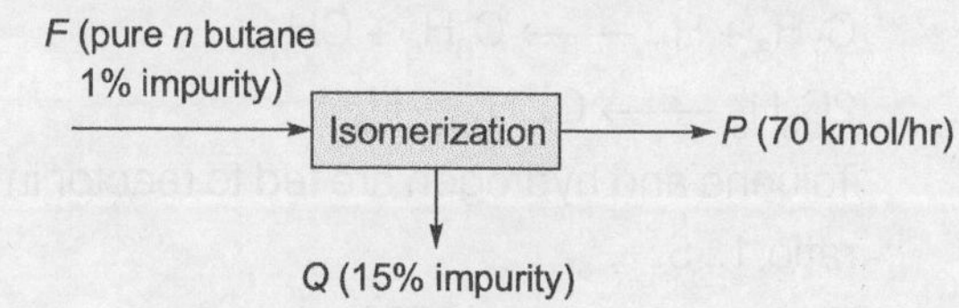

Overall balance :

$$F = P + Q$$

$$F = 70 + Q \quad \text{...(1)}$$

From impurity balance:

$$F \times 0.01 = Q \times 0.15 \quad \text{...(2)}$$

From equation (1) and (2)

$$Q = 5 \text{ kmol/hr}$$

$$F = 75 \text{ kmol/hr}$$

2.3 (d)

Fresh orange juice contains 12% (by weight) solid and rest water.

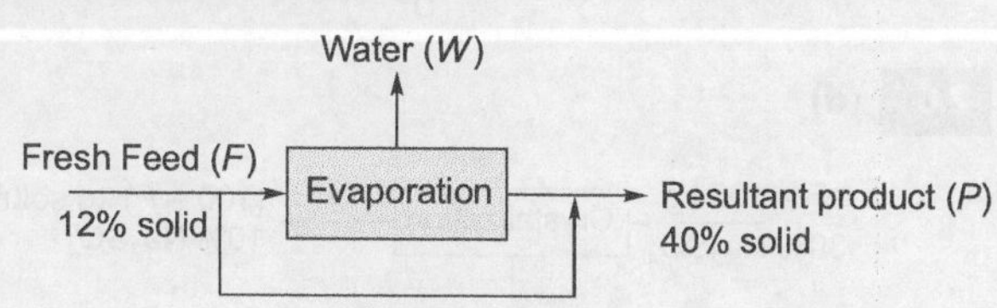

Assuming 100 kg of fresh feed.

From overall solid balance:

$$F \times 0.12 = P \times 0.4$$

$$P = \frac{100 \times 0.12}{0.4} = 30 \text{ kg}$$

From overall balance;

$$F = P + W$$

$$100 = 30 + W \Rightarrow W = 70 \text{ kg}$$

Amount of water removed from 1 kg of fresh feed

$$= \frac{70}{100} = 0.7$$

2.4 (c)

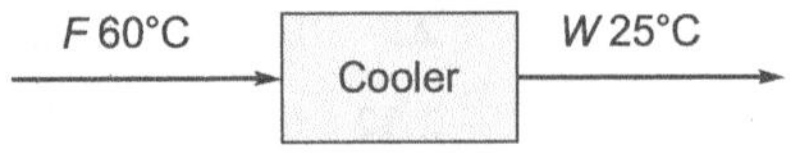

$$\text{Solubility at } 60°\text{C} = \frac{0.6 \text{ kg } A}{\text{kg of water}}$$

$$= \frac{0.6}{1+0.6} = 0.375 \text{ kg of } A/\text{kg solution}$$

Similarly solubility at 25°C = 0.2 kg A/kg water

$$= \frac{0.2}{1+0.2} = 0.167 \text{ kg } A/\text{kg solution}$$

So, crystal formed = 0.375 – 0.167 = 0.2 kg

2.5 (c)

$$C_7H_8 + H_2 \longrightarrow C_6H_6 + CH_4$$

$$2C_6H_6 \longrightarrow C_{12}H_{10} + H_2$$

Toluene and hydrogen are fed to reactor in molar ratio 1 : 5.

Toluene converted is 80%

Selectivity of benzene is 90%

Benzene formed = 0.8 × 0.9 = 0.72 moles

So benzene reacted in 2nd reaction is

$$0.8 - 0.72 = 0.08 \text{ moles}$$

So hydrogen formed in 2nd reaction is

$$\frac{0.08}{2} = 0.04 \text{ moles}$$

Hydrogen reacted in first reactor is 0.8 moles.

So fractional conversion of H_2 is

$$= \frac{0.8 - 0.04}{5} = 0.152$$

2.6 (d)

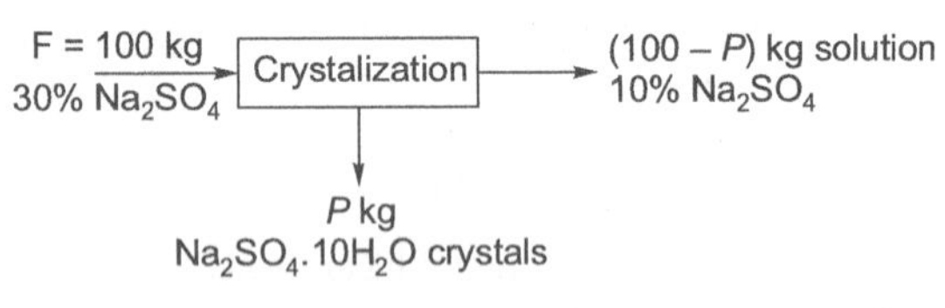

$Na_2SO_4.10H_2O$ crystals are formed by cooling 100 kg of 30% by weight aqueous solution of Na_2SO_4.

From overall balance

$$100 = P + (100 - p) \qquad \ldots(1)$$

From Na_2SO_4 balance

$$100 \times 0.3 = P \times \frac{142}{322} + (100 - P) \times 0.1$$

$$P = 58.65 \text{ kg}$$

2.7 (b)

80 kg of Na_2SO_4 (M.W. = 142) is present in 330 kg of an aqueous solution.

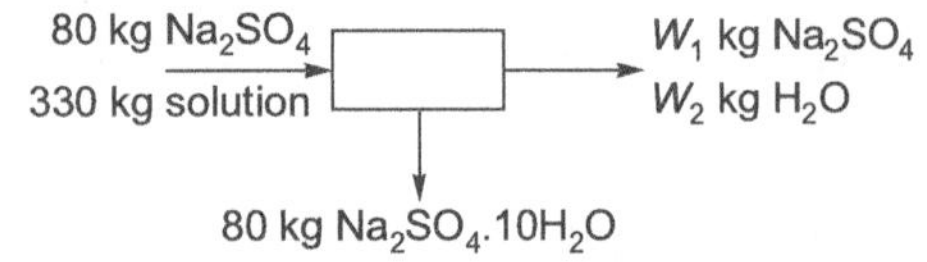

$$80 = 80 \times \frac{142}{(142 + 10 \times 18)} + W_1$$

$$W_1 = 44.72 \text{ kg}$$

$$330 = 80 + W_2 + W_1$$

(Overall material balance)

$$W_2 = 205.92 \text{ kg}$$

Weight fraction of Na_2SO_4 in remaining solution

$$= \frac{44.72}{44.72 + 205.29} = 0.179 \approx 0.18$$

2.8 (b)

A metal recovery unit has intake capacity 5000 kg/hr and recovers 90% of metal in pure form.

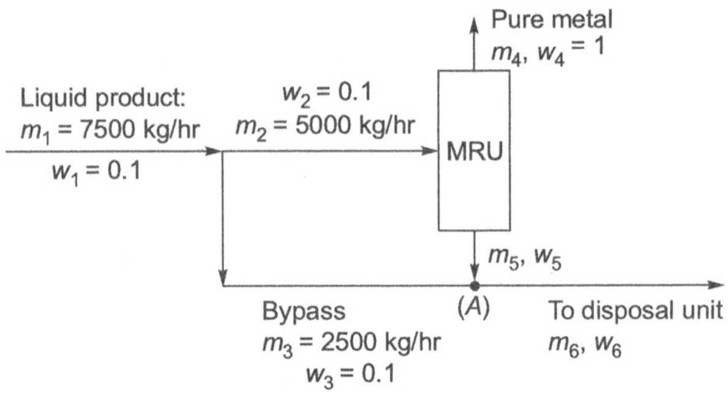

From overall balance

$$m_2 = m_4 + m_5 = 5000 \qquad \ldots(1)$$

From metal material balance

$$m_2 w_2 = m_4 w_4 + m_5 w_5 \qquad \ldots(2)$$

(metal recovery is 90% given in question)

$$m_4 \times 1 = 0.9 \times 5000 \times 0.1$$

$$m_4 = 450 \text{ kg/hr}$$

$$m_5 = 5000 - m_4 \quad \text{(from equation 1)}$$

$$m_5 = 4550 \text{ kg/hr}$$

From equation (2)

$$450 + 4550 \times w_5 = 500$$

$$w_5 = 0.01099$$

Applying material balance at point (A)

$$m_3 w_3 + m_5 w_5 = m_6 w_6$$

$$7050 \times w_6 = 2500 \times 0.1 + 4550 \times 0.01099$$

$$w_6 = 0.04255$$

2.9 (c)

The reaction given in question is $A \rightarrow B$

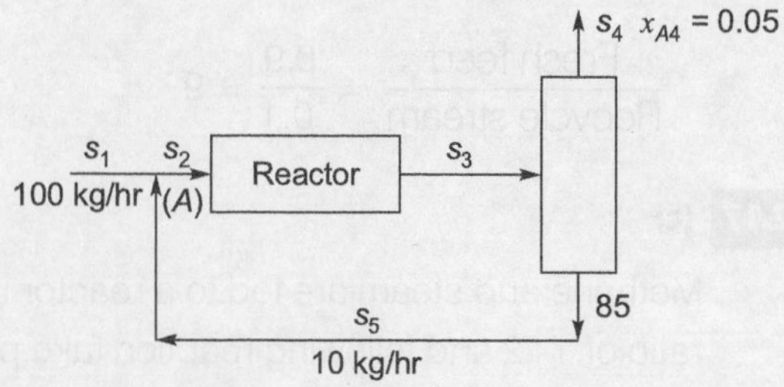

Mass fraction B in s_4 is 0.95.

From material balance at point A

$$s_2 = s_1 + s_5$$

$$s_2 = 100 + 10 = 110 \text{ kg/hr}$$

Separation does not change composition.

So, $$x_{A4} = x_{A3} = x_{A5} = 0.05$$

$$x_{B4} = 0.95$$

Flow rate of A in $s_3 = s_3 x_{A3}$

$$= 110 \times 0.05 = 5.5 \text{ kg/hr}$$

$$\frac{s_3}{s_5} = \frac{110}{10} = 11$$

2.10 (d)

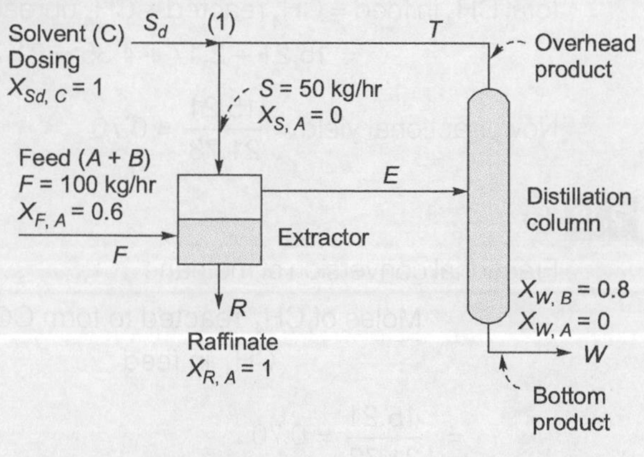

From overall material balance

$$F + S_d = R + W \quad \text{...(1)}$$

Component B balance

$$S_d \times 0 + F \times 0.4 = R \times 0 + W \times 0.8$$

$$W = 100 \times \frac{0.4}{0.8} = 50 \text{ kg/hr}$$

From component C balance

$$F \times 0 + S_d \times 1 = R \times 0 + W \times 0.2$$

$$S_d = 50 \times 0.2 = 10 \text{ kg/hr}$$

2.11 (a)

Applying material balance at point 1

$$S = S_d + T$$

From previous question,

$$S = 50 \text{ kg/hr}$$

$$S_d = 10 \text{ kg/hr}$$

$$T = 50 - 10 = 40 \text{ kg/hr}$$

Applying material balance on distillation column,

$$E = T + W$$

$$= 50 + 40 = 90 \text{ kg/hr}$$

2.12 (c)

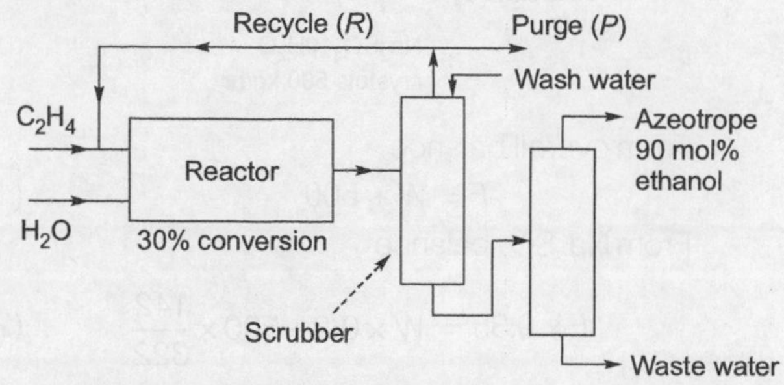

$C_2H_4\,(g) + H_2O \rightarrow C_2H_5OH\,(g)$

Azeotrope product rate is 500 mol/hr which contain 90 mol% ethanol. So, ethanol product rate is

$$= 500 \times 0.9 = 450 \text{ mol/hr}$$

Reactor conversion is 30%.

So, to produce 450 mol/hr ethanol, C_2H_4 required

$$= \frac{450}{0.3} = 1500 \text{ mol/hr}$$

Unreacted C_2H_4 = 1500 − 450

$$= 1050 \text{ mol/hr}$$

Recycle to purge ratio = 34 (given)

$$\frac{R}{P} = 34 \Rightarrow R = 34P \quad \text{...(1)}$$

$$R + P = 1050 \quad \text{...(2)}$$

From eqn. (1) and (2),

$$P = 30 \text{ mol/hr}$$

$$R = 1020 \text{ mol/hr}$$

2.13 (d)

Fresh H_2O feed to reactor = 600 mol/hr

H_2O reacted to produce ethanol (450 mol/hr)

$$= 450 \text{ mol/hr}$$

Water unreacted = 600 − 450 = 150 mol/hr

Wash water flow rate is 20% of condensable coming out of reactor.

Condensable = 450 mol/hr ethanol + 150 mol/hr H_2O + 1050 mol/hr C_2H_4 unreacted = 1650 mol/hr

Wash water flow rate = 0.2 × 1650 = 330 mol/hr

Waste water flow rate

$$= \text{Water unreacted} + \text{Wash water}$$

$$= 150 + 330 = 480 \text{ mol/hr}$$

2.14 (c)

A 35 wt% Na_2SO_4 solution in water initially at 50°C is fed to crystallizer at 20°C as shown in figure below:

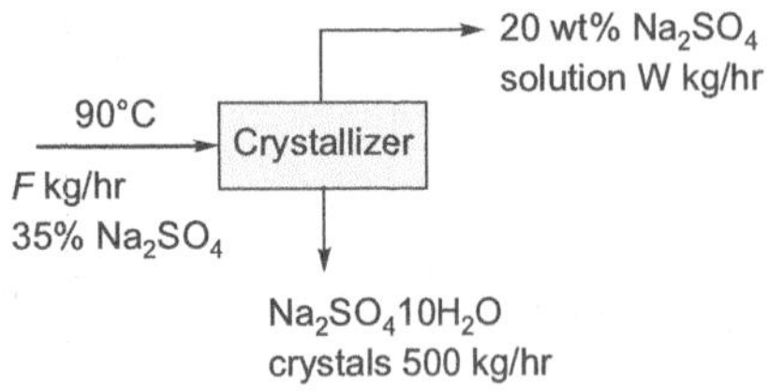

From overall balance

$$F = W + 500 \quad ...(1)$$

From Na_2SO_4 balance

$$F \times 0.35 = W \times 0.2 + 500 \times \frac{142}{322} \quad ...(2)$$

On solving equations (1) and (2)

$$F = 803.3 \text{ kg/hr}$$

2.15 (d)

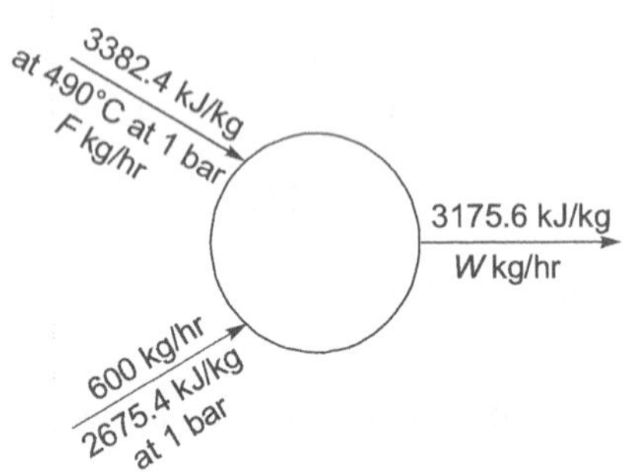

Assuming product stream flow rate is W kg/hr.

Now from overall balance :

$$F + 600 = W \quad ...(1)$$

Now from energy balance

$$F \times 3382.4 + 600 \times 2675.4 = W \times 3175.6 \quad ...(2)$$

On solving equation (1) and (2)

$$F = 1451.257 \text{ kg/hr}$$

$$W = 2051.257 \text{ kg/hr}$$

2.16 (b)

$$CH_4\,(g) \rightarrow C\,(s) + 2H_2\,(g)$$

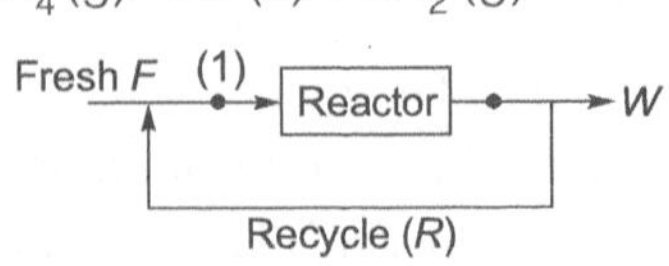

Assuming 1 mol of feed to the reactor at point 1.

∵ Single pass conversion is 0.6

Remaining CH_4 = 1 – 0.6 = 0.4

25% of remaining CH_4 is recycled.

So, 0.4 × 0.25 = 0.1 mol of CH_4 is recycled

Now, Fresh feed = 1 – 0.1 = 0.9

$$\frac{\text{Fresh feed}}{\text{Recycle stream}} = \frac{0.9}{0.1} = 9$$

2.17 (c)

Methane and steam are fed to a reactor in molar ratio of 1 : 2 and following reaction take place:

$$CH_4\,(g) + 2H_2O\,(g) \rightarrow CO_2\,(g) + 4H_2\,(g)$$

$$CH_4\,(g) + H_2O\,(g) \rightarrow CO\,(g) + 3H_2\,(g)$$

$$\text{Selectivity} = \frac{\text{Moles of desired product}}{\text{Moles of undesired product}}$$

$$= \frac{15.21}{2.17} = 7.00$$

2.18 (a)

Basis 100 mole of exit stream.

Moles of desired product CO_2 formed = 15.21 mol

CO_2 is formed when no side reaction is occur and limiting reactant had reacted completely means total CH_4 in feed = CH_4 reacted + CH_4 unreacted

$$= 15.21 + 2.17 + 4.35 = 21.73$$

$$\text{Now, fractional yield} = \frac{15.21}{21.73} = 0.70$$

2.19 (c)

Fractional conversion of methane

$$= \frac{\text{Moles of } CH_4 \text{ reacted to form } CO_2}{CH_4 \text{ in feed}}$$

$$= \frac{15.21}{21.73} = 0.70$$

2.20 (b)

Tie component is defined as component which is used to relate a quantity of one process stream to another process stream and in this question only air is tie component because its mass remains same in feed (wet air) and exit (dry air).

2.21 (b)

Pure water (stream W) is to be obtained from a feed containing 5% salt using a desalination unit as shown below:

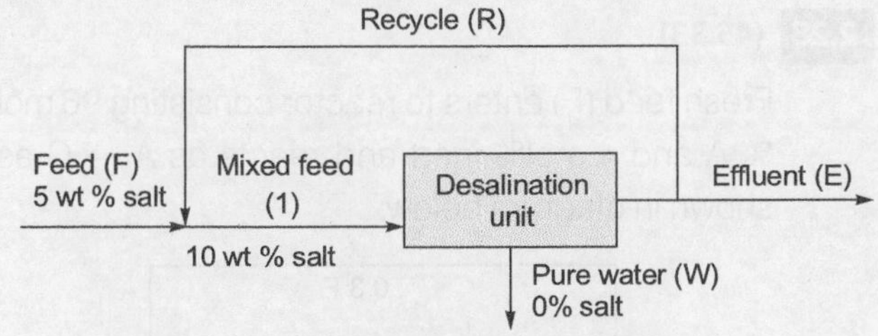

Assuming, $F = 100$ kg for basis

$$F = W + E$$

$$100 = W + E$$

Since overall recovery of pure water is 0.75 kg/kg of feed.

So, $W = 100 \times 0.75 = 75$ kg

$$E = 100 - 75 = 25 \text{ kg}$$

From overall salt balance:

$$F \times 0.05 = W \times 0 + E \times x$$

$$x = 0.2$$

Now, overall balance on point 1

$$M = F + R = 100 + R$$

From salt balance at point 1

$$F \times 0.05 + R \times 0.2 = (R + F) \times 0.1$$

On solving, $R = 50$

$$\therefore \quad R/F = \frac{50}{100} = 0.5$$

2.22 (a)

Ammonia is synthesized at 200 bar and 773 K by the reaction $N_2 + 3H_2 \rightarrow 2NH_3$.

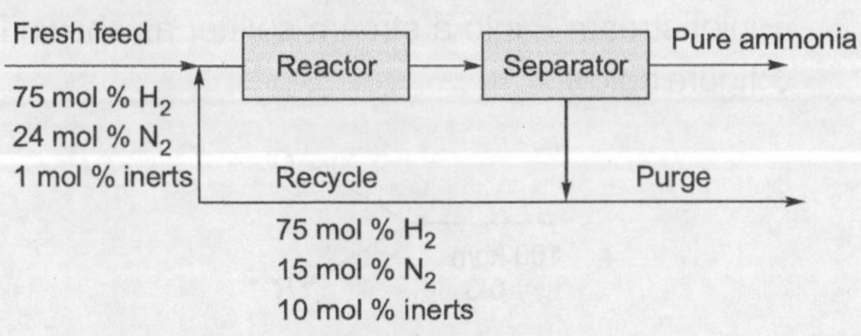

Assuming 100 mol of feed basis

Yield of ammonia = 0.45 mol/mol of fresh feed

Moles of NH_3 produced

$$= 100 \times 0.45 = 45 \text{ mol}$$

From stoichiometry moles of H_2 required

$$= 45 \times \frac{3}{2} = 67.5$$

Moles of H_2 in fresh feed = 75 mole

From inert balance

$$100 \times 0.01 = P \times 0.1$$

$$P = 10$$

H_2 lost in purge = $10 \times 0.75 = 7.5$

$$\frac{\text{Amount of } H_2 \text{ lost in purge}}{H_2 \text{ in fresh feed}} \times 100 = \frac{7.5}{75} \times 100$$

$$= 10\%$$

2.23 (b)

$$A_{(l)} + B_{(g)} \rightarrow C_{(l)} + D_{(g)}$$

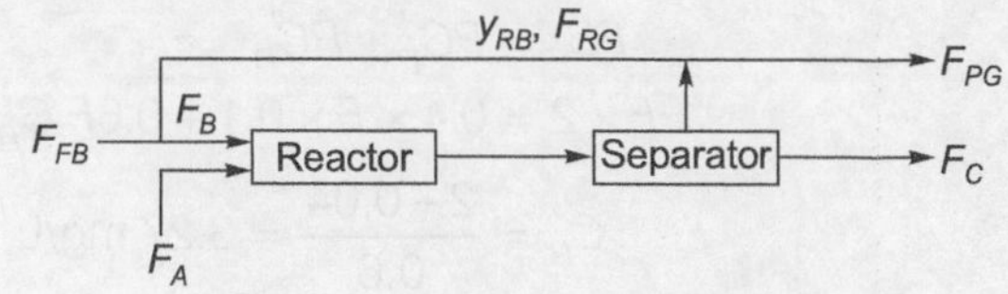

$F_{FB} = 2$ mols, $F_A = 1$ mol/s

$$\frac{F_B}{F_A} = 5$$

$\because \quad X_A = 1$

Hence, $F_C = 1$ mol/s

$$y_{RB} = 0.3$$

From overall material balance

$$F_A + F_{FB} = F_C + F_{PG}$$

$$F_{PG} = 1 + 2 - 1 = 2 \text{ mol/s}$$

Now, from material balance of component B at point 1,

$$F_B = F_{FB} + y_{RB} \times F_{RG}$$

$$F_{RG} = \frac{5-2}{0.3} = 10 \text{ mol/s}$$

$$\frac{F_{RG}}{F_{PG}} = \frac{10}{2} = 5$$

2.24 (a)

Given: $\dfrac{F_{RG}}{F_{PG}} = 4$

$F_{PG} = 2$ (from question 2.23)

So, $F_{RG} = 4 \times 2 = 8$ mol/s

Now from material balance at point 1

$$F_B = F_{FB} + y_{RB} \times F_{RG}$$

$$y_{RB} = \frac{5-2}{8} = \frac{3}{8}$$

2.25 (3.27)

Reverse osmosis plant is treating feed water (F) containing fluoride and its output consists of a permeate stream (P) and a reject stream (R) as shown in diagram below.

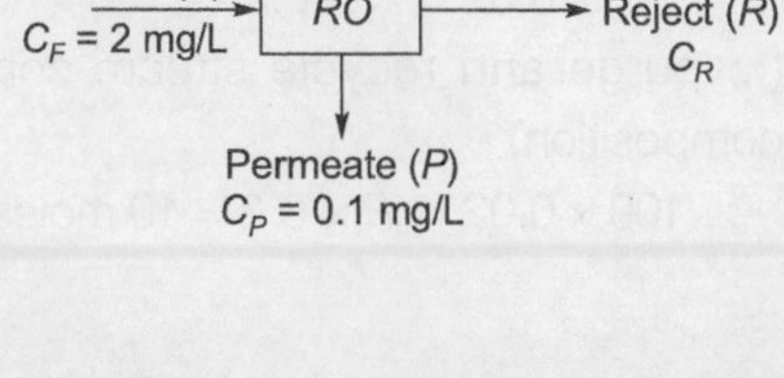

From mass balance:

$$F = P + R$$

Given: $R = 0.6F$

$$F = P + 0.6F$$
$$P = 0.4F$$

From fluoride content balance

$$FC_F = PC_P + RC_R$$
$$F \times 2 = 0.4 \times F \times 0.1 + 0.6F\,C_R$$
$$C_R = \frac{2 - 0.04}{0.6} = 3.27 \text{ mg/L}$$

2.26 (0.264)

The fraction f of the feed is bypassed and mixed with the permeate to obtain treated water having a content of fluoride is 1 mg/L as shown in diagram below.

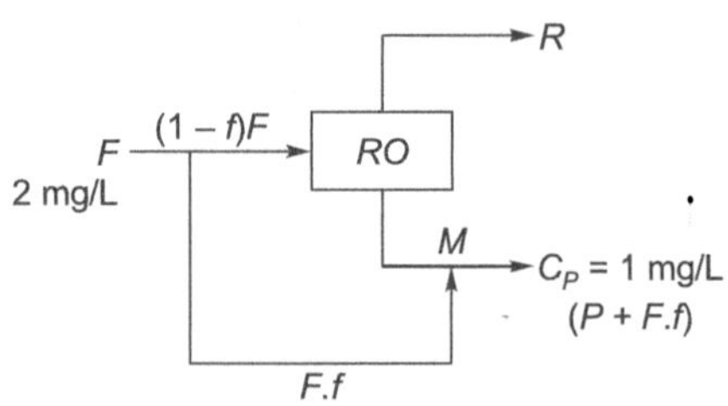

Now from component balance

$$2(f.F) + PC_P = (P + f.F) \times 1$$
$$2(f.F) + P \times 0.1 = P + f.F$$

$\because$ $P = 0.4(1 - f)F$ is given

$$\therefore\ 2f.F + 0.4(1 - f)F \times 0.1 = 0.4(1 - f)F + f.F$$
$$f = 0.264$$

2.27 (10)

Elemental gases (A and B) are reacting to form a liquid (C) in steady state as per the reaction $A + B \rightarrow C$.

Single pass conversion is 20%.

Fresh feed consists 49 mol% A, 49 mol % B and 2 mol % impurities. Maximum allowable impurity in recycle stream is 20% given.

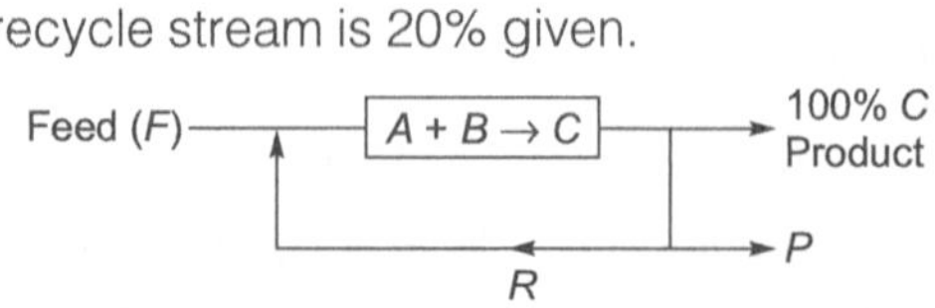

Assuming, $F = 100$ mol

By inert balance

$$F \times 0.02 = P \times 0.2$$

($\because$ purge and recycle stream contains same composition)

$$100 \times 0.02 = P \times 0.2 = 10 \text{ moles}$$

2.28 (45.33)

Fresh feed (F) enters to reactor consisting 96 mol % A and 4 mol% inert and reacts as A → C as shown in diagram below.

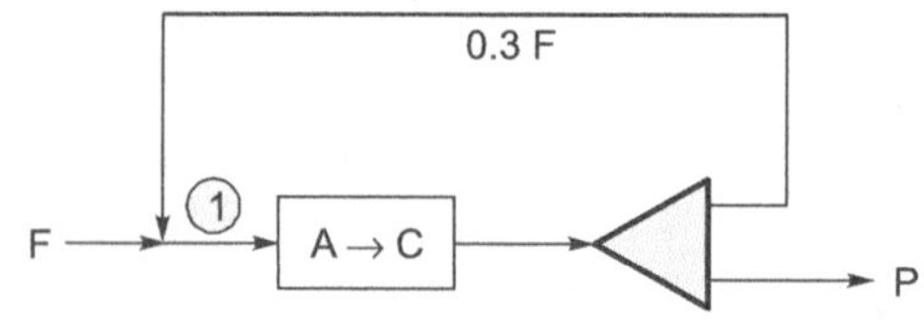

Basis 100 kmol of fresh feed.

From overall steady-state balance :

$$F = P = 100 \text{ kmol}$$

Product stream P contains 50 mol% C (given)

So, P contains 50 kmol of C.

From inert balance:

Inert in feed = Inert in P

= 100 × 0.04 = 4 kmol

A in P = 100 – 4 – 50 = 46%

= 46 kmol

Now A at point 1 = 96 + 0.3 × 46 = 109.8 kmol

A in outlet of reaction

= 46 + 0.3 × 46 = 59.8 kmol

% conversion on the basis of point 1

$$= \frac{109.8 - 59.8}{109.8} \times 100 = 45.33\%$$

2.29 (b)

A liquid mixture of ethanol and water is flowing as inlet stream P into a stream splitter as shown in figure below.

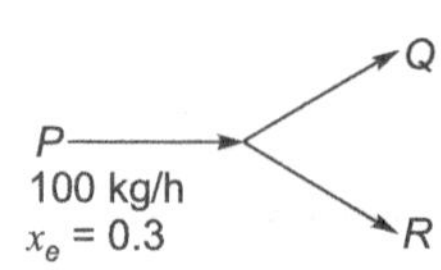

Mass % of ethanol in P stream = 30.

From overall balance :

$$P = Q + R \quad \text{...(1)}$$

From ethanol balance

$$P \cdot x_P = Q \cdot x_Q + R \cdot x_R$$

where, x_Q = Ethanol mass fraction in stream Q

x_R = Ethanol mass fraction in stream R

Since, it is a simple splitting process,

So, $x_P = x_Q = x_R = 0.3$

Now, from equation (1)

$$100 = Q + R$$

For CSTR

$$\frac{V}{F_{AO}} = \frac{X_A}{-r_A}$$

$$\frac{1}{F_{AO}} = \frac{X_A}{k x_A}$$

$$\frac{1}{F_{AO}} = \frac{X_A}{164 \times \frac{F_A}{310}}$$

$$\frac{1}{F_{AO}} = \frac{X_A}{164 \frac{F_{AO}(1-X_A)}{310}}$$

$$X_A = 0.3459$$

$A \to B$

$$F_B = F_{BO} + F_{AO}X_A = F_{AO}X_A$$

$$100 = F_{AO} \times 0.346$$

$$F_{AO} = 289$$

$$F_A = F_{AO}(1-X_A)$$

$$F_A = 289\,(1-0.346) = 189$$

$$\text{Inert} = 1$$

2.33 (0.61)

Let for basis: Moles = 100

$$N_2 = 65\%, \quad O_2 = 10\%$$

$$CO_2 = 15\%, \quad CO = 10\%$$

Moles of $N_2 = 65$

$$CO_2 = 15, \quad CO = 10$$

$$O_2 = 10$$

Let, m = moles of H_2O

$$\frac{m}{100+m} = 0.07$$

$$m = 7.527 \text{ moles}$$

N_2 = on wet bases

$$\frac{65}{100+7.527} = 0.6045 \approx 0.61$$

2.34 (40)

Consider the following reaction in a continuous flow reactor under steady state condition

$$N_2(g) + 3H_2(g) \rightleftharpoons 2NH_3(g)$$

Consider the following inlet flow condition

$$F^0_{N_2} = 100 \text{ mol/s}$$

$$F^0_{H_2} = 300 \text{ mol/s}$$

$$F^0_{inert} = 1 \text{ mol/s}$$

Fractional conversion of $H_2 = 0.6$

Moles of H_2 converted/sec = 300 × 0.6 = 180

Moles of N_2 converted/sec = $\frac{180}{3} = 60$

Outlet moles flow rate of N_2 = 100 – 60 = 40 mol/s

2.35 (18.2)

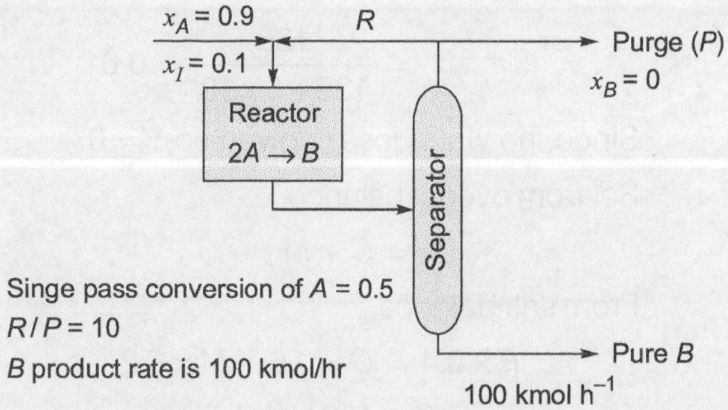

Reaction is given as $2A \to B$ so to form 100 moles of B 200 moles of A should be converted since single pass conversion of A is 50% so feed rate of A to reactor is $\frac{200}{0.5}$ = 400 mol/hr unconverted A either will be in purge as in recycle so applying mass balance for component A.

$$400 \times 0.5 = (R+P)y_A$$

Since, $R = 10P$

So, $200 = 11P\,y_A$

$$Py_n = \frac{200}{11} = 18.18 \simeq 18.2 \text{ kmol/hr}$$

Now, there is two unknown and one independent equation is available. So, additional specification required to known all mass flow rates is 2 – 1 = 1.

2.30 (200)

Aqueous salt-solution enters the crystallizer operating at 25°C as shown in figure below.

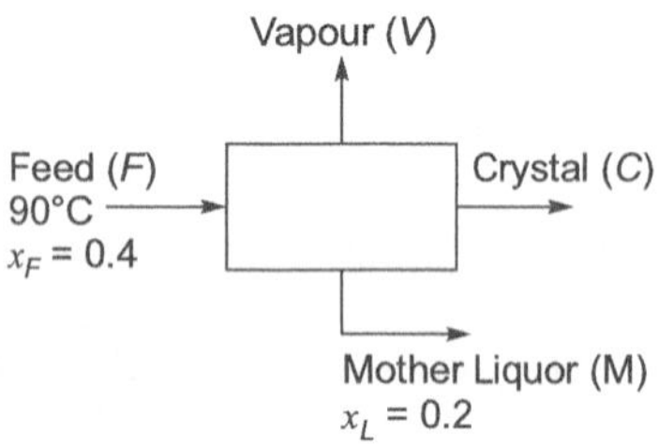

Salt crystallizes as pentahydrate and molecular weight of anhydrous salt is 135.

So, weight fraction of salt in crystal

$$x_c = \frac{135}{135 + 5 \times 18} = 0.6$$

Since, no water loss is given so $V = 0$.

So, from overall balance

$$F = C + M \quad ...(1)$$

From salt balance

$$F \times 0.4 = C \times 0.6 + M \times 0.2$$

$$C = 100 \text{ kg/s (given)}$$

$$F \times 0.4 = 100 \times 0.6 + (F - 100) \times 0.2$$

$$F = \frac{100 \times 0.4}{0.2} = 200 \text{ kg/s}$$

2.31 (c)

Feed stream containing pure species L flows into the reactor and partially converted to M as shown in figure below.

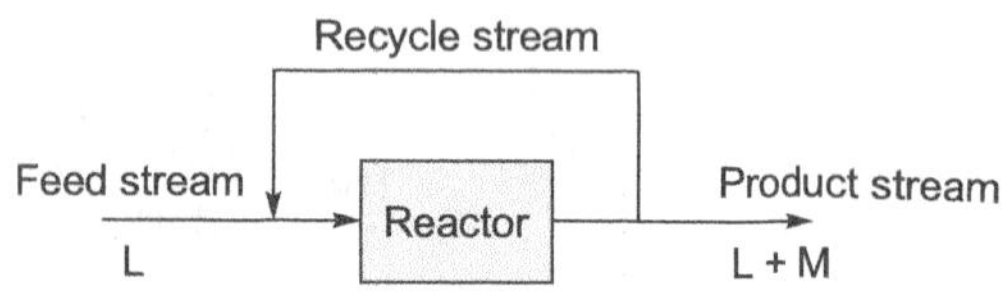

$$R = 0.2P \text{ (given)}$$

So, $$R = 0.2 \times 70 = 14 \text{ mol/s}$$

Overall conversion, $X_0 = 0.3$

Assuming 100 mol/s feed rate of pure L stream.

$$M \text{ in product} = 100 \times 0.3 = 30 \text{ mol/s}$$
$$= 70 \text{ mol/s } (\because \text{ steady state})$$

L in product stream.

Now, assuming X_P is single pass conversion

Pure species 'L' balance around reactor.

$$(100 + 70 \times 0.2)(1 - X_P) = 70 + 70 \times 0.2$$

$$X_P = 26.31\%$$

2.32 (0.99)

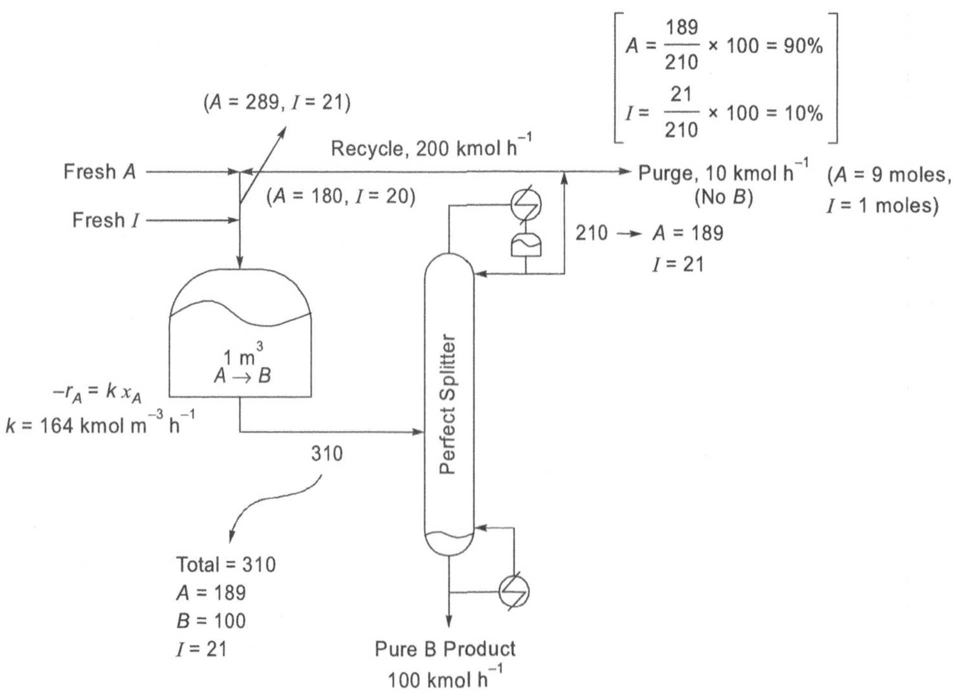

3 Energy Balance Calculations

3.1 Heat capacity of air can be approximately expressed as $C_p = 26.693 + 7.365 \times 10^{-3}$ T where C_p is in J/(mol)(K) and T is in K. The heat given off by 1 mole of air when cooled at 1 atmospheric pressure from 500°C to –100°C is

(a) 10.73 kJ (b) 16.15 kJ
(c) 18.11 kJ (d) 18.33 kJ

[2003 : 2 M]

Common Data for Questions (3.2 & 3.3):

One mole of methane undergoes complete combustion in a stoichiometric amount of air. The reaction proceeds as $CH_4 + 2O_2 \rightarrow CO_2 + 2H_2O$. Both the reactants and the products are in gas phase.

$$\Delta H^0_{298} = -730 \text{ kJ/mol of methane}$$

3.2 Mole fraction of water vapour in the product gases is about

(a) 0.19 (b) 0.33
(c) 0.40 (d) 0.67

[2004 : 2 M]

3.3 If the average specific heat of all the gases/vapour is 40 J/(mol K). The maximum temperature rise of the exhaust gases in °C would be approximately equal to

(a) 1225 (b) 1335
(c) 1525 (d) 1735

[2004 : 2 M]

3.4 Calculate the heat required (in kJ, up to 1 digit after the decimal point) to raise the temperature of 1 mole of a solid material from 100°C to 1000°C. The specific heat (C_p) of the material (in J/mol-K) is expressed as C_p = 20 + 0.005 *T*, where *T* is in *K*. Assume no phase change ________.

[2013 : 2 M]

3.5 The reactant (M) is converted into product (N) in the presence of catalyst in a fixed bed reactor. All the flow rates (*F*, *G*, *H*, *P* and *R*) in mol/s, and mole fraction of reactant (M) in these streams (x_F, x_G, x_H, x_P and x_R) are shown in the diagram.

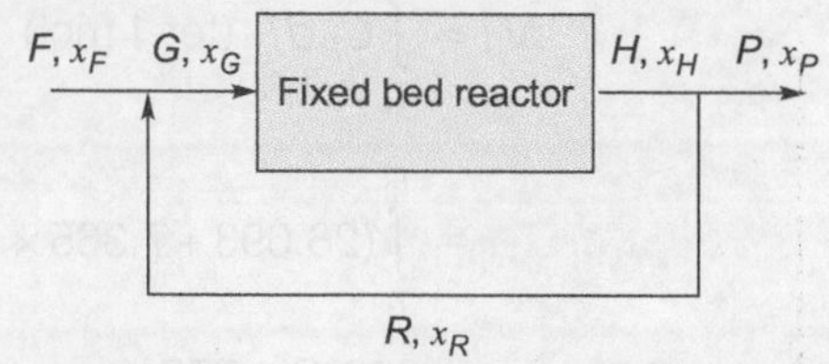

The overall fractional conversion is

(a) $\dfrac{x_G G - x_H H}{x_G G}$ (b) $\dfrac{x_G G - x_P P}{x_G G}$

(c) $\dfrac{x_F F - x_H H}{x_F F}$ (d) $\dfrac{x_F F - x_P P}{x_F F}$

[2018 : 1 M]

3.6 Methane combusts with air in a furnace as $CH_4 + 2O_2 + 2H_2O$. The heat of reaction ΔH_{rxn} = –880 kJ per mol CH_4 and is assumed to be constant. The furnace is well-insulated and no other side reactions occur. All components behave as ideal gases with a constant molar heat capacity of 44 J mol^{-1} $°C^{-1}$. Air may be considered as 20 mol% O_2 and 80 mol% N_2. The air-fuel mixture enters the furnace at 50°C. The methane conversion *X* varies with the air-to-methane mole ratio, *r*, as

$$X = 1 - 0.1e^{-2(r-r_s)} \text{ with } 0.9r_s \le r \le 1.1r_s$$

where r_s is the stoichiometric air-to-methane mole ratio. For $r = 1.05\, r_s$, the exit flue gas temperature in °C, rounded off to 1 decimal place, is ________.

[2024 : 2 M]

Answers Energy Balance Calculations

3.1	(c)	3.2	(a)	3.3	(d)	3.4	(21.7)	3.5	(d)	3.6	(1720 to 1730)

Explanations Energy Balance Calculations

3.1 (c)

Heat capacity of air,

$$C_p = 26.693 + 7.365 \times 10^{-3}\,T \text{ J/molK}$$

where T is in K

$$\Delta H = \int_{T_1}^{T_2} C_P dT \text{ (for 1 mol)}$$

$$= \int_{T_1}^{T_2} (26.693 + 7.365 \times 10^{-3} T) dT$$

$$T_1 = 500°\text{C} = 773 \text{ K}$$
$$T_2 = -100°\text{C} = 173 \text{ K}$$

Now,

$$\Delta H = 26.693(173-773) + \frac{7.365 \times 10^{-3}}{2} \times (173^2 - 773^2)$$

$$\Delta H = -18105.987 \text{ J} = -18.11 \text{ kJ}$$

Negative sign means heat is rejected.

3.2 (a)

1 mole of methane undergoes complete combustion as following reaction

$$CH_4 + 2O_2 \rightarrow CO_2 + 2H_2O$$
$$\Delta H° = -730 \text{ kJ/mol of methane}$$

1 mol of CH_4 required 2 mol of O_2. So, mol of N2

$$\text{supplied} = 2 \times \frac{0.79}{0.21} = 7.52$$

Total moles of product = Moles of CO_2 + Moles of H_2O + Moles of N_2

$$1 + 2 + 7.52 = 10.52$$

Mol fraction of H_2O in product gas

$$= \frac{2}{10.52} = 0.19$$

3.3 (d)

Average specific heat of all the gases/vapour is 40 J/mol.K

$$\not{Q}^0 = (\Sigma n_i h_i)_{out} - (\Sigma n_i h_i)_{in} + \Sigma \Delta H^o_{298}$$

$$O = nC_P\Delta T + \Delta H^o_{298}$$

$$O = 10.524 \times 40 \times \Delta T + (-730 \times 1000)$$

$$O = 420.96 \times \Delta T + (-730000)$$

$$\Delta T = 1734.13°\text{C}$$

3.4 (21.7)

C_p of material is given as

$$C_p = 20 + 0.005T$$

Energy required to raise temperature of 1 mole of material from 100°C to1000°C is given by

$$Q = \int_{T_1}^{T_2} C_p dT = \int_{373}^{1273} (20 + 0.005T) dT$$

$$Q = 20(1273 - 373) + \frac{0.005}{2}(1273^2 - 373^2)$$

$$= 21.7 \text{ kJ}$$

3.5 (d)

Overall fractional conversion = (Amount of reactant M in fresh feed – Amount of reactant M in final product stream)/Amount of reactant M in fresh feed.

So overall fractional conversion = $\dfrac{F \cdot x_F - P \cdot x_P}{F \cdot x_F}$.

■■■■

4 Combustion

4.1 For the case of a fuel gas undergoing combustion with air, if the air/fuel ratio is increased, the adiabatic flame temperature will
(a) increase
(b) decrease
(c) increase or decrease depending on the fuel type
(d) not change

[2001 : 1 M]

4.2 Methane is mixed with stoichiometric proportion of oxygen and completely combusted. The number of additional specifications required to determine the product flow rate and composition is
(a) 0 (b) 1
(c) 2 (d) 3

[2002 : 1 M]

4.3 6 g of carbon is burnt with an amount of air containing 18 g oxygen. The product contains 16.5 g CO_2 and 2.8 g CO besides other constituents. What is the degree conversion on the basis of disappearance of the limiting reactant?
(a) 100% (b) 95%
(c) 75% (d) 20%

[2003 : 2 M]

4.4 A sample of natural gas containing 80% Methane (CH_4) and the rest Nitrogen (N_2) is burnt with 20% excess air. With 80% of the combustibles producing CO_2 and the remainder going to CO, the Orsat analysis in volume percent is
(a) CO_2 : 6.26, CO : 1.56, O_2 : 3.91, H_2O : 15.66, N_2 : 72.60
(b) CO_2 : 7.42, CO : 1.86, O_2 : 4.64, N_2 : 86.08
(c) CO_2 : 6.39, CO : 1.60, O_2 : 3.99, H_2O : 15.96, N_2 : 72.06
(d) CO_2 : 7.60, CO : 1.90, O_2 : 4.75, N_2 : 85.74

[2003 : 2 M]

Statement for Linked Answer Question (4.5 & 4.6):

4.5 44 kg of C_3H_8 is burnt with 1160 kg of air (Mol. Wt. = 29) to produce 88 kg of CO_2 and 14 kg of CO

$$C_3H_8 + 5O_2 = 3CO_2 + 4H_2O$$

What is the percent excess air used?
(a) 55 (b) 60
(c) 65 (d) 68

[2007 : 2 M]

4.6 What is the % carbon burnt?
(a) 63.3 (b) 73.3
(c) 83.3 (d) 93.3

[2007 : 2 M]

4.7 Air (79 mole % nitrogen and 21 mole % oxygen) is passed over a catalyst at high temperature. Oxygen completely reacts with nitrogen as shown below:

$$0.5N_2\,(g) + 0.5\,O_2\,(g) \rightarrow NO\,(g)$$
$$0.5N_2\,(g) + O_2\,(g) \rightarrow NO_2\,(g)$$

The molar ratio of NO to NO_2 in the product stream is 2 : 1. The fractional conversion of nitrogen is
(a) 0.13 (b) 0.20
(c) 0.27 (d) 0.40

[2008 : 2 M]

4.8 The products of combustion of methane in atmospheric air (21% O_2 and 79% N_2) have the following composition on a dry basis :

Products	Mole %
CO_2	10.00
O_2	2.37
CO	0.53
N_2	87.10

The ratio of the moles of CH_4 to the moles of O_2 in the feed stream is
(a) 1.05 (b) 0.60
(c) 0.51 (d) 0.45

[2010 : 2 M]

4.9 The following combustion reaction occur when methane is burnt.

$$CH_4 + 2O_2 \rightarrow CO_2 + 2H_2O$$
$$2CH_4 + 3O_2 \rightarrow 2CO + 4H_2O$$

20% excess air is supplied to the combustor. The conversion of methane is 80% and the molar ratio of CO to CO_2 in the flue gas is 1 : 3. Assume air to have 80 mol % N_2 and rest O_2. The O_2 consumed as a percentage of O_2 entering the combustor is

(a) 20 (b) 62.5
(c) 80 (d) 83.3

[2011 : 2 M]

4.10 Carbon monoxide (CO) is burnt in presence of 200% excess pure oxygen and the flame temperature achieved is 2298 K. The inlet streams are at 25°C. The standard heat of formation (at 25°C) of CO and CO_2 are –110 kJ mol^{-1} and –390 kJ mol^{-1}, respectively. The heat capacities (in J mol^{-1} K^{-1}) of the components are

$$C_{PO_2} = 25 + 14 \times 10^{-3}\,T$$
$$C_{PCO_2} = 25 + 42 \times 10^{-3}\,T$$

where, T is the temperature in K. The heat loss (in kJ) per mole of CO burnt is ______.

[2014 : 2 M]

4.11 Determine the correctness or otherwise of the following Assertion [a] and Reason [r].

Assertion: Significant combustion of coke takes place only if it is heated at higher temperature in presence of air.

Reason: $C + O_2 \rightarrow CO_2$ is an exothermic reaction.

(a) Both [a] and [r] are true and [r] is the correct reason for [a]
(b) Both [a] and [r] are true but [r] is not the correct reason for [a]
(c) [a] is correct but [r] is false
(d) Both [a] and [r] are false

[2014 : 2 M]

4.12 A catalytic reforming plant produces hydrogen and benzene from cyclohexane by de-hydro aromatisation. In order to increase the production of hydrogen, the owner plans to change the process to steam reforming of the same feedstock that produces hydrogen and carbon dioxide. Stoichiometrically, what is the maximum ratio of pure hydrogen produced in the proposed process to that in the existing process?

(a) 1 (b) 2
(c) 5 (d) 6

[2016 : 2 M]

4.13 The ammonia (NH_3) oxidation process occurs over a catalyst as

$$4NH_3 + 5O_2 \rightarrow 6H_2O + 4NO$$

Air is supplied such that oxygen (O_2) is 20% in excess of that required for complete conversion of NH_3. The mole fraction of O_2 in inlet gas mixture (NH_3 + air) is ____ (rounded off to third decimal place).

[2018 : 1 M]

4.14 The heat of combustion of methane, carbon monoxide and hydrogen are P, Q and R respectively. For the reaction below,

$$CH_4 + H_2O \rightarrow CO + 3H_2$$

The heat of reaction is given by

(a) $P - Q - 3R$ (b) $Q + 3R - P$
(c) $P - Q - R$ (d) $Q + R - P$

[2021 : 1 M]

4.15 Formaldehyde is produced by the oxidation of methane in a reactor. The following two parallel reactions occur.

$$CH_4 + O_2 \rightarrow HCHO + H_2O$$
$$CH_4 + 2O_2 \rightarrow CO_2 + 2H_2O$$

Methane and oxygen are fed to the reactor. The product gases leaving the reactor include methane, oxygen, formaldehyde, carbon dioxide and water vapor.

60 mol s^{-1} of methane enters the reactor. The molar flow rate (in mol s^{-1}) of CH_4, O_2 and CO_2 leaving the reactor are 26, 2 and 4, respectively. The molar flowrate of oxygen entering the reactor is ______ mol s^{-1}.

[2021 : 2 M]

4.16 Burning of methane in a combustor yields carbon monoxide, carbon dioxide, and water vapor. Methane is fed to the combustor at 100 mol.hr^{-1}, of which 50% reacts. The theoretical oxygen requirement (in mol.hr^{-1}) is ______ (rounded off to one decimal place).

[2023 : 1 M]

■■■■

Answers Combustion

4.1	(b)	**4.2**	(a)	**4.3**	(b)	**4.4**	(b)	**4.5**	(d)	**4.6**	(c)	**4.7**	(b)
4.8	(d)	**4.9**	(b)	**4.10**	(34.62)	**4.11**	(b)	**4.12**	(d)	**4.13**	(0.184)	**4.14**	(a)
4.15	(40)	**4.16**	(200)										

Explanations Combustion

4.1 (b)

If air/fuel ratio is increased, N_2 content is increased which does not take part in reaction. Some amount of heat will be used for raising the sensible heat content of N_2, thereby reducing the adiabatic flame temperature.

4.2 (a)

Methane is mixed with stoichiometric proportion of oxygen and completely combusted.

$$CH_4 + 2O_2 \rightarrow CO_2 + 2H_2O$$

Combustion of 1 mol CH_4 uses 2 mol O_2 and produces 1 mol of CO_2 and 2 mol of H_2O. So, no more information is required to know flow rate if feed is in stoichiometric proportion.

4.3 (b)

6 g carbon is burnt with an amount of air containing 18 gram oxygen. Product contains 16.5 gram CO_2 and 2.8 gram CO.

$$\text{Mole of carbon in feed} = \frac{6}{12} = 0.5$$

$$\text{Mole of oxygen in feed} = \frac{18}{32} = 0.5625$$

So, carbon is limiting reactant. In product:

$$\text{mol of } CO_2 = \frac{16.5}{44} = 0.375$$

$$\text{mol of CO} = \frac{2.8}{28} = 0.1$$

Reaction involve can be written as

$$C + O_2 \rightarrow CO_2$$

$$C + \frac{1}{2}O_2 \rightarrow CO$$

Total moles of carbon reacted

$$= 0.375 + 0.1 = 0.475$$

So, conversion on the basis of limiting reactant

$$= \frac{0.475}{0.5} = 0.95$$

4.4 (b)

Natural gas containing 80% methane and rest nitrogen is burnt with 20% excess air.

Reaction involve are:

$$CH_4 + 2O_2 \rightarrow CO_2 + 2H_2O$$

$$CH_4 + \frac{3}{2}O_2 \rightarrow CO + 2H_2O$$

Assuming 100 mol of gas

Moles of CH_4 = 80

Moles of N_2 = 20

Moles of CO_2 produced (80% of CH_4 used in first reaction) = 80 × 0.8 = 64

Moles of CO produced = 80 × 0.2 = 16

Moles of O_2 required = 80 × 2 = 160

Moles of O_2 feed = 160 × 1.2 = 192

$$\text{Moles of } O_2 \text{ consumed} = 2 \times 64 + \frac{3}{2} \times 16 = 152$$

O_2 remaining = 192 − 152 = 40

$$\text{Total moles of N2} = 20 + 192 \times \frac{0.79}{0.21} = 742.29$$

Component	Mol	Volume %
CO_2	64	7.42
CO	16	1.86
O_2	40	4.64
N_2	742.29	86.08

Total = 862.29

4.5 (d)

44 kg of C_3H_8 is burnt with 1160 kg of air (Mol. Wt. = 29) to produce 88 kg of CO_2 and 14 kg of CO.

$$C_3H_8 + 5O_2 \rightarrow 3CO_2 + 4H_2O$$

$$\text{Number of moles of } C_3H_8 = \frac{44}{44} = 1 \text{ kmol}$$

$$\text{Number of moles of } CO_2 = \frac{88}{44} = 2 \text{ kmol}$$

From stoichiometry of reactions for 1 kmol of C_3H_8, we require 5 kmol of O_2

$$\text{Air required} = \frac{5}{0.21} = 23.8 \text{ kmol}$$

$$\text{Feed air} = \frac{1160}{29} = 40 \text{ kmol}$$

$$\text{\% of excess air} = \frac{\text{Feed air} - \text{Air required}}{\text{Air required}} \times 100$$

$$= \frac{40 - 23.8}{23.8} \times 100 = 68.06\%$$

4.6 (c)

Carbon in feed = 3 × 12 = 36 kg

$$\text{Carbon burnt to produce } CO_2 = \frac{80}{44} \times 12 = 24 \text{ kg}$$

$$\text{Carbon burnt to produce CO} = \frac{14}{28} \times 12 = 6 \text{ kg}$$

Total carbon burnt = 24 + 6 = 30 kg

$$\text{\% Carbon burnt} = \frac{30}{36} \times 100 = 83.3\%$$

4.7 (b)

Air (79 mol % N_2 and O_2 21 mol%) is passed over a catalyst bed and oxygen completely reacts with nitrogen as following :

$0.5N_2(g) + 0.5O_2(g) \rightarrow NO(g)$

$0.5N_2(g) + O_2(g) \rightarrow NO_2(g)$

Assuming 100 mol of air basis means 79 moles N_2 and 21 moles O_2.

Let x mol of NO_2 is formed.

So, moles of NO formed = 2 × x (given)

From reaction total number of moles O_2 consumed

$$= x + 2x \times 0.5 = 2x$$

So, $2x = 21$

$x = 10.5$

Now from stoichiometry total number of moles N_2 consumed

$$= x + \frac{x}{2}$$

$$\frac{3x}{2} = \frac{3 \times 10.5}{2} = 15.75$$

$$\text{Fractional conversion of } N_2 = \frac{15.75}{79}$$

$$= 0.1993 \simeq 0.20$$

4.8 (d)

The products of combustion of methane in atmospheric air (21% O_2 and 79% N_2) have the following composition on dry basis.

Products	Mole %
CO_2	10.00
O_2	2.37
CO	0.53
N_2	87.10

Reactions of combustion are below:

$$CH_4 + 2O_2 \rightarrow CO_2 + 2H_2O$$

$$CH_4 + \frac{3}{2}O_2 \rightarrow CO + 2H_2O$$

Moles of methane burnt = Moles of CO_2 + Moles of CO = 10 + 0.53 = 10.53

$$\text{Moles of } O_2 \text{ entered} = \text{Moles of } N_2 \times \frac{0.21}{0.79}$$

$$= 87.10 \times \frac{0.21}{0.79} = 23.15$$

Ratio of moles of CH_4 to the moles of O_2 in feed

$$\text{stream} = \frac{10.53}{23.15} = 0.45$$

4.9 (b)

Following combustion reactions occur when methane is burnt.

$$CH_4 + 2O_2 \rightarrow CO_2 + 2H_2O$$

$$2CH_4 + 3O_2 \rightarrow 2CO + 4H_2O$$

Assuming 100 mole of methane supplied in feed.

O_2 required = 2 × 100 = 200 moles

O_2 supplied in 20% excess.

So, O_2 supplied = 200 × 1.2 = 240 moles

80% methane is converted so methane out of reactor = 100 × 0.2 = 20 moles

$$N_2 \text{ supplied} = \frac{240}{20} \times 80 = 960 \text{ moles}$$

CH_4 reacted = 100 − 20 = 80 moles

Assuming number of CH_4 moles reacted in first reaction is a. So, CH_4 reacted in reaction (2) is $(80 - a)$.

$$\frac{\text{Moles of } CO_2}{\text{Moles of CO}} = 3 \text{ (given)}$$

$$\frac{a}{(80 - a)} = 3 \Rightarrow a = 60$$

Now,

$$\frac{O_2 \text{ consumed}}{O_2 \text{ entering}} \times 100 = \frac{2\times 60 + \frac{3}{2}(80-60)}{240} = 62.5\%$$

4.10 (34.62)

Carbon monoxide is burnt in presence of 200% excess pure oxygen and flame temperature 2298 K is achieved.

$$C + \frac{1}{2}O_2 \rightarrow CO\ (\Delta H_1 = -110 \text{ kJ}) \quad ...(1)$$

$$C + O_2 \rightarrow CO_2\ (\Delta H_2 = -390 \text{ kJ}) \quad ...(2)$$

$$CO + \frac{1}{2}O_2 \rightarrow CO_2 \quad ...(3)$$

Now from reaction (1) and reaction (2)

$$\Delta H_3 = -390 - (-110) = -280 \text{ kJ/mol}$$

For 1 mol of CO burnt O_2 supplied = 1.5 mole

O_2 in product = 1 mol

$$\text{Heat of product} = \int_{298}^{2298} (25 + 0.014T)dT + \int_{298}^{2298} (25 + 0.042T)dT = 86344 + 159032$$

$$= 245376 \text{ J/mol} = 245.376 \text{ kJ/mol}$$

So, change in heat = –(280 – 245.376) = –34.624 kJ

Here, negative sign indicates loss in heat.

4.11 (b)

[r] is not the correct reason of [a]. However both [a] and [r] are true.

4.12 (d)

In process-1, cyclohexane is converting into benzene by dehydro aromatization process

$$C_6H_{12} \rightarrow C_6H_6 + 3H_2$$

In process-2, cyclohexane is converting into CO_2 and H_2

$$C_6H_{12} + 2H_2O \rightarrow 6CO_2 + 18H_2$$

$$\text{Now,} \quad \frac{\dfrac{\text{Moles of } H_2 \text{ in process-2}}{\text{Moles of } C_6H_{12}}}{\dfrac{\text{Moles of } H_2 \text{ in process-1}}{\text{Moles of } C_6H_{12}}} = \frac{18}{3} = 6$$

4.13 (0.184)

Ammonia oxidation process occurs over a catalyst as

$$4NH_3 + 5O_2 \rightarrow 6H_2O + 4NO$$

O_2 is supplied 20% in excess.

Assuming 1 mol of ammonia in feed stream. O_2 required for complete combustion of 1 mol of NH_3

$$= \frac{5}{4} = 1.25 \text{ mol}$$

Since O_2 supplied in 20% excess so O_2 supplied

$$= 1.25 \times 1.2 = 1.5$$

$$N_2 \text{ supplied} = \frac{1.5}{0.21} \times 0.79 = 5.64 \text{ mol}$$

Total moles in feed = 1 + 1.5 + 5.64 = 8.14

$$O_2 \text{ mol fraction in feed} = \frac{1.5}{8.14} = 0.184$$

4.14 (a)

$$CH_4 + 2O_2 \rightarrow CO_2 + 2H_2O;\ \Delta H_1 = P$$

$$CO + \frac{1}{2}O_2 \rightarrow CO_2;\ \Delta H_2 = Q$$

$$H_2 + \frac{1}{2}O_2 \rightarrow H_2O;\ \Delta H_3 = R$$

For reaction $CH_4 + H_2O \rightarrow CO + 3H_2$

$$(\Delta H) = \Delta H_1 - 3\Delta H_3 - \Delta H_2$$

$$= P - Q - 3R$$

4.15 (40)

$$CH_4 + O_2 \rightarrow HCHO + H_2O$$

$$CH_4 + 2O_2 \rightarrow CO_2 + 2H_2O$$

It is given that 60 mol/s CH_4 is entering the reactor and 26 mol/s CH_4 is leaving.

CH_4 reacted = 60 – 26 = 34 mol/s

CO_2 leaving = 4 mol/s

So, CH_4 reacted for HCHO formation

= 34 – 4 = 30 mol/s

O_2 consumed for HCHO formation = 30 mol/s

O_2 consumed in CO_2 formation = 2 × 4 = 8 mol/s

O_2 entering = O_2 consumed + O_2 leaving

= 30 + 8 + 2 = 40 mol/s

4.16 (200)

Burning of methane is given as

$$CH_4 + 2CO_2 \rightarrow CO_2 + 2H_2O$$

$$CH_4 + \frac{3}{2}O_2 \rightarrow CO + 2H_2O$$

Given: CH_4 supply = 100 mol/hr

For theoretical oxygen requirement, complete combustion is only considered, i.e.

For 100 mole $CH_4 \rightarrow$ 200 mole O_2 is require.

Thermodynamics

UNIT II

CONTENTS

Thermodynamics

Syllabus

First and Second Laws of Thermodynamics. Applications of First Law to Close and Open Systems. Second Law and Entropy. Thermodynamic Properties of Pure Substances : Equation of State and Residual Properties, Properties of Mixtures : Partial Molar Properties, Fugacity, Excess Properties and Activity Coefficients; Phase Equilibria, Predicting VLE of Systems; Chemical Reaction Equilibriums.

Analysis of Previous GATE Papers

Exam Year	1 Mark Ques.	2 Marks Ques.	5 Marks Ques.	Total Marks
2000	2	2	1	**11**
2001	2	3	–	**8**
2002	4	3	–	**10**
2003	6	6	–	**18**
2004	2	6	–	**14**
2005	3	5	–	**13**
2006	3	3	–	**9**
2007	5	8	–	**21**
2008	2	4	–	**10**
2009	2	3	–	**8**
2010	3	1	–	**5**
2011	2	2	–	**6**
2012	4	2	–	**8**
2013	3	2	–	**7**
2014	3	2	–	**7**
2015	4	3	–	**10**
2016	2	3	–	**8**
2017	2	4	–	**10**
2018	1	3	–	**7**
2019	4	4	–	**12**
2020	2	4	–	**10**
2021	2	3	–	**8**
2022	2	4	–	**10**
2023	2	2	–	**6**
2024	2	3	–	**8**

1 Basic Concepts and Zeroth Law of Thermodynamics

1.1 For the two paths as shown in the figure, one reversible and one irreversible, to change the state of the system from *a* to *b*,

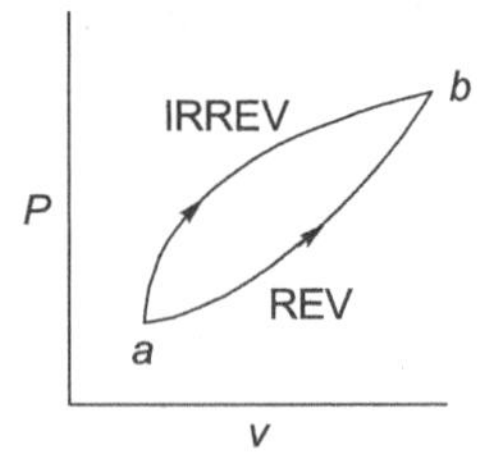

(a) ΔU, Q, W are same
(b) ΔU is same
(c) Q, W are same
(d) ΔU, Q are different

[2007 : 2 M]

1.2 For a pure liquid, the rate of change of vapour pressure with temperature is 0.1 bar/K in the temperature range of 300 to 350 K. If the boiling point of the liquid at 2 bar is 320 K, the temperature (in K) at which it will boil at 1 bar (up to one decimal place) is _______.

[2015 : 1 M]

1.3 Water is heated at atmospheric pressure from 40°C to 80°C using two different processes, In process-I, the heating is done by a source at 80°C. In process-II, the water is first heated from 40°C to 60°C by a source at 60°C. and then from 60°C to 80°C by another source at 80°C.
Identify the correct statement.
(a) Enthalpy change of water in process-I is greater than enthalpy change in process-II
(b) Enthalpy change of water in process-II is greater than enthalpy change in process-I
(c) Process-I is closer to reversibility
(d) Process-II is closer to reversibility

[2017 : 1 M]

1.4 Consider the following properties:
P. temperature
Q. specific gravity
R. chemical potential
S. volume
The option which lists ALL the intensive properties is
(a) P
(b) P and Q
(c) P, Q and R
(d) P, Q, R and S

[2018 : 1 M]

■■■■

Answers Basic Concepts and Zeroth Law of Thermodynamics

1.1 (b) 1.2 (310) 1.3 (d) 1.4 (c)

Explanations Basic Concepts and Zeroth Law of Thermodynamics

1.1 (b)

Internal energy is point function so does not depend upon path.

Heat transfer (Q) and work done (W) are path function so ΔU is same for both process Q and W are different.

1.2 (310)

For pure liquid, the rate of change of vapour pressure with temperature is 0.1 bar/K in temperature range of 300 to 350 K.

$$\frac{dP}{dT} = 0.1$$

On integrating,

$$P = 0.1\,T + C \quad \text{...(i)}$$

At $P = 2$ bar, $T = 320$ K

So, $2 = 0.1 \times 320 + C$

$C = -30$

So, $P = 0.1\,T - 30$

At $P = 1$ bar, $T = ?$

$1 = 0.1 \times T - 30$

$T = 310$ K

1.3 (d)

Since initial and final temperature is same for both process.

Hence enthalpy change of water is same in both process.

Process-II is closer to reversibility because heating is break into two sub paths. If process-II is break into infinitely small temperature difference paths then process will be almost reversible.

1.4 (c)

Temperature, specific gravity and chemical potential are intensive properties.

Volume is extensive properties.

■■■■

2 Energy and Energy Interactions

2.1 A rigid vessel, containing three moles of nitrogen gas at 30°C, is heated to 250°C. Assume the average heat capacities of nitrogen to be C_p = 29.1 J/mol°C and C_v = 20.8 J/mol °C. The heat required, neglecting the heat capacity of the vessel, is

(a) 13728 J
(b) 19206 J
(c) 4576 J
(d) 12712 J

[2002 : 2 M]

2.2 For estimation of heat capacity of a solid compound, one can use

(a) Chaperon's equation
(b) Gibb's equation
(c) Kopp's rule
(d) Trouton's rule

[2003 : 1 M]

2.3 In Joule's experiments, an insulated container contains 20 kg of water initially at 25°C. It is stirred by an agitator, which is made to turn by a slowly falling body weighing 40 kg through a height of 4 m. The process is repeated 500 times. The acceleration due to gravity is 9.8 ms^{-2}. Neglecting the heat capacity of agitator, the temperature of water (in °C) is

(a) 40.5
(b) 34.4
(c) 26.8
(d) 25

[2003 : 1 M]

2.4 The vapour pressure of water is given by $\ln P^{sat} = A - \frac{5000}{T}$ where A is a constant, P^{sat} is vapour pressure in atm, and T is temperature in K. The vapour pressure of water in atm at 50 °C is approximately

(a) 0.07
(b) 0.09
(c) 0.11
(d) 0.13

[2004 : 2 M]

2.5 The work done by one mole of a van der Waals fluid undergoing reversible isothermal expansion from initial volume V_i to final volume V_f is

(a) $RT\ln\left(\frac{V_f}{V_i}\right)$

(b) $RT\ln\left(\frac{V_f - b}{V_i - b}\right)$

(c) $RT\ln\left(\frac{V_f - b}{V_i - b}\right) - a\left(\frac{1}{V_f} - \frac{1}{V_i}\right)$

(d) $RT\ln\left(\frac{V_f - b}{V_i - b}\right) + a\left(\frac{1}{V_f} - \frac{1}{V_i}\right)$

[2008 : 1 M]

2.6 A new linear temperature scale, denoted by °S, has been developed, where the freezing point of water is 200°S and the boiling point is 400°S corresponds in degrees Celsius, to

(a) 100°C
(b) 125°C
(c) 150°C
(d) 300°C

[2010 : 1 M]

2.7 The thermodynamic state of a closed system containing a pure fluid changes from (T_1, p_1) to (T_2, p_2), where T and p denote the temperature and pressure, respectively. Let Q denote the heat absorbed (> 0 if absorbed by the system) and W the work done (> 0 if done by the system). Neglect changes in kinetic and potential energies. Which one of the following is CORRECT?

(a) Q is path-independent and W is path-dependent
(b) Q is path-dependent and W is path-independent
(c) (Q – W) is path-independent
(d) (Q + W) is path-independent

[2013 : 1 M]

2.8 Three identical closed systems of a pure gas are taken from an initial temperature and pressure (T_1, P_1) to a final state (T_2, P_2), each by a different path. Which of the following is ALWAYS TRUE for the three systems? (Δ represents the change between the initial and final states; U, S, G, Q and W are internal energy, entropy, Gibbs free energy, heat added and work done, respectively.)

(a) ΔU, ΔS, Q are same
(b) W, ΔU, ΔG are same
(c) ΔS, ΔW, Q are same
(d) ΔG, ΔU, ΔS are same

[2015 : 1 M]

2.9 An ideal gas is initially at a pressure of 0.1 MPa and a total volume of 2 m^3. It is first compressed to 1 MPa by a reversible adiabatic process and then cooled at constant pressure to a final volume of 0.2 m^3. The total work done (in kJ) on the gas for the entire process (up to one decimal place) is ______.

Data: R = 8.314 J/mol.K; heat capacity at constant pressure (C_P) = 2.5 R

[2015 : 2 M]

2.10 Consider one mole of an ideal gas in a closed system. It undergoes a change in state from L to N through two different non-isothermal processes, as shown in the P-V diagram (where P is the pressure and V is the molar volume of the gas). Process I is the carried out in a single step, namely LN, whereas process II is carried out in two steps, namely LM and MN. All the steps are reversible.

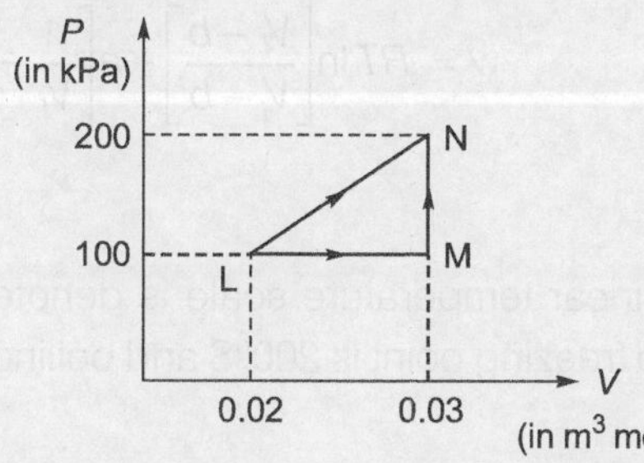

The net heat flowing into the system for process I is Q_I and that for process II is Q_{II}. The value of $Q_I - Q_{II}$ (in J) is

(a) 250 (b) 500
(c) 1000 (d) 1500

[2020 : 2 M]

2.11 The combustion of carbon monoxide is carried out in a closed, rigid and insulated vessel. 1 mol of CO, 0.5 mol of O_2 and 2 mol of N_2 are taken initially at 1 bar and 298 K, and the combustion is carried out to completion.

The standard molar internal energy change of reaction (Δu_R^o) for the combustion of carbon monoxide at 298 K = –282 kJ mol^{-1}. At constant pressure, the molar heat capacities of N_2 and CO_2 are 33.314 J mol^{-1} K^{-1} and 58.314 J mol^{-1} K^{-1}, respectively. Assume the heat capacities to be independent of temperature, and the gases are ideal. Take R = 8.314 J mol^{-1} K^{-1}.

The final pressure in the vessel at the completion of the reaction is ____ bar (round off to 1 decimal place).

[2021 : 2 M]

■■■■

Answers Energy and Energy Interactions

2.1	(a)	2.2	(c)	2.3	(b)	2.4	(d)	2.5	(d)	2.6	(c)	2.7	(c)
2.8	(d)	2.9	(749.97)	2.10	(b)	2.11	(8.93)						

Explanations Energy and Energy Interactions

2.1 (a)

This is the case of heating at constant volume (use C_v).

$$Q = nC_v(T_f - T_i)$$
$$Q = 3 \times 20.8 \times (250 - 30)$$
$$= 13728 \text{ J}$$

2.2 (c)

Kopp's rule states that the molecular heat capacity of a solid compound is the sum of the atomic heat capacities of elements composing it.

2.3 (b)

From energy conservation, all potential energy will be utilized to rise temperature of water because tank is insulated

$$n \cdot mgh = m' C_p (\Delta T)$$
$$500 \times 40 \times 9.8 \times 4 = 20 \times 4180 (T - 25)$$
$$T = 34.4°\text{C}$$

2.4 (d)

$$\ln P^{sat} = A - \frac{5000}{T}$$

T is temperature in Kelvin:

$$P^{sat} = \text{Vapour pressure in atm}$$

$$\ln P^{sat} = A - \frac{5000}{T}$$

At $T = 373$ K

$$P^{sat} = 1 \text{ atm}$$

So, $A = \dfrac{5000}{373} = 13.4048$

Now at, $T = 50°$C

$$\ln P^{sat} = 13.4048 - \frac{5000}{(50 + 273)}$$
$$P^{sat} = 0.13 \text{ atm}$$

2.5 (d)

Vander-Waals equation is written below:

$$\left(P + \frac{a}{v^2}\right)(v - b) = RT \qquad \text{...(i)}$$

For isothermal process $\left(P + \dfrac{a}{v^2}\right)(v - b)$ = constant

Work done from initial volume V_i to final volume V_f can be calculated as

$$w = \int_{V_i}^{V_f} P\,dv$$

By putting value of P from equation (i)

$$w = \int_{V_i}^{V_f} \left(\frac{RT}{v - b} - \frac{a}{v^2}\right) dv$$

$$w = RT \ln\left[\frac{V_f - b}{V_i - b}\right] + a\left[\frac{1}{V_f} - \frac{1}{V_i}\right]$$

2.6 (c)

Now linear temperature scale is denoted by °S where freezing point is 200°S and boiling point is 400°S

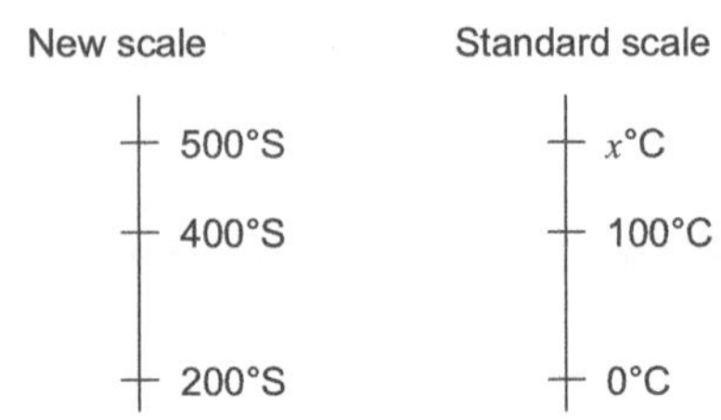

Assume that 500°S on new scale corresponds to x°C on standard scale.

Now performing some operation on both scales

$$\frac{x - 0}{100 - 0} = \frac{500 - 200}{400 - 200}$$
$$x = 150°\text{C}$$

2.7 (c)

From first law of thermodynamics

$$Q - W = \Delta U$$

So Q and W are path dependent and $Q - W$ is path independent as internal energy is property of system (or point function).

2.8 (d)

ΔG, ΔU and ΔS are point function. They are not dependent on path. They depends only upon initial and final point of process.

2.9 (749.97)

Ideal gas is initially at 0.1 MPa and a total volume of 2 m^3 is compressed to 1 MPa by reversible adiabatic process and then cooled at constant pressure to final volume of 0.2 m^3.

Above process can be better understand by following diagram:

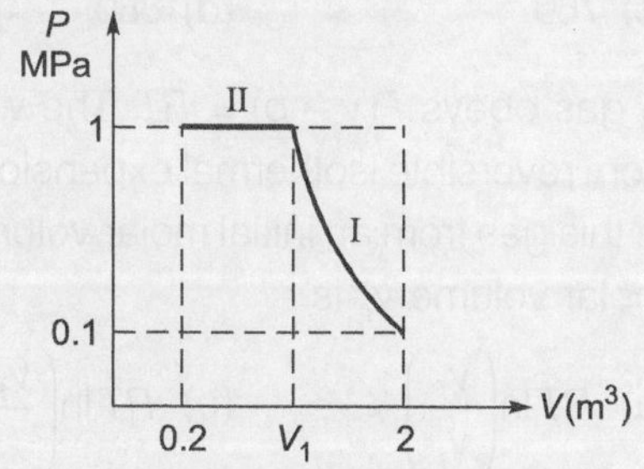

Process I is reversible adiabatic

$$P_1V_1^{\gamma} = P_2V_2^{\gamma}$$

$$\gamma = \frac{C_p}{C_p - R} = \frac{5}{3}$$

$$0.1 \times 2^{1.667} = 1 \times V_1^{1.667}$$

$$V_1 = 0.5 \text{ m}^3$$

Work done for process I

$$W_1 = \frac{P_1V_1 - P_2V_2}{\gamma - 1} = \frac{0.1 \times 2 - 1 \times 0.5}{1.6667 - 1}$$

$$= -0.4499 \times 10^6 \text{ Joule}$$

For process II

$$W_2 = \int_{V_1}^{V_2} P\,dV = 10^6\,(0.2 - 0.5)$$

$$= -0.3 \times 10^6 \text{ Joule}$$

Total work done on the gas = $W_1 + W_2$ = 749.97 kJ

2.10 (b)

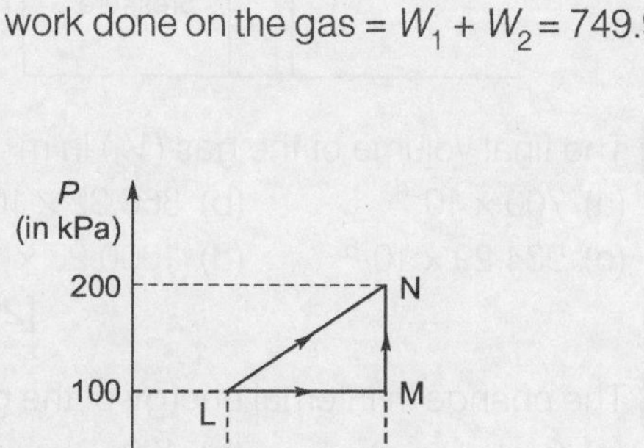

Process 1 (single step process) L to N

Process 2 (double step process) L to M to N

For process 1:

$$Q_1 - W_1 = \Delta U_1 \quad \text{(from first law)}$$

Here works done by system and heat added to the system is possible.

For process 2:

$$Q_2 - W_2 = \Delta U_2$$

$$W_1 = \int_{V_1}^{V_2} P\,dV = \text{Area under curve}$$

$$= \frac{1}{2}(100 + 200) \times 0.01 = 1.5 \text{ kJ}$$

$$W_2 = (0.03 - 0.02) \times 100 = 1 \text{ kJ}$$

Since internal energy is a point function (does not depends upon path) so

$$\Delta U_1 = \Delta U_2$$

$$Q_1 - W_1 = Q_2 - W_2$$

$$Q_1 - Q_2 = W_2 - W_2$$

$$= 1.5 - 1 = 0.5 \text{ kJ} = 500 \text{ Joule}$$

2.11 (8.93)

Since vessel is insulated and rigid so heat loss and volume change can be assumed to zero.

$$C_{P_{N2}} = 33.314 \text{ J/mol.K}$$

$$C_{V_{N2}} = C_P - R = 25 \text{ J/mol.K}$$

$$C_{P_{CO2}} = 58.314 \text{ J/mol.K}$$

$$C_{V_{CO2}} = 50.0 \text{ J/mol.K}$$

$$CO + \frac{1}{2}O_2 \rightarrow CO_2$$

1	0.5	0
0	0	1

ΔU = Change in internal energy

= Heat gain by product

$$282 \times 10^3 = (n_1C_{VN_2} + n_2C_{VCO_2})\Delta T$$

$$\Rightarrow \quad n_1 = 2,\ n_2 = 1$$

$$\Delta T = 2820$$

$$T_2 = 3118 \text{ K}$$

As gases are ideal

$$\frac{P_1V}{P_2V} = \frac{N_1RT_1}{N_2RT_2}$$

$$N_1 = 1 + 0.5 + 2 = 3.5$$

$$N_2 = 1 + 2 = 3$$

$$P_2 = P_1\left(\frac{N_2}{N_1}\right)\left(\frac{T_2}{T_1}\right)$$

$$= 1 \times \frac{3}{3.5} \times \frac{3118}{298} = 8.93 \text{ bar}$$

■■■■

3 First Law of Thermodynamics

3.1 Air enters an adiabatic compressor at 300 K. The exit temperature for a compression ratio of 3, assuming air to be an ideal gas $\left(\gamma = \frac{C_p}{C_v} = \frac{7}{5}\right)$ and the process to be reversible, is

(a) $300(3^{2/7})$ (b) $300(3^{3/5})$
(c) $300(3^{3/7})$ (d) $300(3^{5/7})$

[2001 : 2 M]

3.2 The partial molar enthalpy of a component in an ideal binary gas mixture of composition *z*, at a temperature *T* and pressure *P* is a function only of

(a) *T* (b) *T* and *P*
(c) *T*, *P* and *z* (d) *T* and *z*

[2002 : 1 M]

3.3 1 m^3 of an ideal gas at 500 K and 1000 kPa expands reversibly to 5 times its initial volume in an insulated container. If the specific heat capacity (at constant pressure) of the gas is 21 J/mol K, the final temperature will be

(a) 35 K (b) 174 K
(c) 274 K (d) 154 K

[2002 : 2 M]

3.4 One mole of Nitrogen at 8 bar and 600 K is contained in a piston-cylinder arrangement. It is brought to 1 bar isothermally against a resisting pressure of 1 bar. The work done (in Joules) by the gas is

(a) 30554 (b) 10373
(c) 4988.4 (d) 4364.9

[2003 : 1 M]

3.5 A perfectly insulated container of volume *V* is divided into two equal halves by a partition. One side is under vacuum while the other side has one mole of an ideal gas (with constant heat capacity) at 298 K. If the partition is broken, the final temperature of the gas in the container

(a) will be greater than 298 K
(b) will be 298 K
(c) will be less than 298 K
(d) cannot be determined

[2004 : 1 M]

3.6 A car tyre of volume 0.057 m^3 is inflated to 300 kPa at 300 K. After the car is driven for ten hours, the pressure in the tyre increases to 330 kPa. Assume air is an ideal gas and C_v for air is 21 J/(mol K). The change in the internal energy of air in the tyre in J/mol is

(a) 380 (b) 630
(c) 760 (d) 880

[2004 : 2 M]

3.7 A gas obeys $P(v - b) = RT$. The work obtained from reversible isothermal expansion of one mole of this gas from an initial molar volume v_i to a final molar volume v_f, is

(a) $RT\ln\left(\frac{v_f}{v_i}\right)$ (b) $RT\ln\left(\frac{v_f - b}{v_i}\right)$

(c) $RT\ln\left(\frac{v_f}{v_i - b}\right)$ (d) $RT\ln\left(\frac{v_f - b}{v_i - b}\right)$

[2004 : 2 M]

Statement for Linked Answer Question (3.8 to 3.9):

A frictionless cylinder piston assembly contains an ideal gas. Initially pressure (P_1) = 100 kPa, temperature (T_1) = 500 K and volume (V_1) = 700 × 10^{-6} m^3. This system is supplied with 100 J of heat and pressure is maintained constant at 100 kPa. The enthalpy variation is given by

$$h\,(\text{J/mol}) = 30000 + 50\,T$$

where *T* is the temperature in K, and the universal gas constant *R* = 8.314 J/(mol K).

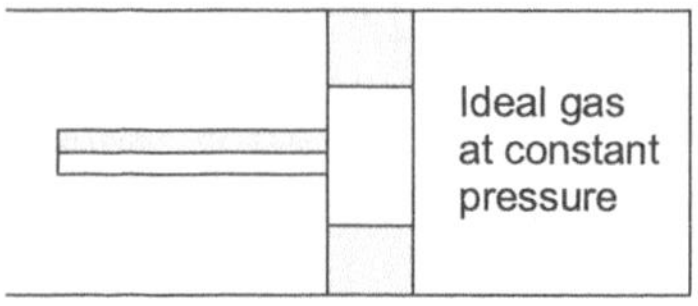

3.8 The final volume of the gas (V_2) in m^3 is

(a) 700×10^{-6} (b) 866.32×10^{-6}
(c) 934.29×10^{-6} (d) 1000.23×10^{-6}

[2005 : 2 M]

3.9 The change in internal energy of the gas in J is

(a) 0 (b) 100
(c) 23.43 (d) 83.37

[2005 : 2 M]

3.10 The state of an ideal gas is changed from (T_1, P_1) to (T_2, P_2) in a constant volume process. To calculate the change in enthalpy, Δh, all of the following properties/variables are required.
(a) C_v, P_1, P_2
(b) C_p, T_1, T_2
(c) C_p, T_1, T_2, P_1, P_2
(d) C_v, P_1, P_2, T_1, T_2

[2007 : 1 M]

Statement for Linked Answer Question (3.11 & 3.12):

3.11 A perfectly insulated cylinder of volume 0.6 m^3 is initially divided into two parts by a thin, frictionless piston, as shown in the figure. The smaller part of volume 0.2 m^3 has ideal gas at 6 bar pressure and 100°C. The other part is evacuated.

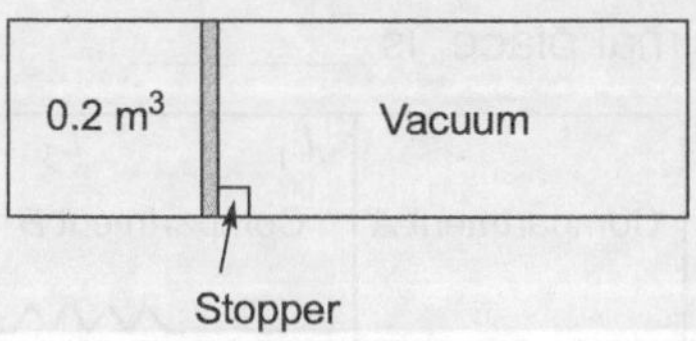

At certain instant of time t, the stopper is removed and the piston moves out freely to the other end. The final temperature is
(a) –149°C (b) –33°C
(c) 33°C (d) 100°C

[2007 : 2 M]

3.12 The cylinder insulation is now removed and the piston is pushed back to restore the system to its initial state. If this is to be achieved only by doing work on the system (no heat addition, only heat removal allowed), what is the minimum work required?
(a) 3.4 kJ (b) 107 kJ
(c) 132 kJ (d) 240 kJ

[2007 : 2 M]

3.13 A saturated liquid at 1500 kPa and 500 K, with an enthalpy of 750 kJ/kg, is throttled to a liquid-vapour mixture at 150 kPa and 300 K. At the exit conditions, the enthalpy of the saturated liquid is 500 kJ/kg and the enthalpy of the saturated vapour is 2500 kJ/kg. The percentage of the original liquid, which vaporizes, is
(a) 87.5% (b) 67%
(c) 12.5% (d) 10%

[2010 : 1 M]

3.14 One mole of methane is contained in a leak proof piston-cylinder assembly at 8 bar and 1000 K. The gas undergoes isothermal expansion to 4 bar under reversible conditions. Methane can be considered as an ideal gas under these conditions. The value of universal gas constant is 8.314 J mol^{-1} K^{-1}. The heat transferred (in kJ) during the process is
(a) 11.52 (b) 5.76
(c) 4.15 (d) 2.38

[2011 : 1 M]

3.15 In a closed system, the isentropic expansion of an ideal gas with constant specific heats is represented by

(a) 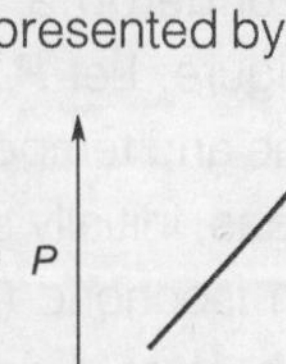

(b) 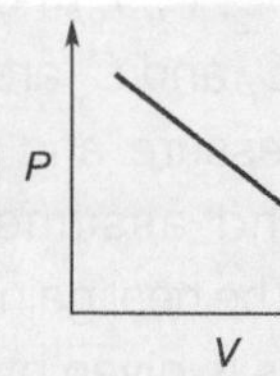

(c) 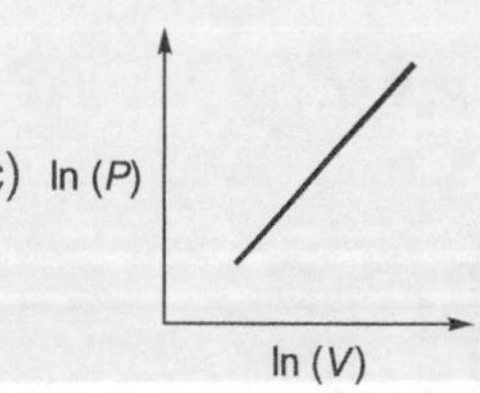

(d) 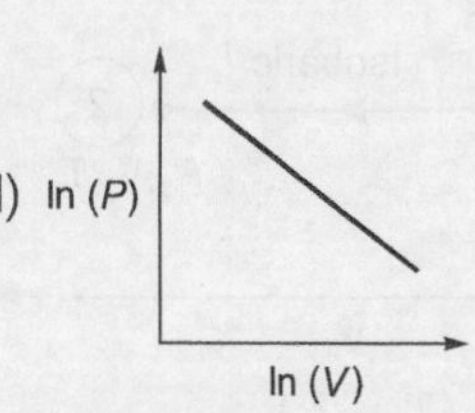

[2014 : 1 M]

3.16 Consider a rigid, perfectly insulated, container partitioned into two unequal parts by a thin membrane (see figure). One part contains one mole of an ideal gas at pressure P_i and temperature T_i while the other part is evacuated. The membrane ruptures, the gas fills the entire volume and the equilibrium pressure is $P_f = \frac{P_i}{4}$. If C_p (molar specific heat capacity at constant pressure), C_V (molar specific heat capacity at constant volume) and R (universal gas constant) have the same units as molar entropy, the change in molar entropy $(S_f - S_i)$ is

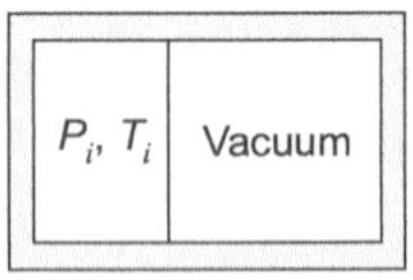

(a) $C_p \ln 2 + R \ln 4$
(b) $-C_v \ln 2 + R \ln 4$
(c) $R \ln 4$
(d) $C_p \ln 2$

[2019 : 1 M]

3.17 N moles of an ideal gas undergo a two-step process as shown in the figure. Let P, V and T denote the pressure, volume and temperature of the gas, respectively. The gas, initially at state-1 (P_1, V_1, T_1), undergoes an isochoric (constant volume) process to reach state-A, and then undergoes an isobaric (constant pressure) expansion to reach state-2 (P_2, V_2, T_2). For an ideal gas, $C_P - C_V = NR$, where C_P and C_V are the heat capacities at constant pressure and constant volume, respectively, and assumed to be temperature independent. The heat gained by the gas in the two-step process is given by

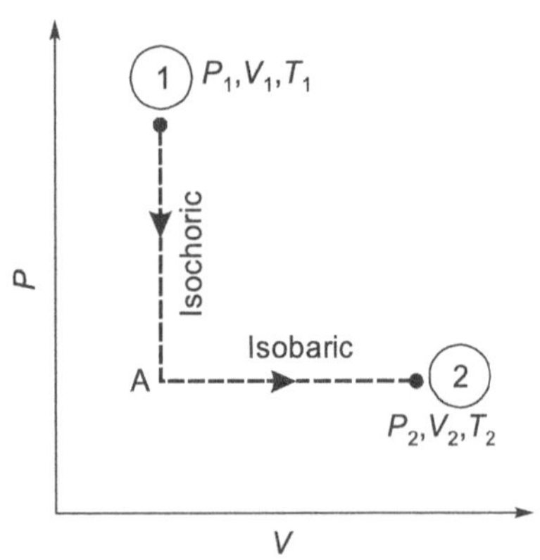

(a) $P_2(V_2 - V_1) + C_V(T_2 - T_1)$
(b) $P_2(V_2 - V_1) + C_P(T_2 - T_1)$
(c) $C_P(T_2 - T_1) + C_V(T_2 - T_1)$
(d) $P_2V_2 - P_1V_1$

[2022 : 2 M]

3.18 An isolated system consists of two perfectly sealed cuboidal compartments A and B separated by a movable rigid wall of cross-sectional area 0.1 m^2 as shown in the figure. Initially, the movable wall is held in place by latches L_1 and L_2 such that the volume of compartment A is 0.1 m^3. Compartment A contains a monoatomic ideal gas at 5 bar and 400 K. Compartment B is perfectly evacuated and contains a massless Hookean spring of force constant 0.3 N m^{-1} at its equilibrium length (stored elastic energy is zero). The latches L_1 and L_2 are released, the wall moves to the right by 0.2 m, where it is held at the new position by latches L_3 and L_4. Assume all the walls and latches are massless. The final equilibrium temperature, in K, of the gas in compartment A, rounded off to 1 decimal place, is ________.

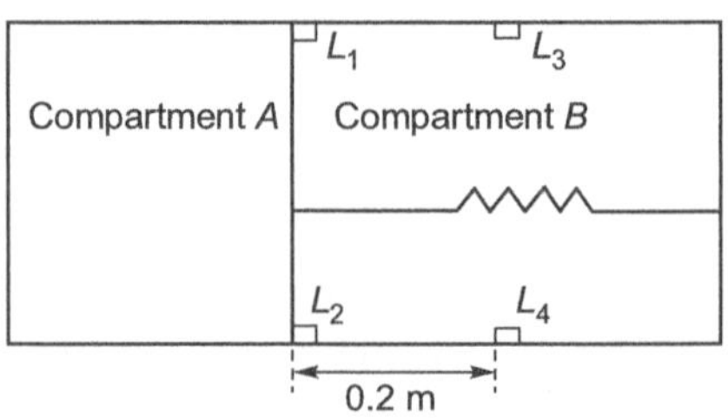

[2024 : 2 M]

■■■■

Answers First Law of Thermodynamics

3.1	(a)	3.2	(a)	3.3	(b)	3.4	(b)	3.5	(b)	3.6	(b)	3.7	(d)
3.8	(b)	3.9	(d)	3.10	(b)	3.11	(d)	3.12	(c)	3.13	(c)	3.14	(b)
3.15	(d)	3.16	(c)	3.17	(a)	3.18	(396.8)						

Explanations First Law of Thermodynamics

3.1 (a)

Given;

$$\left(\frac{P_2}{P_1}\right) = 3$$

$$\gamma = \frac{7}{5}$$

$$\therefore \quad TP^{\frac{1-\gamma}{\gamma}} = \text{Constant}; \; T_1 = 300 \text{ K}$$

$$\therefore \quad T_2 (P_2)^{\frac{1-\gamma}{\gamma}} = T_1 P_1^{\frac{1-\gamma}{\gamma}}$$

$$\left(\frac{T_2}{T_1}\right) = \left(\frac{P_1}{P_2}\right)^{\frac{1-\gamma}{\gamma}} = \left(\frac{1}{3}\right)^{-\frac{2}{7}}$$

$$T_2 = T_1 (3)^{2/7} = 300\,(3)^{2/7}$$

3.2 (a)

$$\bar{h}_i^{ig} = \left[\frac{\partial(nh)}{\partial n_i}\right]_{T,P,n_j}^{ig}$$

For an ideal gas, we know

$$u = u(T)$$

$$\therefore \quad h_i = h_i(T) + PV = h_i(T) + RT$$

$$h_i = f(T)$$

i.e. molar enthalpy of an ideal gas is a function of T only.

3.3 (b)

$$\therefore \quad \left(\frac{T_2}{T_1}\right) = \left(\frac{V_1}{V_2}\right)^{\gamma-1}$$

Data given:

$$T_1 = 500 \text{ K}, \quad V_1 = 1 \text{ m}^3$$

$$V_2 = 5 \text{ m}^3, \quad P_1 = 1000 \text{ Pa}$$

$$T_2 = ?$$

$$\gamma = \left(\frac{C_P}{C_V}\right) = \frac{C_P}{C_P - R}$$

$$= \frac{21}{21 - 8.314} = 1.665$$

$$\therefore \quad T_2 = 500\left[\frac{1}{5}\right]^{1.665-1} = 174.13 \text{ K}$$

3.4 (b)

Work done in an isothermal process given by

$$w = nRT\ln\left(\frac{P_1}{P_2}\right)$$

$$w = 1 \times 8.314 \times 600 \ln\left(\frac{8}{1}\right)$$

$$= 10373 \text{ Joules}$$

3.5 (b)

Given condition is the case of free expansion as we know that in free expansion initial and final temperature is always same.

3.6 (b)

A car tyre of volume 0.057 m^3 is inflated to 300 kPa at 300 K.

Pressure increases to 330 kPa

Change in internal energy is given by

$$\Delta U = C_V \Delta T$$

$$\Delta U = C_V (T_2 - T_1)$$

For ideal gas

$$\frac{P_1}{P_2} = \frac{T_1}{T_2}$$

$$\frac{300}{330} = \frac{300}{T_2}$$

$$T_2 = 330 \text{ K}$$

So, $\Delta U = 21\,(330 - 300) = 630$ J/mol

3.7 (d)

Gas obeys $P(v - b) = RT$

Work obtained for reversible isothermal expansion, from volume V_i to V_f is given by

$$w = \int_{V_i}^{V_f} P\,dV = RT\int_{V_i}^{V_f} \frac{dv}{v-b}$$

$$w = RT\ln\left(\frac{V_f - b}{V_i - b}\right)$$

3.8 (b)

A frictionless cylinder piston assembly contains an ideal gas as shown in figure below:

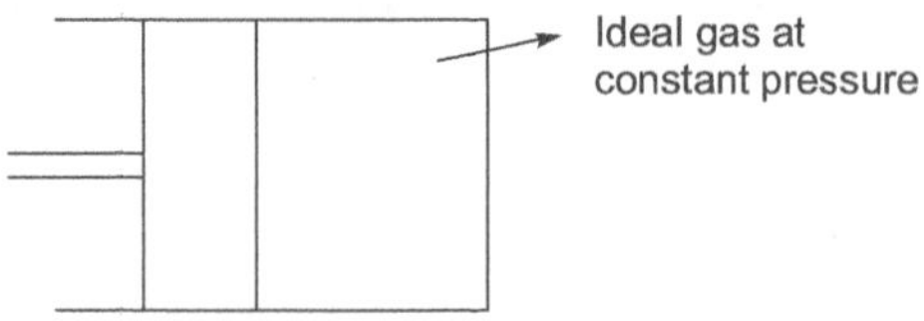

$$P_1 = 100 \text{ kPa}, \; T_1 = 500 \text{ K}$$
$$V_1 = 700 \times 10^{-6} \text{ m}^3$$

System is supplied 100 J of heat at constant P enthalpy variation is given by

$$h\left(\frac{\text{Joule}}{\text{mol}}\right) = 3000 + 50T \quad \text{...(i)}$$

From ideal gas equation

$$PV = nRT$$

$$n = \frac{100000 \times 700 \times 10^{-6}}{8.314 \times 500} = 0.01684$$

From equation (i)

$$\Delta h = n \times 50 \Delta T$$

$$\frac{100}{0.01684} = 50 \times \Delta T$$

$$\Delta T = 118.764 \text{ K}$$

$$T_2 = 118.764 + 500 = 618.76 \text{ K}$$

$$\frac{V_2}{V_1} = \frac{T_2}{T_1} \quad \text{(at constant } P\text{)}$$

$$V_2 = \left(\frac{618.76}{500}\right) \times 700 \times 10^{-6}$$

$$= 866.27 \times 10^{-6} \text{ m}^3$$

3.9 (d)

$$\Delta H = \Delta U + P\Delta V$$
$$100 = \Delta U + 100 \times 10^3 (866.27 - 700) \times 10^{-6}$$
$$\Delta U = 83.37 \text{ Joule}$$

3.10 (b)

For an ideal gas change in enthalpy is given by following relation:

$$\Delta H = C_p \Delta T = C_p (T_2 - T_1)$$

3.11 (d)

After the stopper is removed, the piston moves freely to the end.

$$du = dq - dw$$

For adiabatic, $dq = 0$

For free expansion, $dw = 0$

u = constant (before and after expansion)

$$u = C_v dT \text{ and } du = 0$$
$$dT = 0, \; T \text{ remains constant}$$
$$\text{Final temp.} = 100°\text{C}$$

3.12 (c)

The insulation is now removed.

$$w_{min} = P_2 V_2 \ln\left(\frac{V_2}{V_1}\right)$$

In this case, $V_1 = 0.6 \text{ m}^3$

and $V_2 = 0.2 \text{ m}^3$ at $P_2 = 6$ bar

$$w_{min} = 6 \times 100 \times 0.2 \ln\left(\frac{0.2}{0.6}\right)$$

$$w_{min} = -131.83 \text{ kJ}$$

$\simeq -132$ kJ (–ve sign indicates work done on this system)

3.13 (c)

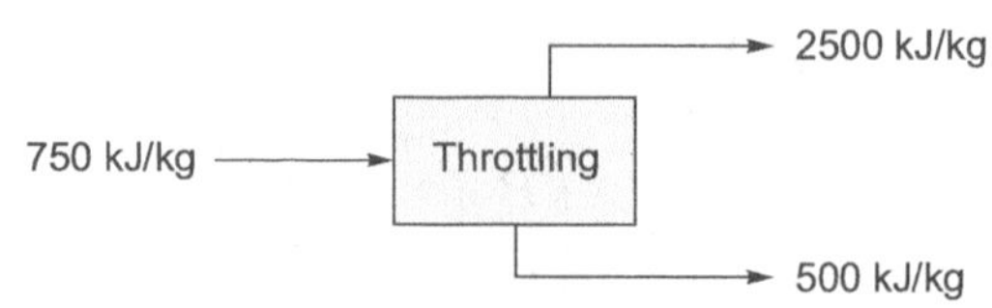

Assume inlet flow rate of stream is 1 kJ/sec

Fraction that vaporizes is x

Now as we know that throttling is constant enthalpy process so,

Enthalpy at inlet = Enthalpy at outlet

$$750 \times 1 = x \times 2500 + (1 - x) \times 500$$
$$x = 0.125$$

So percentage of original liquid that vaporizes is 12.5%.

3.14 (b)

One mole of methane undergoes isothermal expansion from 8 bar and 1000 K to 4 bar.

$$R = 8.314 \text{ Joule / mole K}$$

From first law of thermodynamics

$$dQ = dw + du$$

Here heat added to system is positive work done by the system is positive.

Since isothermal expansion is occurring so $du = 0$

$$dQ = dw$$

Work done for isothermal process is given as

$$w = nRT \ln\left(\frac{P_1}{P_2}\right)$$

$$w = 1 \times 8.314 \times 1000 \ln\left(\frac{8}{4}\right)$$

$$Q = w = 5.76 \text{ kJ}$$

3.15 (d)

Isentropic expansion of ideal gas means reversible adiabatic process.

For adiabatic process PV^{γ} = constant

$$P = V^{-\gamma}$$

Taking log both side

$$\ln P = -\gamma \ln V$$

So graph between $\ln P$ and $\ln V$ will have a negative slope.

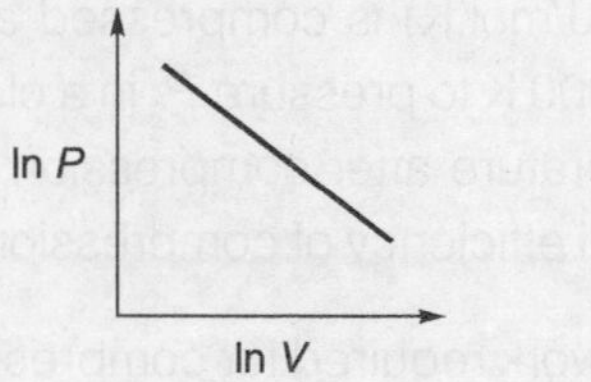

3.16 (c)

Change in molar entropy $(S_f - S_i)$ is given by following formula

$$\Delta S = nC_p \ln\left(\frac{T_f}{T_i}\right) - nR \ln\left(\frac{P_f}{P_i}\right)$$

$$\frac{P_f}{P_i} = \frac{1}{4} \text{ (given)}$$

$$\frac{T_f}{T_i} = 1$$

(Since it is a gas of free expansion so $T_f = T_i$)

Now, $$\Delta S = -nR \ln\left(\frac{1}{4}\right)$$

$$n = 1 \text{ (given)}$$

$$\Delta S = R \ln (4)$$

3.17 (a)

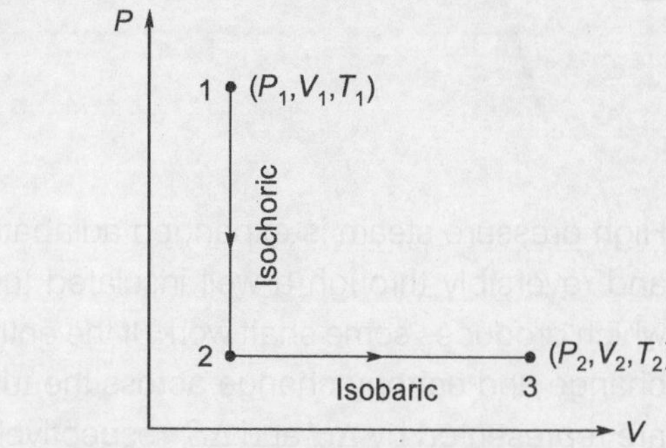

Assume at point '2' temperature is T.

In process 1 to 2 volume is constant so

$$du = dQ - dw$$

$$du_1 = dQ_1$$

$$du_1 = C_v(T - T_1)$$

In process 2-3 pressure is constant so

$$du_2 = dQ_2 - dw$$

Here, $$dw = P_2(V_2 - V_1)$$

$$du_2 = C_v(T_2 - T)$$

So, heat gained by process

$$dQ_1 + dQ_2 = du_1 + du_2 + dw$$

$$= P_2(V_2 - V_1) + C_v(T_2 - T) + C_v(T - T_1)$$

$$P_2(V_2 - V_1) + C_v(T_2 - T_1)$$

Correct option is (a).

■■■■

4 Open System Analysis by First Law

4.1 High pressure steam is expanded adiabatically and reversibly through a well insulated turbine which produces some shaft work. If the enthalpy change and entropy change across the turbine are represented by ΔH and ΔS, respectively, for this process :

(a) $\Delta H = 0$ and $\Delta S = 0$
(b) $\Delta H \neq 0$ and $\Delta S = 0$
(c) $\Delta H \neq 0$ and $\Delta S \neq 0$
(d) $\Delta H = 0$ and $\Delta S \neq 0$

[2001 : 1 M]

4.2 A steam turbine operates with a superheated steam flowing at 1 kg s^{-1}. This steam is supplied at 41 bar and 500°C, and discharges at 1.01325 bar and 100°C

Data: 41 bar, 500°C
Enthalpy: 3443.9 kJ kg^{-1}
Entropy: 7.0785 kJ $kg^{-1}K^{-1}$
41 bar, 251.8 °C
Enthalpy of saturated steam: 2799.9 kJ kg^{-1}
Entropy of saturated steam: 6.0583 kJ kg^{-1} K^{-1}
1.01325 bar 100°C
Enthalpy of saturated vapour: 2676 kJ kg^{-1}
Enthalpy of saturated liquid: 419.1 kJ kg^{-1}
Entropy of saturated vapour: 7.3554 kJ kg^{-1} K^{-1}
Entropy of saturated liquid: 1.3069 kJ kg^{-1} K^{-1}

The maximum power output (in kW) will be

(a) 644.0 (b) 767.9
(c) 871.3 (d) 3024.8

[2003 : 2 M]

4.3 What is the actual power required to drive a reciprocating air compressor which has to compress 34 m^3 of air per minute from 1.013 × 10^5 N/m^2 to 4.052 × 10^5 N/m^2? Assume that $PV^{1.25}$ is constant, where P is the pressure and V is the volume, and the efficiency of the compressor is 85%.

(a) 107.9 kW (b) 200 kW
(c) 82.6 kW (d) 91.7 kW

[2005 : 2 M]

Common Data for Questions (4.4 & 4.5):

An ideal gas with molar heat capacity $C_p = \frac{5}{2}R$ (where R = 8.314 J/mol.K) is compressed adiabatically from 1 bar and 300 K to pressure P_2 in a closed system. The final temperature after compression is 600 K and the mechanical efficiency of compression is 50%.

4.4 The work required for compression (in kJ/mol) is

(a) 3.74 (b) 6.24
(c) 7.48 (d) 12.48

[2009 : 2 M]

4.5 The final pressure P_2 (in bar) is

(a) $2^{3/4}$ (b) $2^{5/4}$
(c) $2^{3/2}$ (d) $2^{5/2}$ [2009 : 2 M]

4.6 Two systems are available for compressing 6 m^3/hr of ambient air to 10 bar. The first one uses a single stage compressor (K1) and the second one uses a multistage compressor with inter-stage cooling (K2). Which one of the following statements is incorrect?

(a) K2 will have knock out pots in between the stages.
(b) Discharge temperature of K1 will be higher than that of K2.
(c) K2 will consume more power than K1.
(d) Cost of K2 will be more than that of K1.

[2011 : 2 M]

4.7 An insulated, evacuated container is connected to a supply line of an ideal gas at pressure P_s, temperature T_s and specific volume v_s. The container is filled with the gas until the pressure in the container reaches P_s. There is no heat transfer between the supply line to the container, and kinetic and potential energies are negligible. If C_P and C_V are the heat capacities at constant pressure and constant volume, respectively $\left(\gamma = \frac{C_P}{C_V}\right)$, then the final temperature of the gas in the container is

(a) γT_s (b) T_s

(c) $(\gamma - 1)T_s$ (d) $\frac{(\gamma - 1)T_s}{\gamma}$

[2012 : 2 M]

4.8 An ideal gas is adiabatically and irreversibly compressed from 3 bar and 300 K to 6 bar in a closed system. The work required for the irreversible compression is 1.5 times the work that is required for reversible compression from the same initial temperature and pressure to the same final pressure. The molar heat capacity of the gas at constant volume is 30 J mol^{-1} K^{-1} (assumed to be independent of temperature); universal gas constant, R is 8.314 J mol^{-1} K^{-1}; ratio of molar heat capacities is 1.277. The temperature (in K, rounded off to the first decimal place) of the gas at the final state in the irreversible compression case is ________.

[2016 : 2 M]

4.9 A tank initially contains a gas mixture with 21 mol % oxygen and 79 mol % nitrogen. Pure nitrogen enters the tank, and a gas mixture of nitrogen and oxygen exits the tank. The molar flow rate of both the inlet and exit streams is 8 mol s^{-1}.

In addition, use the following data and assumptions:

- Assume the tank contents to be well mixed
- Assume ideal gas behaviour
- The temperature and pressure inside the tank are held constant
- Molar density of the gas mixture in the tank is constant at 40 mol m^{-3}.

If the volume of the tank is 20 m^3, then the time (in seconds) required for oxygen content in the tank to decrease to 1 mol % is

(a) 100.45 (b) 304.45

(c) 3.445 (d) 10

[2020 : 2 M]

■■■■

Answers Open System Analysis by First Law

4.1	(b)	**4.2**	(c)	**4.3**	(a)	**4.4**	(c)	**4.5**	(d)	**4.6**	(c)	**4.7**	(a)
4.8	(373)	**4.9**	(b)										

Explanations Open System Analysis by First Law

4.1 (b)

Given adiabatic and reversible ⇒ Isentropic ($\Delta S = 0$)

$$\dot{m}(h_{out} - h_{in}) = \dot{w}_s$$

⇒ Change in enthalpy $\Delta H \neq 0$

4.2 (c)

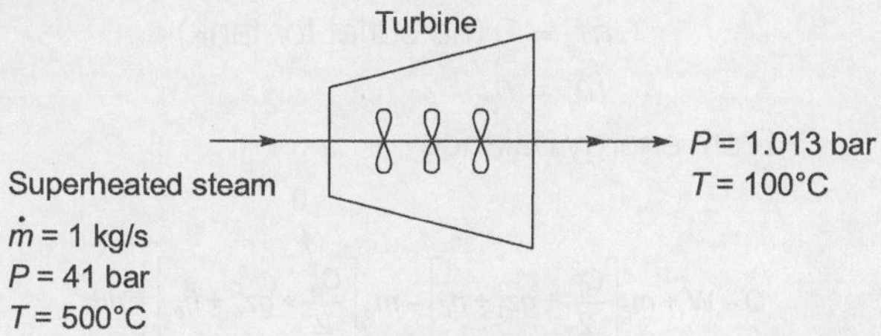

Max power output from turbine ⇒ Under isentropic condition

$\because \quad s_{in} = s_{out}$

From energy balance

$$W_s = -\dot{m}[\Delta h]$$

Thus, for quality of exit steam, we have to use entropy terms

$$x = \frac{s - s_f}{s_g - s_f}$$

$$= \frac{7.0785 - 1.3069}{7.3554 - 1.3069} = 0.954$$

Now,

For $\quad h_{out} = (1 - 0.954)\, 419.1 + 0.954 \times 2676$

$h_{out} = 2572.183$ kJ/kg

$$\therefore \quad W_s = -\dot{m}[\Delta h] = -1\,[3443.9 - 2572.183]$$

$$W_s = -871.717 \text{ kW}$$

4.3 (a)

Given reciprocating air compressor follows the relation $PV^{1.25}$ = constant

This shows an adiabatic compression. As we know for adiabatic compression work can be found by following relation:

$$W_{\text{ideal}} = \frac{\gamma \cdot P_1V_1\left[1 - \left(\frac{P_2}{P_1}\right)^{\frac{\gamma-1}{\gamma}}\right]}{(\gamma - 1)}$$

$$W_{\text{ideal}} = \frac{1.25 \times 1.013 \times 10^5 \times 34\left[1 - \left(\frac{1.052 \times 10^5}{1.013 \times 10^5}\right)^{\frac{1.25-1}{1.25}}\right]}{(1.25 - 1)}$$

$W_{\text{ideal}} = -55.022 \times 10^5$ J/minute $= -0.917 \times 10^5$ J/s

–ve sign indicates that work is done on system.

$$W_{\text{actual}} = \frac{W_{\text{ideal}}}{\text{Compressor efficiency}}$$

$$= \frac{0.917 \times 10^5}{0.85}$$

$$= 1.0788 \times 10^5 \text{ watt} = 107.88 \text{ kW}$$

4.4 (c)

Ideal gas $\left(C_p = \frac{5}{2}R\right)$ is compressed adiabatically from 1 bar 300 K to 600 K.

Mechanical efficiency $\eta_m = 0.5$ (given)

Ideal work done in an adiabatic compression process is given by

$$W_{\text{ideal}} = \frac{nR(T_1 - T_2)}{(\gamma_1 - 1)}$$

$$\gamma = \frac{C_p}{C_v} = \frac{C_p}{C_p - R} = \frac{5}{3}$$

$$W_{\text{ideal}} = \frac{n \times 8.314(300 - 600)}{\left(\frac{5}{3} - 1\right)}$$

$$\left(\frac{W_{\text{ideal}}}{n}\right) = 3.741 \text{ kJ/mol}$$

$$W_{\text{actual}} = \frac{W_{\text{ideal}}}{\eta_m} = \frac{3.471}{0.5} = 7.48 \text{ kJ/mol}$$

4.5 (d)

For adiabatic compression

$$TP^{\frac{\gamma}{1-\gamma}} = \text{Constant}$$

$$\frac{P_2}{P_1} = \left(\frac{T_1}{T_2}\right)^{\frac{\gamma}{1-\gamma}}$$

$$P_2 = 1 \times \left(\frac{300}{600}\right)^{\frac{5/3}{-2/3}} = 2^{5/2}$$

4.6 (c)

K_1 is single stage compression and K_2 is multistage compressor with inter stage cooling: Following conclusion can be made for given condition

(a) K_2 will have knockout pots in between stages since it has inter stage cooling facility.

(b) Discharge temperature of K_1 will be higher than that of K_2.

(c) K_1 will consume more power than K_2 since multistage compression leads to less consumption of power.

(d) Multistage compression unit is costly than the single stage compression unit.

4.7 (a)

An insulated evacuated container is connected to supply line of an ideal gas at temperature T_s and pressure P_s.

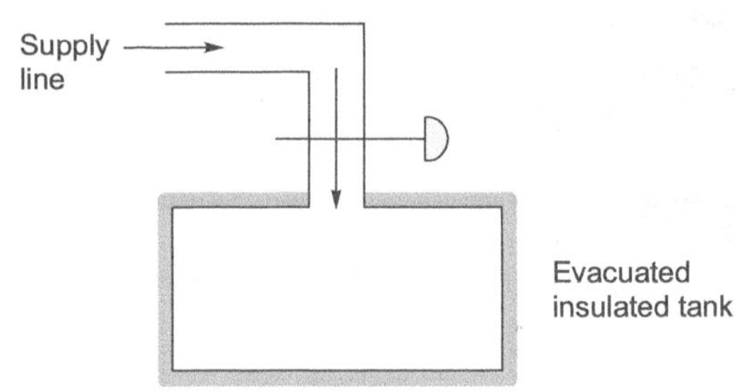

Mass balance for above system

$$m_{\text{in}} - m_{\text{out}} = (\Delta m)_{\text{system}}$$

$$m_i - m_e = (m_2 - m_1)_{\text{C.V.}}$$

$m_1 = 0$ (Initially evacuated tank is used)

$m_e = 0$ (no outlet for tank)

$m_i = m_2$ ...(i)

From energy balance

$$Q - W + m_i\left[\frac{C_i^2}{2} + gz_i + h_i\right] - m_e\left[\frac{C_e^2}{2} + gz_e + h_e\right]^{0} = du$$

$$Q = 0 \text{ (Insulated tank)}$$

$$W = 0 \text{ (Evacuated tank)}$$

Potential and kinetic energy is negligible so

$$gz_1 \text{ and } \frac{C_i^2}{2} = 0$$

So, $$m_i h_i = m_2 u_2 - \overset{0}{\cancel{m_1}} u_1$$

From equation (i) $h_i = u_2$

$$\Rightarrow \quad C_p T_s = C_v T_2$$

So, $$T_2 = \frac{C_p}{C_v} T_s$$

$$\Rightarrow \quad T_2 = \gamma T_s$$

4.8 (373)

Ideal gas is adiabatically and irreversibly compressed from 3 bar and 300 K to 6 bar.

Work irreversible = 1.5 times work reversible

$$C_p = 30 \text{ J/mol K}$$

$$R = 8.314 \text{ J/mol K}$$

$$\frac{C_p}{C_v} = \gamma = 1.277$$

Work done in adiabatic reversible process is given by

$$W = \frac{P_1V_1 - P_2V_2}{\gamma - 1}$$

$$W = \frac{P_1V_1}{\gamma - 1}\left[1 - \left(\frac{P_2}{P_1}\right)^{\frac{\gamma-1}{\gamma}}\right]$$

So, $$W_{irr} = \frac{1.5 \times nRT_1\left[1 - \left(\frac{P_2}{P_1}\right)^{\frac{\gamma-1}{\gamma}}\right]}{(\gamma - 1)} \quad \text{...(i)}$$

Now from first law of thermodynamics

$$dQ = du + dw$$

Since process is adiabatic.

$$dQ = 0$$

$$-du = dw \quad \text{...(ii)}$$

Here, $$du = nC_v \Delta T$$

From equation (i) and (ii)

$$\frac{1.5 \times nR \times 300\left[1 - \left(\frac{6}{3}\right)^{\frac{1.277-1}{1.277}}\right]}{(1.277 - 1)} = -n \times 30 \times (T_2 - 300)$$

$$T_2 = 373 \text{ K}$$

4.9 (b)

Tank initially contains 21 mol% oxygen and 79 mol% nitrogen. Pure nitrogen enters the tank and mixture of nitrogen and oxygen exit the tank as shown in figure below.

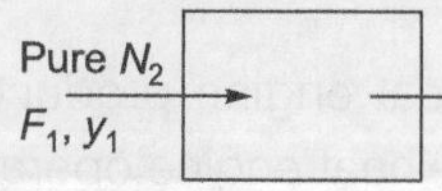

Assuming y_1 is mol fraction of O_2 in entry and y_2 is mol fraction of O_2 in exit respectively.

Molar density of gas mixture in tank (ρ)

$$= 40 \text{ mol/m}^3$$

Volume of tank, $V = 20 \text{ m}^3$

$$y_1 = 0$$

Applying oxygen mass balance in the tank:

$$V\rho \frac{dy}{dt} = F_1y_1 - F_2y_2$$

$$V\rho \frac{dy}{dt} = -F_2y_2$$

$$V\rho \int_{0.21}^{0.01} \frac{dy}{y} = -F_2 \int_0^t dt$$

$$20 \times 40 \times \ln\left(\frac{0.01}{0.21}\right) = -8 \times t$$

$$t = 304.45 \text{ s}$$

■■■■

5 Second Law of Thermodynamics

5.1 The thermal efficiency of a reversible heat engine operating between two given thermal reservoirs is 0.4. The device is used either as a refrigerator or as a heat pump between the same reservoirs. Then the coefficient of performance as a refrigerator $(COP)_R$ and the coefficient of performance as a heat pump $(COP)_{HP}$ are

(a) $(COP)_R = (COP)_{HP} = 0.6$

(b) $(COP)_R = 2.5$; $(COP)_{HP} = 1.5$

(c) $(COP)_R = 1.5$; $(COP)_{HP} = 2.5$

(d) $(COP)_R = (COP)_{HP} = 2.5$

[2000 : 2 M]

5.2 The following heat engine produces power of 100,000 kW. The heat engine operates between 800 K and 300 K. It has a thermal efficiency equal to 50% of that of the Carnot engine for the same temperatures. The rate at which heat is absorbed from the hot reservoir is

(a) 100,000 kW

(b) 160,000 kW

(c) 200,000 kW

(d) 320,000 kW

[2003 : 2 M]

5.3 A cyclic engine exchanges heat with two reservoirs maintained at 100 and 300 °C, respectively. The maximum work (in J) that can be obtained from 1000 J of heat extracted from the hot reservoir is

(a) 349 (b) 651

(c) 667 (d) 1000

[2004 : 2 M]

5.4 Which one of the following statements is TRUE?

(a) Heat can be fully converted into work.

(b) Work cannot be fully converted into heat.

(c) The efficiency of a heat engine increases as the temperature of the heat source is increased while keeping the temperature of the heat sink fixed.

(d) A cyclic process can be devised whose sole effect is to transfer heat from a lower temperature to a higher temperature.

[2005 : 1 M]

5.5 A Carnot heat engine cycle is working with an ideal gas. The work performed by the gas during the adiabatic expansion and compression steps, W_1 and W_2 respectively, are related as

(a) $|W_1| > |W_2|$ (b) $|W_1| < |W_2|$

(c) $W_1 = W_2$ (d) $W_1 = -W_2$

[2005 : 2 M]

5.6 A heat engine operates at 75% of the maximum possible efficiency. The ratio of the heat source temperature (in K) to the heat sink temperature (in K) is $\frac{5}{3}$. The fraction of the heat supplied that is converted in to work is

(a) 0.2 (b) 0.3

(c) 0.4 (d) 0.6

[2006 : 1 M]

5.7 Which of the following represents the Carnot cycle (ideal engine)?

(a) 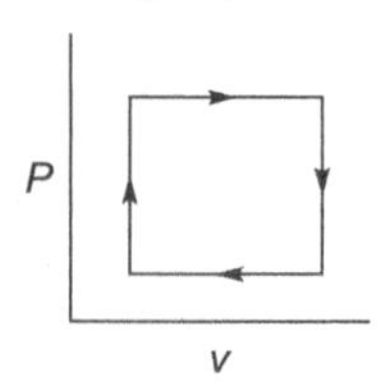

(b) 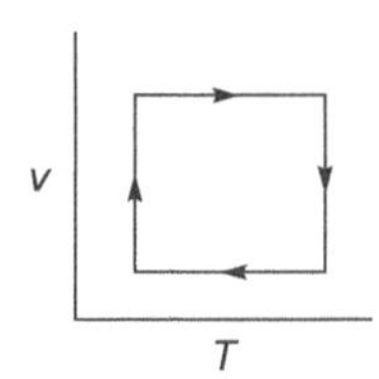

(c) 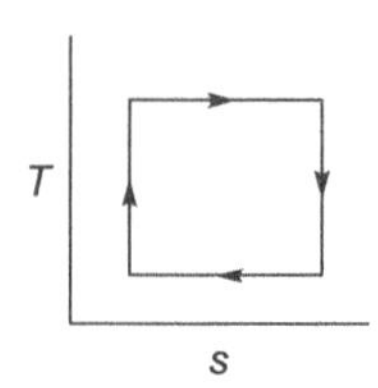

(d) 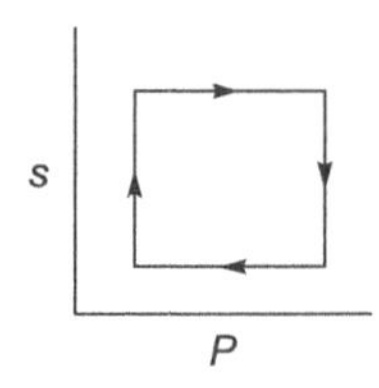

[2007 : 2 M]

5.8 For a Carnot refrigerator operating between 40°C and 25°C, the coefficient of performance is

(a) 1 (b) 1.67

(c) 19.88 (d) 39.74

[2008 : 1 M]

5.9 Consider two Carnot engines C_1 and C_2 as shown in the figure.

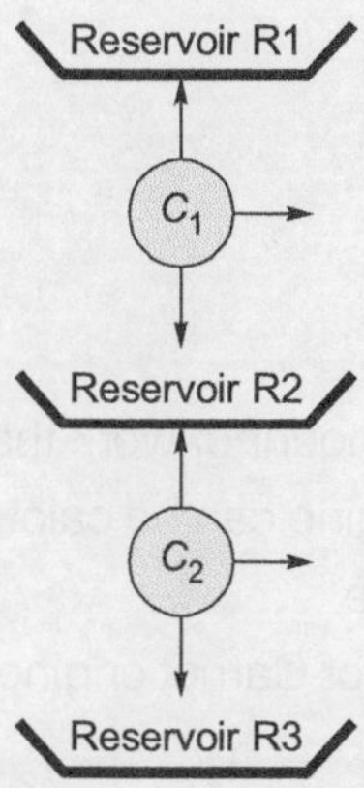

The efficiencies of the engines C_1 and C_2 are 0.40 and 0.35, respectively. If the temperature of Reservoir **R1** is 800 K, then the temperature (in K) of Reservoir **R3** is _____ (round off to nearest integer). **[2020 : 1 M]**

5.10 John and Jane independently performed a thermodynamic experiment, in which X and Y represent the initial and final thermodynamic states of the system, respectively. John performed the experiment under reversible conditions, for which the change in entropy of the system was ΔS_{rev}. Jane performed the experiment under irreversible conditions, for which the change in entropy of the system was ΔS_{irr}. Which one of the following relationships is CORRECT?.

(a) $\Delta S_{rev} = \Delta S_{irr}$ (b) $\Delta S_{rev} > \Delta S_{irr}$
(c) $\Delta S_{rev} < \Delta S_{irr}$ (d) $\Delta S_{rev} = 2\Delta S_{irr}$

[2023 : 1 M]

5.11 Consider a system where a Carnot engine is operating between a source and a sink. Which of the following statements about this system is/are NOT correct?

(a) This engine is reversible.
(b) The engine efficiency is independent of the source and sink temperatures.
(c) This engine has the highest efficiency among all engines that operate between the same source and sink.
(d) The total entropy of this system increases at the completion of each cycle of the engine.

[2023 : 1 M]

5.12 Heat is available at a rate of 2 kW from a thermal reservoir at 400 K. A two-stage process harnesses this heat to produce power. Stages 1 and 2 reject heat at 360 K and 300 K, respectively. Stage 2 is driven by the heat rejected by Stage 1. If the overall process efficiency is 50% of the corresponding Carnot efficiency, the power delivered by the process, in kW, rounded off to 2 decimal places, is _____.

[2024 : 2 M]

■■■■

Answers Second Law of Thermodynamics

5.1	(c)	**5.2**	(d)	**5.3**	(a)	**5.4**	(c)	**5.5**	(d)	**5.6**	(b)	**5.7**	(c)
5.8	(c)	**5.9**	(312)	**5.10**	(a)	**5.11**	(b, d)	**5.12**	(0.25)				

Explanations Second Law of Thermodynamics

5.1 (c)

Assuming T_H is temperature of heat source and T_C is temperature of sink.

Efficiency of heat engine,

$$\eta = 1-\frac{T_C}{T_H}$$

$$0.4 = 1-\frac{T_C}{T_H}$$

$$\frac{T_C}{T_H} = 0.6$$

or $$T_C = 0.6T_H$$

Now, C.O.P. of heat pump

$$= \frac{T_H}{T_H - T_C} = \frac{T_H}{T_H - 0.6T_H} = \frac{1}{0.4} = 2.5$$

C.O.P. of refrigeration

$$= \frac{T_C}{T_H - T_C} = \frac{0.6T_H}{T_H - 0.6T_H} = \frac{0.6}{0.4} = 1.5$$

5.2 (d)

Efficiency of Carnot engine is given by following equation

$$\eta_c = \frac{W}{Q_1} = 1-\frac{T_2}{T_1}$$

Here, W = Work produced

Q_1 = Heat absorbed by engine from source

T_2 = Sink temperature

T_1 = Source temperature

$$\eta_c = 1-\frac{300}{800} = 0.625$$

Thermal efficiency of given engine is equal to 50% of Carnot engine

$$\eta = 0.5 \times \eta_c$$

$$\eta = 0.625 \times 0.5 = 0.3125$$

$$\frac{W}{Q_1} = 0.3125$$

$$Q_1 = \frac{100000}{0.3125} = 320000 \text{ kW}$$

5.3 (a)

Maximum amount of work that can be obtained by cyclic engine can be calculated by assuming Carnot engine.

As we know for Carnot engine:

$$\eta = \frac{W}{Q_1} = 1-\frac{T_2}{T_1}$$

$$= \frac{W}{1000} = \left(1-\frac{373}{573}\right)$$

$$W = 349 \text{ Joule}$$

5.4 (c)

(i) Heat can not be fully converted to work.

(ii) Work can be fully converted to heat.

(iii) Efficiency of heat engine can be defined as

$$\eta = 1-\frac{T_2}{T_1}$$

T_2 = Sink temperature

T_1 = Source temperature

So efficiency can be increased by increasing heat source temperature.

(iv) Option (d) violates 2nd law of thermodynamics so it cannot be true.

5.5 (d)

Carnot cycle is shown as below:

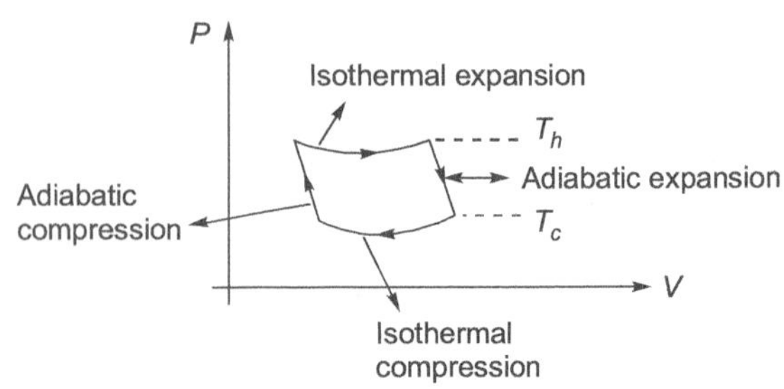

For adiabatic expansion

$$w_1 = \frac{nR(T_h - T_c)}{\gamma - 1}$$

For adiabatic compression

$$w_2 = \frac{nR(T_c - T_h)}{\gamma - 1}$$

So on comparing w_1 and w_2 we can conclude that

$$w_1 = -w_2$$

5.6 (b)

Efficiency of Carnot engine is given by

$$\eta_c = 1 - \frac{T_2}{T_1}$$

$$\eta_c = 1 - \frac{3}{5} = 0.4$$

Efficiency of given heat engine is 75% of Carnot engine so $\eta = 0.75 \times \eta_c$

$$\eta = \frac{W}{Q_1} = 0.75 \times 0.4 = 0.3$$

5.7 (c)

Carnot cycle represents two isentropic (constant entropy) and two isothermal process (constant temperature).

So option (c) is correct option.

5.8 (c)

Coefficient of refrigerator is given as $\frac{T_2}{T_1 - T_2}$

T_2 = Sink temperature

T_1 = Source temperature

$$\text{C.O.P.} = \frac{25 + 273}{40 - 25} = 19.87$$

5.9 (312)

Efficiencies of engine C_1 and C_2 is given as 0.40 and 0.35 respectively. Temperature of R_1 is 800 K.

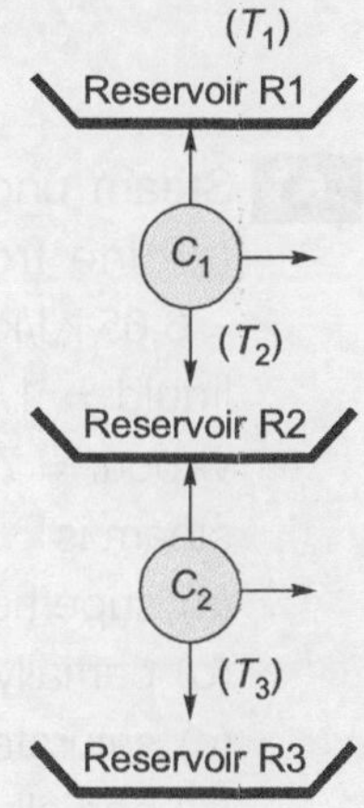

$$\eta_1 = 1 - \frac{T_2}{T_1} = 0.4$$

$$1 - \frac{T_2}{800} = 0.4$$

$$T_2 = 480 \text{ K}$$

Similarly, $\eta_2 = 1 - \frac{T_3}{T_2} = 0.35$

$$1 - \frac{T_3}{480} = 0.35$$

$$T_3 = 312 \text{ K}$$

5.10 (a)

Since entropy is a point function so that is why, it only depends on the initial and final state either performed by reversible path or irreversible path. Therefore the change in entropy in both the cases will be same.

5.12 (0.25)

Overall efficiency is 50% of corresponding Carnot efficiency.

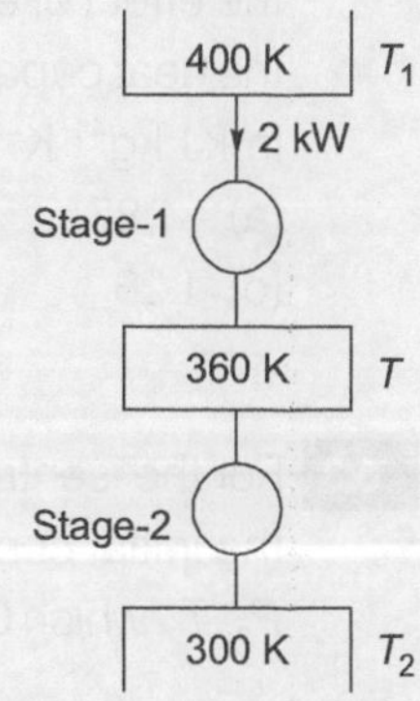

$$\text{Carnot efficiency} = 1 - \frac{T_2}{T_1}$$

$$= 1 - \frac{300}{400} = \frac{1}{4}$$

Overall process efficiency

$$= \frac{1}{4} \times \frac{1}{2} = \frac{1}{8}$$

Power delivered = Efficiency × Heat supplied

$$= 2 \times \frac{1}{8} = 0.25 \text{ kW}$$

■■■■

6 Entropy

6.1 Steam undergoes isentropic expansion in a turbine from 5000 kPa and 400°C (entropy = 6.65 kJ/kg K) to 150 kPa (entropy of saturated liquid = 1.4336 kJ/kg K, entropy of saturated vapour = 7.2234 kJ/kg K). The exit condition of steam is
(a) superheated vapour
(b) partially condensed vapour with quality of 0.9
(c) saturated vapour
(d) partially condensed vapour with quality of 0.1

[2002 : 2 M]

6.2 A solid metallic block weighing 5 kg has an initial temperature of 500°C; 40 kg of water initially at 25°C is contained in a perfectly insulated tank. The metallic block is brought into contact with water. Both of them come to equilibrium. Specific heat of block material is 0.4 kJ kg^{-1} K^{-1}. Ignoring the effect of expansion and contraction, and also the heat capacity of tank, the total entropy change in kJ kg^{-1} K^{-1} is
(a) –1.87 (b) 0.0
(c) 1.26 (d) 3.91

[2003 : 2 M]

6.3 For the isentropic expansion of an ideal gas from the initial conditions P_1, T_1 to the final conditions P_2, T_2, which ONE of the following relations is valid?

Where $\gamma = \frac{C_p}{C_v}$

(a) $\left(\frac{P_1}{P_2}\right) = \left(\frac{T_2}{T_1}\right)^{\gamma}$ (b) $\left(\frac{P_1}{P_2}\right) = \left(\frac{T_1}{T_2}\right)^{\frac{\gamma}{\gamma-1}}$

(c) $\left(\frac{P_1}{P_2}\right) = \left(\frac{T_1}{T_2}\right)$ (d) $\left(\frac{P_1}{P_2}\right) = \left(\frac{T_1}{T_2}\right)^{\frac{\gamma-1}{\gamma}}$

[2006 : 1 M]

6.4 The change in entropy of the system, ΔS_{sys}, undergoing a cyclic irreversible process is
(a) greater than 0 (b) equal to zero
(c) less than zero
(d) equal to the $\Delta S_{surroundings}$

[2007 : 1 M]

6.5 2 kg of steam in a piston-cylinder device at 400 kPa and 175°C undergoes a mechanically reversible, isothermal compression to a final pressure such that the steam becomes just saturated. What is the work, *W*, required for the process. Data :
T = 175°C, P = 400 kPa – v = 0.503 m^3/kg, u = 2606 kJ/kg, s = 7.055 kJ/kg-K
T = 175°C, Satd. vapour – v = 0.216 m^3/kg, u = 2579 kJ/kg, s = 6.622 kJ/kg-K
(a) 0 kJ (b) 230 kJ
(c) 334 kJ (d) 388 kJ

[2007 : 2 M]

6.6 If the temperature of saturated water is increased infinitesimally at constant entropy, the resulting state of water will be
(a) Liquid
(b) Liquid – vapor coexistence
(c) Saturated vapor
(d) Solid

[2012 : 1 M]

6.7 In a parallel flow heat exchanger operating under steady state, hot liquid enters at a temperature $T_{h,in}$ and leaves at a temperature $T_{h,out}$. Cold liquid enters at a temperature $T_{c,in}$ and leaves at a temperature $T_{c,out}$. Neglect any heat loss from the heat exchanger to the surrounding. If $T_{h,in} >> T_{c,in}$, then for a given time interval, which ONE of the following statements is true?
(a) Entropy gained by the cold stream is GREATER than entropy lost by the hot stream
(b) Entropy gained by the cold stream is EQUAL to the entropy lost by the hot stream
(c) Entropy gained by the cold stream is LESS than the entropy lost by the hot stream
(d) Entropy gained by the cold stream is ZERO

[2012 : 1 M]

6.8 In a process occurring in a closed system *F*, the heat transferred from *F* to the surroundings *E* is 600 J. If the temperature of *E* is 300 K and that of F is in the range 380 - 400 K, the entropy changes of the surroundings (ΔS_E) and system (ΔS_F), in J/K, are given by

(a) $\Delta S_E = 2, \Delta S_F = -2$ (b) $\Delta S_E = -2, \Delta S_F = 2$

(c) $\Delta S_E = 2, \Delta S_F < -2$ (d) $\Delta S_E = 2, \Delta S_F > -2$

[2013 : 2 M]

6.9 The molar heat capacity at constant pressure C_p (in J mol^{-1} K^{-1}) for *n*-pentane as a function of temperature (*T* in K) is given by

$$\frac{C_p}{R} = 2.46 + 45.4 \times 10^{-3}\, T - 14.1 \times 10^{-6}\, T^2$$

Take $R = 8.314$ J mol^{-1} K^{-1}

At 1000 K, the rate of change of molar entropy of *n*-pentane with respect to temperature at constant pressure is ______ J mol^{-1} K^{-2} (round off to 2 decimal places).

[2021 : 1 M]

6.10 A gaseous mixture at 1 bar and 300 K consists of 20 mol % CO_2 and 80 mol % inert gas. Assume the gases to be ideal.

Take $R = 8.314$ J mol^{-1} K^{-1}.

The magnitude of minimum work required to separate 100 mol of this mixture at 1 bar and 300 K into pure CO_2 and inert gas at the same temperature and pressure is ______ kJ (round off to nearest integer).

[2021 : 2 M]

6.11 5 moles of liquid benzene, 8 moles of liquid toluene and 7 moles of liquid xylene are mixed at 25 °C and 1 bar. Assuming the formation of an ideal solution and using the universal gas constant $R = 8.314$ J mol^{-1} K^{-1}, the total entropy change is ______ J K^{-1} (rounded off to one decimal place).

[2022 : 1 M]

■■■■

Answers Entropy

6.1	(b)	**6.2**	(c)	**6.3**	(b)	**6.4**	(b)	**6.5**	(c)	**6.6**	(a)	**6.7**	(a)
6.8	(d)	**6.9**	(0.28)	**6.10**	(125)	**6.11**	(179.67)						

Explanations Entropy

6.1 (b)

Steam undergoes isentropic expansion i.e. adiabatic and reversible.

$$\therefore \quad s_{in} = s_{out}$$
$$6.65 = s_f + x(s_g - s_f)$$
$$= 1.4336 + x(7.2234 - 1.4336)$$
$$x = 0.90$$

[Mixture of saturated liquid and vapour with quality 0.9].

6.2 (c)

Entropy charge for sensible heating is given by

$$\Delta S = mC_p \ln\left(\frac{T_f}{T_i}\right)$$

For

$$Q = (m\,C_p\,\Delta T)_{water} = (m\,C_p\,\Delta T)_{metal\ block}$$
$$5 \times 0.4\,(500 - T) = 40 \times 4.18 \times (T - 25)$$
$$T = 30.61°C = 303.61\text{ K}$$
$$(\Delta S)_{metal\ block} = 5\times0.4 \ln\left(\frac{303.61}{773}\right)$$
$$= -1.87\text{ kJ/K}$$
$$\Delta S_{water} = 40\times4.18 \ln\left(\frac{303.61}{298}\right)$$
$$= 3.12\text{ kJ/K}$$
$$\Delta S_{total} = (\Delta S)_{water} + (\Delta S)_{metal\ block}$$
$$= 3.12 - 1.87 = 1.25\text{ kJ/K}$$

6.3 (b)

Isentropic expansion means reversible adiabatic process.

For adiabatic process

$$PV^{\gamma} = \text{Constant} \quad \ldots(i)$$

For ideal gas

$$PV = nRT$$

By putting value of V in equation (i)

$$P\cdot\left(\frac{T}{P}nR\right)^{\gamma} = \text{Constant}$$
$$T^{\gamma}\,P^{1-\gamma} = \text{Constant}$$
$$\left(\frac{P_1}{P_2}\right) = \left(\frac{T_1}{T_2}\right)^{\frac{\gamma}{\gamma-1}}$$

6.4 (b)

Change in entropy of the system undergoing a cyclic process is always zero because entropy is property of system it depends only upon final points, not on path of process.

6.5 (c)

Work required for mechanically reversible process is given as

$$W = \dot{m}(T\Delta s - \Delta u)$$
$$W = 2[(175 + 273)(6.622 - 7.055) - (2579 - 2606)]$$
$$W = -333.97\text{ kJ}$$

–ve sign indicates that work done on system.

6.6 (a)

Given process can be understood by T-S diagram given below:

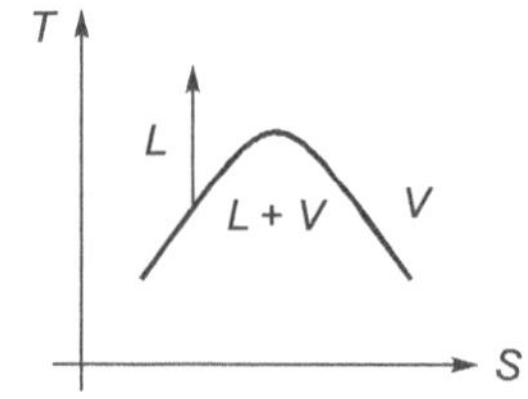

So if we increase temperature of saturated liquid of constant entropy resulting state will always sub-cooled/compressed liquid.

6.7 (a)

Entropy can be define as

$$ds = \frac{dQ}{T}$$

In heat exchanger dQ is same for both hot and cold fluid.

In parallel flow heat exchanger temperature of cold fluid is always less than temperature of hot fluid. So entropy gained by cold stream is greater than entropy lost by hot stream.

6.8 (d)

Amount of heat transferred from system *F* to surrounding *E* is 600 Joule.
Surrounding temperature (T_E) = 300 K
System temperature (T_F) ranges from 380-400 K

$$\Delta S_E = \frac{\delta Q}{T_E} = \frac{600}{300} = 2 \text{ Joule/K}$$

ΔS_E can be calculated by using 2nd law of thermodynamics:

$$\Delta S_{\text{Total}} = \Delta S_{\text{System}} + \Delta S_{\text{Surrounding}} > 0$$
$$\Delta S_F + 2 > 0$$
$$\Delta S_F > -2$$

6.9 (0.28)

$$\frac{C_P}{R} = 2.46 + 45.4 \times 10^{-3} T - 14.1 \times 10^{-6} T^2$$

We have to find $\left.\frac{dS}{dT}\right|_P$

$$dS = \oint \frac{dQ}{T}$$

$$dQ|_P = dH = C_P dT$$

$$\therefore \quad dS = \oint \frac{dQ}{T} = \oint \frac{C_P dT}{T}$$

Hence, $\left.\frac{dS}{dT}\right|_P = \frac{C_P}{T}$

$$= \frac{R(2.46 + 45.4 \times 10^{-3} T - 14.1 \times 10^{-6} T^2)}{T}$$

$$= 8.314\left(\frac{2.46}{1000} + 45.4 \times 10^{-3} - 14.1 \times 10^{-6} T\right)$$

$$\left.\frac{dS}{dT}\right|_P = 8.314\left(\frac{2.46}{1000} + 45.4 \times 10^{-3} - 14.1 \times 10^{-6} \times 10^3\right)$$

$$= 0.28 \text{ J/mol K}$$

6.10 (125)

$$W_{\min} = -\Delta G_{\text{mix}} = -RT\,\Sigma y_i \ln y_i$$

Putting the values

$= -8.314 \times 300 \times (0.2 \ln 0.2 + 0.8 \ln 0.8)$

$= -1248$ J/mol

For 100 moles of mixture work done

$= 1248 \times 100$ J

$= 124.8$ kJ

≈ 125 kJ (to nearest integer)

6.11 (179.67)

5 mol (A) benzene, 7 mol (B) xylene, 8 mol (C) toluene at 25°C and 1 bar are mixed.

$$\Delta S_{\text{mix}}^{\text{id}} = -nR \sum x_i \ln x_i$$

$$x_A = \frac{5}{20} = 0.25$$

$$x_B = \frac{7}{20} = 0.35$$

$$x_C = \frac{8}{20} = 0.4$$

$$\Delta S_{\text{mix}}^{\text{id}} = -20 \times 8.314\,(0.25 \ln 0.25 + 0.35 \ln 0.35 + 0.4 \ln 0.4)$$

$$= 179.67 \text{ (J/K)}$$

■■■■

7 Properties of Pure Substances

7.1 Which of the following conditions are satisfied at the critical point by the P-V-T relation of a real fluid?

(a) $\left(\dfrac{\partial^2 P}{\partial V^2}\right)_T = \left(\dfrac{\partial P}{\partial V}\right)_T = 0$

(b) $\left(\dfrac{\partial^2 P}{\partial V^2}\right)_T > 0, \left(\dfrac{\partial P}{\partial V}\right)_T = 0$

(c) $\left(\dfrac{\partial^2 P}{\partial V^2}\right)_T < 0, \left(\dfrac{\partial P}{\partial V}\right)_T = 0$

(d) $\left(\dfrac{\partial^2 P}{\partial V^2}\right)_T > 0, \left(\dfrac{\partial P}{\partial V}\right)_T > 0$

[2002 : 1 M]

7.2 Which of the following identities can be most easily used to verify steam table data for superheated steam?

(a) $\left(\dfrac{\partial T}{\partial V}\right)_V = -\left(\dfrac{\partial P}{\partial S}\right)_V$ (b) $\left(\dfrac{\partial T}{\partial P}\right)_S = \left(\dfrac{\partial V}{\partial S}\right)_P$

(c) $\left(\dfrac{\partial P}{\partial T}\right)_V = \left(\dfrac{\partial S}{\partial V}\right)_T$ (d) $\left(\dfrac{\partial V}{\partial T}\right)_P = -\left(\dfrac{\partial S}{\partial P}\right)_T$

[2002 : 1 M]

7.3 For organic compounds, Group Contribution Method can be used for estimation of

(a) critical properties
(b) specific gravity
(c) specific volume
(d) thermal conductivity

[2003 : 1 M]

Q.7.4 - 7.5 are based on the data supplied in the paragraph below

One kg of saturated steam at 100°C and 1.01325 bar is contained in a rigid walled vessel. It has a volume 1.673 m^3. It cools to 98°C, the saturation pressure is 0.943 bar; one kg of water vapour under these conditions has a volume of 1.789 m^3.

7.4 The amount of water vapour condensed (in kg) is

(a) 0.0 (b) 0.065
(c) 0.1 (d) 1.0

[2003 : 2 M]

7.5 The latent heat of condensation (kJ kg^{-1}) under these conditions is

(a) 40732 (b) 2676
(c) 2263 (d) 540

[2003 : 2 M]

7.6 In Vander Waals equation of state, what are the criteria applied at the critical point to determine the parameters '*a*' and '*b*'?

(a) $\left(\dfrac{\partial P}{\partial V}\right)_T = 0; \left(\dfrac{\partial^2 P}{\partial V^2}\right)_T = 0$

(b) $\left(\dfrac{\partial V}{\partial P}\right)_T = 0; \left(\dfrac{\partial^2 V}{\partial P^2}\right)_T = 0$

(c) $\left(\dfrac{\partial P}{\partial T}\right)_V = 0; \left(\dfrac{\partial^2 P}{\partial T^2}\right)_V = 0$

(d) $\left(\dfrac{\partial V}{\partial T}\right)_P = 0; \left(\dfrac{\partial^2 V}{\partial T^2}\right)_P = 0$

[2005 : 1 M]

7.7 An ideal gas at temperature T_1 and pressure P_1 is compressed isothermally to pressure $P_2 (> P_1)$ in a closed system. Which one of the following is TRUE for internal energy (U) and Gibbs free energy (G) of the gas at the two states?

(a) $U_1 = U_2, G_1 > G_2$
(b) $U_1 = U_2, G_1 < G_2$
(c) $U_1 > U_2, G_1 = G_2$
(d) $U_1 < U_2, G_1 = G_2$

[2009 : 1 M]

7.8 Consider a vessel containing steam at 180°C. The initial steam quality is 0.5 and the initial volume of the vessel is 1 m^3. The vessel loses heat at a constant rate $\dot{q}$ under isobaric conditions so that the quality of steam reduces to 0.1 after 10 hours. The thermodynamic properties of water at 180°C are (subscript *g* : vapor phase; subscript *f*: liquid phase) :

Specific volume :
$v_g = 0.19405$ m^3/kg; $v_f = 0.001127$ m^3/kg
Specific internal energy :
$u_g = 2583.7$ kJ/kg, $u_f = 762.08$ kJ/kg

Specific enthalpy :

$h_g = 2778.2$ kJ/kg, $h_f = 763.21$ kJ/kg

The rate of heat loss $\dot{q}$ (in kJ/hour) is _____ (rounded off to the nearest integer).

[2019 : 2 M]

7.9 A substance at 4 °C has thermal expansion coefficient $\beta = \frac{1}{v}\left(\frac{\partial v}{\partial T}\right)_P = 0\ \text{K}^{-1}$, an isothermal compressibility, $\kappa_T = -\frac{1}{v}\left(\frac{\partial v}{\partial P}\right)_T = 5 \times 10^{-4}\ \text{Pa}^{-1}$ and a molar volume $v = 18 \times 10^{-6}\ \text{m}^3\ \text{mol}^{-1}$. If s is the molar entropy, then at 4°C, the quantity $\left[v\left(\frac{\partial s}{\partial v}\right)_T\right]$ evaluated for the substance is _____ J mol^{-1} K^{-1} (rounded off to the nearest integer).

[2022 : 2 M]

7.10 For a pure substance, the following data at saturated conditions are given:

ln P^{sat} (bar)	T(K)
0.693	350
1.386	370

Assume that the vapor phase behaves ideally, the molar volume of the liquid is negligible, and the latent heat of vaporization is constant over the given temperature range. The universal gas constant, $R = 8.314$ J K^{-1} mol^{-1}. From the above data, the estimated latent heat of vaporization at 360 K is _____ kJ/mol (rounded off to one decimal place).

[2022 : 2 M]

7.11 Ethylene obeys the truncated virial equation-of-state

$$\frac{PV}{RT} = 1 + \frac{BP}{RT}$$

where P is the pressure, V is the molar volume, T is the absolute temperature and B is the second virial coefficient. The universal gas constant $R = 83.14$ bar cm^3 mol^{-1} K^{-1}. At 340 K, the slope of the compressibility factor vs. pressure curve is -3.538×10^{-3} bar^{-1}. Let G^R denote the molar residual Gibbs free energy. At these conditions, the value of $\left(\frac{\partial G^R}{\partial P}\right)_T$, in cm^3 mol^{-1}, rounded off to 1 decimal place, is _____.

[2024 : 2 M]

■■■■

Answers Properties of Pure Substances

7.1	(a)	7.2	(d)	7.3	(a)	7.4	(b)	7.5	(c)	7.6	(a)	7.7	(b)
7.8	(825.91)	7.9	(0)	7.10	(37.306)	7.11	(–100.0)						

Explanations Properties of Pure Substances

7.1 (a)

At critical point,

$$\left(\frac{\partial^2 P}{\partial V^2}\right)_T = \left(\frac{\partial P}{\partial V}\right)_T = 0$$

Critical point is a state at which no phase boundaries exist.

7.2 (d)

In steam table for superheated steam, values of v, u, h and s are given at constant pressure for different temperatures. Therefore the following identity of Maxwell relations used to verify steam table data for super heated steam.

$$\left(\frac{\partial V}{\partial T}\right)_P = -\left(\frac{\partial S}{\partial P}\right)_T$$

7.4 (b)

At initial condition,

$$M = 1\text{ kg},\quad T = 100°\text{C}$$

$$P = 1.013\text{ bar},\quad V = 1.673\text{ m}^3$$

$$v_{\text{initial}} = \frac{V}{M} = \frac{1.673\text{ m}^3}{1\text{ kg}} = 1.673\text{ m}^3/\text{kg}$$

$$v_g = 1.789\text{ m}^3/\text{kg (given) at } 98°\text{C}$$

$$v_f = \frac{1}{1000}\text{ m}^3/\text{kg [at } 98°\text{C] assume}$$

$$\text{Thus quality of steam} = \frac{1.673 - 0.001}{1.789 - 0.001} = 0.935$$

∴ kg water vapour @ 98°C after condensation = 0.935

∴ Amount of water vapour condensed

$$= 1 - 0.935 = 0.065\text{ kg}$$

7.5 (c)

$$\therefore \quad \ln\left[\frac{P_2}{P_1}\right]_{\text{sat}} = \frac{\lambda}{R}\left[\frac{1}{T_1} - \frac{1}{T_2}\right]$$

Data given:

$$T_1 = 100°\text{C} = 373.15\text{ K}$$

$$T_2 = 98°\text{C} = 371.15\text{ K}$$

$$P_1^{\text{sat}} = 1.01325\text{ bar}$$

$$P_2^{\text{sat}} = 0.943\text{ bar}$$

$$\ln\left|\frac{0.943}{1.01325}\right| = \frac{\lambda}{8.314}\left[\frac{1}{373.15} - \frac{1}{371.15}\right]$$

Solving we get,

$$\lambda = 41368.78\left(\frac{\text{kJ}}{\text{Kmol}}\right)$$

$$\therefore \quad \lambda = 41368.78 \times \frac{1}{18}$$

$$\lambda = 2298.15\text{ kJ/kg}$$

7.6 (a)

Vander-Waals equations of state shows a point of inflection (neither maxima, nor-minima) at critical point on P-V diagram.

So first and 2nd derivative of pressure with respect to volume is equal to zero at constant temperature.

7.7 (b)

For ideal gas internal energy depends upon temperature only so for isothermal process:

$$U_1 = U_2$$

Decrease in Gibbs free energy represents work done by system. Here work done on system in compression so Gibbs free energy increases.

So, $G_2 > G_1$

7.8 (825.91)

Here for both the initial and final condition ($P_I = P_{II}$) i.e. isobaric condition [closed system]; also

$$T_I = T_{II} = 180°\text{C}$$

i.e. $du = dq - dw$

$$du + PdV = dq$$

or ($dh = dq$) at const. P

Heat transferred = Change in enthalpy of steam

Mass of steam in piston cylinder

$$m = \frac{V}{v} = \frac{\text{Total volume}}{\text{Sp. volume}}$$

Initially,

$$T = 180°\text{C}$$

$$v_f = 0.001127\text{ m}^3/\text{kg},\ x = 0.5$$

$$v_g = 0.19405$$

$\therefore \quad v_I = (1-0.5)\,0.001127 + 0.5 \times 0.19405$

$v_I = 0.0976\ m^3/kg$

$\therefore$ Mass of steam inside the vessel

$$M = \frac{V}{v} = \frac{1\ m^3}{0.0976\ m^3/kg} = 10.247\ kg$$

Specific enthalpy at state-I

$$h_I = (1-x)h_f + x\,h_g$$
$$= (1-0.5)\times 763.21 + 0.5\times 2778.2$$
$$h_I = 1770.705\ kJ/kg$$

State-2

$$h_{II} = (1-x)h_f + x\,h_g$$
$$h_{II} = (1-0.1)\,763.21 + 0.1\times 2778.2$$
$$h_{II} = 964.709\ kJ/kg$$

Energy balance

$$dq = dh$$
$$\Delta q = h_{II} - h_I = (964.709 - 1770.705)$$
$$= -805.996\ kJ/kg$$
$$\Delta Q = m\cdot\Delta q$$
$$= 10.247\times(-805.996) = -8259.121\ kJ$$

$\therefore$ Rate of heat transfer $= -8259.121 \times \dfrac{1}{10h}$

$$= -825.91\ kJ/h$$

(–ve sign indicates that heat has been removed from the vessel).

7.9 (0)

Given at 4°C, $\beta = 0\ K^{-1}$

$\alpha = 5\times10^{-4}\ Pa^{-1}$

$\kappa_T = 18\times10^{-6}\ m^3/mol$

Form Maxwell equation

$$\left.\frac{\partial S}{\partial V}\right|_T = \left.\frac{\partial P}{\partial T}\right|_V$$

Given: $\beta = \left.\dfrac{dv}{vdT}\right|_P$, $\kappa_T = -\left.\dfrac{dv}{vdP}\right|_T$

$$\frac{dv}{v} = \beta dT - \kappa_T dP \text{ (at constant volume)}$$

$$\left.\frac{\partial P}{\partial T}\right|_V = 0$$

So value of $v\left(\dfrac{\partial S}{\partial V}\right)_T = v\left(\dfrac{\partial P}{\partial T}\right)_V = 0$

7.10 (37.306)

$$\ln\left(\frac{P_2^{sat}}{P_1^{sat}}\right) = \frac{\Delta H_{vap}}{R}\left(\frac{1}{T_1}-\frac{1}{T_2}\right)$$

$$\ln P_2^{sat} - \ln P_1^{sat} = \frac{\Delta H_{vap}}{R}\left[\frac{1}{350}-\frac{1}{370}\right]$$

$$1.386 - 0.693 = \frac{\Delta H_{vap}}{R}\left[\frac{1}{350}-\frac{1}{370}\right]$$

$$\Delta H_{vap} = \frac{0.693\times8.314\times350\times370}{20}$$

$$= 37.306\ kJ/mol$$

7.11 (–100.0)

Given virial equation is

$$\frac{PV}{RT} = 1+\frac{BP}{RT} \quad \text{...(i)}$$

$R = 83.14\ bar\ cm^3\ mol^{-1}\ K^{-1}$

At 340 K slope of compressibility factor v/s pressure curve is $-3.538\times10^{-3}\ bar^{-1}$

So, $$\frac{\partial G}{\partial P} = -3.538\times10^{-3} \quad \text{...(ii)}$$

From equation (i)

$$Z = 1+\frac{BP}{RT}$$

$$\frac{\partial Z}{\partial P} = \frac{B}{RT}$$

By putting values from equation (ii)

$$\frac{B}{RT} = -3.538\times10^{-3}$$

$$B = -3.538\times10^{-3}\times83.14\times340 = -100.010$$

Now, $$\frac{G^R}{RT} = \int_0^P (Z-1)\frac{dP}{P}$$

$$\frac{G^R}{RT} = \int_0^P \frac{BdP}{RT} \text{ from equation (i)}$$

$$G^R = BP$$

$$\left(\frac{\partial G^R}{\partial P}\right)_T = B = -100.010\ cm^3/mol$$

■■■■

8 Thermodynamic Relations

8.1 On a P-V diagram of an ideal gas, suppose a reversible adiabatic line intersects a reversible isothermal line at point A. Then at point A, the slope of the reversible adiabatic line $\left(\frac{\partial P}{\partial V}\right)_S$ and the slope of the reversible isothermal line $\left(\frac{\partial P}{\partial V}\right)_T$ are related as $\left(\text{where } \gamma = \frac{C_p}{C_v}\right)$

(a) $\left(\frac{\partial P}{\partial V}\right)_S = \left(\frac{\partial P}{\partial V}\right)_T$ (b) $\left(\frac{\partial P}{\partial V}\right)_S = \left[\left(\frac{\partial P}{\partial V}\right)_T\right]^\gamma$

(c) $\left(\frac{\partial P}{\partial V}\right)_S = \gamma\left(\frac{\partial P}{\partial V}\right)_T$ (d) $\left(\frac{\partial P}{\partial V}\right)_S = \frac{1}{\gamma}\left(\frac{\partial P}{\partial V}\right)_T$

[2000 : 1 M]

8.2 The Maxwell relation derived from the differential expression for the Helmholtz free energy (dA) is

(a) $\left(\frac{\partial T}{\partial V}\right)_S = -\left(\frac{\partial P}{\partial S}\right)_V$ (b) $\left(\frac{\partial S}{\partial P}\right)_T = -\left(\frac{\partial V}{\partial T}\right)_P$

(c) $\left(\frac{\partial V}{\partial S}\right)_P = \left(\frac{\partial T}{\partial P}\right)_S$ (d) $\left(\frac{\partial S}{\partial V}\right)_T = \left(\frac{\partial P}{\partial T}\right)_V$

[2001 : 2 M]

8.3 At 100°C, water and methyl cyclohexane both have vapour pressures of 1.0 atm. Also at 100°C, the latent heats of vapourization of these compounds are 40.63 kJ/mol for water and 31.55 kJ/mol for methyl cyclohexane. The vapour pressure of water at 150°C is 4.69 atm. At 150°C, the vapour pressure of methylcyclohexane would be expected to be :

(a) significantly less than 4.69 atm
(b) nearly equal to 4.69 atm
(c) significantly more than 4.69 atm
(d) indeterminate due to a lack of data

[2001 : 2 M]

8.4 The molar density of water vapor at the normal boiling point of water is 33 mol/m^3. The compressibility factor under these conditions is close to which ONE of the following? $R = 8.314$ J/(mol K)

(a) 0.75 (b) 1
(c) 1.25 (d) 1.5

[2006 : 2 M]

8.5 Parameters a and b in the van der Waals and other cubic equations of state represent

(a) a – molecular weight; b – molecular polarity
(b) a – molecular size; b – molecular attraction
(c) a – molecular size; b – molecular speed
(d) a – molecular attraction; b – molecular size

[2007 : 1 M]

8.6 For a pure substance, the Maxwell's relation obtained from the fundamental property relation $du = Tds - Pdv$ is

(a) $\left(\frac{\partial T}{\partial v}\right)_s = -\left(\frac{\partial P}{\partial s}\right)_v$ (b) $\left(\frac{\partial P}{\partial T}\right)_v = -\left(\frac{\partial s}{\partial v}\right)_T$

(c) $\left(\frac{\partial T}{\partial P}\right)_s = -\left(\frac{\partial v}{\partial s}\right)_P$ (d) $\left(\frac{\partial v}{\partial T}\right)_P = -\left(\frac{\partial s}{\partial P}\right)_T$

[2007 : 2 M]

8.7 The standard Gibbs free energy change and enthalpy change at 25°C for the liquid phase reaction

$$CH_3COOH\,(l) + C_2H_5OH\,(l) \rightarrow CH_3COOC_2H_5\,(l) + H_2O\,(l)$$

are given as $\Delta G^o_{298} = -4650$ J/mol and $\Delta H^o_{298} = -3640$ J/mol. If the solution is ideal and enthalpy change is assumed to be constant, the equilibrium constant at 95° is

(a) 0.65 (b) 4.94
(c) 6.54 (d) 8.65

[2008 : 2 M]

8.8 In a throttling process, the pressure of an ideal gas reduces by 50 %. If C_P and C_V are the heat capacities at constant pressure and constant volume, respectively ($\gamma = C_P/C_V$), the specific volume will change by a factor of

(a) 2 (b) $2^{1/\gamma}$
(c) $2^{(\gamma-1)/\gamma}$ (d) 0.5

[2012 : 1 M]

8.9 For an exothermic reversible reaction, which one of the following correctly describes the dependence of the equilibrium constant (K) with temperature (T) and pressure (P)?

(a) K is independent of T and P
(b) K increases with an increase in T and P
(c) K increases with T and decreases with P
(d) K decreases with an increase in T and is independent of P

[2012 : 1 M]

8.10 An equation of state is explicit in pressure p and cubic in the specific volume v. At the critical point 'c', the isotherm passing through 'c' satisfies

(a) $\frac{\partial p}{\partial v} < 0, \frac{\partial^2 p}{\partial v^2} = 0$ (b) $\frac{\partial p}{\partial v} > 0, \frac{\partial^2 p}{\partial v^2} < 0$

(c) $\frac{\partial p}{\partial v} = 0, \frac{\partial^2 p}{\partial v^2} > 0$ (d) $\frac{\partial p}{\partial v} = 0, \frac{\partial^2 p}{\partial v^2} = 0$

[2013 : 1 M]

8.11 Which ONE of the following is CORRECT for an ideal gas in a closed system?

(a) $\left(\frac{\partial U}{\partial V}\right)_s V = nR\left(\frac{\partial U}{\partial S}\right)_V$

(b) $-\left(\frac{\partial H}{\partial P}\right)_s P = nR\left(\frac{\partial H}{\partial S}\right)_P$

(c) $\left(\frac{\partial U}{\partial V}\right)_s V = nR\left(\frac{\partial H}{\partial S}\right)_P$

(d) $\left(\frac{\partial H}{\partial P}\right)_s P = nR\left(\frac{\partial U}{\partial S}\right)_V$

[2014 : 2 M]

8.12 If v, u, s and g represent respectively the molar volume, molar internal energy, molar entropy and molar Gibbs free energy, then match the entries in the left and right columns below and choose the correct option.

P. $-\left(\frac{\partial u}{\partial v}\right)_s$ I. Temperature

Q. $\left(\frac{\partial g}{\partial P}\right)_T$ II. Pressure

R. $-\left(\frac{\partial g}{\partial T}\right)_P$ III. v

S. $-\left(\frac{\partial u}{\partial s}\right)_v$ IV. s

(a) P-II, Q-III, R-IV, S-I
(b) P-II, Q-IV, R-II, S-III
(c) P-I, Q-IV, R-II, S-III
(d) P-III, Q-II, R-IV, S-I

[2015 : 1 M]

8.13 A gas obeying the Clausius equation of state is isothermally compressed from 5 MPa to 15 MPa in a closed system at 400 K. The Clausius equation of state is $P = \frac{RT}{v - b(T)}$ where P is the pressure, T is the temperature, v is the molar volume and R is the universal gas constant. The parameter b in the above equation varies with temperature as $b(T) = b_0 + b_1 T$ with $b_0 = 4 \times 10^{-5}$ m^3 mol^{-1} and $b_1 = 1.35 \times 10^{-7}$ m^3 mol^{-1} K^{-1}. The effect of pressure on the molar enthalpy (h) at a constant temperature is given by $\left(\frac{\partial h}{\partial P}\right)_T = v - T\left(\frac{\partial v}{\partial T}\right)_P$.

Let h_i and h_f denote the initial and final molar enthalpies, respectively. The change in the molar enthalpy $h_f - h_i$ (in J mol^{-1}, rounded off to be first decimal place) for this process is ______.

[2016 : 2 M]

8.14 The volumetric properties of two gases M and N are described by the generalized compressibility chart which expresses the compressibility factor (Z) as a function of reduced pressure and reduced temperature only. The operating pressure (P) and temperature (T) of two gases M and N along with their critical properties (P_c, T_c) are given in the table below.

Gas	P (bar)	T(K)	P_C (bar)	T_C (K)
M	25	300	75	150
N	75	1000	225	500

Z_M and Z_N are the compressibility factor of the gases M and N under the given operating conditions, respectively.

The relation between Z_M and Z_N is

(a) $Z_M = 8\, Z_N$
(b) $Z_M = 3\, Z_N$
(c) $Z_M = Z_N$
(d) $Z_M = 0.333\, Z_N$

[2017 : 1 M]

8.15 In a closed piston-cylinder system, methane was observed to obey the following equation of state

$$P(V - nb) = nRT$$

where $b = 0.029$ m^3/mol. The temperature and volume are 500°C and 5 m^3 respectively for 100 moles of methane. At this state of the system, the isobaric rate of change of temperature with volume (in °C/m^3) is _______ (rounded off to second decimal place).

[2018 : 2 M]

8.16 The Vander Waals equation of state is given by

$$P_r = \frac{8T_r}{3v_r - 1} - \frac{3}{v_r^2}$$

where P_r, T_r and v_r represent reduced pressure, reduced temperature and reduced molar volume, respectively. The compressibility factor at critical point (z_c) is $\frac{3}{8}$.

If $v_r = 3$ and $T_r = \frac{4}{3}$, then the compressibility factor based on the van der Waals equation of state is _______ (round off to 2 decimal places).

[2021 : 1 M]

8.17 For a single component system at vapor-liquid equilibrium, the extensive variables A, V, S and N denote the Helmholtz free energy, volume, entropy, and number of moles, respectively, in a given phase. If superscripts (v) and (l) denote the vapor and liquid phase, respectively, the relation that is **NOT CORRECT** is

(a) $\left(\frac{\partial A^{(l)}}{\partial V^{(l)}}\right)_{T,N^{(l)}} = \left(\frac{\partial A^{(v)}}{\partial V^{(v)}}\right)_{T,N^{(v)}}$

(b) $\left(\frac{\partial A^{(l)}}{\partial N^{(l)}}\right)_{T,N^{(l)}} = \left(\frac{\partial A^{(v)}}{\partial N^{(v)}}\right)_{T,N^{(v)}}$

(c) $\left(\frac{A+PV}{N}\right)^{(l)} = \left(\frac{A+PV}{N}\right)^{(v)}$

(d) $\left(\frac{A+TS}{N}\right)^{(l)} = \left(\frac{A+TS}{N}\right)^{(v)}$

[2022 : 1 M]

■■■■

Answers Thermodynamic Relations

8.1	(c)	**8.2**	(d)	**8.3**	(a)	**8.4**	(b)	**8.5**	(d)	**8.6**	(a)	**8.7**	(b)
8.8	(b)	**8.9**	(d)	**8.10**	(d)	**8.11**	(d)	**8.12**	(a)	**8.13**	(400)	**8.14**	(c)
8.15	(368.09)	**8.16**	(0.84)	**8.17**	(d)								

Explanations Thermodynamic Relations

8.1 (c)

Slope of adiabatic lines are γ times higher than slope of isothermal lines on P-V diagram.

Slope of adiabatic lines = $\gamma \times$ Slope of isothermal line.

8.2 (d)

By definition

$$A = U - TS$$

$$\therefore \quad dA = dU - TdS - SdT \text{ (for closed system)}$$

$$\therefore \quad dA = -PdV - SdT$$

$$\therefore \quad \left(\frac{\partial A}{\partial V}\right)_T = -P \text{ and } \left(\frac{\partial A}{\partial T}\right)_V = -S$$

$$\therefore \quad \frac{\partial}{\partial T}\bigg|_V \left(\frac{\partial A}{\partial V}\right)_T = -\left(\frac{\partial P}{\partial T}\right)_V$$

$$\text{and} \quad \frac{\partial}{\partial V}\bigg|_T \left(\frac{\partial A}{\partial T}\right)_V = -\left(\frac{\partial S}{\partial V}\right)_T$$

A is a state function

$$\frac{\partial}{\partial T}\bigg|_V \left(\frac{\partial A}{\partial V}\right)_T = \frac{\partial}{\partial V}\bigg|_T \left(\frac{\partial A}{\partial T}\right)_V$$

$$-\left(\frac{\partial P}{\partial T}\right)_V = -\left(\frac{\partial S}{\partial V}\right)_T$$

$$\left(\frac{\partial S}{\partial V}\right)_T = \left(\frac{\partial P}{\partial T}\right)_V$$

8.3 (a)

From Clausius–Clapeyron equation;

$$\ln\left|\frac{P^{vap}, 150°C}{P^{vap}, 100°C}\right| = \frac{\Delta H}{R}\left[\frac{1}{373} - \frac{1}{423}\right]$$

Given,

$$P^{vap}_{100\text{ MCH}} = P^{vap}_{100\text{ Water}} = 1 \text{ atm}$$

$$\Delta H^{vap}_{MCH} = 31.55 \text{ kJ/mol}$$

$$\Delta H^{vap}_{W} = 40.63 \text{ kJ/mol}$$

$$(P^{vap})_{150} = 4.69 \text{ atm}$$

$$\ln(P^{vap})_{150\text{ MCH}} = \ln(P^{vap})_{150\text{ Water}} \times \frac{31.55}{40.63}$$

$$\ln(P^{vap})_{150\text{ MCH}} = \ln|4.69| \times \frac{31.55}{40.63}$$

$$P^{vap} = 3.320 \text{ atm}$$

8.4 (b)

Molar density of water vapour at normal boiling point is given as 33 mol/m³

$$Z = \frac{\text{Actual volume}}{\text{Ideal volume}} = \frac{v}{\frac{RT}{P}}$$

$$Z = \frac{P \cdot v}{RT}$$

$$P = 101325$$

$$v = \frac{1}{\rho} = \frac{1}{33} \text{ m}^3\text{/mol}$$

$$R = 8.314 \text{ J/mol K}$$

$$T = 100 + 273 = 373 \text{ K}$$

$$z = \frac{101325}{33 \times 8.314 \times 373} = 0.99 \approx 1$$

8.6 (a)

If a fundamental property is written as

$$dp = Mdx + Ndy$$

Then from Maxwell's relation

$$\left(\frac{\partial M}{\partial y}\right)_x = \left(\frac{\partial N}{\partial x}\right)_y$$

So from $du = Tds - Pdv$ Maxwell's relation obtained is

$$\left(\frac{\partial T}{\partial v}\right)_s = -\left(\frac{\partial P}{\partial s}\right)_v$$

8.7 (b)

$$\Delta G°_{298} = -4650 \text{ J/mol}$$

$$\Delta G = -RT \ln k_{eq}$$

$$-4650 = -8.314 \times 298 \ln k_{eq}$$

$$k_{eq} = 6.53$$

Now, $$\ln\left(\frac{k_{2eq}}{k_{1eq}}\right) = \frac{\Delta H}{R}\left[\frac{1}{T_1}-\frac{1}{T_2}\right]$$

$$\ln\left(\frac{k_{2eq}}{6.53}\right) = -\frac{3640}{8.314}\left[\frac{1}{298}-\frac{1}{368}\right]$$

$$k_{2,eq} = 4.94$$

8.8 (b)

Throttling process is always isentropic process. So throttling process can be considered as reversible adiabatic process.

For adiabatic process ⇒ PV^{γ} = Constant

$$P_1V_1^{\gamma} = P_2V_2^{\gamma}$$

$$P_2 = 0.5P_1 \text{ (given)}$$

So, $$\frac{V_2}{V_1} = \left(\frac{P_1}{P_2}\right)^{\frac{1}{\gamma}} = 2^{\frac{1}{\gamma}}$$

8.9 (d)

$$\left(\frac{\partial(\ln K_{eq})}{\partial T}\right)_P = \frac{\Delta H_{Rxn}(T)}{RT^2}$$

For exothermic reaction $\Delta H_{R\times n}$ is negative

So, $$\frac{\partial \ln K_{eq}}{\partial T} = -\text{ve}$$

So with increase in temperature K_{eq} decreases and at low pressure K_{eq} is independent of P.

8.10 (d)

At critical point p-v curves indicates "a point of inflection" means no maxima and no minima so first and 2nd derivative is zero.

8.11 (d)

$$du = TdS - Pdv \quad \text{...(i)}$$

$$dH = TdS + VdP \quad \text{...(ii)}$$

From equation (ii)

$$\left(\frac{\partial H}{\partial P}\right)_S = V$$

From equation (i)

$$\left(\frac{\partial U}{\partial S}\right)_V = T$$

So from given option (d)

$$\left(\frac{\partial H}{\partial P}\right)_S P = nR\left(\frac{\partial U}{\partial S}\right)_V$$

$$PV = nRT$$

So option (d) is correct for an ideal gas system.

8.12 (a)

$$du = Tds - Pdv \quad \text{...(i)}$$

$$dg = vdP - sdT \quad \text{...(ii)}$$

From equation (i) and (ii)

$$\left.\frac{\partial u}{\partial v}\right|_s = -P, \quad \left.\frac{\partial g}{\partial p}\right|_T = V$$

$$\left.\frac{\partial g}{\partial T}\right|_P = -S, \quad \left.\frac{\partial u}{\partial s}\right|_v = T$$

8.13 (400)

Gas obeying Clausius equation of state is isothermally compressed from 5 MPa to 15 MPa at 400 K.

Clausius equation: $P = \dfrac{RT}{v-b(T)}$ ...(i)

where,
P = Pressure
T = Temperature
v = Molar volume
R = Universal gas constant
$b(T) = b_0 + b_1T$
$b_0 = 4 \times 10^{-5}$ m³/mol
$b_1 = 1.35 \times 10^{-7}$ m³/mol K

The effect of pressure on molar enthalpy (h) at constant temperature is given by

$$\left(\frac{\partial h}{\partial P}\right)_T = v - T\left(\frac{\partial v}{\partial T}\right)_P$$

From equation (i)

$$v - b(T) = \frac{RT}{P}$$

$$\left(\frac{\partial v}{\partial T}\right)_P = \frac{R}{P} + b_1$$

$$T\left(\frac{\partial v}{\partial T}\right)_P = \frac{RT}{P} + b_1T$$

$$v - T\left(\frac{\partial v}{\partial T}\right)_P = \frac{RT}{P} + b_0 + b_1T - \frac{RT}{P} + b_1T$$

$$v - T\left(\frac{\partial v}{\partial T}\right)_P = b_0$$

By putting this value in equation (ii)

$$\left(\frac{\partial h}{\partial P}\right)_T = b_0$$

$$\Delta h = b_0(\Delta P)$$

$$= 4 \times 10^{-5}(15-5) \times 10^6$$

$$= 400 \text{ J/mol}$$

8.14 (c)

According to law of corresponding state all fluids when compared to same reduced temperature and reduced pressure, have approximately same compressibility factor.

Here, reduced temperature $(T_R) = \dfrac{T}{T_C}$

Reduced pressure $(P_R) = \dfrac{P}{P_C}$

For gas M : $P_R = \dfrac{25}{75} = \dfrac{1}{3}$, $T_R = \dfrac{300}{150} = 2$

For gas N : $P_R = \dfrac{75}{225} = \dfrac{1}{3}$, $T_R = \dfrac{1000}{500} = 2$

Since P_R and T_R is same for both gases M and N so $Z_M = Z_N$.

8.15 (368.09)

In closed piston-cylinder system methane was observed to obey the following equation

$$P(V - nb) = nRT \qquad \text{...(i)}$$

where, $b = 0.029$ m^3/mol

Temperature and volume are 500°C and 5 m^3 for 100 moles of methane.

$$\left.\frac{dT}{dV}\right|_{\text{constant } P} = ?$$

From equation (i)

$$T = \frac{P}{nR}(V - nb)$$

$$\left.\frac{dT}{dV}\right|_P = \frac{P}{nR}$$

$$\left.\frac{dT}{dV}\right|_P = \frac{T}{V - nb} = \frac{500 + 273}{5 - 100 \times 0.029}$$

$$\left.\frac{dT}{dV}\right|_P = 368.09 \text{ K/m}^3 = 368.09 \text{ °C/m}^3$$

8.16 (0.84)

The Vander Waals equation is

$$P_r = \frac{8T_r}{(3V_r - 1)} - \frac{3}{V_r^2}$$

Given: $V_r = 3$ and $T_r = \dfrac{4}{3}$

Compressibility factor at critical point

$$Z_c = \frac{3}{8}$$

Now, $Z_c = \dfrac{P_C V_C}{RT_C}$

Also, $V_r = \dfrac{V}{V_C}$

$$P_r = \frac{P}{P_C}$$

$$T_r = \frac{T}{T_C}$$

Putting values of V_r and T_r in given expression

$$P_r = \frac{8 \times \frac{4}{3}}{3 \times 3 - 1} - \frac{3}{3^2} = 1$$

$$\frac{P}{P_C} = 1 \Rightarrow P = P_C$$

Compressibility factor $Z = \dfrac{PV}{RT}$

$$\frac{Z}{Z_C} = \frac{\dfrac{PV}{RT}}{\left(\dfrac{PV}{RT}\right)_C} = \frac{\dfrac{V}{V_C}}{\dfrac{T}{T_C}}$$

$$= \frac{V_r}{T_r} = \frac{9}{4}$$

$$Z = \frac{9}{4} \times \frac{3}{8} = \frac{27}{32} = 0.84$$

8.17 (d)

At equilibrium of vapor and liquid phases

$$G^L = G^V$$

$$T^L = T^V$$

$$P^L = P^V$$

We know, $A = U - TS$

$$A + TS = U$$

So internal energy is not equal at equilibrium condition.

Option (d) is incorrect.

$$dA = -SdT - PdV$$

$\left.\dfrac{\partial A}{\partial V}\right|_{T,N} = -P$ at equilibrium it is equal in both phase

$\left.\dfrac{\partial A}{\partial N}\right|_{T,V} = u$ (chemical potential also equal in both phase at equilibrium)

■■■■

9 Solution Thermodynamics

9.1 A reasonably general expression for vapour-liquid phase equilibrium at low to moderate pressures is $\phi_i y_i P = \gamma_i x_i f_i^0$ where ϕ_i is a vapour fugacity coefficient, γ_i is the liquid activity coefficient, and f_i^0 is the fugacity of pure component i. The K_i value ($y_i = K_i x_i$) is therefore, in general, a function of

(a) temperature only
(b) temperature and pressure only
(c) temperature, pressure, and liquid composition x_i only
(d) temperature, pressure, liquid composition x_i, and vapour composition y_i. [2001 : 1 M]

9.2 For water at 300°C, it has a vapour pressure 8592.7 kPa and fugacity 6738.9 kPa. Under these conditions, one mole of water in liquid phase has a volume 25.28 cm^3, and that in vapour phase 391.1 cm^3. Fugacity of water (in kPa) at 9000 kPa will be

(a) 6738.9 (b) 6753.5
(c) 7058.3 (d) 9000 [2003 : 1 M]

9.3 As pressure approaches zero, the ratio of fugacity to pressure (f/P) for a gas approaches

(a) zero (b) unity
(c) infinity (d) an indeterminate value [2004 : 1 M]

9.4 If $m_i, \bar{m}_i, m_i^R, m_i^E$ are molar, partial molar, residual and excess properties respectively for a pure species i, the maximum property M of a binary non-ideal mixture of components 1 and 2, is given by

(a) $x_1\bar{m}_1 + x_2\bar{m}_2$ (b) $x_1 m_1^R + x_2 m_2^R$
(c) $x_1 m_1 + x_2 m_2$ (d) $x_1 m_1^E + x_2 m_2^E$ [2007 : 1 M]

9.5 The molar volume (v) of a binary mixture, of species 1 and 2 having mole fractions x_1 and x_2 respectively is given by

$$v = 220x_1 + 180x_2 + x_1x_2(90x_1 + 50x_2)$$

The partial molar volume of species 2 at $x_2 = 0.3$ is

(a) 183.06 (b) 212.34
(c) 229.54 (d) 256.26 [2008 : 2 M]

9.6 An equimolar liquid mixture of species 1 and 2 is in equilibrium with its vapour at 400 K. At this temperature, the vapour pressures of the species are $P_1^{sat} = 180$ kPa and $P_2^{sat} = 120$ kPa. Assuming that Raoult's law is valid, the value of y_1 is

(a) 0.30 (b) 0.41
(c) 0.50 (d) 0.60 [2010 : 1 M]

9.7 At constant T and P, the molar density of a binary mixture is given by $\rho = 1 + x_2$, where x_2 is the mole fraction of component 2. The partial molar volume at infinite dilution for component 1, $\bar{V}_1^{\infty}$ is

(a) 0.75 (b) 1.0
(c) 2.0 (d) 4.0 [2010 : 2 M]

9.8 Minimum work (W) required to separate a binary gas mixture at a temperature T_0 and pressure P_0 is

$$W = -RT_0\left[y_1 \ln\left(\frac{\hat{f}_1}{f_{pure,1}}\right) + y_2 \ln\left(\frac{\hat{f}_2}{f_{pure,2}}\right)\right]$$

where y_1 and y_2 are mole fractions, $f_{pure,1}$ and $f_{pure,2}$ are fugacities of pure species at T_0 and P_0, and $\hat{f}_1$ and $\hat{f}_2$ are fugacities of species in the mixture at T_0, P_0 and y_1. If the mixture is ideal then W is

(a) 0
(b) $-RT_0[y_1 \ln y_1 + y_2 \ln y_2]$
(c) $RT_0[y_1 \ln y_1 + y_2 \ln y_2]$
(d) RT_0 [2011 : 1 M]

9.9 The partial molar enthalpies of mixing (in J/mol) for benzene (component 1) and cyclohexane (component 2) at 300 K and 1 bar are given by $\Delta\bar{H}_1 = 3600x_2^2$ and $\Delta\bar{H}_2 = 3600x_1^2$, where x_1 and x_2 are the mole fractions. When one mole of benzene is added to two moles of cyclohexane, the enthalpy change (in J) is

(a) 3600 (b) 2400
(c) 2000 (d) 800 [2011 : 1 M]

9.10 Consider a binary mixture of methyl ethyl ketone (component 1) and toluene (component 2). At 323 K the activity coefficients γ_1 and γ_2 are given by

$$\ln\gamma_1 = x_2^2(\psi_1 - \psi_2 + 4\psi_2 x_1),$$
$$\ln\gamma_2 = x_1^2(\psi_1 + \psi_2 - 4\psi_2 x_2)$$

where x_1 and x_2 are the mole fractions in the liquid mixture, and ψ_1 and ψ_2 are parameters independent of composition. At the same temperature, the infinite dilution activity coefficients, γ_1^∞ and γ_2^∞ are given by $\ln \gamma_1^\infty = 0.4$ and $\ln \gamma_2^\infty = 0.2$. The vapour pressures of methyl ethyl ketone and toluene at 323 K are 36.9 and 12.3 kPa respectively. Assuming that the vapour phase is ideal, the equilibrium pressure (in kPa) of a liquid mixture containing 90 mol % toluene is

(a) 19 (b) 18
(c) 16 (d) 15 [2011 : 2 M]

9.11 Consider a binary liquid mixture at constant temperature T and pressure P. If the enthalpy change of mixing, $\Delta H = 5\,x_1x_2$, where x_1 and x_2 are the mole fraction of species 1 and 2 respectively, and the entropy change of mixing $\Delta S = -R[x_1 \ln x_1 + x_2 \ln x_2]$ (with $R = 8.314$ J/mol.K), then the minimum value of the Gibbs free energy change of mixing at 300 K occurs when

(a) $x_1 = 0$ (b) $x_1 = 0.2$
(c) $x_1 = 4$ (d) $x_1 = 0.5$ [2012 : 2 M]

9.12 A binary liquid mixture is in equilibrium with its vapor at a temperature $T = 300$ K. The liquid mole fraction x_1 of species 1 is 0.4 and the molar excess Gibbs free energy is 200 J/mol. The value of the universal gas constant is 8.314 J/mol-K, and γ_i denotes the liquid-phase activity coefficient of species i. If $\ln(\gamma_1) = 0.09$, then the value of $\ln(\gamma_2)$, up to 2 digits after the decimal point, is ______. [2013 : 2 M]

9.13 Given that molar residual Gibbs free energy, g^R, and molar residual volume, v^R, are related as $\frac{g^R}{RT} = \int_0^P \left(\frac{v^R}{RT}\right) dP$, find g^R at $T = 27°$C and $P = 0.2$ MPa. The gas may be assumed to follow the virial equation of state, $Z = 1 + \frac{BP}{RT}$, where $B = -10^{-4}$ m^3/mol at the given conditions. ($R = 8.314$ J/mol.K). The value of g^R in J/mol is:

(a) 0.008 (b) −2.4
(c) 20 (d) −20 [2015 : 2 M]

9.14 The partial molar enthalpy (in kJ/mol) of species 1 in a binary mixture is given by $\bar{h}_1 = 2 - 60x_2^2 + 100x_1x_2^2$ where x_1 and x_2 are the mole fractions of species 1 and 2, respectively. The partial molar enthalpy (in kJ/mol, rounded off to the first decimal place) of species 1 at infinite dilution is ______. [2016 : 1 M]

9.15 A binary liquid mixture of benzene and toluene contains 20 mol% of benzene. At 350 K the vapour pressures of pure benzene and pure toluene are 92 kPa and 35 kPa, respectively. The mixture follows Raoult's law. The equilibrium vapour phase mole fraction (rounded off to the second decimal place) of benzene in contact with this liquid mixture at 350 K is ______. [2016 : 1 M]

9.16 A binary system at a constant pressure with species '1' and '2' is described by the two-suffix Margules equation, $\frac{g^E}{RT} = 3x_1x_2$, where g^E is the molar excess Gibbs free energy, R is the universal gas constant, T is the temperature and x_1, x_2 are the mole fractions of species 1 and 2 respectively. At a temperature T, $\frac{g_1}{RT} = 1$ and $\frac{g_2}{RT} = 2$, where g_1 and g_2 are the molar Gibbs free energies of pure species 1 and 2 respectively. At the same temperature, g represents the molar Gibbs free energy of the mixture. For a binary mixture with 40 mole % of species 1, the value (rounded off to the second decimal place) of $\frac{g}{RT}$ is ______. [2016 : 2 M]

9.17 The pressure of a liquid is increased isothermally. The molar volume of the liquid decreases from 50.45×10^{-6} m^3/mol to 48×10^{-6} m^3/mol during this process. The isothermal compressibility of the liquid is 10^{-9} Pa^{-1}, which can be assumed to be independent of pressure.
The change in the molar Gibbs free energy of the liquid, rounded to nearest integer, is ______ J/mol. [2017 : 2 M]

9.18 The vapour pressure of a pure substance at a temperature T is 30 bar. The actual and ideal gas values of $\frac{g}{RT}$ for the saturated vapour at this temperature T and 30 bar are 7,0 and 7,7. respectively. Here g is the molar Gibbs free energy and R is the universal gas constant.
The fugacity of the saturated liquid at these conditions, rounded to 1 decimal place, is ______ bar. [2017 : 2 M]

9.19 Consider a sealed rigid bottle containing CO_2 and H_2O at 10 bar and ambient temperature. Assume that the gas phase in the bottle is pure CO_2 and follows the ideal gas law. The liquid phase in the bottle contains CO_2 dissolved in H_2O and is an ideal solution. The Henry's constant at the system pressure and temperature is H_{CO_2} = 1000 bar. The equilibrium mole fraction of CO_2 dissolved in H_2O is ______ (rounded off to three decimal places).

[2019 : 1 M]

9.20 For a given binary system at constant temperature and pressure, the molar volume (in m^3/ mol) is given by : $\upsilon = 30x_A + 20x_B + x_A x_B(15x_A - 7x_B)$, where x_A and x_B are the mole fractions of components *A* and *B*, respectively. The volume change of mixing Δv_{mix} (in m^3/mol) at $x_A = 0.5$ is _____ (rounded off to one decimal place).

[2019 : 2 M]

9.21 A fractionator recovers 95 mol % *n*-propane as the distillate from an equimolar mixture of *n*-propane and *n*-butane. The condensate is a saturated liquid at 55°C. The Antoine equation is of the form $\ln\left(P^{sat}[\text{in bar}]\right) = A - \frac{B}{T[\text{in K}] + C}$, and the constants are provided below:

	A	B	C
n-propane	9.1058	1872.46	–25.16
n-butane	9.0580	2154.90	–34.42

Assuming Raoult's law, the condenser pressure (in bar) is_____ (rounded off to one decimal place).

[2019 : 2 M]

9.22 Mole fraction and activity coefficient of component 1 in a binary liquid mixture are x_1 and γ_1, respectively. G^E is excess molar Gibbs energy of the mixture, *R* is universal gas constant and *T* is absolute temperature of the mixture. Which one of the following is always true?

(a) $\lim_{x_1 \to 1} \frac{G^E}{RT} = 0$ (b) $\lim_{x_1 \to 1} \frac{G^E}{RT} = 0.5$

(c) $\lim_{x_1 \to 1} \gamma_1 = 0$ (d) $\lim_{x_1 \to 1} \gamma_1 = 0.5$

[2020 : 1 M]

9.23 A pure gas obeys the equation of the state given by

$$\frac{PV}{RT} = 1 + \frac{BP}{RT}$$

where *P* is the pressure, *T* is the absolute temperature, *V* is the molar volume of the gas, *R* is the universal gas constant, and *B* is a parameter independent of *T* and *P*. The residual molar Gibbs energy, G^R, of the gas is given by the relation

$$\frac{G^R}{RT} = \int_0^P (Z-1)\frac{dP}{P}$$

where *Z* is the compressibility factor that the integral is evaluated at constant *T*. If the value of B is 1×10^{-4} m^3 mol^{-1}, the residual molar enthalpy (in J mol^{-1} of the gas at 1000 kPa and 300 K is

(a) 100 (b) 300

(c) 2494 (d) 30000 [2020 : 2 M]

9.24 The molar excess Gibbs free energy (g^E) of a liquid mixture of *A* and *B* is given by

$$\frac{g^E}{RT} = x_A x_B [C_1 + C_2 (x_A - x_B)]$$

where x_A and x_B are the mole fraction of *A* and *B*, respectively, the universal gas constant, $R = 8.314$ J K^{-1} mol^{-1}, *T* is the temperature in *K*, and C_1, C_2 are temperature-dependent parameters. At 300 K, $C_1 = 0.45$ and $C_2 = -0.018$. If γ_A and γ_B are the activity coefficients of *A* and *B*, respectively, the value of $\int_0^1 \ln\left(\frac{\gamma_A}{\gamma_B}\right) dx_A$ at 300 K and 1 bar is ________ (rounded off to the nearest integer).

[2022 : 2 M]

9.25 The enthalpy (*H*, in J.mol.$^{-1}$) of a binary liquid system at constant temperature and pressure is given as

$$H = 40x_1 + 60x_2 + x_1x_2(4x_1 + 2x_2),$$

where x_1 and x_2 represent the mole fractions of species 1 and 2 in the liquid, respectively. Which one of the following is the CORRECT value of the partial molar enthalpy of species 1 at infinite dilution, $\bar{H}_1^\infty$ (in J.mol^{-1})?

(a) 100 (b) 42

(c) 64 (d) 40 [2023 : 2 M]

9.26 Consider a binary mixture of components *A* and *B* at temperature *T* and pressure *P*. Let $\bar{V}_A$ and $\bar{V}_B$ be the partial molar volumes of *A* and *B*, respectively. At a certain mole fraction of *A*, x_A

$$\left(\frac{\partial \bar{V}_A}{\partial x_A}\right)_{T,P} = 22 \text{ cm}^3 \text{ mol}^{-1} \text{ and } \left(\frac{\partial \bar{V}_B}{\partial x_A}\right)_{T,P} = -18 \text{ cm}^3 \text{ mol}^{-1}$$

The value of x_A, rounded off to 2 decimal places, is ________.

[2024 : 1 M]

Answers Solution Thermodynamics

9.1	(c)	**9.2**	(b)	**9.3**	(b)	**9.4**	(a)	**9.5**	(b)	**9.6**	(d)	**9.7**	(a)
9.8	(b)	**9.9**	(d)	**9.10**	(c)	**9.11**	(d)	**9.12**	(0.073)	**9.13**	(d)	**9.14**	(–58)
9.15	(0.40)	**9.16**	(1.6471)	**9.17**	(2450)	**9.18**	(14.89)	**9.19**	(0.01)	**9.20**	(1)	**9.21**	(17.94)
9.22	(a)	**9.23**	(a)	**9.24**	(0)	**9.25**	(b)	**9.26**	(0.45)				

Explanations Solution Thermodynamics

9.1 (c)

$$\therefore \quad \phi_i\, y_i\, P = \gamma_i\, x_i\, f_i^o$$

$$k_i = \frac{y_i}{x_i} = \frac{\gamma_i f_i^o}{\phi_i P}$$

$$\Rightarrow \quad \gamma_i = f(T, P, x_i)$$

$$f_i^o = f(T, P)$$

$$\phi_i = \phi_i(T, P, y_i) \equiv \phi_i(T, P) \text{ at low pressure}$$

Thus, $k_i = f(T, P, x_i)$

9.2 (b)

Vapour pressure of water at 300°C is given as

$$P_1^{sat} = 8592.7 \text{ kPa}$$

Fugacity, $F^{l\,sat} = 6738.9$ kPa

1 mole of water has liquid phase volume

$$V^l = 25.28 \text{ cm}^3$$

Fugacity at 9000 kPa can be calculated by

$$F^l = F^{l\,sat} \exp\left[\frac{V^l}{RT}(P - P^{sat})\right]$$

$$F^l = 6738.9 \times \exp\left[\frac{25.28\times10^{-6}}{8.314\times573}(9000 - 8592.7)\times10^3\right]$$

$$F^l = 6753.5 \text{ kPa}$$

9.5 (b)

Molar volume of a binary mixture is given by

$$\begin{aligned} v &= 220x_1 + 180x_2 + x_1x_2(90x_1 + 50x_2) \\ &= 220x_1 + 180(1-x_1) + x_1(1-x_1)[(90x_1 + 50(1-x_1)] \\ &= 220x_1 + 180 - 180x_1 + (x_1 - x_1^2)(90x_1 + 50 - 50x_1) \\ &= 40x_1 + 180 + 50x_1 + 40x_1^2 - 50x_1^2 - 40x_1^3 \\ &= 180 + 90x_1 - 10x_1^2 - 40x_1^3 \end{aligned}$$

$$\bar{v}_2 = v - x_1\frac{\partial v}{\partial x_1} \quad \text{...(i)}$$

$$\frac{\partial v}{\partial x_1} = 90 - 20x_1 - 120x_1^2$$

$$x_1\frac{\partial v}{\partial x_1} = 90x_1 - 20x_1^2 - 120x_1^3$$

By putting these values in equation (i)

$$\bar{v}_2 = 180 + 10x_1^2 + 80x_1^3$$

$\bar{v}_2$ at $x_2 = 0.3$ or $x_1 = 0.7$ is given by

$$\bar{v}_2\big|_{x_1=0.7} = 180 + 10\times0.7^2 + 80\times0.7^3 = 212.34$$

9.6 (d)

Equimolar mixture of two components is given

So, $x_1 = x_2 = 0.5$

$$P_1^{sat} = 180 \text{ kPa}$$

$$P_2^{sat} = 120 \text{ kPa}$$

From Raoult's law

$$P_T = x_1 P_1^{sat} + x_2 P_2^{sat}$$

$$P_T = 0.5\times180 + 0.5\times120$$

$$P_T = 150 \text{ kPa}$$

Now, $y_1 P_T = x_1 P_1^{sat}$

$$y_1 = \frac{0.5\times180}{150} = 0.6$$

9.7 (a)

At constant T and P molar density of a binary mixture is given by

$$\rho = 1 + x_2$$

Here x_2 is mole fraction of component-2.

Partial molar volume of component-1 at infinite dilution $\bar{V}_1^\infty = ?$

$$\text{Molar density} = \rho = 1 + x_2$$

$$\text{Molar volume } (V) = \frac{1}{1+x_2}$$

$$\bar{V}_1 = V - x_2\frac{\partial V}{\partial x_2}$$

$$\frac{\partial V}{\partial x_2} = -\frac{1}{(1+x_2)^2}$$

$$\bar{V}_1 = \frac{1}{(1+x_2)} + \frac{x_2}{(1+x_2)^2}$$

$$\bar{V}_1^\infty = \bar{V}_1\big|_{\text{at } x_1=0} = \frac{1}{2} + \frac{1}{2^2} = 0.75$$

9.8 (b)

Minimum work required to separate a binary gas mixture at a temperature T_0 and pressure P_0 is given as

$$w = -RT_0\left[y_1 \ln\left(\frac{\hat{f}_1}{f_{\text{pure, 1}}}\right) + y_2 \ln\left(\frac{\hat{f}_2}{f_{\text{pure, 2}}}\right)\right]$$

From Lewis Randall rule

$$\hat{f}_1 = y_1 f_{\text{pure 1}}$$

Similarly, $\hat{f}_2 = y_2 f_{\text{pure 2}}$

So, $w = -RT_0 [y_1 \ln y_1 + y_2 \ln y_2]$

9.9 (d)

Partial molar enthalpies of mixing is given as

$$\overline{\Delta H_1} = 3600\, x_2^2, \quad \overline{\Delta H_2} = 3600\, x_1^2$$

1 mole of benzene is added to 2 moles of cyclohexane.

So $x_1 = \frac{1}{1+2} = \frac{1}{3}$, $x_2 = \frac{2}{2+1} = \frac{2}{3}$

$$\Delta H_{\text{mixing}} = \Sigma x_1 \cdot \overline{\Delta H_i} = x_1\overline{\Delta H_1} + x_2\overline{\Delta H_2}$$

$$= \frac{1}{3}\times 3600\times\left(\frac{2}{3}\right)^2 + \frac{2}{3}\times 3600\times\left(\frac{1}{3}\right)^2$$

$$= 800 \text{ Joule}$$

9.10 (c)

Binary mixture of methyl ethyl ketone (component-1) and toluene (component-2) is given in question:

$$\ln\gamma_1 = x_2^2(\Psi_1 - \Psi_2 + 4\Psi_2 x_1) \quad \text{...(i)}$$

$$\ln\gamma_2 = x_1^2(\Psi_1 + \Psi_2 - 4\Psi_2 x_2) \quad \text{...(ii)}$$

Infinite dilution activity coefficients are given as

$$\ln\gamma_1^\infty = 0.4$$

$$\ln\gamma_2^\infty = 0.2$$

From equation (i) and (ii)

$$\ln\gamma_1^\infty = 0.4 = \Psi_1 - \Psi_2 \quad (\text{at } x_1 = 0)$$

$$\ln\gamma_2^\infty = 0.2 = \Psi_1 + \Psi_2 \quad (\text{at } x_2 = 0)$$

On solving,

$$\Psi_1 = 0.3, \quad \Psi_2 = -0.1$$

Now γ_1 and γ_2 at $x_2 = 0.9$ should be calculated.

$$\ln\gamma_1 = 0.9^2 [(0.3 + 0.1 + 4(-0.4)\times 0.1)]$$

$$\gamma_1 = 1.34$$

$$\ln\gamma_2 = 0.1^2 (0.3 - 0.1 - 4(-0.1)\times 0.9)$$

$$\gamma_2 = 1.005$$

Now from modified Roult's law

$$P_T = x_1\gamma_1 P_1^{\text{sat}} + x_2\gamma_2 P_2^{\text{sat}}$$

$$P_T = 0.1\times 1.314\times 36.9 + 0.9\times 1.005\times 12.3$$

$$P_T = 16.07 \text{ kPa}$$

9.11 (d)

$$G = H - TS, \quad dG = dH - TdS - SdT$$

At constant temperature and pressure

$$dG = dH - TdS, \quad \Delta G = \Delta H - T\Delta S$$

From given data in question,

$$\Delta G = 5x_1x_2 - RT[x_1 \ln x_1 + x_2 \ln x_2]$$

$$\Delta G = 5x_1(1-x_1) - RT[x_1 \ln x_1 + (1-x_1)\ln(1-x_1)]$$

For minimum Gibbs free energy change of mixing

$$\frac{dG}{dx_1} = 0$$

$$\frac{dG}{dx_1} = 5[(1-x_1) + x_1(-1)] - RT[1 + \ln x_1 + (-1) + \ln(1-x_1)(-1)]$$

On solving

$$\frac{dG}{dx_1} = 0 \text{ at } x_1 = 0.5$$

9.12 (0.073)

Liquid mixture is in equilibrium with its vapour at a temperature $T = 300$ K.

Liquid mole fraction of species 1 $(x_1) = 0.4$

Molar Gibbs free energy $G^E = 200$ J/mol

$$\ln\gamma_1 = 0.09 \text{ (given)}$$

$$\frac{G^E}{RT} = \Sigma x_i \ln\gamma_i = x_1 \ln\gamma_1 + x_2 \ln\gamma_2$$

$$\frac{200}{8.314\times 300} = 0.09\times 0.4 + 0.6 \ln\gamma_2$$

$$\ln\gamma_2 = 0.073$$

9.13 (d)

Molar residual Gibbs free energy (g^R) and molar residual volume V^R, are related as

$$\frac{g^R}{RT} = \int_0^P \frac{V^R}{RT} dP$$

At $T = 27°\text{C}$ and $P = 0.2$ MPa

$$\frac{g^R}{RT} = \int_0^P (Z-1)\frac{dP}{P} \quad \text{...(i)}$$

$$Z = 1 + \frac{BP}{RT}$$

where, $B = -10^{-4}\ \text{m}^3/\text{mol}$

By putting this value in equation (i)

$$\frac{g^R}{RT} = \int_0^P \frac{B}{RT} dP = \frac{BP}{RT}$$

So, $g^R = BP$

$$= -10^{-4}\times 0.2\times 10^6 = -20 \text{ J./mol}$$

9.14 (–58)

Partial molar enthalpy of species 1 in a binary mixture is given as

$$\bar{h}_1 = 2 - 60\,x_2^2 + 100\,x_1\,x_2^2$$

Partial molar enthalpy of species 1 at infinite dilution

$$(\bar{h}_1^{\infty}) = \bar{h}_1\Big|_{\text{At } x_1=0} = 2 - 60 = -58$$

9.15 (0.40)

A binary liquid mixture of benzene and toluene containing 20 mol% benzene.

Assuming benzene as component 1 and toluene as component 2.

$$P_1^{\text{sat}} = 92 \text{ kPa}, \quad P_2^{\text{sat}} = 35 \text{ kPa}$$

Now from Raoult's law:

$$P_T = x_1 P_1^{\text{sat}} + x_2 P_2^{\text{sat}}$$

$$P_T = 0.2 \times 92 + 0.8 \times 35 = 46.4 \text{ kPa}$$

Now, $y_1 \times P_T = x_1 P_1^{\text{sat}}$

$$y_1 \times 46.4 = 0.2 \times 92$$

$$y_1 = 0.396 \simeq 0.40$$

9.16 (1.6471)

$$x_1 = 0.4, \quad \frac{g_1}{RT} = 1$$

$$x_2 = 0.6, \quad \frac{g_2}{RT} = 2$$

$$\therefore \; g = (x_1 g_1 + x_2 g_2) + \Delta g^{\text{mix}} + g^E$$

$$\Delta_g^{\text{mix}} = RT \sum x_i \ln x_i = RT(x_1 \ln x_1 + x_2 \ln x_2)$$

$$\therefore \; g = (x_1 g_1 + x_2 g_2) + RT(x_1 \ln x_1 + x_2 \ln x_2) + g^E$$

$$\frac{g}{RT} = \left[x_1 \frac{g_1}{RT} + x_2 \frac{g_2}{RT}\right] + [x_1 \ln x_1 + x_2 \ln x_2] + \frac{g^E}{RT}$$

$$\frac{g}{RT} = [0.4 + 0.6 \times 2] + [0.4 \ln 0.4 + 0.6 \ln 0.6] + 3 \times 0.4 \times 0.6$$

$$\frac{g}{RT} = 1.6 - 0.6729 + 0.72 = 1.6471$$

9.17 (2450)

Isothermal compressibility of liquid is 10^{-9} Pa^{-1}

$$\beta = -\frac{1}{V}\frac{\partial V}{\partial P}\Big|_{\text{At constant temperature}} \quad \text{...(i)}$$

$$dG = VdP - SdT$$

For isothermal process $SdT = 0$

$$dG = VdP$$

From equation (i)

$$VdP = -\frac{\partial V}{\beta}$$

$$= \frac{(50.49 - 48) \times 10^{-6}}{10^{-9}} = 2450$$

So, $dG = 2450$ J/mol

9.18 (14.89)

$$P^{\text{sat}} = 30 \text{ bar (given)}$$

$$\frac{g^E}{RT} = \left(\frac{g}{RT}\right)_{\text{Actual}} - \left(\frac{g}{RT}\right)_{\text{Ideal}}$$

$$\frac{g^E}{RT} = 7.0 - 7.7 = -0.7$$

$$\frac{g^E}{RT} = \ln \phi = -0.7$$

$$\phi = 0.4965 = \frac{F^l}{P^{\text{sat}}}$$

Here, F^l = Fugacity of saturated liquid

$= 14.89$ bar

9.19 (0.01)

Rigid bottle containing CO_2 and H_2O at 10 bar

From Henry's law

$$P_A = HX_A$$

Here, H = Henry law constant

H_{CO_2} = 1000 bar (given)

X_A = Mole fraction of CO_2 dissolved in H_2O

P_A = Partial pressure of CO_2

So, $10 = 1000X_A$

$$X_A = 0.01$$

9.20 (1)

Molar volume $\upsilon = 30x_A + 20x_B + x_A x_B(15x_A - 7x_B)$

Molar volume of pure A, υ_A at $x_A = 1$

$$\upsilon_A = 30 \times 1 + 20 \times 0 + 0\,(15 \times 1 - 7 \times 0)$$

$$\upsilon_A = 30 \text{ m}^3/\text{mol}$$

Similarly molar volume of pure B, υ_B at $x_B = 1$

$$\upsilon_B = 30 \times 0 + 20 \times 1 + 0 = 20 \text{ m}^3/\text{mol}$$

Now property change of mixing is given by

$$\Delta M_{\text{mix}} = M - \sum x_i M_i \quad \text{...(i)}$$

M = Molar property of mixture

$\sum x_i M_i$ = Molar properties of pure components

Molar volume of mixture at $x_A = 0.5$ is given

$$\upsilon_{\text{mix}} = 30 \times 0.5 + 20 \times 0.5 + 0.5 \times 0.5\,[15 \times 0.5 - 7 \times 0.5] = 26 \text{ m}^3/\text{mol}$$

Now from equation (i) volume change of mixing

$$\Delta\upsilon_{\text{mix}} = 26 - (0.5 \times 30 + 0.5 \times 20) = 1 \text{ m}^3/\text{mol}$$

9.21 (17.94)

Fractionator recovers 95 mol % *n*-propane as distillate from equimolar mixture of *n*-propane and *n*-butane.

Condensate is saturated liquid at 55°C.

Antoine equation is given as

$$\ln P^{sat} \text{ (in bar)} = A - \frac{B}{T[\text{in K}] + C}$$

$$\ln P_1^{sat} = 9.1058 - \frac{1872.46}{328 - 25.16}$$

$$P_1^{sat} = 18.593 \text{ bar}$$

Similarly, $\ln P_2^{sat} = 9.0580 - \dfrac{2154.90}{328 - 34.42}$

$$P_2^{sat} = 5.57 \text{ bar}$$

Condenser pressure

$$P_T = \Sigma x_i P_1^{sat}$$

$$= 0.95 \times 18.593 + 0.05 \times 5.51 = 17.94 \text{ bar}$$

9.22 (a)

$$\frac{G^E}{RT} = x_1 \ln \gamma_1 + x_2 \ln \gamma_2$$

at $x_1 = 1,\ x_2 = 0$ and $\gamma_1 = 1$

By putting these to values in above expression

$$\lim_{x_1 \to 1} \frac{G^E}{RT} = 0$$

9.23 (a)

Pure gas obeys the following equation

$$\frac{PV}{RT} = 1 + \frac{BP}{RT}$$

Residual Gibbs energy $\dfrac{G^R}{RT}$ is given by

$$\frac{G^R}{RT} = \int_0^P (Z-1)\frac{dP}{P}$$

where, Z = Compressibility factor

Value of B is 1×10^{-4} m^3/mol

$$\frac{G^R}{RT} = \int_0^P \frac{BP}{RT}\cdot\frac{dP}{P} = \frac{BP}{RT}$$

$$G^R = 1 \times 10^{-4} \times 10^6 = 100 \text{ J/mol}$$

9.24 (0)

Given: $\dfrac{g^E}{RT} = x_A x_B\left[C_1 + C_2(x_A - x_B)\right]$

where, $C_1 = 0.45$ and $C_2 = -0.018$

$$\frac{g^E}{RT} = x_A x_B\left[0.45 - 0.018(x_A - x_B)\right]$$

$$\frac{g^E}{RT} = x_A x_B[0.45 - 0.018x_A + 0.018x_B]$$

$$\frac{d}{dx_A}\left[\frac{g^E}{RT}\right] = \ln\left[\frac{\gamma_A}{\gamma_B}\right]$$

$$\frac{g^E}{RT} = x_A(1 - x_A)[0.45 - 0.018x_A + 0.018 - 0.018x_A]$$

$$(x_A - x_A^2)[0.468 - 0.036x_A]$$

$$= 0.468x_A - 0.036x_A^2 - 0.468x_A^2 + 0.036x_A^3$$

$$\frac{d\left(\frac{g^E}{RT}\right)}{dx_A} = 0.468 - 0.072x_A - 0.936x_A + 0.108x_A^2$$

$$\ln\left(\frac{\gamma_A}{\gamma_B}\right) = 0.108x_A^2 - 1.008x_A + 0.468$$

$$\int_0^1 \ln\left(\frac{\gamma_A}{\gamma_B}\right)dx_A = \int_0^1 \left(0.108x_A^2 - 1.008x_A + 0.468\right)dx_A$$

$$= \left[\frac{0.108x_A^3}{3} - \frac{1.008x_A^2}{2} + 0.468x_A\right]_0^1$$

$$= [0.036 - 0.504 + 0.468] = 0$$

9.25 (b)

For $\overline{H}_1^\infty$, we have

$$\overline{H}_1 = H + X_2\frac{dH}{dX_1}$$

$$H = 40X_1 + 60X_2 + X_1X_2[4X_1 + 2X_2]$$

$$H = -20X_1 + 60 + 2X_1 - 2X_1^3 \text{ \{In terms of } X_1\text{\}}$$

Now, $\overline{H}_1 = 40X_1 + 60X_2 + X_1X_2[4X_1 + 2X_2] + X_2\dfrac{dH}{dX_1}$

$$\frac{dH}{dX_1} = -20 + 2 - 6X_1^2 = -18 - 6X_1^2$$

$$\therefore\ \overline{H}_1^\infty = 60 - 18 = 42 \qquad \{X_1 \to 0, X_2 \to 1\}$$

$$\overline{H}_1^\infty = 42$$

9.26 (0.45)

The partial molar volume of A and B in a binary mixture at temperature T and pressure P is given as

$$\left(\frac{\partial \overline{V}_A}{\partial x_A}\right)_{T,P} = 22 \text{ cm}^3\text{/mol and } \left(\frac{\partial \overline{V}_B}{\partial x_A}\right)_{T,P} = -18 \text{ cm}^3\text{/mol}$$

From Gibbs Duhem equation

$$\Sigma x_i d\overline{M}_i = 0$$

$$x_A\frac{\partial \overline{V}_A}{\partial x_A} + x_B\frac{\partial \overline{V}_B}{\partial x_A} = 0$$

$$x_A \times 22 + (1 - x_A)(-18) = 0$$

$$x_A = \frac{18}{40} = 0.45$$

■■■■

10 VLE

10.1 In a binary liquid solution of components A and B, if component A exhibits positive deviation from Raoult's law, then component B

(a) exhibits positive deviation from Raoult's law.
(b) exhibits negative deviation from Raoult's law.
(c) obeys Raoult's law.
(d) may exhibit either positive or negative deviation from Raoult's law. **[2000 : 1 M]**

10.2 The number of degrees of freedom for an azeotropic mixture of ethanol and water in vapour-liquid equilibrium, is

(a) 3 (b) 1
(c) 2 (d) 0 **[2002 : 1 M]**

10.3 When dilute aqueous solutions of two salts are mixed, the process is associated with

(a) decrease in temperature
(b) increase in temperature
(c) no change in temperature
(d) change in temperature which is a function of composition **[2003 : 1 M]**

10.4 At 60°C, vapour pressures of methanol and water are 84.562 kPa and 19.953 kPa respectively. An aqueous solution of methanol at 60°C exerts a pressure of 39.223 kPa; the liquid phase and vapour phase mole fractions of methanol are 0.1686 and 0.5714 respectively. Activity coefficient of methanol is

(a) 1.572 (b) 1.9398
(c) 3.389 (d) 4.238 **[2003 : 2 M]**

10.5 The number of degrees of freedom for an azeotropic mixture in a two component vapour-liquid equilibria is/are

(a) zero (b) one
(c) two (d) three **[2004 : 2 M]**

10.6 A process flow sheet analysis results in the degrees of freedom having a value of –2. Which one of the following steps must be next carried out?

(a) Identify and add two new independent equations from process model
(b) Remove two equations that have been wrongly assumed to be independent
(c) Assign values to two variables in the process
(d) Assign value to one variable and remove one equation that was wrongly assumed to be independent **[2005 : 1 M]**

10.7 The Van Laar activity coefficient model for a binary mixture is given by the form. Given $\gamma_1 = 1.40$, $\gamma_2 = 1.25$, $x_1 = 0.25$, $x_2 = 0.75$, determine the constants A^* and B^*.

(a) $A^* = 0.5$, $B^* = 0.3$
(b) $A^* = 3$, $B^* = 0.5$
(c) $A^* = 0.333$, $B^* = 0.2$
(d) $A^* = 2$, $B^* = 0.333$ **[2005 : 2 M]**

10.8 At a given temperature and pressure, a liquid mixture of benzene and toluene is in equilibrium with its vapor. The available degree(s) of freedom is (are)

(a) Zero (b) 1
(c) 2 (d) 3 **[2006 : 1 M]**

10.9 For a binary mixture of A and B at 400 K and 1 atm, which ONE of the following equilibrium states deviates significantly from ideality?

Given: $\ln(P_A^{sat}) = 6.2 - \dfrac{2758}{T}$

where, P_A^{sat} = vapour pressure of A, atm;
T = Temperature, K
x_A = mole fraction of A in liquid;
y_A = mole fraction of A in vapor

(a) $x_A = 0.5$; $y_A = 0.25$
(b) $x_A = 0.5$; $p_A = 0.25$
(c) $x_A = 0.5$; $p_A = 0.5$
(d) $x_A = 0.6$; $y_A = 0.3$ **[2006 : 2 M]**

10.10 A methanol-water vapor liquid system is at equilibrium at 60°C and 60 kPa. The mole fraction of methanol in liquid is 0.5 and in vapor is 0.8. Vapor pressure of methanol and water at 60°C are 85 kPa and 20 kPa respectively. Assuming vapor phase to be an ideal gas mixture, what is the activity coefficient of water in the liquid phase?

(a) 0.3 (b) 1.2
(c) 1.6 (d) 7.5 **[2007 : 2 M]**

10.11 What is the excess Gibbs free energy (g^E, in J/mol) of the liquid mixture?

(a) 9.7 (b) 388
(c) 422 (d) 3227

[2007 : 2 M]

Statement for Linked Answer Question (10.12 to 10.13):

A binary mixture containing species 1 and 2 forms an azeotrope at 105.4°C and 1.013 bar. The liquid phase mole fraction of component 1 (x_1) of this azeotrope is 0.62. At 105.4°C, the pure component vapor pressures for species 1 and 2 are 0.878 bar and 0.665 bar, respectively. Assume that the vapour phase is an ideal gas mixture. The van Laar constants, A and B, are given by the expressions:

$$A = \left[1+\frac{x_2 \ln\gamma_2}{x_1 \ln\gamma_1}\right]^2 \ln\gamma_1,\ B = \left[1+\frac{x_1 \ln\gamma_1}{x_2 \ln\gamma_2}\right]^2 \ln\gamma_2$$

10.12 The activity coefficients (γ_1, γ_2) under these conditions are

(a) (0.88, 0.66) (b) (1.15, 1.52)
(c) (1.52, 1.15) (d) (1.52, 0.88)

[2008 : 2 M]

10.13 The Van Laar constants (A, B) are

(a) (0.92, 0.87) (b) (1.00, 1.21)
(c) (1.12, 1.00) (d) (1.52, 1.15)

[2008 : 2 M]

10.14 Dehydrogenation of ethane, $C_2H_6\,(g) \rightarrow C_2H_4\,(g) + H_2\,(g)$, is carried out in a continuous stirred tank reactor (CSTR). The feed is pure ethane. If the reactor exit stream contains unconverted ethane along with the products, then the number of degrees of freedom for the CSTR is

(a) 1 (b) 2
(c) 3 (d) 4

[2009 : 1 M]

10.15 For a binary mixture at constant temperature and pressure, which one of the following relations between activity coefficient (γ_i) and mole fraction (x_i) is thermodynamically consistent?

(a) $\ln\gamma_1 = -1+2x_1 - x_1^2, \ln\gamma_2 = \frac{1}{2}x_1^2$

(b) $\ln\gamma_1 = -1+2x_1 - x_1^2, \ln\gamma_2 = x_1^2$

(c) $\ln\gamma_1 = -1+2x_1 - x_1^2, \ln\gamma_2 = -\frac{1}{2}x_1^2$

(d) $\ln\gamma_1 = -1+2x_1 - x_1^2, \ln\gamma_2 = -x_1^2$

[2009 : 2 M]

10.16 A gaseous system contains H_2, I_2, and HI, which participate in the gas-phase reaction

$$2HI \rightleftharpoons H_2 + I_2$$

At a state of reaction equilibrium, the number of thermodynamic degrees of freedom is ______ .

[2013 : 1 M]

10.17 From the following list, identify the properties which are equal in both vapour and liquid phases at equilibrium:

P. Density
Q. Temperature
R. Chemical potential
S. Enthalpy

(a) P and Q only (b) Q and R only
(c) R and S only (d) P and S only

[2014 : 1 M]

10.18 Consider a binary liquid mixture at equilibrium with its vapour at 25° C.

Antoine equation for this system is given as

$\log_{10} p_i^{sat} = A - \frac{B}{t+C}$ where t is in °C and p in torr. The Antoine constants (A, B and C) for the system are given in the following table.

Component	A	B	C
1	7.0	1210	230
2	6.5	1206	223

The vapour phase is assumed to be ideal and the activity coefficients (γ_i) for the non-ideal liquid phase are given by

$\ln(\gamma_1) = x_2^2[2 - 0.6x_1]$, $\ln(\gamma_2) = x_1^2[1.7 - 0.6x_2]$

If the mole fraction of component 1 in liquid phase (x_1) is 0.11, then the mole fraction of component 1 in vapour phase (y_1) is ______.

[2014 : 2 M]

10.19 A binary mixture of components (1) and (2) forms an azeotrope at 130°C and $x_1 = 0.3$. The liquid phase non-ideality is described by $\ln\gamma_1 = Ax_2^2$ and $\ln\gamma_2 = Ax_1^2$, where γ_1, γ_2 are the activity coefficients, and x_1, x_2 are the liquid phase mole fractions. For both components, the fugacity coefficients are 0.9 at the azeotropic composition. Saturated vapor pressures at 130° C are P_1^{sat} = 70 bar and P_2^{sat} = 30 bar.

The total pressure in bars for the above azeotropic system (up to two decimal places) is ______.

[2015 : 2 M]

10.20 A sparingly soluble gas (solute) is in equilibrium with a solvent at 10 bar. The mole fraction of the solvent in the gas phase is 0.01, At the operating temperature and pressure, the fugacity coefficient of the solute in the gas phase and the Henry's law constant are 0.92 and 1000 bar, respectively. Assume that the liquid phase obeys Henry's law.

The MOLE PERCENTAGE of the solute in the liquid phase, rounded to 2 decimal places, is ______. **[2017 : 2 M]**

10.21 G denotes the Gibbs free energy of a binary mixture, n_T denotes the total number of moles present in the system, μ_i is the chemical potential of the ith component ($\mu_i \neq 0$ and $\mu_1 > \mu_2$) and x_i is the mole fraction of the ith component. The correct variation of $\frac{G}{nT}$ (in J/mol) at constant temperature and pressure is given by

(a) 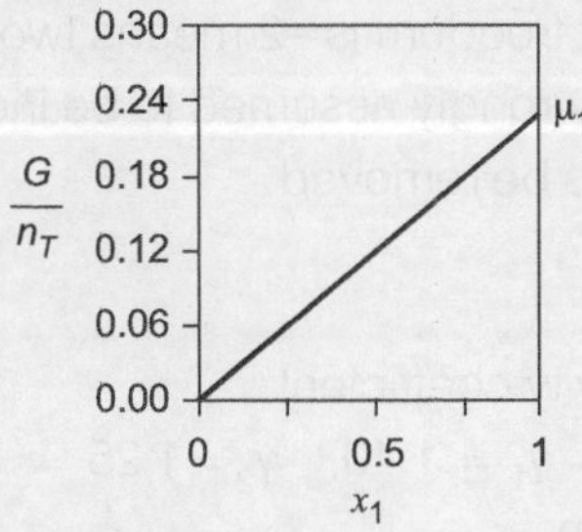

(b) 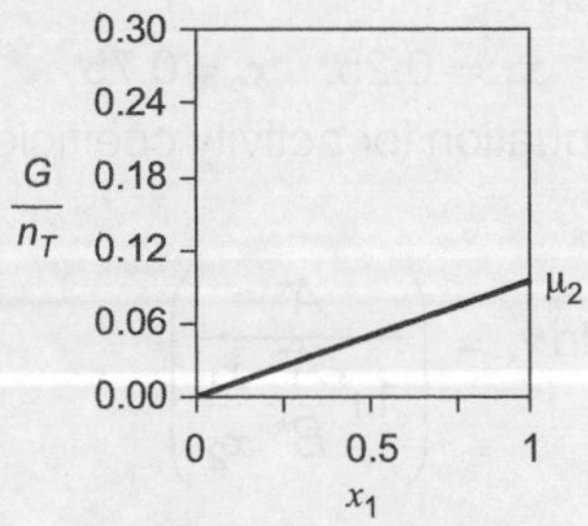

(c) 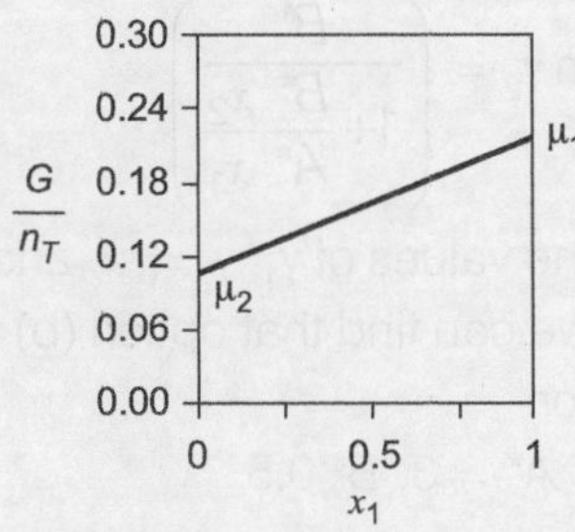

(d) 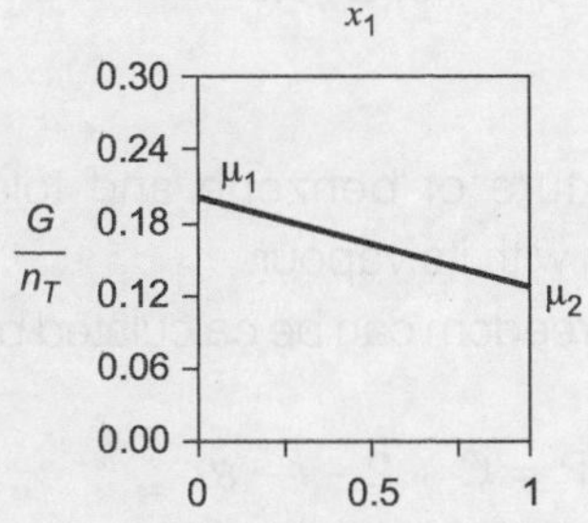

[2018 : 2 M]

10.22 An azeotropic mixture of ethanol and water is to be separated in a distillation column using benzene as an entrainer. At the column operating conditions, two liquid phases are formed on a tray. The degree(s) of freedom of the system for the choice of intensive properties at equilibrium is/are ______. **[2018 : 2 M]**

10.23 For a binary non-ideal A-B mixture exhibiting a minimum boiling azeotrope, the activity coefficients, γ_i ($i = A, B$), must satisfy

(a) $\gamma_A > 1, \gamma_B > 1$
(b) $\gamma_A < 1, \gamma_B > 1$
(c) $\gamma_A = 1, \gamma_B = 1$
(d) $\gamma_A < 1, \gamma_B < 1$

[2019 : 1 M]

10.24 A binary liquid mixture consists of two species 1 and 2. Let γ and x represent the activity coefficient and the mole fraction of the species, respectively. Using a molar excess Gibbs free energy model, $\ln \gamma_1$ vs. x_1 and $\ln \gamma_2$ vs. x_1 are plotted. A tangent drawn to the $\ln \gamma_1$ vs. x_1 curve at a mole fraction of $x_1 = 0.2$ has a slope $= -1.728$.

The slope of the tangent drawn to the $\ln \gamma_2$ vs. x_1 curve at the same mole fraction is ______ (correct to 3 decimal places). **[2021 : 2 M]**

10.25 Water in a container at 290 K is exposed to air containing 3 % CO_2 by volume. Air behaves like an ideal gas and is maintained at 100 kPa pressure. The liquid phase comprising of dissolved CO_2 in water behaves like an ideal solution. Use Henry's constant of CO_2 dissolved in water at 290 K as 12 MPa. Under equilibrium conditions, which one of the following is the CORRECT value of the mole fraction of CO_2 dissolved in water?

(a) 2.9×10^{-4}
(b) 0.9×10^{-4}
(c) 2.5×10^{-4}
(d) 0.5×10^{-4}

[2023 : 2 M]

10.26 Consider a vapour-liquid mixture of components A and B that obeys Raoult's law. The vapour pressure of A is half that of B. The vapour phase concentrations of A and B are 3 mol m^{-3} and 6 mol m^{-3}, respectively. At equilibrium, the ratio of the liquid phase concentration of A to that of B is

(a) 1.0
(b) 0.5
(c) 2.0
(d) 1.5

[2024 : 1 M]

Answers VLE

10.1	(a)	**10.2**	(b)	**10.3**	(d)	**10.4**	(a)	**10.5**	(b)	**10.6**	(b)	**10.7**	(b)
10.8	(c)	**10.9**	(c)	**10.10**	(b)	**10.11**	(c)	**10.12**	(b)	**10.13**	(c)	**10.14**	(c)
10.15	(d)	**10.16**	(3)	**10.17**	(b)	**10.18**	(0.657)	**10.19**	(27.54)	**10.20**	(0.91)	**10.21**	(c)
10.22	(2)	**10.23**	(a)	**10.24**	(0.432)	**10.25**	(c)	**10.26**	(a)				

Explanations VLE

10.1 (a)

In a binary liquid solution if one component exhibits positive deviation from Raoult's law, then other component also exhibits positive deviation from Raoult's law.

10.2 (b)

For phase rule we have

[F = C – P + 2 – *r* – *s*]

C = Number of components
= 2 (water and ethanol)
P = Number of phases
= 2 (liquid and vapour)
r = 0 (No reaction)
s = Special constraints (azeotropes)
F = 2 – 2 + 2 – 1 = 1

10.3 (d)

Mixing causes change in temperature which is a function of composition.

10.4 (a)

Vapour pressure of methanol (P_1^{sat}) = 84.562 kPa
Vapour pressure of water (P_2^{sat}) = 19.953 kPa
Mole fraction of methanol in solution (x_1) = 0.1686
Mole fraction of methanol in vapour (y_1) = 0.5714
From modified Roult's law:

$$y_1 \times P_T = \gamma_1 x_1 P_1^{sat}$$

$$0.5714 \times 39.223 = \gamma_1 \times 0.1686 \times 84.562$$

$$\gamma_1 = 1.572$$

10.5 (b)

Given mixture forms an azeotropes in two component vapour liquid equilibrium.
Degree of freedom can be found by following equation.

$$F + P = C + 2 - r - s$$

F = Degree of freedom
C and *P* are number of component and phases respectively

$$s = 1 \text{ (azeotropes)}$$

$$F = 2 - 2 + 2 - 1 = 1$$

10.6 (b)

Degree of freedom can be define as the difference between number of independent equations and variables that have known values.
If degree of freedom is –2 means two equations have been wrongly assumed to be independent which has to be removed.

10.7 (b)

Given, activity coefficient

$$\gamma_1 = 1.40, \quad \gamma_2 = 1.25$$

Mole fraction

$$x_1 = 0.25, \quad x_2 = 0.75$$

Van Laar equation for activity coefficient is given as

$$\ln\gamma_1 = \left(\frac{A^*}{1+\frac{A^*}{B^*}\frac{x_1}{x_2}}\right)$$

$$\ln\gamma_2 = \left(\frac{B^*}{1+\frac{B^*}{A^*}\frac{x_2}{x_1}}\right)$$

By putting the values of γ_1, $\gamma_2 x_1$, x_2 and checking all options we can find that option (b) is Correct option

$$A^* = 3, B^* 0.5$$

10.8 (c)

Liquid mixture of benzene and toluene is in equilibrium with its vapour.
Degree of freedom can be calculated by following equation:

$$F + P = C + 2 - r - s$$

F = Degree of freedom
P = Number of phases
C = Number of components
r = Number of independent reaction
s = Special constrains

So, $F = 2 + 2 - 2$

$F = 2$

10.9 (c)

Given,

$$\ln(P_A^{sat}) = 6.2 - \frac{2758}{T}$$

$$T = 400\text{ K}$$

So, $$\ln(P_A^{sat}) = 6.2 - \frac{2758}{400}$$

$$P_A^{sat} = 0.5\text{ atm}$$

From Roult's law:

$$P_A = x_A P_A^{sat}$$

$$y_A \times P_T = x_A \cdot P_A^{sat}$$

$$P_T = 1\text{ atm } P_A^{sat} = 0.5\text{ atm}$$

$$y_A = 0.5x_A$$

From given options, option (c) $x_A = 0.5$ and $P_A = 0.5$ deviates significantly from ideality.

10.10 (b)

Methanol and water are equilibrium at 60°C and 60 kPa

Methanol mole fraction $(x_1) = 0.5$

Water mole fraction $(x_2) = 0.5$

Mole fraction of methanol in vapour $y_1 = 0.8$

$$P_1^{sat} = 85\text{ kPa},\quad P_2^{sat} = 20\text{ kPa}$$

From Roult's modified law:

$$y_2 P_T = x_2 \gamma_2 P_2^{sat}$$

$$(1-0.8)\times 60 = 0.5\times\gamma_2\times 20$$

$$\gamma_2 = 1.2$$

Similarly for methanol

$$0.8\times 60 = 0.5\times\gamma_1\times 85$$

$$\gamma_1 = 1.129$$

10.11 (c)

$$\frac{G^E}{RT} = \Sigma x_i \ln\gamma_i$$

$$\frac{G^E}{8.314\times(60+273)} = 0.5\ln 1.2 + 0.5\ln 1.129$$

$$G^E = 421\text{ J/mol}$$

10.12 (b)

Binary mixture forms an azeotropes at 105.4°C and 1.013 bar

$$x_1 = 0.62$$

$$P_1^{sat} = 0.878\text{ bar}$$

$$P_2^{sat} = 0.665\text{ bar}$$

From Roult's modified law,

$$y_1 P_T = x_1 P_1^{sat}\cdot\gamma_1$$

For azeotropes

$$x_1 = y_1$$

So, $$\gamma_1 = \frac{P_T}{P_1^{sat}} = \frac{1.013}{0.878} = 1.15$$

Similarly, $$\gamma_2 = \frac{P_T}{P_2^{sat}} = \frac{1.013}{0.665} = 1.52$$

10.13 (c)

Van Laar constants A and B are given by expressions:

$$A = \left[1 + \frac{x_2\ln\gamma_2}{x_1\ln\gamma_1}\right]^2 \ln\gamma_1$$

$$B = \left[1 + \frac{x_1\ln\gamma_1}{x_2\ln\gamma_2}\right]^2 \ln\gamma_2$$

$$x_1 = 0.62,\quad x_2 = 1 - x_1 = 0.38$$

$$\gamma_1 = 1.15,\quad \gamma_2 = 1.52$$

By putting all these values

$$A = 1.12,\quad B = 1.00$$

10.14 (c)

Degree of freedom can be calculated by following formula :

$$F + P = C + 2 - r - s$$

The reaction that is occurring in CSTR is

$$C_2H_6\,(g) \rightarrow C_2H_4\,(g) + H_2\,(g)$$

Number of species $(C) = 3$

r = Number of independent reaction = 1

From design equation of CSTR

$$V = \frac{F_{Ao}\times X_A}{-r_A}$$

Number of unknown variable = 4

p = Number of phases = 1

So, degree of freedom,

$$F = 3 + 2 - 1 - 1$$

$$F = 3$$

10.15 (d)

Thermodynamic consistency of system can be check by following equation:

$$\Sigma x_i \, d\ln\gamma_i = 0$$

$$x_1 \frac{d\ln\gamma_1}{dx_1} + x_2 \frac{d\ln\gamma_2}{dx_1} = 0 \qquad \text{...(i)}$$

From option (d),

$$\ln\gamma_1 = -1 + 2x_1 - x_1^2$$

$$\frac{d\ln\gamma_1}{dx_1} = 2 - 2x_1$$

$$\frac{\ln\gamma_2}{dx_1} = -2x_1$$

Now by putting these values in equation (i)

$$x_1(2-2x_1) + x_2(-2x_1) = 0$$

$$2x_1 - 2x_1^2 + (1 - x_1)(-2x_1) = 0$$

$$2x_1 - 2x_1^2 - 2x_1 + 2x_1^2 = 0$$

$$0 = 0$$

$$\text{LHS} = \text{RHS}$$

So, option (d) is thermodynamically consistent.

10.16 (3)

A gaseous system contains H_2, I_2 and HI and following reaction occurs

$$2\,HI \rightleftharpoons H_2 + I_2$$

Thermodynamic degree of freedom can be found by following relation

$$F + P = C + 2 - r - s$$

Here, F = Degree of freedom

P = Number of phases

C = Number of components

r = Number of independent reactions

s = Special constrains

For given question

$$F + 1 = 3 + 2 - 1$$

$$F = 3$$

10.17 (b)

When vapour and liquid phases are at equilibrium, then temperature and chemical potential of two phases are equal.

10.18 (0.657)

Binary liquid mixture at equilibrium with vapour pressure at 25°C.

Antoine equation is given as

$$\log_{10} P_1^{sat} = A - \frac{B}{t+C}$$

For component 1:

$$\log_{10} P_1^{sat} = 7 - \frac{1210}{25+230}$$

$$P_1^{sat} = 179.846 \text{ torr.}$$

For component 2:

$$\log_{10} P_2^{sat} = 6.5 - \frac{1206}{25+223}$$

$$P_2^{sat} = 43.36 \text{ torr.}$$

$$\ln\gamma_1 = x_2^2(2 - 0.6\,x_1)$$

$$x_1 = 0.11 \text{ (given)}$$

So, $x_2 = 1 - 0.11 = 0.89$

$$\ln\gamma_1 = 0.89^2(2 - 0.6 \times 0.11) = 1.531$$

$$\gamma_1 = 4.62$$

$$\ln\gamma_2 = x_1^2(1.7 + 0.6\,x_2)$$

$$\ln\gamma_2 = 0.11^2(1.7 + 0.6 \times 0.89)$$

$$\gamma_2 = 1.027$$

Now, $P_T = x_1\gamma_1 P_1^{sat} + x_2\gamma_2 P_2^{sat}$

From modified Roult's law

$$P_T = 0.11 \times 4.62 \times 179.846 + 0.89 \times 1.027 \times 43.36$$

$$= 131.03 \text{ torr}$$

Now again from modified Roult law

$$y_1 P_T = x_1\gamma_1 P_1^{sat}$$

$$y_1 \times 131.03 = 0.11 \times 4.62 \times 179.846$$

$$y_1 = 0.657$$

10.19 (27.54)

Binary mixture of components (1) and (2) forms an azeotrope at 130°C and $x_1 = 0.3$. Liquid phase non ideality is described by

$$\ln\gamma_1 = A\,x_2^2$$

$$\ln\gamma_2 = A\,x_1^2$$

For both components fugacity coefficient are 0.9

$$P_1^{sat} = 70 \text{ bar}$$

$$P_2^{sat} = 30 \text{ bar}$$

From Roult's law

$$\phi_1 y_1 P_T = \gamma_1 x_1 P_1^{sat}$$

For azeotropes

$$x_1 = y_1$$

So, $\gamma_1 \propto \frac{1}{P_1^{sat}}$

$$\frac{\gamma_1}{\gamma_2} = \frac{30}{70}$$

$$\ln\left(\frac{\gamma_1}{\gamma_2}\right) = \ln\left(\frac{3}{7}\right) = A(x_2^2 - x_1^2)$$

$$\ln\left(\frac{3}{7}\right) = A(0.7^2 - 0.3^2)$$

$$A = -2.118$$
$$\ln \gamma_1 = A x_2^2$$
$$\gamma_1 = 0.354$$
$$\ln \gamma_2 = A x_1^2$$
$$\gamma_2 = 0.8264$$

Now, $\phi_1 P_T = x_1 \gamma_1 P_1^{sat} + x_2 \gamma_2 P_2^{sat}$

$$0.9 P_T = 0.3 \times 0.354 \times 70 + 0.7 \times 0.8264 \times 30$$

So, $P_T = 27.54$ bar

10.20 (0.91)

Sparingly soluble gas is in equilibrium with solvent at 10 bar.

Mole fraction of solvent in gas phase = 0.01

So mole fraction of solute in gas phase = 1 – 0.01 = 0.99

Fugacity coefficient = 0.92

Henry law constant = 1000 bar

From Henry's law

$$\phi P_A = x_A H_A$$
$$0.92 \times 0.99 \times 10 = x_A \times 1000$$
$$x_A = 9.108 \times 10^{-3}$$

Mole percentage $= 100 x_A = 0.91$ mol %

10.21 (c)

$$\frac{G}{n_T} = \Sigma \mu_i x_i + RT \Sigma x_i \ln x_i + g^E$$

(g^E) Molar excess Gibbs free energy.

For ideal mix = 0

$$\frac{G}{n_T} = \mu_1 x_1 + \mu_2 x_2 + RT(x_1 \ln x_1 + x_2 \ln x_2)$$

when, $x_1 = 0, \ \frac{G}{n_T} = \mu_2$

when, $x_1 = 1, \ \frac{G}{n_T} = \mu_1$

and ($\mu_1 > \mu_2$) (given)

Only option (c) follows these criteria.

10.22 (2)

Degree of freedom $F = C + 2 - P - r - s$

Component, $C = 3$

Phase, $P = 2$

$s = 1$ due to azeotrope mix

Then, $F = 3 + 2 - 2 - 1 = 2$

10.23 (a)

For minimum boiling azeotrope both component must show positive deviation from Roult's law so $\gamma_A > 1, \gamma_B > 1$.

10.24 (0.432)

From Gibbs Duhem equation

$$\Sigma x_i d \ln \gamma_i = 0$$
$$\Rightarrow \quad x_1 d \ln \gamma_1 + x_2 d \ln \gamma_2 = 0$$
$$x_1 \frac{d \ln \gamma_1}{dx_1} + x_2 \frac{d \ln \gamma_2}{dx_1} = 0$$
$$\because \quad \frac{d \ln \gamma_1}{dx_1} = -1.728 \text{ at } x_1 = 0.2$$
$$\therefore \quad x_2 \frac{d \ln \gamma_2}{dx_1} = -0.2 \times (-1.728)$$
$$\frac{d \ln \gamma_2}{dx_1} = 0.432$$

10.25 (c)

Given: $y_{CO_2} = 0.03, \ P_T = 100$ kPa

$$H_{CO_2} = 12 \times 10^3 \text{ kPa}$$

From Henry law, we have

$$0.03 \times 100 = 12 \times 10^3 \times x_{CO_2}$$
$$x_{CO_2} = 2.5 \times 10^{-4}$$

10.26 (a)

Vapour phase concentration of $A = 3$ mol/m^3

Vapour phase concentration of $B = 6$ mol/m^3

Vapour phase mole fraction of

$$A(y_{A_1}) = \frac{3}{3+6} = \frac{1}{3}$$

Vapour phase mole fraction of $B \ y_B = \frac{2}{3}$

From Raoult's law,

$$y_1 P_T = x_1 P_1^{sat} \quad \text{...(i)}$$

and $$y_2 P_T = x_2 P_2^{sat} \quad \text{...(ii)}$$

$$P_1^{sat} = \frac{1}{2} P_2^{sat}$$

From equation (i) / (ii)

$$\frac{y_1}{y_2} = \left(\frac{x_1}{x_2}\right) \frac{P_1^{sat}}{P_2^{sat}}$$
$$\frac{\frac{1}{3}}{\frac{2}{3}} = \left(\frac{x_1}{x_2}\right)\left(\frac{1}{2}\right)$$
$$\frac{x_1}{x_2} = 1$$

11 Chemical Reaction Equilibria

11.1 At a given temperature, K_1, K_2 and K_3 are the equilibrium constants for the following reactions 1, 2, 3 respectively:

$$CH_4\,(g) + H_2O \rightleftharpoons CO\,(g) + 3H_2\,(g) \quad ...(1)$$
$$CO\,(g) + H_2O\,(g) \rightleftharpoons CO_2\,(g) + H_2\,(g) \quad ...(2)$$
$$CH_4\,(g) + 2H_2O\,(g) \rightleftharpoons CO_2\,(g) + 4H_2\,(g) ...(3)$$

Then K_1, K_2 and K_3 are related as

(a) $K_3 = K_1K_2$ (b) $K_3 = (K_1K_2)^{0.5}$

(c) $K_3 = \frac{(K_1K_2)}{2}$ (d) $K_3 = (K_1K_2)^2$

[2000 : 2 M]

11.2 Industrial grade methanol can be produced according to the reaction

$$CO\,(g) + 2H_2\,(g) \rightleftarrows CH_3OH\,(g)$$

For this reaction $\Delta G^\circ_{400} = -1.3484$ kJ. If an equimolar mixture of CO and H_2 is fed to a reactor maintained at 400 K and 10 bar, determine the fraction of CO that is converted into CH_3OH at equilibrium. Assume that the reaction mixture behaves like an ideal gas.

[2000 : 5 M]

11.3 At standard conditions,

$$N_2 + 2O_2 \Leftrightarrow 2NO_2;\ \Delta G^0 = 100 \text{ kJ/mol}$$

$$NO + \frac{1}{2}O_2 \Leftrightarrow NO_2;\ \Delta G^0 = -35 \text{ kJ/mol}$$

The standard free energy of formation of NO in kJ/mol is

(a) 15 (b) 30

(c) 85 (d) 170

[2004 : 2 M]

11.4 For a reversible exothermic gas phase reaction, $A + B \rightleftharpoons C$, the equilibrium conversion will increase with

(a) increase in pressure and increase in temperature

(b) decrease in pressure and increase in temperature

(c) increase in pressure and decrease in temperature

(d) decrease in pressure and decrease in temperature

[2006 : 2 M]

11.5 Vapor phase hydration of C_2H_4 to ethanol by the following reaction

$$C_2H_4\,(g) + H_2O\,(g) \leftrightarrow C_2H_5OH\,(g)$$

attains equilibrium at 400 K and 3 bar. The standard Gibbs free energy change of reaction at these conditions is $\Delta g^\circ = 4000$ J/mol. For 2 moles of an equimolar feed of ethylene and steam, the equation in terms of the extent of reaction ε (in mols) at equilibrium is

(a) $\frac{\varepsilon(2-\varepsilon)}{(1-\varepsilon)^2} - 0.3 = 0$ (b) $\frac{(1-\varepsilon)^2}{\varepsilon(2-\varepsilon)} - 0.9 = 0$

(c) $\frac{\varepsilon}{(1-\varepsilon)^2} - 0.3 = 0$ (d) $\frac{\varepsilon(2-\varepsilon)}{(1-\varepsilon)^2} - 0.9 = 0$

[2007 : 2 M]

11.6 Match the following:

Group-1	Group-2
P. $\left(\frac{\partial G}{\partial n_i}\right)_{T,P,n_{j\neq i}}$	I. Arrhenius equation
Q. $\left(\frac{\partial G}{\partial n_i}\right)_{S,V,n_{j\neq i}}$	II. Reaction equilibrium constant
R. $\exp\left(\frac{-\Delta G^0_{reaction}}{RT}\right)$	III. Chemical potential
S. $\sum(n_i\, d\mu_i)_{T,P} = 0$	IV. Gibbs-Duhem equation

(a) Q-III, R-I, S-II (b) Q-III, R-II, S-IV

(c) P-III, R-II, S-IV (d) P-III, R-IV, S-I

[2014 : 1 M]

11.7 For a gas phase cracking reaction $A \rightarrow B + C$ at 300°C, the Gibbs free energy of the reaction at this temperature is $\Delta G^\circ = -2750$ J/mol. The pressure is 1 bar and the gas phase can be assumed to be ideal. The universal gas constant $R = 8.314$ J/mol.K. The fractional molar conversion of A at equilibrium is:

(a) 0.44 (b) 0.50
(c) 0.64 (d) 0.80

[2015 : 1 M]

11.8 The reversible reaction of t-butyl alcohol (TBA) and ethanol (EtOH) to ethyl t-butyl ether (ETBE) is

$$\text{TBA + EtOH} \rightleftharpoons \text{ETBE + Water}$$

The equilibrium constant for this reaction is $K_c = 1$. Initially 74 g of TBA is mixed with 100 g of aqueous solution containing 46 weight % ethanol. The molecular weights are: 74 g/mol for TBA, 46 g/mol for EtOH, 102 g/mol for ETBE, and 18 g/mol for water.

The mass of ETBE at equilibrium, rounded to one decimal place, is ______ g.

[2017 : 2 M]

11.9 For a single component system, vapor (subscript *g*) and liquid (subscript *f*) coexist in mechanical, thermal and phase equilibrium when

(a) $u_g = u_f$ (equality of specific internal energy)
(b) $h_g = h_f$ (equality of specific enthalpy)
(c) $s_g = s_f$ (equality of specific entropy)
(d) $g_g = g_f$ (equality of specific Gibbs free energy)

[2019 : 1 M]

11.10 Carbon monoxide (CO) reacts with hydrogen sulphide (H_2S) at a constant temperature of 800 K and a constant pressure of 2 bar as:

$$CO + H_2S \rightleftarrows COS + H_2$$

The Gibbs free energy of the reaction Δg°_{rxn} = 22972.3 J/mol and universal gas constant R = 8.314 J/(mol K). Both the reactants and products can be assumed to be ideal gases. if initially only 4 mol of H_2S and 1 mol of CO are present, the extent of the reaction (in mol) at equilibrium is ______ (rounded off to two decimal places).

[2019 : 2 M]

11.11 Consider the gas phase reaction $N_2O_4 \rightleftarrows 2NO_2$ occurring in an isothermal and isobaric reactor maintained at 298 K and 1.0 bar. The standard Gibbs energy change of the reaction at 298 K is ΔG^{o}_{298} = 5253 J mol^{-1}. The standard states are those of pure ideal gases at 1.0 bar. The equilibrium mixture in the reactor behaves as an ideal gas. The value of the universal gas constant is 8.314 J mol^{-1} K^{-1}. If one mole of pure N_2O_4 is initially charged to the reactor, the fraction of N_2O_4 that decomposes into NO_2 at equilibrium is

(a) 0 (b) 0.17
(c) 0.38 (d) 1

[2020 : 2 M]

■■■■

Answers Chemical Reaction Equilibria

11.1	(a)	**11.2**	(0.439)	**11.3**	(c)	**11.4**	(c)	**11.5**	(d)	**11.6**	(c)	**11.7**	(d)
11.8	(20.4)	**11.9**	(d)	**11.10**	(0.29)	**11.11**	(b)						

Explanations Chemical Reaction Equilibria

11.1 (a)

$$CH_4\,(g) + H_2O\,(g) \rightleftharpoons CO\,(g) + 3H_2\,(g)$$

$$K_1 = \frac{[CO]\,[H_2]^3}{[H_2O][CH_4]} \quad ...(1)$$

$$CO\,(g) + H_2O\,(g) \rightleftharpoons CO_2\,(g) + H_2\,(g)$$

$$K_2 = \frac{[CO_2]\,[H_2]}{[H_2O][CO]} \quad ...(2)$$

$$CH_4\,(g) + 2H_2O\,(g) \rightleftharpoons CO_2\,(g) + 4H_2\,(g)$$

$$K_3 = \frac{[H_2]^4\,[CO_2]}{[H_2O]^2[CH_4]} \quad ...(3)$$

From equation (1), (2) and (3),

$$K_3 = K_2K_1$$

11.2 (0.439)

$$\ln K = -\frac{\Delta G}{RT}$$

$$\ln K = \frac{-1.3484\times 1000}{8.314\times 400}$$

$$K = 0.667$$

$$CO\,(g) + 2H_2\,(g) \rightleftharpoons CH_3OH\,(g)$$

Initially	1	1	0
At equilibrium	$1-\epsilon$	$1-2\epsilon$	ϵ

Total number of moles at equilibrium

$$= (1-\epsilon) + (1-2\epsilon) + \epsilon$$

$$= 2(1-\epsilon)$$

$$K_y.P^{\Sigma v_i} = K$$

$$\frac{\dfrac{\epsilon}{2(1-\epsilon)}}{\left(\dfrac{1-2\epsilon}{2(1-\epsilon)}\right)^2\times\dfrac{1-\epsilon}{2(1-\epsilon)}}\times 10^{-2} = 0.667$$

$$\frac{4\epsilon(1-\epsilon)}{1-4\epsilon+4\epsilon^2} = 6.67$$

$$270.8\epsilon^2 - 270.8\epsilon + 66.7 = 0$$

$$\epsilon = 0.439$$

So, fractional conversion of CO at equilibrium is 0.439.

11.3 (c)

At standard conditions:

$$N_2 + 2O_2 \rightleftharpoons 2NO_2 \quad ...(A)$$

$$\Delta G^\circ = 100 \text{ kJ/mol}$$

$$NO + \frac{1}{2}O_2 \rightleftharpoons NO_2 \quad ...(B)$$

$$\Delta G^\circ = -35 \text{ kJ/mol}$$

Reaction (A) – 2 × Reaction (B) results in

$$N_2 + O_2 \rightleftharpoons 2\,NO$$

$$\Delta G^\circ = 100 - (-2\times 35) = 170 \text{ KJ/K mol}$$

For 1 mol NO

$$\frac{1}{2}N_2 + \frac{1}{2}O_2 \rightleftharpoons NO$$

$$\Delta G^\circ = \frac{170}{2} = 85 \text{ KJ/K mol}$$

11.4 (c)

For reversible exothermic reaction given as $A + B \rightleftharpoons C$ equilibrium conversion will decrease with temperature because reaction is exothermic and equilibrium conversion will increase with increase in pressure because number of moles are decreasing.

Above conclusion are given by Le-Chatelier principle.

11.5 (d)

Hydration of C_2H_4 to ethanol is occurring by following reaction at 400 K and 3 bar

Initially 2 moles of equimolar feed is present

	C_2H_4	$+ H_2O$	$\rightleftharpoons C_2H_5OH$
Initial	1	1	0
At equilibrium	$1-\epsilon$	$1-\epsilon$	ϵ

Total number of moles at equilibrium is

$$= (1-\epsilon) + (1-\epsilon) + \epsilon = 2-\epsilon$$

$$y_{C_2H_4} = \frac{1-\epsilon}{2-\epsilon}$$

$$y_{H_2O} = \frac{1-\epsilon}{2-\epsilon}$$

$$y_{C_2H_5OH} = \frac{\epsilon}{2-\epsilon}$$

$$-\Delta G = RT \ln k_{eq}$$

$$-4000 = 8.314 \times 400 \ln k_{eq}$$

$$k_{eq} = 0.3$$

$$k_{eq} = k_y \cdot k_\phi \cdot k_p \quad \text{...(i)}$$

$$k_p = P^{\Sigma \upsilon_i} = P^{1-2} = P^{-1}$$

$$k_\phi = 1$$

$$k_y = \frac{\dfrac{\epsilon}{2-\epsilon}}{\dfrac{1-\epsilon}{2-\epsilon} \cdot \dfrac{1-\epsilon}{2-\epsilon}} = \frac{\epsilon(2-\epsilon)}{(1-\epsilon)^2}$$

By putting all the values in equation (i)

$$\frac{\epsilon(2-\epsilon)}{(1-\epsilon)^2} \times 3^{-1} = 0.3$$

$$\frac{\epsilon(2-\epsilon)}{(1-\epsilon)^2} = 0.9$$

11.6 (c)

$$\Sigma(n_i \, d\mu_i)_{T,P} = 0$$

This represents Gibbs-Duhem equation at constant temperature and pressure.

$$-\Delta G = RT \ln K_{eq}$$

This represent reaction equilibrium constant.

$$\left(\frac{\partial G}{\partial n_i}\right)_{T,P,n_{j \neq i}} = \mu_i \text{ (Chemical potential)}$$

11.7 (d)

$$-\Delta G = 2750 \text{ J/mol}$$

$$R = 8.314 \text{ J/mol K}$$

$$\text{Pressure } (P) = 1 \text{ bar}$$

Now equilibrium constant can be calculated by

$$\Delta G = -RT \ln K_{eq}$$

$$2750 = 8.314 \times (300 + 273) \ln K_{eq}$$

$$K_{eq} = 1.781$$

Gas phase cracking reaction is given as

	A	→	B	+	C
Initially	1		0		0
At equilibrium	$1-\epsilon$		ϵ		ϵ

(Initially assume 1 mole of A)

Total number of moles at equilibrium is

$$= (1-\epsilon) + (\epsilon) + (\epsilon) = (1+\epsilon)$$

$$y_A = \frac{1-\epsilon}{1+\epsilon}$$

$$y_B = y_C = \frac{\epsilon}{1+\epsilon}$$

$$K_{eq} = K_y \cdot K_p \, K_\phi$$

$$K_\phi = 1$$

$$K_p = P^{\Sigma \upsilon_i} = P^{2-1} = P$$

$$\frac{\left(\dfrac{\epsilon}{1+\epsilon}\right)\left(\dfrac{\epsilon}{1+\epsilon}\right)}{\left(\dfrac{1-\epsilon}{1+\epsilon}\right)} . P = K_{eq}$$

$$\frac{\epsilon^2}{1-\epsilon^2} \times 1 = 1.781 \Rightarrow \epsilon = 0.8$$

Fractional conversion of A at equilibrium $= \frac{\epsilon}{1} = 0.8$

11.8 (20.4)

Reversible reaction of *t*-butyl alcohol (TBA) and ethanol (EtOH) to ethyl *t*-butyl ether (ETBE) is as

$$\text{TBA + EtOH} \rightleftharpoons \text{ETBE + Water}$$

Initially 74 gram of TBA is mixed with 100 g of aqueous solution contains 46 weight % ethanol.

$$\text{Initial moles of TBA} = \frac{\text{Weight}}{M_{wt}} = \frac{74}{74} = 1 \text{ moles}$$

$$\text{Similarly initial moles of alcohol} = \frac{46}{46} = 1 \text{ moles}$$

$$\text{Initial moles of water} = \frac{100-46}{18} = 3 \text{ moles}$$

	TBA +	EtOH	⇌ ETBE +	Water
Initially	1	1	0	3
At equilibrium	$1-\epsilon$	$1-\epsilon$	ϵ	$3+\epsilon$

Total number of moles at equilibrium

$$= (1-\epsilon) + (1-\epsilon) + (\epsilon) + (3+\epsilon) = 5$$

$$y_{EtOH} = y_{TBA} = \frac{1-\epsilon}{5}$$

$$y_{ETBE} = \frac{\epsilon}{5}$$

$$y_{water} = \frac{3+\epsilon}{5}$$

Equilibrium constant for reaction $K_c = 1$

$$K_c = K_y \cdot K_p \cdot K\phi$$

$$K\phi = 1$$

$$K_p = 1$$

Moles are not changing so $P^{\Sigma\upsilon_i} = 1$

$$1 = \frac{\left(\frac{1-\epsilon}{5}\right)^2}{\frac{\epsilon}{5}\frac{(3+\epsilon)}{5}} = \frac{(1+\epsilon)^2}{\epsilon(3+\epsilon)}$$

$$\epsilon^2 + 3\epsilon = 1 + \epsilon^2 - 2\epsilon$$

$$\epsilon = 0.2$$

So mass of ETBE at equilibrium

$$= \text{Moles at equilibrium} \times M_{wt}$$

$$= 0.2 \times 102 = 20.4 \text{ gram}$$

11.9 (d)

Vapour and liquid coexist in thermodynamic equilibrium (mechanical, thermal and phase equilibrium) when chemical potential or Gibbs free energy of both phase becomes equal.

11.10 (0.29)

CO reacts with H_2S at 800 K and 2 bar as

$CO + H_2S \rightleftharpoons COS + H_2$

$$\Delta g^\circ_{rxn} = 22972.3 \text{ J/mol}$$

$$R = 8.314 \text{ J/mol K}$$

Initially 4 moles of H_2S and 1 mole CO are present

	CO +	H_2S ⇌	COS +	H_2
Initially	1	4	0	0
At equilibrium	$1-\epsilon$	$4-\epsilon$	ϵ	ϵ

Total number of moles at equilibrium is

$$= (1-\epsilon) + (4-\epsilon) + (\epsilon) + (\epsilon) = 5$$

$$y_{CO} = \frac{1-\epsilon}{5}$$

$$y_{H_2S} = \frac{4-\epsilon}{5}$$

$$y_{H_2} = y_{COS} = \frac{\epsilon}{5}$$

$$K_{eq} = K_y \cdot K_p$$

Here, $K_p = P^{\Sigma\nu_i}$

Since number of moles are not changing

So, $K_p = 1$

K_{eq} can be found by following relation

$$-RT \ln K_{eq} = \Delta G$$

$$-8.314 \times 800 \times \ln K_{eq} = 22972.3$$

$$K_{eq} = 0.03162$$

$$\text{So, } 0.03162 = \frac{\left(\frac{\epsilon}{5}\right)^2}{\left(\frac{1-\epsilon}{5}\right)\left(\frac{4-\epsilon}{5}\right)} = \frac{\epsilon^2}{(1-\epsilon)(4-\epsilon)}$$

$$\epsilon = 0.29$$

11.11 (b)

Gas phase reaction $N_2O_4 \rightleftharpoons 2NO_2$ occurring in isothermal and isobaric reactor maintained at 298 K and 1 bar.

$$\Delta G^\circ_{298} = 5253 \text{ J/mol}$$

$$R = 8.314 \text{ J/mol K}$$

1 mole of N_2O_4 is initially charged

	N_2O_4 ⇌	$2NO_2$
Initially	1	0
At equilibrium	$1-\epsilon$	2ϵ

Total number of moles at equilibrium

$$= 1 - \epsilon + 2\epsilon = 1 + \epsilon$$

Mole fraction of N_2O_4 at equilibrium

$$y_{N_2O_4} = \frac{1-\epsilon}{1+\epsilon}$$

Similarly mole fraction of NO_2 at equilibrium

$$y_{NO_2} = \frac{2\epsilon}{1+\epsilon}$$

Now equilibrium constant can be found as

$$\ln K_{eq} = -\frac{\Delta G}{RT}$$

$$K_{eq} = 0.12$$

$$K_{eq} = K_y \cdot K_p$$

Here $K_p = P^{\Sigma\nu_i}$

Since number of moles one not changing so

$$P^{\Sigma\nu_i} = 1$$

$$\text{So, } 0.12 = \frac{\left(\frac{2\epsilon}{1+\epsilon}\right)^2}{\frac{1-\epsilon}{(1+\epsilon)}} = \frac{4\epsilon^2}{1-\epsilon^2}$$

$$\frac{\epsilon^2}{1-\epsilon^2} = 0.03$$

$$\epsilon = 0.17$$

So fraction of N_2O_4 decomposes in NO_2 is

$$\frac{\epsilon}{1} = 0.17$$

■■■■

Fluid Mechanics

UNIT

III

CONTENTS

Fluid Mechanics

Syllabus

Fluid Statics, Surface Tension, Newtonian and Non-Newtonian Fluids, Transport Properties, Shell-Balances including Differential Form of Bernoulli Equation and Energy Balance, Equation of Continuity, Equation of Motion, Equation of Mechanical Energy, Macroscopic Friction Factors, Dimensional Analysis and Similitude, Flow through Pipe Systems, Velocity Profiles, Flow Meters, Pumps and Compressors, Elementary Boundary Layer Theory, Flow Past Immersed Bodies including Packed and Fluidized Beds, Turbulent Flow: Fluctuating Velocity, Universal Velocity Profile and Pressure Drop

Analysis of Previous GATE Papers

Exam Year	1 Mark Ques.	2 Marks Ques.	5 Marks Ques.	Total Marks
2000	3	1	2	**15**
2001	2	2	–	**6**
2002	3	–	1	**8**
2003	3	4	–	**11**
2004	1	4	–	**9**
2005	1	5	–	**11**
2006	2	4	–	**10**
2007	2	5	–	**12**
2008	4	6	–	**16**
2009	–	4	–	**8**
2010	1	4	–	**9**
2011	2	3	–	**8**
2012	3	2	–	**7**
2013	6	1	–	**8**
2014	3	4	–	**11**
2015	2	2	–	**6**
2016	2	3	–	**8**
2017	2	3	–	**8**
2018	3	2	–	**7**
2019	2	2	–	**6**
2020	1	4	–	**9**
2021	2	3	–	**8**
2022	1	4	–	**9**
2023	3	2	–	**8**
2024	3	3	–	**9**

1 Fluid Properties

1.1 A Bingham fluid of viscosity $\mu = 10$ Pa.s, and yield stress $\tau_0 = 10$ kPa, is sheared between flat parallel plates separated by a distance 10^{-3} m. The top plate is moving with a velocity of 1 m/s. The shear stress on the plate is

(a) 10 kPa (b) 20 kPa
(c) 30 kPa (d) 40 kPa

[2001 : 2 M]

1.2 A lubricant 100 times more viscous than water would have a viscosity (in Pa-s)

(a) 0.01 (b) 0.1
(c) 1 (d) 10

[2003 : 1 M]

1.3 The velocity profile for a Bingham plastic fluid flowing (under laminar conditions) in a pipe is

(a) parabolic
(b) flat
(c) flat near the wall and parabolic in the middle
(d) parabolic near the wall and flat in the middle

[2003 : 1 M]

1.4 Viscosity of water at 40°C lies in the range of

(a) $1 \times 10^{-3} - 2 \times 10^{-3}$ kg/(ms)
(b) $0.5 \times 10^{-3} - 1 \times 10^{-3}$ kg/(ms)
(c) 1 – 2 kg/(ms)
(d) 0.5 – 1 kg/(ms)

[2004 : 2 M]

1.5 Match the following types of fluid (in Group-I) with their respective constitutive relations (in Group-II), where τ is the stress and $\dot{\gamma}$ is the strain rate,

Group-I	Group-II
A. Psedoplastic	1. $\tau = \mu\dot{\gamma}$
B. Bingham plastic	2. $\tau = \tau_0 + K\dot{\gamma}$
	3. $\tau = K\lvert\dot{\gamma}\rvert^n; n<1$
	4. $\tau = K\lvert\dot{\gamma}\rvert^n; n>1$

Codes:

	A	B
(a)	1	4
(b)	4	1
(c)	2	3
(d)	3	2

[2005 : 1 M]

1.6 A fluid obeying the constitutive equation

$$\tau = \tau_0 + K\left(\frac{dv_x}{dy}\right)^{1/2}, \tau > \tau_0$$

is held between two parallel plates a distance d apart. If the stress applied to the top plate is $3\tau_0$, then the velocity with which the top plate moves relative to the bottom plate would be

(a) $2\left(\frac{\tau_0}{K}\right)^2 d$ (b) $3\left(\frac{\tau_0}{K}\right)^2 d$
(c) $4\left(\frac{\tau_0}{K}\right)^2 d$ (d) $9\left(\frac{\tau_0}{K}\right)^2 d$

[2006 : 2 M]

1.7 Consider a soap film bubble of diameter D. If the external pressure is P_o and the surface tension of the soap film is σ, the expression for the pressure inside the bubble is

(a) P_o (b) $P_o + \frac{2\sigma}{D}$
(c) $P_o + \frac{4\sigma}{D}$ (d) $P_o + \frac{8\sigma}{D}$

[2007 : 1 M]

1.8 The units of the isothermal compressibility are

(a) m^{-3} (b) Pa^{-1}
(c) $m^3\,Pa^{-1}$ (d) $m^{-3}\,Pa^{-1}$

[2013 : 1 M]

1.9 The apparent viscosity of a fluid is given by $0.007\left|\frac{dV}{dy}\right|^{0.3}$ where $\left(\frac{dV}{dy}\right)$ is the velocity gradient. The fluid is

(a) Bingham plastic (b) dilatant
(c) pseudoplastic (d) thixotropic

[2013 : 1 M]

1.10 Which of the following statements are CORRECT?

P : For a rheopectic fluid, the apparent viscosity increases with time under a constant applied shear stress.

Q : For a pseudoplastic fluid, the apparent viscosity decreases with time under a constant applied shear stress.

R : For a Bingham plastic, the apparent viscosity increases exponentially with the deformation rate.

S : For a dilatant fluid, the apparent viscosity increases with increasing deformation rate.

(a) P and Q only
(b) Q and R only
(c) R and S only
(d) P and S only

[2014 : 1 M]

1.11 At a shear rate of 10 s^{-1}, the apparent viscosity of a non-Newtonian liquid was found to be 1 Pa s. At a shear rate of 100 s^{-1}, the apparent viscosity of the same liquid was found to be 0.5 Pa s. If the liquid follows power law behavior, the apparent viscosity (in Pa s) at a shear stress of 10 N m^{-2} is ________ (rounded off to two decimal place).

[2018 : 2 M]

1.12 Consider a fluid confined between two horizontal parallel plates and subjected to shear flow.

In the first experiment, the plates are separated by a distance of 1 mm. It is found that a shear stress of 2 N m^{-2} has to be applied to keep the top plate moving with a velocity of 2 m s^{-1}, while the other plate is fixed.

In the second experiment, the plates are separated by a distance of 0.25 mm. It is found that a shear stress of 3 N m^{-2} has to be applied to keep the top plate moving with a velocity of 1 m s^{-1}, while the other plate is fixed.

In the range of shear rates studied, the rheological character of the fluid is

(a) Newtonian
(b) Pseudoplastic
(c) Dilatant
(d) Ideal and inviscid

[2021 : 2 M]

1.13 An infinitely long cylindrical water filament of radius R is surrounded by air. Assume water and air to be static. The pressure outside the filament is P_{out} and the pressure inside is P_{in}. If γ is the surface tension of the water-air interface, then $P_{in} - P_{out}$ is

(a) $\frac{2\gamma}{R}$
(b) 0
(c) $\frac{\gamma}{R}$
(d) $\frac{4\gamma}{R}$

[2024 : 1 M]

■■■■

Answers Fluid Properties

1.1	(b)	1.2	(b)	1.3	(d)	1.4	(b)	1.5	(d)	1.6	(c)	1.7	(d)
1.8	(b)	1.9	(b)	1.10	(d)	1.11	(1.002)	1.12	(b)	1.13	(c)		

Explanations Fluid Properties

1.1 (b)

For Bingham fluid

$$\tau = \tau_0 + \frac{\mu dV}{dy}$$

{Given, $\tau_0 = 10$ kPa, $\mu = 10$ Pa.s}

$$\frac{dV}{dy} = \frac{1-0}{10^{-3}} = 10^3 \text{s}^{-1}$$

$$\tau = 10^4 + 10 \times 10^3 = 20 \text{ kPa}$$

1.2 (b)

$$\mu_{\text{water}} = 10^{-3} \text{ Pa.s}$$

100 times more viscous, i.e.

$$= 100 \times 10^{-3} = 0.1 \text{Pa.S}$$

1.3 (d)

For Bingham plastic fluid

$$\tau_{yx} = -\mu_0 \frac{dV_x}{dy} \pm \tau_0, \text{ if } |\tau_{yx}| > \tau_0$$

$$\frac{dV_x}{dy} = 0, \text{ if } |\tau_{yx}| < \tau_0$$

Momentum flux distribution in pipe is given by

$$\tau_{rz} = \left(\frac{P_0 - P_L}{2L}\right) r$$

∴ Parabolic near the wall and flat in the middle.

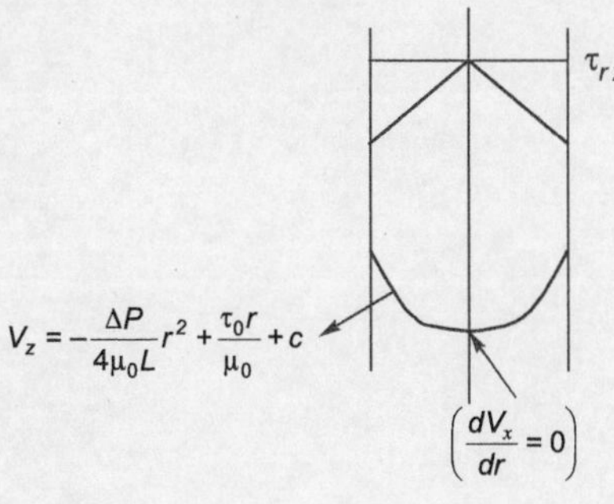

1.4 (b)

$$\mu_{\text{water}} = 1\ c_p = \frac{0.1}{100} \text{ poise}$$

$$= \frac{1}{100}\left(\frac{g}{\text{cm.s}}\right) \times \frac{1}{1000}\frac{\text{kg}}{g} \times 100\frac{\text{cm}}{m}$$

$$= 10^{-3} \text{ kg/m.s}$$

1.5 (d)

Pseudoplastic fluid $\rightarrow \tau = k|Y|^n; n < 1$

Bingham plastic fluid $\rightarrow \tau = \tau_0 + kY$

1.6 (c)

$$\tau = \tau_0 + k\left(\frac{dV_x}{dy}\right)^{1/2} \quad \tau = \tau_0$$

$$\tau = 3\tau_0 \text{ (given)}$$

$$\therefore \quad 3\tau_0 = \tau_0 + k\left(\frac{dV_x}{dy}\right)^{1/2}$$

$$\left(\frac{2\tau_0}{k}\right)^2 = \frac{dV_x}{dy} \rightarrow \int_0^d 4\left(\frac{\tau_0}{k}\right)^2 dy = \int_0^d dV_x$$

Velocity V_x is integrated 0 to V because velocity has to be found w.r.t. bottom plate

$$\therefore \quad V_{\text{bottom}} = 0 \text{ (for calculation)}$$

$$V = 4\left(\frac{\tau_0}{k}\right)^2 d$$

1.7 (d)

For soap bubble :

$$\Delta P = \frac{8\sigma}{d} \Rightarrow P_i - P_o = \frac{8\sigma}{d}$$

1.8 (b)

Isothermal compressibility is define as

$$\beta = -\frac{1}{v}\left(\frac{\partial v}{dp}\right)_T$$

So unit of β is 1 m^3 / m^3.Pa = Pa^{-1}

1.9 (b)

$$\text{Apparent viscosity} = m\left|\frac{dV_x}{dy}\right|^{n-1}$$

$$m = 0.007, \ n - 1 = 0.3$$

$$n = 1.3 \rightarrow (n > 1) \text{ Dilatant}$$

1.10 (d)

- For a rheopectic fluid, the apparent viscosity increases with time under a constant applied shear stress. (True)
- For a dilatant fluid, the apparent viscosity increases with increasing deformation rate. (True)

Others two option (Q) and (R) are incorrect.

1.11 (1.002)

"From power law model [Ostwald-de-Waele equation] for non-Newtonian fluid"

$$\tau = m\left(\frac{du}{dy}\right)^n = m\left(\frac{du}{dy}\right)^{n-1}\left(\frac{du}{dy}\right)$$

$$\mu_{app} = m\left(\frac{du}{dy}\right)^{n-1}$$

Given,

$$\mu_{app} = 1, \quad \left(\frac{du}{dy}\right) = 10$$

$$\mu_{app} = 0.5, \quad \left(\frac{du}{dy}\right) = 100$$

$$\frac{1}{0.5} = \frac{m\,(10)^{n-1}}{m\,(100)^{n-1}}$$

$$\Rightarrow \quad n = 0.698 \cong 0.7$$

$$1 = m(10)^{0.7-1}$$

$$m = 1.995 \cong 2$$

For $\tau = 10\ \text{N/m}^2$

$$\left(\frac{du}{dy}\right) = ?$$

$$10 = 2\left(\frac{du}{dy}\right)^{0.7}$$

$$\frac{du}{dy} \cong 10$$

Now, $\mu_{app} = 2[10]^{0.7-1} = 1.002$

1.12 (b)

$$\therefore \quad \mu_1 = \frac{\tau_1}{\left(\frac{du}{dy}\right)_L} = \frac{2}{\left(\frac{2}{10^{-3}}\right)} = 10^{-3}\ \text{Pa.s}$$

$$\mu_2 = \frac{3}{\left(\frac{1}{0.25\times10^{-3}}\right)} = 0.75\times10^{-3}\ \text{Pa.s}$$

Thus, $\mu_1 \neq \mu_2$ (Non-newtonian fluid)

$$\tau = m\left(\frac{du}{dy}\right)^n \quad (\because \text{for non-newtonian fluid})$$

$$\frac{2}{3} = \frac{m\left(\frac{2}{10^{-3}}\right)^n}{m\left[\frac{1}{0.25\times10^{-3}}\right]^n} = (0.50)^n$$

Take ln both sides,

$$\ln\left|\frac{2}{3}\right| = n\ln|0.50|$$

$$n = 0.585 < 1$$

Thus, Pseudoplastic fluid.

1.13 (c)

Infinitely long cylindrical water filament of radius R is surrounded by air

$$P_{in} - P_{out} = \Delta P = \frac{2\gamma}{D} = \frac{2\gamma}{2R} = \frac{\gamma}{R}$$

where, γ = Surface tension

2 Fluid Pressure and Its Measurement

2.1 For the manometer setup shown in the figure, the pressure difference $P_A - P_B$ is given by

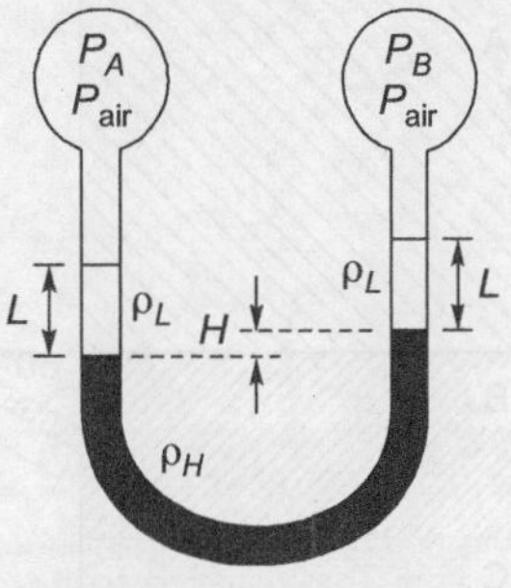

(a) $(\rho_H - \rho_{air})gH$
(b) $(\rho_H - \rho_L)gH$
(c) $(\rho_H - \rho_L)gH + (\rho_L - \rho_{air})g(L - H)$
(d) $(\rho_H - \rho_L)gL + (\rho_L - \rho_{air})gH$ **[2004 : 2 M]**

2.2 Three containers are filled with water upto the same height as shown. The pressures at the bottom of the containers are denoted as P_1, P_2 and P_3. Which one of the following relationship is true?

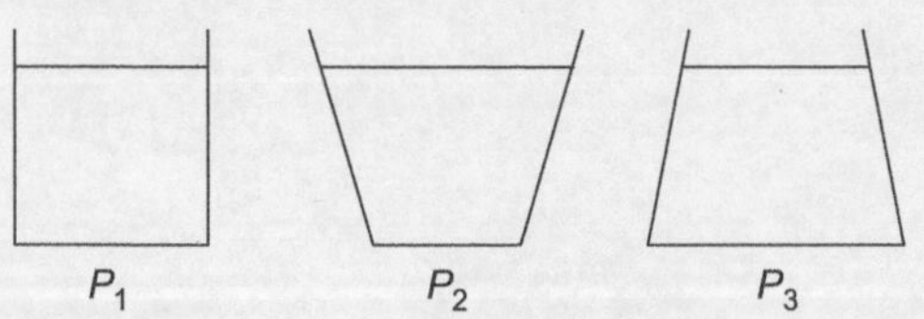

(a) $P_3 > P_1 > P_2$ (b) $P_2 > P_1 > P_3$
(c) $P_1 > P_2 = P_3$ (d) $P_1 = P_2 = P_3$

[2008 : 1 M]

2.3 An open tank contains two immiscible liquids of densities (800 kg/m^3 and 1000 kg/m^3) as shown in the figure. If g = 10 m/s^2, under static conditions, the gauge pressure at the bottom of the tank in Pa is _______ .

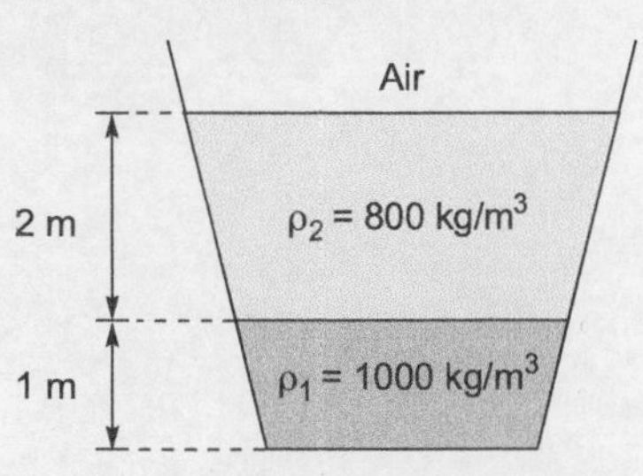

[2013 : 1 M]

2.4 A vertical cylindrical vessel has a layer of kerosene (of density 800 kg/m^3) over a layer of water (of density 1000 kg/m^3). L-shaped glass tubes are connected to the column 30 cm apart. The interface between the two layers lies between the two points at which the L-tubes are connected. The levels (in cm) to which the liquids rise in the respective tubes are shown in the figure below.

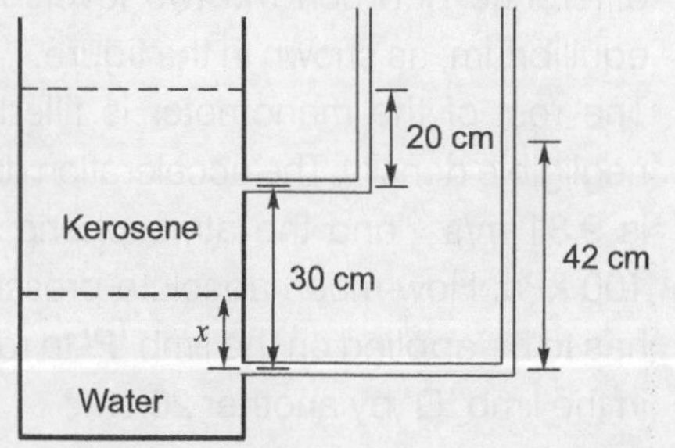

The distance (x in cm, rounded off to the first decimal place) of the interface from the point at which the lower L-tube is connected is _______.

[2016 : 1 M]

2.5 Under isothermal condition, a vertical tube of length L = 100 m contains a gas of molecular weight equal to 60. The pressure and temperature at the top of the tube are 100 kPa and 25°C respectively. Consider the universal gas constant and acceleration due to gravity as 8.314 J mol^{-1} K^{-1} and 9.81 m s^{-2} respectively. If the gas is ideal, the pressure (in kPa) at the bottom of the tube will be _______ (rounded off to third decimal place).

[2018 : 2 M]

2.6 For a hydraulic lift with dimensions shown in figure, assuming g = 10 m/s^2, the maximum diameter D_{left} (in m) that lifts a vehicle of mass 1000 kg using a force of 100 N is ______ (rounded off to two decimal places).

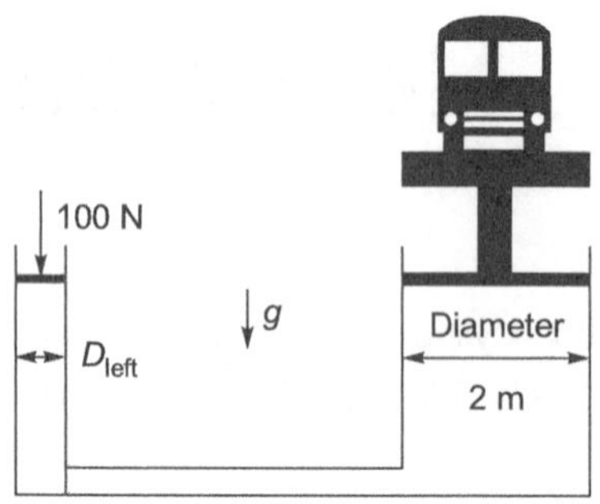

[2019 : 1 M]

2.7 A U-tube manometer contains two manometric fluids of densities 1000 kg m^{-3} and 600 kg m^{-3}. When both the limbs are open to atmosphere, the difference between the two levels is 10 cm at equilibrium, as shown in the figure.

The rest of the manometer is filled with air of negligible density. The acceleration due to gravity is 9.81 m/s^{-2} and the atmospheric pressure is 100 kPa. How much absolute pressure (in kPa) has to be applied on the limb **'P'** to raise the fluid in the limb **'Q'** by another 20 cm?

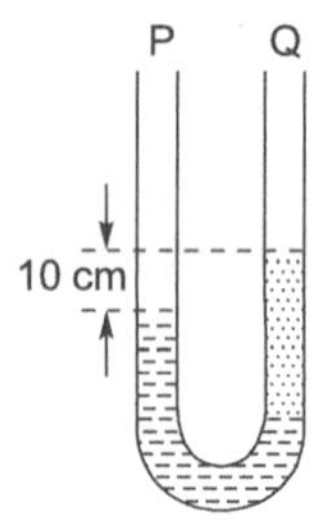

(a) 100.175 (b) 103.924
(c) 547.231 (d) 833.206

[2020 : 2 M]

2.8 Consider a tank filled with 3 immiscible liquids *A*, *B* and *C* at static equilibrium, as shown in the figure. At 2 cm below the liquid *A* – liquid *B* interface, a tube is connected from the side of the tank. Both the tank and the tube are open to the atmosphere.

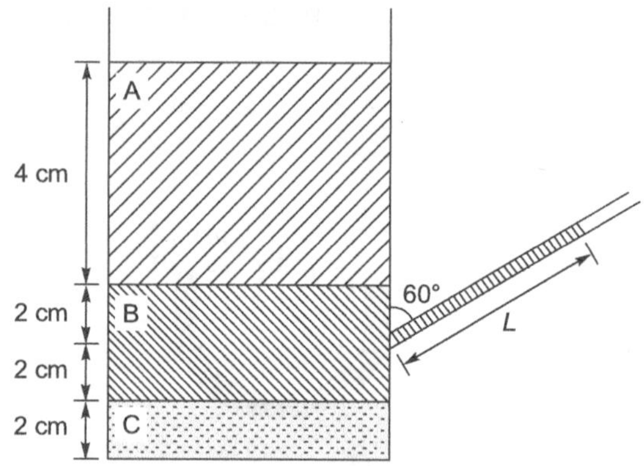

At the operating temperature and pressure, the specific gravities of liquids *A*, *B* and *C* are 1, 2 and 4, respectively. Neglect any surface tension effects in the calculations. The length of the tube *L* that is wetted by liquid *B* is _____ cm.

[2021 : 2 M]

■■■■

Answers Fluid Pressure and Its Measurement

2.1 (a) **2.2** (d) **2.3** (26000) **2.4** (10) **2.5** (102.376) **2.6** (0.2) **2.7** (b)
2.8 (8)

Explanations Fluid Pressure and Its Measurement

2.1 (a)

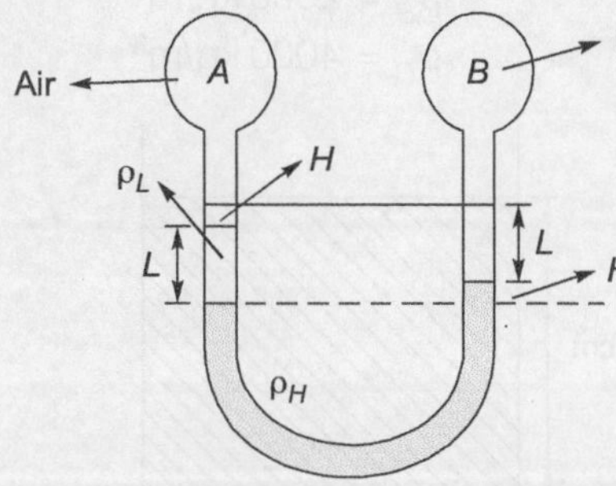

Manometric balance on the datum line,

$$P_A + \rho_{air} gH + \rho_L gL = P_B + \rho_L gL + \rho_H gH$$

$$P_A - P_B = (\rho_H - \rho_{air})\, gH$$

2.2 (d)

From hydrostatic law :

$$P = \rho \times g \times h$$

Since, H is same for all the three containers $(P_1 = P_2 = P_3)$

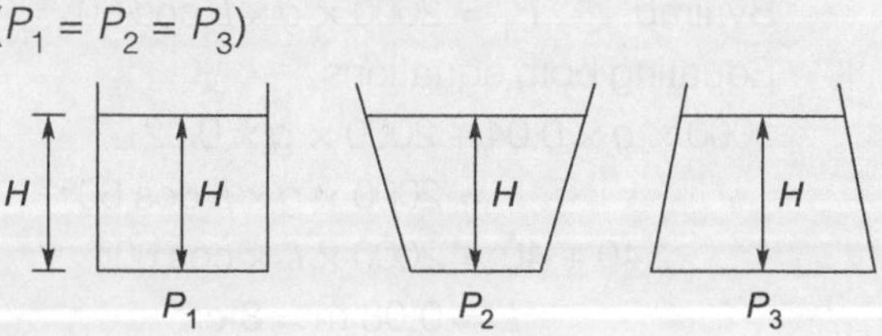

2.3 (26000)

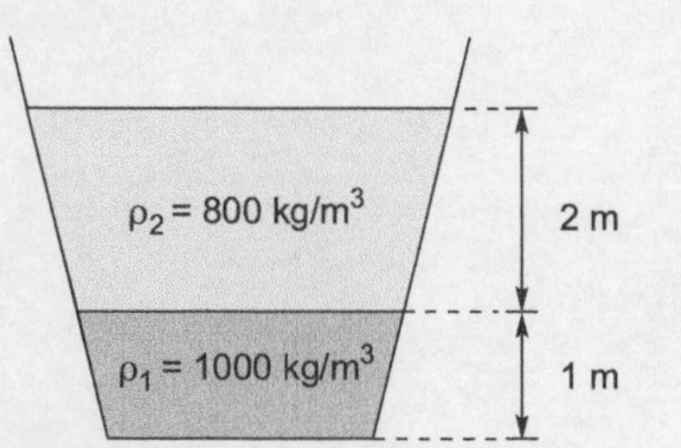

Pressure at bottom is,

$$P_B = P_1 + P_2$$

$$P_B = 800 \times 10 \times 2 + 1000 \times 10 \times 1$$

$$P_B = 26000 \text{ Pascal}$$

2.4 (10)

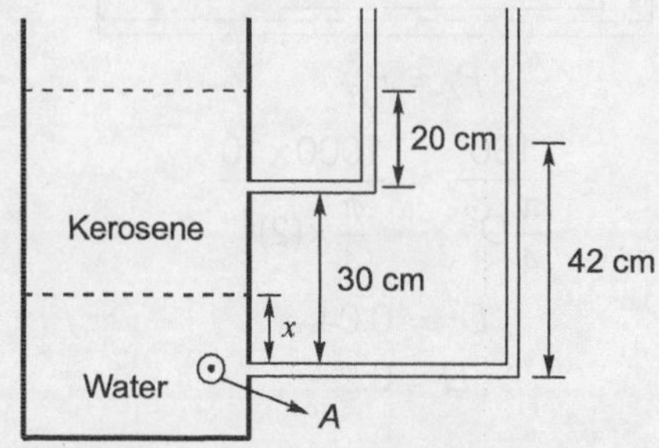

Pressure at point A will be $= 1000 \times 0.42 \times g$

Again at the same point A,

$$P = 800\,(0.30 - x)\, g + 800 \times g \times 0.20 + 1000 \times g \times x$$

Equating both pressure, we get

$$x = 10 \text{ cm}$$

2.5 (102.376)

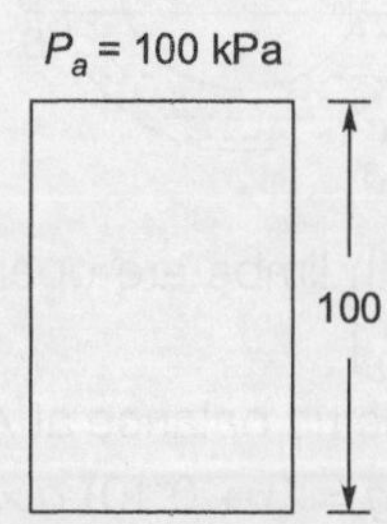

Molecular weight of gas $M_{wt} = 60$

$$R = 8.314 \text{ J/mol K}$$

$$g = 9.81 \text{ m/s}^2$$

$$PV = nRT$$

$$PV = \frac{m}{m_{wt}} RT, \quad P = \frac{\rho}{m_{wt}} RT$$

$$\rho = \frac{P \cdot M_{wt}}{RT}$$

$$= \frac{100 \times 60}{8.314 \times 298} = 2.42 \text{ kg/m}^3$$

$$P_2 = P_1 + \rho gh$$

$$= 100 + 2.42 \times 9.81 \times 100 \times 10^{-3}$$

$$= 102.376 \text{ kPa}$$

2.6 (0.2)

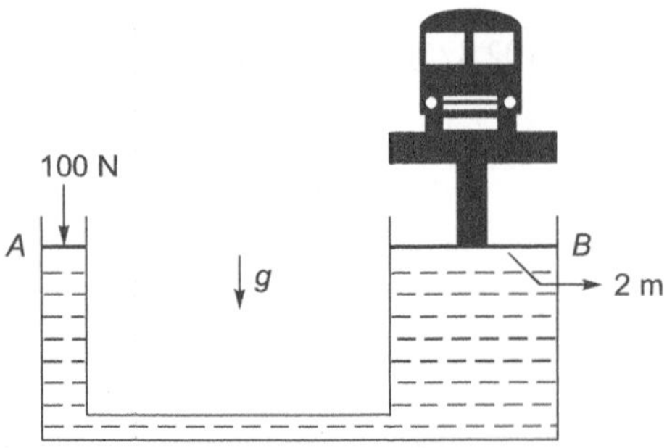

$$P_A = P_B$$

$$\frac{100}{\frac{\pi}{4}D^2} = \frac{1000 \times 10}{\frac{\pi}{4} \times (2)^2}$$

$$D^2 = 0.04$$

$$D = 0.2$$

2.7 (b)

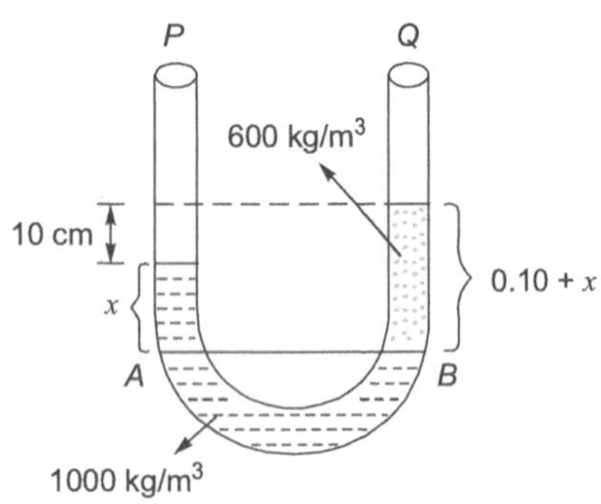

Since, the both limbs are open to atmosphere, that is $[P_A = P_B]$

Apply manometric balance at A and B, i.e.

$$1000 \times x \times g = (x + 0.10)\, g \times 600$$

$$x = 0.15 \text{ m} = 15 \text{ cm}$$

Total height in RHS = 15 + 10 = 25 cm

Now, if 20 cm has to rise in right side limb at the same time 20 cm has to down in left side limb.

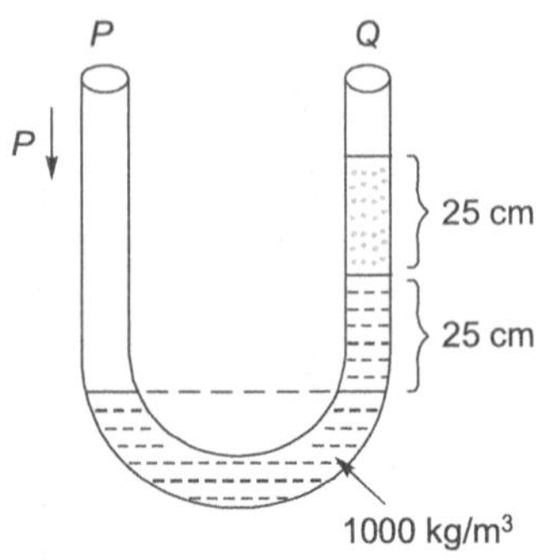

$$g = 9.81 \text{ m/s}^2$$

$$P = 1000 \times g \times 0.25 + 600 \times g \times 0.25$$

$$= 3924 \text{ pascal}$$

i.e. 3.924 kPa

Absolute pressure = 100 kPa + 3.924

= 103.924 kPa

2.8 (8)

$$\rho_A = 1000 \text{ kg/m}^3$$

$$\rho_B = 2000 \text{ kg/m}^3$$

$$\rho_C = 4000 \text{ kg/m}^3$$

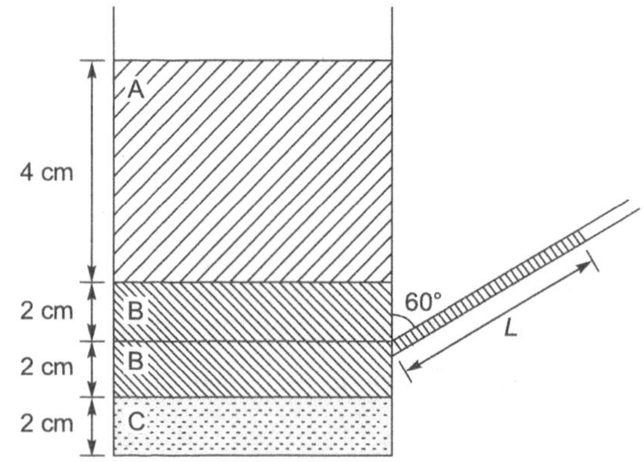

Now, pressure at point (1) will be equivalent to pressure shown by limb. Thus, by pressure balance,

$P_1 = 1000 \times g \times 0.04 + 2000 \times g \times 0.02$ ...(i)

By limb, $P_1 = 2000 \times g \times L \cos 60°$ ...(ii)

Equating both equations,

$$1000 \times g \times 0.04 + 2000 \times g \times 0.02 = 2000 \times g \times L \cos 60°$$

$$40 + 40 = 2000 \times L \times \cos 60°$$

$$L = 0.08 \text{ m} = 8 \text{ cm}$$

■■■■

3 Buoyancy and Floatation

3.1 What is the force required (in Newtons) to hold a spherical balloon stationary in water at a depth of H from the air-water interface? The balloon is of radius 0.1 m and is filled with air.

(a) $\frac{4\pi g}{3}$ (b) $\frac{0.1\pi gH}{4}$

(c) $\frac{0.1\pi gH}{8}$ (d) $\frac{0.04\pi gH}{3}$

[2004 : 1 M]

3.2 A conical tank with a bottom opening of cross-sectional area A is filled with water and is mounted on supports, as shown in the figure. What is the force F with which plate X must be pushed up to prevent water from leaking? Assume that the density of air is negligible as compared to the density of water ρ_L.

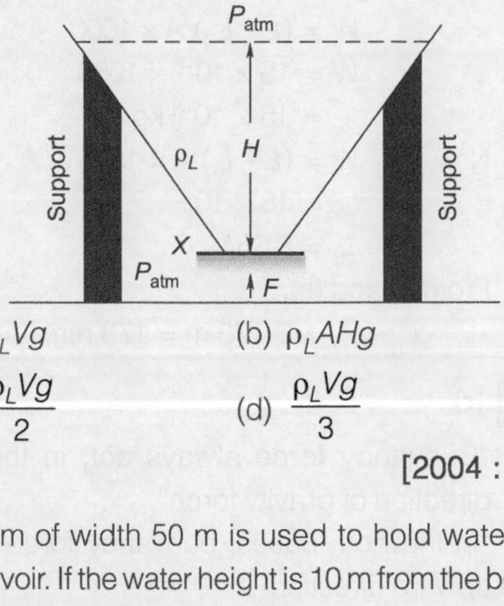

(a) $\rho_L Vg$ (b) $\rho_L AHg$

(c) $\frac{\rho_L Vg}{2}$ (d) $\frac{\rho_L Vg}{3}$

[2004 : 2 M]

3.3 A dam of width 50 m is used to hold water in a reservoir. If the water height is 10 m from the bottom of the dam, what is the total force F acting on the dam due to the water? Assume $g = 10$ m/s^2, and the fluid density is 1000 kg/m^3.

(a) $F = 12.5 \times 10^6$ N (b) $F = 25 \times 10^6$ N

(c) $F = 50 \times 10^6$ N (d) $F = 5 \times 10^6$ N

[2005 : 2 M]

3.4 The magnitude of the force (in N) required to hold a body of volume 0.05 m^3 and mass 40 kg in water (density 1000 kg/m^3) at a depth of 0.1 m is ($g = 9.81$ m/s^2)

(a) Zero (b) 98.1

(c) 490.5 (d) 882.9 [2006 : 1 M]

3.5 A hydrometer, with stem cross-sectional area of 2.82×10^{-5} m^2, is immersed in a very large vessel containing water as shown in the figure. The immersed volume is 15×10^{-6} m^3 and the length of the stem above water surface is L_w. If the entire volume of water is replaced by a liquid with specific gravity 1.5 and if the length of the stem above the liquid surface is L_l, then the difference, $L_l - L_w$ is

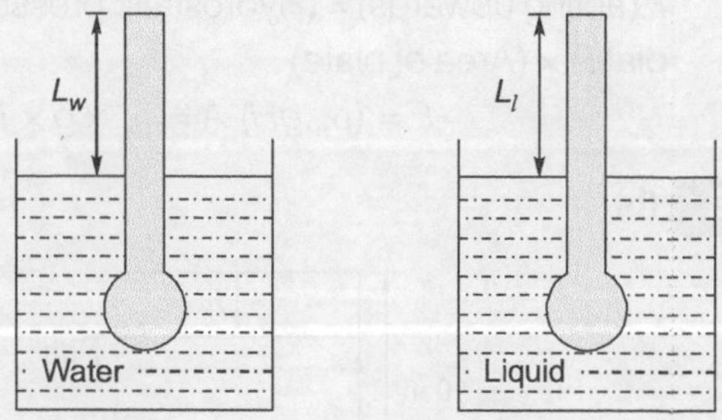

(a) –177 mm (b) 177 mm

(c) –266 mm (d) 266 mm

[2010 : 2 M]

3.6 Consider the following two cases of movement of particles. In case I, the particle moves along the positive y-direction and in case II, the particle moves along negative y-direction. Gravity acts along the positive y-direction. Which one of the following options corresponds to the correct directions of buoyancy acting on the particles?

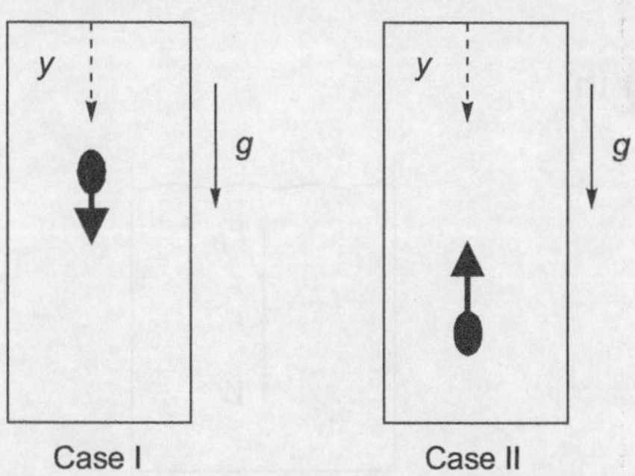

(a) Positive y-direction for both the cases.

(b) Negative y-direction for case I, positive y-direction for case II.

(c) Negative y-direction for both the cases.

(d) Positive y-direction for Case I, negative y-direction for case II. [2011 : 1 M]

Answers Buoyancy and Floatation

3.1	(a)	3.2	(b)	3.3	(b)	3.4	(b)	3.5	(b)	3.6	(c)

Explanations Buoyancy and Floatation

3.1 (a)

$$\begin{aligned}\text{Force required} &= \text{Buoyancy force}\\ &= \rho_f(V_{\text{submerged}})\cdot g\\ &= \text{wt. of water displaced}\\ &= \frac{4}{3}\pi(0.1)^3\times 1000\times g = \frac{4}{3}\pi g\end{aligned}$$

3.2 (b)

F(acting upwards) = (Hydrostatic pressure on the plate) × (Area of plate)

$$F = (\rho_L\, gH)\, A = \rho_L\times g\times H\times A$$

3.3 (b)

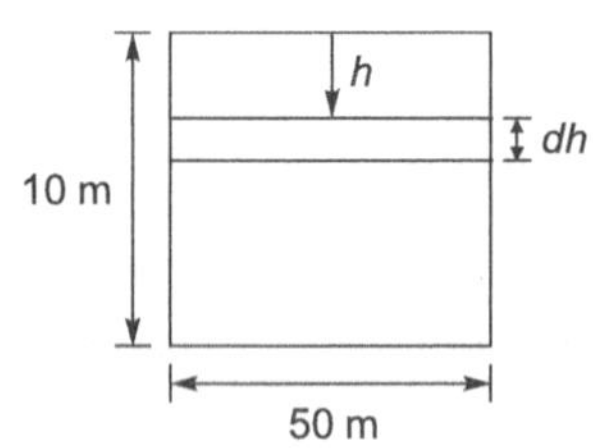

$$F = \int_0^{10} \underbrace{(h\rho g)}_{\text{Pressure}}\underbrace{(50\times dh)}_{\text{Area}}$$

$$F = 50\rho g\left.\frac{h^2}{2}\right|_0^{10} = 50\times 1000\times 10\times\frac{10^2}{2}$$

$$F = 25\times 10^6\ \text{N}$$

3.4 (b)

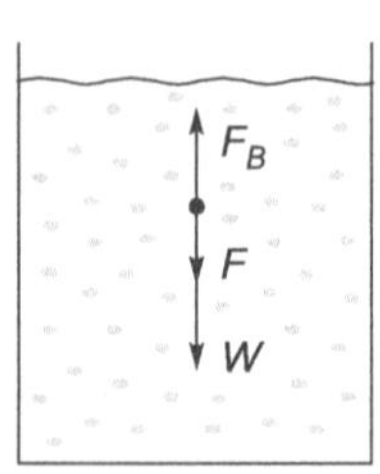

Applied force + weight = Buoyancy force

$$F + W = F_B$$

$$F + 40\times 9.81 = 0.05\times 1000\times 9.81$$

$$F = 98.1\ \text{N}$$

3.5 (b)

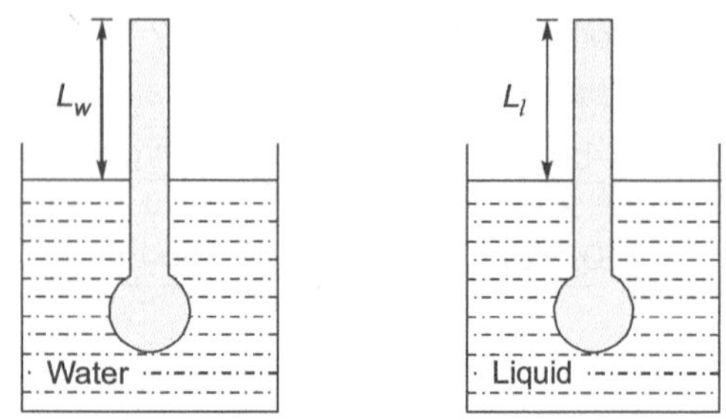

Neglect volume of bulb so that we can assume the hydrometer as a cylinder.

$$(L - L_w)\,A = 15\times 10^{-6}\ \text{m}^3$$

$$L - L_w = \frac{15\times 10^{-6}}{2.82\times 10^{-5}} = 0.532 \qquad \text{...(i)}$$

Also, weight of bulb = weight of displaced liquid

$$\begin{aligned}W &= (L - L_w)\,A\times 1000\\ W &= 15\times 10^{-6}\times 1000\\ &= 15\times 10^{-3}\ \text{kg}\end{aligned}$$

Now,

$$\begin{aligned}W &= (L - L_l)\,A\times 1500\\ &= 15\times 10^{-3}\\ L - L_l &= 0.354 \qquad \text{...(ii)}\end{aligned}$$

From (i) and (ii),

$$L_l - L_w = 0.178\ \text{m} = 178\ \text{mm}$$

3.6 (c)

"Buoyancy force always act, in the opposite direction of gravity force".

"In the both cases, buoyancy force acts in the upward direction".

■■■■

4 Fluid Kinematics

4.1 A steady flow field of an incompressible fluid is given by $\vec{V} = (Ax + By)\hat{i} - Ay\hat{j}$, where $A = 1\ s^{-1}$, $B = 1\ s^{-1}$, and x, y are in meters. The magnitude of the acceleration (in m/s²) of a fluid particle at (1, 2) is

(a) 1 (b) $\sqrt{2}$
(c) $\sqrt{5}$ (d) $\sqrt{10}$

[2008 : 2 M]

4.2 For an incompressible flow, the x- and y-components of the velocity vector are

$$v_x = 2(x + y);\ v_y = 3(y + z)$$

where x, y, z are in metres and velocities are in m/s. Then the z-component of the velocity vector (v_z) of the flow for the boundary condition $v_z = 0$ at $z = 0$ is

(a) $5z$ (b) $-5z$
(c) $2x + 3z$ (d) $-2x - 3z$

[2009 : 2 M]

4.3 The stream function in a xy-plane is given below:

$$\psi = \frac{1}{2}x^2y^3$$

The velocity vector for this stream function is

(a) $xy^3\hat{i} - \frac{3}{2}x^2y^2\hat{j}$ (b) $\frac{3}{2}x^2y^2\hat{i} - xy^3\hat{j}$

(c) $\frac{3}{2}x^2y^2\hat{i} + xy^3\hat{j}$ (d) $xy^3\hat{i} + \frac{3}{2}x^2y^2\hat{j}$

[2010 : 1 M]

4.4 An incompressible fluid is flowing through a contraction section of length L and has a 1-D (x-direction) steady state velocity distribution, $u = u_0\left(1 + \frac{2x}{L}\right)$. If $u_0 = 2$ m/s and $L = 3$ m, the convective acceleration (in m/s²) of the fluid at L is ________. [2014 : 2 M]

4.5 In a steady incompressible flow, the velocity distribution is given by $\vec{V} = 3x\hat{i} - Py\hat{j} + 5z\hat{k}$, where, V is in m/s and x, y and z are in m. In order to satisfy the mass conservation, the value of the constant P(in s^{-1}) is ________. [2014 : 2 M]

4.6 A three-dimensional velocity field is given by $V = 5x^2yi + Cyj - 10xyzk$, where i, j, k are the unit vectors in x, y, z directions, respectively, describing a Cartesian coordinate system. The coefficient C is a constant. If V describes an incompressible fluid flow, the value of C is

(a) –1 (b) 0
(c) 1 (d) 5 [2021 : 1 M]

4.7 The velocity field in an incompressible flow is $v = \alpha xy\hat{i} + v_y\hat{j} + \beta\hat{k}$, where $\hat{i}$, $\hat{j}$ and $\hat{k}$ are unit-vectors in the (x, y, z) Cartesian coordinate system. Given that α and β are constants, and $v_y = 0$ at $y = 0$, the correct expression for v_y is

(a) $\frac{-\alpha xy}{2}$ (b) $\frac{-\alpha y^2}{2}$

(c) $\frac{\alpha y^2}{2}$ (d) $\frac{\alpha xy}{2}$ [2024 : 1 M]

■■■■

Answers Fluid Kinematics

4.1 (c) **4.2** (b) **4.3** (b) **4.4** (8) **4.5** (8) **4.6** (b) **4.7** (b)

Explanations Fluid Kinematics

4.1 (c)

$$\vec{V} = (Ax+By)\hat{i} - Ay\hat{j}$$

$$\vec{a} = \frac{\partial \vec{V}}{\partial t} + V_x\frac{\partial \vec{V}}{\partial x} + V_y\frac{\partial \vec{V}}{\partial y} + V_z\frac{\partial \vec{V}}{\partial z}$$

$$\frac{\partial \vec{V}}{\partial t} = \frac{\partial \vec{V}}{\partial z} = 0$$

$\because$ $\vec{V}$ does not have z, and 't' component

$$a_x = V_x\frac{\partial V_x}{\partial x} + V_y\frac{\partial V_x}{\partial y}$$

$$a_y = V_x\frac{\partial V_y}{\partial x} + V_y\frac{\partial V_y}{\partial y}$$

Here, $V_x = Ax + By$ and $V_y = -Ay$

$$a_x = (Ax + By)\,A + (-Ay)\,B$$

$$a_y = (Ax + By)\,0 + (-Ay)(-A) = A^2 y$$

Putting values,

$$A = 1,\ B = 1,\ x = 1\ y = 2$$

$$a_x = 1,\ a_y = 2$$

$$|\vec{a}| = \sqrt{1+4} = \sqrt{5}$$

4.2 (b)

$\therefore$ From continuity equation

$$\Delta\vec{V} = 0$$

$$\frac{\partial u}{\partial x} + \frac{\partial v}{\partial y} + \frac{\partial w}{\partial z} = 0$$

$$2+3+\frac{\partial w}{\partial z} = 0$$

$$w = -5z + C$$

$\because$ $z = 0,\quad w = 0$

$$w = -5z$$

4.3 (b)

$$\vec{v} = v_x\hat{i} + v_y\hat{j}$$

$$= \frac{\partial \psi}{\partial y}\hat{i} - \frac{\partial \psi}{\partial x}\hat{j}$$

Given, $\psi = \frac{1}{2}x^2y^3$

$$\therefore \quad \frac{\partial \psi}{\partial y} = \frac{1}{2}x^2(3y^2) = \frac{3}{2}x^2y^2$$

$$\frac{\partial \psi}{\partial x} = \frac{1}{2}(2x)y^3 = xy^3$$

$$\therefore \quad \vec{v} = \frac{3}{2}x^2y^2\hat{i} - xy^3\hat{j}$$

4.4 (8)

$$u = u_0\left[1+\frac{2x}{L}\right]$$

$$a_x = \frac{du}{dt} = \frac{\partial u}{\partial x}\cdot\frac{\partial u}{\partial t} = u\cdot\frac{\partial u}{\partial x}$$

$$a_x = u_0\left[1+\frac{2x}{L}\right]\frac{\partial}{\partial x}\left[u_0\left(1+\frac{2x}{L}\right)\right]$$

$$\therefore \quad a\big|_{x=L} = 4\left(\frac{2}{3}\right)3 = 8\text{ m/s}^2$$

4.5 (8)

$$\frac{D\rho}{Dt} = -(\nabla\cdot\rho V)$$

For steady flow

$$\frac{D\rho}{Dt} = 0$$

For steady incompressible flow,

$$(\nabla\cdot\vec{V}) = 0$$

$$\frac{\partial u_x}{\partial y} + \frac{\partial u_y}{\partial y} + \frac{\partial u_z}{\partial z} = 0$$

$$3 - P + 5 = 0$$

$$P = 8\text{ s}^{-1}$$

4.6 (b)

For incompressible fluid flow,

$$\frac{\partial u}{\partial x} + \frac{\partial v}{\partial y} + \frac{\partial w}{\partial z} = 0$$

$$10xy + C - 10xy = 0$$

$$C = 0$$

4.7 (b)

Velocity field, $v = \alpha xy\hat{i} + v_y\hat{j} + \beta\hat{k}$ where, α and β are constant

$$v_y = 0 \text{ at } y = 0 \text{ (given)}$$

For incompressible flow $\nabla\cdot\vec{v} = 0$

$$\frac{\partial v_y}{\partial y} + \frac{\partial}{\partial x}(\alpha xy) + \frac{\partial \beta}{\partial z} = 0$$

$$\frac{\partial v_y}{\partial y} = -\alpha y$$

$$v_y = -\frac{\alpha y^2}{2} + c$$

$\because$ $v_y = 0$ at $y = 0$ so $c = 0$

$$v_y = \frac{-\alpha y^2}{2}$$

■■■■

5 Fluid Dynamics

5.1 A free jet of water of cross-sectional area 0.01 m^2 and velocity of 20 m/s strikes a plate and then flows in the plane parallel to the plate as shown in the figure below. The horizontal component of the force on the support is

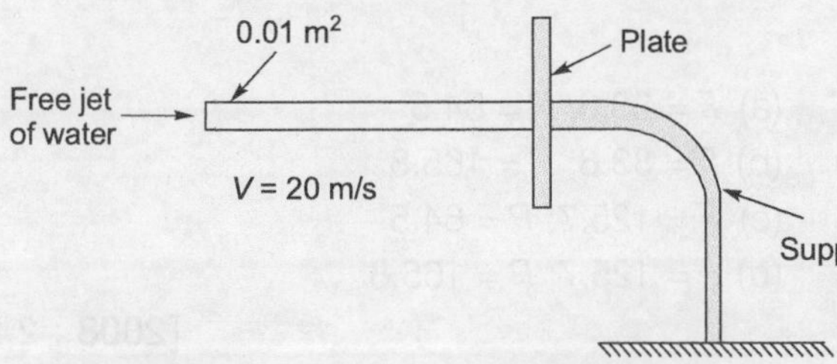

(a) 200 N (b) 400 N
(c) 2000 N (d) 4000 N

[2000 : 2 M]

5.2 A free jet of water is produced using a 75 mm diameter nozzle attached to a 200 mm diameter pipe, as shown in the figure. If the average velocity of water at plane B is 3.8 m/s, calculate the velocity of water at point A in the free jet. Neglect friction losses in the nozzle and pipe.

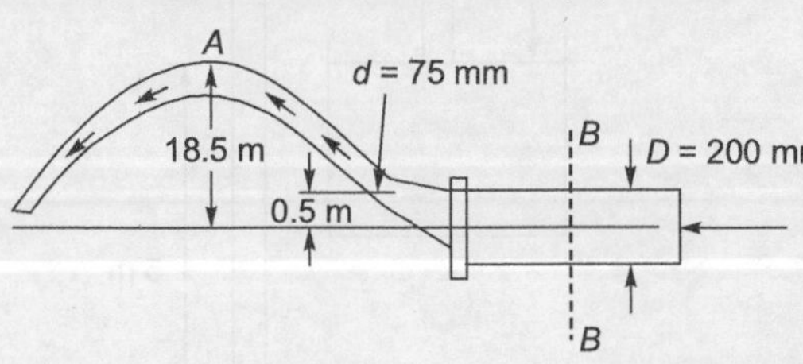

[2000 : 5 M]

5.3 The inlet velocity of water (ρ = 1000 kg/m^3) in a right angled bend-reducer is v_1 = 1 m/s, as shown below. The inlet diameter is D_1 = 0.8 m and the outlet diameter is D_2 = 0.4 m. The flow is turbulent and the velocity profiles at the inlet and outlet are flat (plug flow). Gravitational forces are negligible.

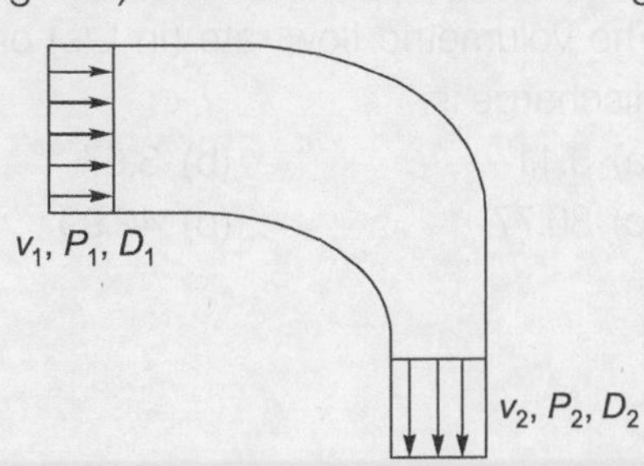

(a) Find the pressure drop ($P_1 - P_2$) across the bend assuming negligible friction losses.

(b) If the actual pressure drop is ($P_1 - P_2$) = 8.25 kPa, find the friction loss factor (K_f) based on the velocity v_1.

[2001 : 2 M]

5.4 Consider the flow in a liquid film of constant thickness (δ) along a vertical wall as shown in the figure.

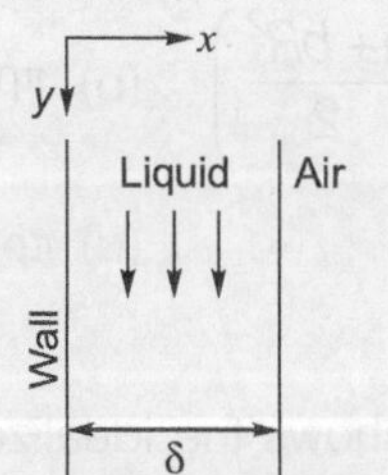

Assuming laminar, one-dimensional, fully developed flow, the y-direction Navier Stokes equation reduces to

$$\mu \frac{d^2 v_y}{dx^2} + \rho g = 0$$

(where v_y is the velocity in y direction, μ is the viscosity and ρ is the density of the liquid.)

(a) State the boundary conditions to be used for the solution of velocity profile?

(b) Solve for the velocity profile.

(c) If Q is the volumetric flow rate per unit width of the wall, how is it related to the film thickness δ.

[2002 : 5 M]

5.5 A pipe has a porous section of length L as shown in the figure. Velocity at the start of this section is V_o. If fluid leaks into the pipe through the porous section at a volumetric rate per unit area $q\,(x/L)^2$, what will be the axial velocity in the pipe at any x? Assume incompressible, one-dimensional flow, i.e., no gradients in the radial direction

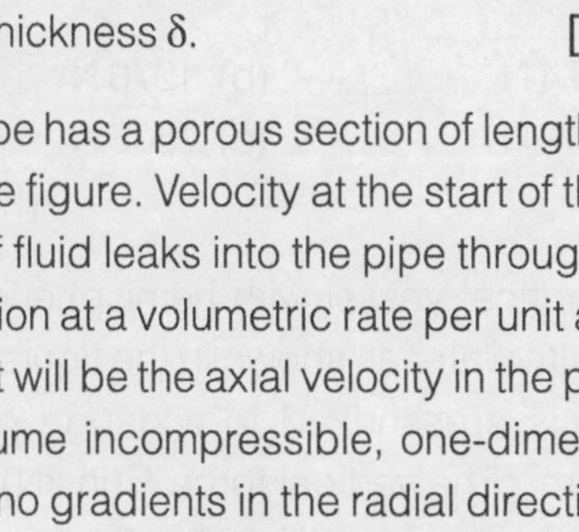

(a) $V_x = V_o + q\dfrac{x^3}{L^2D}$ (b) $V_x = V_o + \dfrac{1}{3}q\dfrac{x^3}{L^2}$

(c) $V_x = V_o + 2q\dfrac{x^2}{LD}$ (d) $V_x = V_o + \dfrac{4}{3}q\dfrac{x^3}{L^2D}$

[2003 : 2 M]

5.6 The liquid surface in a cylindrical bucket of radius R rotating about its axis acquires a parabolic profile given by the equation $y = a + br^2$, where y is the height of the liquid surface from the bottom of the bucket at a radial distance r from the bucket axis. If the liquid has density ρ, then the mass of the liquid in the bucket is

(a) $\pi\rho R^2\left(\dfrac{a+bR^2}{2}\right)$ (b) $\pi\rho R^2\left(a+\dfrac{bR^2}{2}\right)$

(c) $\pi\rho R^2 a$ (d) $\pi\rho R^2(a + bR^2)$

[2006 : 2 M]

5.7 The figure shows the idealized view of a return elbow or U bend, which is connected to two pipes by flexible hoses that transmit no force. Water with density 1000 kg/m^3 flows at velocity of 10 m/s through the pipe, which has a uniform ID of 0.1 m. The gauge pressure at points 1 and 2 are 304 kPa and 253 kPa respectively. The horizontal force F required to keep the elbow in position is

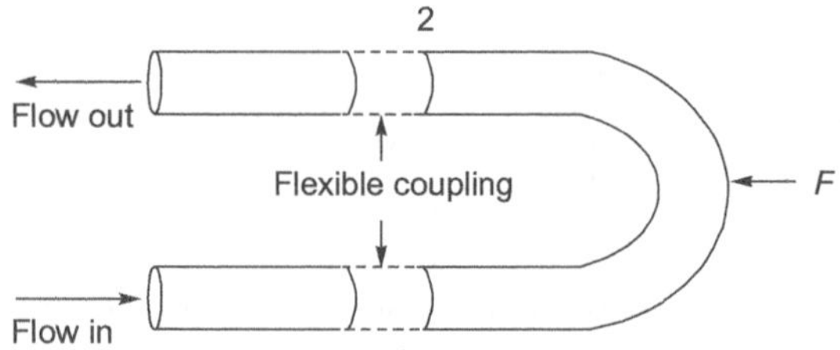

(a) 1574 N (b) 1970 N
(c) 5942 N (d) 7533 N

[2007 : 2 M]

5.8 A cylindrical vessel with hemispherical ends is filled with water as shown in the figure. The head space is pressurized to a gauge pressure of 40 kN/m^2. The vertical force F (in kN) tending to lift the top dome and the absolute pressure P (in kN/m^2) at the bottom of the vessel are ($g = 9.8$ m/s^2, density of water = 1000 kg/m^3)

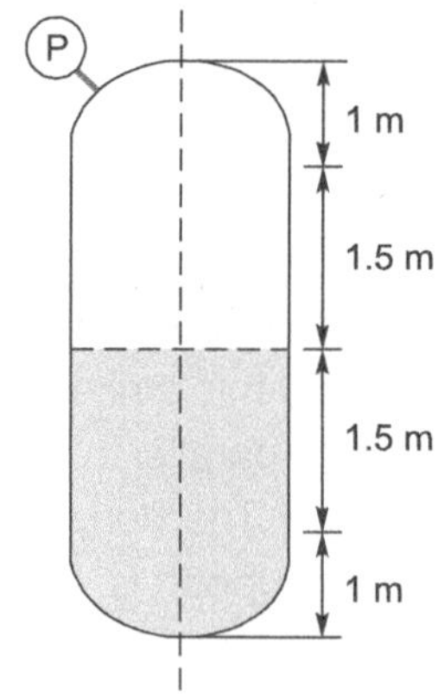

(a) $F = 83.6$; $P = 64.5$
(b) $F = 83.6$; $P = 165.8$
(c) $F = 125.7$; $P = 64.5$
(d) $F = 125.7$; $P = 165.8$

[2008 : 2 M]

Statement for Linked Answer Question (5.9 & 5.10) :
A siphon tube having a diameter of 2 cm draws water from a large open reservoir and discharges into the open atmosphere as shown in the figure. Assume incompressible fluid and neglect frictional losses ($g = 9.8$ m/s^2)

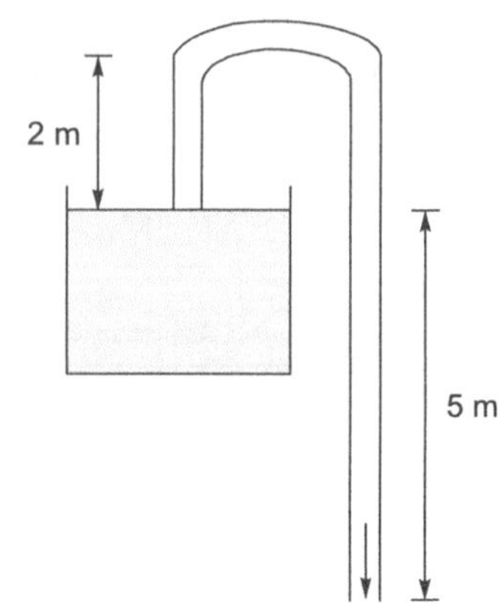

5.9 The velocity (in m/s) at the discharge point is
(a) 9.9 (b) 11.7
(c) 98 (d) 136.9

[2008 : 2 M]

5.10 The volumetric flow rate (in L/s) of water at the discharge is
(a) 3.11 (b) 3.67
(c) 30.77 (d) 42.99

[2008 : 2 M]

Statement for Linked Answer Questions 5.11 and 5.12 :

A free jet of water is emerging from a nozzle (diameter 75 mm) attached to a pipe (diameter 225 mm) as shown below.

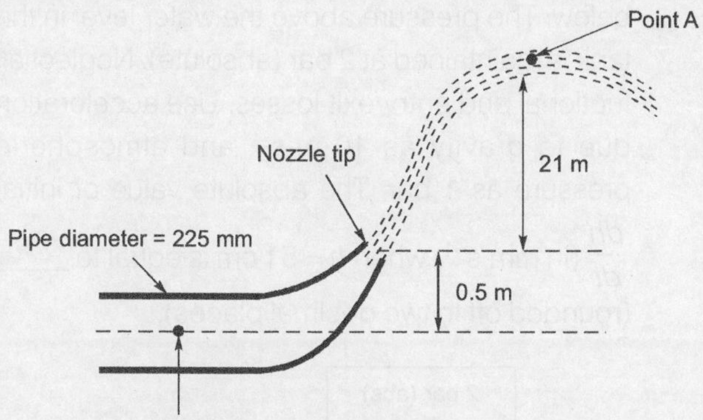

The velocity of water at point A is 18 m/s. Neglect friction in the pipe and nozzle. Use $g = 9.81$ m/s^2 and density of water = 1000 kg/m^3.

5.11 The velocity of water at the tip of the nozzle (in m/s) is

(a) 13.4 (b) 18.0
(c) 23.2 (d) 27.1

[2009 : 2 M]

5.12 The gauge pressure (in kPa) at point B is

(a) 80.0 (b) 100.0
(c) 239.3 (d) 367.6

[2009 : 2 M]

Statement for Linked Answer Questions 5.13 & 5.14 :

A Newtonian fluid of viscosity μ flows between two parallel plates due to the motion of the bottom plate (as shown below), which is moved with a velocity V. The top plate is stationary.

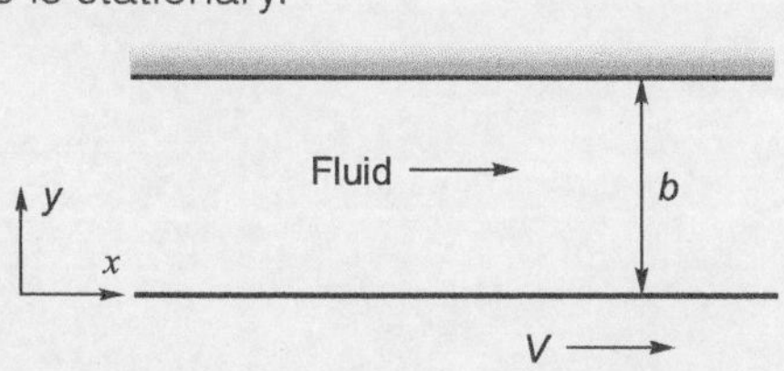

5.13 The steady, laminar velocity profile in the x-direction is

(a) $V\left[\frac{y}{b}\right]$ (b) $V\left[\left(\frac{y}{b}\right)^2 - 1\right]$

(c) $V\left[1 - \left(\frac{y}{b}\right)^2\right]$ (d) $V\left[1 - \frac{y}{b}\right]$

[2012 : 2 M]

5.14 The force per unit area (in the x-direction) that must be exerted on the bottom plate to maintain the flow is

(a) $\frac{\mu V}{b}$ (b) $\frac{-\mu V}{b}$

(c) $\frac{2\mu V}{b}$ (d) $\frac{-2\mu V}{b}$

[2012 : 2 M]

5.15 The mass balance for a fluid with density (ρ) and velocity vector ($\vec{V}$) is

(a) $\frac{\partial \rho}{\partial t} + \nabla \cdot (\rho \vec{V}) = 0$ (b) $\frac{\partial \rho}{\partial t} + \vec{V} \cdot (\nabla \rho) = 0$

(c) $\frac{\partial \rho}{\partial t} + \rho(\nabla \cdot \vec{V}) = 0$ (d) $\frac{\partial \rho}{\partial t} - \vec{V} \cdot (\nabla \rho) = 0$

[2013 : 1 M]

5.16 An incompressible Newtonian fluid, filled in an annular gap between two concentric cylinders of radii R_1 and R_2 as shown in the figure, is flowing under steady state conditions. The outer cylinder is rotating with an angular velocity of Ω while the inner cylinder is stationary. Given that $(R_2 - R_1) << R_1$, the profile of the θ-component of the velocity V_θ can be approximated by,

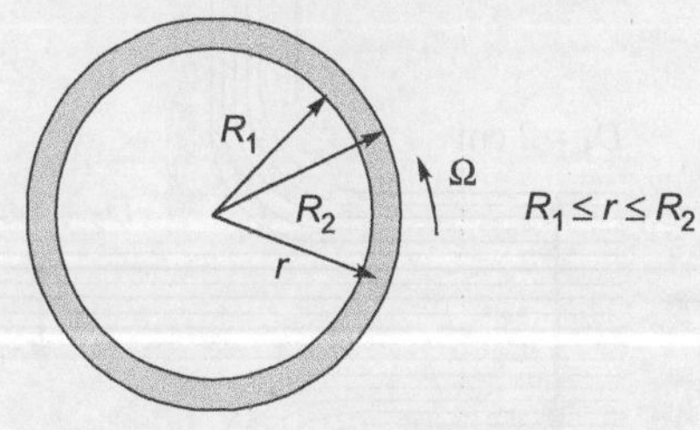

(a) $R_2 \Omega$ (b) $\frac{(r - R_2)}{(R_2 - R_1)} r\Omega$

(c) $\frac{(r + R_2)}{(R_2 + R_1)} R_1 \Omega$ (d) $\frac{(r - R_1)}{(R_2 - R_1)} R_2 \Omega$

[2013 : 1 M]

5.17 Water (density 1000 kg/m^3) is flowing through a nozzle, as shown below and exiting to the atmosphere. The relationship between the diameters of the nozzle at locations 1 and 2 is $D_1 = 4\,D_2$. The average velocity of the stream at location 2 is 16 m/s and the frictional loss between location 1 and location 2 is 10000 Pa. Assuming steady state and turbulent flow, the gauge pressure in Pa, at location 1 is ______.

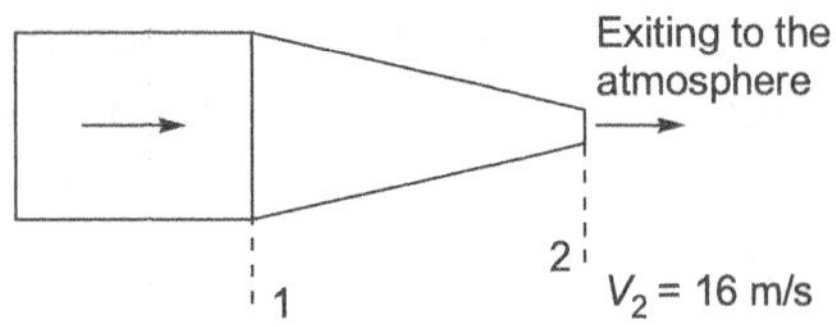

[2013 : 2 M]

5.18 The initial water level in a tank is 4 m. When the valve located at the bottom is opened, the rate of change of water level (L) with respect to time (t) is, $\frac{dL}{dt} = -k\sqrt{t}$, where $k = 0.6\ ms^{-3/2}$. The level of water (in m) in the tank at time 0.5 s after opening the valve is ________ (rounded off to second decimal place).

[2018 : 1 M]

5.19 A horizontal cylindrical water jet of diameter $D_1 = 2$ cm strikes a vertical solid plate with a hole of diameter $D_2 = 1$ cm, as shown in the figure. A part of the jet passes through the hole and the rest is deflected along the plate. The density of water is 1000 kg m^{-3}. If the speed of the jet is 20 m s^{-1}, the magnitude of the horizontal force, in N, required to hold the plate stationary is

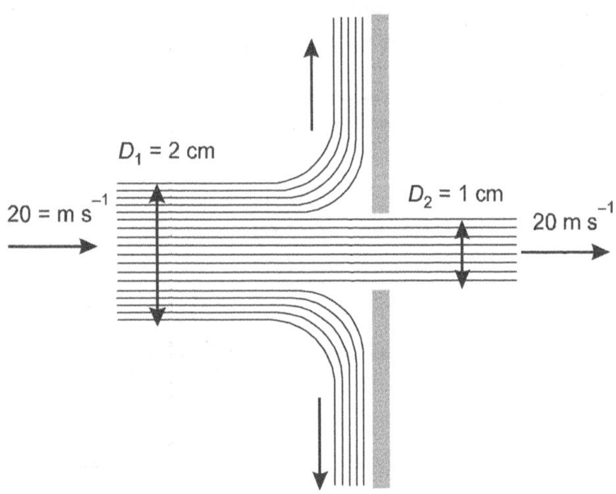

(a) 30π
(b) 10π
(c) 20π
(d) 5π

[2022 : 2 M]

5.20 A cylindrical tank with a diameter of 500 mm contains water (density =1 g.cm^{-3}) upto a height h. A 5 mm diameter round nozzle, whose center is 1 cm above the base of the tank, has its exit open to the atmosphere as shown in the schematic below. The pressure above the water level in the tank is maintained at 2 bar (absolute). Neglect all frictional and entry/exit losses. Use acceleration due to gravity as 10 m.s^{-2} and atmospheric pressure as 1 bar. The absolute value of initial $\frac{dh}{dt}$ (in mm.s^{-1}) when $h = 51$ cm is equal to _____ (rounded off to two decimal places).

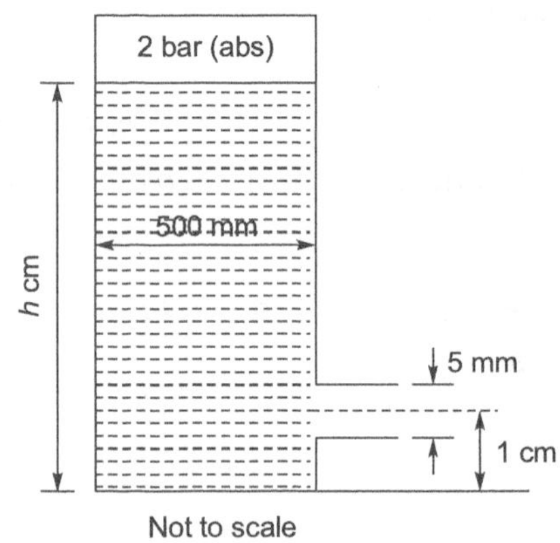

[2023 : 2 M]

Answers Fluid Dynamics

5.1	(d)	**5.2**	(19.36)	**5.3**	(7500,1.5)	**5.4**	(Sol)	**5.5**	(d)	**5.6**	(b)	**5.7**	(c)
5.8	(d)	**5.9**	(a)	**5.10**	(a)	**5.11**	(d)	**5.12**	(d)	**5.13**	(d)	**5.14**	(a)
5.15	(a)	**5.16**	(d)	**5.17**	(137500)	**5.18**	(3.858)	**5.19**	(a)	**5.20**	(1.45)		

Explanations Fluid Dynamics

5.1 (d)

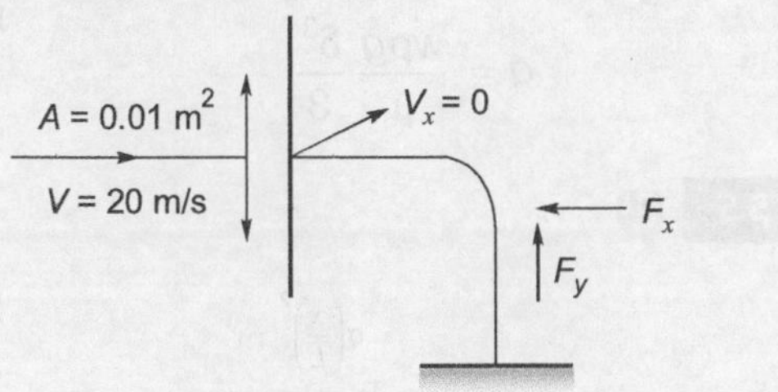

$$F_x = \iint V_x \rho V \cos\alpha \, dA + \frac{\partial}{\partial t}\iiint \rho V_x \, dV$$

($\alpha = 0$; second term $= 0$ (Steady state))

$$F_x = V_x \rho V A = 20 \times 1000 \times 20 \times 0.01 = 4000 \text{ N}$$

↓

Horizontal component of force on the support.

5.2 (19.36)

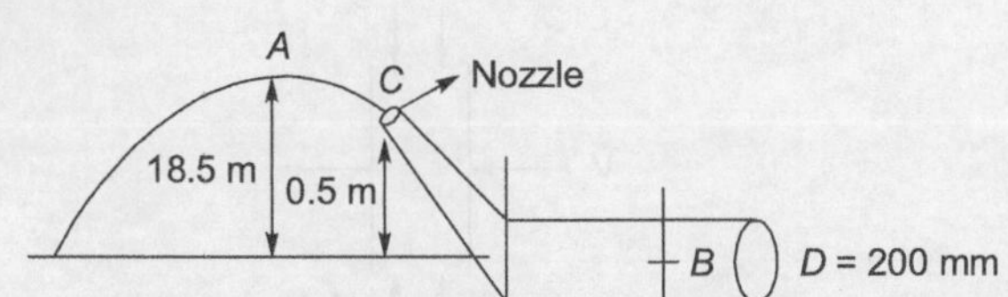

Diameter of nozzle = 75 mm

Pipe diameter = 200 mm

$$\bar{V}_B = 3.8 \text{ m/s}$$

From $[A_C V_C = A_B V_B]$

$\therefore$ $\bar{V}_C = 27$ m/s

Apply Bernoulli's theorem between A and C

$$\frac{P_C}{\rho} + g z_C + \frac{V_C^2}{2} = \frac{P_A}{\rho} + g z_A + \frac{V_A^2}{2} + h_f$$

$P_A = P_C =$ Atmospheric pressure

$Z_C = 0.5$ m, $Z_A = 18.5$ m

($h_f = 0$) (Neglect frictional loss)

$$g(0.5) + \frac{V_C^2}{2} = g(18.5) + \frac{V_A^2}{2}$$

$$V_A^2 = V_C^2 + 2g(-18) \because (V_C = 27 \text{ m/s})$$

$$V_A = \sqrt{375.84} = 19.36 \text{ m/s}$$

5.3 (7500, 1.5)

(a)

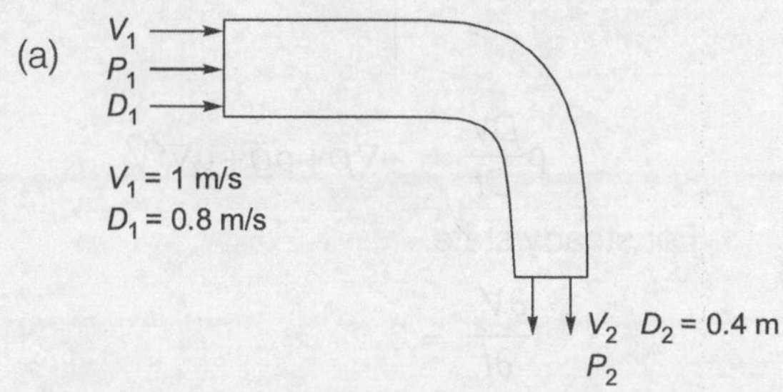

From $[A_1 V_1 = A_2 V_2]$ $V_2 = 4$ m/s

Negligible friction loss

$$\frac{P_1}{\rho} + g z_1 + \frac{V_1^2}{2} = \frac{P_2}{\rho} + g z_2 + \frac{V_2^2}{2}$$

$$Z_1 \cong Z_2$$

$$\frac{P_1 - P_2}{\rho} = \frac{1}{2}[4^2 - 1^2] = 7.5$$

$$P_1 - P_2 = 7.5 \times 1000$$

$$P_1 - P_2 = 7500 \text{ Pascal}$$

(b) In the II case, after apply Bernoulli's equation

$$\therefore \quad \frac{P_1}{\rho} + \frac{V_1^2}{2} = \frac{P_2}{\rho} + \frac{V_2^2}{2} + h_f$$

$$\therefore \quad h_f = \frac{P_1 - P_2}{\rho} + \frac{V_1^2 - V_2^2}{2}$$

$$h_f = \frac{8.25 \times 1000}{1000} + \frac{1^2 - 4^2}{2}$$

$$h_f = 8.25 - 7.5 = 0.75$$

$$h_f = k_f \cdot \frac{V_1^2}{2}$$

Thus, $0.75 = k_f \dfrac{1^2}{2}$

$\Rightarrow$ $k_f = 1.5$

5.4 (Sol)

(a) From Navier Stokes equation,

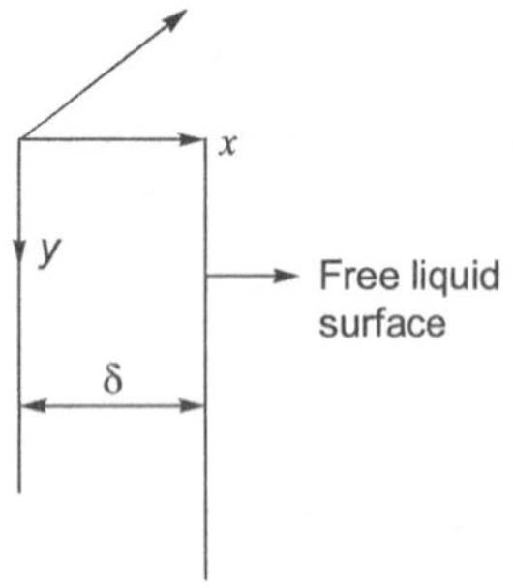

$$\rho\frac{DV}{Dt} = -\nabla p+\rho\vec{g}+\mu\nabla^2\vec{V}$$

For steady state,

$$\frac{\partial V_y}{\partial t} = 0$$

$$V_y \neq V_y(y),\ V_z = 0,\ V_x = 0$$

$$\rho\left[\cancelto{0}{\frac{\partial V_y}{\partial t}} + \cancelto{0}{V_x}\frac{\partial V_y}{\partial x} + V_y\cancelto{0}{\frac{\partial V_y}{\partial y}} + \cancelto{0}{V_z}\frac{\partial V_y}{\partial z}\right]$$

$$= \mu\left[\frac{\partial^2 V_y}{\partial x^2} + \cancelto{0}{\frac{\partial^2 V_y}{\partial y^2}} + \cancelto{0}{\frac{\partial^2 V_y}{\partial z^2}}\right] - \cancelto{0}{\frac{\partial p}{\partial y}} + \rho g_y$$

$$0 = \mu\frac{\partial^2 V_y}{\partial x^2}+\rho g \qquad [g_y = g]$$

Boundary condition:

$x = 0,\quad V_y = 0$ (No slip)

$x = \delta$

$\tau_{yx} = 0$ (free liquid surface)

$$-\mu\frac{dV_y}{dx} = 0$$

$$\frac{dV_y}{dx} = 0$$

(b) $$\mu\frac{\partial^2 V_y}{\partial x^2} = -\rho g$$

$$\frac{dV_y}{dx} = -\left(\frac{\rho g}{\mu}\right)x + C_1$$

at $$x = \delta,\ \frac{dV_y}{dx} = 0$$

$$C_1 = \frac{\rho g\delta}{\mu}$$

Again integrate using BC: at $x = 0$, $V_y = 0$

$$V_y = \frac{\rho g}{\mu}\left[\delta x - \frac{x^2}{2}\right]$$

(c) $$q = \iint V_y dA = \int_0^w\int_0^\delta V_y dx\,dz$$

$$= \frac{w\rho g}{\mu}\left[\frac{\delta x^2}{2} - \frac{x^3}{6}\right]_0^\delta$$

$$q = \frac{w\rho g}{\mu}\frac{\delta^3}{3}$$

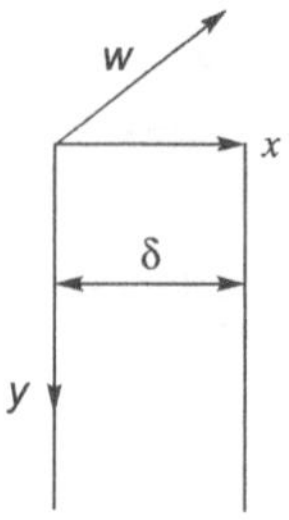

5.5 (d)

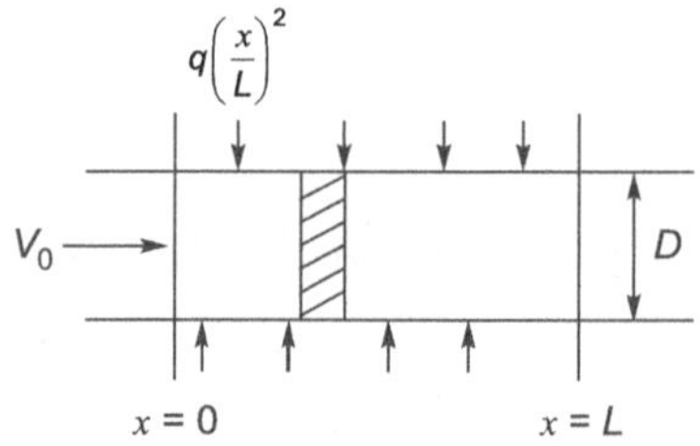

Incompressible fluid, constant density (ρ), steady state.

Mass balance at the differential element between x and $x + \Delta x$

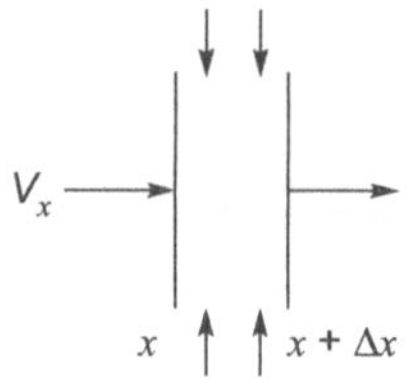

(Input at x) – [Output at $x + \Delta x$] + Mass of fluid leaking through porous section of Δx thickness.

At steady state:

$$\left(\frac{\pi}{4}D^2\right)V_x\rho\Big|_x - \left(\frac{\pi}{4}D^2\right)V_x\rho\Big|_{x+\Delta x} + \pi D\Delta x\cdot q\left(\frac{x}{L}\right)^2\rho = 0$$

$$\frac{D}{4}\left[V_x\big|_{x+\Delta x} - V_x\big|_x\right] = q\left(\frac{x}{L}\right)^2\Delta x$$

$$\lim_{\Delta x\to 0}\frac{V_x\big|_{x+dx} - V\big|_x}{\Delta x} = \left(\frac{4q}{L^2 D}\right)x^2$$

$$\int V_x = \left(\frac{4q}{L^2 D}\right)\int x^2 dx + c$$

B.C: $x = 0,\quad V_x = V_0$

$V_0 = c$

$$\therefore \qquad V_x = V_0 + \frac{4}{3}q\frac{x^3}{L^2 D}$$

5.6 (b)

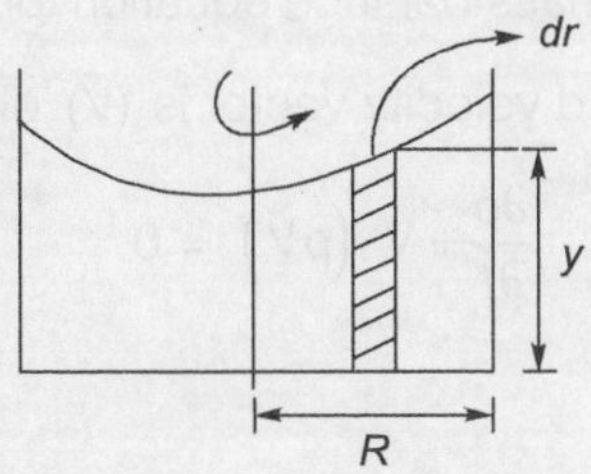

$$\int_0^M dM = \int_0^R \rho\, dV$$

$dV = 2\pi r\, dr\, y = dA \cdot y$ (differential volume)

Total mass in the bucket,

$$M = \rho 2\pi \int_0^R r(a + br^2)dr = \rho 2\pi \left[\frac{aR^2}{2} + \frac{bR^4}{4}\right]$$

$$M = R^2 \pi \rho \left[a + \frac{bR^2}{2}\right]$$

5.7 (c)

$$F_x = \rho Q\,[V_2 \cos\theta - V_1] + P_2 A_2 \cos\theta - P_1 A_1$$

Here, $\theta = 180°$

$$F_x = \rho Q\,[V_2 \cos 180° - V_1] + P_2 A_2\,[\cos 180°] - P_1 A_1$$

$$F_x = 0.0785 \times 1000\,[10 \cos 180 - 10] + 253 \times 10^3 \times \frac{\pi}{4}(0.1)^2 \cos 180 - 304 \times 10^3 \times \frac{\pi}{4}(0.1)^2$$

$$F_x = 5945 \text{ N}$$

Closer value $\rightarrow$ [5942 N]

5.8 (d)

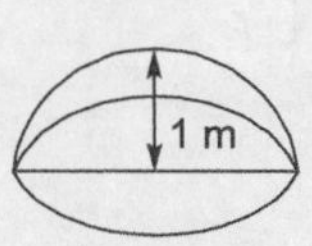

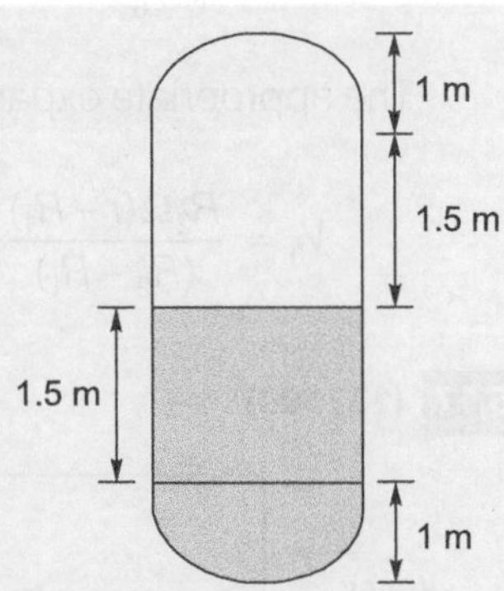

(Given) effective pressure tending to lift the dome $= 40$ kN/m^2

$$\text{Force} = \text{Pressure} \times \text{Area}$$

$$= 40 \times \frac{\pi}{4} D^2 = 125.7 \text{ kN}$$

Absolute pressure at bottom = Head space pressure + atm pressure + $\rho g h$

$$P_B = 40 + 101.3 + \frac{1000 \times 9.8 \times 2.5}{1000}$$

$$= 165.8 \text{ kN/m}^2$$

5.9 (a)

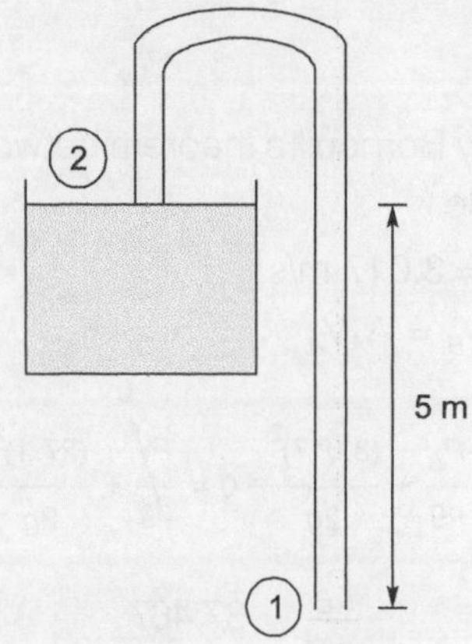

Apply Bernoulli's equation

$$\frac{P_1}{\rho g} + \frac{V_1^2}{2g} + z_1 = \frac{P_2}{\rho g} + \frac{V_2^2}{2g} + z_2$$

$$P_1 = P_2 = 1 \text{ atm}$$

$$V_2 = 0 \text{ m/s}$$

$$\frac{V_1^2}{2g} + z_1 = z_2$$

$$\therefore \quad V_1 = \sqrt{2g(z_2 - z_1)} = \sqrt{2 \times g \times 5}$$

$$V_1 = 9.899 \text{ m/s} \cong 9.9 \text{ m/s}$$

5.10 (a)

$$Q = AV$$

$$Q = \frac{\pi}{4} D^2 \times 9.9 = \frac{\pi (0.02)^2 \times 9.9}{4}$$

$$= 3.10 \times 10^{-3} \text{ m}^3/\text{s}$$

$$= 3.10 \text{ L/s}$$

5.11 (d)

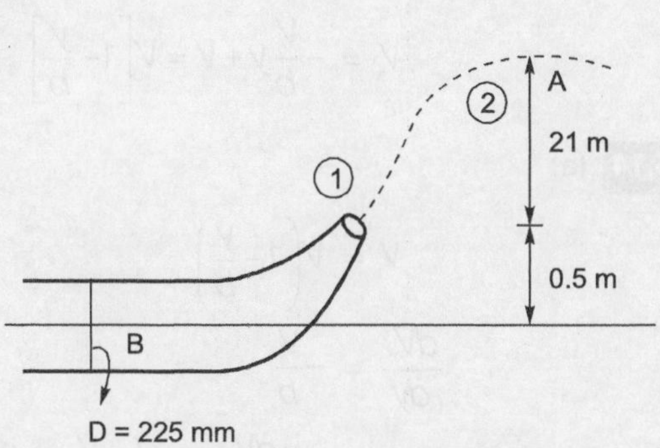

Apply energy equation between at (1) and (2)

$$\frac{P_1}{\rho g}+\frac{V_1^2}{2g}+z_1=\frac{P_2}{\rho g}+\frac{V_2^2}{2g}+z_2+h_f$$

$$\frac{V_1^2}{2g}+0.5 = \frac{(18)^2}{2g}+21.5+0$$

$$V_1 = 27.1 \text{ m/s}$$

5.12 (d)

Apply Bernoulli's theorem between (B) and tip of nozzle.

$$\begin{Bmatrix} V_B = 3.017 \text{ m/s} \\ A_BV_B = A_AV_A \end{Bmatrix}$$

$$\frac{P_B}{\rho g}+\frac{(3.017)^2}{2g}+0=\frac{P}{\rho g}+\frac{(27.1)^2}{2g}+0.5$$

$$\frac{P_B}{\rho g} = 37.467$$

$$P_B = 367.58 \cong 367.6 \text{ kPa}$$

5.13 (d)

The steady, laminar velocity profile in x-direction is

From Navier Stokes equations:

Steady state, Horizontal flow

$$\rho\frac{DV}{Dt} = -\nabla p+\mu\nabla^2 V+\rho g$$

$$0 = 0+\mu\nabla^2+0$$

$$\frac{d^2V_x}{dy^2} = 0$$

Integrating, $V_x = C_1 y + C_2$

Boundary condition,

$$y = 0, \quad V_x = V$$
$$y = b, \quad V_x = 0$$

Hence, $C_2 = V$

$$C_1 = -\frac{V}{b}$$

$$V_x = -\frac{V}{b}y+V=V\left[1-\frac{y}{b}\right]$$

5.14 (a)

$$V_x = V\left(1-\frac{y}{b}\right)$$

$$\frac{dV_x}{dy} = -\frac{V}{b}$$

$$\tau_{yx} = -\frac{\mu dV_x}{dy}=-\mu\left(-\frac{V}{b}\right)=\frac{\mu V}{b}$$

5.15 (a)

The mass balance equation for a fluid of density 'ρ' and velocity vector is $(\vec{V})$ given by

$$\therefore \quad \frac{\partial\rho}{\partial t}+\nabla\cdot\left(\rho\vec{V}\right) = 0$$

5.16 (d)

Applying equation of continuity for this flow

$$\frac{\partial}{\partial r}\left[\frac{1}{r}\frac{\partial}{\partial r}(rV_\theta)\right] = 0$$

Integrating $\frac{\partial}{\partial r}(rV_\theta) = C_1 r$

Integrating again, $rV_\theta = C_1\frac{r^2}{2}+C_2$ ...(i)

At $r = R_1,\ V_\theta = 0$

$$0 = C_1\frac{R_1^2}{2}+C_2$$

$$C_2 = -\frac{C_1R_1^2}{2}$$

At, $r = R_2,\ V_\theta = R_2\,\Omega$

$$\therefore \quad R_2^2\,\Omega = C_1\frac{R_2^2}{2}+C_2 = C_1\frac{R_2^2}{2}-\frac{C_1R_1^2}{2}$$

$$R_2^2\,\Omega = \frac{C_1}{2}[R_2^2-R_1^2]$$

From equation (i),

$$rV_\theta = C_1\frac{r^2}{2}-\frac{C_1R_1^2}{2} = \frac{C_1}{2}[r^2-R_1^2]$$

$$rV_\theta = \frac{R_2^2\Omega(r-R_1)(r+R_1)}{(R_2^2-R_1^2)}$$

The appropriate expansion is

$$V_\theta = \frac{R_2\Omega(r-R_1)}{(R_2-R_1)} \qquad \begin{bmatrix} \because\ r+R_1=R_2+R \\ \frac{R_2}{r}\cong 1 \end{bmatrix}$$

5.17 (137500)

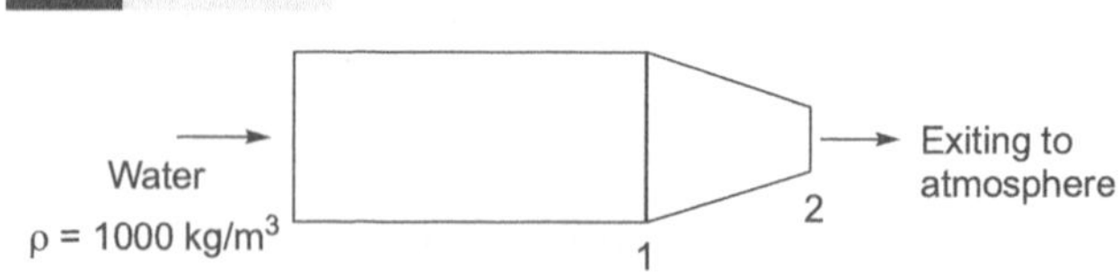

$$P_2 = 1 \text{ atm}$$
$$V_2 = 16 \text{ m/s}$$

$(D_1 = 4D_2)$ given

Friction loss = 10000 Pa

From continuity equation,

$$Q = A_1V_1 = A_2V_2$$

$$V_2 = \left(\frac{D_2^2}{D_1^2}\right)\cdot 16$$

$$\Rightarrow \quad V_1 = 1 \text{ m/s}$$

Apply energy equation between 1 and 2,

$$\frac{P_1}{\rho g}+\frac{V_1^2}{2g} = \frac{P_2}{\rho g}+\frac{V_2^2}{2g}+\frac{10000}{1000\times 9.81}$$

↓
Open to atmosphere

$$\frac{P_1}{1000}+\frac{1}{2} = \frac{(16)^2}{2}+10$$

$$P_1 = 137500 \text{ pascal}$$

5.18 (3.858)

Given, $\left\{\frac{dL}{dt} = -k\sqrt{t}\right\}$ the rate of change of water level (L) with respect to time (t).

At, $t = 0, \quad L = 4$ m

$t = 0.5$ sec $\quad (L = ?)$

($k = 0.6$) given

$$\frac{dL}{dt} = -k\sqrt{t} \rightarrow \int_4^L dL = -\int_0^{0.5} k\sqrt{t}\,dt$$

$$L-4 = \left[-0.6\times\frac{2}{3}\times(t)^{3/2}\right]$$

$$L = 3.858 \text{ m}$$

5.19 (a)

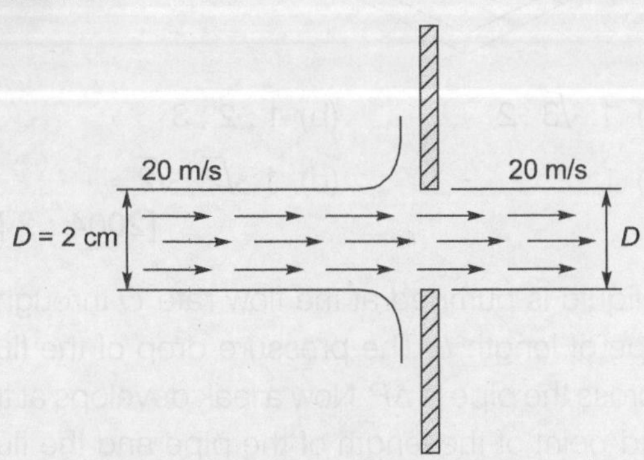

Applying momentum balance, we get

$$\Sigma F_x = \dot{m}v|_{\text{out}} - \dot{m}v|_{\text{in}}$$

$$= \rho Q_{\text{out}}v_{\text{out}} - \rho Q_{\text{in}}v_{\text{in}}$$

$$= 1000\left[\frac{\pi}{4}(0.01)^2\times 400-\frac{\pi}{4}(0.02)^2\times 400\right]$$

$$= \frac{1000\times 400\times \pi}{4}\left[(0.01)^2-(0.02)^2\right] = 30\pi$$

5.20 (1.45)

From mass balance, we get

$$\rho A_N V_N = \frac{dm}{dt} = \rho A_T\frac{dh}{dt}$$

$$\left(\frac{A_N}{A_T}\right)V_N = \frac{dh}{dt}$$

$$\frac{dh}{dt} = \left(\frac{d_N}{d_T}\right)^2 V_N \quad \text{...(i)}$$

where, d_N = Nozzle diameter

d_T = Tank diameter

V_N = Nozzle velocity

Applying Bernoulli's theorem between tank water surface and nozzle exit, we get

$$\frac{P_1}{\rho g}+\frac{V_1^2}{2g}+\frac{h}{100} = \frac{P_2}{\rho g}+\frac{V_N^2}{2g}+\frac{1}{100}$$

$$V_1 \cong 0 \text{ (surface velocity)}$$

$$\frac{2\times 10^5}{10^3\times 10}+\frac{h}{100} = \frac{10^5}{10^4}+\frac{(V_N)^2}{20}+\frac{1}{100}$$

$$20+\frac{h}{100}-\frac{1}{100} = 10+\frac{(V_N)^2}{20}$$

$$10+\frac{h-1}{100} = \frac{V_N^2}{20}$$

$$V_N = \sqrt{20\left[10+\frac{h-1}{100}\right]}$$

Putting V_N in equation (i), we get

$$\frac{dh}{dt} = \left(\frac{5}{500}\right)^2\sqrt{20\left[10+\frac{51-1}{100}\right]}$$

$$\frac{dh}{dt} = 1.45\times 10^{-3} \text{ m/s} = 1.45 \text{ mm/sec}$$

■■■■

6 Flow Through Pipes

6.1 In a fully turbulent flow (Re > 10^5) in a pipe of diameter d, for a constant pressure gradient, the dependence of volumetric flow rate of an incompressible fluid is

(a) d (b) d^2

(c) $d^{2.5}$ (d) d^4

[2000 : 1 M]

6.2 A hydrocarbon oil (viscosity 0.025 Pa s and density 900 kg/m^3) is transported using a 0.6 m diameter, 10 km long pipe The maximum allowable pressure drop across the pipe length is 1 MPa. Due to a maintenance schedule on this pipeline, it is required to use a 0.4 m diameter, 10 km long pipe to pump the oil at the same volumetric flow rate as in the previous case. Estimate the pressure drop for the 0.4 m diameter pipe. Assume both pipes to be hydro-dynamically smooth and in the range of operating conditions, the Fanning friction factor is given by:

$$f = 0.079\ \text{Re}^{-0.25}$$

[2000 : 5 M]

6.3 The operation of a Rotameter is based on

(a) variable flow area

(b) rotation of a turbine

(c) pressure drop across a nozzle

(d) pressure at a stagnation point

[2001 : 1 M]

6.4 The pressure differential across a vertical venturimeter (shown in figure) is measured with the help of a mercury manometer to estimate flow rate of water flowing through it. The expression for the velocity of water at the throat is

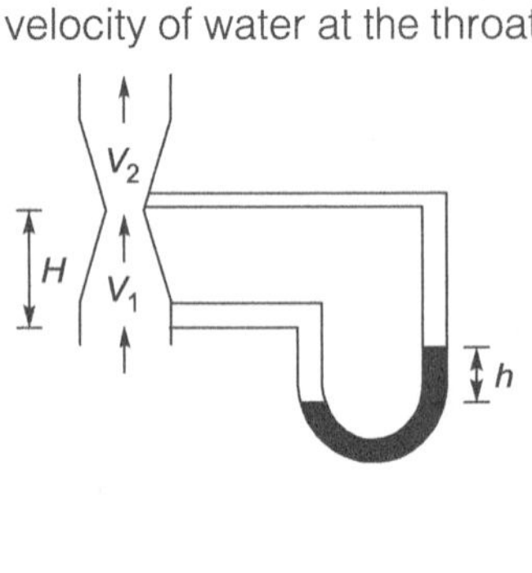

(a) $\dfrac{V_2^2 - V_1^2}{2g} = \dfrac{h\rho_m}{\rho_w}$

(b) $\dfrac{V_2^2 - V_1^2}{2g} = \dfrac{h(\rho_m - \rho_w)}{\rho_w}$

(c) $\dfrac{V_2^2 - V_1^2}{2g} = \dfrac{H + h(\rho_m - \rho_w)}{\rho_w}$

(d) $\dfrac{V_2^2}{g} = \dfrac{h(\rho_m - \rho_w)}{\rho_w}$

[2003 : 2 M]

6.5 Three piping networks, as shown in the figure, are placed horizontally. They are made using identical pipe segments and are subjected to the same pressure drop across them.

Assuming no pressure losses at junctions, the flow rates across the three networks are related as $Q_1 : Q_2 : Q_3$:

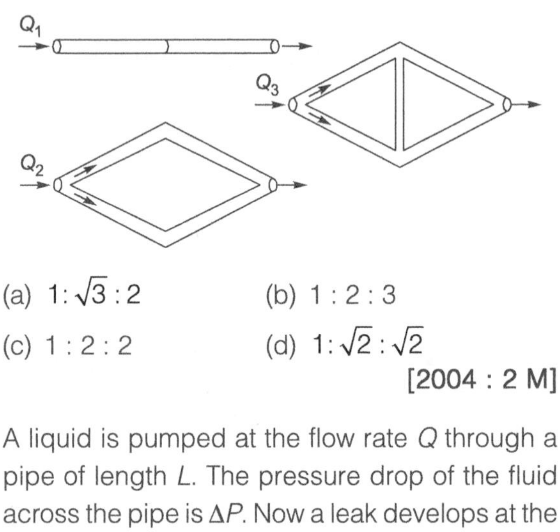

(a) $1 : \sqrt{3} : 2$ (b) 1 : 2 : 3

(c) 1 : 2 : 2 (d) $1 : \sqrt{2} : \sqrt{2}$

[2004 : 2 M]

6.6 A liquid is pumped at the flow rate Q through a pipe of length L. The pressure drop of the fluid across the pipe is ΔP. Now a leak develops at the mid-point of the length of the pipe and the fluid leaks at the rate of $Q/2$. Assuming that the friction factor in the pipe remains unchanged, the new pressure drop across the pipe for the same inlet flow rate (Q) will be

(a) $\left(\dfrac{1}{2}\right)\Delta P$ (b) $\left(\dfrac{5}{8}\right)\Delta P$

(c) $\left(\dfrac{3}{4}\right)\Delta P$ (d) ΔP [2006 : 2 M]

6.7 A pipeline system carries crude oil of density 800 kg/m^3. The volumetric flow rate at point 1 is 0.28 m^3/s. The cross sectional areas of the branches 1, 2 and 3 are 0.012, 0.008 and 0.004 m^2 respectively. All the three branches are in a horizontal plane and the friction is negligible. If the pressure at the points 1 and 3 are 270 kPa and 240 kPa respectively, then the pressure at point 2 is

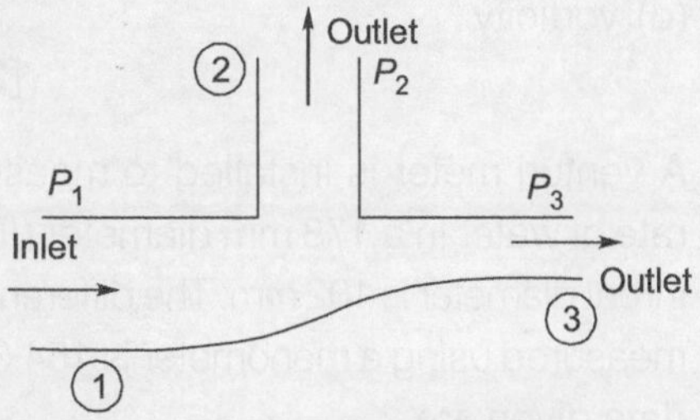

(a) 202 kPa (b) 240 kPa
(c) 284 kPa (d) 355 kPa

[2007 : 2 M]

6.8 The pressure differential across a venturimeter, inclined at 45° to the vertical (as shown in the figure) is measured with the help of a manometer to estimate the flowrate of a fluid through it. If the density of the flowing fluid is ρ and the density of the manometer fluid is ρ_m, the velocity of the fluid at the throat can be obtained from the expression

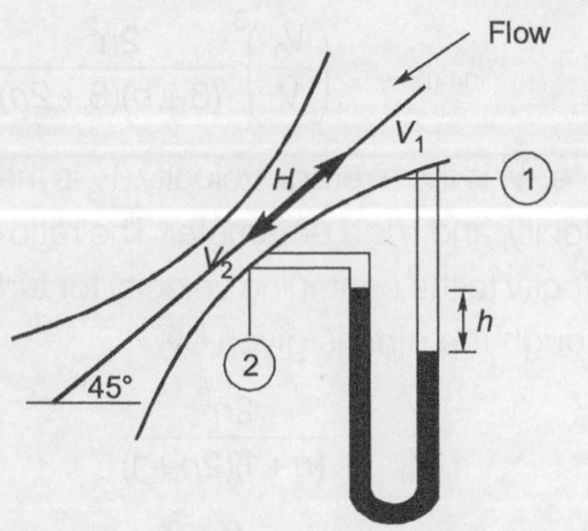

(a) $\dfrac{V_2^2 - V_1^2}{2g} = \dfrac{h(\rho_m - \rho)}{\rho} + H\sin 45°$

(b) $\dfrac{V_2^2 - V_1^2}{2g} = \dfrac{h\rho_m}{\rho} + H\sin 45°$

(c) $\dfrac{V_2^2 - V_1^2}{2g} = \dfrac{h\rho_m}{\rho}$

(d) $\dfrac{V_2^2 - V_1^2}{2g} = \dfrac{h(\rho_m - \rho)}{\rho}$

[2007 : 2 M]

6.9 Losses for flow through valves and fittings are expressed in terms of
(a) drag coefficient
(b) equivalent length of a straight pipe
(c) shape factor
(d) roughness factor

[2008 : 1 M]

6.10 In Hagen-Poiseuille flow through a cylindrical tube, the radial profile of shear stress is
(a) constant (b) cubic
(c) parabolic (d) linear

[2010 : 2 M]

6.11 In an orifice meter, if the pressure drops across the orifice is overestimated by 5%, then the percentage error in the measured flow rate is
(a) 2.47 (b) 5
(c) –2.47 (d) –5

[2011 : 2 M]

6.12 Two liquids (*P* and *Q*) having same viscosity are flowing through a double pipe heat exchanger as shown in the schematic below.

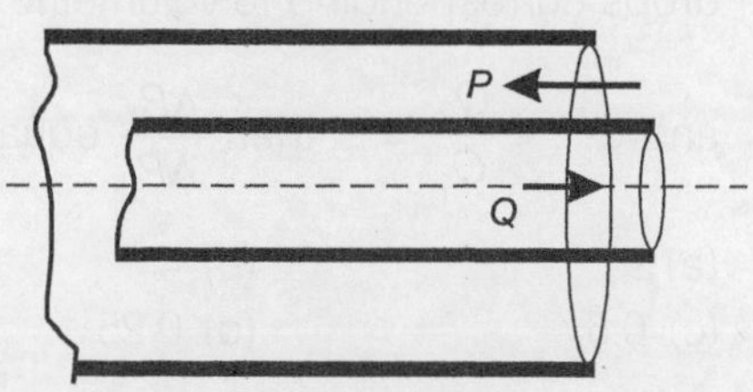

Densities of *P* and *Q* are 1000 and 800 kg/m^3 respectively. The average velocities of the liquids *P* and *Q* are 1 and 2.5 m/s respectively. The inner diameters of the pipes are 0.31 and 0.1 m. Both pipes are 5 mm thick. The ratio of the Reynolds numbers Re_P to Re_Q is
(a) 2.5 (b) 1.55
(c) 1 (d) 4

[2011 : 2 M]

6.13 The local velocity of a fluid along a streamline can be measured by
(a) Pitot tube (b) Venturi meter
(c) Rotameter (d) Orifice meter

[2012 : 1 M]

6.14 In case of a pressure driven laminar flow of a Newtonian fluid of viscosity (μ) through a horizontal circular pipe, the velocity of the fluid is proportional to
(a) μ (b) $\mu^{0.5}$
(c) μ^{-1} (d) $\mu^{-0.5}$

[2014 : 1 M]

6.15 In a steady and incompressible flow of a fluid (density = 1.25 kg m^{-3}), the difference between stagnation and static pressures at the same location in the flow is 30 mm of mercury (density = 13600 kg m^{-3}). Considering gravitational acceleration as 10 m s^{-2}, the fluid speed (in m s^{-1}) is ________.

[2014 : 2 M]

6.16 Two different liquids are flowing through different pipes of the same diameter. In the first pipe, the flow is laminar with centerline velocity, $V_{max,1}$, whereas in the second pipe, the flow is turbulent. For turbulent flow, the average velocity is 0.82 times the centerline velocity, $V_{max\,2}$. For equal volumetric flow rates in both the pipes, the ratio $\frac{V_{max\,1}}{V_{max\,2}}$ (up to two decimal places) is ________.

[2015 : 1 M]

6.17 In a venturi meter, ΔP_1 and ΔP_2 are the pressure drops corresponding to volumetric flowrates Q_1 and Q_2. If $\frac{Q_2}{Q_1} = 2$, then $\frac{\Delta P_2}{\Delta P_1}$ equals

(a) 2 (b) 4
(c) 0.5 (d) 0.25

[2017 : 1 M]

6.18 Oil is being delivered at a steady flowrate through a circular pipe of radius 1.25×10^{-2} m and length 10 m. The pressure drop across the pipe is 500 Pa. The shear stress at the pipe wall rounded to 2 decimal places, is ________ Pa.

[2017 : 2 M]

6.19 The following table provides four sets of Fanning friction factor data, for different values of Reynolds number (Re) and roughness factor $\left(\frac{k}{D}\right)$.

	Re	10^2	10^3	10^5	10^5
	$\left(\frac{k}{D}\right)$	0	0.001	0	0.001
Set I	f	0.16	0.016	16×10^{-5}	16×10^{-5}
Set II	f	0.016	0.16	0.0055	0.0045
Set III	f	0.16	0.016	0.0045	0.0055
Set IV	f	0.0045	0.0055	0.016	0.16

Which of the above sets of friction factor data is correct?

(a) Set I (b) Set II
(c) Set III (d) Set IV

[2017 : 2 M]

6.20 Pitot tube is used to measure

(a) liquid level in a tank
(b) flow velocity at a point
(c) angular deformation
(d) vorticity

[2018 : 1 M]

6.21 A venturi meter is installed to measure the flow rate of water in a 178 mm diameter (ID) pipe. The throat diameter is 102 mm. The differential pressure measured using a manometer is 154.3 kN/m^2. The data given are :
discharge coefficient = 0.98;
water density = 1000 kg/m^3.
The volumetric flow rate of water (in m^3/s) is ____.

[2018 : 1 M]

6.22 Consider a steady flow of an incompressible, Newtonian fluid through a smooth circular pipe. Let $\alpha_{laminar}$ and $\alpha_{turbulent}$ denote the kinetic energy correction factors for laminar and turbulent flow through the pipe, respectively. For turbulent flow through the pipe

$$\alpha_{turbulent} = \left(\frac{V_0}{\bar{V}}\right)^3 \frac{2n^2}{(3+n)(3+2n)}$$

Here, $\bar{V}$ is the average velocity, V_0 is the centerline velocity, and n is a parameter. The ratio of average velocity to the centerline velocity for turbulent flow through the pipe is given by

$$\frac{\bar{V}}{V_0} = \frac{2n^2}{(n+1)(2n+1)}$$

For $n = 7$, the value of $\frac{\alpha_{turbulent}}{\alpha_{laminar}}$ is _____ (round off to 2 decimal places).

[2021 : 1 M]

6.23 Water of density 1000 kg m^{-3} flows in a horizontal pipe of 10 cm diameter at an average velocity of 0.5 m s^{-1}. The following plot shows the pressure measured at various distances from the pipe entrance.

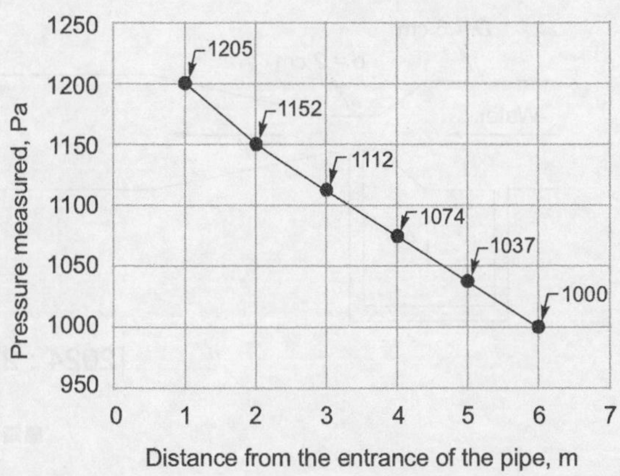

Using the data shown in the figure, the Fanning friction factor in the pipe when the flow is FULLY DEVELOPED is

(a) 0.0012 (b) 0.0074
(c) 0.0082 (d) 0.0106

[2021 : 2 M]

6.24 Consider a horizontal rod of radius aR ($a < 1$) in a stationary pipe of radius R. The rod is pulled coaxially at a constant velocity V as shown in the figure. The annular region is filled with a Newtonian incompressible fluid of viscosity μ. The steady state fully developed axial velocity profile in the fluid is given by $u(r) = V\dfrac{\ln\left(\dfrac{r}{R}\right)}{\ln(a)}$, where r is the radial coordinate. Ignoring end effects, the magnitude of the pulling force per unit rod length is

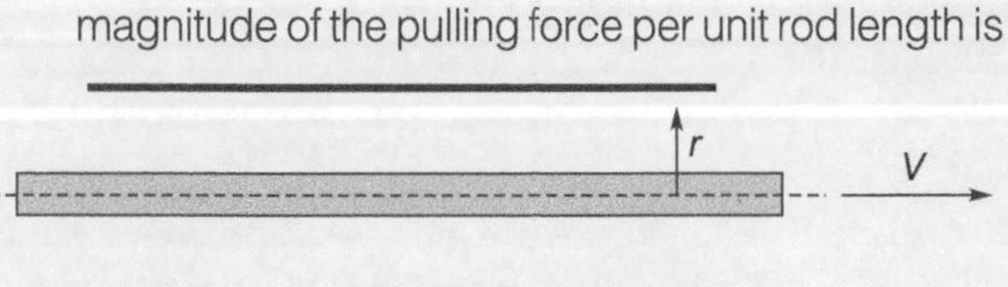

(a) $\pi\mu V$ (b) $-\dfrac{2\pi\mu V}{\ln(a)}$

(c) 0 (d) $-\dfrac{\pi\mu V}{\ln(a)}$ [2022 : 2 M]

6.25 A venturi meter (venturi coefficient, $C_v = 0.98$) is connected to a pipe of inner diameter 50 mm. Water (density 1000 kg m^{-3}) is flowing through the pipe. The pressure-drop measured across the venturi meter is 50 kPa. If the venturi throat diameter is 20 mm, the estimated flow rate of water is _______ × 10^{-3} m^3 s^{-1} (rounded off to two decimal places).

[2022 : 2 M]

6.26 Which one of the following quantities of a flowing fluid is measured using a rotameter?

(a) Static pressure
(b) Dynamic pressure
(c) Volumetric flow rate
(d) Viscosity

[2023 : 1 M]

6.27 Consider the steady, uni-directional, fully-developed, pressure-driven laminar flow of an incompressible Newtonian fluid through a circular pipe of inner radius 5.0 cm. The magnitude of shear stress at the inner wall of the pipe is 0.1 N m^{-2}. At a radial distance of 1.0 cm from the pipe axis, the magnitude of the shear stress, in N m^{-2}, rounded off to 3 decimal places, is _______.

[2024 : 1 M]

6.28 Consider a steady, fully-developed, uni-directional laminar flow of an incompressible Newtonian fluid (viscosity μ) between two infinitely long horizontal plates separated by a distance $2H$ as shown in the figure. The flow is driven by the combined action of a pressure gradient and the motion of the bottom plate at $y = -H$ in the negative x direction. Given that $\dfrac{\Delta P}{L} = \dfrac{(P_1 - P_2)}{L} > 0$, where P_1 and P_2 are the pressures at two x locations separated by a distance L. The bottom plate has a velocity of magnitude V with respect to the stationary top plate at $y = H$. Which one of the following represents the x-component of the fluid velocity vector?

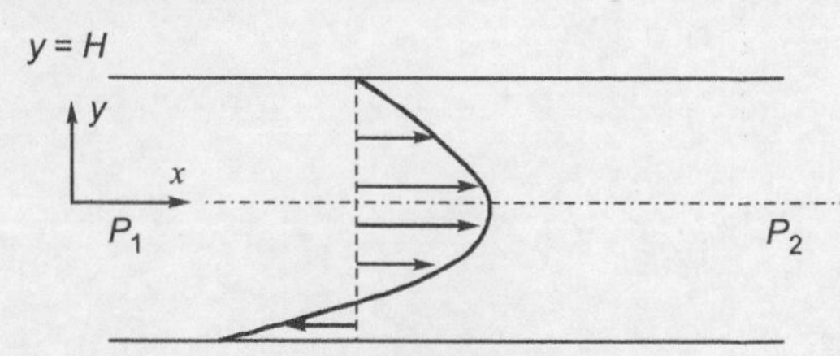

(a) $\dfrac{\Delta P}{L}\dfrac{H^2}{2\mu}\left(1 - \dfrac{y^2}{H^2}\right) + \dfrac{V}{2}\left(\dfrac{y}{H} - 1\right)$

(b) $\dfrac{\Delta P}{L}\dfrac{H^2}{2\mu}\left(\dfrac{y^2}{H^2} - 1\right) + \dfrac{V}{2}\left(\dfrac{y}{H} - 1\right)$

(c) $\dfrac{\Delta P}{L}\dfrac{H^2}{2\mu}\left(\dfrac{y^2}{H^2} - 1\right) - \dfrac{V}{2}\left(\dfrac{y}{H} - 1\right)$

(d) $\frac{\Delta P}{L}\frac{H^2}{2\mu}\left(1-\frac{y^2}{H^2}\right)-\frac{V}{2}\left(\frac{y}{H}-1\right)$

[2024 : 2 M]

6.29 A Venturi meter with a throat diameter $d = 2$ cm measures the flow rate in a pipe of diameter $D = 6$ cm, as shown in the figure. A U-tube manometer is connected to measure the pressure drop. Assume the discharge coefficient is independent of the Reynolds number and geometric ratios. If the volumetric flow rate through the pipe is doubled $Q_2 = 2Q_1$, the corresponding ratio of the manometer readings $\frac{\Delta h_2}{\Delta h_1}$, rounded off to the nearest integer, is ________.

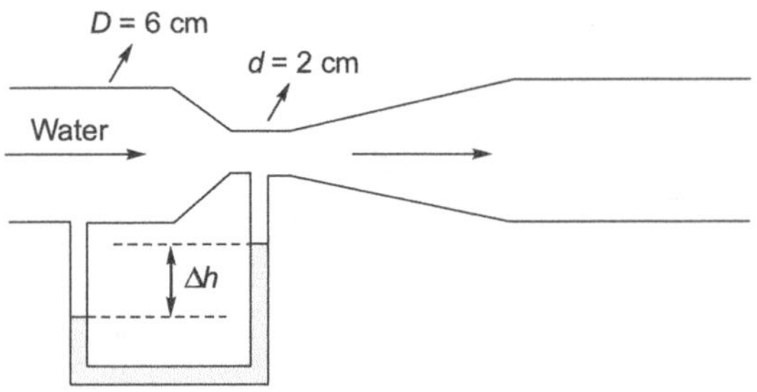

[2024 : 2 M]

■■■■

Answers Flow Through Pipes

6.1	(c)	**6.2**	(6.861)	**6.3**	(a)	**6.4**	(b)	**6.5**	(c)	**6.6**	(b)	**6.7**	(c)
6.8	(d)	**6.9**	(b)	**6.10**	(d)	**6.11**	(a)	**6.12**	(c)	**6.13**	(a)	**6.14**	(c)
6.15	(80.8)	**6.16**	(1.64)	**6.17**	(b)	**6.18**	(0.3125)	**6.19**	(c)	**6.20**	(b)	**6.21**	(0.1488)
6.22	(0.53)	**6.23**	(b)	**6.24**	(b)	**6.25**	(3.12)	**6.26**	(c)	**6.27**	(0.02)	**6.28**	(a)
6.29	(4.0)												

Explanations Flow Through Pipes

6.1 (c)

$$\frac{\Delta p_s}{L} = \frac{2f\rho\bar{V}^2}{D}$$

At Re > 10^5 $f \cong$ constant

$$\frac{\Delta P}{L} = \text{Press gradient} = \text{Constant (given)}$$

$$\therefore \quad (\bar{V})^2 = \left(\frac{\Delta P}{L}\right)\frac{D}{2f\rho}$$

$$\bar{V} = \sqrt{\left(\frac{\Delta P}{L}\right)\frac{1}{2f\rho}}\sqrt{D}$$

$$Q = \left(\frac{\pi}{4}D^2\right)\bar{V}$$

$$= \left(\frac{\pi}{4}D^2\right)\sqrt{\left(\frac{\Delta P}{L}\right)\frac{1}{2f\rho}}\sqrt{D}$$

$$\therefore \quad Q \propto D^{2.5}$$

6.2 (6.861)

Case-I:

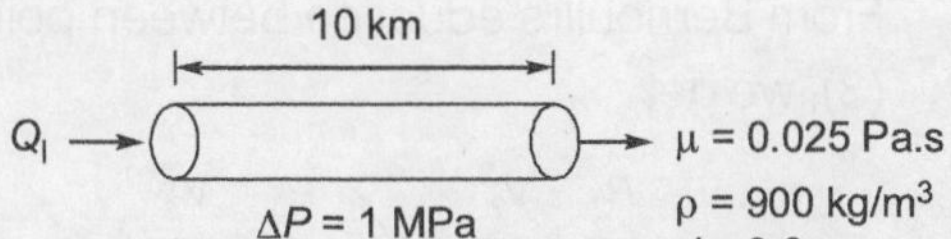

Case-II:

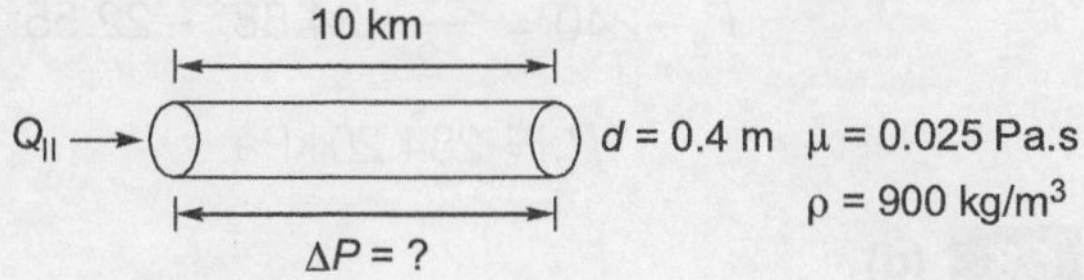

$[Q_I = Q_{II}]$ given

$$\frac{\Delta P}{L} = \frac{2f\rho\bar{V}^2}{D} \qquad \left\{f = \frac{0.079}{\text{Re}^{0.25}}\right\}$$

$$\Delta P = 0.158\, D^{-1.25}\, V^{-1.75}\, \rho^{0.75}\, L\, \mu^{0.25} \quad \text{...(i)}$$

Equation in terms of flow rate, then

$$\Delta P = \left[\frac{0.158 \times 4^{1.75}}{\pi^{1.75}}\rho^{0.75} L \mu^{0.25} Q^{1.75}\right] D^{-4.75}$$

$Q_I = Q_{II}$, $\rho_I = \rho_{II}$, $L_I = L_{II}$, $\mu_I = \mu_{II}$

$$\frac{\Delta P(0.6)_m}{(\Delta P)_{0.4}} = \left(\frac{0.6}{0.4}\right)^{-4.75}$$

$$\therefore \quad (\Delta P)_{0.4} = 6.861 \text{ MPa}$$

6.3 (a)

"Rotameter is a device which measures the flow rate of liquid or gas in closed tube. It works on variable area principle".

6.4 (b)

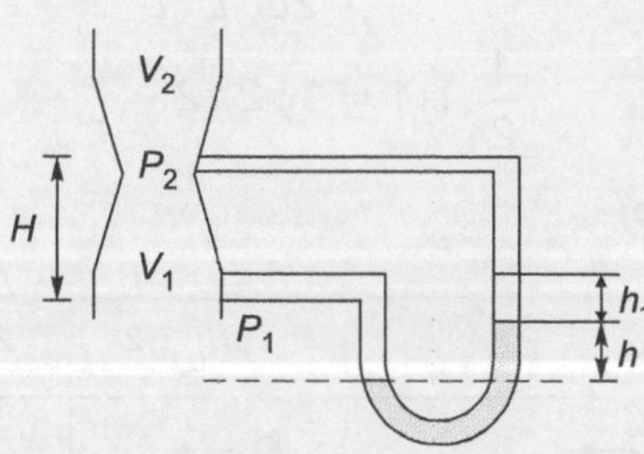

Force balance across the datum line

$$P_1 + (h + h_1)\rho_w g = (H + h_1)\rho_w g + h\rho_m g + P_2$$

$$(P_1 - P_2) = h\rho_m g + \rho_w g(H - h) \quad \text{...(i)}$$

By Bernoulli equation;

$$\frac{P_1}{\rho_w g} + \frac{V_1^2}{2g} + Z_1 = \frac{P_2}{\rho_w g} + \frac{V_2^2}{2g} + Z_2$$

$$\frac{V_2^2 - V_1^2}{2g} = \frac{P_1 - P_2}{\rho_w g} - H$$

From equation (i),

$$\frac{V_2^2 - V_1^2}{2g} = \frac{\rho_m gh + \rho_w g(H - h)}{\rho_w g} - H$$

$$\frac{V_2^2 - V_1^2}{2g} = \frac{(\rho_m - \rho_w)h}{\rho_w}$$

6.5 (c)

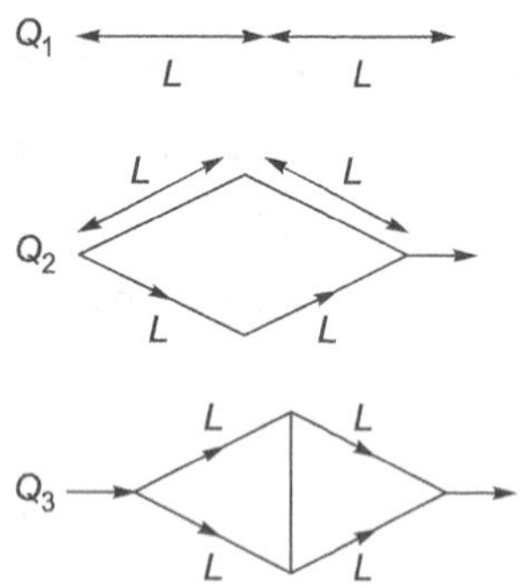

From Hagen–Poiseuille's equation,

$$Q = \frac{\pi \Delta P D^4}{128 \mu L}$$

Now, (ΔP = constant) given

$$\therefore \quad Q = \frac{k}{L} \text{ where } k = \frac{\pi \Delta P D^4}{128\mu} = \text{constant}$$

$$Q_1 = \frac{k}{2L}$$

$$\frac{Q_2}{2} = \frac{k}{2L} \rightarrow Q_2 = \frac{k}{L} \quad (\because \text{ flow is halved})$$

$$\frac{Q_3}{2} = \frac{k}{2L} \rightarrow Q_3 = \frac{k}{L}$$

{$\because$ middle branching does not matter}

$$\therefore \; Q_1 : Q_2 : Q_3 = \frac{k}{2L} : \frac{k}{L} : \frac{k}{L}$$

$$\frac{1}{2} : 1 : 1 = 1 : 2 : 2$$

6.6 (b)

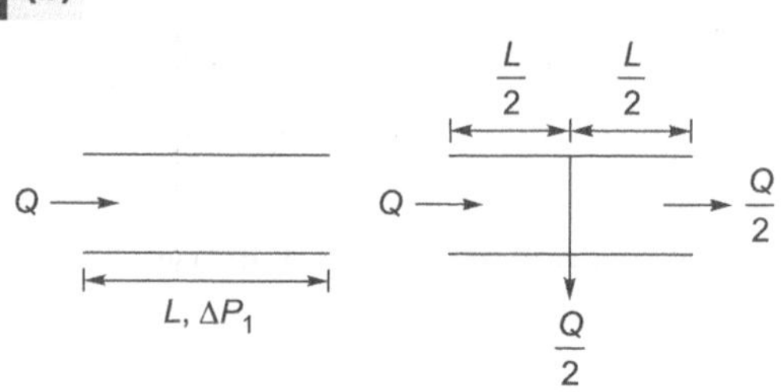

$$\frac{\Delta P}{\rho g} = \frac{f l Q^2}{3 d^5}$$

[for both laminar and turbulent]

f = remains unchanged

d = same

Hence, $\Delta P \propto lQ^2 \rightarrow \Delta P = klQ^2$

$$\Delta P_1 = klQ^2 \qquad \text{...(i)}$$

$$\Delta P_2 = k\left(\frac{l}{2}\right)Q^2 + k\left(\frac{l}{2}\right)\left(\frac{Q}{2}\right)^2 = klQ^2 \frac{5}{8}$$

$$\Delta P_2 = \Delta P_1 \cdot \left(\frac{5}{8}\right)$$

6.7 (c)

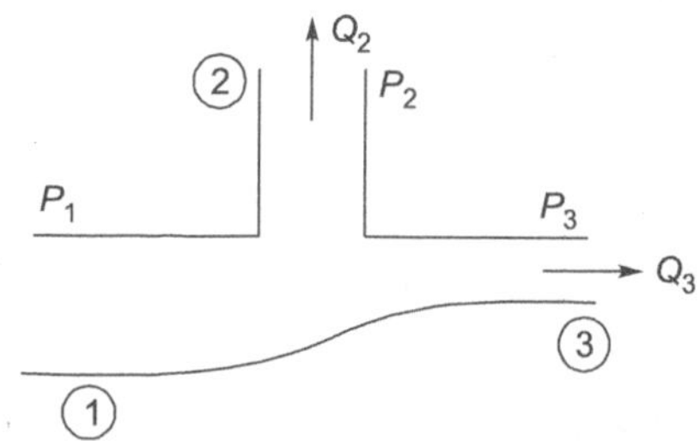

$$Q_1 = Q_2 + Q_3$$

$$0.28 = A_2V_2 + A_3V_3 \qquad \text{...(i)}$$

$$0.28 = 0.008V_2 + 0.004V_3$$

$$V_2 = 35 - \frac{V_3}{2} \qquad \text{...(ii)}$$

Bernoulli's equation between point (1) and (3), we get

$$\frac{P_1}{\rho g} + \frac{V_1^2}{2g} + z_1 = \frac{P_3}{\rho g} + \frac{V_3^2}{2g} + z_3$$

$$P_1 - P_3 = \frac{\rho}{2}\left[V_3^2 - V_1^2\right]$$

$\{z_1 = z_3\}$ horizontal plate

$$V_1 = \frac{0.28}{0.012} = 23.33$$

$$1000[270 - 240] = \frac{800}{2}\left[V_3^2 - (23.33)^2\right]$$

$$V_3 = 24.88 \text{ m/s}$$

From equation (ii),

$$V_2 = 35 - \frac{24.88}{2} = 22.55 \text{ m/s}$$

From Bernoulli's equation between point (2) and (3), we get

$$\frac{P_2}{\rho g} + \frac{V_2^2}{2g} + \cancel{z_2} = \frac{P_3}{\rho g} + \frac{V_3^2}{2g} + \cancel{z_3}$$

$$P_2 - 240 = \frac{800}{2}[24.88^2 - 22.55^2]$$

$$P_2 = 284.20 \text{ kPa}$$

6.8 (d)

Apply Bernoulli's equation for point (1) and (2)

$$\frac{P_1}{\rho g} + \frac{V_1^2}{2g} = \frac{P_2}{\rho g} + \frac{V_2^2}{2g} + H\sin 45°$$

$$\frac{V_2^2 - V_1^2}{2g} = \frac{P_1 - P_2}{\rho g} - H\sin 45° \qquad \text{...(i)}$$

$$\frac{V_2^2 - V_1^2}{2} = \frac{(\rho_m - \rho)gh}{\rho}$$

$\because$ $P_1 + (H \sin 45° + h)\rho g = P_2 + h P_m g$

$$P_1 - P_2 = (\rho_m - \rho)\, gh - H \sin 45° \rho g \quad ...(ii)$$

From equation (i) and (ii),

$$\therefore \quad \frac{V_2^2 - V_1^2}{2g} = \frac{(\rho_m - \rho)gh}{g\rho} = \frac{(\rho_m - \rho)h}{\rho}$$

6.9 (b)

"Losses for flow through valves and fittings are expressed in terms of equivalent length of a straight pipe".

6.10 (d)

$$\tau_{rx} = \left(\frac{\Delta P}{2L}\right) r$$

$\therefore$ $\tau_{rx} \propto r$ (linear dependence)

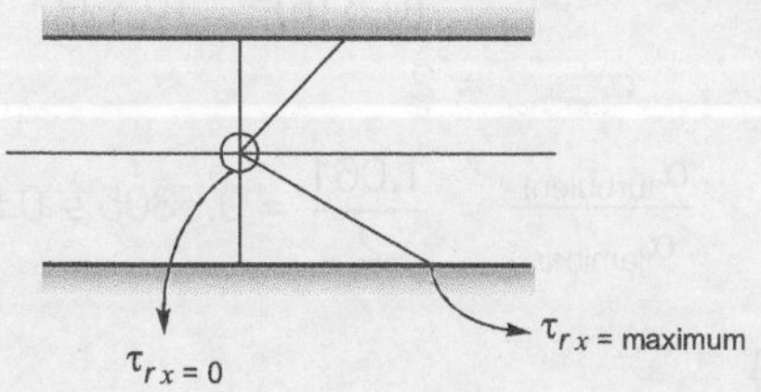

6.11 (a)

For an orifice meter,

$$Q = C_d A_2 \sqrt{\frac{2\Delta P}{\rho_f (1 - \beta^4)}}$$

$$A_2 = \text{Area of throat}, \; \beta = \frac{d_2}{d_1}$$

$$Q_1 \propto \sqrt{\Delta P}$$

$$Q_2 \propto \sqrt{1.05P}$$

$\therefore$ % error in measured flow rate

$$= \left(\frac{Q_2 - Q_1}{Q_1}\right) \times 100 = 2.47\%$$

6.12 (c)

Internal diameter of outer pipe = 0.31 m

Equivalent diameter of an annular $\Rightarrow D = 4\, r_H$

$$r_H = \frac{\text{Cross-sectional area of a chennel}}{\text{Wetted perimeter of a channel}}$$

i.e. $$D = 4\frac{\frac{\pi}{4}D_1^2 - \frac{\pi}{4}D_2^2}{\pi D_1 + \pi D_2} = D_1 - D_2$$

$\therefore$ $D_P = 0.31 - 0.1 = 0.2$ m

$$\frac{R_{ep}}{R_{eq}} = \frac{\dfrac{(0.20 \times 1 \times 1000)}{\mu_P}}{\dfrac{(0.1 \times 2.5 \times 800)}{\mu_Q}} \quad \{\mu_P = \mu_Q\} \text{ given}$$

$$= 1.0$$

6.13 (a)

The local velocity of a fluid along a stream line can be measured by Pitot tube.

6.14 (c)

From Hagen-Poiseuille equation

$$u_{av} = \frac{\Delta P \cdot D^2}{32\mu L}$$

$\therefore$ $u_{av} \propto \mu^{-1}$

6.15 (80.8)

Velocity through Pitot tube is given by

$$V = \sqrt{\frac{2\Delta P}{\rho_f}}$$

$$\Delta P = 13600 \times 0.03 \times 10 = 4080 \text{ Pa}$$

$$V = \sqrt{\frac{2 \times 4080}{1.25}} = 80.79 \cong 80.8 \text{ m/s}$$

6.16 (1.64)

For laminar flow $\Rightarrow V_{avg} = 0.5\, V_{max}$

For turbulent flow $\Rightarrow V_{avg} = 0.82\, V_{max}$

Equal volumetric flow rate

$$\left(\frac{\pi}{4}D_1^2\right)V_{avg1} = \left(\frac{\pi}{4}D_2^2\right)V_{avg2} \quad (D_1 = D_2) \text{ given}$$

$$0.5\, V_{max1} = 0.82\, V_{max2}$$

$$\frac{V_{max1}}{V_{max2}} = \frac{0.82}{0.5} = 1.64$$

6.17 (b)

In the case of venturimeter,

$\therefore$ $$Q \propto \sqrt{\Delta P} \quad \left\{\frac{Q_2}{Q_1} = 2\right\} \text{ given}$$

i.e. $$\frac{Q_2}{Q_1} = \frac{\sqrt{\Delta P_2}}{\sqrt{\Delta P_1}}$$

$$\Rightarrow \quad \left(\frac{Q_2}{Q_1}\right)^2 = \frac{\Delta P_2}{\Delta P_1}$$

$$\frac{\Delta P_2}{\Delta P_1} = (2)^2 = 4$$

6.18 (0.3125)

For flow through pipes,

$$\tau = \left(-\frac{\partial P}{\partial x}\right)\frac{r}{2}$$

$$\tau = \left(-\frac{\partial P}{\partial x}\right)\cdot\frac{r}{2}$$

$$\tau_w = \left(\frac{500}{10}\right)\times\frac{1.25\times10^{-2}}{2} = 0.3125 \text{ Pa}$$

6.19 (c)

For laminar flow, $f = \frac{16}{\text{Re}}$

For Re = 10^2, $f = \frac{16}{100} = 0.16$

For Re = 10^3, $f = \frac{16}{1000} = 0.016$

i.e. the set I and III satisfy

Now,

For Re = 10^5 i.e. [Turbulent regime] as the value of $\left(\frac{K}{D}\right)$ increases, f also increases.

Hence, option (c) is correct [f increases from 0.0045 to 0.0055].

6.20 (b)

Pitot tube is used to measure "point velocity".

6.21 (0.1488)

For flow through the venturi meter, flow rate is given by

$$Q = C_d \times A_2 \times \sqrt{\frac{2\Delta P}{\rho_f \cdot (1-\beta^4)}}$$

$$\beta = \frac{\text{Diameter of throat}}{\text{Diameter of pipe}}$$

$$= \frac{102}{178} = 0.5730$$

C_d = 0.98 (discharge coefficient)

$$\Delta P = 154.3 \text{ kN/m}^2$$

$$Q = 0.98\times\frac{\pi}{4}\times(0.102)^2\times\sqrt{\frac{2\times154.3\times1000}{1000\times(1-0.573^4)}}$$

$$Q = 0.1488 \text{ m}^3/\text{s}$$

6.22 (0.53)

Kinetic energy correction factor for turbulent flow through pipe is given as

$$\alpha_{\text{turbulent}} = \left(\frac{V_o}{\bar{V}}\right)^3\cdot\frac{2n^2}{(3+n)(3+2n)}$$

$$\left(\frac{\bar{V}}{V_o}\right) = \frac{2n^2}{(n+1)(2n+1)} \text{ (given)}$$

For $n = 7$, $\left(\frac{\bar{V}}{V_o}\right) = \frac{2\times49}{8\times15} = 0.816$

$$\alpha_{\text{turbulent}} = \frac{1}{(0.816)^3}\times\frac{2\times49}{10\times17} = 1.061$$

$\because \quad \alpha_{\text{laminar}} = 2$

So, $\frac{\alpha_{\text{turbulent}}}{\alpha_{\text{laminar}}} = \frac{1.061}{2} = 0.5305 \cong 0.53$

6.23 (b)

In fully developed flow, friction factor becomes constant. So, only constant slope will be considered from graph.

$$\Rightarrow \quad m_4 = m_5 = 37$$

$$= \frac{\Delta P}{l} = \text{Constant (from graph)}$$

$$\therefore \quad \frac{\Delta P}{\rho g} = \frac{4flV^2}{2gd} \Rightarrow \left(\frac{\Delta P}{l}\right) = \frac{4fV^2\rho}{2d}$$

$$37 = \frac{4\times5\times0.25\times1000}{2\times0.10}$$

$$f = 0.0074$$

6.24 (b)

From Newton law of viscosity, we have

$$\tau = -\mu\frac{du}{dr}$$

$$F = -A\,.\,\mu\,.\,\frac{du}{dr}$$

$$u(r) = V\frac{\ln\left(\frac{r}{R}\right)}{\ln a} \qquad \because \; r = aR$$

$$A = 2\pi aR.1$$

$$\frac{du}{dr} = \frac{V}{r\ln a}$$

$$F = -2\pi aR\,.\,\mu\frac{V}{r\ln a} = -2\pi aR\mu\frac{V}{aR\,.\ln a}$$

$$F = \frac{-2\pi\mu V}{\ln(a)}$$

6.25 (3.12)

Data given: $C_d = 0.98$

Pipe diameter, $D_o = 50$ mm, $d_o = 20$ mm (throat diameter)

$$\Delta P = 50 \text{ kPa}$$

$$\rho_w = 1000 \text{ kg/m}^3$$

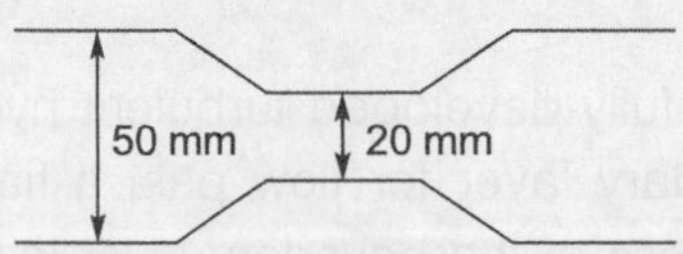

For venturimeter, the flow rate is given by

$$Q = C_d A_2\sqrt{\frac{2\Delta P}{\rho_f\left(1-\beta^4\right)}}$$

$$\beta = \frac{d_2}{d_1} = \frac{20}{50} = 0.4$$

$$A_2 = \frac{\pi}{4}(0.0.2)^2 = 0.000314$$

$$Q = 0.98\times 0.000314\sqrt{\frac{2\times 50\times 1000}{1000\left(1-(0.4)^4\right)}}$$

$$Q = 3.12\times 10^{-3}\text{ m}^3/\text{s}$$

6.27 (0.02)

For steady, uni-directional, fully-developed, pressure-driven laminar flow of an incompressible Newtonian fluid through circular pipe

$$\left(\frac{\partial P}{\partial x}\right) = \frac{2\tau}{r}$$

$$\tau = \left(\frac{\partial P}{\partial x}\right)\frac{r}{2}$$

At $r = 5.0$ cm, $\tau = 0.1$ N/m^2

At $r = 1.0$ cm, $\tau = ?$

$$\frac{\tau}{0.1} = \frac{1}{5}$$

$$\tau = 0.02 \text{ N/m}^2$$

6.28 (a)

From the given data and conditions in question boundary conditions are given as below:

At $y = -H,\ \ u = -V$

At $y = +H,\ \ u = 0$ ($\because$ top plate is stationary)

This question can be done by elimination method by applying above two boundary conditions options (c) and (d) can be eliminated easily now we have to check option (a) and option (b).

For option (b):

$$u = \frac{\Delta P}{L}\frac{H^2}{2\mu}\left(\frac{y^2}{H^2}-1\right)+\frac{V}{2}\left(\frac{y}{H}-1\right)$$

At $y = H,\ \ u = 0$ (satisfied)

At $y = -H,\ \ -V$ (satisfied)

Now from figure it is clear that at $y = +\frac{H}{2}$

$$u > 0$$

$$u = \frac{\Delta P}{L}\frac{H^2}{2\mu}\left[\frac{1}{4}-1\right]+\frac{V}{2}\left[\frac{1}{2}-1\right]$$

$$u = \frac{\Delta P}{L}\frac{H^2}{2\mu}\left[-\frac{3}{4}\right]+\frac{V}{2}\left[-\frac{1}{2}\right]$$

From option (b) it is clear that at $y = \frac{H}{2}$ we are getting negative velocity instead of getting positive.

Similarly we can judge that option (a) is correct option as we will get $V > 0$ at $y = \frac{H}{2}$.

6.29 (4.0)

Volumetric flow rate through a venturi meter is given by

$$Q = \frac{C_D A_1 A_2\sqrt{2gh}}{\sqrt{A_1^2 - A_2^2}}$$

$$Q \propto \sqrt{\Delta h_1}$$

$$Q_2 = 2Q_1 \text{ (given)}$$

$$\frac{\Delta h_2}{\Delta h_1} = \left(\frac{Q_2}{Q_1}\right)^2 = 2^2 = 4$$

■■■■

7 Boundary Layer Theory

7.1 In the laminar boundary layer flow over a flat plate, the ratio $\left(\frac{\delta}{x}\right)$ varies as :

(a) Re (b) $\sqrt{Re}$

(c) $\frac{1}{Re}$ (d) $Re^{-1/2}$

[2000 : 1 M]

7.2 For flow past a flat plate, if x is the distance along the plate in the direction of flow, the boundary layer thickness is proportional to

(a) $\sqrt{x}$ (b) $\frac{1}{\sqrt{x}}$

(c) x (d) $\frac{1}{x}$

[2002 : 1 M]

7.3 For uniform laminar flow (in the x-direction) past a flat plate at high Reynolds number, the local boundary layer thickness (δ) varies with the distance along the plate (x) as

(a) $\delta \propto x^{1/4}$ (b) $\delta \propto x^{1/3}$

(c) $\delta \propto x^{1/2}$ (d) $\delta \propto x$

[2012 : 1 M]

7.4 For uniform laminar flow over a flat plate, the thickness of the boundary layer, δ at a distance x from the leading edge of the plate follows the relation:

(a) $\delta(x) \propto x^{-1}$ (b) $\delta(x) \propto x$

(c) $\delta(x) \propto x^{1/2}$ (d) $\delta(x) \propto x^{-1/2}$

[2015 : 1 M]

7.5 The thickness of laminar boundary layer over a flat plate varies along the distance from the leading edge of the plate. As the distance increases, the boundary layer thickness

(a) increases

(b) decreases

(c) initially increases and then decreases

(d) initially decreases and then increases

[2017 : 1 M]

7.6 For a fully-developed turbulent hydrodynamic boundary layer for flow past a flat plate, the thickness of the boundary layer increases with distance x from the leading edge of the plate, along the free-stream flow direction, as

(a) $x^{0.5}$ (b) $x^{1.5}$

(c) $x^{0.4}$ (d) $x^{0.8}$

[2019 : 1 M]

7.7 Consider an incompressible flow of a constant property fluid over a smooth, thin and wide flat plate. The free stream flows parallel to the surface of the plate along its length and its velocity is constant. Value of the Reynolds number at a distance of 2.0 m from the leading edge of the plate is 8000. The flow within the boundary layer at a distance of 1.0 m from the leading edge of the plate is

(a) laminar

(b) turbulent

(c) transitioning from laminar to turbulent

(d) inviscid

[2020 : 1 M]

■■■■

Answers Boundary Layer Theory

7.1 (d) 7.2 (a) 7.3 (c) 7.4 (c) 7.5 (a) 7.6 (d) 7.7 (a)

Explanations Boundary Layer Theory

7.1 (d)

In laminar region;

$$\frac{\delta}{x} = \frac{5.0}{\sqrt{Re_x}}$$

$$\frac{\delta}{x} \propto \frac{1}{\sqrt{Re_x}} \propto (Re)^{-1/2}$$

7.2 (a)

$$\frac{\delta}{x} = \frac{5.0}{\sqrt{Re_x}} = \frac{5.0\sqrt{\mu}}{\sqrt{\rho V x}}$$

$$\delta \propto \frac{x}{\sqrt{x}} \propto \sqrt{x}$$

7.3 (c)

From Blasius reaction, the boundary layer thickness is given by

$$\delta = 5.0\sqrt{\frac{\mu \cdot x}{\rho V_\infty}}$$

$\therefore$ $\delta \propto x^{1/2}$ [For laminar flow]

7.4 (c)

For laminar flow over a flat plate

Boundary layer thickness is given by

$$\delta(x) = \frac{5.0x}{Re_x^{1/2}}$$

$$\delta(x) \propto x^{1/2}$$

7.5 (a)

Boundary layer thickness $\propto$ distance from leading edge i.e.

$$\frac{\delta}{x} = \frac{4.98}{\sqrt{Re_x}} \text{ [laminar region]}$$

[$\delta \propto \sqrt{x}$] so, as the distance increases, the boundary layer thickness increases.

7.6 (d)

For fully developed turbulent flow, over flow past a flat plate, the value of thickness comes out to be, from integral momentum analysis

$$\frac{\delta}{x} = \frac{0.376}{(Re_x)^{1/5}}$$

$$\frac{\delta}{x} = \frac{0.376}{\left[\dfrac{\rho \times V \times d_{eq}}{\mu}\right]^{1/5}} = \frac{0.376}{\left(\dfrac{\rho \times V \times x}{\mu}\right)^{0.2}}$$

$$\delta \propto \frac{x}{x^{0.2}} \propto x^{0.8}$$

7.7 (a)

For a flat plate,

If value of Reynold's number is less than or equal to 5×10^5, i.e. considered as laminar flow.

[$Re \le 5 \times 10^5$] [Re number $\propto x$]

$\therefore$ If at distance of 2 m $\rightarrow$ Re is 8000 therefore at distance of 1 m. It will be less than 8000, hence laminar flow.

$$\text{Reynold's number} = \frac{\rho \times V \times d_{eq}}{\mu} = \frac{\rho \times V \times x}{\mu}$$

■■■■

8 Laminar Flow

8.1 For laminar flow of a shear-thinning liquid in a pipe, if the volumetric flow rate is doubled, the pressure gradient will increase by a factor of

(a) 2 (b) < 2

(c) > 2 (d) $\frac{1}{2}$ [2000 : 1 M]

8.2 Applying a pressure drop across capillary results in a volumetric flow rate Q under laminar flow conditions. The flow rate, for the same pressure drop, in a capillary of the same length but half the radius is

(a) $\frac{Q}{2}$ (b) $\frac{Q}{4}$

(c) $\frac{Q}{8}$ (d) $\frac{Q}{16}$ [2001 : 1 M]

Common Data for Questions (8.3 & 8.4):

Two tanks, A and B, of cross sectional area 1 m^2 each, contain a fluid of density 1000 kg/m^3 and viscosity 1 kg/m.s. The tanks are connected by a pipe of diameter 0.02 m and length 1 m, and a check valve, at the bottom. Assume that the flow is laminar, and there is no friction in the check valve. In the initial state, the height of the fluid in the tank A is 6 m, and the height of the fluid in tank B is 2 m (as shown in the figure below). The check valve is opened, and the fluid flows from tank A to tank B till the levels in the two tanks are equal in the final state. Assume g = 10 m/s^2 in the calculations.

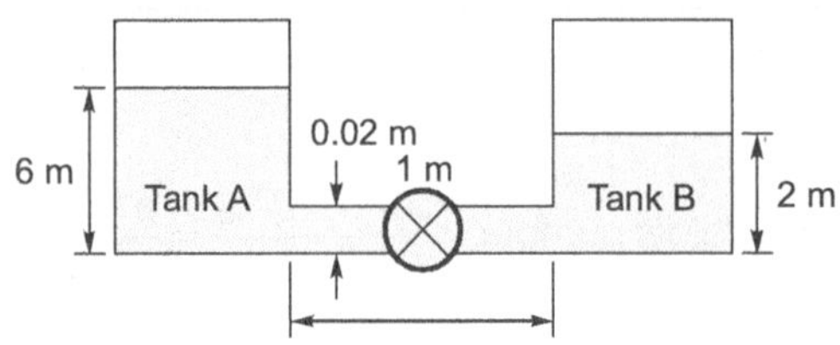

8.3 What is the average fluid velocity in the pipe as soon as the valve is opened?

(a) 0.25 m/s (b) 0.5 m/s

(c) 1 m/s (d) 2 m/s [2005 : 2 M]

8.4 What is the total energy loss between the initial and final states due to the fluid flow?

(a) 2×10^4 J (b) 16×10^4 J

(c) 8×10^4 J (d) 4×10^4 J

[2005 : 2 M]

8.5 In a laminar flow through a pipe of radius R, the fraction of the total fluid flowing through a circular cross-section of radius $\frac{R}{2}$ centered at the pipe axis is

(a) $\frac{3}{8}$ (b) $\frac{7}{16}$

(c) $\frac{1}{2}$ (d) $\frac{3}{4}$ [2006 : 2 M]

8.6 A tube of diameter D and length L is initially filled with a liquid of density ρ and viscosity μ. It is then pushed out by the application of a constant force F to the plunger as shown in the figure. Assuming laminar flow and pseudo steady state, the time required to expel one half of the liquid out of the tube is

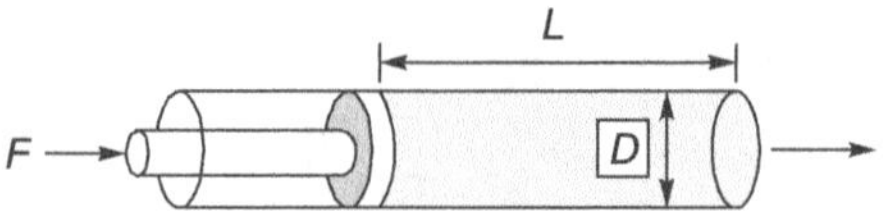

(a) $\frac{3\pi\mu L^2}{F}$ (b) $\frac{3\pi\mu D^2}{F}$

(c) $\frac{6\pi\mu L^2}{F}$ (d) $\frac{6\pi\mu DL}{F}$ [2007 : 2 M]

8.7 Given a pipe of diameter D, the entrance length necessary to achieve fully developed laminar flow is proportional to (N_{Re} is Reynolds number)

(a) DN_{Re} (b) $\frac{D}{N_{Re}}$

(c) DN_{Re}^2 (d) $\frac{D}{N_{Re}^2}$ [2008 : 1 M]

8.8 Two identical reservoirs, open at the top, are drained through pipes attached to the bottom of the tanks as shown below. The two drain pipes are of the same length, but of different diameters ($D_1 > D_2$).

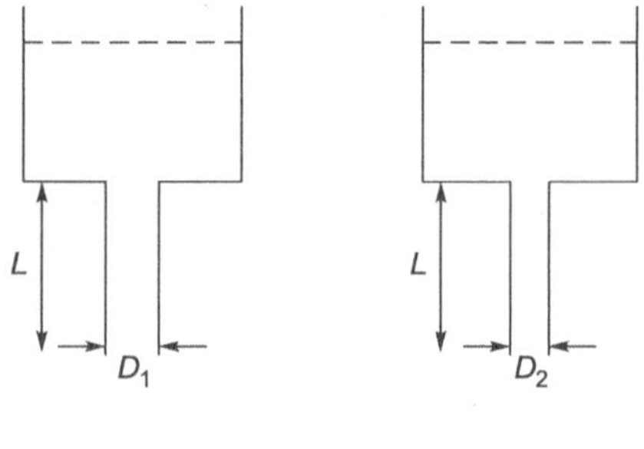

Assume the flow to be steady and laminar in both drain pipes, if the volumetric flow rate in the larger pipe is 16 times of that in the smaller pipe, the ratio $\frac{D_1}{D_2}$ is

(a) 2 (b) 4
(c) 8 (d) 16 [2009 : 2 M]

8.9 A liquid is flowing through the following piping network. The length of pipe sections *P*, *Q*, *R* and *S* shown in the schematic are equal. The diameters of the sections *P* and *R* are equal and the diameter of the section *Q* is twice that of *S*. The flow is steady and laminar. Neglecting curvature and entrance effects, the ratio of the volumetric flow rate in the pipe section *Q* to that in *S* is

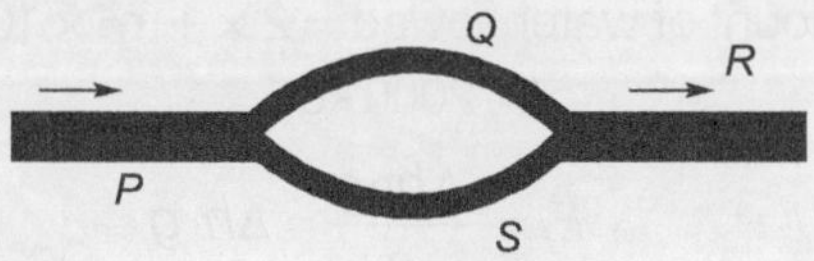

(a) 16 (b) 8
(c) 2 (d) 1

[2011 : 2 M]

8.10 Water is flowing under laminar conditions in a pipe of length *L*. If the diameter of the pipe is doubled, for a constant volumetric flow rate, the pressure drop across the pipe

(a) decreases 2 times (b) decreases 16 times
(c) increases 2 times (d) increases 16 times

[2012 : 1 M]

8.11 For a Newtonian fluid flowing in a circular pipe under steady state conditions in fully developed laminar flow, the Fanning friction factor is

(a) $0.046\,\text{Re}^{-0.2}$ (b) $0.0014 + \frac{0.125}{\text{Re}^{0.32}}$

(c) $\frac{16}{\text{Re}}$ (d) $\frac{24}{\text{Re}}$ [2013 : 1 M]

8.12 An incompressible Newtonian fluid flows in a pipe of diameter D_1 at volumetric flow rate *Q*. Fluid with same properties flows in another pipe of diameter $D_2 = \frac{D_1}{2}$ at the same flow rate *Q*. The transition length required for achieving fully-developed flow is l_1 for the tube of diameter D_1, while it is l_2 for the tube of diameter D_2. Assuming steady laminar flow in both cases, the ratio $\frac{l_1}{l_2}$ is

(a) $\frac{1}{4}$ (b) 1
(c) 2 (d) 4

[2019 : 2 M]

8.13 Consider steady, laminar, fully developed flow of an incompressible Newtonian fluid through two horizontal straight pipes, I and II, of circular cross section. The volumetric flow rates in both the pipes are the same. The diameter of pipe II is twice the diameter of pipe I, i.e., $d_{\text{II}} = 2d_{\text{I}}$. The ratio of the shear stress at the wall of pipe I to the shear stress at the wall of pipe II is

(a) 0.5 (b) 2
(c) 4 (d) 8

[2020 : 2 M]

8.14 A student performs a flow experiment with Bingham Plastic under fully developed laminar flow conditions in a tube of radius 0.01 m with a pressure drop (ΔP) of 10 kPa over tube length (L) of 1.0 m. The velocity profile is flat for $r < r_c$ and parabolic for $r \geq r_c$, as shown in the figure.

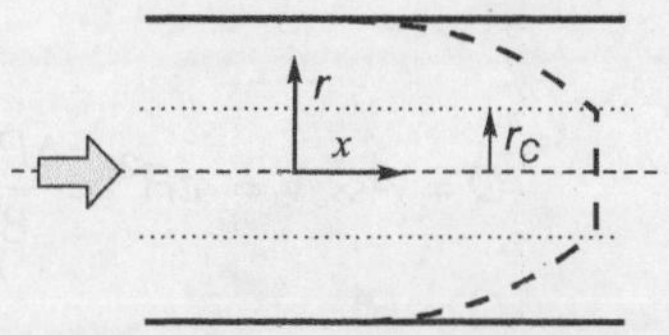

Consider *r* and *x* as the radial and axial directions, and the shear stress is finite as *r* approaches zero. A force balance results in the following equation

$$\frac{d(r\tau_{rx})}{dr} = r\frac{(-\Delta P)}{L}$$

where τ_{rx} is the shear stress. If r_c is 0.001 m, then the magnitude of yield stress for this Bingham Plastic (in Pa) is

(a) 1 (b) 5
(c) 8 (d) 12 [2020 : 2 M]

8.15 The viscosity of an incompressible Newtonian fluid is measured using a capillary tube of diameter 0.5 mm and length 1.5 m. The fluid flow is laminar, steady and fully developed. For a flow rate of 1 $cm^3\,s^{-1}$, the pressure drop across the length of the tube is 1 MPa. If the viscosity of the fluid is $k \times 10^{-3}$ Pa.s, the value of *k* is _____ (rounded off to two decimal places).

[2023 : 1 M]

■■■■

Answers Laminar Flow

8.1	(b)	**8.2**	(d)	**8.3**	(b)	**8.4**	(d)	**8.5**	(b)	**8.6**	(a)	**8.7**	(a)
8.8	(b)	**8.9**	(a)	**8.10**	(b)	**8.11**	(c)	**8.12**	(b)	**8.13**	(d)	**8.14**	(b)
8.15	(1.02)												

Explanations Laminar Flow

8.1 (b)

For laminar flow (in terms of flow rate).

$$\left(\frac{\Delta P}{L}\right) = \frac{128\mu Q}{\pi d^4}$$

↓ Pressure gradient ↓ Volumetric flow rate

- For a Newtonian fluid, μ = constant

$\therefore$ If Q is doubled $\rightarrow \left(\frac{\Delta P}{L}\right)$ will double

- For shear thinning fluid, as Q increases μ decreases.

$\therefore$ If Q is doubled then $\left(\frac{\Delta P}{L}\right)$ will increase by a factor less than 2.

8.2 (d)

$$Q = A \times V = \pi R^2 \times \frac{\Delta P \cdot R^2}{8\mu L}$$

$\therefore \quad Q \propto R^4$

$\because \quad R_2 = \frac{R_1}{2}$ (given)

$$\frac{Q_2}{Q_1} = \left(\frac{R_2}{R_1}\right)^4$$

Hence, $\quad Q_2 = \frac{Q_1}{16}$

8.3 (b)

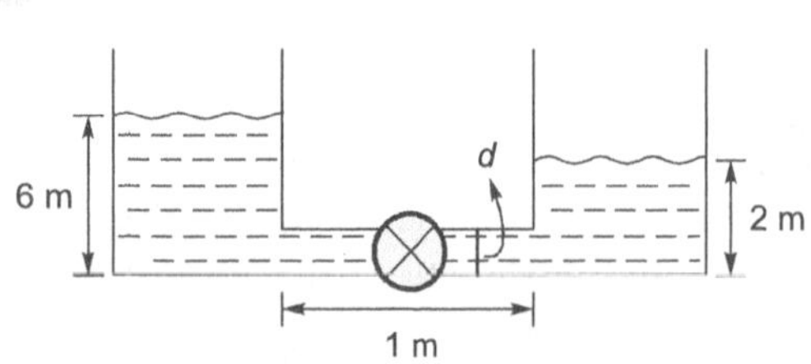

Laminar flow; Hagen-Poiseuille equation

$$V = \frac{\Delta P D^2}{32 L\mu}, \; d = 0.02 \text{ m}$$

Initially, $\quad \Delta P = (6 \text{ m} - 2 \text{ m}) \times 1000 \times 10$
$= 4 \times 10^4$ Pa

$$V = \frac{4 \times 10^4 \times (0.02)^2}{32 \times 1 \times 1} = 0.5 \text{ m/s}$$

8.4 (d)

Mechanical loss, in energy due to skin friction in the pipe is given by

$$F_f = \frac{\Delta P}{\rho}$$

$\therefore$ Water in the tank loses level by 6 – 4 = 2 m

Amount of water flowed = 2 × 1 m² × 1000
= 2000 kg

$$F_f = \frac{\Delta h \rho g}{\rho} = \Delta h \cdot g$$

$= 2 \times 10 = 20 \text{ m}^2/\text{s}^2 = 20$ J/kg

$\therefore$ Total energy loss = 2000 × 20 (J) = 4×10^4 J

8.5 (b)

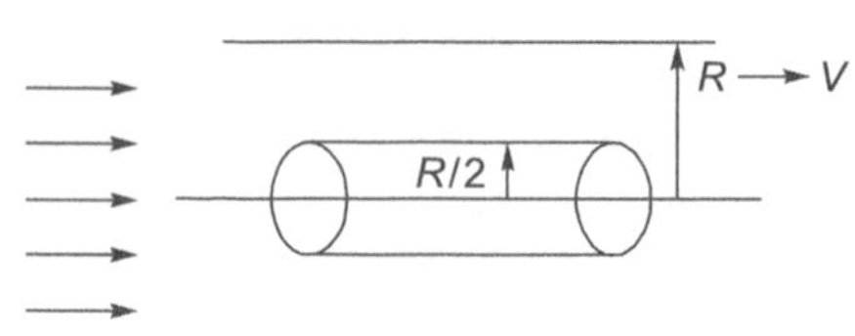

Laminar flow

Let us consider a differential section centered, area of the differential section.

$$= \pi[(r + dr)^2 - r^2] = 2\pi r\, dr$$

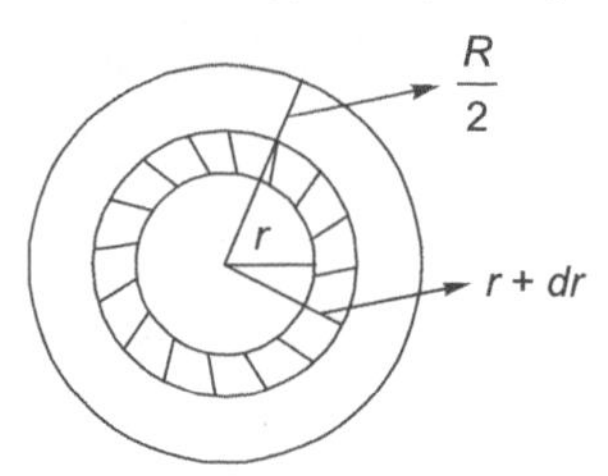

Volumetric flow rate of fluid through the differential section.

$$dq = u \cdot dA$$

$$\int dq = \int u_{max}\left[1 - \left(\frac{r}{R}\right)^2\right] 2\pi r\, dr$$

$$Q_{R/2} = 2\pi u_{max} \int_0^{\frac{R}{2}} \left[r\, dr - \frac{r^3}{R^2} dr\right]$$

$$Q_{R/2} = (\pi R^2 u_{max})\left(\frac{7}{32}\right)$$

Now, $$Q_R = (\pi R^2)\frac{u_{max}}{2} \text{ (for laminar flow)}$$

$$\therefore \quad \frac{Q_{R/2}}{Q_R} = \frac{\pi R^2 u_{max}\left(\frac{7}{32}\right)}{(\pi R^2 u_{max})\left(\frac{1}{2}\right)} = \frac{7}{16}$$

8.6 (a)

Hagen-Poiseuille equation,

$$\Delta P = \frac{32\mu v L}{D^2} = \frac{F}{A} \qquad \because v = \frac{dL}{dt}$$

$$\frac{F\times 4}{\pi D^2} = \frac{32\mu L}{D^2}\times\frac{dL}{dt}$$

$$F = 8\mu L\frac{\pi dL}{dt}$$

$$\int_0^t dt = \frac{8\pi\mu}{F}\int_L^{L/2} L\,dL$$

$$\Rightarrow \quad t = \frac{8\pi\mu}{F}\frac{3L^2}{8}$$

$$t = \frac{3\pi\mu L^2}{F}$$

8.7 (a)

To achieve fully developed laminar flow,

$\therefore$ Re = Reynold's number

$$L_e = 0.05 \text{ Re } D$$

$$L_e \propto D\times(\text{Re})$$

8.8 (b)

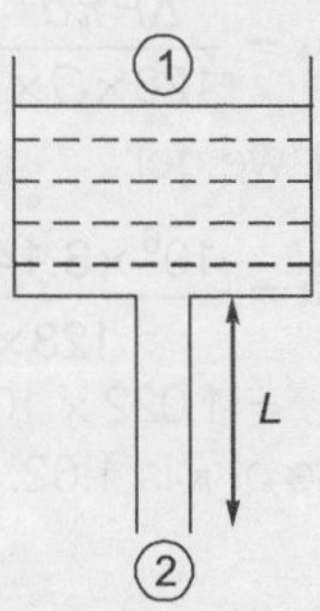

"Velocity of discharge will remain same for both tank".

Bernoulli's equation,

$$\frac{P_1}{\rho} + \frac{V_1^2}{2} + g z_1 = \frac{P_2}{\rho} + \frac{V_2^2}{2} + g z_2 \quad (P_1 = 0,\ V_1 = 0,\ P_2 = 0)$$

$$V_2 = \sqrt{2g(z_1 - z_2)} = \sqrt{2gh} = \sqrt{2gL}$$

So, $$Q = A\times V$$

$$Q \propto A$$

$$\frac{Q_2}{Q_1} = \frac{A_2}{A_1} \Rightarrow \frac{Q_2}{Q_1} = \frac{\frac{\pi}{4}D_2^2}{\frac{\pi}{4}D_1^2}$$

$$\frac{1}{16} = \left(\frac{D_2}{D_1}\right)^2 \Rightarrow \frac{D_1}{D_2} = 4$$

8.9 (a)

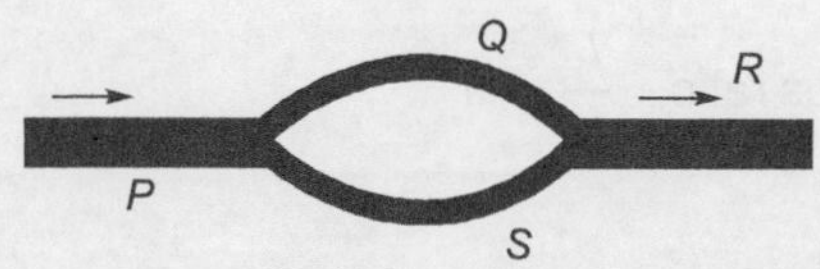

$$l_P = l_R, \quad l_Q = l_S$$

$$d_P = d_R, \quad d_Q = 2d_S$$

From Hagen-Poiseuille equation

$$V_{ag} = \frac{\Delta P D^2}{32\mu L} \text{ and } q = \left(\frac{\pi}{4}D^2\right)V_{avg}$$

$$q = \frac{\pi\Delta P D^4}{128\mu L} \rightarrow q \propto D^4$$

$$\frac{q_Q}{q_S} = \left(\frac{D_q}{D_s}\right)^4 = 2^4 = 16 \qquad \{D_q = 2D_s\}$$

8.10 (b)

For laminar flow with constant flow rate,

$$\frac{\Delta P}{L} = \frac{128\mu Q}{\pi d^4}$$

$$\Rightarrow \quad \Delta P \propto \frac{1}{d^4}$$

$$\frac{\Delta P_2}{\Delta P_1} = \left(\frac{d_1}{d_2}\right)^4 = \left(\frac{d_1}{2d_1}\right)^4 = \frac{1}{16}$$

$$\Delta P_2 = \frac{\Delta P_1}{16} \qquad \text{[Decreases 16 times]}$$

8.11 (c)

For fully developed laminar flow, the fanning friction factor is given by $f = \frac{16}{\text{Re}}$

whereas Re = Reynold's number

8.12 (b)

The transition length for fully developed flow in laminar regime is

$$L_e = 0.05\times\text{Re}\times D$$

$$\text{Re} = \frac{\rho \bar{V} d}{\mu} = \frac{\rho \times Q \times d}{\frac{\pi}{4} d^2 \times \mu}$$

Putting value of Re number in (L_e), we get

$$L_e = \frac{0.05 \times \rho \times Q \times d \times d}{\frac{\pi}{4} d^2 \times \mu} = \text{const.}$$

So, the expression for transition length does not depend on the diameter of pipe.

Therefore, for both the pipe, l_1 and l_2 will be same.

Thus ratio, $\frac{l_1}{l_2} = 1$

8.13 (d)

From Hagen–Poiseuille equation, we know,

For laminar flow, $\Delta P = \frac{32\mu \bar{V} l}{d^2}$ as flow rate is const. in both pipes.

So, $$\Delta P = \frac{128\mu Q \times l}{\pi d^4} \quad \text{...(i)}$$

For flow through pipes,

$$\frac{dP}{dx} = \frac{d\tau}{dy} \text{ and } = -\frac{dP}{dx} = \frac{2\tau}{r}$$

i.e. $$\tau = \left(-\frac{dP}{dx}\right)\frac{r}{2} = \left(-\frac{dP}{dx}\right)\frac{d}{4}$$

As for pipe, from equation (i)

$$\frac{dP}{dx} \propto \frac{128\mu Q}{\pi d^4} \text{ i.e. } \left(\frac{dP}{dx} \propto \frac{1}{d^4}\right)$$

keeping other parameter const.

$$\tau \propto \frac{d}{d^4}$$

So, $\left[\tau \propto \frac{1}{d^3}\right]$ $d_{\text{II}} = 2d_{\text{I}}$ ($\because$ given)

$$\frac{\tau_{\text{I}}}{\tau_{\text{II}}} = \frac{d_{\text{II}}^3}{d_{\text{I}}^3} = \frac{(2d_{\text{I}})^3}{(d_{\text{I}})^3} = 8$$

8.14 (b)

For Bingham plastic fluid, the profile comes out to be,

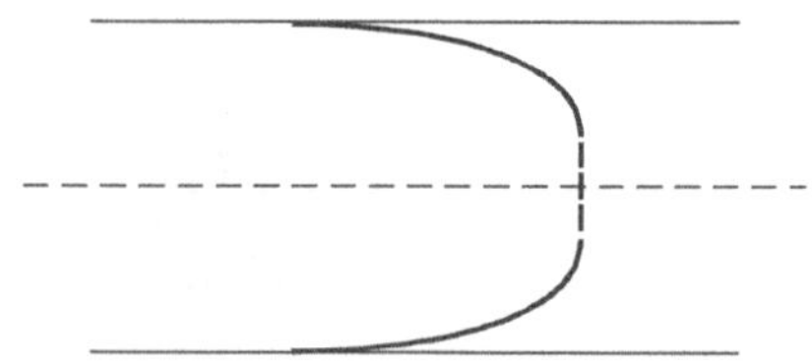

i.e. parabolic near the wall and flat in the middle.

For flow through pipe

$$\therefore \quad \tau_{rx} = \left(-\frac{dP}{dx}\right)\frac{r_c}{2}$$

and for Bingham plastic, at $r = r_c$, $\tau_{rx} = \tau_0$

$$\therefore \quad \tau_0 = \left(-\frac{dP}{dx}\right)\left(\frac{r_c}{2}\right)$$

$-\Delta P = 10$ kPa, $r_c = 0.001$ m

$L = 1$ m

$$\tau_0 = \left(10 \text{ kPa} \times 1000 \frac{\text{Pa}}{\text{kPa}}\right) \times \left(\frac{1}{1.0 \text{ m}}\right)\left(\frac{0.001}{2}\right) \text{m}$$

$\tau_0 = 5$ Pa yield stress

8.15 (1.02)

Given: $d = 0.5$ mm, $L = 1.5$ m

$\Delta P = 1$ MPa, $Q = 1$ cm^3/s

$\mu = k \times 10^{-3}$ Pa.s

Given the flow is laminar, steady and fully developed, thus we have

$$\frac{\Delta P}{L} = \frac{128\mu Q}{\pi d^4}$$

$$\mu = \frac{\Delta P \pi d^4}{128 \times Q \times L}$$

Putting values, we get

$$\mu = \frac{10^6 \times 3.14 \times (0.5 \times 10^{-3})^4}{128 \times 10^{-6} \times 1.5}$$

$= 1.022 \times 10^{-3}$ Pa.s

Thus, the value of k is 1.02.

■■■■

9 Turbulent Flow in Pipe

9.1 For turbulent flow of an incompressible fluid through a pipe, the flow rate Q is proportional to $(\Delta P)^n$, where ΔP is the pressure drop. The value of exponent n is

(a) 1 (b) 0

(c) < 1 (d) > 1

[2002 : 1 M]

9.2 For turbulent flow past a flat plate, when no form drag is present, the friction factor f and the Chilton-Colburn factor j_D air related as

(a) f and j_D cannot he related

(b) f is equal to j_D

(c) f is greater than J_D

(d) f is less than J_D

[2005 : 2 M]

9.3 For a flow through a smooth pipe, the Fanning friction factor (f) is given by $f = m\text{Re}^{-0.2}$ in the turbulent flow regime, where Re is the Reynolds number and m is a constant. Water flowing through a section of this pipe with a velocity 1 m/s results in a frictional pressure drop of 10 kPa. What will be the pressure drop across this section (in kPa), when the velocity of water is 2 m/s?

(a) 11.5 (b) 20

(c) 34.8 (d) 40

[2016 : 1 M]

9.4 Water flows through a smooth circular pipe under turbulent conditions. In the viscous sub-layer, the velocity varies linearly with the distance from the wall. The Fanning friction factor is defined as,

$$f = \frac{\tau_w}{\frac{\rho \bar{u}^2}{2}}$$ where τ_w is the shear stress at the wall of the pipe, ρ is the density of the fluid and $\bar{u}$ is the average velocity in the pipe. Water (density = 1000 kg m^{-3}, viscosity = 1×10^{-3} kg m^{-1} s^{-1}) flows at an average velocity of 1 m s^{-1} through the pipe. For this flow condition, the friction factor f is 0.005. At a distance of 0.05 mm from the wall of the pipe (in the viscous sub-layer), the velocity (in m s^{-1}, rounded off to the third decimal place), is ______.

[2016 : 2 M]

9.5 Consider turbulent flow in a pipe under isothermal conditions. Let r denote the radial coordinate and z denote the axial flow direction. On moving away from the wall towards the center of the pipe, the rz-component of the Reynolds stress

(a) Increases and then decreases

(b) Decreases and then increases

(c) Remains unchanged

(d) Only increases

[2022 : 1 M]

9.6 For a fully developed turbulent flow of an incompressible Newtonian fluid through a pipe of constant diameter, which of the following statements is/are CORRECT?

(a) Reynolds stress, averaged over a sufficiently long time, is zero everywhere inside the pipe.

(b) Reynolds stress at the pipe wall is zero.

(c) Average velocity of the fluid is half of its center-line velocity.

(d) Average pressure gradient in the flow direction is constant.

[2023 : 1 M]

■■■■

Answers Turbulent Flow in Pipe

9.1	(c)	9.2	(c)	9.3	(c)	9.4	(0.125)	9.5	(a)	9.6	(b, d)

Explanations Turbulent Flow in Pipe

9.1 (c)

$$\frac{\Delta P}{\rho g} = \frac{4fIV^2}{2gd} \qquad \because \ \bar{u} = \frac{4Q}{\pi D^2}$$

For high Reynold's number $f \to 0.0014$

$$\therefore \qquad \Delta P = (2\times 0.0014)\frac{L\rho}{D}\cdot\left(\frac{4Q}{\pi D^2}\right)^2$$

or, simply we can write

$$\frac{\Delta P}{\rho g} = \frac{fIQ^2}{3d^5}$$

$$Q \propto (\Delta P)^{1/2}$$

$$\therefore \qquad n = 0.5 < 1$$

For turbulent flow, $Q \propto \sqrt{\Delta P}$

9.2 (c)

When no form drag is present for turbulent flow past a flat plate;

i.e. Chilton-Colburn analogy $J_D = \frac{f}{2}$

$$\therefore \qquad f > J_D$$

9.3 (c)

$$\frac{\Delta P}{\rho g} = \frac{4flv^2}{2gd} \qquad \{\text{given } f = m\,\text{Re}^{-0.2}\}$$

$$\Delta P = \frac{4flv^2\rho g}{2gd} = kfv^2 \qquad \left\{k = \frac{4l\rho}{2d}\right\}$$

$$\Delta P = \frac{kmv^2}{\left(\frac{\rho v d}{\mu}\right)^{0.2}} = k''v^{1.8} \qquad \left\{k'' = \frac{km}{\left(\frac{\rho d}{\mu}\right)^{0.2}}\right\}$$

$$\Delta P = k''\,v^{1.8} \Rightarrow \frac{10}{x} = \frac{k''(1)^{1.8}}{k''(2)^{1.8}}$$

New, $\Delta P = 34.8$ kPa

9.4 (0.125)

$$\tau_w = \frac{f\rho u^2}{2} = \frac{0.005\times 1000\times 1}{2} = 2.5$$

$$\frac{du}{dy} = \frac{\tau_w}{\mu} = \frac{2.5}{10^{-3}} = 2500$$

Integrating we get,

$$u = 2500\,y + c$$

Using boundary condition

at $y = 0,\ u = 0, \quad \therefore c = 0$

$$u = 2500\,y$$

at $y = 0.05 \times 10^{-3}$ m

$$u = 0.125 \text{ m/s}$$

9.5 (a)

Reynolds stress remain unchanged for fully developed flow.

If flow is not fully developed than first increase and than decrease near center.

9.6 (b, d)

For the case of fully developed turbulent flow,

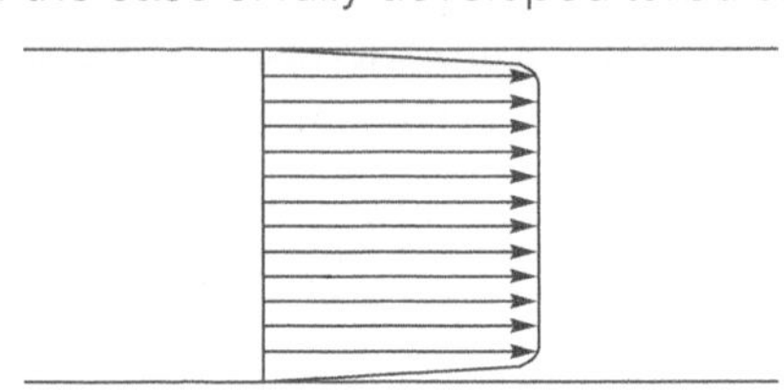

$\bar{V}_{avg}$ is found to be 0.817 times of the centre line velocity.

10 Dimensional Analysis

10.1 Match the following dimensionless numbers with the appropriate ratio of forces

List-I (Dimensionless Number)

A. Froude Number
B. Reynolds Number
C. Friction factor
D. Nusselt Number

List-II (Ratio of forces)

1. Shear force/inertial force
2. Convective heat transfer/conductive heat transfer
3. Gravitational force/viscous force
4. Inertial force/viscous force
5. Inertial force/gravitational force

Codes:

	A	B	C	D
(a)	1	2	5	3
(b)	5	4	3	2
(c)	5	4	1	2
(d)	3	4	5	1

[2003 : 2 M]

10.2 The relation between the stress τ and the strain rate $\left(\frac{du_x}{dy}\right)$ for the rapid flow of a granular material is given by

$$\tau = B\left(\frac{du_x}{dy}\right)^2$$

where B is a constant. If M, L and T are the mass, length and time dimension respectively, what is the dimension of the constant B?

(a) $ML^{-1}T^{-1}$
(b) $ML^{-1}T^{-2}$
(c) MT^{-1}
(d) ML^{-1}

[2005 : 2 M]

10.3 Match the following:

Group-1

P. Euler number
Q. Froude number
R. Weber number

Group-2

1. Viscous force/Inertial force
2. Pressure force/Inertial force
3. Inertial force/Gravitational force
4. Inertial force/Surface tension force

(a) P-1, Q-2, R-3
(b) P-2, Q-3, R-4
(c) P-3, Q-2, R-1
(d) P-4, Q-3, R-2

[2008 : 2 M]

10.4 A propeller (diameter $D = 15$ m) rotates at $N = 1$ revolution per second (rps). To understand the flow around the propeller, a lab-scale model is made. Important parameters to study the flow are velocity of the propeller tip ($V = \pi ND$), diameter D and acceleration due to gravity (g). The lab-scale model is $\frac{1}{100^{th}}$ of the size of the actual propeller.

The rotation speed of the lab-scale model, to the nearest integer, should be ______ rps.

[2017 : 2 M]

■■■■

Answers Dimensional Analysis

10.1 (c) **10.2** (d) **10.3** (b) **10.4** (10 rps)

Explanations Dimensional Analysis

10.1 (c)

$$\text{Froude number} = \frac{\text{Inertial force}}{\text{Gravitational force}}$$

$$\text{Reynold's number} = \frac{\text{Inertial force}}{\text{Viscous force}}$$

$$\text{Friction factor} = \frac{\text{Shear force}}{\text{Inertial force}}$$

$$\text{Nusselt number} = \frac{\text{Convective heat transfer}}{\text{Conductive heat transfer}}$$

10.2 (d)

$$\tau = B\left(\frac{du_x}{dy}\right)^2$$

$$\tau = \frac{F}{A} = \frac{MLT^{-2}}{L^2} = ML^{-1}T^{-2}$$

$$ML^{-1}T^{-2} = B\left[\frac{LT^{-1}}{L}\right]^2$$

$$B = L^{-1}M$$

10.3 (b)

$$\text{Euler number} = \frac{P}{\rho V^2} = \frac{\text{Pressure force}}{\text{Inertial force}}$$

$$\text{Froude number} = \frac{V}{\sqrt{lg}} = \frac{\text{Inertial force}}{\text{Gravity force}}$$

$$\text{Weber number} = \frac{V^2 L\rho}{\sigma\cdot l} = \frac{\text{Inertial force}}{\text{Surface tension force}}$$

10.4 (10)

Assuming geometrical similarity and Froude number to be constant.

$$\text{Fr} = \frac{N_i^2 D_i}{g}$$

$$\therefore \quad N_1^2 D_1 = D_2 N_2^2$$

$$D_1 = 15 \text{ m}, N_1 = 1 \text{ rps}$$

$$D_2 = \frac{15}{100} = 0.15 \text{ m}$$

$$\left(\frac{N_2}{N_1}\right)^2 = \frac{D_1}{D_2} = 100$$

$$N_2 = \sqrt{100} = 10 \text{ rps}$$

■■■■

11 Drag and Lift

11.1 The diameter of a drop of liquid fuel changes with time, due to combustion, according to the relationship, $D = D_0\left(1-\frac{t}{t_b}\right)$. While burning, the drop falls at its terminal velocity under Stokes' flow regime. The distance it will travel before complete combustion, is given by

(a) $\frac{D_o^2 \Delta\rho t_b g}{18\mu}$ (b) $\frac{D_o^2 \Delta\rho t_b g}{36\mu}$

(c) $\frac{D_o^2 \Delta\rho t_b g}{54\mu}$ (d) $\frac{D_o^2 \Delta\rho t_b g}{108\mu}$

[2010 : 2 M]

11.2 For Fanning friction factor f (for flow in pipes) and drag coefficient C_D (for flow over immersed bodies), which of the following statements are true?

P : f accounts only for the skin friction

Q : C_D accounts only for the form friction

R : C_D accounts for both skin friction and form friction

S : Both f and C_D depend on the Reynolds number

T : For laminar flow through a pipe, f doubles on doubling the volumetric flow rate.

(a) R, S, T (b) P, Q, S

(c) P, R, S (d) P, Q, S, T

[2015 : 2 M]

■■■■

Answers Drag and Lift

11.1 (c) **11.2** (c)

Explanations Drag and Lift

11.1 (c)

$$\therefore \quad D = D_0\left(1-\frac{t}{t_b}\right)$$

At $t = 0, \quad D = D_0$

At $t = t_b, \quad D = 0$

From Stokes law,

$$V_t = \frac{D^2(\rho_p - \rho_f)g}{18\mu_f} = \frac{ds}{dt}$$

$$\int_0^s ds = \frac{D_0^2(\rho_p - \rho_f)g}{18\mu}\int_0^{t_b}\left[1-\frac{t}{t_b}\right]^2 dt$$

$$s = \frac{D_0^2 \Delta\rho g}{18\mu}\left[t + \frac{t^3}{3t_b^2} - \frac{t^2}{t_b}\right]_0^{t_b}$$

$$\therefore \quad s = \frac{D_0^2 \Delta\rho g}{18\mu} \times \frac{t_b}{3} = \frac{D_0^2 \Delta\rho g t_b}{54\mu}$$

11.2 (c)

$f \rightarrow$ skin friction and C_D accounts not only for form friction, but for skin and form both.

f and C_D both depend on Reynold's number.

■■■■

12 Hydraulic Pumps

12.1 With increasing flow rate, the hydraulic efficiency of a centrifugal pump
(a) monotonically decreases
(b) decreases and then increases
(c) remains constant
(d) increases and then decreases
[2002 : 1 M]

12.2 The variables required to be known in correlations used for estimating the horse power of a centrifugal gas compressor and hence its cost are
1. inlet pressure
2. compressor RPM
3. delivery pressure
4. volumetric flow rate at inlet
(a) 1, 2 and 3 (b) 1 and 3
(c) 3 and 4 (d) 1, 3 and 4
[2003 : 1 M]

12.3 A centrifugal pump is used to pump water through a horizontal distance of 150 m and then raised to an overhead tank 10 m above. The pipe is smooth with an I.D. of 50 mm. What head (m of water) must the pump generate at its exit (E) to deliver water at a flow rate of 0.001 m^3/s? The Fanning friction factor, f is 0.0062

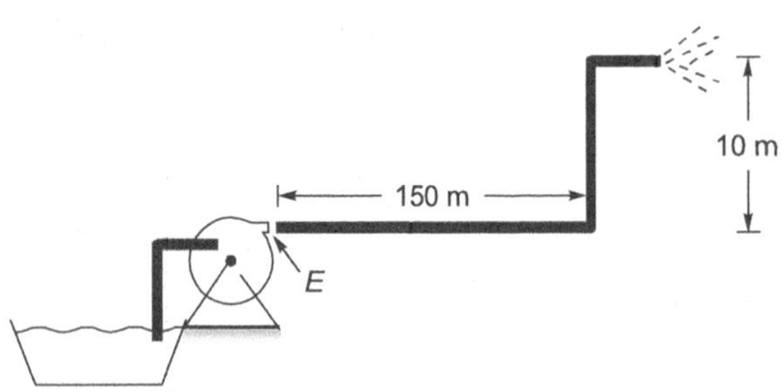

(a) 10 m (b) 11 m
(c) 12 m (d) 20 m
[2003 : 2 M]

12.4 Match the following for a centrifugal pump with impeller speed n

List-I	List-II
A. Capacity	1. proportional to n
B. Head	2. proportional to n^2
	3. proportional to n^3

Codes:

	A	B
(a)	2	1
(b)	1	3
(c)	2	3
(d)	1	2

[2006 : 1 M]

12.5 Select the most appropriate pump from Group-2 to handle each fluid flow given in Group-1.
Group-1
P. Highly viscous fluid flow
Q. Fluid containing large amount of abrasive solids
Group-2
1. Piston pump
2. Gear pump
3. Plunger pump
4. Centrifugal pump
(a) P-2, Q-1 (b) P-2, Q-4
(c) P-3, Q-4 (d) P-4, Q-3 [2007 : 1 M]

12.6 The figure shows a series-parallel configuration of three identical centrifugal pumps. The head increases ΔH across a single such pump varies with flow rate Q according to $\Delta H = a - bQ^2$. The expression for the total head increases $\Delta H = H_2 - H_1$ in terms of a and b and the total flow rate Q_1 for this configuration is given by

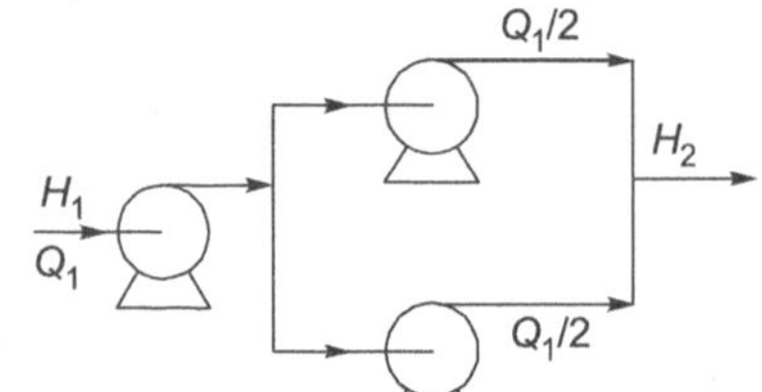

(a) $2a - \frac{5}{4}bQ_1^2$ (b) $2a - bQ_1^2$
(c) $2a - 2bQ_1^2$ (d) $a - bQ_1^2$ [2007 : 2 M]

12.7 To determine the performance of a compressor, a standardized test is performed. In the testing process, when the compressor is under operation, "shut off" term signifies
(a) maximum flow (b) zero flow
(c) steady flow (d) intermittent flow
[2008 : 1 M]

12.8 A pump draws oil (specific gravity 0.8) from a storage tank and discharges it to an overhead tank. The mechanical energy delivered by the pump to the fluid is 50 J/kg. The velocities at the suction and the discharge points of the pump are 1 m/s and 7 m/s, respectively. Neglecting friction losses and assuming kinetic energy correction factor to be unity, the pressure developed by the pump (in kN/m^2) is

(a) 19.2 (b) 20.8
(c) 40 (d) 80 [2008 : 2 M]

12.9 A storage vessel exposed to atmosphere (absolute pressure = 10.3 m of water) has a diameter of 3 m and is initially filled with water to a height of 2 m. The pump draws water from the vessel and is located at an elevation of 5 m above the bottom of the vessel. The frictional head loss in the suction pipe is 2 m of water. If the vapour pressure of the liquid at the temperature of operation is 3 m of water, then the available NPSH is

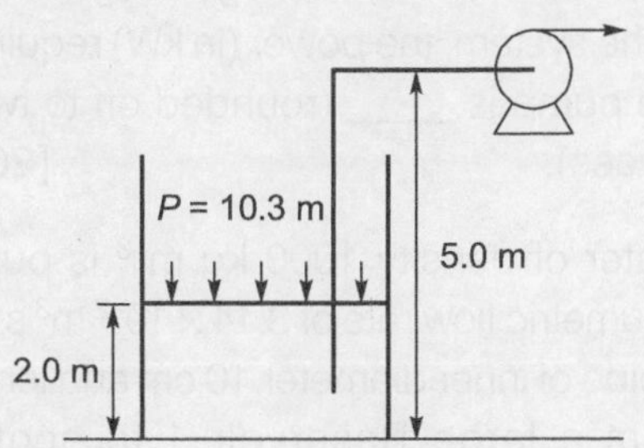

(a) 2.3 m (b) 5.3 m
(c) 6.3 m (d) 8.3 m
[2010 : 2 M]

12.10 Match the pumps in Group-I with the corresponding fluids in Group-II.

Group-I
P. Gear pump
Q. Peristaltic pump

Group-II
1. Highly viscous liquid
2. Aqueous sterile liquid
3. Slurry

(a) P-3, Q-1 (b) P-2, Q-1
(c) P-3, Q-2 (d) P-1, Q-2
[2011 : 1 M]

12.11 Slurries are most conveniently pumped by a

(a) syringe pump (b) diaphragm pump
(c) vacuum pump (d) gear pump
[2014 : 1 M]

12.12 Match the following:

Group-1	Group-2
P. Turbulence	I. Reciprocating pump
Q. NPSH	II. Packed bed
R. Ergun equation	III. Fluctuating velocity
S. Rotameter	IV. Impeller
T. Power number	V. Vena contact

(a) P-III, R-II, T-IV (b) Q-V, R-II, S-III
(c) P-III, R-IV, T-II (d) Q-III, S-V, T-IV
[2014 : 2 M]

12.13 A centrifugal pump delivers water at the rate of 0.22 m^3/s from a reservoir at ground level to another reservoir at a height H, through a vertical pipe of 0.2 m diameter. Both the reservoirs are open to atmosphere. The power input to the pump is 90 kW and it operates with an efficiency of 75%.

Data: Fanning friction factor for pipe flow is f = 0.004. Neglect other head losses. Take gravitational acceleration, g = 9.8 m/s^2 and density of water is 1000 kg/m^3.

The height H, in meters, to which the water can be delivered (up to one decimal place) is _____.
[2015 : 2 M]

12.14 Water (density = 1000 kg m^{-3}) is pumped at a rate of 36 m^3/h, from a tank 2 m below the pump, to an overhead pressurized vessel 10 m above the pump. The pressure values at the point of suction from the bottom tank and at the discharge point to the overhead vessel are 120 kPa and 240 kPa, respectively. All pipes in the system have the same diameter. Take acceleration due to gravity, g = 10 m s^{-2}. Neglecting frictional losses, what is the power (in kW) required to deliver the fluid?

(a) 1.2 (b) 2.4
(c) 3.6 (d) 4.8 [2016 : 2 M]

12.15 The characteristics curve (Head - Capacity relationship) of a centrifugal pump is represented by the equation ΔH_{pump} = 43.8 – 0.19Q, where ΔH_{pump} is the head developed by the pump (in m) and Q is the flowrate (in m^3/h) through the pump. This pump is to be used for pumping water through a horizontal pipeline. The frictional head loss ΔH_{piping} (in m) is related to the water flowrate Q_L (in m^3/h) by the equation $\Delta H_{piping} = 0.0135\, Q_L^2 + 0.045\, Q_L$. The flowrate (in m^3/h, rounded off to the first decimal place) of water pumped through the above pipeline, is _____. [2016 : 2 M]

12.16 A centrifugal pump is used to pump water (density 1000 kg/m^3) from an inlet pressure of 10^5 Pa to an exit pressure of 2×10^5 Pa. The exit is at an elevation of 10 m above the pump. The average velocity of the fluid is 10 m/s. The cross-sectional area of the pipes at the pump inlet and outlet is 10^{-3} m^2 and acceleration due to gravity is $g = 10$ m/s^2. Neglecting losses in the system, the power (in Watts) delivered by the pump is _______ (rounded off to the nearest integer).

[2019 : 2 M]

12.17 Liquid water is pumped at a volumetric flow rate of 0.02 m^3 s^{-1} from Tank-I to Tank-II, as shown in the figure.

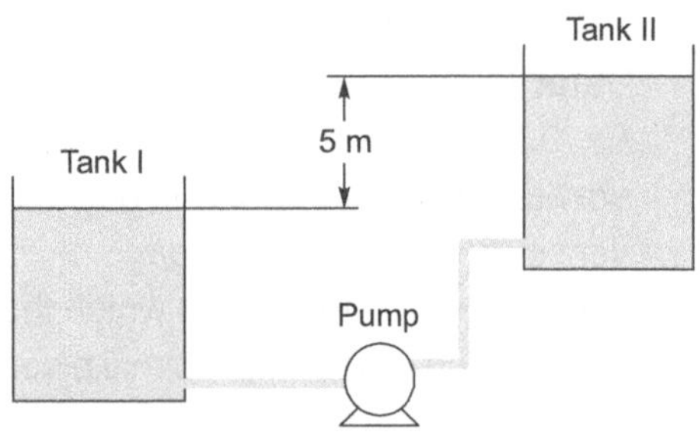

Both the tanks are open to the atmosphere. The total frictional head loss for the pipe system is 1.0 m of water.

In addition, use the following data and assumptions:

- Density of water is 1000 kg m^{-3}.
- Acceleration due to gravity is 9.81 m/s^{-2}.
- Efficiency of the pump is 100%.
- The liquid surfaces in the tanks have negligible velocities.

The power supplied (in W) by the pump to lift the water is _____ (round off to 1 decimal place).

[2020 : 2 M]

12.18 Two reservoirs located at the same altitude are connected by a straight horizontal pipe of length 120 m and inner diameter 0.5 m, as shown in the figure. A pump transfers the liquid of density 800 kg m^{-3} at a flow rate of 1 m^3 s^{-1} from Reservoir-1 to Reservoir-2. The liquid levels in Reservoir-1 and Reservoir-2 are 2 m and 10 m, respectively. Assume that the reservoirs' cross-section areas are large enough to neglect the liquid velocity at the top of the reservoirs. All minor losses can be ignored. The acceleration due to gravity is 9.8 m s^{-2}. If the friction factor for the pipe-flow is 0.01, the required power of the pump is ________ kW (rounded off to one decimal place).

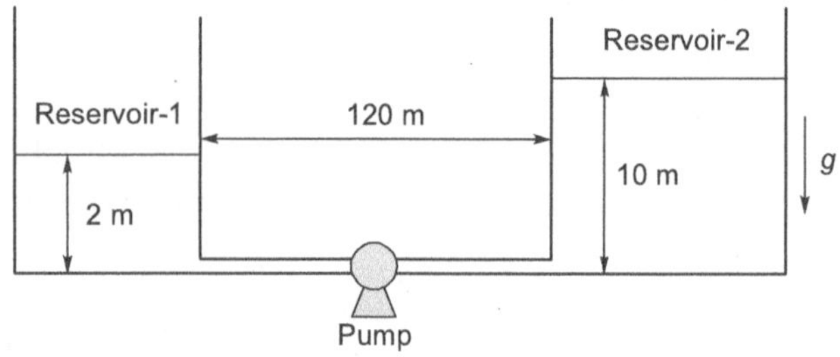

[2022 : 2 M]

12.19 A pump draws water (density =1000 kg.m^{-3}) at a steady rate of 10 kg.s^{-1}. The pressures at the suction and discharge sides of the pump are –20 kPa (gauge) and 350 kPa (gauge), respectively. The pipe diameters at the suction and discharge side are 70 mm and 50 mm, respectively. The suction and discharge lines are at the same elevation, and the pump operates at an efficiency of 80 %. Neglecting frictional losses in the system, the power (in kW) required to drive the pump is _____ (rounded off to two decimal places).

[2023 : 2 M]

12.20 Water of density 1000 kg m^{-3} is pumped at a volumetric flow rate of 3.14×10^{-2} m^3 s^{-1}, through a pipe of inner diameter 10 cm and length 100 m, from a large Reservoir 1 to another large Reservoir 2 at a height 50 m above Reservoir 1, as shown in the figure. The flow in the pipe is in the turbulent regime with a Darcy friction factor $f = 0.06$, and a kinetic energy correction factor $\alpha = 1$. Take $g = 9.8$ m s^{-2}. If all minor losses are negligible, and the pump efficiency is 100%, the pump power, in kW, rounded off to 2 decimal places, is ________.

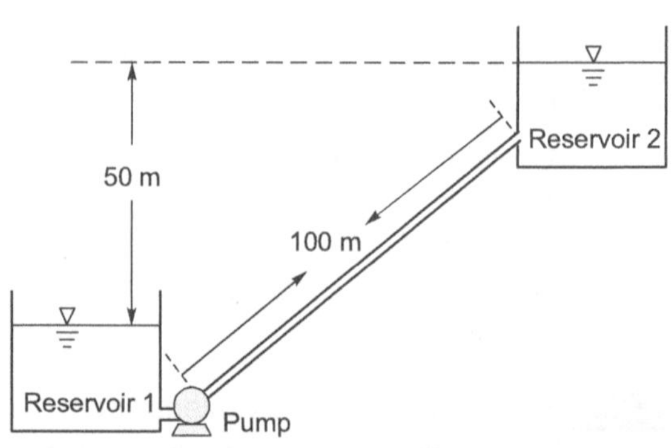

[2024 : 2 M]

Answers Hydraulic Pumps

12.1	(d)	**12.2**	(d)	**12.3**	(b)	**12.4**	(d)	**12.5**	(b)	**12.6**	(a)	**12.7**	(b)
12.8	(b)	**12.9**	(a)	**12.10**	(d)	**12.11**	(b)	**12.12**	(a)	**12.13**	(24)	**12.14**	(b)
12.15	(48.92)	**12.16**	(2000)	**12.17**	(1177.2)	**12.18**	(87.75)	**12.19**	(4.74)	**12.20**	(30.44)		

Explanations Hydraulic Pumps

12.1 (d)

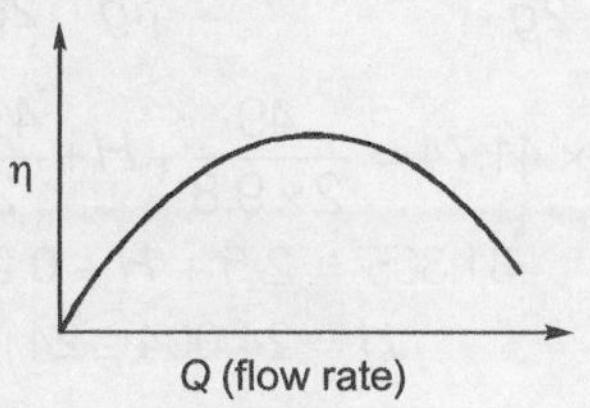

$\eta \rightarrow$ Efficiency

$Q \rightarrow$ Flow rate

$\therefore$ with increasing flow rate η first increases then decreases.

12.3 (b)

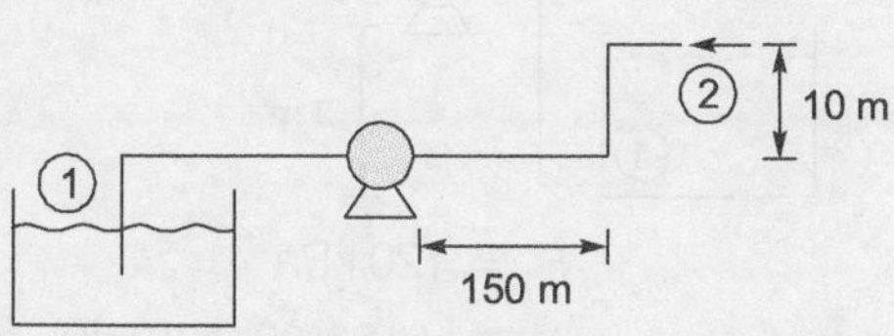

$$Q = A \times \bar{V}$$

$$\bar{V} = \frac{0.001}{\left(\frac{\pi}{4}\right)D^2} = 0.5093 \text{ m/s}$$

Apply Bernoulli theorem between (1) and (2)

$$\frac{P_1}{\rho g} + \frac{V_1^2}{2g} + z_1 + H_P = \frac{P_2}{\rho g} + \frac{V_2^2}{2g} + z_2 + h_f$$

$$H_P = \frac{4 \times 0.0062 \times 160 \times (0.5093)^2}{2 \times 9.81 \times 0.05} + 10$$

$$H_P = 11.049 \cong 11 \text{ m}$$

12.4 (d)

Affinity laws for pump

Characteristic	Capacity	Head	Power
Constant D	$q \propto n$	$\Delta H \propto n^2$	$P \propto n^3$
Constant n	$q \propto D$	$q \propto D^2$	$q \propto D^3$

$n \rightarrow$ Impeller speed

$D \rightarrow$ Diameter of pump impeller

12.5 (b)

- Gear pumps are positive displacement pump. The pump can handle high viscosity fluids efficiently.
- Piston pumps handle slurries and sludge with moderate abrasively.
- Plunger pumps can be used at higher pressure.

12.6 (a)

$$\Delta H = a - bQ^2 \text{ (given)}$$

$\therefore$ for pump (1)

$$H^1 = a - bQ_1^2 + H_1 \quad ...(i)$$

For pump (2)

$$H_2 - H^1 = a - b(Q_2)^2 \qquad \because Q_2 = \frac{Q_1}{2}$$

$$H_2 = H^1 + a - \frac{bQ_1^2}{4}$$

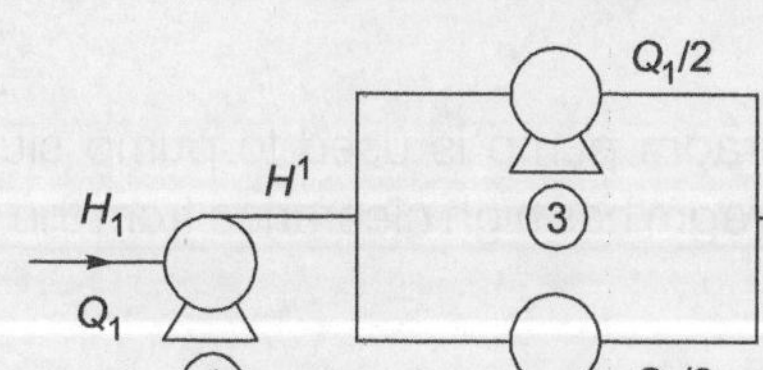

Put H^1 from equation (i)

$$H_2 = a - bQ_1^2 + H_1 + a - \frac{bQ_1^2}{4}$$

$$H_2 - H_1 = 2a - \frac{5}{4}bQ_1^2$$

12.7 (b)

When the valve is shut, the mass flow rate is zero.

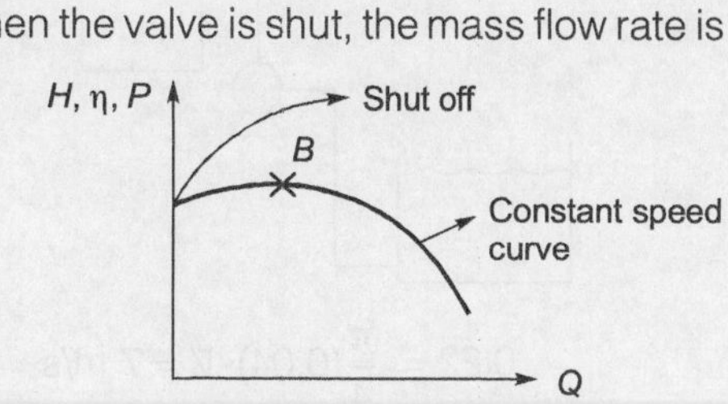

12.8 (b)

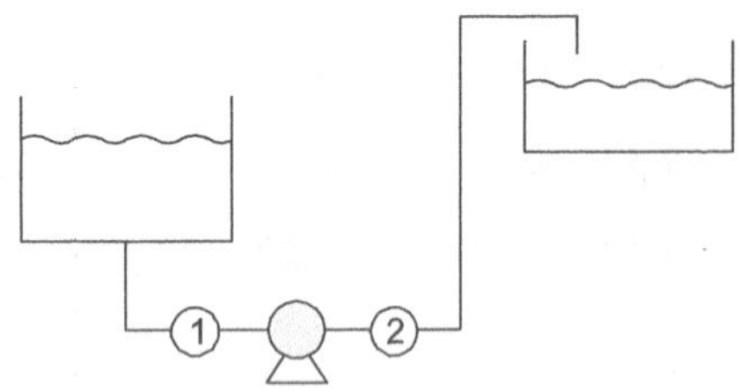

Apply Bernoulli's equation between points (1) and (2)

$$\frac{P_1}{\rho}+\frac{V_1^2}{2}+g\cancel{z_1}+h_p=\frac{P_2}{\rho}+\frac{V_2^2}{2}+g\cancel{z_2}+h_f$$

$$(h_f = 0)$$

$$V_1 = 1\text{ m/s},\ V_2 = 7\text{ m/s},\ h_p = 50\text{ J/kg}$$

$$\frac{P_1}{\rho}+\frac{1}{2}+50 = \frac{P_2}{\rho}+\frac{49}{2}+0 \quad \{\rho = 800\text{ kg/m}^3\}$$

$$= 20800\text{ N/m}^2 = 20.8\text{ kN/m}^2$$

12.9 (a)

The NPSH of a pump is given by

$$\text{NPSH} = \text{Helping head} - \text{Losing head}$$

$$\text{NPSH} = \frac{P}{\rho g}-\left[\frac{P^{\text{vapour}}}{\rho g}+h_f+h_s\right]$$

$$\text{NPSH} = 10.3-[3+2+3] = 2.3\text{ m}$$

12.10 (d)

Gear pump → Highly viscous liquid
Peristaltic pump → Aqueous sterile liquid

12.11 (b)

Diaphragm pump is used to pump slurries b/c diaphragm has high clearance from the wall.

12.12 (a)

Turbulence → Fluctuating velocity
Ergun equation → Packed bed
Power number → Impeller

12.13 (24)

$$Q = 0.22\text{ m}^3/\text{s}$$

Given, $\text{Power} = 90\text{ kW} = \rho \times g \times H_p \times Q$

$$90\times 10^3 = 1000\times 9.8\times H_p\times 0.22$$

$$H_p = 41.77\text{ m}$$

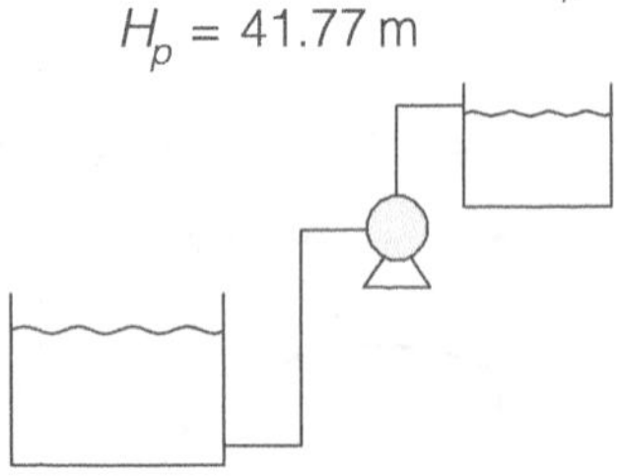

$$0.22 = \frac{\pi}{4}(0.04)\cdot V = 7\text{ m/s}$$

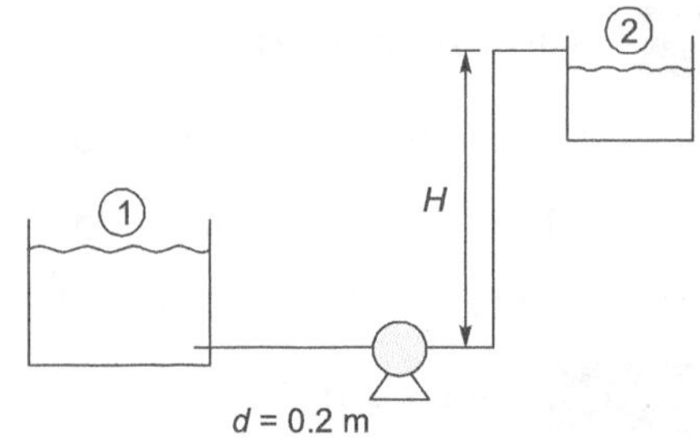

Apply energy balance between (1) and (2), we get

$$\frac{P_1}{\rho g}+\frac{V_1^2}{2g}+z_1+n\,H_p = \frac{P_2}{\rho g}+\frac{V_2^2}{2g}+z_2+h_f$$

$$0.75\times 41.74 = \frac{49}{2\times 9.8}+H+\frac{4\times f\times H\times(7)^2}{2\times 9.8\times 0.2}$$

$$31.305 = 2.5+H+0.2\,H$$

$$H = 24.004 \cong 24\text{ m}$$

12.14 (b)

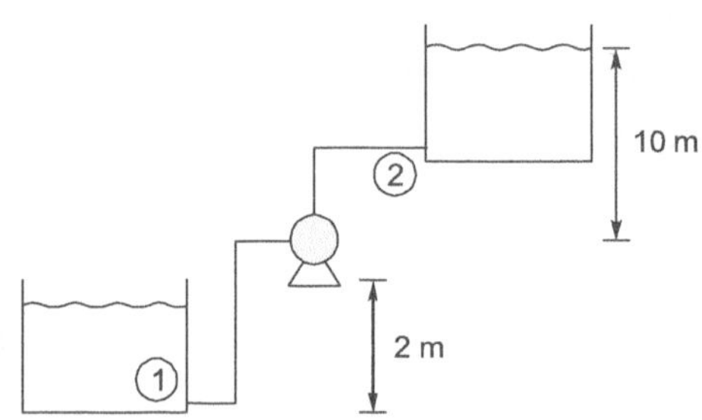

$$P_1 = 120\text{ kPa}$$

$(u_1 = u_2)$ same diameter

Apply Bernoulli's theorem between point (1) and (2),

$$\frac{P_1}{\rho g}+\frac{V_1^2}{2g}+z_1+\eta\,H_p = \frac{P_2}{\rho g}+\frac{V_2^2}{2g}+z_2+h_f$$

$$\frac{120\times 10^3}{10^3\times 10}+0+\eta H_p = \frac{240\times 10^3}{10^3\times 10}+12$$

$$12+H_p = 24+12$$

$$H_p = 24 \quad \{\eta \Rightarrow 100\%\}\text{ if not given}$$

$$\text{Power} = \frac{1000\times g\times 24}{1000}\times\frac{36}{3600} = 2.4\text{ kW}$$

12.15 (48.92)

$$\Delta H_{\text{pump}} = 43.8-0.19\,Q$$

$$\Delta H_{\text{piping}} = 0.0135\,Q_L^2+0.045\,Q_L$$

The pump head to compensate for head loss due to friction, equating both equation

$$43.8-0.19\,Q = 0.0135\,Q^2+0.045\,Q$$

Solving for Q, we get $Q = 48.92\text{ m}^3/\text{h}$

12.16 (2000)

Power delivered by the pump is $= \rho\times g\times h_p\times Q$
i.e. Power = Density of fluid $\times\, g\times h_p\times Q$

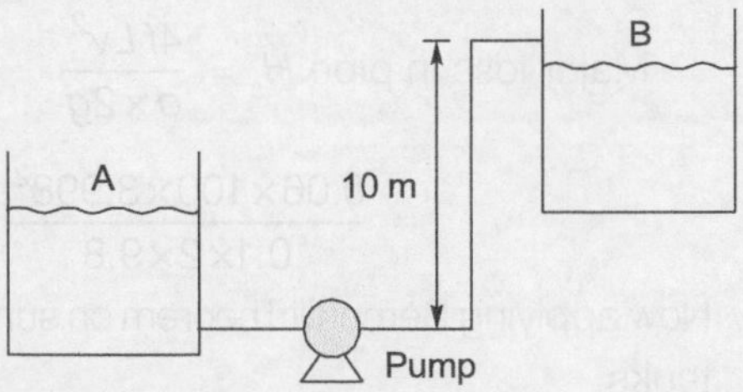

Apply Bernoulli's equation between A and B, [Neglect frictional loss]

$$\left[\frac{P_A}{\rho g}+\frac{V_A^2}{2g}+Z_A+\eta H_p=\frac{V_B^2}{2g}+\frac{P_B}{\rho g}+Z_B\right]$$

At diameter of pipe is uniform so, $V_A = V_B$

$$\frac{10^5}{1000\times 10}+H_p=\frac{2\times 10^5}{1000\times 10}+10$$

$$H_p = 20 \text{ m}$$

Thus [Power delivered = $\rho_f \times g \times H_p \times Q$]

$$Q = A\times\bar{V} = 10^{-3} \text{ m}^2 \times 10 \text{ m/s} = 10^{-2} \text{ m}^3/\text{s}$$

$$\text{Power} = 1000 \times 10 \times 20 \times 10^{-2} = 2000 \text{ watt}$$

12.17 (1177.2)

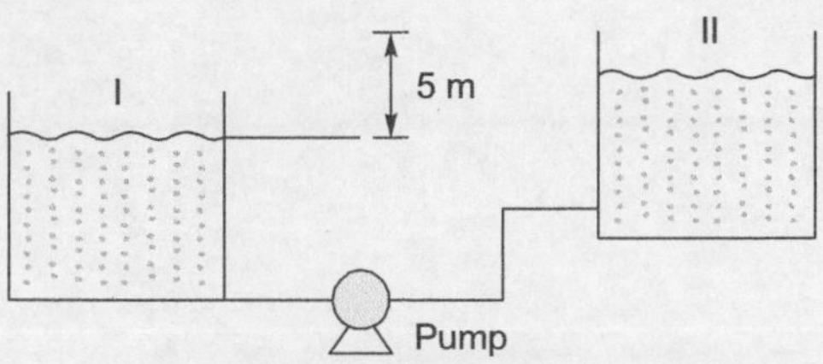

$$\rho_w = 1000 \text{ kg/m}^3,\ g = 9.81 \text{ m/s}^2$$

$$\eta = 100\%$$

[$Q = 0.02$ m^3/s]

($h_f = 1$ m of water) frictional head loss

Now, apply Bernoulli's theorem between I and II,

$$\frac{P_A}{\rho_g}+\frac{V_A^2}{2g}+z_A+\eta\cdot h_p = \frac{P_B}{\rho_g}+\frac{V_B^2}{2g}+z_B+h_f$$

As both are open to atmosphere,

$$P_A = P_B = 0,\ V_A = V_B \cong \text{negligible}$$

$$h_p = 5 + 1 = 6 \text{ m}$$

Power supplied by pump = $\rho \cdot g \cdot h_p \cdot Q$

$$= 1000 \times 9.81 \times 6 \times 0.02$$

$$P = 1177.2 \text{ watt}$$

12.18 (87.75)

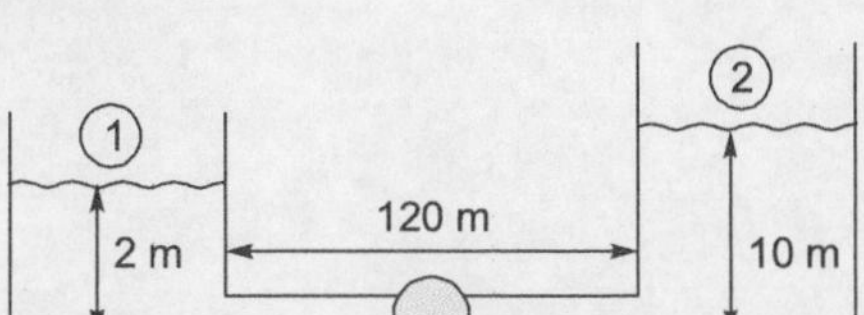

Given data: $d = 0.5$ m

$4f$ = F (friction factor) = 0.01

$\rho = 800$ kg/m^3, $Q = 1$ m^3/s

Now (neglecting minor losses) apply Bernoulli's equation (between 1 and 2), we get

$$\frac{P_1}{\rho g}+\frac{V_1^2}{2g}+Z_1+H_P = \frac{P_2}{\rho g}+\frac{V_2^2}{2g}+Z_2+h_f$$

As open to atmosphere and having large surface area of tank, so [$v_1 = v_2 = 0$]; $p_1 = p_2$

$$2 + h_p = 10 + h_f$$

$$h_f = \frac{4flv^2}{2gd} = \frac{0.01\times 120\times v^2}{2\times 9.81\times 0.5} \quad \left\{\begin{aligned}1 &= \frac{\pi}{4}(0.25)V\\ V &= 5.1 \text{ m/s}\end{aligned}\right\}$$

$$h_f = 3.18 \text{ m},\ H_P = 8 + 3.18 = 11.18 \text{ m}$$

$$\text{Power} = 800 \times 9.81 \times 11.18 \times 1 = 87.75 \text{ kW}$$

12.19 (4.74)

Given: $\dot{m} = 10$ kg/s, $\rho_w = 1000$ kg/m^3

$P_S = -20$ kPa (gauge), $P_D = 350$ kPa (gauge)

$Z_1 = Z_2$ (same), $g = 9.81$ m/s^2

$d_S = 70$ mm, $d_D = 50$ mm, $\eta = 80\%$

Now, from continuity equation, we have

$$A_1V_1 = A_2V_2$$

$$Q = \frac{10}{1000} = 0.01 \text{ m}^3/\text{s}$$

$$0.01 = \frac{\pi}{4}(0.07)^2 V_s \rightarrow V_s = 2.6 \text{ m/s}$$

$$0.01 = \frac{\pi}{4}(0.05)^2 \cdot V_D \rightarrow V_D = 5.1 \text{ m/s}$$

Now, applying Bernoulli's theorem between suction and discharge, we get

$$\frac{P_S}{\rho g}+\frac{V_S^2}{2g}+Z_1+H_P = \frac{P_D}{\rho g}+\frac{V_D^2}{2g}+Z_2+h_f$$

$$h_f \cong 0 \text{ (given)}$$

$$(Z_1 = Z_2)$$

Putting values, we get

$$\frac{-20\times 10^3}{10^3\times g}+\frac{(2.6)^2}{2\times g}+H_P = \frac{35\times 10^3}{10^3\times g}+\frac{(5.1)^2}{2\times g}$$

$$-20+\frac{(2.6)^2}{2}+gH_P = 350+\frac{(5.1)^2}{2}$$

$$gH_P = 383.005-\frac{(2.6)^2}{2} = 379.625$$

$$\text{Power} = \frac{\rho g H_P\cdot Q}{\eta} = \frac{1000\times 379.625\times 0.01}{1000\times 0.8}$$

$$P = 4.74 \text{ kW}$$

12.20 (30.44)

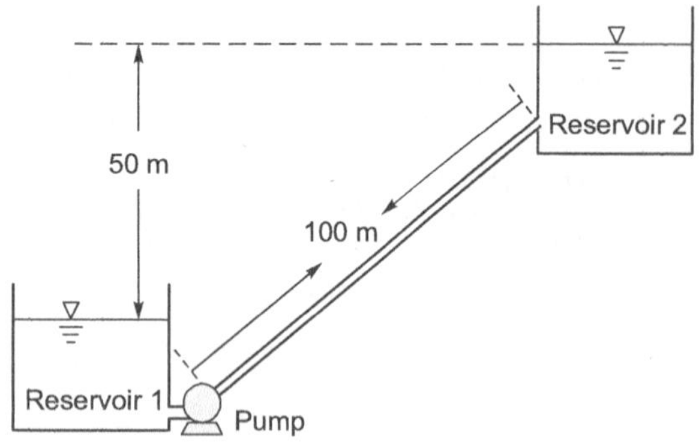

Volumetric flow rate $Q = 3.14 \times 10^{-2}$ m³/s

Diameter of pipe = 0.1 m

Darcy friction factor $4f = 0.06$

Kinetic energy connection factor $\alpha = 1$

$$g = 9.8 \text{ m/s}^2$$

$$Q = \frac{\pi}{4} d^2 \cdot v$$

$$3.14 \times 10^{-2} = \frac{\pi}{4} \times 0.1^2 \times v$$

$$v = 3.998 \text{ m/s}$$

$$\text{Major loss in pipe } H_s = \frac{4fLv^2}{d \times 2g}$$

$$= \frac{0.06 \times 100 \times 3.998^2}{0.1 \times 2 \times 9.8} = 48.9299$$

Now applying Bernoulli Theorem on surface of both tanks

$$\frac{P_1}{\rho g} + \frac{V_1^2}{2g} + Z_1 + H_p = \frac{P_2}{\rho g} + \frac{V_2^2}{2g} + Z_2 + H_s$$

$$P_1 = P_2 \text{ and } V_1 = V_2$$

$$H_p = 50 + 48.9299 = 98.9299$$

Power supplied by pump = $\rho g H Q$

$$= 10^3 \times 9.8 \times 98.9299 \times 3.14 \times 10^{-2}$$

$$= 30.44 \times 10^3 \text{ W} = 30.44 \text{ kW}$$

■■■■

Mechanical Operations

UNIT IV

CONTENTS

Mechanical Operations

Syllabus

Particle Size and Shape, Particle Size Distribution, Size Reduction and Classification of Solid Particles; Free and Hindered Settling; Centrifuge and Cyclones; Thickening and Classification, Filtration, Agitation and Mixing; Conveying of Solids.

Analysis of Previous GATE Papers

Exam Year	1 Mark Ques.	2 Marks Ques.	5 Marks Ques.	Total Marks
2000	1	3	–	**7**
2001	1	2	–	**5**
2002	2	4	–	**10**
2003	1	3	–	**7**
2004	4	5	–	**14**
2005	2	2	–	**6**
2006	–	6	–	**12**
2007	4	4	–	**12**
2008	2	2	–	**6**
2009	2	1	–	**4**
2010	2	–	–	**2**
2011	–	3	–	**6**
2012	1	1	–	**3**
2013	2	2	–	**6**
2014	1	–	–	**1**
2015	1	3	–	**7**
2016	1	2	–	**5**
2017	2	1	–	**4**
2018	2	2	–	**6**
2019	1	1	–	**3**
2020	2	1	–	**4**
2021	1	1	–	**3**
2022	–	1	–	**2**
2023	2	1	–	**4**
2024	–	1	–	**2**

1 Properties and Handling of Particulate Solids

1.1 The sphericity of a solid particle of cubical shape is

(a) π (b) $\left(\frac{\pi}{6}\right)^{\frac{1}{3}}$

(c) $\left(\frac{\pi}{6}\right)^{\frac{1}{2}}$ (d) $\left(\frac{\pi}{3}\right)$

[2000 : 2 M]

1.2 If d_p is the equivalent diameter of a non-spherical particle, v_p its volume and s_p its surface area, then its sphericity ϕ is defined by

(a) $\varphi_s = \frac{6v_p}{d_p s_p}$ (b) $\varphi_s = \frac{v_p}{d_p s_p}$

(c) $\varphi_s = \frac{6d_s s_p}{v_p}$ (d) $\varphi_s = \frac{d_s s_p}{v_p}$

[2002 : 1 M]

1.3 A sand mixture was screened through a standard 10-mesh screen. The mass fraction of the oversize material in feed, overflow and underflow were found to be 0.38, 0.79 and 0.22, respectively. The screen effectiveness based on the oversize is

(a) 0.50 (b) 0.58
(c) 0.68 (d) 0.62

[2002 : 2 M]

1.4 The cumulative mass fraction of particles smaller than size d_j for a collection of N_i particles of diameter d_i, and mass m_i (i = 1, 2, 3, ..., ∞) is given by

(a) $\frac{\sum_{i=1}^{j} N_i d_i^3}{\sum_{i=1}^{\infty} N_i d_i^3}$ (b) $\frac{\sum_{i=1}^{j} N_i m_i d_i^3}{\sum_{i=1}^{\infty} N_i m_i d_i^3}$

(c) $\frac{\sum_{i=1}^{j} N_i m_i d_i^2}{\sum_{i=1}^{\infty} N_i m_i d_i^2}$ (d) $\frac{\sum_{i=1}^{j} N_i m_i d_i}{\sum_{i=1}^{\infty} N_i m_i d_i}$

[2004 : 2 M]

1.5 In Tyler series, the ratio of the aperture size of a screen to that of the next smaller screen is

(a) $\frac{1}{\sqrt{2}}$ (b) $\sqrt{2}$
(c) 1.5 (d) 2 [2007 : 1 M]

1.6 In the Tyler standard screen scale series, when the mesh number increases from 3 mesh to 10 mesh, then

(a) the clear opening decreases
(b) the clear opening increases
(c) the clear opening is unchanged
(d) the wire diameter increases

[2013 : 1 M]

1.7 The diameters of sand particles in a sample range from 50 to 150 microns. The number of particles of diameter x in the sample is proportional to $\frac{1}{50 + x}$. The average diameter, in microns, (up to one decimal place) is ______.

[2015 : 2 M]

1.8 Size analysis was earned out on a sample of gravel. The data for mass fraction (x_i) and average particle diameter (D_{pi}) of the fraction is given in the table below:

x_i	D_{p_i} (mm)
0.2	5
0.4	10
0.4	20

The mass mean diameter of the sample, to the nearest integer, is ______ mm.

[2017 : 2 M]

1.9 Consider a cylinder (diameter D and length D), a sphere (diameter D) and a cube (side length D). Which of the following statements concerning the sphericity (ϕ) of the above objects is true :

(a) $\phi_{sphere} > \phi_{cylinder} > \phi_{cube}$
(b) $\phi_{sphere} = \phi_{cylinder} = \phi_{cube}$
(c) $\phi_{sphere} < \phi_{cylinder} < \phi_{cube}$
(d) $\phi_{cylinder} > \phi_{sphere} > \phi_{cube}$

[2019 : 1 M]

Answers Properties and Handling of Particulate Solids

1.1 (b)	1.2 (a)	1.3 (b)	1.4 (b)	1.5 (b)	1.6 (a)	1.7 (94.3)
1.8 (13)	1.9 (a)					

Explanations Properties and Handling of Particulate Solids

1.1 (b)

Sphericity (ϕ_s) is given by $= \dfrac{6}{dp}\dfrac{V_p}{S_p}$

$$\frac{\pi}{6}dp^3 = a^3$$

$$D_p = \left(\frac{6}{\pi}\right)^{1/3} a$$

Now, surface area of sphere $= \pi D_s^2 = 6^{2/3}\pi^{1/3}a^2$

Surface area of the cube $= 6a^2$

$$\therefore \qquad \phi_s = \frac{\pi\left[\left(\frac{6}{\pi}\right)^{1/3} a\right]^2}{6a^2} = \left(\frac{\pi}{6}\right)^{1/3}$$

1.2 (a)

Sphericity is given by, $\left[\phi_s = \dfrac{6}{d_p}\dfrac{v_p}{s_p}\right]$.

1.3 (b)

$$F = W + Z$$

$$0.38\,F = 0.22\,W + 0.79\,Z = 0.22\,(F - Z) + 0.79\,Z$$

$$\left(\frac{Z}{F}\right) = \frac{16}{57}$$

$$\text{Screen effectiveness} = \frac{0.79 \times Z}{0.38 \times F} = \frac{0.79}{0.38} \times \frac{16}{57} = 0.583$$

1.4 (b)

Cumulative mass fraction of particle = Fraction of total weight of particles having a size smaller than a given diameter of the screen.

$$\text{i.e.} \quad \frac{\sum_{i=1}^{I} N_i m_i d_i^3}{\sum_{i=1}^{\infty} N_i m_i d_i^3}$$

1.5 (b)

The ratio of the aperture size of a screen to that of the next smaller screen is given by

$$\frac{d_1}{d_2} = \sqrt{2}$$

1.6 (a)

Mesh number increases, the clear opening decreases.

1.7 (94.3)

$$D_{avg} = \frac{\int_a^b \left[x \cdot \frac{1}{50 + x}\right] dx}{\int_a^b \left(\frac{1}{50 + x}\right) dx} \qquad \text{...(i)}$$

Range of $(a, b) = (50, 150)$

$$D_{avg} = \frac{\int_{50}^{150} \frac{x}{50 + x} dx}{\int_{50}^{150} \frac{1}{50 + x} dx} = \frac{(150 - 50) - 50\ln\left|\frac{200}{100}\right|}{\ln\left|\frac{200}{100}\right|}$$

$$D_{avg} = 94.3$$

1.8 (13)

For mass mean diameter ($\overline{D_m}$) is given by

$$\overline{D_m} = \sum_{i=1}^{n} X_i \overline{D_{pi}}$$

$$= 0.2 \times 5 + 0.4 \times 10 + 0.4 \times 20 = 13 \text{ mm}$$

1.9 (a)

Sphericity = Surface area of a hypothetical sphere having the same volume as that of the given particle/surface area of the particle

$$\left[\phi_s = \frac{6}{d_p}\frac{V_p}{s_p}\right]$$

$$\therefore \qquad \frac{\pi}{6}d_p^3 = \text{Volume of particle}$$

The values of sphericity are:

∴ For a sphere $\phi_s = 1$ (always)

∴ For a cylinder $\phi_s = 0.874$ if $(L = D)$

∴ For a cube $\phi_s = 0.806$

For the given problem $\phi_{sphere} > \phi_{cylinder} > \phi_{cube}$

■■■■

2 Size Reduction Laws and Equipments

2.1 The energy required per unit mass to grind limestone particles of very large size to 100 μm is 12.7 kWh/ton. An estimate (using Bond's Law) of the energy to grind the particles from a very large size to 50 μm is
(a) 6.35 kWh/ton (b) 9.0 kWh/ton
(c) 18 kWh/ton (d) 25.4 kWh/ton

[2001 : 2 M]

2.2 Arrange the following size reduction equipment in the decreasing order of the average particle size produced by each of them.
(a) Jaw crushers, Ball mills, Fluid energy mills
(b) Ball mills, Jaw crushers. Fluid energy mills
(c) Fluid energy mills, Jaw crushers, Ball mills
(d) Fluid energy mills, Ball mills, Jaw crushers

[2002 : 1 M]

2.3 What is the critical rotational speed, in revolutions per second, for a ball mill of 1.2 m diameter charged with 70 mm diameter balls?
(a) 0.5 (b) 1.0
(c) 2.76 (d) 0.66

[2002 : 2 M]

2.4 Energy requirement (per unit mass of material crushed/ground) is highest for
(a) Jaw crusher (b) Rod mill
(c) Ball mill (d) Fluid energy mill

[2003 : 1 M]

2.5 The critical speed of the ball mill of radius R, which contains balls of radius r, is proportional to
(a) $(R-r)^{-0.5}$ (b) $(R-r)^{-1}$
(c) $(R-r)$ (d) $(R-r)^2$

[2005 : 1 M]

Statement for Linked Answer Question (2.6 & 2.7):
A continuous grinder obeying the Bond crushing law grinds a solid at the rate of 1000 kg/hr from the initial diameter of 10 mm to the final diameter of 1 mm.

2.6 If the market now demands particles of size 0.5 mm, the output rate of the grinder (in kg/hr) for the same power input would be reduced to
(a) 227 (b) 474
(c) 623 (d) 856

[2006 : 2 M]

2.7 In order to restore the output back to 1000 kg/hr, an additional grinder was installed. The two grinders can be operated either in series (configuration-1) or parallel (configuration-2). Compare the two configurations in terms of the additional power consumption over the case above.
(a) configuration-1 consumes less power than configuration-2
(b) configuration-2 consumes less power than configuration-1
(c) Both configurations consume the same power
(d) configuration-2 consumes less or more power than configuration-1 depending on how the feed is distributed between the two grinders in configuration-2 (the parallel configuration)

[2006 : 2 M]

2.8 Size reduction of coarse hard solids using a crusher is accompanied by
(a) attrition
(b) compression
(c) cutting
(d) impact

[2007 : 1 M]

2.9 The power required for size reduction in crushing is
(a) proportional to $\dfrac{1}{\text{Surface energy of the material}}$
(b) proportional to $\sqrt{\dfrac{1}{\text{Surface energy of the material}}}$
(c) proportional to Surface energy of the material
(d) independent of the Surface energy of the material

[2008 : 1 M]

2.10 The critical speed (revolutions per unit time) of a ball mill of radius R, which uses balls of radius r, is

(a) $\frac{1}{2\pi}\sqrt{\frac{g}{\sqrt{Rr}}}$ (b) $\frac{1}{2\pi}\sqrt{\frac{g}{R}}$

(c) $\frac{1}{2\pi}\sqrt{\frac{g}{r}}$ (d) $\frac{1}{2\pi}\sqrt{\frac{g}{R-r}}$

[2010 : 1 M]

2.11 100 ton/h of a rock feed, of which 80% passed through a mesh size of 2.54 mm, were reduced in size such that 80% of the crushed product passed through a mesh size of 1.27 mm. The power consumption was 100 kW. If 100 ton/h of the same material is similarly crushed from a mesh size of 5.08 mm to a mesh size of 2.54 mm, the power consumption (in kW, to the nearest integer) using Bond's law, is ________.

[2013 : 2 M]

2.12 In order to produce fine solid particles between 5 and 10 μm, the appropriate size reducing equipment is

(a) fluid energy mill
(b) hammer mill
(c) jaw crusher
(d) smooth roll crusher

[2014 : 1 M]

2.13 Which of the following is the correct sequence of equipment for size reduction of solids?

(a) Fluid energy mill → Jaw crusher → Hammer mill

(b) Hammer mill → Jaw crusher → Fluid energy mill

(c) Fluid energy mill → Hammer mill → Jaw crusher

(d) Jaw crusher → Hammer mill → Fluid energy mill

[2017 : 1 M]

2.14 Critical speed of a ball mill depends on

(a) the radius of the mill (shell) and the radius of the particles
(b) the radius of the mill (shell) and the density of the particles
(c) the radius of the balls and the radius of the particles
(d) the radius of the balls and the radius of the mill (shell)

[2018 : 1 M]

2.15 In a roll crusher, rolls of diameter 1 m each are set in such a manner that minimum clearance between the crushing surfaces is 15 mm. If the angle of nip is 31°, the maximum diameter of the particle (in mm) which can be crushed is ________ (rounded off to third decimal place).

[2018 : 2 M]

■■■■

Answers Size Reduction Laws and Equipments

2.1	(c)	**2.2**	(a)	**2.3**	(d)	**2.4**	(d)	**2.5**	(a)	**2.6**	(c)	**2.7**	(c)
2.8	(b)	**2.9**	(c)	**2.10**	(d)	**2.11**	(71)	**2.12**	(a)	**2.13**	(d)	**2.14**	(d)
2.15	(53.3)												

Explanations Size Reduction Laws and Equipments

2.1 (c)

From Bond's law:

$$\frac{P}{\dot{m}} = \frac{k_b}{\sqrt{D_p}}$$

where, d_p = Product size

$$\frac{\left(\frac{P}{\dot{m}}\right)_2}{\left(\frac{P}{\dot{m}}\right)_1} = \sqrt{\frac{D_{P_1}}{D_{P_2}}} = \sqrt{\frac{100}{50}}$$

$$\therefore \quad \left(\frac{P}{\dot{m}}\right)_2 = 12.7\sqrt{2}$$

$$= 17.96 \cong 18 \text{ kWh/tonne}$$

2.2 (a)

The correct sequence of size reduction is
Jaw crusher → Ball mills → Fluid energy mills

2.3 (d)

Critical speed for ball mill is given as

$$N_c = \frac{1}{2\pi}\sqrt{\frac{g}{R-r}}$$

$$= \frac{1}{2\pi}\sqrt{\frac{9.81}{0.6-35\times10^{-3}}} = 0.663 \text{ rps}$$

2.4 (d)

Fluid energy mills are used for producing very fine particles (1 – 20 μm) from large particles $\left(\frac{1}{2}\text{ inch}\right)$. As the energy required per unit mass of feed is proportional to the new surface area produced, large amount of energy is consumed by the fluid energy mills.

2.5 (a)

Critical speed of the ball mill is given as y,

$$N_c = \frac{1}{2\pi}\sqrt{\frac{g}{R-r}}$$

$$\therefore \quad N_c \propto \frac{1}{\sqrt{R-r}}$$

2.6 (c)

From Bond's law:

$$\frac{P}{\dot{m}} = k_b\left[\frac{1}{\sqrt{D_P}} - \frac{1}{\sqrt{D_F}}\right]$$

$$\frac{P}{\dot{m}_1} = k_b\left[\frac{1}{\sqrt{1}} - \frac{1}{\sqrt{10}}\right]$$

$$\frac{P}{\dot{m}_2} = k_b\left[\frac{1}{\sqrt{0.5}} - \frac{1}{\sqrt{10}}\right]$$

$$\therefore \quad \frac{\dot{m}_2}{\dot{m}_1} = \frac{0.684}{1.098} = 0.623$$

$$\therefore \quad \dot{m}_2 = (0.623)\times(1000) = 623 \text{ kg/h}$$

2.7 (c)

Both the configuration consume the same power.

2.8 (b)

Size reduction of coarse hard solids crushers utilizes the compressive force between two solid surface.

2.9 (c)

Energy required in size reduction is proportional to the surface energy of the material.

2.10 (d)

Critical speed is given as

$$N_c = \frac{1}{2\pi}\sqrt{\frac{g}{R-r}}$$

where, N_c = Critical speed
r = Radius of ball
R = Radius of ball mill

2.11 (70.7)

Given, $\dot{m}$ = 100 ton/hr

Fond Bond law:

$$\frac{P}{\dot{m}} = 0.3162 w_i \left[\frac{1}{\sqrt{D_P}} - \frac{1}{\sqrt{D_F}}\right]$$

Case-I:

D_P = 1.27 mm, D_F = 2.54 mm

Case-II:

D_P = 2.54 mm, D_F = 5.08 mm

$$\frac{P}{\dot{m}} \propto \frac{1}{\sqrt{D_p}} - \frac{1}{\sqrt{D_F}}$$

$$\frac{P_2}{P_1} = \frac{\frac{1}{\sqrt{2.54}} - \frac{1}{\sqrt{5.08}}}{\frac{1}{\sqrt{1.27}} - \frac{1}{\sqrt{2.54}}} = 0.7071 = \frac{P_2}{100}$$

$\because$ (P_1 = 100) thus P_2 = 70.71 kW ≅ 71 kW

2.12 (a)

The appropriate size reducing equipment to produce fine particles between 5 to 10 μm is fluid energy mill.

2.13 (d)

The correct sequence of size reduction are:
Jaw crusher → Hammer mill → Fluid energy mill

2.14 (d)

Critical speed of a ball mill is given by,

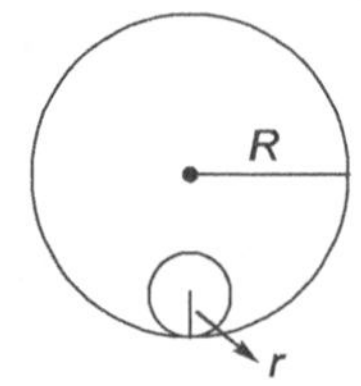

$$N_c = \frac{1}{2\pi}\sqrt{\frac{g}{R-r}}$$

$$N_c \propto \frac{1}{\sqrt{R-r}}$$

R = Radius of shell

r = Radius of balls

Hence, N_c depends on R and r respectively.

2.15 (53.3)

For a roll crusher:

$$\cos\left(\frac{\alpha}{2}\right) = \frac{D + D_P}{D + D_F}$$

α = Angle of Nip = 31°

D = Rolls diameter = 1 m

D_P = 0.015 m

D_F = ?

$$\cos[15.5] = \frac{1 + 0.015}{1 + D_F}$$

D_F = 53.31 mm

■■■■

3 Sedimentation and Fluidization

3.1 For a sphere falling in the constant drag coefficient regime, its terminal velocity depends on its diameter (d) as

(a) d (b) $\sqrt{d}$
(c) d^2 (d) $\frac{1}{d}$

[2000 : 1 M]

3.2 A 30% (by volume) suspension of spherical sand particles in a viscous oil has hindered settling velocity of 4.44 μm/s. If the Richardson-Zaki hindered setting index is 4.5, then the terminal velocity of a sand grain is

(a) 0.90 μm/s (b) 1 mm/s
(c) 22.1 μm/s (d) 0.02 μm/s

[2000 : 2 M]

3.3 A 1 m high bed made up of a mm particles is to be fluidised by an oil (density 900 kg/m^3; viscosity 0.01 Pa-s). If at the point of incipient fluidization, the bed voidage is 39% and the pressure drop across the bed is 10 kPa, then the density of particles is

(a) 2574 kg/m^3 (b) 3514 kg/m^3
(c) 4000 kg/m^3 (d) 4350 kg/m^3

[2000 : 2 M]

3.4 The drag coefficient for a bacterium moving in water at 1 mm/s, will be of the following order of magnitude (assume size of the bacterium to be 1 micron and kinematic viscosity of water to be 10^{-6} m^2/s)

(a) 24000 (b) 24
(c) 0.24 (d) 0.44

[2002 : 2 M]

3.5 A gas (density = 1.5 kg/m^3, viscosity = 2×10^{-5} kg/m.s) flowing through a packed bed (particle size = 0.5 cm, porosity = 0.5) at a superficial velocity of 2 m/s causes a pressure drop of 8400 Pa/m. The pressure drop for another gas, with density of 1.5 kg/m^3 and viscosity of 3×10^{-5} kg/m s, flowing at 3 m/s will be

(a) 8400 Pa/m (b) 18900 Pa/m
(c) 12600 Pa/m (d) 16800 Pa/m

[2002 : 2 M]

Q.3.6 - Q.3.7 are based on the data supplied in the paragraph below:

A bed of spherical particles (sp. gravity 2.5) of uniform size 1500 μm is 0.5 m in diameter and 0.5 m high. In the packed bed state, the porosity may be taken as 0.4. Ergun's equation for the above particle-fluid system (in SI units) is given below:

$$\frac{\Delta P}{L} = 375 \times 10^3 V_{OM} + 10.94 \times 10^6 V_{OM}^2 \text{ (SI Units)}$$

3.6 If water is to be used as the fluidizing medium, the minimum fluidization velocity, V_{OM} is

(a) 12 mm/s (b) 16 mm/s
(c) 24 mm/s (d) 28 mm/s

[2003 : 2 M]

3.7 In actual operation, the above bed has a height = 1 m. What is the porosity of the fluidized bed?

(a) 0.2 (b) 0.5
(c) 0.7 (d) 0.8

[2003 : 2 M]

3.8 Match the systems in **Group-I** with equipment used to separate them in **Group-II**.

Group-I
A. Gas–solid
B. Liquid–liquid

Group-II
1. Filter press
2. Cyclone
3. Decanter
4. Thickener

Codes:

	A	B
(a)	1	2
(b)	2	3
(c)	3	4
(d)	4	1

[2004 : 1 M]

3.9 For a cyclone of diameter 0.2 m with a tangential velocity of 15 m/s at the wall, the separation factor is

(a) 2250 (b) 1125
(c) 460 (d) 230

[2004 : 1 M]

3.10 For a particle settling in water at its terminal settling velocity, which of the following is true?

(a) buoyancy = weight + drag
(b) weight = buoyancy + drag
(c) drag = buoyancy + weight
(d) drag = weight

[2004 : 1 M]

3.11 A centrifugal filtration unit operating at a rotational speed of ω has inner surface of the liquid (density ρ_L) located at a radial distance R from the axis of rotation. The thickness of the liquid film is δ and no cake is formed. The initial pressure drop during filtration is

(a) $\frac{1}{2}\omega^2 R^2 \rho_L$ (b) $\frac{1}{2}\omega^2 \delta^2 \rho_L$
(c) $\frac{1}{2}\omega^2 \delta \rho_L (2R+\delta)$ (d) $\frac{1}{2}\omega^2 R \rho_L (R+2\delta)$

[2004 : 2 M]

3.12 U_{mf} is the minimum fluidisation velocity for a bed of particles. An increase in the superficial gas velocity from $2U_{mf}$ to $2.5U_{mf}$ results in (all velocities are smaller than the entrainment velocity of the particles) no change in

(a) drag on particles (b) drag on column walls
(c) the bed height (d) the bed void age

[2004 : 2 M]

3.13 The Kozeny-Carman equation, rewritten in terms of non-dimensional numbers, gives $\left(\frac{\Delta P}{\rho u^2}\right)$ proportional to

(a) $\frac{\left(\frac{L}{D_P}\right)}{\text{Re}}$ (b) $\frac{\text{Re}}{\left(\frac{D_P}{L}\right)}$
(c) $\frac{\left(\frac{L}{D_P}\right)}{\text{Re}^2}$ (d) $\frac{\text{Re}^2}{\left(\frac{D_P}{L}\right)}$

[2004 : 2 M]

3.14 A centrifuge of diameter 0.2 m in a pilot plant rotates at a speed of 50 Hz in order to achieve effective separation. If this centrifuge is scaled up to a diameter of 1 m in the chemical plant, and the same separation factor is to be achieved, what is the rotational speed of the scaled up centrifuge?

(a) 15 Hz (b) 22.36 Hz
(c) 30 Hz (d) 44.72 Hz

[2005 : 2 M]

3.15 What is the terminal velocity in m/s, calculated from Stokes law, for a particle of diameter 0.1×10^{-3} m, density 2800 kg/m^3 settling in water of density 1000 kg/m^3 and viscosity 10^{-3} kg/m.s? (Assume $g = 10$ m/s^2)

(a) 2×10^{-2} (b) 4×10^{-3}
(c) 10^{-2} (d) 8×10^{-3}

[2005 : 2 M]

3.16 A bed fluidized by water is used for cleaning sand contaminated with salt. The particles of sand and salt have the same shape and size but different densities (ρ_{sand} = 2500 kg/m^3 and ρ_{salt} = 2000 kg/m^3). If the initial volume fraction of the salt in the mixture is 0.3 and if the initial value of the minimum fluidization velocity (U_{mf}) is 0.9 m/s, find the final value of the U_{mf} (in m/s) when the sand is washed free of the salt. Assume that the bed characteristics (bed porosity and solid surface area per unit volume) do not change during the operation and that the pressure drop per unit length is directly proportional to the fluid velocity

(a) 0.70 (b) 0.90
(c) 1.00 (d) 1.46

[2006 : 2 M]

3.17 Two spherical particles have the same outer diameter but are made of different materials The first one (with material density ρ_1) is solid, whereas the second (with material density ρ_2) is a hollow sphere with the inner shell diameter equal to half the outer diameter. If both the spheres have the same terminal velocity in any fluid, then the ratio of their material densities, $\frac{\rho_2}{\rho_1}$, is

(a) 1 (b) $\frac{8}{7}$
(c) 2 (d) 8

[2006 : 2 M]

3.18 In the Stokes regime, the terminal velocity of particles for centrifugal sedimentation is given by

$$U_t = \frac{\omega^2 r(\rho_p - \rho)d_p^2}{18\mu}$$

where, ω : angular velocity; r : distance of the particle from the axis of rotation; ρ_p : density of the particle; ρ : density of the fluid; d_p : diameter of the particle and μ : viscosity of the fluid.

In a Bowl centrifugal classifier operating at 60 rpm with water (μ = 0.001 kg/m.s), the time taken for a particle (d_p = 0.0001 m, sp.gr = 2.5) in seconds to traverse a distance of 0.05 m from the liquid surface is

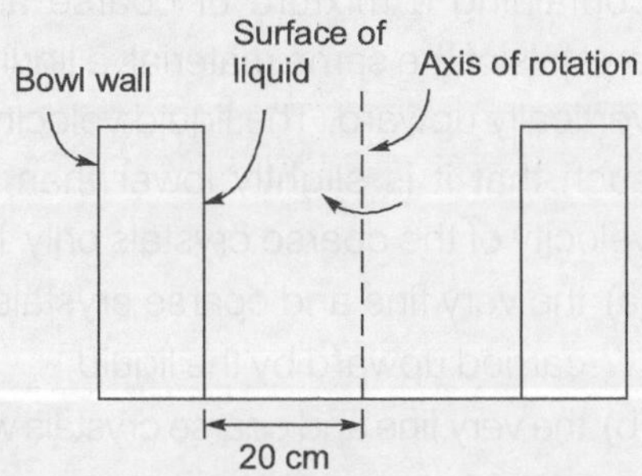

(a) 4.8 (b) 5.8
(c) 6.8 (d) 7.8

[2007 : 2 M]

Statement for Linked Answer Question (3.19 & 3.20):

3.19 A fluidized bed (0.5 m dia, 0.5 high) of spherical particles (diameter = 2000 μm, specific gravity = 2.5) uses water as the medium. The porosity of the bed is 0.4. The Ergun equation for the system is

$$\frac{\Delta P}{L} = 4 \times 10^5\, U_{mf} + 1 \times 10^7\, U_{mf}^{\,2}$$

(SI unit, U_{mf} in m/s).

$\frac{\Delta P}{L}$ (SI unit) at minimum fluidization condition is

($g = 9.8$ m/sec^2)

(a) 900 (b) 8820
(c) 12400 (d) 17640

[2007 : 2 M]

3.20 The minimum fluidization velocity (mm/sec) is

(a) 12.8 (b) 15.8
(c) 24.8 (d) 28.8

[2007 : 2 M]

3.21 Two identically sized spherical particles A and B having densities ρ_A and ρ_B, respectively, are settling in a fluid of density ρ. Assuming free settling under turbulent flow conditions, the ratio of the terminal settling velocity of particle A to that of particle B is given by

(a) $\sqrt{\frac{(\rho_A - \rho)}{(\rho_B - \rho)}}$ (b) $\sqrt{\frac{(\rho_B - \rho)}{(\rho_A - \rho)}}$

(c) $\frac{(\rho_A - \rho)}{(\rho_B - \rho)}$ (d) $\frac{(\rho_B - \rho)}{(\rho_A - \rho)}$

[2008 : 2 M]

3.22 Under fully turbulent flow conditions, the frictional pressure drop across a packed bed varies with the superficial velocity (V) of the fluid as

(a) V^{-1} (b) V
(c) $V^{3/2}$ (d) V^2

[2009 : 1 M]

3.23 The terminal settling velocity of a 6 mm diameter glass sphere (density: 2500 kg/m^3) in a viscous Newtonian liquid (density : 1500 kg/m^3) is 100 μm/s. If the particle Reynolds number is small and the value of acceleration due to gravity is 9.81 m/s^2, then the viscosity of the liquid (in Pa.s) is

(a) 100 (b) 196.2
(c) 245.3 (d) 490.5

[2009 : 2 M]

3.24 The height of a fluidized bed at incipient fluidization is 0.075 m, and the corresponding voidage is 0.38. If the voidage of the bed increases to 0.5, then the height of the bed would be

(a) 0.058 m (b) 0.061 m
(c) 0.075 m (d) 0.093 m

[2010 : 1 M]

Common Data for Questions 3.25 and 3.26:

For a liquid flowing through a packed bed, the pressure drop per unit length of the bed $\left(\frac{\Delta P}{L}\right)$ is given as

$$\frac{\Delta P}{L} = \frac{150\mu_f \overline{V}_0}{(\phi_s \overline{d}_p)^2}\frac{(1-\varepsilon)^2}{\varepsilon^3} + \frac{1.75\rho_f \overline{V}_0^2 (1-\varepsilon)}{\phi_s \overline{d}_p \varepsilon^3}$$

where $\bar{V}_0$ is the superficial liquid velocity, ε is the bed porosity, $\bar{d}_p$ is average particle size, ϕ_s is particle sphericity, ρ_f is liquid density and μ_f is liquid viscosity.

Given data: $\bar{d}_p = 1 \times 10^{-3}$ m, $\phi_s = 0.8$, $\rho_f = 1000$ kg/m^3, $\mu_f = 1 \times 10^{-3}$ kg m^{-1} s^{-1}, particle density $\rho_p = 2500$ kg/m^3 and acceleration due to gravity, $g = 9.8$ m/s^2.

3.25 When $\bar{V}_0$ is 0.005 m/s and $\varepsilon = 0.5$, which one of the following is the correct value for the ratio of the viscous loss to the kinetic energy loss?

(a) 0.09 (b) 1.07
(c) 10.71 (d) 93

[2011 : 2 M]

3.26 On further increasing $\bar{V}_0$, incipient fluidisation is achieved. Assuming that the porosity of the bed remains unaltered, the pressure drop per unit length (in Pa/m) under incipient fluidisation condition is

(a) 3675 (b) 7350
(c) 14700 (d) 73501

[2011 : 2 M]

3.27 The particle size distributions of the feed and collected solids (sampled for same duration) for a gas cyclone are given below:

Size range (μm)	1-5	5-10	10-15	15-20	20-25	25-30
Weight of feed in the size range (g)	2.0	3.0	5.0	6.0	3.0	1.0
Weight of collected solids in the size range (g)	0.1	0.7	3.6	5.5	2.9	1.0

What is the collection efficiency (in percentage) of the gas cyclone?

(a) 31 (b) 60
(c) 65 (d) 69

[2011 : 2 M]

3.28 A bed of spherical glass beads (density 3000 kg/m^3, diameter 1 mm, bed porosity 0.5) is to be fluidized by a liquid of density 1000 kg/m^3 and viscosity 0.1 Pa.s. Assume that the Reynolds number based on particle diameter is very small compared to one. If $g = 10$ m/s^2, then the minimum velocity (in m/s) required to fluidize the bed is

(a) 3.33×10^{-4} (b) 3.33×10^{-1}
(c) 3 (d) 30

[2012 : 2 M]

3.29 Taking the acceleration due to gravity to be 10 m/s^2, the separation factor of a cyclone 0.5 m in diameter and having a tangential velocity of 20 m/s near the wall is ________.

[2013 : 1 M]

3.30 In the elutriation leg of a commercial crystallizer containing a mixture of coarse and very fine crystals of the same material, a liquid is pumped vertically upward. The liquid velocity is adjusted such that it is slightly lower than the terminal velocity of the coarse crystals only. Hence

(a) the very fine and coarse crystals will both be carried upward by the liquid
(b) the very fine and coarse crystals will both settle at the bottom of the tube
(c) the very fine crystals will be carried upward and the coarse crystals will settle
(d) the coarse crystals will be carried upward and the very fine crystals will settle

[2013 : 2 M]

3.31 A cylindrical packed bed of height 1 m is filled with equal sized spherical particles. The particles are nonporous and have a density of 1500 kg/m^3. The void fraction of the bed is 0.45. The bed is fluidized using air (density 1 kg/m^3). If the acceleration due to gravity is 9.8 m/s^2, the pressure drop (in Pa) across the bed at incipient fluidization (up to one decimal place) is ______.

[2015 : 1 M]

3.32 A spherical solid particle of 1 mm diameter is falling with a downward velocity of 1.7 mm/s through a liquid (viscosity 0.04 Pa.s) at a low Reynolds number (Stokes regime). The liquid is flowing upward at a velocity of 1 mm/s. All velocities are with respect to a stationary reference frame. Neglecting the wall effects, the drag force per unit projected area of the particle, in Pa, (up to two decimal places) is ________.

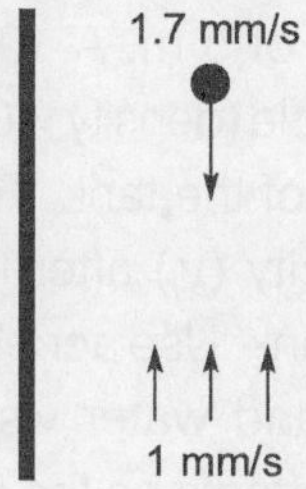

[2015 : 2 M]

3.33 In a cyclone separator used for separation of solid particles from a dust laden gas, the separation factor is defined as the ratio of the centrifugal force to the gravitational force acting on the particle. S_r denotes the separation factor at a location (near the wall) that is at a radial distance r from the centre of the cyclone. Which one of the following statements is INCORRECT?
(a) S_r depends on mass of the particle
(b) S_r depends on the acceleration due to gravity
(c) S_r depends on tangential velocity of the particle
(d) S_r depends on the radial location (r) of the particle

[2016 : 1 M]

3.34 Consider a rigid solid sphere falling with a constant velocity in a fluid. The following data are known at the conditions of interest: viscosity of the fluid = 0.1 Pa s, acceleration due to gravity = 10 m s^{-2}, density of the particle = 1180 kg m^{-3} and density of the fluid = 1000 kg m^{-3}. The diameter (in mm, rounded off to the second decimal place) of the largest sphere that settles in the Stokes's law regime (Reynolds number $\leq$ 0.1), is ______ .

[2016 : 2 M]

3.35 A gas bubble (gas density ρ_g = 2 kg/m^3 ; bubble diameter D = 10^{-4} m) is rising vertically through water (density ρ = 1000 kg/m^3 ; viscosity μ = 0.001 Pa.s). Force balance on the bubble leads to the following equation,

$$\frac{dv}{dt} = -g\frac{\rho_g - \rho}{\rho_g} - \frac{18\mu}{\rho_g D^2}v$$

where v is the velocity of the bubble at any given time t. Assume that the volume of the rising bubble does not change. The value of g = 9.81 m/s^2. The terminal rising velocity of the bubble (in cm/s). Rounded to 2 decimal places, is ______ cm/s.

[2017 : 1 M]

3.36 The terminal velocity of a spherical particle in gravitational settling under Stokes' regime varies
(a) linearly with the particle diameter
(b) linearly with the viscosity of the liquid
(c) directly with the square of particle diameter
(d) inversely with the density of particle

[2018 : 1 M]

3.37 Match the equipment in **Column A** with the corresponding process in **Column B**:

Column A	Column B
(P) Centrifugal sifter	(1) Mixing
(Q) Bowl mill	(2) Sedimentation
(R) Gravity thickner	(3) Screening
(S) Two-arm kneader	(4) Grinding

(a) P-1, Q-4, R-2, S-3 (b) P-3, Q-4, R-2, S-1
(c) P-4, Q-1, R-2, S-3 (d) P-4, Q-3, R-1, S-2

[2018 : 2 M]

3.38 A rigid spherical particle undergoes free settling in a liquid of density 750 kg m^{-3} and viscosity 9.81 × 10^{-3} Pa.s. Density of the particle is 3000 kg m^{-3} and the particle diameter is 2 × 10^{-4} m. Acceleration due to gravity is 9.81 m/s^{-2}. Assuming Stokes' law to be valid, the terminal settling velocity (in m/s) of the particle is
(a) 2 × 10^{-3} (b) 3 × 10^{-3}
(c) 4 × 10^{-3} (d) 5 × 10^{-3}

[2020 : 1 M]

3.39 A vertically held packed bed has a height of 1 m, and a void fraction of 0.1, when there is no flow through the bed. The incipient (minimum) fluidization is set in by injection of a fluid of density 1 kg m^{-3}. The particle density (ρ_p) of the solids is 3000 kg m^{-3}. Acceleration due to gravity is 9.81 m s^{-2}. The pressure drop (in Pa) across the height of the bed is _____ (rounded off to nearest integer).

[2020 : 2 M]

3.40 A batch settling experiment is performed in a long column using a dilute dispersion containing equal number of particles of type A and type B in water (density 1000 kg m^{-3}) at room temperature.
Type A are spherical particles of diameter 30 μm and density 1100 kg m^{-3}.
Type B are spherical particles of diameter 10 μm and density 1900 kg m^{-3}.

Assuming that Stokes' law is valid throughout the duration of the experiment, the settled bed would

(a) consist of a homogeneous mixture of type A and type B particles.
(b) consist of type B particles only.
(c) be completely segregated with type B particles on top of type A particles.
(d) be completely segregated with type A particles on top of type B particles.

[2021 : 1 M]

3.41 Sea water is passed through a column containing a bed of resin beads.

Density of sea water = 1025 kg m^{-3}
Density of resin beads = 1330 kg m^{-3}
Diameter of resin beads = 50 μm
Void fraction of the bed at the onset of fluidizatoin = 0.4
Acceleration due to gravity = 9.81 m s^{-2}

The pressure drop per unit length of the bed at the onset of fluidization is _____ Pa m^{-1} (round off to nearest integer). [2021 : 2 M]

3.42 For a packed-bed comprising of uniform-sized spherical particles of diameter D_p, the pressure drop across the bed is given by the Kozeny-Carman equation when the particle Reynolds number $Re_p < 1$. Under this condition, minimum fluidization velocity is proportional to D_p^n. Which one of the following is the CORRECT value of exponent n?

(a) 2 (b) −1
(c) −2 (d) 1 [2023 : 1 M]

3.43 A large tank is filled with water (density = 1 $g.cm^{-3}$) upto a height of 5 m. A 100 μm diameter solid spherical particle (density = 0.8 $g.cm^{-3}$) is released at the bottom of the tank. The particle attains its terminal velocity (v_t) after traveling to a certain height in the tank. Use acceleration due to gravity as 10 $m.s^{-2}$ and water viscosity as 10^{-3} Pa.s. Neglect wall effects on the particle. If Stokes law is applicable, the absolute value of v_t (in $mm.s^{-1}$) is _____ (rounded off to two decimal places).

[2023 : 2 M]

3.44 A metallic spherical particle of density 7001 kg m^{-3} and diameter 1 mm is settling steadily due to gravity in a stagnant gas of density 1 kg m^{-3} and viscosity 10^{-5} kg m^{-1} s^{-1}. Take g = 9.8 ms^{-2}. Assume that the settling occurs in the regime where the drag coefficient C_D is independent of the Reynolds number, and equals 0.44. The terminal settling velocity of the particle, in m s^{-1}, rounded off to 2 decimal places, is _______.

[2024 : 2 M]

■■■■

Answers Sedimentation and Fluidization

3.1	(b)	**3.2**	(c)	**3.3**	(a)	**3.4**	(a)	**3.5**	(b)	**3.6**	(b)	**3.7**	(c)
3.8	(b)	**3.9**	(d)	**3.10**	(b)	**3.11**	(c)	**3.12**	(c)	**3.13**	(a)	**3.14**	(b)
3.15	(c)	**3.16**	(c)	**3.17**	(b)	**3.18**	(d)	**3.19**	(b)	**3.20**	(b)	**3.21**	(a)
3.22	(d)	**3.23**	(b)	**3.24**	(d)	**3.25**	(c)	**3.26**	(b)	**3.27**	(d)	**3.28**	(a)
3.29	(160)	**3.30**	(c)	**3.31**	(8079.6)	**3.32**	(1.296)	**3.33**	(a)	**3.34**	(2.15)	**3.35**	(0.543)
3.36	(c)	**3.37**	(b)	**3.38**	(d)	**3.39**	(26478.17)	**3.40**	(a)	**3.41**	(1795)	**3.42**	(a)
3.43	(1.11)	**3.44**	(14.42)										

Explanations Sedimentation and Fluidization

3.1 (b)

For constant drag regime, i.e. $C_D = 0.44$
[Turbulent regime will be valid]
Terminal velocity is given by

$$V_t = \sqrt{\frac{3.03 d_p(\rho_p - \rho_f)\cdot g}{\rho_f}}$$

$$V_t \propto \sqrt{D_P}$$

3.2 (c)

Given, $\in = 1 - 0.3 = 0.7$

$$\because \quad \in = \frac{\text{Volume of void}}{\text{Total volume}}$$

For hindered velocity,

$$u_h = V_t \cdot \in^k$$

$\downarrow \quad \downarrow$

(Hindered velocity) (Terminal velocity)

$$4.44 = V_t \times (0.7)^{4.5}$$

$$V_t = 22.1\ \mu\text{m/s}$$

3.3 (a)

For incipient fluidization

$$\frac{\Delta P}{L} = g(1-\in)(\rho_p - \rho_f)$$

$$\Rightarrow \quad \frac{10^4}{1} = (1-0.39)(\rho_p - 900)\, 9.8$$

$$\rho_p = 2572.80\ \text{kg/m}^3 \text{ close (A)}$$

3.4 (a)

Size of bacterium = 1 μm
"Re number is very small i.e. stokes law will be valid".

$$d_p = 10^{-6}\ \text{m},\ V = 10^{-3}\ \text{m/s}$$

$$\nu = 10^{-6}\ \text{m}^2/\text{s}$$

$$\therefore \quad (R_{ep}) = \frac{d\cdot u}{\nu} = \frac{10^{-6}\times 10^{-3}}{10^{-6}} = 10^{-3}$$

For low Reynold's number

$$\Rightarrow \quad C_D = \frac{24}{R_{ep}}$$

$$C_D = \frac{24}{10^{-3}} = 24000$$

3.5 (b)

Check Reynolds number

Case I:

$$\text{Re}_\text{I} = \frac{1.5\times 2\times 0.5\times 10^{-2}}{2\times 10^{-5}\times(1-0.5)} = 1500$$

Case II:

$$\text{Re}_\text{II} = \frac{1.5\times 3\times 0.5\times 10^{-2}}{3\times 10^{-5}\times 1-0.5} = 1500$$

As Reynold's number > 1000, so Burke-Plummer equation is applicable

$$\left(\frac{\Delta P}{L}\right) = \frac{1.75\rho_f V_0^2(1-\in)}{(\phi_s d_p)\in^3}$$

$$\left(\frac{\Delta P}{L}\right) \propto V_0^2$$

$$\frac{\left(\frac{\Delta P}{L}\right)_1}{\left(\frac{\Delta P}{L}\right)_2} = \frac{(V_0^2)_1}{(V_0^2)_2} = \left(\frac{2}{3}\right)^2 = \frac{8400}{\left(\frac{\Delta P}{L}\right)_2}$$

$$\left(\frac{\Delta P}{L}\right)_2 = \frac{9}{4}\times 8400 = 18900\ \text{Pa./m}$$

3.6 (b)

At minimum fluidization velocity

$$\left(\frac{\Delta P}{L}\right) = (1-\epsilon)(\rho_p - \rho_f)g$$

Given, $$\left(\frac{\Delta P}{L}\right) = 375\times10^3 V_{0m} + 10.94\times10^6 V_{0m}^2$$

$$(1-\epsilon)(\rho_p-\rho_f)g = 375\times10^3 V_{0m} + 10.94\times10^6 V_{0m}^2$$

$\epsilon = 0.4$, $\rho_p = 2.5\times1000 = 2500\ \text{kg/m}^3$

$\rho = 1000\ \text{kg/m}^3$, $g = 9.8$

$$(1-0.4)(1500)9.8 = 375\times10^3 V_{0m} + 10.94\times10^6 V_{0m}^2$$

Solving, $V_{0m} = 0.01604$ m/s $= 16.04$ mm/s

3.7 (c)

For fluidization,

$$\left(\frac{\Delta P}{L}\right) = (1-\epsilon)g(\rho_p-\rho_f)$$

$$\frac{1-\epsilon_2}{1-\epsilon_1} = \frac{L_1}{L_2}$$

$\because$ $\epsilon_1 = 0.4$, $L_1 = 0.5$ m, $L_2 = 1$ m

$\therefore$ $$\frac{1-\epsilon_2}{1-0.4} = \frac{0.5}{1}$$

$\Rightarrow$ $\epsilon_2 = 0.7$

3.8 (b)

"Liquid-liquid separation is done in decanter and gas-solid separation is done in cyclone".

3.9 (d)

Separation factor:

$\Rightarrow$ $$S = \frac{V_0^2}{r\cdot g}$$

$$S = \frac{(15)^2}{(0.1)(9.81)} = 229.36 \cong 230$$

3.10 (b)

For particle setting, in water

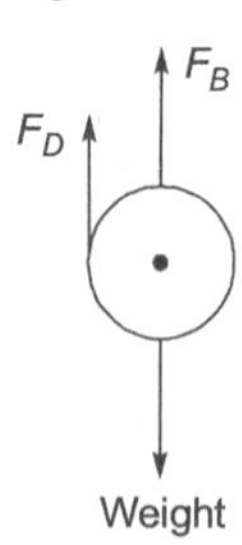

i.e. Weight = Buoyancy + Drag force

3.11 (c)

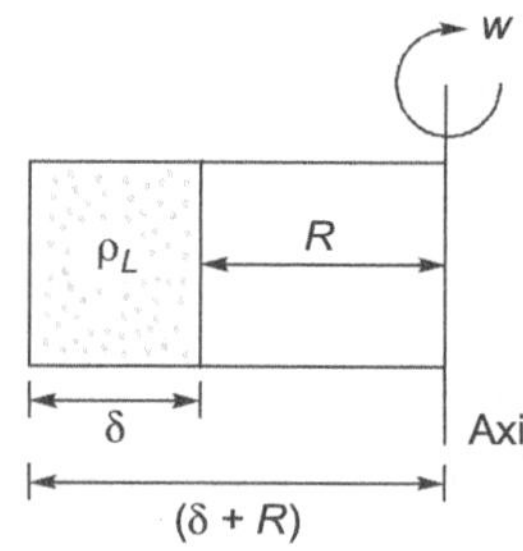

As we know,

$$\Delta P = \frac{\rho(v_2^2 - v_1^2)}{2}$$

$\because$ $v = \omega r$

$$\Delta P = \frac{\rho\omega^2(r_2^2 - r_1^2)}{2}$$

$$r_2 = r_1 + \delta = R + \delta$$
$$r_1 = r_1 = R$$

$$\Delta P = \frac{1}{2}\rho_L\omega^2\left[(\delta+R)^2 - R^2\right]$$

$$= \frac{1}{2}\rho_L\omega^2 R(\delta + 2R)$$

3.12 (c)

"With increase in superficial velocity the bed height remains same but the pressure drop increase".

3.13 (a)

From **Kozeny-Carman** equation:

$$\left(\frac{\Delta P}{L}\right) = \frac{150\mu_f V_0}{(\phi_s d_p)^2}\frac{(1-\epsilon)^2}{\epsilon^3}$$

$$\left(\frac{\Delta P}{L}\right) \propto \frac{\bar{V}_0\mu_f}{D_P^2}$$

$$\frac{\Delta P}{\rho u^2} \propto \frac{\mu\bar{V}_0 L}{D_P^2\rho u^2} \propto \left(\frac{\mu}{\rho v D_P}\right)\left(\frac{L}{D_P}\right)$$

$$\frac{\Delta P}{\rho u^2} \propto \frac{\frac{L}{D_P}}{\text{Re}}$$

3.14 (b)

Separation factor:

$\Rightarrow$ $$S = \frac{V^2}{Rg}$$

$\because$ $[V = \omega . r]$

$$S = \frac{\omega^2 R^2}{Rg} = \frac{\omega^2 R}{g}$$

As same separation factor

$$(50)^2 \times \frac{0.2}{2} = (\omega)^2 \times \left(\frac{1}{2}\right)$$

$$\omega = 22.36 \text{ Hz}$$

3.15 (c)

Terminal velocity under Stokes regime is given by

$$V_t = \frac{d_p^2 g(\rho_p - \rho_f)}{18\mu_f}$$

Putting values, we get

$$\therefore \quad V_t = \frac{(0.1\times10^{-3})\times10\times(2800-1000)}{18\times10^{-3}}$$

$$V_t = 10^{-2} \text{ m/s}$$

3.16 (c)

Pressure drop at the start of fluidization can be obtained as

Pressure drop × Cross sectional area = Weight of the bed – Buoyant force

Initially [sand is 70% and salt is 30%]

$$\Delta P_1 \times A = L\,A(1-\epsilon)g\,[\rho_{sand} \times 0.7 + \rho_{salt} \times 0.30] - L\,A(1-\epsilon)g\,\rho_{water}$$

$$\rho_{sand} = 2500 \text{ kg/m}^3, \rho_{salt} = 2000 \text{ kg/m}^3,$$

$$\rho_{water} = 1000 \text{ kg/m}^3$$

$$\therefore \quad \frac{\Delta P_1}{L} = (1-\epsilon)g \times 1350 \qquad \text{...(i)}$$

At final condition (No salt is there)

$$\Delta P_2 \times A = L\,A(1-\epsilon)g\,\rho_{sand} - L\,A(1-\epsilon)g\,\rho_{water}$$

$$\frac{\Delta P_2}{L} = (1-\epsilon)g \times 1500 \qquad \text{...(ii)}$$

$$\therefore \quad \frac{\Delta P_1}{\Delta P_2} = 0.9$$

$$\because \quad \Delta P \propto U_m$$

$$\therefore \quad \frac{U_{m1}}{U_{m2}} = 0.9$$

$$\because \quad U_{m1} = 0.9 \text{ m/s}$$

$$\text{Thus,} \quad U_{m2} = 1 \text{ m/s}$$

3.17 (b)

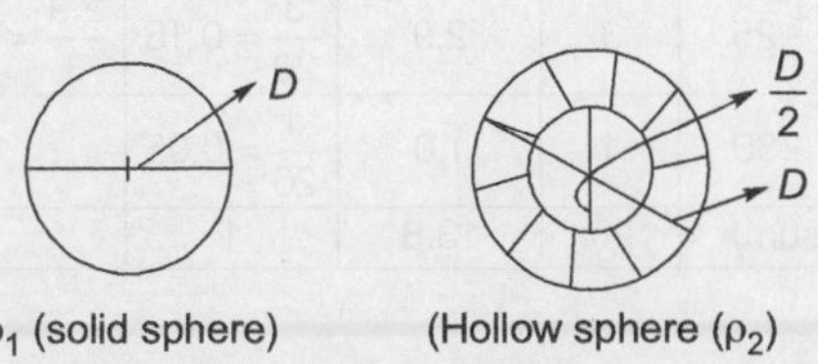

ρ_1 (solid sphere) (Hollow sphere (ρ_2))

$$\text{Now,} \quad V_t = \sqrt{\frac{2g(\rho_p - \rho_f)m}{A C_D \rho_f \rho_p}}$$

∵ Outer diameter is same, we assume the significant area A and C_D are same for both spheres.

$$\text{Also,} \quad \frac{\rho_p - \rho_f}{\rho_p} = \frac{\rho_p}{\rho_p} \cong 1$$

∴ Fluid density is assumed to be much smaller than ρ_p.

$$\therefore \quad u_t \propto \sqrt{m}$$

$$\frac{U_{t1}}{U_{t2}} = \sqrt{\frac{m_1}{m_2}}$$

$$\because \quad U_{t1} = U_{t2}$$

$$\therefore \quad m_1 = m_2$$

$$\frac{\pi}{6}D^3\rho_1 = \frac{\pi}{6}D^3\left(\frac{7}{8}\right)\rho_2$$

$$\therefore \quad \left(\frac{\rho_2}{\rho_1}\right) = \frac{8}{7}$$

3.18 (d)

Given,

$$U_t = \frac{\omega^2 r(\rho_p - \rho)d_p^2}{18\mu}$$

$$\because \quad \omega = 60 \text{ rpm}$$

$$N = \frac{2\pi\times60}{60} = 2\pi \text{ rad/s}$$

$$r = 0.2 \text{ m}$$

$$\rho_p = 2500 \text{ kg/m}^3, \rho = 1000 \text{ kg/m}^3$$

$$d_p = 0.0001 \text{ m}, \mu = 0.001 \text{ kg/m.s}$$

$$V_t = \frac{(6.294)^2\times0.2\times1500\times(10^{-4})^2}{18\times0.001}$$

$$V_t = 6.602\times10^{-3} \text{ m/s}$$

Time taken for a particle to traverse 0.05 m

$$t = \frac{0.05}{6.602\times10^{-3}} = 7.572$$

Close value [7.8] seconds.

3.19 (b)

After minimum fluidization condition,

$$\frac{\Delta P}{L} = g(1-\epsilon)(\rho_p - \rho_f)$$

$$\frac{\Delta P}{L} = 9.81\times(1-0.4)(2.5-1)\times1000$$

$$\frac{\Delta P}{L} = 8829 \text{ (Pa/m)}$$

3.20 (b)

$$\left(\frac{\Delta P}{L}\right)_{min} = 4\times10^5 U_{mf} + 1\times10^7 U_{mf}^2$$

$$8820 = 4\times10^5 U_{mf} + 10^7 U_{mf}^2$$

$$\left[\because \frac{\Delta P}{L} = 8820\right]$$

Solving, we get

$$U_{mf} = 0.0158 \text{ m/s} = 15.8 \text{ mm/sec}$$

3.21 (a)

For turbulent condition,

$$V_t = \sqrt{\frac{\frac{40g(\rho_p-\rho_f)D_P}{3}}{\rho_f}}$$

$\therefore$ For two identically sized particles,

$$\frac{V_{t_A}}{V_{t_B}} = \sqrt{\frac{\rho_A-\rho}{\rho_B-\rho}}$$

3.22 (d)

For turbulent conditions,

$$\frac{\Delta P}{l} = \frac{1.75\rho_f V_0^2(1-\epsilon)}{(\phi_s d_p)\epsilon^3}$$

$$\Delta P \propto V_0^2$$

3.23 (b)

As the particle Reynold's number is small thus Stokes law is valid

$$V_t = \frac{d_p^2(\rho_f-\rho_f)g}{18\mu_f}$$

$$g = 9.81 \text{ m/s}^2$$

$$V_t = 100\times10^{-6} \text{ m/s}$$

$$\rho_p = 2500 \text{ kg/m}^3, \quad \rho_f = 1500 \text{ kg/m}^3$$

Putting values,

$$100\times10^{-6} = \frac{(0.006)^2(2500-1500)9.81}{18\mu_f}$$

$$\mu_f = 196.2 \text{ Pa.s}$$

3.24 (d)

Total volume of solid = $L.A(1-\epsilon)$ = Constant

L = Height of bed

ϵ = Void age

$$L \propto \frac{1}{(1-\epsilon)}$$

$$\frac{L_2}{L_1} = \frac{1-\epsilon_1}{1-\epsilon_2}$$

$$L_2 = L_1\left[\frac{1-\epsilon_1}{1-\epsilon_2}\right]$$

$$L_2 = 0.075\left|\frac{1-0.38}{1-0.5}\right| = 0.093 \text{ m}$$

3.25 (c)

From Ergun equation,

$$\therefore \quad \frac{\Delta P}{l} = \text{Viscous loss + Kinetic energy loss}$$

$$\frac{\text{Viscous loss}}{\text{K.E loss}} = \frac{150\mu_f V_o(1-\epsilon)^2}{(\phi_s d_p)^2\epsilon^3}\times\frac{(\phi_s d_p)\epsilon^3}{1.75\rho_f V_0^2(1-\epsilon)}$$

$$\frac{\text{Viscous loss}}{\text{K.E loss}} = \frac{150\times10^{-3}\times0.5}{1.75\times1000\times0.005\times0.8\times10^{-3}}$$

$$= 10.71$$

3.26 (b)

At incipient fluidization condition,

$$\frac{\Delta P}{l} = (1-\epsilon)(\rho_p-\rho_f)g$$

$$= 0.5\times1500\times9.8 = 7350 \text{ Pa/m}$$

3.27 (d)

$$n_i = \frac{\text{wt. of collected solid}}{\text{wt. of feed}}$$

$$\Delta w_{fi} = \frac{\text{Weight of the feed}}{\text{Total wt. of feed}}$$

$$\eta = \sum_{i=1}^{n} n_i \Delta wf_i$$

Size range (μm)	wt. of feed	wt. of collected solid	Δwf_i	η_i	$\eta_i\Delta wf_i$
1–5	2	0.1	$\frac{2}{20}=0.1$	$\frac{0.1}{2.0}=0.05$	0.005
5–10	3	0.7	$\frac{3}{20}=0.15$	$\frac{0.7}{3}=0.233$	0.035
10–15	5	3.6	$\frac{5}{20}=0.25$	$\frac{3.6}{5}=0.72$	0.18
15–20	6	5.5	$\frac{6}{20}=0.3$	$\frac{5.5}{6}=0.917$	0.275
20–25	3	2.9	$\frac{3}{20}=0.15$	$\frac{2.9}{3}=0.967$	0.145
25–30	1	1.0	$\frac{1}{20}=0.05$	1	0.05
Σ(sum)	20	13.8	1		0.69

3.28 (a)

For minimum velocity, required to fluidize the bed would be,

$$\frac{\Delta P}{L} = g(\rho_p - \rho_f)(1-\epsilon) = \frac{150\mu_f V_0(1-\epsilon)^2}{(\phi_s d_p)^2 \times (\epsilon)^3}$$

$\because$ As Reynold's number is very small so assuming stokes (laminar regime).

$$10\times(3000-1000) = \frac{150\times 0.1\times V_0 \times (1-0.5)}{(10^{-3})^2\times(0.5)^3}$$

$$10\times 2000 = \frac{150\times 0.1\times V_0\times 0.5}{10^{-6}\times(0.5)^3}$$

$$V_0 = 3.33\times 10^{-4}\text{ m/s}$$

3.29 (160)

$$\text{Separation factor} = \frac{u_{\tan}^2}{rg} = \frac{(20)^2}{0.25\times 10} = 160$$

3.30 (c)

$\because$ Liquid velocity is slightly below the terminal velocity of coarse particles, it is expected that liquid velocity is higher than terminal velocity of fine crystals i.e. fine crystals will be carried upward and coarse will be settled down.

3.31 (8079.6)

At incipient fluidization condition

$$\left(\frac{\Delta P}{L}\right) = g\times(\rho_p - \rho_f)(1-\epsilon)$$

$$\frac{(\Delta P)}{1} = 9.8\times(1500-1)\times(1-0.45)$$

$$\Delta P = 8079.6\text{ Pascal}$$

3.32 (1.296)

Given: $d_p = 10^{-3}$ m, $v = 1.7\times 10^{-3}$ m/s

$\mu = 0.04$ Pa.s

$$\text{Drag force per unit projected area} = \frac{F_D}{A_P}$$

$$\text{Since,}\quad C_D = \frac{\frac{F_D}{A_P}}{\frac{1}{2}\rho u^2}$$

$$\therefore\quad C_D = \frac{24}{\text{Re}}\text{ (Stokes regime)}$$

$$\text{Thus,}\quad \frac{F_D}{A_P} = \frac{1}{2}\rho u^2\times\frac{24\mu}{\rho v d} = 12\mu\frac{v}{d}$$

u = Relative velocity of solid particle

$u = 2.7\times 10^{-3}$ m/s

$$\frac{F_D}{A_P} = \frac{12\times 0.04\times 2.7\times 10^{-3}}{10^{-3}} = 1.296\text{ Pa}$$

3.33 (a)

$$\text{Separation factor} = \frac{F_{\text{centrifugal}}}{F_{\text{gravity}}} = \frac{\frac{mv^2}{r}}{mg}$$

$\therefore$ Separation factor does not depend on mass of particle.

3.34 (2.15)

For Stoke's law regime

$$R_{ep} = 0.1$$

$$R_{ep} = \frac{\rho u_t d_p}{\mu} = 0.1 \qquad \text{...(i)}$$

In stokes law regime,

$$V_t = \frac{g d_p^2(\rho_p - \rho_f)}{18\mu_f}$$

$$\text{Hence,}\quad V_t = \frac{g d_p^2(\rho_p - \rho_f)}{18\mu}$$

$$= \frac{10\times d_p^2\times 180}{18\times 0.1} = 1000 d_p^2$$

Now, placing the value of V_t in equation (i)

$$\text{We get,}\quad R_{ep} = \frac{1000\times 1000 d_p^3}{0.1} = 0.1$$

$$d_p^3 = 10^{-8}$$

$$d_p = 2.15\times 10^{-3}\text{ m} = 2.15\text{ mm}$$

3.35 (0.543)

At terminal velocity condition,

$$\frac{dv}{dt} = 0$$

$$g\left[\frac{\rho_g - \rho}{\rho g}\right] = \frac{-18\mu}{\rho_g D^2}v$$

$$v = 0.543\times 10^{-3}\text{ m/s} = 0.543\text{ cm/s}$$

3.36 (c)

Terminal velocity of a spherical particle, under stokes law regime, is given by

$$V_t = \frac{d_p^2(\rho_p - \rho_f)g}{18\mu_f}$$

$$\therefore\quad V_t \propto d_p^2$$

3.37 (b)

P. Centrifugal sifter → It is a powder screening machine designed for screening.

Q. Bowl mill is a pulverizer or grinder used for grinding different materials.

R. Gravity thickener → generally bio solids are condensed to produce a concentrated solids product and a relatively solids-free supernatant.

S. Two-air kneader is used for mixing and kneading substances, particularly highly viscous materials.

3.38 (d)

The terminal velocity under stokes law regime is given by:

$$V_t = \frac{d_p^2(\rho_p - \rho_f)g}{18\mu_f}$$

$$V_t = \frac{(2\times10^{-4})^2(3000-750)9.81}{18\times9.81\times10^{-3}}$$

$$V_t = 5\times10^{-3}\text{ m/s}$$

3.39 (26478.17)

At the minimum fluidization condition, inside the packed bed,

$$\frac{\Delta P}{l} = g(\rho_p - \rho_f)\times(1-\epsilon)$$

Void fraction = 0.1

Fluid density = 1 kg/m^3

Particle density = 3000 kg/m^3

Height = 1 m

Therefore, pressure drop across the bed is

$$\Delta P = 9.81\times(1-0.1)\times(3000-1)\times1$$

$$\Delta P = 26478.17\text{ Pa}$$

3.40 (a)

Given: Stokes law regime

Particle (A) ⇒ Density = 1100 kg/m^3, d = 30 μm

Particle (B) ⇒ Density = 1900 kg/m^3, d = 10 μm

$$(V_t)_A = \frac{dp^2(\rho_p - \rho f)g}{18\mu_f} \text{ (in Stokes law region)}$$

$$(V_t)_A = \frac{(30\times10^{-6})^2(1100-1000)\times9.81}{18\times10^{-3}}$$

$$= 49050\times10^{-9}\text{ m/s}$$

$$(V_t)_B = \frac{(10\times10^{-6})^2(1900-1000)\times9.81}{18\times10^{-3}}$$

$$= 49050\times10^{-9}\text{ m/s}$$

Since terminal velocity of both the particles are same, so there will be homogeneous mixture of type A and type B particles.

3.41 (1795)

At the onset of fluidization, the pressure drop would be

$$\frac{\Delta P}{l} = g(\rho_P - \rho_f)(1-\epsilon)$$

Data given:

$$\rho_{\text{sea water}} = 1025\text{ kg/m}^3,\quad \rho_P = 1330\text{ kg/m}^3$$

$$\epsilon = 0.4$$

Putting values, we get

$$\frac{\Delta P}{l} = 9.81(1330-1025)(1-0.4)$$

$$\frac{\Delta P}{l} = 1795.23 \cong 1795\text{ Pa/m}$$

3.42 (a)

From Carman-Kozeny equation, we have

$$\frac{\Delta P}{L} = \frac{150\mu_f V_0(1-\epsilon)^2}{(\phi_s d_p)^2\epsilon^3} \quad \{\text{when } Re_p < 1\}$$

V_0 = Fluidization velocity

Under minimum fluidization condition, we have

$$g(\rho_p - \rho_f)(1-\epsilon) = \frac{150\mu_f V_0(1-\epsilon)^2}{(\phi_s d_p)^2\epsilon^3}$$

i.e. $V_0 \propto (D_p)^2$

$n = 2$

3.43 (1.11)

Given: ρ_f = 1 g/cm^3, ρ_p = 0.8 g/cm^3

$R_{ep} < 1$ (Stokes regime)

d_p = 100 μm, $\mu = 10^{-3}$ Pa.s

V_t = ?

From Stokes regime, we have

$$V_t = \frac{d_p^2(\rho_p - \rho_f)g}{18\mu_f}$$

$$V_t = \frac{(100\times10^{-6})^2(1000-800)\times9.81}{18\times10^{-3}}$$

$$V_t = \frac{10^{-8}\times200\times9.81}{18\times10^{-3}} = 1.09\text{ mm/s} = 1.1\text{ mm/s}$$

3.44 (14.42)

Terminal settling velocity is given by

$$V_T = \sqrt{\frac{4}{3}\frac{(\rho_P - \rho_F)g\,dp}{P_F C_D}}$$

C_D = 0.44 (given, ρ_P = 7001 kg/m^3

ρ_F = 1 kg/m^3, g = 9.8 m/s^2

d = 0.001 m

$$V_T = \sqrt{\frac{4}{3}\times\frac{(7001-1)\times9.8\times0.001}{1\times0.44}} = 14.42\text{ m/s}$$

4 Filtration

4.1 The volumetric flow rate during constant pressure filtration is $\frac{dV}{dt} = \frac{1}{K_c V + \beta}$ where V is the total volume of filtrate collected in time t and K_c and β are constants.

(a) Integrate the above equation to obtain a relation between V and t.

(b) Make a sketch of $\frac{t}{V}$ versus V from your results.

(c) Given $V = 1.0$ litre at $t = 41.3$ s and $V = 2.0$ liter at $t = 108.3$ s, find K_c.

[2001 : 2 M]

4.2 The basic filtration equation is given as

$$\frac{dt}{dV} = \frac{\mu}{A\Delta P}\left(\frac{\alpha CV}{A} + R_m\right)$$

where V is volume of the filtrate; A is the filtration area; α is specific cake resistance, μ is viscosity of the filtrate, and C is the concentration of solids in the feed slurry. In a 20 min, constant rate filtration, 5 m^3 of filtrate was obtained. If this is followed by a constant pressure filtration, how much more time in minutes will it take for another 5 m^3 of filtrate to be produced? Neglect filter medium resistance, R_m; assume incompressible cake

(a) 10
(b) 20
(c) 25
(d) 30

[2003 : 2 M]

4.3 In constant pressure filtration,

(a) resistance decreases with time
(b) rate of filtration is constant
(c) rate of filtration increases with time
(d) rate of filtration decreases with time

[2004 : 1 M]

4.4 A filtration is conducted at constant pressure to recover solids from dilute slurry. To reduce the time of filtration, the solids concentration in the feed slurry is increased by evaporating half the solvent. If the resistance of the filter medium is negligible, the filtration time will be reduced by a factor of

(a) 1
(b) 2
(c) 4
(d) 8

[2006 : 2 M]

4.5 In constant pressure filtration, the rate of filtration follows the relation (v : filtration volume, t : time, k and c : constants).

(a) $\frac{dv}{dt} = kv + c$

(b) $\frac{dv}{dt} = \frac{1}{kv + c}$

(c) $\frac{dv}{dt} = kv$

(d) $\frac{dv}{dt} = kv^2$

[2007 : 1 M]

4.6 A fluidized bed (0.5 m dia, 0.5 m high) of spherical particles (diameter = 2000 μm, specific gravity = 2.5) uses water as the medium. The porosity of the bed is 0.4. The Ergun equation for the system is

$$\frac{\Delta P}{L} = 4 \times 10^5\, U_{mf} + 1 \times 10^7\, U_{mf}^2$$

(SI unit, U_{mf} in m/s)

$\frac{\Delta P}{L}$ (SI unit) at minimum fluidization condition is _______ ($g = 9.8$ m/sec^2)

(a) 900
(b) 8820
(c) 12400
(d) 17640

[2007 : 2 M]

4.7 For laminar flow conditions, the relationship between the pressure drop (ΔP_c) across an incompressible filter cake and the specific surface area (S_o) of the particles being filtered is given by which one of the following?

(a) ΔP_c is proportional to S_o
(b) ΔP_c is proportional to $1/S_o$
(c) ΔP_c is proportional to S_o^2
(d) ΔP_c is proportional to $1/S_o^2$

[2008 : 1 M]

4.8 A typical batch filtration cycle consists of filtration followed by washing. One such filtration unit operating at constant pressure difference first filters a slurry during which 5 liters of filtrate is collected in 100 s. This is followed by washing, which is done for t_W seconds and uses 1 liter of wash water. Assume the following relation to be applicable between the applied pressure drop ΔP, cake thickness L at time t, and volume of liquid V collected in time t:

$$\frac{\Delta P}{L} = k_1 \frac{dV}{dt}; \quad L = k_2 V, \text{ if } L \text{ is changing}$$

k_1 and k_2 can be taken to be constant during filtration and washing. The wash time t_W, in seconds (up to one decimal place), is ______ .

[2015 : 2 M]

4.9 In a constant-pressure cake filtration with an incompressible cake layer, volume of the filtrate (V) is measured as a function of time t. The plot of $\frac{t}{V}$ versus V results in a straight line with an intercept of 10^4 s m^{-3}. Area of the filter is 0.05 m^2, viscosity of the filtrate is 10^{-3} Pa s, and the overall pressure drop across the filter is 200 kPa. The value of the filter-medium resistance (in m^{-1}) is

(a) 1×10^9 (b) 1×10^{10}
(c) 1×10^{11} (d) 1×10^{12}

[2020 : 1 M]

4.10 In a constant-rate cake filtration operation, the collected filtrate volumes are 120 m^3 and 240 m^3 at 1 min and 2 min, respectively. Assume the cake resistance to be constant and the filter medium resistance to be negligible. If the pressure-drop across the cake is 10 kPa at 1 min, its value at 2 min is ________ kPa (rounded off to the nearest integer).

[2022 : 2 M]

■■■■

Answers Filtration

4.1 (25.70) 4.2 (a) 4.3 (d) 4.4 (b) 4.5 (b) 4.6 (b) 4.7 (c)

4.8 (40) 4.9 (c) 4.10 (20)

Explanations Filtration

4.1 (25.70)

$$\frac{dv}{dt} = \frac{1}{k_c V + \beta}$$

Integrating the above equation, we get

$$t = k\cdot\frac{V^2}{2} + \beta V$$

(a) $$\left(\frac{t}{v}\right) = \frac{k_c}{2}.V + \beta$$

(b)

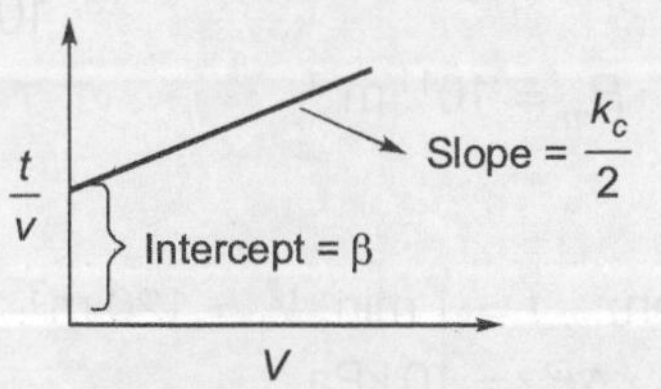

(c) $$\left(\frac{t}{v}\right) = \left(\frac{k_c}{2}\right)V + \beta$$

$$\frac{108.3}{2} = \left(\frac{k_c}{2}\right)2 + \beta \quad \text{...(i)}$$

$$\frac{41.3}{1} = \left(\frac{k_c}{2}\right)(1) + \beta \quad \text{...(ii)}$$

From equation (i) and (ii)

$$k_c = 25.70 \text{ s}/L^2,\ \beta = 0.035 \text{ s}/L$$

4.2 (a)

Stage I: [Constant rate filtration]

$$\frac{dt}{dv} = k\cdot v + \beta \nearrow 0$$

$$\Rightarrow \quad \frac{t}{v} = k\cdot v \rightarrow k = \frac{t}{V^2}$$

$$\left(k = \frac{20}{25}\right)$$

Stage II: [Constant pressure process]

$$\frac{dt}{dv} = k\cdot v$$

$$\Rightarrow \quad \int_0^t dt = k\int_0^5 v\,dv = k\left(\frac{v^2}{2}\right)_0^5$$

$$t = \frac{20}{25}\times\frac{25}{2} = 10 \text{ min}$$

4.3 (d)

$$\frac{dv}{dt} = \frac{\Delta P}{\mu\left[\frac{\alpha CV}{A} + R_m\right]}$$

$\therefore$ Volume of filtrate collected $\propto$ Time

As time passes, resistance increases so rate of filtration decreases.

4.4 (b)

From filtration equation, we have:

$$\frac{dt}{dv} = \frac{\mu}{A\Delta P}\left[\frac{\alpha CV}{A} + R_m\right]$$

Given, $R_m \equiv 0$

$$\Rightarrow \quad \frac{dt}{dv} = \frac{\mu\alpha CV}{A^2\Delta P}$$

$$t = \frac{\mu\alpha C}{A^2\Delta P}\frac{V^2}{2} + c$$

$\because$ C become $2C$ on evaporating half the solvent.

$$\Rightarrow \quad t \ \alpha \ CV^2$$

$$\frac{t_1}{t_2} = \frac{CV^2}{(2C)\left(\frac{V}{2}\right)^2} = 2 \qquad \left(t_2 = \frac{t_1}{2}\right)$$

$\because$ C becomes $2C$ on evaporating half the solvent.

4.5 (b)

From filtration equation, we have

$$\frac{dt}{dv} = \frac{\mu}{A\Delta P}\left[\frac{\alpha CV}{A} + R_m\right]$$

$$\frac{dt}{dv} = kv + \beta$$

$$\left(\frac{dV}{dt}\right) = \frac{1}{kv + \beta}$$

$$k = \frac{\mu\alpha C}{A^2\Delta P}, \quad \beta = \frac{\mu R_m}{A\Delta P}$$

4.6 (b)

For minimum fluidization,

$$\frac{\Delta P}{L} = g(1-\in)(\rho_p - \rho_f)$$

$$\left(\frac{\Delta P}{L}\right) = 9.8\,(1-0.4)\,(2500-1000)$$

$$= 8820 \text{ kg/m}^2\text{ s}^2$$

4.7 (c)

From filtration equation,

$$\frac{dt}{dv} = \frac{\mu}{A\Delta P}\left[\frac{\alpha C \cdot V}{A} + R_m\right]$$

where, $\alpha = \dfrac{k_L(1-\in)S_o^2}{\rho_p \in^3}$

i.e. $\Delta P \;\alpha\; \text{'}\alpha\text{'}$

$\alpha\; S_o^2$

4.8 (40)

$$\frac{\Delta P}{L} = k_1\frac{dV}{dt} \text{ and } L = k_2 V$$

So, $\dfrac{dV}{dt} = \dfrac{1}{k_1}\dfrac{\Delta P}{k_2 V}$

$$\int_0^5 V\,dV = \frac{\Delta P}{k_1 k_2}\int_0^{100} dt$$

$$\frac{\Delta P}{k_1 k_2} = \frac{1}{8}$$

Final rate of filtration is

$$\frac{dV}{dt} = \frac{\Delta P}{k_1 k_2 V} = \frac{1}{40} \quad \because (V = 5)$$

$$\text{Washing time} = \frac{\text{Volume of wash water}}{\text{Final rate of filtration}}$$

$$= \frac{1}{\frac{1}{40}} = 40 \text{ seconds}$$

4.9 (c)

From filtration equation,

$$\frac{dt}{dV} = \frac{\mu}{A\Delta P}\left[\frac{\alpha C V}{A} + R_m\right] = k_p V + \beta$$

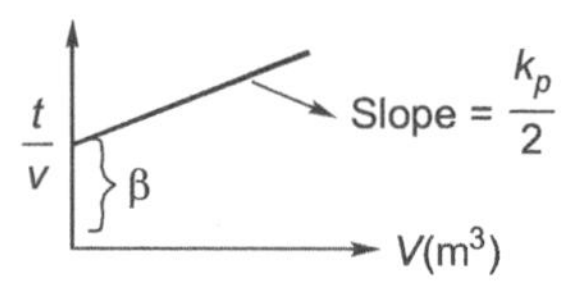

Intercept $\Rightarrow \beta = 10^4$ s/m^3 (given)

$$\therefore \quad \frac{\mu}{A\Delta P}R_m = 10^4$$

$A = 0.05$ m^2, $\mu = 10^{-3}$ Pa.s

$\Delta P = 200$ kPa

$$\therefore \quad R_m = 0.05 \times 200 \times 1000 \times \frac{1}{10^{-3}} \times 10^4$$

$$R_m = 10^{11} \text{ m}^{-1}$$

4.10 (20)

Data given: $t = 1$ min, $V_1 = 120$ m^3,

$\Delta P_1 = 10$ kPa

$t = 2$ min, $V_2 = 240$ m^3, $\Delta P_2 = ?$

$\propto$ = Constant, $R_m \cong 0$

For constant rate filtration, we have

$$\frac{dt}{dV} = \frac{t}{V} = k_c V \text{, where } k_c = \frac{\mu \propto c}{A_1^2 \Delta P}$$

$$k_c = \frac{t}{V^2} \quad \because k_c \propto \frac{1}{\Delta P}$$

Thus, $\Delta P \propto \dfrac{V^2}{t} \rightarrow \dfrac{\Delta P_1}{\Delta P_2} = \dfrac{V_1^2}{t_1}\dfrac{t_2}{V_2^2}$

$$\frac{\Delta P_1}{\Delta P_2} = \frac{(120)^2 2}{1(240)^2} = \frac{1}{2}$$

$$\frac{10}{\Delta P_2} = \frac{1}{2} \rightarrow \Delta P_2 = 20 \text{ kPa}$$

■■■■

5 Mixing and Agitation

5.1 The Power number for a stirred tank becomes constant at high Reynolds number. In this limit, the variation of power input with impeller rotational speed (N) is proportional to

(a) N^0 (b) N^1
(c) N^2 (d) N^3

[2001 : 1 M]

5.2 To keep the power input constant for a stirred vessel operating under fully developed turbulent flow conditions (constant power number), if the impeller diameter is increased by 20%, the impeller speed should be decreased by a factor of

(a) $(1.2)^{3/2}$ (b) $(1.2)^{3/5}$
(c) $(1.2)^{2/3}$ (d) $(1.2)^{5/3}$

[2004 : 2 M]

5.3 If the frequency of the stirrer in a mixing tank is increased by a factor of 2 while all other parameters are kept constant, by what factor is the power requirement increased at high Reynolds number?

(a) 4 (b) 8
(c) 16 (d) 32

[2005 : 1 M]

5.4 The mixing of rubber latex solution was studied in an unbaffled mixer in the laboratory. The mixer was equipped with a six blade turbine impeller. A tyre company scales this process up using a baffled tank. The baffled tank has 3 times the diameter of the lab scale mixer. It uses the same type of impeller operated at the same speed. The relevant shape factors are also the same. Assuming that laminar conditions prevail in both cases, the power requirement in the industrial scale mixer

(a) is 3 times that of the lab scale mixer
(b) is 9 times that of the lab scale mixer
(c) is 27 times that of the lab scale mixer
(d) cannot be estimated reliably due to the presence of baffles

[2006 : 2 M]

5.5 Consider the scale-up of a cylindrical baffled vessel configured to have the standard geometry (i.e., Height = Diameter). In order to maintain an equal rate of mass transfer under turbulent conditions for a Newtonian fluid, the ratio of the agitator speeds should be
(Given : N_1, D_1 are agitator speed and vessel diameter before scale-up; N_2, D_2 are agitator speed and vessel diameter after scale up)

(a) $\frac{N_1}{N_2}=\frac{D_1}{D_2}$ (b) $\frac{N_1}{N_2}=\frac{D_2}{D_1}$

(c) $\frac{N_1}{N_2}=\left(\frac{D_1}{D_2}\right)^{2/3}$ (d) $\frac{N_1}{N_2}=\left(\frac{D_2}{D_1}\right)^{2/3}$

[2008 : 2 M]

5.6 For a mixing tank operating in the laminar regime, the power number varies with the Reynolds number (Re) as

(a) $Re^{-1/2}$ (b) $Re^{1/2}$
(c) Re (d) Re^{-1}

[2009 : 1 M]

5.7 In a mixing tank operating at very high Reynolds number ($> 10^4$), if the diameter of the impeller is doubled (other conditions remaining constant), the power required increases by a factor of

(a) $\frac{1}{32}$ (b) $\frac{1}{4}$
(c) 4 (d) 32

[2012 : 1 M]

5.8 An agitated cylindrical vessel is fitted with baffles and flat blade impellers. The power number for this system is given by $N_p = \frac{P}{\rho n^3 D^5}$ where P is the power consumed for the mixing, ρ is the density of the fluid, n is the speed of the impeller and D is the diameter of the impeller. The diameter of the impeller is $\frac{1}{3}$ rd the diameter of the tank and the height of liquid level is equal to the tank diameter. The impeller speed to achieve the desired degree of mixing is 4 rpm. In a scaled up design, the linear dimensions of the equipment are to be doubled, holding the power input per unit volume constant. Assuming the liquid to be Newtonian and N_p to be independent of Reynolds number, what is the impeller speed (in rpm) to achieve the same degree of mixing in the scaled up vessel?

(a) 0.13 (b) 1.26
(c) 2.52 (d) 3.82

[2016 : 2 M]

5.9 A disk turbine is used to stir a liquid in a baffled tank. To design the agitator, experiments are performed in a lab-scale model with a turbine diameter of 0.05 m and a turbine impeller speed of 600 rpm. The liquid viscosity is 0.001 Pa-s, while the liquid density is 1000 kg/m^3. The actual application has a turbine diameter of 0.5 m, an impeller speed of 600 rpm, a liquid viscosity of 0.1 Pa-s and a liquid density of 1000 kg/m^3. The effect of gravity is negligible. If the power required in the lab-scale model is P_1 and the estimated power for the actual application is P_2, then the ratio $\frac{P_2}{P_1}$ is

(a) 10^3 (b) 10^4
(c) 10^5 (d) 10^6

[2019 : 2 M]

Answers Mixing and Agitation

5.1	(d)	5.2	(d)	5.3	(b)	5.4	(c)	5.5	(d)	5.6	(d)	5.7	(d)
5.8	(c)	5.9	(c)										

Explanations Mixing and Agitation

5.1 (d)

Power number

$$\Rightarrow \quad N_P = \frac{P}{\rho N^3 D^5}$$

At high Reynolds number (N_P = constant)

$$\therefore \quad P \propto N^3$$

5.2 (d)

For high Reynold's number

$$N_P(\text{Power number}) = \frac{P}{\rho N^3 D^5} = \text{Constant}$$

$$P = \text{Constant}, \quad \rho = \text{Constant}$$

$$N_1^3 D_1^5 = N_2^3 D_2^5$$

$$\therefore \quad \left(\frac{N_2}{N_1}\right)^3 = \left(\frac{D_1}{D_2}\right)^3 = \left(\frac{1}{1.2}\right)^5$$

$$N_2 = \frac{N_1}{(1.2)^{5/3}}$$

5.3 (b)

At high Reynold's number, N_p = Constant

$$N_p = \frac{P}{\rho N^3 D^5} = \text{Constant}$$

$$P \propto N^3$$

$$\therefore \quad \frac{P_1}{P_2} = \frac{N^3}{(2N)^3}$$

$$P_2 = 8P_1$$

5.5 (d)

In turbulent condition,

N_p (Power number) = Constant

$$N_p = \frac{P}{\rho N^3 D^5}$$

To keep equal rate of mass transfer, power per unit volume must be constant

For scale up,

$$\left(\frac{P}{V}\right) = \frac{N_p \rho N^3 D^5}{\frac{\pi}{4} D^3} = \text{Constant}$$

For scale up,

$$N^3 D^2 = \text{Constant}$$

$$\left(\frac{N_1}{N_2}\right) = \left(\frac{D_2}{D_1}\right)^{2/3}$$

5.6 (d)

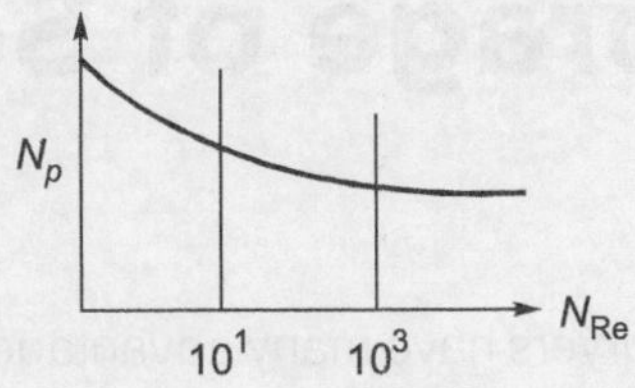

In laminar region, [$N_p \times$ Re = constant]

$$N_p \propto \frac{1}{\text{Re}}$$

5.7 (d)

At high Reynold's, the power number is nearly constant.

$$N_p = \frac{P}{\rho N_i^3 D_i^5} = \text{const.}$$

$$\therefore \quad \frac{P_1}{\rho N_1^3 D_1^5} = \frac{P_2}{\rho N_2^3 D_2^5}$$

$$N_P = \text{Constant}$$

$$N_1 = N_2$$

$$\therefore \quad \frac{P_2}{P_1} = (2)^5 = 32$$

5.8 (c)

$$N_p = \frac{P}{\rho N_i^3 D_i^5}$$

Since $\quad N_p \neq f(\text{Re}) = \text{constant}$

$$P = K\rho N_i^3 D_i^5$$

Given, power per unit volume is constant.

$$\text{Volume of tank} = \frac{\pi D_T^2 L}{4}$$

$$D = \frac{1}{3} D_T$$

Hence, $\quad D_T = 3D,\ L = D_T = 3D$

$$\text{Volume} = \frac{27}{4}\pi D^3$$

$$\left(\frac{P}{V}\right) = \frac{K\rho N_i^3 D_i^5}{\frac{27}{4}\pi D^3}$$

$$\therefore \quad \frac{P}{V} \propto N_i^3 D^2 = \text{const.}$$

$$\left(\frac{N_1}{N_2}\right)^3 = \left(\frac{D_2}{D_1}\right)^2$$

$$\because \quad D_2 = 2D_1$$

Thus, $\quad N_2 = 2.53$ rpm

5.9 (c)

For lab scale model, the value of Reynold's number comes out to be,

$$\text{Re}_{\text{I}} = \frac{\rho N_i D_i^2}{\mu} = \frac{600 \times (0.05)^2 \times 1000}{60 \times 0.001}$$

$$\text{Re}_{\text{I}} = 25000$$

$$\text{Re}_{\text{II}}\ (\text{Actual application}) = \frac{600 \times (0.5)^2 \times 1000}{60 \times 0.1}$$

$$= 25000$$

The power number for the impellers is constant, when Re > 10^4

$$N_{P\text{I}} = N_{P\text{II}}$$

$$\frac{P_1}{N_i^3 D_i^5 \rho_{\text{I}}} = \frac{P_2}{N_{\text{II}}^3 D_{\text{II}}^5 \rho_{\text{II}}}$$

Putting values, we get

$$\frac{P_1}{(0.05)^5} = \frac{P_2}{(0.5)^5}$$

$$\frac{P_2}{P_1} = 10^5$$

■■■■

6 Transportation and Storage of Solids

6.1 Sticky materials are transported by

(a) apron conveyor

(b) screw conveyor

(c) belt conveyor

(d) hydraulic conveyor

[2007 : 1 M]

6.2 Spray dryers have many advantages. Which one of the following is NOT an advantage of a typical spray dryer?

(a) has short drying time

(b) produces hollow spherical particles

(c) has high heat efficiency

(d) is suitable for heat sensitive materials

[2023 : 1 M]

■■■■

Answers Transportation and Storage of Solids

6.1 (b) **6.2** (c)

Explanations Transportation and Storage of Solids

6.1 (b)

Screw conveyor uses a rotating helical screw blade to move the liquid or granular material. For sticky material we use shaftless screw conveyor.

■■■■

Heat Transfer

UNIT

V

CONTENTS

Heat Transfer

Syllabus

Equation of Energy, Steady and Unsteady Heat Conduction, Convection and Radiation, Thermal Boundary Layer and Heat Transfer Coefficients, Boiling, Condensation and Evaporation; Types of Heat Exchangers and Evaporators and Their Process Calculations; Design of Double Pipe, Shell and Tube Heat Exchangers and Single and Multiple Effect Evaporators.

Analysis of Previous GATE Papers

Exam Year	1 Mark Ques.	2 Marks Ques.	5 Marks Ques.	Total Marks
2000	2	3	3	**23**
2001	1	5	–	**11**
2002	1	3	1	**12**
2003	3	6	–	**15**
2004	3	4	–	**11**
2005	4	8	–	**20**
2006	2	7	–	**16**
2007	1	7	–	**15**
2008	1	5	–	**11**
2009	2	6	–	**14**
2010	2	4	–	**10**
2011	2	3	–	**8**
2012	3	5	–	**13**
2013	2	2	–	**6**
2014	2	3	–	**8**
2015	1	3	–	**7**
2016	3	3	–	**9**
2017	2	3	–	**8**
2018	2	4	–	**10**
2019	3	3	–	**9**
2020	2	4	–	**10**
2021	4	3	–	**10**
2022	2	5	–	**12**
2023	4	3	–	**10**
2024	3	3	–	**9**

1 Introduction and Basic Concepts

Common Data for Question 1.1 and 1.2:
A slab of thickness L with one side ($x = 0$) insulated and the other side ($x = L$) maintained at a constant temperature T_o is shown below:

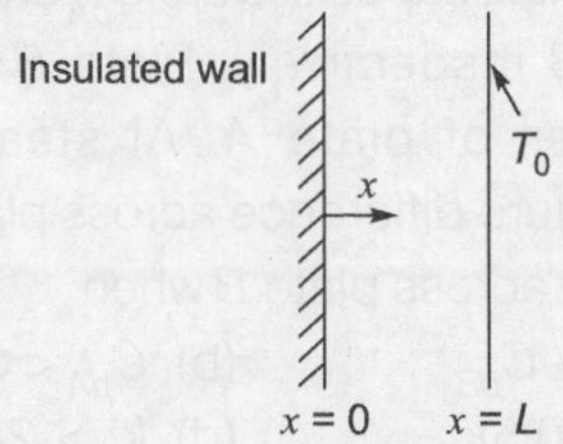

A uniformly distributed internal heat source produces heat in the slab at the rate of S W/m^3. Assume the heat conduction to be steady and I-D along the x-direction.

1.1 The maximum temperature in the slab occurs at x equal to

(a) 0 (b) $\frac{L}{4}$

(c) $\frac{L}{2}$ (d) L [2009 : 2 M]

1.2 The heat flux at $x = L$ is

(a) 0 (b) $\frac{SL}{4}$

(c) $\frac{SL}{2}$ (d) SL [2009 : 2 M]

1.3 Ratio of momentum diffusivity to thermal diffusivity is

(a) Reynolds number (b) Prandtl number
(c) Nusselt number (d) Peclet number

[2020 : 1 M]

1.4 Match the quantities in Group 1 with their units in Group 2 listed in the table below:

Group 1
P. Thermal conductivity
Q. Convective heat transfer coefficient
R. Stefan-Boltzmann constant
S. Heat capacity rate

Group-2
I. W.m^{-2} K^{-1}
II. W.m^{-1} K^{-1}
III. W.K^{-1}
IV. W.m^{-2} K^{-4}

(a) P - II, Q - I, R - IV, S II
(b) P - I, Q - II, R - III, S - IV
(c) P - III, Q - IV, R - II, S - I
(d) P - IV, Q - I, R - III, S - IV [2023 : 1 M]

1.5 Which of the following statements is/are CORRECT?

(a) Bond number includes surface tension.
(b) Jakob number includes latent heat.
(c) Prandtl number includes liquid-vapor density difference.
(d) Biot number includes gravity.

[2023 : 2 M]

■■■■

Answers Introduction and Basic Concepts

1.1 (a) 1.2 (d) 1.3 (b) 1.4 (a) 1.5 (a, b)

Explanations Introduction and Basic Concepts

1.1 (a)

Maximum temperature in the slab will occur at left wall as this insulated i.e. at the insulated surface

$\left(\frac{dT}{dX} = 0\right)$ and temperature will be max.

1.2 (d)

Heat flux at $x = L$.
Heat is conducted in x direction and steady state.
In case of steady state, rate of heat generation is equal to rate of heat conduction.
Heat generation rate = S W/m^3
So, heat flux at $x = L$ will be

$$q = S \times L = S \cdot L \text{ (W/m}^2\text{)}$$

■■■■

2 Steady State Heat Conduction

2.1 A composite flat wall of a furnace is made of two materials *A* and *B*. The thermal conductivity of *A* is twice of that of material *B*, while the thickness of layer of *A* is half of that of *B*. If the temperatures at the two sides of the wall are 400 and 1200 K, then the temperature drop (in K) across the layer of material *A* is

(a) 125 (b) 133

(c) 150 (d) 160

[2000 : 2 M]

2.2 The outside surface temperature of a pipe (radius = 0.1 m) is 400 K. The pipe is losing heat to atmosphere, which is at 300 K. The film heat transfer coefficient is 10 W/m^2 K. To reduce the rate of heat loss, the pipe is insulated by a 50 mm thick layer of asbestos (k = 0.5 W/m K). Calculate the percentage reduction in the rate of heat loss.

[2000 : 5 M]

2.3 The heat flux (from outside to inside) across an insulating wall with thermal conductivity k = 0.04 W/m K and thickness 0.16 m is 10 W/m^2. The temperature of the inside wall is –5°C. The outside wall temperature is

(a) 25°C (b) 30°C

(c) 35°C (d) 40°C

[2001 : 2 M]

2.4 A 10 cm diameter steam pipe, carrying steam at 180°C, is covered with an insulation (conductivity = 0.6 W/m°C). It loses heat to the surroundings at 30°C. Assume a heat transfer coefficient of 8.0 W/m^2 °C for heat transfer from surface to the surroundings. Neglect wall resistance of the pipe and film resistance of steam. If the insulation thickness is 2 cm, the rate of heat loss from this insulated pipe will be

(a) greater than that of the un-insulated steam pipe

(b) less than that of the un-insulated steam pipe

(c) equal to that of the un-insulated steam pipe

(d) less than the steam pipe with 5 cm insulation

[2002 : 2 M]

2.5 A composite wall consists of two plates *A* and *B* placed in series normal to the flow of heat. The thermal conductivities are k_A and k_B and the specific heat capacities are C_{pA} and C_{pB}, for plates *A* and *B* respectively. Plate *B* has twice the thickness of plate *A*. At steady state, the temperature difference across plate *A* is greater than that across plate *B* when

(a) $C_{pA} > C_{pB}$ (b) $C_{pA} < C_{pB}$

(c) $k_A < 0.5k_B$ (d) $k_A > 2k_B$

[2002 : 2 M]

2.6 The units of resistance to heat transfer are

(a) $Jm^{-2} K^{-1}$ (b) $Jm^{-1} K^{-1}$

(c) $Wm^{-2} K^{-1}$ (d) $W^{-1} m^2 K$

[2003 : 1 M]

2.7 The inner wall of a furnace is at a temperature of 700°C. The composite wall is made of two substances, 10 and 20 cm thick with thermal conductivities of 0.05 and 0.1 $Wm^{-1}°C^{-1}$ respectively. The ambient air is at 30°C and the heat transfer coefficient between the outer surface of wall and air is 20 W m^{-2} °C^{-1}. The rate of heat loss from the outer surface in W/m^2 is

(a) 165.4 (b) 167.5

(c) 172.8 (d) 175 [2003 : 2 M]

2.8 For a given ambient air temperature with increase in the thickness of insulation of a hot cylindrical pipe, the rate of heat loss from the surface would

(a) decrease

(b) increase

(c) first decrease and then increase

(d) first increase and then decrease

[2003 : 2 M]

2.9 The left face of a one dimensional slab of thickness 0.2 m is maintained at 80°C and the right face is exposed to air at 30°C. The thermal conductivity of the slab is 1.2 W/(mK) and the heat transfer coefficient from the right face is 10 W/(m^{2}K). At steady state, the temperature of the right face in °C is

(a) 77.2 (b) 71.2
(c) 63.8 (d) 48.7

[2004 : 2 M]

2.10 A circular tube of outer diameter 5 cm and inner diameter 4 cm is used to convey hot fluid. The inner surface of the wall of the tube is at a temperature of 80°C, while the outer surface of the wall of the tube is at 25°C. What is the rate of heat transport across the tube wall per meter length of the tube at steady state, if the thermal conductivity of the tube wall is 10 W/(mK)?

(a) 13823 W/m (b) 15487 W/m
(c) 17279 W/m (d) 27646 W/m

[2005 : 2 M]

2.11 A stagnant liquid film of 0.4 mm thickness is held between two parallel plates. The top plate is maintained at 40°C and the bottom plate is maintained at 30°C. If the thermal conductivity of the liquid is 0.14 W/(m K), then the steady state heat flux (in W/m^2) assuming one dimensional heat transfer is

(a) 3.5 (b) 350
(c) 3500 (d) 7000

[2006 : 1 M]

2.12 One dimensional steady state heat transfer occurs from a flat vertical wall of length 0.1 m into the adjacent fluid. The heat flux into this fluid is 21 W/m^2. The wall thermal conductivity is 1.73 W/(mK). If the heat transfer coefficient is 30 W/(m^2 K) and the Nusselt number based on the wall length is 20, then the magnitude of the temperature gradient at the wall on the fluid side (in K/m) is

(a) 0.7 (b) 12.14
(c) 120 (d) 140

[2006 : 2 M]

2.13 The composite wall of an oven consists of three materials A, B and C. Under steady state operating conditions, the outer surface temperature T_{so} is 20°C, the inner surface temperature T_{si} is 600°C and the oven air temperature is T_∞ = 800°C. For the following data:

Thermal conductivities k_A = 20 W/(mK) and k_C = 50 W/(mK),

Thickness L_A = 0.3 m, L_B = 0.15 m and L_C = 0.15 m,

Inner-wall heat transfer coefficient h = 25 W/(m^{2}K),

The thermal conductivity k_B (W/(mK) of the material B, is calculated as

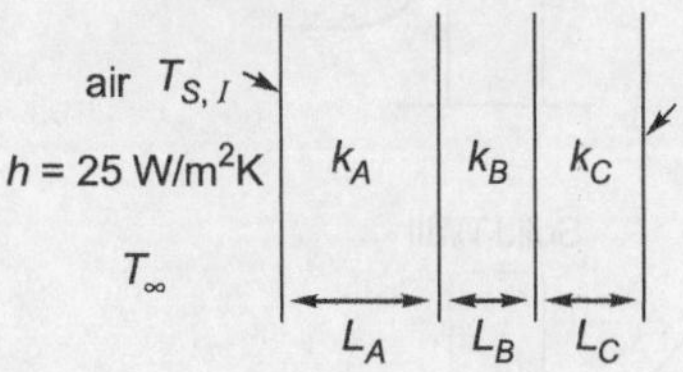

(a) 35 (b) 1.53
(c) 0.66 (d) 0.03

[2007 : 2 M]

2.14 Two plates of equal thickness (t) and cross-sectional area, are joined together to form a composite as shown in the figure. If the thermal conductivities of the plates are k and $2k$ then, the effective thermal conductivity of the composite is

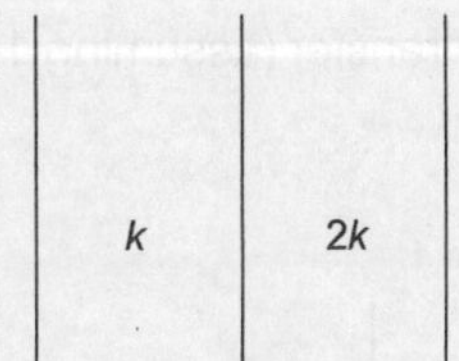

(a) $\frac{3k}{2}$ (b) $\frac{4k}{3}$
(c) $\frac{3k}{4}$ (d) $\frac{2k}{3}$

[2008 : 2 M]

2.15 The temperature profile for heat transfer from one fluid to another separated by a solid wall is

(a)

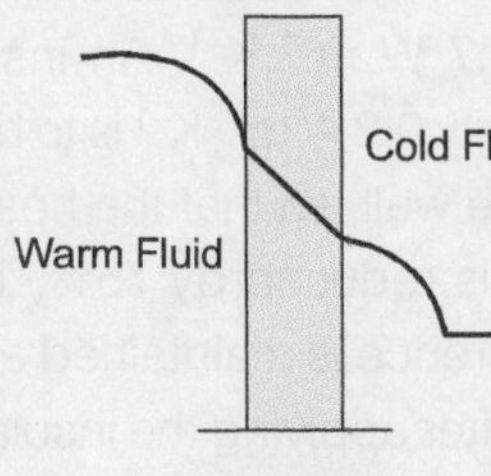

(b)

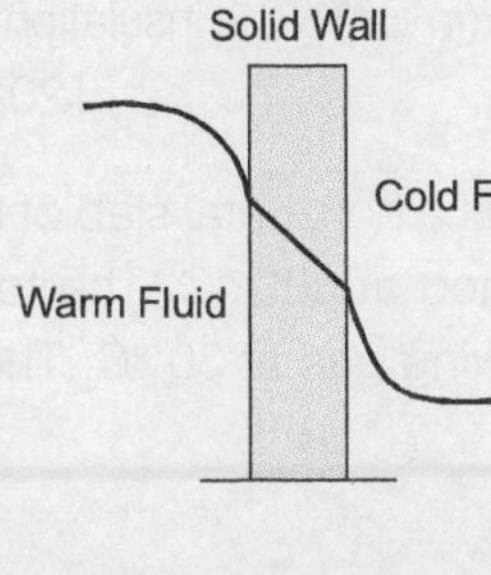

(c)

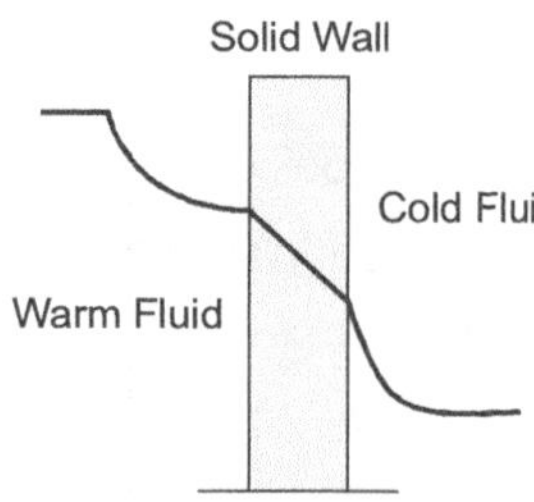

(d)

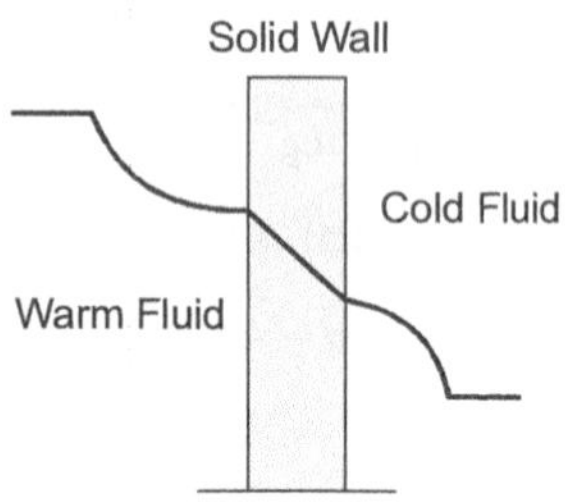

[2008 : 2 M]

2.16 For the composite wall shown below (Case 1), the steady state interface temperature is 180°C. If the thickness of layer *P* is doubled (Case 2), then the rate of heat transfer (assuming 1-D conduction) is reduced by

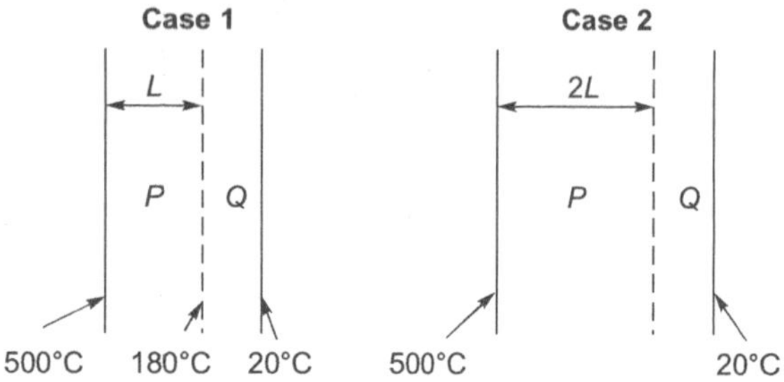

(a) 20%
(b) 40%
(c) 50%
(d) 70%

[2009 : 2 M]

2.17 A brick wall of 20 cm thickness has thermal conductivity of 0.7 W m^{-1} K^{-1}. An insulation of thermal conductivity 0.2 W m^{-1} K^{-1} is to be applied on one side of the wall, so that the heat transfer through the wall is reduced by 75%. The same temperature difference is maintained across the wall before and after applying the insulation. The required thickness (in cm) of the insulation is ______.

[2014 : 2 M]

2.18 The bottom face of a horizontal slab of thickness 6 mm is maintained at 300 °C. The top face is exposed to a flowing gas at 30 °C. The thermal conductivity of the slab is 1.5 W m^{-1} K^{-1} and the convective heat transfer coefficient is 30 W m^{-2} K^{-1}. At steady state, the temperature (in °C) of the top face is ______.

[2014 : 2 M]

2.19 A composite wall is made of four different materials of construction in the fashion shown below. The resistance (in K/W) of each of the sections of the wall is indicated in the diagram.

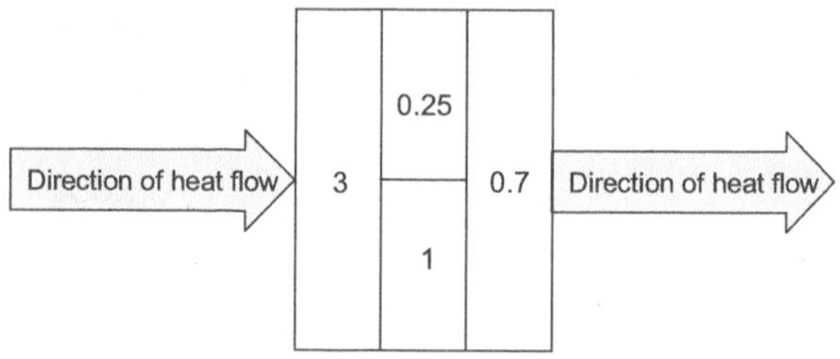

The overall resistance (in K/W, rounded off to the first decimal place) of the composite wall, in the direction of heat flow, is ______.

[2016 : 1 M]

2.20 In an experimental setup, mineral oil is filled in between the narrow gap of two horizontal smooth plates. The setup has arrangements to maintain the plates at desired uniform temperatures. At these temperatures, ONLY the radiative heat flux is negligible. The thermal conductivity of the oil does not vary perceptibly in this temperature range. Consider four experiments at steady state under different experimental conditions, as shown in the figure below. The figure shows plate temperatures and the heat fluxes in the vertical direction.

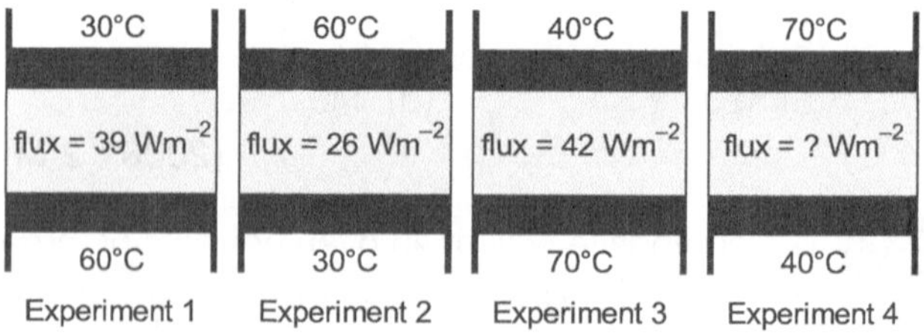

What is the steady state heat flux (in Wm^{-2}) with the top plate at 70°C and the bottom plate at 40°C?

(a) 26
(b) 39
(c) 42
(d) 63

[2016 : 2 M]

2.21 The wall of a pipe of radius 1 m is at a uniform temperature of 200°C, and is covered by insulation of thickness 0.1 m. The ambient air outside the insulated pipe is at 20°C and has heat transfer coefficient of 10 W m^{-2} K^{-1}. The thermal conductivity of the insulation material is 0.05 W m^{-1} K^{-1}. If the heat transfer occurs at steady state, the temperature (in °C) of the outer surface of insulation is ________ (rounded off to 2nd decimal place). **[2018 : 2 M]**

2.22 Consider a bare long copper wire of 1 mm diameter. Its surface temperature is T_s and the ambient temperature is $T_a (T_s > T_a)$. The wire is to be coated with a 2 mm thick insulation. The convective heat transfer coefficient is 20 W m^{-2} K^{-1}. Assume that T_s and T_s remain unchanged. To reduce heat loss from the wire, the maximum allowed thermal conductivity of the insulating material, in W m^{-1} K^{-1}, rounded off to two decimal places, is

(a) 0.02 (b) 0.04

(c) 0.10 (d) 0.01 **[2022 : 2 M]**

2.23 A slab of thickness L, as shown in the figure below, has cross-sectional area A and constant thermal conductivity k. T_1 and T_1 are the temperatures at $x = 0$ and $x = L$, respectively. Which one of the following options is the CORRECT expression of the thermal resistance for steady-state one-dimensional heat conduction?

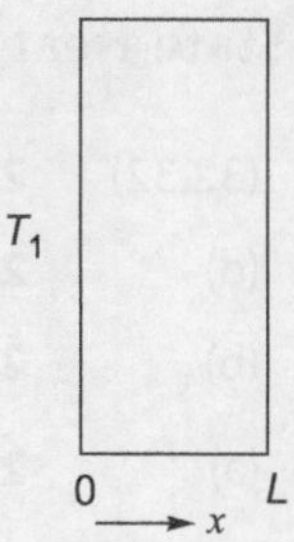

(a) $\frac{L}{kA}$ (b) $\frac{k}{LA}$

(c) $\frac{kA(T_1 - T_2)}{L}$ (d) $\frac{A}{Lk}$ **[2023 : 1 M]**

2.24 The opposite faces of a metal slab of thickness 5 cm and thermal conductivity 400 W m^{-1} °C^{-1} are maintained at 500°C and 200°C. The area of each face is 0.02 m^2. Assume that the heat transfer is steady and occurs only in the direction perpendicular to the faces. The magnitude of the heat transfer rate, in kW, rounded off to the nearest integer, is ________. **[2024 : 1 M]**

2.25 The capital cost of a distillation column is Rs. 90 lakhs. The cost is to be fully depreciated (salvage value is zero) using the double-declining balance method over 10 years. At the end of two years of continuous operation, the book-value of the column, in lakhs of rupees, rounded off to 1 decimal place, is ________. **[2024 : 1 M]**

■■■■

Answers Steady State Heat Conduction

2.1	(d)	**2.2**	(32.32)	**2.3**	(c)	**2.4**	(a)	**2.5**	(c)	**2.6**	(d)	**2.7**	(a)
2.8	(d)	**2.9**	(d)	**2.10**	(b)	**2.11**	(c)	**2.12**	(d)	**2.13**	(b)	**2.14**	(b)
2.15	(b)	**2.16**	(b)	**2.17**	(17.14)	**2.18**	(271.07)	**2.19**	(3.9)	**2.20**	(a)	**2.21**	(28.2)
2.22	(a)	**2.23**	(a)	**2.24**	(48)	**2.25**	(57.6)						

Explanations Steady State Heat Conduction

2.1 (d)

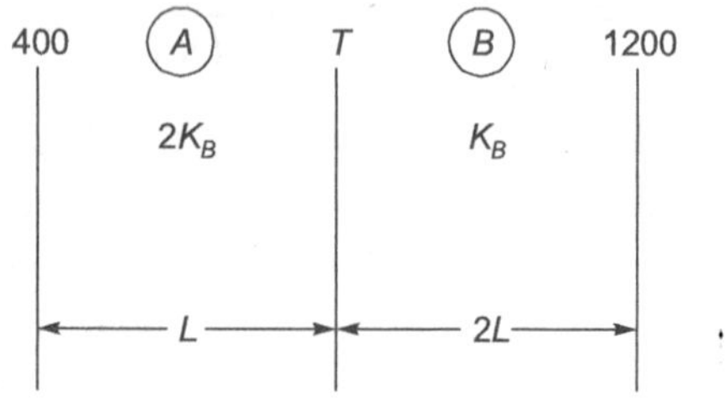

At steady state

$$\frac{1200-400}{\frac{2L}{K_B}+\frac{L}{2K_B}} = \frac{1200-T}{\frac{2L}{K_B}} = \frac{T-400}{\frac{L}{2K_B}}$$

$$1200 - T = 4T - 1600 = 560 \text{ K}$$

So, temperature drop across material A is $560 - 400 = 160$ K.

2.2 (28.73)

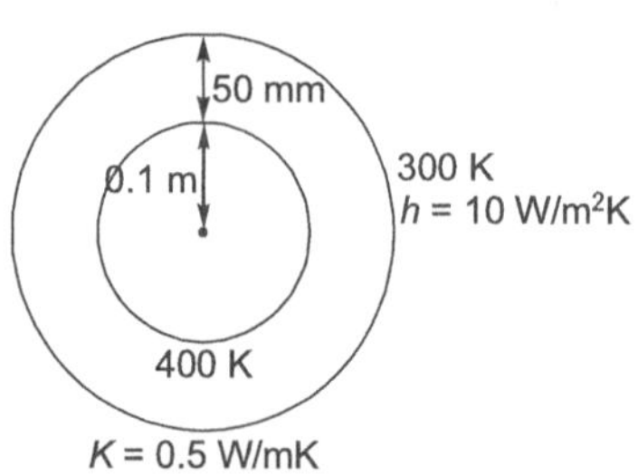

$$Q_1 = hA\Delta T$$
$$= 10 \times \pi \times 0.2 \times L \times (400 - 300)$$
$$\frac{Q_1}{L} = 628.32 \text{ W/m}$$
$$Q_2 = \frac{400-300}{\frac{\ln\left(\frac{0.15}{0.1}\right)}{2\pi\times 0.5L}+\frac{1}{10\pi\times 0.30\times L}}$$
$$\frac{Q_2}{L} = 425.25$$

$$\% \text{ reduction heat loss} = \frac{\frac{Q_1}{L}-\frac{Q_2}{L}}{\frac{Q_1}{L}}\times 100$$

$$= \frac{628.32-425.25}{628.32}\times 100 = 32.32\%$$

2.3 (c)

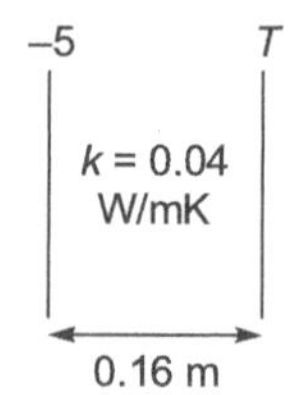

$$Q = \frac{\Delta T}{\frac{b}{KA}}$$
$$\frac{Q}{A} = \frac{T+5}{\frac{0.16}{0.04}} = 10$$
$$T = 35°\text{C}$$

2.4 (a)

Critical radius of insulation for cylinder is given as

$$r_c = \frac{k_{\text{insulation}}}{h_o}$$
$$= \frac{0.6}{8} = 7.5 \text{ cm}$$

Since, radius of pipe (5 cm) is less than r_c. So, initially heat flux will increase then decrease.

Radius after insulation is (5 + 2) = 7 cm.

So, option (a) is correct.

2.5 (c)

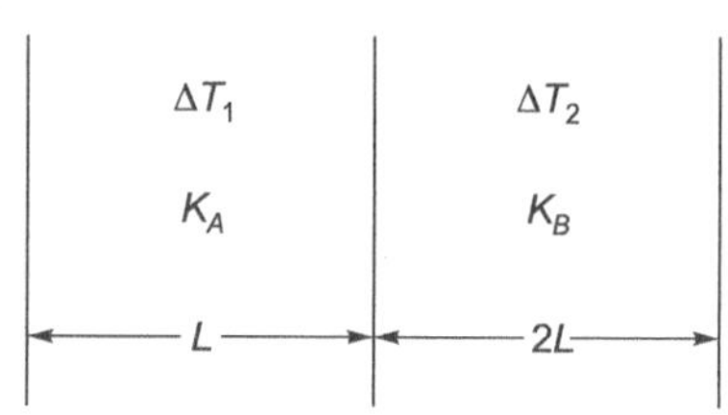

At steady state

$$\frac{\Delta T_1}{\frac{L}{K_A}} = \frac{\Delta T_2}{\frac{2L}{K_B}}$$
$$\frac{\Delta T_1}{\Delta T_2} = \frac{K_B}{2K_A}$$
$$\Delta T_1 > \Delta T_2 \text{ (given)}$$

So, $K_B > 2K_A$

$$K_A < 0.5K_B$$

2.7 (a)

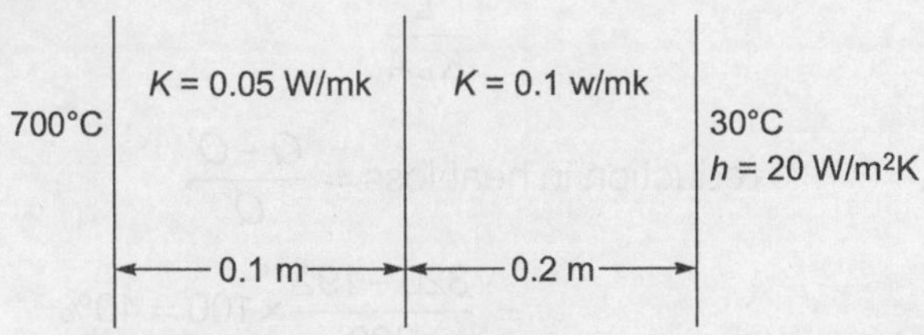

Rate of heat loss

$$Q = \frac{\Delta T}{\Sigma R} = \frac{\Delta T}{\frac{b_1}{K_1 A} + \frac{b_2}{K_2 A} + \frac{1}{hA}}$$

$$\frac{Q}{A} = \frac{700 - 30}{\frac{0.1}{0.05} + \frac{0.2}{0.1} + \frac{1}{20}} = 165.43 \text{ W/m}^2$$

2.8 (d)

For cylinder on adding insulation heat loss first increases upto critical thickness of insulation and then decreases.

2.9 (d)

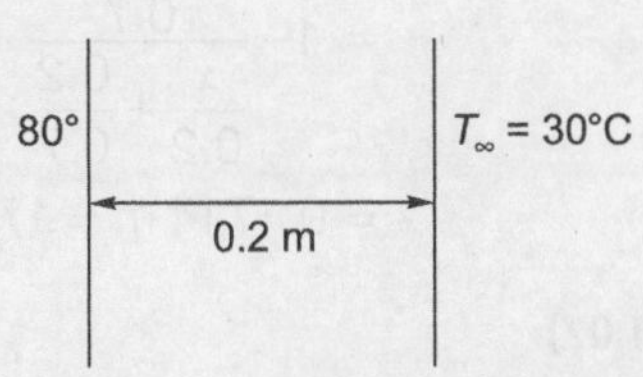

$K = 1.2$ W/mK, $h = 10$ W/m²K

At steady state

$$\frac{80 - T}{\left(\frac{0.2}{1.2A}\right)} = \frac{80 - 30}{\frac{0.2}{1.2A} + \frac{1}{10A}}$$

Solving, we get $T = 48.7°C$

2.10 (b)

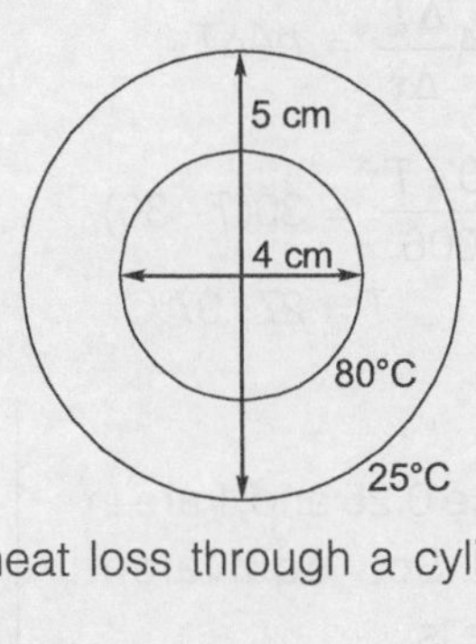

Rate of heat loss through a cylindrical body is given by

$$Q = \frac{\Delta T}{\frac{\ln\left(\frac{r_2}{r_1}\right)}{2\pi kL}}$$

$$\frac{Q}{L} = \frac{80 - 25}{\frac{\ln\left(\frac{2.5}{2}\right)}{2\pi \times 10}} = 15486.67 \text{ W/m}$$

2.11 (c)

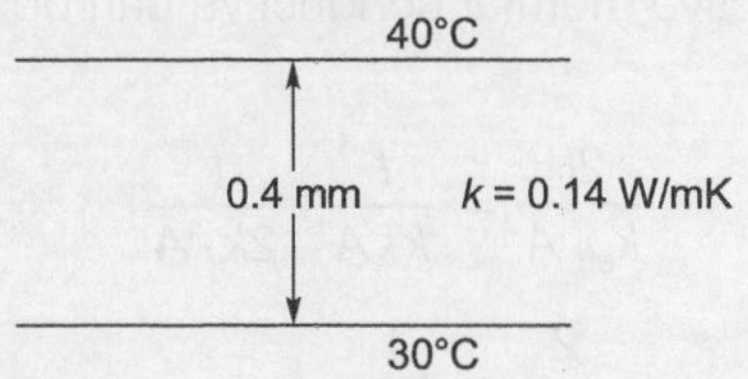

Heat flux, $Q = \dfrac{\Delta T}{\frac{b}{kA}}$

$$\left(\frac{Q}{A}\right) = \frac{40 - 30}{\frac{0.0004}{0.14}} = 3500 \text{ W/m}^2$$

2.12 (d)

Nusselt number based on wall length is 20.

$$\frac{hL}{k_F} = 20$$

$$\frac{30 \times 0.1}{k_F} = 20$$

$$k_F = 0.15 \text{ W/mK}$$

Now, $Q = -kA\dfrac{dT}{dx} \rightarrow \left(\dfrac{Q}{A}\right) = -k_f \dfrac{dT}{dx}$

$$21 = -0.15 \times \frac{dT}{dx}$$

$$\frac{dT}{dx} = -140 \text{ K/m}$$

Magnitude of $\dfrac{dT}{dx} = 140$ K/m

2.13 (b)

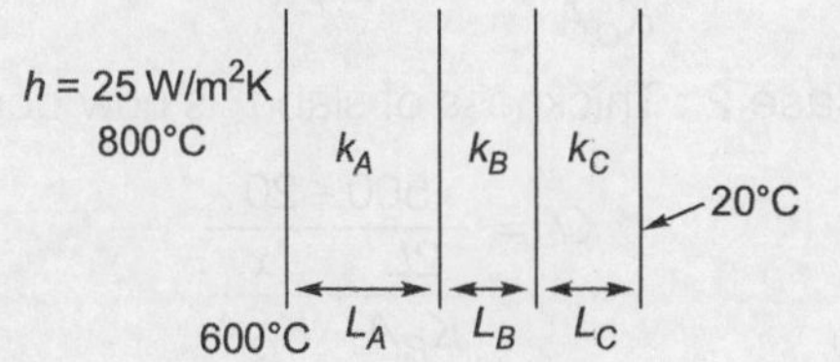

At steady state,

$$hA\Delta T_1 = \frac{\Delta T_2}{\frac{L_A}{k_A \cdot A} + \frac{L_B}{k_B \cdot A} + \frac{L_C}{k_C \cdot A}}$$

$$25(800-600) = \frac{600-20}{\frac{0.3}{20}+\frac{0.15}{k_B}+\frac{0.15}{50}}$$

$$k_B = 1.53 \text{ W/mK}$$

2.14 (b)

Effective thermal conductive can be calculated by

$$\frac{2t}{k_{\text{eff}}A} = \frac{t}{k\cdot A}+\frac{t}{2k\cdot A}$$

$$\frac{2}{k_{\text{eff}}} = \frac{3}{2k}$$

$$k_{\text{eff}} = \frac{4k}{3}$$

2.15 (b)

In slab heat transfer occurs by conduction and near the wall heat transfer will be by convection. In all given options, only option (b) has non-zero slope of temperature at the born end of wall. So, option (b) is the correct option.

2.16 (b)

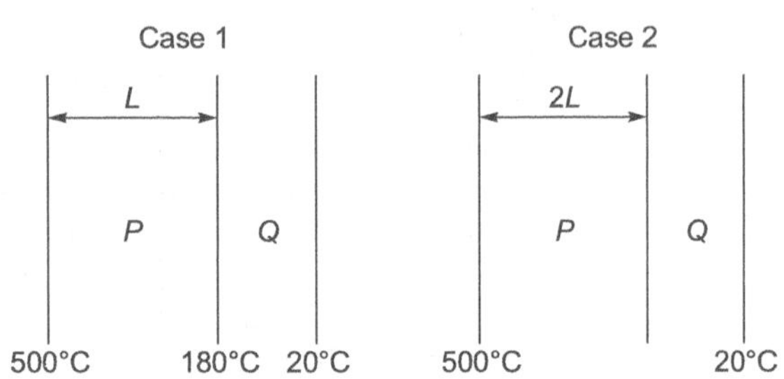

Case 1 : For steady state :

Heat transfer by slab P = Heat transfer by slab Q

$$Q = \frac{500-180}{\frac{L}{K_PA}} = \frac{180-20}{\frac{x}{K_QA}}$$

$$\frac{x}{K_QA} = 0.5\, L/K_PA \qquad \text{...(1)}$$

Case 2 : Thickness of slab P is now doubled

$$Q' = \frac{500-20}{\frac{2L}{K_PA}+\frac{x}{K_QA}}$$

From equation (1)

$$Q' = \frac{480}{2.5\frac{L}{K_PA}}$$

$$Q' = \frac{192}{\frac{L}{K_PA}} \qquad \text{...(ii)}$$

$$\% \text{ reduction in heat loss} = \frac{Q-Q'}{Q'}$$

$$= \frac{320-192}{320}\times 100 = 40\%$$

2.17 (17.14)

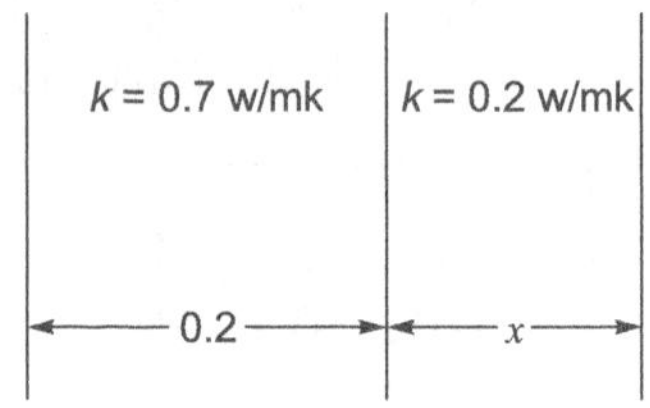

Assume thickness of insulation of x meters.

% reduction in heat loss

$$= \frac{Q_1-Q_2}{Q_1} = 1-\frac{Q_2}{Q_1} = 1-\frac{R_1}{R_2}$$

$$= 1-\frac{\frac{0.2}{0.7}}{\frac{x}{0.2}+\frac{0.2}{0.7}} = 0.75$$

$$x = 0.1714 \text{ m} = 17.14 \text{ cm}$$

2.18 (271.07)

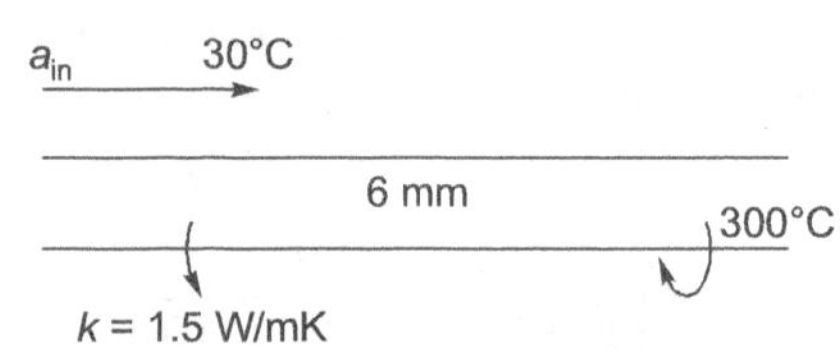

At steady state

$$Q_{\text{conduction}} = Q_{\text{convection}}$$

$$kA\frac{\Delta T}{\Delta x} = hA\Delta T$$

$$1.5\times\frac{300-T}{0.006} = 30(T-30)$$

$$T = 271.07°\text{C}$$

2.19 (3.9)

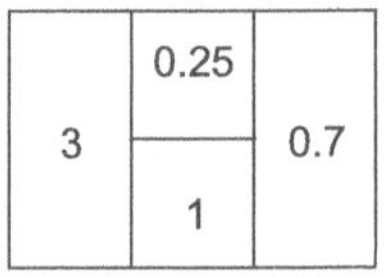

Resistance 0.25 and 1 are in parallel. So, equivalent resistance is

$$\frac{1}{R} = \frac{1}{R_1}+\frac{1}{R_2}$$

$$\frac{1}{R} = \frac{1}{0.25}+\frac{1}{1}$$

$R = 0.2$

Now, above three resistances are in series. So, overall resistance

3	0.2	0.7

$$R_0 = R_1 + R_2 + R_3$$
$$R_0 = 3 + 0.2 + 0.7 = 3.9 \text{ K/watt}$$

2.20 (a)

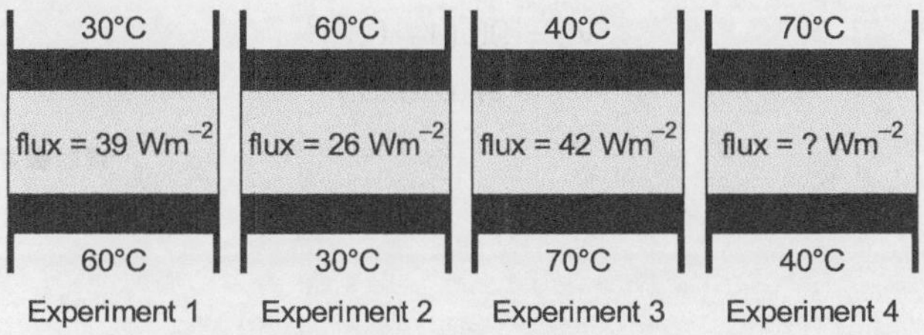

Radiation heat flux is negligible. (Given)
Thermal conductivity does not vary.
In given setup in experiment-2 and experiment-4, upper plate is maintained at higher temperature and lower plates is at low temperature. So, convection cannot occur i.e. 26 W/m^2 in these two cases because in convection heat transfer is due to density difference.
So, heat flux is only caused by conduction and will be same because temperature difference is same.

2.21 (28.2)

$$r_o = 1 \text{ m}$$
$$T_s = 200°\text{C}$$

Insulation thickness = 0.1 m

$$T_\infty = 20°\text{C}$$
$$h = 10 \text{ W/m}^2\text{K}$$

At steady state

$$\begin{pmatrix}\text{Heat transfer rate}\\ \text{by conduction}\end{pmatrix} = \begin{pmatrix}\text{Heat transfer rate}\\ \text{by convection}\end{pmatrix}$$

$$\frac{(200 - T_0)}{\dfrac{\ln\left(\dfrac{1.1}{1}\right)}{2\pi \times 0.05}} = \frac{T_0 - 20}{\dfrac{1}{10 \times 2\pi \times 1.1}}$$

$$T_0 = 28.19°\text{C}$$

2.22 (a)

Given: $r_0 = \dfrac{1}{2} = 0.5 \text{ mm} = 0.5 \times 10^{-3} \text{ m}$

$$R_0 = 0.5 + 2 = 2.5 \text{ mm} = 2.5 \times 10^{-3} \text{ m}$$
$$h = 20 \text{ W/m}^2 \text{ K}$$

$$q_{\text{without insulation}} \geq q_{\text{with insulation}}$$

For maximum limit of thermal conductivity

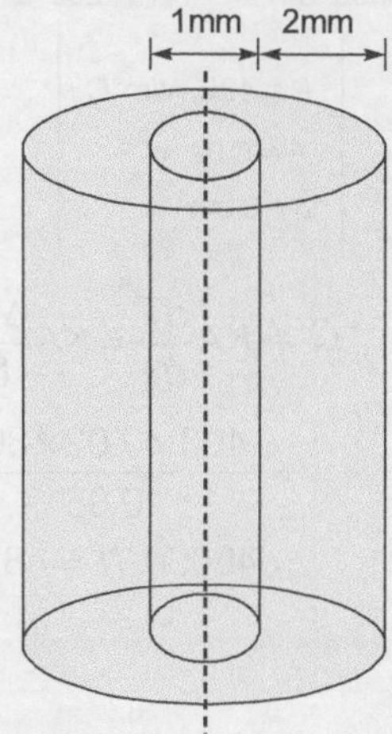

$$q_{\text{without insulation}} = q_{\text{with insulation}}$$

$$\frac{(T_s - T_\infty)}{\dfrac{1}{hA_1}} = \frac{(T_s - T_\infty)}{\dfrac{1}{hA_2} + \dfrac{\ln\left(\dfrac{R_0}{r_0}\right)}{2\pi KL}}$$

$$\Rightarrow \quad \frac{1}{hA_1} = \frac{1}{hA_2} + \frac{\ln\left(\dfrac{R_0}{r_0}\right)}{2\pi KL}$$

$$= \frac{1}{20 \times 2\pi \times 0.5 \times 10^{-3} \times L}$$

$$= \frac{1}{20 \times 2\pi \times 2.5 \times 10^{-3} \times L} + \frac{\ln\left(\dfrac{2.5}{0.5}\right)}{2\pi \times K \times L}$$

$$\Rightarrow \quad 100 = 20 + \frac{\ln 5}{K}$$

$$\Rightarrow \quad K = \frac{\ln(5)}{80} = 0.02 \text{ W/mK}$$

2.23 (a)

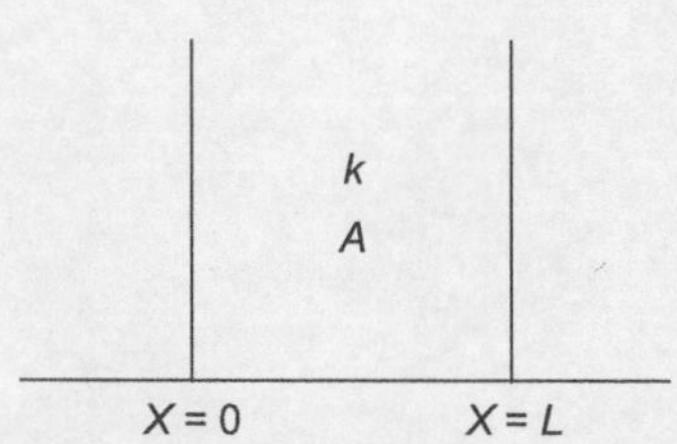

From, conduction we have

$$\Sigma R = \frac{b}{kA}, \quad R = \frac{L}{kA}$$

where, R = Resistance, k = Thermal conductivity
A = Cross-sectional area
L = Thickness of slab

2.24 (48)

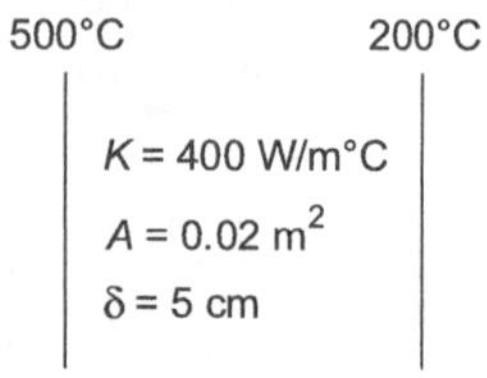

$$Q = KA\frac{dT}{dx} = KA\frac{\Delta T}{\delta}$$

$$= \frac{400 \times 0.02 \times 300}{0.05}$$

$$= 48000 \text{ W} = 48 \text{ kW}$$

2.25 (57.6)

Capital cost of a distillation column = Rs. 90 lakhs

For double-decline balance method when salvage value is zero.

$$F = \frac{2}{n} = \frac{2}{10} = 0.2$$

$$V_a = V(1 - F)^n$$

For $n = 2$ year

$$V_a = 90\,(1 - 0.2)^2$$

$$= 57.6 \text{ lakh}$$

■■■■

3 Steady-State Conduction with Internal Heat Generation

3.1 A fluid flows through a cylindrical pipe under fully developed, steady state laminar flow conditions. The tube wall is maintained at constant temperature. Assuming constant physical properties and negligible viscous heat dissipation, the governing equation for the temperature profile is (z, axial direction; r, radial direction)

(a) $u_{max}\left[1-\left(\frac{r}{R}\right)^2\right]\left(\frac{\partial T}{\partial z}\right)=\frac{k}{\rho C_p}\left[\frac{1}{r}\frac{\partial}{\partial r}\left(r\frac{\partial T}{\partial r}\right)+\frac{\partial^2 T}{\partial z^2}\right]$

(b) $u_{max}\left[1-\left(\frac{r}{R}\right)^2\right]\left(\frac{\partial T}{\partial r}\right)=\frac{k}{\rho C_p}\left[\frac{1}{r}\frac{\partial}{\partial r}\left(r\frac{\partial T}{\partial r}\right)+\frac{\partial^2 T}{\partial r^2}\right]$

(c) $2u_{max}\left[1-\left(\frac{r}{R}\right)^2\right]\left(\frac{\partial^2 T}{\partial z^2}\right)=\frac{k}{\rho C_p}\left[\frac{1}{r}\frac{\partial}{\partial r}\left(r\frac{\partial T}{\partial r}\right)+\frac{\partial^2 T}{\partial r^2}\right]$

(d) $u_{max}\left[1-\left(\frac{r}{R}\right)^2\right]\left(\frac{\partial T}{\partial z}\right)=\frac{k}{\rho C_p}\left[\frac{1}{z}\frac{\partial}{\partial z}\left(z\frac{\partial T}{\partial z}\right)+\frac{\partial^2 T}{\partial r^2}\right]$

[2006 : 2 M]

3.2 A process fluid has to be cooled from 22°C to 2°C using brine in a 2-4 shell-and-tube heat exchanger shown below. The brine enters at –3°C and leaves at 7°C. The overall heat transfer coefficient is 500 W/(m^{2}K). The design heat load is 30 kW. The brine flows on the tube side and the process fluid on the shell side. The heat transfer area in (m^2) is

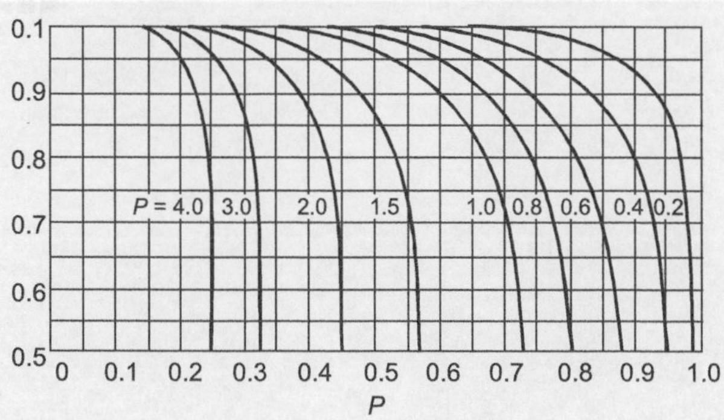

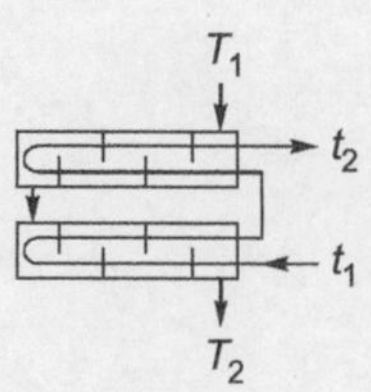

$$P=\frac{t_2-t_1}{T_1-t_1}$$

$$R=\frac{T_1-T_2}{t_2-t_1}$$

(a) 1.1 (b) 5.77
(c) 6.59 (d) 7.53

[2006 : 2 M]

3.3 The figure below shows steady state temperature profiles for one dimensional heat transfer within a solid slab for the following cases:

P : uniform heat generation with left surface perfectly insulated

Q : uniform heat generation with right surface perfectly insulated

R : uniform heat consumption with left surface perfectly insulated

S : uniform heat consumption with right surface perfectly insulated

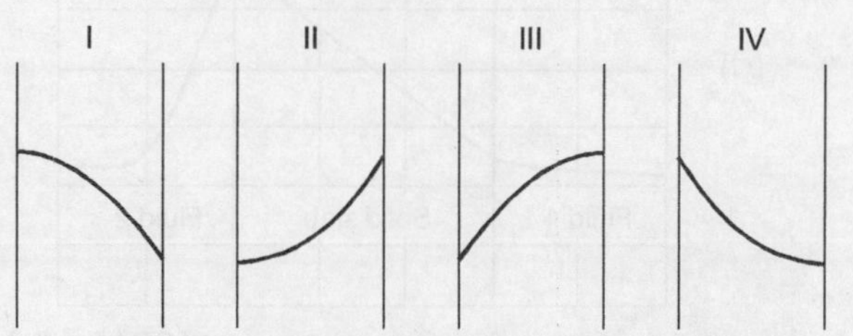

Match the profiles with appropriate cases :

(a) P-1, Q-3, R-2, S-4 (b) P-2, Q-3, R-1, S-4
(c) P-1, Q-4, R-2, S-3 (d) P-2, Q-4, R-1, S-3

[2010 : 2 M]

3.4 Heat is generated uniformly within a solid slab. The slab separates fluid 1 from fluid 2. The heat transfer coefficients between the solid slab and the fluids are h_1 and h_2 ($h_2 > h_1$) respectively. The steady state temperature profile (T vs. x) for one-dimensional heat transfer is correctly shown by

(a)

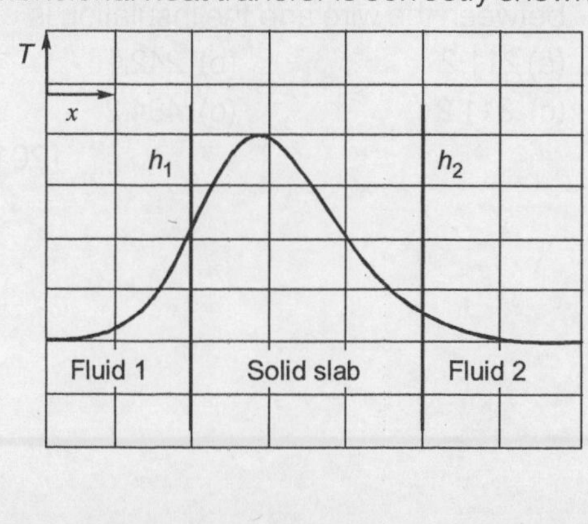

(b)

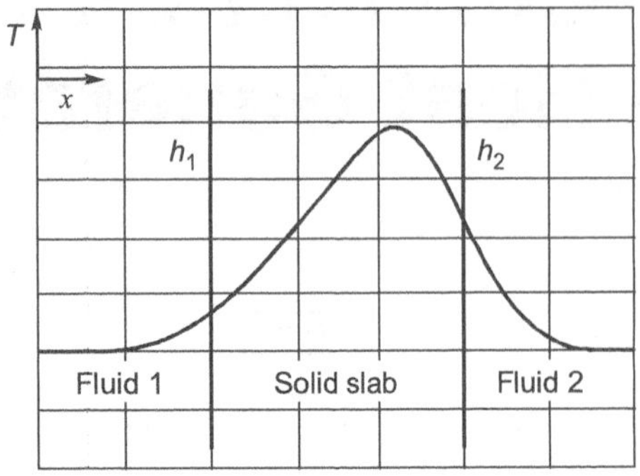

(c)

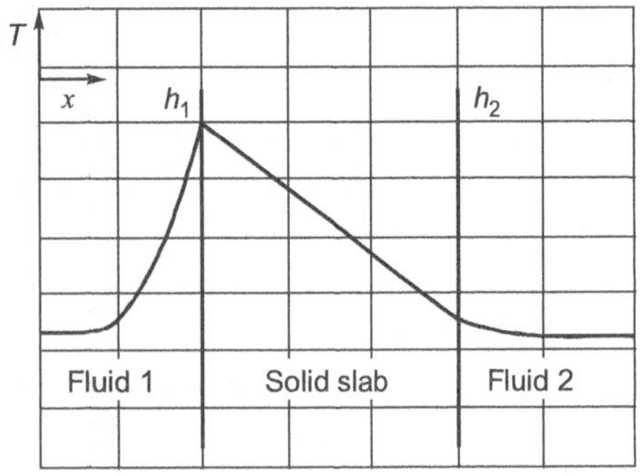

(d)

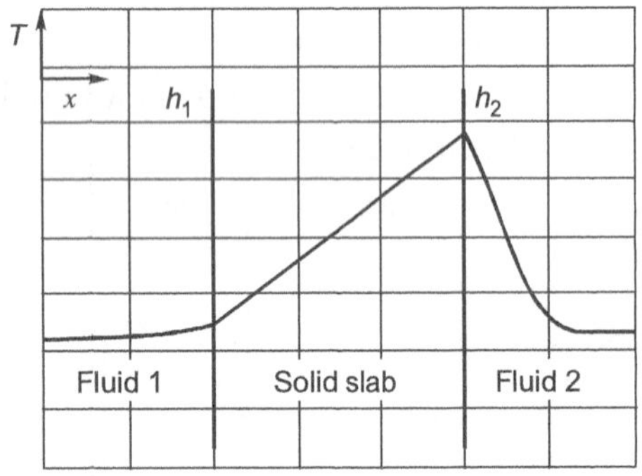

[2011 : 2 M]

3.5 Heat is generated at a steady rate of 100 W due to resistance heating in a long wire (length = 5 m, diameter = 2 mm). This wire is wrapped with an insulation of thickness 1 mm that has a thermal conductivity of 0.1 W/m.K. The insulated wire is exposed to air at 30°C. The convective heat transfer between the wire and surrounding air is characterized by a heat transfer coefficient of 10 W/m^2.K. The temperature (in °C) at the interface between the wire and the insulation is

(a) 211.2 (b) 242.1
(c) 311.2 (d) 484.2

[2012 : 2 M]

3.6 A 20 cm diameter cylindrical solid pellet of a nuclear fuel with density 6000 kg/m^3 and conductivity of 300 W/(mK) generates heat by nuclear fission at a spatially uniform rate of 10^4 W/kg. The heat from the fuel pellet is transferred to the surrounding coolant by convection such that the pellet wall temperature remains constant at 300°C. Neglecting the axial and azimuthal dependence, the maximum temperature (in °C) in the pellet at steady state is _____ (rounded off to the nearest integer).

[2019 : 2 M]

3.7 Consider a solid slab of thickness $2L$ and uniform cross section A. The volumetric rate of heat generation within the slab is $\dot{g}$ (W m^{-3}). The slab loses heat by convection at both the ends to air with heat transfer coefficient h. Assuming steady state, one-dimensional heat transfer, the temperature profile within the slab along the thickness is given by:

$$T(x) = \frac{\dot{g}L^2}{2k}\left[1-\left(\frac{x}{L}\right)^2\right]+T_s \text{ for } -L \le x \le L$$

where k is the thermal conductivity of the slab and T_s is the surface temperature. If T_s = 350 K, ambient air temperature T_∞ = 300 K, and Biot number (based on L as the characteristic length) is 0.5, the maximum temperature in the slab is _______ K (round off to nearest integer).

[2021 : 2 M]

■■■■

Answers Steady-State Conduction with Internal Heat Generation

3.1 (a) **3.2** (d) **3.3** (a) **3.4** (a) **3.5** (a) **3.6** (800) **3.7** (362.5)

Explanations Steady-State Conduction with Internal Heat Generation

3.2 (d)

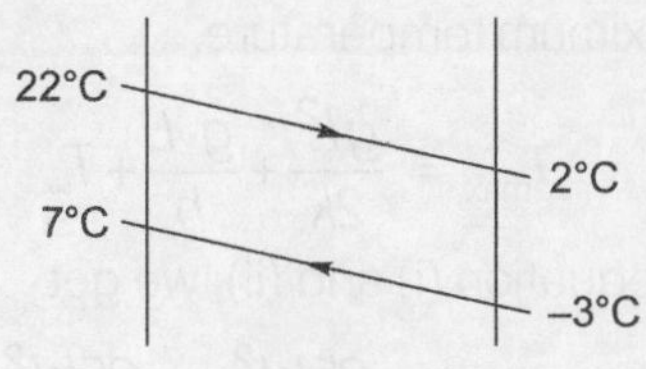

$$U = 500 \text{ W/m}^2\text{K}$$
$$Q = 30 \text{ kW}$$
$$P = \frac{t_2 - t_1}{T_2 - t_1} = \frac{7-(-3)}{22-(-3)}$$
$$= \frac{10}{5} = 0.4$$
$$R = \frac{T_1 - T_2}{t_2 - t} = \frac{20}{10} = 2$$

From figure given in question, $F = 0.85$

$$\Delta T_{\text{LMTD}} = \frac{\Delta T_1 - \Delta T_2}{\ln\left(\frac{\Delta T_1}{\Delta T_2}\right)} = \frac{15-5}{\ln\left(\frac{15}{5}\right)} = 9.102$$

Now, $$Q = UA\Delta T_{\text{LMTD}} \times F$$
$$30 \times 10^3 = 500 \times A \times 9.102 \times 0.85$$
$$A = 7.75 \text{ m}^2$$

So, closest option is (d).

3.3 (a)

Perfectly insulated means heat loss is zero.

$$Q = 0$$
$$-kA\frac{dT}{dx} = 0$$
$$\frac{dT}{dx} = 0$$

So, slope at perfect insulated surface will be zero.

3.4 (a)

This is the case of heat generation. So, temperature profile will not be linear and $h_2 > h_1$. So maximum temperature within slab will be more closer to left side.

If $h_2 > h_1$ then $\left(\frac{dT}{dx}\right)_{\text{fluid 1}} > \left(\frac{dT}{dx}\right)_{\text{fluid 2}}$,

Therefore option (a) will be correct one, as $\frac{dT}{dx}$ of fluid 1 is greater than fluid 2.

3.5 (a)

$$T_\infty = 30^\circ\text{C}$$

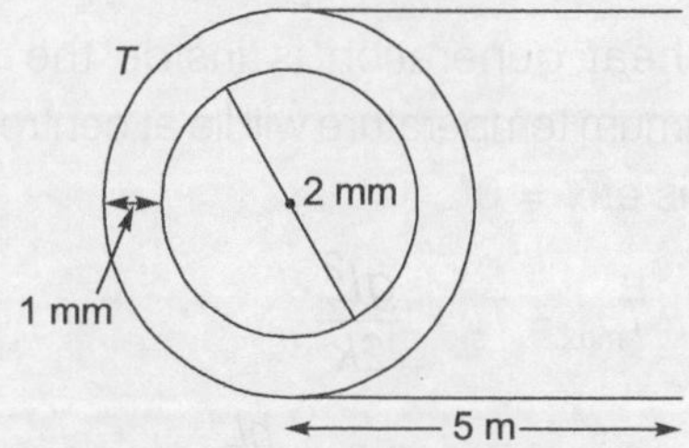

$$k_{\text{insulation}} = 0.1 \text{ W/mK}$$

At steady state

$$Q = hA\Delta T$$
$$100 = 10 \times \pi \times 0.004 \times 5 \times (T - 30)$$
$$T = 189.15^\circ\text{C}$$

Now, $$100 = \frac{T_s - 189.15}{\frac{\ln\left(\frac{0.002}{0.001}\right)}{2\pi \times 0.1 \times 5}}$$
$$T_s = 211.21^\circ\text{C}$$

3.6 (800)

$$\text{Dia} = 0.2 \text{ m}, \quad P = 6000 \text{ kg/m}^3$$
$$K = 300 \text{ W/mK}, \quad q = 10^4 \text{ W/kg}$$
$$T_w = 300^\circ\text{C}$$
$$q\,(\text{W/m}^3) = 10^4 \times 6000 = 6 \times 10^7 \text{ W/m}^3$$

For cylindrical solid with heat generation

$$T = \frac{-qr^2}{4K} + C_1 \ln r + C_2$$

at $r = 0$, $T = T_{\max}$, so $C_1 = 0$

$$T = \frac{-qr^2}{4K} + C_2$$

at $r = 0.1$ m, $T = 300^\circ\text{C}$

$$300 = \frac{-6 \times 10^7 \times (0.1)^2}{4 \times 300} + C_2$$
$$C_2 = 800^\circ\text{C}$$

at $r = 0$, $T = T_{\max}$

So, $T_{\max} = C_2 = 800^\circ\text{C}$

3.7 (362.5)

Given, $T_s = 350$ K, $T_\infty = 300$ K

Biot number = 0.5 (based on length L)

Temperature profile is given as

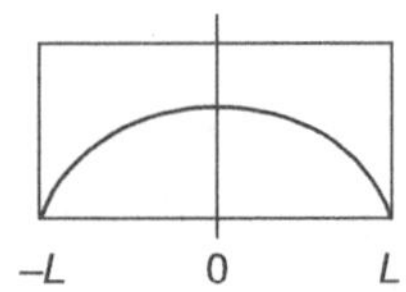

$$T(x) = \frac{\dot{g}L^2}{2k}\left[1-\left(\frac{x}{L}\right)^2\right]+T_s$$

"As heat generation is inside the slab, thus maximum temperature will lie at centre of the slab that is at $x = 0$".

$$T_{max} - T_s = \frac{\dot{g}L^2}{2k}$$

Now, $\quad \text{Bi} = 0.5 = \frac{hL}{k} \Rightarrow h = \frac{0.5k}{L} \quad ...(i)$

From boundary condition,

$$h(T_s - T_\infty)(2A) = \dot{g}\cdot V$$

$$T_s - T_\infty = \frac{\dot{g}\cdot L}{h} = 350 - 300 = 50$$

$$\dot{g} = \frac{50h}{L} \quad ...(i)$$

For maximum temperature,

$$T_{max} = \frac{\dot{g}L^2}{2k} + \frac{g\cdot L}{h} + T_\infty$$

Using equation (i) and (ii), we get

$$T_{max} = \frac{25kL^2}{L^2\cdot k(2)} + \frac{25kL^2}{L^2(0.5)k} + 300$$

$$T_{max} = 12.5 + 50 + 30 = 362.5°C$$

■■■■

4 Heat Transfer from Extended Surfaces (FINS)

4.1 Consider an infinitely long rectangular fin exposed to a surrounding fluid at a constant temperature $T_a = 27°C$.

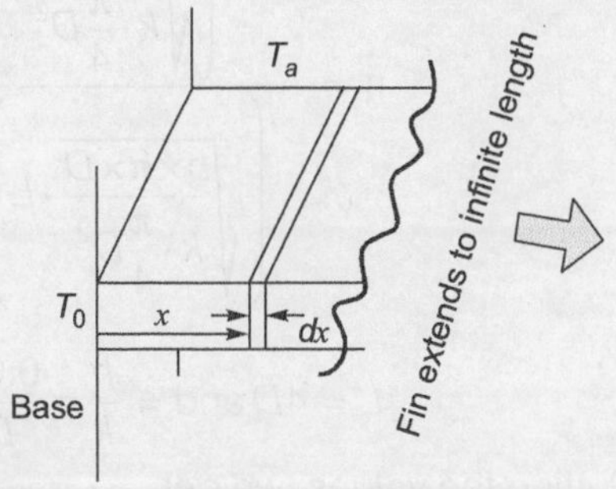

The steady state one dimensional energy balance on an element of the fin of thickness dx at a distance x from its base yields

$$\frac{d^2\theta}{dx^2} = m^2\theta$$

where $\theta = T_x - T_a$, T_x is the temperature of the fin at the distance x from its base in °C. The value of m is 0.04 cm^{-1} and the temperature at the base is $T_0 = 227°C$. The temperature (in °C) at $x = 25$ cm is _____ (round off to 1 decimal place).

[2020 : 2 M]

4.2 A straight fin of uniform circular cross section and adiabatic tip has an aspect ratio (length/diameter) of 4. If the biot number (based on radius of the fin as the characteristic length) is 0.04, the fin efficiency is ________ % (round off to nearest integer).

[2021 : 2 M]

4.3 A cylindrical fin of diameter 24 mm is attached horizontally to a vertical planar wall. The heat transfer rate from the fin to the surrounding air is 60% of the heat transfer rate if the entire fin were at the wall temperature. If the fin effectiveness is 10, its length is ________ mm (rounded off to the nearest integer).

[2022 : 2 M]

■■■■

Answers Heat Transfer from Extended Surfaces (FINS)

4.1 (100.6) **4.2** (43%) **4.3** (94)

Explanations Heat Transfer from Extended Surfaces (FINS)

4.1 (100.6)

Infinitely long rectangular fin is exposed to surrounding at temperature $T_a = 27°C$.

$$\frac{d^2\theta}{dx^2} = m^2\theta \quad ...(1)$$

$$\theta = T_x - T_a$$

$$m = 0.04 \text{ cm}^{-1}$$

$$T_o = 227°C$$

On solving equation (1)

$$\frac{\theta}{\theta_b} = e^{-mx}$$

where, $\theta_b = T_o - T_\theta$

$$= 227 - 27 = 200°C$$

Temperature at $x = 0.25$ m is

$$T|_{x=0.25m} = e^{-0.04 \times 29} \times 200 + 27°C$$

$$= 100.6°C$$

4.2 (43)

Given, $\frac{L}{D} = 4$

Bi = 0.04

[Based on radius as characteristic length]

∴ Fin is a uniform cross-section and adiabatic tip, its efficiency is given by

$$\eta = \frac{\tanh(mL)}{(mL)}$$

where, m is given as, $m = \sqrt{\frac{hP}{k \cdot A_c}}$

P = Perimeter

A_C = Cross-sectional area

Now, $\text{Bi} = 0.04 = \frac{h_D}{k(2)} \Rightarrow \frac{h_D}{k} = 0.08$

Now, efficiency,

$$\eta = \frac{\tanh\left(\sqrt{\frac{h \times \pi D}{k \cdot \frac{\pi}{4} D^2}} \cdot L\right)}{\left(\sqrt{\frac{h \times \pi \times D}{k \cdot \frac{\pi}{4} D^2}} \cdot L\right)}$$

∵ $L = 4D$ and $= \frac{h}{k} = \frac{0.08}{D}$

Putting the values, we get

$$\eta = \frac{\tanh\left(4 \times \sqrt{4 \times 0.08}\right)}{\left(4 \times \sqrt{4 \times 0.08}\right)} = 0.432$$

$$\eta = 43.2\%$$

4.3 (94)

Data given: d = 24 mm

η = 60% (efficiency)

ξ = 10 [effectiveness]

From the relation between η and ξ, we have

is $$\xi = \eta \times \frac{A_s}{A_c}$$

Here, $A_s = \pi dl + \frac{\pi}{4}d^2$ {Total surface area, here tip area is also included}

$$A_c = \frac{\pi}{4}d^2$$

Thus, $$10 = 0.60 \frac{\pi dl + \frac{\pi}{4}d^2}{\frac{\pi}{4}d^2}$$

$$0.25 \times 0.024 \times 16.667 = l + 0.25 \times 0.024$$

$$l = 0.094 = 94 \text{ mm}$$

■■■■

5 Transient Conduction

5.1 A steel sphere of radius 0.1 m at 400 K is immersed in an oil at 390 K. If the centre of the sphere reaches 350 K in 20 minutes, how long will it take for a 0.05 m radius steel sphere to reach the same temperature (at the centre) under identical conditions? Assume that the convective heat transfer coefficient is infinitely large.

(a) 5 min (b) 10 min
(c) 20 min (d) 40 min

[2000 : 2 M]

5.2 1000 kg of liquid at 30°C in a well-stirred vessel has to be heated to 120°C, using immersed coils carrying condensing steam at 150°C. The area of the steam coils is 1.2 m^2 and overall heat transfer coefficient to the liquid is 1500 W/m^2 °C. Assuming negligible heat loss to surrounding and specific heat capacity of the liquid to be 4 kJ/kg °C, the time taken for the liquid to reach desired temperature will be

(a) 15 min (b) 22 min
(c) 44 min (d) 51 min

[2002 : 2 M]

5.3 Three solid objects of the same material and of equal mass - a sphere, a cylinder (length = diameter) and a cube - are at 500°C initially. These are dropped in a quenching bath containing a large volume of cooling oil each attaining the bath temperature eventually. The time required for 90% change of temperature is smallest for

(a) cube (b) cylinder
(c) sphere (d) equal for all the three

[2003 : 1 M]

5.4 A metal ball of radius 0.1 m at a uniform temperature of 90°C is left in air at 30°C. The density and the specific heat of the metal are 3000 kg/min and 0.4 kJ/(kg K), respectively. The heat transfer coefficient is 50 W/(m^2K). Neglecting the temperature gradients inside the ball, the time taken (in hours) for the ball to cool to 60 °C is

(a) 555 (b) 55.5
(c) 0.55 (d) 0.15 [2004 : 2 M]

5.5 A semi-infinite slab occupying the region $x = 0$ and $x = \infty$ is at an initial temperature T_0. At time $t = 0$, the surface of the slab at $x = 0$ is brought into contact with a heat bath at a temperature T_H. The temperature $T(x, t)$ of the slab rises according to the equation

$$\frac{T_H - T(x,t)}{T_H - T_0} = \frac{2}{\sqrt{\pi}} \int_0^{x/\sqrt{4\alpha}} e^{-\eta^2} d\eta$$

where x is position and t is time. The heat flux at the surface $x = 0$ is proportional to

(a) $t^{-1/2}$ (b) $t^{1/2}$
(c) t (d) $t^{3/2}$

[2005 : 2 M]

Statement for Linked Answer Question (5.6 and 5.7):

A liquid of mass 7 kg and specific heat 4 kJ/(kg °C) is contained in a cylindrical heater of diameter 0.15 m and height 0.40 m. The cylindrical surface of the heater is exposed to air at 25°C while the end caps are insulated, so that heat transfer takes place only through the cylindrical surface.

The thickness of the wall of the heater = 2 mm
The wall thermal conductivity = 10 W/(mK),
The heat transfer coefficient in the liquid = 100 W/(m^2K)
The heat transfer coefficient in air = 10 W/(m^2K)
The liquid is initially maintained at a temperature of 75°C. At time $t = 0$, the heater is switched off, and the temperature of the liquid in the heater decreases due to heat loss across. The cylindrical surface.

5.6 What is the overall heat transfer coefficient in W/(m^2K)?

(a) 1 (b) 4.04
(c) 9.07 (d) 10 [2005 : 2 M]

5.7 What is the time required for the temperature of the liquid to reduce to 50°C after the heater is switched off, assuming lumped system analysis is valid?

(a) 7.874×10^3 s (b) 11.346×10^3 s
(c) 16.828×10^3 s (d) 23.213×10^3 s

[2005 : 2 M]

5.8 Transient three-dimensional heat conduction is governed by one of the following differential equations (α-thermal diffusivity, k-thermal conductivity and Ψ-volumetric rate of heat generation)

(a) $\frac{1}{\alpha}\cdot\frac{\partial T}{\partial t} = \nabla T + \psi k$

(b) $\frac{1}{\alpha}\cdot\frac{\partial T}{\partial t} = \nabla T + \frac{\psi}{k}$

(c) $\frac{1}{\alpha}\cdot\frac{\partial T}{\partial t} = \nabla^2 T + \psi k$

(d) $\frac{1}{\alpha}\cdot\frac{\partial T}{\partial t} = \nabla^2 T + \frac{\psi}{k}$

[2008 : 1 M]

5.9 A metallic ball (ρ = 2700 kg/m^3 and C_p = 0.9 kJ/kg°C) of diameter 7.5 cm is allowed to cool in air at 25°C. When the temperature of the ball is 125°C, it is found to cool at the rate of 4°C per minute. If thermal gradients inside the ball are neglected, the heat transfer coefficient (in W/m^2 °C) is

(a) 2.034 (b) 20.34
(c) 81.36 (d) 203.4

[2008 : 2 M]

5.10 During the transient convective cooling of a solid object, biot number → 0 indicates

(a) uniform temperature throughout the object
(b) negligible convection at the surface of the object
(c) significant thermal resistance within the object
(d) significant temperature gradient within the object

[2009 : 1 M]

5.11 For heat transfer across a solid-fluid interface, which one of the following statements is NOT true when the Biot number is very small compared to 1?

(a) Conduction resistance in the solid is very small compared to convection resistance in the fluid
(b) Temperature profile within the solid is nearly uniform
(c) Temperature drop in the fluid is significant
(d) Temperature drop in the solid is significant

[2012 : 1 M]

5.12 A solid sphere with an initial temperature T_i is immersed in a large thermal reservoir of temperature T_0. The sphere reaches a steady temperature after a certain time t_1. If the radius of the sphere is doubled, the time required to reach steady-state will be

(a) $\frac{t_1}{4}$ (b) $\frac{t_1}{2}$
(c) $2t_1$ (d) $4t_1$

[2012 : 1 M]

5.13 The one-dimensional unsteady state heat conduction equation in a hollow cylinder with a constant heat source q is $\frac{\partial T}{\partial t} = \frac{1}{r}\frac{\partial}{\partial r}\left(r\frac{\partial T}{\partial r}\right) + q$.

If A and B are arbitrary constants, then the steady state solution to the above equation is

(a) $T(r) = -\frac{qr^2}{2} + \frac{A}{r} + B$

(b) $T(r) = -\frac{qr^2}{4} + A\ln r + B$

(c) $T(r) = A\ln r + B$

(d) $T(r) = \frac{qr^2}{4} + A\ln r + B$

[2012 : 2 Mark]

5.14 A heated solid copper sphere (of surface area A and volume V) is immersed in a large body of cold fluid. Assume the resistance to heat transfer inside the sphere to be negligible and heat transfer coefficient (h), density (ρ), heat capacity (C), and thermal conductivity (k) to be constant. Then, at time t, the temperature difference between the sphere and the fluid is proportional to:

(a) $\exp\left[-\frac{hA}{\rho VC}t\right]$ (b) $\exp\left[-\frac{\rho VC}{hA}t\right]$

(c) $\exp\left[-\frac{4\pi k}{\rho CA}t\right]$ (d) $\exp\left[-\frac{\rho CA}{4\pi k}t\right]$

[2015 : 2 M]

5.15 The one-dimensional unsteady heat conduction equation is

$$\rho C_p\frac{\partial T}{\partial t} = \frac{1}{r^n}\frac{\partial}{\partial r}\left(r^n k\frac{\partial T}{\partial r}\right)$$

where T - temperature, t - time, r - radial position, k - thermal conductivity, ρ - density and C_p - specific heat.

For the cylindrical coordinate system, the value of n in the above equation is

(a) 0 (b) 1
(c) 2 (d) 3 [2017 : 1 M]

5.16 An insulated storage tank contains 1000 kg liquid of specific heat 10 kJ kg^{-1} K^{-1}. The liquid is heated by saturated steam, condensing in a helical coil at a temperature of 180°C. The heat transfer area of the coil is 0.1 m^2. If the overall heat transfer coefficient is constant at 1000 Wm^{-2} K^{-1}, then the time (in hours) required to raise the temperature of the liquid in the tank from 20°C to 80°C is _____ (rounded off to second decimal place).

[2018 : 2 M]

5.17 A solid sphere of radius 1 cm and initial temperature of 25°C is exposed to a gas stream at 100°C. For the solid sphere, the density is 10^4 kg/m^3 and the specific heat capacity is 500 J/(kg K). The density of the gas is 0.6 kg/m^3 and its specific heat capacity is 10^3 J/(kg K). The solid sphere is approximated as a lumped system (Biot number ≪ 1) and all specific heats are constant. If the heat transfer coefficient between the solid and gas is 50 W/(m^2 K), the time (in seconds) needed for the sphere to reach 95°C is _____ (rounded off to the nearest integer).

[2019 : 2 M]

5.18 A solid slab of thickness H_1 is initially at a uniform temperature T_0. At time $t = 0$, the temperature of the top surface at $y = H_1$ is increased to T_1, while the bottom surface at $y = 0$ is maintained at T_0 for $t \geq 0$. Assume heat transfer occurs only in the y-direction, and all thermal properties of the slab are constant. The time required for the temperature at $y = \frac{H_1}{2}$ to reach 99% of its final steady value is τ_1. If the thickness of the slab is doubled to $H_2 = 2H_1$, and the time required for the temperature at $y = \frac{H_2}{2}$ to reach 99% of its final steady value is τ_2, then $\frac{\tau_2}{\tau_1}$ is

(a) 2
(b) $\frac{1}{4}$
(c) 4
(d) $\frac{1}{2}$

[2024 : 2 M]

■■■■

Answers Transient Conduction

5.1	(a)	5.2	(d)	5.3	(a)	5.4	(d)	5.5	(a)	5.6	(c)	5.7	(b)
5.8	(d)	5.9	(b)	5.10	(a)	5.11	(d)	5.12	(c)	5.13	(b)	5.14	(a)
5.15	(b)	5.16	(13.06)	5.17	(903)	5.18	(c)						

Explanations Transient Conduction

5.1 (a)

In the case of $h \to \infty$, i.e. unsteady state heat transfer

$$t \propto x^2 \text{ or } r^2$$

So,
$$\frac{20}{(0.1)^2} = \frac{t_2}{(0.05)^2}$$

$$t_2 = 5 \text{ minutes}$$

5.2 (d)

Time required for heating is given by

$$\frac{T - T_\infty}{T_i - T_\infty} = e^{-\frac{hAt}{\rho V C_P}}$$

Putting the values, we get

$$\frac{120 - 150}{30 - 150} = e^{-\frac{1500 \times 1.2 \times t}{1000 \times 4000}}$$

$$t = 51.34 \text{ minutes}$$

5.3 (a)

For the given material of equal mass, cube has highest area available for heat transfer than cylinder and sphere. So, time required for 90% change is smaller for cube.

5.4 (d)

For lumped heat capacity system:

$$\frac{T - T_\infty}{T_i - T_\infty} = e^{-\frac{hAt}{\rho V C_P}}$$

Putting values,

$$\frac{60 - 30}{90 - 30} = e^{-\frac{50 \times 3 \times t}{3000 \times 400 \times 0.1}}$$

$$t = 554.4 \text{ sec} = 0.15 \text{ hours}$$

5.5 (a)

$$\frac{T_H - T(x,t)}{T_H - T_o} = \frac{2}{\sqrt{\pi}} \int_0^\eta e^{-x^2} dx$$

where, $\eta = \dfrac{x}{\sqrt{4\alpha t}}$

$$\left.\frac{dT}{dx}\right|_{x=0} = \frac{1}{(T_H - T_o)} \cdot \frac{2}{\sqrt{\pi}} \frac{d}{dx}\left.\left(\int_0^\eta e^{-x^2} dx\right)\right|_{x=0}$$

$$= \frac{1}{(T_H - T_o)} \cdot \frac{2}{\sqrt{\pi}} \frac{d}{dx}\left(\int_0^\eta e^{-x^2} dx\right) \cdot \left.\frac{dx}{dx}\right|_{x=0}$$

$$= \frac{1}{(T_H - T_o)} \cdot \frac{2}{\sqrt{\pi}} e^{-x^2} \left.\frac{dh}{dx}\right|_{x=0}$$

$$\left(x = \frac{x}{\sqrt{4\alpha t}}\right)$$

On solving,

$$\left.\frac{dT}{dx}\right|_{x=0} = \frac{-1}{(T_H - T_o)\sqrt{\pi \propto t}}$$

$$q = -k\left.\frac{dT}{dx}\right|_{x=0} = \frac{k}{(T_H - T_o)} \cdot \frac{1}{\sqrt{\pi \propto t}}$$

So, $q \propto t^{-1/2}$

5.6 (c)

Overall heat transfer coefficient can be found by

$$\frac{1}{U_o A_o} = \frac{1}{h_i A_i} + \frac{\ln\left(\frac{r_2}{r_1}\right)}{2\pi k l} + \frac{1}{h_o A_o}$$

$$\frac{1}{U_o} = \frac{1}{h_i d_i} d_o + \frac{\ln\left(\frac{r_2}{r_1}\right)}{k} r_o + \frac{1}{h_o}$$

$$\frac{1}{U_o} = \frac{1}{100}\left(\frac{0.154}{0.15}\right) + \frac{\ln\left(\frac{0.154}{0.15}\right) \times \frac{0.154}{2}}{10} + \frac{1}{10}$$

$$U_o = 9.07 \text{ W/m}^2\text{K}$$

5.7 (b)

Applied lumped capacity system:

$$\frac{T - T_\infty}{T_i - T_\infty} = e^{-\frac{hAt}{mC_P}}$$

$$\frac{50-25}{75-25} = e^{-\frac{9.07\times\pi\times0.154\times0.4t}{7\times4000}}$$

$$t = 11.1 \times 10^3 \text{ sec}$$

5.9 (b)

Thermal gradients inside the ball are neglected means it is case of lumped heat capacity system. For lumped heat capacity system :

$$\frac{T-T_\infty}{T_i-T_\infty} = e^{\frac{-hAt}{\rho VC_P}}$$

$$\frac{dT}{dt} = (T_i - T_\infty)\cdot\frac{-hA}{\rho VC_P}e^{\frac{-hAt}{\rho VC_P}}$$

Assume at $t = 0$, $T = 125°C$

So, $$\frac{dT}{dt} = \frac{4°C}{\text{minute}} \text{ (Given)} = \frac{4}{60}\left(\frac{°C}{s}\right)$$

So, $$\frac{-4}{60} = (125-25)\times\left(\frac{-h\times6}{2700\times0.075\times900}\right)$$

$$h = 20.25 \text{ W/m}^2\text{K}$$

5.10 (a)

$$\text{Biot No.} = \frac{\text{Conductive resistance within solid}}{\text{Convective resistance in fluid}}$$

Biot Number → 0 means conductive resistance within solid is negligible. So, uniform temperature throughout the object.

5.11 (d)

$$\text{Biot Number} = \frac{\text{Conductive resistance in solid}}{\text{Convective resistance in fluid}}$$

$$\text{Biot Number} = \frac{hL}{K}$$

It is less than 1 given in question. So, temperature difference in solid will be minimum as conductive resistance is very less.

5.12 (c)

$$\frac{T-T_\infty}{T_i-T_\infty} = \exp\left(\frac{-hAt}{\rho VC_P}\right)$$

Since final temperature is same for both sphere.

So, $$\exp\left(\frac{-hA_1t_1}{\rho V_1C_P}\right) = \exp\left(\frac{-hA_2t_2}{\rho V_2C_P}\right)$$

$$\frac{t_2}{t_1} = \left(\frac{A_1}{A_2}\right)\left(\frac{V_2}{V_1}\right) \quad (\because r_2 = 2r_1)$$

$$\frac{t_2}{t_1} = \frac{1}{4}\times 8 = 2$$

$$\Rightarrow \quad t_2 = 2t_1$$

5.13 (b)

$$\frac{1}{r}\left[\frac{\partial}{\partial r}\left(Tr\frac{\partial T}{\partial r}\right)+q\right] = \frac{\partial T}{\partial t}$$

At steady state $\frac{\partial T}{\partial t} = 0$

$$\frac{1}{n}\left[\frac{\partial}{\partial r}\left(n\frac{\partial T}{\partial r}\right)\right] = -q$$

$$\frac{\partial}{\partial y}\left(r\frac{\partial T}{\partial r}\right) = -qr$$

$$r\frac{\partial T}{\partial r} = \frac{-qr^2}{2} + C_1$$

$$\frac{\partial T}{\partial r} = \frac{-qr}{2} + \frac{C_1}{r}$$

$$T = \frac{-qr^2}{4} + C_1 \ln r + C_2$$

5.15 (b)

$$\frac{1}{r^n}\cdot\frac{\partial}{\partial r}\left(kr^n\frac{\partial T}{\partial r}\right) + \dot{q}_g = \rho C_P\frac{\partial T}{\partial t}$$

This is general equation for heat conduction.

For slab $n = 0$; $r \to x$

For cylinder $n = 1$

For sphere $n = 2$

5.16 (13.09)

Time required for lumped heat capacity system to get a particular temperature T is given by

$$\frac{T-T_\infty}{T_i-T_\infty} = e^{\frac{-hAt}{\rho VC_P}}$$

$$\frac{80-180}{20-180} = e^{-\frac{1000\times0.1\times t}{1000\times10\times10^3}}$$

$$t = 47000.36 \text{ sec}$$

$$= 13.06 \text{ hours}$$

5.17 (903)

For lumped capacity system, time required to reach the system at a particular temperature is given by

$$\frac{T - T_\infty}{T_i - T_\infty} = e^{-\frac{hAt}{\rho V C_p}}$$

$$\frac{95 - 100}{29 - 100} = e^{\frac{-90 \times 6 \times t}{10^4 \times 0.01 \times 10^3}}$$

$$t = 902.68 \text{ sec} \approx 903 \text{ sec}$$

5.18 (c)

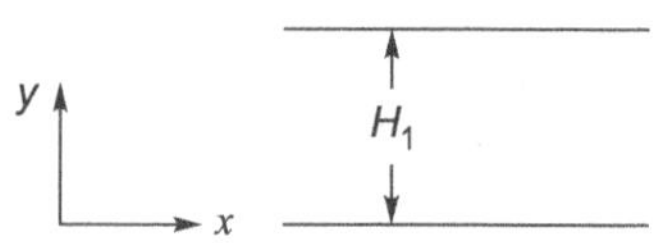

For unsteady state but not lumped heat transfer time required to reach a certain temperature is given by

$$t \propto x^2$$

where t is time and x is thickness.

For case-1 $\quad \tau_1 = k\left(\frac{H_1}{2}\right)^2 \quad$ ...(i)

For case-2 $\quad \tau_2 = k\left(\frac{H_2}{2}\right)^2 \quad$ ...(ii)

For equation (i) and (ii)

$$\frac{\tau_2}{\tau_1} = \left(\frac{H_2}{H_1}\right)^2$$

$$H_2 = 2H_1 \text{ (given)}$$

So, $$\frac{\tau_2}{\tau_1} = (2)^2 = 4$$

■■■■

6 Forced Convection

6.1 For turbulent flow in a tube the heat transfer coefficient is obtained from the Dittus-Boelter correlation. If the tube diameter is halved and the flow rate is doubled, then the heat transfer coefficient will change by a factor of

(a) 1 (b) 1.74
(c) 6.1 (d) 37

[2000 : 2 M]

6.2 The Sieder-Tate correlation for heat transfer in turbulent flow in a pipe gives $Nu \propto Re^{0.8}$, where Nu is the Nusselt number and Re is the Reynolds number for the flow. Assuming that this relation is valid, the heat transfer coefficient varies with pipe diameter (D) as

(a) $D^{-1.8}$ (b) $D^{-0.2}$
(c) $D^{0.2}$ (d) $D^{1.8}$

[2001 : 2 M]

6.3 A 200 W heater has a spherical casing of diameter 0.2 m. The heat transfer coefficient for conduction and convection from the casing to the ambient air is obtained from $Nu = 2 + 0.6Re^{1/2} Pr^{1/3}$ with $Re = 10^4$ and $Pr = 0.69$. The temperature of the ambient air is 30°C and the thermal conductivity of air is $k = 0.02$ W/m K.

(a) Find the heat flux from the surface at steady state.
(b) Find the steady state surface temperature of the casing.
(c) Find the temperature of the casing at steady state for stagnant air. Why is this situation physically infeasible?

[2001 : 2 M]

6.4 A fluid is flowing inside the inner tube of a double pipe heat exchanger with diameter 'd'. For a fixed mass flow rate, the tube side heat transfer coefficient for turbulent flow conditions is proportional to

(a) $d^{0.8}$ (b) $d^{-0.2}$
(c) d^{-1} (d) $d^{-1.8}$

[2003 : 2 M]

6.5 In forced convection, the Nusselt number Nu is a function of

(a) Re and Pr (b) Re and Gr
(c) Pr and Gr (d) Re and Sc

[2004 : 1 M]

6.6 The thermal boundary layer is significantly thicker than the hydrodynamic boundary layer for

(a) Newtonian liquids
(b) polymeric liquids
(c) liquid metals
(d) gases

[2005 : 1 M]

6.7 Consider the flow of a gas with density 1 kg/m^3, viscosity 1.5×10^{-5} kg/m.s, specific heat C_p = 846 J/(kg K) and thermal conductivity k = 0.01665 W/(mK), in a pipe of diameter D = 0.01 m and length L = 1 m, and assume the viscosity does not change with temperature. The Nusselt number for a pipe with (L/D) ratio greater than 10 and Reynolds number greater than 20000 is given by

$$Nu = 0.026\, Re^{0.8}\, Pr^{1/3}$$

while the Nusselt number for a laminar flow for Reynolds number less than 2100 and (Re Pr D/L) < 10 is

$$Nu = 1.86\, [Re\, Pr\, (D/L)]^{1/3}$$

If the gas flows through the pipe with an average velocity of 0.1 m/s, the heat transfer coefficient is

(a) 0.68 W/(m^2 K) (b) 1.14 W/(m^2 K)
(c) 2.47 W/(m^2 K) (d) 24.7 W/(m^2 K)

[2005 : 2 M]

6.8 Let d_h be the hydrodynamic entrance length for mercury in laminar flow in a pipe under isothermal conditions. Let d_t be its thermal entrance length under fully developed hydrodynamic conditions. Which ONE of the following is TRUE?

(a) $d_h > d_t$
(b) $d_h < d_t$
(c) $d_h = d_t$
(d) $d_h < d_t$ only if the pipe is vertical

[2006 : 1 M]

6.9 Experiments conducted with a sparingly dissolving cylinder wall in a flowing liquid yielded the following correlation for the Sherwood number

$$Sh = 0.023\,(Re)^{0.83}\,(Sc)^{1/3}$$

Assuming the applicability of the Chilton-Colburn analogy, the corresponding correlation for heat transfer is

(a) $Sh = 0.023\,(Gr)^{0.83}\,(Pr)^{1/3}$
(b) $Nu = 0.023\,(Re)^{0.83}\,(Pr)^{1/3}$
(c) $j_H = 0.023\,(Re)^{0.83}\,(Pr)^{2/3}$
(d) $Nu = 0.069\,(We)^{0.5}\,(Pr)^{4/3}$

[2006 : 2 M]

6.10 Water enters a thin walled tube (L = 1 m, D = 3 mm) at an inlet temperature of 97°C and mass flow rate 0.015 kg/s. The tube wall is maintained at a constant temperature of 27°C. Given the following data for water

Density, ρ = 1000 kg/m^3
Viscosity, μ = 48910^{-6} Ns/m^2
Specific heat,
C_p = 4184 J/kgK

Inside heat transfer coefficient h = 12978 W/(m^{2}K), the outlet temperature of water in °C is

(a) 28 (b) 37
(c) 62 (d) 96

[2007 : 2 M]

6.11 A hot fluid entering a well-stirred vessel is cooled by feeding cold water through a jacket around the vessel. Assume the jacket is well-mixed. For the following data,

mass flow rates of the hot fluid = 0.25 kg/s,
mass flow rate of cold water = 0.4 kg/s,
specific heats of oil = 6000 J/kgK
specific heat of cold water = 4184 J/kgK
The inlet and exit temperature of the hot fluid is 150°C and 100°C respectively.
Inlet temperature of cold water = 20°C
The overall heat transfer coefficient is 500 W/m^{2}K.
The heat transfer area in m^2, is

(a) 1.82 (b) 2.1
(c) 3 (d) 4.26 [2007 : 2 M]

6.12 Hot liquid is flowing at a velocity of 2 m/s through a metallic pipe having an inner diameter of 3.5 cm and length 20 m. The temperature at the inlet of the pipe is 90°C. Following data is given for liquid at 90°C.

Density = 950 kg/m^3;
Specific heat = 4.23 kJ/kg °C;
Viscosity = 2.55 × 10^{-4} kg/m.s;
Thermal conductivity = 0.685 W/m °C

The heat transfer coefficient (in W/m^2 °C) inside the tube is

(a) 222.22 (b) 111.11
(c) 22.22 (d) 11.11

[2008 : 2 M]

6.13 The Prandtl number of a fluid is the ratio of

(a) thermal diffusivity to momentum diffusivity
(b) momentum diffusivity to thermal diffusivity
(c) conductive resistance to convective resistance
(d) thermal diffusivity to kinematic viscosity

[2009 : 1 M]

6.14 The ratio of Nusselt number to Biot number is

(a) $\dfrac{\text{Conductive resistance of fluid}}{\text{Conductive resistance of solid}}$

(b) $\dfrac{\text{Conductive resistance of fluid}}{\text{Convective resistance of fluid}}$

(c) $\dfrac{\text{Conductive resistance of solid}}{\text{Conductive resistance of fluid}}$

(d) Unity

[2010 : 1 M]

6.15 If the Nusselt number (Nu) for heat transfer in a pipe varies with Reynolds number (Re) as 0.8, $Nu \propto Re^{0.8}$, then for constant average velocity in the pipe, the heat transfer coefficient varies with the pipe diameter D as

(a) $D^{-1.8}$ (b) $D^{-0.2}$
(c) $D^{0.2}$ (d) $D^{1.8}$ [2012 : 1 M]

6.16 Air is flowing at a velocity of 3 m/s perpendicular to a long pipe as shown in the figure below. The outer diameter of the pipe is d = 6 cm and temperature at the outside surface of the pipe is maintained at 100 °C. The temperature of the air far from the tube is 30 °C.

Data for air: Kinematic viscosity, v = 18 × 10^{-6} m^2/s; Thermal conductivity, k = 0.03 W/(m.K) Using the Nusselt number correlation:

$Nu = \dfrac{hd}{k} = 0.024 \times Re^{0.8}$, the rate of heat loss per unit length (W/m) from the pipe to air (up to one decimal place) is _______.

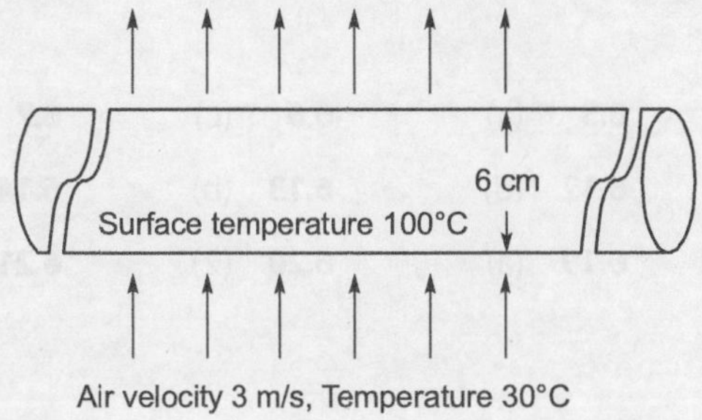

[2015 : 2 M]

6.17 Match the dimensionless numbers in **Group-1** with the ratios in **Group-2**.

Group-1

P. Biot number
Q. Schmidt number
R. Grashof number

Group-2

I. $\dfrac{\text{Buoyancy force}}{\text{Viscous force}}$

II. $\dfrac{\text{Internal thermal resistance of a solid}}{\text{Boundary layer thermal resistance}}$

III. $\dfrac{\text{Momentum diffusivity}}{\text{Mass diffusivity}}$

(a) P-II, Q-I, R-III (b) P-I, Q-III, R-II
(c) P-III, Q-I, R-II (d) P-II, Q-III, R-I

[2016 : 1 M]

6.18 A fluid flows over a heated horizontal plate maintained at temperature T_w. The bulk temperature of the fluid is T_∞. The temperature profile in the thermal boundary layer is given by:

$$T = T_w + (T_w - T_\infty)\left[\frac{1}{2}\left(\frac{y}{\delta_t}\right)^3 - \frac{3}{2}\left(\frac{y}{\delta_t}\right)\right], 0 \le y \le \delta_t$$

Here, y is the vertical distance from the plate, δ_t is the thickness of the thermal boundary layer and k is the thermal conductivity of the fluid.

The local heat transfer coefficient is given by

(a) $\dfrac{k}{2\delta_t}$ (b) $\dfrac{k}{\delta_t}$

(c) $\dfrac{3}{2}\dfrac{k}{\delta_t}$ (d) $2\dfrac{k}{\delta_t}$ [2017 : 2 M]

6.19 Prandtl number signifies the ratio of

(a) $\dfrac{\text{Momentum Diffusivity}}{\text{Thermal Diffusivity}}$

(b) $\dfrac{\text{Mass Diffusivity}}{\text{Thermal Diffusivity}}$

(c) $\dfrac{\text{Thermal Diffusivity}}{\text{Momentum Diffusivity}}$

(d) $\dfrac{\text{Thermal Diffusivity}}{\text{Mass Diffusivity}}$

[2019 : 1 M]

6.20 A fluid is flowing steadily under laminar conditions over a thin rectangular plate at temperature T_s as shown in the figure below. The velocity and temperature of the free stream are u_∞ and T_∞, respectively. When the fluid flow is only in the x-direction, h_x is the local heat transfer coefficient. Similarly, when the fluid flow is only in the y-direction, h_y is the corresponding local heat transfer coefficient. Use the correlation Nu = 0.332 $(\text{Re})^{1/2}(\text{Pr})^{1/3}$ for the local heat transfer coefficient, where, Nu, Re, and Pr, respectively are the appropriate Nusselt, Reynolds and Prandtl numbers. The average heat transfer coefficients are defined as $\bar{h}_l = \frac{1}{l}\int_0^l h_x\,dx$ and $\bar{h}_w = \frac{1}{w}\int_0^w h_y\,dy$. If $w = 1$ and $l = 4$ m, the value of the ratio of $\bar{h}_l$ is _____ (in integer).

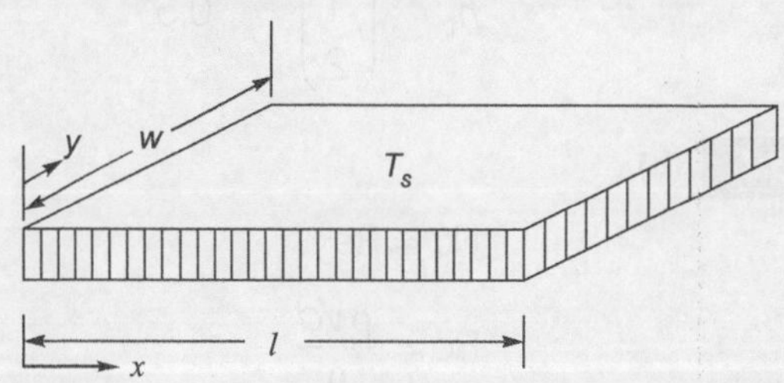

[2023 : 2 M]

6.21 If the Prandtl number Pr = 0.01, which of the following statements is/are correct?

(a) The momentum diffusivity is much larger than the thermal diffusivity.
(b) The thickness of the momentum boundary layer is much smaller than that of the thermal boundary layer.
(c) The thickness of the momentum boundary layer is much larger than that of the thermal boundary layer.
(d) The momentum diffusivity is much smaller than the thermal diffusivity.

[2024 : 1 M]

■■■■

Answers Forced Convection

6.1	(c)	**6.2**	(b)	**6.3**	(Sol)	**6.4**	(d)	**6.5**	(a)	**6.6**	(c)	**6.7**	(c)
6.8	(a)	**6.9**	(b)	**6.10**	(b)	**6.11**	(d)	**6.12**	(d)	**6.13**	(b)	**6.14**	(a)
6.15	(b)	**6.16**	(250.83)	**6.17**	(d)	**6.18**	(c)	**6.19**	(a)	**6.20**	(2)	**6.21**	(b, d)

Explanations Forced Convection

6.1 (c)

From Dittus-Boelter equation, we have

$$Nu = 0.023\, Re^{0.8}\, Pr^{0.33}$$

$$Re = \frac{\rho d V}{\mu} = \frac{\rho d Q}{A\mu}$$

$$Nu \propto Re^{0.8}$$

$$\frac{hd}{K} \propto \left(\frac{Qd}{d^2}\right)^{0.8}$$

$$h \propto \left(\frac{Q}{d}\right)^{0.8} \cdot \frac{1}{d} \propto \frac{Q^{0.8}}{d^{1.8}}$$

Now, $Q_2 = 2Q_1$ and $d_2 = 0.5d_1$

$$\frac{h_2}{h_1} = \left(\frac{2}{\frac{1}{2}}\right)^{0..8} \times \frac{1}{0.5} = 6.06$$

6.2 (b)

$$Nu \propto Re^{0.8}$$

$$Re = \frac{\rho V d}{\mu}$$

$$\frac{hd}{k} \propto d^{0.8} \Rightarrow h \propto d^{-0.2}$$

6.3 (Sol)

(a) At steady state,

Heat generated = Heat transfer by conduction

= Heat transfer by convection

So, $$\frac{q}{A} = \frac{200}{4\pi \times (0.1)^2} = 1592.35 \text{ W/m}^2$$

(b) $$Nu = 2 + 0.6\, Re^{0.5}\, Pr^{1/3}$$

$$Nu = 2 + 0.6 \times (10^4)^{0.5} \times (0.69)^{1/3}$$

$$\frac{h \times 0.2}{0.02} = 55.02 \Rightarrow 5.502 \text{ W/m}^2\text{K}$$

Surface temperature T_s can be found by

$$200 = 5.502 \times 4\pi \times (0.1)^2 \times (T_s - 30)$$

$$T_s = 319.27°C$$

(c) For stagnant air Re = 0 as $V = 0$

$$Nu = 2.0$$

$$\frac{hd}{k} = 2.0$$

$$h = 2.0 \times 0.02 \times \frac{1}{0.2}$$

$$= 0.2 \text{ W/m}^2\text{K}$$

Now, $$q = hA\Delta T$$

$$200 = 0.2 \times 4\pi \times (0.1)^2 \times (T_s - 30°)$$

$$T_s = 7987.75°C$$

This situation is physically unattainable as high temperature gradient starts a strong convection current which reduces casing temperature.

6.4 (d)

For turbulent flow

$$Nu = 0.023\, Re^{0.8}\, Py^{0.33}$$

$$\frac{hd}{k} \propto \left(\frac{\rho d V}{\mu}\right)^{0.8} \propto \left(\frac{\dot{m}}{d}\right)^{0.8} \cdot \frac{1}{d}$$

For fixed mass flow rate

$$h \propto \left(\frac{\rho d Q}{\frac{\pi}{4} d^2 \mu}\right)^{0.8} \cdot \frac{1}{d}$$

$$h \propto d^{-1.8}$$

6.5 (a)

Under forced convection:

$$Nu = f(Re, Pr)$$

6.6 (c)

For liquid metals, thermal diffusivity (α) is much higher than momentum diffusivity (ν). So for liquid metals Prandtl number is less than 1 and thermal boundary layer thickness is much thicker than hydrodynamic boundary layer thickness.

6.7 (c)

$$Re = \frac{\rho du}{\mu} = \frac{1\times0.01\times0.1}{1.5\times10^{-5}} = 66.667$$

$$Pr = \frac{\mu C_P}{k} = \frac{1.5\times10^{-5}\times846}{0.01665}$$

$$= 0.7621$$

$$\because \quad Nu = 1.86\left[Re\cdot Pr\left(\frac{D}{L}\right)\right]^{1/3}$$

$$\frac{h\times0.01}{0.01665} = 1.86\left[66.667\times0.7621\times\left(\frac{0.01}{1}\right)\right]^{1/3}$$

$$h = 2.47 \text{ W/m}^2\text{ K}$$

6.8 (a)

$L_t = L_H \cdot Pr$

L_t = Thermal entrance length

L_H = Hydrodynamic entrance length

For mercury $Pr < 1$ {Liquid metal}

So, $L_t < L_H$

6.9 (b)

In heat transfer Prandtl number plays same role as Schmidt number plays in mass transfer.

6.10 (b)

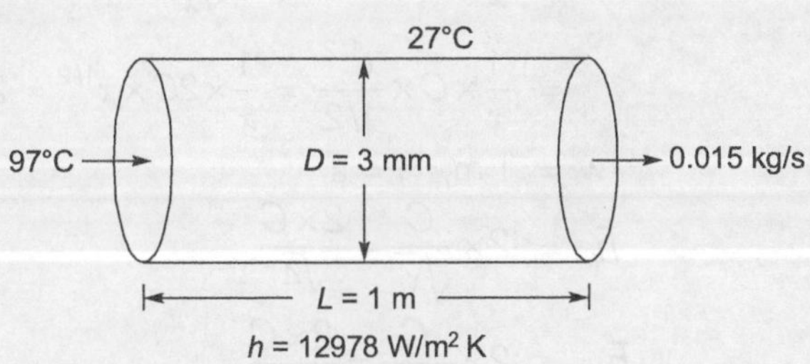

Let outlet temperature is T.

$$LMTD = \frac{\Delta T_1 - \Delta T_2}{\ln\left(\frac{\Delta T_1}{\Delta T_2}\right)}$$

$$= \frac{70-(T-27)}{\ln\left(\frac{70}{T-27}\right)}$$

$$mC_P\Delta T = UA\Delta T_{LMTD}$$

$$= 0.015\times4184(97-T)$$

$$= \frac{(97-T)}{\ln\left(\frac{70}{T-27}\right)}\times12978\times\pi\times0.003$$

$$T = 36.97 \cong 37°C$$

6.11 (d)

Well stirred vessel and jacket is also well-mixed. So, instead of ΔT_{LMTD} we will use here the outlet temperature difference because well mixing means temperature will be uniform in stirred vessel and coolant jacket.

From energy balance:

$$(mC_P\Delta T)_{\text{cold fluid}} = (mC_P\Delta T)_{\text{hot fluid}}$$

So, $0.4 \times 4184(T-20) = 0.25 \times 6000(150-100)$

$$T = 64.81$$

$$Q = mC_P\Delta T = UA\Delta T_{LMTD}$$

For stirred vessel ΔT_{emtd} is equal to difference of exit temperature of both fluids.

$$= 0.25 \times 6000 \times (150-100)$$

$$= 500 \times A \times (100-64.81)$$

$$A = 4.26 \text{ m}^2$$

6.12 (d)

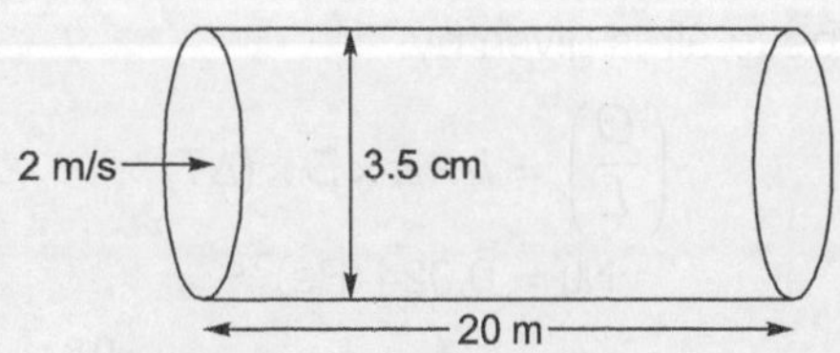

$$T_{inlet} = 90°C$$

$$\rho = 950 \text{ kg/m}^3$$

$$C_P = 4.23 \text{ kJ/kg °C}$$

$$\mu = 2.55 \times 10^{-4} \text{ Kg/m.s}$$

$$k = 0.685 \text{ W/m°C}$$

$$Re = \frac{\rho du}{\mu} = \frac{950\times2\times0.035}{2.55\times10^{-4}}$$

$$= 260784.31$$

$$Pr = \frac{2.55\times10^{-4}\times4230}{0.685} = 1.57$$

$$\therefore \quad Nu = 0.023\, Re^{0.8}\, Pr^{0.33}$$

$$\frac{h\times0.035}{0.685} = (260784.31)^{0.8}\times(1.57)^{0.33}\times0.023$$

$$h = 11.12 \text{ kW}$$

6.13 (b)

$$\text{Prandtl number} = \frac{\text{Momentum diffusivity}}{\text{Thermal diffusivity}}$$

6.14 (a)

$$\therefore \text{Nusselt number} = \frac{\text{Conductive resistance in fluid}}{\text{Convective resistance in fluid}}$$

$$\therefore \frac{(\text{Nu})}{(\text{Bi})} = \frac{\text{Conductive resistance in fluid}}{\text{Conductive resistance in solid}}$$

$$\text{Biot Number} = \frac{\text{Conductive resistance in solid}}{\text{Convective resistance in fluid}}$$

6.15 (b)

$$\text{Nu} \propto \text{Re}^{0.8}$$

$$\text{Nu} = \frac{hd}{K} \text{ and Re} = \frac{\rho V d}{\mu}$$

$$\frac{hd}{K} \propto d^{0.8}$$

$$h \propto d^{-0.2}$$

6.16 (250.83)

Given, $d = 6\text{ cm} = 0.06\text{ m}$, $V = 3\text{ m/s}$

$T_s = 100°C$, $T_{air} = 30°C$

$v = 18 \times 10^{-6}\text{ m}^2\text{/s}$, $k = 0.03\text{ W/m.K}$

$\therefore$ For rate of heat loss per unit length from the pipe to air is given by

$$\left(\frac{Q}{L}\right) = h \times \pi \times D \times (\Delta T)$$

$$\therefore \quad \text{Nu} = 0.024\,(\text{Re})^{0.8}$$

$$\frac{h \times 0.06}{0.03} = 0.024\left[\frac{3 \times 0.06}{18 \times 10^{-6}}\right]^{0.8}$$

$$h = 19.02\text{ W/m}^2\text{K}$$

Now, $$\left(\frac{Q}{L}\right) = 19.02 \times \pi \times 0.06 \times (100 - 30)$$

$$\left(\frac{Q}{L}\right) = 250.83\text{ (W/m)}$$

6.18 (c)

Local heat transfer coefficient can be found by following co-relation:

$$h_x A(T_w - T_\infty) = -kA\left.\frac{dT}{dx}\right|_{x=0}$$

$$h_x = \frac{-k\left.\frac{dT}{dx}\right|_{x=0}}{(T_w - T_\infty)} \quad \ldots(1)$$

From given temperature profile

$$T = T_w + (T_w - T_\infty)\left[\frac{1}{2}\left(\frac{y}{\delta_t}\right)^3 - \frac{3}{2}\left(\frac{y}{\delta_t}\right)\right]$$

$$\left.\frac{dT}{dy}\right|_{y=0} = -(T_w - T_\infty)\left(\frac{3}{\delta_t}\right)$$

By putting this value in equation (1)

$$h|_{y=0} = \frac{-k(T_w - T_\infty)\left(\frac{-3}{\delta_t}\right)}{(T_w - T_\infty)} = \frac{3k}{\delta_t}$$

6.19 (a)

$$\text{Prandtl number} = \frac{\nu}{\alpha} = \frac{\mu C_P}{k}$$

$$\text{Prandtl number} = \frac{\text{Momentum diffusivity}}{\text{Thermal diffusivity}}$$

6.20 (2)

For the given case of plate:

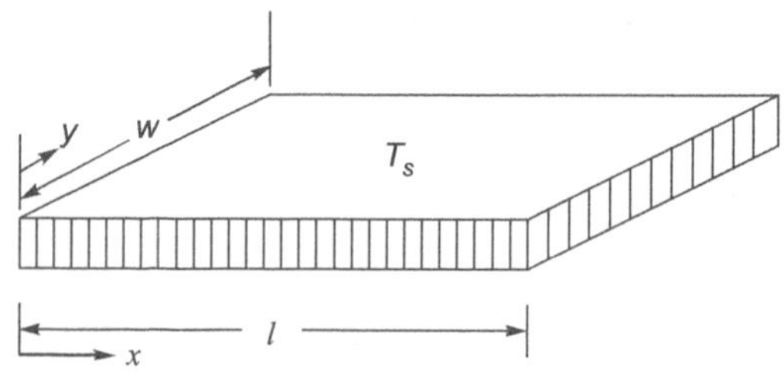

Steady flow

Laminar flow

$$\text{Nu}_x = 0.332\,\text{Re}_x^{1/2}\,\text{Pr}^{1/3}$$

$$\frac{h_x x}{k} = 0.332\left(\frac{V_\infty x}{\nu}\right)^{1/2}\text{Pr}^{1/3}$$

$$h_x = \frac{\text{Constant}}{\sqrt{x}} = \frac{C}{\sqrt{x}}$$

$$\bar{h}_x = \frac{1}{x}\int_0^x h_x\,dx = \frac{1}{x}\int_0^x \frac{C}{\sqrt{x}}\,dx$$

$$= \frac{1}{x} \times C \times \frac{x^{1/2}}{1/2} = \frac{1}{x} \times 2C \times x^{1/2} = 2\frac{C}{\sqrt{x}}$$

$$w = 1\text{ m}, \quad l = \text{s}$$

$$\bar{h}_l = 2 \times \frac{C}{\sqrt{x_l}} = \frac{2 \times C}{\sqrt{4}}$$

$$\bar{h}_w = 2 \times \frac{C}{\sqrt{x_w}} = \frac{2 \times C}{\sqrt{4}}$$

$$\frac{\bar{h}_w}{\bar{h}_l} = \frac{\frac{2 \times C}{\sqrt{1}}}{\frac{2 \times C}{\sqrt{4}}} = 2$$

6.21 (b, d)

$$\text{Prandtl number (Pr)} = \frac{\text{Momentum diffusivity }(\nu)}{\text{Thermal diffusivity }(\alpha)}$$

Given Pr = 0.01

It means momentum diffusivity is much smaller than thermal diffusivity and thickness of momentum boundary layer is much smaller than that of the thermal boundary layer.

7 Boiling and Condensation

7.1 An electrically heated element is submerged in a pool of water at its saturation temperature. As the temperature of the element increases, the maximum heat transfer coefficient is observed

(a) in the free convection regime
(b) between the nucleate boiling and partial nucleate boiling mixed with unstable film boiling regimes
(c) in the incipient nucleate boiling regime
(d) in the stable film boiling regime without significant radiation effects

[2005 : 1 M]

7.2 In film type condensation of liquid along a vertical tube, the thickness of the condensate layer increases towards the bottom. This implies that the local heat transfer coefficient

(a) increases from top to bottom
(b) decreases from top to bottom
(c) remains constant from top to bottom
(d) first increases and then decreases from top to bottom

[2005 : 1 M]

Statement for Linked Answer Question (7.3 and 7.4):

In film condensation on a vertical plane surface, the x-directional velocity distribution is given by

$$u(y) = \frac{g(\rho_l - \rho_v)}{\mu_l}\left(\delta y - \frac{1}{2}y^2\right)$$

where δ is the film thickness at any x.

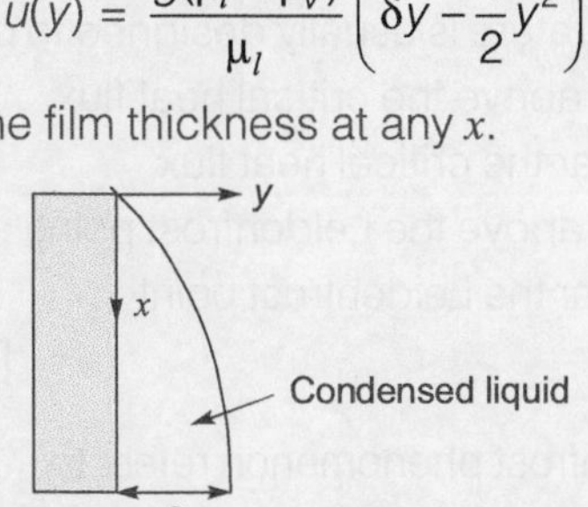

7.3 The mass flow rate of the condensate $m(x)$ through any axial position x per unit width of the plate is given by

(a) $m(x) = \dfrac{g\rho_l(\rho_l - \rho_v)\delta^3}{3\mu_l}$

(b) $m(x) = \dfrac{g(\rho_l - \rho_v)\delta^2}{3\mu_l}$

(c) $m(x) = \dfrac{g\rho_v^2 - \delta^3}{\mu_l}$

(d) $m(x) = \dfrac{g\rho_l(\rho_v)\delta^3}{3\mu_l}$

[2006 : 2 M]

7.4 Differentiate $m(x)$ with respect to δ to get the differential increase in condensate mass dm with film thickness i.e., $\dfrac{dm}{d\delta}$. Then obtain $\dfrac{dm}{dx}$ assuming heat flux through the film to be due to conduction based on a linear temperature profile between the vapor and wall. Hence determine $\dfrac{d\delta}{dx}$.

Here μ_l is liquid viscosity, k_l is thermal conductivity, and λ is latent heat of condensation. T_v is the vapor temperature and T_W is the wall temperature.

(a) $\dfrac{d\delta}{dx} = \dfrac{\mu_l k_l (T_v - T_w)}{g\rho_1(\rho_1 - \rho_v)\lambda}\cdot\dfrac{1}{\delta^2}$

(b) $\dfrac{d\delta}{dx} = \dfrac{\mu_l k_l (T_v - T_w)}{g\rho_v(\rho_1 - \rho_v)\lambda}\cdot\dfrac{1}{\delta^3}$

(c) $\dfrac{d\delta}{dx} = \dfrac{\mu_l (T_v - T_w)}{g\rho_v k_l(\rho_1 - \rho_v)\lambda}\cdot\dfrac{1}{\delta^2}$

(d) $\dfrac{d\delta}{dx} = \dfrac{\mu_l k_l (T_v - T_w)}{g\rho_l(\rho_l - \rho_v)\lambda}\cdot\dfrac{1}{\delta^3}$

[2006 : 2 M]

7.5 Consider a liquid stored in a container exposed to its saturated vapor at constant temperature T_{sat}. The bottom surface of the container is maintained at a constant temperature $T_s < T_{sat}$ while its side walls are insulated. The thermal conductivity k_l of the liquid, its latent heat of vaporisation λ and density ρ_l are known. Assuming a linear temperature distribution in the liquid, the expression for the growth of the liquid layer δ as a function of time t is given by

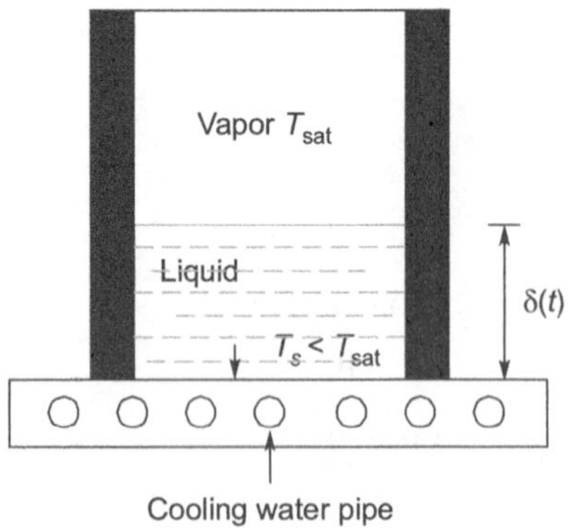

(a) $\delta(t) = \left[\frac{4k_l(T_{sat} - T_s)}{\rho_l\lambda}t\right]^{1/2}$

(b) $\delta(t) = \left[\frac{k_l(T_{sat} - T_s)}{2\rho_l\lambda}t\right]^{1/2}$

(c) $\delta(t) = \left[\frac{2k_l(T_{sat} - T_s)}{\rho_l\lambda}t\right]^{1/2}$

(d) $\delta(t) = \left[\frac{k_l(T_{sat} - T_s)}{\rho_l\lambda}t\right]^{1/2}$

[2007 : 2 M]

7.6 In film type condensation over a vertical tube, local heat transfer coefficient is
(a) Inversely proportional to local film thickness
(b) directly proportional to local film thickness
(c) equal to local film thickness
(d) independent of local film thickness

[2011 : 1 M]

7.7 In a pool boiling experiment, the following phenomena were observed.
P. Natural convection
Q. Film boiling
R. Transition boiling
S. Nucleate boiling

What was the CORRECT sequence of their occurrence?
(a) P, Q, R, S (b) S, R, Q, P
(c) Q, R, P, S (d) P, S, R, Q

[2013 : 1 M]

7.8 Steam at 100°C is condensing on a vertical steel plate. The condensate flow is laminar. The average Nusselt numbers are Nu_1 and Nu_2, when the plate temperatures are 10° C and 55°C, respectively. Assume the physical properties of the fluid and steel to remain constant within the temperature range of interest. Using Nusselt equations for film-type condensation, what is the value of the ratio $\frac{Nu_2}{Nu_1}$?

(a) 0.5 (b) 0.84
(c) 1.19 (d) 1.41

[2016 : 1 M]

7.9 In nucleate boiling, the pressure inside a bubble is higher than the pressure of the surrounding liquid. Assuming that both the liquid and vapour are saturated, the temperature of the liquid will ALWAYS be
(a) at 100°C
(b) lower than the temperature of the vapour
(c) equal to the temperature of the vapour
(d) higher than the temperature of the vapour

[2017 : 2 M]

7.10 Vapour bubbles are formed in the nucleate boiling regime at a frequency of 10 bubbles per second per nucleation site. There are 100 nucleation sites per m^2 of heating area. The latent heat of vapourization and the density of vapour under the operating condition are 1000 kJ/kg and 1 kg/m^3 respectively. The diameter of each bubble is 10^{-3} m. Assume that the entire heat supplied is used for vapour generation. The heat flux (in Watt per m^2 of heating area) is _______ (rounded off to third decimal place).

[2018 : 2 M]

7.11 Pool boiling equipment operating above ambient temperature is usually designed to operate
(a) far above the critical heat flux
(b) near the critical heat flux
(c) far above the Leidenfrost point
(d) near the Leidenfrost point

[2019 : 1 M]

7.12 Leidenfrost phenomenon refers to
(a) the condensation of vapour on a cold surface.
(b) the exchange of heat between two solids.
(c) the melting of frost.
(d) film boiling and evaporation of liquid droplets falling on a very hot surface.

[2020 : 1 M]

7.13 Heat transfer coefficient for a vapor condensing as a film on a vertical surface is given by
(a) Dittus-Boelter equation
(b) Nusselt theory
(c) Chilton-Colburn analogy
(d) Sieder-Tate equation

[2021 : 1 M]

7.14 Saturated steam condenses on a vertical plate maintained at a constant wall temperature. If x is the vertical distance from the top edge of the plate, then the local heat transfer coefficient $h(x) \propto \Gamma(x)^{-1/3}$, where $\Gamma(x)$ is the local mass flow rate of the condensate per unit plate width. The ratio of the average heat transfer coefficient over the entire plate to the heat transfer coefficient at the bottom of the plate is

(a) 4 (b) $\frac{4}{3}$
(c) $\frac{3}{4}$ (d) 3

[2022 : 2 M]

Answers Boiling and Condensation

7.1	(b)	**7.2**	(b)	**7.3**	(a)	**7.4**	(d)	**7.5**	(c)	**7.6**	(a)	**7.7**	(d)
7.8	(c)	**7.9**	(b)	**7.10**	(0.523)	**7.11**	(b)	**7.12**	(d)	**7.13**	(b)	**7.14**	(b)

Explanations Boiling and Condensation

7.2 (b)

$$h(x) \propto \frac{1}{\delta}$$

Here, $h(x)$ is local heat transfer coefficient in film type condensation and δ is film thickness.
So, from top to bottom δ increases and $h(x)$ decreases.

7.3 (a)

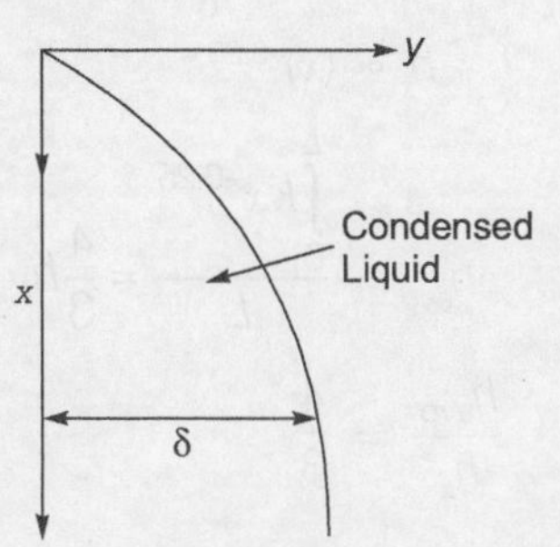

$$\because \quad u(y) = \frac{g(\rho_l - \rho_V)}{\mu_l}\left(\delta y - \frac{1}{2}y^2\right) \quad \text{\{given\}}$$

$$m(x) = \int_0^{\delta} \rho_l\, u(y) w \cdot dy \quad \{w = \text{width}\}$$

$$= \frac{g\rho_l(\rho_l - \rho_V)w}{\mu_l}\int_0^{\delta}\left(\delta y - \frac{1}{2}y^2\right)dy$$

$$\frac{m(x)}{w} = \frac{g(\rho_l - \rho_V)\rho_l\delta^3}{3\mu_l} \qquad (\because w = 1)$$

7.4 (d)

$$m(\delta) = \frac{g\rho_l(\rho_l - \rho_V)\delta^3 W}{3\mu_i}$$

$$\frac{d(m)}{d\delta} = \frac{g\rho_l(\rho_l - \rho_V)\delta^2 W}{\mu_1} \quad \text{...(1)}$$

From energy balance

$$q_x = -k_l(dx \cdot \omega)\left.\frac{dT}{dy}\right|_{y=0} = dm.\lambda$$

$$dm \cdot \lambda = k_l(dx\omega)\frac{T_V - T_W}{\delta} \quad \text{...(2)}$$

From equations (1) and (2)

$$\frac{g\rho_l(\rho_l - \rho_V)\delta^2 \omega d\delta}{\mu} = k_l(dx\omega)\left(\frac{T_V - T_W}{\delta}\right)$$

$$\frac{d\delta}{dx} = \frac{\mu_l k_l(T_V - T_W)}{g\rho_l(\rho_l - \rho_V)\lambda\delta^3}$$

7.5 (c)

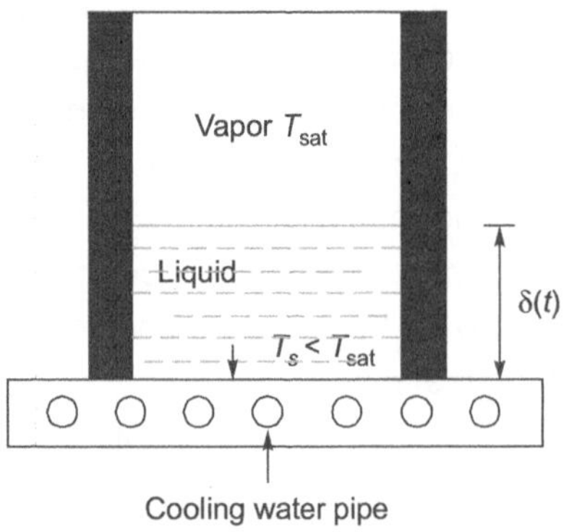

At steady state,

Heat supplied by conduction = Latent heat of condensation

$$Q = -k_l A\frac{dT}{d\delta} = \frac{m\lambda}{t}$$

$$-k_l A\frac{dT}{d\delta} = \frac{\rho V\lambda}{t}$$

$$-k_l A\frac{dT}{d\delta} = \frac{\rho A\delta(t)\lambda}{t}$$

$$-k_l At\int_{T_{sat}}^{T_s} dT = \rho A\lambda\int_0^{\delta} \delta(t)d\delta$$

$$k_l t(T_{sat} - T_s) = \rho\lambda\frac{\delta^2}{2}$$

$$\delta(t) = \left[\frac{2k_l(T^{sat} - T^s)t}{\rho_l\lambda}\right]^{1/2}$$

7.6 (a)

In film wise condensation

$$h(x) = \frac{k(x)}{\delta}$$

where, δ = Film thickness

7.8 (c)

For film-wise condensation

$$N_u \propto (\Delta T)^{-1/4}$$

So,
$$\frac{N_{u2}}{N_{u1}} = \left(\frac{\Delta T_1}{\Delta T_2}\right)^{1/4}$$

$$= \left(\frac{100-10}{100-55}\right)^{1/4} = 2^{0.25}$$

$$\frac{N_{u2}}{N_{u1}} = 1.19$$

7.9 (b)

Pressure inside bubble is higher than pressure of surrounding liquid. So, temperature inside bubble of vapour will be always greater than saturated liquid temperature because temperature is always proportional to pressure.

7.10 (0.523)

10 bubbles are formed per nucleation site and 100 nucleation site per m^2 of heating area.

So, total bubble produced per m^2 is

$$10 \times 100 = 1000$$

$$\text{Heat flux} = m\lambda = \rho v\lambda$$

$$= 1 \times 1000 \times \frac{4}{3}\pi\left(\frac{10^{-3}}{2}\right)^3 \times 1000 \times 1000$$

$$= 0.523 \text{ watt/m}^2$$

7.11 (b)

Pool boiling equipment operating above ambient temperature is usually designed to operate near the critical heat flux.

7.12 (d)

Leidenfrost phenomenon refers to film boiling. In film boiling, Leidenfrost point is known as the point where heat flux is minimum.

7.14 (b)

From Nusselt theory, we have

$$h_x \propto (\dot{m})^{-1/3}$$

$$\dot{m} \propto \delta^3$$

$$h_x \propto \delta^{-1} \propto \frac{1}{\delta}$$

$$h_x \propto (x)^{-0.25}$$

$$h_{avg} = \frac{\int_0^L kx^{-0.25}}{L} = \frac{4}{3}h_x$$

$$\frac{h_{avg}}{h_x} = \frac{4}{3}$$

■■■■

8 Natural Convection

8.1 The Grashof number is defined as the ratio of
(a) buoyancy to inertial forces.
(b) buoyancy to viscous forces.
(c) inertial to viscous forces.
(d) buoyancy to surface tension forces.
[2000 : 1 M]

8.2 Heat transfer by natural convection is enhanced in systems with
(a) high viscosity
(b) high coefficient of thermal expansion
(c) low temperature gradients
(d) low density changes with temperature
[2001 : 2 M]

8.3 The Grashof number is
(a) thermal diffusivity/mass diffusivity
(b) inertial force/surface tension force
(c) sensible heat/latent heat
(d) buoyancy force/viscous force
[2007 : 1 M]

Answers Natural Convection

8.1 (b) **8.2** (b) **8.3** (d)

Explanations Natural Convection

8.2 (b)

High coefficient of thermal expansion will cause high change in density and this will enhance natural convection.

8.3 (d)

$$\text{Grashof number} = \frac{\rho^2 g\beta\Delta T L^3}{\mu^2}$$

$$= \frac{\text{Buoyancy force}}{\text{Viscous force}}$$

9 Radiation Heat Transfer

9.1 A sphere of radius R_1 is enclosed in a sphere of radius R_2. The view (or shape) factor for radiative heat transfer of the outer sphere with respect to the inner sphere is

(a) 0 (b) $\dfrac{R_2}{R_1 + R_2}$

(c) 1 (d) $\left(\dfrac{R_1}{R_2}\right)^2$

[2000 : 1 M]

9.2 The heat transfer by radiation from a mild steel surface is to be reduced by reducing the emissivity of the surface. This can be best achieved by

(a) painting the surface black
(b) painting the surface white
(c) giving the surface a mirror finish
(d) roughening the surface

[2001 : 1 M]

9.3 For an ideal black body

(a) absorptivity = 1 (b) reflectivity = 1
(c) emissivity = 0 (d) transmissivity = 1

[2004 : 1 M]

9.4 A black body at a higher temperature T_H transfers energy by radiation to a black body at a lower temperature T_L. Initially, T_H = 1850°C, T_L = 500°C, and the net rate of energy transfer is 25 W. After some time, when T_H = 1500°C and T_L = 750°C, what is the net rate of energy transfer?

(a) 8.73 W (b) 9.60 W
(c) 13.89 W (d) 11.01 W

[2005 : 2 M]

9.5 An insulated cylindrical pipe of 0.2 m diameter has a surface temperature of 45°C. It is exposed to black body surroundings at 25°C. The emissivity and absorptivity of the insulation surface are 0.96 and 0.93, respectively. The convective heat transfer coefficient outside the insulation surface is 3.25 W/(m²K). The Stefan-Boltzmann constant is 5.67×10^{-8} W/(m² K⁴). The surrounding fluid may be assumed to be transparent. Find the percentage contribution from radiation to the total heat transfer rate to the surroundings

(a) 30.9 (b) 50.0
(c) 57.6 (d) 68.4

[2006 : 2 M]

9.6 For the two long concentric cylinders with surface areas A_1 and A_2, the view factor F_{22} is given by

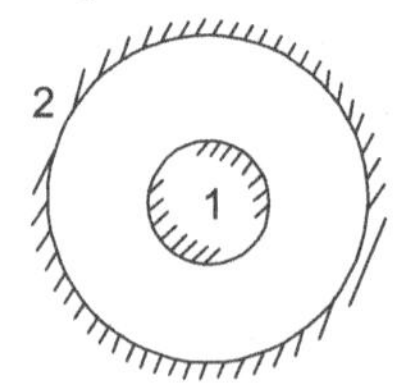

(a) 0 (b) 1

(c) $1-\dfrac{A_1}{A_2}$ (d) $\dfrac{A_1}{A_2}$

[2007 : 2 M]

9.7 A well-insulated hemispherical furnace (radius = 1 m) is shown below :

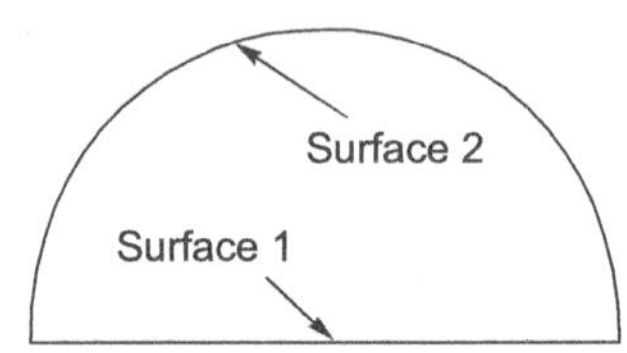

The self-view factor of radiation for the curved surface 2 is

(a) $\dfrac{1}{4}$ (b) $\dfrac{1}{2}$

(c) $\dfrac{2}{3}$ (d) $\dfrac{3}{4}$

[2009 : 2 M]

9.8 The view factor matrix for two infinitely long co-axial cylinders, shown in the figure below, is

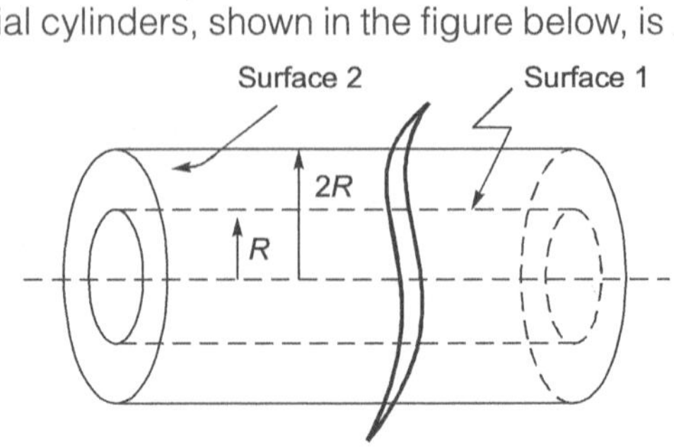

(a) $\begin{bmatrix} 0 & 1 \\ 0.5 & 0.5 \end{bmatrix}$ (b) $\begin{bmatrix} 0 & 1 \\ 1 & 0 \end{bmatrix}$

(c) $\begin{bmatrix} 1 & 0 \\ 0 & 1 \end{bmatrix}$ (d) $\begin{bmatrix} 0.5 & 0.5 \\ 0 & 1 \end{bmatrix}$

[2010 : 2 M]

9.9 Consider two black bodies with surfaces S_1 (area = 1 m^2) and S_2 (area = 4 m^2). They exchange heat only by radiation. 40% of the energy emitted by S_1 is received by S_2. The fraction of energy emitted by S_2 that is received by S_1 is

(a) 0.05 (b) 0.1
(c) 0.4 (d) 0.6 [2011 : 1 M]

9.10 For the enclosure formed between two concentric spheres as shown below ($R_2 = 2R_1$), the fraction of radiation leaving the surface area A_2 that strikes itself is

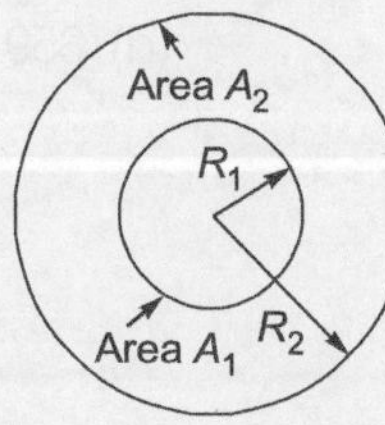

(a) $\frac{1}{4}$ (b) $\frac{1}{2}$
(c) $\frac{1}{\sqrt{2}}$ (d) $\frac{3}{4}$ [2012 : 2 M]

9.11 A hole of area 1 cm^2 is opened on the surface of a large spherical cavity whose inside temperature is maintained at 727°C. The value of Stefan-Boltzmann constant is 5.67×10^{-8} W/m^2-K^4. Assuming black body radiation, the rate at which the energy is emitted (in W) by the cavity through the hole, up to 3 digits after the decimal point, is ________.

[2013 : 1 M]

9.12 In a completely opaque medium, if 50% of the incident monochromatic radiation is absorbed, then which of the following statements are CORRECT?

P. 50% of the incident radiation is reflected.
Q. 25% of the incident radiation is reflected.
R. 25% of the incident radiation is transmitted.
S. No incident radiation is transmitted.

(a) P and S only (b) Q and R only
(c) P and Q only (d) R and S only

[2014 : 1 M]

9.13 Two infinitely large parallel plates (I and II) are held at temperatures T_I and T_{II} ($T_I > T_{II}$) respectively, and placed at a distance $2d$ apart in vacuum. An infinitely large flat radiation shield (III) is placed in parallel in between I and II. The emissivities of all the plates are equal. The ratio of the steady state radiative heat fluxes with and without the shield is

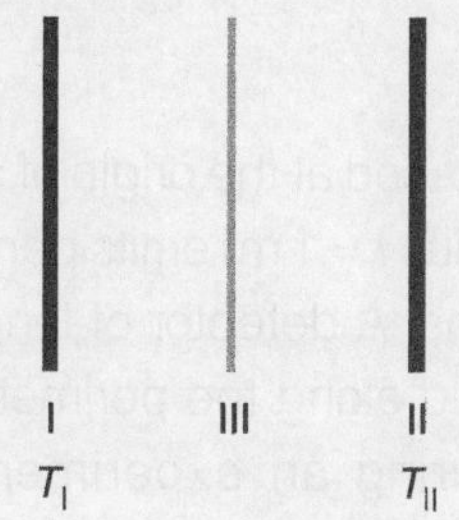

(a) 0.5 (b) 0.75
(c) 0.25 (d) 0 [2015 : 1 M]

9.14 The space between two hollow concentric spheres of radii 0.1 m and 0.2 m is under vacuum. Exchange of radiation (uniform in all directions) occurs only between the outer surface (S_1) of the smaller sphere and the inner surface (S_2) of the larger sphere. The fraction (rounded off to the second decimal place) of the radiation energy leaving S_2, which reaches S_1 is ________.

[2016 : 2 M]

9.15 Let $I_{b\lambda}$ be the spectral blackbody radiation intensity per unit wavelength about the wavelength λ. The blackbody radiation intensity emitted by a blackbody over all wavelengths is

(a) $\frac{dI_{b\lambda}}{d\lambda}$ (b) $\frac{d^2 I_{b\lambda}}{d\lambda^2}$

(c) $\int_0^\infty I_{b\lambda} d\lambda$ (d) $\int_0^\infty \lambda I_{b\lambda} d\lambda$

[2017 : 2 M]

9.16 A hollow cylinder of equal length and inner diameter (i.e., $L = D$) is sealed at both ends with flat plate, as shown in the figure. Its inner surfaces, A_1, A_2 and A_3 radiate energy.

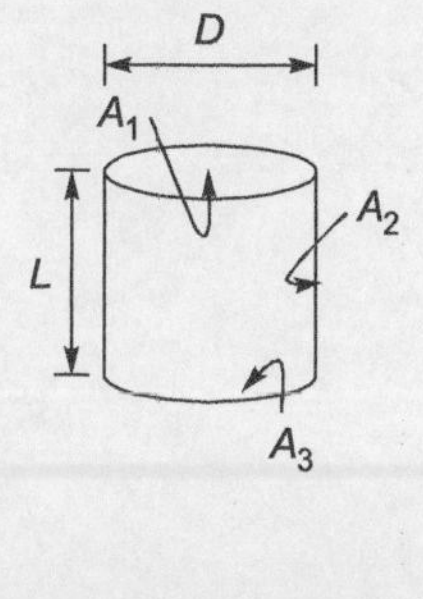

F_{ij} denotes the fraction of radiation energy leaving the surface A_i which reaches the surface A_j. It is also known that $F_{13} = 3-2\sqrt{2}$. Which one of the following is correct?

(a) $F_{21} = \sqrt{2}-1$ (b) $F_{21} = \frac{\sqrt{2}-1}{2}$

(c) $F_{21} = \frac{\sqrt{2}-1}{4}$ (d) $F_{21} = \frac{\sqrt{2}-1}{8}$

[2020 : 2 M]

9.17 A source placed at the origin of a circular sample holder (radius $r = 1$ m) emits particles uniformly in all directions. A detector of length $l = 1$ cm has been placed along the perimeter of the sample holder. During an experiment, the detector registers 14 particles. The total number of particles emitted during the experiment is ________.

[2021 : 1 M]

9.18 Two large parallel planar walls are maintained at 1000 K and 500 K. Parallel radiation shields are to be installed between the two walls. Assume that the emissivities of the walls and the shields are equal. If the melting temperature of the shields is 900 K, the maximum number of shield(s) that can be installed between the walls is (are)

(a) 1 (b) 0

(c) 2 (d) 3 [2022 : 2 M]

9.19 Given that E (in W.m^{-2}) is the total hemispherical emissive power of a surface maintained at a certain temperature, which of the following statements is/are CORRECT?

(a) E does not depend on the direction of the emission.

(b) E depends on the viewfactor.

(c) E depends on the wavelength of the emission.

(d) E does not depend on the frequency of the emission. [2023 : 1 M]

9.20 The temperatures of two large parallel plates of equal emissivity are 900 K and 300 K. A reflection radiation shield of low emissivity and negligible conductive resistance is placed parallelly between them. The steady-state temperature of the shield, in K, is

(a) 759 (b) 559

(c) 659 (d) 859

[2024 : 2 M]

■■■■

Answers Radiation Heat Transfer

9.1	(d)	**9.2**	(b)	**9.3**	(a)	**9.4**	(d)	**9.5**	(d)	**9.6**	(c)	**9.7**	(b)
9.8	(a)	**9.9**	(b)	**9.10**	(d)	**9.11**	(5.67)	**9.12**	(a)	**9.13**	(a)	**9.14**	(0.25)
9.15	(c)	**9.16**	(b)	**9.17**	(8792)	**9.18**	(a)	**9.19**	(a, d)	**9.20**	(a)		

Explanations Radiation Heat Transfer

9.1 (d)

$$F_{12} = 1$$

From reciprocity theorem

$$A_1 F_{12} = A_2 F_{21}$$

$$F_{21} = \frac{A_1}{A_2} = \left(\frac{R_1}{R_2}\right)^2$$

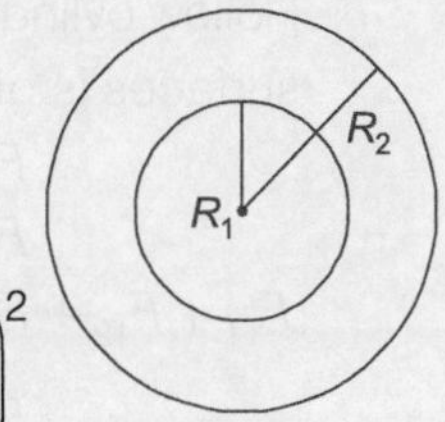

9.2 (b)

Rough surface has more emissivity than smooth surface. By painting the surface white, we can reduce the thermal radiation.

9.4 (d)

$$Q_{\text{radiation}} = \sigma A(T_1^4 - T_2^4)$$

(from Stefan Boltzmann law)

So, $$Q \propto (T_1^4 - T_2^4)$$

$$\frac{Q_2}{Q_1} = \frac{(1773)^4 - (1023)^4}{(2123)^4 - (773)^4} \rightarrow \{Q_1 = 11\}$$

$$Q_2 = 11.01 \text{ W}$$

9.5 (d)

Heat transfer due to convection = $hA\Delta T$

$$Q_1 = 3.25A(45 - 25)$$

Heat transfer due to radiation

$$= \sigma\,\varepsilon A(T_1^4) - \sigma A\,\alpha T_2^4$$

where, α = Absorptivity of surface

$$Q_2 = 5.67 \times 10^{-8} \times 0.96 \times A \times 318^4 - 5.67 \times 10^{-8} \times A \times 0.93 \times (298)^4$$

% contribution from radiation is given as

$$= \frac{Q_2}{Q_1 + Q_2}$$

Putting values, we get = 68.4%

9.6 (c)

$$F_{12} = 1$$

$$A_1 F_{12} = A_2 F_{21} \text{ (from reciprocity theorem)}$$

$$F_{21} = \frac{A_1}{A_2}$$

$$F_{21} + F_{22} = 1 \qquad \{\text{Summation rule}\}$$

$$F_{22} = 1 - \frac{A_1}{A_2}$$

9.7 (b)

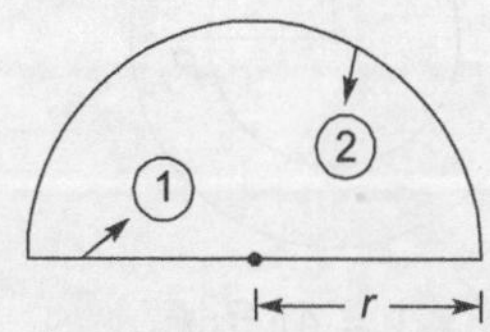

Radius = r

$$F_{12} = 1$$

$$A_1 F_{12} = A_2 F_{21} \text{ (from reciprocity theorem)}$$

$$\pi r^2 \times 1 = 2\pi r^2 \times F_{21}$$

$$F_{21} = \frac{1}{2}$$

$$F_{21} + F_{22} = 1$$

$$F_{22} = 1 - \frac{1}{2} = \frac{1}{2}$$

9.8 (a)

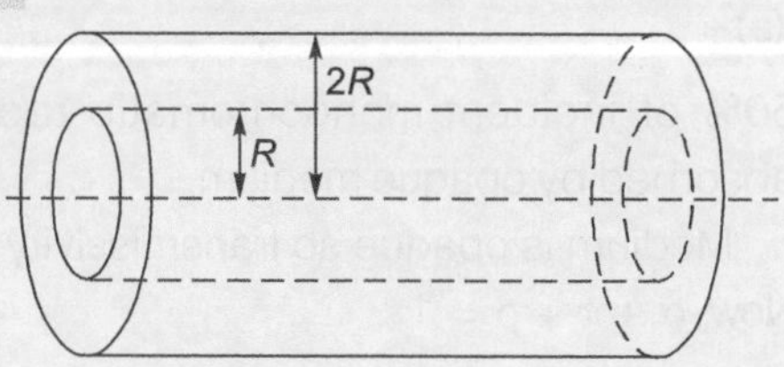

$$F_{11} = 0$$

$$F_{12} = 1$$

$$A_1 F_{12} = A_2 F_{21} \text{ (Reciprocity theorem)}$$

So, $$F_{21} = \frac{2\pi RL}{2\pi(2R)L} \times 1$$

$$F_{21} = 0.5$$

$$F_{21} + F_{22} = 1$$

$$F_{22} = 1 - 0.5 = 0.5$$

So, $$\text{Matrix} = \begin{bmatrix} F_{11} & F_{12} \\ F_{21} & F_{22} \end{bmatrix} = \begin{bmatrix} 0 & 1 \\ 0.5 & 0.5 \end{bmatrix}$$

9.9 (b)

From reciprocity theorem

$$s_1 f_{12} = s_2 f_{21}$$
$$f_{12} = 0.4 \text{ (given)}$$
$$s_1 = 1 \text{ m}^2, s_2 = 4 \text{ m}^2$$
$$f_{21} = \frac{s_1}{s_2} \times f_{12} = \frac{1}{4} \times 0.4 = 0.1$$

9.10 (d)

$$F_{12} = 1$$
$$A_1 F_{12} = A_2 F_{21} \text{ (reciprocity theorem)}$$

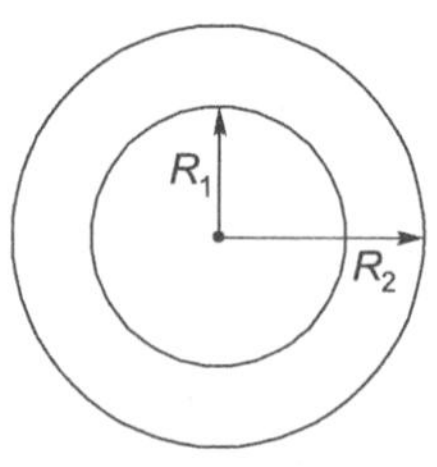

$$4\pi \times R_1^2 \times 1 = 4\pi R_2^2 F_{21}$$
$$F_{21} = \frac{1}{4} = 0.25$$
$$F_{21} + F_{22} = 1$$
So, $F_{22} = 1 - 0.25 = 0.75$

9.11 (5.67)

Energy emitted by black body is given by $Q = \sigma A(T_1^4)$ (Stefan-Boltzmann law)

$$Q = 5.67 \times 10^{-8} \times 10^{-4} \times (727 + 273)^4$$
$$= 5.67 \text{ watt}$$

9.12 (a)

50% of incident monochromatic radiation is absorbed by opaque medium.

$\because$ Medium is opaque so transmissivity (τ) = 0

Now, $\alpha + \tau + \rho = 1$

$$\alpha + \rho = 1$$

Given :

$$\alpha = 0.5$$

So, $\rho = 1 - 0.5 = 0.5$

So, option P and S is correct.

9.13 (a)

For parallel plates with inserted shield and with equal areas and emissivities:

$$\frac{Q_{\text{with}}}{Q_{\text{without}}} = \frac{1}{n+1} = \frac{1}{1+1} = \frac{1}{2}$$

9.14 (0.25)

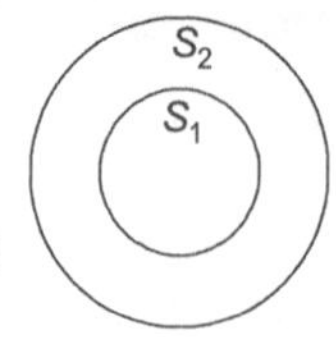

$$F_{12} = 1$$
$$A_1 F_{12} = A_2 F_{21}$$
(reciprocity theorem)
$$4\pi \times 0.1^2 F_{12} = 4\pi \times 0.2^2 F_{21}$$
$$F_{21} = 0.25$$

9.16 (b)

Hollow cylinders of length = diameter with sealed surfaces is given.

$$F_{13} = 3 - 2\sqrt{2}$$
$$F_{11} = 0 \text{ (from figure)}$$
$$F_{11} + F_{12} + F_{13} = 1$$
$$F_{12} = 1 - F_{13} = 2\sqrt{2} - 2$$
$$A_1 F_{12} = A_2 F_{21}$$
$$\frac{\pi}{4} D^2 F_{12} = \pi d L F_{21} \qquad \text{(Given : } L = d\text{)}$$
$$F_{21} = \frac{F_{12}}{4}$$
$$F_{21} = \frac{2\sqrt{2} - 2}{4} = \frac{\sqrt{2} - 1}{2}$$

9.17 (8792)

Source placed at the origin of circular sample holder (radius r = 1 m) emits particle in all directions.

Number of particles detected by detector is proportional to length of detector.

$\therefore$ $L \propto N_d$ ...(1)

(N_d = Number of particle detected)

Number of particles emitted will be proportional to the perimeter of circular sample holder.

$P \propto N_{Te}$ ...(2)

(N_{Te} = Number of particle emitted)

From (1) / (2)

$$\frac{L}{P} = \frac{N_d}{N_{Te}}$$

Here, $L = 1$ cm, $P = \pi D = 2\pi$, $N_d = 14$

So, $$N_{Te} = \frac{2\pi}{0.01} \times 14 = 8792$$

9.18 (a)

Data given:

Emissivities of walls and the shield are equal.

Now, melting point of shield is given as 900 K.

$$q_{\text{without}} = \frac{A\sigma\left[1000^4 - 500^4\right]}{\frac{2}{\epsilon} - 1} \quad \text{...(i)}$$

$$q_{\text{with shield}} = \frac{A\sigma\left[1000^4 - 900^4\right]}{\frac{2}{\epsilon} - 1}$$

Now, $\frac{q_{\text{with}}}{q_{\text{without}}} = \frac{1}{n+1}$

$$n + 1 = 2.72$$

$$n = 1.72$$

Now, the maximum number of shield will be 1.72 so, its integer value will be one.

$$n = 1$$

9.20 (a)

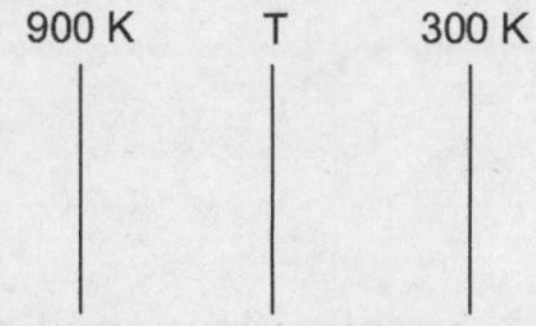

The steady state temperature of third plate inserted between two plates of temperature T_1 and T_2 is given by

$$T = \left[\frac{T_1^4 + T_2^4}{2}\right]^{1/4}$$

$$= \left[\frac{900^4 + 300^4}{2}\right]^{1/4} = 759.13 \text{ K}$$

■■■■

10 Heat Exchangers

10.1 In a 1 – 1 counter flow shell and tube heat exchanger, a process stream (C_p = 4.2 kJ/kg K) is cooled from 450 to 350 K using water (C_p = 4.2 kJ/kg K) at 300 K. The process stream flows on the shell-side at a rate of 1 kg/s and the water on the tube-side at a rate of 5 kg/s. If heat transfer, coefficients on the shell and tube sides are 1000 W/m^2 K and 1500 W/m^2 K, respectively, determine

(a) the required heat transfer area.

(b) by what factor will the required area change if the flow is cocurrent?

Neglect tube wall resistance and fouling resistances. [2000 : 5 M]

10.2 An aqueous solution of a solute is concentrated from 5% to 20% (mass basis) in a single-effect short-tube evaporator. The feed enters the evaporator at a rate of 10 kg/s and at a temperature of 300 K. Steam is available at a saturation pressure of 1.3 bar. The pressure in the vapour space of the evaporator is 0.13 bar and the corresponding saturation temperature of steam is 320 K. If the overall heat transfer coefficient is 5000 W/m^2 K, calculate the

(a) steam economy

(b) heat transfer surface area.

Data:

	Enthalpy (kJ/kg)	Heat of vaporization (kJ/kg)
Saturated steam (1.3 bar; 380 K)	–	2000
Saturated steam (0.13 bar; 320 K)	2200	–
Feed (5%; 300 K)	80	–
Concentrated liquor (20%; 325 K)	400	–
Boiling point elevation is 5 K		

[2000 : 5 M]

10.3 The overall heat transfer coefficient for a shell and tube heat exchanger for clean surfaces is U_0 = 400 W/m^2 K. The fouling factor after one year of operation is found to be h_{do} = 2000 W/m^2 K. The overall heat transfer coefficient at this time is

(a) 1200 W/m^2 K (b) 894 W/m^2 K

(c) 333 W/m^2 K (d) 287 W/m^2 K

[2001 : 2 M]

10.4 If the baffle spacing in a shell and tube heat exchanger increases, then the Reynolds number of the shell side fluid

(a) Remains unchanged

(b) Increases

(c) Increases or decreases depending on number of shell passes

(d) Decreases [2002 : 1 M]

10.5 A double pipe countercurrent heat exchanger is designed to cool 3500 kg/hr of benzene flowing in the inner pipe from 80°C to 35°C. Water enters at 20°C and exits at 37°C in the annular space. The inside pipe has an inner diameter of 3.5 cm and wall thickness of 3.56 mm. The outer pipe has an inner diameter of 5.25 cm and is insulated. Neglecting the wall resistance to heat transfer from the inner pipe, and assuming the individual film heat transfer coefficient for water to be 6600 W/m^2 °C. Calculate:

(a) the individual heat transfer coefficient for benzene flowing in the inner pipe.

(b) the overall heat transfer coefficient based on inside diameter of inner pipe.

(c) the total length required for the heat exchanger.
$Nu = 0.023 (Re)^{0.8} (Pr)^{0.3}$
Where Nu is the Nusselt Number and Pr is the Prandtl Number.

Average properties of benzene: Viscosity = 4×10^{-4} kg/m s, thermal conductivity = 0.147 W/m °C, specific heat capacity = 1880 J/kg °C, density = 837 kg/m^3.

[2002 : 5 M]

10.6 A dilute aqueous solution is to be concentrated in an evaporator system. High pressure steam is available. Multiple effect evaporator system is employed because

(a) total heat transfer area of all the effects is less than that in a single effect evaporator system
(b) total amount of vapour produced per kg of feed steam in a multi-effect system is much higher than in a single effect
(c) boiling point elevation in a single effect system is much higher than that in any effect in a multi-effect system
(d) heat transfer coefficient in a single effect is much lower than that in any effect in a multi-effect system

[2003 : 1 M]

10.7 A process stream of dilute aqueous solution flowing at the rate of 10 kg s^{-1} is to be heated. Steam condensate at 95°C is available for heating purpose, also at a rate of 10 kg s^{-1}. A1-1 shell and tube heat exchanger is available. The best arrangement is
(a) counterflow with process stream on shell side
(b) counterflow with process stream on tube side
(c) parallel flow with process stream on shell side
(d) parallel flow with process stream on tube side

[2003 : 2 M]

10.8 Steam is to be condensed in a shell and tube heat exchanger; 5 m long with a shell diameter of 1 m. Cooling water is to be used for removing the heat. Heat transfer coefficient for the cooling water, whether on shell side or tube side, is same. The best arrangement is
(a) vertical heat exchanger with steam on tube side
(b) vertical heat exchanger with steam on shell side
(c) horizontal heat exchanger with steam on tube side
(d) horizontal heat exchanger with steam on shell side

[2003 : 2 M]

10.9 Air is to be heated by condensing steam. Two heat exchangers are available : (i) a shell and tube heat exchanger, and (ii) a finned tube heat exchanger. Tube side heat transfer area is equal in both cases. The recommended arrangement is
(a) finned tube heat exchanger with air inside and steam outside
(b) finned tube heat exchanger with air outside and steam inside
(c) shell and tube heat exchanger with air inside tubes and steam shell side
(d) shell and tube heat exchanger with air on shell side and steam inside tubes

[2003 : 2 M]

10.10 The equivalent diameter for flow through a rectangular duct of width B and height H is
(a) $\dfrac{HB}{2(H+B)}$ (b) $\dfrac{HB}{(H+B)}$
(c) $\dfrac{2HB}{(H+B)}$ (d) $\dfrac{4HB}{(H+B)}$

[2004 : 1 M]

10.11 It is desired to concentrate a 20% salt solution (20 kg of salt in 100 kg of solution) to a 30% salt solution in an evaporator. Consider a feed of 300 kg/min at 30°C. The boiling point of the solution is 110°C, the latent heat of vaporization is 2100 kJ/kg, and the specific heat of the solution is 4 kJ/(kg K). The rate at which heat has to be supplied (in kJ/min) to the evaporator is
(a) 3.06×10^5 (b) 6.12×10^5
(c) 7.24×10^5 (d) 9.08×10^5

[2004 : 2 M]

10.12 Hot water (0.01 m^3/min) enters the tube side of a cocurrent shell and tube heat exchanger at 80°C and leaves at 50°C. Cold oil (0.05 m^3/min) of density 800 kg/m^3 and specific heat of 2 kJ/(kg K) enters at 20°C. The log mean temperature difference in °C is approximately
(a) 32 (b) 37
(c) 45 (d) 50

[2004 : 2 M]

10.13 Baffles are used in heat exchangers in order to
(a) increase the tube side fluid's heat transfer coefficient
(b) promote vibration in the heat exchanger
(c) promote cross flow and turbulence in the shell side Fluid
(d) to prevent shell expansion due to thermal effects.

[2005 : 1 M]

10.14 A countercurrent flow double pipe heat exchanger is used to heat water flowing at 1 kg/s from 40°C to 80°C. Oil is used for heating and its temperature changes from 100°C to

70°C. The overall heat transfer coefficient is 300 W/(m²°C).

If it is replaced by a 1-2 shell and tube heat exchanger with countercurrent flow configuration with water flowing in shell and oil flowing in the tube, what is the excess area required with respect to the double pipe heat exchanger? The correction factor, F_t for LMTD (log mean temperature difference) based on the above double pipe heat exchanger is 0.5. The heat transfer coefficient remains unchanged, and the same inlet and outlet conditions are maintained.

$C_{p,\text{ water}}$ = 4180 J/kg °C, $C_{p,\text{ oil}}$ = 2000 J/kg °C,

(a) 0 m^2
(b) –20.15 m^2
(c) 22.6 m^2
(d) 9.69 m^2

[2005 : 2 M]

10.15 Fluid flows in an annulus of inner diameter 0.8 m and outer diameter 1 m. Heat is transferred to the fluid from inner tube surface of the annulus. What is the equivalent diameter For heat transfer in m?

(a) 0.45
(b) 0.20
(c) 1.64
(d) 0.90

[2005 : 2 M]

10.16 The following list of options P, Q, R and S are some of the important considerations in the design of a shell and tube heat exchanger.

P. Square pitch permits the use of more tubes in a given shell diameter.
Q. The tube side clearance should not be less than one fourth of the tube diameter.
R. Baffle spacing is not greater than the diameter of the shell or less than one-fifth of the shell diameter.
S. The pressure drop on the tube side is less than 10 psi.

Pick out the correct combination of TRUE statements from the following:

(a) P, Q and R
(b) Q, R and S
(c) R, S and P
(d) P, Q, R and S

[2007 : 2 M]

10.17 In a double pipe heat exchanger the ID and OD of the inner pipe are 4 cm and 5 cm respectively. The ID of the outer pipe is 10 cm with a wall thickness of 1 cm. Then, the equivalent diameters (in cm) of the annulus for heat transfer and pressure drop respectively are

(a) 15, 5
(b) 21, 6
(c) 6, 19
(d) 15, 21

[2007 : 2 M]

10.18 In a shell and tube heat exchanger, if the shell length is L_S, the baffle spacing is L_B and the thickness of baffle is t_b, the number of baffles on the shell side, N_B, is,

(a) $\dfrac{L_S}{L_B + t_b}$
(b) $\dfrac{L_S}{L_B + t_b} - 1$
(c) $\dfrac{L_S}{L_B + t_b} + 1$
(d) $\dfrac{L_S}{L_B + t_b} + 2$

[2008 : 2 M]

10.19 A double-pipe heat exchanger is to be designed to heat 4 kg/s of a cold feed from 20 to 40°C using a hot stream available at 160°C and a flow rate of 1 kg/s. The two streams have equal specific heat capacities and the overall heat transfer coefficient of the heat exchanger is 640 W/m².K. Then the ratio of the heat transfer areas required for the co-current to counter-current modes of operation is

(a) 0.73
(b) 0.92
(c) 1.085
(d) 1.25

[2009 : 2 M]

10.20 Match the equipment in **Group-I** to the internals in **Group-II**.

Group-I	Group-II
P. Centrifugal pump	1. Baffle
Q. Distillation column	2. Impeller
R. Heat exchanger	3. Tray
	4. Volute

(a) P-2, Q-1, R-4
(b) P-4, Q-4, R-3
(c) P-1, Q-3, R-4
(d) P-4, Q-3, R-1

[2009 : 2 M]

10.21 Which one of the following statements about baffles in a shell and tube heat exchanger is false? Baffles

(a) act as a support to the tube bundle
(b) reduce the pressure drop on the shell-side
(c) alter the shell-side flow pattern
(d) help in increasing the shell-side heat transfer coefficient

[2010 : 1 M]

Common Data for Question 10.22 and 10.23:

Hot oil at 150°C is used to preheat a cold fluid at 30°C in a 1:1 shell and tube heat exchanger. The exit temperature of the hot oil is 110°C. Heat capacities (product of mass flow rate and specific heat capacity) of both the streams are equal. The heat duty is 2 kW.

10.22 Under co-current flow conditions, the overall heat transfer resistance (1/UA) is

(a) 0.4 °C/W (b) 0.04 °C/W
(c) 0.36 °C/W (d) 0.036 °C/W

[2010 : 2 M]

10.23 Under counter-current flow conditions, the overall heat transfer resistance (1/*UA*) is

(a) 0.4 °C/W (b) 0.04 °C/W
(c) 0.36 °C/W (d) 0.036 °C/W

[2010 : 2 M]

10.24 Oil at 120°C is used to heat water at 30°C in a 1-1 co-current shell and tube heat exchanger. The available heat exchange area is S_1. The exit temperatures of the oil and the water streams are 90°C and 60°C respectively. The co-current heat exchanger is replaced by a 1-1 counter-current heat exchanger having heat exchange area S_2. If the exit temperatures and the overall heat transfer coefficients are same, the ratio of S_1 to S_2 is

(a) ∞ (b) 1.1
(c) 0.91 (d) 0

[2011 : 2 M]

10.25 An aqueous sodium chloride solution (10 wt %) is fed into a single effect evaporator at a rate of 10000 kg/hr. It is concentrated to a 20 wt% sodium chloride solution. The rate of consumption of steam in the evaporator is 8000 kg/hr. The evaporator capacity (kg/hr) and economy are

(a) 5000, 0.625
(b) 10000, 0.625
(c) 5000, 1.6
(d) 10000, 1.6

[2011 : 2 M]

10.26 In a counter-flow double pipe heat exchanger, oil ($\dot{m}$ = 2 kg/s, C_P = 2.1 kJ/kg.°C) is cooled from 90°C to 40 °C by water ($\dot{m}$ = 1 kg/s, C_P = 4.2 kJ/kg.°C) which enters the inner tube at 10 °C. The radius of the inner tube is 3 cm and its length is 5 m. Neglecting the wall resistance, the overall heat transfer coefficient based on the inner radius, in kW/ m^2.K, is

(a) 0.743 (b) 7.43
(c) 74.3 (d) 2475

[2012 : 2 M]

10.27 In a 1-1 pass floating head type shell and tube heat exchanger, the tubes (od = 25 mm; id = 21 mm) are arranged in a square pitch. The tube pitch is 32 mm. The thermal conductivity of the shell side fluid is 0.19 W/m.K, and the Nusselt number is 200. The shell-side heat transfer coefficient (in W/m^2.K), rounded off to the nearest integer, is

(a) 1100 (b) 1400
(c) 1800 (d) 2100

[2012 : 2 M]

10.28 The effectiveness of a heat exchanger in the ε-NTU method is defined as

(a) $\dfrac{\text{Increase in temperature of the cold fluid}}{\text{Decrease in temperature of the hot fluid}}$

(b) $\dfrac{\text{Actual exit temperature attained by the cold fluid}}{\text{Maximum exit temperature attainable by the cold fluid}}$

(c) $\dfrac{\text{Actual exit temperature attained by the hot fluid}}{\text{Minimum exit temperature attainable by the hot fluid}}$

(d) $\dfrac{\text{Actual heat transfer rate}}{\text{Maximum possible heat transfer rate from hot fluid to cold fluid}}$

[2013 : 1 M]

10.29 In a double pipe counter-current heat exchanger, the temperature profiles shown in the figure were observed. During operation, due to fouling inside the pipe, the heat transfer rate reduces to half of the original value. Assuming that the flow rates and the physical properties of the fluids do not change, the LMTD (in °C) in the new situation is

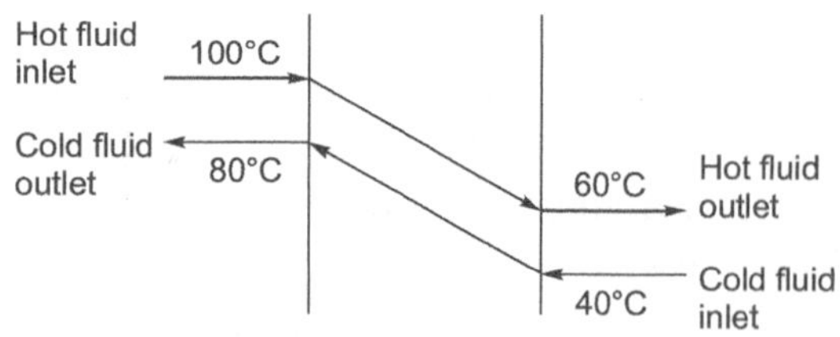

(a) 0
(b) 20
(c) 40
(d) indeterminate

[2013 : 2 M]

10.30 Steam economy of a multiple effect evaporator system is defined as

(a) kilogram of steam used per hour
(b) kilogram of steam consumed in all the effects for each kilogram of steam fed
(c) kilogram of steam used in all the effects for each kilogram of water vaporized per hour
(d) kilogram of water vaporized from all the effects for each kilogram of steam fed to the first effect

[2014 : 1 M]

10.31 An oil with a flow rate of 1000 kg/h is to be cooled using water in a double-pipe counter-flow heat exchanger from a temperature of 70 °C to 40 °C. Water enters the exchanger at 25 °C and leaves at 40 °C. The specific heats of oil and water are 2 kJ kg^{-1} K^{-1} and 4.2 kJ kg-1 K-1, respectively. The overall heat transfer coefficient is 0.2 kW m^{-2} K-1. The minimum heat exchanger area (in m^2) required for this operation is ______.

[2014 : 2 M]

10.32 In the figure below, the temperature profiles of cold and hot fluids in counter current double pipe heat exchangers (in different modes of operation) are shown on the left. For each case, match the heat exchange process for the fluid represented by the bold curve with the options given on the right.

Group-1

I.
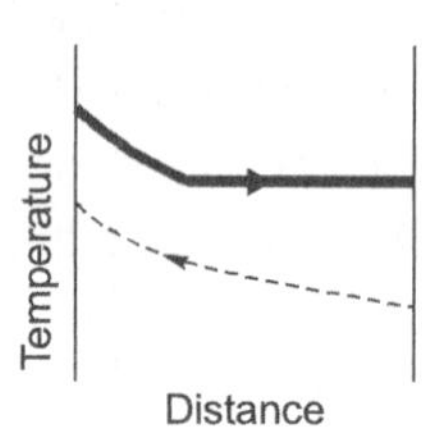

II.
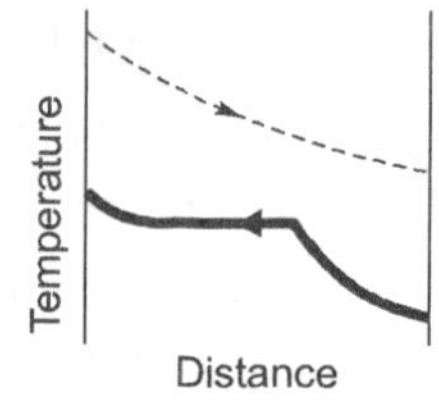

III.
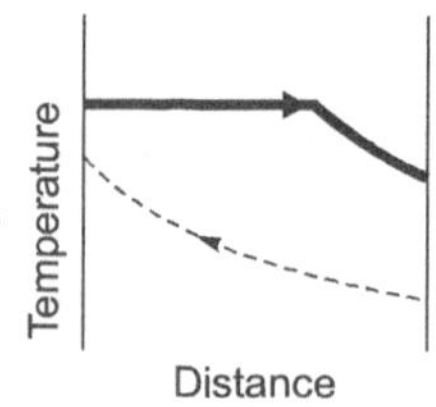

IV.
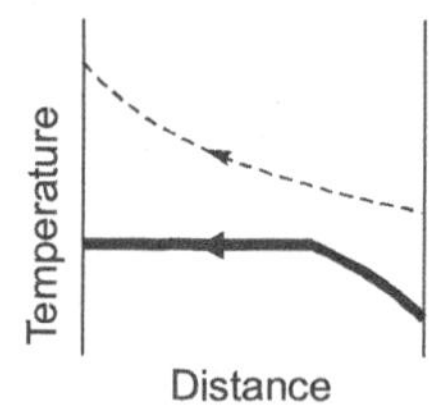

Group-2

P. Heating of sub-cooled feed to super heated vapour.
Q. Condensation of super heated vapour.
R. Boiling of sub-cooled liquid.
S. Condensation of saturated vapour followed by sub-cooling.

(a) I - P, II - Q, III - R, IV - S
(b) I - P, II - Q, III - S, IV - R
(c) I - Q, II - P, III - S, IV - R
(d) I - Q, II - S, III - P, IV - R

[2015 : 2 M]

10.33 In a 1-1 pass shell and tube exchanger, steam is condensing in the shell side at a temperature (T_s) of 135°C and the cold fluid is heated from a temperature (T_1) of 20°C to a temperature (T_2) of 90° C. The energy balance equation for this heat exchanger is

$$\ln\frac{T_s - T_1}{T_s - T_2} = \frac{UA}{\dot{m}c_p}$$

where U is the overall heat transfer coefficient, A is the heat transfer area, $\dot{m}$ is the mass flow rate of the cold fluid and c_p is its specific heat capacity. Tube side fluid is in a turbulent flow and the heat transfer coefficient can be estimated from the following equation:

$$\text{Nu} = 0.023\,(\text{Re})^{0.8}\,(\text{Pr})^{1/3}$$

where Nu is the Nusselt number, Re is the Reynold's number and Pr is the Prandtl number. The condensing heat transfer coefficient in the shell side is significantly higher than the tube side heat transfer coefficient. The resistance of the wall to heat transfer is negligible. If only the mass flow rate of the cold fluid is doubled, what is the outlet temperature (in °C) of the cold fluid at steady state?

(a) 80.2 (b) 84.2
(c) 87.4 (d) 88.6

[2016 : 2 M]

10.34 In a heat exchanger, the inner diameter of a tube is 25 mm and its outer diameter is 30 mm, The overall heat transfer coefficient based on the inner area is 360 W/m^2 °C. Then, the overall heat transfer coefficient based on the outer area, rounded to the nearest integer, is _____ W/m^2 °C.

[2017 : 1 M]

10.35 Economy of evaporators used for concentrating sugarcane juice is

(a) $\dfrac{\text{kg of concentrated juice produced}}{\text{kg of steam supplied}}$

(b) $\dfrac{\text{kg of steam supplied}}{\text{kg of sugarcane juice fed}}$

(c) $\dfrac{\text{kg of water vaporized}}{\text{kg of steam supplied}}$

(d) $\dfrac{\text{kg of sugarcane juice fed}}{\text{kg of water vaporized}}$

[2018 : 1 M]

10.36 Segmental baffles in a 2-4 shell and tube heat exchanger

(a) change the flow pattern of the tube side fluid and increase the overall heat transfer coefficient.
(b) increase the heat transfer coefficient in the shell side and support the tubes.
(c) help to reduce the thermal expansion of the tubes and increase the heat transfer coefficient in the tube side.
(d) increase the number of passes in the shell side and increase the heat transfer coefficient in the tube side

[2018 : 1 M]

10.37 A hot liquid is to be cooled in a 1-1 shell and tube heat exchanger from 80°C to 50°C. Cooling water enters the tube side at 30°C, and exits at 45°C. The properties of the liquids are constant. Also, the overall heat transfer coefficient is same for counter-current and cocurrent modes. The percentage saving in heat transfer area for counter-current option with respect to the area of co-current option is _____ (rounded off to third decimal place).

[2018 : 2 M]

10.38 Consider the two countercurrent heat exchanger designs for heating a cold stream from t_{in} to t_{out}, as shown in figure. The hot process stream is available at T_{in}. The inlet stream conditions and overall heat transfer coefficients are identical in both the designs. The heat transfer area in Design-I and Design-II are respectively A^I_{HX} and A^{II}_{HX}.

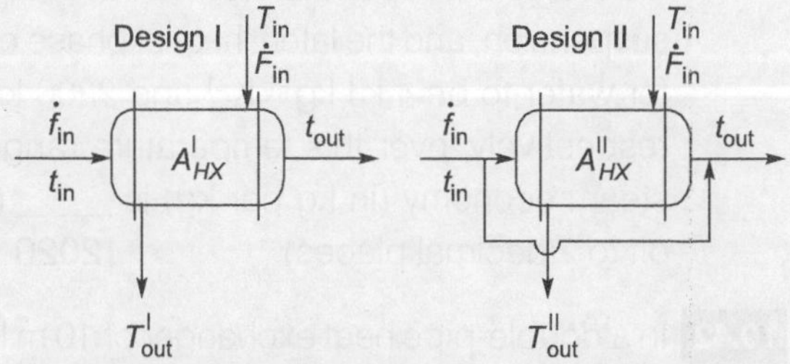

If heat losses are neglected, and if both the designs are feasible, which of the following statements holds true:

(a) $A^I_{HX} > A^{II}_{HX}$; $T^I_{out} < T^{II}_{out}$
(b) $A^I_{HX} = A^{II}_{HX}$; $T^I_{out} = T^{II}_{out}$
(c) $A^I_{HX} < A^{II}_{HX}$; $T^I_{out} > T^{II}_{out}$
(d) $A^I_{HX} < A^{II}_{HX}$; $T^I_{out} = T^{II}_{out}$

[2019 : 1 M]

10.39 Stream A with specific heat capacity C_{PA} = 2000 J/(kg K) is cooled from 90°C to 45°C in a concentric double pipe counter current heat exchanger having a heat transfer area of 8 m^2. The cold stream *B* of specific heat capacity C_{PB} = 1000 J/(kg K) enters the exchanger at a flow rate 1 kg/s and 40°C. The overall heat transfer coefficient *U* = 250 W/(m^{2}K). Assume that the mean driving force is based on the arithmetic mean temperature difference, that is

$$[\Delta T]_{AMTD} = \left(\frac{T_{A,in} + T_{A,out}}{2}\right) - \left(\frac{T_{B,in} + T_{B,out}}{2}\right)$$

where $T_{i,in}$ and $T_{i,out}$ refer to the temperature of the ith stream (i = A, B) at the inlet and exit, respectively. The mass flow rate of stream A (in kg/s), is ______ (rounded off to two decimal places). [2019 : 2 M]

10.40 A fluid is heated from 40°C to 60°C in a counter-current, double pipe heat exchanger. Hot fluid enters at 100°C and exits at 70°C. The log mean temperature difference, i.e., LMTD (in °C), is _____ (round off to 2 decimal places). [2020 : 2 M]

10.41 An aqueous suspension at 60°C is fed to the first effect of a double effect forward feed evaporator with a mass flow rate of 1.25 kg s^{-1}. The sum of the rates of water evaporated from the first and second effects is 1.0 kg s^{-1}. Temperatures of the exit streams from the first and the second effects are 100°C and 60°C, respectively. Consider the specific heat of the aqueous suspension, and the latent heat of phase change for water to be 4 kJ kg^{-1} K^{-1} and 2200 kJ kg^{-1}, respectively, over this temperature range. The steam economy (in kg per kg) is _____ (round off to 2 decimal places). [2020 : 2 M]

10.42 In a double-pipe heat exchanger of 10 m length, a hot fluid flows in the annulus and a cold fluid flows in the inner pipe. The temperature profiles of the hot (T_h) and cold (T_c) fluids along the length of the heat exchanger (x, such that $x \geq 0$), are given by

$$T_h(x) = 80 - 3x$$
$$T_c(x) = 20 + 2x$$

where T_h and T_c are in °C, and x is in meter.
The logarithmic mean temperature difference (in °C) is

(a) 24.6 (b) 27.9
(c) 30.0 (d) 50.0 [2021 : 1 M]

10.43 For a shell-and-tube heat exchanger, the clean overall heat transfer coefficient is calculated as 250 W m^{-2} K^{-1} for a specific process condition. It is expected that the heat exchanger may be fouled during the operation, and a fouling resistance of 0.001 m^2 KW^{-1} is prescribed. The dirt overall heat transfer coefficient is ______ Wm^{-2} K^{-1}.

(a) 100 (b) 150
(c) 200 (d) 250 [2021 : 1 M]

10.44 A double-effect evaporator is used to concentrate a solution. Steam is sent to the first effect at 110 °C and the boiling point of the solution in the second effect is 63.3 °C. The overall heat transfer coefficient in the first effect and second effect are 2000 W m^{-2} K^{-1} and 1500 W m^{-2} K^{-1}, respectively. The heat required to raise the temperature of the feed to the boiling point can be neglected. The heat flux in the two evaporators can be assumed to be equal.
The temperature at which the solution boils in the first effect is _____ °C (round off to nearest integer).

[2021 : 2 M]

10.45 Liquid flowing through a heat exchanger (HX) is heated. A bypass stream is provided to control the temperature of the heated exit stream. From the given plumbing options, the one that provides the most effective temperature control for large disturbances while avoiding vaporization in the heat exchanger

(a)
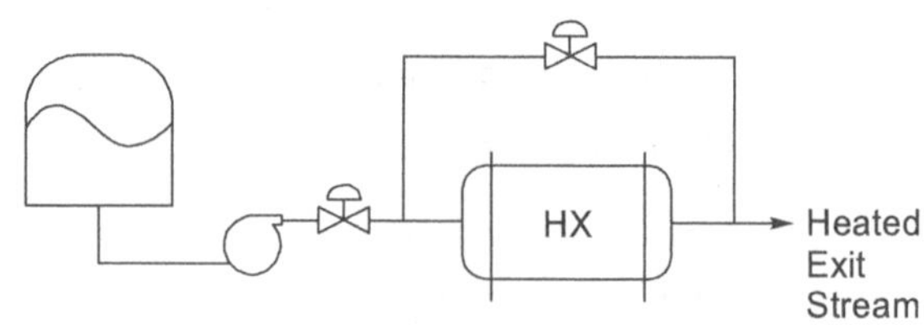

(b)
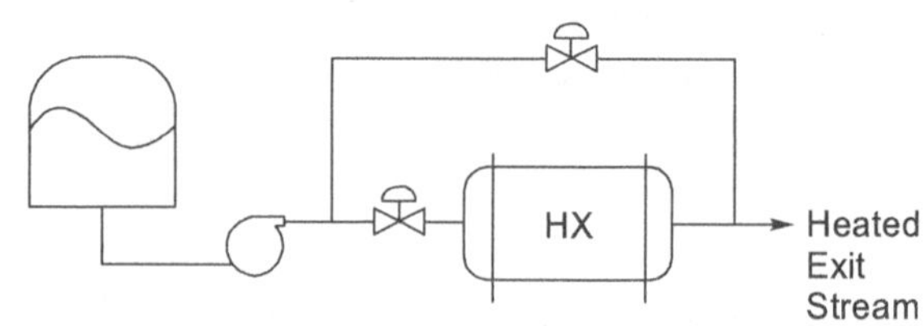

(c)
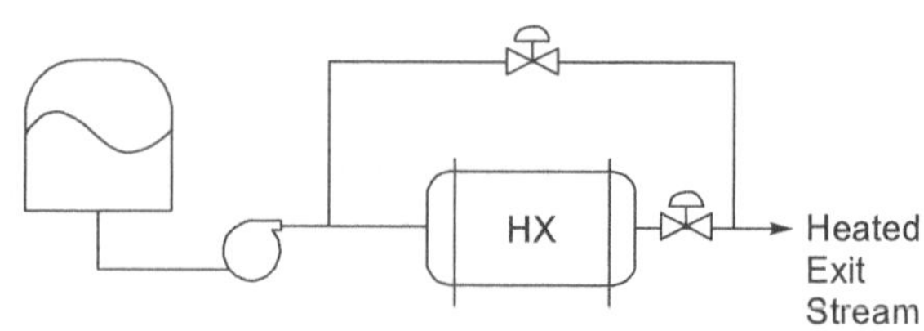

(d)
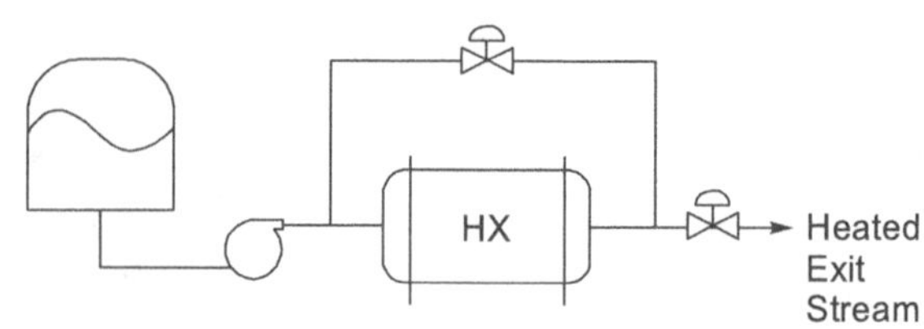

[2022 : 1 M]

10.46 A perfectly insulated double pipe heat exchanger is operating at steady state. Saturated steam enters the inner pipe at 100°C and leaves as saturated water at 100°C. Cooling water enters the outer pipe at 75°C and exits at 95°C. The overall heat transfer coefficient is 1 kW m^{-2} K^{-1} and the heat transfer area is 1 m^2. The average specific heat capacity of water at constant pressure is 4.2 kJ kg^{-1} K^{-1}. The required cooling water flow rate is ________ kg s^{-1} (rounded off to two decimal places). **[2022 : 1 M]**

10.47 A single-effect evaporator with a heat transfer area of 70 m^2 concentrates a salt solution using steam. The salt solution feed rate and temperature are 10000 kg h^{-1} and 40°C, respectively. The saturated steam feed rate and temperature are 7500 kg h^{-1} and 150°C, respectively. The boiling temperature of the solution in the evaporator is 80°C. The average specific heat of the solution is 0.8 kcal kg^{-1} K^{-1}. The latent heat of vaporization is 500 kcal kg^{-1}. If the steam-economy is 0.8, the overall heat transfer coefficient is ________ kcal h^{-1} m^{-1} K^{-1} (rounded off to the nearest integer). **[2022 : 2 M]**

10.48 An isothermal jacketed continuous stirred tank reactor (CSTR) operating at 150°C is shown in the figure below. The cold feed entering the system at 30°C is preheated to a temperature T (T < 150°C) using a heat exchanger HX_1. This preheated feed is further heated to 150°C using the utility heater HX_2. The mass flow rate and heat capacity are same for all the process streams, and the overall heat transfer coefficient is independent of temperature. Which one of the following statements is the CORRECT action to take if it is desired to increase the value of T?

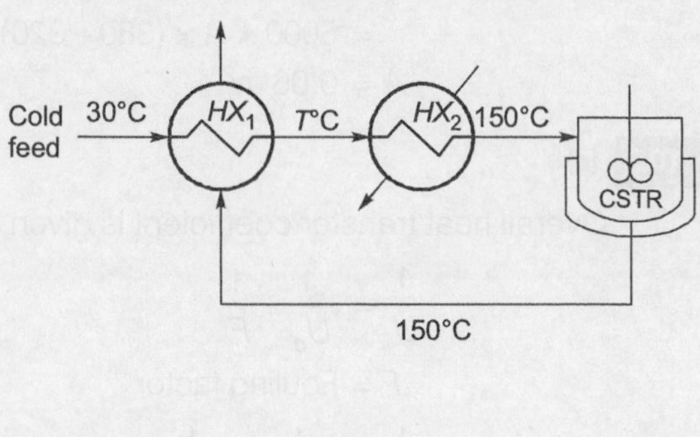

(a) Increase both heat transfer area of HX_1 and heat duty of HX_2.
(b) Decrease both heat transfer area of HX_1 and heat duty of HX_2.
(c) Increase the heat transfer area of HX_1 and decrease the heat duty of HX_2.
(d) Decrease the heat transfer area of HX_1 and increase the heat duty of HX_2.

[2023 : 1 M]

10.49 A perfectly insulated, concentric tube countercurrent heat exchanger is used to cool lubricating oil using water as a coolant (see figure below). Oil enters the outer annulus at a mass flow rate of 2 kg.s^{-1} with a temperature of 100 °C and leaves at 40 °C. Water enters the inner tube at a mass flow rate of 1 kg.s^{-1} with a temperature of 20 °C and leaves at 80 °C. Use specific heats of oil and water as 2089 J.$kg^{-1}K^{-1}$ and 4178 J.$kg^{-1}K^{-1}$, respectively. There is no phase change in both the streams. Under steady-state conditions, the number of transfer units (NTU) is ______ (in integer).

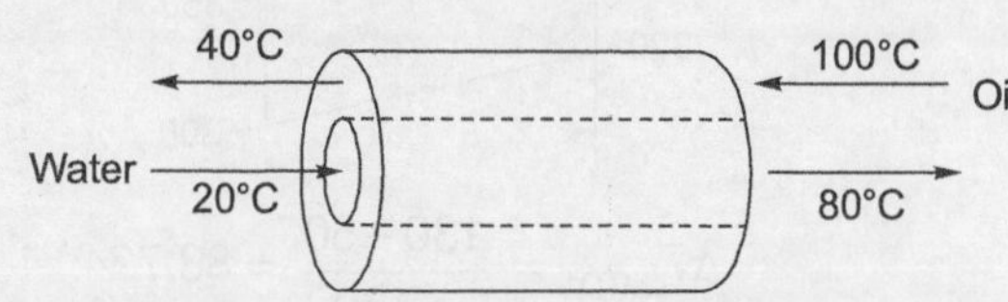

[2023 : 2 M]

10.50 Hot oil at 110°C heats water from 30°C to 70°C in a counter-current double-pipe heat exchanger. The flow rates of water and oil are 50 kg min^{-1} and 100 kg min^{-1}, respectively and their specific heat capacities are 4.2 kJ kg^{-1} °C^{-1} and 2.0 kJ kg^{-1} °C^{-1}, respectively. Assume the heat exchanger is at steady state. If the overall heat transfer coefficient is 200 W m^{-2} °C^{-1}, the heat transfer area in m^2 is

(a) 17.9
(b) 1.1
(c) 5.2
(d) 35.2

[2024 : 2 M]

■■■■

Answers Heat Exchangers

10.1 (8.36 m^2, 1.12) **10.2** (0.898, 0.06) **10.3** (c) **10.4** (d)

10.5 (1428, 1210, 22.45) **10.6** (b) **10.7** (b) **10.8** (d) **10.9** (b) **10.10** (c)

10.11 (a) **10.12** (a) **10.13** (c) **10.14** (c) **10.15** (b) **10.16** (b) **10.17** (a)

10.18 (b) **10.19** (c) **10.20** (d) **10.21** (b) **10.22** (d) **10.23** (b) **10.24** (b)

10.25 (a) **10.26** (b) **10.27** (b) **10.28** (d) **10.29** (c) **10.30** (d) **10.31** (3.85)

10.32 (c) **10.33** (b) **10.34** (300) **10.35** (c) **10.36** (b) **10.37** (27.08) **10.38** (d)

10.39 (0.305) **10.40** (34.76) **10.41** (1.78) **10.42** (b) **10.43** (c) **10.44** (90) **10.45** (c)

10.46 (0.148) **10.47** (678) **10.48** (c) **10.49** (3) **10.50** (a)

Explanations Heat Exchangers

10.1 (8.36 m^2)

(a) $Q = (mC_P\Delta T)_{hot} = (mC_P\Delta T)_{cold} = UA\Delta T_{LMTD}$ ...(1)

$1 \times 4.2 \times (450 - 350) = 5 \times 4.2 \times (T - 300)$

$T = 320$ K

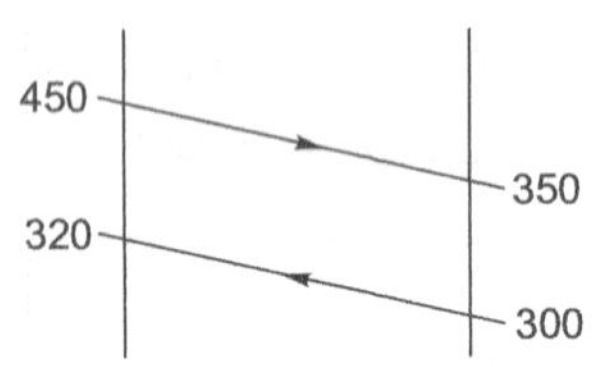

$$\Delta T_{LMTD1} = \frac{130-50}{\ln\left(\frac{130}{50}\right)} = 83.724$$

$$\Delta T_{LMTD2} = \frac{150-30}{\ln\left(\frac{150}{30}\right)} = 74.56$$

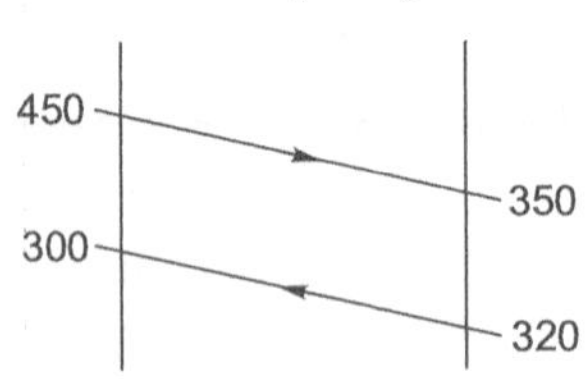

Overall heat transfer coefficient L can be calculated by

$$\frac{1}{U} = \frac{1}{h_i} + \frac{1}{h_o}$$

$$\frac{1}{U} = \frac{1}{1000} + \frac{1}{1500}$$

$U = 600$

From eqn. (1)

$$A = \frac{4.2 \times (450-350) \times 1000}{600 \times 83.724} = 8.36 \text{ m}^2$$

(b) Area will change by a factor of $\left(\frac{A_2}{A_1}\right)$ if we use co-current

$$\frac{A_2}{A_1} = \frac{\Delta T_{LMTD1}}{\Delta T_{LMTD2}} = 1.12$$

10.2 (0.898, 0.06)

From overall balance

$F = L + V,\; 10 = L + V$

From solute balance

$Fx_L = Lx_L$

$10 \times 0.05 = L \times 0.2,\; L = 2.5$ kg/s

$V = 10 - 2.5 = 7.5$ kg/s

$T_F = 300$ K, $T_{sat} = 320$ K

$U_o = 5000$ W/m^2 K

Energy balance $\Rightarrow FH_F + s\lambda_s = VH_L + LH_L$

$10 \times 80 + s \times 2000 = 7.5 \times 2200 + 2.5 \times 400$

$s = 8.35$ kg/s

(a) Steam economy

$$\frac{V}{s} = \frac{7.5}{8.35} = 0.898$$

(b) $s\lambda_s = UA\Delta T = 8.39 \times 2000$

$= 5000 \times A \times (380 - 320)$

$A = 0.06$ m^2

10.3 (c)

Overall heat transfer coefficient is given by

$$\frac{1}{U} = \frac{1}{U_o} + \frac{1}{F}$$

F = Fouling factor

$$\frac{1}{U} = \frac{1}{400} + \frac{1}{2000}$$

$U = 333.33$ W/m^2 K

10.4 (d)

Baffle spacing is increased means area available for flow of shell side fluid is increased. This decreases velocity of shell side fluid. Hence, Reynolds number is decreased.

10.5 (1428, 1210, 22.45)

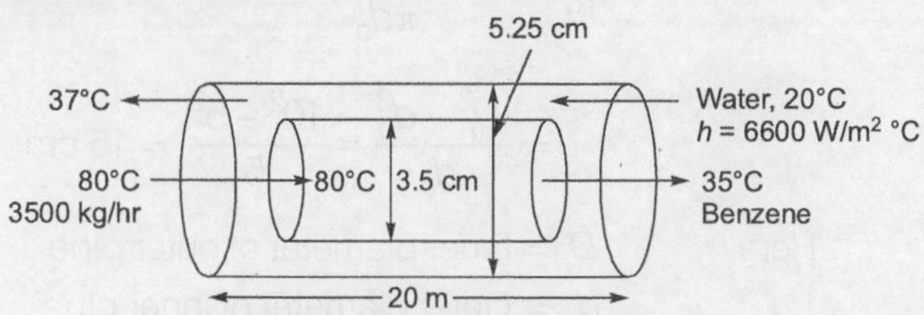

Inner pipe wall thickness is 3.56 mm.

$$\rho = 837 \text{ kg/m}^3$$
$$\mu = 4 \times 10^{-4} \text{ kg/ms},$$
$$k = 0.147 \text{ W/m °C}$$
$$C_P = 1880 \text{ J/kgC}$$

(a)
$$\text{Re} = \frac{\rho d V}{\mu} = \frac{d\dot{m}}{\mu A}$$
$$= \frac{0.035 \times 3500}{3600 \times 4 \times 10^{-4} \times \frac{\pi}{4} \times 0.035^2}$$
$$\text{Re} = 88416.13$$
$$\text{Pr} = \frac{\mu C_P}{K} = \frac{4 \times 10^{-4} \times 1880}{0.147} = 5.11$$

Now,
$$\text{Nu} = 0.023\,\text{Re}^{0.8}\,\text{Pr}^{0.3}$$
$$\frac{h_i \times 0.035}{0.147} = 0.023(88416.13)^{0.8} \times (5.11)^{0.3}$$
$$h_i = 1428 \text{ W/m}^2 \text{°C}$$

(b)
$$\frac{1}{U} = \frac{1}{h_i} = \frac{1}{h_o d_{io}} di$$

Here, $d_{io} = 3.5 + 2 \times 0.356 = 4.212$ cm

U = Overall heat transfer coefficient based on inner side.

$$\frac{1}{U} = \frac{1}{1428} + \frac{1}{6600} + \frac{3.5}{4.212}$$
$$U = 1210 \text{ W/m}^2 \text{°C}$$

(c)
$$\Delta T_{\text{LMTD}} = \frac{\Delta T_1 - \Delta T_2}{\ln\left(\frac{\Delta T_1}{\Delta T_2}\right)} = \frac{50 - 17}{\ln\left(\frac{50}{17}\right)} = 30.6\text{°C}$$

Now,
$$Q = mC_P\,\Delta T = UA\,\Delta T_{\text{LMTD}}$$
$$= \frac{3500}{3600} \times 1880 \times 50$$
$$= 1210 \times \pi \times 0.035 \times L \times 30.6$$
$$L = 22.45 \text{ m}$$

10.6 (b)

In multiple effect evaporator steam economy is increased while capacity is decreased.

10.7 (b)

Counter flow gives better LMTD and minimizes the area of heat exchanger. Process stream in question is dilute aqueous solution which may have a fouling tendency so it will be in tube side. So, option (b) is correct option.

10.8 (d)

Cooling water have more tendency to foul as compared to clean steam. So, it will be in tube side. Horizontal heat exchanger have better heat transfers characteristics than vertical heat exchanger. So, correct option is (d).
Vertical heat exchanger with steam on shell side favours drop wise condensation.

10.9 (b)

For air heat transfer coefficient is very low. So we must use finned type heat exchanger with air outside.

10.10 (c)

Equivalent diameter = $4\,r_H\{r_H$ = Hydraulic radius$\}$

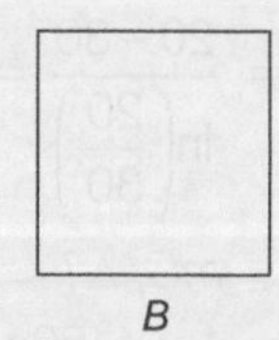

$$r_H = \frac{\text{Cross-sectional area}}{\text{Wetted perimeter}} = \frac{BH}{2(B+H)}$$
$$\therefore \quad d_{eq} = 4\left[\frac{BH}{2(B+H)}\right] = \frac{2BH}{B+H}$$

10.11 (a)

$$F = L + V$$
$$300 = L + V$$

On solute balance
$$300 \times 0.2 = L \times 0.3$$
$$L = 200 \text{ kg/min}$$
$$V = 300 - 200 = 100 \text{ kg/min}$$
$$Q = FC_P(T^{\text{sat}} - T_F) + V\lambda_V$$
$$= 300 \times 4 \times (80) + 100 \times 2100$$
$$= 3.06 \times 10^5 \text{ kJ/min}$$

10.12 (a)

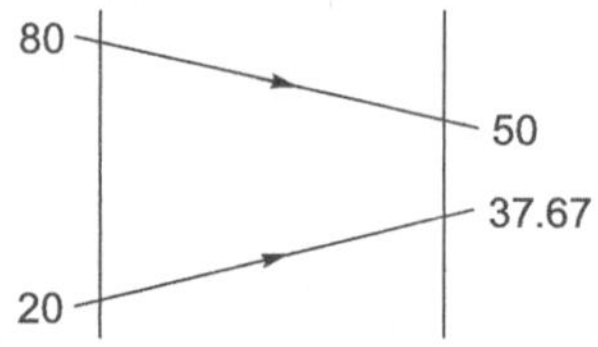

From energy balance

$(mC_P\Delta T)_{\text{cold fluid}} = (mC_P\Delta T)_{\text{hot fluid}}$

$0.01 \times 1000 \times 4.18 \times (80-50)$

$= 0.05 \times 800 \times 2 \times (T-20)$

$T = 35.67°C$

$$\Delta T_{LMTD} = \frac{\Delta T_1 - \Delta T_2}{\ln\left(\dfrac{\Delta T_1}{\Delta T_2}\right)} = \frac{60-14.325}{\ln\left(\dfrac{60}{14.325}\right)} = 31.9$$

10.13 (c)

Baffles creates turbulence and increases heat transfer coefficient in shell side fluid.

10.14 (c)

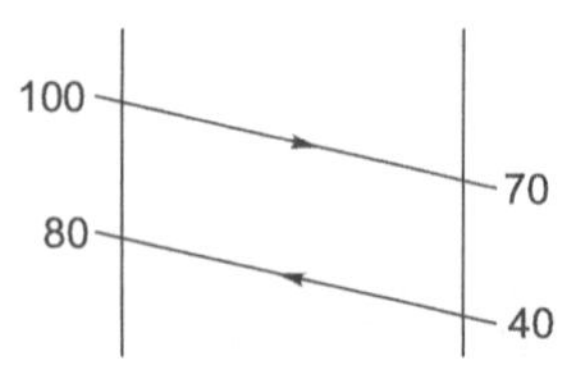

$$\Delta T_{LMTD} = \frac{20-30}{\ln\left(\dfrac{20}{30}\right)} = 24.66°C$$

$Q = mC_P\Delta T = UA\,\Delta T_{LMTD}$

$= 1 \times 4180 \times 40 = 300 \times A \times 24.66$

$A = 22.6 \text{ m}^2$

For shell and tube heat exchanger

$Q = UA\,\Delta T_{LMTD} \times F_t = mC_P\Delta T$

Correction factor $F_t = 0.5$

For same flow rate and temperature limits, now area will be doubled because $F_t = 0.5$.

So, excess area = 22.6 m^2

10.15 (b)

Triangular pitch permits maximum numbers of tubes in given shell diameters.

$C_L \le 0.25d_t$ (d_t = Diameter of tube)

$P_t \le 1.25d_t$

Baffles spacing

$\frac{1}{5}D \le L_B \le D$, L_B = Baffle spacing

Here, D = Shell diameter

10.17 (a)

$$\text{Equivalent dia} = \frac{4 \times \text{Area for flow}}{\text{Wetted perimeter}}$$

For heat transfer

$$d_{eq} = \frac{4 \times \frac{\pi}{4}(D_i^2 - d_o^2)}{\pi d_o}$$

$$= \frac{D_i^2 - d_o^2}{d_o} = \frac{10^2 - 5^2}{5} = 15 \text{ cm}$$

Here, D_i = Inner diameter of outer pipe

d_o = Outer diameter of inner pipe

For pressure drop

$$d_{eq} = \frac{4\frac{\pi}{4}(10^2 - 5^2)}{\pi(10+5)} = \frac{75}{15} = 5 \text{ cm}$$

10.19 (c)

$\dot{m}_{\text{cold fluid}} = 4$ kg/s

$T_{C1} = 20°C$, $T_{C2} = 40°C$

$\dot{m}_{\text{hot fluid}} = 1$ kg/s

$T_{H1} = 160°C$

C_P value is same for both fluid.

$V_o = 640 \text{ W/m}^2\text{K}$

From energy balance:

$(mC_P\Delta T)_{\text{cold fluid}} = (mC_P\Delta T)_{\text{hot fluid}}$

$4C_P(40-20) = 1C_P(160 - T_{H2})$

$T_{H2} = 80°C$

Co-current :

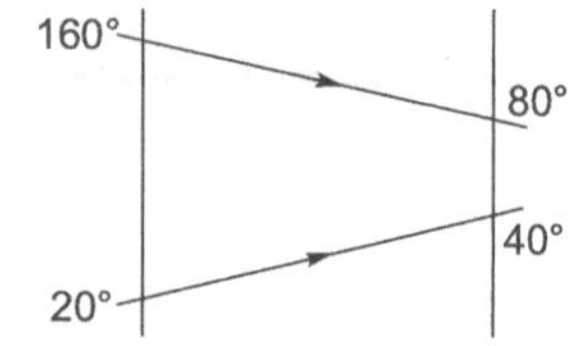

$$\Delta T_{LMTD} = \frac{\Delta T_1 - \Delta T_2}{\ln\left(\dfrac{\Delta T_1}{\Delta T_2}\right)} = \frac{140-40}{\ln\left(\dfrac{140}{40}\right)} = 79.82°C$$

Counter-current :

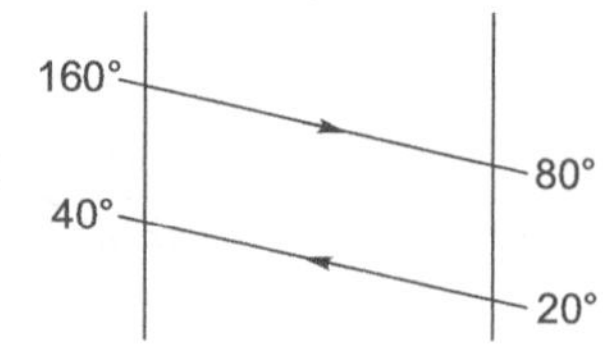

$$\Delta T_{LMTD} = \frac{120-60}{\ln\left(\frac{120}{60}\right)} = 86.56°C$$

$$A \propto \frac{1}{\Delta T_{LMTD}}$$

$$\frac{A_1}{A_2} = \frac{86.56}{79.82} = 1.085$$

10.21 (b)

Baffles reduces area available for flow in shell side and increases flow velocity. So, shell side pressure drop will be increased by using Baffles.

10.22 (d)

mC_P is same for both fluids.

From energy balance:

$$(mC_P\Delta T)_{\text{hot fluid}} = (mC_P\Delta T)_{\text{cold fluid}}$$

$$150 - 110 = T - 30°$$

$$T = 70°C$$

Cold fluid outlet temperature is 70°C.

Heat duty, $Q = 2\text{ kW} = 2000\text{ W}$

For co-current,

$$\Delta T_{LMTD} = \frac{\Delta T_1 - \Delta T_2}{\ln\left(\frac{\Delta T_1}{\Delta T_2}\right)} = \frac{120-40}{\ln\left(\frac{120}{40}\right)} = 72.81$$

$$Q = UA\Delta T_{LMTD} \Rightarrow \frac{1}{UA} = \frac{\Delta T_{LMTD}}{Q}$$

$$= \frac{72.81}{2000} = 0.036\ °C/\text{Watt}$$

10.23 (b)

For counter-current

$$\Delta T_{LMTD} = \frac{\Delta T_1 - \Delta T_2}{\ln\left(\frac{\Delta T_1}{\Delta T_2}\right)}$$

mC_P of both fluid is same.

So, $\Delta T_{LMTD} = 110 - 30 = 80°C$

$$\frac{1}{UA} = \frac{\Delta T}{Q} = \frac{80}{2000} = 0.04\ °C/W$$

10.24 (b)

$$Q = UA\Delta T_{LMTD}$$

$$A \propto \frac{1}{\Delta T_{LMTD}} \quad \text{...(1)}$$

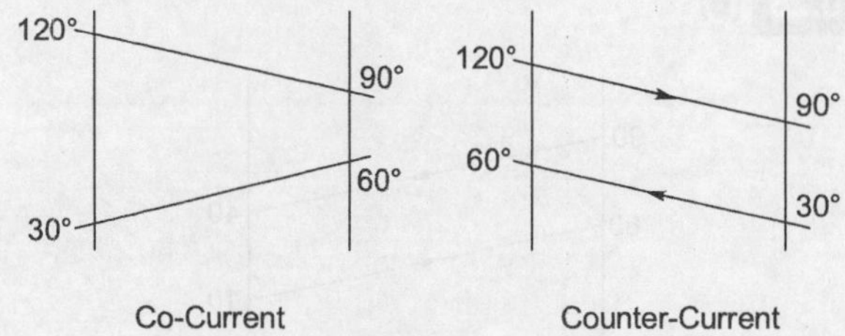

ΔT_{LMTD1}, co-current

$$= \frac{\Delta T_1 - \Delta T_2}{\ln\left(\frac{\Delta T_1}{\Delta T_2}\right)} = \frac{90-30}{\ln\left(\frac{90}{30}\right)} = 54.614°C$$

ΔT_{LMTD2} counter-current $= 120 - 60 = 60°C$

From relation (1)

$$\frac{A_1}{A_2} = \frac{\Delta T_{LMTD2}}{\Delta T_{LMTD1}}$$

$$= \frac{60}{54.614} = 1.1$$

Note: Area of counter-current heat exchanger is always less than area of co-current heat exchanger with same temperature limit. So, by options we can choose 1.1 without solving it.

10.25 (a)

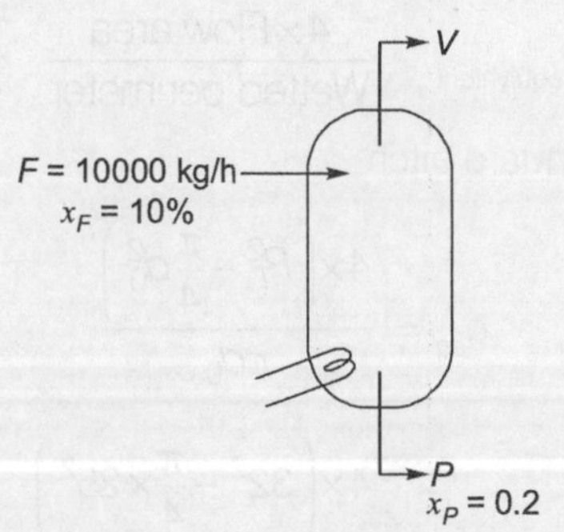

Steam consumption (s) = 8000 kg/h

From balance,

$$F = P + V$$

$$10000 = P + V \quad \text{...(i)}$$

By solute balance

$$Fx_F = Px_P$$

$$10000 \times 0.1 = P \times 0.2$$

$$P = 5000$$

From eqn. (i)

$$V = 10000 - 5000$$

$$= 5000\text{ kg/h (capacity)}$$

$$\text{Steam economy} = \frac{\text{Water evaporated}}{\text{Steam consumption}} = \frac{V}{s}$$

$$= \frac{5000}{8000} = 0.625$$

10.26 (b)

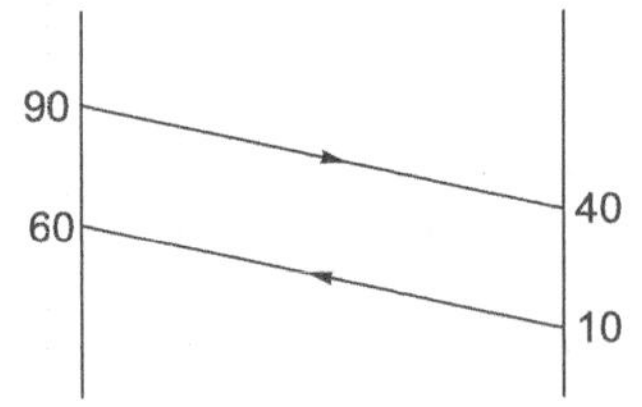

$$(\dot{m}C_P\Delta T)_{\text{cold fluid}} = (\dot{m}C_P\Delta T)_{\text{hot fluid}}$$

$$2\times 2.1\times(90-40) = 1\times 4.2\times(T-10)$$

Now, $\quad Q = mC_P\Delta T = UA\Delta T_{\text{LMTD}}$

mC_P is same for hot fluid.

So, $\quad \Delta T_{\text{LMTD}} = 90-60 = 30°\text{C}.$

Now $\quad Q = mC_P\Delta T = UA\Delta T_{\text{LMTD}}$

$$2\times 2.1\times(90-40) = U\times\pi\times 0.06\times 5\times 30$$

$$U = 7.43\ \text{kW/m}^2\,\text{K}$$

10.27 (b)

$$\text{Nu} = 200$$

Shell side fluid,

$$k = 0.19\ \text{W/mK}$$

$$\text{Tube pitch} = 32\ \text{mm}$$

$$\text{Outer diameter} = 0.025\ \text{m}$$

$$d_{\text{equivalent}} = \frac{4\times\text{Flow area}}{\text{Wetted perimeter}}$$

For square pitch

$$d_{eq} = \frac{4\times\left(P_T^2-\frac{\pi}{4}d_0^2\right)}{\pi d_0}$$

$$= \frac{4\times\left(32^2-\frac{\pi}{4}\times 25^2\right)}{\pi\times 25^2} = 27.1\ \text{mm}$$

$$\text{Nu} = \frac{hd_{eq}}{k} = 200$$

$$\frac{h\times 0.0271}{0.19} = 200$$

$$h = 1400\ \text{W/m}^2\,\text{K}$$

10.29 (c)

Due to fouling heat transfer rate reduces to half

$$Q = mC_P(100-60) \qquad \text{...(i)}$$

$$\frac{Q}{2} = mC_P(100-T_1) \qquad \text{...(ii)}$$

From equation (i) and (ii)

$$T_1 = 80°\text{C}$$

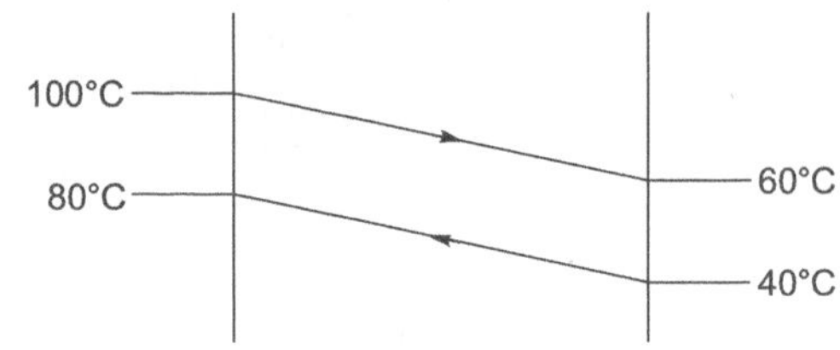

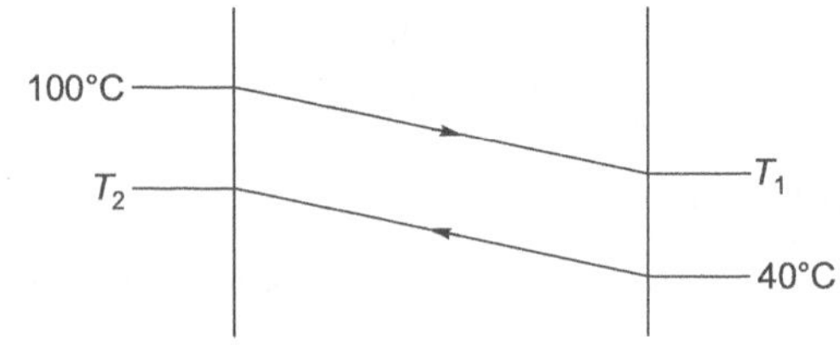

Similar for cold fluid,

$$Q = mC_P(80-40) \qquad \text{...(iii)}$$

$$\frac{Q}{2} = mC_P(T_2-40) \qquad \text{...(iv)}$$

From equation (iii) and (iv)

$$T_2 = 60°\text{C}$$

Now new temperature profile is

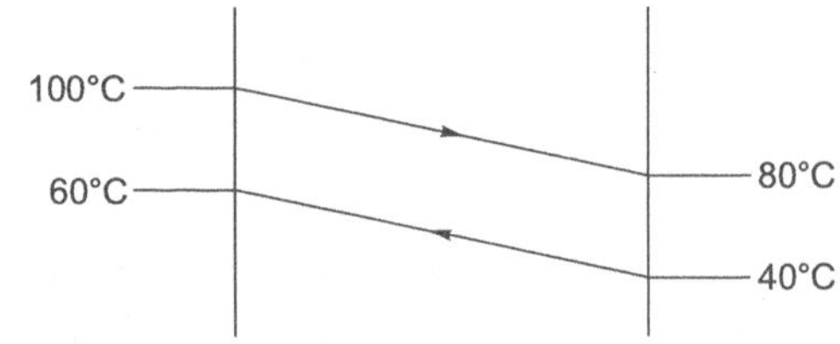

Now new $\Delta T_{\text{LMTD}} = 100-60 = 40°\text{C}$

10.31 (3.85)

Area of heat exchanger is minimum for counter-current heat exchanger

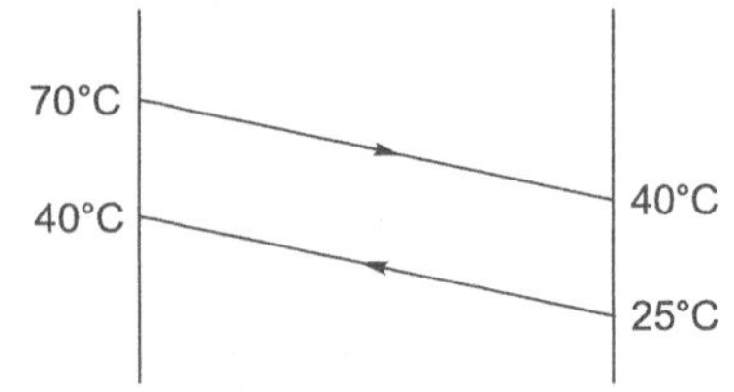

$$\Delta T_{\text{LMTD}} = \frac{\Delta T_1-\Delta T_2}{\ln\left(\frac{\Delta T_1}{\Delta T_2}\right)}$$

$$= \frac{30-15}{\ln\left(\frac{30}{15}\right)} = 21.64°\text{C}$$

$$Q = UA\Delta T_{\text{LMTD}} = mC_P\Delta T$$

$$= 200\times A\times 21.64$$

$$= \frac{1000}{3600}\times 2000\times(70-40)$$

$$A = 3.85\ \text{m}^2$$

10.32 (c)

During sensible heating temperature will rise and sensible cooling will cause down in temperature in phase change occur at constant temperature.

10.33 (b)

In 1.1 pass shell and tube heat exchanger system is condensing at shell side at 135°C.

Energy balance equation results in

$$\ln\left(\frac{T_s - T_1}{T_s - T_2}\right) = \frac{UA}{\dot{m}C_P} \quad ...(1)$$

Heat transfer coefficient is given by

$$N_u = 0.023R_e^{0.8}Pr^{1/3}$$

So, $U \propto m^{0.8}$

Now, by putting this relation in equation (1)

$$\ln\left(\frac{T_s - T_1}{T_s - T_2}\right) = \frac{m^{-0.2}A}{C_P}$$

$$\ln\left(\frac{T_s - T_1}{T_s - T_2}\right) \propto m^{-0.2}$$

It is given that mass flow rate of cold fluid doubles.

So,

$$\frac{\ln\left(\frac{135-20}{135°C - T_2}\right)}{\ln\left(\frac{135-20}{135-90}\right)} = \frac{(2\,m)^{-0.2}}{(m)^{-0.2}} = (2)^{-0.2}$$

Solving for, $T_2 = 84.2°C$

10.34 (300)

$$U_oA_o = U_iA_i$$

Here, A_o = Outer Area

A_i = Inner Area

U_i = Inner Heat Transfer Coefficient

U_o = Outer Heat Transfer Coefficient

$$360 \times \pi \times 0.025L = U_o \times \pi \times 0.03 \times L$$

$$U_o = 300\ W/m^2\,°C$$

10.36 (b)

Use of segmental baffles in a 2-4 shell and tube heat exchanger:

(i) To change the direction of shell side fluid.

(ii) To support tubes.

(iii) To prevent vibration in tubes.

(iv) To increase heat transfer coefficient in shell side.

10.37 (27.08)

$$\Delta T_{LMTD} = \frac{\Delta T_1 - \Delta T_2}{\ln\left(\frac{\Delta T_1}{\Delta T_2}\right)}$$

$$Q = UA\,\Delta T_{LMTD}$$

$$A \propto \frac{1}{\Delta T\ LMTD}$$

$$LMTD|_{counter\text{-}current} = \frac{35-20}{\ln\left(\frac{35}{20}\right)} = 26.8°C$$

$$LMTD|_{co\text{-}current} = \frac{50-5}{\ln\left(\frac{50}{5}\right)} = 19.54°C$$

% saving in area

$$= \frac{A_1 - A_2}{A_1} = 1 - \frac{A_2}{A_1}$$

$$\left\{\begin{matrix} A_1 = \text{Area of co-current} \\ A_2 = \text{Area of counter-current} \end{matrix}\right\}$$

$$= 1 - \frac{\Delta T_{LMTD}\ \text{co-current}}{\Delta T_{LMTD}\ \text{counter-current}}$$

$$= 1 - \frac{19.54}{26.8} = 0.2708 = 27.08\%$$

10.38 (d)

Energy balance for I design:

$f_{in}\,C_{Pt}\,(t_{out} - t_{in}) = F_{in}\,C_{Ps}[T_{in} - T^{I}_{out}]$.

Energy balance for II design:

$f_{in}\,C_{Pt}(t_{out} - t_{in}) = F_{in}\,C_{Ps}(T_{in} - T^{II}_{out})$

According to given conditions of question f_{in}, F_{in}, t_{in}, t_{out}, T_{in} are same for both designs.

So $T^{I}_{out} = T^{II}_{out}$ and to produce higher temperature for design II $A^{II}_{HX} > A^{I}_{HX}$.

10.39 (0.305)

$$T_{A,\,in} = 90°C, \quad T_{A,\,out} = 45°C$$

$$A = 8\ m^2;\ U = 290\ W/m^2\,K$$

$$C_{PA} = 2000\ J/KgK$$

$$C_{PB} = 1000\ J/KgK$$

$$\dot{m}_B = 1\ Kg/s, \quad T_{B,\,in} = 40°C$$

$$\Delta T_{AMTD} = \left(\frac{T_{A,in} + T_{A,out}}{2}\right) - \left(\frac{T_{B,in} + T_{B,out}}{2}\right)$$

$$2 = UA\,\Delta T_{AMTD} = \dot{m}C_P\Delta T$$

$$290 \times 8 \times \left[\left(\frac{90+45}{2}\right) - \left(\frac{40+T_{B,out}}{2}\right)\right] = [1 \times 1000 \times (T_B - 40)]$$

$$T_{B,\,out} = 67.5°C$$

Now, $Q = (mC_P\Delta T)_A = (mC_P\Delta T)_B$

$$\dot{m}_A \times 2000 \times 45 = 1 \times 1000 \times (67.5 - 40)$$

$$\dot{m}_A = 0.305 \text{ (kg/s)}$$

10.40 (34.76)

Counter-current double pipe heat exchanger

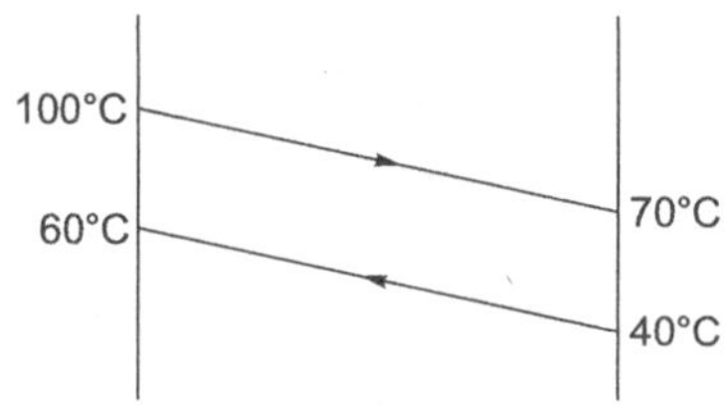

$$\text{LMTD} = \frac{\Delta T_1 - \Delta T_2}{\ln\left(\frac{\Delta T_1}{\Delta T_2}\right)} = \frac{40-30}{\ln\left(\frac{40}{30}\right)} = 34.76°C$$

10.41 (1.78)

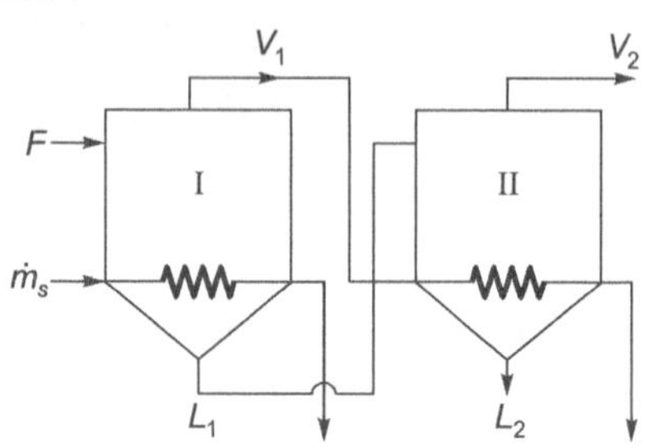

Mass balance on evaporator I

$$F = L_1 + V_1$$

Similarly mass balance on evaporator II

$$L_1 = V_2 + L_2$$

$$F = (V_1 + V_2) + L_2$$

$$F = 1 + L_2 \qquad (\because V_1 + V_2 = 1)$$

By energy balance on first evaporator

$$\dot{Q} = \dot{m}\lambda_s = FC_P\Delta T + V\lambda_V$$

$$\dot{m}_s \times 2200 = 1.25 \times 4(100 - 60) + V_1 \times 2200$$

$$(\dot{m}_s - V_1) = \frac{200}{2200} \qquad \text{...(1)}$$

Similarly energy on evaporator II

$$V_1 \times 2200 = L_1 \times 4 \times (60 - 100) + V_2 \times 2200$$

$$2200(V_1 - V_2) = (F - V_1) \times (-160)$$

Here, $V_1 + V_2 = 1$ and $F = 1.25$ (given)

So,

$$2200(2V_1 - 1) = (1.25 - V_1)(-160)$$

$$V_1 = 0.4717$$

By putting value of V_1 in equation (1)

$$(\dot{m}_s - 0.4717) = \frac{200}{2200}$$

$$\dot{m}_s = 0.5626 \text{ kg/s}$$

$$\text{Steam economy} = \frac{\text{kg of vapour evaporated}}{\text{kg of steam used}} = \frac{V_1 + V_2}{\dot{m}_s} = \frac{1}{0.5626} = 1.777 \cong 1.78 \text{ kg/s}$$

10.42 (b)

Length of heat exchanger is 10 m.

$$T_h(x) = 80 - 3x, \quad T_c(x) = 20 + 2x$$

$$T_h\big|_{x=0} = 80; \quad T_c\big|_{x=0} = 20°C$$

$$T_h\big|_{x=10} = 50; \quad T_c\big|_{x=10} = 40°C$$

Above data shows that flow of fluids is co-current in heat exchanger.

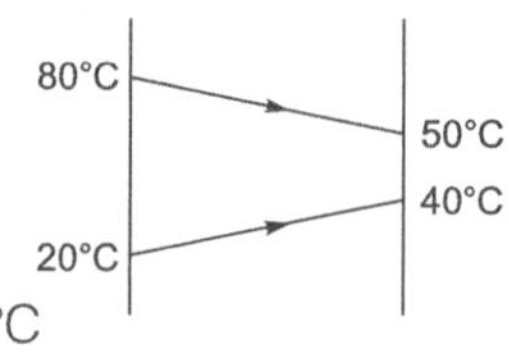

$$\Delta T_1 = 80 - 20 = 60°C$$

$$\Delta T_2 = 50 - 40 = 10°C$$

$$\Delta T_{LMTD} = \frac{60-10}{\ln\left(\frac{60}{10}\right)} = 27.9°C$$

10.43 (c)

Clean overall heat transfer coefficient

$$U_c = 250 \text{ w/m}^2\text{K}$$

Fouling resistance,

$$d_F = 0.001 \text{ m}^2\text{k/w}$$

Overall heat transfer coefficient can be found by

$$\frac{1}{U_o} = \frac{1}{U_c} + d_F = \frac{1}{250} + 0.001$$

$$U_o = 200 \text{ w/m}^2\text{K}$$

10.44 (90)

Heat flux is same in both the evaporators (given in question).

$$Q_1 = U_1 A\Delta T_1, \quad Q_2 = U_2 A\Delta T_2$$

$$2000(110 - T_1) = 1500\,(T_1 - 63.3)$$

$$T_1 = 89.9°C \approx 90°C$$

10.45 (c)

We have design effective temperature control for large disturbances.

Option (a) → O/P is not controller.

Option (b) → It is controlling only I/P.

Option (c) → As valve is in recycle it will control I/P and valve at exit of H.E. will control O/P.

Option (d) → As it is controlling only final O/P.

Option (c) is correct.

10.46 (0.148)

Data given:

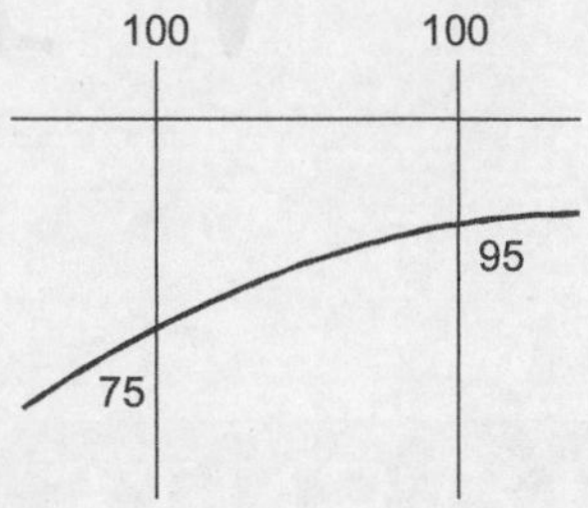

$U = 1\ \text{kW/m}^2\ \text{K}$

$A = 1\ \text{m}^2$

$t_{c_1} = 75°\text{C},$

$C_p = 4.2\ \text{KJ/kg K}$

$t_{c_2} = 95°\text{C}$

$\dot{m}_c = ?$

Here hot fluid is saturated water at 100°C.

$$Q = U \cdot A(\Delta T)_{\text{lmtd}} = \dot{m}_c C_{pc}(95-75)$$

$$\Delta T_{\text{lmtd}} = \frac{25-5}{\ln\left|\frac{25}{5}\right|} = 12.43$$

$$Q = U \cdot A(\Delta T)_{\text{lmtd}}$$

$$= 1 \times 1 \times 12.43 = 12.43\ \text{kW}$$

$$1 \times 1 \times 12.43 = m \times 4.2 \times 20$$

$$\dot{m}_c = 0.148\ \text{kg/sec}$$

10.47 (678)

Given, Area = 70 m²

Latent heat of vaporization (λ_v) = 500 Kcal/kg

Steam economy = 0.8

Feed flow rate = 10000 kg/h at T = 40°C

Steam rate = 7500 kg/h at T = 150°C

Boiling point of solution, T = 80°C

Applying energy balance,

$$F\,C_P(T - T_F) + V\lambda_v = S\lambda_s = u_0 A_0 (T_S - T)$$

$$\frac{V}{S} = 0.8$$

$$V = 0.8 \times 7500 = 6000\ \text{kg}/n$$

$$10000 \times 0.8\ (80-40) + 6000 \times 500 = u_0 \times 70 \times (150-80)$$

$$u_0 = 677.55\ \text{Kcal/m}^2.\text{h.k}$$

10.49 (3)

Given: $\dot{m}_w$ = 1 kg/s (mass flow rate of water)

C_{pw} = 4178 J/kg.K (specific heat of water)

For NTU, we have,

$$\text{NTU} = \frac{U \cdot A}{(mC_p)_{\min}}$$

From energy balance,

$$(\dot{m}C_p)_{\text{hot}}[100-40] = (\dot{m}C_p)_{\text{cold}}[80-20]$$

$$(\dot{m}C_p)_{\text{hot}} = (\dot{m}C_p)_{\text{cold}}$$

Again, $1 \times 4178 \times (80-20) = U \cdot A\,\Delta T_{\text{lmtd}}$

$$\because \quad \Delta T_{\text{lmtd}} = 20°\text{C} \quad \{\Delta T_1 = \Delta T_2 = 20°\text{C}\}$$

$$UA = \frac{4178 \times 60}{20}$$

$$\therefore \quad (\text{NTU}) = \frac{UA}{(mC_p)_{\min}} = \frac{4178 \times 60}{20 \times 1 \times 4178} = 3$$

10.50 (a)

Heat balance equation for a double pipe heat exchanger is given by

$$\dot{m}_H C_{P_H}(T_{H\text{ in}} - T_{H\text{ out}}) = m_c C_{P_c}(T_{c\text{ out}} - T_{c\text{ in}}) = UAT_{\text{LMTD}}$$

$$\frac{100}{60} \times 2.0 \times 10^3 \times (110 - \text{T}) = \frac{50}{60} \times 4.2 \times 10^3 \times (70-30)$$

$$= 14 \times 10^4\ \text{W}$$

$$\text{T} = 68°\text{C}$$

110 ⟶ 68

70 ⟵ 90

$$\text{LMTD} = \frac{(110-70)-(68-30)}{\ln\left(\frac{110-70}{68-30}\right)} = 38.99°\text{C}$$

$$U = 200\ \text{W/m}^2\ °\text{C (given)}$$

Now from equation (i)

$$14 \times 10^4 = 200 \times A \times 38.99$$

$$A = 17.95\ \text{m}^2$$

Mass Transfer

UNIT

VI

CONTENTS

Mass Transfer

Syllabus

Fick's Laws, Molecular Diffusion in Fluids, Mass Transfer Coefficients, Film, Penetration and Surface Renewal Theories; Momentum, Heat and Mass Transfer Analogies; Stage-wise and Continuous Contacting and Stage Efficiencies; HTU & NTU Concepts; Design and Operation of Equipment for Distillation, Absorption, Leaching, Liquid-Liquid Extraction, Drying, Humidification, Dehumidification and Adsorption, Membrane Separations (Micro-Filtration, Ultra-Filtration, Nano-Filtration and Reverse Osmosis).

Analysis of Previous GATE Papers

Exam Year	1 Mark Ques.	2 Marks Ques.	5 Marks Ques.	Total Marks
2000	5	2	2	**19**
2001	3	5	–	**13**
2002	2	2	2	**16**
2003	3	9	–	**21**
2004	7	10	–	**27**
2005	3	8	–	**19**
2006	1	7	–	**15**
2007	–	7	–	**14**
2008	1	11	–	**23**
2009	2	6	–	**14**
2010	1	5	–	**11**
2011	2	4	–	**10**
2012	3	3	–	**9**
2013	2	5	–	**12**
2014	4	3	–	**10**
2015	4	4	–	**12**
2016	1	4	–	**9**
2017	3	4	–	**11**
2018	3	4	–	**11**
2019	3	4	–	**11**
2020	2	6	–	**14**
2021	2	4	–	**10**
2022	5	2	–	**9**
2023	3	4	–	**11**
2024	2	6	–	**14**

1 Introduction and Basic Concepts

1.1 The weight fraction of methanol in an aqueous solution is 0.64. The mole fraction of methanol x_M satisfies

(a) $x_M = 0.5$ (b) $x_M < 0.5$
(c) $0.5 < x_M < 0.64$ (d) $x_M \geq 0.64$

[2004 : 1 M]

Answers Introduction and Basic Concepts

1.1 (a)

Explanations Introduction and Basic Concepts

1.1 (a)

Weight fraction of methanol in an aqueous solution is 0.64.

Molecular weight of methanol = 32 kg/kmol

Molecular weight of water = 18 kg/kmol

Assuming 100 kg of aqueous solution

$$\text{Methanol} = 100 \times 0.64 = 64 \text{ kg}$$

$$\text{Water} = 100 - 64 = 36 \text{ kg}$$

Mole fraction of methanol

$$= \frac{\frac{64}{32}}{\frac{64}{32}+\frac{36}{18}} = \frac{2}{2+2} = 0.5$$

■■■■

2 Diffusion

2.1 At 750 K and 1 atm, the approximate value of the Schmidt number for air is

(a) 0.01 (b) 0.1
(c) 1 (d) 10 [2000 : 1 M]

2.2 The individual mass transfer coefficients (mol/m^2 s) for absorption of a solute from a gas mixture into a liquid solvent are k_L = 4.5 and k_G = 1.5. The slope of the equilibrium line is 3. Which of the following resistance(s) is/are controlling?

(a) liquid – side (b) gas – side
(c) interfacial (d) both (a) and (b)

[2000 : 2 M]

2.3 The surface renewal frequency in Danckwerts' model of mass transfer is given by (k_L : mass transfer coefficient, m/s)

(a) $\sqrt{k_L^2 D_A}$ (b) $k_L^2 D_A$
(c) $\dfrac{k_L^2}{D_A}$ (d) $\dfrac{k_L}{D_A^2}$

[2001 : 1 M]

2.4 A sugary substance A is added to a pot of milk (initially containing no A) and stirred vigorously by a spoon so that the concentration of A, C_A, is uniform everywhere. The mass transfer coefficient for the transfer of A into the liquid is $k_{sl} = 1 \times 10^{-4}$ m/s. Solid A is added in great excess compared to the saturation capacity of milk to dissolve A. Assume that the solid-liquid interfacial area stays constant throughout the dissolution process and is given by $a = 1000$ cm^2. Derive the expression for C_A versus time, t. Find the time taken for $\dfrac{C_A}{C_A^*} = 0.95$. $C_A^* = 5 \times 10^{-2}$ kmol/m^3; $V_L = 1000$ cm^3.

[2001 : 2 M]

2.5 The dimensionless group in mass transfer that is equivalent to Prandtl number in heat transfer is

(a) Nusselt number (b) Sherwood number
(c) Schmidt number (d) Stanton number

[2002 : 1 M]

2.6 The Reynolds analogy for momentum, heat and mass transfer is best applicable for

(a) Gases in turbulent flow
(b) Gases in laminar flow
(c) Liquids in turbulent flow
(d) Liquids and gases in laminar flow

[2002 : 1 M]

2.7 The mass flux from a 5 cm diameter naphthalene ball, placed in stagnant air at 40°C and atmospheric pressure, is 1.47×10^{-3} mol/m^2 s. Assume the vapour pressure of naphthalene to be 0.15 atm at 40°C and negligible bulk concentration of naphthalene in air. If air starts blowing across the surface of naphthalene ball at 3 m/s by what factor will the mass transfer rate increase, all other conditions remaining the same? For spheres: Sh = 2.0 + 0.6 (Re)$^{0.5}$ (Sc)$^{0.33}$ where Sh is the Sherwood number and Sc is the Schmidt number. The viscosity and density of air are 1.8×10^{-5} kg/m s and 1.123 kg/m^3, respectively and the gas constant is 82.06 cm^3 atm/mol K.

[2002 : 5 M]

2.8 The diffusion coefficient, in m^2/s, of Acetic acid in Benzene (liquid in liquid) is

(a) 2.09×10^{-4} (b) 2.09×10^{-5}
(c) 2.09×10^{-9} (d) 2.09×10^{-12}

[2003 : 1 M]

2.9 Component A is diffusing in a medium B. The flux NA relative to a stationary point is equal to the flux due to molecular diffusion if

(a) mass transfer is accompanied by reaction
(b) diffusion of A is in stagnant medium B
(c) molecular mean free path is high
(d) there is equimolar counter-diffusion

[2003 : 1 M]

2.10 Experiments were conducted to determine the flux of a species A in a stagnant medium across a gas-liquid interface. The overall mass transfer coefficient based on the liquid side for dilute systems for the above was estimated to be 4×10^{-3} kg mol/m²s. The equilibrium data for the system is given as $y = 2x$. The flux across the interface (in kg mol/m²s) for bulk concentrations of A in gas phase and liquid phase as $y = 0.4$ and $x = 0.01$, respectively is

(a) 3.6×10^{-4} (b) 8.5×10^{-4}
(c) 5.6×10^{-3} (d) 8.5×10^{-3}

[2003 : 2 M]

2.11 The Reynolds Number of the liquid was increased 100 fold for a laminar falling film used for gas-liquid contacting. Assuming penetration theory is applicable, the fold-increase in the mass transfer coefficient (k_c) for the same system is

(a) 100 (b) 10
(c) 5 (d) 1

[2003 : 2 M]

2.12 A pure drug is administered as a sphere and as a cube. The amount of drug is the same in the two tablets. Assuming that the shape and size do not influence the mass transfer, the ratio of rate of dissolution in water at $t = 0$ for the cubic to spherical tablet is

(a) 0.54 (b) 1.04
(c) 1.24 (d) 1.94

[2003 : 2 M]

Common Data for Questions (2.13 & 2.14):

Pure aniline is evaporating through a stagnant air film of 1 mm thickness at 300 K and a total pressure of 100 kPa. The vapour pressure of aniline at 300 K is 0.1 kPa. The total molar concentration under these conditions is 40.1 mol/m³. The diffusivity of aniline in air is 0.74×10^{-5} m²/s.

2.13 The numerical value of the mass transfer coefficient is 7.4×10^{-3}. Its units are

(a) m/s (b) cm/s
(c) mol/(m² s Pa) (d) kmol/(m² s Pa)

[2004 : 1 M]

2.14 The rate of evaporation of aniline is 2.97×10^{-4}. Its units are

(a) mol/s (b) mol/(cm² s)
(c) mol/(m² s) (d) kmol/(m² s)

[2004 : 1 M]

2.15 A long cylinder and a sphere both of 5 cm diameter are made from the same porous material. The flat ends of cylinder are sealed. Both the cylinder and sphere are saturated with the same solution of sodium chloride. Later, both the objects are immersed for a short and equal interval of time in a large tank of water, which is well agitated. The fractions of salt remaining in the cylinder and the sphere are X_C and X_S, respectively. Which of the following statements is correct?

(a) $X_C > X_S$
(b) $X_C = X_S$
(c) $X_C < X_S$
(d) X_C is greater/less than X_S depending on the length of the cylinder

[2004 : 2 M]

2.16 The ratio of the diffusion coefficient in a gas to that in a liquid is of the order of

(a) 10^5 (b) 10^{-5}
(c) 10^{-2} (d) 10^2

[2005 : 1 M]

2.17 Two solid discs of benzoic acid (molecular weight = 122) of equal dimensions are spinning separately in large volumes of water and air at 300 K. The mass transfer coefficients for benzoic acid in water and air are 0.9×10^{-5} and 0.47×10^{-2} m/s respectively. The solubility of benzoic acid in water is 3 kg/m³ and the equilibrium vapor pressure of benzoic acid in air is 0.04 kPa. Then the disc

(a) dissolves faster in air than in water
(b) dissolves faster in water than in air
(c) dissolves at the same rate in both air and water
(d) does not dissolve either in water or in air

[2005 : 2 M]

2.18 Match the variation of mass transfer coefficient given by the theory in **Group-I** with the appropriate variation in **Group-II**:

Group-I

A. Film Theory
B. Penetration theory
C. Boundary layer theory

Group-II

1. $\propto D_{AB}$
2. $\propto D_{AB}^{2/3}$
3. $\propto D_{AB}^{1/2}$

Codes:

	A	B	C
(a)	1	2	3
(b)	2	1	3
(c)	1	3	2
(d)	3	2	1

[2005 : 2 M]

2.19 A drug tablet of mass M_0 administered orally at time $t = 0$ reaches the intestine at time $t = \tau$ without losing any mass. From the intestine, the drug is absorbed into blood. The rate of absorption is found to be proportional to the mass of the drug in the intestine with the proportionality constant k. Assuming no drug is lost from the blood, the total mass of the drug in the blood, M_b, at time $t \geq \tau$ is given by

(a) $M_b = M_0[1 - \exp\{-k(t-\tau)\}]$
(b) $M_b = M_0[1 - \exp\{-kt\}]$
(c) $M_b = M_0 \exp\{-k(t-\tau)\}$
(d) $M_b = M_0[1 - \exp\{-k(t+\tau)\}]$

[2006 : 2 M]

2.20 The following figure depicts steady one-dimensional diffusion of water vapour from the surface of water taken in a conical flask at room temperature. Derive the governing equation for determining the concentration profile of water vapour in the gas medium. Neglect change of level of water due to condensation. The temperatures of the gas and the liquid media are identical and constant.

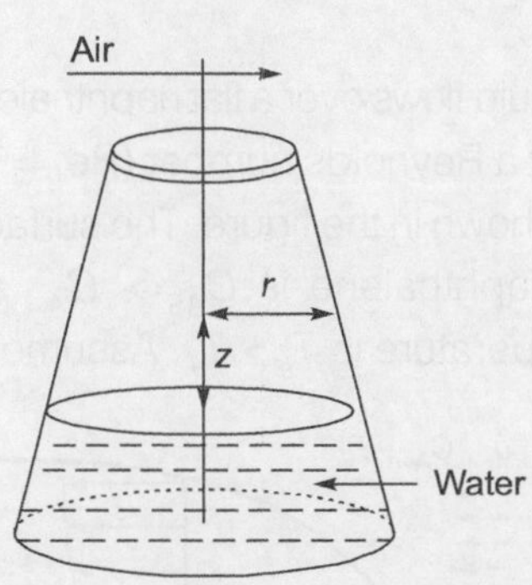

(a) $\dfrac{d}{dz}\left[r^2 \dfrac{CD_{WA}}{1-x_w} \cdot \dfrac{dx_w}{dz}\right] = 0$

(b) $\dfrac{d}{dz}\left[\dfrac{CD_{WA}}{1-x_w} \cdot \dfrac{dx_w}{dz}\right] = 0$

(c) $\dfrac{d}{dz}\left[r \dfrac{CD_{WA}}{1-x_w} \cdot \dfrac{dx_w}{dz}\right] = 0$

(d) $\dfrac{d}{dz}\left[rCD_{WA} \dfrac{dx_w}{dz}\right]$

[2007 : 2 M]

2.21 A rectangular slab of thickness $2b$ along the x-axis and extending to infinity along the other directions is initially at concentration c_{AO}. At time $t = 0$, both surfaces of the slab ($x = \pm b$) have their concentrations increased to c_{AW} and maintained at that value. Solute A diffuses into the solid. The dimensionless concentration C is defined as

$$C = \frac{c_A - c_{AO}}{c_{AW} - c_{AO}}$$

The diffusivity of A inside the solid is assumed constant. At a certain time instant, which one of the following is the correct representation of the concentration profile?

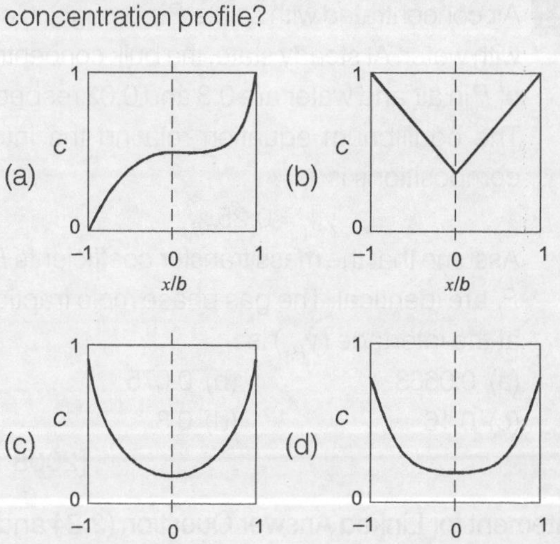

[2008 : 2 M]

2.22 A sparingly soluble solute in the form of a circular disk is dissolved in an organic solvent as shown in the figure. The area available for mass transfer from the disk in A and the volume of the initially pure organic solvent is V. The disk is rotated along the horizontal plane at a fixed rpm to produce a uniform concentration of the dissolving solute in the liquid.

The convective mass transfer coefficient under these conditions is k_c. The equilibrium concentration of the solute in the solvent is C^*. The time required for the concentration to reach 1% of the saturation value is given by

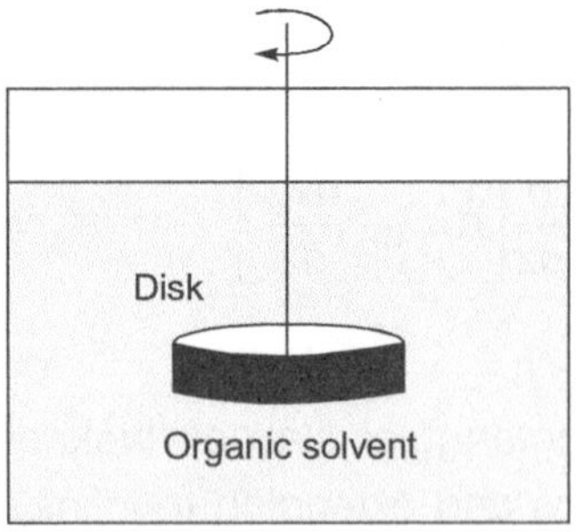

(a) $\exp\left(-\frac{k_c A}{V}t\right) = 0.99$

(b) $\exp\left(-\frac{k_c A}{V}t\right) = 0.01$

(c) $\frac{V}{Ak_c}\exp(-0.99) = t$

(d) $\frac{V}{Ak_c}\exp(0.01) = t$

[2008 : 2 M]

2.23 Air concentrated with solute P is brought in contact with water. At steady state, the bulk concentrations of P in air and water are 0.3 and 0.02 respectively. The equilibrium equation relating the interface compositions is

$$y_{P,i} = 0.25x_{P,i}$$

Assume that the mass transfer coefficients F_G and F_L are identical. The gas phase mole fraction of P at the interface ($y_{P,i}$) is

(a) 0.0663 (b) 0.075
(c) 0.16 (d) 0.3

[2008 : 2 M]

Statement for Linked Answer Question (2.24 and 2.25)

A thin liquid film flows at steady state along a vertical surface as shown in the figure. The average velocity of the liquid film is 0.05 m/s. The viscosity of the liquid is 1 cP and its density is 1000 kg/m^3. The initially pure liquid absorbs a sparingly soluble gas from air as it flows down. The length of the wall is 2 m and its width is 0.5 m. The solubility of the gas in the liquid is 3.4×10^{-2} kmol/m^3 and isothermal conditions may be assumed.

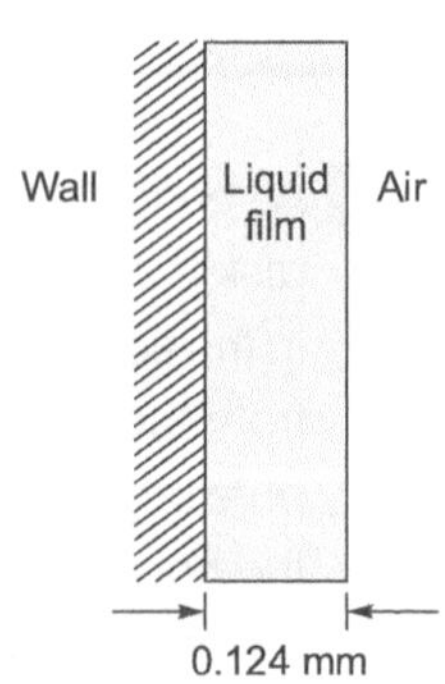

2.24 If the exit average concentration in the liquid is measured to be 1.4×10^{-2} kmol/m^3, the total mass transfer rate (in kmol/s) of the sparingly soluble gas into the liquid is

(a) 0.133×10^{-4} (b) 0.434×10^{-7}
(c) 3.4×10^{-2} (d) 17×10^{-2}

[2008 : 2 M]

2.25 The mass transfer coefficient $k_{c,avg}$ (in m/s), averaged along the length of the vertical surface is

(a) 2.94×10^{-6} (b) 2.27×10^{-6}
(c) 1.94×10^{-6} (d) 1.65×10^{-6}

[2008 : 2 M]

2.26 According to the penetration theory of mass transfer, the mass transfer coefficient (k) varies with diffusion coefficient (D) of the diffusing species as

(a) D (b) $D^{-1/2}$
(c) $D^{1/2}$ (d) $D^{3/2}$

[2009 : 1 M]

2.27 Species A is diffusing at steady state from the surface of a sphere (radius = 1 cm) into a stagnant fluid. If the diffusive flux at a distance r = 3 cm from the center of the sphere is 27 mol/cm^2.s, the diffusive flux (in mol/cm^2.s) at a distance r = 9 cm is

(a) 1 (b) 3
(c) 9 (d) 27

[2009 : 2 M]

2.28 The ratio of the thermal boundary layer thickness to the concentration boundary layer thickness is proportional to

(a) Nu (b) Le
(c) Sh (d) Pr

[2010 : 1 M]

2.29 A liquid flows over a flat naphthalene plate of length L, at a Reynolds number ($Re_L = L\rho u_\infty/\mu$) of 1500, as shown in the figure. The surface concentration of naphthalene is $C_{AS} > C_{A\infty}$ and the surface temperature is $T_s > T_\infty$. Assume Pr = Sc = 1.

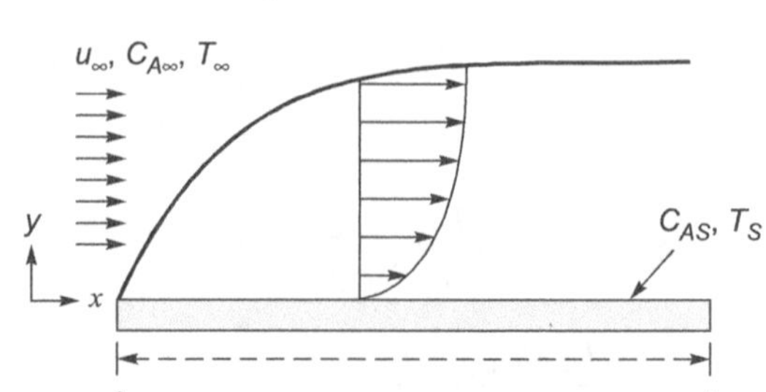

If at $x = L, \left.\dfrac{\partial C_A^*}{\partial y^*}\right|_{y^*=0} = 10$ where $C_A^* = \dfrac{C_A - C_{AS}}{C_{A\infty} - C_{AS}}$ and $y^* = \dfrac{y}{L}$, then the Nusselt number and the friction coefficient at $x = L$, are

(a) $10, \dfrac{1}{75}$ (b) 10, 10

(c) 20, 10 (d) $\dfrac{1}{75}, 5$

[2010 : 2 M]

2.30 Ammonia (component 1) is evaporating from a partially filled bottle into surrounding air (component 2). The liquid level in the bottle and the concentration of ammonia at the top of the bottle are maintained constant. N_i is the molar flux relative to a fixed location in space and J_i is the molar flux with respect to the average molar velocity of the constituent species in the gas phase. Assume that air in the bottle is stagnant. Which one of the following is correct?

(a) N_1 = constant, $N_2 = 0$, $J_1 + J_2 = 0$
(b) $N_1 + N_2 = 0$, $J_1 + J_2 = 0$
(c) $N_1 + N_2 = 0$, J_1 = constant, $J_2 = 0$
(d) N_1 = constant, $N_2 = 0$, J_1 = constant, $J_2 = 0$

[2011 : 1 M]

2.31 Simultaneous heat and mass transfer is occurring in a fluid flowing over a flat plate. The flow is laminar. The concentration boundary layer will coincide with the thermal boundary layer, when

(a) Sc = Nu (b) Sh = Nu
(c) Sh = Pr (d) Sc = Pr

[2011 : 1 M]

2.32 A gas mixture is in contact with a liquid. Component in the gas mixture is highly soluble in the liquid. Possible concentration profiles during absorption of *P* are shown in the choices, where

x : mole fraction of *P* in bulk liquid
y : mole fraction of *P* in bulk gas
x_i : mole fraction of *P* at the interface in liquid
y_i : mole fraction of *P* at the interface in gas
y^* : equilibrium gas phase mole fraction corresponding to x_i

The correct profile is

(a)
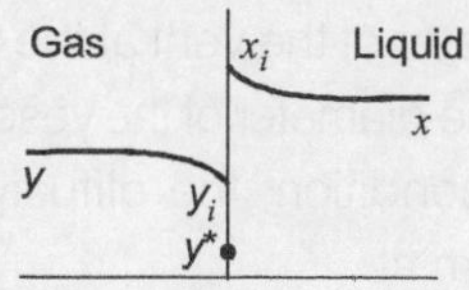

(b)
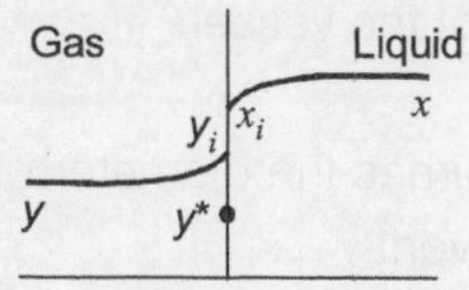

(c)
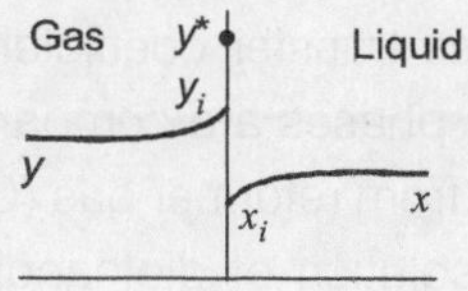

(d)
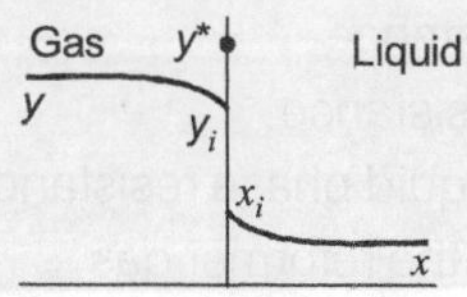

[2011 : 2 M]

2.33 For which of the following combinations, does the absorption operation become gas-film controlled?

P. The solubility of gas in the liquid is very high
Q. The solubility of gas in the liquid is very low
R. The liquid-side mass transfer coefficient is much higher than the gas-side mass transfer coefficient
S. The liquid-side mass transfer coefficient is much lower than the gas-side mass transfer coefficient

(a) P and Q (b) P and R
(c) P and S (d) Q and R

[2012 : 1 M]

2.34 A study was conducted in which water was pumped through cylindrical pipes made of a sparingly soluble solid. For a given pipe and certain flow conditions, the mass transfer coefficient k_c has been calculated as 1 mm/s using the correlation

$$Sh = 0.025\, Re^{0.6}\, Sc^{0.33}$$

If the velocity of the fluid and the diameter of the pipe are both doubled, what is the new value of k_c in mm/s, up to 2 digits after the decimal point?

[2013 : 2 M]

2.35 A spherical storage vessel is quarter–filled with toluene. The diameter of the vent at the top of the vessel is 1/20th of the diameter of the vessel. Under the steady state condition, the diffusive flux of toluene is maximum at

(a) the surface of the liquid
(b) the mid-plane of the vessel
(c) the vent
(d) a distance 20 times the diameter of the vent away from the vent

[2014 : 1 M]

2.36 Assuming the mass transfer coefficients in the gas and the liquid phases are comparable, the absorption of CO_2 from reformer gas ($CO_2 + H_2$) into an aqueous solution of diethanolamine is controlled by

(a) gas phase resistance
(b) liquid phase resistance
(c) both gas and liquid phase resistances
(d) composition of the reformer gas

[2014 : 1 M]

2.37 Which **ONE** of the following statements is **CORRECT** for the surface renewal theory?

(a) Mass transfer takes place at steady state
(b) Mass transfer takes place at unsteady state
(c) Contact time is same for all the liquid elements
(d) Mass transfer depends only on the film resistance

[2014 : 1 M]

2.38 A spherical ball of benzoic acid (diameter = 1.5 cm) is submerged in a pool of still water. The solubility and diffusivity of benzoic acid in water are 0.03 kmol/m^3 and 1.25×10^{-9} m^2/s respectively. Sherwood number is given as $Sh = 2.0 + 0.6\ Re^{0.5}\ Sc^{0.33}$. The initial rate of dissolution (in kmol/s) of benzoic acid approximately is

(a) 3.54×10^{-11} (b) 3.54×10^{-12}
(c) 3.54×10^{-13} (d) 3.54×10^{-14}

[2014 : 2 M]

2.39 For a binary mixture of components *A* and *B*, N_A and N_B denote the total molar fluxes of components *A* and *B*, respectively. J_A and J_B are the corresponding molar diffusive fluxes. Which of the following is true for equimolar counter-diffusion in the binary mixture?

(a) $N_A + N_B = 0$ and $J_A + J_B \neq 0$
(b) $N_A + N_B \neq 0$ and $J_A + J_B = 0$
(c) $N_A + N_B \neq 0$ and $J_A + J_B \neq 0$
(d) $N_A + N_B = 0$ and $J_A + J_B = 0$

[2015 : 1 M]

2.40 A spherical naphthalene ball of 2 mm diameter is subliming very slowly in stagnant air at 25°C. The change in the size of the ball during the sublimation can be neglected. The diffusivity of naphthalene in air at 25°C is 1.1×10^{-6} m^2/s.

The value of mass transfer coefficient is $B \times 10^{-3}$ m/s, where B (up to one decimal place) is ______

[2015 : 1 M]

2.41 Consider a steady state mass transfer process between well-mixed liquid and vapour phases of a binary mixture comprising of components *A* and *B*. The mole fractions of component *A* in the bulk liquid (x_A) and bulk vapour (y_A) phases are 0.36 and 0.16, respectively. The mass transfer coefficients for component A in liquid and vapour phases are 0.1 mol/(m^2.s) and 0.05 mol/(m^2.s), respectively. The vapour-liquid equilibrium can be approximated as $y_A^* = 2x_A$, for x_A less than 0.4. The mole fraction of *A* in the liquid at the interface (up to two decimal places) is ______ .

[2015 : 2 M]

2.42 Consider the following two cases for a binary mixture of ideal gases A and B under steady state conditions. In Case 1, the diffusion of A occurs through non-diffusing B. In Case 2, equimolal counter diffusion of A and B occurs. In both the cases, the total pressure is 100 kPa and the partial pressures of A at two points separated by a distance of 10 mm are 10 kPa and 5 kPa. Assume that the Fick's first law of diffusion is applicable. What is the ratio of molar flux of A in Case 1 to that in Case 2?

(a) 0.58 (b) 1.08
(c) 1.58 (d) 2.18

[2016 : 2 M]

2.43 Consider steady state mass transfer of a solute A from a gas phase to a liquid phase. The gas phase bulk and interface mole fractions are $y_{A,g}$ and $y_{A,i}$ respectively. The liquid phase bulk and interface mole fractions are $x_{A,L}$ and $x_{A,i}$

respectively. The ratio $\frac{(x_{A,t} - x_{A,L})}{(y_{A,G} - y_{A,i})}$ is very close to zero.

This implies that mass transfer resistance is

(a) negligible in the gas phase only
(b) negligible in the liquid phase only
(c) negligible in both the phases
(d) considerable in both the phases

[2017 : 1 M]

2.44 The Sherwood number (Sh_L) correlation for laminar flow over a flat plate of length L is given by

$$Sh_L = 0.664 Re_L^{0.5} Sc^{1/3}$$

where Re_L and Sc represent Reynolds number and Schmidt number, respectively. This correlation, expressed in the form of Chilton-Colburn j_D factor, is

(a) $j_D = 0.664$
(b) $j_D = 0.664\ Re_L^{-0.5}$
(c) $j_D = 0.664\ Re_L$
(d) $j_D = 0.664\ Re_L^{-0.5}\ Sc^{2/3}$

[2017 : 2 M]

2.45 According to the surface renewal theory, the unit of fractional rate of surface renewal is

(a) $m^2\ s^{-2}$ (b) $m^2\ s^{-1}$
(c) $m\ s^{-1}$ (d) s^{-1}

[2018 : 1 M]

2.46 For absorption of H_2S from a mixture with hydrocarbon vapour into an aqueous alkanolamine solution, the liquid phase mass transfer resistance is

(a) significantly higher than that of the gas phase
(b) negligible compared to that of the gas phase
(c) equal to that of the gas phase
(d) dependent on the gas phase mass transfer resistance

[2018 : 1 M]

2.47 The correct expression for the Colburn j-factor for mass transfer that relates Sherwood number (Sh), Reynolds number (Re) and Schmidt number (Sc) is

(a) $\frac{Sh}{(Re)(Sc)^{1/3}}$ (b) $\frac{Sh}{(Re)^{1/2}(Sc)}$

(c) $\frac{Sh}{(Re)^{1/2}(Sc)^{1/3}}$ (d) $\frac{Sh}{(Re)(Sc)}$

[2019 : 1 M]

2.48 Two spherical camphor particles of radii 20 cm and 5 cm, far away from each other, are undergoing sublimation in a stream of air. The mass transfer coefficient is proportional to $\frac{1}{\sqrt{r(t)}}$, where $r(t)$ is the radius of the sphere at time t. Assume that the partial pressure of camphor far away from the surface of the particle is zero. Also, assume quasi-steady state, identical ambient conditions, and negligible heat effects. If t_1 and t_2 are the times required for complete sublimation of the 20 cm and 5 cm camphor particles, respectively, the ratio $\frac{t_1}{t_2}$ is ______ (rounded off to one decimal place).

[2019 : 2 M]

2.49 Pure gas P is being absorbed into a liquid. The dissolved P undergoes an irreversible reaction in the liquid film. The reaction is first order with respect to P. Which one of the following represents the concentration profile of P in the liquid film at steady state?

(a)
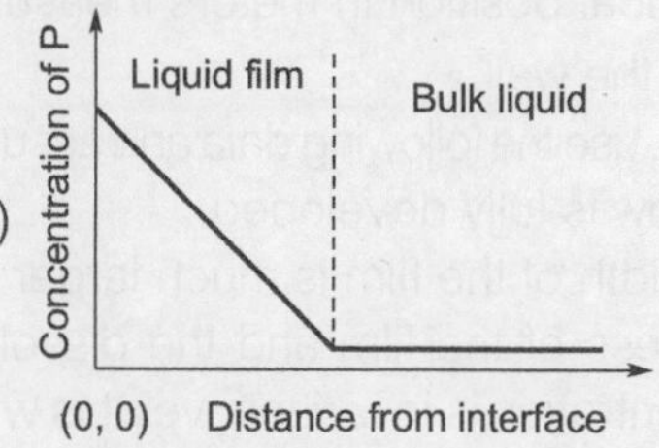

(b)
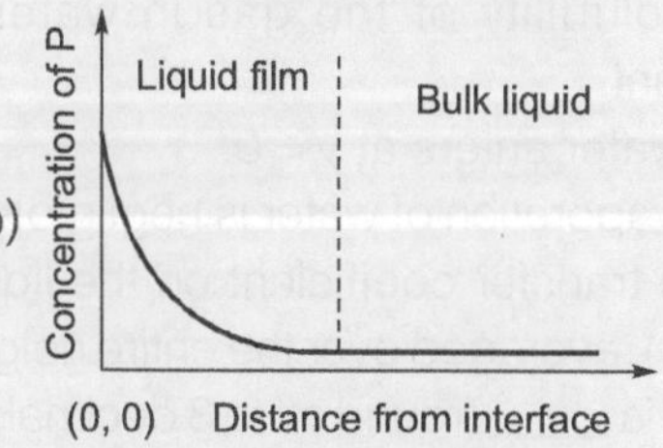

(c)
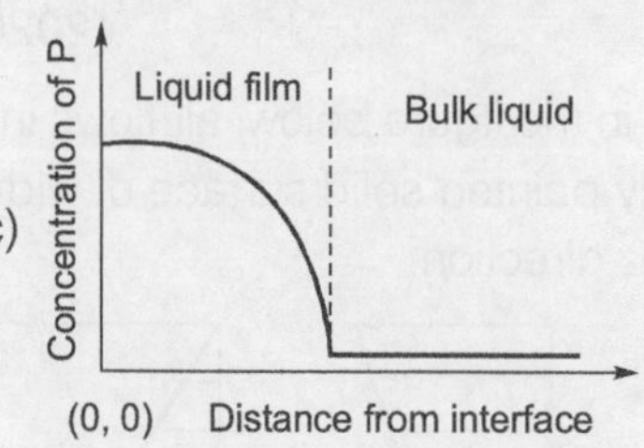

(d)
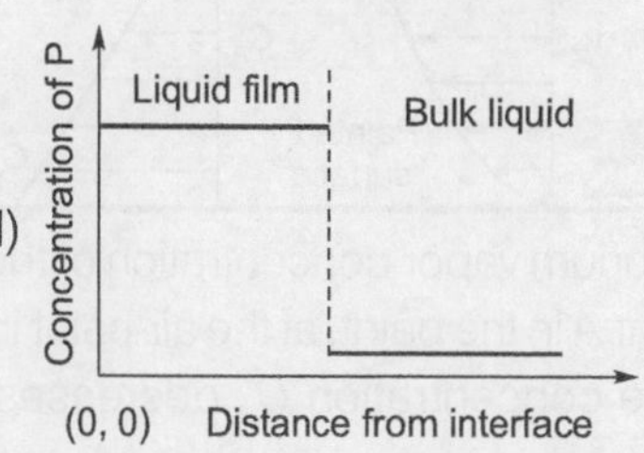

[2020 : 1 M]

2.50 Two film theory applies for absorption of a solute from a gas mixture into a liquid solvent. The interfacial mass transfer coefficient (in mol m^{-2} s^{-1}) for the gas side is 0.1 and for the liquid side is 3. The equilibrium relationship is $y^* = 2x$, where x and y^* are mole fractions of the solute in the liquid and gas phases, respectively. The ratio of the mass transfer resistance in the liquid film to the overall resistance is

(a) 0.0161
(b) 0.0322
(c) 0.0625
(d) 0.0645 [2020 : 2 M]

2.51 Consider a vertically falling film of water over an impermeable wall. The film is in contact with a static pool of non-reactive pure gas. The gas diffuses into the water film over the entire height of the falling film. The height of the film is 1.0 m, and its thickness is 10^{-4} m. The velocity of water, averaged over the film thickness, is 0.01 m s^{-1}. The gas concentration (in kg m^{-3}), averaged over the film thickness is $\overline{C_{A_y}} = C_{A_i}(1 - e^{-30y})$ where y is the vertical position in meters measured from the top of the wall.

In addition, use the following data and assumptions:

- The flow is fully developed.
- The width of the film is much larger than the thickness of the film and the dissolved gas concentration is invariant over this width.
- The solubility of the gas in water, C_{A_i} is constant.
- Pure water enters at $y = 0$.
- The evaporation of water is negligible.

The mass transfer coefficient on the liquid side (in mm s^{-1}), averaged over the entire height of the falling film is _____ (round off to 3 decimal places).

[2020 : 2 M]

2.52 As shown in the figure below, air flows in parallel to a freshly painted solid surface of width 10 m, along the z-direction.

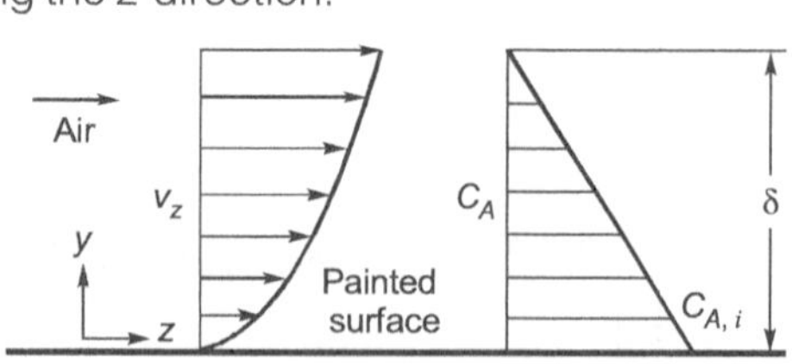

The equilibrium vapor concentration of the volatile component A in the paint, at the air-paint interface, is $C_{A,i}$. The concentration C_A decreases linearly from this value to zero along the y-direction over a distance δ of 0.1 m in the air phase. Over this distance, the average velocity of the air stream is 0.033 m s^{-1} and its velocity profile (in m s^{-1}) is given by $v_z(y) = 10y^2$ where y is in meter.

Let $C_{A,m}$ represent the flow averaged concentration. The ratio of $C_{A,m}$ to $C_{A,i}$, is _____ (round off to 2 decimal places).

[2021 : 2 M]

2.53 Consider two stationary spherical pure water droplets of diameters D_1 and $2D_1$. CO_2 diffuses into the droplets from the surroundings. If the rate of diffusion of CO_2 into the smaller droplet is W_1 mol s^{-1}, the rate of diffusion of CO_2 into the larger droplet is

(a) $2W_1$
(b) $4W_1$
(c) W_1
(d) $0.5W_1$ [2022 : 1 M]

2.54 Consider steady-state diffusion in a binary A-B liquid at constant temperature and pressure. The mole-fraction of A at two different locations is 0.8 and 0.1. Let N_{A1} be the diffusive flux of A calculated assuming B to be non-diffusing, and N_{A2} be the diffusive flux of A calculated assuming equimolar counter-diffusion. The quantity $\frac{(N_{A1} - N_{A2})}{N_{A1}} \times 100$ is _________ (rounded off to one decimal place).

[2022 : 1 M]

2.55 Consider interphase mass transfer of a species S between two immiscible liquids A and B. The interfacial mass transfer coefficient of S in liquid A is twice of that in liquid B. The equilibrium distribution of S between the liquids is given by $y_S^A = 0.5\, y_S^B$, where y_S^A and y_S^B are the mole-fractions of S in A and B, respectively. The bulk phase mole-fraction of S in A and B is 0.10 and 0.02, respectively. If the steady-state flux of S is estimated to be 10 kmol h^{-1} m^{-2}, the mass transfer coefficient of S in A is _______ kmol h^{-1} m^{-2} (rounded off to one decimal place).

[2022 : 1 M]

2.56 In a binary gas-liquid system, $N_{A,EMD}$ is the molar flux of a gas A for equimolar counter diffusion with a liquid B. $N_{A,UMD}$ is the molar flux of A for steady one-component diffusion through stagnant B. Using the mole fraction of A in the bulk of the gas phase as 0.2 and that at the gas-liquid interface as 0.1 for both the modes of diffusion, the ratio of $N_{A,UMD}$ to $N_{A,EMD}$ is equal to _____ (rounded off to two decimal places).

[2023 : 1 M]

2.57 Consider the steady, uni-directional diffusion of a binary mixture of *A* and *B* across a vertical slab of dimensions 0.2 m × 0.1 m × 0.02 m as shown in the figure. The total molar concentration of *A* and *B* is constant at 100 mol m^{-3}. The mole fraction of *A* on the left and right faces of the slab are maintained at 0.8 and 0.2, respectively. If the binary diffusion coefficient $D_{AB} = 1 \times 10^{-5}$ m^2 s^{-1}, the molar flow rate of *A* in mol s^{-1}, along the horizontal *x* direction is

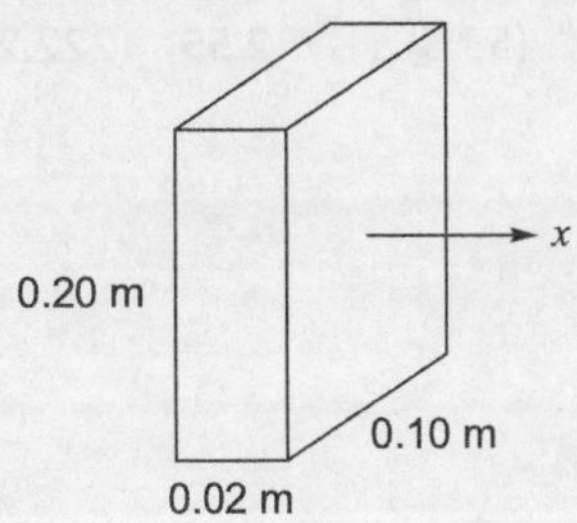

(a) 6×10^{-4} (b) 6×10^{-6}
(c) 3×10^{-6} (d) 3×10^{-4}

[2024 : 1 M]

2.58 Sulfur dioxide (SO_2) gas diffuses through a stagnant air-film of thickness 2 mm at 1 bar and 30°C. The diffusion coefficient of SO_2 in air is 1×10^{-5} m^2 s^{-1}. The SO_2 partial pressures at the opposite sides of the film are 0.15 bar and 0.05 bar. The universal gas constant is 8.314 J mol^{-1} K^{-1}. Assuming ideal gas behavior, the steady-state flux of SO_2 in mol m^{-2} s^{-1} through the air-film is

(a) 0.077 (b) 0.022
(c) 0.085 (d) 0.057

[2024 : 2 M]

2.59 Let *r* and θ be the polar coordinates defined by $x = r\cos\theta$ and $y = r\sin\theta$. The area of the cardioid $r = a(1 - \cos\theta)$, $0 \le \theta \le 2\pi$, is

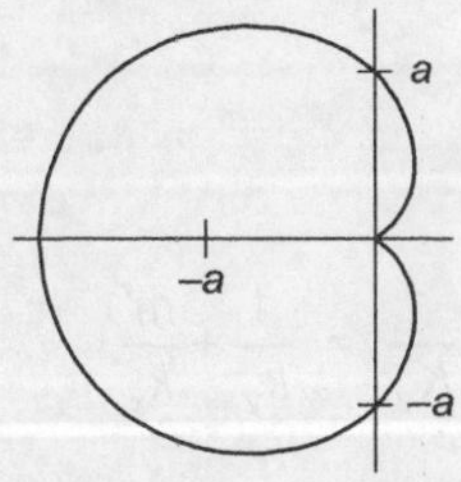

(a) $\frac{3\pi a^2}{2}$ (b) $\frac{2\pi a^2}{3}$
(c) $3\pi a^2$ (d) $2\pi a^2$

[2024 : 2 M]

■■■■

Answers Diffusion

2.1	(c)	**2.2**	(d)	**2.3**	(c)	**2.4**	(300)	**2.5**	(c)	**2.6**	(a)	**2.7**	(39.729)
2.8	(c)	**2.9**	(d)	**2.10**	(b)	**2.11**	(b)	**2.12**	(c)	**2.13**	(a)	**2.14**	(d)
2.15	(c)	**2.16**	(a)	**2.17**	(b)	**2.18**	(c)	**2.19**	(a)	**2.20**	(a)	**2.21**	(c)
2.22	(a)	**2.23**	(a)	**2.24**	(b)	**2.25**	(d)	**2.26**	(c)	**2.27**	(b)	**2.28**	(b)
2.29	(a)	**2.30**	(a)	**2.31**	(d)	**2.32**	(d)	**2.33**	(b)	**2.34**	(1.148)	**2.35**	(c)
2.36	(a)	**2.37**	(b)	**2.38**	(b)	**2.39**	(d)	**2.40**	(1.1)	**2.41**	(0.22)	**2.42**	(b)
2.43	(b)	**2.44**	(b)	**2.45**	(d)	**2.46**	(b)	**2.47**	(a)	**2.48**	(8)	**2.49**	(b)
2.50	(c)	**2.51**	(0.03)	**2.52**	(0.25)	**2.53**	(a)	**2.54**	(53.5)	**2.55**	(222.22)	**2.56**	(1.17)
2.57	(a)	**2.58**	(b)	**2.59**	(a)								

Explanations Diffusion

2.2 (d)

We know,

$$\frac{1}{K_y} = \frac{1}{k_y} + \frac{m'}{k_x} \quad ...(1)$$

where $\frac{1}{k_y}$ = gas phase resistance

$\frac{m'}{k_x}$ = liquid phase resistance

Given: $k_y = 1.5$ mil/m²-sec

$k_x = 4.5$ mol/m²-sec

$m' = 3$

The ratio of resistance in the liquid to gas-phase

$$= \frac{\left(\frac{m'}{k_x}\right)}{\left(\frac{1}{k_y}\right)} = \frac{\left(\frac{3}{4.5}\right)}{\left(\frac{1}{1.5}\right)} = \frac{1.5}{1.5} = 1$$

Thus, the ratio of resistance in the liquid to gas-phase is 1 : 1. Hence, the liquid and gas-phase resistances contribute equally to the overall mass transfer resistance.

2.3 (c)

Danckwert's surface renewal theory

$$k_c \propto \sqrt{D_{AB}S}$$

$$\frac{k_L^2}{D_{AB}} \propto S \text{ (surface renewal frequency)}$$

2.4 (300)

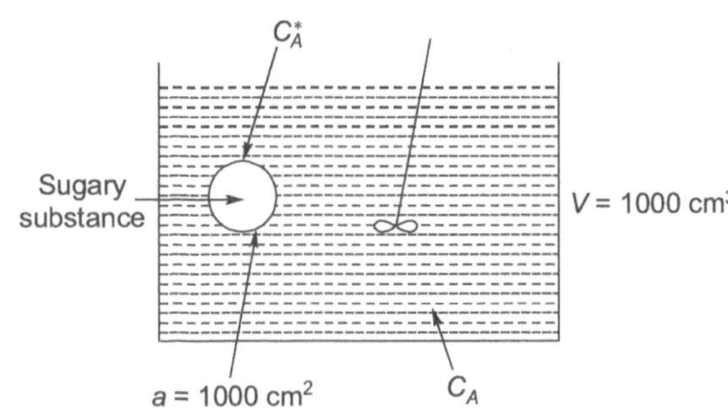

Transient mass balance for sugar

$$V\frac{dC_A}{dt} = k_m \cdot a(C_A^* - C_A)$$

C_A = Concentration at any time t

C_A^* = Surface concentration

$$\int \frac{dC_A}{C_A^* - C_A} = \int \frac{k_m\, a}{V} dt$$

$$-\ln(C_A^* - C_A) = \frac{k_m\, a}{V} t + C$$

at $t = 0$, $C_A = 0$

Then, $C = -\ln C_A^*$

$$\ln\left(\frac{C_A^* - C_A}{C_A^*}\right) = \frac{-k_m a}{V} t$$

$$1 - \frac{C_A}{C_A^*} = \exp\left[-\frac{k_m a}{V} t\right]$$

$$C_A = C_A^*\left[1 - \exp\left(\frac{-k_m a}{V} t\right)\right]$$

$$\ln\left[1 - \frac{C_A}{C_A^*}\right] = \frac{-k_m a}{V} t$$

$$\frac{C_A}{C_A^*} = 0.95 \text{ (given)}$$

$$\ln(1-0.95) = -1\times10^{-4}\times\frac{1000}{1000}\times t\times100$$

$$\frac{-t}{100} = -2.996$$

$$t = 299.6 \simeq 300 \text{ s}$$

2.6 (a)

Reynold analogy is best applicable for

(i) Gases in turbulent flow.

(ii) When Prandtl and Schmidt number is 1.

(iii) Only friction drag (no form drag is available).

2.7 (39.729)

For stagnant air

$$N_A = k_{cs}(C_s - \cancel{C_\infty}^{0})$$

$$N_A = 1.47 \times 10^{-3} \text{ mol/m}^2\text{ s}$$

C_s = Concentration of naphthalene at the surface of ball

$$C_s = \frac{P_A^{sat}}{RT} = \frac{0.15\times101.3}{8.314\times(40+273)}$$

$$= 5.84 \times 10^{-3} \text{ kmol/m}^3$$

$$1.47 \times 10^{-3} = k_{cs} \times 5.84 \times 10^{-3} \times 1000$$

$$k_{cs} = 0.2517 \times 10^{-3} \text{ m/s}$$

$$\text{Sh} = 2 + 0.6\,\cancel{\text{Re}^{0.5}}_{0}\,\text{Sc}^{0.33}$$

(for stagnant air)

$$\text{Sh} = 2.0$$

$$\frac{k_{cs}\cdot d}{D_{AB}} = 2$$

$$D_{AB} = \frac{2.517\times10^{-4}\times0.05}{2}$$

$$= 6.295 \times 10^{-6} \text{ m}^2\text{/s}$$

For air blowing across the surface of naphthalene ball

$$V_\infty = 3 \text{ m/s}$$

$$\text{Re} = \frac{\rho V_\infty d}{\mu} = \frac{0.05\times3\times1.123}{1.8\times10^{-5}}$$

$$= 9358.33$$

$$S_c = \frac{\mu}{\rho_{air}D_{AB}}$$

$$= \frac{1.8\times10^{-5}}{1.123\times6.295\times10^{-6}}$$

$$S_c = 2.5462$$

$$\text{Sh} = 2 + 0.6\times(9358.33)^{0.5}\times(2.5462)^{0.33}$$

$$\text{Sh} = 81$$

$$\frac{k_{CB}\cdot d}{D_{AB}} = 81$$

where k_{CB} is mass transfer coefficient for air blowing.

$$k_{CB} = \frac{81\times6.295\times10^{-6}}{0.05}$$

$$k_{CB} = 0.01 \text{ m/s}$$

Rate of mass transfer (stagnant air)

$$N_{CS} = k_{CS}(C_s - \cancel{C_\infty}^{0})$$

Rate of mass transfer (air blowing)

$$N_{CB} = k_{CB}(C_s - \cancel{C_\infty}^{0})$$

$$\frac{N_{CB}}{N_{CS}} = \frac{k_{CB}}{k_{CS}} = \frac{0.01}{2.517\times10^{-4}}$$

$$= 39.729$$

2.8 (c)

Liquid in liquid diffusion coefficient is order of 10^{-9} m²/s.

2.9 (d)

$$N_A = J_A^* + \frac{C_A}{C}(N_A + N_B)$$

Given: Flux, $N_A = J_A^*$

From equation,

$$N_A + N_B = 0$$

$$N_A = -N_B \text{ (equimolar diffusion)}$$

2.10 (b)

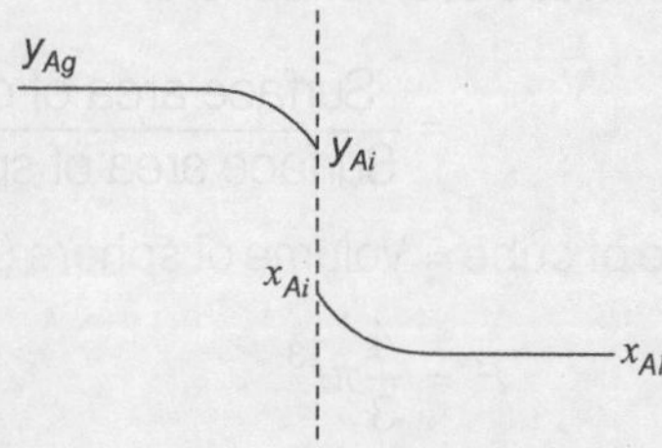

$$N_A = k_{Ox}\left[\frac{(x_A^* - x_{Al})}{(1-x_A)_{lm}}\right]$$

$$\text{where, } (1-x_A)_{lm} = \frac{(1-x_{Al})-(1-x_A^*)}{\ln\left(\frac{1-x_{Al}}{1-x_A^*}\right)}$$

Given that $y_{Ag} = 0.4$, $x_{Al} = 0.01$ and

$y^* = 2x$, $k_{ox} = 4 \times 10^{-3}$ kmol/m²s

y_A^* and x_{Al} are in equilibrium and x_A^* and y_{Ag} are in equilibrium. Hence,

$$x_A^* = \frac{y_{Ag}}{2} = \frac{0.4}{2} = 0.2$$

then, $$(1-x_A)_{lm} = \frac{(1-0.01)-(1-0.2)}{\ln\left(\frac{1-0.01}{1-0.2}\right)} = 0.891$$

$$N_A = 4\times10^{-3}\times\frac{(0.2-0.01)}{0.891}$$

$$= 8.51 \times 10^{-4} \text{ kmol/m}^2\text{s}$$

2.11 (b)

$$k_c = \sqrt{\frac{4D_{AB}}{\pi\theta}}$$

where, $$\theta = \frac{L}{u}$$

$$k_c = \sqrt{\frac{4D_{AB}u}{\pi L}}$$

$$k_c \propto \sqrt{u}$$

$$\frac{k_{c2}}{k_{c1}} = \sqrt{\frac{\rho u_2 d}{\mu}\times\frac{\mu}{\rho u_1 d}}$$

$$\frac{k_{c2}}{k_{c1}} = \sqrt{\frac{\text{Re}_2}{\text{Re}_1}} = \sqrt{100}$$

$$\frac{k_{c2}}{k_{c1}} = 10$$

2.12 (c)

$$\frac{\text{Rate of dissolution of cube}}{\text{Rate of dissolution of sphere}} = \frac{\text{Surface area of cube}}{\text{Surface area of sphere}}$$

Volume of cube = Volume of sphere (given)

$$l^3 = \frac{4}{3}\pi r^3$$

$$l^3 = \frac{\pi}{6}d^3$$

$$\left(\frac{l}{d}\right)^3 = \frac{\pi}{6}$$

$$\frac{l}{d} = \left(\frac{\pi}{6}\right)^{1/3} \quad \text{...(i)}$$

$$\frac{(\text{Rate of dissolution})_{\text{cube}}}{(\text{Rate of dissolution})_{\text{sphere}}} = \frac{6l^2}{\pi d^2}$$

$$= \frac{6}{\pi}\left(\frac{l}{d}\right)^2$$

$$= \frac{6}{\pi}\left(\frac{\pi}{6}\right)^{2/3}$$

$$= \left(\frac{6}{\pi}\right)^{1/3} = 1.24$$

2.13 (a)

Mass transfer coefficient,

$$K_c = \frac{D_{AB}}{z}$$

Here, $z = 1$ mm

$$K_c = \frac{0.74\times10^{-5}\ \text{m}^2/\text{s}}{1\times10^{-3}\ \text{m}}$$

$$= 7.4 \times 10^{-3} \text{ m/s}$$

2.14 (d)

Evaporation rate,

$$N_A = k_C(C_A - C_{AS})$$

$$C_{AS} = 0 \text{ (due to bulk concentration)}$$

$$N_A = k_C \cdot C_A$$

$$N_A = 7.4 \times 10^{-3} \times 40.1$$

$$= 0.297 \text{ mol/m}^2\text{.s}$$

$$N_A = 2.97 \times 10^{-4} \text{ kmol/(m}^2\text{s)}$$

2.15 (c)

For cylinder :

Area available for diffusion (A_1) = πDl

For sphere :

Area available for diffusion (A_2) = πD^2

$$l > D \text{ (given)}$$

$$A_1 > A_2$$

Diffusion in cylinder will be more. So, salt remain in cylinder will be less as compared to sphere.

$$X_C < X_S$$

2.16 (a)

Diffusion coefficient in gas = 10^{-4} order

Diffusion coefficient in liquid = 10^{-9} order

$$\frac{D_{\text{gas}}}{D_{\text{liquid}}} = \frac{10^{-4}}{10^{-9}} = 10^5$$

2.17 (b)

Mass transfer rate in water per unit area

$$N_w = K_m(C_s - C_\infty). \text{ Here } C_\infty = 0$$

Given that $k_m = 0.9 \times 10^{-5}$ m/s

$$C_s = 3 \text{ kg/m}^3$$
$$N_w = 0.9 \times 10^{-5}(3 - 0)$$
$$= 2.7 \times 10^{-5} \text{ kg/m}^2\text{.s}$$

Mass transfer rate in air per unit area

$$N_{air} = K'_m(C_s - C_\infty). \text{ Here } C_\infty = 0$$

Given : $P_A^{sat} = 0.04$ kPa, $K'_m = 0.47 \times 10^{-2}$ m/s

Concentration in kg/m^3

$$= \frac{PM_{air}}{RT}$$

$P = 0.04$ kPa, $M_{air} = 122$, $T = 300$ K, $R = 8.314$ (given)

Hence, $$N_{air} = k'_m \frac{PM_{air}}{RT}$$
$$= 0.47 \times 10^{-2}\left(\frac{0.04 \times 122}{8.314 \times 300}\right)$$
$$= 0.919 \times 10^{-5} \text{ kg/m}^2\text{s}$$

Hence, $N_W > N_{air}$. It means disc dissolves faster in water than in air.

2.18 (c)

$$K \propto D_{AB} \text{ (Film theory)}$$
$$K \propto D_{AB}^{1/2} \text{ (Penetration theory)}$$
$$K \propto D_{AB}^{2/3} \text{ (Boundary theory)}$$

2.19 (a)

$$\frac{dm}{dt} \propto (m_o - m)$$
$$\frac{dm}{dt} = k(m_o - m)$$
$$\int_0^{m_b} \frac{dm}{m_o - m} = k\int_\tau^t dt$$
$$[-\ln(m_o - m)]_0^{m_b} = k(t - \tau)$$
$$\ln\left(\frac{m_o}{m_o - m_b}\right) = k(t - \tau)$$
$$\frac{m_o}{m_o - m_b} = e^{k(t-\tau)}$$
$$m_o - m_b = m_o e^{-k(t-\tau)}$$
$$m_b = m_o \times [1 - e^{-k(t-\tau)}]$$

2.20 (a)

At steady state

$$\pi(r + \Delta r)^2 N_A\big|_z - \pi r^2 N_A\big|_{z+\Delta z} = 0$$
$$\lim_{\Delta z \to 0} \frac{\pi(r + \Delta r)^2 N_A\big|_z - \pi r^2 N_A\big|_{z+\Delta z}}{\Delta z} = 0$$

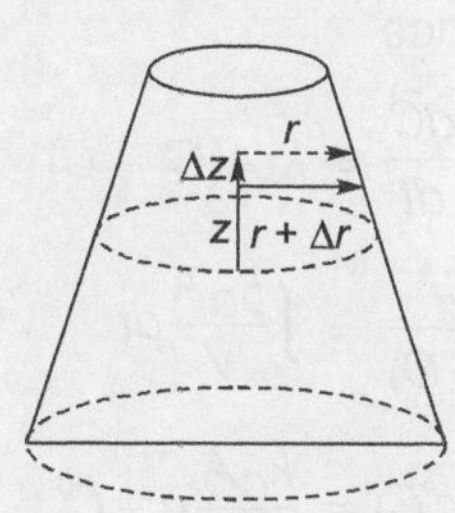

$$\frac{d}{dz}(r^2 N_A) = 0 \quad \text{...(i)}$$

Since $$N_A = (N_A + N_B)x_A - D_{AB}\frac{dC_A}{dz} \quad \text{...(ii)}$$

Here, $N_B = 0$

Hence, $$N_A(1 - x_A) = -D_{AB}\frac{dC_A}{dz}$$

Since, $$x_A = \frac{C_A}{C_T}$$

then, $$N_A = \frac{-D_{AB}C_T}{(1 - x_A)}\frac{dx_A}{dz} \quad \text{...(iii)}$$

Now, putting equation (iii) in equation (i), we get

$$\frac{d}{dz}\left(\frac{r^2 D_{AB}C_T}{(1 - x_A)} \cdot \frac{dx_A}{dz}\right) = 0$$

Here, $D_{AB} = D_{WA}$ and $x_A = x_W$

2.21 (c)

At initial (in figure $\frac{x}{b} = -1$), $C = 1$ and near the centre, concentration profile should be exponential.

Option (a) at $\frac{x}{b} = -1$, $C = 0$, not correct.

Option (b) at $\frac{x}{b} = -1$, $C = 1$ at centre concentration profile is linear, not correct.

Option (c) at $\frac{x}{b} = -1$, C = 1 at centre, concentration profile is exponential, correct.

Option (d) at $\frac{x}{b} = -1$, C = 1, but at centre, concentration profile is flat-not correct.

2.22 (a)

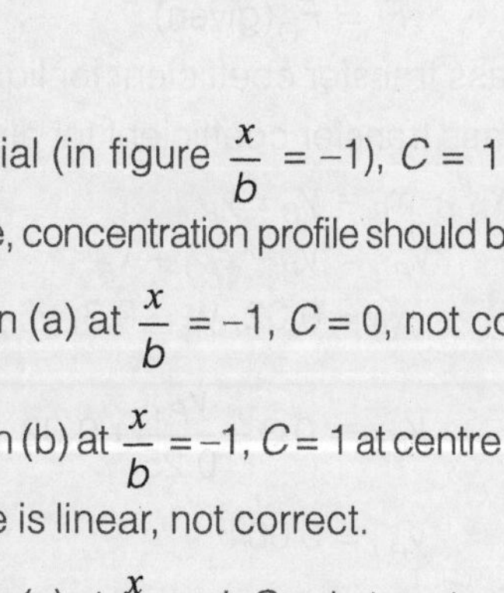

Solute balance

$$V\frac{dC}{dt} = k_c A(C^* - C)$$

$$\int \frac{dC}{(C^* - C)} = \int \frac{k_C A}{V} dt$$

$$-\ln(C^* - C) = \frac{k_C A}{V} \cdot t + C'$$

at $t = 0$, $C = 0$ (given)

$$C' = -\ln C^*$$

$$-\ln(C^* - C) = \frac{k_C A}{V} \cdot t - \ln C^*$$

$$\ln \frac{C^*}{C^* - C} = \frac{k_C A}{V} \cdot t$$

$$\frac{C^* - C}{C^*} = \exp\left[\frac{-k_C A}{V} \cdot t\right]$$

$$C = 0.01C^*$$

$$0.99 = \exp\left[\frac{-k_C A}{V} \cdot t\right]$$

2.23 (a)

$$N_A = F_L(x_{P,i} - x_B) = F_G(y_B - y_{P,i})$$

$$F_L = F_G \text{(given)}$$

$F_L \Rightarrow$ mass transfer coefficient for liquid

$F_G \Rightarrow$ mass transfer coefficient for air

$$x_{P,i} - x_B = y_B - y_{P,i}$$

$$y_{P,i} = y_B - x_{P,i} + x_B$$

Given : $x_B = 0.02$, $y_B = 0.3$, $y_{P,i} = 0.25x_{P,i}$

$$y_{P,i} = 0.3 - \frac{y_{P,i}}{0.25} + 0.02$$

$$y_{P,i} = 0.064$$

2.24 (b)

$$N_A = K_{avg} A(C_A^* - C_A)_{LMTD}$$

where A is area exposed to air.

$$N_A = 1.645 \times 10^{-6} \times \frac{(C_A^* - C_{in}) - (C_A^* - C_{out})}{\ln\left(\frac{C_A^* - C_{in}}{C_A^* - C_{out}}\right)}$$

$C_{in} = 0$ (Pure liquid)

$$N_A = 1.645 \times 10^{-6} \times \frac{(3.4 \times 10^{-2}) - (3.4 \times 10^{-2} - 1.4 \times 10^{-2})}{\ln\left(\frac{3.4 \times 10^{-2}}{(3.4 - 1.4) \times 10^{-2}}\right)}$$

$N_A = 0.434 \times 10^{-7}$ kmol/s

2.25 (d)

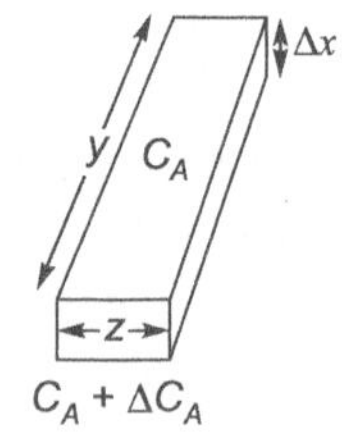

$$N_A = Q\Delta C_A$$

(where Q is volumetric flow rate)

$$= k_L(y \cdot \Delta x)(C_A^* - C_A)$$

$$N_A = Vzy\Delta C_A = k_L(y\Delta x)(C_A^* - C_A)$$

$$\frac{\Delta C_A}{\Delta x} = \frac{k_L}{Vz}(C_A^* - C_A)$$

$$\int \frac{dC_A}{C_A^* - C_A} = \frac{k_L}{Vz}\int dx$$

$$-\ln(C_A^* - C_A) = \frac{k_L}{Vz}x + C$$

at $x = 0$, $C_A = 0$ (pure liquid)

$$C' = -\ln C_A^*$$

$$-\ln(C_A^* - C_A) = \frac{k_L}{Vz}x + (-\ln C_A^*)$$

$$\ln\left(\frac{C_A^*}{C_A^* - C_A}\right) = \frac{k_L}{Vz} \cdot x$$

Given : $z = 0.124 \times 10^{-3}$ m

$C_A = 1.4 \times 10^{-2}$ kmol/m^3

$x = 2$ m $= L$

$C_A^* = 3.4 \times 10^{-2}$ kmol/m^3

$V = 0.05$ m/s

$$\ln\left(\frac{3.4 \times 10^{-2}}{3.4 \times 10^{-2} - 1.4 \times 10^{-2}}\right) = \frac{k_L \times 2}{0.05 \times 0.124 \times 10^{-3}}$$

$$k_{C,\,avg} = 1.645 \times 10^{-6} \text{ m/s}$$

2.26 (c)

According to penetration theory

$$k \propto \sqrt{\frac{4D_{AB}}{\pi\theta}}$$

where D_{AB} is diffusivity, θ = time of penetration

$$k \propto D_{AB}^{1/2}$$

2.27 (b)

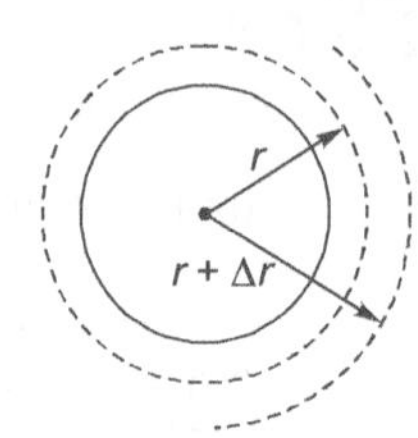

At steady state

$$4\pi r^2 N_A|_r - 4\pi r^2 N_A|_{r+\Delta r} = 0$$

$$-4\pi[r^2 N_A|_{r+\Delta r} - r^2 N_A|_r] = 0$$

$$4\pi\frac{d}{dr}(r^2N_A) = 0$$

$$r^2N_A = \text{Constant}$$

$$\frac{N_{A_2}}{N_{A_1}} = \frac{r_1^2}{r_2^2}$$

$$\frac{N_{A_2}}{27} = \left(\frac{3}{9}\right)^2$$

$$N_{A2} = \frac{27}{9} = 3 \text{ mol/cm}^2\text{.s}$$

2.28 (b)

Thermal boundary layer is signified by Prandtl number. Similarly concentration boundary layer is signified by Schmidt number.

$$N_{Pr} = \frac{\nu}{\alpha}$$

$$N_{Sc} = \frac{\nu}{D_{AB}}$$

So, ratio of thermal boundary layer thickness to concentration boundary layer thickness is given by $\frac{\alpha}{D_{AB}}$ (Lewis Number).

2.29 (a)

Given : $\left.\frac{\partial C_A^*}{\partial y^*}\right|_{y^*=0} = 10 = \text{Sh}$

at Pr = Sc = 1 then

Sh = Nusselt Number (Nu) = 10

$$\frac{\text{Nu}}{\text{Re.Pr}} = \frac{f}{2}$$

$$\frac{10}{1500\times 1} = \frac{f}{2}$$

Friction coefficient, $f = \frac{2}{150} = \frac{1}{75}$

2.30 (a)

Here, N_1 = Constant and $N_2 = 0$

Diffusion of one specie through non-diffusing species.

$$N_1 = J_1 + x_A(N_1 + \cancel{N_2}^0) \quad \text{...(i)}$$

$$\cancel{N_2}^0 = J_2 + x_B(N_1 + \cancel{N_2}^0) \quad \text{...(ii)}$$

Eqn. (i) + (ii)

$$N_1 = J_1 + J_2 + N_1(x_A + x_B)$$

$$N_1 = J_1 + J_2 + N_1$$

$$J_1 + J_2 = 0$$

2.31 (d)

Prandtl number signifies thermal boundary layer growth. Similarly, Schmidt number signifies concentration boundary layers. When both will coincide then, Sc= Pr.

2.32 (d)

Component P is highly soluble in liquid and diffuses from gas to liquid phase.

y must be > y_i and x_i must be > x. So, B and C are not possible.

A and D are two possible diagrams.

$$y_i = Hx_i \text{ (Henry's law)}$$

For highly soluble systems, $y_i > x_i$. So, option (d) is correct.

2.33 (b)

$$\frac{1}{k_{og}} = \frac{1}{k_g} + \frac{m}{k_l}$$

For a highly soluble gas, m is small and if $k_l >> k_g$ then

$$\frac{1}{k_{og}} \simeq \frac{1}{k_g} \text{ (gas phase is controlling)}$$

2.34 (1.148)

$$\text{Sh} = 0.025\text{Re}^{0.6}\text{Sc}^{0.33}$$

$$\frac{k_c d}{D_{AB}} = 0.025\left(\frac{\rho V d}{\mu}\right)^{0.6}\left(\frac{\mu}{\rho D_{AB}}\right)^{0.33}$$

$$k_c \propto (V)^{0.6}(d)^{-0.4}$$

$$\frac{k_{c2}}{k_{c1}} = \left(\frac{V_2}{V_1}\right)^{0.6}\left(\frac{d_1}{d_2}\right)^{0.4}$$

$$\frac{k_{c2}}{1 \text{ mm/s}} = (2)^{0.6}\cdot\frac{1}{(2)^{0.4}} = (2)^{0.2}$$

$$k_{c2} = 1.148 \text{ mm/s}$$

2.35 (c)

At steady state diffusive molar flow rate at the liquid surface is the same as that through the vent.

$$\bar{N}_A = \text{(Flux).(Area)}$$

If area is minimum then flux is maximum. So, diameter of the vent at the top of vessel is minimum then diffusive flux is maximum at the vent.

2.36 (a)

$$\frac{1}{k_{oy}} = \frac{1}{k_y} + \frac{m}{k_x}$$

Since CO_2 is highly soluble in diethanolamine hence m is very small.

So, $\dfrac{1}{k_{oy}} \simeq \dfrac{1}{k_y}$

It means all resistances are in gas phase only.

2.37 (b)

Option (a) is correct for film theory.
Option (b) is correct for penetration theory and surface renewal theory.
Option (c) is only correct for penetration theory.
Option (d) is correct for film theory.

2.38 (b)

Mass transfer of benzoic acid from a spherical ball to still water.

Given : $Sh = 2.0 + 0.6\, Re^{0.5} Sc^{0.33}$

$Sh = 2.0$ for still water

$$\frac{k_c \cdot d}{D_{AB}} = 2$$

$$\frac{k_c \times 1.5 \times 10^{-2}}{1.25 \times 10^{-9}} = 2$$

$$k_c = 1.67 \times 10^{-7} \text{ m/s}$$

Initial rate of dissolution,

$$\bar{N}_A = k_c \cdot A(C_s - C_\infty)$$

At time $t = 0$, $C_\infty = 0$

$C_s = 0.03$ kmol/m^3 (given)

$$\bar{N}_A = 1.67 \times 10^{-7} \times A \times (0.03 - 0)$$

Initial area of sphere $(A) = \pi d^2$

$$A = 3.14 \times (1.5 \times 10^{-2})^2 = 7.065 \times 10^{-4} \text{ m}^2$$

$$\bar{N}_A = 1.67 \times 10^{-7} \times (0.03 - 0) \times 7.065 \times 10^{-4}$$

$$\bar{N}_A = 3.539 \times 10^{-12} \text{ kmol/s}$$

2.39 (d)

For equimolar counter diffusion

Only equimolar ($N_A = -N_B$) and $j_A + j_B = 0$ for both type of diffusion

$$N_A = j_A + x_A(N_A + N_B) \quad \text{...(i)}$$

$$N_B = j_B + x_B(N_A + N_B) \quad \text{...(ii)}$$

Adding eqn. (i) and (ii)

$$j_A + j_B = 0$$

So for equimolar counter diffusion

$$N_A + N_B = 0 \text{ and } j_A + j_B = 0$$

2.40 (1.1)

For naphthalene ball in stagnant air

$$Sh = 2$$

$$\frac{K_m d}{D_{eff}} = 2$$

$$\frac{K_m \times 2 \times 10^{-3}}{1.1 \times 10^{-6}} = 2$$

$$K_m = 1.1 \times 10^{-3} \text{ m/sec}$$

2.41 (0.22)

Given : $x_A = 0.36$, $y_A = 0.16$

$k_L = 0.1$ mol/m^{2}s, $k_g = 0.05$ mol/m^{2}s

$y_A^* = 2x_A$ for $x_A < 0.4$

So, at interface,

$$y_{Ai} = 2x_{Ai}$$

At steady state, mass flux is constant.

$$k_l(x_A - x_{Ai}) = k_g(y_{Ai} - y_A)$$

$$k_l(x_A - x_{Ai}) = k_g(2x_{Ai} - y_A)$$

$$0.1(0.36 - x_{Ai}) = 0.05(2x_{Ai} - 0.16)$$

$$x_{Ai} = 0.22$$

2.42 (b)

Case 1 : Diffusion of *A* occurs through non-diffusing *B*

Case 2 : Equimolar counter diffusion of *A* and *B*

For case 1 :

$$N_{A1} = \frac{D_{AB} \cdot P_T}{RTZ} \cdot \frac{(P_{A1} - P_{A2})}{P_{Blm}}$$

$$P_{A1} + P_{B1} = P_T$$

$P_{A1} = 10$ kpas

$P_{B1} = 90$ kpas

$P_{A2} = 5$ kpas

$P_{B2} = 95$ kpas

For case 2 :

$$N_{A2} = \frac{D_{AB}}{RTZ} \cdot (P_{A1} - P_{A2})$$

$$\frac{N_{A1}}{N_{A2}} = \frac{P_T}{P_{Blm}} = \frac{100}{95-90}\ln\left(\frac{95}{90}\right) = 1.081$$

2.43 (b)

At steady state mass transfer from gas to liquid phase

$$k_y(y_{Ag} - y_{Ai}) = k_x(x_{Ai} - x_{Al})$$

$$\frac{k_y}{k_x} = \frac{x_{Ai} - x_{AL}}{y_{Ag} - y_{Ai}}$$

It is very close to zero (given).

Then, $k_y \simeq 0$ and $\frac{1}{k_y} \sim \infty$

It shows that mass transfer resistance negligible in the liquid phase only.

2.44 (b)

$$\text{Sh}_L = 0.664\,\text{Re}_L^{0.5}\text{Sc}^{1/3}$$

For $j_D = \frac{\text{Sh}_L}{\text{Re}_L\,\text{Sc}}\text{Sc}^{2/3}$

$$j_D = \frac{0.664}{\text{Re}_L\,\text{Sc}}\text{Re}_L^{0.5}\,\text{Sc}^{1/3}\text{Sc}^{2/3}$$

$$= 0.664\text{Re}_L^{-0.5}$$

2.45 (d)

According to surface renewal theory, the unit of fractional rate of surface renewal is

$$k_{L,\text{avg}} \propto \sqrt{D_{AB}\cdot s}$$

D_{AB} unit m^2/sec and $k_{L,\text{avg}}$ unit m/sec, then unit of s is sec^{-1}.

2.46 (b)

The alkanolamine are very reactive and remove H_2S by absorption and chemical reaction.
For highly soluble in case of H_2S with alkanolamines the value of slope is very small.

$$\frac{1}{K_{oy}} = \frac{1}{k_y} + \frac{m}{k_x}^{\,0}$$

So, $\left(\frac{m}{k_x}\right)$ is negligible as compared to that of gas phase.

2.47 (a)

Stanton number for mass transfer from Reynold analogy.

$$(\text{St})_m = \frac{f}{2} = \frac{Sh}{\text{Re.Sc}}$$

where Colburn j-factor is

$$j_H = (\text{St})_m.\text{Sc}^{2/3} = \frac{Sh}{\text{Re.Sc}}\cdot\text{Sc}^{2/3}$$

$$j_H = \frac{\text{Sh}}{(\text{Re})(\text{Sc})^{1/3}}$$

2.48 (8)

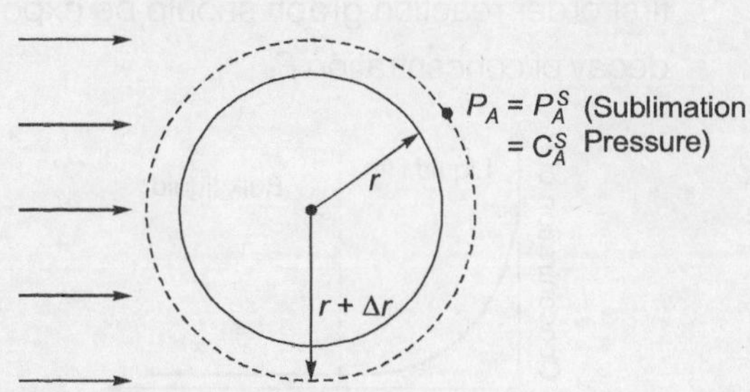

Assume in time Δt the radius of sphere decreases by Δr.
Volume of camphor sublimated

$$= \frac{4}{3}\pi\left[\underset{\text{(final)}}{r^3} - \underset{\text{(initial)}}{(r+\Delta r)^3}\right]$$

$$= -4\pi r^2\Delta r$$

Mass of camphor sublimated in time Δt

$$= -(4\pi r^2\Delta r)\rho$$

Moles of camphor sublimated in time Δt

$$= \frac{-(4\pi r^2\Delta r)\rho}{(\text{M.W})} \quad \text{...(i)}$$

Mass transferred during time interval Δt

$$k_c(C_{As} - C_{A\infty})4\pi r^2\Delta t \quad \text{...(ii)}$$

Equating eqn. (i) and (ii)

$$-(4\pi r^2\Delta r)\frac{\rho}{M} = k_c\cdot C_{As}4\pi r^2\Delta t$$

$$\lim_{\Delta t\to 0}\frac{-\Delta r}{\Delta t} = \frac{k_c C_{As} M}{\rho}$$

$$\left[k_c \propto \frac{1}{\sqrt{r(t)}} \text{ given}\right]$$

$$\frac{dr}{dt} = \frac{k'}{r^{1/2}}$$

where $k' = \frac{k_c C_{As} M}{\rho}$

$$\int_R^0 r^{1/2}dr = k'\int_0^{t_C} dt$$

From this, $t_C \propto R^{3/2}$

So for complete sublimation of 2 camphor spheres of diameter 20 cm and 5 cm is

$$\frac{t_{c_1}}{t_{c_2}} = \left(\frac{20}{5}\right)^{3/2} = 8$$

2.49 (b)

The reaction is first order with respect to *P*. For a first order reaction graph should be exponential decay of concentration *P*.

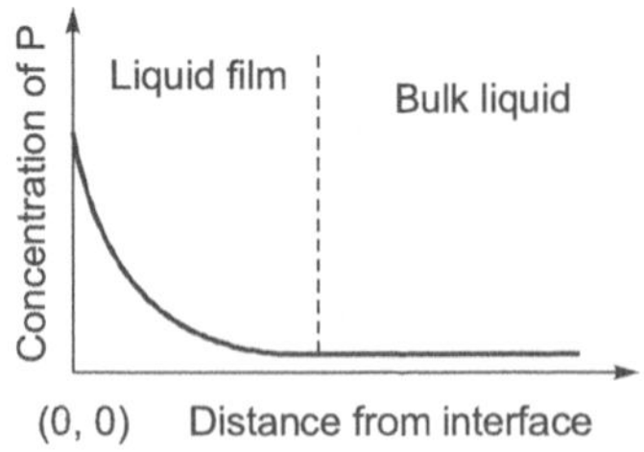

2.50 (c)

In film theory, the interfacial mass transfer coefficient.

For gas side:

$$k_y = 0.1 \text{ mol/m}^2\text{-sec}$$

For liquid side:

$$k_x = 3 \text{ mol/m}^2\text{-sec}$$

Equilibrium relation $y^* = 2x$

where x and y^* are mole fraction of solute in liquid and gas phase respectively.

From film theory,

$$\frac{1}{k_{oy}} = \frac{1}{k_y} + \frac{m}{k_x}$$

$$\frac{1}{k_{oy}} = \frac{1}{0.1} + \frac{2}{3} = 10.66$$

We have to find out

$$\frac{m/k_x}{1/k_{oy}} = \frac{2/3}{10.66} = 0.0625$$

where $\frac{m}{k_x}$ is mass transfer resistance in liquid film and $\frac{1}{k_{oy}}$ is overall mass transfer resistance.

2.51 (0.03)

Basis 1 meter width of film.

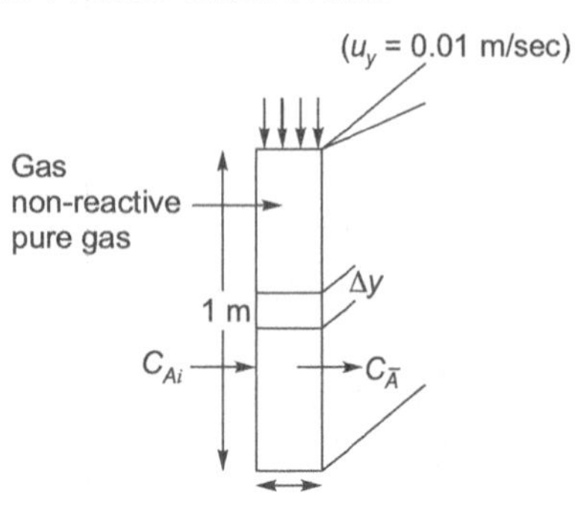

kmol/sec of pure gas transferred from gas phase to the control volume of liquid.

Change in concentration of transferred gas component in liquid phase during its transport in the control volume.

$$k_L(C_{A_i} - \bar{C}_A)(\Delta y \times 1) = u_y(\delta \times 1)\Delta \bar{C}_A$$

$$\frac{d\bar{C}_A}{dy} = \frac{k_l}{u_y \cdot \delta}(C_{A_i} - \bar{C}_A)$$

$$\int_{\bar{C}_A = C_{A_0}}^{\bar{C}_A = \bar{C}_{A_l}} \frac{d\bar{C}_A}{(C_{A_i} - \bar{C}_A)} = \int_{y=0}^{y=l} \left(\frac{k_l}{u_y \cdot \delta}\right) dy$$

$$\ln \frac{C_{A_i} - C_{A_0}}{C_{A_i} - \bar{C}_{A,L}} = \frac{k_l L}{u_y \cdot \delta}$$

at $y = 0$, $C_{A_0} = 0$ (pure water falls at the entrance of wall), then

$$C_{A_i} - C_{A_0} \simeq C_{A_i}$$

at $y = L \Rightarrow (C_{A_i} - \bar{C}_{A,L})$

$$= C_{A_i} - C_{A_i}(1 - e^{-30L})$$

$$= C_{A_i} e^{-30 \times L} \qquad \text{(where } L = 1 \text{ m)}$$

$$\ln\left(\frac{C_{A_i}}{C_{A_i} e^{-30}}\right) = \frac{(k_{Lavg}) : L}{u_y \delta}$$

$$30 = \frac{(k_{Lavg}) : L}{u_y \delta}$$

$$k_{L,\,avg} = \frac{30 u_y \cdot \delta}{L} = \frac{30 \times 0.01 \times 10^{-4}}{1}$$

$$= 3 \times 10^{-5} \text{ m/sec}$$

$$k_{L,\,avg} = 0.03 \text{ mm/sec}$$

2.52 (0.25)

Given : $C_A = F(y)$

$$V_z(y) = 10y^2$$

$$\bar{V} = 0.033 \text{ m/sec}, \delta = 0.1 \text{ m}$$

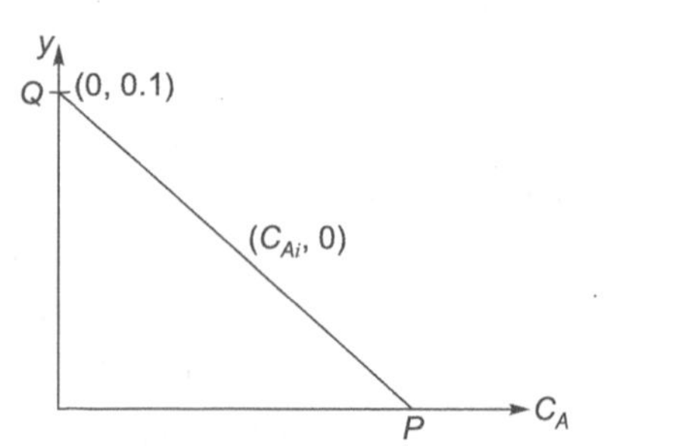

From the above diagram, the slope of PQ line is

$$m = \frac{0.1-0}{0-C_{A,i}}$$

We can also write the equation of line PQ

$$y-0 = \frac{0.1}{-C_{A,i}}(C_A - C_{Ai})$$

$$C_A = C_{Ai}(1-10y) = F(y)$$

For unit width of plate area, average concentration is given by

$$\dot{m} = A\bar{V}C_{A,m} = \int_0^{\delta} V_z(y)\cdot C_A(y)(dy\cdot 1)$$

$$= (\delta \times 1)\times 0.033C_{A,m}$$

$$= \int_0^{0.1} 10y^2 C_{A,i}(1-10y)dy$$

$$0.0033C_{A,m} = C_{Ai}\int_0^{0.1}(10y^2-100y^3)dy$$

After solving, we get $\frac{C_{A,m}}{C_{A,i}} = 0.25$

2.53 (a)

From definition we know that for a spherical droplet which is stationary.

Sherwood number = 2

$$N_{sh} = 2$$

From definition of Sherwood number

$$N_{sh} = \frac{k_c' D}{D_{AB}}$$

where, k_c' = Mass transfer coefficient
D = Diameter of droplet
D_{AB} = Mass diffusivity

For this condition,

$$N_{sh} = 2$$

$$2 = \frac{k_c' D}{D_{AB}} \text{ or } k_c' = \frac{2D_{AB}}{D}$$

For small droplet,

$$k_{c_1}' = \frac{2D_{AB}}{D_1}$$

For large droplet,

$$k_{c_2}' = \frac{2D_{AB}}{D_2}$$

We also know that,

$$\text{Rate} = \text{Flux} \times \text{Area}$$

$$\text{Rate }(W) = N_A \times \text{Area}$$

$$W = k_c'(C_{A_1} - C_{A_2}) \times \text{Area}$$

For small droplet,

$$A_1 = \pi D_1^2$$

$$A_2 = \pi(2D_1)^2 = 4\pi D_1^2$$

So, $W_1 = k_{c_1}'(C_{A_1} - C_{A_2}) \times \pi D_1^2$

($\because$ Dilute $k_c = k_c'$)

$$W_1 \propto k_{c_1}' \times \pi D_1^2$$

In similar manner

$$W_2 \propto k_{c_2}' \times 4\pi D_1^2$$

Put the values of k_{c_1}' and k_{c_2}'

$$W_1 \propto \frac{2D_{AB}}{D_1}\times \pi D_1^2$$

where, $D_2 = 2D_1$

$$W_2 \propto \frac{2D_{AB}}{D_2}\times 4\pi D_1^2$$

On comparing these two equations

$$W_2 = 2W_1$$

2.54 (53.5)

For A diffusing in non-diffusing B

$$N_{A_1} = \frac{D_{AB}P_T}{RTZP_{Blm}}(P_{A_1} - P_{A_2})$$

For equimolar diffusion

$$N_{A_2} = \frac{D_{AB}}{RTZ}(P_{A_1} - P_{A_2})$$

$$\frac{N_{A_1}}{N_{A_2}} = \frac{P_T}{P_{Blm}} = \frac{1}{y_{Blm}}$$

Given, $y_{A_1} = 0.8$

$$y_{A_2} = 0.1$$

$$y_{B_1} = 1 - y_{A_1} = 0.2$$

$$y_{B_2} = 1 - y_{A_2} = 0.9$$

$$y_{Blm} = \frac{0.9-0.2}{\ln\left(\frac{9}{2}\right)} = \frac{0.7}{\ln(4.5)} = 0.465$$

$$\frac{N_{A_2}}{N_{A_1}} = y_{Blm}$$

$$\frac{N_{A_1} - N_{A_2}}{N_{A_1}} \times 100 = 53.45$$

2.55 (222.22)

Liquid	Liquid
(A)	(B)
k_a	k_b
$(y_{Ag})0.10$	$0.02(y_{Bl})$

Given: $k_a = 2k_b$

$$N_A = k_A(y_{Ag} - y_S^A)$$
$$= k_B(y_S^B - y_{Bl})$$
$$2k_B(0.10 - y_S^A) = k_b(y_S^B - 0.02)$$
$$0.20 - 2y_S^A = y_S^B - 0.02$$

Equilibrium relation is

$$y_S^A = 0.5y_S^B$$
$$0.20 - y_S^B = y_S^B - 0.02$$
$$0.22 = 2y_S^B$$
$$y_S^B = 0.11$$

So, $N_A = k_b(0.11 - 0.02)$

$$k_b = \frac{10}{0.11 - 0.02} = 111.11$$
$$k_a = 2 \times 111.11 = 222.22 \text{ kmol/m}^2\text{.hr}$$

2.56 (1.17)

The ratio of flux for the case of A non-diffusing B to equimolar counter diffusion is given by:

$$\frac{(N_A)_{ANB}}{(N_A)_{EMCD}} = \frac{D_{AB}P_T(y_{A_1} - y_{A_2})R_T\Delta z}{R_T\Delta z\, y_{Blm} \cdot D_{AB}P_T(y_{A_1} - y_{A_2})}$$

$$\frac{(N_A)_{ANB}}{(N_A)_{EMCD}} = \frac{1}{y_{Blm}} = \frac{1}{\dfrac{y_{B_1} - y_{B_2}}{\ln\left|\dfrac{y_{B_1}}{y_{B_2}}\right|}}$$

$$y_{B_1} = 1 - 0.2 = 0.8$$
$$y_{B_2} = 1 - 0.1 = 0.9$$
$$Y_{Blm} = \frac{0.8 - 0.9}{\ln\left|\dfrac{0.8}{0.9}\right|} = 0.85$$

$$\therefore \quad \frac{(N_A)_{ANB}}{(N_A)_{EMCD}} = \frac{1}{0.85} = 1.176$$

2.57 (a)

Molar flux N_A is given by

$$N_A = \frac{D_{AB}C_T\left(X_{A_1} - X_{A_2}\right)}{(Z_2 - Z_1)}$$

Here, $D_{AB} = 10^{-5}$ m²/s

$C_T = 100$ mol/m³

$X_{A_1} = 0.8, \quad X_{A_2} = 0.2$

$Z_2 - Z_1 = 0.02$ m

So,
$$N_A = \frac{10^{-5} \times 100 \times (0.8 - 0.2)}{0.02}$$
$$\overline{N_A} = N_A \cdot A$$
$$= \frac{0.6 \times 100 \times 10^{-5}}{0.02} \times 0.02$$
$$= 6 \times 10^{-4} \text{ mol/s}$$

2.58 (b)

For A non-diffusing B steady state molar flux is given by

$$N_A = \frac{D_{AB}P_T}{RT(Z_2 - Z_1)} \ln\left[\frac{1 - y_{A_2}}{1 - y_{A_1}}\right]$$
$$= \frac{10^{-5} \times 10^5}{8.314 \times 303 \times 0.002} \ln\left[\frac{1 - 0.05}{1 - 0.15}\right]$$
$$= 0.022 \text{ mol/m}^2\text{.s}$$

3 Absorption and Stripping

3.1 The absorption factor is defined as

(a) $\frac{L}{mG}$ (b) $\frac{G}{mL}$

(c) $\frac{mL}{G}$ (d) $\frac{LG}{m}$

where L = liquid flow rate, G = gas flow rate and m = slope of the equilibrium line.

[2000 : 1 M]

3.2 For gas absorption the height of a transfer unit, based on the gas phase, is given by (G : superficial molar gas velocity; L : superficial molar liquid velocity; F_G: mass transfer coefficient, mol/m^2 s; a : interfacial area per unit volume of tower)

(a) $\frac{G}{F_G a}$ (b) $\frac{F_G}{Ga}$

(c) $\frac{Ga}{F_G}$ (d) $\frac{L}{F_G G}$

[2001 : 1 M]

3.3 The interfacial area per unit volume of dispersion, in a gas-liquid contactor, for fractional hold-up of gas = 0.1 and gas bubble diameter = 0.5 mm is given by (in m^2/m^3)

(a) 500 (b) 1200

(c) 900 (d) 800

[2001 : 2 M]

3.4 H_2S is being absorbed in a gas absorber unit. The height of the transfer unit based on the overall mass transfer coefficient on the gas side is 0.4 m. The equilibrium data given by $y = 1.5x$. The bulk concentration of H_2S has to be reduced from 0.05 to 0.001 mole fraction in the gas side. The height of the tower (in meters) corresponding to an operating line given by $y = 5x + 0.001$ is

(a) 2.0 (b) 1.56

(c) 1.0 (d) 0.56

[2003 : 2 M]

3.5 Acetone is to be removed from air in an isothermal dilute absorber using pure water as solvent. The incoming air contains 5 mol% of acetone (y_{in} = 0.05). The design equation to be used for obtaining the number of trays (N) of the absorber is

$$N + 2 = 6\log\left(\frac{y_{in}}{y_{out}}\right)$$

For 98% recovery of acetone, the number of trays required is/are

(a) 1 (b) 8

(c) 9 (d) 10

[2004 : 1 M]

Statement for Linked Answer Question (3.6 to 3.7):
A binary gas mixture of a solute and a carrier gas is treated in a countercurrent gas absorption column, containing ideal trays, using a solvent. The compositions y and x (see figure below) are the mole fractions of the solute in the gas and liquid respectively. Also, V and L are the molar flow rates of the gas and liquid respectively. Assume that the carrier gas is insoluble in the solvent and that the vapour pressure of the solvent is very low at the given conditions of the column. Further, the gas and liquid streams are sufficiently dilute that L and V may be assumed to be constant throughout the column. The equilibrium relation is given by $y^* = mx$, where m is a positive constant.

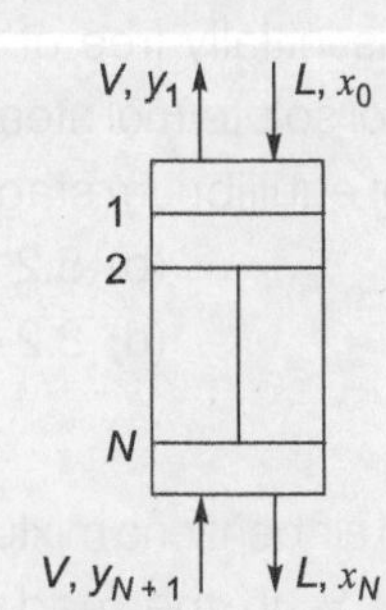

3.6 For any value of m, the change in liquid composition across a tray is independent of the tray location if

(a) $\frac{L}{V} = 1$ (b) $\frac{mL}{V} = 1$

(c) $\frac{m}{(LV)} = 1$ (d) $\frac{L}{(mV)} = 1$

[2005 : 2 M]

3.7 Under the correct condition corresponding to part (a), the number of ideal trays in the column is given by

(a) $N = \dfrac{x_0 - x_N}{x_0 - (y_{N+1}/m)}$

(b) $N = \dfrac{x_0 - (y_{N+1}/m)}{x_0 - x_N}$

(c) $N = \dfrac{x_N - x_0}{(y_{N+1}/m) - x_N}$

(d) $N = \dfrac{(y_{N+1}/m) - x_0}{x_N - x_0}$

[2005 : 2 M]

3.8 In a multistage countercurrent isothermal stripping column, feed containing 0.05 mol of solute/mol of solute free oil is treated with steam. The absorption factor $A = 0.65$. The equilibrium relation is given by $Y^* = 2X$, where Y^* and X refer to the equilibrium mole ratio in the steam and oil phases respectively. The Kremser equation is given as follows ('0' refers to liquid inlet at the top, 'N_p' refers to the last stage at the bottom).

$$N_p = \frac{\log\left[\left(\dfrac{X_0 - \dfrac{Y_{N_{P+1}}}{m}}{X_{N_P} - \dfrac{Y_{N_{P+1}}}{m}}\right)(1-A) + A\right]}{\log\left[\dfrac{1}{A}\right]}$$

If the steam is initially free of solute and its exit mole ratio (mol solute/mol steam) is 0.0624, then the number of equilibrium stages required is

(a) 4.2 (b) 5.2
(c) 7.2 (d) 8.2

[2006 : 2 M]

3.9 Benzene in an air-benzene mixture is to be reduced from 5.2 mol % in the feed to 0.5 mol % by contacting with wash oil in a multistage countercurrent gas absorber. The inlet flowrate of air-benzene mixture is 10 mol/s while benzene free wash oil comes in at 9.5 mol/s. If the equilibrium curve is given as $Y^* = X$, where Y^* and X are equilibrium mole ratios of benzene in air and benzene in oil, the number of equilibrium stages required to achieve the above separation is

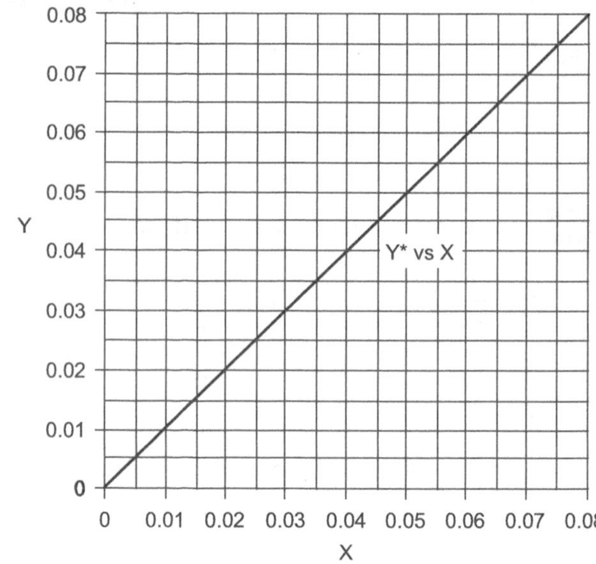

(a) 12 (b) 10
(c) 8 (d) 6

[2007 : 2 M]

3.10 Given the following statements listed from P to T, select the correct combination of TRUE statements from the choices that follow this list.

P : Plate columns are preferred when the operation involves liquids containing suspended solids.

Q : Packed towers are preferred if the liquids have a large foaming tendency.

R : The pressure drop through packed towers is more than the pressure drop through plate columns designed for the same duty.

S : Packed columns are preferred when large temperature changes are involved in distillation operations.

T : Packed towers are cheaper than plate towers if highly corrosive fluids must be handled.

(a) T, S, P (b) P, Q, T
(c) S, R, T (d) R, Q, S

[2007 : 2 M]

3.11 In a countercurrent gas absorber, both the operating and equilibrium relations are linear. The inlet liquid composition and the exit gas composition are maintained constant. In order to increase the absorption factor

(a) the liquid flow rate should decrease
(b) the gas flow rate should increase
(c) the slope of the equilibrium line should increase
(d) the slope of the equilibrium line should decrease

[2008 : 1 M]

3.12 A packed tower containing Berl saddles is operated with a gas-liquid system in the countercurrent mode. Keeping the gas flow rate constant, if the liquid flow rate is continuously increased,
(a) the void fraction available for the gas to flow will decrease beyond the loading point.
(b) the gas pressure drop will decrease.
(c) liquid will continue to flow freely down the tower beyond the loading point.
(d) the entrainment of liquid in the gas will considerably decrease near the flooding point.

[2008 : 2 M]

3.13 The ratio of the liquid to gas flow rate in a counter-current gas absorption column is increased, at otherwise identical conditions. Which one of the following statements is TRUE?
(a) The operating line shifts towards the equilibrium curve.
(b) The operating line shifts away from the equilibrium curve.
(c) The concentration of the absorbed species increases in the exit liquid stream.
(d) The operating line does not shift.

[2009 : 1 M]

Common Data for Questions 3.14 and 3.15:

A counter-current extraction column is designed to remove 99% of solute *C* from a solution of solvent *A* and solute *C* using pure solvent *B*. The initial concentration of solute in the solution of *A* + *C* is 20 wt%, and the total flow of solution is 1000 kg/h. If the equilibrium relationship is $Y = 2X$, where Y = mass of *C* /mass of *A* and X = mass of C/mass of *B*.

3.14 The minimum flow rate of solvent *B* required (in kg/h) is
(a) 1454 (b) 1584
(c) 1676 (d) 1874

[2012 : 2 M]

3.15 If the flow rate of *B* is 2400 kg/h, then the theoretical number of stages in the column, using Kremser's equation (adjusted to the next integer) is
(a) 5 (b) 9
(c) 11 (d) 13

[2012 : 2 M]

3.16 The packing of an existing absorption tower is replaced with a new type of packing. The height of the packing and the inlet conditions are maintained the same as before. Tests reveal that the number of transfer units is lower than before. This indicates that the tower with the new packing, when compared to that with the old packing, will
(a) have a higher rate of absorption of the solute from the gas stream
(b) have a lower rate of absorption of the solute from the gas stream
(c) have the same rate of absorption of the solute from the gas stream
(d) have a lower height of transfer unit

[2013 : 1 M]

3.17 Packed towers are preferred for gas-liquid mass transfer operations with foaming liquids because
(a) in packed towers, high liquid to gas ratios are best handled
(b) in packed towers, continuous contact of gas and liquid takes place
(c) packed towers are packed with random packings
(d) in packed towers, the gas is not bubbled through the liquid pool

[2014 : 1 M]

3.18 Benzene is removed from air by absorbing it in a non-volatile wash-oil at 100 kPa in a counter-current gas absorber. Gas flow rate is 100 mol/min, which includes 2 mol/min of benzene. The flow rate of wash-oil is 50 mol/min. Vapor pressure of benzene at the column conditions is 50 kPa. Benzene forms an ideal solution with the wash-oil and the column is operating at steady state. Gas phase can be assumed to follow ideal gas law. Neglect the change in molar flow rates of liquid and gas phases inside the column.

For this process, the value of the absorption factor (up to two decimal places) is ______.

[2015 : 1 M]

3.19 In a countercurrent stripping operation using pure steam, the mole ratio of a solute in the liquid stream is reduced from 0.25 to 0.05. The liquid feed flow rate on a solute-free basis, is 3 mol/s. The equilibrium line for the system is given in the figure below.

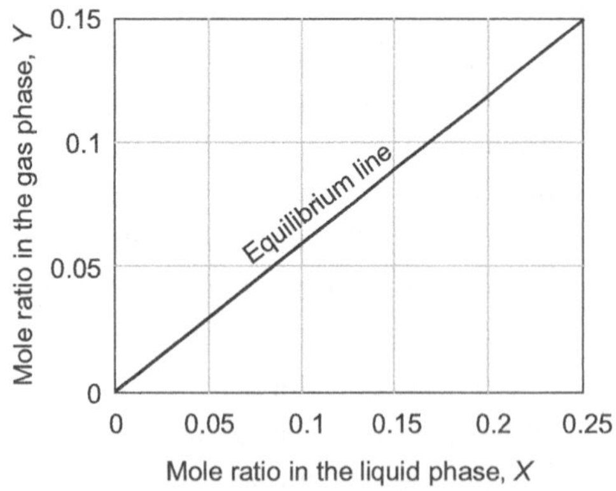

The MINIMUM flowrate of pure steam for this process, rounded to 1 decimal place, is _____ mol/s. [2017 : 2 M]

3.20 A counter-current absorption tower is designed to remove 95% of component *A* from an incoming binary gas mixture using pure solvent *B*. The mole ratio of A in the inlet gas is 0.02. The carrier gas flow rate is 50 kmol/h. The equilibrium relation is given by $Y = 2X$, where Y and X are the mole ratios of *A* in the gas and liquid phases, respectively. If the tower is operated at twice the minimum solvent flow rate, the mole ratio of *A* in the exit liquid stream is ______ (rounded off to three decimal places). [2019 : 2 M]

3.21 SO_2 from air is absorbed by pure water in a counter current packed column operating at constant pressure. The compositions and the flow rates of the streams are shown in the figure.

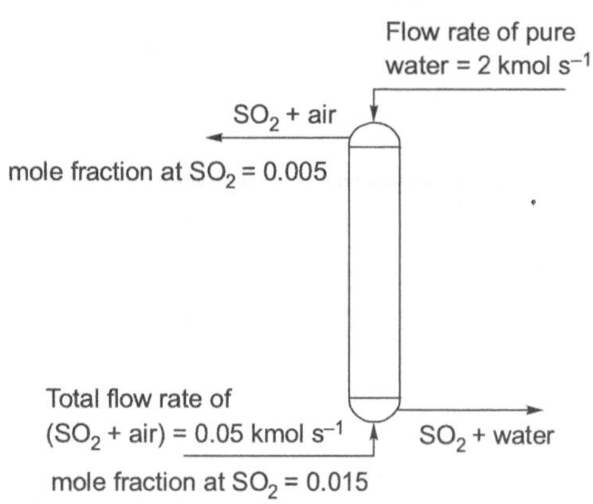

In addition, use the following data and assumptions :

- Column operates under isothermal conditions.
- At the operating temperature of the column, $y^* = 40x$, where y^* is the mole fraction of SO_2 in the gas that is in equilibrium with water containing SO_2 at a mole fraction of x.
- Solution is dilute and the operating line is linear.
- Negligible amount of water evaporates.

The number of transfer units (NTU) for this column is

(a) 0.5 (b) 1.0
(c) 1.5 (d) 2.0 [2020 : 2 M]

3.22 In a solvent regeneration process, a gas is used to strip a solute from a liquid in a counter-current packed tower operating under isothermal condition. Pure gas is used in this stripping operation. All solutions are dilute and Henry's law, $y^* = mx$, is applicable. Here, y^* is the mole fraction of the solute in the gas phase in equilibrium with the liquid phase of solute mole fraction x, and m is the Henry's law constant.

Let x_1 be the mole fraction of the solute in the leaving liquid, and x_2 be the mole fraction of solute in the entering liquid.

When the value of the ratio of the liquid-to-gas molar flow rates is equal to m, the overall liquid phase Number of Transfer Units, NTU_{OL}, is given by

(a) $\frac{x_2 - x_1}{x_1}$ (b) $\frac{x_2 + x_1}{x_2 - x_1}$

(c) $\ln\left(\frac{x_2}{x_1}\right)$ (d) $\ln\left(\frac{x_2 + x_1}{x_2 - x_1}\right)$

[2021 : 2 M]

3.23 A packed distillation column, with vapor having an average molecular weight of 45 kg.kmol^{-1}, density of 2 kg.m^{-3} and a molar flow rate of 0.1 kmol.s^{-1}, has a flooding velocity of 0.15 m.s^{-1}. The column is designed to operate at 60% of the flooding velocity. Which one of the following is the CORRECT value for the column diameter (in m)?

(a) $\frac{5}{\sqrt{\pi}}$ (b) $5\sqrt{\pi}$

(c) 4π (d) $\frac{10}{\sqrt{\pi}}$ [2023 : 1 M]

3.24 A liquid *L* containing a dissolved gas *S* is stripped in a countercurrent operation using a pure carrier gas *V*. The liquid phase inlet and outlet mole fractions of *S* are 0.1 and 0.01, respectively. The equilibrium distribution of *S* between *V* and *L* is governed by $y_e = x_e$, where y_e and x_e are the mole fractions of *S* in *V* and *L*, respectively. The molar feed rate of the carrier gas stream is twice as that of the liquid stream. Under dilute solution conditions, the minimum number of ideal stages required is _____ (in integer). [2023 : 1 M]

3.25 Which one of the given statements is correct with reference to gas-liquid contactors for mass transfer applications?

(a) A tray tower is more suitable for foaming systems than a packed tower.
(b) Tray towers are preferred over packed towers for systems requiring frequent cleaning.
(c) For a given liquid flow rate, the gas flow rate in the loading region is greater than that in the flooding region.
(d) Flooding can never occur for counter-current contact.

[2024 : 1 M]

3.26 A gas stream containing 95 mol% CO_2 and 5 mol% ethanol is to be scrubbed with pure water in a counter-current, isothermal absorption column to remove ethanol. The desired composition of ethanol in the exit gas stream is 0.5 mol%. The equilibrium mole fraction of ethanol in the gas phase, y^*, is related to that in the liquid phase, x, as $y^* = 2x$. Assume CO_2 is insoluble in water and neglect evaporation of water. If the water flow rate is twice the minimum, the mole fraction of ethanol in the spent water is

(a) 0.0225
(b) 0.0126
(c) 0.0428
(d) 0.0316

[2024 : 2 M]

■■■■

Answers Absorption and Stripping

3.1	(a)	**3.2**	(a)	**3.3**	(b)	**3.4**	(a)	**3.5**	(b)	**3.6**	(d)	**3.7**	(c)
3.8	(b)	**3.9**	(b)	**3.10**	(b)	**3.11**	(d)	**3.12**	(a)	**3.13**	(b)	**3.14**	(b)
3.15	(b)	**3.16**	(b)	**3.17**	(d)	**3.18**	(1.02)	**3.19**	(4.0)	**3.20**	(0.005)	**3.21**	(d)
3.22	(a)	**3.23**	(d)	**3.24**	(3)	**3.25**	(b)	**3.26**	(b)				

Explanations Absorption and Stripping

3.1 (a)

Absorption factor is defined as,

$$A = \frac{L_S}{mG_S}$$

where, $\frac{L_S}{G_S}$ = Slope of operating line

m = Slope of equilibrium line

3.2 (a)

For packed bed absorption tower

$$\text{HTU} = \frac{G}{k_G P \cdot a}$$

where $k_G P = F_G$

$$\text{HTU} = \frac{G}{F_G a}$$

3.3 (b)

Fractional hold up (ε)

$$= \frac{\text{Total volume of bubbles}}{\text{Total volume of gas ? liquid contactor}}$$

Basis = 1 m^3 of gas-liquid contactor

$$\varepsilon = \frac{n \times \frac{4}{3}\pi r^3}{1}$$

$$n = \frac{3\varepsilon}{4\pi r^3}$$

Interfacial area of n bubbles

$$a = 4\pi r^2 \times n$$

$$a = 4\pi r^2 \times \frac{3\varepsilon}{4\pi r^3}$$

$$a = \frac{6\varepsilon}{d_P} = \frac{6 \times 0.1}{0.5} \times 1000$$

$$= 1200 \text{ m}^2/\text{m}^3$$

3.4 (a)

Height of tower (z) = HTU × NTU

$$\text{NTU} = \frac{y_1 - y_2}{(y - y_i)_{lm}}$$

where $(y - y_i)_{lm} = \frac{(y_1 - y_{i1}) - (y_2 - y_{i2})}{\ln\left(\frac{y_1 - y_{i1}}{y_2 - y_{i2}}\right)}$

where 1 and 2 indicates bottom and top composition respectively.

Given : $y_1 = 0.05$, $y_2 = 0.001$

The operating line is given as $y = 5x + 0.001$

Equilibrium relation, $y = 1.5x$

At $y_1 = 0.05, x_1 = \frac{0.05 - 0.001}{5} = 0.0098$

and at $y_2 = 0.001$, $x_2 = 0$

and from equilibrium relation at $x_1 = 0.0098$,

$$y_{i1} = 1.5 \times 0.0098 = 0.0147$$

and at $x_2 = 0$, $y_{i2} = 0$

$$(y - y_i)_{lm} = \frac{(0.05 - 0.0147) - (0.001 - 0)}{\ln\left(\frac{0.05 - 0.0147}{0.001}\right)}$$

$$= 0.009624$$

Hence, $\text{NTU} = \frac{y_1 - y_2}{(y - y_i)_{lm}}$

$$= \frac{0.05 - 0.001}{0.009624} = 5.09$$

Since, HTU = 0.4 (given)

then, Z = HTU × NTU = 5.09 × 0.4

= 2.036 m

3.5 (b)

Basis : 100 mole of incoming feed contains 5 mole % of acetone.

Mole of acetone = 5 mole

Mole of air = 95 mole

98% recovery of acetone required.

Acetone in vent gas = 5 × (1 – 0.98)

= 0.1 mole

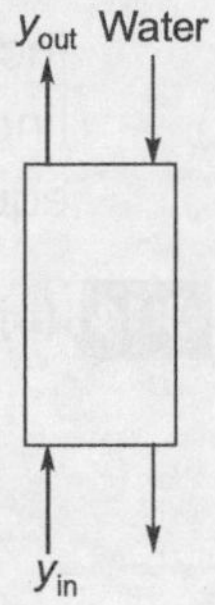

Mole fraction of acetone in vent gas

$$y_{out} = \frac{0.1}{95+0.1} = 1.0515 \times 10^{-3}$$

$$\text{Number of trays} = N + 2 = 6\log\left(\frac{y_{in}}{y_{out}}\right)$$

$$N + 2 = 6\log\left(\frac{0.05}{1.0515\times10^{-3}}\right)$$

$$N + 2 = 10.062$$

$$N = 8.062 \simeq 8$$

3.6 (d)

When difference in liquid composition across a tray is same for each tray in the absorber means when change in liquid composition across a tray is independent of tray location, the equilibrium line and operating lines are parallel. Therefore, slope of operating line = Slope of equilibrium line

$$\frac{L}{V} = m$$

$$\frac{L}{mV} = 1$$

3.7 (c)

Number of ideal trays in the column,

$$N = \frac{x_N - x_0}{(y_{N+1}/m) - x_N}$$

3.8 (b)

$$N_P = \frac{\log\left[\left(\dfrac{X_o - \dfrac{Y_{N_P+1}}{m}}{X_{N_P} - \dfrac{Y_{N_P+1}}{m}}\right)(1-A) + A\right]}{\log\left(\dfrac{1}{A}\right)}$$

Given : $X_o = 0.05,\ Y_{N_P+1} = 0$

$Y_1 = 0.0624,\ Y^* = 2X$

$$A = \frac{L}{mV} = \frac{L}{V} = 2 \times 0.65 = 1.3$$

Apply solute balance

$$L(X_o - X_{N_P}) = V(Y_1 - Y_{N_P+1})$$

$$\frac{L}{V}(0.05 - X_{N_P}) = 0.0624 - 0$$

$$X_{N_P} = 0.002$$

$$N_P = \frac{\log\left[\left(\dfrac{0.05-0}{0.002-0}\right)(1-0.65)+0.65\right]}{\log\left[\dfrac{1}{0.65}\right]}$$

$$N_P = 5.201$$

3.9 (b)

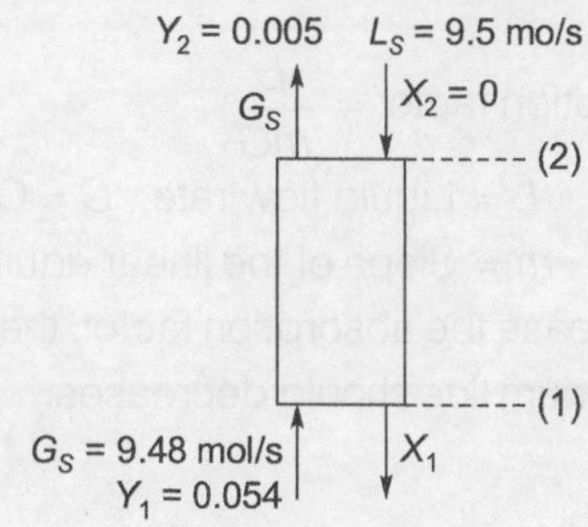

Given : $L = 9.5$ mol/s

$$y_1 = 0.052,\ Y_1 = \frac{y_1}{1-y_1} = \frac{0.052}{1-0.052} = 0.054$$

$$y_2 = 0.005,\ Y_2 = \frac{y_2}{1-y_2} = \frac{0.005}{1-0.005} = 0.005$$

$X_2 = 0$ and $m = 1$ (given)

$$A = \text{Absorption factor} = \frac{L_s}{mG_s}$$

where L_s and G_s are solute free solvent and gas flow rate.

$$L_s = 9.5 \text{ mol/s and } G_s = G(1 - y_1)$$
$$= 10(1 - 0.052) = 9.48 \text{ mol/s}$$

Hence, $$A = \frac{L_s}{mG_s} = \frac{9.5}{9.48\times1} = 1.00$$

For absorption factor A unity the number of stages

$$N = \left[\frac{Y_1 - Y_2}{Y_2 - Y_2^*}\right]$$

where, $Y_2^* = mX_2 = 0$

$$N = \frac{0.054 - 0.005}{0.005 - 0} = 9.8 \simeq 10$$

3.10 (b)

If solid particle are present in liquid the cleaning of a packing section is much too complicated and expensive therefore plate columns are preferred. Handling of foaming liquid in packed column is more preferred because of the relatively low degree of agitation by the gas. In packed column pressure drop is lower which is important in vacuum distillation, plate columns are preferred at high temperature because presence of thermal or mechanical stress due to large temperature changes which might lead to cracked packings. In plate columns tray are made of metals or alloys, which are not highly corrosion resistance as ceramic and plastic materials are. Therefore, packed tower are cheaper if we want to handle highly corrosive fluids.

3.11 (d)

$$\text{Absorption factor} = \frac{L}{mG}$$

where, L = Liquid flow rate, G = Gas flow rate
m = Slope of the linear equilibrium line

To increase the absorption factor, the slope of the equilibrium line should decreases.

3.12 (a)

Keeping the gas flow rate constant if the liquid flow rate is continuously increases, the void fraction available for the gas to flow will decrease beyond the loading point.

3.13 (b)

For a counter-current gas absorption column

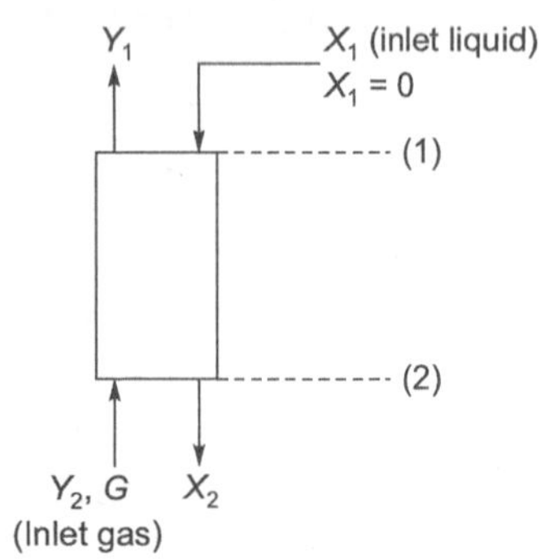

$$Y_2 > Y_1$$

(due to absorption transfer of solute from gas to liquid)

$$X_2 > X_1$$

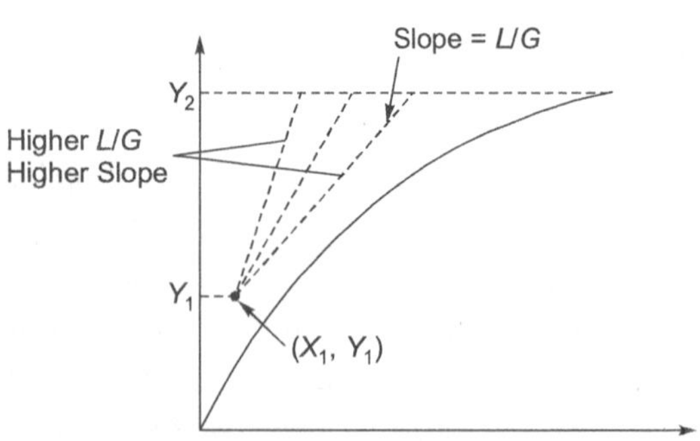

$$G(Y_2 - Y_1) = L(X_2 - X_1)$$

$$Y_2 = Y_1 + \frac{L}{G}(X_2 - X_1)$$

Increasing L result in higher slope of operating line and the operating line shifts away from equilibrium line.

3.14 (b)

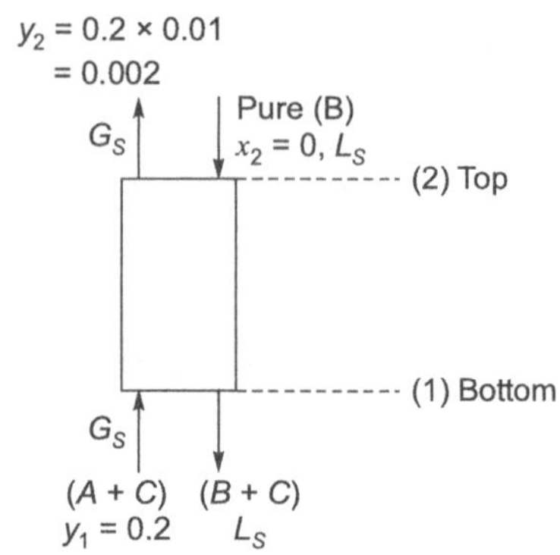

Given:
$$G = 1000 \text{ kg/h}$$
$$G_s = 1000(1 - 0.2)$$
$$G_s = 800 \text{ kg/h}$$
$$Y_1 = \frac{0.2}{1-0.2} = \frac{0.2}{0.8}$$
$$Y_1 = 0.25$$
$$Y_2 = 0.0025$$

Overall solute balance

$$G_s(Y_1 - Y_2) = L_s(X_1 - X_2)$$

For minimum flow rate of solvent ($L_{s,min}$) bottom component is in equilibrium for counter-current extraction column.

$$Y_1 = mX_1$$
$$0.25 = 2X_1$$
$$X_1 = 0.125$$
$$800(0.25 - 0.0025) = L_{s,min}(0.125 - 0)$$
$$L_{s,min} = 1584 \text{ kg/h}$$

3.15 (b)

Kremser's equation

$$N = \frac{\ln\left[\left(\frac{Y_1 - mX_2}{Y_2 - mX_2}\right)\left(1 - \frac{1}{A}\right) + \frac{1}{A}\right]}{\ln(A)}$$

where, $$A = \frac{2400}{2 \times 800} = 1.5$$

$$N = \frac{\ln\left[\left(\frac{0.25 - 0}{0.0025 - 0}\right)\left(1 - \frac{1}{1.5}\right) + \frac{1}{1.5}\right]}{\ln(1.5)}$$

$$= \frac{\ln(34)}{\ln(1.5)} = 8.697 \simeq 9$$

3.16 (b)

Height of Packing,

$$Z = (\text{HTU})(\text{NTU})$$

where HTU is measure of separation effectiveness of a packing and NTU is measure of difficulty of separation.

For this problem,

$$Z = \text{Constant}$$

Lower NTU $\Rightarrow$ Higher HTU

If HTU is high then less efficient mass transfer.

3.17 (d)

In packed towers, gas is not bubbled through the liquid pool.

3.18 (1.02)

Counter-current gas absorber

$$G = 100 \text{ mol/min}$$

and $G_s = 98$ mol/min, $L_s = 50$ mol/min

$$P_A^{\text{Sat}} = 50 \text{ kpas}$$

Absorption factor,

$$A = \frac{L_s}{mG_s}$$

$$P_A^{\text{Sat}} = mP_T \Rightarrow m = \frac{50}{100} = 0.5$$

$$A = \frac{50}{0.5 \times 98} \simeq 1.02$$

3.19 (4.0)

Counter-current stripping :

X_o to X_N is given 0.25 to 0.05 respectively.

$$L_s = 3 \text{ mol/sec}$$

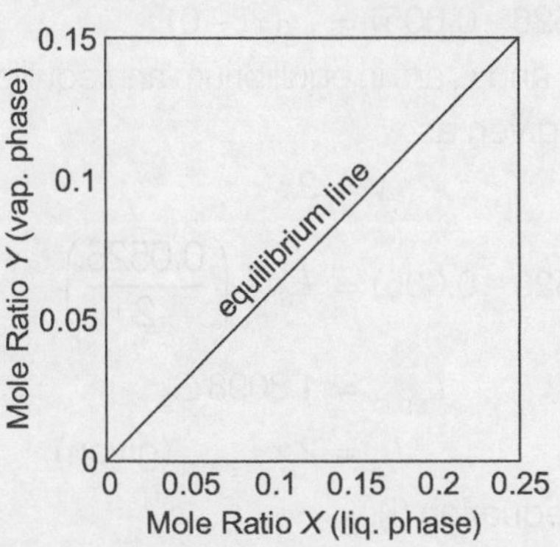

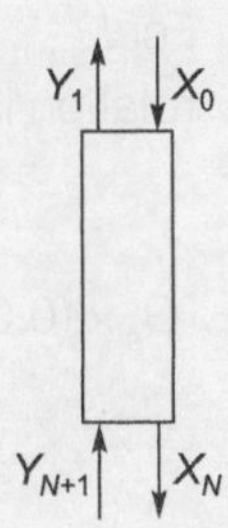

$$\text{Slope} = \frac{0.15-0}{0.25-0} = 0.6$$

$$G_{s,\min} = ?$$

$$G_{s,\min}(Y_{N+1} - Y_1) = L_S(X_N - X_0)$$

$$G_{s,\min}(0 - Y_1) = 3(0.05 - 0.25)$$

For counter-current stripping, top component will be in equilibrium.

$$Y_1 = mX_0$$

$$Y_1 = 0.6 \times 0.25 = 0.15$$

$$G_{s,\min} \times 0.15 = 3 \times 0.2$$

$$G_{s,\min} = 4.0 \text{ mol/sec}$$

3.20 (0.005)

Counter-current absorption tower

Mole ratio of A in inlet gas

$$Y_{A,\text{in}} = 0.02$$

$$Y_{A,\text{out}} = 0.02 \times 0.05$$

$$G_s = 50 \text{ kmol/h (carrier gas flow rate)}$$

Gas phase, $Y = 2X$ (given $L_s = 2\,L_{s,\min}$)

$$G_s(Y_{A,\text{in}} - Y_{A,\text{out}}) = L_s(X_{A,\text{out}} - X_{A,\text{in}})$$

$$50(0.02 - 0.02 \times 0.05) = L_s X_{\text{out}}$$

At min. solvent flow rate

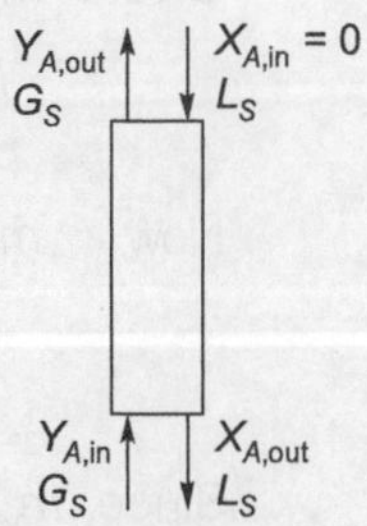

$$Y_{A,\text{in}} = mX_{\text{out}}$$

$$50 \times 0.019 = L_{S,\min} \times \frac{0.02}{2}$$

$$L_{S,\min} = 95$$

then $L_S = 2 \times 95$

$$= 190 \text{ kmol/h}$$

$$50(0.02 - 0.02 \times 0.05) = 190X_{\text{out}}$$

$$X_{\text{out}} = 0.005$$

3.21 (d)

SO_2 from air is absorbed by pure water in a counter current packed column operating at constant pressure and temperature.

(G) Total flow rate (SO_2 + air) = 0.05 kmol/sec

Mole fraction of SO_2 = 0.015 at inlet

Mole fraction of SO_2 = 0.005 at outlet

(L_S) Flow rate of pure water at inlet = 2 kmol/sec

$$\text{Absorption factor} = \frac{L_s}{mG_s}$$

$$y^* = 40x \text{ where } m = 40$$

where solution is dilute then $G_s = G$ and $L_s = L$.

$$A = \frac{2}{40 \times 0.05} = 1$$

If absorption factor is one then number of plate

$$N = \frac{y_{\text{in}} - y_{\text{out}}}{y_{\text{out}} - y^*_{\text{out}}}$$

where $y^*_{\text{out}} = mx_{\text{in}} \simeq 0$ (pure water)

$$N = \frac{0.015 - 0.005}{0.005 - 0} = 2$$

Number of transfer units is 2.

3.22 (a)

Given : $y^* = mx$

In stripping $S = \frac{mG_s}{L_s}$

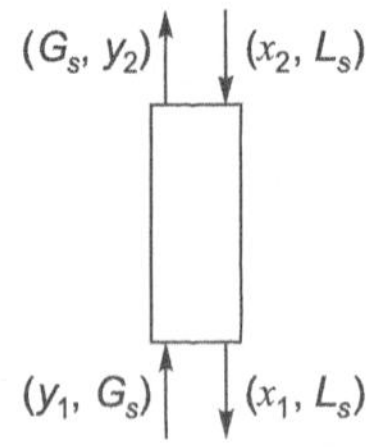

$$\frac{L_s}{G_s} = m$$

So $S = 1$

For this condition

$$(NTU)_{OL} = \frac{x_2 - x_1}{x_1 - \frac{y_1}{m}} \text{ where } y_1 = 0 \text{ (given)}$$

So, $(NTU)_{OL} = \frac{x_2 - x_1}{x_1}$

3.23 (d)

Given: $\dot{m}_v = 0.1$ kmol/s

$M = 45$ kg/kmol, $V_f = 0.15$ m/s

$\rho_v = 2$ kg/m^3, $V_{operating} = 60\%$ of $V_{flooding}$

Now, $\dot{m}_v = \rho A V_{operating}$

$$A = \frac{\dot{m}_v}{\rho V_{operating}}$$

Since, $\dot{m}_v$ is in kmol/s, thus expression becomes

$$A = \frac{\dot{m}_v M}{\rho V_{operating}}$$

$$\frac{\pi}{4}d^2 = \frac{0.1 \times 45}{2 \times 0.60 \times 0.15}$$

$$d = \sqrt{\frac{100}{3.14}} = 5.64 \text{ m} = \frac{10}{\sqrt{\pi}} \text{ m}$$

3.24 (3)

Given: $X_0 = 0.1$, $X_{N+1} = 0.01$

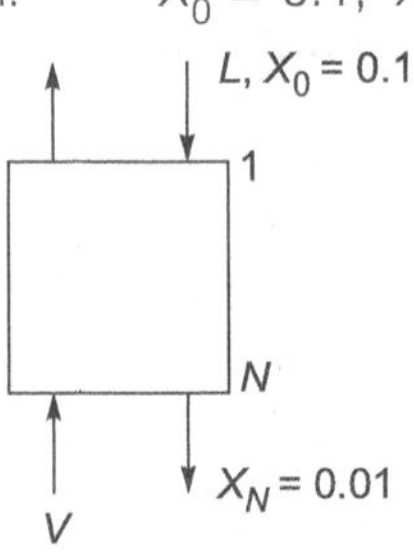

$y_{n+1} = 0$ (pure gas), $L = \frac{V}{2}$, $y_e = x_e$

Now, for stripping, the number of stages is given as

$$N_P = \frac{\ln\left[\left(\frac{X_0 - \frac{Y_{N+1}}{m}}{X_N - \frac{Y_{N+1}}{m}}\right)\left(1 - \frac{1}{s}\right) + \frac{1}{s}\right]}{\ln s}$$

where, S = Stripping factor = 2

Putting values, we get

$$N_P = \frac{\ln\left[\left(\frac{0.1}{0.01}\right)(1 - 0.5) + 0.5\right]}{\ln 2}$$

$$N_P = 2.46 \cong 3 \text{ stage}$$

3.25 (b)

- Packed towers are suitable for foaming because of less bubbling.
- Tray tower is more suitable for cleaning.
- Gas flow rate in flooding region is higher than loading region.
- Flooding occurs in counter-current contact operation.

3.26 (b)

The counter current absorption column along with the data is given as below:

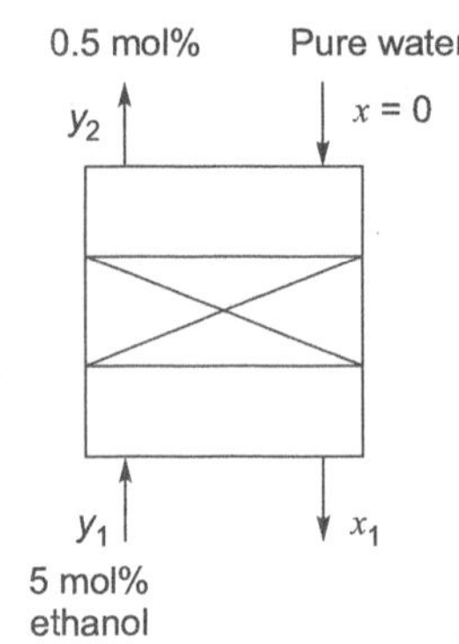

$$5 \text{ mol\%} = \frac{0.05}{0.95} = 0.0526 \text{ mole ratio}$$

Now from component balance.

$$G_s(0.0526 - 0.005) = L_s(x_1 - 0) \quad ...(i)$$

For L_{smin} y_1 and x_1 are in equilibrium and equilibrium relation is given as

$$y = 2x$$

$$G_s \times (0.0526 - 0.005) = L_{smin}\left(\frac{0.0526}{2}\right)$$

$$L_{smin} = 1.8098\, G_s$$

$$L_s = 2 \times L_{smin} \text{ (given)}$$

Now from equation (i)

$$G_s(0.0526 - 0.005) = 1.8098 \times 2 \times G_s \times (x_1 - 0)$$

$$x_1 = 0.013$$

$$\text{mole fraction} = \frac{0.013}{1 + 0.013} = 0.012$$

4 Distillation

4.1 Assume that benzene is insoluble in water. The normal boiling points of benzene and water are 80.1°C and 100°C, respectively. At a pressure of 1-atm, the boiling point of a mixture of benzene and water is

(a) 80.1 °C
(b) less than 80.1°C
(c) 100°C
(d) greater than 80.1°C but less than 100°C

[2000 : 1 M]

4.2 For the nth tray (counted from the bottom of a distillation column), the Murphree tray efficiency is given by

(a) $\dfrac{y_{n+1} - y_n}{y_n^* - y_{n-1}}$ (b) $\dfrac{y_n - y_{n-1}}{y_n^* - y_{n-1}}$

(c) $\dfrac{y_{n-1} - y_n}{y_{n+1} - y_n}$ (d) $\dfrac{y_n^* - y_{n-1}}{y_n^* - y_{n+1}}$

[2000 : 1 M]

4.3 The McCabe ΔL law states that the

(a) molar heats of vaporization of components are nearly equal.
(b) linear crystal growth rate depends on the degree of super saturation.
(c) linear crystal growth rate does not depend on the crystal size.
(d) linear crystal growth rate depends on the crystal size.

[2000 : 1 M]

4.4 A packed gas liquid contactor employs ceramic Intalox saddles (38 mm). The gas flow rate, G = 1.5 kg/s and gas density, ρ_G = 1.5 kg/m^3. The liquid flow rate L = 30 kg/s and liquid density ρ_L = 1000 kg/m^3. The column is to be designed for a pressure drop of 42 mm H_2O. Norton's correlation for this pressure drop is

$$K = \frac{13.1 G_W^2 F_P \left(\dfrac{\mu_L}{\rho_L}\right)^{0.1}}{\rho_G(\rho_L - \rho_G)}$$

where K = flow coefficient, μ_L, liquid viscosity = 10^{-3} Ns/m^2, G_w = gas flow rate per unit cross sectional area, kg/(m^{2}s), F_p = packing factor = 170 m^{-1}. K is given by the equation: $K = 0.62 - 0.5\,(m - 0.5)$ for $0.5 \le m \le 1.0$ where

$$m = \frac{L}{G}\sqrt{\frac{\rho_g}{\rho_L}}$$

Calculate the diameter of the contactor.

[2001 : 2 M]

4.5 According to the Fenske equation, what will be the minimum number of plates required in a distillation column to separate an equimolar binary mixture of components A and B into an overhead fraction containing 99 mole % A and a bottoms fraction containing 98 mole % B? [Assume that the relative volatility (α_{AB} = 2) does not change appreciably in the column]

(a) 5 (b) 9
(c) 12 (d) 28

[2002 : 2 M]

4.6 Minimum reflux ratio in a distillation column results in

(a) optimum number of trays
(b) minimum reboiler size
(c) maximum condenser size
(d) minimum number of trays

[2003 : 1 M]

4.7 In distillation column sizing calculations by short cut methods, match the following:

A. Underwood's equation
B. Fenske's equation
C. Gilliland's equation
D. Vapour velocity at flooding

1. Number of real trays
2. Column diameter
3. Minimum number of ideal trays
4. Actual number of ideal trays
5. Minimum reflux ratio
6. Tray efficiency

Codes:

	A	B	C	D
(a)	1	3	4	6
(b)	2	5	1	3
(c)	5	3	6	2
(d)	5	3	4	2

[2003 : 2 M]

4.8 A distillation column separates 10,000 kg/h of a benzene-toluene mixture as shown in the figure below. In the figure x_F, x_D, and x_w represent the weight fraction of benzene in the feed, distillate, and residue, respectively.

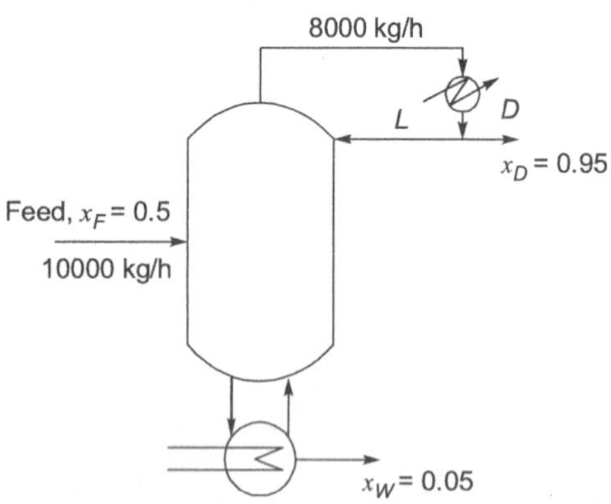

The reflux ratio is

(a) 0.5 (b) 0.6
(c) 1.0 (d) 2.0

[2004 : 1 M]

Common Data for Questions (4.9 & 4.10) :

The boiling points for pure water and pure toluene are 100°C and 110.6°C, respectively. Toluene and water are completely immiscible in each other. A well agitated equimolar mixture of toluene and water is prepared.

4.9 The temperature at which the above mixture will exert a pressure of one standard atm is

(a) less than 100°C
(b) 100°C
(c) between 100 and 110°C
(d) 110.6°C

[2004 : 2 M]

4.10 At a total pressure of one standard atm exerted by the vapours of water and toluene, the mole fraction of water x_w in the vapour phase satisfies

(a) $0 < x_w < 0.5$ (b) $x_w = 0.5$
(c) $0.5 < x_w < 1.0$ (d) $x_w = 1.0$

[2004 : 2 M]

4.11 In a distillation operation, what is the effect of the temperature of the reflux stream (given below) on the condenser and reboiler loads?

Reflux conditions:

(i) reflux stream is completely liquid and is at its bubble point
(ii) reflux stream is below its bubble point

(a) condenser and reboiler loads are the same in both the cases
(b) reboiler load is the same in both the cases but condenser load is higher in case (ii)
(c) condenser load is the same in both the cases but reboiler load is higher in case (ii)
(d) both condenser and reboiler loads are higher in case (ii) as compared to case (i)

[2004 : 2 M]

4.12 A distillation column with *N* plates is being operated under normal conditions. At some point in time, the operation is shifted to total reflux condition (i.e., no product and residue are being withdrawn and feed to the column is stopped). At the new steady state,

(a) composition of vapours and that of liquid do not vary throughout the column
(b) reboiler load and condenser load are minimum
(c) the top and bottom compositions are unchanged with and without total reflux
(d) the top and bottom compositions correspond to the maximum enrichment achievable

[2004 : 2 M]

4.13 An aqueous solution of methanol is to be distilled in a tray column. High pressure steam is available as a source of heat. For a given reflux ratio and overhead composition two options are being explored :

(i) a reboiler is used, and
(ii) no reboiler is used but steam is fed directly to the bottom of the column.

As compared to option (i), in option (ii)

(a) less number of trays are required
(b) composition of the residue remains unchanged
(c) more number of trays are required but the residue composition remains unchanged
(d) more number of trays are required and the residue composition is more dilute in methanol

[2004 : 2 M]

4.14 A distillation column at a pilot plant is sealed up by 3 times for industrial use at steady state. After scaling up

(a) the number of theoretical trays increases by 3 times
(b) the minimum reflux ratio is increased by three times
(c) the feed flow rate and product flow rates are increased by three times
(d) the feed composition and product compositions are increased by three times

[2005 : 1 M]

4.15 In a tray column, separating a binary mixture, with non-ideal stages, ONE of the following statements is TRUE.

(a) Point efficiency can exceed 100%
(b) Murphree efficiency cannot exceed 100%
(c) Murphree efficiency can exceed 100%
(d) Both Murphree and point efficiencies can exceed 100%

[2005 : 1 M]

4.16 For a two-phase feed, where 80% of the feed is vaporized under column conditions, the feed line slope in the McCabe-Thiele method for distillation column design, is

(a) $-\frac{1}{4}$ (b) $\frac{1}{4}$
(c) 4 (d) –4

[2005 : 2 M]

4.17 A liquid mixture of benzene and toluene is in equilibrium with its vapour at 101 kPa and 373 K. The vapor pressures of benzene and toluene at 373 K respectively are 156 and 63 kPa respectively. Assuming that the system obeys Raoult's law, the mole fraction of benzene in the liquid phase is

(a) 0.65
(b) 0.41
(c) 0.065
(d) 0.04

[2005 : 2 M]

4.18 A separation column for vapour-liquid contact processes 200 kmol/hr of vapour. The flooding velocity is 3 m/s. If the column operates at 85% of flooding velocity and the downcomer area is 10% of the total cross sectional area, what is the diameter of the column.? Average density of vapour = 2 kg/m^3 and its molecular weight = 44.

(a) 0.82 m (b) 0.72 m
(c) 0.78 m (d) 1 m

[2005 : 2 M]

4.19 100 moles of a binary mixture F containing 60 mol% A (more volatile) and 40 mol% B is treated in a batch distillation still. After 1 hour, 70 moles of the distillate D is collected leaving behind the residue W. Relative volatility α is 2. The governing equation is

$$\log\frac{Fx_F}{Wx_w} = \alpha\log\frac{F(1-x_F)}{W(1-x_w)}$$

The average mole fraction of A in the distillate is

(a) 0.43 (b) 0.61
(c) 0.69 (d) 0.73

[2006 : 2 M]

4.20 A saturated vapor is fed to a distillation column at 180 kmol/hr. Both the rectifying and stripping sections of the column operate at 60% of their respective flooding velocities. The flooding velocity of the rectifying section is twice that of the stripping section. The assumptions of constant molar overflow and constant molar vapor density throughout the column are valid. If the boil-up rate is 60 kmol/hr, then the relationship between the diameters of the rectifying section (d_r) and the stripping section (d_s) is

(a) $d_r = \sqrt{2}d_s$ (b) $d_r = \sqrt{3}d_s$
(c) $d_r = 2d_s$ (d) $d_r = 3d_s$

[2006 : 2 M]

Common Data for Questions (4.21, 4.22 & 4.23) :

A binary distillation column separates 100 mol/hr of a feed mixture into distillate D and residue W. The McCabe-Thiele diagram for this process is given below. The relative volatility for the binary system is constant at 2.4.

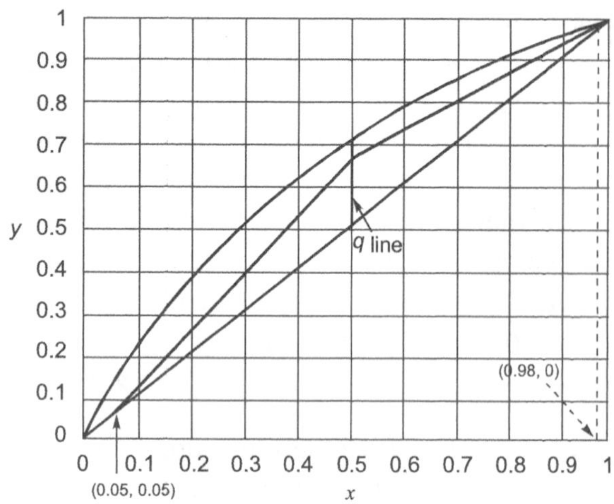

4.21 The distillate and residue flow rates (in mol/hr) are
(a) $D = 48.4$, $W = 51.6$
(b) $D = 51.6$, $W = 48.4$
(c) $D = 54.7$, $W = 45.3$
(d) $D = 45.3$, $W = 54.7$

[2006 : 2 M]

4.22 The ratio of liquid to vapor molar flow rates in the rectifying section is
(a) 0.64 (b) 1.00
(c) 1.55 (d) 1.80

[2006 : 2 M]

4.23 The minimum number of theoretical stages (inclusive of reboiler) for this process is
(a) 5.2 (b) 6.1
(c) 7.8 (d) infinite

[2006 : 2 M]

4.24 In a distillation operation, it is desired to have a very high purity bottom product. Initially, a kettle-type reboiler is used at the bottom of the column and the following analytical equation is used to obtain the equilibrium trays in the exhausting section of the column

$$N_p - m + 1 = \frac{\log\left[\dfrac{x_m - x_w/\alpha}{x_w - x_w/\alpha}(1-\bar{A}) + \bar{A}\right]}{\log\left(\dfrac{1}{\bar{A}}\right)}$$

where x_m is the composition of the liquid leaving tray m. Tray m is the last equilibrium tray obtained by a McCabe Thiele graph of the exhausting section. If the kettle-type reboiler is replaced by a thermo-syphon reboiler, the analytical equation, for the exhausting section will be

(a) $N_p - m + 1 = \dfrac{\log\left[\dfrac{x_m - x_w/\alpha}{x_w - x_w/\alpha}(1-\bar{A}) + \bar{A}\right]}{\log\left(\dfrac{1}{\bar{A}}\right)}$

(b) $N_p + 1 = \dfrac{\log\left[\dfrac{x_m - x_w/\alpha}{x_w - x_w/\alpha}(1-\bar{A}) + \bar{A}\right]}{\log\left(\dfrac{1}{\bar{A}}\right)}$

(c) $N_p - m = \dfrac{\log\left[\dfrac{x_m - x_w/\alpha}{x_w - x_w/\alpha}(1-\bar{A}) + \bar{A}\right]}{\log\left(\dfrac{1}{\bar{A}}\right)}$

(d) $N_p - m + 2 = \dfrac{\log\left[\dfrac{x_m - x_w/\alpha}{x_w - x_w/\alpha}(1-\bar{A}) + \bar{A}\right]}{\log\left(\dfrac{1}{\bar{A}}\right)}$

[2007 : 2 M]

4.25 In a binary mixture containing components *A* and *B*, the relative volatility of *A* with respect to *B* is 2.5 when mole fractions are used. The molecular weights of *A* and *B* are 78 and 92 respectively. If the compositions are however expressed in mass fractions the relative volatility will then be
(a) 1.18 (b) 2.12
(c) 2.5 (d) 2.95

[2008 : 2 M]

4.26 An ideal flash vaporization is carried out with a binary mixture at constant temperature and pressure. A process upset leads to an increase in the mole fraction of the heavy component in the feed. The flash vessel continues to operate at the previous temperature and pressure and still produces liquid and vapor. After steady state is reestablished,
(a) the amount of vapor produced will increase
(b) the amount of liquid produced will decrease
(c) the new equilibrium compositions of the vapor and liquid products will be different.
(d) the new equilibrium compositions of the vapor and liquid products will remain as they were before the upset occurred. [2008 : 2 M]

4.27 A batch distillation operation is carried out to separate a feed containing 100 moles of a binary mixture of *A* and *B*. The mole fraction of *A* in the feed is 0.7. The distillation progresses until the mole fraction of *A* in the residue decreases to 0.6. The equilibrium curve in this composition range may be linearized to $y^* = 0.7353x + 0.3088$. Here, x and y are the mole fractions of the more volatile component *A* in the liquid and vapor phases respectively. The number of moles of residue is

(a) 73.53 (b) 48.02
(c) 40 (d) 30.24

[2008 : 2 M]

4.28 The feed to a binary distillation column has 40 mol % vapor and 60 mol % liquid. Then, the slope of the *q*-line in the McCabe-Thiele plot is

(a) –1.5 (b) –0.6
(c) 0.6 (d) 1.5

[2009 : 2 M]

4.29 An equi-molar mixture of four hydrocarbons (1, 2, 3, 4) is to be separated into high purity individual components using a sequence of simple distillation columns (one overhead and one bottom stream). Four possible schemes are shown below :

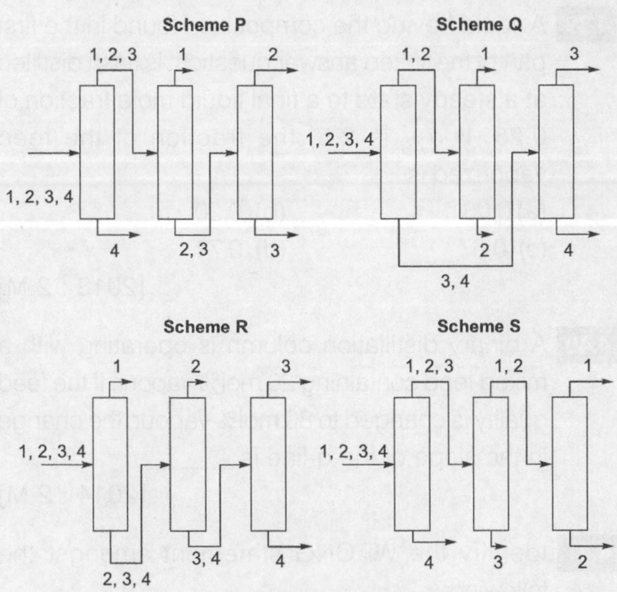

Component	K_i
1	6
2	3
3	2.5
4	1.1

Using the $K_i\left(=\frac{y_i^*}{x_i}\right)$ values given above, the optimal scheme is

(a) *P* (b) *Q*
(c) *R* (d) *S*

[2009 : 2 M]

Common Data for Questions 4.30 and 4.31 :

A flash distillation drum (see fig. below) is used to separate a methanol-water mixture. The mole fraction of methanol in the feed is 0.5, and the feed flow rate is 1000 kmol/hr. The feed is preheated in a heater with heat duty Q_h and is subsequently flashed in the drum. The flash drum can be assumed to be an equilibrium stage, operating adiabatically. The equilibrium relation between the mole fractions of methanol in the vapor and liquid phases is $y = 4x$. The ratio of distillate to feed flow rate is 0.5.

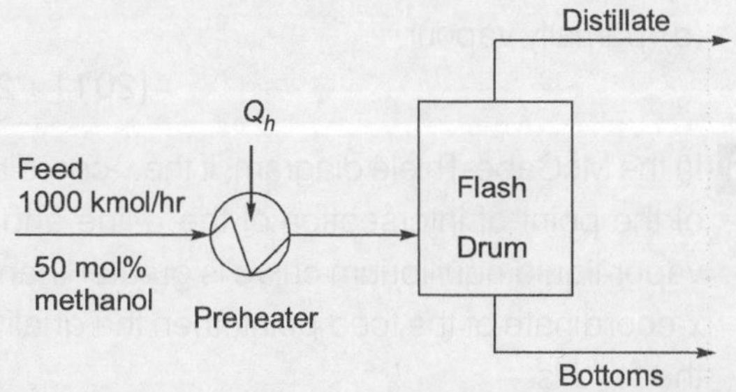

4.30 The mole fraction of methanol in the distillate is

(a) 0.2 (b) 0.7
(c) 0.8 (d) 0.9

[2009 : 2 M]

4.31 If the enthalpy of the distillate with reference to the feed is 3000 kJ/kmol, and the enthalpy of the bottoms with reference to the feed is –1000 kJ/kmol, the heat duty of the preheater (Q_h in kJ/hr) is

(a) -2×10^6 (b) -1×10^6
(c) 1×10^6 (d) 2×10^6

[2009 : 2 M]

4.32 The flooding velocity in a plate column, operating at 1 atm pressure, is 3 m/s. If the column is operated at 2 atm pressure, under otherwise identical conditions, the flooding velocity will be

(a) $\frac{3}{\sqrt{2}}$ (b) $\frac{3}{2}$

(c) 1 (d) $\frac{3}{4}$

[2010 : 2 M]

Common Data for Question 4.33 and 4.34 :

A binary feed mixture containing equimolar quantities of components *S* and *T* is to be distilled in a fractionating tower at atmospheric pressure. The distillate contains 96 mol % *S*. The *q*-line (feed line) intersects the equilibrium line at $x' = 0.46$ and $y' = 0.66$, where x' and y' are mole fractions. Assume that the McCabe-Thiele method is applicable and the relative volatility is constant.

4.33 The Minimum reflux ratio is

(a) 1.6 (b) 1.5
(c) 0.66 (d) 0.6

[2011 : 2 M]

4.34 The feed is

(a) at dew point
(b) at bubble point
(c) superheated vapour
(d) partially vapour

[2011 : 2 M]

4.35 In the McCabe-Thiele diagram, if the x-coordinate of the point of intersection of the q-line and the vapor-liquid equilibrium curve is greater than the x-coordinate of the feed point, then the quality of the feed is

(a) super-heated vapor
(b) liquid below bubble point
(c) saturated vapor
(d) saturated liquid

[2012 : 1 M]

4.36 An equimolar mixture of A and B (A being more volatile) is flash distilled continuously at a feed rate of 100 kmol/h, such that the liquid product contains 40 mol% of A. If the relative volatility is 6, then the vapor product, in kmol/h, is

(a) 10 (b) 20
(c) 25 (d) 45

[2012 : 2 M]

4.37 The vapor-liquid equilibrium curve of a binary mixture A-B, may be approximated by a linear equation over a narrow range of liquid mole fractions $(0.2 < x_A < 0.3)$ as follows

$$y_A^* = 1.325x_A + 0.121$$

Here y_A^* is the mole fraction of A in the vapor. 100 moles of a feed ($x_{A,F} = 0.28$) is batch distilled to a final residue ($x_{A,W} = 0.2$). Using the Rayleigh equation, the number of moles of the residue left behind in the distillation unit, up to 2 digits after the decimal point, is ________ .

[2013 : 2 M]

Statement for Linked Answer Questions 4.38 and 4.39:

The vapor liquid equilibrium relation for an ideal binary system is given by

$$y_A^* = \frac{\alpha_{AB}x_A}{1+(\alpha_{AB}-1)x_A}$$

Here x_A and y_A^* are the mole fractions of species A in the liquid and vapor, respectively. The relative volatility (α_{AB}) is greater than unity.

4.38 The liquid mole fraction x_A at which the maximum difference between the equilibrium vapor mole fraction and liquid mole fraction occurs is

(a) $\dfrac{1}{(1+\sqrt{\alpha_{AB}})}$ (b) $\dfrac{0.75}{(1+\sqrt{\alpha_{AB}})}$

(c) $\dfrac{0.5}{(\sqrt{\alpha_{AB}}+1)}$ (d) $\dfrac{0.75}{(\sqrt{\alpha_{AB}}+1)}$

[2013 : 2 M]

4.39 A liquid having the composition found in the first part of the linked answer question, is flash distilled at a steady state to a final liquid mole fraction of 0.25. If α_{AB} is 2.5, the fraction of the feed vaporized is

(a) 0.08 (b) 0.20
(c) 0.67 (d) 0.74

[2013 : 2 M]

4.40 A binary distillation column is operating with a mixed feed containing 20 mol% vapour. If the feed quality is changed to 80 mol% vapour, the change in the slope of the q-line is ______.

[2014 : 2 M]

4.41 Identify the WRONG statement amongst the following:

(a) Steam distillation is used for mixtures that are immiscible with water.
(b) Vacuum distillation is used for mixtures that are miscible with water.
(c) Steam distillation is used for mixtures that are miscible with water.

(d) Vacuum distillation columns have larger diameters as compared to atmospheric columns for the same throughput.

[2015 : 1 M]

4.42 A binary feed consisting of 25 mol% liquid and 75 mol% vapour is separated in a staged distillation column. The mole fraction of the more volatile component in the distillate product is 0.95. The molar flow rate of distillate is 50% of the feed flow rate and the McCabe-Thiele method can be used to analyze the column. The *q*-line intersects the operating line of the enriching section at (0.35, 0.5) on the x-y diagram. The slope of the stripping section operating line (up to one decimal place) is ________.

[2015 : 2 M]

4.43 A binary distillation column is to be designed using McCabe Thiele method. The distillate contains 90 mol% of the more volatile component. The point of intersection of the *q*-line with the equilibrium curve is (0.5, 0.7). The minimum reflux ratio (rounded off to the first decimal place) for this operation is ________.

[2016 : 2 M]

4.44 The composition of vapour entering a tray in a distillation column is 0.47. The average composition of the vapour leaving the tray is 0.53. The equilibrium composition of the vapour corresponding to the liquid leaving this tray is 0.52. All the compositions are expressed in mole fraction of the more volatile component.
The Murphree efficiency based on the vapour phase, rounded to the nearest integer, is ____%.

[2017 : 1 M]

4.45 The vapor phase composition and relative volatilities (with respect to *n*-propane) on an ideal tray of a distillation column are

Component	Methane	Ethane	*n*-Propane
Mole fraction in vapour	0.12	0.28	0.60
Relative volatility	10	4	1

The mole fraction of *n*-propane in the liquid phase, rounded to 2 decimal places, is ______.

[2017 : 2 M]

4.46 Two packed towers are designed for the same mass velocity of the gas. The first has liquid and gas flow rates of 30 kg/s and 1.2 kg/s, respectively, while the corresponding flow rates in the second tower are 67.5 kg/s and 1.8 kg/s. The ratio of the design diameter of the wider tower to that of the narrower tower is

(a) 2 (b) 1.8
(c) 1.5 (d) 1.225

[2018 : 1 M]

4.47 A binary distillation column is designed by McCabe-Thiele method to get a distillate mole fraction of 0.9. The enriching section operating line has an intercept with *y*-axis at 0.3 mole fraction. The ratio of liquid to vapour molar flow rate in the enriching section is ________ (rounded off to third decimal place).

[2018 : 2 M]

4.48 Three distillation schemes for separating an equimolar, constant relative volatility ABC mixture into nearly pure components are shown. The usual simplifying assumptions such as constant molal overflow, negligible heat loss, ideal trays are valid. All the schemes are designed for minimum total reboiler duty. Given that the relative volatilities are in the ratio $\alpha_A : \alpha_B : \alpha_C \equiv 8:2:1$, the correct option that arranges the optimally-designed schemes in ascending order of total reboiler duty is

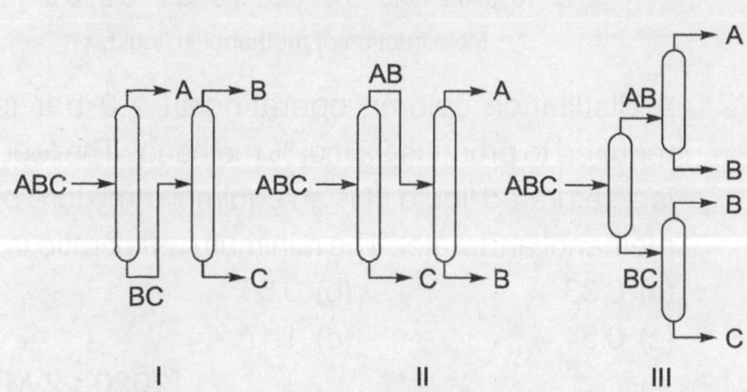

(a) I, II, III (b) III, I, II
(c) II, I, III (d) III, II, I

[2019 : 1 M]

4.49 A binary mixture with component *A* and *B* is to be separated in a distillation column to obtain 95 mol % A as the top product. The binary mixture has a constant relative volatility $\alpha_{AB} = 2$. The column feed is a saturated liquid containing 50 mol % A. Under the usual simplifying assumptions such as constant molal overflow, negligible heat loss, ideal trays, the minimum reflux ratio for this separation is _____ (rounded off to one decimal place).

[2019 : 2 M]

4.50 Consider a batch distillation process for an equimolar mixture of benzene and toluene at atmospheric pressure. The mole fraction of benzene in the distillate collected after 10 minutes is 0.6. The process is further continued for additional 10 minutes. The mole fraction of benzene in the total distillate collected after 20 minutes of operation is

(a) less than 0.6
(b) exactly equal to 0.6
(c) exactly equal to 0.7
(d) greater than 0.7

[2020 : 1 M]

4.51 Consider the equilibrium data for methanol-water system at 1 bar given in the figure below:

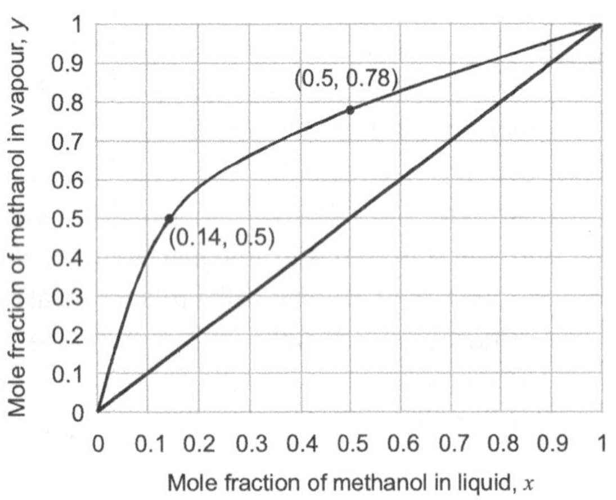

A distillation column operating at 1.0 bar is required to produce 92 mol % methanol. The feed is a saturated liquid. It is an equimolar mixture of methanol and water. The minimum reflux ratio is

(a) 0.33
(b) 0.50
(c) 0.54
(d) 1.17

[2020 : 2 M]

4.52 Equilibrium data for a binary mixture of E and F at two different pressures is shown in the figure.

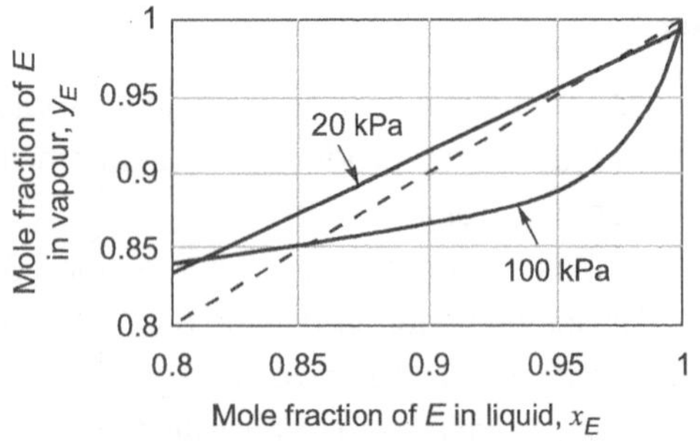

It is desired to process a feed containing 80 mol % E and 20 mol F, and obtain a product with a purity of 99.5 mol % E. A sequence of two distillation columns, one operating at pressure P_1 and another at P_2, is employed for this operations, as shown below.

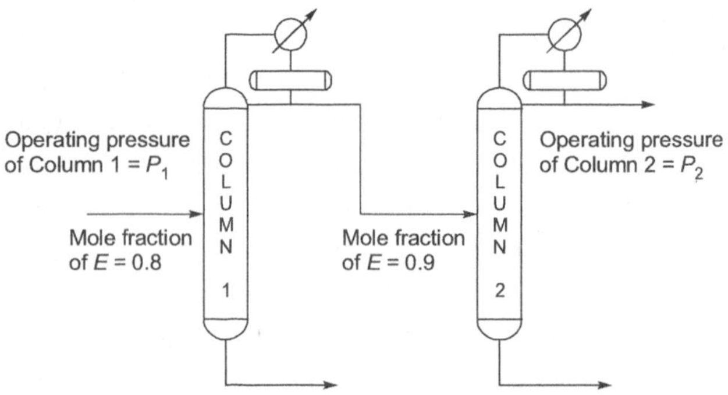

Mole fraction of E in the distillate obtained from column 1 is 0.9. If the column pressures P_1 and P_2 are in kPa, which one of the following is correct?

(a) P_1 = 100, P_2 = 20 and high purity E is recovered from the top of column 2.
(b) P_1 = 100, P_2 = 20 and high purity E is recovered from the bottom of column 2.
(c) P_1 = 20, P_2 = 100 and high purity E is recovered from the top of column 2.
(d) P_1 = 20, P_2 = 100 and high purity E is recovered from the bottom of column 2.

[2020 : 2 M]

4.53 Two ideal cross-current stages operate to extract P from a feed containing P and Q, as shown below.

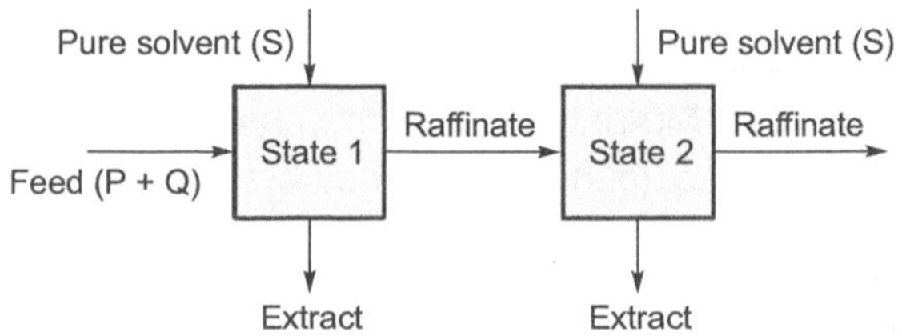

The mass flow rate of P and Q fed to Stage 1 are 1,000 kg h^{-1} and 10,000 kg h^{-1}, respectively. Pure solvent (S) is injected at mass flow rates of 5,000 kg h^{-1} and 15,000 kg h^{-1} to Stages 1 and 2, respectively. The components Q and S are immiscible. The equilibrium relation is given by $Y^* = 1.5X$, where X is the mass of P per unit mass of Q in the raffinate and Y^* is the mass of P per unit mass of S is the extract, which is in equilibrium with the raffinate. The mass flow rate of P (in kg h^{-1}) in the raffinate from Stage 2 is ______ (round off to nearest integer).

[2020 : 2 M]

4.54 A distillation column handling a binary mixture of A and B is operating at total reflux. It has two ideal stages including the reboiler. The mole fraction of the more volatile component in the residue (x_W) is 0.1. The average relative volatility α_{AB} is 4. The mole fraction of A in the distillate (x_D) is ______ (round off to 2 decimal places).

[2021 : 2 M]

4.55 A simple distillation column is designed to separate an ideal binary mixture to specified distillate and bottoms purities at a given column pressure. If RR_{min} is the minimum reflux ratio for this separation, select the statement that is **NOT CORRECT** with regard to the variation in the total annualized cost (TAC) of the column with reflux ratio (RR).

(a) TAC has a minimum with respect to RR
(b) The sharpest rise in TAC occurs as RR approaches RR_{min} from above
(c) The sharpest decrease in TAC occurs as RR approaches RR_{min} from above
(d) TAC increases with RR for RR >> RR_{min}

[2022 : 1 M]

4.56 An equimolar binary mixture is to be separated in a simple tray-distillation column. The feed rate is 50 kmol min^{-1}. The mole fractions of the more volatile component in the top and bottom products are 0.90 and 0.01, respectively. The feed as well as the reflux stream are saturated liquids. On application of the McCabe-Thiele method, the operating line for the stripping section is obtained as

$$y = 1.5\,x - 0.005$$

where y and x are the mole fractions of the more volatile component in the vapor and liquid phases, respectively. The reflux ratio is _______ (rounded off to two decimal places).

[2022 : 2 M]

4.57 Distillation of a non-reactive binary mixture with components A and B is carried out in a batch still as shown in the figure below. The initial charge of the mixture in the still is 1 kmol. The initial and final amounts of A in the still are 0.1 kmol and 0.01 kmol, respectively. Use a constant relative volatility of 4.5. The mole fraction of B remaining in the vessel is ________ (rounded off to three decimal places).

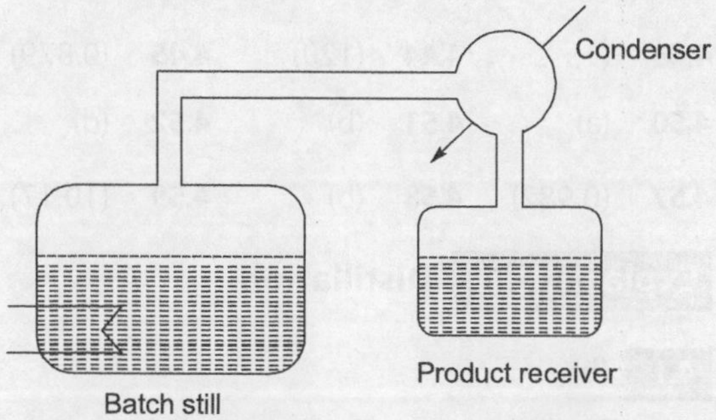

[2023 : 2 M]

4.58 A simple distillation column separates a binary mixture of A and B. The relative volatility of A with respect to B is 2. The steady-state composition of A in the vapour leaving the 1st, 2nd and 3rd trays in the rectifying section are 94, 90 and 85 mol%, respectively. For ideal trays and constant molal overflow, the reflux-to-distillate ratio is

(a) 1.9 (b) 2.7
(c) 1.2 (d) 1.1

[2024 : 2 M]

4.59 Consider a tray-column of diameter 120 cm. Each downcomer has a cross-sectional area of 575 cm^2. For a tray, the percentage column cross-sectional area not available for vapour flow due to the downcomers, rounded off to 1 decimal place, is __________.

[2024 : 2 M]

■■■■

Answers Distillation

4.1	(b)	**4.2**	(b)	**4.3**	(c)	**4.4**	(1.320)	**4.5**	(c)	**4.6**	(b)	**4.7**	(d)
4.8	(b)	**4.9**	(a)	**4.10**	(c)	**4.11**	(d)	**4.12**	(d)	**4.13**	(a)	**4.14**	(c)
4.15	(c)	**4.16**	(a)	**4.17**	(b)	**4.18**	(a)	**4.19**	(c)	**4.20**	(a)	**4.21**	(a)
4.22	(a)	**4.23**	(c)	**4.24**	(c)	**4.25**	(c)	**4.26**	(d)	**4.27**	(b)	**4.28**	(a)
4.29	(d)	**4.30**	(c)	**4.31**	(c)	**4.32**	(a)	**4.33**	(b)	**4.34**	(d)	**4.35**	(b)
4.36	(c)	**4.37**	(66.62)	**4.38**	(a)	**4.39**	(c)	**4.40**	(3.75)	**4.41**	(c)	**4.42**	(1.40)
4.43	(1)	**4.44**	(120)	**4.45**	(0.879)	**4.46**	(d)	**4.47**	(0.67)	**4.48**	(b)	**4.49**	(1.69)
4.50	(a)	**4.51**	(b)	**4.52**	(d)	**4.53**	(175.82)	**4.54**	(0.64)	**4.55**	(c)	**4.56**	(0.632)
4.57	(0.982)	**4.58**	(b)	**4.59**	(10.17)								

Explanations Distillation

4.1 (b)

The bubble point temperature of a mixture is lower than boiling point of either *A* or *B*. This property of an immiscible liquid mixture forms the basis of steam distillation. Therefore, the boiling point of a mixture of benzene and water is less than 80.1°C.

4.2 (b)

Murphree tray efficiency is defined as the ratio of change in actual concentration to the change in predicted by equilibrium.

$$\eta_M = \frac{\text{Change in actual concentration}}{\text{Change predicted by equilibrium condition}}$$

or
$$\eta_M = \frac{y_n - y_{n+1}}{y_n^* - y_{n+1}}$$

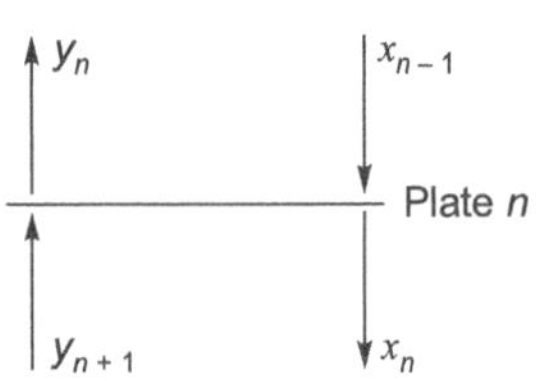

As given, counting is from bottom of a distillation column, thus, Murphree tray efficiency is

$$\eta_M = \frac{y_n - y_{n-1}}{y_n^* - y_{n-1}}$$

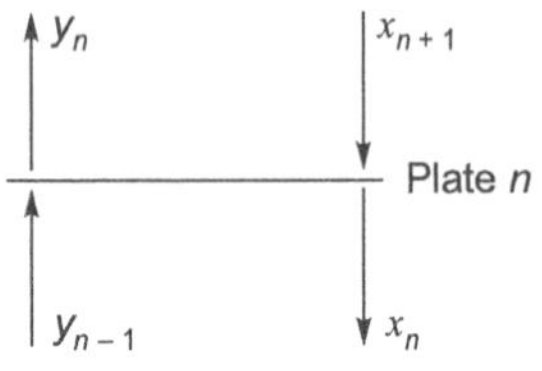

4.3 (c)

McCabe ΔL law, if all crystals in magma grow in a uniform supersaturation field and at the constant temperature and if all crystals grow from birth at a rate determined by the supersaturation, then the crystals are not only invariant but also have the identical growth that is independent of size. This generalization is called the McCabe ΔL law.

4.4 (1.320)

$$m = \frac{L}{G}\sqrt{\frac{\rho_g}{\rho_l}}$$

$$m = \frac{30}{1.5}\sqrt{\frac{1.5}{1000}} = 0.775$$

m is within the range $(0.5 \le m \le 1.0)$

$$k = 0.62 - 0.5(m - 0.5)$$
$$= 0.62 - 0.5(0.775 - 0.5) = 0.4825$$

$$k = \frac{13.1\, G_w^2 F_p \left(\frac{\mu_l}{\rho_l}\right)^{0.1}}{\rho_g(\rho_l - \rho_g)}$$

where, $G_w = \frac{G}{A_c}$ (A_c is cross-sectional area)

$$0.4825 = \frac{13.1\, G_w^2 \times 170 \left(\frac{10^{-3}}{1000}\right)^{0.1}}{1.5(1000 - 1.5)}$$

$$G_w^2 = 1.2929,\ G_w = 1.136$$

$$A_c = \frac{G}{G_w} = \frac{1.5}{1.136} = 1.320\ \text{m}^2$$

4.5 (c)

Given: $\alpha = 2$

Fenske equation,

$$N_{min} = \frac{\log\left[\left(\frac{x_D}{1-x_D}\right)\left(\frac{1-x_w}{x_w}\right)\right]}{\log \alpha_{avg}}$$

$$x_D = 0.99,\ x_w = (1-0.98) = 0.02,\ \alpha_{avg} = 2$$

$$N_{min} = \frac{\log\left(\frac{0.99}{0.01}\times\frac{0.98}{0.02}\right)}{\log 2} = 12.24 = 3 \text{ plates}$$

Minimum number of plates required in distillation column is $(N_{min} - 1) = 13 - 1 = 12$

4.6 (b)

Minimum reflux ratio in distillation column results in minimum reboiler size.

4.7 (d)

Underwood's equation → Minimum reflux ratio

Fenske's equation:

$$N_{min} = \frac{\log\left[\left(\frac{x_D}{1-x_D}\right)\left(\frac{1-x_W}{x_W}\right)\right]}{\log \alpha_{avg}}$$

N_{min} = Minimum number of theoretical plates required

(Gilliland's equation → Actual number of ideal trays).

$$\frac{N - N_{min}}{N+1} = f\left(\frac{R_D - R_{Dm}}{R_D + 1}\right)$$

where R_{Dm} is calculated by underwood equation.

N_{min} is calculated by Fenske's equation.

Vapour velocity at flooding → Column diameter.

4.8 (b)

A distillation column separates 10000 kg/hr of benzene-toluene as shown in figure.

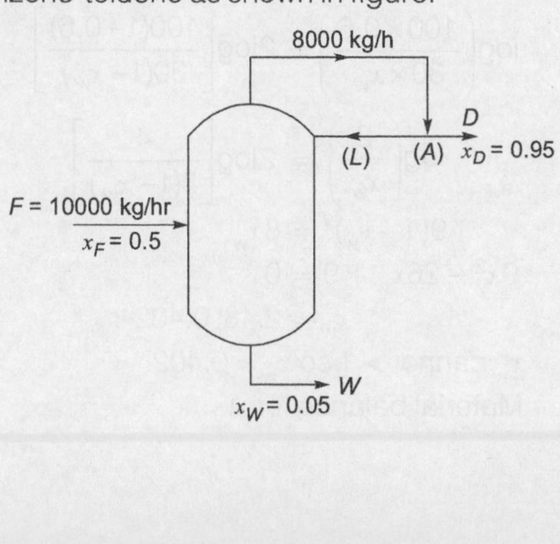

$$F = D + W \text{(from overall balance)}$$

$$10000 = D + W \quad ...(1)$$

$$F \cdot x_F = D \cdot x_D + W \cdot x_W$$

$$10000 \times 0.5 = D \times 0.95 + W \times 0.05 \quad ...(2)$$

From eqn. (1) and (2)

$$D = 5000 \text{ kg/hr}$$

$$W = 5000 \text{ kg/hr}$$

From material balance at point (A)

$$8000 = D + L$$

$$L = 8000 - 5000 = 3000 \text{ kg/hr}$$

$$\text{Reflux ratio} = \frac{L}{D} = \frac{3000}{5000} = 0.6$$

4.9 (a)

The boiling point of a mixture of two immiscible liquids is always less than the boiling point of each individual component such that the boiling point of a mixture of water and toluene will be always less than 100°C.

4.10 (c)

As water has a lower boiling point than toluene at 1 std. atm. pressure (and for all temperatures), water has a higher vapor pressure than toluene. So, the vapor phase will always be richer in water but can never be equal to one if it is a mixture of toluene and water.

$$X_w > 0.5 \text{ but less than } 1.0$$

4.11 (d)

Reflux stream below its bubble point requires more heating means more boiler load and requires more cooling means more condenser load. Therefore, both condenser load and the reboiler load will be higher for case (ii) as compared to case (i).

4.12 (d)

In the distillation column as we increase reflux ratio then overhead product purity is increased. At total reflux condition the top and bottom composition correspond to the maximum enrichment achievable.

4.13 (a)

Reboiler is considered as an extra tray therefore as compared to option (i), in option (ii) less number of trays are required.

4.14 (c)

After scaling 3 times a distillation column the feed flow rate and product flow rates are increased by three times.

4.15 (c)

Murphree efficiency,

$$\eta_m = \frac{y_n - y_{n+1}}{y_n^* - y_{n+1}}$$

where, y_n = Actual concentration of vapor leaving plate n

y_{n+1} = Actual concentration of vapor entering plate n

y_n^* = Concentration of vapor in equilibrium

When local efficiency is high, sometimes the presence of concentration gradients in the liquid gives an average vapor concentration greater than y_n^*, then Murphree efficiency goes higher than 100%.

4.16 (a)

Feed line equation :

$$y = \frac{q}{q-1}x - \frac{x_f}{q-1}$$

$$\text{Slope} = \frac{q}{q-1}$$

where q is defined as the mole fraction of the liquid flow in the stripping section that result from the introduction of each moles of feed.

$$q = 0.2 \text{ (80\% feed is vaporized)}$$

$$\text{Slope} = \frac{0.2}{0.2-1} = -\frac{1}{4}$$

4.17 (b)

$$P = x_A P_A^{sat} + x_B P_B^{sat}$$

$$x_A + x_B = 1$$

$$101 = x_A \times 156 + (1 - x_A) \times 63$$

$$x_A = 0.409$$

4.18 (a)

Volumetric flow rate (m³/s)

$$Q = \frac{\text{Mass flow rate}}{\text{Density}}$$

$$= \frac{\text{Molar flow rate} \times \text{Molecular wt.}}{\text{Density}}$$

$$Q = \frac{\dot{m} \times M_w}{\rho_V}$$

Given that :

Vapour velocity = 85% of flooding velocity

= 0.85 × 3 = 2.55 m/s

Vapor flow rate ($\dot{m}$)

$$= 200 \text{ kmol/h} = \frac{200}{3600}$$

$$= 0.055 \text{ kmol/s}$$

$$\rho_v = 2 \text{ kg/m}^3, M_w = 44 \text{ kg/kmol}$$

Hence, $$Q = \frac{0.055 \times 44}{2} = 1.21 \text{ m}^3\text{/s}$$

Since, $$Q = \text{Area} \times \text{Vapour velocity}$$

$$\text{Area} = \frac{Q}{\text{Vapor velocity}}$$

$$= \frac{1.21}{2.55} = 0.48 \text{ m}^2$$

Given that, downcomer area = 10% of total area

Area available for vapor flow = 90% of plate area

Hence, Plate area = $\frac{0.48}{0.9} = 0.527 \text{ m}^2$

$$\frac{\pi}{4}d^2 = 0.527$$

$$d = 0.82 \text{ m}$$

4.19 (c)

$$\log\left[\frac{Fx_F}{Wx_w}\right] = \alpha \log\left[\frac{F(1-x_F)}{W(1-x_w)}\right] \quad \text{...(i)}$$

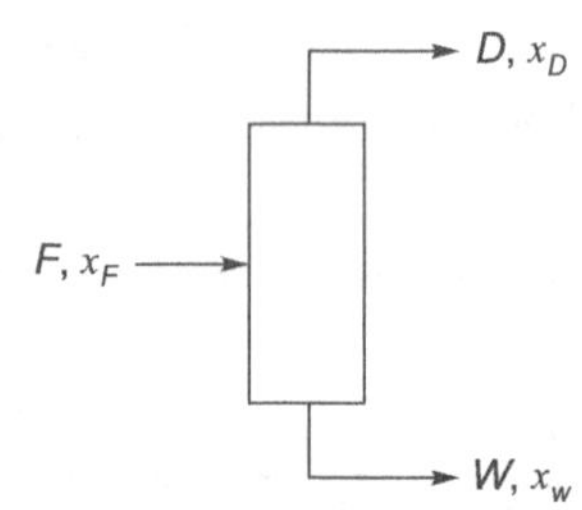

Given : $F = 100$ mol, $X_F = 0.6$

$D = 70$ mol, $\alpha = 2$

Overall material balance

$$F = D + W$$

$$100 = 70 + W$$

$$W = 30 \text{ mol}$$

Put all values in equation (i)

$$\log\left(\frac{100 \times 0.6}{30 \times x_w}\right) = 2\log\left[\frac{100(1-0.6)}{30(1-x_w)}\right]$$

$$\log\left(\frac{2}{x_w}\right) = 2\log\left[\frac{4}{3(1-x_w)}\right]$$

$$9(1 - x_w)^2 = 8x_w$$

$$9x_w^2 - 26x_w + 9 = 0$$

$$x_w = 2.48, 0.402$$

x_w cannot > 1 so $x_w = 0.402$

Material balance for A

$$Fx_F = Dx_D + Wx_w$$
$$100 \times 0.6 = 70x_D + 30 \times 0.4$$
$$x_D \simeq 0.69$$

4.20 (a)

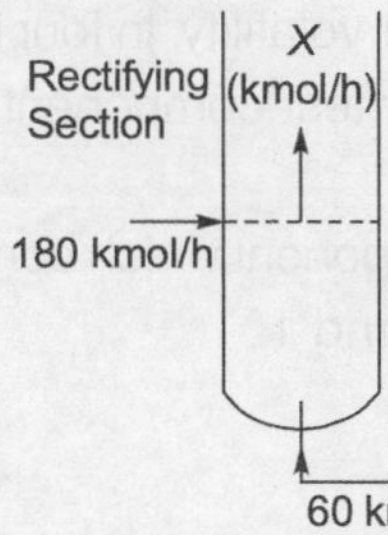

For system,

$$180 + 60 = X$$
$$X = 240 \text{ kmol/h}$$

For stripping section

$$\text{Feeding rate} = \text{Area} \times \text{Velocity} \times \frac{\text{Density}}{\text{Molecular Wt.}}$$

$$60 = \frac{\pi d_s^2}{4} \times u_s \times \frac{\rho}{m} \quad \text{...(i)}$$

For rectifying section

$$240 = \frac{\pi d_r^2}{4} \times u_r \times \frac{\rho}{m} \quad \text{...(ii)}$$

Equation (ii) divided by equation (i)

$$4 = \left(\frac{d_r}{d_s}\right)^2 \times \frac{u_r}{u_s}$$

Given : $u_r = 0.6 \times u_{fr}$, $u_s = 0.6 \times u_{fs}$

$$4 = \left(\frac{d_r}{d_s}\right)^2 \times \frac{u_{fr}}{u_{fs}}$$
$$u_{fr} = 2u_{fs} \text{ (given)}$$
$$4 = \left(\frac{d_r}{d_s}\right)^2 \times 2$$
$$d_r = \sqrt{2} d_s$$

4.21 (a)

$$F = D + W$$
$$100 = D + W$$
$$Fx_F = Dx_D + Wx_W$$

From fig., $x_W = 0.05$, $x_F = 0.5$, $x_D = 0.98$

$$100 \times 0.5 = D \times 9.8 + W \times 0.05$$
$$50 = D \times 9.8 + (100 - D) \times 0.05$$
$$50 = D \times 0.98 + 5 - 0.05D$$
$$45 = 0.93D$$
$$D = 48.38 \text{ mol/h}$$
$$W = 100 - 48.38 = 51.61 \text{ mol/h}$$

4.22 (a)

Slope of upper line = Ratio of liquid to vapor molar Flow rates in rectifying section

$$\frac{L}{V} = \frac{0.98 - 0.68}{0.98 - 0.5} = 0.625$$

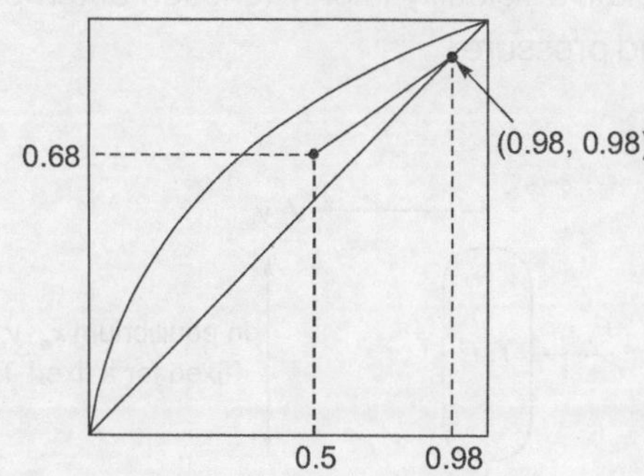

4.23 (c)

Minimum number of theoretical stages

$$N_m = \frac{\log\left[\left(\frac{x_D}{1-x_D}\right)\left(\frac{1-x_W}{x_W}\right)\right]}{\log \alpha_{avg}}$$

Given : $x_D = 0.98$, $x_W = 0.05$, $\alpha_{avg} = 2.4$

$$N_m = \frac{\log\left[\left(\frac{0.98}{1-0.98}\right)\left(\frac{1-0.05}{0.05}\right)\right]}{\log(2.4)} = 7.8$$

4.24 (c)

Pressure drop of boiling fluid side in kettle type reboiler is negligible and high in thermo-syphon reboiler. Kettle type reboiler works as additional equilibrium stage in distillation column.

For thermo-syphon reboiler

$$N_P - m = \frac{\log\left[\frac{x_m - \frac{x_w}{\alpha}}{x_w - \frac{x_w}{\alpha}}(1-\overline{A}) + \overline{A}\right]}{\log\left(\frac{1}{\overline{A}}\right)}$$

4.25 (c)

$$\text{Relative volatility } (\alpha_{AB}) = \frac{\left(\frac{y_A}{x_A}\right)}{\left(\frac{y_B}{x_B}\right)}$$

If composition are expressed in mass fractions

$$\alpha'_{AB} = \left(\frac{y_A \cdot m_A}{x_A \cdot m_A}\right)\left(\frac{x_B \cdot m_B}{y_B \cdot m_B}\right)$$

$$\alpha'_{AB} = \frac{\left(\frac{y_A}{x_A}\right)}{\left(\frac{y_B}{x_B}\right)} = \alpha_{AB}$$

Relative volatility is only function of temperature and pressure.

4.26 (d)

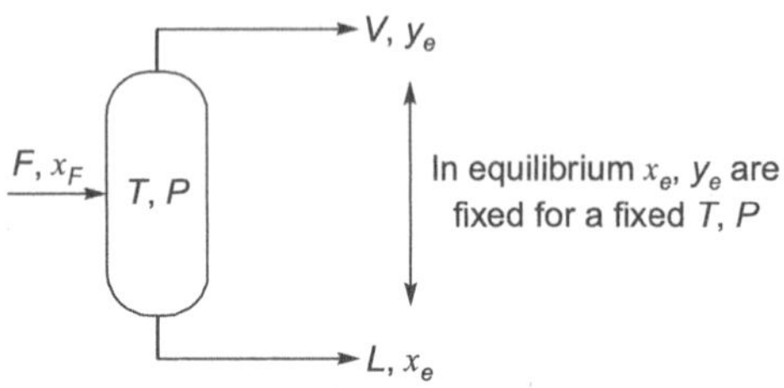

4.27 (b)

For batch distillation

$$\ln\left(\frac{F}{W}\right) = \int_{x_W}^{x_F} \frac{dx}{y^* - x}$$

where, W = Moles of residue
F = Moles of feed
x_W = Final composition
x_F = Initial composition
$y^* = 0.7353x + 0.3088$

$$\ln\left(\frac{F}{W}\right) = \int_{x_W}^{x_F} \frac{dx}{(0.7353x + 0.3088 - x)}$$

F = 100 mole
$x_F = 0.7$, $x_W = 0.6$

$$\ln\left(\frac{100}{W}\right) = \int_{0.6}^{0.7} \frac{dx}{-0.2647x + 0.3088}$$

$$\ln\left(\frac{100}{W}\right) = \frac{1}{-0.2647}\ln\left[\frac{-0.2647 \times 0.7 + 0.3088}{-0.2647 \times 0.6 + 0.3088}\right]$$

$$\ln\left(\frac{100}{W}\right) = 0.7336$$

$$\frac{100}{W} = e^{0.7336} = 2.082$$

$$W = \frac{100}{2.082} = 48.018 \text{ mole}$$

4.28 (a)

Given: $q = 0.6$

$$\text{Slope} = \frac{q}{q-1} = \frac{0.6}{0.6-1} = -\frac{0.6}{0.4} = -1.5$$

4.29 (d)

$$K_i = \frac{y_i^*}{x_i}$$

Therefore, K_i is showing volatility here. Higher K_i means higher volatility. In four hydrocarbon (1, 2, 3, 4), 1 is lightest component and 4 is heaviest component.

Order of component which comes out will be 4, 3, 2 and at the end 1.

4.30 (c)

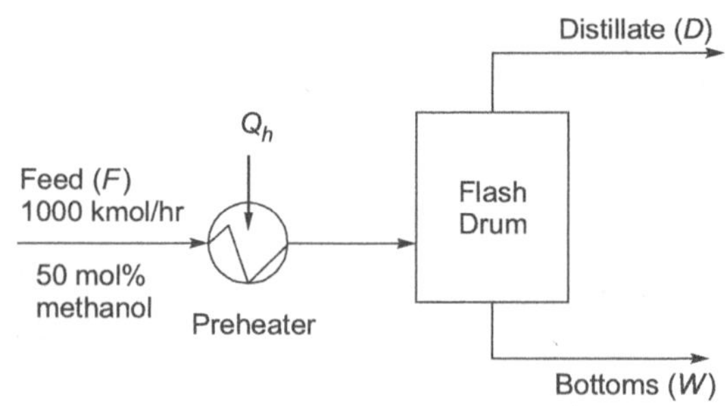

From overall balance

$$F = D + W$$

$$\frac{D}{F} = 0.5 \text{ (given)}$$

and $F = 1000$ kmol/hr (given)

So, $D = 0.5 \times 1000 = 500$ kmol/hr

$W = 1000 - 500 = 500$ kmol/hr

Now, from methanol balance,

$$F \cdot x_F = Dy + Wx$$

Here, $y = 4x$ (given)

So, $1000 \times 0.5 = 500 \times 4x + 500x$

On solving, $x = 0.2$

So, $y = 4x = 4 \times 0.2 = 0.8$

4.31 (c)

From overall energy balance

$$F \cdot H_F + Q_h = D \cdot H_D + W \cdot H_w$$

Enthalpies are given with the reference of feed enthalpy.

So, $Q_H = 500 \times 3000 + 500(-1000)$
$= 10^6$ kJ/hr

4.32 (a)

Flooding velocity is given by

$$u_f = k_l\sqrt{\frac{\rho_l - \rho_v}{\rho_v}}$$

k_l is coefficient obtained from chart.

Assume $\rho_l \gg \rho_v$

$$u_f = k_l\sqrt{\frac{\rho_l}{\rho_v}}$$

where, ρ_l = Density of liquid

ρ_v = Density of vapor

$$\rho_v = \frac{P}{RT}$$

$$u_f \propto \frac{1}{\sqrt{P}}$$

$$\sqrt{\frac{P_1}{P_2}} = \frac{u_{f2}}{u_{f1}} = \sqrt{\frac{1}{2}} = \frac{1}{\sqrt{2}}$$

$$u_{f2} = \frac{3}{\sqrt{2}} \text{ m/s}$$

4.33 (b)

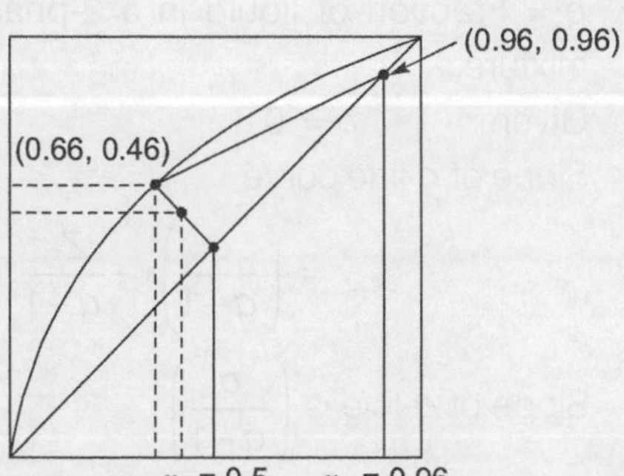

At minimum reflux, the point of contact of the operating and equilibrium lines is at the intersection of feed line with equilibrium curve.

$$\text{Slope} = \frac{0.96 - 0.66}{0.96 - 0.46} = 0.6$$

$$\frac{R_{\min}}{1 + R_{\min}} = 0.6$$

$$R_{\min} = \frac{0.6}{0.4} = 1.5$$

4.34 (d)

$$\text{Slope of } q\text{-line} = \frac{0.66 - 0.5}{0.46 - 0.5} = -4$$

$$\frac{q}{q-1} = -4$$

$$q = \frac{4}{5} = 0.8$$

Feed is partially vapor ($0 < q < 1$)

4.35 (b)

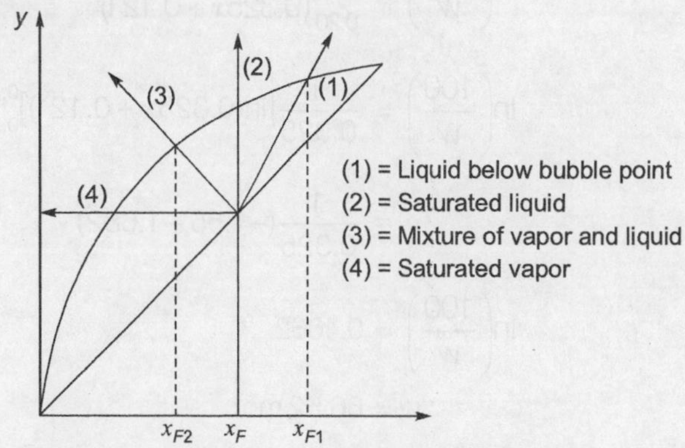

Here $x_{F1} > x_F$. So present case falls under the case 1.

4.36 (c)

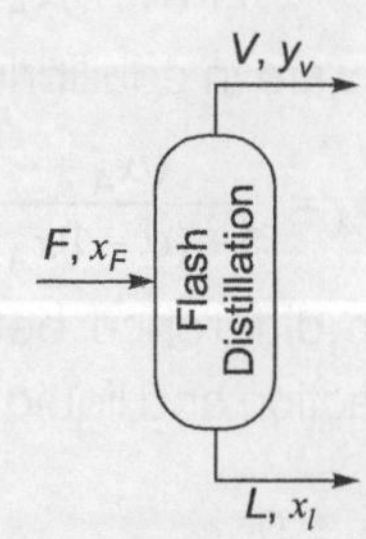

Given : $\alpha = 6$

$$y_V = \frac{\alpha x_l}{1 + (\alpha - 1)x_l}$$

$$y_V = \frac{6 \times 0.4}{1 + 5 \times 0.4} = 0.8$$

$$F = L + V$$

$$Fx_F = Lx_l + Vy_V$$

$$100 \times 0.5 = (100 - V) \times 0.4 + V \times 0.8$$

$$50 = 40 + 0.4\text{ V}$$

$$V = \frac{10}{0.4} = 25 \text{ kmol/h}$$

4.37 (66.62)

Rayleigh equation for batch distillation

$$\ln\left(\frac{F}{w}\right) = \int_{x_w}^{x_f} \frac{dx}{y^* - x}$$

where F = 100 mol

$x_f = 0.28$

$x_w = 0.20$

$y^* = 1.325x + 0.121$

$$\ln\left(\frac{100}{w}\right) = \int_{0.20}^{0.28} \frac{dx}{(1.325x + 0.121) - x}$$

$$\ln\left(\frac{100}{W}\right) = \int_{0.20}^{0.28} \frac{dx}{(0.325x + 0.121)}$$

$$\ln\left(\frac{100}{W}\right) = \frac{1}{0.325}[\ln(0.325x + 0.121)]_{0.20}^{0.28}$$

$$= \frac{1}{0.325}(-1.55 + 1.682)$$

$$\ln\left(\frac{100}{W}\right) = 0.4062$$

$$W = 66.62 \text{ mol}$$

4.38 (a)

Binary VLE

$$y_A^* = \frac{\alpha x_A}{1 + (\alpha - 1)x_A}$$

where y^* and x are in equilibrium.

$$y_A^* - x_A = \frac{\alpha x_A}{1 + (\alpha - 1)x_A} - x_A$$

For maximum difference between equilibrium vapor mole fraction and liquid mole fraction

$$\frac{d}{dx_A}(y_A^* - x_A) = 0$$

$$\frac{\alpha[1 + (\alpha - 1)x_A] - (\alpha - 1)\alpha x_A}{(1 + (\alpha - 1)x_A)^2} - 1 = 0$$

$$\alpha + \alpha(\alpha - 1)x_A - \alpha^2 x_A + \alpha x_A - [1 + (\alpha - 1)x_A]^2 = 0$$

$$\alpha + \alpha^2 x_A - \alpha x_A - \alpha^2 x_A + \alpha x_A - [1 + (\alpha - 1)x_A]^2 = 0$$

$$\sqrt{\alpha} = 1 + (\alpha - 1)x_A$$

$$\frac{\sqrt{\alpha} - 1}{(\alpha - 1)} = x_A$$

$$x_A = \frac{\sqrt{\alpha} - 1}{(\sqrt{\alpha} - 1)(\sqrt{\alpha} + 1)} = \frac{1}{\sqrt{\alpha} + 1}$$

where $\alpha = \alpha_{AB}$

So, $$x_A = \frac{1}{\sqrt{\alpha_{AB}} + 1}$$

4.39 (c)

$$x_F = \frac{1}{\sqrt{\alpha_{AB}} + 1}$$

$$= \frac{1}{\sqrt{2.5} + 1} = \frac{1}{2.582}$$

$$x_F = 0.387$$

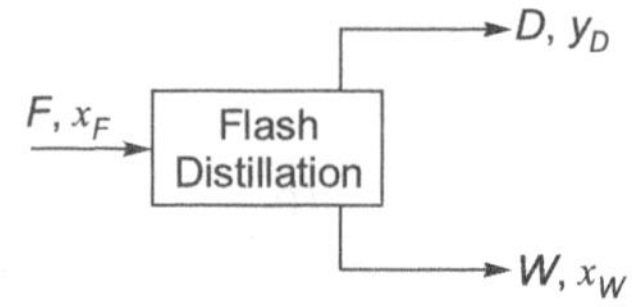

where $x_W = 0.25$

So, $$y_o = \frac{\alpha x_W}{1 + (\alpha - 1)x_W} = \frac{25 \times 0.25}{1 + 1.5 \times 0.25}$$

$$y_o = 0.4545$$

Component 'A' balance

$$Fx_F = Dy_D + Wx_W$$

$$Fx_F = Dy_D + (F - D)x_W$$

$$F \times 0.387 = D \times 0.4545 + (F - D) \times 0.25$$

$$F \times 0.1374 = D \times 0.2045$$

$$\frac{D}{F} = \frac{0.1374}{0.2045} = 0.6718$$

4.40 (3.75)

q = Fraction of liquid in a 2-phase vap-liquid mixture.

Given: $q = 0.8$

Slope of q-line curve

$$y = \left(\frac{q}{q-1}\right)x - \frac{Z_F}{q-1}$$

Slope of q-line is $\left(\frac{q}{q-1}\right)$

$$\text{Initial slope} = \frac{0.8}{0.8 - 1} = \frac{0.8}{-0.2} = -4$$

Now, feed quality is changed to 80 mole % vapour

$$q = 0.2$$

$$\text{Slope} = \frac{0.2}{0.2 - 1} = -0.25$$

Change in slope = Final slope – Initial slope

= (–0.25) – (–4)

= 4 – 0.25 = 3.75

4.41 (c)

In steam distillation, steam must be inert for the mixtures. If steam is miscible with mixtures then vacuum distillation is used.

4.42 (1.40)

Given: $q = 0.25, x_D = 0.95$

Now by equation of rectifying section

$$y_{n+1} = \frac{R}{R+1}x_n + \frac{x_D}{R+1}$$

Since this line passes through (0.35, 0.50), so we get

$$0.5 = \frac{R}{R+1} \times 0.35 + \frac{0.95}{R+1}$$

$$R = 3$$

So, reflux ratio,

$$R = \frac{L}{D} = 3$$

$$L = 3D$$

Molal flow rate of distillate is 50% of the feed.

So, $D = 0.5F$

$$F = D + W$$

$$W = D$$

Now for stripping section

$$y_{m+1} = \frac{L_m}{V_{m+1}} x_m - \frac{W_{x_w}}{V_{m+1}}$$

In stripping, $L_m = L + qF$

$$L_m = 3D + 0.25 \times 2D$$

$$L_m = 3.5D$$

and $V_{m+1} = L_m - W = 3D - 0.5D$

$$V_{m+1} = 2.5D$$

Now, slope of operating line of stripping section

$$= \frac{L_m}{V_{m+1}}$$

$$\frac{L_m}{V_{m+1}} = \frac{3.5D}{2.5D} = 1.4$$

4.43 (1)

$x_D = 0.9$, intersection point of q-line with the equilibrium curve is (0.5, 0.7).

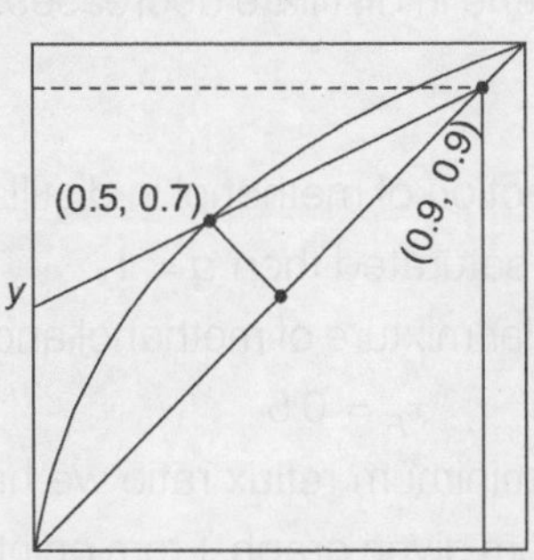

For minimum reflux ratio, slope of enriching section

$$\text{is equal} = \frac{R_m}{1+R_m} = \frac{0.9-0.7}{0.9-0.5}$$

$$1 + R_m = 2R_m$$

$$R_m = 1$$

4.44 (120)

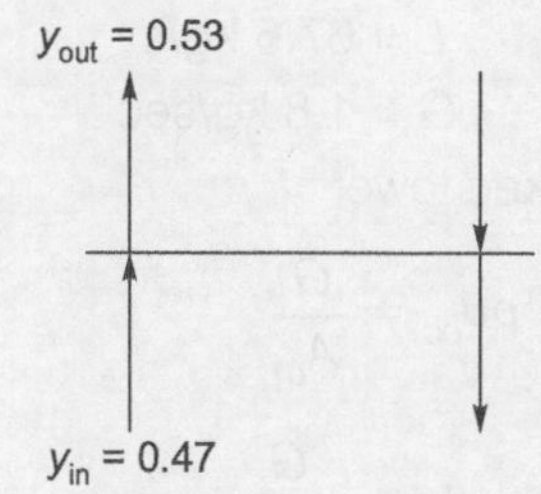

Given : $y^* = 0.52$

Murphree efficiency based on vapour phase

$$= \frac{y_{out} - y_{in}}{y^*_{out} - y_{in}}$$

$$= \frac{0.53-0.47}{0.52-0.47} = 1.2$$

Percentage murphree efficiency

$$= 1.2 \times 100$$

$$= 120\%$$

4.45 (0.879)

Vapour phase composition and relative volatilities on an ideal tray of a distillation column are :

Component	Methane	Ethane	n-Propane
Mole fraction in vapor	0.12	0.28	0.60
Relative volatility	10	4	1

Mole fraction in liquid phase,

$$x_i = \frac{\frac{y_i}{\alpha_i}}{\sum \frac{y_i}{\alpha_i}}$$

$$x(n\text{-propane}) = \frac{y_{n\text{-propane}} / \alpha_{n\text{-propane}}}{\Sigma\left(\frac{y_i}{\alpha_i}\right)}$$

$$= \frac{\frac{0.6}{1}}{\frac{0.6}{1} + \frac{0.28}{4} + \frac{0.12}{10}}$$

$$= \frac{0.6}{0.682} = 0.879$$

4.46 (d)

Mass velocity of gas is same.

For first tower,

$$L = 30 \text{ kg/s}$$

$$G = 1.2 \text{ kg/sec}$$

For second tower,

$$L = 67.5 \text{ kg/s}$$
$$G = 1.8 \text{ kg/sec}$$

For a packed tower

$$\rho u_{op} = \frac{G}{A_c}$$

$$A_c = \frac{G}{\rho u_{op}}$$

where mass velocity = $\rho \mu_{op}$

$$D_{c_1}^2 \propto G_1$$

$$\left(\frac{D_{c_2}}{D_{c_1}}\right)^2 = \frac{G_2}{G_1} = \frac{1.8}{1.2}$$

$$\frac{D_{c_2}}{D_{c_1}} = 1.224$$

4.47 (0.67)

$$x_D = 0.9$$

Enriching section operating line intercept

$$\frac{x_D}{R+1} = 0.3 \text{ (given)}$$

$\left(\frac{L}{V}\right)$ in enriching section = ?

We know, $\frac{L}{V} = \frac{R}{R+1}$

$$\frac{x_D}{R+1} = 0.3 = \frac{0.9}{R+1}$$
$$R + 1 = 3$$
$$R = 2$$
$$\frac{L}{V} = \frac{2}{3} = 0.67$$

4.48 (b)

Remove final products one by one as overhead distillate. Select the sequence that favours near equimolar amounts of overhead and bottoms in each column (optimal energy cost). On the basis of several heuristics, the sequence of distillate column in terms of energy requirement is III < I < II.

4.49 (1.69)

$x_D = 0.95$, $\alpha_{AB} = 2$ and $x_F = 0.5$

(Feed is saturated liquid, $q = 1$)

Assumption: Constant molal overflow negligible heat, loss, ideal trays.

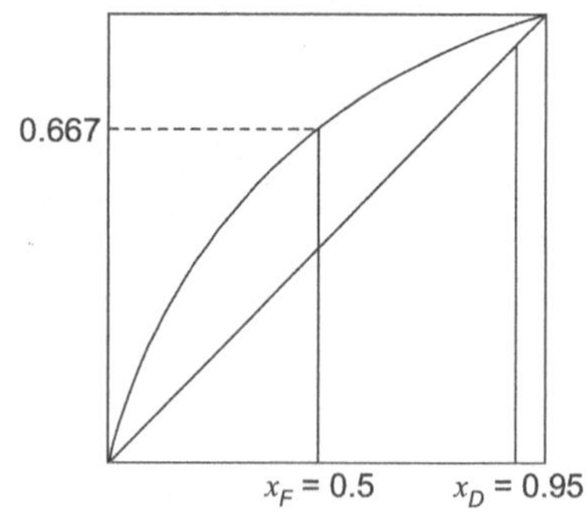

$$y = \frac{\alpha x_F}{1+(\alpha-1)x_F}$$

$$y = \frac{2\times 0.5}{1+0.5} = 0.667$$

$$\frac{R_m}{R_{m+1}} = \frac{0.95-0.667}{0.95-0.5} = 0.628$$

$$1.6R_m = R_m + 1$$
$$R_m = 1.69$$

4.50 (a)

At $t \geq 0$ when batch distillation is started, mole fraction of benzene (y_B) in the vapor at equilibrium is highest (given by benzene-toluene VLE relationship).

As time progresses, mole fraction of benzene in the liquid (in batch distillation still) decreases and correspondingly mole fraction of toluene in the liquid increases and then mole fraction of benzene in the vapor (correspondingly in the distillate collected) decreases, while mole fraction of toluene increases so as time passes mole fraction of benzene in distillate decreases.

4.51 (b)

Mole fraction of methanol in distillate is 0.92.

Feed is saturated then $q = 1$.

Equimolar mixture of methanol and water then

$$x_F = 0.5$$

To find minimum reflux ratio we have to find out slope from given graph. From point (0.92, 0.92) to (0.5, 0.78). We know

$$\text{Slope} = \frac{R_{min}}{1+R_{min}}$$

$$\frac{0.92-0.78}{0.92-0.5} = \frac{R_m}{1+R_m}$$

$$\frac{1}{3} = \frac{R_m}{R_m+1}$$

$$R_m + 1 = 3R_m$$

$$1 = 2R_m$$

$$R_{min} = 0.5$$

4.52 (d)

It is desired to process a feed containing 80 mol% E and 20 mol% F.

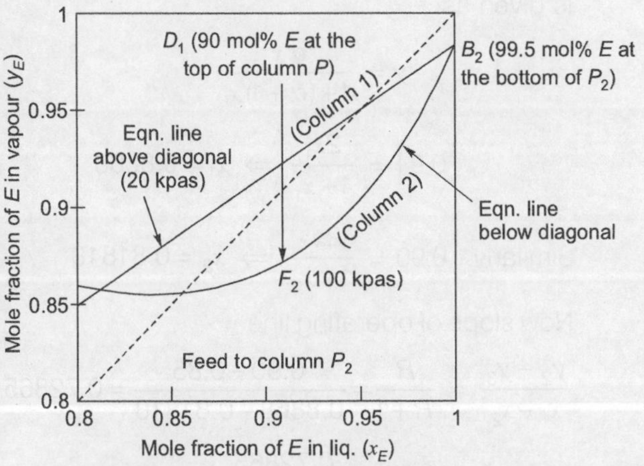

From graph we can say that P_1 = 20, P_2 = 100 and high purity *E* is recovered from the bottom of column 2.

4.53 (175.82)

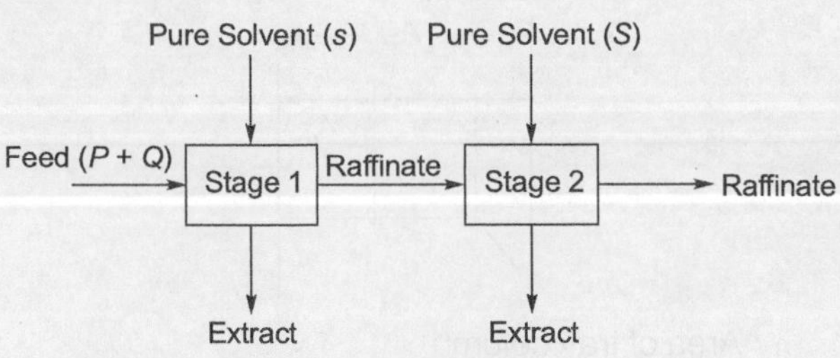

Given : F_P = 1000 kg/hr

F_Q = 10000 kg/hr

and F_{S1} = 5000 kg/hr

F_{S2} = 15000 kg/hr

Q and *S* are immiscible.

Y^* = 1.5*X* where *X* is mass of *P* per unit mass of *Q* in Raffinate phase and Y^* is mass of *P* per unit mass of *S* in extract phase.

Mass balance at stage 1 on solute *P*

$$1000 + 0 = 5000Y1 + 10000X_1$$

$$1000 = 5000 \times 1.5X_1 + 10000X_1$$

$$X_1 = 0.0571$$

Mass balance at stage 2 on solute *P*

$$571.42 + 0 = 10000X_2 + 15000Y_2$$

$$571.42 = 32500X_2$$

$$X_2 = 0.01758$$

Mass flow rate of *P* (in kg h^{-1}) in Raffinate from stage 2 is

$$= F_Q X_2 = 175.82$$

4.54 (0.64)

At total reflux, we require minimum number of ideal stages. From Fenske equation

$$N_m + 1 = \frac{\log\left[\left(\frac{x_D}{1-x_D}\right)\left(\frac{1-x_w}{x_w}\right)\right]}{\log(\alpha_{AB})}$$

Given : $N_m + 1 = 2$ and $x_w = 0.1$

$$2 = \frac{\log\left(\frac{x_D}{1-x_D} \times \frac{0.9}{0.1}\right)}{\log(4)}$$

$$1.78 = \frac{x_D}{1-x_D}$$

$$1 - x_D = 0.5625x_D$$

$$x_D = 0.64$$

4.55 (c)

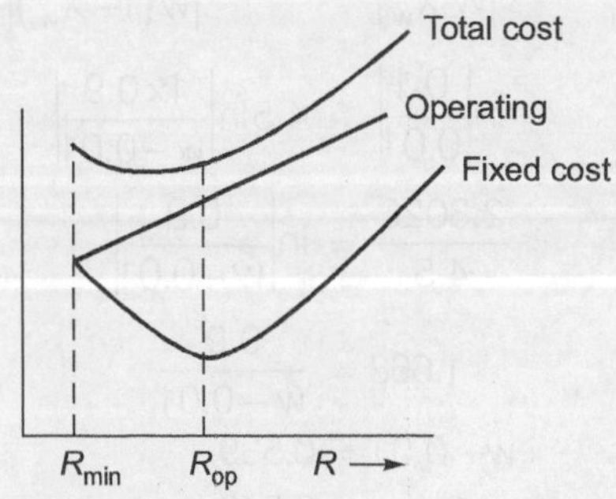

Not correct option (c).

Because sharpest increase in TAC occurs as RR approaches RR_{min} from above.

4.56 (0.632)

Equimolar binary mixture, *x* = 0.5

Feed flow rate = 50 kmol/min

$x_D = 0.9$, $x_w = 0.01$

Stripping line equation:

$$y = 1.5x - 0.005$$

and saturated feed (*q*) = 1

$$\frac{\bar{L}}{\bar{V}} = 1.5$$

$$Fx_F = Dx_D + Wx_W$$
$$50 \times 0.5 = D \times 0.9 + w \times 0.01$$
$$25 = D \times 0.9 + (50 - D) \times 0.01$$
$$25 - 0.5 = D(0.9 - 0.01)$$
$$D = 27.52,\ w = 22.47$$
$$\frac{Wx_W}{\bar{V}} = 0.005$$
$$\frac{22.47 \times 0.01}{0.005} = \bar{V} = 44.94 \text{ kmol/min}$$

For saturated feed

$$V = \bar{V} + F(1-q) = \bar{V}$$
$$= 44.94 \text{ kmol/min}$$

For enriching section

$$V = L + D$$
$$44.94 - 27.52 = L = 17.42 \text{ kmol/min}$$

So, $$R = \frac{L}{D} = \frac{17.42}{(27.53)} = 0.632$$

4.57 (0.982)

Given: $F = 1$ kmol

$$FX_F = 0.1 \text{ kmol}$$
$$WX_w = 0.01 \text{ kmole}$$
$$\alpha_{AB} = 4.5$$

From batch distillation, we have

$$\ln\left|\frac{FX_F}{WX_w}\right| = \alpha_{AB} \ln\left|\frac{F(1-X_F)}{w(1-X_w)}\right|$$
$$\ln\left|\frac{0.1}{0.01}\right| = 4.5\ln\left|\frac{1 \times 0.9}{w - 0.01}\right|$$
$$\frac{2.3026}{4.5} = \ln\left|\frac{0.9}{w - 0.01}\right|$$
$$1.668 = \frac{0.9}{w - 0.01}$$
$$w - 0.01 = 0.539$$
$$w = 0.549$$

$\because$ $$w \times X_w = 0.01$$
$$X_w = \frac{0.01}{0.549} = 0.0182$$

The mole fraction of *B* remaining in the vessel is

$$X_B = 1 - X_w = 0.982$$

4.58 (b)

Relative volatility of component *A* with respect to *B* is given $(\alpha_{AB}) = 2$

The vapour composition in rectifying section is given as below:

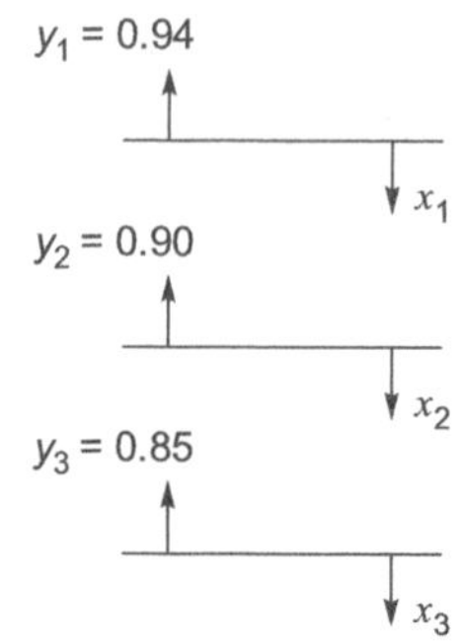

y_1 and x_1 are in equilibrium and equilibrium relation is given as

$$y = \frac{\alpha x}{1 + (\alpha + 1)x}$$
$$0.94 = \frac{2x_1}{1 + x_1} \Rightarrow x_1 = 0.8868$$

Similarly, $$0.90 = \frac{2x_2}{1 + x_2} \Rightarrow x_2 = 0.81818$$

Now slope of operating line

$$\frac{y_2 - y_3}{x_1 - x_2} = \frac{R}{R+1} = \frac{0.90 - 0.85}{0.8868 - 0.81818} = 0.72865$$
$$R = \frac{0.72865}{1 - 0.72865} = 2.68$$

4.59 (10.17)

Given, Diameter of tray column = 120 cm

Downcomer area $(A_d) = 575\ \text{cm}^2$

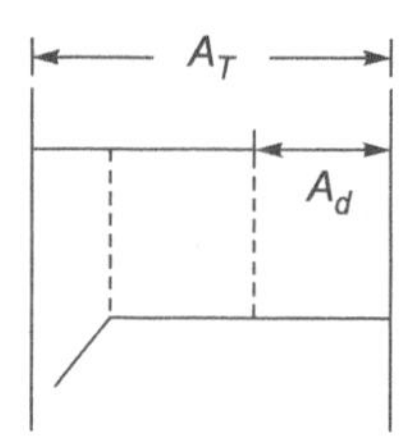

Area of tray column

$$A_T = \frac{\pi}{4} \times d^2 = \frac{\pi}{4} \times 120^2$$
$$= 11309.7335\ \text{cm}^2$$

Area not available for vapour flow = $2 \times A_d$

$$= 2 \times 575 = 1150$$

% area not available for vapour flow

$$= \frac{1150 \times 100}{11309.7335} = 10.17\%$$

■■■■

5 Liquid-Liquid Extraction, Leaching and Adsorption

5.1 A continuous – contact extraction column is used to extract a solute from an aqueous stream (F) using an organic solvent (S). The distribution coefficient (y/x) is 1.0, where x and y are the mass fractions of solute in raffinate and extract phases, respectively. The height of transfer unit based on the extract phase is 1.0 m. The rest of the data are given in the figure. Assuming that the phase flow rates are constant, find the height of the tower.

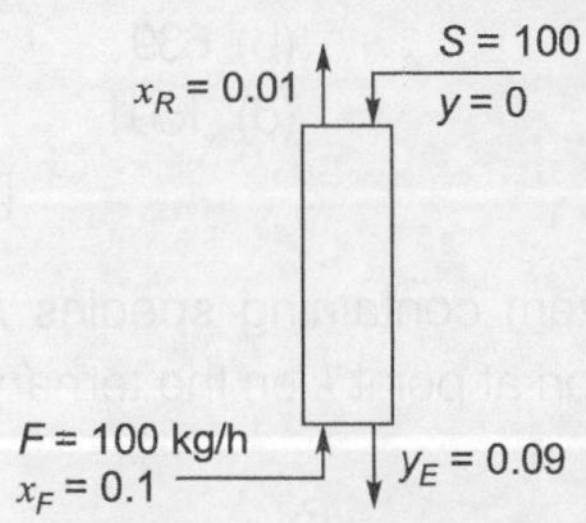

[2000 : 5 M]

5.2 In a single stage extraction process, 10 kg of pure solvent S (containing no solute A) is mixed with 30 kg of feed F containing A at a mass fraction $x_f = 0.2$. The mixture splits into an extract phase E and a raffinate phase R, containing A at $x_E = 0.5$ and $x_R = 0.05$, respectively. The total mass of the extract phase is (in kg)

(a) 6.89 (b) 8.89
(c) 10 (d) 8.25

[2001 : 2 M]

5.3 It is desired to extract acetone from a feed containing acetone and water, using chloroform as the solvent, in two cross current extraction stages as shown below:

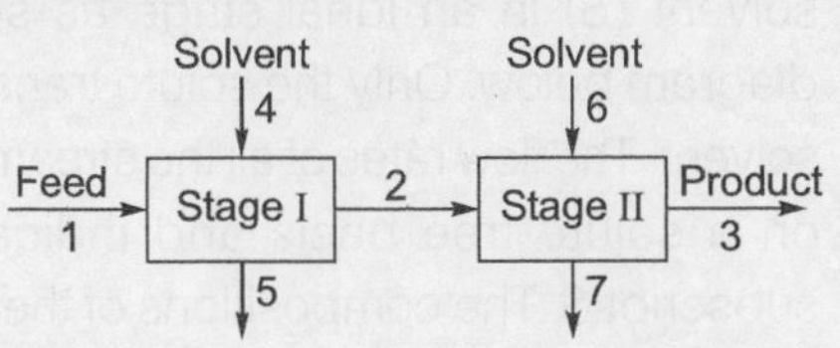

Assume that water and chloroform are immiscible. The following data are given for the process:

(i) The feed is an equimolar mixture of acetone and water.

(ii) The quantities of chloroform used in the two stages are equal.

(iii) 60 mole % of the acetone in the feed is extracted in stage 1.

(iv) The extract and raffinate phases exiting from each stage are in equilibrium. The equilibrium relation for the distribution of acetone is given by

$$\frac{\text{Mole of acetone in water rich phase}}{\text{Moles of water in water rich phase}} = 2\times\frac{\text{Mole of acetone in chloroform rich phase}}{\text{Moles of chloroform in choloroform rich phase}}$$

(a) Indicate (with the help of the above diagram) the components in each stream.

(b) Determine the quantity of chloroform used in each stage per mol of feed.

(c) Determine the mole fraction of acetone in final product stream.

[2002 : 5 M]

5.4 Compound A is extracted from a solution of $A + B$ into a pure solvent S. A co-current unit is used for the liquid-liquid extraction. The inlet rate of the solution containing A is 200 moles of B/hr m^2 and the solvent flow rate is 400 moles of S/hr m^2. The equilibrium data is represented by $Y = 3X^2$ where Y is in moles of A/moles of B and X is in moles of A/moles of S. The maximum percentage extraction achieved in the unit is

(a) 25% (b) 50%
(c) 70% (d) 90%

[2003 : 2 M]

5.5 In liquid-liquid extraction 10 kg of a solution containing 2 kg of solute C and 8 kg of solvent A is brought into contact with 10 kg of solvent B. Solvents A and B are completely immiscible in each other whereas solute C is soluble in both the solvents. The extraction process attains equilibrium. The equilibrium relationship between the two phases is $Y^* = 0.9X$ where Y^* is kg of C/kg of B and X is kg of C/kg of A. Choose the correct answer.

(a) the entire amount of C is transferred to solvent B.
(b) less than 2 kg but more than 1 kg of C is transferred to solvent B.
(c) less than 1 kg of C is transferred to B.
(d) no amount of C is transferred to B.

[2004 : 2 M]

5.6 At equilibrium, the concentration of water in vapor phase (C^*) in kg/m^3 of air space and the amount of water (m) absorbed per kg of dry silica gel are related by $C^* = 0.0667$ m. To maintain dry conditions in a room of air space 100 m^3 containing 2.2 kg of water vapour initially, 10 kg of dry silica gel is kept in the room. The fraction of initial water remaining in the air space after a long time (during which the temperature is maintained constant) is

(a) 0.0 (b) 0.2
(c) 0.4 (d) 1.0

[2004 : 2 M]

5.7 In the triangular diagram represented below for a batch separation process a stream F, is mixed with a solvent B to produce products R and E. Substance A is the carrier liquid and C is the solute to be extracted. The amounts of B and E are 1 kg and 1.20 kg respectively. The length FM is 3.1 and length FB is 8.5 units on the figure. The ratio R/E is estimated to be

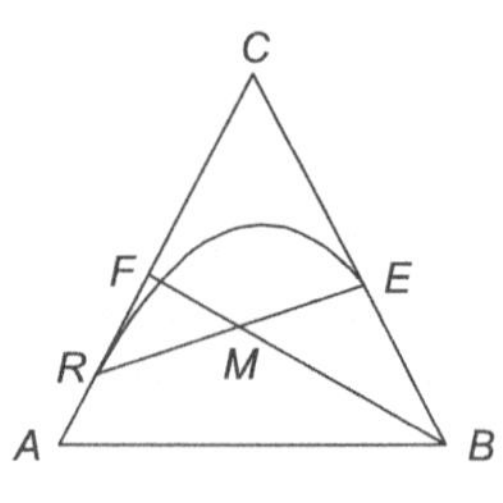

(a) 1.285 (b) 2
(c) 0.751 (d) 2.5

[2005 : 2 M]

5.8 An ideal single stage extraction process is used to treat 100 mol/s of an organic feed solution. The solution concentration in this solution is to be reduced from 0.5 mol% to 0.1 mol%. A pure solvent S is used. To reduce the solvent requirement by half for the same separation,

(a) add one more ideal co-current stage
(b) use another pure solvent S^* whose partition coefficient is twice that of S
(c) use solvent S containing 0.02 mole fraction of the solute
(d) double the residence time of the solvent S in the contactor

[2006 : 1 M]

5.9 It is desired to reduce the concentration of pyridine in 500 kg of aqueous solution from 20 weight percent to 5 wt percent in a single batch extraction using chloro-benzene as solvent. Equilibrium compositions (end points of the tie line) in terms of weight percent of pyridine-water-chlorobenzene are (5, 95, 0) and (11, 0, 89).
The amount of pure solvent required in kg for the operation is

(a) 607 (b) 639
(c) 931 (d) 1501

[2007 : 2 M]

5.10 For a system containing species P, Q and R, composition at point k on the ternary plot is

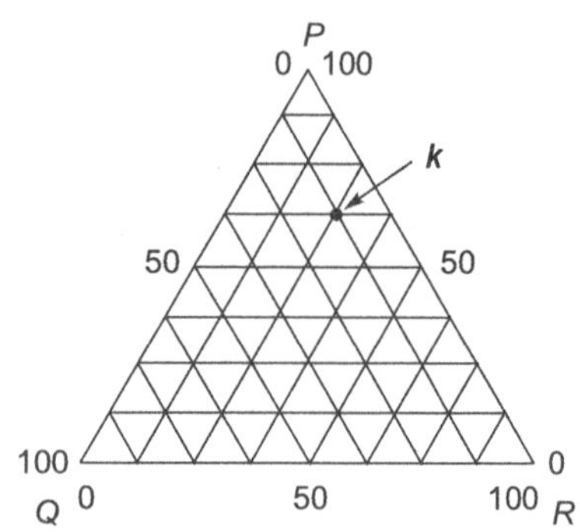

(a) 62.5% P, 12.5 Q, 25% R
(b) 25% P, 62.5% Q, 12.5% R
(c) 12.5% P, 62.5% Q, 12.5% R
(d) 12.5% P, 25% Q, 62.5% R

[2008 : 1 M]

5.11 A feed (F) containing a solute is contacted with a solvent (S) in an ideal stage as shown in the diagram below. Only the solute transfers into the solvent. The flow rates of all the streams are shown on a solute free basis and indicated by the subscript S. The compositions of the streams are expressed on a mole ratio basis. The extract leaving the contactor is divided into two equal parts, one part collected as the product (P) and the other stream is recycled to join the solvent. The equilibrium relationship is

$$Y^* = 2X$$

The product flow rate (P_S) and composition (Y_{out}) are

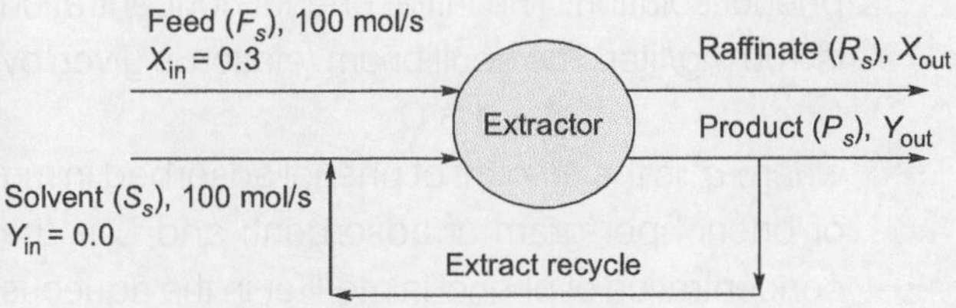

(a) $P_S = 50$ mol/s, $Y_{out} = 0.3$
(b) $P_S = 100$ mol/s, $Y_{out} = 0.2$
(c) $P_S = 200$ mol/s, $Y_{out} = 0.1$
(d) $P_S = 100$ mol/s, $Y_{out} = 0.4$

[2008 : 2 M]

Linked Answer Questions :

Statement for Linked Answer Questions 5.12 and 5.13:
Water is used to absorb ammonia from a gas mixture in a single separation stage contactor. The process is schematically represented in the figure below.

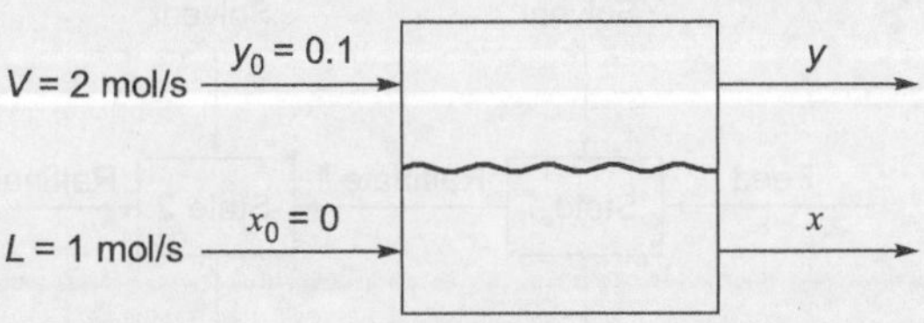

The molar gas and liquid flow rates, and the inlet mole fractions are given in the figure. Both the liquid and the gas phases are well mixed, and the equilibrium relation between y and x is given by $y^* = x$.

5.12 If the stage is ideal, then the value of y is

(a) $\frac{1}{15}$ (b) $\frac{1}{10}$
(c) $\frac{2}{15}$ (d) $\frac{1}{6}$

[2010 : 2 M]

5.13 If the stage efficiency is 50%, then the value of y is

(a) $\frac{1}{12}$ (b) $\frac{1}{6}$
(c) $\frac{1}{4}$ (d) $\frac{1}{3}$

[2010 : 2 M]

5.14 A cross-current cascade of N ideal stages is used to treat a feed stream of molar flow rate E. The feed stream contains a solute which is to be recovered by a pure solvent having a molar flow rate S. The solvent is divided equally between these N stages. The linear equilibrium curve relating the mole fractions x and y^* of the solute in the raffinate and the extract respectively, is given by $y^* = mx$. Assume dilute conditions. The ratio of the solute mole fraction in the original feed to that in the exit raffinate stream i.e. $\left(\frac{x_0}{x_N}\right)$ is given by

(a) $\left[1+\left(\frac{mS}{NE}\right)\right]^N$ (b) $\left[1+\left(\frac{NS}{mE}\right)\right]^N$
(c) $\left[1+\left(\frac{NE}{mS}\right)\right]^N$ (d) $\left[1+\left(\frac{mE}{NS}\right)\right]^N$

[2013 : 2 M]

5.15 A multi-stage, counter-current liquid-liquid extractor is used to separate solute C from a binary mixture (F) of A and C using solvent B. Pure solvent B is recovered from the raffinate R by distillation, as shown in the schematic diagram below.

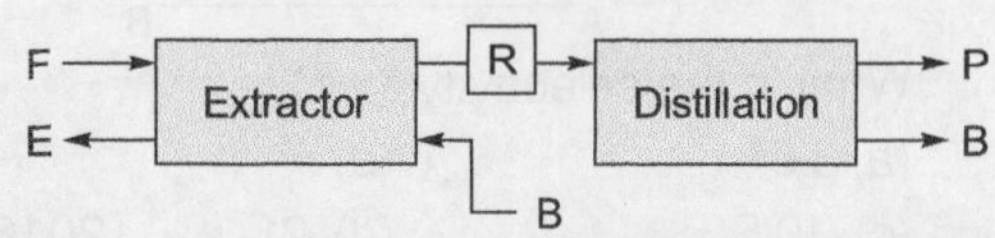

Locations of different mixtures for this process are indicated on the triangular diagram below. P is the solvent-free raffinate, E is the extract, F is the feed and Δ is the difference point from which the mass balance lines originate. The line PB intersects the binodal curve at U and T. The lines PΔ and FB intersect the binodal at V and W respectively.

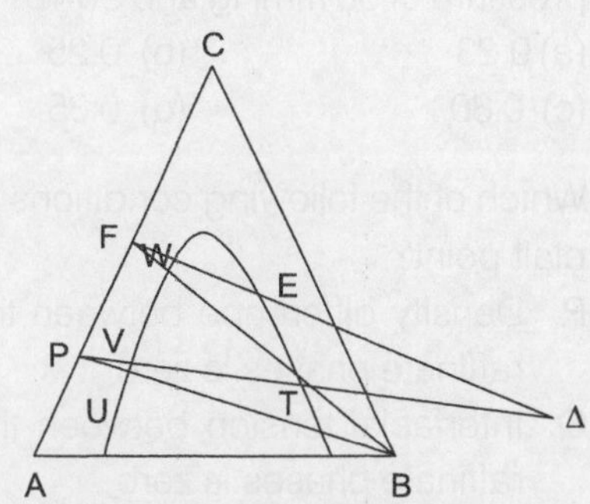

The raffinate coming out of the extractor is represented in the diagram by the point:

(a) T (b) U
(c) V (d) W

[2015 : 2 M]

5.16 Adsorption on activated carbon is to be used for reducing phenol concentration in wastewater from 0.04 mol/*l* to 0.008 mol/*l*. The adsorption isotherm at the operating temperature can be expressed as $q = 0.025C^{1/3}$; where q is the phenol concentration in solid (mol/g solid) and C is the phenol concentration in water (mol/l). The minimum amount of solid (in grams) required per liter of wastewater (up to one decimal place) is ______.

[2015 : 2 M]

5.17 Solute C is extracted in a batch process from its homogenous solution of A and C, using solvent B. The combined composition of the feed and the extracting solvent is shown in the figure below as point M, along with the tie line passing through it. The ends of the tie line are on the equilibrium curve.

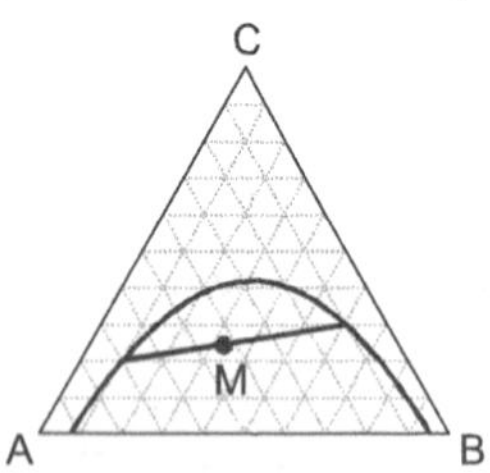

What is the selectivity for C?

(a) 3.5 (b) 7
(c) 10.5 (d) 21 [2016 : 2 M]

5.18 At 30°C, the amounts of acetone adsorbed at partial pressures of 10 and 100 mmHg are 0.1 and 0.4 kg acetone/kg activated carbon, respectively. Assume Langmuir isotherm describes the adsorption of acetone on activated carbon. What is the amount of acetone adsorbed (in kg per kg of activated carbon) at a partial pressure of 50 mmHg and 30° C?

(a) 0.23 (b) 0.25
(c) 0.30 (d) 0.35 [2016 : 2 M]

5.19 Which of the following conditions are valid at the plait point?

P. Density difference between the extract and raffinate phases is zero
Q. Interfacial tension between the extract and raffinate phases is zero
R. Composition difference between the extract and raffinate phases is zero

(a) P and Q only (b) Q and R only
(c) P and R only (d) P, Q and R

[2017 : 1 M]

5.20 In a batch adsorption process. 5 g of fresh adsorbent is used to treat 1 liter of an aqueous phenol solution. The initial phenol concentration is 100 mg/liter. The equilibrium relation is given by

$$q^* = 1.3\ C$$

where q^* is the amount of phenol adsorbed in mg of phenol per gram of adsorbent: and C is the concentration of phenol in mg/liter in the aqueous solution.

When equilibrium is attained between the adsorbent and the solution, the concentration of phenol in the solution, rounded to 1 decimal place, is ________ mg/liter.

[2017 : 2 M]

5.21 It is decided to extract A from a feed containing 20 mol% A and 80 mol% B in two ideal cross-current stages as shown below, using equal amount of pure solvent C in each stage.

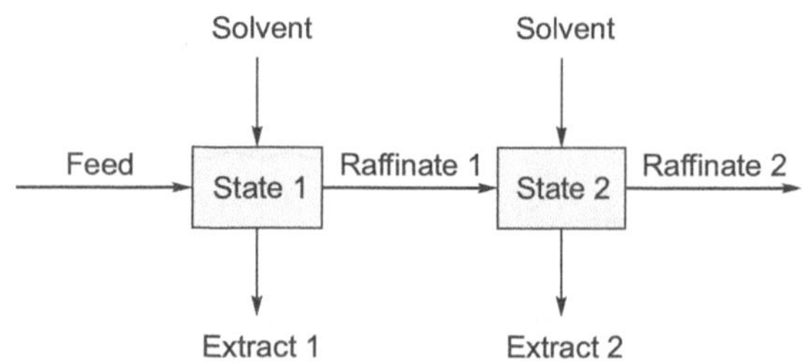

Components B and C are immiscible. 60% of A in the feed is extracted in Stage 1. The equilibrium relation is given by $Y^* = 1.5X$ where,

X = moles of A per mole of B in raffinate

Y^* = moles of A per mole of C in extract in equilibrium with raffinate

The mol % of A in raffinate from Stage 2 is _______ (rounded off to second decimal place).

[2018 : 2 M]

5.22 100 kg of a feed containing 50 wt% of a solute C is contacted with 80 kg of a solvent containing 0.5 wt. % of C in a mixer-settler unit. From this operation, the resultant extract and raffinate phases contain 40 wt. % and 20 wt. % of C, respectively. If E and R denote the mass of the extract and raffinate phases, respectively, the ratio E/R is

(a) $\frac{1}{4}$ (b) $\frac{1}{2}$
(c) $\frac{2}{3}$ (d) 1

[2019 : 2 M]

5.23 Feed solution F is contacted with solvent B in an extraction process. Carrier liquid in the feed is A and the solute is C. The ternary diagram depicting a single ideal stage extraction is given below. The dashed lines represent the tie-lines.

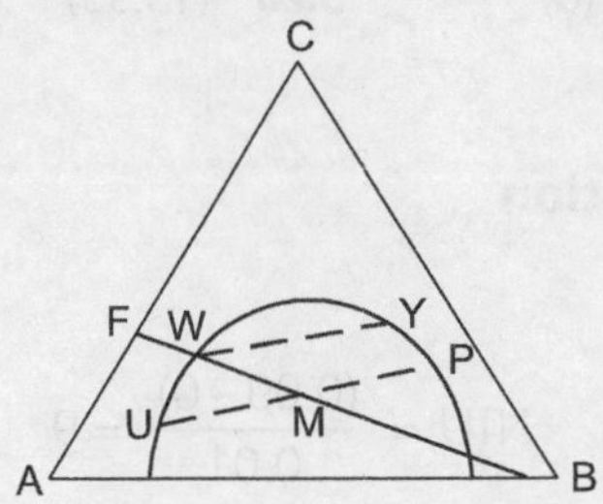

The correct option(s) is/are

(a) For the tie-lines shown, concentration of solute in the extract is higher than that in the raffinate.

(b) Maximum amount of solvent is required if the mixture composition is at W.

(c) Y represents the composition of extract when minimum amount of solvent is used.

(d) U represents the raffinate composition if the mixture composition is at M.

[2021 : 1 M]

■■■■

Answers Liquid-Liquid Extraction, Leaching and Adsorption

5.1	(9)	**5.2**	(b)	**5.3**	(0.138)	**5.4**	(b)	**5.5**	(b)	**5.6**	(c)	**5.7**	(a)
5.8	(b)	**5.9**	(b)	**5.10**	(a)	**5.11**	(b)	**5.12**	(a)	**5.13**	(a)	**5.14**	(a)
5.15	(b)	**5.16**	(6.4)	**5.17**	(c)	**5.18**	(c)	**5.19**	(d)	**5.20**	(13.33)	**5.21**	(3.85)
5.22	(c)	**5.23**	(a, c, d)										

Explanations Liquid-Liquid Extraction, Leaching and Adsorption

5.1 (9)

We know,

(Height of tower) = (NTU) × (HTU) ...(1)

Given: HTU = 1 m

$$F = S = 100 \text{ kg/h}$$

$$\frac{y^*}{x} = 1$$

or $y^* = x$

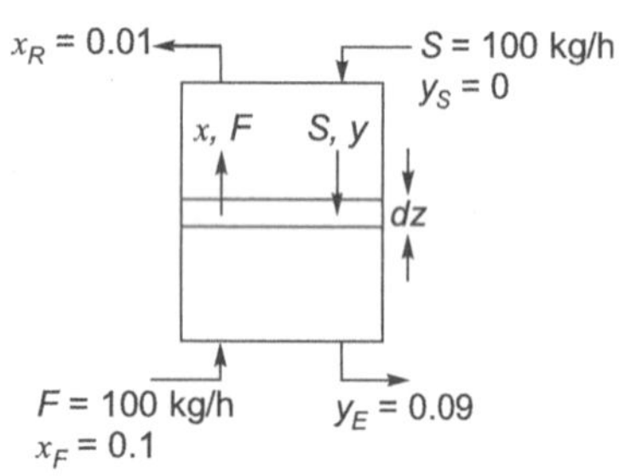

From the definition of NTU, we have

$$\text{NTU} = \int_{y_1}^{y_2} \frac{dy}{y^* - y}$$

$$\Rightarrow \quad \text{NTU} = \int_{y_1}^{y_2} \frac{dy}{x - y} \quad (\text{Given } y^* = x)...(2)$$

From equation of operating line, we have

$$Fx_F + Sy = Fx + Sy_E$$

$$x_F + y = x + y_E \quad (\text{Given : } F = S)$$

$$\Rightarrow (x_F - y_E) + y = x \quad ...(3)$$

Substituting the values in Eq. (3), we get

$(0.1 - 0.09) + y = x$

$$\Rightarrow \quad x = y + 0.01$$

Substituting the value of x in Eq. (2), we get

$$\text{NTU} = \int_{y_1}^{y_2} \frac{dy}{y + 0.01 - y} = \int_{y_1}^{y_2} \frac{dy}{0.01}$$

$$\text{NTU} = \frac{(y_2 - y_1)}{0.01}$$

or
$$\text{NTU} = \frac{(y_E - y_S)}{0.01}$$

$$\Rightarrow \quad \text{NTU} = \frac{(0.09 - 0)}{0.01} = 9$$

Substituting the value of NTU in Eq. (1), we get

Height of tower = (NTU) × (HTU)

= 9 × 1 = 9 m

Thus, the height of tower is 9 m.

5.2 (b)

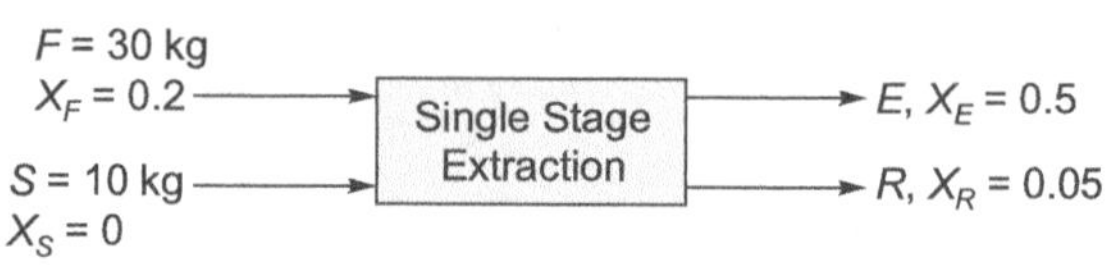

Overall balance

$$F + S = E + R$$

$$E + R = 40 \text{ kg}$$

Balance for 'A'

$$FX_F + SX_S = EX_E + RX_R$$

$$30 \times 0.2 + S \times 0 = E \times 0.5 + R \times 0.05$$

$$6 = E \times 0.5 + (40 - E) \times 0.05$$

$$E = 8.89 \text{ kg}$$

5.3 (0.138)

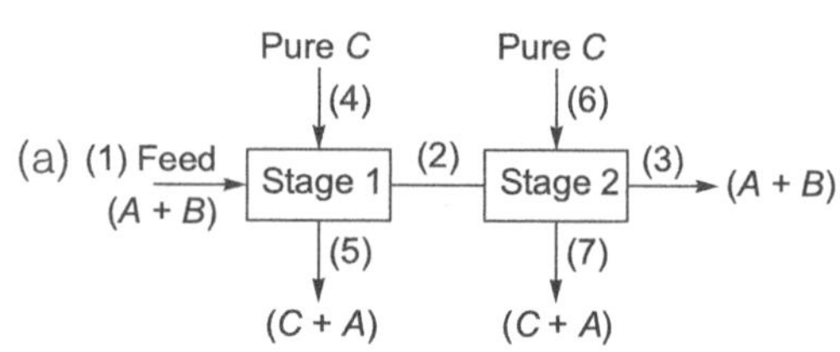

where, $A \Rightarrow$ Acetone (solute)

$B \Rightarrow$ Water (diluent)

$C \Rightarrow$ Chloroform (solvent)

Equilibrium relationship

$$X = 2Y$$

where, $X = \dfrac{\text{Mole of acetone in water rich phase}}{\text{Mole of water in water rich phase}}$

$$Y = \frac{\text{Mole of acetone in chloroform rich phase}}{\text{Moles of chloroform in chloroform rich phase}}$$

Basis : 100 mole feed

$$X_1 = \frac{50 \text{ mole acetone}}{50 \text{ mole water}} = 1$$

$$Y_0 = 0 \text{ (pure solvent)}$$

$$FX_1 + SY_0 = FX_2 + SY_1$$

60 mole% acetone extracted in stage - I[st]

$$X_2 = \frac{50(1-0.6)}{50} = \frac{20}{50} = 0.4$$

$$F \times 1 + S \times 0 = F \times 0.4 + S \times Y_1$$

$$Y_1 = \frac{X_2}{2} = \frac{0.4}{2} = 0.2$$

where, $F = 50$ mole

So,
$$S = \frac{50 \times 0.6}{0.2}$$
$$= 150 \text{ mole in each stage}$$

(b)
$$\frac{\text{Mole of acetone added in each stage } (S)}{\text{Mole of feed}} = \frac{150}{100} = 1.5$$

(c) Acetone balance over stage II

$$FX_2 + SY_0 = FX_3 + SY_2$$

where, $F = 50$ mole, $S = 150$ mole

$$X_2 = 0.4,\ Y_0 = 0$$
$$X_3 = 2Y_2$$

$$50 \times 0.4 + 0 = 50X_3 + 150\frac{X_3}{2}$$

$$20 = X_3\left(50 + \frac{150}{2}\right)$$

$$20 = X_3(125)$$

$$X_3 = \frac{20}{125} = 0.16$$

$$= \frac{\text{Mole of acetone in product}}{\text{Mole of water}}$$

$$X_3 = \frac{\text{Mole of acetone}}{\text{Mole of acetone + mole of water}}$$

$$X_3 = \frac{0.16}{1+0.16} = 0.138$$

Mole fraction of acetone in product $(X_3) = 0.138$

5.4 (b)

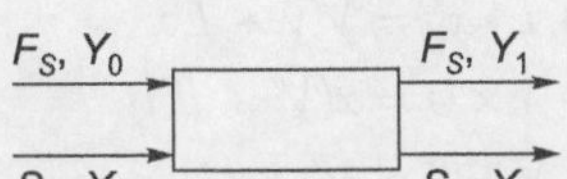

Given :

Molar flux of B in inlet (F_s) = 200 mol/m²h

Molar flux of solvent in inlet (S_s) = 400 mol/m²h

Equilibrium data, $Y = 2X$

Apply component C balance

$$F_sY_o + S_sX_o^{\,0} = F_sY_1 + S_sX_1$$

$$200Y_0 = 200Y_1 + 400\frac{Y_1}{2}$$

$$200Y_0 = 400Y_1$$

$$\frac{Y_0}{Y_1} = 2$$

$$\% \text{ of extraction} = \frac{\text{Amount of } C \text{ removed}}{\text{Mole of } C \text{ in feed}} \times 100$$

$$= \frac{F_s(Y_0 - Y_1)}{F_sY_0} \times 100$$

$$= \frac{Y_0 - \frac{Y_0}{2}}{Y_0} \times 100 = 50\%$$

5.5 (b)

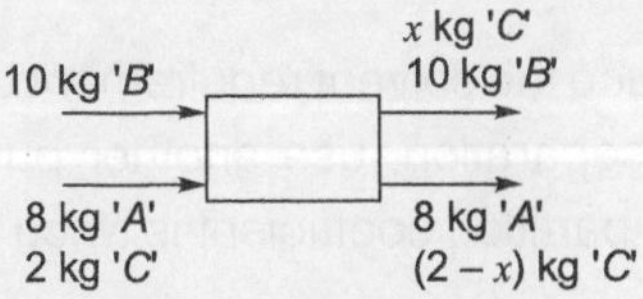

Given : $Y^* = 0.9X$

$$\frac{\text{(kg of } C)}{\text{(kg of } B)} = 0.9\frac{\text{(kg of } C)}{\text{(kg of } A)}$$

$$\frac{x}{10} = 0.9\frac{(2-x)}{8}$$

$$x = 1.058 \text{ kg}$$
$$1 < x < 2$$

5.6 (c)

Given : $C^* = 0.0667\,m$

Let x = kg of water absorbed by 10 kg of dry silica

C^* = Concentration of water in vapor phase in kg/m³

$$m = \frac{\text{Amount of water adsorbed}}{\text{kg of dry silica gel}} = \frac{x}{10}$$

$$\frac{2.2 - x}{100} = 0.0667 \times \frac{x}{10}$$

$$x \simeq 1.32 \text{ kg}$$

Fraction of initial water remaining in air space

$$= \frac{2.2 - x}{2.2} = \frac{2.2 - 1.32}{2.2} = 0.4$$

5.7 (a)

We know,

$$\frac{F}{B} = \frac{BM}{FM} = \frac{FB - FM}{FM}$$

$$\frac{F}{B} = \frac{8.5 - 3.1}{3.1} = 1.742$$

$$F = 1.742 \times 1$$
$$= 1.742 \text{ kg}$$

From material balance, we have

$$F + B = R + 1.2$$
$$R = F + B - 1.2$$
$$R = 1.742 + 1 - 1.2$$
$$R = 1.542$$

Now, the ratio is

$$\frac{R}{E} = \frac{1.542}{1.20} = 1.285$$

5.8 (b)

To reduce the solvent requirement by half, for the same separation, use another pure solvent S^* whose partition coefficient is twice of that of S.

5.9 (b)

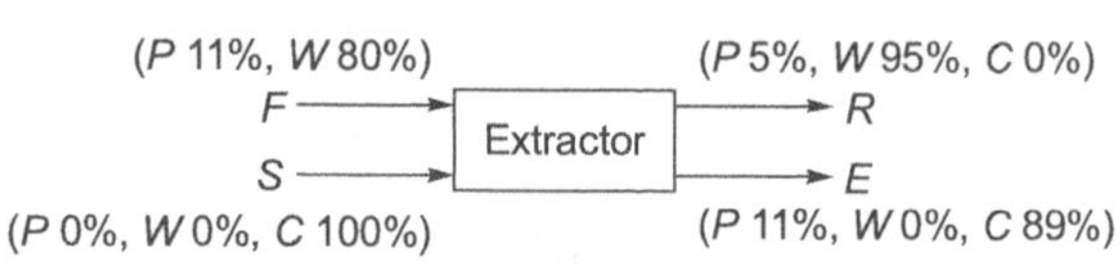

where, F = Feed; P = Pyridine
R = Raffinate
W = Water
S = Solvent
C = Chlorobenzene
E = Extract

Water balance over extractor

$$F \cdot x_{PW} = R \cdot x_{RW}$$
$$500 \times 0.80 = R \times 0.95$$
$$R = 421.05 \text{ kg}$$

Pyridine balance

$$F \cdot x_{FP} + S \cdot x_{SP} = R \cdot x_{RP} + E \cdot x_{EP}$$
$$500 \times 0.2 + S \times 0 = 421.05 \times 0.05 + E \times 0.11$$
$$E = 717.70 \text{ kg}$$

Overall balance :

$$F + S = E + R$$
$$S = 717.70 + 421.05 - 500$$
$$S = 638.75 \text{ kg} \simeq 639 \text{ kg}$$

5.10 (a)

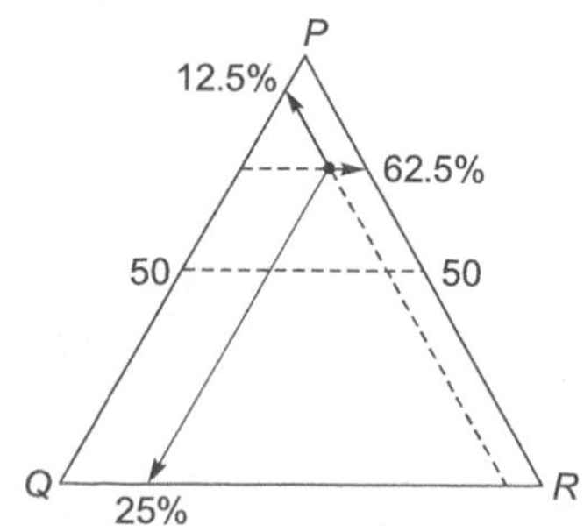

5.11 (b)

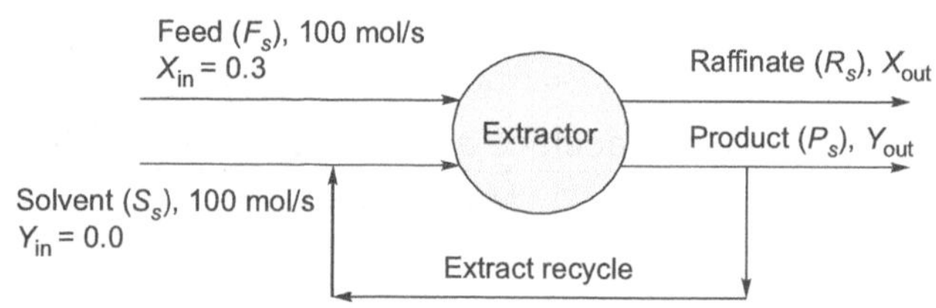

Overall material balance

$$F_s + S_s = R_s + P_s$$
$$R_s = P_s \text{ (given)}$$

So, $R_s = 100$ mol/s
$P_s = 100$ mol/s

Given : $F_s = 100$ mol/s
$S_s = 100$ mol/s

Solute balance:

$$F_s X_{in} + S_s Y_{in} = R_s X_{out} + P_s Y_{out}$$
$$Y_{out} = 2X_{out} \text{ (given)}$$

$$100 \times 0.3 + 100 \times 0 = 100\frac{Y_{out}}{2} + 100 Y_{out}$$

$$Y_{out} = 0.2$$

5.12 (a)

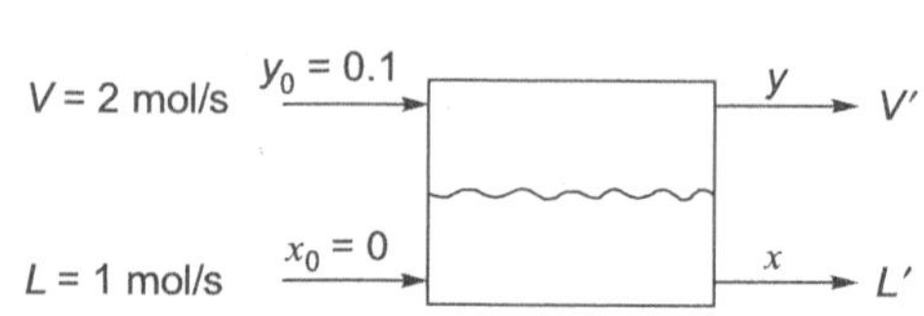

where $y^* = x$ (given)

$$V \cdot y_0 + L \cdot x_0 = V'y + L'x$$
$$2 \times 0.1 + 1 \times 0 = x(V' + L')$$
$$V + L = V' + L'$$
$$0.2 = x \times 3$$
$$x = \frac{2}{30} = \frac{1}{15}$$

5.13 (a)

Stag efficiency 50%.

Mole transferred,

$$= \frac{1}{2} \times \frac{1}{15} = \frac{1}{30} \text{ mol/s}$$

$$\frac{1}{30} = 2(0.1 - y')$$

$$\frac{1}{60} = \frac{1}{10} - y'$$

$$y' = \frac{5}{60} = \frac{1}{12}$$

5.14 (a)

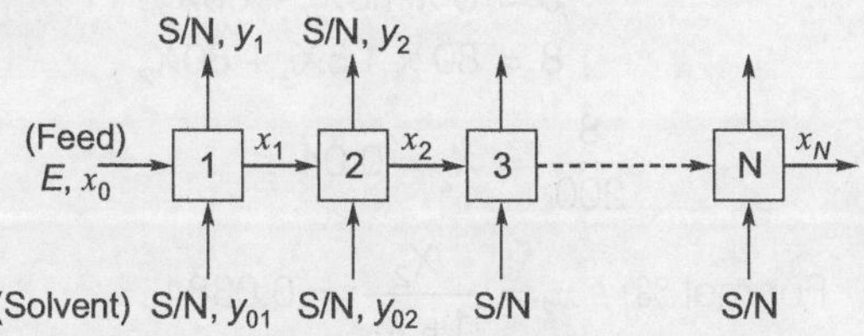

Solute balance over stage 1st

$$Ex_0 + 0 = \frac{S}{N} y_1 + Ex_1$$

$$y_1 = mx_1$$

$$Ex_0 = \frac{mS}{N} x_1 + Ex_1$$

$$Ex_0 = \left(\frac{mS}{N} + E\right) x_1$$

$$\frac{x_0}{x_1} = \left(\frac{\frac{mS}{N} + E}{E}\right) \quad \text{...(i)}$$

Solute balance over stage 2nd

$$\frac{x_1}{x_2} = \left(\frac{\frac{mS}{N} + E}{E}\right) \quad \text{...(ii)}$$

Solute balance over stage N^{th}

$$\frac{x_{N-1}}{x_N} = \left(\frac{\frac{mS}{N} + E}{E}\right) \quad \text{...(iii)}$$

Multiply equation (i), (ii) and (iii)

$$\frac{x_o}{x_N} = \left(\frac{\frac{mS}{N} + E}{E}\right)^N$$

$$\frac{x_o}{x_N} = \left[\left(\frac{mS}{NE}\right) + 1\right]^N$$

5.15 (b)

Comparing the given diagram with standard graph, the raffinate coming out the extractor is represented by point U.

5.16 (6.4)

Phenol removed from waste water

$$= 0.04 - 0.008 = 0.032 \text{ mol}$$

Let q be the mol of phenol in 1 g solid in saturation

$$q \text{ mol phenol} = 1 \text{ g solid}$$

$$0.032 \text{ mol phenol} = \frac{0.032}{q} \text{ g solid required} \quad \text{...(i)}$$

$$q = 0.025C^{1/3}$$

$$q = 0.025(0.008)^{1/3} = 0.005$$

By eqn. (i), solid required

$$= \frac{0.032}{q} = \frac{0.032}{0.005} = 6.4 \text{ gm}$$

5.17 (c)

Raffinate phase,

$$x_C = 0.2$$
$$x_A = 0.7$$
$$x_B = 0.1$$

Extract phase,

$$y_C = 0.3$$
$$y_A = 0.1$$
$$y_B = 0.6$$

Selectivity of C

$$= \frac{\frac{y_C}{y_A}}{\frac{x_c}{x_A}} = \frac{\frac{0.3}{0.1}}{\frac{0.2}{0.7}} = \frac{21}{2} = 10.5$$

5.18 (c)

Langmuir isotherm, where q is amount of acetone absorbed.

$$\frac{q}{q_{max}} = \frac{kP}{1+kP}$$

$$\frac{q_1}{q_2} = \frac{kP_1}{1+kP_1} \times \frac{1+kP_2}{kP_2}$$

$$\frac{0.1}{0.4} = \frac{10k}{1+10k} \times \frac{1+100k}{100k}$$

$$\frac{10}{4} = \frac{1+100k}{1+10k}$$

$$10 + 100k = 4 + 400k$$

$$6 = 300k$$

$$k = \frac{1}{50}$$

$$\frac{q_3}{q_2} = \frac{50k}{1+50k} \times \frac{1+100k}{100k}$$

$$\frac{q_3}{q_2} = \frac{1+2}{2\times 2} = \frac{3}{4}$$

$$q_3 = \frac{3}{4} \times 0.4$$

$$= 0.3 \text{ kg/kg of activated carbon}$$

5.19 (d)

At plait point density difference, interfacial tension and composition difference between the extract and Raffinate phases is zero.

5.20 (13.33)

$$q^* = 1.3C$$

q^* = amount of phenol adsorbed in mg of phenol per gram of adsorbent

C = concentration of phenol in mg/lit. in aq. solution.

At equilibrium,

$$V(C_i - C_F) = M(q_F - q_i)$$

$$q_F = 1.3C_F \text{ where } q_i = 0$$

$$V(C_i - C_F) = 1.3C_F \times M$$

$$V = 1 \text{ lit.}, M = 5 \text{ gm}$$

$$1(100 - C_F) = 1.3 \times 5 \times C_F$$

$$100 = 7.5C_F$$

$$C_F = 13.33 \text{ mg/lit.}$$

5.21 (3.85)

Assume feed is 100 mol/hr.

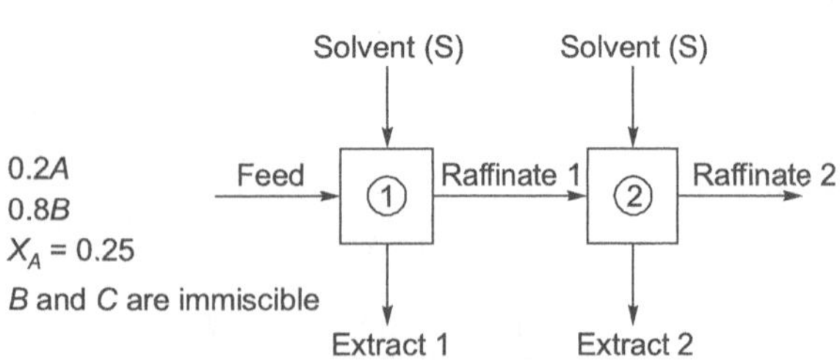

$$Y^* = 1.5X$$

where X = moles of A per mole of B in raffinate.

Y^* = Moles of A per mole of C in extract in equilibrium with raffinate.

Component balance of A over stage 1st

$$80 \times 0.25 + 0 = SY_1 + 80X_1$$

$$20 = S \times 0.15 + 8$$

$$S = 80 \text{ mol/hr}$$

60% of A in feed is extracted in stage 1st then in stage 2nd. 8 mole entering, then

$$X_1 = \frac{8}{80} = 0.1$$

$$8 + 0 = SY_2 + 80X_2$$

$$8 = S \times 1.5X_2 + 80X_2$$

$$8 = 80 \times 1.5X_2 + 80X_2$$

$$\frac{8}{200} = X_2 = 0.04$$

For mol %, $x_2 = \dfrac{X_2}{1+X_2} = 0.0384$

$$0.0384 \times 100 = 3.85$$

5.22 (c)

Mixer settler unit

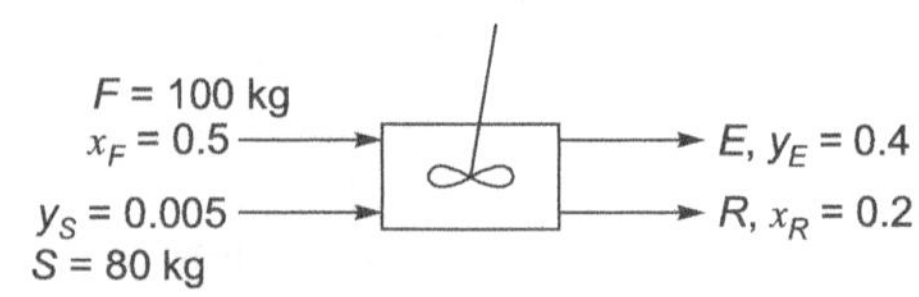

Overall material balance

$$F + S = E + R$$

$$E + R = 100 + 80 = 180 \quad \text{...(i)}$$

Component balance of C

$$100 \times 0.5 + 80 \times 0.005 = E \times 0.4 + R \times 0.2$$

$$50.4 = 0.4E + 0.2R \quad \text{...(ii)}$$

$$50.4 = 0.4(180 - R) + 0.2R$$

$$50.4 = -0.2R + 72$$

$$R = \frac{21.6}{0.2} = 108 \text{ kg}$$

$$E = 180 - 108 = 72 \text{ kg}$$

$$\frac{E}{R} = \frac{72}{108} = \frac{8}{12} = \frac{2}{3}$$

■■■■

6 Humidification

6.1 Air at 1 atm is blown past the bulb of a mercury thermometer. The bulb is covered with a wick. The wick is immersed in an organic liquid (molecular weight = 58). The reading of the thermometer is 7.6 °C. At this temperature, the vapour pressure of the liquid is 5 kPa. Find the air temperature, given that the ratio of the heat transfer coefficients to the mass transfer coefficient (psychrometric ratio) is 2 kJ/(kg K) and the latent heat of vapourization of the liquid is 360 kJ/kg. Assume that the air, which is blown, is free from the organic vapour.

[2000 : 5 M]

6.2 The Lewis relation for air-water humidification is given by (k_Y: mass transfer coefficient of moisture in air; h_G : heat transfer coefficient; C_S : heat capacity of vapour-gas mixture)

(a) $\frac{h_G^2}{k_y C_s} = 1$ (b) $\frac{k_y C_s^2}{h_G} = 1$

(c) $\frac{h_G}{k_y C_s} = 1$ (d) $\frac{k_y^2 h_G}{C_s} = 1$

[2001 : 1 M]

6.3 Air at a temperature of 20°C and 750 mm Hg pressure has a relative humidity of 80. What is its percentage humidity? Vapour pressure of water at 20°C is 17.5 mm Hg.

(a) 80.38 (b) 80

(c) 79.62 (d) 78.51

[2003 : 2 M]

6.4 The following plot gives the saturated humidity (He) versus Temperature (T)

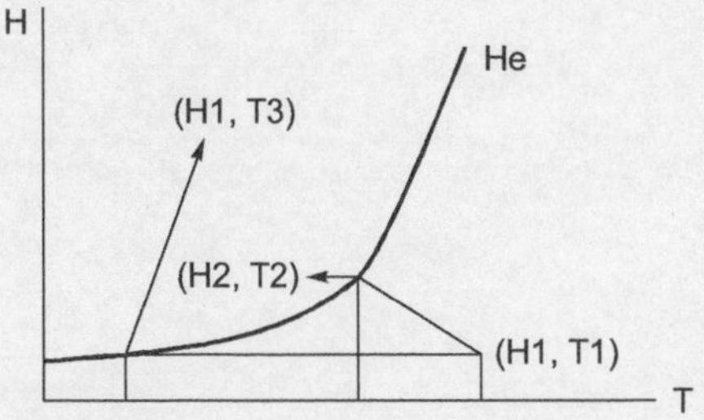

Line joining (H1, T1) and (H2, T2) is the constant enthalpy line. Choose the correct one from among the alternatives A, B, C and D

(a) T1 – Dew Point Temp.; T2 – Dry Bulb Temp.; T3 – Wet Bulb Temp.

(b) T1 – Dew Point Temp.; T2 – Wet Bulb Temp.; T3 – Dry Bulb Temp.

(c) T1 – Wet Bulb Temp.; T2 – Dry Bulb Temp.; T3 – Dew Point Temp.

(d) T1 – Dry Bulb Temp.; T2 – Wet Bulb Temp.; T3 – Dew Point Temp.

[2003 : 2 M]

Common Data for Questions (6.5 & 6.6):

An air-water vapour mixture has a dry bulb temperature of 60 °C and a dew point temperature of 40 °C. The total pressure is 101.3 kPa and the vapour pressures of water at 40°C and 60°C are 7.30 kPa and 19.91 kPa, respectively.

6.5 The humidity of air sample expressed as kg of water vapour/kg of dry air is

(a) 0.048 (b) 0.079

(c) 0.122 (d) 0.152

[2004 : 1 M]

6.6 The wet bulb temperature T_w for the above mixture would be

(a) less than 40°C

(b) 40°C

(c) 40°C < T_w < 60°C

(d) 60°C

[2004 : 1 M]

6.7 A vessel of volume 1000 m^3 contains air which is saturated with water vapour. The total pressure and temperature are 100 kPa and 20°C, respectively. Assuming that the vapour pressure of water at 20°C is 2.34 kPa, the amount of water vapour (in kg) in the vessel is approximately

(a) 17 (b) 20

(c) 25 (d) 34

[2004 : 2 M]

6.8 If the percent humidity of air (30°C, total pressure 100 kPa) is 24% and the saturation pressure of water vapor at that temperature is 4 kPa, the percent relative humidity and the absolute humidity of air are

(a) 25.2, 0.0062 (b) 25, 0.0035
(c) 20.7, 0.0055 (d) 18.2, 0.0035

[2007 : 2 M]

6.9 At 25°C and 90% relative humidity, water evaporates from the surface of a lake at the rate of 1.0 kg/(m^2.h). The relative humidity that will lead to an evaporation rate of 3.0 kg/(m^2.h), with other conditions remaining the same, is

(a) 30% (b) 50%
(c) 60% (d) 70%

[2010 : 2 M]

6.10 For what value of Lewis number, the wet-bulb temperature and adiabatic saturation temperature are nearly equal?

(a) 0.33 (b) 0.5
(c) 1 (d) 2

[2016 : 1 M]

6.11 The humidity of air at a dry-bulb temperature of 65°C is 0.025 kg water/kg dry air. The latent heat of vaporization of water at 0°C is 2500 kJ/kg. The psychrometric ratio of air is 0.95 kJ (kg dry air)$^{-1}$ K^{-1}. Considering 0°C as reference temperature, the enthalpy of air (in kJ/kg) at its adiabatic saturation temperature of 35°C is ______ (rounded off to two decimal places).

[2018 : 2 M]

6.12 The dry-bulb temperature of air in a room is 30°C. The Antoine equation for water is given as

$$\ln P^{sat} = 12.00 - \frac{4000}{T - 40}$$

where T is the temperature in K and P^{sat} is the saturation vapor pressure in bar. The latent heat of vaporization of water is 2000 kJ kg^{-1}, the humid heat is 1.0 kJ kg^{-1} K^{-1}, and the molecular weights of air and water are 28 kg kmol^{-1} and 18 kg kmol^{-1}, respectively. If the absolute humidity of air is Y' kg moisture per kg dry air, then for a wet-bulb depression of 9 °C, $1000 \times Y'$ = ________ (rounded off to one decimal place).

[2022 : 2 M]

6.13 Partially saturated air at 1 bar and 50°C is contacted with water in an adiabatic saturator. The air is cooled and humidified to saturation, and exits at 25°C with an absolute humidity of 0.02 kg water per kg dry air. Use latent heat of vaporization of water as 2450 kJ.kg^{-1}, and average specific heat capacity for dry air and water, respectively as 1.01 kJ.kg^{-1}K^{-1} and 4.18 kJ.kg^{-1}K^{-1}. If the absolute humidity of air entering the adiabatic saturator is $H \times 10^{-3}$ water per kg dry air, the value of H is ________ (rounded off to two decimal places).

[2023 : 2 M]

■■■■

Answers Humidification

6.1	(26.3)	**6.2**	(c)	**6.3**	(c)	**6.4**	(d)	**6.5**	(d)	**6.6**	(c)	**6.7**	(a)
6.8	(a)	**6.9**	(d)	**6.10**	(c)	**6.11**	(124.25)	**6.12**	(11.3)	**6.13**	(9.3)		

Explanations Humidification

6.1 (26.3)

Given: $p_w^{sat} = 5$ kPa

$p_t = 1$ atm $= 101.325$ kPa

$$\frac{h_G}{k_y} = 2 \text{ kJ/kg-K}$$

$\lambda_w = 360$ kJ/kg

$t_w = 7.6°C$

$Y' = 0$

(As air is free from vapors of acetone)

We know

$$t_G - t_w = \frac{\lambda_w(Y'_w - Y')}{\left(\frac{h_G}{k_y}\right)} \quad ...(1)$$

where, t_G = Temperature of entering gas mixture

t_w = Wet bulb temperature

λ_w = Latent heat of vaporization at t_w

Y'_w = Absolute humidity at pressure p_t and wet-bulb temperature, t_w

Y' = Absolute humidity at pressure p_t, and dry-bulb temperature, t_G

$\frac{h_G}{k_y}$ = Psychrometric ratio

The absolute humidity at pressure p_t is given by

$$Y'_w = \frac{p_w^{sat}}{p_t - p_w^{sat}} \times \frac{M_{acetone}}{M_{air}} \quad ...(2)$$

Substituting the values in Eq. (2), we get

$$Y'_w = \left(\frac{5}{101.325 - 5}\right) \times \left(\frac{58}{29}\right)$$

$Y'_w = 0.1038$ kg of vapor-dry mixture

Substituting the given values and Y'_w in Eq. (1), we get

$$t_G - 7.6 = \frac{350(0.1038 - 0)}{2}$$

$$\Rightarrow \quad t_G = 26.3°C$$

6.2 (c)

For air-water vapor mixture

$$\frac{h_G}{k_y C_s} = 1$$

where, h_G = Heat transfer coefficient

k_y = Mass transfer coefficient

C_S = Molar specific heat

6.3 (c)

$$\text{Relative Humidity } (R) = \frac{p_A}{p_A^{sat}} \times 100$$

$$\therefore \quad p_A = \frac{R \times p_A^{sat}}{100} \quad ...(1)$$

$$\% \text{ humidity} = \frac{p_A(P_T - p_A^{sat})}{p_A^{sat}(P_T - P_A)} \times 100$$

By putting value of p_A from equation (1)

$$\% \text{ humidity} = \frac{R(P_T - p_A^{sat})}{\left(P_T - \frac{R \cdot p_A^{sat}}{100}\right)}$$

$$= 80 \times \frac{(750 - 17.5)}{\left(750 - \frac{80 \times 17.5}{100}\right)}$$

$$= 80 \times 0.995 = 79.62\%$$

6.4 (d)

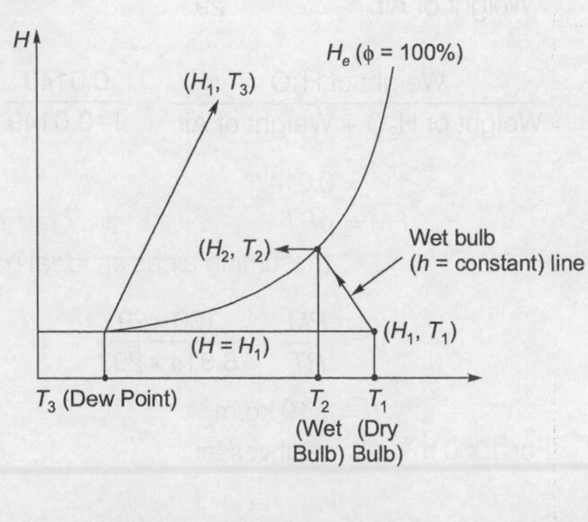

6.5 (d)

$$\text{Humidity} = \left(\frac{p_a}{P_T - p_a}\right)\frac{M_w}{M_{air}}$$

$$= \frac{19.91}{101.3 - 19.91} \times \frac{18}{29}$$

$$= 0.152 \frac{\text{kg water vapor}}{\text{kg dry air}}$$

6.6 (c)

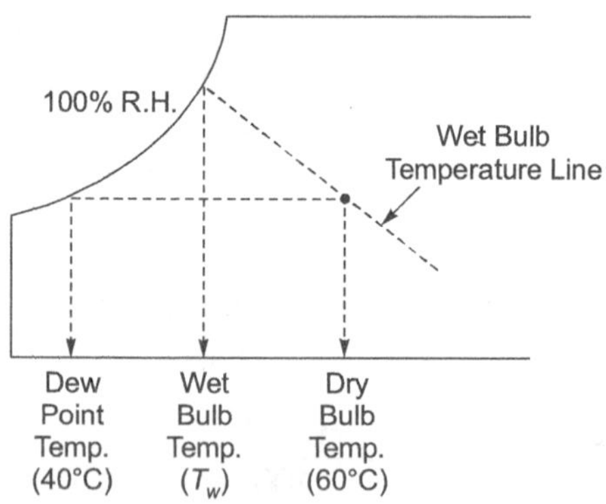

From fig., 40°C < T_w < 60°C

6.7 (a)

Volume of vessel, $V = 1000\ m^3$

$$P_T = 100 \text{ kPa}$$
$$T = 20°C$$

Vapour pressure of water at 20°C is 2.34 kPa

$$\frac{\text{Moles of } H_2O}{\text{Moles of Air}} = \frac{P_w^{sat}}{P_T - P_W^{sat}}$$

$$= \frac{2.34}{100 - 2.34} = 0.024$$

$$\frac{\text{Weight of } H_2O}{\text{Weight of Air}} = 0.024 \times \frac{18}{29} = 0.0149$$

$$\frac{\text{Weight of } H_2O}{\text{Weight of } H_2O + \text{Weight of Air}} = \frac{0.0149}{1+0.0149}$$

$$= 0.0147$$

$\because$ $PM = \rho RT$

(assuming air as an ideal gas)

$$\rho = \frac{PM}{RT} = \frac{100 \times 29}{8.314 \times 293}$$

$$= 1.19 \text{ kg/m}^3$$

For 1000 m^3 volume of vessel

Weight of H_2O + Air = 1.19 × 1000 = 1190 kg

Weight of H_2O = 1190 × 0.0147 = 17.49 kg

6.8 (a)

Percentage humidity = 24 (given)

$$= \frac{p_A}{P - p_A} \times \frac{P - p_A^{sat}}{p_A^{sat}} \times 100$$

where $P = 100$ kPa, $p_A^{sat} = 4$ kPa

$$\left(\frac{p_A}{100 - p_A}\right) \times \left(\frac{100-4}{4}\right) \times 100 = 24$$

$$100p_A = 100 - p_A$$
$$P_A = 0.990 \text{ kPa}$$

$$\% \text{ relative humidity} = \frac{p_A}{p_A^{sat}} \times 100$$

$$= \frac{0.990}{4} \times 100 = 24.75$$

Absolute humidity

$$= \frac{p_A}{P - p_A} \times \frac{18}{29}$$

$$= \frac{18 \times 0.990}{29(100 - 0.990)} = 0.0062$$

6.9 (d)

$$N_A = k_g(P_A^{sat} - P_A)$$

$$N_A = \frac{k_g}{P_A^{sat}}\left(1 - \frac{P_A}{P_A^{sat}}\right)$$

$$N_A = \frac{k_g}{P_A^{sat}}(1 - \text{R.H.})$$

$$N_A \propto (1 - \text{R.H.})$$

$$\frac{N_{A2}}{N_{A1}} = \frac{1-(R.H.)_2}{1-(R.H.)_1}$$

$$\frac{3}{1} = \frac{1-(R.H.)_2}{1-0.9}$$

$$1 - (\text{R.H.})_2 = 0.3$$
$$(\text{R.H.})_2 = 0.7 = 70\%$$

6.10 (c)

For air-water vapour system the wet bulb temperature and adiabatic saturation temperature are nearly equal when

$$C_s = \frac{h_g}{k_y}$$

$$\frac{h_g}{k_y \cdot C_s} = 1 = \text{Lewis number}$$

6.11 (124.25)

Humidity of air at dry bulb temperature at 65°C

$$= 0.025 \text{ kg water/kg dry air}$$

$$\lambda_{\text{at }0°C} = 2500 \text{ kJ/kg}$$

For water-air system

$$C_s = \frac{h}{k_y} = 0.95 \text{ kJ/kg dry air.K}$$

Reference temperature = 0°C

$$\text{Enthalpy of air} = C_s(T_{dry} - T_{ref}) + \lambda_{\text{at Ref.}} Y_{air}$$

$$= 0.95(65 - 0) + 2500 \times 0.025$$

$$= 124.25 \text{ kJ/kg}$$

6.12 (11.3)

We have the equation for wet bulb depression

$$T_D - T_W = \frac{\lambda[Y_s' - Y']}{c_s}$$

T_D = Dry bulb temperature

T_W = Wet bulb temperature

Y_s' = Saturation humidity

c_s = Humid heat

For Y_s', we have

$$\ln P_s = 12 - \frac{4000}{294 - 40}$$

$$P_s = 0.024 \text{ bar}$$

$$Y_s' = \left(\frac{0.024}{1 - 0.024}\right)\frac{18}{28} = 0.0158$$

Now,
$$9 = \frac{2000(0.0158 - Y')}{1}$$

$$Y' = 0.0113$$

$$1000 \times Y' = 11.3$$

6.13 (9.3)

Given:

T_{DB} = 50°C (Dry bulb temperature)

T_W = 25°C (Wet bulb temperature)

L = 2450 kJ/kg

C_{Pw} = 4.18 kJ/kgK {Water}

$(C_P)_{air}$ = 1.01 kJ/kgK {Air}

γ_w' = 0.02 kg H_2O/kg dry air

For the humid heat, we have

$$C_S = [(C_P)_{air} + \gamma(C_P)_{water}]$$

$$C_S = 1.01 + 4.18\gamma' \quad ...(i)$$

From the energy balance, we have

$$T_D - T_w = \frac{\lambda(\gamma_w' - \gamma')}{C_S}$$

$$(50 - 25) = \frac{2450[0.02 - \gamma']}{1.01 + 4.18\gamma'}$$

Solving, we get $\gamma' = 9.29 \times 10^{-3}$

Thus, $H = 9.29$

■■■■

7 Drying

7.1 In a laboratory test run, the rate of drying was found to be 0.5×10^{-3} kg/m^{2}s when the moisture content reduced from 0.4 to 0.1 on a dry basis. The critical moisture content of the material is 0.08 on a dry basis. A tray drier is used to dry 100 kg (dry basis) of the same material under identical conditions. The surface area of the material is 0.04 m^2/kg of dry solid. The time required (in seconds) to reduce the moisture content of the solids from 0.3 to 0.2 (dry basis) is

(a) 2000 (b) 4000
(c) 5000 (d) 6000 [2000 : 2 M]

7.2 200 kg of solid (on dry basis) is subjected to a drying process for a period of 5000 s. The drying occurs in the constant rate period with the drying rate as $N_C = 0.5 \times 10^{-3}$ kg/m^{2}s. The initial moisture content of the solid is 0.2 kg moisture/kg dry solid. The interfacial area available for drying is 4 m^2/100 kg of dry solid. The moisture content at the end of the drying period is (in kg moisture/kg dry solid)

(a) 0.5 (b) 0.05
(c) 0.1 (d) 0.15 [2001 : 2 M]

7.3 It takes 6 hours to dry a wet solid from 50% moisture content to the critical moisture content of 15%. How much longer will it take to dry the solid to 10% moisture content, under the same drying conditions (the equilibrium moisture content of the solid is 5%)

(a) 15 min (b) 51 min
(c) 71 min (d) 94 min

[2002 : 2 M]

7.4 A solid is being dried in the linear drying rate regime from moisture content X_0 to X_F. The drying rate is zero at $X = 0$ and the critical moisture content is the same as the initial moisture, X_0. The drying time for $M = L_S/AR_C$ is : (where L_S: total mass of dry solids, A: total surface area for drying, R_C: constant maximum drying rate per unit area, and X : moisture content (in mass of water/mass of dry solids)

(a) $M(X_0 - X_F)$ (b) $M\left(\frac{X_0}{X_F}\right)$
(c) $M\ln\left(\frac{X_0}{X_F}\right)$ (d) $MX_0\ln\left(\frac{X_0}{X_F}\right)$

[2003 : 2 M]

7.5 A 25 cm × 25 cm × 1 cm flat sheet weighing 1.2 kg initially was dried from both sides under constant drying rate conditions. It took 1500 seconds for the weight of the sheet to reduce to 1.05 kg. Another 1 m × 1 m × 1 cm flat sheet of the same material is to be dried from one side only. Under the same constant drying rate conditions, the time required for drying (in seconds) from its initial weight of 19.2 kg to 17.6 kg is

(a) 1000 (b) 1500
(c) 2000 (d) 2500

[2004 : 2 M]

7.6 A 50 cm × 50 cm × 1 cm flat wet sheet weighing 2 kg initially, was dried from both the sides under constant drying rate period. It took 1000 secs for the weight of the sheet to reduce to 1.75 kg. Another 1 m × 1 m × 1 cm flat sheet is to be dried from one side only. Under the same drying rate and other conditions, time required for drying (in secs) from initial weight of 4 kg to 3 kg is

(a) 1000 (b) 1500
(c) 2000 (d) 2500

[2007 : 2 M]

7.7 The equilibrium moisture curve for a solid is shown below :

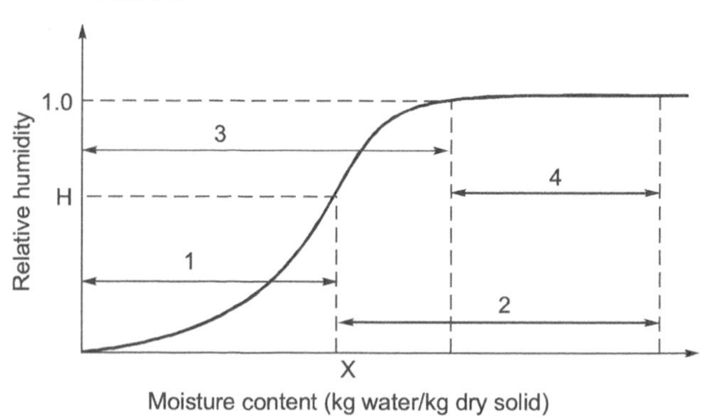

The total moisture content of the solid is X and it is exposed to air of relative humidity H. In the table below, Group-I lists the types of moisture, and Group-II represents the region in the graph above.

	Group-I	Group-II
P.	Equilibrium moisture	1
Q.	Bound moisture	2
R.	Unbound moisture	3
S.	Free moisture	4

Which one of the following is the correct match?

(a) P-1, Q-2, R-3, S-4 (b) P-1, Q-3, R-4, S-2
(c) P-1, Q-4, R-2, S-3 (d) P-1, Q-2, R-4, S-3

[2009 : 2 M]

7.8 A batch of 120 kg wet solid has initial moisture content of 0.2 kg water/kg dry solid. The exposed area for drying is 0.05 m^2/kg dry solid. The rate of drying follows the curve given below.

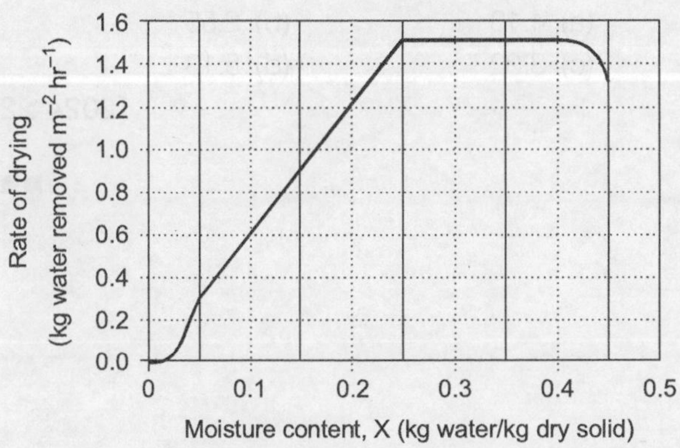

The time required (in hours) for drying this batch to a moisture content of 0.1 kg water/kg dry solid is

(a) 0.033 (b) 0.43
(c) 0.6 (d) 2.31

[2011 : 2 M]

7.9 Consider the drying operation shown in the figure below for a solid loading (dry basis) of 50 kg/m^2 with a constant drying rate of 5 kg/m^2.h. The falling rate of drying is linear with moisture content.

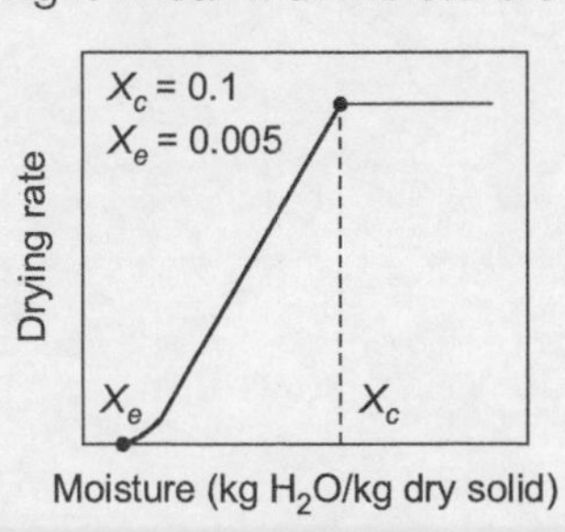

The drying time (in hrs) required to reduce an initial moisture content of 25% to a final moisture content of 2% is

(a) 1.55 (b) 1.75
(c) 3.25 (d) 4.55

[2012 : 2 M]

7.10 A wet solid is dried over a long period of time by unsaturated air of nonzero constant relative humidity. The moisture content eventually attained by the solid is termed as the

(a) unbound moisture content
(b) bound moisture content
(c) free moisture content
(d) equilibrium moisture content

[2013 : 1 M]

7.11 A wet solid of 100 kg is dried from a moisture content of 40 wt% to 10 wt%. The critical moisture content is 15 wt% and the equilibrium moisture content is negligible. All moisture contents are on dry basis. The falling rate is considered to be linear. It takes 5 hours to dry the material in the constant rate period. The duration (in hours) of the falling rate period is ________.

[2014 : 2 M]

7.12 A fiberboard sheet (1.5 m × 2.0 m × 15 mm) is being dried by suspending it horizontally in a current of hot, dry air. The edges are insulated so that drying takes place only from the top and bottom surfaces. The wet sheet weighing 16 kg with initial moisture content of 60% loses moisture at a constant rate of 1.25×10^{-5} kg m^{-2} s^{-1} until the moisture content falls to 30%. All moisture contents are on dry basis. The time required for drying during constant rate period (in hour) is ________ (rounded off to third decimal place).

[2018 : 2 M]

7.13 In the drying of non-dissolving solids at constant drying conditions, the internal movement of moisture in the solids has a dominant effect on the drying rate during

(a) the initial adjustment period only
(b) the constant rate period only
(c) the falling rate period only
(d) both the initial adjustment and constant rate periods

[2019 : 1 M]

7.14 In a batch drying experiment, a solid with a critical moisture content of 0.2 kg H_2O/kg dry solid is dried from an initial moisture content of 0.35 kg H_2O/kg dry solid to a final moisture content of 0.1 kg H_2O/kg dry solid in 5 h. In the constant rate regime, the rate of drying is 2 kg $H_2O/(m^2.h)$ The entire falling rate regime is assumed to be uniformly linear. The equilibrium moisture content is assumed to be zero.

The mass of the dry solid per unit area is _____ kg/m^2 (round off to nearest integer).

[2021 : 2 M]

7.15 A wet solid containing 20% (w/w) moisture (based on mass of bone-dry solid) is dried in a tray-dryer. The critical moisture content of the solid is 10% (w/w). The drying rate (kg m^{-2} s^{-1}) is constant for the first 4 hours, and then decreases linearly to half the initial value in the next 1 hour. At the end of 5 hours of drying, the percentage moisture content of the solid is ________% (w/w) (rounded off to one decimal place).

[2022 : 1 M]

7.16 Wet air containing 10 mole percent water vapor is dried by continuously passing it through a column of $CaCl_2$ pellets. The pellets remove 50 percent of water from wet air entering the column. The mole percent of water vapor in the product stream exiting the column is _______ (rounded off to two decimal places).

[2023 : 2 M]

7.17 Alumina particles with an initial moisture content of 5 kg per kg dry solid are dried in a batch dryer. For the first two hours, the measured drying rate is constant at 2 kg m^{-2} h^{-1}. Thereafter, in the falling-rate period, the rate decreases linearly with the moisture content. The equilibrium moisture content is 0.05 kg per kg dry solid and the drying area of the particles is 0.5 m^2 per kg dry solid. The total drying time, in h, to reduce the moisture content to half its initial value is

(a) 4.13 (b) 2.55
(c) 3.22 (d) 5.13

[2024 : 2 M]

■■■■

Answers Drying

7.1	(c)	**7.2**	(c)	**7.3**	(c)	**7.4**	(d)	**7.5**	(c)	**7.6**	(c)	**7.7**	(b)
7.8	(d)	**7.9**	(c)	**7.10**	(d)	**7.11**	(1.22)	**7.12**	(11.11)	**7.13**	(c)	**7.14**	(34.64)
7.15	(8.2)	**7.16**	(5.26%)	**7.17**	(b)								

Explanations Drying

7.1 (c)

In the lab test,

$$X_c = 0.08 \text{ kg moisture/kg dry solid}$$
$$X_1 = 0.4 \text{ kg moisture/kg dry solid}$$
$$X_2 = 0.1 \text{ kg moisture/kg dry solid}$$

Since, $X_2 > X_c$, it is the problem of constant rate period. Time required for drying in a constant rate period is given by

$$t_C = \frac{W_S}{AN_C}(X_1 - X_2) \quad \text{...(1)}$$

Given, $\frac{A}{W_S} = 0.04 \text{ m}^2\text{/kg}$

$$X_1 = 0.3 \text{ kg moisture/kg dry solid}$$
$$N_c = 0.5 \times 10^{-3} \text{ kg/m}^2\text{-s}$$
$$X_2 = 0.2 \text{ kg moisture/kg dry solid}$$

Substituting the values in eqn. (1), we get

$$t_C = \frac{0.3 - 0.2}{0.04 \times 0.5 \times 10^{-3}}$$

$$\Rightarrow \quad t_C = 5000 \text{ s}$$

7.2 (c)

Drying rate at constant rate period is

$$N_c = \frac{W_s}{A} \cdot \frac{(X_1 - X_2)}{t}$$

$$0.5 \times 10^{-3} = \frac{200}{8} \cdot \frac{(0.2 - X_2)}{5000}$$

$$X_2 = 0.1$$

7.3 (c)

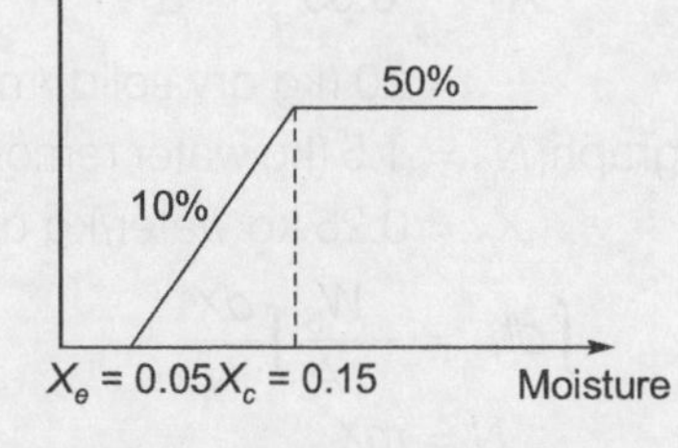

Given : Constant rate period = 6 hours

$$X_1 = 0.5, X_c = 0.15$$
$$X_e = 0.05, X_F = 0.10$$

For constant rate period

$$t_1 = \frac{W_s}{AR_c}(X_1 - X_c)$$

$$t_1 = \frac{W_s}{AR_c}(0.5 - 0.15)$$

$$t_1 = 0.35\frac{W_s}{AR_c}$$

$$\frac{6}{0.35} = \frac{W_s}{AR_c} = 17.14 \frac{\text{kg } H_2O}{\text{kg solid}}.\text{h}$$

Now for falling rate period

$$t_2 = \frac{W_s}{AR_c}(X_c - X_e)\ln\left(\frac{X_c - X_e}{X_f - X_e}\right)$$

$$t_2 = 17.14(0.15 - 0.05)\ln\left(\frac{0.15 - 0.05}{0.10 - 0.05}\right)$$

$$t_2 = 1.18 \text{ hr} = 70.80 \text{ min}$$

7.4 (d)

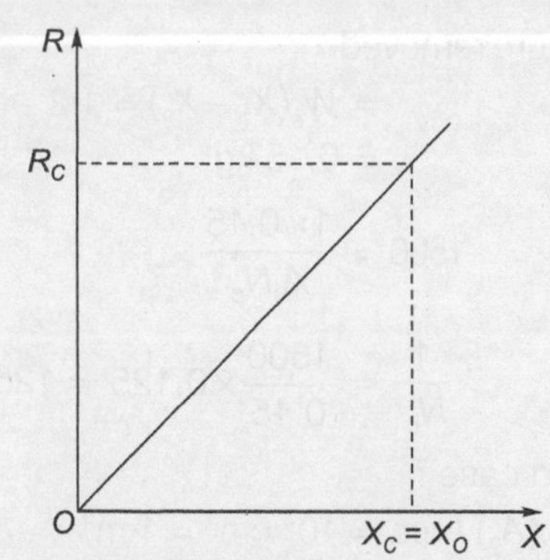

Finding the equation of this line

$$R = mX$$

where Slope, $m = \frac{R_c}{X_o}$

Since drying rate,

$$R = \frac{-W_s}{A} \cdot \frac{dX}{d\theta}$$

$$d\theta = \frac{-W_s}{AR}dX$$

Putting value of R

$$d\theta = \frac{-W_s X_o}{AR_c}\cdot\frac{dX}{X}$$

On integrating above equation

$$\int_0^\theta d\theta = \frac{-W_s X_o}{AR_c}\int_{X_o}^{X_F}\frac{dX}{X}$$

$$\theta = \frac{W_s X_o}{AR_c}\ln\left(\frac{X_o}{X_F}\right)$$

where, $M = \frac{W_s}{AR_c}$ (given)

Then, $\theta = MX_o \ln\left(\frac{X_o}{X_F}\right)$

7.5 (c)

In 1st case :

Area of drying surface

$= 2(25 \times 25) = 1250\ cm^2$

$= 0.125\ m^2$

For constant drying rate

$$N_c = \frac{-W_s}{A}\times\frac{dX}{d\theta}$$

$$\int_0^\theta d\theta = \frac{-W_s}{AN_c}\int_{X_1}^{X_2}dX$$

$$\theta_1 = \frac{W_s}{A_1 N_c}(X_1 - X_2)$$

Moisture removed

$= W_s(X_1 - X_2) = 1.2 - 1.05$

$= 0.15$ kg

$$1500 = \frac{1\times 0.15}{A_1 N_c}$$

$$\frac{1}{N_c} = \frac{1500}{0.15}\times 0.125 = 1250$$

In 2nd case :

(A_2) Area $= 10^4\ cm^2 = 1\ m^2$

Moisture removed

$= W_s(X_1 - X_2)$

$= 19.2 - 17.6 = 1.6$ KG

$$\theta_2 = \frac{1}{A_2 N_c}\times 1.6$$

$$\theta_2 = 1250 \times 1.6 = 2000\ s$$

7.6 (c)

For the constant rate period

$$t = \frac{W_s}{AN_c}(X_1 - X_2)$$

where, t = Time of drying

W_s = kg dry solid

N_c = Constant rate of drying

X_1 and X_2 is initial and final moisture content.

Case 1 : $A = 2(50 \times 50)\times 10^{-4}$

$A = 0.50\ m^2$ (both side)

Moisture removed from sheet = Initial weight of sheet – final weight of sheet

$X_1 - X_2 = (2 - 1.75)$ kg $= 0.25$ kg

$$1000 = \frac{W_s}{AN_c}\times 0.25$$

$$1000 = \frac{W_s}{0.50N_c}\times 0.25$$

$$\frac{W_s}{N_c} = 2000$$

Case 2 : Area of drying = $1\ m^2$

Moisture removed from sheet = $(4 - 3)$ kg = 1 kg

$$t_2 = \frac{W_s}{N_c A}(X_1 - X_2)$$

$$t_2 = \frac{2000}{1}\times 1 = 2000\ s$$

7.7 (b)

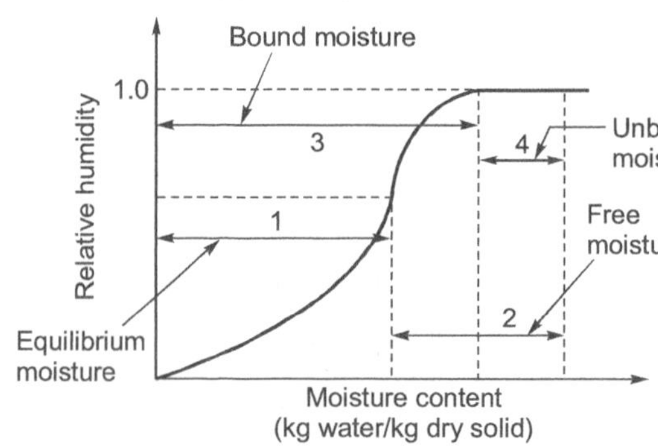

7.8 (d)

Given : $\frac{W_S}{A} = \frac{1}{0.05} = \frac{100}{5}$

$= 20$ (kg dry solid)/m^2

From graph, $N_C = 1.5$ (kg water removed)/m^2.h

$X_C = 0.25$ kg water/kg dry solid

$$\int dt_f = -\frac{W_S}{A}\int\frac{dX}{N}$$

$N = mX$

$$\text{Slope } m = \frac{1.2-0.6}{0.2-0.1} = 6$$

$$t_f = -\frac{W_S}{A\cdot m}[\ln X]_{0.2}^{0.1}$$

$$t_f = \frac{20}{6}[\ln 0.2 - \ln 0.1] = 2.31 \text{ hr}$$

7.9 (c)

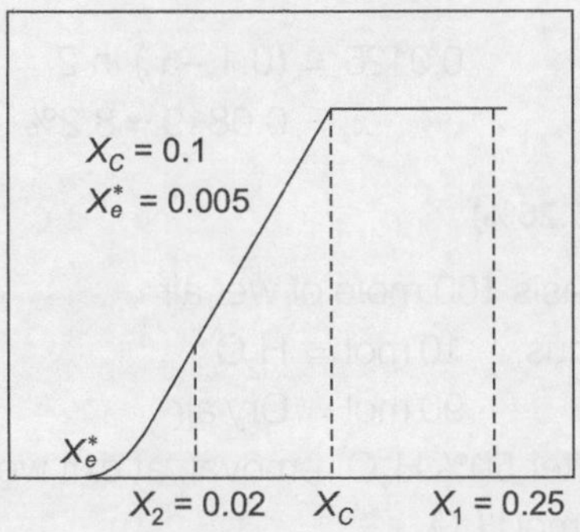

Total time $= t_c + t_f$

t_c = Time required for the constant rate period

t_f = Time required for the falling rate period

$$t_c = \frac{W_s}{AN_c}(X_1 - X_C)$$

$$\frac{W_s}{A} = 50 \text{ kg/m}^2$$

$$N_c = 5.0 \text{ (kg H}_2\text{O)/m}^2\text{h}$$

$$t_c = \frac{50}{5}(0.25-0.1) = 1.5 \text{ hr}$$

$$t_f = \frac{W_s}{AN_c}(X_c - X_e^*)\ln\left(\frac{X_c - X_e^*}{X_f - X_e^*}\right)$$

$$t_f = \frac{50}{5}(0.10-0.005)\ln\left(\frac{0.10-0.005}{0.02-0.005}\right)$$

$$t_f = 10 \times 0.175 = 1.75 \text{ hr.}$$

Total time $= t_c + t_f = 1.5 + 1.75 = 3.25$ hr.

7.11 (1.22)

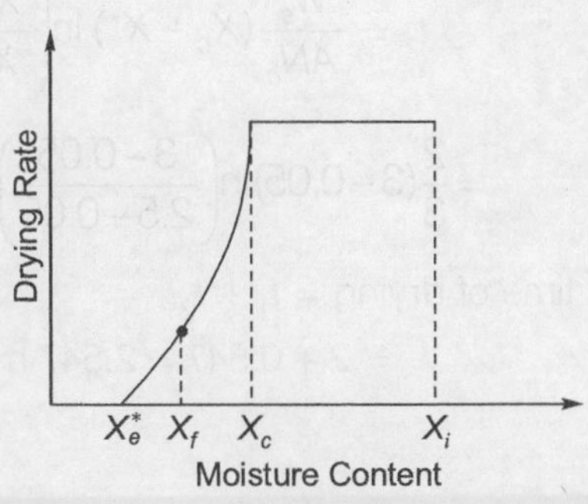

Given :

Wet solid = 100 kg

$X_i = 0.40,\ X_f = 0.10$

$X_c = 0.15$

$X_e^* = 0$

Moisture content is always expressed as

$$= \frac{\text{kg moisture}}{\text{kg dry solid}}$$

In constant rate period

$$t_c = \frac{W_s}{aN_c}(X_i - X_c)$$

$$5 = \frac{W_s}{aN_c}(0.4-0.15)$$

$$\frac{W_s}{aN_c} = \frac{5}{0.4-0.15}$$

$$= 20\ \frac{\text{(kg dry solid).(hr)}}{\text{(kg moisture)}}$$

In falling rate period

$$t_f = \frac{W_s}{a\cdot N_c}(X_c - X_e^*)\ln\frac{(X_c - X_e^*)}{(X_f - X_e^*)}$$

$$t_f = 20(0.15-0)\ln\left(\frac{(0.15-0)}{(0.10-0)}\right)$$

$$t_f = 3 \times 0.4054 = 1.216 \text{ hr.}$$

7.12 (11.11)

Fiberboard sheet (1.5 m × 2.0 m × 1.5 mm) drying only from top and bottom surfaces.

Wet sheet weight = 16 kg

$$(S_s) \text{ dry weight} = 16\left(1-\frac{0.6}{1.6}\right) = 10 \text{ kg}$$

$$N_c = 1.25 \times 10^{-5} \text{ kg/m}^2\text{sec}$$

$$X_i = 0.6,\ X_F = 0.3$$

Time required for constant rate period

$$t_c = \frac{-S_s}{N_c\cdot A}(X_F - X_i)$$

$$= \frac{10}{1.25\times10^{-5}\times2\times1.5\times2}\times(0.6-0.3)$$

$$= 40{,}000 \text{ s}$$

$$t_c = 11.11 \text{ hr.}$$

7.13 (c)

In the non-dissolving solids drying, the internal movement of moisture in the solid has a dominant effect on the drying rate during the falling rate period only.

7.14 (34.64)

Given: $X_c = 0.2$ kg H_2O/kg dry solid
$X_i = 0.35$ kg H_2O/kg dry solid
$X_f = 0.1$ kg H_2O/kg dry solid
Total time, $t_T = 5$ h
$N_c = 2$ kg H_2O/m^2h, $X^* = 0$

$$t_T = \frac{W_s}{A_cN_c}\left[(X_i - X_c) + (X_c - X^*)\ln\left(\frac{X_c - X^*}{X_f - X^*}\right)\right]$$

$$5 = \frac{W_s}{A_c \times 2}\left[(0.35-0.2) + 0.2\ln\left(\frac{0.2}{0.1}\right)\right]$$

$$\frac{W_s}{A_c} = 34.64 \text{ kg/m}^2$$

7.15 (8.2)

Data given:
Initial moisture, $x_i = 0.20$
Critical moisture, $x_c = 0.10$
For constant drying rate period, we have

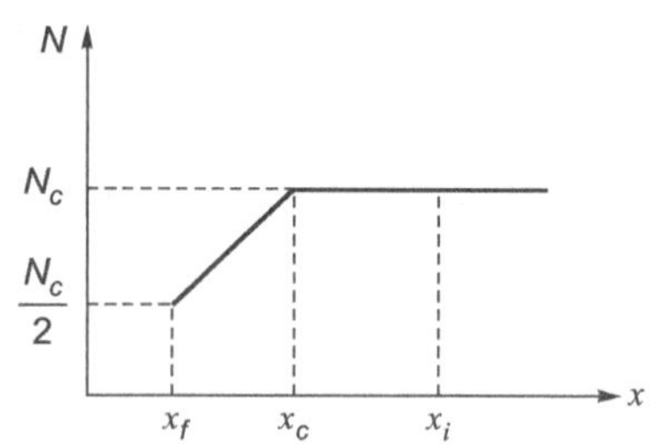

$$\theta_c = \frac{S_s}{AN_c}[x_i - x_c]$$

$$4 \times 3600 = \frac{S_s}{AN_c}(0.20 - 0.10)$$

$$\frac{S_s}{AN_c} = 144000 \qquad \text{(i)}$$

For falling rate period, the rate expression will be:

$$N - \frac{N_c}{2} = \left(\frac{N_c - \frac{N_c}{2}}{x_c - x_f}\right)(x - x_f)$$

$$N = \frac{N_c}{2} + \frac{N_c}{2}\left(\frac{x - x_f}{0.10 - x_f}\right)$$

$$\theta_f = \frac{S_s}{A}\int_{x_f}^{0.10}\frac{dx}{N}$$

$$\Rightarrow \quad 3600 \times 1 = \frac{S_s}{A}\int_{x_f}^{0.10}\frac{dx}{\frac{N_c}{2} + \frac{N_c}{2}\frac{(x - x_f)}{(0.10 - x_f)}}$$

$$3600 = 2\left(\frac{S_s}{AN_c}\right)\int_{x_f}^{0.1}\frac{(0.10 - x_f)dx}{[0.10 - 2x_f + x]}$$

$$0.0125 = \int\frac{(0.1 - x_f)dx}{(0.10 - 2x_f + x)}$$

Solving we get,

$$0.0125 = (0.1 - x_f)\ln|0.1 - 2x_f + x|_{x_f}^{0.2}$$

$$0.0125 = (0.1 - x_f)\ln 2$$

$$x_f = 0.0819 \simeq 8.2\%$$

7.16 (5.26%)

Basis 100 mole of wet air
Thus, 10 mol = H_2O
90 mol = Dry air
After 50% H_2O removal, at exit we have
5 mol H_2O
90 mol dry air

$$X_{\text{water}} = \frac{5}{95} = 0.053 = 5.3\%$$

7.17 (b)

Initial moisture constant $X_I = 5$ kg/kg of dry solid

Constant drying rate $N_c = \frac{2 \text{ kg}}{\text{m}^2\text{.h}}$

Equilibrium moisture constant $X^* = 0.05$ kg/kg dry solid

For constant rate of drying

$$t_1 = \frac{W_s}{AN_c}(X_I - X_c)$$

$$\frac{W_s}{AN_c} = 2 \text{ kg of dry solid/m}^2 \text{ (given)}$$

X_c = Critical moisture conent

$$2 = \frac{2}{2}(5 - X_c)$$

$$X_c = 3 \text{ kg/kg of dry solid}$$

For falling rate of drying

$$t_2 = \frac{W_s}{AN_c}(X_c - X^*)\ln\left(\frac{X_c - X^*}{X_I - X^*}\right)$$

$$= \frac{2}{3}(3 - 0.05)\ln\left(\frac{3 - 0.05}{2.5 - 0.05}\right) = 0.547 \text{ h}$$

Total time of drying = $t_1 + t_2$
= 2 + 0.547 = 2.547 h

■■■■

8 Membrane Separation

8.1 In reverse osmosis, the hydraulic pressure and osmotic pressure at the feed side of the membrane are P_1 and π_1, respectively. The corresponding values are P_2 and π_2 at the permeate side. The membrane, feed, and permeate are at the same temperature. For equilibrium to prevail, the general criterion that should be satisfied is

(a) $\pi_1 = \pi_2$
(b) $P_1 = P_2$
(c) $P_1 + \pi_1 = P_2 + \pi_2$
(d) $P_1 - \pi_1 = P_2 - \pi_2$

[2021 : 1 M]

8.2 Which one of the following represents the CORRECT effects of concentration polarization in a reverse osmosis process?

(a) Reduced water flux and reduced solute rejection
(b) Increased water flux and increased solute rejection
(c) Reduced water flux and increased solute rejection
(d) Increased water flux and reduced solute rejection

[2023 : 2 M]

Answers Membrane Separation

8.1 (d) **8.2** (a)

Explanations Membrane Separation

8.1 (d)

In reverse osmosis

Hydraulic pressure drop = Osmotic pressure drop

$$P_1 - P_2 = \pi_1 - \pi_2$$
$$P_1 - \pi_1 = P_2 - \pi_2$$

8.2 (a)

Concentration polarization reduces the flux of water because the increase in osmotic pressure reduces the driving force for water transport and the solute rejection decreases because of lower water flux.

Chemical Reaction Engineering

UNIT **VII**

CONTENTS

Chemical Reaction Engineering

Syllabus

Theories of Reaction Rates; Kinetics of Homogeneous Reactions, Interpretation of Kinetic Data, Single and Multiple Reactions in Ideal Reactors, Kinetics of Enzyme Reactions (Michaelis-Menten and Monod Models), Non-Ideal Reactors; Residence Time Distribution, Single Parameter Model; Non-Isothermal Reactors; Kinetics of Heterogeneous Catalytic Reactions; Diffusion Effects in Catalysis; Rate and Performance Equations for Catalyst Deactivation.

Analysis of Previous GATE Papers

Exam Year	1 Mark Ques.	2 Marks Ques.	5 Marks Ques.	Total Marks
2000	4	5	–	14
2001	3	4	–	11
2002	3	3	2	19
2003	3	5	–	13
2004	6	8	–	22
2005	–	8	–	16
2006	3	8	–	19
2007	–	10	–	20
2008	1	9	–	19
2009	3	4	–	11
2010	1	4	–	9
2011	–	6	–	12
2012	2	4	–	10
2013	3	4	–	11
2014	2	3	–	8
2015	3	4	–	11
2016	3	5	–	13
2017	3	4	–	11
2018	2	5	–	12
2019	2	4	–	10
2020	3	4	–	11
2021	2	3	–	8
2022	2	4	–	10
2023	1	4	–	9
2024	4	4	–	12

1 Kinetics of Homogeneous Reactions

1.1 The experimentally determined overall order for the reaction $A + B \rightarrow C + D$ is two. Then the

(a) reaction is elementary with a molecularity of 2.
(b) molecularity of the reaction is 2, but the reaction may not be elementary.
(c) reaction may be elementary with a molecularity of 2.
(d) reaction is elementary but the molecularity may not be 2.

[2000 : 1 M]

1.2 The reaction rate constants at two different temperatures T_1 and T_2 are related by

(a) $\ln\left(\frac{k_2}{k_1}\right) = \frac{E}{R}\left(\frac{1}{T_2} - \frac{1}{T_1}\right)$

(b) $\ln\left(\frac{k_2}{k_1}\right) = \frac{E}{R}\left(\frac{1}{T_1} - \frac{1}{T_2}\right)$

(c) $\exp\left(\frac{k_2}{k_1}\right) = \frac{E}{R}\left(\frac{1}{T_1} - \frac{1}{T_2}\right)$

(d) $\exp\left(\frac{k_2}{k_1}\right) = \frac{E}{R}\left(\frac{1}{T_2} - \frac{1}{T_1}\right)$

[2001 : 1 M]

1.3 The extent of a reaction is

(a) different for reactants and products
(b) dimensionless
(c) dependent on the stoichiometric coefficients
(d) all of the above

[2002 : 1 M]

1.4 For a series of reactions $A \xrightarrow{k_1} B \xrightarrow{k_2} C$ having $k_1 << k_2$, the reaction system can be approximated as

(a) $A \xrightarrow{k_1} B$ (b) $A \xrightarrow{k_2} B$

(c) $A \xrightarrow{k_2} C$ (d) $A \xrightarrow{k_1} C$

[2003 : 1 M]

1.5 Find a mechanism that is consistent with the rate equation and reaction given below:

$2A + B \rightarrow A_2B \; (-r_A) = kC_AC_B$

(a) $A + B \rightleftharpoons AB;\; AB + A \rightarrow A_2B$
(b) $A + B \rightarrow AB;\; AB + A \rightarrow A_2B$
(c) $A + A \rightarrow AA;\; AA + B \rightarrow A_2B$
(d) $A + A \rightleftharpoons AA;\; AA + B \rightarrow A_2B$

[2003 : 1 M]

1.6 The rate expression for the gaseous phase reaction $CO + 2H_2 \Leftrightarrow CH_3OH$ is given by $r = k_1 p^{\alpha}_{CO}\, p^{\beta}_{H_2} - k_2 p^{\gamma}_{CH_2OH}$. Which of the following is NOT possible?

(a) $\alpha = 1, \beta = 1, \gamma = 1$
(b) $\alpha = 1, \beta = 2, \gamma = 1$
(c) $\alpha = \frac{1}{3}, \beta = \frac{2}{3}, \gamma = \frac{1}{3}$
(d) $\alpha = \frac{1}{2}, \beta = 1, \gamma = \frac{1}{2}$

[2004 : 1 M]

1.7 The rate of ammonia synthesis for the reaction $N_2 + 3H_2 \Leftrightarrow 2NH_3$ is given by $r = 0.8p_{N_2}\,p^3_{H_2} - 0.6p^2_{NH_2}$. If the reaction is represented as, $\frac{1}{2}N_2 + \frac{3}{2}H_2 \Leftrightarrow NH_3$ the rate of ammonia synthesis is

(a) $r = 0.8p^{0.5}_{N_2}p^{1.5}_{H_2} - 0.6p_{NH_3}$

(b) $r = 0.8p_{N_2}p^3_{H_2} - 0.6p^2_{NH_3}$

(c) $r = 0.5(0.8p_{N_2}p^3_{H_2} - 0.6p^2_{NH_3})$

(d) $r = 0.5(0.8p^{0.5}_{N_2}p^{1.5}_{H_2} - 0.6p^2_{NH_3})$

[2004 : 1 M]

1.8 For the reaction $2R + S \rightarrow T$, the rates of formation, r_R, r_S and r_T of the substances R, S and T respectively, are related by

(a) $2r_R = r_S = r_T$ (b) $2r_R = r_S = -r_T$
(c) $r_R = 2r_S = 2r_T$ (d) $r_R = 2r_S = -2r_T$

[2005 : 2 M]

1.9 Which is the correct statement from the following statements on the Arrhenius model of the rate constant $k = -Ae^{-E/RT}$?

(a) A is always dimensionless.
(b) For two reactions 1 and 2, if $A_1 = A_2$ and $E_1 > E_2$, then $k_1(T) > k_2(T)$
(c) For a given reaction, the % change of k with respect to temperature is higher at lower temperatures.
(d) The % change of k with respect to temperature is higher for higher A. [2005 : 2 M]

1.10 The rate expression for the reaction of A is given by

$$-r_A = \frac{k_1 c_A^2}{1 + k_2 c_A^{1/2}}$$

The units of k_1 and k_2 are, respectively,
(a) $(mol^{-1}\ m^3\ s^{-1})$, $(mol^{-1/2}\ m^{3/2})$
(b) $(mol^{-1}\ m^3\ s^{-1})$, $(mol^{1/2}\ m^{-3/2})$
(c) $(mol\ m^{-3}\ s^{-1})$, $(mol^{1/2}\ m^{3/2}\ s^{-1})$
(d) $(mol^{-1}\ m^3\ s^{-1})$, $(mol^{-1/2}\ m^{3/2}\ s^{-1/2})$
[2005 : 2 M]

1.11 The reaction $2A + B \rightarrow 2C$ occurs on a catalyst surface. The reactants A and B diffuse to the catalyst surface and get converted completely to the product C, which diffuses back. The steady state molar fluxes of A, B and C are related by
(a) $N_A = 2N_B = N_C$
(b) $N_A = -\left(\frac{1}{2}\right)N_B = -N_C$
(c) $N_A = 2N_B = -N_C$
(d) $N_A = \left(\frac{1}{2}\right)N_B = N_C$
[2006 : 1 M]

1.12 For a homogeneous reaction system, where
C_j is the concentration of j at time t
N_j is the number of moles of j at time t
V is the reaction volume at time t
t is the reaction time
The rate of reaction for species j is defined as
(a) $\frac{dC_j}{dt}$ (b) $-\left(\frac{dC_j}{dt}\right)$
(c) $\frac{1}{V}\cdot\frac{dN_j}{dt}$ (d) $-\left(\frac{1}{V}\cdot\frac{dN_j}{dt}\right)$
[2009 : 1 M]

1.13 Which of the following can change if only the catalyst is changed for a reaction system?
(a) Enthalpy of reaction
(b) Activation energy
(c) Free energy of the reaction
(d) Equilibrium constant
[2015 : 1 M]

1.14 For which reaction order, the half-life of the reactant is half of the full lifetime (time for 100 % conversion) of the reactant?
(a) Zero order (b) Half order
(c) First order (d) Second order
[2015 : 1 M]

1.15 An irreversible, homogeneous reaction $A \rightarrow$ products, has the rate expression:

$$\text{Rate} = \frac{2C_A^2 + 0.1C_A}{1 + 50C_A},$$

where C_A is the concentration of A.
C_A varies in the range 0.5 - 50 mol/m^3.
For very high concentrations of A, the reaction order tends to:
(a) 0 (b) 1
(c) 1.5 (d) 2
[2015 : 1 M]

1.16 For a non-catalytic homogeneous reaction $A \rightarrow B$, the rate expression at 300 K is $-r_A(mol\ m^{-3}\ s^{-1})$ $= \frac{10C_A}{1 + 5C_A}$, where C_A is the concentration of A (in mol/m^3). Theoretically, the upper limit for the magnitude of the reaction rate ($-r_A$ in mol m^{-3} s^{-1}, rounded off to the first decimal place) at 300 K is ________.
[2016 : 1 M]

1.17 Hydrogen iodide decomposes through the reaction $2HI \rightleftharpoons H_2 + I_2$. The value of the universal gas constant R is 8.314 J mol^{-1} K^{-1}. The activation energy for the forward reaction is 184000 J mol^{-1}. The ratio (rounded off to the first decimal place) of the forward reaction rate at 600 K to that at 550 K is ________. [2016 : 1 M]

1.18 For a chemical reaction, the ratio of rate constant at 500 K to that at 400 K is 2.5. Given R = 8.314 J mol^{-1} K^{-1}, the value of activation energy (in kJ/mol) is
(a) 10.5 (b) 12.0
(c) 15.2 (d) 18.4
[2018 : 1 M]

1.19 The following homogeneous liquid phase reactions are at equilibrium:

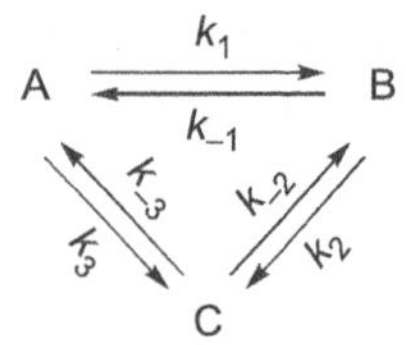

The values of rate constants are given by: $k_1 = 0.1\ s^{-1}$, $k_{-1} = 0.2\ s^{-1}$, $k_2 = 1\ s^{-1}$, $k_{-2} = 10\ s^{-1}$, $k_3 = 10\ s^{-1}$.

The value of rate constant k_{-3} is _____ s^{-1} (round off to 1 decimal place).

[2021 : 1 M]

1.20 An elementary irreversible gas-phase reaction, $A \rightarrow B + C$, is carried out at fixed temperature and pressure in two separate ideal reactors: (i) a 10 m^3 plug flow reactor (PFR), (ii) a 10 m^3 continuous-stirred tank reactor (CSTR). If pure A is fed at 5 $m^3\ h^{-1}$ to the PFR operating at 400 K, the conversion is 80%. If a mixture of 50 mol% of A and 50 mol% of an inert is fed at 5 $m^3\ h^{-1}$ to the CSTR operating at 425 K, the conversion is 80%. The universal gas constant $R = 8.314$ J $mol^{-1}\ K^{-1}$. Assuming the Arrhenius rate law, the estimated activation energy is _______ kJ mol^{-1} (rounded off to one decimal place).

[2022 : 2 M]

1.21 For a reversible endothermic chemical reaction with constant heat of reaction over the operating temperature range, K is the thermodynamic equilibrium constant. Which one of the following figures shows the CORRECT dependence of K on temperature T?

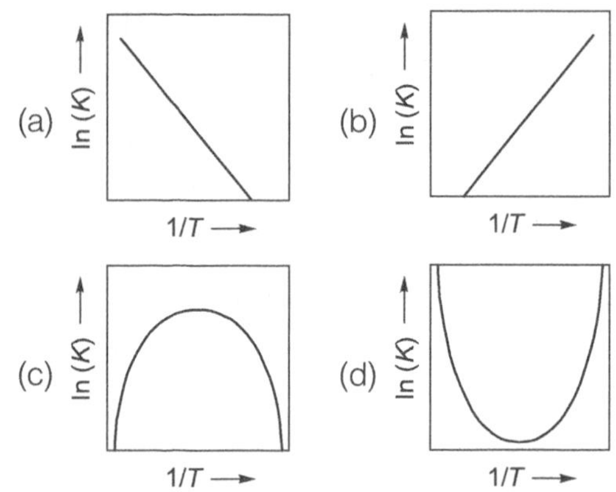

[2023 : 1 M]

1.22 The ratio of the activation energy of a chemical reaction to the universal gas constant is 1000 K. The temperature-dependence of the reaction rate constant follows the collision theory. The ratio of the rate constant at 600 K to that at 400 K is

(a) 2.818
(b) 4.323
(c) 1.502
(d) 1.000

[2024 : 1 M]

Answers Kinetics of Homogeneous Reactions

1.1 (c)	**1.2** (b)	**1.3** (d)	**1.4** (d)	**1.5** (b)	**1.6** (a)	**1.7** (b)
1.8 (d)	**1.9** (c)	**1.10** (a)	**1.11** (c)	**1.12** (c)	**1.13** (b)	**1.14** (a)
1.15 (b)	**1.16** (2)	**1.17** (28.6)	**1.18** (c)	**1.19** (200)	**1.20** (47.432)	**1.21** (a)
1.22 (a)						

Explanations Kinetics of Homogeneous Reactions

1.1 (c)

Reaction given is $A + B \rightarrow C + D$ now molecularity is define for only elementary reaction and for elementary reaction order of reaction is same as stoichiometric coefficients.

So reaction may be elementary with molecularity of 2 for given condition.

1.2 (b)

From Arrhenius law:

$$k_1 = k_0 e^{-E/RT_1}$$

$$k_2 = k_0 e^{-E/RT_2}$$

On dividing

$$\frac{k_2}{k_1} = e^{-\frac{E}{R}\left[\frac{1}{T_2} - \frac{1}{T_1}\right]}$$

By taking in both side

$$\ln\left(\frac{k_2}{k_1}\right) = \frac{E}{R}\left[\frac{1}{T_1} - \frac{1}{T_2}\right]$$

1.5 (b)

$$A + B \xrightarrow{k_1} AB$$

$$AB + A \xrightarrow{k_2} A_2B$$

$$-r_{AB} = \frac{dC_{AB}}{dt}$$

$$= k_1 C_A C_B - k_2 C_{AB} C_A$$

For Pseudo steady state

$$\frac{dC_{AB}}{dt} = 0$$

$$k_1 C_A C_B = k_2 C_{AB} C_A \quad \text{...(i)}$$

$$-r_A = \frac{dC_A}{dt}$$

$$= k_1 C_A C_B + k_2 C_{AB} C_A$$

From equation (i)

$$-r_A = 2k_1 C_A C_B \quad \text{...(ii)}$$

Given reaction $2A + B \rightarrow A_2B$

$$-r_A = kC_A C_B \quad \text{...(iii)}$$

Rate equation (ii) matches with rate equation (iii).

1.7 (b)

Rate of reaction will be same for both the reaction as it is found experimentally, it does not have any effect by multiply or divide by a constant value.

$$N_2 + 3H_2 \rightleftharpoons 2NH_3 \text{ and } \frac{1}{2}N_2 + \frac{3}{2}H_2 \rightleftharpoons NH_3$$

1.8 (d)

$2R + S \rightarrow T$

$$\frac{-r_R}{2} = \frac{-r_S}{1} = \frac{r_T}{1}$$

$$r_R = 2r_S = -2r_T$$

1.9 (c)

% change in rate constant is higher at lower temperature in comparison of high temperature.

1.10 (a)

$$-r_A = \frac{k_1 C_A^2}{1 + k_2 C_A^{1/2}}$$

$$\text{Unit of } k_1 = (\text{conc.})^{1-n} \text{ time}^{-1}$$

$$\text{mol}^{-1}\text{ m}^3\text{ s}^{-1}$$

$$\text{Unit of } k_2 = \text{mol}^{-1/2}\text{ m}^{3/2}$$

1.11 (c)

$2A + B \rightarrow 2C$

At steady state

$$\frac{-N_A}{2} = \frac{-N_B}{1} = \frac{N_C}{2}$$

$$N_A = 2N_B = -N_C$$

1.12 (c)

Rate of reaction for any species is define as

$$r_j = \frac{1}{V}\frac{dN_j}{dt}$$

In question it is not given that species j is a reactant so option (d) is incorrect because it includes negative sign.

1.13 (b)

Catalyst follows an alternative path of reaction which have comparatively lower activation energy.

1.14 (a)

For zero order reaction:

$$\text{Half life} = \frac{C_{A_0}}{2K}$$

Time required for full conversion of reactant is $\frac{C_{A_0}}{K}$.

1.15 (b)

Given, $$\text{Rate} = \frac{2C_A^2 + 0.1C_A}{1+50C_A}$$

For very high concentration of A

$$\text{Rate} = \frac{2C_A^2}{50C_A} = \frac{1}{25}C_A$$

So at very high concentration of A order tends to 1.

1.16 (2)

Given rate, $$-r_A = \frac{10C_A}{1+5C_A}$$

$$-r_A = \frac{10C_A+2}{1+5C_A} - \frac{2}{1+5C_A}$$

$$-r_A = 2 - \frac{2}{1+5C_A}$$

We can see that the second term is negative for all possible values of C_A. So upper limit of rate of reaction for magnitude is 2.

1.17 (28.6)

From Arrhenius law:

$$\ln\left(\frac{r_2}{r_1}\right) = \frac{E}{R}\left[\frac{1}{T_1} - \frac{1}{T_2}\right]$$

$$\ln\left(\frac{r_2}{r_1}\right) = \frac{184000}{8.314}\left[\frac{1}{550} - \frac{1}{600}\right]$$

$$\frac{r_2}{r_1} = 28.6$$

1.18 (c)

From Arrhenius equation:

$$\ln\left(\frac{k_2}{k_1}\right) = \frac{E}{R}\left[\frac{1}{T_1} - \frac{1}{T_2}\right]$$

Putting the values, we get

$$\ln(2.5) = \frac{E}{8.314}\left[\frac{1}{400} - \frac{1}{500}\right]$$

$$E = 15.23 \text{ kJ/mol}$$

1.19 (200)

$$A \underset{k_{-1}}{\overset{k_1}{\rightleftharpoons}} B$$

$$A \underset{k_{-3}}{\overset{k_3}{\rightleftharpoons}} C$$

$$B \underset{k_{-2}}{\overset{k_2}{\rightleftharpoons}} C$$

Since, they are in equilibrium. Hence,

$$k_1C_A - k_{-1}C_B = 0 \quad \text{...(1)}$$

$$k_3C_A - k_{-3}C_C = 0 \quad \text{...(2)}$$

$$k_2C_B - k_{-2}C_C = 0 \quad \text{...(3)}$$

From (1), $$\frac{C_B}{C_A} = \frac{k_1}{k_{-1}} = \frac{0.1}{0.2} = 0.5 \quad \text{...(4)}$$

From (2), $$\frac{C_C}{C_A} = \frac{k_3}{k_{-3}} = k_3 \quad \text{...(5)}$$

From (3), $$\frac{C_C}{C_B} = \frac{k_2}{k_{-2}} = k_2 = \frac{1}{10} = 0.1 \quad \text{...(6)}$$

Multiplying equation (4) and (6)

$$\frac{C_C}{C_A} = \frac{C_B}{C_A} \times \frac{C_C}{C_B} = k_1 \times k_2$$

Again from (5),

$$\frac{C_C}{C_A} = k_3$$

$$k_3 = k_1 \times k_2$$
$$= 0.5 \times 0.1 = 0.05$$

$$\frac{k_3}{k_{-3}} = 0.05$$

$$k_{-3} = \frac{10}{0.05} = 200 s^{-1}$$

1.20 (47.432)

PFR:

$$A \rightarrow B + C \quad \varepsilon_A = \left(\frac{2-1}{1}\right) = 1$$

$$\frac{\tau}{C_{A0}} = \int_o^{x_A} \frac{dx_A}{(-r_A)} \Rightarrow \frac{10}{5C_{A0}} = \int_o^{0.80} \frac{dx_A}{\frac{k_1C_{A0}(1-x)}{(1+\varepsilon_{x_A})}}$$

$$2k_1 = \int_o^{0.80}\left(\frac{1+x}{1-x}\right)dx = \int \frac{2dx}{1-x} - \int dx = -2\ln|1-x| - x$$

$$-2k_1 = \left[2\ln|1-x| + x\right]_0^{0.80}$$

$$k_1 = 1.21 \text{ at } 400 \text{ K}$$

Now for CSTR,

$$\tau = \frac{10}{5} = \frac{C_{A0}X_A(1+\varepsilon_A \times A)}{k_2 C_{A0}(1-x)}$$

$$2 = \frac{0.80(1+0.50\times 0.80)}{k_2(0.20)}$$

$$k_2 = 2.8 \text{ at } 425 \text{ K}$$

Using Arrhenius equation,

$$\ln\left|\frac{k_2}{k_1}\right| = \frac{E}{8.314}\left(\frac{1}{400}-\frac{1}{425}\right)$$

$$\ln\left|\frac{2.8}{1.21}\right| = \frac{E}{8.314}\left(\frac{1}{400}-\frac{1}{425}\right)$$

$$E = 47.432 \text{ kJ/mol}$$

1.21 (a)

For reversible endothermic reactions, the value of equilibrium constant (K) increases as the value of operating temperature increases, thus

K increases, as T increases

K increases, as $\frac{1}{T}$ decreases

Thus, option (a) is the correct one.

1.22 (a)

$$\frac{E}{R} = 1000 \text{ K (given)}$$

From Collision theory

$$\ln\left(\frac{k_2}{k_1}\right) = \frac{1}{2}\ln\left(\frac{T_2}{T_1}\right) + \frac{E}{R}\left[\frac{1}{T_1}-\frac{1}{T_2}\right]$$

$$T_1 = 400 \text{ K and } T_2 = 600 \text{ K (given)}$$

So,
$$\ln\left(\frac{k_2}{k_1}\right) = \frac{1}{2}\ln\left(\frac{600}{400}\right) + 1000\left(\frac{1}{400}-\frac{1}{600}\right)$$

$$\frac{k_2}{k_1} = 2.818$$

■■■■

2 Interpretation of Batch Reactor Data

2.1 The reaction $A \rightarrow B$ is conducted in an isothermal batch reactor. If the conversion of A increases linearly with holding time, then the order of the reaction is

(a) 0 (b) 1
(c) 1.5 (d) 2

[2000 : 1 M]

2.2 The following half-life data are available for the irreversible liquid phase reaction, $A \rightarrow$ products:

Initial concentration (kmol/m^3)	Half - life (min)
2	2
8	1

The overall order of the reaction is

(a) 0.5 (b) 1
(c) 1.5 (d) 2

[2000 : 2 M]

2.3 The first order series reaction $A \xrightarrow{k_1} B \xrightarrow{k_2} C$ is conducted in a batch reactor. The initial concentrations of A, B and C (C_{A0}, C_{B0}, C_{C0} respectively) are all non-zero. The variation of C_B with reaction time will not show a maximum if

(a) $k_2 C_{B0} > k_1 C_{A0}$ (b) $k_1 C_{A0} > k_2 C_{B0}$
(c) $C_{B0} > C_{A0}$ (d) $C_{A0} > C_{B0}$

[2000 : 2 M]

2.4 The conversion for a second order, irreversible reaction (constant volume), $A \xrightarrow{k_2} B$, in batch mode is given by

(a) $\dfrac{1}{1+k_2 C_{A0} t}$ (b) $\dfrac{k_2 C_{A0} t}{1+k_2 C_{A0} t}$

(c) $\dfrac{(k_2 C_{A0} t)^2}{1+k_2 C_{A0} t}$ (d) $\dfrac{k_2 C_{A0} t}{(1+k_2 C_{A0} t)^2}$

[2001 : 1 M]

2.5 The first-order, gas phase reaction $A \xrightarrow{k_I} 2B$ is conducted isothermally in batch mode. The rate of change of conversion with time is given by

(a) $\dfrac{dX_A}{dt} = k_1 (1-X_A)^2 (1+2X_A)$

(b) $\dfrac{dX_A}{dt} = k_1 (1-X_A)^2 (1+0.5X_A)$

(c) $\dfrac{dX_A}{dt} = k_1 (1-X_A)$

(d) $\dfrac{dX_A}{dt} = k_1 (1-X_A)/(1+X_A)$

[2001 : 2 M]

2.6 For an isothermal second order aqueous phase reaction $A \rightarrow B$, the ratio of the time required for 90% conversion to the time required for 45% conversion is

(a) 2 (b) 4
(c) 11 (d) 22

[2004 : 2 M]

2.7 For the liquid phase reaction $A \rightarrow P$, in a series of experiments in a batch reactor, the half-life $t_{1/2}$ was found to be inversely proportional to the square root of the initial concentration of A. The order of the reaction is

(a) $\dfrac{3}{2}$ (b) 1

(c) $\dfrac{1}{2}$ (d) $-\dfrac{1}{2}$

[2005 : 2 M]

2.8 An irreversible gas phase reaction $A \rightarrow S$ is conducted in an isothermal batch reactor at constant pressure in the presence of an inert. The feed contains no B. If the volume of the gas at complete conversion must not exceed three times the initial volume, the minimum mole percent of the inert in the feed must be

(a) 0 (b) 20
(c) 33 (d) 50 [2006 : 1 M]

2.9 A first order reversible reaction $A \underset{k_2}{\overset{k_1}{\rightleftharpoons}} B$ occurs in a batch reactor. The exponential decay of the concentration of A has the time constant

(a) $\dfrac{1}{k_1}$ (b) $\dfrac{1}{k_2}$

(c) $\dfrac{1}{k_1 - k_2}$ (d) $\dfrac{1}{k_1 + k_2}$

[2006 : 1 M]

Common Data for Questions (2.10 & 2.11) :

2.10 The following liquid phase reaction is taking place in an isothermal batch reactor

$$A \xrightarrow{k_1 \text{ (first order)}} B \xrightarrow{k_2 \text{ (zero order)}} C$$

Feed concentration = 1 mol/litre

The time at which the concentration of *B* will reach its maximum value is given by

(a) $t = \dfrac{1}{k_1}\ln\left(\dfrac{k_1}{k_2}\right)$ (b) $t = \dfrac{1}{k_2 - k_1}\ln\left(\dfrac{k_2}{k_1}\right)$

(c) $t = \dfrac{1}{k_2}\ln\left(\dfrac{k_2}{k_1}\right)$ (d) $t = \dfrac{1}{k_2}\ln\left(\dfrac{k_1}{k_2}\right)$

[2007 : 2 M]

2.11 The time at which the concentration of *B* will become zero is given by the following equation :

(a) $(1 - e^{k_1 t}) = k_2 t$ (b) $t = \infty$

(c) $t = \dfrac{1}{k_2}$ (d) $t = \dfrac{1}{k_1}$

[2007 : 2 M]

2.12 The liquid phase reaction $A \rightarrow$ Products is governed by the kinetics

$$-r_A = kC_A^{1/2}$$

If the reaction undergoes 75% conversion of *A* in 10 minutes in an isothermal batch reactor, the time (in minutes) for complete conversion of *A* is

(a) $\dfrac{40}{3}$ (b) 20

(c) 30 (d) ∞

[2008 : 2 M]

2.13 The half-life of a first order liquid phase reaction is 30 seconds. Then the rate constant, in min^{-1}, is

(a) 0.0231 (b) 0.602

(c) 1.386 (d) 2.0

[2009 : 1 M]

2.14 An autocatalytic liquid phase reaction, $A + R \rightarrow 2R$ is conducted in an isothermal batch reactor with a small initial concentration of *R*. Assume that the order of reaction with respect to both reactants is positive. The rate of reaction $(-r_A)$ versus concentration, C_A, as the reaction proceeds, is depicted by

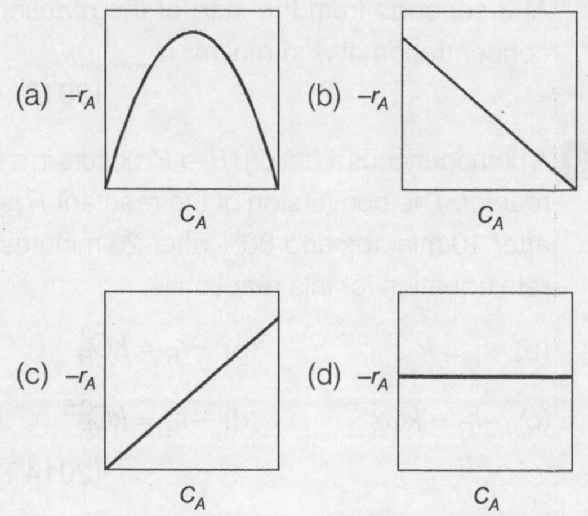

[2010 : 2 M]

2.15 Reactant *R* forms three products *X*, *Y* and *Z* irreversibly, as shown below.

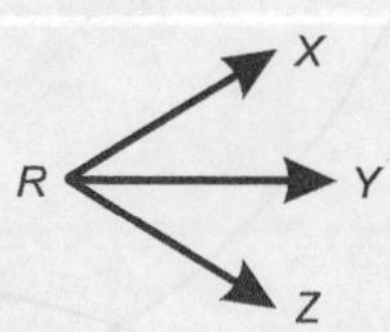

The reaction rates are given by $r_X = k_X C_R$, $r_Y = k_Y C_R^{1.5}$ and $r_Z = k_Z C_R$. The activation energies for formation of *X*, *Y* and *Z* are 40, 40 and 5 kJ/mol respectively. The pre-exponential factors for all reactions are nearly same. The desired conditions for maximizing the yield of *X* are

(a) high temperature, high concentration of *R*
(b) high temperature, low concentration of *R*
(c) low temperature, high concentration of *R*
(d) low temperature, low concentration of *R*

[2011 : 2 M]

2.16 The half-life of an nth order reaction in a batch reactor depends on

(a) only the rate constant
(b) only the rate constant and the order of the reaction
(c) only the rate constant and the initial reactant concentration
(d) the rate constant, initial reactant concentration, and the order of the reaction

[2012 : 1 M]

2.17 An isothermal liquid phase zero order reaction $A \rightarrow B$ ($k = 0.5$ mol/m^3-s) is carried out in a batch reactor. The initial concentration of A is 2 mol/m^3. At 3 seconds from the start of the reaction, the concentration of A in mol/m^3 is ________.

[2013 : 1 M]

2.18 A homogeneous reaction ($R \rightarrow P$) occurs in a batch reactor. The conversion of the reactant R is 67% after 10 minutes and 80% after 20 minutes. The rate equation for this reaction is

(a) $-r_R = k$ (b) $-r_R = kC_R^2$

(c) $-r_R = kC_R^3$ (d) $-r_R = kC_R^{0.5}$

[2014 : 2 M]

2.19 The variations of the concentrations (C_A, C_R and C_S) for three species (A, R and S) with time, in an isothermal homogeneous batch reactor are shown in the figure below.

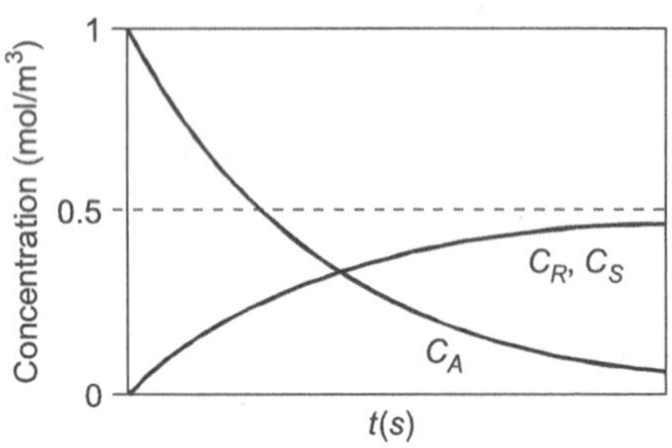

Select the reaction scheme that correctly represents the above plot. The numbers in the reaction schemes shown below, represent the first order rate constants in unit of s^{-1}.

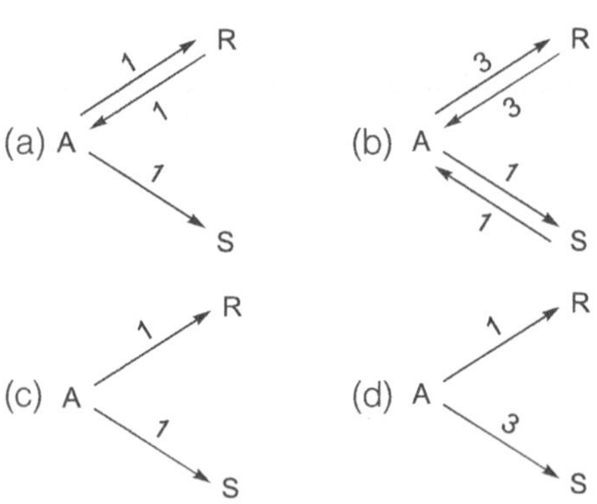

[2016 : 1 M]

2.20 The following gas-phase reaction is carried out in a constant-volume isothermal batch reactor

$$A + B \rightarrow R + S$$

The reactants A and B as well as the product S are non-condensable. At the operating temperature, the saturation pressure of the product R is 40 kPa. Initially, the batch reactor contains equimolar amounts of A and B (and no products) at a total pressure of 100 kPa. The initial concentrations of the reactants are $C_{A,0} = C_{B,0} = 12.5$ mol/m^3. The rate of reaction is given by $(-r_A) = 0.08\, C_A C_B$ mol/m^3.s.

The time at which R just starts condensing, rounded to 1 decimal place, is ________ s.

[2017 : 2 M]

2.21 An irreversible gas phase reaction $2P \rightarrow 4Q + R$ is conducted in an isothermal and isobaric batch reactor. Assume ideal gas behaviour. The feed is an equimolar mixture of the reactant P and an inert gas. After complete conversion of P, the fractional change in volume is _____ (round off to 2 decimal places).

[2020 : 1 M]

2.22 The rate of a reaction $A \rightarrow B$ is 0.2 mol m^{-3} s^{-1} at a particular concentration C_{A1}. The rate constant of the reaction at a given temperature is 0.1 m^3 mol^{-1} s^{-1}. If the reactant concentration is increased to 10 C_{A1} at the same temperature, the reaction rate, in mol m^{-3} s^{-1}, is

(a) 20 (b) 10

(c) 100 (d) 50

[2024 : 1 M]

2.23 Two parallel first-order liquid phase reactions $A \xrightarrow{k_1} B$ and $A \xrightarrow{k_2} C$ are carried out in a well-mixed isothermal batch reactor. The initial concentration of A in the reactor is 1 kmol m^{-3}, while that of B and C is zero. After 2 hours, the concentration of A reduces to half its initial value, and the concentration of B is twice that of C. The rate constants k_1 and k_2, in h^{-1}, are, respectively

(a) 0.40, 0.20 (b) 0.23, 0.12

(c) 0.50, 0.25 (d) 0.36, 0.18

[2024 : 1 M]

■■■■

Answers Interpretation of Batch Reactor Data

2.1	(a)	2.2	(c)	2.3	(b)	2.4	(b)	2.5	(c)	2.6	(c)	2.7	(a)
2.8	(d)	2.9	(d)	2.10	(a)	2.11	(a)	2.12	(b)	2.13	(a)	2.14	(a)
2.15	(b)	2.16	(d)	2.17	(0.5)	2.18	(b)	2.19	(c)	2.20	(4)	2.21	(0.75)
2.22	(a)	2.23	(b)										

Explanations Interpretation of Batch Reactor Data

2.1 (a)

For zero-order reaction:

$$C_{A_0}X_A = kt \text{ i.e. } X_A \propto t$$

where, t = Holding time

2.2 (c)

$$t_{1/2} = \frac{C_{A_0}^{1-n}\left[0.5^{1-n}-1\right]}{k(n-1)}$$

$$t_{1/2} \propto C_{A_0}^{1-n}$$

$$\left(\frac{2}{1}\right) = \left(\frac{2}{8}\right)^{1-n}$$

$$1 = 2n-2$$

$$n = 1.5$$

2.3 (b)

Concentration of B will show a maxima only if its consumption rate $(k_2 C_{B_0})$ will be less than its formation rate $k_1 C_{A_0}$.

2.4 (b)

For 2nd order reaction

$$-\frac{dC_A}{dt} = kC_A^2$$

$$-\int_{C_{A_0}}^{C_A} \frac{dC_A}{C_A^2} = k\int_0^t dt$$

$$\frac{1}{C_A}-\frac{1}{C_{A_0}} = kt$$

$$\frac{1}{1-X_A}-1 = kC_{A_0}t$$

$$X_A = \frac{kC_{A_0}t}{1+kC_{A_0}t}$$

2.5 (c)

Consider following gas phase first order reaction occurring in batch reactor:

$$A \xrightarrow{k_1} 2B$$

For first order reaction

$$X_A = 1-e^{-k_1 t} \quad \text{...(i)}$$

$$\frac{dX_A}{dt} = k_1 e^{-k_1 t} \quad \text{...(ii)}$$

On arranging equation (i) and (ii)

$$\frac{dX_A}{dt} = k_1(1-X_A)$$

2.6 (c)

Time required for fractional conversion in 2nd order reaction is given by

$$kC_{A_0}t = \frac{X_A}{1-X_A} \text{ i.e. } t \propto \frac{X_A}{1-X_A}$$

$$\frac{t_{90\%}}{t_{45\%}} = \frac{\dfrac{0.9}{1-0.9}}{\dfrac{0.45}{1-0.45}} = 11$$

2.7 (a)

For half life of nth order reaction:

$$t_{1/2} \propto C_{A_0}^{1-n}$$

According to question $t_{1/2} \propto \dfrac{1}{\sqrt{C_{A_0}}}$

$$1-n = -\frac{1}{2}$$

$$n = \frac{3}{2},\ n = \text{order}$$

2.8 (d)

$$A \to 5B, \quad \epsilon_A = y_{A_0}\delta_A = y_{A_0}\cdot 4$$

$$V = V_0(1+\epsilon_A X_A)$$

At $X_A = 1,\ V = 3V_0$ (given)

$$3V_0 = V_0(1+\epsilon_A)$$

$$\epsilon_A = 2$$

$$y_{A_0}\cdot 4 = 2$$

$$y_{A_0} = 0.5$$

So mole percent of inert = 50%

2.9 (d)

$$A \underset{k_2}{\overset{k_1}{\rightleftharpoons}} B$$

$$-\frac{dC_A}{dt} = \frac{dC_B}{dt} = k_1 C_A - k_2 C_B$$

$$C_B = C_{B_0} + (C_{A_0} - C_A)$$

So, $$-\frac{dC_A}{dt} = k_1C_A - k_2(C_{B_0} + C_{A_0} - C_A)$$

$$-\frac{dC_A}{dt} = (k_1+k_2)C_A - k_2(C_{B_0} + C_{A_0}) \quad ...(i)$$

At equilibrium $\frac{dC_A}{dt} = 0$ and $C_A = C_{A_e}$

$$0 = (k_1+k_2)C_{A_e} - k_2(C_{B_0} + C_{A_0})$$

$$k_2 = \frac{(k_1+k_2)C_{A_e}}{(C_{B_0} + C_{A_0})} \quad ...(ii)$$

From equation (i) and (ii)

$$-\frac{dC_A}{dt} = (k_1+k_2)C_A - \frac{(k_1+k_2)}{C_{B_0}+C_{A_0}}C_{A_e}(C_{B_0}+C_{A_0})$$

$$-\frac{dC_A}{dt} = (k_1+k_2)C_A - (k_1+k_2)C_{A_e}$$

$$\frac{dC_A}{dt} = (k_1+k_2)(C_{A_e} - C_A)$$

$$\int_{C_{A_0}}^{C_A} \frac{dC_A}{C_{A_e} - C_A} = \int_0^t (k_1+k_2)dt$$

On integrating

$$\frac{C_A - C_{A_e}}{C_{A_0} - C_{A_e}} = e^{(k_1+k_2)t} = e^{-t/\tau}$$

So, $$\tau = \frac{1}{(k_1+k_2)}$$

2.11 (a)

From equation (i)

$$\frac{dC_B}{dt} = k_1C_{A_0}e^{-k_1t} - k_2$$

$$C_B = \int_0^t (k_1e^{-k_1t} - k_2)dt = 0$$

$$1 - e^{-k_1t} = k_2t$$

2.12 (b)

$$-r_A = kC_A^{1/2}$$

In 10 minutes reaction goes 75% conversion from the relation between time required and fractional conversion

$$t_F = \frac{C_{A_0}^{1-n}\left[F^{1-n}-1\right]}{k(n-1)}$$

$$10 = \frac{C_{A_0}^{1-n}\left[0.25^{1-0.5}-1\right]}{k(n-1)}$$

$$\frac{C_{A_0}^{1-n}}{k(1-n)} = 20 \text{ minutes}$$

$$t_{100}\% = \frac{C_{A_0}^{1-n}}{k(1-n)} = 20 \text{ minutes}$$

2.13 (a)

Half life of a first order liquid phase reaction can be found by following:

$$t_{1/2} = \frac{\ln(2)}{k}$$

$$k = \frac{\ln(2)}{t_{1/2}} = 0.0231 \text{ sec}^{-1}$$

2.14 (a)

Rate of reaction concentration curve for a autocatalytic reaction passes through a maxima so option (a) is correct.

2.15 (b)

Reactant R forms three products X, Y and Z

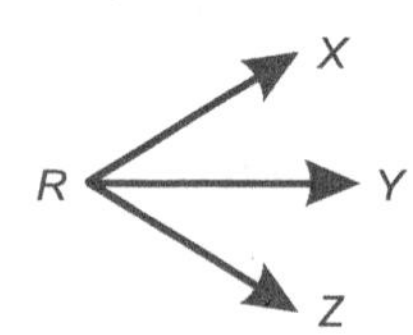

$r_x = k_xC_R$ $\quad$ $E_x = 40$ kJ/mol

$r_y = k_yC_R^{1.5}$ $\quad$ $E_y = 40$ kJ/mol

$r_z = k_zC_R$ $\quad$ $E_z = 5$ kJ/mol

For maximizing yield of X temperature must be high because high activation energy favours high temperature and concentration must be low because high concentration will favours formation of Y strongly.

2.16 (d)

Half-life of nth order reaction is given by

$$t_{(1/2)} = \frac{C_{A_0}^{1-n}\left[0.5^{1-n}-1\right]}{k(n-1)}$$

So $t_{1/2}$ depends upon rate constant, initial concentration and order of reaction.

2.17 (0.5)

Liquid phase zero order reaction $A \rightarrow B$ is given

$$k = 0.5 \text{ mol/m}^3\text{.s}$$

$$C_{A_0} = 2 \text{ mol/m}^3$$

$$C_A = C_{A_0} - kt = 2 - 0.5 \times 3 = 0.5 \text{ mol/m}^3$$

2.18 (b)

Homogeneous reaction ($R \rightarrow P$) occurs in a batch reaction.

Time required for fractional conversion of any n^{th} order reaction is given by

$$t_F = \frac{C_{A_0}^{1-n}[F^{1-n} - 1]}{k(n-1)}$$

Here F denotes fraction of reactant unconverted

$$\frac{10}{20} = \frac{[0.33^{1-n} - 1]}{[0.2^{1-n} - 1]}$$

Now, solving for n, we get $n = 2$ or this can also solved by cross-checking of options.

2.19 (c)

From graph we can see that rate of formation of R and S is same for all the time.

So option (c) is only who fulfills this criteria.

2.20 (4)

	A	+	$B \longrightarrow R$	+ S
	A		B	R	S

	A	B	R	S
Concentration (mol/m^3) at $t = 10$	12.5	12.5	0	0
Concentration (mol/m^3) at $t = t$	$12.5 - x$	$12.5 - x$	x	x
Mole fraction at $t = t$	$\frac{12.5 - x}{25}$	$\frac{12.5 - x}{25}$	$\frac{x}{25}$	$\frac{x}{25}$

R will start condensing when $P_R = 40$ kPa

$$\frac{x}{25} \times P_T = 40 \text{ where } P_T = 100 \text{ kPa}$$

$$x = \frac{40 \times 25}{100} = 10 \text{ mol/m}^3$$

$$\text{Conversion: } X_A = \frac{10}{12.5} = 0.8$$

$$-r_A = 0.08 C_A C_B$$

$$-r_A = 0.08\, C_A^2 \qquad (\because C_A = C_B)$$

Now for 2nd order reaction

$$\frac{X_A}{1 - X_A} = kC_{A_0}t$$

$$\frac{0.8}{0.2} = 0.08 \times 12.5t$$

$$t = 4 \text{ second}$$

2.21 (0.75)

Gas phase irreversible reaction

$$2P \rightarrow 4Q + R$$

Equimolar feed is given with inert means 50% P and 50% inert.

$$\in_P = y_{A0}\delta_A$$

$$\in_P = \frac{5-2}{2} \times 0.5 = 0.75$$

After complete conversion of P fraction change in volume can be found from following coordination.

$$V = V_o(1 + \in_A X_A)\left(\frac{P_o}{P}\right)\left(\frac{T}{T_o}\right)$$

$$\frac{V}{V_o} = (1 + 0.75), \ \{X_A = 1\}$$

$$\text{Fractional change in volume} = \frac{V - V_o}{V_o} = 0.75$$

2.22 (a)

The rate of reaction $A \rightarrow B$ is

$$-r_A = kC_A^n \qquad \text{...(i)}$$

$-r_A = 0.2$ mol/m^{3}s and $k = 0.1$ m^3/mol.s (given)

$$\text{Unit of } k = \text{con}^{(1-n)} \cdot \text{time}^{-1}$$

$$1 - n = -1$$

$$n = 2$$

$$-r_A = kC_A^2$$

So, from equation (i)

$$0.2 = 0.1C_{A1}^2 \qquad \text{...(ii)}$$

Now, $-r_A$ at $10\,C_{A1}$

$$-r_A = 0.1\,(10C_{A1})^2 \qquad \text{...(iii)}$$

From equation (ii) and (iii)

$$\frac{-r_A}{0.2} = \frac{100C_{A1}^2 k_1}{k_1 C_{A1}^2}$$

$$-r_A = 20 \text{ mol/m}^3\text{.s}$$

2.23 (b)

Following two parallel first order liquid phase reaction carried out in a wall mixed isothermal batch reactor

$$A \xrightarrow{k_1} B \qquad -r_{A_1} = k_1 C_A$$

$$A \xrightarrow{k_2} C \qquad -r_{A_2} = k_2 C_A$$

$$C_{A_0} = 1 \text{ kmol/m}^3$$

In two hours.

$$C_A = 0.5C_{A_0}, \ C_B = 2C_A$$

So,
$$k_1 = 2k_2 \qquad \text{...(i)}$$

and
$$C_A = C_{A_0}e^{-(k_1+k_2)t}$$

$$0.5 = e^{-(k_1+k_2)2}$$

$$k_1 + k_2 = 0.34657 \qquad \text{...(ii)}$$

From equation (i) and (ii)

$$k_2 = 0.12, \quad k_1 = 0.23$$

3 Design of Isothermal Ideal Reactors

3.1 For the liquid phase parallel reactions

$$A \rightarrow R \quad r_R = k_1 C_A^2; \quad E_1 = 80 \text{ kJ/mol}$$
$$A \rightarrow S \quad r_S = k_2 C_A; \quad E_2 = 120 \text{ kJ/mol}$$

the desired product is R. A higher selectivity of R will be achieved if the reaction is conducted at

(a) low temperature in a CSTR.
(b) high temperature in a CSTR.
(c) low temperature in a PFR.
(d) high temperature in a PFR.

[2000 : 1 M]

3.2 The conversion for a first-order liquid-phase reaction $A \rightarrow B$ in a CSTR is 50%. If another CSTR of the same volume is connected in series, then the % conversion at the exit of the second reactor will be

(a) 60 (b) 75
(c) 90 (d) 100

[2000 : 2 M]

3.3 A CSTR and a PFR of equal volume V (each) are given for the conduct of a second-order, isothermal, liquid phase reaction. The reactors are to be arranged sequentially (in series). Find the values of the conversion for the two possible reactor arrangements. $A \xrightarrow{k_2} B$, k_2 = 1 m³/kmol s, C_{A_0} = 0.1 kmol/m³s and τ = 5 s (for volume V).

[2001 : 2 M]

3.4 For an ideal plug flow reactor the value of the Peclet number is

(a) 0 (b) ∞
(c) 1 (d) 10

[2002 : 1 M]

3.5 A simple chemical reaction $A \rightarrow B$ is being carried out in two CSTRs connected in series. The volume of the first reactor is 1.5 times that of the second reactor. The temperatures of the reactors are such that the rate constant in the first reactor is 1.5 times the rate constant in the second reactor.

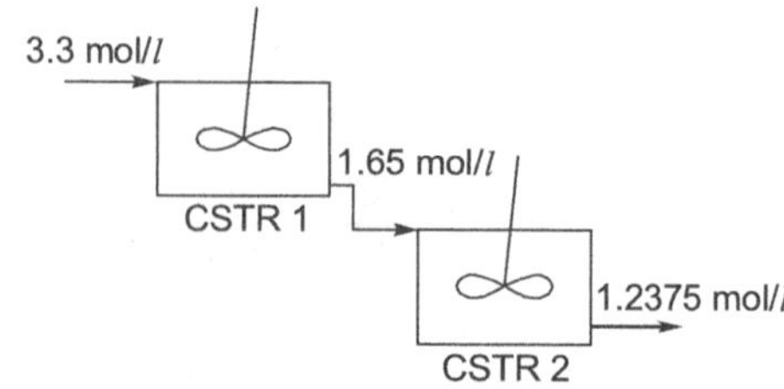

(a) is the data given above consistent for a first order reaction kinetics? Justify.
(b) is the data given above consistent for a second order reaction kinetics? Justify.

[2002 : 5 M]

3.6 An elementary liquid phase decomposition reaction $A \xrightarrow{k} 2B$ is to be carried out in a CSTR. The design equation is

(a) $k\tau =$ [illegible]

(b) $k\tau = \dfrac{X_A(1+X_A)}{(1-X_A)}$

(c) $k\tau = \dfrac{X_A}{(1-X_A)^2}$

(d) $k\tau C_{A0} = \dfrac{\dfrac{X_A}{(1+X_A)^2}}{(1-X_A)^2}$

[2003 : 1 M]

3.7 A liquid phase reaction is to be carried out under isothermal conditions. The reaction rate as a function of conversion has been determined experimentally and is shown in figure given below. What choice of reactor or combination of reactors will require the minimum overall reactor volume, if a conversion of 0.9 is desired?

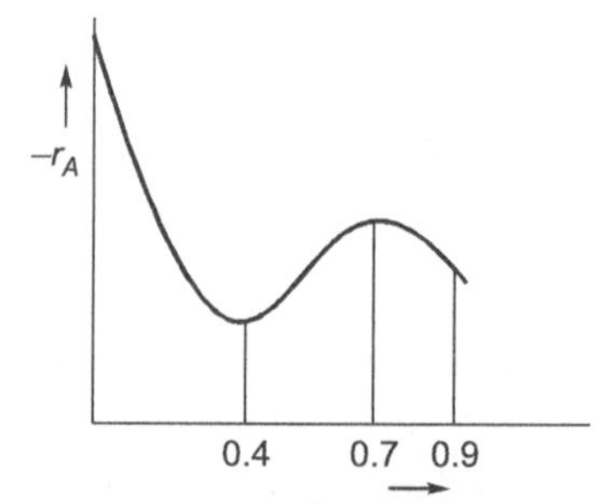

(a) CSTR followed by a PFR
(b) PFR followed by a CSTR
(c) CSTR followed by a PFR followed by CSTR.
(d) PFR followed by a CSTR followed by a PFR

[2003 : 2 M]

Q.3.8 - 3.9 are based on the data supplied in the paragraph below:

The following gas phase reactions are carried out isothermally in a CSTR

$A. \to 2R$; $r_1 = k_1\,p_A$; $k_1 = 20$ mol/(sec.m^3 bar)
$A \to 3S$; $r_2 = k_2\,p_A$; $k_2 = 40$ mol/(sec.m^3 bar)
Total pressure = 1 bar; $F_{Ao} = 1$ mol/sec; feed is pure A

3.8 What is the maximum possible value of F_R (mol/sec)?

(a) $\frac{1}{3}$ (b) $\frac{1}{2}$
(c) $\frac{2}{3}$ (d) 2

[2003 : 2 M]

3.9 The volume of a CSTR required for a fractional conversion of A equal to 0.3 due to the first reaction is

(a) 0.011 (b) 0.21
(c) 0.275 (d) 0.375 [2003 : 2 M]

3.10 The following gas phase reaction is taking place in a plug flow reactor.

$$A + \frac{1}{2}B \to C$$

A stoichiometric mixture of A and B at 300 K is fed? to the reactor. At 1 m along the length of the reactor, the temperature is 360 K. The pressure drop is negligible and an ideal gas behavior can be assumed. Identify the correct expression relating the concentration of A at the inlet (C_{A0}), concentration of A at 1 m (C_A) and the corresponding conversion of A is X.

(a) $C_A = 1.2C_{A0}\dfrac{(1-X)}{(1-0.33X)}$

(b) $C_A = 1.2C_{A0}\dfrac{(1-X)}{(1-0.5X)}$

(c) $C_A = 0.83C_{A0}\dfrac{(1-X)}{(1-0.33X)}$

(d) $C_A = 0.83C_{A0}\dfrac{(1-X)}{(1-0.5X)}$

[2004 : 2 M]

3.11 A second order liquid phase reaction $A \to B$ is carried out in a mixed flow reactor operated in semi-batch mode (no exit stream). The reactant A at concentration C_{AF} is fed to the reactor at a volumetric flow rate of F. The volume of the reacting mixture is V and the density of the liquid mixture is constant. The mass balance for A is

(a) $\dfrac{d(VC_A)}{dt} = -F(C_{AF} - C_A) - kC_A^2V$

(b) $\dfrac{d(VC_A)}{dt} = F(C_{AF} - C_A) - kC_A^2V$

(c) $\dfrac{d(VC_A)}{dt} = -FC_A - kC_A^2V$

(d) $\dfrac{d(VC_A)}{dt} = FC_{AF} - kC_A^2V$

[2004 : 2 M]

3.12 An isothermal aqueous phase reversible reaction $P \Leftrightarrow R$ is to be carried out in a mixed flow reactor. The reaction rate in kmol/(m^{3}h) is given by $r = 0.5C_P - 0.125C_R$.
A stream containing only P enters the reactor. The residence time required (in hours) for 40% conversion of P is

(a) 0.80 (b) 1.33
(c) 1.60 (d) 2.67

[2004 : 2 M]

3.13 The first order liquid phase reaction $A \to P$ is to be carried out isothermally in the following ideal reactor configurations

P. A 1 m^3 CSTR followed by a 1 m^3 PFR.
Q. A 2 m^3 CSTR
R. A 1 m^3 PFR followed by a 1 m^3 CSTR.
S. A 1 m^3 CSTR followed by a 1 m^3 CSTR.

The overall exit conversions, X, for the above configurations P, Q, R and S, assuming identical inlet conditions and temperature, are related as

(a) $X_P > X_R > X_S > X_Q$
(b) $X_P = X_R > X_S > X_Q$
(c) $X_P = X_S = X_Q = X_R$
(d) $X_Q > X_P > X_R > X_S$

[2005 : 2 M]

3.14 The gas phase reaction $A \to B + C$ is carried out in an ideal PFR achieving 40% conversion of A. The feed has 70 mol% A and 30 mol% inerts. The inlet temperature is 300 K and the outlet temperature is 400 K. The ratio of the outlet to inlet molar concentration of A (assuming ideal gas mixture and uniform pressure) is

(a) 0.60 (b) 0.30
(c) 0.47 (d) 0.35

[2005 : 2 M]

3.15 The rate of the liquid phase reversible reaction $A \Leftrightarrow 2B$ in (kmol m^{-3} min^{-1}) at 298 K, is

$$-r_A = 0.02c_A - 0.01c_B$$

where the concentrations c_A and c_B are expressed in (kmol m^{-3}). What is the maximum limiting conversion of A achievable in an isothermal CSTR at 298 K, assuming pure A is fed at the inlet?

(a) 1 (b) $\frac{2}{3}$
(c) $\frac{1}{2}$ (d) $\frac{1}{3}$

[2005 : 2 M]

3.16 A reaction $A \rightarrow B$ is to be conducted in two CSTRs in series. The steady state conversion desired is X_f. The reaction rate as a function of conversion is given by $r = \frac{-1}{(1+X)}$. If the feed contains no B, then the conversion in the first reactor that minimizes the total volume of the two reactors is

(a) $1 - X_f$ (b) $0.2X_f$
(c) $0.5X_f$ (d) $0.5(1 - X_f)$

[2006 : 2 M]

3.17 Consider the following elementary reaction network

$$\begin{array}{ccc} A & \xrightarrow{1} & B \\ 2\downarrow & & \downarrow 3 \\ C & \xrightarrow{4} & D \end{array}$$

The activation energies for the individual reactions are E_1 = 100 kJ/mol, E_2 = 150 kJ/mol, E_3 = 100 kJ/mol, and E_4 = 200 kJ/mol. If the feed is pure A and the desired product is C, then the desired temperature profile in a plug flow reactor in the direction of flow should be

(a) constant at low temperature
(b) constant at high temperature
(c) increasing
(d) decreasing [2006 : 2 M]

3.18 A well-stirred reaction vessel is operated as a semi-batch reactor in which it is proposed to conduct a liquid phase first order reaction of the type $A \rightarrow B$. The reactor is fed with the reactant A at a constant rate of 1 liter/min having feed concentration equal to 1 mol/liter. The reactor is initially empty. Given k = 1 min^{-1}, the conversion of reactant A based on moles of A fed at t = 2 min is

(a) 0.136 (b) 0.43
(c) 0.57 (d) 0.864

[2007 : 2 M]

3.19 The following liquid phase reaction is taken placing in an isothermal CSTR

$$A \xrightarrow{k_1} B \xrightarrow{k_2} C$$
$$2A \xrightarrow{k_3} D$$

Reaction mechanism is same as the stoichiometry given above. Given k_1 = 1 min^{-1}; k_2 = 1 min^{-1}; k_3 = 0.5 lit/(mol(min); C_{A0} = 10 mol/litre, C_{B0} = 0 mol/litre and C_B = 1 mol/litre, the solution for F/V (flow rate/reactor volume in min^{-1}) yields

(a) 6.7 (b) 6 and 0.5
(c) 2 and $\frac{4}{3}$ (d) 8 [2007 : 2 M]

3.20 Determine the level of C_{A0} (high, low, intermediate), temperature profile (high, low, increasing, decreasing) which will favor the formation of the desired product indicated in the reaction scheme given below.

$$A \xrightarrow{1} R \xrightarrow{3} S_{\text{desired}}$$
$$A \xrightarrow{2} U$$

n_1	E_1	n_2	E_2	n_3	E_3
2	25	1	35	3	45

(a) High C_{A0}, increasing T, plug flow reactor
(b) Low C_{A0}, increasing T, plug flow reactor
(c) High C_{A0}, decreasing T, mixed flow reactor
(d) High C_{A0}, decreasing T, plug flow reactor

[2007 : 2 M]

3.21 The gas phase reaction $A + 3B \rightarrow 2C$ is conducted in a PFR at constant temperature and pressure. The PFR achieves a conversion of 20% of A. The feed is a mixture of A, B and an inert I. It is found that the concentration of A remains the same throughout the reactor.

Which one of the following ratios of inlet molar rates ($F_{A,in}$: $F_{B,in}$: $F_{I,in}$) is consistent with this observation? Assume the reaction mixture is an ideal gas mixture.

(a) 2 : 3 : 0 (b) 2 : 2 : 1
(c) 3 : 2 : 1 (d) 1 : 2 : 1

[2008 : 2 M]

3.22 The elementary liquid phase series-parallel reaction scheme

$$A \to B \to C$$
$$A \to R$$

is to be carried out in an isothermal CSTR. The rate laws are given by

$$r_R = k'C_A$$
$$r_B = kC_A - kC_B$$

Feed is pure A. The space time of the CSTR which results in the maximum exit concentration of B is given by

(a) $\frac{1}{\sqrt{kk'}}$ (b) $\frac{1}{\sqrt{k'(k+k')}}$

(c) $\frac{1}{(k+k')}$ (d) $\frac{1}{\sqrt{k(k+k')}}$

[2008 : 2 M]

Statement for Linked Answer Questions (3.23 & 3.24) :

The liquid phase reaction $A \to$ Products is to be carried out at constant temperature in a CSTR followed by a PFR in series. The overall conversion of A achieved by the reactor system (CSTR + PFR) is 95%. The CSTR has a volume of 75 litres. Pure A is fed to the CSTR at a concentration C_{A_0} = 2 mol/liter and a volumetric flow rate of 4 liters/min. The kinetics of the reaction is given by

$$-r_A = 0.1C_A^2 \text{ mol/liter/min}$$

3.23 The conversion achieved by the CSTR is

(a) 40% (b) 50%
(c) 60% (d) 80%

[2008 : 2 M]

3.24 The volume of the PFR required (in liters) is

(a) 380 (b) 350
(c) 75 (d) 35

[2008 : 2 M]

Statement for Linked Answer Questions (3.25 & 3.26) :

The liquid phase reaction $A \to B + C$ is conducted isothermally at 50°C in a continuous stirred tank reactor (CSTR). The inlet concentration of A is 8.0 gmol/liter. At a space time of 5 minutes, the concentration of A at the exit of CSTR is 4.0 gmol/liter. The kinetics of the reaction is

$$-r_A = kC_A^{0.5} \text{ gmol/liter.min}$$

A plug flow reactor of the same volume is added in series the existing CSTR.

3.25 The rate constant (k) for this reaction at 50°C is

(a) $0.2\left(\frac{\text{gmol}}{\text{liter}}\right)^{0.5}\cdot\text{min}^{-1}$

(b) $0.2\left(\frac{\text{liter}}{\text{gmol}}\right)^{0.5}\cdot\text{min}^{-1}$

(c) $0.4\left(\frac{\text{gmol}}{\text{liter}}\right)^{0.5}\cdot\text{min}^{-1}$

(d) $0.4\left(\frac{\text{liter}}{\text{liter}}\right)^{0.5}\cdot\text{min}^{-1}$

[2009 : 2 M]

3.26 The concentration of A (in gmol/liter) at the exit of the plug flow reactor is

(a) 0.5 (b) 1.0
(c) 2.0 (d) 2.5

[2009 : 2 M]

Common Data for Questions 3.27 and 3.28 :

A liquid phase reaction, $A \to B$, is conducted isothermally in a CSTR having a residence time of 2 s. The inlet concentration of species A is 2 moles/litre, and the outlet concentration is 1 mole/litre. The rate law for the reaction is $-r_A = \frac{kC_A}{K+C_A}$ where k = 5 moles/litre/s.

3.27 The value of K, in moles/litre, is

(a) 11 (b) 9
(c) 5 (d) 2

[2010 : 2 M]

3.28 If the same reaction is conducted in a series of two CSTRs with residence time 1 s and 0.2 s, then the inlet concentration of A, in moles/litre, required to attain an outlet concentration of A of 1 mole/litre, is

(a) 2.64 (b) 2.00
(c) 1.64 (d) 0.54

[2010 : 2 M]

Linked Answer Questions 3.29 and 3.30:

3.29 In an aqueous solution, reaction $P \rightarrow Q$ occurs under isothermal conditions following first order kinetics. The feed rate is 500 cm^3/min and concentration of P in the feed is 1.5×10^{-4} mol/cm^3. The reaction is carried out in a 5 litre CSTR. At steady state, 60% conversion is observed. The rate constant (in min^{-1}) is

(a) 0.06 (b) 0.15
(c) 0.21 (d) 0.28

[2011 : 2 M]

3.30 The 5 litre CSTR is replaced by five CSTRs in series. If the capacity of each new CSTR is 1 litre, then the overall conversion (in %) is

(a) 65 (b) 67
(c) 73 (d) 81

[2011 : 2 M]

3.31 Consider the reaction scheme shown below:

$$A \xrightarrow{k_1} B \xrightarrow{k_2} C$$

Both the reactions are first-order. The activation energies for k_1 and k_2 are 80 and 20 kJ/mol, respectively. To maximize the yield of B, it is preferable to use

(a) CSTR and high temperature
(b) PFR and high temperature
(c) CSTR and low temperature
(d) PFR and low temperature

[2012 : 1 M]

Statement for Linked Answer Questions 3.32 & 3.33:

The first order liquid phase reaction $A \rightarrow P$ is conducted isothermally in a plug flow reactor of 5 liter volume. The inlet volumetric flow rate is 1 liter/min and the inlet concentration of A is 2 mole/liter.

3.32 If the exit concentration of A is 0.5 mole /liter, then the rate constant, in min^{-1}, is

(a) 0.06 (b) 0.28
(c) 0.42 (d) 0.64

[2012 : 2 M]

3.33 The plug flow reactor is replaced by 3 mixed flow reactors in series, each of 2.0 liters volume. The exit conversion of A (in %) is

(a) 35.9 (b) 52.5
(c) 73.7 (d) 94.8

[2012 : 2 M]

3.34 A first order liquid phase reaction is carried out isothermally at a steady state in a CSTR and 90% conversion is attained. With the same inlet conditions and for the same overall conversion, if the CSTR is replaced by two smaller and identical isothermal CSTRs in series, the % reduction in total volume, to the nearest integer, is _______ .

[2013 : 2 M]

Common Data for Questions 3.35 and 3.36:

Liquid reactant A decomposes as follows:

$A \rightarrow R \quad r_R = k_1 C_A^2 \quad k_1 = 0.5$ m^3/mol-s

$A \rightarrow S \quad r_S = k_2 C_A \quad k_2 = 1\ s^{-1}$

An aqueous feed of composition $C_{A0} = 30$ mol/m^3, $C_{R0} = 2$ mol/m^3, and $C_{S0} = 1$ mol/m^3 enters a CSTR in which the above reactions occur. Assume isothermal and steady state conditions.

3.35 If the conversion of A is 80 %, the concentration of R in the exit stream in mol/m^3, to the nearest integer, is _______.

[2013 : 2 M]

3.36 What is the % conversion of A, to the nearest integer, so that the concentration of S in the exit stream is 11.8 mol/m^3 ? _______

[2013 : 2 M]

3.37 In order to achieve the same conversion under identical reaction conditions and feed flow rate for a non-autocatalytic reaction of positive order, the volume of an ideal CSTR is

(a) always greater than that of an ideal PFR
(b) always smaller than that of an ideal PFR
(c) same as that of an ideal PFR
(d) smaller than that of an ideal PFR only for first order reaction

[2014 : 1 M]

3.38 Consider two steady isothermal flow configurations shown schematically as Case-I and Case-II below. In Case-I, a CSTR of volume V_1 is followed by a PFR of volume V_2, while in Case-II a PFR of volume V_2 is followed by a CSTR of volume V_1. In each case, a volumetric flow rate Q of liquid reactant is flowing through the two units in series. An irreversible reaction $A \rightarrow$ products (order n) takes place in both cases, with a reactant concentration C_{A0} being fed into the first unit.

Case I:

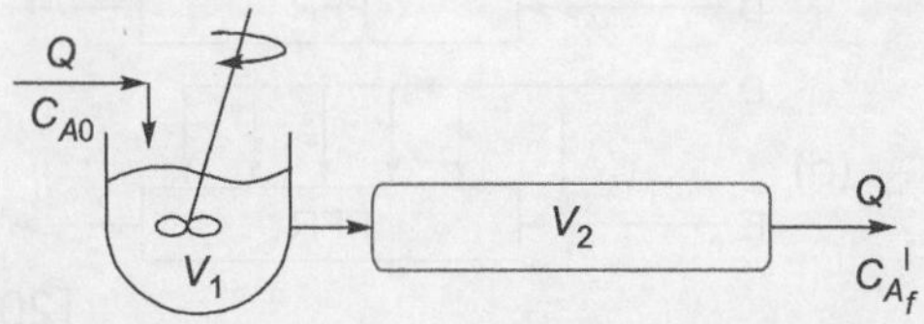

Case II:

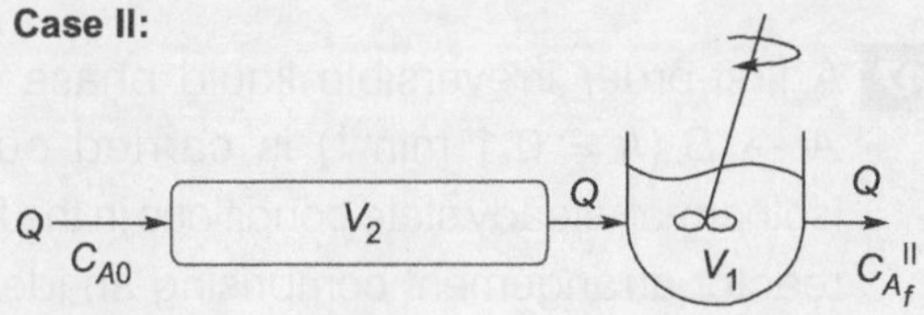

Choose the correct option:

(a) $\dfrac{C_{A_f}^{I}}{C_{A_f}^{II}} > 1$ for $n = 1$ (b) $\dfrac{C_{A_f}^{I}}{C_{A_f}^{II}} = 1$ for $n = 1$

(c) $\dfrac{C_{A_f}^{I}}{C_{A_f}^{II}} < 1$ for $n = 1$ (d) $\dfrac{C_{A_f}^{I}}{C_{A_f}^{II}} = 1$ for $n > 1$

[2015 : 2 M]

3.39 An isothermal steady state mixed flow reactor (CSTR) of 1 m³ volume is used to carry out the first order liquid-phase reaction $A \rightarrow$ products. Fresh feed at a volumetric flow rate of Q containing reactant A at a concentration C_{A0} mixes with the recycle stream at a volumetric flow rate RQ as shown in the figure below.

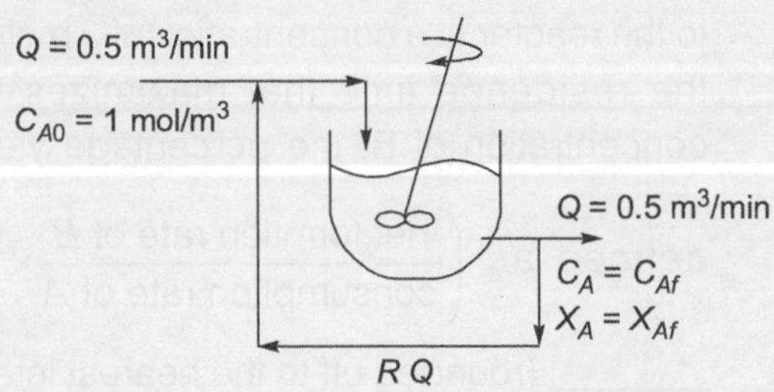

It is observed that when the recycle ratio $R = 0.5$, the exit conversion $X_{Af} = 50\%$. When the recycle ratio is increased to $R = 2$, the new exit conversion (in percent) will be:

(a) 50.0 (b) 54.3

(c) 58.7 (d) 63.2

[2015 : 2 M]

3.40 The liquid phase reversible reaction $A \rightleftharpoons B$ is carried out in an isothermal CSTR operating under steady state conditions. The inlet stream does not contain B and the concentration of A in the inlet stream is 10 mol/lit. The concentrations of A at the reactor exit, for residence times of 1 s and 5 s are 8 mol/lit and 5 mol/lit, respectively. Assume the forward and backward reactions are elementary following the first order rate law. Also assume that the system has constant molar density. The rate constant of the forward reaction (in s^{-1}, rounded off to the third decimal place) is ______.

[2016 : 2 M]

3.41 The following reaction rate curve is shown for a reaction $A \rightarrow P$. Here, $(-r_A)$ and X_A represent reaction rate and conversion, respectively. The feed is pure A and 90% conversion is desired.

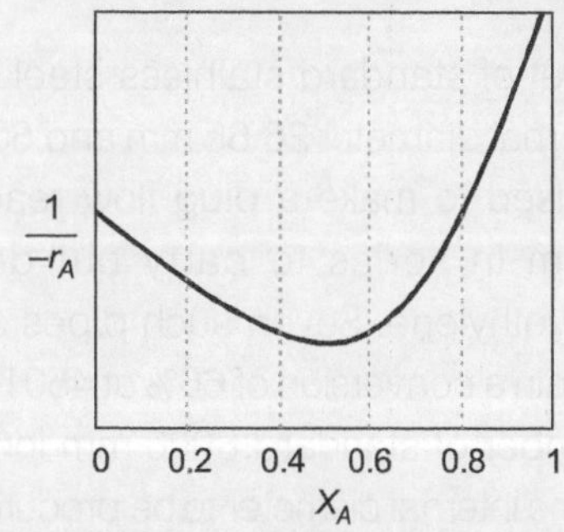

Which amongst the following reactor configurations gives the lowest total volume of the reactor(s)?

(a) CSTR followed by PFR

(b) Two CSTRs in series

(c) PFR followed by CSTR

(d) A single PFR

[2017 : 1 M]

3.42 The following liquid phase second-order reaction is carried out in an isothermal CSTR at steady state

$$A \rightarrow R \qquad (-r_A) = 0.005\, C_A^2 \text{ mol/m}^3\text{.hr}$$

where C_A is the concentration of the reactant in the CSTR. The reactor volume is 2 m³. The inlet flow rate is 0.5 m³/hr and the inlet concentration of the reactant is 1000 mol/m³.

The fractional conversion, rounded to 2 decimal places, is ______.

[2017 : 2 M]

3.43 A CSTR and a PFR of equal volume are connected in series as shown below to carry out a first-order, isothermal, liquid phase reaction $A \rightarrow P$

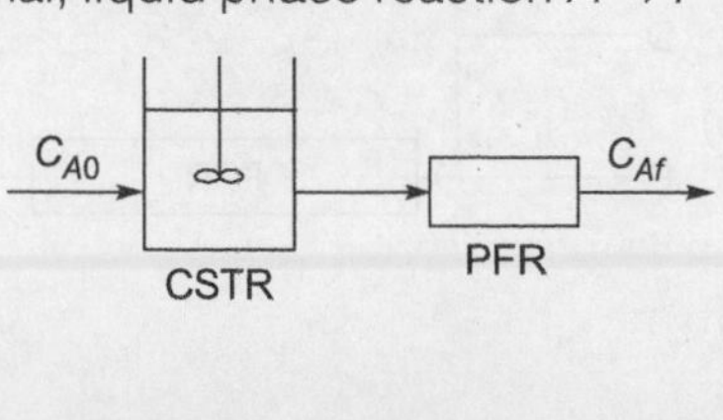

The rate constant is 0.2 s^{-1}. The space-time is 5 s for both the reactors. The overall fractional conversion of A is _______ (rounded off to third decimal place).

[2018 : 2 M]

3.44 The elementary second-order liquid phase reaction $A + B \rightarrow C + D$ is carried out in an isothermal plug flow reactor of 2 m^3 volume. The inlet volumetric flow rate is 10 m^3/h. The initial concentrations of both A and B are 2 kmol/m^3. The rate constant is given as 2.5 m^3 $kmol^{-1}$ h^{-1}. The percentage conversion of A is _______.

[2018 : 2 M]

3.45 A set of standard stainless steel pipes, each of internal diameter 26.65 mm and 6000 mm length, is used to make a plug flow reactor by joining them in series to carry out degradation of polyethylene. Seven such pipes are required to obtain a conversion of 66% at 450 K. The minimum number of standard 8000 mm long pipes of the same internal diameter to be procured for obtaining at least 66% conversion under the same reaction conditions is _______.

[2018 : 2 M]

3.46 The desired liquid-phase reaction:

$$D + E \xrightarrow{k_1} F;\ r_F = k_1 C_D^2 C_E^{0.3}$$

is accompanied by an undesired side reaction

$$D + E \xrightarrow{k_2} G;\ r_G = k_2 C_D^{0.4} C_E^{1.5}$$

Four isothermal reactor schemes (CSTR : ideal Continuous-Stirred Tank Reactor; PFR: ideal Plug Flow Reactor) for processing equal molar feed rates of D and E are shown in figure. Each scheme is designed for the same conversion. The scheme that gives the most favorable product distribution is

(a)

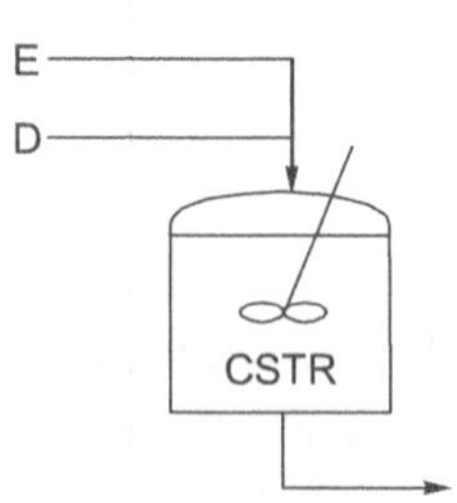

(b)

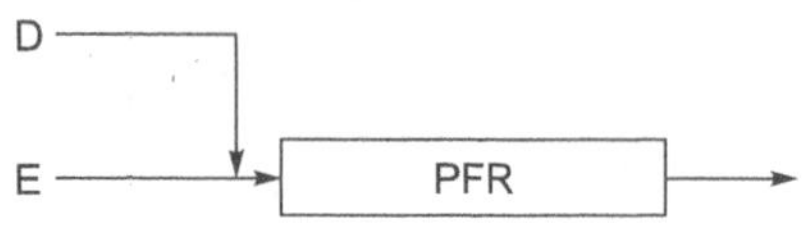

(c)

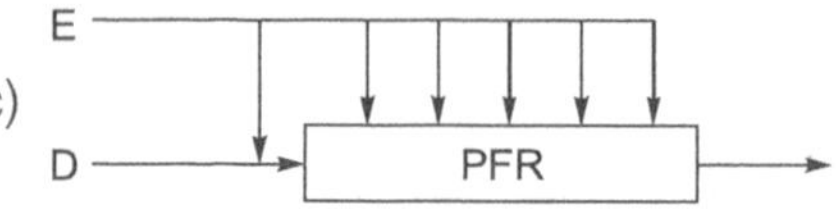

(d)

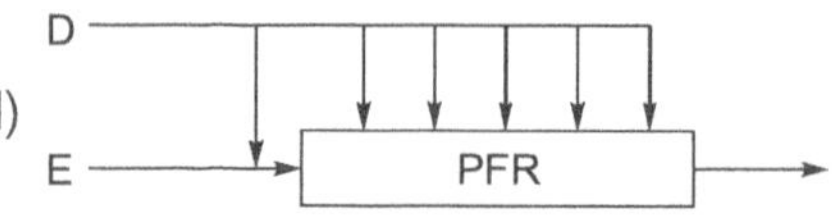

[2019 : 1 M]

3.47 A first-order irreversible liquid phase reaction $A \rightarrow B$ (k = 0.1 min^{-1}) is carried out under isothermal, steady state conditions in the following reactor arrangement comprising an ideal CSTR (Continuous-Stirred Tank Reactor) and two ideal PFRs (Plug Flow Reactors). From the information in the figure, the volume of the CSTR (in liters) is _______ (rounded off to the nearest integer).

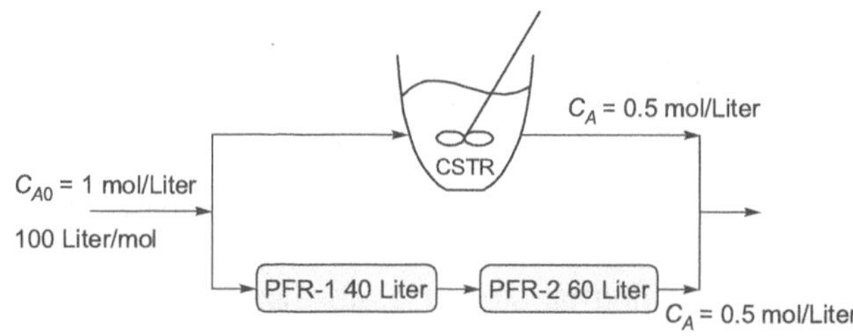

[2019 : 2 M]

3.48 The elementary liquid-phase irreversible reactions

$$A \xrightarrow{k_1 = 0.4\ min^{?\ 1}} B \xrightarrow{k_2 = 0.1\ min^{?\ 1}} C$$

take place in an isothermal ideal CSTR (Continuous-Stirred Tank Reactor). Pure A is fed to the reactor at a concentration of 2 mol/liter. For the residence time that maximizes the exit concentration of B, the percentage yield of B, defined as $\left(\dfrac{\text{net formtion rate of } B}{\text{consumption rate of } A} \times 100\right)$ is _______ (rounded off to the nearest integer).

[2019 : 2 M]

3.49 An elementary liquid phase reversible reaction $P \rightleftarrows Q$ is carried out in an ideal continuous stirred tank reactor (CSTR) operated at steady state. The rate of consumption of P, $-r_P$ (in mol liter^{-1} min^{-1}), is given by

$$-r_P = C_P - 0.5C_Q$$

where C_P and C_Q are the concentrations (in mol litre^{-1}) of P and Q, respectively. The feed contains only the reactant P at a concentration of 1 mol liter^{-1}, and the conversion of P at the exit of the CSTR is 75% of the equilibrium conversion.

Assume that there is no volume change associated with the reaction, and the temperature of the reaction mixture is constant throughout the operation. The space time (in minutes) of the CSTR is ______ (round off to 1 decimal place).

[2020 : 2 M]

3.50 The following isothermal autocatalytic reaction,

$$A + B \rightarrow 2B$$

$$(-r_A) = 0.1C_AC_B \text{ (mol L}^{-1}\text{ s}^{-1})$$

is carried out in an ideal continuous stirred tank reactor (CSTR) operating at steady state. Pure A at 1 mol L^{-1} is fed, and 90% of A is converted in the CSTR. The space time of the CSTR is ______ seconds.

[2021 : 2 M]

3.51 The following homogeneous, irreversible reaction involving ideal gases,

$$A \rightarrow B + C$$

$$(-r_A) = 0.5C_A \text{ (mol L}^{-1}\text{ s}^{-1})$$

is carried out in a steady state ideal plug flow reactor (PFR) operating at isothermal and isobaric conditions. The feed stream consists of pure *A*, entering at 2 m s^{-1}.

In order to achieve 50% conversion of A, the required length of the PFR is ______ meter (round off to 2 decimal places).

[2021 : 2 M]

3.52 Consider the process in the figure. The liquid phase elementary reactions

$A + B \rightarrow P \qquad -r_{B1} = k_1 x_A x_B$

$P + B \rightarrow S \qquad -r_{B2} = k_2 x_P x_B$

$S + A \rightarrow 2P \qquad -r_{S3} = k_3 x_S x_A$

occur in the continuous stirred tank reactor (CSTR), where x_j is the mole fraction of the jth component (j = *A*, *B*, *P*, *S*) in the CSTR. It is given that $k_2 = k_3$. All process feed, process exit and recycle streams are pure. At steady state, the net generation rate of the undesired product, *S*, in the CSTR is zero.

As $q = \dfrac{x_A}{x_B}$ is varied at constant reactor temperature, the reactor volume is adjusted to maintain a constant single-pass conversion of *B*. For a fixed product rate and 90% conversion of *B* in the reactor, the value of *q* that minimizes the sum of the molar flow rates of the *A* and *S* recycle streams is ______ (round off to one decimal place).

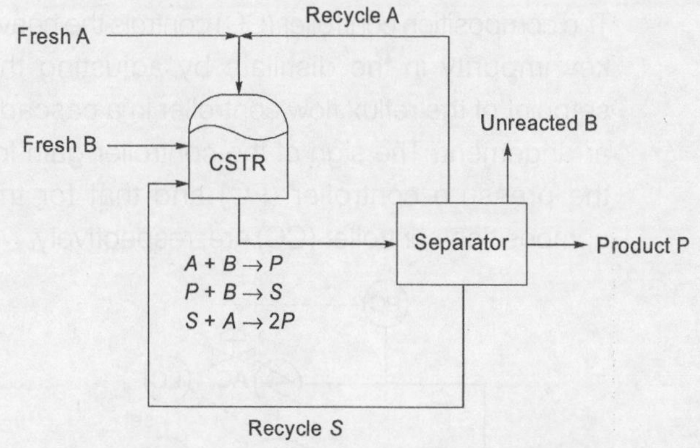

All fresh feeds, process exit streams and recycle streams are pure

[2022 : 2 M]

3.53 An elementary irreversible liquid-phase reaction, $2A \rightarrow B$, is carried out under isothermal conditions in a 1 m^3 ideal plug flow reactor (PFR) as shown in the figure. The volumetric flow rate of fresh *A*, v_1 = 10 m^3 h^{-1}, and its concentration C_{A1} = 2 kmol m^{-3}. For a recycle ratio R = 0, the conversion of *A* at location 2 with respect to the fresh feed (location 1) is 50%. For $R \rightarrow \infty$, the corresponding conversion of *A* is ______% (rounded off to one decimal place).

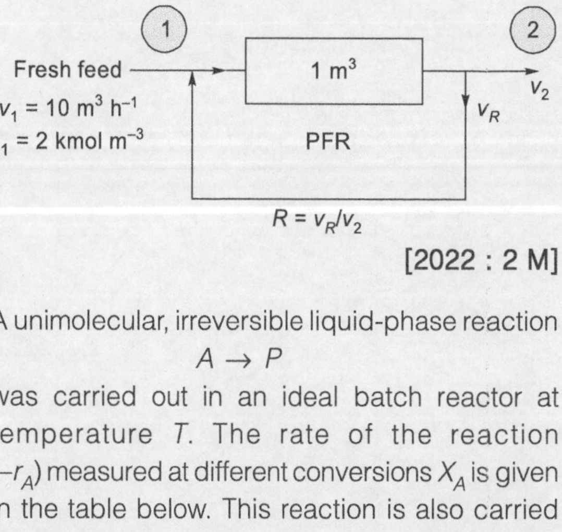

[2022 : 2 M]

3.54 A unimolecular, irreversible liquid-phase reaction

$$A \rightarrow P$$

was carried out in an ideal batch reactor at temperature *T*. The rate of the reaction $(-r_A)$ measured at different conversions X_A is given in the table below. This reaction is also carried out in an ideal continuous stirred tank reactor (CSTR) at the same temperature *T* with a feed concentration of 1 mol.m^{-3}, under steady-state conditions. For a conversion of 0.8, the space time (in s) of the CSTR is ______ (in integer).

X_A	0	0.1	0.2	0.4	0.6	0.8
$-r_A$(mol.$m^{-3}s^{-1}$)	0.45	0.35	0.31	0.18	0.11	0.05

[2023 : 2 M]

3.55 Consider the control structure for the overhead section of a distillation column shown in the figure. The composition controller (CC) controls the heavy key impurity in the distillate by adjusting the setpoint of the reflux flow controller in a cascade arrangement. The sign of the controller gain for the pressure controller (PC) and that for the composition controller (CC) are, respectively,

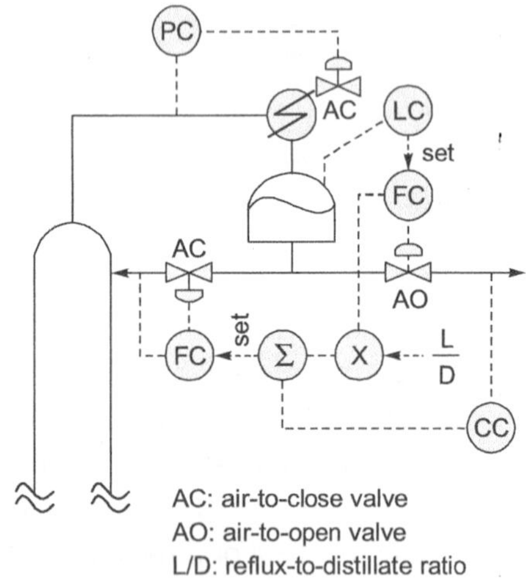

(a) negative, negative (b) negative, positive
(c) positive, positive (d) positive, negative

[2024 : 1 M]

3.56 A first-order liquid phase reaction $A \rightarrow B$ is carried out in two isothermal plug flow reactors (PFRs) of volume 1 m^3 each, connected in series. The feed flow rate and concentration of A to the first reactor are 10 m^3 h^{-1} and 1 kmol m^{-3}, respectively. At steady-state, the concentration of A at the exit of the second reactor is 0.2 kmol m^{-3}. If the two PFRs are replaced by two equal-volume continuously stirred tank reactors (CSTRs) to achieve the same overall steady-state conversion, the volume of each CSTR, in m^3, is

(a) 1.54 (b) 3.84
(c) 7.28 (d) 1.98

[2024 : 2 M]

Answers Design of Isothermal Ideal Reactors

3.1	(c)	**3.2**	(b)	**3.3**	(Sol)	**3.4**	(b)	**3.5**	(Sol)	**3.6**	(a)	**3.7**	(d)
3.8	(c)	**3.9**	(d)	**3.10**	(c)	**3.11**	(d)	**3.12**	(c)	**3.13**	(b)	**3.14**	(d)
3.15	(c)	**3.16**	(c)	**3.17**	(d)	**3.18**	(c)	**3.19**	(b)	**3.20**	(a)	**3.21**	(c)
3.22	(d)	**3.23**	(c)	**3.24**	(b)	**3.25**	(c)	**3.26**	(b)	**3.27**	(b)	**3.28**	(c)
3.29	(b)	**3.30**	(c)	**3.31**	(b)	**3.32**	(b)	**3.33**	(c)	**3.34**	(51.95)	**3.35**	(20)
3.36	(90)	**3.37**	(a)	**3.38**	(b)	**3.39**	(a)	**3.40**	(0.267)	**3.41**	(a)	**3.42**	(0.80)
3.43	(0.816)	**3.44**	(50)	**3.45**	(6)	**3.46**	(c)	**3.47**	(855.73)	**3.48**	(67)	**3.49**	(2)
3.50	(100)	**3.51**	(3.54)	**3.52**	(3)	**3.53**	(38.2%)	**3.54**	(16)	**3.55**	(d)	**3.56**	(a)

Explanations Design of Isothermal Ideal Reactors

3.1 (c)

Given paralleled reaction is

$$A \rightarrow R \quad r_R = k_1 C_A^2 \quad E_1 = 80 \text{ kJ/mol}$$
$$A \rightarrow S \quad r_S = k_2 C_A \quad E_2 = 120 \text{ kJ/mol}$$

Desired product of R

$$\text{Selectivity of } R = \frac{k_1 C_A^2}{k_2 C_A} = \left(\frac{k_1}{k_2}\right) C_A$$

k_2 has high activation energy so we have to maintain low temperature.

So desired condition is low temperature and PFR for high C_A.

3.2 (b)

For CSTR
$$\frac{\tau}{C_{A_0}} = \frac{V}{F_{A_0}} = \frac{X_A}{-r_A}$$

$$\frac{\tau}{C_{A_0}} = \frac{X_A}{kC_{A_0}(1-X_A)}$$

$$k\tau = \frac{X_A}{1-X_A} = \frac{0.5}{1-0.5} = 1$$

Another CSTR of same volume is added in series

Then
$$\frac{\tau}{C_{A_0}} = \frac{X_{A_2} - X_{A_1}}{kC_{A_0}(1-X_{A_2})}$$

$$k\tau = \frac{X_{A_2} - X_{A_1}}{1-X_{A_2}} \quad \{X_{A_1} = 50\}$$

$$1 = \frac{X_{A_2} - 0.5}{1-X_{A_2}}$$

$$X_{A_2} = 0.75$$

3.3 (Sol)

Second order liquid phase reaction

$$A \xrightarrow{k_2} B \qquad k_2 = \frac{1\text{ m}^3}{\text{kmol.s}}$$

$$C_{A_0} = 0.1 \text{ kmol/m}^3$$

$$\tau = 5s \text{ for both reactor}$$

Case-1: PFR followed by CSTR

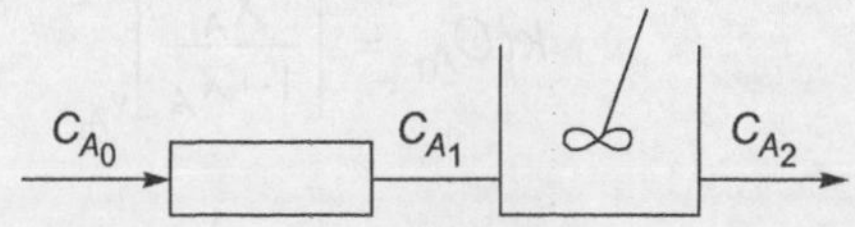

For PFR
$$\frac{\tau}{C_{A_0}} = \int_0^{X_A} \frac{dX_A}{-r_A}$$

$$\frac{\tau}{C_{A_0}} = \int_0^{X_A} \frac{dX_A}{kC_{A_0}^2(1-X_A)^2}$$

$$kC_{A0}\tau = \frac{X_A}{1-X_A}$$

So,
$$X_{A_1} = \frac{1\times 0.1\times 5}{1+1\times 0.1\times 5} = 0.333$$

For CSTR
$$\frac{\tau}{C_{A_0}} = \frac{X_{A_2} - X_{A_1}}{kC_{A_0}^2(1-X_{A_2})^2}$$

$$kC_{A_0}\tau = \frac{X_{A_2} - X_{A_1}}{(1-X_{A_2})^2}$$

$$1\times 0.1\times 5 = \frac{X_{A_2} - 0.333}{(1-X_{A_2})^2}$$

$$0.5X_{A_2}^2 - X_{A_2} + 0.5 = X_{A_2} - 0.333$$

$$X_{A_2} = 0.4724$$

Solving for X_{A_2},

Case-2: CSTR followed by PFR

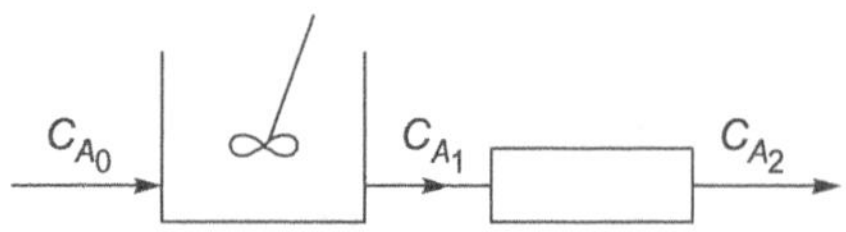

For CSTR
$$\frac{\tau}{C_{A_0}} = \frac{X_A}{-r_A}$$

$$\frac{\tau}{C_{A_0}} = \frac{X_{A1}}{kC_{A_0}^2(1-X_{A_1}^2)}$$

$$k\tau C_{A_0} = \frac{X_{A_1}}{(1-X_{A_1})^2}$$

$$0.5 = \frac{X_{A_1}}{(1-X_{A_1})^2}$$

$$X_{A_1} = 0.268$$

For PFR
$$\frac{\tau}{C_{A_0}} = \int_{X_{A_1}}^{X_{A2}} \frac{dX_A}{-r_A}$$

$$\frac{\tau}{C_{A_0}} = \int_{X_{A_1}}^{X_{A2}} \frac{dX_A}{kC_{A_0}^2(1-X_A)^2}$$

$$k\tau C_{A_0} = \left[\frac{X_A}{1-X_A}\right]_{X_{A_1}}^{X_{A2}}$$

$$0.5 = \left[\frac{1}{1-X_{A_2}} - \frac{1}{1-0.268}\right]$$

$$X_{A_2} = 0.4641$$

Here we can also conclude that CSTR and PFR in series for order greater than 1 conversion obtained in PFR followed by CSTR is always greater than conversion obtained in CSTR followed by PFR.

3.4 (b)

Peclet number is define as ratio of transfer rate due to bulk or convection flow to the diffusion flow. So for PFR, Peclet number is infinite because transfer rate due to diffusion is zero in PFR.

3.5 (Sol)

$A \rightarrow B$

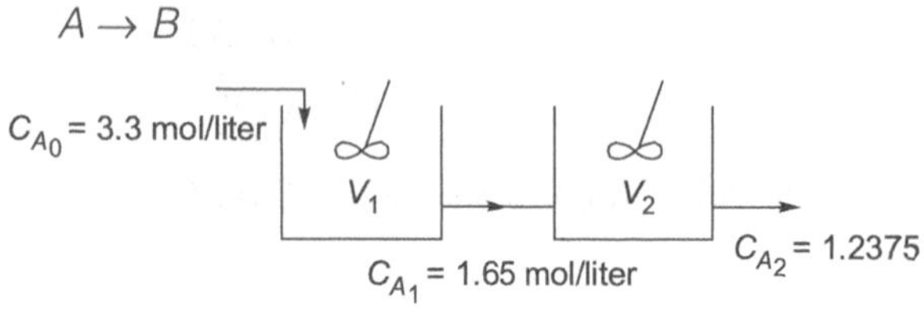

$$\frac{V_1}{V_2} = 1.5, \quad \frac{k_1}{k_2} = 1.5$$

(a) For first order reaction

$$\tau_1 = \frac{C_{A_0} - C_A}{k_1 C_A}$$

$$k_1\tau_1 = \frac{3.3-1.65}{1.65} = 1$$

Similarly for 2nd CSTR

$$k_2\tau_2 = \frac{1.65-1.2375}{1.2375} = 0.333$$

From given data in question

$$\frac{k_1\tau_1}{k_2\tau_2} = \left(\frac{k_1}{k_2}\right)\left(\frac{\tau_1}{\tau_2}\right) = 1.5 \times 1.5 = 2.25$$

So data is not consistent for first order reaction.

(b) For second order reaction

For first CSTR

$$k_1\tau_1 = \frac{C_{A_0} - C_A}{C_A^2}$$

$$= \frac{3.3-1.65}{1.65^2} = 0.606$$

For second CSTR

$$k_2\tau_2 = \frac{1.65-1.2375}{1.2375^2} = 0.2693$$

$$\left(\frac{k_1}{k_2}\right)\left(\frac{\tau_1}{\tau_2}\right) = \frac{0.606}{0.2693} = 2.25$$

So given data is consistent for 2nd order reaction.

3.6 (a)

$A \xrightarrow{k} 2B$

For CSTR

$$\frac{\tau}{C_{A_0}} = \frac{V}{F_{A_0}} = \frac{X_A}{-r_A}$$

$$\frac{\tau}{C_{A_0}} = \frac{X_A}{kC_{A_0}(1-X_A)}$$

$$k\tau = \frac{X_A}{1-X_A}$$

3.7 (d)

Graph is between $-r_A$ v/s conversion so for minimum overall reactor volume area under curve must be maximized so correct arrangement is PFR followed by a CSTR followed by a PFR.

3.8 (c)

$$\frac{F_{A_0}-F_{A_1}}{(-r_A)} = \frac{F_R-F_{R_0}}{r_R}$$

$$-r_A = r_1 + r_2 = 60\text{ pa}$$

$(\because r_1 = 20p_A \text{ and } r_2 = 40p_A)$

$$F_{A_0} = 1\text{ mol/s and } F_{R_0} = F_{S_0} = 0$$

and $$\frac{r_1}{-r_A} = \frac{20p_A}{60p_A} = \frac{1}{3}$$

and $$r_R = 2(r_1) = \frac{2}{3}(-r_A)$$

Hence,

$$\frac{F_{A_0}X_{A_1}}{(k_1+k_2)p_A} = \frac{F_R}{\frac{2}{3}(k_1+k_2)p_A}$$

$$\Rightarrow \quad F_{A_0}X_{A_1} = \frac{2}{3}F_R$$

$(F_R)_{max}$ when $X_{A_1} = 1$

So, $$(F_R)_{max} = \frac{2F_{A_0}}{3} = \frac{2}{3}$$

3.9 (d)

For CSTR:

$$\frac{V}{F_{A_0}} = \frac{X_A}{-r_A}$$

$$\because \quad -r_A = 20p_A + 40p_A = 60p_A$$

$$\because \quad \tau = \frac{F_R-F_{R_0}}{40p_A} = \frac{F_S-F_{S_0}}{120p_A}$$

$$= \frac{2\times0.3\times F_{A_0}}{40} = \frac{X_{A_2}\times3\times F_{A_0}}{120}$$

$$X_{A_2} = 0.6$$

Now, $-r_A = 60p_A$

For $p_A = y_A P_T$

$\because \quad P_T = 1$ bar

$$y_A = \frac{F_A}{F_A+F_R+F_S}$$

$$= \frac{F_{A_0}(1-X_{A_1}-X_{A_2})}{F_{A_1}+2X_{A_1}F_{A_0}+3F_{A_0}X_{A_2}}$$

$$y_A = \frac{0.1}{0.1+2\times0.3+3\times0.6} = 0.04$$

$$\because \quad \frac{V}{F_{A0}} = \frac{X_A}{-r_A}$$

Thus, $$\frac{V}{1} = \frac{0.9}{60\times0.04}$$

$$V = 0.375\text{ m}^3$$

3.10 (c)

$$A+\frac{B}{2}\rightarrow C$$

Stoichiometric feed is used

$$T_0 = 300\text{ K}, \quad T = 360\text{ K}$$

$$y_{A_0} = \frac{1}{1+\frac{1}{2}} = \frac{2}{3}$$

$$\in_A = y_{A_0}\delta_A = \frac{2}{3}\frac{\left(1-\frac{3}{2}\right)}{1} = -\frac{1}{3}$$

$$C_A = \frac{C_{A_0}(1-X_A)}{(1+\in_A X_A)\left(\frac{P_0}{P}\right)\left(\frac{T}{T_0}\right)}$$

$$C_A = \frac{0.833C_{A_0}(1-X_A)}{(1+\in_A X_A)}$$

$$C_A = \frac{0.833C_{A_0}(1-X_A)}{(1-0.33X_A)}$$

3.11 (d)

2nd order liquid phase

Reaction $A\rightarrow B$ carried out in MFR operated in semi batch (no exit).

Initial concentration = C_{A_F}

Volumetric flow rate = F

Volume of reacting mixture = V

Mass balance over A

Rate of A in – Rate of A out ± Generation/consumption rate = Accumulation rate of A

$$FC_{A_F} - 0 - r_AV = \frac{d(VC_A)}{dt}, \qquad \{-r_A = kC_A^2\}$$

$$\frac{d(VC_A)}{dt} = FC_{A_F} - kC_A^2V$$

3.12 (c)

Isothermal aqueous phase reversible reaction $P \rightleftharpoons R$ carried out of MFR.

$$-r_P = 0.5C_P - 0.125C_R\text{ kmol/m}^3\text{.hr}$$

$$X_P = 0.4$$

For CSTR,

$$\frac{\tau}{C_{A_0}} = \frac{X_A}{-r_A}$$

$$\frac{\tau}{C_{R_0}} = \frac{0.4}{0.5C_{R_0}(1-0.4)-0.125(C_{R_0}+C_{R_0}\times 0.4)}$$

$$\tau = 1.6 \text{ hr}$$

3.13 (b)

For first order liquid phase reaction CSTR and PFR in series in any sequence will result in same overall conversion.

For first order reaction two equal volume CSTR in series will always have a greater conversion than a single CSTR.

3.14 (d)

$A \rightarrow B + C$

Conversion obtain of A in PFR is 40%

Feed has 30% inert and 70 mole% A

Inlet temperature = 300 K

Outlet temperature = 400 K

$$\epsilon_A = y_{A_0}\delta_A = \frac{2-1}{1}\times 0.7 = 0.7$$

$$C_A = \frac{C_{A_0}(1-X_A)}{(1+\epsilon_A X_A)\left(\frac{P_0}{P}\right)\left(\frac{T}{T_0}\right)}$$

$$\frac{C_A}{C_{A_0}} = \frac{1-0.4}{1+0.7\times 0.4\left(\frac{400}{300}\right)} = 0.351$$

3.15 (c)

$A \rightleftharpoons 2B$

$$-r_A = 0.02C_A - 0.01C_B$$

Maximum achievable conversion is equilibrium conversion in CSTR.

At equilibrium conversion

$$-r_A = 0$$

$$0.02C_A = 0.01C_B$$

$$0.02C_{A_0}(1-X_{A_e}) = 0.01\left(\cancelto{0}{C_{B_0}} + \frac{2}{1}C_{A_0}X_{A_e}\right)$$

$$0.02 - 0.02X_{A_e} = 0.02X_{A_e}$$

$$X_{A_e} = \frac{1}{2}$$

3.16 (c)

Reaction $A \rightarrow B$ is conducted in two CSTR in series. Overall conversion is X_F. Rate of reaction is given as:

$$r = \frac{-1}{(1+X)}$$

Feed is pure A

Now from design equation of CSTR

$$\frac{\tau}{C_{A_0}} = \frac{V}{F_{A_0}} = \frac{X_A}{-r_A}$$

$$\frac{V_1}{F_{A_0}} = -\frac{X_{A_1}}{\frac{1}{1+X_{A_1}}}$$

$$\frac{V_1}{F_{A_0}} = -X_{A_1}(1+X_{A_1})$$

Similar for 2nd CSTR in series

$$\frac{V_2}{F_{A_0}} = \frac{X_{A_F}-X_{A_1}}{-\frac{1}{(1+X_{A_F})}}$$

$$\frac{V_2}{F_{A_0}} = -(X_{A_F}-X_{A_1})(1+X_{A_F})$$

For minimum volume $\dfrac{d\left(\frac{V_1+V_2}{F_{A_0}}\right)}{dX_{A_1}} = 0$

So,

$$\frac{d}{dX_{A_1}}\left[-X_{A_1}(1+X_{A_1})-(X_{A_F}-X_{A_1})(1+X_{A_F})\right] = 0$$

On solving $X_{A_1} = 0.5X_{A_F}$

3.18 (c)

For semi batch reactor with first order reaction if reactor is initially empty, the concentration of A is given as

$$C_A = \frac{C_{A_0}}{kt}(1-e^{-kt})$$

$$\frac{C_A}{C_{A_0}} = \frac{1}{2\times 1}\left[1-e^{-2\times 1}\right]$$

$$\frac{C_A}{C_{A_0}} = 0.4323$$

$$X_A = 1-\frac{C_A}{C_{A_0}} = 0.57$$

3.19 (b)

For CSTR:

$$\frac{\tau}{C_{A_0}} = \frac{V}{F_{A_0}} = \frac{X_A}{-r_A}$$

On $$\tau = \frac{C_{A_0} - C_A}{-r_A}$$

$$A \xrightarrow{k_1} B \xrightarrow{k_2} C$$

$$2A \xrightarrow{k_3} D$$

$$k_1 = 1\text{ min}^{-1}, \quad k_2 = 1\text{ min}^{-1}$$

$$k_3 = \frac{0.5\text{ lit}}{\text{mol.min}}$$

$$C_{A_0} = \frac{10\text{ mol}}{\text{liter}}, \; C_{B_0} = 0$$

$$C_B = \frac{1\text{ mol}}{\text{liter}}$$

$$\tau = \frac{10 - C_A}{C_A + 0.5C_A^2} \quad \text{...(i)}$$

Similarly for component (*B*)

$$\tau = \frac{C_B - C_{B_0}}{r_B}$$

$$\tau = \frac{1-0}{C_A - 1} \quad \text{...(ii)}$$

From equation (i) and (ii)

$$\frac{10 - C_A}{C_A + 0.5C_A^2} = \frac{1}{C_A - 1}$$

On solving $C_A = 5.44$ and 1.23

From equation (ii)

$$\tau = \frac{V}{F} = \frac{1}{C_A - 1}$$

$$\frac{F}{V} = C_A - 1 = 4.44 \text{ and } 1.23$$

Closest option is option (b).

3.20 (a)

High activation energy favours high temperature and for high C_A use PFR.

3.21 (c)

Following gas phase reaction is conducted in a PFR

$$A + 3B \rightarrow 2C$$

$$X_A = 0.2$$

Feed contains *A*, *B* and inert *I*

It is given in the question that concentration of *A* remains same throughout the reaction so

$$\frac{F_{A_{in}}}{F_{A_{in}} + F_{B_{in}} + F_{I_{in}}} = \frac{F_A}{F_A + F_B + F_C + F_{I_0}}$$

$$\frac{F_{A_0}}{F_{A_0} + F_{B_0} + F_{I_0}} = \frac{F_{A_0}(1 - X_A)}{F_{A_0}(1 - X_A) + \left(F_{B_0} - \frac{b}{a}F_{A_0}X_A\right) + \left(F_{C_0} + \frac{c}{a}F_{A_0}X_A\right) + F_{I_0}}$$

$$\frac{F_{A_0}}{F_{A_0} + F_{B_0} + F_{I_0}} = \frac{F_{A_0} \times 0.8}{F_{A_0} \times 0.8 + (F_{B_0} - 3F_{A_0} \times 0.2) + 2F_{A_0} \times 0.2 + F_{I_0}}$$

$$0.8F_{A_0} + F_{B_0} - 0.6F_{A_0} + 0.4F_{A_0} + F_{I_0} = 0.8F_{A_0} + 0.8F_{B_0} + 0.8F_{I_0}$$

$$0.2F_{B_0} + 0.2F_{I_0} = 0.2F_{A_0}$$

$$F_{A_0} = F_{B_0} + F_{I_0}$$

Only option (c) follows this criteria.

3.22 (d)

Following liquid phase series-parallel reaction is carried out in isothermal CSTR.

$$A \rightarrow B \rightarrow C \qquad r_B = kC_A - kC_B$$

$$A \rightarrow R \qquad r_R = k'C_A$$

From design equation of CSTR:

$$\tau = \frac{C_{A_0} - C_A}{-r_A} = \frac{C_{A_0} - C_A}{(k + k')C_A}$$

$$C_A = \frac{C_{A_0}}{1 + (k + k')\tau} \quad \text{...(i)}$$

Now for component *B*.

$$\tau = \frac{C_B - C_{B_0}}{r_B}$$

Here $C_{B_0} = 0$ (feed is pure *A*)

$$\tau = \frac{C_B}{k(C_A - C_B)}$$

$$C_B = \frac{k\tau C_A}{1 + k\tau}$$

By putting value of C_A from equation (i)

$$C_B = \frac{k\tau C_{A_0}}{(1 + k\tau)[1 + (k + k')\tau]}$$

For maximum C_B

$$\frac{dC_B}{d\tau} = 0$$

$$\frac{dC_B}{d\tau} = (1 + k\tau)\,(1 + (k + k')\,\tau)\,k - k\tau\,[k\,(1 + (k + k')\,\tau) + (1 + k\tau)\,(k + k')] / [(1 + k\tau)\,[1 + (k + k')\,\tau]]^2$$

On solving above equal to zero, we get

$$\tau = \frac{1}{\sqrt{k(k+k')}}$$

3.23 (c)

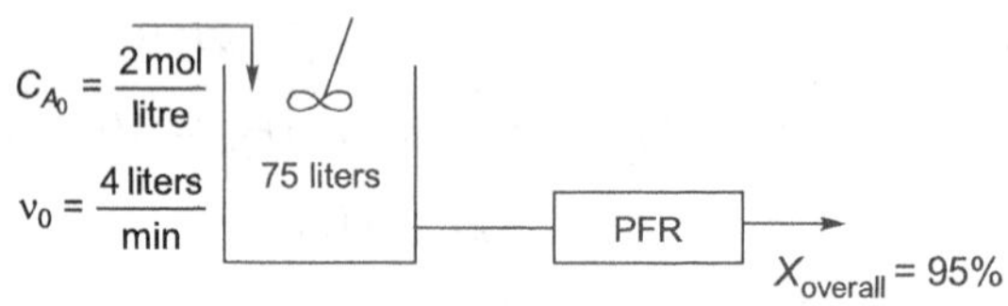

$$-r_A = 0.1C_A^2$$

From design equation of CSTR

$$\frac{\tau}{C_{A_0}} = \frac{V}{F_{A_0}} = \frac{X_A}{-r_A}$$

$$\frac{75}{4C_{A_0}} = \frac{X_A}{0.1C_{A_0}^2(1-X_A)^2}$$

$$\frac{75\times 2}{4} = \frac{X_A}{0.1(1-X_A)^2}$$

Solving for X_A, we get = 60%

3.24 (b)

From design equation of PFR

$$\frac{\tau}{C_{A_0}} = \frac{V}{F_{A_0}} = \int_{X_{A_1}}^{X_{A_2}} \frac{dX_A}{-r_A}$$

$$\frac{V}{4\times C_{A_0}} = \int_{0.6}^{X_{A_2}} \frac{dX_A}{0.1C_{A_0}^2(1-X_A)^2}$$

$$\frac{V\times 0.1\times 2}{4} = +\left[\frac{1}{(1-X_A)}\right]_{0.6}^{0.95}$$

$$V = 350 \text{ liters}$$

3.25 (c)

From design equation of CSTR:

τ = 5 minute, C_{A_0} = 8 gmol/liter

C_A = 4 gmol/liter

$$\tau = \frac{C_{A_0} - C_A}{-r_A}$$

$$5 = \frac{8-4}{k\times 4^{0.5}}$$

$$k = 0.4\left(\frac{\text{gmol}}{\text{liter}}\right).\text{min}^{-1}$$

3.26 (b)

For PFR $\quad \tau$ = 5 minute

C_{A_0} = 4 mol/litre

From design equation of PFR

$$\tau = -\int_{C_{A_0}}^{C_A} \frac{dC_A}{-r_A}$$

$$\tau = \int_{C_A}^{C_{A_0}} \frac{dC_A}{-r_A}$$

$$5 = \int_{C_A}^{4} \frac{dC_A}{0.4C_A^{0.5}}$$

$$= \frac{1}{0.4\times 0.5}\left[C_A^{0.5}\right]$$

$$5 = \frac{1}{0.4\times 0.5}\left[4^{0.5} - C_A^{0.5}\right]$$

C_A = 1 gmol/litre

3.27 (b)

Given liquid phase reaction $A \rightarrow B$ conducted in CSTR of residence time 2 sec.

C_{A_0} = 2 mol/liter

C_A = 1 mol/liter

Rate of reaction ($-r_A$) is given as

$$-r_A = \frac{kC_A}{K + C_A}$$

Here, $\quad k = \frac{5 \text{ mole}}{\text{litre.sec}}$

From design equation of CSTR

$$\tau = \frac{C_{A_0} - C_A}{-r_A}$$

$$2 = \frac{2-1}{\dfrac{5\times 1}{K+1}}$$

$K + 1 = 10$

K = 9 mol/litre

3.28 (c)

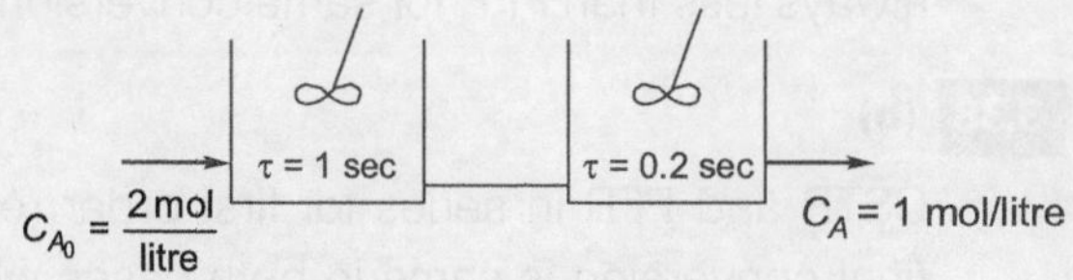

For 2nd CSTR from design equation:

$$\tau = \frac{C_{A_0} - C_A}{-r_A}$$

$$0.2 = \frac{C_{A_1} - 1}{\frac{5\times 1}{9+1}}$$

$$C_{A_1} = 1.1 \text{ mol/liter}$$

C_{A_0} for 2nd MFR will be C_{A_1} of first MFR so by design equation of MFR

$$\tau = \frac{C_{A_0} - C_{A_1}}{-r_A}$$

$$1 = \frac{C_{A_0} - 1.1}{\frac{5\times 1.1}{9+1.1}}$$

$$C_{A_0} = 1.64 \text{ mol/liter}$$

3.29 (b)

From design equation of CSTR

$$\frac{\tau}{C_{A_0}} = \frac{V}{F_{A_0}} = \frac{X_A}{-r_A}$$

$$\frac{5000}{500 C_{A_0}} = \frac{0.6}{k C_{A_0}(1-0.6)}$$

$$k = 0.15 \text{ minute}^{-1}$$

3.30 (c)

For n equal volume CSTR in series and first order reaction overall conversion can be found by

$$X_A = 1 - \frac{1}{(1+k\tau)^n}$$

$$X_A = 1 - \frac{1}{(1+0.15\times 2)^5} = 0.7306$$

% conversion = 73.06%

3.31 (b)

To maximize yield of any intermediate component we must prefer PFR instead of CSTR and high activation energy favours high temperature because ($k = k_0 e^{-E/RT}$). At high temperature k is comparatively large if E is large.

3.32 (b)

For first order liquid phase reaction $A \rightarrow P$

Volume of PFR V_P = 5 liter

$$\upsilon_0 = 1 \text{ liter/min}$$

$$C_{A_0} = 2 \text{ mol/liter}$$

$$C_A = 0.5 \text{ mol/liter}$$

From design equation of PFR:

$$\tau = -\int_{C_{A_0}}^{C_A} \frac{dC_A}{-r_A}$$

$$\frac{5}{1} = -\int_{2}^{0.5} \frac{dC_A}{kC_A}$$

$$k \times 5 = \ln\left(\frac{2}{0.5}\right)$$

$$k = 0.28 \text{ min}^{-1}$$

3.33 (c)

For n equal volume CSTR in series for first order reaction conversion can be found by

$$X_A = 1 - \frac{1}{(1+k\tau)^n}$$

Putting values, we get

$$X_A = 1 - \frac{1}{(1+0.28\times 2)^3} = 0.7366$$

% conversion = 73.66%

3.34 (51.95)

From design equation of CSTR:

$$\frac{\tau}{C_{A_0}} = \frac{X_A}{-r_A}$$

Here first order liquid phase reaction is given in question

So, $$\frac{\tau}{C_{A_0}} = \frac{X_A}{kC_{A_0}(1-X_A)}$$

$$X_A = \frac{k\tau}{1+k\tau}$$

$$X_A = 0.9 \text{ given}$$

$$k\tau = \frac{0.9}{1-0.9} = 9$$

CSTR is replaced by two identical CSTR in series.

$$X_A = 1 - \frac{1}{(1+k\tau')^2}$$

For same value of X_A

$$k\tau' = 2.162$$

$$\tau = \frac{V}{v_0}$$

So $\qquad \tau \propto V$

% reduction in total volume is given by

$$\frac{\tau k - 2\tau' k}{\tau k} \times 100 = \frac{9 - 2\times 2.162}{9} \times 100 = 51.95\%$$

3.35 (20)

Liquid reactant A decomposes as follows:

$A \rightarrow R \qquad r_R = k_1 C_A^2, \quad k_1 = 0.5\ \text{m}^3/\text{mol.s}$

$A \rightarrow S \qquad r_S = k_2 C_A, \quad k_2 = 1\ \text{s}^{-1}$

$C_{A_0} = 30\ \text{mol/m}^3 \qquad X_A = 0.8$

$C_{R_0} = 2\ \text{mol/m}^3 \qquad C_A = C_{A_0}(1 - X_A)$

$C_{S_0} = 1\ \text{mol/m}^3 \qquad C_A = 6\ \text{mol/m}^3$

From material balance in CSTR

$$\tau = \frac{C_{A_0} - C_A}{-r_A} = \frac{C_R - C_{R_0}}{r_R}$$

$$\frac{30-6}{0.5\times 6^2 + 6} = \frac{C_R - 2}{0.5\times 6^2}$$

$$C_R = 20\ \text{mol/m}^3$$

3.36 (90)

$C_S = 11.8\ \text{mol/m}^3$ (given)

From material balance equation on CSTR

$$\tau = \frac{C_{A_0} - C_A}{-r_A} = \frac{C_S - C_{S_0}}{r_S}$$

$$\frac{30 - C_A}{0.5C_A^2 + C_A} = \frac{11.8 - 1}{C_A}$$

$$\frac{30 - C_A}{0.5C_A + 1} = 10.8$$

$$C_A = 3\ \text{mol/m}^3$$

$$X_A = 1 - \frac{C_A}{C_{A_0}} = 1 - \frac{3}{30} = 0.9$$

% conversion of A is 90%.

3.37 (a)

In order to achieve the same conversion volume required by CSTR is always greater than PFR for positive order.

For negative order volume required by CSTR is always less than PFR for same conversion.

3.38 (b)

CSTR and PFR in series for first order reaction final conversion is same in both cases whether CSTR followed by PFR or PFR followed by CSTR.

3.39 (a)

In ideal CSTR back mixing is assume to be infinite. So in CSTR there is no effect of recycling on value of recycle ratio.

3.40 (0.267)

Liquid phase reversible reaction $A \rightleftharpoons B$

$$C_{A_0} = 10\ \text{mol/liter}$$

For, $\quad \tau = 1$ sec $\quad C_A = 8$ mol/liter $\quad X_A = 0.2$

$\qquad \tau = 5$ sec $\quad C_A = 5$ mol/liter $\quad X_A = 0.5$

For CSTR

$$\tau = \frac{C_{A_0} - C_A}{-r_A}$$

$$1 = \frac{10-8}{8k_1 - 2k_2} \Rightarrow 4k_1 - k_2 = 1 \qquad \text{...(i)}$$

$$5 = \frac{10-5}{5k_1 - 5k_2} \Rightarrow 5k_1 - 5k_2 = 1 \qquad \text{...(ii)}$$

On solving equation (i) and (ii), we get

$$k_1 = 0.267\ \text{sec}^{-1}$$

3.41 (a)

The graph shown in figure is drawn between $\frac{1}{-r_A}$ and X_A. So for lowest volume we have to minimize area.

So correct option is CSTR followed by PFR.

3.42 (0.80)

$A \rightarrow R \quad -r_A = 0.05\, C_A^2$

From design equation of CSTR

$$\frac{\tau}{C_{A_0}} = \frac{V}{F_{A_0}} = \frac{X_A}{-r_A}$$

$$\frac{2}{0.5\times 1000} = \frac{X_A}{(1-X_A)^2 \times 0.005 \times 1000^2}$$

$$20 = \frac{X_A}{(1-X_A)^2}$$

$$20X_A^2 - 41X_A + 20 = 0$$
$$X_A = 0.80$$

3.43 (0.816)

$$A \rightarrow P$$

First-order, isothermal, liquid phase reaction.
From design equation of CSTR

$$\frac{\tau}{C_{A_0}} = \frac{X_A}{-r_A}$$

$$\frac{5}{C_{A_0}} = \frac{X_A}{0.2C_{A_0}(1-X_A)}$$

$$X_A = \frac{1}{1+1} = 0.5$$

From design equation of PFR

$$\frac{\tau}{C_{A_0}} = \int_{X_{A_1}}^{X_{A_2}} \frac{dX_A}{-r_A}$$

$$\frac{5}{C_{A_0}} = \int_{0.5}^{X_{A_2}} \frac{dX_A}{0.2C_{A_0}(1-X_A)}$$

$$1 = -[\ln(1-X_A)]_{0.5}^{X_{A_2}}$$

$$X_{A_2} = 0.816$$

3.44 (50%)

Elementary second order liquid phase reaction

$$A + B \rightarrow C + D$$
$$-r_A = kC_AC_B$$

Initial concentration of A and B is same and equal to 2 kmol/m^3 so concentration of A and B will be ($C_A = C_B$) same throughout the reactor

So, $-r_A = kC_A^2$

Now from design equation of PFR

$$\frac{V}{F_{A_0}} = \int_0^{X_A} \frac{dX_A}{kC_{A_0}^2(1-X_A)^2}$$

$$\frac{2}{20} = \int_0^{X_A} \frac{dX_A}{2.5\times 2^2(1-X_A)^2}$$

$$\frac{2}{20}\times 2.5\times 2^2 = \frac{X_A}{1-X_A}$$

$$X_A = 0.5 = 50\%$$

3.45 (6)

For plug flow reactor for same conversion of polyethylene both the plug flow reactors must have same volume.

So, $7\times \frac{\pi}{4}\times 26.65^2\times 6000 = n\times \frac{\pi}{4}\times 26.65^2\times 8000$

$$n = 5.26$$

Number of pipes must be integer because all the pipes are identical.
So numbers of pipes are 6.

3.46 (c)

From given rate 2 kinetics:

$$D + E \xrightarrow{k_1} F \quad r_F = k_1 C_D^2 C_E^{0.3}$$

$$D + E \xrightarrow{k_2} G \quad r_G = k_2 C_D^{0.4} C_E^{1.5}$$

Formation of F is desired and formation of G is undesired.

$$\frac{r_F}{r_G} = \left(\frac{k_1}{k_2}\right) C_D^{1.6} C_E^{-1.2}$$

So we have to keep concentration of D high and concentration of E lower for favourable product distribution.
Thus option (c) would be best one.

3.47 (855.73)

For PFR in series both the PFR of 40 liter and 60 liter can be replaced by PFR of 100 liter.
From design equation of PFR

$$\frac{V}{F_{A_0}} = \int_0^{X_A} \frac{dX_A}{-r_A}$$

$$\frac{100}{v_0 \times C_{A_0}} = \frac{1}{kC_{A_0}} \int_0^{X_A} \frac{dX_A}{(1-X_A)}$$

$$\because \quad X_A = 1 - \frac{C_A}{C_{A_0}} = 1 - \frac{0.5}{1} = 0.5$$

$$\frac{10}{v_0} = -\ln(1-0.5)$$

$$v_0 = 14.427 \text{ liter/min}$$

So volumetric flow rate for CSTR is
$(100 - 14.247) = 85.573$ mol/liter
From design equation of CSTR

$$\frac{V}{F_{A_0}} = \frac{X_A}{-r_A}$$

$$\frac{V}{1\times 85.573} = \frac{0.5}{0.1\times 1(0-0.5)}$$

$$V = 855.73 \text{ liter}$$

3.48 (67)

Elementary liquid-phase irreversible reactions

$$A \xrightarrow{k_1 = 0.4\,\text{min}^{-1}} B \xrightarrow{k_2 = 0.1\,\text{min}^{-1}} C$$

Pure A fed to reactor with $C_{A_0} = 2$ mol/liter

From design equation of CSTR

$$\tau = \frac{C_{A_0} - C_A}{-r_A} = \frac{C_{A_0} - C_A}{k_1 C_A}$$

$$C_A = \left(\frac{C_{A_0}}{1 + k_1\tau}\right) \qquad \text{...(i)}$$

Similarly for component B

$$\tau = \frac{C_B - C_{B_0}}{+r_B}$$

$$\tau = \frac{C_B}{k_1 C_A - k_2 C_B} \qquad C_{B_0} = 0$$

$k_1 \tau C_A - k_2 \tau C_B = C_B$

$$C_B = \frac{k_1 \tau C_A}{(1 + k_2\tau)}$$

By putting value of C_A from equation (i)

$$C_B = \frac{k_1 \tau C_{A_0}}{(1 + k_1\tau)(1 + k_2\tau)} \qquad \text{...(ii)}$$

From maximum C_B

$$\frac{dC_B}{d\tau} = 0$$

On solving, $\tau = \dfrac{1}{\sqrt{k_1 k_2}} = \dfrac{1}{\sqrt{0.04}} = 5$ minute

$$= \left(\frac{\text{Net formation rate of } B}{\text{Consumption rate of } A} \times 100\right)$$

$$= \frac{k_1 C_A - k_2 C_B}{k_1 C_A} \times 100 \qquad \text{...(iii)}$$

From equation (i) $C_A = \left(\dfrac{2}{3}\right)$

From equation (ii) $C_B = 0.889$

By putting values in term (iii)

$$\frac{k_1 C_A - k_2 C_B}{k_1 C_A} \times 100 = 66.66\% \cong 67\%$$

3.49 (2)

Given elementary phase reaction $P \rightleftharpoons Q$

Feed is pure P

Conversion obtained of P in CSTR is 75% of the equilibrium conversion.

The rate equation given in the question is

$$-r_P = C_P - 0.5\, C_Q$$

At equilibrium

$$0 = C_{P_0}(1 - X_{P_e}) - 0.5(C_{Q_0} + C_{P_0} X_{P_e})$$

$$C_{P_0}(1 - X_{P_e}) = 0.5 C_{P_0} X_{P_e} \qquad \{C_{Q_0} = 0\}$$

$$X_{P_e} = 0.667$$

$$X_P = X_{P_e} \times 0.75 = 0.5$$

From design equation of CSTR

$$\frac{\tau}{C_{A_0}} = \frac{V}{F_{A_0}} = \frac{X_A}{-r_A}$$

$$\frac{\tau}{C_{A_0}} = \frac{0.5}{C_{P_0}(1 - 0.5) - 0.5(C_{Q_0} + C_{P_0} \times 0.5)}$$

$$\tau = 2.0 \text{ min}$$

3.50 (100)

The isothermal autocatalytic reaction is

$$A + B \rightarrow 2B$$

$$-r_A = 0.1 C_A C_B$$

For autocatalytic reaction

$$C_A + C_B = C_0 = C_{A_0} + C_{B_0}$$

Given: $C_{A_0} = 1$ mol/l and $X_A = 0.9$

For CSTR, $\tau_m = \dfrac{C_{A_0} - C_A}{(-r_A)_{\text{exit}}}$

$$= \frac{C_{A_0} - C_{A_f}}{0.1 C_{A_f}(C_0 - C_{A_f})}$$

$\because$ For C_{A_f}, $C_{A_f} = C_{A_0}(1 - X_{A_f})$

$$= 1(1 - 0.9) = 0.1 \text{ mol/L}$$

$$\tau = \frac{1 - 0.1}{0.1 \times 0.1 \times 0.9} = 100 \text{ sec}$$

3.51 (3.54)

$$A \rightarrow B + C$$

$$-r_A = 0.5 C_A$$

If area of PFR be A, length be L and volume is V_P, then

$$V_P = AL \qquad \text{...(1)}$$

If velocity is v, then volumetric flow rate

$$Q = Av$$

$$\tau_{\text{plug}} = C_{A_0} \int_0^{X_A} \frac{dX_A}{(-r_A)}$$

$$\tau_{\text{plug}} = \frac{V_P}{Q} = \frac{AL}{Av} = \frac{L}{v} \qquad \text{...(2)}$$

For gas phase reaction,

$$\epsilon_A = \frac{2 - 1}{1} = 1$$

$$\therefore \quad -r_A = 0.5C_{A_0}\frac{(1-X_A)}{(1+X_A)}$$

From equation (2)

$$\frac{L}{v} = C_{A_0}\int_0^{0.5}\frac{dX_A}{\dfrac{0.5C_{A_0}(1-X_A)}{(1+X_A)}}$$

$$= 2\int_0^{0.5}\left(\frac{1+X_A}{1-X_A}\right)dX_A$$

On integrating,

$$\frac{L}{v} = 1.772$$

$$L = 2\times 1.772 = 3.544 \text{ m}$$

3.53 (38.2%)

Now for PFR $\quad \dfrac{X}{1-X} = kC_{A_0}t$

$$\frac{0.50}{0.50} = 5\times 2\times\frac{1}{10}$$

$$\Rightarrow \quad k = 5 \text{ m}^3\text{/kmol.hr}$$

Now, $\quad \dfrac{1}{10\times C_{A_0}} = \dfrac{x_A}{5\times C_{A_0}^2(1-x)^2}$

$$x^2 - 3x + 1 = 0$$

Solving, $\quad x = 0.382,\ 38.2\%$

3.54 (16)

From the performance equation of CSTR we have,

$$\frac{\tau}{C_{A_0}} = \frac{V}{F_{A0}} = \left.\frac{X_A}{-r_A}\right|_{exit}$$

$$C_{A_0} = 1 \text{ mol/m}^3$$

$$X_A = 0.8$$

Since in CSTR the rate at exit conversion is only considered, therefore we have,

$$\frac{\tau}{1} = \frac{0.8}{0.05}$$

$\therefore$ Space time $= \tau = 16$ sec

3.55 (d)

The control valve attached to pressure controller is air to close.

Suppose pressure inside distillation column increases then pressure controller have to open the valve to release the over pressure.

It is case of increasing input (error to pressure controller) and to open the valve (since it is air to close) pressure controller have to reduce its output. So pressure controller is an indirect controller so gain of PC controller will be positive.

Similarly for composition controller if purity of distillate increases flow controller have to minimize reflux flow. It means as error (input) to composition controller increases its output increases to close the flow control valve since this valve is air to close. It is a case of direct controller so gain will be negative.

Direct controller → Negative gain

Indirect controller → Positive gain

3.56 (a)

A first-order liquid phase reaction $A \rightarrow B$ is carried out in two isothermal PFR of volume 1 m^3 each along with following data

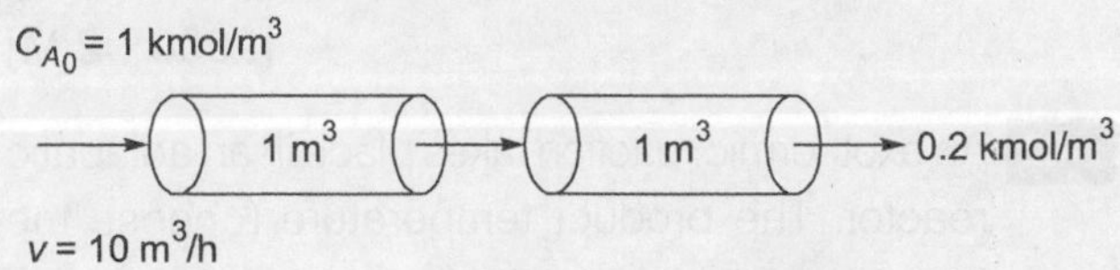

$$X_A = 1-\frac{C_A}{C_{A_0}} = 1-\frac{0.2}{1} = 0.8$$

From design equation of PFR

$$\frac{\tau}{C_{A_0}} = \frac{V}{F_{A_0}} = \int_0^{X_A}\frac{dX_A}{(-r_A)}$$

$$k\tau = -\ln(1-X_A)$$

$$k\tau = -\ln(0.2) = 1.6094$$

$$k\times\frac{2}{10} = 1.60944$$

$$k = 8.0471$$

For equal size MFR connected in series

$$X_A = 1-\left(\frac{1}{1+k\tau}\right)^n$$

$$0.8 = 1-\left(\frac{1}{1+k\tau}\right)^2$$

$$k\tau = 1.236$$

$$8.0471\times\frac{V}{10} = 1.236$$

$$V = 1.54 \text{ m}^3$$

■■■■

4 Design of Non-Isothermal Ideal Reactors

4.1 The reaction $A \rightarrow B$ is conducted in an adiabatic plug flow reactor (PFR). Pure A at a concentration of 2 kmol/m^3 is fed to the reactor at the rate of 0.01 m^3/s and at a temperature of 500 K. If the exit conversion is 20%, then the exit temperature (in K) is

(a) 400 (b) 500

(c) 600 (d) 1000

Data:

Heat of reaction at 298 K = –50,000 kJ/kmol of A reacted.

Heat capacities, $C_{PA} = C_{PB} = 100$ kJ/kmol K (may be assumed to be independent of temperature.)

[2000 : 2 M]

4.2 An exothermic reaction takes place in an adiabatic reactor. The product temperature (choose the correct option) the reactor feed temperature

(a) is always equal to

(b) is always greater than

(c) is always less than

(d) may be greater or less than

[2002 : 1 M]

4.3 A batch adiabatic reactor at an initial temperature of 373 K is being used for the reaction $A \rightarrow B$. Assume the heat of reaction is –1 kJ/mol at 373 K and the heat capacity of both A and B to be constant and equal to 50 J/mol K. The temperature rise after a conversion of 0.5 will be

(a) 5°C (b) 10°C

(c) 20°C (d) 100°C

[2002 : 2 M]

4.4 Ammonia is produced by the following reaction

$$N_2 + 3H_2 \Leftrightarrow 2NH_3$$

In a commercial process for ammonia production, the feed to an adiabatic reactor contains 1 kmol/s of nitrogen and stoichiometric amount of hydrogen at 700 K. What is the maximum allowable conversion in the reactor, if the adiabatic temperature rise across the reactor should not exceed 100 K?

Assume the feed and product streams to be ideal gas mixtures. The heat of reaction at 700 K for the above reaction is calculated to be –94.2 kJ/mol. Mean molar heat capacities (C_p) in the range 700-800 K, are 0.03, 0.0289 and 0.0492 kJ/mol K for nitrogen, hydrogen and ammonia, respectively.

[2002 : 5 M]

4.5 A CSTR is to be designed in which an exothermic liquid phase first order reaction of the type $A \rightarrow R$ is taking place. The reactor is to be provided with a jacket in which coolant is flowing. Following data is given C_{Ao} = 5 kmol/m^3; X_A = 0.5; feed temperature = reactor temperature = 40°C; rate constant at 40°C = 1 min^{-1}; (ΔH) = –40 kJ/mol; ρ = 1000 kg/m^3; C_p = 4 J/gm°C; $q = 10^{-3}$ m^3/min (ρ and C_p are same for the reactant and product streams). The amount of heat to be removed is

(a) 2/3 kW (b) 1 kW

(c) 5/3 kW (d) 4 kW

[2003 : 2 M]

4.6 For an ideal gas mixture undergoing a reversible gaseous phase chemical reaction, the equilibrium constant

(a) is independent of pressure

(b) increases with pressure

(c) decreases with pressure

(d) increases/decreases with pressure depending on the stoichiometric coefficients of the reaction

[2004 : 1 M]

4.7 An endothermic aqueous phase first order irreversible reaction is carried out in an adiabatic plug flow reactor. The rate of reaction

(a) is maximum at the inlet of the reactor

(b) goes through a maximum along the length of the reactor

(c) goes through a minimum along the length of the reactor

(d) is maximum at the exit of the reactor

[2004 : 1 M]

4.8 Consider a reversible exothermic reaction in a plug flow reactor. The maximum and minimum permissible temperatures are T_{max} and T_{min}, respectively. Which of the following temperature (T) profiles will require the shortest residence time to achieve the desired conversion?

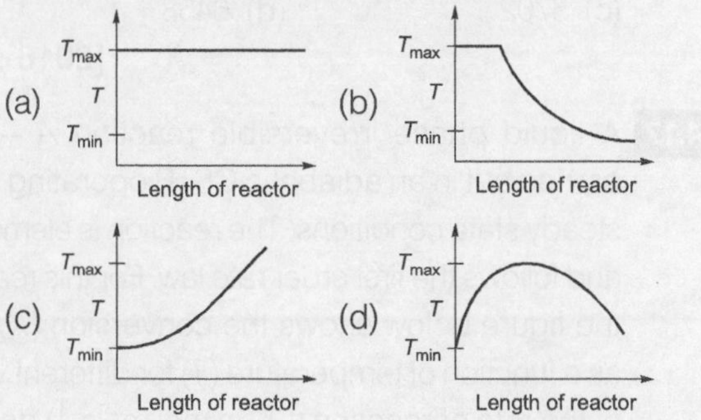

[2004 : 2 M]

4.9 An irreversible aqueous phase reaction $A + B \rightarrow P$ is carried out in an adiabatic mixed flow reactor. A feed containing 4 kmol/m^3 of each A and B enters the reactor at 8 m^3/h. If the temperature of the exit stream is never to exceed 390 K, what is the maximum feed inlet temperature allowed?
Data: Heat of reaction = –50 kJ/mol, density of the reacting mixture = 1000 kg/m^3, specific heat of reacting mixture = 2 kJ/kg K.
The above data can be assumed to be independent of composition and temperature.

(a) 190 (b) 290
(c) 390 (d) 490

[2004 : 2 M]

4.10 Pick the WRONG design guideline for a reactor function which the reactions $A \rightarrow R$ (desired) and $A \rightarrow S$ (undesired) are to take place. The ratio of the reaction rates is $\frac{r_R}{r_S} = \left(\frac{k_1}{k_2}\right) C_A^{a-b}$.

(a) use high pressure and eliminate inerts when $a > b$
(b) avoid recycle when $a > b$
(c) use batch reactor or plug flow reactor when $a > b$
(d) use CSTR with a high conversion when $a > b$

[2004 : 2 M]

4.11 Pure A at 200°C is fed to a steady slate adiabatic continuous reactor at the rate of 100 kg/hr where it undergoes an exothermic reaction to give its isomer B. The product stream is at temperature 500°C. The heat of reaction is 21 kJ/mol of A and the specific heat of the reaction mixture is constant at 35 J/(mol °C). The conversion in the reactor is

(a) 25% (b) 50%
(c) 75% (d) 100%

[2006 : 2 M]

4.12 A liquid phase exothermic first order reaction is being conducted in a batch reactor under isothermal conditions by removing heat generated in the reactor with the help of cooling water. The cooling water flows at a very high rate through a coil immersed in the reactor such that there is negligible rise in its temperature from inlet to outlet of the coil. If the rate constant is given as k, heat of reaction $(-\Delta H)$, volume of the reactor V, initial concentration as C_{A0}, overall heat transfer coefficient U, heat transfer area of the coil is equal to A, the required cooling water inlet temperature, T_{ci} is given by the following equation :

(a) $T_{ci} = T - \frac{(-\Delta H)VkC_{A_0}}{UA}$

(b) $T_{ci} = T - \frac{(-\Delta H)VkC_{A_0}e^{-kt}}{UA}$

(c) $T_{ci} = T - \frac{(-\Delta H)VC_{A_0}e^{-kt}}{UAt}$

(d) $T_{ci} = T - \frac{(-\Delta H)VC_{A_0}}{UAt}$

[2007 : 2 M]

4.13 The homogeneous reaction $A + B \rightarrow C$ is conducted in an adiabatic CSTR at 800 K so as to achieve a 30% conversion of A. The relevant specific heats and enthalpy change of reaction are given by

$$C_{P,A} = 100 \text{ J/(mol K)}$$
$$C_{P,C} = 150 \text{ J/(mol K)}$$
$$C_{P,B} = 50 \text{ J/(mol K)}$$
$$\Delta h^{r\times n} = -100 \text{ kJ/mol}$$

If the feed, a mixture of A and B, is available at 550 K, the mole fraction of A in the feed that is consistent with the above data is

(a) $\frac{5}{7}$ (b) $\frac{1}{4}$
(c) $\frac{1}{2}$ (d) $\frac{2}{7}$

[2008 : 2 M]

4.14 The liquid-phase reaction $A \rightarrow B$ is conducted in an adiabatic plug flow reactor.

Data:

Inlet concentration of $A = 4.0$ kmol/m^3

Density of reaction mixture (independent of temperature) = 1200 kg/m^3

Average heat capacity of feed stream (independent of temperature) = 2000 J/kg.K

Heat of reaction (independent of temperature) = –120 kJ/mol of A reacting

If the maximum allowable temperature in the reactor is 800 K, then the feed temperature (in K) should not exceed

(a) 400 (b) 500

(c) 600 (d) 700

[2009 : 2 M]

4.15 The elementary reversible exothermic gas-phase reaction

$$A + 3B \rightleftharpoons 2C$$

is to be conducted in a non-isothermal, non-adiabatic plug flow reactor. The maximum allowable reactor temperature is T_{max}. To minimize the total reactor volume, the variation of reactor temperature (T) with axial distance from the inlet (z) should be

(a)
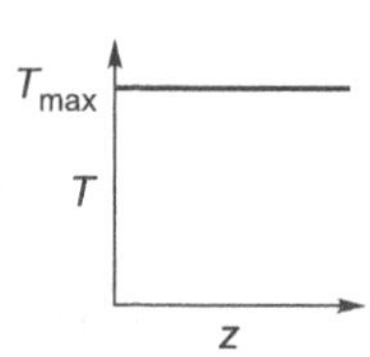

(b)
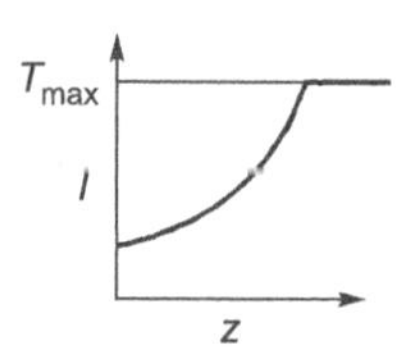

(c)
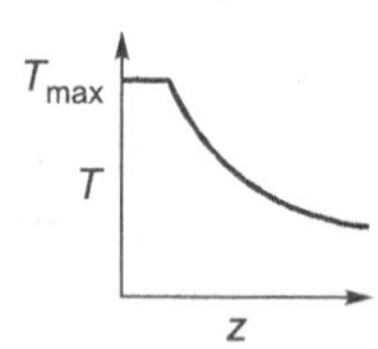

(d)
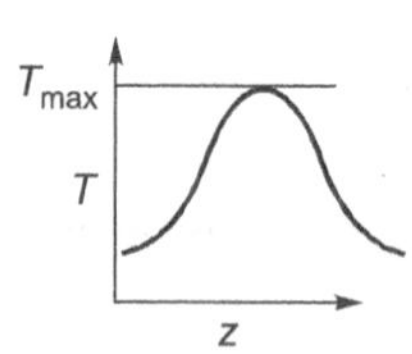

[2012 : 2 M]

4.16 A jacketed stirred tank with a provision for heat removal is used to mix sulphuric acid and water in a steady state flow process. H_2SO_4 (l) enters at a rate of 4 kg/h at 25°C and H_2O (l) enters at a rate of 6 kg/h at 10°C. The following data are available:

Specific heat capacity of water = 4.2 kJ kg^{-1} K^{-1}.

Specific heat capacity of aqueous solution of 40 mass% H_2SO_4 = 2.8 kJ (kg solution)$^{-1}$ K^{-1}.

Assume the specific heat capacities to be independent of temperature. Based on reference states of H_2SO_4 (l) and H_2O (l) at 25°C, the heat of mixing for aqueous solution of 40 mass% H_2SO_4 = –650 kJ (kg H_2SO_4)$^{-1}$.

If the mixed stream leaves at 40°C, what is the rate of heat removal (in kJ/h)?

(a) 1802 (b) 2558

(c) 5702 (d) 6458

[2016 : 2 M]

4.17 A liquid phase irreversible reaction $A \rightarrow B$ is carried out in an adiabatic CSTR operating under steady state conditions. The reaction is elementary and follows the first order rate law. For this reaction, the figure below shows the conversion (X_A) of A as a function of temperature (T) for different values of the rate of reaction ($-r_A$ in mol m^{-3}s^{-1}) denoted by the numbers to the left of each curve. This figure can be used to determine the rate of the reaction at a particular temperature, for a given conversion of A.

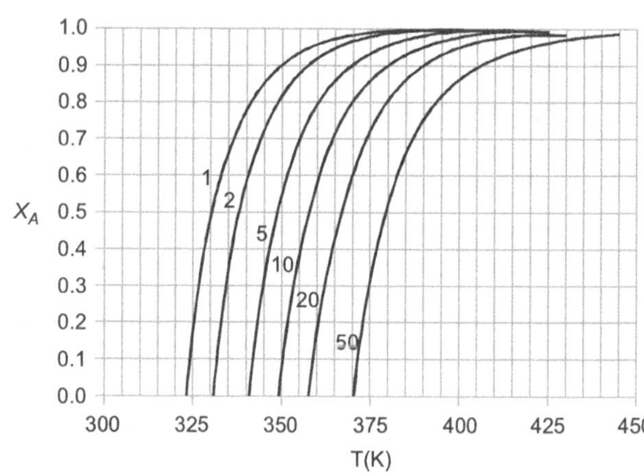

The inlet stream does not contain B and the concentration of A in the inlet stream is 5 mol/m^3. The molar feed rate of A is 100 mol/s. A steady state energy balance for this CSTR results in the following relation: $T = 350 + 25\, X_A$ where T is the temperature (in K) of the exit stream and X_A is the conversion of A in the CSTR. For an exit conversion of 80 % of A, the volume (in m^3, rounded off to the first decimal place) of CSTR required is ________.

[2016 : 2 M]

4.18 Reaction $A \rightarrow B$ is carried out in a reactor operating at steady state and 1 mol/s of pure A at 425°C enters the reactor. The outlet stream leaves the reactor at 325°C. The heat input to the reactor is 17 kW, The heat of reaction at the reference temperature of 25°C is 30 kJ/mol. The specific heat capacities (in kJ/mol.K) of A and B are 0.1 and 0.15, respectively.

The molar flow rate of B leaving the reactor, rounded to 2 decimal places, is _____ mol/s.

[2017 : 2 M]

4.19 The elementary irreversible gas-phase reaction $A \rightarrow B + C$ is carried out adiabatically in an ideal CSTR (Continuous-Stirred Tank Reactor) operating at 10 atm. Pure A enters the CSTR at a flow rate of 10 mol/s and a temperature of 450 K. Assume A, B and C to be ideal gases. The specific heat capacity at constant pressure (C_{Pi}) and heat of formation (H_i^0), of component $i(i = A, B, C)$, are :

$C_{PA} = 30$ J/(mol K); $C_{PB} = 10$ J/(mol K);

$C_{PC} = 20$ J/(mol K)

$H_A^0 = -90$ kJ/mol; $H_B^0 = -54$ kJ/mol;

$H_C^0 = -45$ kJ/mol

The reaction rate constant k (per second) = 0.133 $\exp\left\{\frac{E}{R}\left(\frac{1}{450} - \frac{1}{T}\right)\right\}$, where $E = 31.4$ kJ/mol and universal gas constant $R = 0.082$ L atm/(mol K) = 8.314 J/(mol K). The shaft work may be neglected in the analysis, and specific heat capacities do not vary with temperature. All heats of formation are referenced to 273 K. The reactor volume (in Liters) for 75% conversion is _____ (rounded off to the nearest integer).

[2019 : 2 M]

4.20 An exothermic, aqueous phase, irreversible, first order reaction, $Y \rightarrow Z$ is carried out in an ideal continuous stirred tank reactor (CSTR) operated adiabatically at steady state. Rate of consumption of Y (in mol liter^{-1} minute^{-1}) is given by

$$-r_Y = 10^9 e^{-\frac{6500}{T}} C_Y$$

where C_Y is the concentration of Y (in mol liter^{-1}), and T is the temperature of the reaction mixture (in K). Reactant Y is fed at 50°C. Its inlet concentration is 1.0 mol liter^{-1}, and its volumetric flow rate is 1.0 liter minute^{-1}.

In addition, use the following data and assumptions:

- Heat of the reaction = –42000 J mol^{-1}
- Specific heat capacity of the reaction mixture = 4.2 J g^{-1} K^{-1}
- Density of reaction mixture = 1000 g liter^{-1}
- Heat of the reaction, specific heat capacity and density of the reaction mixture do not vary with temperature.
- Shaft work is negligible.

If the conversion of Y at the exit of the reactor is 90%, the volume of the CSTR (in liter) is _____ (round off to 2 decimal places).

[2020 : 2 M]

■■■■

Answers Design of Non-Isothermal Ideal Reactors

4.1	(c)	**4.2**	(b)	**4.3**	(b)	**4.4**	(12.12)	**4.5**	(c)	**4.6**	(d)	**4.7**	(a)
4.8	(b)	**4.9**	(b)	**4.10**	(d)	**4.11**	(b)	**4.12**	(b)	**4.13**	(a)	**4.14**	(c)
4.15	(c)	**4.16**	(a)	**4.17**	(8.0)	**4.18**	(0.6)	**4.19**	(133.3)	**4.20**	(2.86)		

Explanations Design of Non-Isothermal Ideal Reactors

4.1 (c)

$A \rightarrow B$ is conducted in adiabatic PFR.

$C_{A_0} = 2$ kmol/m^3 $\quad T_F = 500$ K

$v_0 = 0.01$ m^3/s $\quad X_A = 0.2$

From energy balance of adiabatic PFR

$$T - T_F = (-\Delta H_R)F_{A_0}\int_0^{X_A}\frac{dX_A}{\Sigma F_i C_{P_i}}$$

$$\Sigma F_i C_{P_i} = F_A C_{P_A} + F_B C_{P_B}$$

$$= F_{A_0}(1-X_A)C_{P_A} + \left(F_{B_0}^{\nearrow 0} + \frac{b}{a}F_{A_0}X_A\right)C_{P_B}$$

$$F_{A_0}C_{P_A} - F_{A_0}X_A C_{P_A} + F_{A_0}X_A C_{P_B}$$

$$F_{A_0}C_{P_A} - F_{A_0}X_A \times 100 + F_{A_0}X_A \times 100$$

$$= F_{A_0}C_{P_A} \text{ (not function of } X_A)$$

So, $$T - T_F = \frac{-\Delta H_R F_{A_0} X_A}{F_{A_0}C_{P_A}}$$

$$T - 500 = \frac{50000 \times 0.2}{100}$$

$$T = 600 \text{ K}$$

4.2 (b)

For exothermic reaction in adiabatic reaction product temperature is always greater than feed temperature because heat generates in exothermic reaction and total heat is absorbed by product as the reactor is adiabatic.

4.3 (b)

$T_0 = 373$ K

$\Delta H = -1$ kJ/mol at 373 K

$C_{P_A} = C_{P_B} = 50$ J/mol.K

$X_A = 0.5$

$A \rightarrow B$

For adiabatic batch reactor

$$T - T_0 = (-\Delta H_R)N_{A_0}\int_0^{X_A}\frac{dX_A}{\Sigma N_i C_{P_i}}$$

$$\Sigma N_i C_{P_i} = N_A C_{P_A} + N_B C_{P_B}$$

$$N_{A_0}(1-X_A)C_{P_A} + \left(N_{B_0}^{\nearrow 0} + \frac{b}{a}N_{P_0}X_A\right)C_{P_B}$$

$$N_{A_0}C_{P_A} - N_{A_0}C_{P_A}X_A - N_{A_0}C_{P_B}X_A$$

$$\Sigma N_i C_{P_i} = N_{A_0}C_{P_A} \text{ (not function of } X_A)$$

So, $$T - T_0 = \frac{(-\Delta H_R)N_{A_0}\int_0^{X_A} dX_A}{N_{A_0}C_{P_A}}$$

$$\Delta T = \frac{1000}{50.0}\times 0.5 = 10°\text{C}$$

4.5 (c)

Following liquid phase first order reaction $A \rightarrow R$ taking place in CSTR.

$C_{A_0} = 5$ kmol/m^3 $\quad \Delta H_R = -40$ kJ/mol

$X_A = 0.5$ $\quad \rho = 1000$ kg/ m^3

$T_0 = T = 40$°C $\quad C_P = 4$ J/gram mol °C

$k = 1$ min^{-1} $\quad V_0 = 10^{-3}$ m^3/min

From energy balance on CSTR

$$\Sigma F_{i0}(h_{i_0} - h_i) + F_{A_0}X_A(-\Delta H_R) + Q = 0$$

Since, $T_0 - T = 0$

$$\Sigma F_{i0}(h_{i_0} - h_i) = 0$$

So heat removed $= F_{A_0}X_A(\Delta H_R) = Q$

$$= \frac{5\times 10^{-3}}{60}\times 0.5 \times 40 \times 1000$$

$$= \frac{5}{3}\text{ kW}$$

4.6 (d)

According to thee Le Chatelier principle equilibrium constant increases with increase in pressure when number of moles are decreasing and decreases with increase in pressure when number of moles are increasing in reaction.

4.7 (a)

First order exothermic irreversible reaction.

$$-r_A = k\,C_A$$
$$-r_A = k_0\,e^{-E/RT}C_A$$

Reaction rate will be higher at higher C_A and higher temperature.

Since reaction is endothermic and adiabatic PFR is used so both C_A and T will be higher at inlet so rate of reaction is maximum at inlet of the reactor.

4.8 (b)

For reversible exothermic reaction, reaction start at maximum allowable temperature and then temperature decreases along the path at which rate is maximum for a given conversion.

4.9 (b)

Irreversible reaction $A + B \rightarrow P$ carried out in adiabatic MFR.

$$C_{A_0} = 4\ \text{kmol/m}^3 \qquad (-\Delta H_R) = 50\ \text{kJ/mol}$$
$$V_0 = 8\ \text{m}^3/\text{hr} \qquad \rho = 1000\ \text{kg/m}^3$$
$$T_{max} = 390\ \text{K} \qquad C_{P_{mix}} = 2\ \text{kJ/kg K}$$
$$T_F = ?$$
$$T - T_F = \frac{(-\Delta H_R)F_{A_0}X_A}{\rho_{mix}V_{mix}C_{P_{mix}}}$$

For $T_{max} \Rightarrow X_A = 1$

So, $$390 - T_F = \frac{50\times 4\times V_0\times 1000}{1000\times V_0\times 2}$$
$$T_F = 290\ \text{K}$$

4.10 (d)

$A \rightarrow R$ desired

$A \rightarrow S$ undesired

Ratio of reaction rate

$$\frac{r_R}{r_S} = \left(\frac{k_1}{k_2}\right)C_A^{a-b}$$

If $a > b$ means high concentration

high pressure

low inert

use PFR (avoid recycle)

So option (d) is wrong.

4.12 (b)

Liquid phase exothermic first order reaction is being conducted in a batch reactor under isothermal condition.

Energy balance equation for batch reactor is as following:

Isothermal condition

$$Q + (-r_A)(-\Delta H)V = \frac{d[N_iC_{P_i}T]}{dt}$$

Here, $$Q = v_0A_0(T_c - T)$$
$$U_0A_0(T_{ci} - T) = -(-\Delta H)Vk\,C_{A_0}e^{-kt}$$
$$T_{ci} = T - \frac{(-\Delta H)Vk\,C_{A_0}e^{-kt}}{U_0A_0}$$

4.13 (a)

Following reaction is conducted in adiabatic CSTR

$A + B \rightarrow C$

$$X_A = 0.3, \quad F_{C_0} = 0$$

Energy balance equation for adiabatic CSTR is as following:

$$T - T_F = \frac{(-\Delta H_R)F_{A_0}X_A}{\Sigma F_{i_0}C_{P_i}}$$
$$800 - 550 = \frac{(100\times 10^3)F_{A_0}\times 0.3}{F_{A_0}\times 100 + F_{B_0}\times 50}$$
$$10F_{A_0} + 5F_{B_0} = 40F_{A_0}\times 0.3$$
$$10F_{A_0} + 5F_{B_0} = 12F_{A_0}$$
$$5F_{B_0} = 2F_{A_0} \qquad \ldots(i)$$

Mole fraction of A in feed is

$$\frac{F_{A_0}}{F_{A_0}+F_{B_0}} = \frac{F_{A_0}}{F_{A_0}+\frac{2}{5}F_{A_0}} = \frac{5}{7}$$

4.14 (c)

For adiabatic plug flow reactor energy balance results in following equation:

$$T - T_0 = \frac{(-\Delta H_R)F_{A_0}X_A}{\Sigma F_i\,C_{P_i}} = \frac{(-\Delta H_R)F_{A_0}X_A}{\rho_{mix}\,C_{P_{mix}}\,v_{mix}}$$

For maximum allowable temperature $X_A = 1$

So, putting values we get,

$$800 - T_0 = \frac{120\times 10^3\times 4\times v\times 1000}{1200\times 2000\times v}$$
$$T_0 = 600\ \text{K}$$

4.15 (c)

For exothermic reversible reaction to minimize reactor volume reaction should start at highest possible temperature, then temperature should decrease along the path of maximum reaction rate.

4.16 (a)

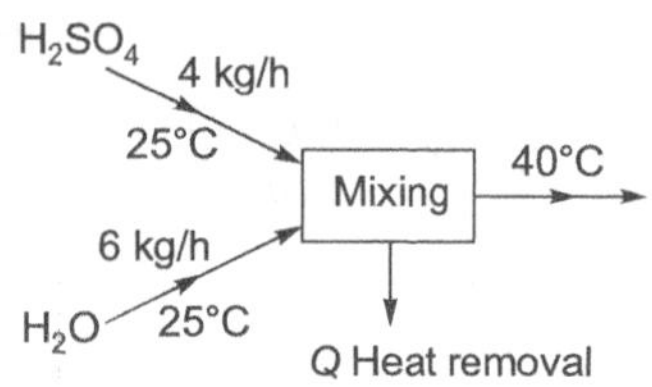

Product stream leaves at 40°C.

On applying energy balance,

Heat removal = Heat of mixing – Heat required by feed stream to reach at product stream temperature

$\therefore\ Q = 650 \times 4 - 10 \times 2.8(40-25) - 6 \times 4.2(25-10)$

$= 1802$ kJ/h

4.17 (8.0)

Liquid phase irreversible reaction is carried out in an adiabatic CSTR

$$A \to B$$

$$C_{A_0} = 5 \text{ mol/m}^3$$

$$F_{A_0} = 100 \text{ mol/sec}$$

Steady state energy balance results in the following relation

$$T = 350 + 25X_A$$

$$X_A = 0.8 \text{ (Given)}$$

So, $\quad T = 350 + 25 \times 0.8 = 370$ K

From graph at $T = 370$ K and $X_A = 0.8$ rate of reaction $(-r_A) = 10$ mol/m^3.s

From design equation of CSTR

$$\frac{\tau}{C_{A_0}} = \frac{V}{F_{A_0}} = \frac{X_A}{-r_A}$$

$$\frac{V}{100} = \frac{0.8}{10}$$

$$V = 8.0 \text{ m}^3$$

4.18 (0.6)

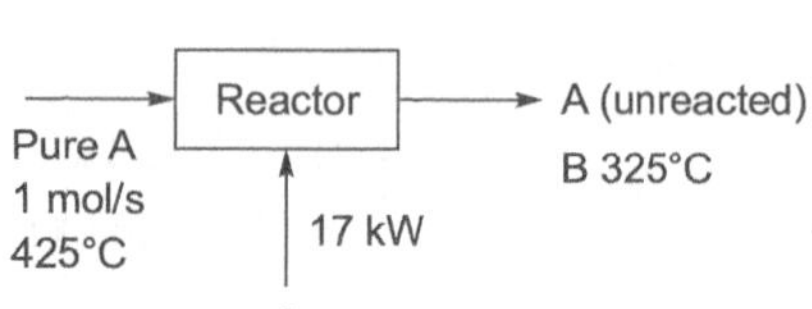

$$\Delta H_R|_{298\text{ K}} = 30 \text{ kJ/mol}$$

$$C_{P_A} = 0.1 \text{ kJ/mol K},\ C_{P_B} = 0.15 \text{ kJ/mol K}$$

Assuming conversion of A is X_A

So moles of B formed = X_A

Moles of A out = $1 - X_A$

From energy balance over reactor, we get

$$Q = \Sigma(n_i h_i)_{out} - \Sigma(n_i h_i)_{in} + (\Delta H_R)F_{A_0}X_A$$

$$17 = [(1-X_A)\times 0.1 + X_A \times 0.15](325-25) - 1 \times 0.1)(425-25) + [1 \times X_A \times 30]$$

$$17 = [0.1 - 0.1X_A + 0.15X_A](300) - 0.1 \times 400 + 30X_A$$

$$X_A = 0.6$$

Molar flow rate of B leaving = 0.6 mol/s

4.19 (133.3)

$A \to B + C$

$$-r_A = kC_A \text{ (elementary reaction)}$$

$$F_{A_0} = 10 \text{ mol/s}$$

Heat of reaction at 273 K

$$= H_B^0 + H_C^0 - H_A^0$$

$$= -45 + (-54) - (-90) = -9 \text{ kJ/mol}$$

$$\Delta C_P = C_{P_C} + C_{P_B} - C_{P_A}$$

$$= 20 + 10 - 30 = 0 \text{ J/mol K}$$

So heat of reaction is not changing with temperature for adiabatic reaction

$$T - T_F = \frac{-\Delta H_R F_{A_0} X_A}{\Sigma F_{i_o} C_{p_i}}$$

$$T - 450 = \frac{9000 \times 10 \times 0.75}{10 \times 30}$$

$$T = 675 \text{ K}$$

Now rate constant at 675 K

$$k_2 = 0.133 \exp\left[\frac{31400}{8.314}\left(\frac{1}{450} - \frac{1}{675}\right)\right]$$

$$k_2 = 2.182 \text{ sec}^{-1}$$

Now for CSTR

$$\frac{V}{F_{A_0}} = \frac{X_A}{-r_A}$$

$$\frac{V}{10} = \frac{0.75}{\dfrac{kC_{A_0}(1-X_A)}{(1+\epsilon_A X_A)}} \quad \text{...(i)}$$

$$\epsilon_A = \frac{2-1}{1} = 1$$

$$C_{A_0} = \frac{P_{A_0}}{RT} = \frac{10\times 101.325}{8.314\times 675}$$

$= 0.1805\ \text{kmol/m}^3$

$= 0.1805\ \text{mol/liter}$

Now from equation (i)

$$\frac{V}{10} = \frac{0.75}{\dfrac{2.182\times 0.1805(1-0.75)}{(1+0.75\times 1)}}$$

$V = 133.3$ liter

4.20 (2.86)

First order reaction $Y \to Z$ is given

$$-r_y = 10^9 e^{-\frac{6500}{T}} C_Y$$

From the equation of energy balance on adiabatic CSTR

$$T - T_o = \frac{(-\Delta H_R)F_{A_0}X_A}{\Sigma F_i C_{P_i}} = \frac{(-\Delta H_R)F_{A_0}X_A}{(\rho V C_p)_{mix}}$$

$$T - 50 = \frac{42000\times 1.0\times 1.0\times 0.9}{1000\times 1\times 4.2}$$

$T = 50 + 9 = 59°\text{C} = 332\ \text{K}$

Rate constant at 332 K is

$$k = 10^9 e^{-\frac{6500}{332}} = 3.142$$

From design equation of CSTR

$$\frac{V}{F_{Y_0}} = \frac{X_Y}{-r_Y}$$

Putting values, we get

$$V = \frac{0.9}{3.142\times 1(1-0.9)} = 2.86\ \text{liter}$$

■■■■

5 Design of Non-Ideal Reactors

5.1 The E-curve for a non-ideal reactor defines the fraction of fluid having age between t and $t + dt$

(a) at the inlet
(b) at the outlet
(c) in the reactor
(d) averaged over the inlet and outlet

[2001 : 1 M]

5.2 The mean conversion in the exit stream, for a second-order, liquid phase reaction in a non-ideal flow reactor is given by

(a) $\int_0^\infty \frac{k_2 C_{A_0} t}{1 + k_2 C_{A_0} t} E(t)\, dt$

(b) $\int_0^\infty \frac{1}{1 + k_2 C_{A_0} t} E(t)\, dt$

(c) $\int_0^\infty \frac{1}{1 + k_2 C_{A_0} t} (1 - E(t)\pi)\, dt$

(d) $\int_0^\infty \frac{\exp\left(-k_2 C_{A_0} t\right)}{1 + k_2 C_{A_0} t} E(t)\, dt$

[2001 : 2 M]

5.3 A pulse tracer is introduced in an ideal CSTR (with a mean residence time τ) at time = 0. The time taken for the exit concentration of the tracer to reach half of its initial value will be

(a) 2τ (b) 0.5τ
(c) $\frac{\tau}{0.693}$ (d) 0.693τ

[2002 : 2 M]

5.4 A step input tracer test is used to explore the flow pattern of fluid through a vessel of total volume equal to 1 m^3 having a feed rate of 1 m^3/min.

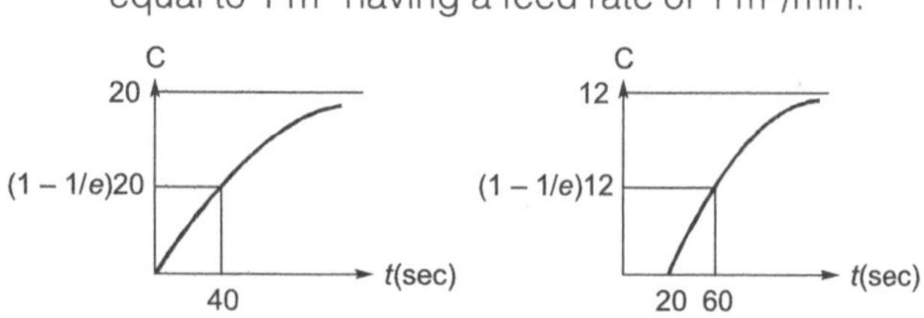

Curve 1 Curve 2

Identify for each curve in **Group-1** a suitable flow model from the list given under **Group-2**:

Group-1
A. Curve 1
B. Curve 2

Group-2
1. PFR and CSTR in series
2. CSTR with dead space
3. PFR in series with a CSTR and dead space
4. CSTR

(a) A-4, B-3 (b) A-4, B-1
(c) A-2, B-3 (d) A-2, B-1

[2003 : 2 M]

5.5 For a packed bed reactor, the presence of a long tail in the residence time distribution curve is an indication of

(a) ideal plug flow (b) bypass
(c) dead zone (d) channeling

[2004 : 1 M]

5.6 The exit age distribution in a stirred reactor is given by

$$E(t) = \frac{1}{\tau} e^{-\frac{1t}{\tau}}$$

Fluid elements e_1 and e_2 enter the reactor at times $t = 0$ and $t = \theta > 0$, respectively. The probability that e_2 exists the reactor before e_1 is

(a) $\frac{1}{2}$ (b) $\frac{1}{2} e^{-\theta/\tau}$
(c) $e^{-\theta/\tau}$ (d) zero

[2006 : 2 M]

5.7 A pulse of concentrated KCl solution is introduced as tracer into the fluid entering a reaction vessel having volume equal to 1 m^3 and flow rate equal to 1 m^3/min. The concentration of tracer measured in the fluid leaving the vessel is showing in the figure given below. The flow model parameters that fit the measured RTD in terms of one or all of the following mixing elements, namely, volume of plug flow reactor V_p, mixed flow volume V_m, and dead space V_d, are

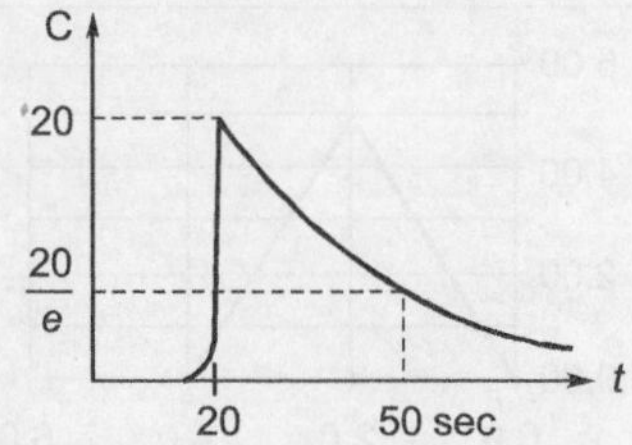

(a) $V_p = \frac{1}{6}$ m^3, $V_m = \frac{1}{2}$ m^3, $V_d = \frac{1}{3}$ m^3

(b) $V_p = V_m = V_d = \frac{1}{3}$ m^3

(c) $V_p = \frac{1}{3}$ m^3, $V_m = \frac{1}{2}$ m^3, $V_d = \frac{1}{6}$ m^3

(d) $V_m = \frac{5}{6}$ m^3, $V_d = \frac{1}{6}$ m^3 [2007 : 2 M]

Common Data for Questions (5.8 & 5.9) :

A liquid is flowing through a reactor at a constant flow rate. A step input of tracer at a molar flow rate of 1 mol/min is given to the reactor at time $t = 0$. The time variation of the concentration (C) of the tracer at the exit of the reactor is as shown in the figure:

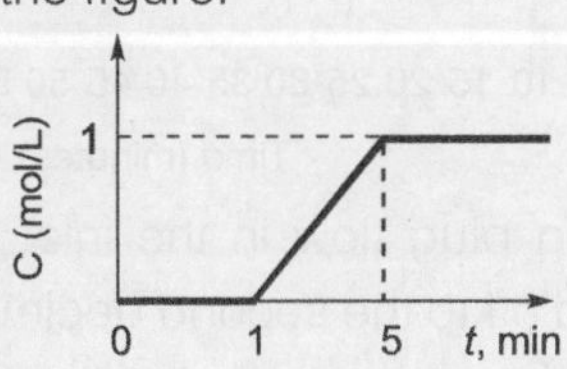

5.8 The volumetric flow rate of the liquid through the reactor (in L/min) is

(a) 1 (b) 2
(c) 1.5 (d) 4 [2008 : 2 M]

5.9 The mean residence time of the fluid in the reactor (in minutes) is

(a) 1 (b) 2
(c) 3 (d) 4 [2008 : 2 M]

5.10 An isothermal pulse test is conducted on a reactor and the variation of the outlet tracer concentration with time is shown below:

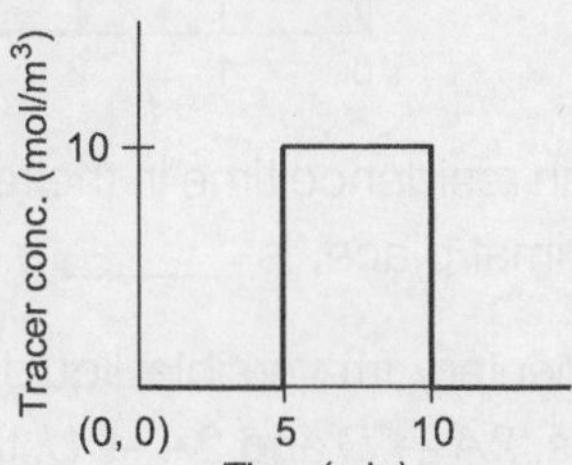

The mean residence time of the fluid in the reactor (in minutes) is

(a) 5.0 (b) 7.5
(c) 10.0 (d) 15.0 [2009 : 2 M]

5.11 Two reactors (reactor 1 and reactor 2) with average residence times, τ_1 and τ_2, respectively, are placed in series. Reactor 1 has zero dispersion and reactor 2 has infinite dispersion. The residence time distribution $E(t)$ of this system is given by

(a) $\begin{cases} 0 & t \le \tau_1 \\ \frac{1}{\tau_2}\exp\left(-\frac{t-\tau_1}{\tau_2}\right) & t > \tau_1 \end{cases}$

(b) $\begin{cases} 0 & t \le \tau_2 \\ \frac{1}{\tau_1}\exp\left(-\frac{t-\tau_2}{\tau_1}\right) & t > \tau_2 \end{cases}$

(c) $\begin{cases} 0 & t \le \tau_1 \\ \frac{1}{\tau_1}\exp\left(-\frac{t-\tau_1}{\tau_2}\right) & t > \tau_1 \end{cases}$

(d) $\begin{cases} 0 & t \le \tau_2 \\ \frac{1}{\tau_2}\exp\left(-\frac{t}{\tau_1}\right) & t > \tau_2 \end{cases}$ [2010 : 2 M]

5.12 The following figures show the outlet tracer concentration profiles (c vs. t) for a pulse input.

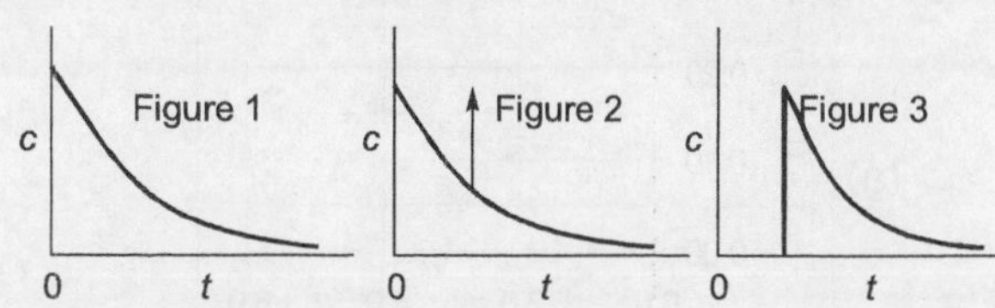

Match the figures in **Group-I** with the reactor configurations in **Group-II**:

Group-I	Group-II
P. Figure 1	1. PFR
Q. Figure 2	2. CSTR
R. Figure 3	3. PFR and CSTR in series
	4. PFR and CSTR in parallel

(a) P-2, Q-4, R-3 (b) P-4, Q-3, R-1
(c) P-3, Q-4, R-2 (d) P-1, Q-3, R-2

[2011 : 2 M]

5.13 The exit age distribution for a reactor is given by $E(t) = \delta(t-4)$, where t is in seconds. A first order liquid phase reaction ($k = 0.25$ s^{-1}) is carried out in this reactor under steady state and isothermal conditions. The mean conversion of the reactant at the exit of the reactor, up to 2 digits after the decimal point, is ______.

[2013 : 1 M]

5.14 The vessel dispersion number for an ideal CSTR is

(a) –1 (b) 0

(c) 1 (d) ∞

[2014 : 1 M]

5.15 The impulse response to a tracer pulse experiment for a flow reactor is given below:

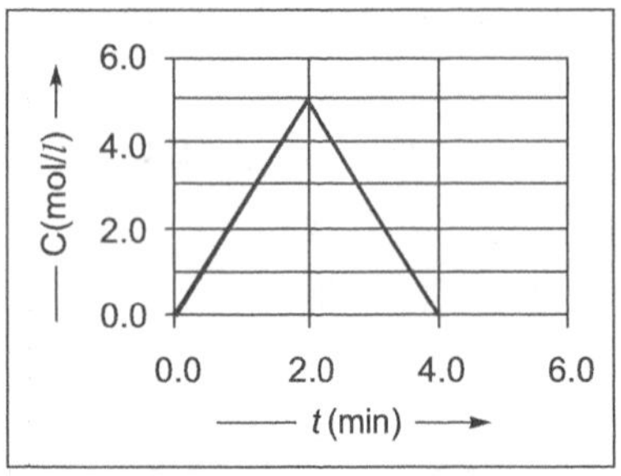

In the above figure, C is the exit tracer concentration. The corresponding E or E_θ (normalized E) curve is correctly represented by which of the following choices? Here, θ is dimensionless time.

(a)

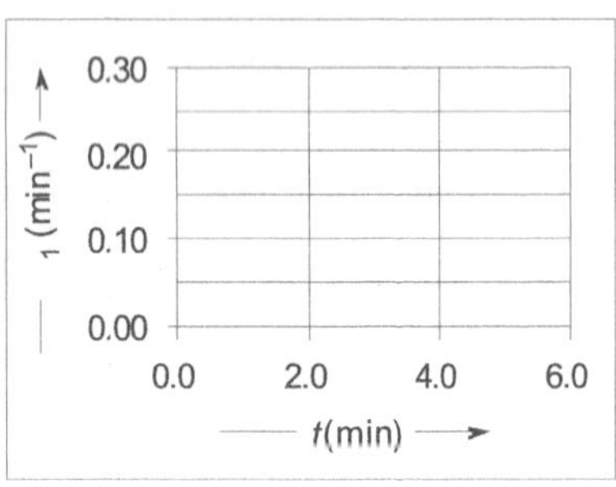

(b)

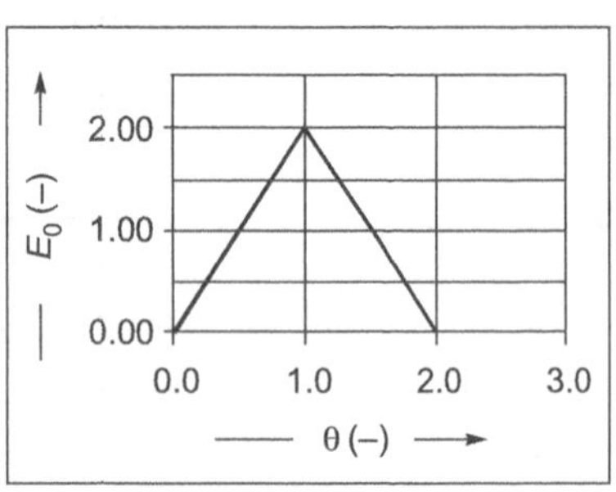

(c)

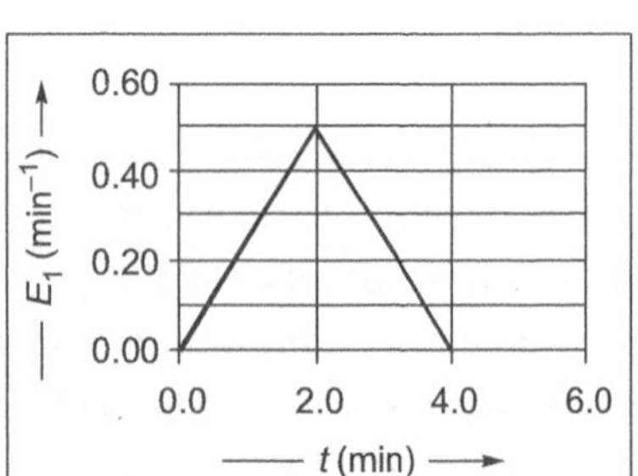

(d)

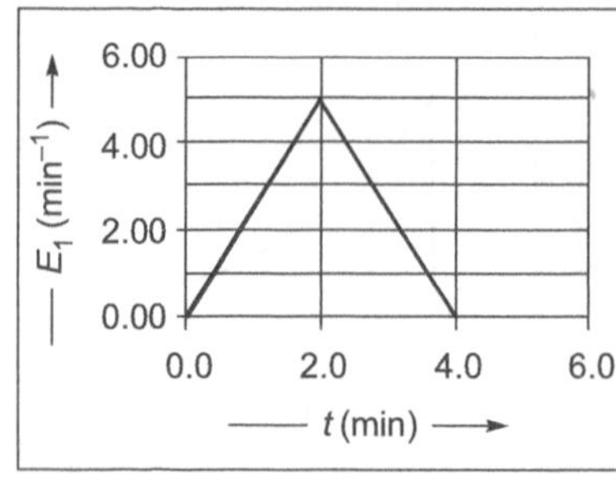

[2015 : 2 M]

5.16 A CSTR has a long inlet pipe. A tracer is injected at the entrance of the pipe. The E-curve obtained at the exit of the CSTR is shown in the figure below.

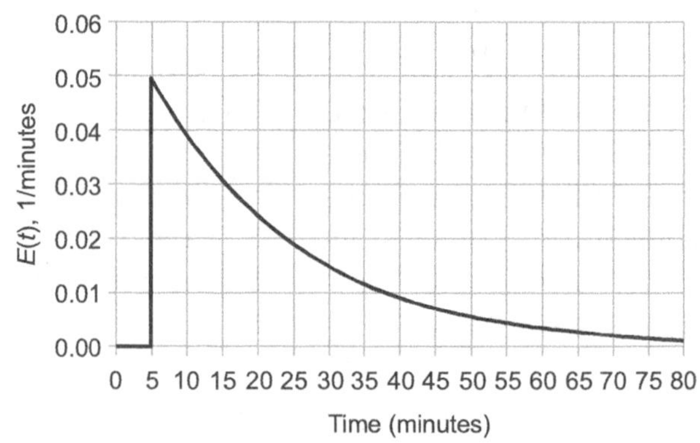

Assuming plug flow in the inlet pipe, the ratio (rounded off to the second decimal place) of the volume of the pipe to that of the CSTR is ______

[2016 : 2 M]

5.17 The C-curve measured during a pulse tracer experiment is shown below, in the figure, $C(t)$ is the concentration of the tracer measured at the reactor exit in mol/liter at time t seconds.

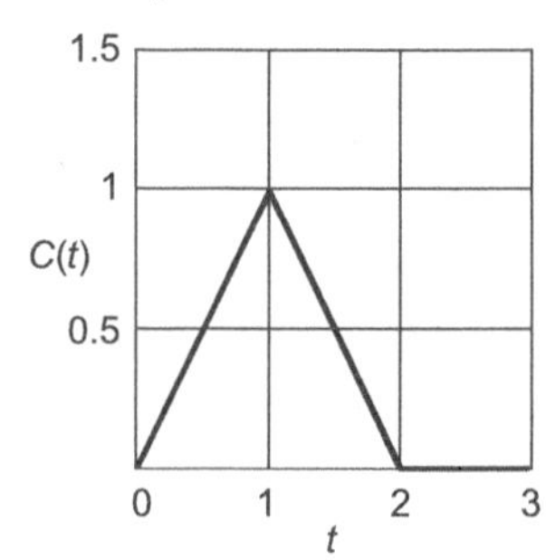

The mean residence time in the reactor, rounded to 1 decimal place, is _______ s. [2017 : 2 M]

5.18 The elementary, irreversible, liquid-phase, parallel reactions, $2A \rightarrow D$ and $2A \rightarrow U$, take place in an isothermal non-ideal reactor. The C-curve measured in a tracer experiment is shown in the figure, where $C(t)$ is the concentration of the tracer in g/m^3 at the reactor exit at time t (in min.)

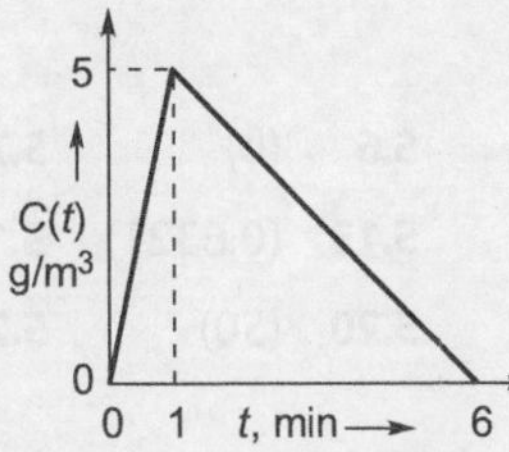

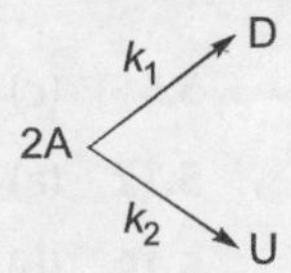

The rate constants are k_1 = 0.2 Liter/(mol min) and k_2 = 0.3 Liter/(mol min). Pure A is fed to the reactor at a concentration of 2 mol/Liter. Using the segregated model, the percentage conversion in the reactor is _____ (rounded off to the nearest integer). [2019 : 2 M]

5.19 In a laboratory experiment, a unit pulse input of tracer is given to an ideal plug flow reactor operating at steady state with a recycle ratio, $R = 1$. The exit age distribution, $E(t)$, of the tracer at the outlet of the reactor is measured. The first four pulses observed at t_1, t_2, t_3 and t_4 are shown below:

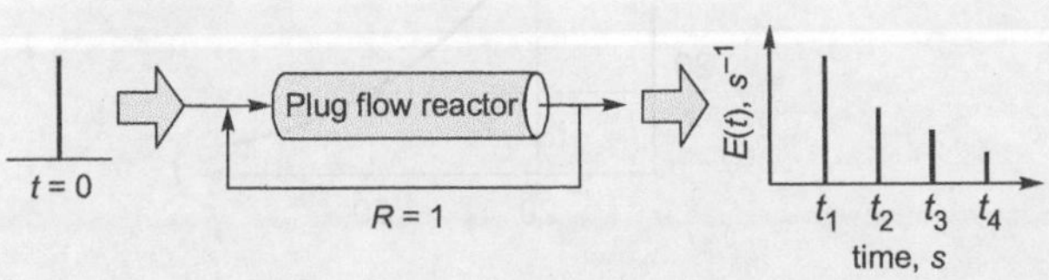

In addition, use the following data and assumptions:

- R is defined as ratio of the volume of fluid returned to the entrance of the reactor to the volume leaving the system.
- No reaction occurs in the reactor.
- Ignore any dead volume in the recycled loop.

If the space time of the plug flow reactor is τ seconds, which one of the following is correct?

(a) $t_1 = \tau$, $t_2 = 2\tau$, $t_3 = 3\tau$, $t_4 = 4\tau$

(b) $t_1 = \frac{\tau}{2}$, $t_2 = \tau$, $t_3 = \frac{3\tau}{2}$, $t_4 = 2\tau$

(c) $t_1 = \frac{\tau}{2}$, $t_2 = \frac{\tau}{4}$, $t_3 = \frac{\tau}{8}$, $t_4 = \frac{\tau}{16}$

(d) $t_1 = \frac{\tau}{3}$, $t_2 = \frac{2\tau}{3}$, $t_3 = \tau$, $t_4 = \frac{4\tau}{3}$

[2020 : 1 M]

5.20 An elementary irreversible liquid-phase reaction, $2P \xrightarrow{k} Q$, where the rate constant $k = 2$ L mol^{-1} min^{-1}, takes place in an isothermal non-ideal reactor. The E-curve in a tracer experiment is shown in the figure. Pure P(2 mol L^{-1}) is fed to the reactor. Using the segregated model, the percentage conversion of P at the exit of the reactor is _________% (rounded off to the nearest integer).

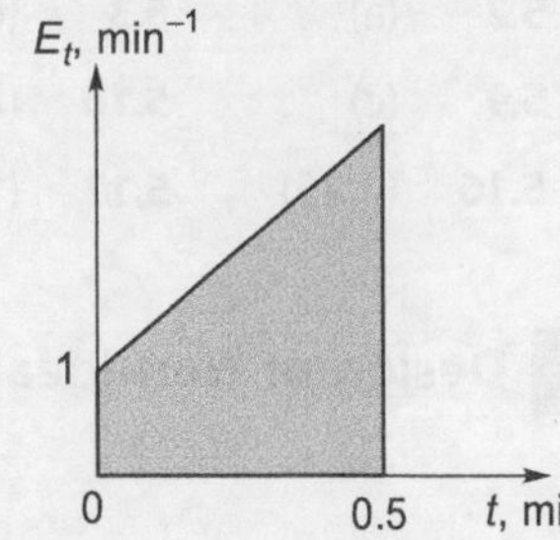

[2022 : 2 M]

5.21 An irreversible liquid-phase second-order reaction

$$A \xrightarrow{k} B$$

with rate constant k = 0.2 liter.mol^{-1}min^{-1}, is carried out in an isothermal non-ideal reactor. A tracer experiment conducted on this reactor resulted in a residence time distribution (E-curve) as shown in the figure below. The areas of the rectangles (i), (ii), and (iii) are equal. Pure A at a concentration of 1.5 mol.liter^{-1} is fed to the reactor. The segregated model mimics the nonideality of this reactor. The percentage conversion of A at the exit of the reactor is _____ (rounded off to the nearest integer).

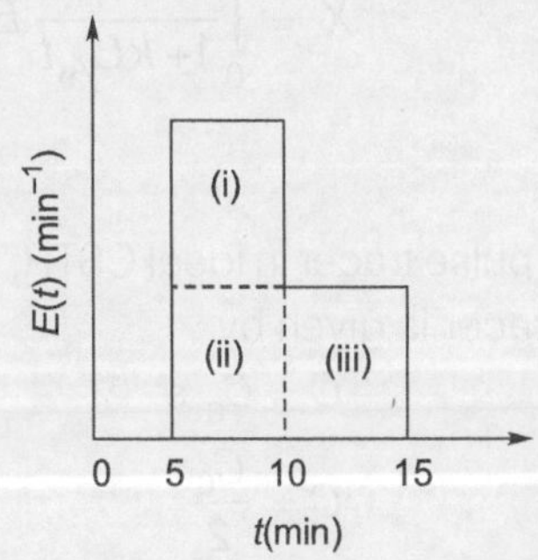

[2023 : 2 M]

5.22 The residence time distribution, E, for a non-ideal flow reactor is given in the figure. A first-order liquid phase reaction with a rate constant 0.2 min^{-1} is carried out in the reactor. For an inlet reactant concentration of 2 mol L^{-1}, the reactant concentration (in mol L^{-1}) in the exit stream is

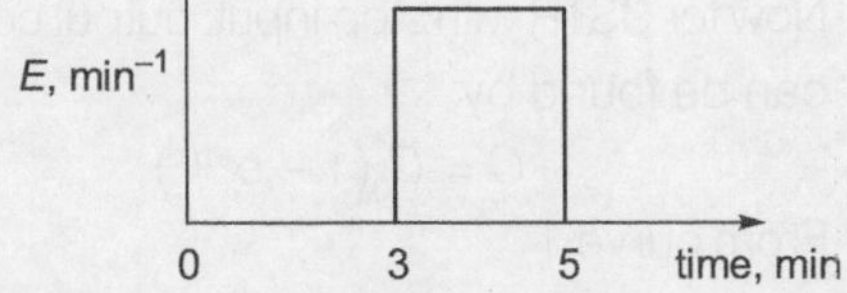

(a) 0.905 (b) 0.452

(c) 1.902 (d) 0.502 [2024 : 2 M]

Answers Design of Non-Ideal Reactors

5.1	(b)	**5.2**	(a)	**5.3**	(d)	**5.4**	(d)	**5.5**	(c)	**5.6**	(b)	**5.7**	(c)
5.8	(a)	**5.9**	(c)	**5.10**	(b)	**5.11**	(a)	**5.12**	(a)	**5.13**	(0.6321)	**5.14**	(d)
5.15	(c)	**5.16**	(0.25)	**5.17**	(1)	**5.18**	(64.64)	**5.19**	(b)	**5.20**	(50)	**5.21**	(72)
5.22	(a)												

Explanations Design of Non-Ideal Reactors

5.1 (b)

All residence time distribution is done at outlet of reactor.

5.2 (a)

Mean conversion in any non-ideal reactor is given by

$$\bar{X} = \int_0^\infty X(t)E(t)dt$$

where $X(t)$ = Conversion in batch reactor

For 2nd order reaction,

$$X(t) = \frac{kC_{A_0}t}{1+kC_{A_0}t}$$

So, $$\bar{X} = \int_0^\infty \frac{kC_{A_0}t}{1+kC_{A_0}t}E(t)dt$$

5.3 (d)

For pulse tracer in ideal CSTR, the concentration of tracer is given by

$$C = C_0\, e^{-t/\tau}$$

$$C = \frac{C_0}{2}$$

$$\frac{1}{2} = e^{-t/\tau}$$

$$t = 0.693\,\tau$$

5.4 (d)

$$\text{Space time of CSTR} = \frac{1\,m^3}{1\,m^3/\min} = 1\,\min$$

$$\tau_m = 60\text{ s}$$

Now for CSTR with step input, output concentration can be found by

$$C = C_0(1 - e^{-t/\tau})$$

From curve 1

$$t_{obs} = 40\text{ s}$$

$t_{obs} < \tau_m$ so CSTR with dead zone. Similarly for curve two.

$$\tau_p = 20\text{ s}$$

$$t_{obs} = 60$$

$$\because \quad t_{obs} = \tau_m$$

So PFR in series with CSTR.

5.7 (c)

Volume of vessel = 1 m^3

Flow rate = 1 m^3/min

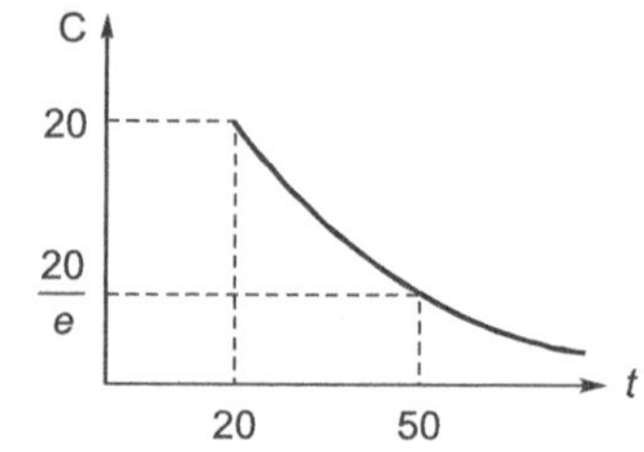

$$\tau_{total} = \frac{V}{V_0} = \frac{1\,m^3}{\frac{1\,m^3}{\min}} = 1\,\min = 60\text{ sec}$$

$$\tau_p = 20\text{ sec}$$

$$\tau_p = \frac{V_p}{V_0},\; V_p = \frac{20}{60} = \frac{1}{3}\,m^3$$

Similarly for mixed flow

$$C = C_0 e^{-t/\tau}$$

From figure, $\tau_m = 50 - 20 = 30$ sec

$$V_m = \frac{30}{60} = \frac{1}{2}\,m^3$$

$$\text{Dead space volume} = 1 - \left(\frac{1}{2}+\frac{1}{3}\right) = \frac{1}{6}\,m^3$$

5.8 (a)

Step tracer input given to reactor

$$F_{A_0} = 1\text{ mol/min}$$

$$C_{A_0} = 1\text{ mol/litre}$$

$$\text{Volumetric flow rate } v_0 = \frac{F_{A_0}}{C_{A_0}} = 1\text{ L/min}$$

5.9 (c)

$$\bar{t} \times C_0 = \text{Area on Y axis}$$

$$\bar{t} \times 1 = \frac{1}{2}(1+5) \times 1 = 3 \text{ minute}$$

5.10 (b)

$$\text{Mean residence time } \bar{t} = \frac{\int_0^\infty t\,C(t)\,dt}{\int_0^\infty C(t)\,dt}$$

$$\bar{t} = \frac{\int_0^\infty t \cdot C(t)\,dt}{\text{Area under C-curve}}$$

$$= \frac{\int_5^{10} t \cdot 10\,dt}{10 \times 5} = \frac{10\left[\frac{t^2}{2}\right]_5^{10}}{10 \times 5} = 7.5$$

5.11 (a)

Zero dispersion means ideal PFR and infinite dispersion stands for ideal CSTR, for CSTR and PFR in series $E(t)$ value is define as

$$E(t) = \frac{1}{\tau_m} \exp\left(-\frac{t-\tau_p}{\tau_m}\right)$$

Here τ_p is average residence time of PFR which corresponds to τ_1 and τ_m is average residence time of CSTR which corresponds to τ_2.

5.13 (0.6321)

Exit age distribution is given by

$$E(t) = \delta(t-4)$$

First order liquid phase reaction with $k = 0.25\ \text{sec}^{-1}$

Mean conversion is given by following formula:

$$\bar{X} = \int_0^\infty X(t)E(t)\,dt$$

Here, $X(t)$ = Conversion in batch reactor

Here, $X(t) = 1 - e^{-kt}$

$$\bar{X} = \int_0^\infty (1-e^{-kt})\delta(t-4)\,dt$$

$$\bar{X} = 1 - e^{-0.25 \times 4} = 0.6321$$

5.14 (d)

Dispersion number is define as the ratio of rate of transport due to dispersion to rate of transport due to convection or bulk.

$$\text{Dispersion number} = \frac{D}{uL}$$

For MFR dispersion is infinite.

It is reciprocal of Peclet numbers.

5.15 (c)

Relation between C-curve and E-curve is given by following relation:

$$E(t) = \frac{C(t)}{\int_0^\infty C(t)\,dt} = \frac{C(t)}{\text{Area under C-curve}}$$

$$\text{Area under C-curve} = \frac{1}{2} \times 4 \times 5 = 10$$

So, shape of E-curve and C-curve will be same only value of Y-axis on C-curve will be divided by 10 to get value of Y-axis on E-curve.

Alternatively, area under the E-curve is one, thus only option (c) fulfills this criteria.

5.16 (0.25)

Plug flow in the inlet pipe is given from graph $\tau_p = 5$ minute.

$$\frac{1}{\tau_m} = 0.05 \quad \text{\{from graph\}}$$

$$\tau_m = 20 \text{ minute}$$

$$\frac{V_p}{V_m} = \frac{\tau_p}{\tau_m} = \frac{5}{20} = 0.25$$

5.17 (1)

$$\text{Area under the curve} = \int_0^\infty C(t)\,dt = \frac{1}{2} \times 2 \times 1 = 1$$

$$E(t) = \frac{C(t)}{\text{Area}}$$

$$\text{Mean residence time} = \int_0^\infty t\,E(t)\,dt$$

$$\Rightarrow \quad \bar{t} = \int_0^1 t \cdot t\,dt + \int_1^2 t(-t+2)\,dt$$

$$\bar{t} = \left[\frac{t^3}{3}\right]_0^1 - \left[\frac{t^3}{3}\right]_1^2 + 2\left[\frac{t^2}{2}\right]_1^2$$

$$\bar{t} = \frac{1}{3} - \frac{8}{3} + \frac{1}{3} + 4 - 1$$

$$\bar{t} = 4 - 3 = 1$$

5.18 (64.64)

$$k_1 = 0.2\frac{L}{\text{mol.min}} \text{ (2}^{\text{nd}}\text{ order)}$$

$$k_2 = 0.3\frac{L}{\text{mol.min}} \text{ (2}^{\text{nd}}\text{ order)}$$

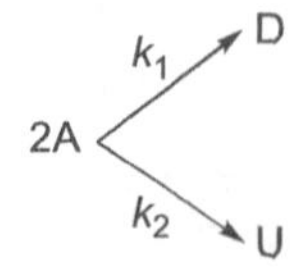

$$-r_A = k_1C_A^2 + k_2C_A^2$$
$$= (0.2 + 0.3)C_A^2 = 0.5C_A^2$$

From segregation model

$$\bar{X} = \int_0^\infty X(t)\cdot E(t)dt$$

$$E(t) = \frac{C(t)}{\int_0^\infty C(t)dt}$$

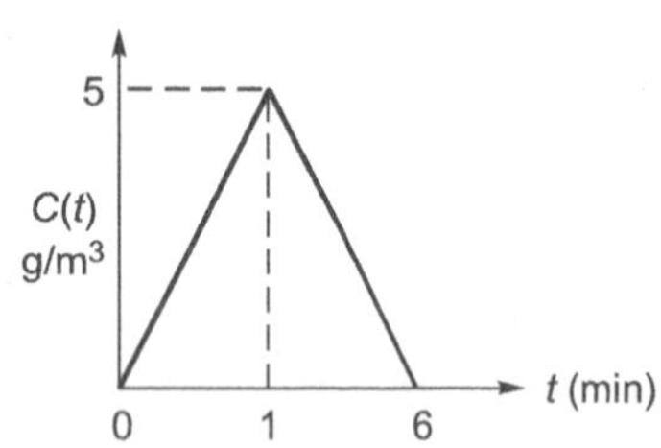

$$\int_0^\infty C(t)dt = \text{Area under the curve}$$

$$= \frac{1}{2}\times 5\times 6 = 15$$

From $0 < t < 1$, $C(t) = 5t$

From $1 < t < 6$, $C(t) = -t + 6$

For 2nd order reaction

$$X(t) = \frac{kC_{A_0}t}{1+kC_{A_0}t}$$

$$\bar{X} = \int_0^\infty X(t)\cdot E(t)dt$$

$$= \int_0^1 \frac{t}{1+t}\frac{5t}{15}dt + \int_1^6 \frac{t}{1+t}\left(\frac{6-t}{15}\right)dt$$

On integrating and applying limits, we get

$$\bar{X} = 0.6464$$

$$\%\text{ conversion} = 64.64\%$$

5.19 (b)

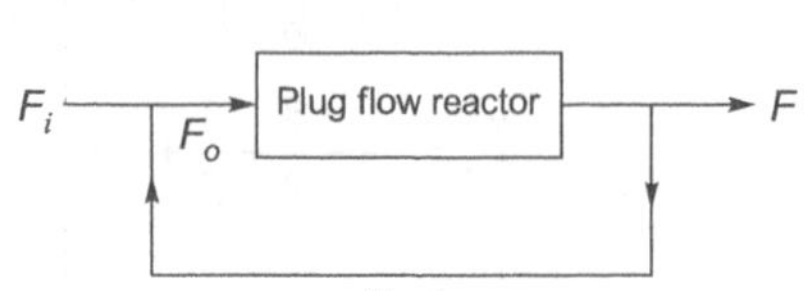

At steady state $F_i = F$

$$F_o = F_i + RF = F_i + RF_i = F_i(R + 1)$$

F_o = Total volumetric flow rate of PFR

$$\text{Space time of PFR} = \frac{\text{Volume of PFR}}{\text{Volumetric flow rate}}$$

$$\tau' = \frac{V_{PFR}}{F_i(R+1)} \text{ (For PFR with recycle)}$$

$$\tau' = \frac{V_{PFR}}{2F_i} \qquad (\because R = 1)$$

$$\tau' = \frac{\tau}{2}$$

where, $\tau = \frac{V_{PFR}}{F_i}$ without recycle.

For an ideal PFR first pulse tracer shows up at exit of PFR at time τ but in this case of recycle PFR it is known at time $\frac{\tau}{2}$.

So, $t_1 = \frac{\tau}{2}$, $t_2 = \tau$

$t_3 = \frac{3\tau}{2}$, $t_4 = 2\tau$

5.20 (50)

Using segregation model, the mean conversion is given by

$$\bar{x} = \int_o^\infty x(t)E(t)dt$$

Now for II order, we know

$$\frac{X}{1-X} = kC_{AO}t$$

$$\Rightarrow \qquad X = \frac{kC_{AO}t}{1+kC_{AO}t} = \frac{4t}{1+4t}$$

From E-curve,

$$E = \int_o^\infty E(t)dt = 1 \quad \text{[Area will be one]}$$

$$1 = \frac{1}{2}[1+x]\times 0.5 \Rightarrow x = 3$$

$$E(t) - 1 = \frac{3-1}{0.5}(t-0)$$

$$\Rightarrow \qquad E(t) = 4t + 1$$

Now for, $$(\bar{x}) = \int_o^{0.5}\frac{4t}{1+4t}\times(1+4t)dt = \left[2t^2\right]_0^{0.5}$$

$$\bar{x} = 0.5 = 50\%$$

5.21 (72)

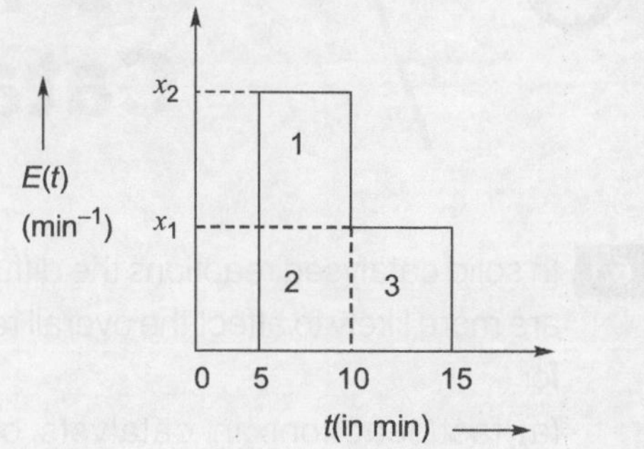

Given: Area under E-curve = 1

$$3\times(10-5)\times x_1 = 1$$

$$3\times5\times x_1 = 1 \text{ or } x_1 = \frac{1}{15}$$

Similarly, $$x_2 = \frac{2}{15}$$

For second-order liquid reaction, we have

$$\therefore \quad \frac{X_A}{1-X_A} = kC_{A_0}t$$

or $$X_A = (1-X_A)kC_{A_0}t$$

$$X_A = kC_{A_0}t - X_A kC_{A_0}t$$

$$X_A + X_A kC_{A_0}t = kC_{A_0}t$$

$$X_A(1+kC_{A_0}t) = kC_{A_0}t$$

$$X_A = \frac{0.3t}{1+0.3t} = \frac{kC_{A_0}t}{1+kC_{A_0}t} = \frac{0.2\times1.5t}{1+0.2\times1.5\times t}$$

$$\therefore \text{ Mean conversion } = \bar{X} = \int_0^{\infty} X(t).E(t)\,dt$$

$$\bar{X} = \int_5^{10}\frac{2}{15}\times\frac{0.3t}{1+0.3t}dt + \int_5^{10}\frac{1}{15}\times\frac{0.3t}{1+0.3t}dt = \int_5^{10}2\times0.02\frac{t}{1+0.3t}dt + \int_5^{10}\frac{1}{15}\times\frac{0.3t}{1+0.3t}dt$$

$$\bar{X} = \frac{2}{15}\int_5^{10}\frac{0.3t}{1-0.3t}dt + \frac{1}{15}\int_{10}^{15}\frac{0.3t}{1-0.3t}dt = \frac{2}{15}\int_5^{10}\frac{0.3t+1-41}{1-0.3t}dt + \frac{1}{15}\int_{10}^{15}\frac{0.3t+1-1}{1+0.3t}dt$$

$$\bar{X} = \frac{2}{15}\int_5^{10}\left(1-\frac{1}{1+0.3t}\right)dt + \frac{1}{15}\int_{10}^{15}\left(1-\frac{1}{1+0.3t}\right)dt$$

$$\bar{X} = \frac{2}{15}\left[t-\frac{1}{0.3}\ln|1+0.3t|\right]_5^{10} + \frac{1}{15}\left[t-\frac{1}{0.3}\ln[1+0.3t]\right]_{10}^{15}$$

$$\bar{X} = \frac{2}{15}[5.38-1.945] + \frac{1}{15}[9.32-5.38] = 0.7206 \cong 72\%$$

5.22 (a)

The residence time distribution for a non-ideal flow reactor is given below:

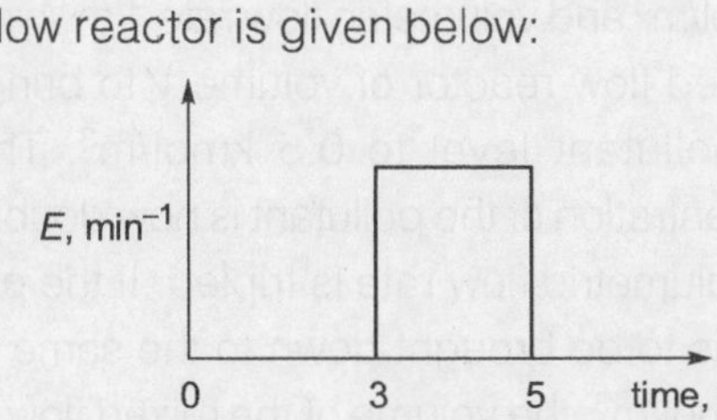

$$C_{A_0} = 2 \text{ mol}/l$$

$$k = 0.2 \text{ min}^{-1}$$

$$\frac{\overline{C_A}}{C_{A_0}} = \int_0^{\infty}\left(\frac{C_A}{C_{A_0}}\right)E(t)\,dt \qquad \text{...(i)}$$

Area under (E) curve = 1

$$(5-3)\times E = 1$$

$$E = 0.5$$

For first order reaction and batch reactor

$$C_A = C_{A_0}e^{-kt}$$

$$\frac{\overline{C_A}}{C_{A_0}} = \int_3^5 0.5e^{-kt}\,dt = 0.5\int_3^5 e^{-0.2t}\,dt$$

$$= -\frac{0.5}{0.2}\left[e^{-1}-e^{-0.6}\right]$$

$$\frac{\overline{C_A}}{C_{A_0}} = 0.452$$

$$\overline{C_A} = 0.905$$

■■■■

6 Heterogeneous Reactions, Rate & Performance Equation for Catalyst Deactivation Boiling

6.1 In solid catalysed reactions the diffusional effects are more likely to affect the overall rate of reaction for

(a) fast reactions in catalysts of small pore diameter.

(b) fast reactions in catalysts of large pore diameter.

(c) slow reactions in catalysts of small pore diameter.

(d) slow reactions in catalysts of large pore diameter.

[2000 : 1 M]

6.2 The rate controlling step for the heterogeneous irreversible catalytic reaction

$$A\,(g) + B\,(g) \rightarrow C\,(g)$$

is the surface reaction of adsorbed A with adsorbed B to give adsorbed C. The rate expression for this reaction can then be written as

[2000 : 2 M]

6.3 For a vapor phase catalytic reaction ($A + B \rightarrow F$) which follows the Rideal mechanism and the reaction step is rate controlling, the rate of reaction is given by (reaction step is irreversible, product also adsorbs)

(a) $-r_A = \dfrac{kp_Ap_B}{1+K_Ap_A+K_pp_p}$

(b) $-r_A = \dfrac{kp_A^2 - p_p}{1+K_Ap_A+K_pp_p}$

(c) $-r_A = \dfrac{kp_Ap_B}{1+K_Ap_A+K_Bp_B+K_pp_p}$

(d) $-r_A = \dfrac{kp_Ap_B}{1+K_Ap_A}$

[2001 : 2 M]

6.4 Following isothermal kinetic data are obtained in a basket type of mixed flow reactor for a porous catalyst. Determine the role of pore diffusion and external mass transfer processes

Pellet diameter	Leaving concentration of the reactant	Spinning rate of basket	$(-r_A^1)$
1	1	High	2
2	1	Low	1
3	1	High	1

(a) Strong pore diffusion control and mass transfer not controlling

(b) Both pore diffusion and mass transfer not controlling

(c) Both pore diffusion and mass transfer controlling

(d) Mass transfer controlling

[2003 : 2 M]

6.5 A first order gaseous phase reaction is catalyzed by a non-porous solid. The kinetic rate constant and the external mass transfer coefficient are k and k_g, respectively. The effective rate constant (k_{eff}) is given by

(a) $k_{eff} = k + k_g$

(b) $k_{eff} = \dfrac{(k+k_g)}{2}$

(c) $k_{eff} = (k\,k_g)^{1/2}$

(d) $\dfrac{1}{k_{eff}} = \dfrac{1}{k} + \dfrac{1}{k_g}$

[2004 : 1 M]

6.6 A pollutant P degrades according to first order kinetics. An aqueous stream containing P at 2 kmol/m^3 and volumetric flow rate 1 m^3/h requires a mixed flow reactor of volume V to bring down the pollutant level to 0.5 kmol/m^3. The inlet concentration of the pollutant is now doubled and the volumetric flow rate is tripled. If the pollutant level is to be brought down to the same level of 0.5 kmol/m^3, the volume of the mixed flow reactor should be increased by a factor of

(a) 7 (b) 6

(c) 3 (d) $\dfrac{7}{3}$

[2004 : 2 M]

6.7 Match the items in **Group-I** with those in **Group-II**:

Group-I
A. Porous catalyst
B. Parallel reactions
C. Non-ideal tubular reactor
D. Gas-solid non-catalytic reaction

Group-II
1. Selectivity
2. Shrinking core model
3. Thiele modulus
4. Dispersion number

Codes:

	A	B	C	D
(a)	3	1	4	2
(b)	1	3	2	4
(c)	1	4	2	3
(d)	3	4	1	2

[2005 : 2 M]

6.8 The rate at which an antiviral drug acts increases with its concentration in the blood, C, according to the equation

$$r = \frac{kC}{C_{50} + C}$$

where C_{50} is the concentration at which the rate is 50% of the maximum rate k. Often, the concentration C_{90}, when the rate is 90% of the maximum, is measured instead of C_{50}. The rate equation then becomes

(a) $r = \dfrac{1.8kC}{(C_{90} + C)}$ (b) $r = \dfrac{kC}{\left(\dfrac{C_{90}}{9} + C\right)}$

(c) $r = \dfrac{kC}{C_{90}}$ (d) $r = \dfrac{0.9kC}{C_{90}}$

[2006 : 2 M]

6.9 Consider the following reactions between gas A and two solid spherical particles, B and C of the same size

$$A + B \xrightarrow{k_1} \text{gaseous product}$$

$$A + C \xrightarrow{k_2} \text{ash}$$

The ash does not leave the particle C. Let t_1 and t_2 be the times required for A to completely consume particles B and C, respectively. If k_1 and k_2 are equal at all temperatures and the gas phase mass transfer resistance is negligible, then

(a) $t_1 = t_2$ at all temperatures
(b) $t_1 = t_2$ at high temperatures
(c) $t_1 > t_2$ at high temperatures
(d) $t_1 < t_2$ at high temperatures

[2006 : 2 M]

Statement for Linked Answer Question (6.10 & 6.11):
Consider the diffusion of a reactant A through a cylindrical catalyst pore of radius R and length $L >> R$. Reactant A undergoes a zeroth order reaction on the cylindrical surface of the pore. The following equation describes changes in the concentration of A within the pore due to the axial diffusion of A and the disappearance of A due to reaction $\dfrac{d^2c_A}{dx^2} = K$ where c_A is the concentration of A at a distance x from the pore entrance, and K is a constant.

6.10 If the concentration of A at the pore entrance ($x = 0$) is C_{AO}, and $x = L$ is a dead end where no reaction occurs, the concentration profile of A in the pore is given by

(a) $c_A(x) = \dfrac{Kx^2}{2} - KLx + c_{A0}$

(b) $c_A(x) = \dfrac{Kx(x-L)}{2} - c_{A0}\dfrac{x}{L} + c_{A0}$

(c) $c_A(x) = c_{A0}$

(d) $c_A(x) = c_{A0}\left(\dfrac{L-x}{L}\right)$

[2006 : 2 M]

6.11 The minimum pore length for A to be completely converted within the pore is

(a) $\left(\dfrac{c_{A0}}{K}\right)^{1/2}$ (b) $\left(\dfrac{c_{A0}}{2K}\right)^{1/2}$

(c) $\left(\dfrac{2c_{A0}}{K}\right)$ (d) $\left(\dfrac{2c_{A0}}{K}\right)^{1/2}$

[2006 : 2 M]

6.12 The first order reaction of A to R is run in an experimental mixed flow reactor. Find the role played by pore diffusion in the run given below. C_{A0} is 100 and W is fixed. Agitation rate was found to have no effect on conversion.

d_p	F_{A0}	X_A
4	2	0.8
6	4	0.4

(a) strong pore diffusion control
(b) diffusion free
(c) intermediate role by pore diffusion
(d) external mass transfer

[2007 : 2 M]

6.13 A packed bed reactor converts A to R by first order reaction with 9 mm pellets in strong pore diffusion regime to 63.2% level. If 18 mm pellets are used what is the conversion?

(a) 0.39 (b) 0.61
(c) 0.632 (d) 0.865

[2007 : 2 M]

6.14 The following rate-concentration data are calculated from experiment. Find the activation energy temperature (E/R) of the first order reaction.

D_p	C_A	$-r_A$	T
1	20	1	480
2	40	2	480
2	40	3	500

(a) 2432.8 (b) 4865.6
(c) 9731.2 (d) 13183.3

[2007 : 2 M]

6.15 A species (A) reacts on a solid catalyst to produce R and S as follows :

1. $A \rightarrow R$ $\quad r_R = k_1 C_A^2$
2. $A \rightarrow S$ $\quad r_S = k_2 C_A^2$

Assume film resistance to mass transfer is negligible. The ratio of instantaneous fractional yield of R in the presence of pore diffusion to that in the absence of pore diffusion is

(a) 1 (b) > 1
(c) < 1 (d) Zero

[2008 : 1 M]

6.16 The irreversible zero order reaction $A \rightarrow B$ takes place in a porous cylindrical catalyst that is sealed at both ends as shown in the figure. Assume dilute concentrations and neglect any variations in the axial direction.

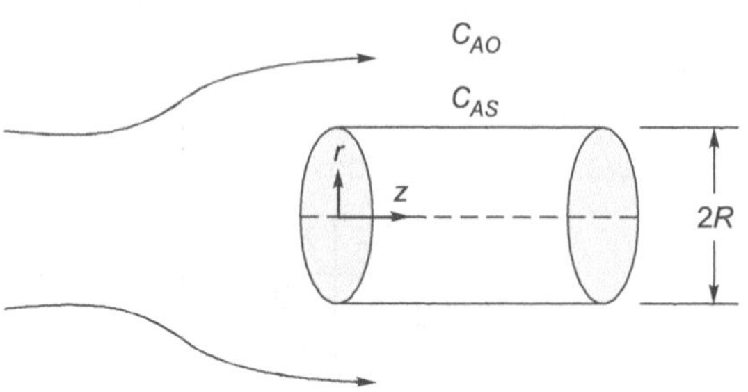

The steady state concentration profile is

$$\frac{C_A}{C_{AS}} = 1 + \frac{\phi_0^2}{4}\left[\left(\frac{r}{R}\right)^2 - 1\right]$$

where ϕ_0 is the Thiele modulus. For $\phi_0 = 4$, the range of r where $C_A = 0$ is

(a) $0 < r < \frac{R}{4}$ (b) $0 < r < \frac{R}{2}$
(c) $0 \le r \le \sqrt{\frac{3}{4}}R$ (d) $0 \le r \le R$

[2008 : 2 M]

6.17 For a solid-catalyzed reaction, the Thiele modulus is proportional to

(a) $\sqrt{\frac{\text{intrinsic reaction rate}}{\text{diffusion rate}}}$

(b) $\sqrt{\frac{\text{diffusion rate}}{\text{intrinsic reaction rate}}}$

(c) $\frac{\text{intrinsic reaction rate}}{\text{diffusion rate}}$

(d) $\frac{\text{diffusion rate}}{\text{intrinsic reaction rate}}$

[2009 : 1 M]

6.18 For a first order isothermal catalytic reaction, $A \rightarrow P$, occurring in an infinitely long cylindrical pore, the relationship between effectiveness factor, ε, and Thiele modulus, ϕ, is

(a) $\varepsilon = \frac{1}{\phi^2}$ (b) $\varepsilon = \phi$
(c) $\varepsilon = 1$ (d) $\varepsilon = \frac{1}{\phi}$

[2010 : 1 M]

6.19 Consider an irreversible, solid catalysed, liquid phase first order reaction. The diffusion and the reaction resistances are comparable. The overall rate constant (k_o) is related to the overall mass transfer coefficient (k_m) and the reaction rate constant (k) as

(a) $k_o = \frac{kk_m}{k + k_m}$ (b) $k_o = \frac{k + k_m}{kk_m}$
(c) $k_o = \frac{k + k_m}{2}$ (d) $k_o = k + k_m$

[2011 : 2 M]

6.20 For a first order catalytic reaction the Thiele modulus (ϕ) of a spherical pellet is defined as

$$\phi = \frac{R_s}{3}\sqrt{\frac{k\rho_p}{D_e}}$$

where, ρ_p = Pellet density
R_s = Pellet radius
D_e = Effective diffusivity
k = First order reaction rate constant

If $\phi > 5$, then the apparent activation energy (E_a) is related to the intrinsic (or true) activation energy (E) as

(a) $E_a = E^{0.5}$ (b) $E_a = 0.5E$
(c) $E_a = 2E$ (d) $E_a = E^2$

[2011 : 2 M]

6.21 The rate-controlling step for the solid-catalyzed irreversible reaction

$$A + B \rightarrow C$$

is known to be the reaction of adsorbed A with adsorbed B to give adsorbed C. If P_i is the partial pressure of component i and K_i is the adsorption equilibrium constant of component i, then the form of the Langmuir-Hinshelwood rate expression will be

(a) rate $\propto \dfrac{P_A P_B}{1 + K_A P_A + K_B P_B + K_C P_C}$

(b) rate $\propto \dfrac{P_A P_B}{(1 + K_A P_A + K_B P_B + K_C P_C)^2}$

(c) rate $\propto \dfrac{P_A P_B}{(1 + K_A P_A + K_B P_B + K_C P_C)^{0.5}}$

(d) rate $\propto \dfrac{P_A P_B}{P_C}$

[2012 : 2 M]

6.22 The overall rates of an isothermal catalytic reaction using spherical catalyst particles of diameters 1 mm and 2 mm are r_{A1} and r_{A2} (in mol (kg-catalyst)$^{-1}$ h^{-1}), respectively. The other physical properties of the catalyst particles are identical. If pore diffusion resistance is very high, the ratio

$\dfrac{r_{A2}}{r_{A1}}$ is ______. [2013 : 1 M]

6.23 The gas phase decomposition of azomethane to give ethane and nitrogen takes place according to the following sequence of elementary reactions.

$$(CH_3)_2N_2 + (CH_3)N_2 \xrightarrow{k_1} (CH_3)_2N_2 + [(CH_3)_2N_2]^*$$

$$[(CH_3)_2N_2]^* + (CH_3)N_2 \xrightarrow{k_2} (CH_3)_2N_2 + [(CH_3)_2N_2]$$

$$[(CH_3)_2N_2]^* \xrightarrow{k_3} C_2H_6 + N_2$$

Using the pseudo-steady-state-approximation for $[(CH_3)_2N_2]^*$, the order with respect to azomethane in the rate expression for the formation of ethane, in the limit of high concentrations of azomethane, is ______.

[2013 : 2 M]

6.24 A vapour phase catalytic reaction on ($Q + R \rightarrow S$) follows Rideal mechanism (R and S are not adsorbed). Initially, the mixture contains only the reactants in equimolar ratio. The surface reaction step is rate controlling. With constants a and b, the initial rate of reaction ($-r_0$) in terms of total pressure (P_T) is given by

(a) $-r_0 = \dfrac{aP_T}{1 + bP_T}$ (b) $-r_0 = \dfrac{aP_T}{1 + bP_T^2}$

(c) $-r_0 = \dfrac{aP_T^2}{1 + bP_T}$ (d) $-r_0 = \dfrac{aP_T^2}{(1 + bP_T)^2}$

[2014 : 2 M]

6.25 Match the following:

Group-1
P. Tank in series model
Q. Liquid-liquid extraction
R. Optimum temperature progression
S. Thiele modulus

Group-2
I. Non-isothermal reaction
II. Mixer-settler
III. PFR with axial mixing
IV. Solid catalyzed reaction

(a) P-II, Q-IV, R-I, S-III (b) P-I, Q-II, R-III, S-IV
(c) P-III, Q-I, R-II, S-IV (d) P-III, Q-II, R-I, S-IV

[2014 : 2 M]

6.26 A catalyst slab of half-thickness L (the width and length of the slab >> L) is used to conduct the first order reaction $A \rightarrow B$. At 450 K, the Thiele modulus for this system is 0.5. The activation energy for the first order rate constant is 100 kJ/mol. The effective diffusivity of the reactant in the slab can be assumed to be independent of temperature, and external mass transfer resistance can be neglected. If the temperature of the reaction is increased to 470 K, then the effectiveness factor at 470 K (up to two decimal places) will be ________.

Value of universal gas constant = 8.314 J/mol. K.

[2015 : 2 M]

6.27 A porous pellet with Pt dispersed in it is used to carry out a catalytic reaction. Following two scenarios are possible.

Scenario 1: Pt present throughout the pores of the pellet is used for catalyzing the reaction.

Scenario 2: Pt present only in the immediate vicinity of the external surface of the pellet is used for catalyzing the reaction.

At a large value of Thiele modulus, which one of the following statements is TRUE?

(a) Since the reaction rate is much greater than the diffusion rate, Scenario 1 occurs
(b) Since the reaction rate is much greater than the diffusion rate , Scenario 2 occurs
(c) Since the reaction rate is much lower than the diffusion rate, Scenario 1 occurs
(d) Since the reaction rate is much lower than the diffusion rate, Scenario 2 occurs

[2016 : 2 M]

6.28 Consider a first order catalytic reaction in a porous catalyst pellet.

Given R - characteristic length of the pellet: D_e - effective diffusivity; k_c - mass transfer coefficient: k_1 - rate constant based on volume ot the catalyst pellet: C_s - concentration of reactant on the pellet surface.

The expression for Thiele modulus is

(a) $\frac{k_C R}{D_e}$
(b) $R\sqrt{\frac{k_1}{D_e}}$
(c) $R\sqrt{\frac{k_1 C_s}{D_e}}$
(d) $R\sqrt{\frac{D_e}{k_1}}$

[2017 : 1 M]

6.29 For a solid-catalyzed gas phase reversible reaction, which of the following statements is ALWAYS TRUE?

(a) Adsorption is rate-limiting
(b) Desorption is rate-limiting
(c) Solid catalyst does not affect equilibrium conversion
(d) Temperature does not affect equilibrium conversion

[2017 : 1 M]

6.30 Liquid phase isomerization of *o*-xylene to *p*-xylene using a zeolite catalyst was carried out in a CSTR. Three sets of kinetic data at different temperatures and stirring speeds were obtained as shown below.

	Set A			Set B			Set C		
Temperature (K)	500	500	500	600	600	600	700	700	700
Stirring speed (rpm)	1000	2000	3000	1000	2000	3000	1000	2000	3000
Reaction rate (mol L^{-1} s^{-1})	0.020	0.025	0.025	0.037	0.047	0.047	0.069	0.078	0.086

The operating condition at which the reaction rate is not controlled by external mass transfer resistance is

(a) T = 500 K; rpm = 3000
(b) T = 600 K; rpm = 1000
(c) T = 700 K; rpm = 1000
(d) T = 700 K; rpm = 2000

[2018 : 1 M]

6.31 In a laboratory batch setup, reaction of P over a catalyst was studied at various temperatures. The reactions occurring are

$$P \rightarrow 2Q;\ P \rightarrow R$$

At the end of one hour of operation, the batch contains x_P, x_Q and x_R mole fractions of P, Q and R components, respectively. The mole fractions of product components (x_Q and x_R) were found to vary linearly with temperature as given in the figure.

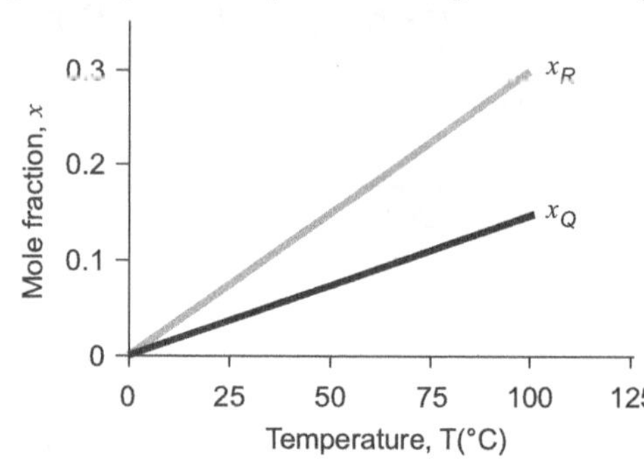

If the yield of Q based on reactant P consumed (Y_Q) at 25°C was found to be 0.40, then the value of Y_Q at 60°C is ________ (rounded off to second decimal place).

[2018 : 2 M]

6.32 Hydrogenation of benzene is to be carried out using Ni (density = 8910 kg/m^3) as catalyst, cast in the form of non-porous hollow cylinders, as shown below. The reaction occurs on all the surfaces of the hollow cylinder. During an experiment, one such cylinder is suspended in the reactant stream. If the observed rate of reaction is 0.39 mol (m^2 of catalyst surface)$^{-1}$ min^{-1}, then the rate of reaction in mol (kg of catalyst)$^{-1}$ min^{-1} is ______ (rounded off to three decimal places).

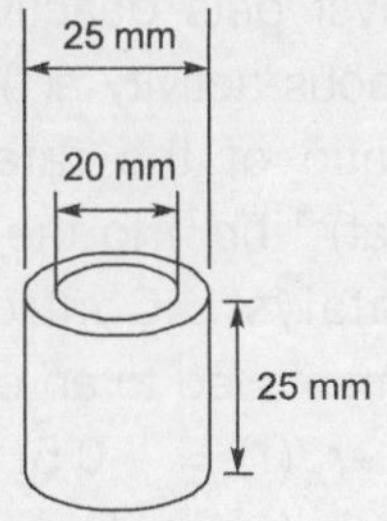

[2018 : 2 M]

6.33 For a first order reaction in a porous spherical catalyst pellet, diffusional effects are most likely to lower the observed rate of reaction for
(a) slow reaction in a pellet of small diameter
(b) slow reaction in a pellet of large diameter
(c) fast reaction in a pellet of small diameter
(d) fast reaction in a pellet of large diameter

[2019 : 1 M]

6.34 The square of Thiele modulus, M_T, is given by $M_T^2 = \frac{L^2 k}{D_{eff}}$, where L is the characteristic length of the catalyst pellet, k is the rate constant of a first order reaction, and D_{eff} is the effective diffusivity of the species in the pores. M_T^2 is a measure of

(a) $\frac{\text{Time scale of pore diffusion}}{\text{Time scale of reaction}}$

(b) $\frac{\text{Rate of pore diffusion}}{\text{Rate of reaction}}$

(c) $\frac{\text{Time scale of reaction}}{\text{Time scale of pore diffusion}}$

(d) $\frac{\text{Rate of reaction}}{\text{Rate of external mass transfer}}$

[2020 : 1 M]

6.35 The liquid phase irreversible reactions, $P \xrightarrow{k_1} Q$ and $P \xrightarrow{k_2} R$ are carried out in an ideal continuous stirred tank reactor (CSTR) operating isothermally at steady state. The space time of the CSTR is 1 minute. Both the reactions are first order with respect to the reactant P, and k_1 and k_2 denote the rate constants of the two reaction. At the exit of the reactor, the conversion of reactant P is 60%, and the selectivity of Q with respect to R is 50%. The value of the first order rate constant k_1 (in minute^{-1}) is ______ (correct upto one decimal place).

[2020 : 2 M]

6.36 A catalytic gas phase reaction $P \rightarrow Q$ is conducted in an isothermal packed bed reactor operated at steady state. The reaction is irreversible and second order with respect to the reactant P. The feed is pure P with a volumetric flow rate of 1.0 liter minute^{-1} and concentration of 2.0 mol liter^{-1}.

In addition, use the following assumptions :

- The reactant and product are ideal gases.
- There is no volume change associated with the reaction.
- Ideal plug flow conditions prevail in the packed bed.

When the mass of catalyst in the reactor is 4 g, the concentration of P measured at the exit is 0.4 mol liter^{-1}. The second order rate constant (in liter^2 $g^{-1}_{catalyst}$ mol^{-1} minute^{-1}) is ______ (correct upto one decimal place).

[2020 : 2 M]

6.37 Ethylene adsorbs on the vacant active sites V of a transition metal catalyst according to the following mechanism.

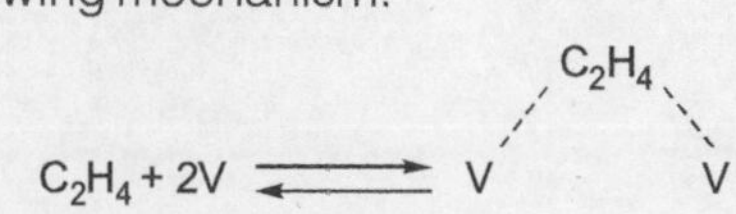

If N_T, N_V and $N_{C_2H_4}$ denote the total number of active sites, number of vacant active sites and number of adsorbed C_2H_4 molecules, respectively, the balance on the total number of active sites is given by
(a) $N_T = N_V + N_{C_2H_4}$
(b) $N_T = N_V + 2N_{C_2H_4}$
(c) $N_T = 2N_V + N_{C_2H_4}$
(d) $N_T = N_V + 0.5N_{C_2H_4}$

[2021 : 1 M]

6.38 The reaction $A \rightarrow B$ is carried out isothermally on a porous catalyst. The intrinsic reaction rate is kC_A^2, where k is the rate constant and C_A is the concentration of A. If the reaction is strongly pore-diffusion controlled, the observed order of the reaction is
(a) 1
(b) 2
(c) $\frac{3}{2}$
(d) $\sqrt{2}$

[2022 : 1 M]

6.39 CO and H_2 participate in a catalytic reaction. The partial pressures (in atm) of the reacting species CO and H_2 in the feed stream are p_{CO} and p_{H_2}, respectively. While CO undergoes molecular adsorption, H_2 adsorbs via dissociative adsorption, that is, as hydrogen atoms. The equilibrium constants (in atm^{-1}) corresponding to adsorption of CO and H_2 to the catalyst sites are K_{CO} and K_{H_2}, respectively. Total molar concentration of active sites per unit mass of the catalyst is C_t (in mol.(g cat)$^{-1}$). Both the adsorption steps are at equilibrium. Which one of the following expressions is the CORRECT ratio of the concentration of catalyst sites occupied by CO to that by hydrogen atoms?

(a) $\frac{K_{CO}p_{CO}}{\sqrt{K_{H_2}p_{H_2}}}$ (b) $\frac{K_{CO}}{\sqrt{K_{H_2}}}$

(c) $\frac{p_{CO}}{\sqrt{p_{H_2}}}$ (d) $\frac{K_{CO}p_{CO}}{K_{H_2}p_{H_2}}$

[2023 : 2 M]

6.40 Fresh catalyst is loaded into a reactor before the start of the following catalytic reaction.

$$A \rightarrow \text{products}$$

The catalyst gets deactivated over time. The instantaneous activity $a(t)$, at time t, is defined as the ratio of the rate of reaction $-r_A'(t)$ (mol.(g cat)$^{-1}$ hr^{-1}) to the rate of reaction with fresh catalyst. Controlled experimental measurements led to an empirical correlation

$$-r_A'(t) = -0.5t + 10$$

where t is in hours. The activity of the catalyst at t = 10 hr is ______ (rounded off to one decimal place). [2023 : 2 M]

6.41 A first-order heterogenous reaction $A \rightarrow B$ is carried out using a porous spherical catalyst. Assume isothermal conditions, and that intraphase diffusion controls the reaction rate. At a bulk A concentration of 0.3 mol L^{-1}, the observed reaction rate in a 3 mm diameter catalyst particle is 0.2 mol s^{-1} L^{-1} catalyst volume. At a bulk A concentration of 0.1 mol L^{-1}, the observed reaction rate, in mol s^{-1} L^{-1} catalyst volume, in a 6 mm diameter catalyst particle, is

(a) 0.011 (b) 0.033

(c) 0.022 (d) 0.005 [2024 : 2 M]

■■■■

Answers Heterogeneous Reactions, Rate & Performance Eqn. for Catalyst Deactivation Boiling

6.1	(a)	**6.2**	(Sol)	**6.3**	(a)	**6.4**	(a)	**6.5**	(d)	**6.6**	(a)	**6.7**	(a)
6.8	(b)	**6.9**	(d)	**6.10**	(b)	**6.11**	(d)	**6.12**	(b)	**6.13**	(a)	**6.14**	(b)
6.15	(a)	**6.16**	(c)	**6.17**	(a)	**6.18**	(d)	**6.19**	(a)	**6.20**	(b)	**6.21**	(b)
6.22	(0.5)	**6.23**	(1)	**6.24**	(c)	**6.25**	(d)	**6.26**	(0.8017)	**6.27**	(b)	**6.28**	(b)
6.29.	(c)	**6.30**	(a)	**6.31**	(0.4)	**6.32**	(0.038)	**6.33**	(d)	**6.34**	(a)	**6.35**	(0.5)
6.36	(0.5)	**6.37**	(b)	**6.38**	(c)	**6.39**	(a)	**6.40**	(0.5)	**6.41**	(b)		

Explanations Heterogeneous Reactions, Rate & Performance Eqn. for Catalyst Deactivation Boiling

6.2 (d)

$A(g) + B(g) \rightarrow C(g)$

Rate controlling step is reaction rate

Reaction of adsorbed A with adsorbed B to give absorbed C occurs.

On writing the reaction mechanism:

$$A + S \underset{k_2}{\overset{k_1}{\rightleftharpoons}} A.S$$

$$B + S \underset{k_4}{\overset{k_3}{\rightleftharpoons}} B.S$$

$$A.S + B.S \xrightarrow{k_5} C.S$$

$$C.S \underset{k_8}{\overset{k_7}{\rightleftharpoons}} C + S$$

On writing their rate expression accordingly:

$$r_1 = k_1 p_A C_V - k_2 C_{A_S}$$

$$r_2 = k_3 p_B C_V - k_4 C_{B_S}$$

$$r_3 = k_5 C_{A_S} \cdot C_{B_S}$$

$$r_4 = k_7 C_{C_S} - k_8 p_C C_V$$

Reaction rate is controlling so r_1, r_2 and r_4 can be equated to zero.

So, $$C_{A_S} = \frac{k_1 p_A C_V}{k_2}$$

$$C_{B_S} = \frac{k_3 p_B C_V}{k_4}$$

$$C_{C_S} = \frac{k_8 p_C C_V}{k_7}$$

$$C_T = C_{A_S} + C_{B_S} + C_{C_S} + C_V$$

$$C_V = \frac{C_T}{\left(1 + \frac{k_1 p_A}{k_2} + \frac{k_3 p_B}{k_4} + \frac{k_8 p_C}{k_7}\right)}$$

By putting the value of C_{A_S} and C_{B_S} in r_3

$$r_3 = \frac{k_5 k_1}{k_2} p_A C_V \cdot \frac{k_3}{k_4} p_B C_V$$

$$r_3 = \frac{k_1 k_5}{k_2} p_A \cdot \frac{k_3}{k_4} p_B C_V^2$$

$$r_3 = \frac{\frac{k_1 k_5}{k_2} p_A \cdot \frac{k_3}{k_4} p_B C_T^2}{\left(1 + \frac{k_1}{k_2} p_A + \frac{k_3}{k_4} p_B + \frac{k_8}{k_7} p_C\right)^2}$$

6.3 (a)

Vapour phase catalytic reaction given in question is $A + B \rightarrow F$

In Eley Rideal mechanism only one components adsorbs on catalyst surface other is present in gaseous form.

Reaction mechanism for above reaction is given below:

$$A + S \underset{k_2}{\overset{k_1}{\rightleftharpoons}} A.S$$

$$A.S + B(g) \xrightarrow{k_3} F.S$$

$$F.S \underset{k_5}{\overset{k_4}{\rightleftharpoons}} F + S$$

On writing the rate equations accordingly for above reactions

$$-r_1 = k_1 p_A C_V - k_2 C_{A_S}$$

$$-r_2 = k_3 C_{A.S} p_B$$

$$-r_3 = k_4 C_{P.S} - k_5 p_P C_V$$

Reaction rate is controlling step so r_1 and r_3 must be equal to zero.

So, $$C_{A.S} = \frac{k_1 p_A C_V}{k_2}$$

$$C_{P_S} = \frac{k_5 p_P C_V}{k_4}$$

$$C_T = C_{A_s} + C_{P_s} + C_V$$

$$C_T = \left(\frac{k_1 p_A}{k_2} + \frac{k_5 p_P}{k_4} + 1\right) C_V$$

$$C_V = \frac{C_T}{\left(\frac{k_1 p_A}{k_2} + \frac{k_5}{k_4} p_P + 1\right)}$$

On putting the value of C_{A_s} and C_V in rate r_2

$$r_2 = \frac{\left(\frac{k_3 k_1 p_A p_B}{k_2}\right) C_T}{\left(\frac{k_1 p_A}{k_2} + \frac{k_5}{k_4} p_P + 1\right)}$$

So, $$-r_A = \frac{k\, p_A p_B}{1 + K_A P_A + K_P P_P}$$

On arranging the constants.

6.4 (a)

Since increasing spinning rate is not affecting reaction rate means it is not mass transfer controlled.

In strong pore diffusion regime.

$$\frac{-r_{A_1}}{-r_{A_2}} = \frac{d_2}{d_1} = 2$$

So strong pore diffusion control and mass transfer not controlling is correct option.

6.5 (d)

For mass transfer $N_A = k_g(C_{A_g} - C_{A_i})$

For reaction rate $-r_A = kC_{A_i}$

At steady state $k_g(C_{A_g} - kC_{Ai}) = kC_{A_i}$

$$C_{A_i} = \frac{k_g}{k_g + k_i}$$

So, $$-r_A = \frac{k k_g}{k_g + k}$$

$$C_{A_g} = \left(\frac{1}{k} + \frac{1}{k_g}\right) C_{A_g}$$

$$\frac{1}{k_{eff}} = \frac{1}{k} + \frac{1}{k_g}$$

6.6 (a)

P degrades according to first order.

$$C_{P_0} = 2 \text{ kmol/m}^3$$

$$v_0 = 1 \text{ m}^3\text{/hr}$$

$$C_P = 0.5 \text{ kmol/m}^3$$

Case-1:

For CSTR

$$\frac{V}{F_{A_0}} = \frac{X_A}{-r_A}$$

$$\frac{V}{v_0} = \frac{C_{A_0} - C_A}{-r_A}$$

$$\frac{V}{1} = \frac{2 - 0.5}{k \times 0.5}$$

$$kV = 3 \quad \text{...(i)}$$

Now in **Case-2:**

$$C_{P_0} = 4 \text{ kmol/m}^3$$

$$V = 3V_0, \; C_P = 0.5 \text{ kmol/m}^3$$

$$\frac{V'}{3} = \frac{4 - 0.5}{k \times 0.5}$$

$$kV' = 7 \times 3 \quad \text{...(ii)}$$

From equation (i) and (ii)

$$\frac{V'}{V} = \frac{7 \times 3}{3} = 7$$

6.8 (b)

Rate at which antiviral drug acts increases with its concentration is given by

$$r = \frac{kC}{C_{50} + C}$$

C_{50} is the concentration at which the rate is 50% of maximum rate k.

Similarly C_{90} is when rate is 90% of maximum rate.

So, $$0.9\,k = \frac{kC_{90}}{C_{50} + C_{90}}$$

$$0.9\,C_{50} = 0.1\,C_{90}$$

$$C_{50} = \frac{C_{90}}{9}$$

By putting this value of C_{50} in rate equation

$$r = \frac{kC}{\frac{C_{90}}{9} + C}$$

6.9 (d)

Reactions between gas *A* and two solid spherical particle *B* and *C* is as following:

$$A + B \xrightarrow{k_1} \text{Gaseous product}$$

$$A + C \xrightarrow{k_2} \text{Ash}$$

Time required for particular A to be completely converted on particle C will more as compared to particle B because formation of Ash (do not leave the surface) will cause resistance for reactants to diffuse.

6.10 (b)

Given : $\frac{d^2C_A}{dx^2} = K$

Integrate,

$$\frac{dC_A}{dx} = Kx + C_1$$

Integrate, $C_A = \frac{Kx^2}{2} + C_1x + C_2$ ...(i)

Boundary conditions at $x = 0$, $C_A = C_{A_0}$

Put in equation (i)

$$C_{A_0} = 0 + 0 + C_2$$

$\therefore$ $C_2 = C_{A_0}$ ...(ii)

From equation (ii) and (i)

$$C_A = \frac{Kx^2}{2} + C_1 x + C_{A_0} \quad ...(iii)$$

2nd B.C. at $x = L$, $C_A = 0$

Put in equation (iii)

$$0 = \frac{KL^2}{2} + C_1 L + C_{A_0}$$

$$C_1 = -\frac{KL}{2} - \frac{C_{A_0}}{L} \quad ...(iv)$$

From equation (iii) and (iv)

$$C_A = \frac{Kx^2}{2} + \left(-\frac{KL}{2} - \frac{C_{A_0}}{L}\right)x + C_{A_0}$$

$$C_A = \frac{Kx(x-L)}{2} + \left(-C_{A_0}\frac{x}{L}\right) + C_{A_0}$$

6.11 (d)

$$C_A = \frac{Kx(x-L)}{2} - C_{A_0}\frac{x}{L} + C_{A_0} \quad \text{(from last question)}$$

For minimum pore length put $\frac{dC_A}{dL} = 0$

$$\therefore \quad \frac{dC_A}{dL} = \frac{-Kx}{2} + C_{A_0}\frac{x}{L^2} = 0$$

On solving, $L = \left(\frac{2C_{A_0}}{K}\right)^{1/2}$

6.12 (b)

First order reaction $A \rightarrow R$ run in mixed flow reactor.

$$\frac{W}{F_{A_0}} = \frac{X_A}{(-r_A)_{obs}}$$

$$(-r_A)_{obs} = \frac{F_{A_0}X_A}{W}$$

$$(-r_A)_{obs_1} = \frac{2\times 0.8}{W} = \frac{1.6}{W}$$

$$(-r_A)_{obs_2} = \frac{4\times 0.4}{W} = \frac{1.6}{W}$$

$$\frac{(-r_A)_{obs_1}}{(-r_A)_{obs_2}} = 1 \quad \text{(diffusion free)}$$

6.13 (a)

$A \rightarrow R$ first order reaction in packed bed reactor

For packed bed reactor

$$\frac{W}{F_{A_0}} = \int_0^{X_A} \frac{dX_A}{(-r_A)_{obs}}$$

$$\eta = \frac{(-r_A)_{obs}}{(-r_A)}$$

So, $\frac{W}{F_{A_0}} = \int_0^{X_A} \frac{dX_A}{\eta(1-r_A)}$

$$\frac{W_\eta}{F_{A_0}} = \int_0^{X_A} \frac{dX_A}{kC_{A_0}(1-X_A)}$$

$$\frac{kC_{A_0}\eta}{F_{A_0}} = -\ln(1-X_A) \quad ...(i)$$

For strong pore diffusion regime

$$\frac{\eta_1}{\eta_2} = \frac{\phi_2}{\phi_1} = \frac{r_2}{r_1}$$

From equation (i)

$$\frac{\eta_1}{\eta_2} = \frac{-\ln(1-X_{A_1})}{-\ln(1-X_{A_2})}$$

$$\frac{\ln(1-0.632)}{\ln(1-X_{A_2})} = \frac{18}{9} = 2$$

$$X_{A_2} = 0.39$$

6.14 (b)

For first order reaction

$$-r_A = kC_A = k_0e^{-E/RT}C_A$$

$$\frac{-r_A}{C_A} = k_0 e^{-E/RT}$$

$$\ln\left(\frac{-r_A}{C_A}\right) = \ln k_0 - \frac{E}{RT}$$

Now from data given in question

C_A	$-r_A$	T	$\ln\left(\frac{-r_A}{C_A}\right)$	$\frac{1}{T}$
20	1	480	−2.99573	0.002083
40	2	480	−2.99573	0.002083
40	3	50	−2.59027	0.002

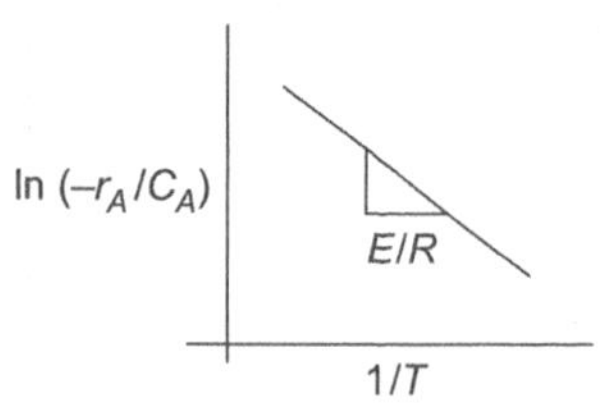

$$\text{Slope} = \frac{-2.99573-(-2.59027)}{0.002083-0.002}$$

$$= -4869.68 = -E/R$$

$$E/R = 4865.68$$

6.15 (a)

Species A reacts on a solid catalyst to produce R and S as following:

1. $A \to R \qquad r_R = k_1 C_A^2$
2. $A \to S \qquad r_S = k_2 C_A^2$

Instantaneous fractional yield is define as

$$\psi = \frac{r_R}{-r_A} = \frac{k_1 C_A^2}{k_1 C_A^2 + k_2 C_A^2}$$

$$\psi\left(\frac{R}{A}\right) = \frac{k_1}{k_1 + k_2}$$

Here we can see that instantaneous fractional yield is independent of C_A.

So ratio of instantaneous fractional yield in presence to absence of pore diffusion will be 1.

6.16 (c)

$$\frac{C_A}{C_{As}} = 1 + \frac{\phi_0^2}{4}\left[\left(\frac{r}{R}\right)^2 - 1\right]$$

Here, $\phi_0 = 4$ (given)

$$C_A = 0 \Rightarrow 1 + \frac{16}{4}\left[\left(\frac{r}{R}\right)^2 - 1\right] = 0$$

$$\left(\frac{r}{R}\right)^2 - 1 = -\frac{1}{4}$$

$$\left(\frac{r}{R}\right)^2 = \frac{3}{4}$$

$$r = \sqrt{\frac{3}{4}}R$$

6.17 (a)

Thiele modulus is define as square root ratio of reaction rate to diffusion rate.

For n^{th} order $\phi = L_c\sqrt{\frac{K}{D_e}\frac{n+1}{2}C_{A_s}^{n-1}}$

6.18 (d)

Infinitely long cylindrical pore will lead a situation of strong pore diffusion resistance. Because in infinitely long cylindrical pore diffusion rate of reactants will be limiting condition.

6.19 (a)

For irreversible, solid catalyzed, liquid phase first order reaction with diffusion rate for external mass transfer with overall mass transfer coefficient (k_m) is written below:

$$N_A = k_m(C_{A_g} - C_{A_i})$$

Similarly for rate of reaction with rate constant k

$$-r_A = kC_{A_i} \qquad \ldots(i)$$

At steady state

$$kC_{A_i} = k_m(C_{A_g} - C_{A_i})$$

$$C_{A_i} = \frac{k_m C_{A_g}}{(k+k_m)}$$

On putting this value of C_{A_i} in equation (i)

$$-r_A = \frac{k k_m C_{Ag}}{k+k_m}$$

On comparing with $-r_A = k_0 C_{A_g}$ we can see that

$$k_0 = \frac{k k_m}{k+k_m}$$

6.20 (b)

For strong pore diffusion resistant regime apparent activation energy E_a is related to the intrinsic activation energy and diffusion activation energy as

$$E_a = \frac{E + E_D}{2}$$

If $E \gg E_D$

$$E_a = 0.5\,E$$

6.21 (b)

Solid catalyzed irreversible reaction is given as

$A + B \rightarrow C$

$$A + S \underset{k_2}{\overset{k_1}{\rightleftharpoons}} A\cdot S \quad \text{...(i)}$$

$$B + S \underset{k_4}{\overset{k_3}{\rightleftharpoons}} B\cdot S \quad \text{...(ii)}$$

$$A\cdot S + B\cdot S \xrightarrow{k_5} C\cdot S \quad \text{...(iii)}$$

$$C\cdot S \underset{k_7}{\overset{k_6}{\rightleftharpoons}} C + S \quad \text{...(iv)}$$

Now writing their rate equations accordingly

$$-r_1 = k_1 p_A C_v - k_2 C_{AS}$$

$$-r_2 = k_3 p_B C_v - k_4 C_{B.S}$$

$$-r_3 = k_5 C_{AS}\, C_{B.S}$$

$$-r_4 = k_6 C_{CS} - k_7 p_C C_v$$

Reaction rate is rate controlling so r_1, r_2 and r_4 must be put as equal to zero.

So, $$C_{AS} = \frac{k_1 p_A C_v}{k_2} \quad \text{...(v)}$$

$$C_{BS} = \frac{k_3 p_B C_v}{k_4} \quad \text{...(vi)}$$

$$C_{CS} = \frac{k_7 p_C C_v}{k_6} \quad \text{...(vi)}$$

Now, $$C_T = C_{AS} + C_{BS} + C_v + C_{C.S}$$

$$C_T = \left[\frac{k_1 p_A}{k_2} + \frac{k_3 p_B}{k_4} + \frac{k_7 p_C}{k_6} + 1\right] C_v$$

$$C_v = \frac{C_T}{\left[1 + \frac{k_1 p_A}{k_2} + \frac{k_3 p_B}{k_4} + \frac{k_7 p_C}{k_6}\right]}$$

Now on putting the values of C_{AS}, C_{BS} and C_v in rate expression (r_3)

$$-r_3 = \frac{k_5 \frac{k_1 p_A}{k_2} \cdot \frac{k_3 p_B}{k_4} C_T^2}{\left[1 + \frac{k_1 p_A}{k_2} + \frac{k_3 p_B}{k_4} + \frac{k_7 p_C}{k_6}\right]^2}$$

$$\text{So rate} \propto \frac{P_A P_B}{(1 + k_A P_A + k_B P_B + k_C P_C)^2}$$

6.22 (0.5)

In the case of strong pore diffusion resistance regime.

$$\eta = \frac{1}{\phi}$$

$$-r_{A_1} \propto \eta \qquad \left(\because \eta = \frac{(-r_A)_{\text{obs}}}{(-r_A)}\right)$$

So, $$\frac{-r_{A_2}}{-r_{A_1}} = \frac{\eta_2}{\eta_1} = \frac{\phi_1}{\phi_2} \quad \text{...(i)}$$

Thiele modulus

$$\phi = L_c \sqrt{\frac{K}{D_e}}$$

$$\phi \propto L_c$$

In case of spherical catalyst $L_c = \frac{R}{3}$ now from equation (i)

$$\frac{(r_{A_2})}{(r_{A_1})} = \frac{R_1}{R_2} = \frac{1}{2} = 0.5$$

6.23 (1)

$$(CH_3)_2\,N_2 + (CH_3)_2\,N_2 \xrightarrow{k_1} (CH_3)_2\,N_2 + [(CH_3)_2\,N_2]^*$$

$$[(CH_3)_2\,N_2]^* + (CH_3)_2\,N_2 \xrightarrow{k_2} (CH_3)_2\,N_2 + (CH_3)_2\,N_2$$

$$[(CH_3)_2\,N_2]^* \xrightarrow{k_3} C_2H_6 + N_2$$

For simplification $(CH_3)_2\,N_2$ is denoted by A

$[(CH_3)_2\,N_2]^*$ is denoted by A*

C_2H_6 is denoted by B

Now net rate of formation of A* is given by

$$+r_A^* = k_1\, C_A^2 - k_2\, C_A^*\, C_A - k_3\, C_A^*$$

According to Pseudo steady state approximation

For $(CH_3)_2\,N_3$ $\quad r_A^* = 0$

$$C_A^* = \frac{k_1 C_A^2}{(k_2 C_A + k_3)} \quad \text{...(i)}$$

Now net rate of formation of ethane (B)

$$+r_B = k_3\, C_A^*$$

By putting value of C_A^* from equation (i)

$$r_B = \frac{k_3 k_1 C_A^2}{(k_2 C_A + k_3)}$$

In limit of high value of C_A

r_B becomes as $\frac{k_3 k_1}{k_2} C_A$

So order = 1

6.24 (c)

In Rideal mechanism only one reactant is adsorbed on catalyst surface other is present in gaseous form.

Rate of reaction $Q + R \rightarrow S$ in equimolar ratio can be given as

$$Q + S \rightleftharpoons Q.S \quad \text{(Adsorption)}$$

$$Q.S + R \longrightarrow S \quad \text{(Chemical reaction)}$$

Rate equation for adsorption mechanism

$$r_{ad} = k_1 P_Q C_V - k_2 C_{QS} \quad \text{...(i)}$$

Similarly for chemical reaction

$$r_{\text{reaction}} = k_r C_{Q.S} P_R \quad \text{...(ii)}$$

Total site balance $C_T = C_V + C_{QS}$ ...(iii)

Rate of reaction is limiting so

$$\frac{r_{ad}}{k_1} = 0$$

$$C_{QS} = \frac{k_1 P_Q C_V}{k_2}$$

By putting this value in equation (iii)

$$C_T = C_V\left(1 + \frac{k_1 P_Q}{k_2}\right)$$

$$C_V = \frac{C_T}{1 + K P_Q}$$

By putting value of C_{QS} and C_V in equation (ii)

$$r_{\text{reaction}} = \frac{K_r k_1 P_Q C_T P_r}{k_2(1 + K P_Q)}$$

$$P_Q = P_r = \frac{P_T}{2}$$

So reaction rate can be optimized in

$$r_{\text{reaction}} = \frac{a P_T^2}{1 + b P_T}$$

where, $a = \dfrac{K_r k_1 C_T}{K_2 \cdot 4}$, $b = \dfrac{K}{2}$

6.26 (0.8017)

First order reaction $A \rightarrow B$

Ratio of rate constant at 470 K to rate constant at 450 K is given by Arrhenius equation.

$$\ln\left(\frac{k_2}{k_1}\right) = \frac{100000}{8.314}\left[\frac{1}{450} - \frac{1}{470}\right]$$

$$\frac{k_2}{k_1} = 3.1186$$

$$\text{Thiele modules} = L_c\sqrt{\frac{k}{D_{eff}}}$$

First order reaction effective diffusivity (D_{eff}) is independent of temperature.

So Thiele modulus $\phi \propto \sqrt{k}$

$$\frac{\phi_2}{\phi_1} = \sqrt{\frac{k_2}{k_1}} = \sqrt{3.1186} = 1.766$$

$$\phi_2 = 1.766 \times \phi_1$$

$$= 1.766 \times 0.5 = 0.883$$

Now effectiveness factor η is given by

$$\eta = \frac{\tanh\phi}{\phi} = \frac{\tanh 0.883}{0.883} = 0.8017$$

6.27 (b)

According to option (b) reaction rate is much greater than diffusion rate. So P_t need to be present only in immediate vicinity of external surface of pellet.

Reaction rate is already fast so P_t need not to present throughout the pores of catalyst.

6.28 (b)

Thiele modulus is ratio of reaction rate to diffusion rate.

For n^{th} order $\phi = L_c\sqrt{\dfrac{n+1}{2}\dfrac{k}{D_o}C_{A_s}^{n-1}}$

For, $n = 1$

$$\phi = L_c\sqrt{\frac{k}{D_e}}$$

L_c = Characteristic length

6.29 (c)

Solid catalyst does not affect equilibrium conversion and equilibrium constant, it speeds-up both forward and backward reaction.

6.30 (a)

In option (a) we can see that the temperature at 500 K and rpm = 3000. From the set A data in 2nd and 3rd column, the temperature is same as 500 K but instead of increasing stirring speed rate from 2000 to 3000, the reaction rate is same as 0.025. So at temperature 500 K and rpm rate 3000 the reaction rate is not controlled by external mass transfer.

6.31 (0.4)

$P \rightarrow 2Q$

$P \rightarrow R$

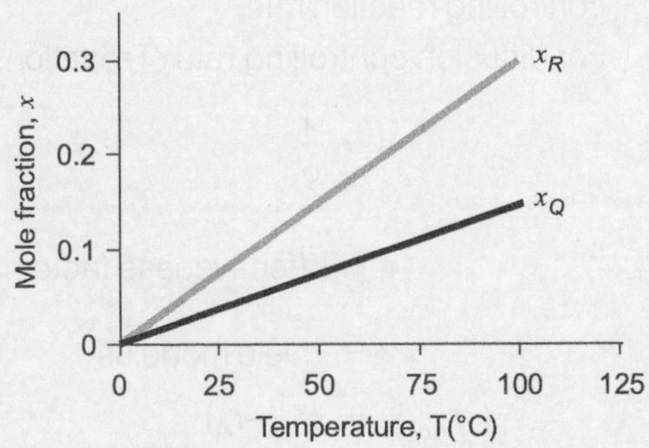

From above plot:

Case 1 : Mole fraction of Q and 25° ($X_Q = 0.035$)

Mole fraction of R and 25°C ($X_R = 0.069$)

$$\text{Yield of } Q \text{ at } 25° = \frac{\text{Moles of } Q \text{ produced}}{\text{Moles of } P \text{ consumed}}$$

$$Y_Q = \frac{0.035}{0.0175 + 0.069} = 0.40$$

Case 2 : At 60°C (basis 1 mol of product):

$$X_Q = 0.0952, \; X_R = 0.177$$

$$\text{Yield of } Q = \frac{0.0952}{0.177 + \frac{0.0952}{2}} = 0.40$$

It means the product yield will remain constant over the temperature range which is given.

6.32 (0.038)

Reaction occurs at all surface of the hollow cylinder. So area available for reaction is

$$A = 2 \times \frac{\pi}{4}(0.025^2 - 0.02^2) + \pi(0.025 + 0.02) \times 0.025$$

$$A = 3.8877 \times 10^{-3}\ \text{m}^2$$

Mass of catalyst $= V \times \rho$

$$= 8910 \times \frac{\pi}{4}(0.025^2 - 0.02^2) \times 0.025$$

$$= 0.03936\ \text{kg}$$

So rate of reaction in $\dfrac{\text{mol}}{\text{kg of catalyst min}}$

$$= \frac{0.39 \times 3.8877 \times 10^{-3}}{0.03936} = 0.038$$

6.33 (d)

For a first order reaction in porous spherical catalyst pellet,

$$\text{Thiele modulus } \phi = L_c\sqrt{\frac{k}{D_e}}$$

Diffusional effects are most likely to lower than observed rate of reaction for the value of ϕ greater than 4.

For the Thiele modulus to be large reaction should be fast in a pellet of large diameter.

6.34 (a)

Thiele modulus measure the relative importance of surface reaction rate over diffusion rate.

$$M_T = L_c\sqrt{\frac{K}{D_{\text{eff}}}}$$

Rate is inversely proportional to time. So M_T^2 is ratio of time scale of pore diffusion to time scale of reaction also.

6.35 (0.5)

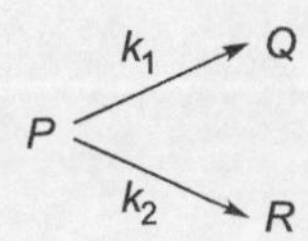

$$-r_P = (k_1 + k_2)\,C_p$$

Given, $\tau_m = 1$ min

$$X_P = 60\%, \; S\left(\frac{Q}{R}\right) = 0.50$$

Now, from selectivity, we have

$$0.50 = \frac{k_1 C_p}{k_2 C_p} = \frac{k_1}{k_2}$$

$$k_2 = 2k_1 \quad \text{...(i)}$$

For CSTR, we have

$$\tau_m = \frac{C_{P_0} X_P}{(-r_P)} = \frac{C_{P_0} \times 0.60}{(k_1 + k_2) C_{P_0}(1 + 0.60)}$$

$$1 = \frac{0.60}{3k_1 \times 0.40}$$

$$k_1 = 0.5\ \text{min}^{-1}$$

6.36 (0.5)

Catalytic gas phase reaction $P \rightarrow Q$ is conducted in isothermal packed bed reactor.

Irreversible 2nd order reaction.

From the design equation of packed bed reaction

$$\frac{W}{F_{A_0}} = \int_0^{X_A} \frac{dX_A}{(-r_A)}$$

$$F_{A_0} = v_0 \times C_{A_0}$$

$$\frac{4}{1 \times 2} = \int_0^{X_A} \frac{dX_A}{kC_{A_0}^2(1 - X_A)^2} \quad \text{...(i)}$$

$$\because \quad X_A = 1 - \frac{C_A}{C_{A_0}} = 1 - \frac{0.4}{2} = 0.8$$

Now integrating equation (i) and applying limits, we get

$$\frac{4}{1\times 2} = \frac{1}{k\times 2^2}\left(\frac{X_A}{1-X_A}\right)$$

$$\frac{4}{2} = \frac{1}{k\times 4}\cdot\frac{0.8}{0.2}$$

$$k = 0.5 \text{ liter}^2\, g_{\text{catalyst}}^{-1} \text{ minute}^{-1}$$

6.38 (c)

For strong pore resistance regime, observed order is given by,

$$n_{obs} = \frac{n_{actual}+1}{2} = \frac{2+1}{2} = \frac{3}{2}$$

6.40 (0.5)

Given: $A \rightarrow$ Products

$$-r_A' = -0.5t + 10 \text{ mol.(g cat)}^{-1}\text{ hr}^{-1})$$

$$a = \frac{-r_A'}{-r_0} = \frac{-0.5t+10}{10}$$

$$= \frac{\text{Rate of reaction at any time } t}{\text{Rate of reaction when analyst is fresh}}$$

$\because \quad -r_0' = 10 \qquad \{\text{at } t = 0\}$

Thus activity of catalyst at time t = 10 hr

$$a = \frac{-0.5\times 10+10}{10} = \frac{5}{10} = 0.5$$

6.41 (b)

First-order heterogenous reaction $A \rightarrow B$ is carried out using a porous spherical catalyst with diffusion controlling reaction rate.

For diffusion controlling rate of reaction.

$$\eta = \frac{1}{\phi}$$

$$\eta = \text{Effectiveness factor} = \frac{-r_{A\text{obs}}}{(-r_A)}$$

ϕ = Thiele modulus

$$(-r_{A\text{ obs}}) = \phi^{-1}(-r_A)$$

$$\frac{(-r_A)_{obs_2}}{(-r_A)_{obs_1}} = \left(\frac{\phi_2}{\phi_1}\right)^{-1}\frac{(-r_A)_2}{(-r_A)_1}$$

$\phi \propto$ dia of catalyst

and $(-r_A) = kC_A$

$$\frac{(-r_A)_{obs_2}}{0.2} = \frac{3}{6}\times\frac{0.1}{0.3}$$

$$(-r_A)_{obs_2} = 0.033 \text{ mol}/l.\text{s}$$

■■■■

7 Kinetics of Enzyme Reactions

7.1 Reactant A decomposes to products B and C in the presence of an enzyme in a well-stirred batch reactor. The kinetic rate expression is given by

$$-r_A = \frac{0.01C_A}{0.05 + C_A} \text{ (mol L}^{-1}\text{ min}^{-1}\text{)}$$

If the initial concentration of A is 0.02 mol L^{-1}, the time taken to achieve 50% conversion of A is _____ min (round off to 2 decimal places).

[2021 : 2 M]

7.2 In an enzymatic reaction, an inhibitor (I) competes with the substrate (S) to bind with the enzyme (E), thereby reducing the rate of product (P) formation. The competitive inhibition follows the reaction mechanism shown below. Let [S] and [I] be the concentration of S and I, respectively, and r_s be the rate of consumption of S. Assuming pseudo-steady state, the correct plot of $\frac{1}{-r_s}$ vs $\frac{1}{[s]}$ is

$$E + S \underset{k_2}{\overset{k_1}{\rightleftharpoons}} E.S \xrightarrow{k_3} E + P$$

$$E + I \underset{k_2}{\overset{k_1}{\rightleftharpoons}} E.I$$

(a) 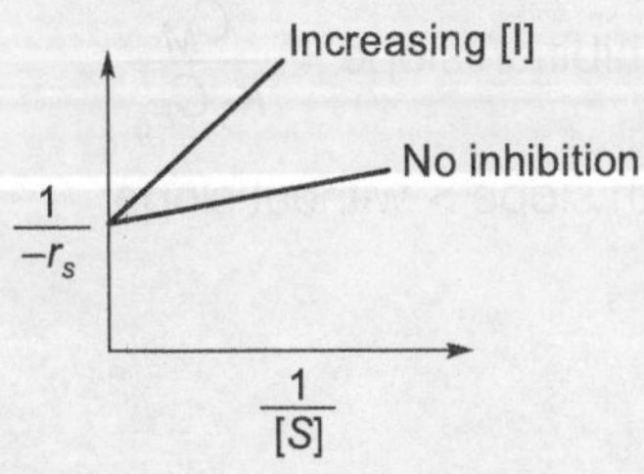

(b) 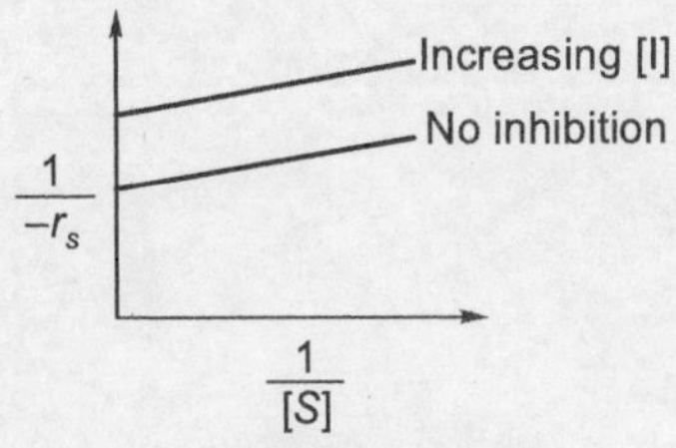

(c) 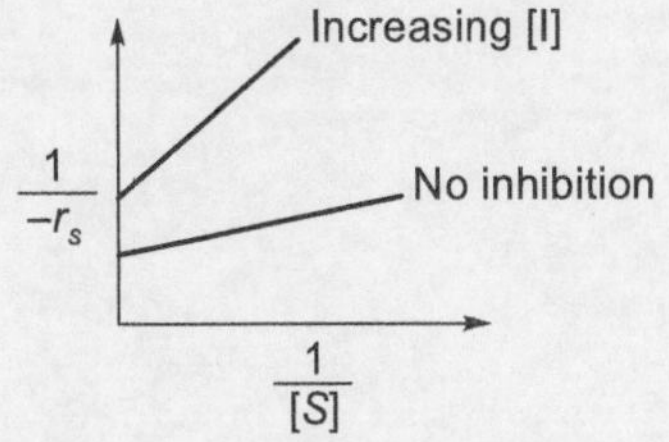

(d) 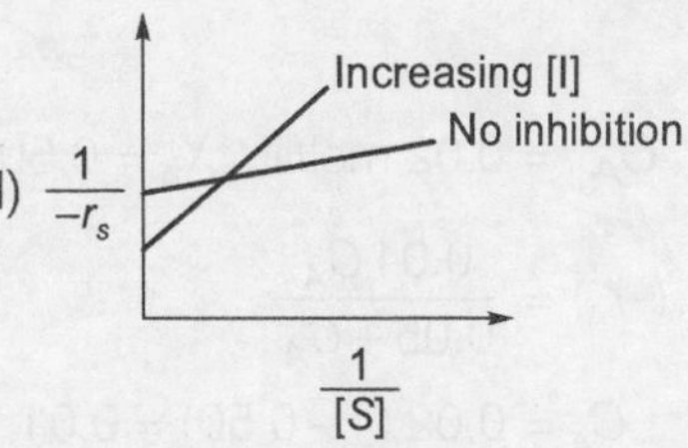

[2022 : 1 M]

7.3 A chemostat with cell recycle is shown in the figure. The feed flow rate and culture volume are F = 75 L h^{-1} and V = 200 L, respectively. The glucose concentration in the feed C_{S0} = 15 g L^{-1}. Assume Monod kinetics with specific cell growth rate $\mu_g = \frac{1}{C_C}\frac{dC_C}{dt} = \frac{\mu_m C_C}{K_S + C_S}$, where $\mu_m = 0.25\, h^{-1}$ and $K_S = 1$ g L^{-1}. Assume maintenance and death rates to be zero, input feed to be sterile ($C_{C0} = 0$) and steady-state operation. The glucose concentration in the recycle stream, C_{S1}, in g L^{-1}, rounded off to 1 decimal place, is ________.

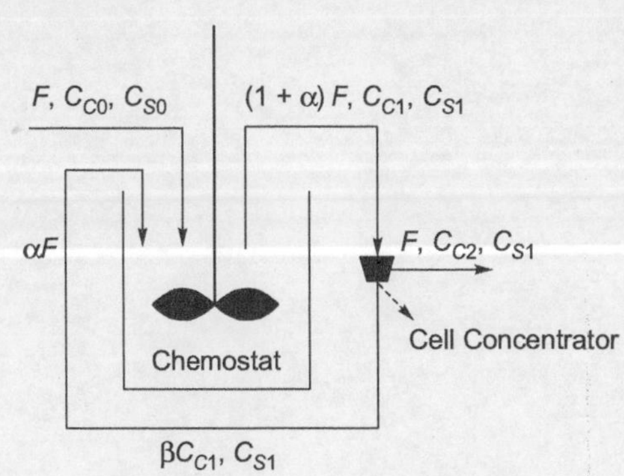

F : Volumetric flow rate
C_C: Cell concentration
C_S: Substrate (glucose) concentration
α : Recycle ratio (α = 0.5)
β : Concentration factor (β = 2.0)

[2024 : 2 M]

■■■■

Answers Kinetics of Enzyme Reactions

7.1 (4.47) **7.2** (a) **7.3** (3.0)

Explanations Kinetics of Enzyme Reactions

7.1 (4.47)

Given: $C_{A_0} = 0.02$ mol/lit, $X_A = 0.50$

$$(-r_A) = \frac{0.01\,C_A}{0.05 + C_A}$$

$\because$ $C_A = 0.02\,(1 - 0.50) = 0.01$

From design equation of batch reactor, we have

$$(-r_A) = -\frac{1}{V}\frac{dN_A}{dt} = -\frac{dC_A}{dt} \quad \{V = \text{constant}\}$$

$$\int dt = -\int_{C_{A0}}^{C_A} \frac{dC_A}{\frac{0.01C_A}{0.05 + C_A}}$$

$$= \int_{C_A}^{C_{A0}} \left[\frac{0.05 + C_A}{0.01C_A}\right] dC_A$$

Integrating, we get

$$t = [5\ln C_A + 100\,C_A]_{0.01}^{0.02}$$

$$t = 4466 \text{ minutes}$$

7.2 (a)

With inhibition, $-r_s = \dfrac{k_3\,C_{E_0}\,C_s}{C_M\left(1 + LC_{B_0}\right) + C_S}$

Without inhibition, $-r_s = \dfrac{k_3\,C_{E_0}\,C_s}{C_M + C_S}$

Now C_S high, $-r_s = k_3\,C_{E_0}$

$-r_s = k_3\,C_{E_0}$

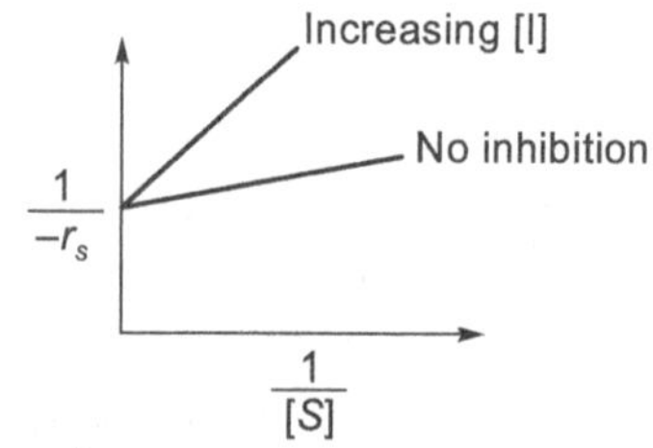

$\dfrac{1}{-r_s}$ vs $\dfrac{1}{[s]}$ same

C_S low $\therefore$ with $-r_s = \dfrac{k_3 C_{E_0}\,C_s}{C_M\left(1 + LC_{B_0}\right)}$

With inhibition slope $= \dfrac{C_M\left(1 + LC_{B_0}\right)}{k_3\,C_{E_0}}$

Without inhibition slope $= \dfrac{C_M}{k_3\,C_{E_0}}$

That it, with slope > without slope.

UNIT VIII

Instrumentation and Process Control

CONTENTS

Instrumentation and Process Control

Syllabus

Measurement of Process Variables; Sensors and Transducers; P&ID Equipment Symbols; Process Modeling and Linearization, Transfer Functions and Dynamic Responses of Various Systems, Systems with Inverse Response, Process Reaction Curve, Controller Modes (P, PI and PID); Control Valves; Transducer Dynamics; Analysis of Closed Loop Systems including Stability, Frequency Response, Controller Tuning, Cascade and Feed Forward Control.

Analysis of Previous GATE Papers

Exam Year	1 Mark Ques.	2 Marks Ques.	5 Marks Ques.	Total Marks
2000	4	–	2	**14**
2001	3	3	–	**9**
2002	2	–	2	**12**
2003	3	6	–	**15**
2004	1	5	–	**11**
2005	4	5	–	**14**
2006	1	9	–	**19**
2007	1	7	–	**15**
2008	–	8	–	**16**
2009	2	4	–	**10**
2010	1	5	–	**11**
2011	2	3	–	**8**
2012	2	3	–	**8**
2013	1	4	–	**9**
2014	3	4	–	**11**
2015	2	3	–	**8**
2016	4	3	–	**10**
2017	2	5	–	**12**
2018	1	4	–	**9**
2019	2	3	–	**8**
2020	2	2	–	**6**
2021	3	4	–	**11**
2022	4	3	–	**10**
2023	1	4	–	**9**
2024	1	3	–	**7**

1 Introduction

1.1 A process is initially at steady state with its output $y = 1$ for an input $u = 1$. The input is suddenly changed to 2 at time $t = 0$. The output response is $y(t) = 1 + 2t$. The transfer function of the process is

(a) $\frac{2}{s}$ (b) $1+\frac{2}{s^2}$

(c) $1+\frac{2}{s}$ (d) $\frac{1}{s}\left(1+\frac{2}{s}\right)$

[2001 : 1 M]

1.2 Find the Laplace transform of the following input function shown in figure

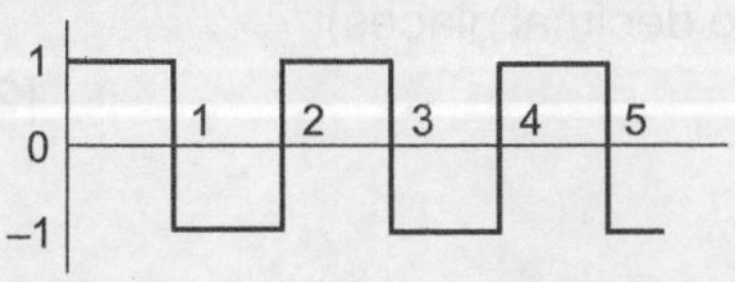

(a) $\frac{(1-e^{-s})}{s(1-e^{-2s})}$ (b) $\frac{2(1-e^{-s})}{s}$

(c) $\frac{(1-e^{-s})}{s(1+e^{-s})}$ (d) $\frac{(1-e^{-s})e^{-s}}{s(1-e^{-2s})}$

[2003 : 2 M]

1.3 For the time domain function $f(t) = t$ the Laplace transform of $\int_0^t f(t)dt$ is given by

(a) $\frac{1}{(2s^3)}$ (b) $\frac{2}{s^3}$

(c) $\frac{1}{s^3}$ (d) $\frac{2}{s^3}$ [2004 : 1 M]

1.4 The Laplace transform of the input function, $X(t)$, given in the figure below:

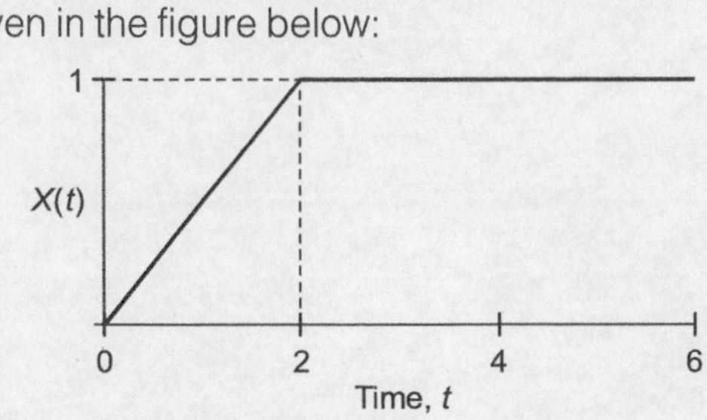

is given by

(a) $\frac{1}{2s^2}(1-e^{-2s})$ (b) $\frac{1}{2s^2}(1+e^{-2s})$

(c) $\frac{1}{s^2}(1+e^{2s})$ (d) $\frac{1}{s^2}(1+e^{-2s})$

[2006 : 2 M]

1.5 The unit impulse response of a first order process is given by $2e^{-0.5t}$. The gain and time constant of the process are, respectively,

(a) 4 and 2 (b) 2 and 2

(c) 2 and 0.5 (d) 1 and 0.5

[2008 : 2 M]

1.6 A unit step input is given to a process that is represented by the transfer function $\frac{(s+2)}{(s+5)}$. The initial value ($t = 0$) of the response of the process to the step input is

(a) 0 (b) $\frac{2}{5}$

(c) 1 (d) ∞

[2008 : 2 M]

1.7 The inverse Laplace transform of $\frac{1}{2s^2+3s+1}$ is

(a) $e^{-t/2} - e^{-t}$

(b) $2e^{-t/2} - e^{-t}$

(c) $e^{-t} - 2e^{-t/2}$

(d) $e^{-t} - e^{-t/2}$

[2009 : 2 M]

1.8 A unit IMPULSE response of a first order system with time constant τ and steady state gain K_p is given by

(a) $\frac{1}{K_p\tau}e^{t/\tau}$ (b) $K_p e^{-t/\tau}$

(c) $\tau K_p e^{-t/\tau}$ (d) $\frac{K_p}{\tau}e^{-t/\tau}$

[2014 : 1 M]

1.9 The Laplace transform of $e^{at}\sin(bt)$ is

(a) $\dfrac{b}{(s-a)^2+b^2}$ (b) $\dfrac{(s-a)}{(s-a)^2+b^2}$

(c) $\dfrac{(s-a)}{(s-a)^2-b^2}$ (d) $\dfrac{b}{(s-a)^2-b^2}$

[2016 : 1 M]

1.10 The Laplace transform of a function is $\dfrac{s+1}{s(s+2)}$.

The initial and final values, respectively, of the function are

(a) 0 and 1 (b) 1 and $\dfrac{1}{2}$

(c) $\dfrac{1}{2}$ and 1 (d) $\dfrac{1}{2}$ and 0

[2017 : 2 M]

1.11 A system has a transfer function $G(s) = \dfrac{3e^{-4s}}{12s+1}$.

When a step change of magnitude M is given to the system input, the final value of the system output is measured to be 120. The value of M is ________.

[2021 : 2 M]

1.12 A process described by the transfer function

$$G_p(s) = \frac{(10s+1)}{(5s+1)}$$

is forced by a unit step input at time $t = 0$. The output value immediately after the step input (at $t = 0^+$) is ________ (rounded off to the nearest integer).

[2022 : 1 M]

1.13 Consider a single-input-single-output (SISO) system with the transfer function

$$G_p(s) = \frac{2(s+1)}{\left(\frac{1}{2}s+1\right)\left(\frac{1}{4}s+1\right)}$$

where the time constants are in minutes. The system is forced by a unit step input at time $\tau = 0$. The time at which the output response reaches the maximum is ________ minutes (rounded off to two decimal places).

[2022 : 2 M]

■■■■

Answers Introduction

1.1	(a)	1.2	(c)	1.3	(c)	1.4	(a)	1.5	(a)	1.6	(c)	1.7	(a)
1.8	(d)	1.9	(a)	1.10	(b)	1.11	(40)	1.12	(2)	1.13	(0.549)		

Explanations Introduction

1.1 (a)

Change in input $X(t) = x(t) - x_{ss} = 2 - 1 = 1$
Change in output $Y(t) = y(t) - y_{ss} = 1 + 2t - 1 = 2t$
Transfer function of from is given by

$$G(s) = \frac{Y(s)}{X(s)} = \frac{\frac{2}{s^2}}{\frac{1}{s}} = \frac{2}{s}$$

1.2 (c)

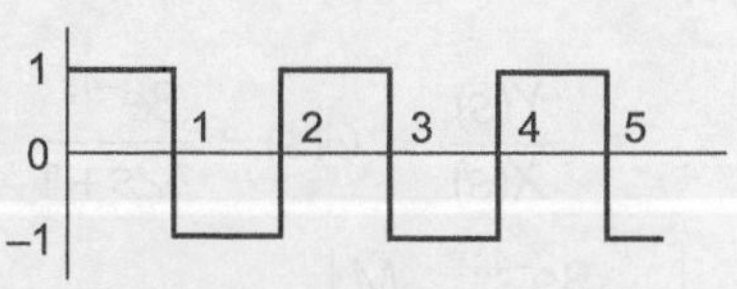

$f(t) = 1\,u(t) - 2 \cdot u(t-1) + 2\,u(t-2) - 2\,u(t-3) + 2\,u(t-4)\ ...$

$$F(s) = \frac{1}{s} - \frac{2}{s}e^{-s} + \frac{2}{s}e^{-2s} - \frac{2}{s}e^{-3s} + \frac{2}{s}e^{-4s}...$$

$$F(s) = \frac{1}{s}\left(1 - 2e^{-s} + 2e^{-2s} - 2e^{-3s} + 2e^{-4s}...\right)$$

$$F(s) = \frac{1}{s}\left[1 - \left(2e^{-s} - 2e^{-2s} + 2e^{-3s} - 2e^{-4s} + ...\right)\right]$$

$$F(s) = \frac{1}{s}\left[1 - \left(\frac{a}{1-r}\right)\right]$$

where, $a = 2e^{-s},\ r = -e^{-s}$

$$F(s) = \frac{1}{s}\left[1 - \frac{2e^{-s}}{1+e^{-s}}\right]$$

$$F(s) = \frac{1}{s}\left(\frac{1+e^{-s}-2e^{-s}}{1+e^{-s}}\right)$$

$$F(s) = \frac{1}{s}\left(\frac{1-e^{-s}}{1+e^{-s}}\right)$$

1.3 (c)

$$F(t) = t$$

$$L[F(t)] = \frac{1}{s^2}$$

Laplace of $\int_0^t F(t)dt$ is given by $\frac{F(s)}{s}$ so Laplace of

$$\int_0^t t\,dt = \frac{1}{s^3}$$

1.4 (a)

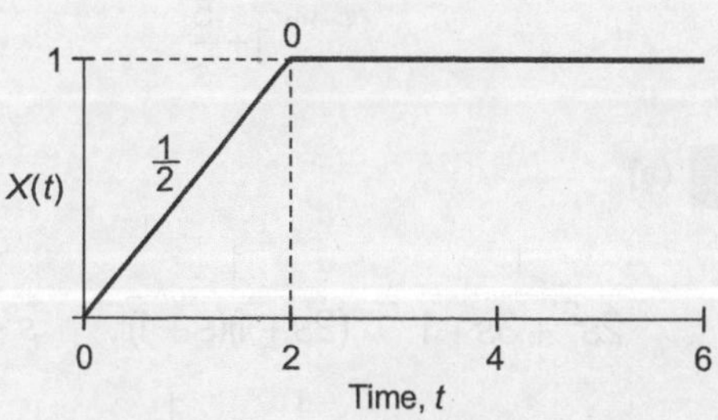

From graph $X(t)$ can be define as

$$X(t) = \frac{1}{2}t \qquad 0 \le t \le 2$$

$$X(t) = 1 \qquad t > 2$$

Laplace of any time domain function $X(t)$ is define as

$$L[X(t)] = \int_0^\infty e^{-st}X(t)dt$$

$$= \int_0^2 e^{-st}\frac{1}{2}(t)dt + \int_0^\infty e^{-st}\,dt$$

On integrating

$$L[X(t)] = \frac{1}{2s^2}(1-e^{-2s})$$

1.5 (a)

Unit impulse response of first order system with gain K_p and time constant τ_p is given by

$$Y(t) = \frac{K_p}{\tau_p}e^{-t/\tau_p}$$

$$Y(t) = 2e^{-0.5t}$$

On comparing

$$\tau_p = 2$$

$$K_p = 4$$

1.6 (c)

Transfer function $G(s) = \frac{s+2}{s+5}$

For unit step input $X(s) = \frac{1}{s}$

$$\frac{Y(s)}{X(s)} = G(s)$$

From initial value theorem

$$\text{Initial value} = \lim_{s\to\infty}[sY(s)]$$

$$= \lim_{s\to\infty} s\frac{(s+2)}{(s+5)}\cdot\frac{1}{s}$$

$$= \lim_{s\to\infty} \frac{1+\frac{2}{s}}{1+\frac{5}{s}} = 1$$

1.7 (a)

$$\frac{1}{2s^2+3s+1} = \frac{1}{(2s+1)(s+1)} = -\frac{1}{s+1}+\frac{2}{2s+1}$$

$$= \frac{1}{s+\frac{1}{2}} - \frac{1}{s+1}$$

Taking inverse Laplace

$\Rightarrow$ $e^{(-1/2)t} - e^{-t}$

1.8 (d)

Impulse response of first order system with time constant τ and steady state gain K_p can be found by differentiating corresponding step response

Step input response

$$Y(t) = K_p(1-e^{-t/\tau_p})$$

Impulse input response

$$\frac{dY(t)}{dt} = \frac{K_p}{\tau_p}e^{-t/\tau_p}$$

1.9 (a)

Laplace transform of sin (bt)

$$L[\sin(bt)] = \frac{b}{s^2+b^2}$$

According to first shifting theorem

$$L[e^{-at}F(t)] = F(s+a)$$

So, $$L[e^{at}\sin(bt)] = \frac{b}{(s-a)^2+b^2}$$

1.10 (b)

Laplace transform of function is given as $\frac{s+1}{s(s+2)}$

From final value theorem,

$$\text{Final value} = \lim_{s\to 0}[sY(s)]$$

$$= \lim_{s\to 0}\frac{s(s+1)}{s(s+2)} = \frac{1}{2}$$

From initial value theorem,

$$\text{Initial value} = \lim_{s\to\infty}[sY(s)] = \lim_{s\to\infty}\frac{s(s+1)}{s(s+2)}$$

$$= \lim_{s\to\infty}\frac{1+\frac{1}{s}}{1+\frac{1}{s}} = 1$$

1.11 (40)

$$\frac{Y(s)}{X(s)} = G(s) = \frac{3e^{-4s}}{12s+1}$$

$$\lim_{s\to 0}\left[s\cdot\frac{3e^{-4s}}{12s+1}\times\frac{M}{s}\right] = 120$$

$$\frac{3M}{1} = 120$$

$$M = \frac{120}{3} = 40$$

1.12 (2)

$$G_p(s) = \frac{(10s+1)}{(5s+1)}$$

From initial value theorem:

$$\frac{y(s)}{x(s)} = \frac{(10s+1)}{(5s+1)}$$

$$y(s) = \frac{(10s+1)}{s(5s+1)}$$

$$y(t)\big|_{t\to 0} = \lim_{s\to\infty} s.y(s)$$

$$= \lim_{s\to\infty}\left(\frac{10s+1}{5s+1}\right)$$

$$= \lim_{s\to\infty}\left(\frac{10s+\frac{1}{s}}{5s+\frac{1}{s}}\right) = 2$$

1.13 **(0.549)**

$$G_p = \frac{2(s+1)}{\left(\frac{1}{2}s+1\right)\left(\frac{1}{4}s+1\right)}$$

After step unit, we have

$$y = \frac{1}{s}\frac{2(s+1)\times 8}{(s+2)(s+4)}$$

$$\frac{16(s+1)}{s(s+2)(s+4)} = \frac{A}{s} + \frac{B}{s+2} + \frac{C}{s+4}$$

$$16s + 16 = s^2(A + B + C) + s(6A + 4B + 2C) + 8A$$

Solving,

$$A = 2$$

$$B = 4$$

$$C = -6$$

$$y(t) = 2 + 4e^{-2t} - 6e^{-4t}$$

$$\frac{dy}{dt} = 0 - 8e^{-2t} + 24e^{-4t}$$

$$= 0 \quad \left\{\text{for mox } \frac{dy}{dt} = 0\right\}$$

$$e^{-2t} + 4t = 3$$

$$e^{2t} = 3$$

$$t = \frac{\ln 3}{2} = 0.549$$

■■■■

2 Dynamic Behaviour of First and Second Order System

2.1 The unit step response of the transfer function $\frac{2s-1}{(3s+1)(4s+1)}$ reaches its final steady state asymptotically after

(a) a monotonic increase.
(b) a monotonic decrease.
(c) initially increasing and then decreasing.
(d) initially decreasing and then increasing.

[2000 : 1 M]

2.2 The unit step response of the transfer function $\frac{1}{s^2+2s+3}$

(a) has a non-zero slope at the origin.
(b) has a damped oscillatory characteristic.
(c) is overdamped.
(d) is unstable.

[2000 : 1 M]

2.3 The response of a thermocouple can be modelled as a first order process to changes in the temperature of the environment. If such a thermocouple at 25°C is immersed suddenly in a fluid at 80°C and held there, it is found that the thermocouple reading (in °C) reaches 63.2% of the final steady state value in 40 seconds. Find the time constant of the thermocouple.

[2000 : 5 M]

2.4 A second order system can be obtained by connecting two first order systems $\frac{1}{(\tau_1 s+1)}$ and $\frac{1}{(\tau_2 s+1)}$ in series. The damping ratio of the resultant second order system for the case $\tau_1 \neq \tau_2$ will be

(a) > 1 (b) = 1

(c) < 1 (d) $= \frac{\tau_2}{\tau_1}$

[2001 : 2 M]

2.5 A mercury thermometer can be used to measure the body temperature by placing it either in the mouth or in the armpit of a patient. The true body temperature can be taken to be the temperature inside the mouth, which is usually higher than the temperature in the armpit by 0.5 K. Assume that the true body temperature of a patient is 312 K and the thermometer is initially at 300 K. Also assume that the thermometer behaves like a first-order system with a time constant of 40 seconds.

(a) Obtain a relation for the thermometer reading $T(t)$ as a function of time in terms of its initial temperature $T(0)$, and body temperature T_B.
(b) How long should the thermometer be placed in the patient's mouth in order to ensure that the error in the measurement is not greater than 0.05%.
(c) Since the body temperature in the armpit is less, the measurement made here using the thermometer is corrected by adding 0.5 K. How long should the thermometer be placed in the armpit in order to ensure that the error in the corrected measurement is not greater than 0.05% of the true body temperature.

[2002 : 5 M]

2.6 Consider a system of two tanks in series as shown below:

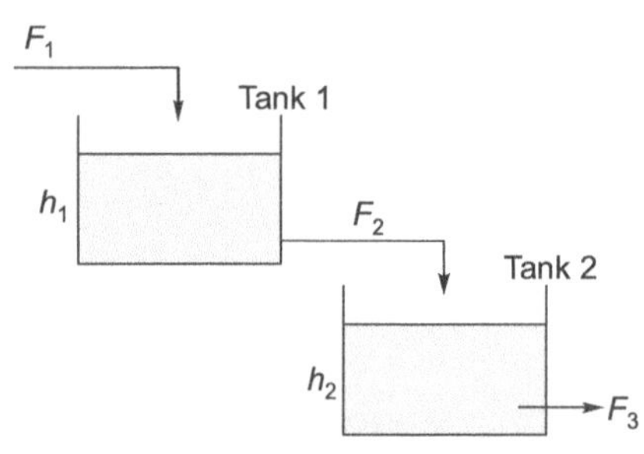

The level h_2 in Tank II is measured and has to be controlled by manipulating the flow rate F_1. It is given that $F_2 = 0.005h_1$ and $F_3 = 0.0025h_2$ m³/s, where h is in m. The cross- sectional areas of Tank I and II are both equal to 1 m².

(a) Determine the transfer function of the process.
(b) Compute the time constants of the process. Is the open loop process over damped, under damped or critically damped?
(c) If proportional control is used with constant $K_c > 0$, determine the value of K_c for which the closed loop response becomes critically damped.

[2002 : 5 M]

2.7 Suppose that the gain, time constant, and dead time of a process with the following transfer function $G_c(s) = 10\exp(-0.1\ s)/(0.5\ s + 1)$ are known with a possible error of ±20% of their values. The largest permissible gain K_c of a proportional controller needs to be calculated by taking the values of process gain, time constant and dead time as
(a) 8, 0.6, 0.08
(b) 12, 0.6, 0.12
(c) 8, 0.6, 0.12
(d) 12, 0.4, 0.08

[2003 : 1 M]

2.8 Water is entering a storage tank at a temperature T_0 and flow rate Q_0 and leaving at a flow rate Q and temperature T. There are negligible heat losses in the tank. The area of cross section of the tank is A_c. The model that describes the dynamic variation of water temperature in the tank with time is given as

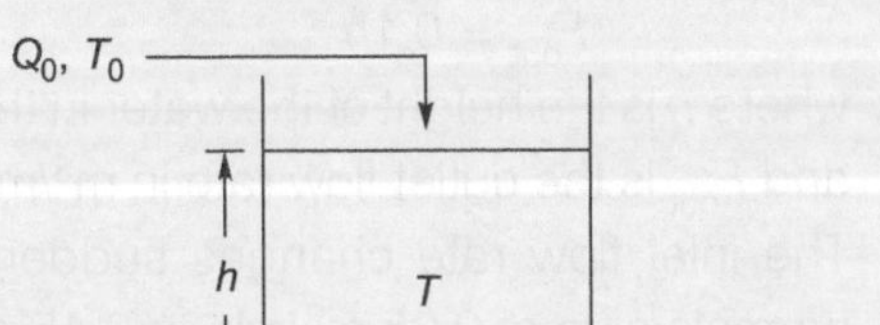

(a) $Q_0(T_0 - T) = A_c h\dfrac{dT}{dt}$

(b) $Q_0T_0 - QT = A_c h\dfrac{dT}{dt}$

(c) $Q(T_0 - T) = A_c h\dfrac{dT}{dt}$

(d) $Q(T_0 - T) = A_c\dfrac{d(Th)}{dt}$

[2003 : 2 M]

2.9 Match first order system given in **Group-I** with the appropriate time constant in **Group-II**.

Group-I	Group-II
A. Thermometer	1. $\dfrac{(mC_p)}{(hA)}$
B. Mixing	2. $\dfrac{q}{V}$
	3. $\dfrac{V}{q}$
	4. $\dfrac{(hA)}{(mC_p)}$

	A	B
(a)	4	2
(b)	4	3
(c)	1	2
(d)	1	3

[2004 : 2 M]

2.10 The unit step response of a first order system with time constant τ and steady state gain K_p is given by
(a) $K_p(1 - e^{-t/\tau})$
(b) $K_p(1 + e^{-t/\tau})$
(c) $K_p(1 - e^{-2t/\tau})$
(d) $K_p e^{-t/\tau}/\tau$

[2005 : 1 M]

2.11 An example of an open-loop second order under damped system is
(a) liquid level in a tank
(b) U-tube manometer
(c) thermocouple in a thermo-well
(d) two non-interacting first order systems in series

[2005 : 1 M]

2.12 A 2-input, 2-output process can be described in the Laplace transform domain as given below:

$$(\tau_1 s + 1)Y_1(s) = K_1U_1(s) + K_2U_2(s)$$
$$(\tau_2 s + 1)Y_2(s) = K_3U_2(s) + K_4Y_1(s)$$

where U_1 and U_2 are the inputs and Y_1 and Y_2 are the outputs. The gains of the transfer functions $Y_1(s)/U_2(s)$ and $Y_2(s)/U_2(s)$, respectively, are
(a) K_2 and K_3
(b) K_1 and $K_3 + K_2K_4$
(c) K_2 and $K_3 + K_1K_4$
(d) K_2 and $K_3 + K_2K_4$

[2006 : 2 M]

2.13 A process is perturbed by a sinusoidal input, $u(t) = A \sin \omega t$. The resulting process output is

$Y(s) = \dfrac{kA\omega}{(\tau s+1)(s^2+\omega^2)}$. If $y(0) = 0$, the differential equation representing the process is

(a) $\dfrac{dy(t)}{dt} + \tau y(t) = Ku(t)$

(b) $\tau \dfrac{dy(t)}{dt} + y(t) = KAu(t)$

(c) $\tau \dfrac{dy(t)}{dt} + y(t) = Ku(t)$

(d) $\tau \left(\dfrac{dy(t)}{dt} + y(t) \right) = KAu(t)$

[2006 : 2 M]

2.14 The dynamic model for a mixing tank open to atmosphere at its top as shown below is to be written. The objective of mixing is to cool the hot water stream entering the tank at a flow rate q_2 and feed temperature of T_s with a cold water feed stream entering the tank at a flow rate q_1 and feed temperature of T_0. A water stream is drawn from the tank bottom at a flow rate of q_4 by a pump and the level in the tank is proposed to be controlled by drawing another water stream at a flow rate q_3. Neglect evaporation and other heat losses from the tank.

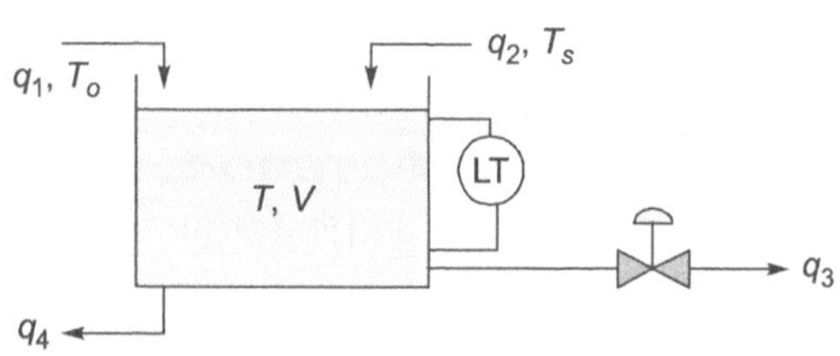

The dynamic model for the tank is given as

(a) $\dfrac{dV}{dt} = q_1 + q_2 - q_3,\ V\dfrac{dV}{dt} = q_1T_0 + q_2T_3 - q_3T$

(b) $\dfrac{dV}{dt} = q_1 - q_4,\ \dfrac{d(VT)}{dt} = q_1T_s + q_4T$

(c) $\dfrac{dV}{dt} = q_1 + q_2 - q_4,$

$\dfrac{d(VT)}{dt} = q_1T_0 + q_2T_s - q_4T$

(d) $\dfrac{dV}{dt} = q_1 + q_2 - q_3 - q_4,$

$V\dfrac{dV}{dt} = q_1(T_0 - T) + q_2(T_s - T)$

[2007 : 2 M]

2.15 Match the transfer functions with the response to a unit step input shown in the figure:

(i) $\dfrac{-2.5(-4s+1)}{4s^2+4s+1}$ (ii) $\dfrac{-2e^{-10s}}{10s+1}$

(iii) $\dfrac{-5}{-20s+1}$ (iv) $\dfrac{-0.1}{s}$

(v) $\dfrac{4s+3}{2s+1}$

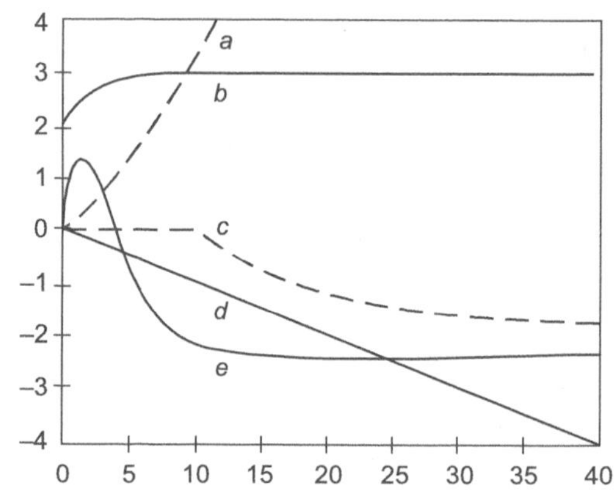

(a) (i)–(e), (ii)–(c), (iii)-(a), (iv)-(d), (v)-(b)
(b) (i)–(a), (ii)–(b), (iii)-(c), (iv)-(d), (v)-(e)
(c) (i)–(b), (ii)–(a), (iii)-(c), (iv)-(e), (v)-(d)
(d) (i)–(e), (ii)–(a), (iii)-(c), (iv)-(b), (v)-(d)

[2007 : 2 M]

2.16 A tank of volume 0.25 m³ and height 1 m has water flowing in at 0.05 m³/min. The outlet flow rate is governed by the relation

$$F_{out} = 0.1h$$

where h is the height of the water in the tank in m and F_{out} is the outlet flow rate in m³/min.
The inlet flow rate changes suddenly from its normal value of 0.05 m³/min to 0.15 m³/min and remains there. The time (in minutes) at which the tank will begin to overflow is given by

(a) 0.28 (b) 1.01
(c) 1.73 (d) ∞

[2008 : 2 M]

2.17 Which one of the following transfer funtions corresponds to an inverse response process with a positive gain?

(a) $\dfrac{1}{2s+1} - \dfrac{2}{3s+1}$ (b) $\dfrac{2}{s+1} - \dfrac{5}{s+10}$

(c) $\dfrac{3(0.5s-1)}{(2s+1)(s+1)}$ (d) $\dfrac{5}{s+1} - \dfrac{3}{2s+1}$

[2008 : 2 M]

2.18 The roots of the characteristic equation of an underdamped second order system are
(a) real, negative and equal
(b) real, negative and unequal
(c) real, positive and unequal
(d) complex conjugates

[2009 : 1 M]

2.19 For a tank of cross-sectional area 100 cm^2 and inlet flow rate (Q_i in cm^3/s), the outlet flow rate (Q_o in cm^3/s) is related to the liquid height (H in cm) as $Q_o = 3\sqrt{H}$ (see figure)

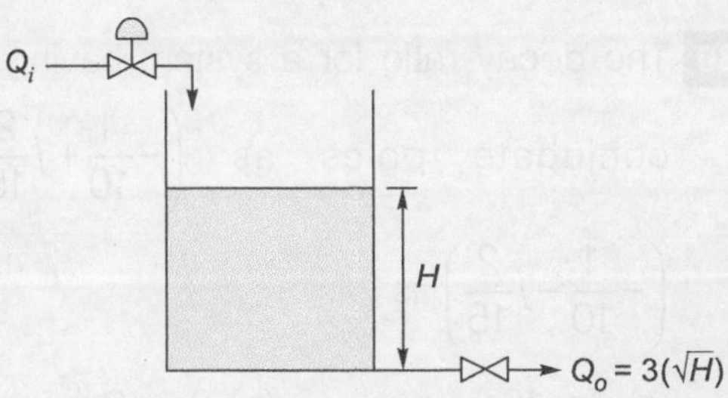

Then the transfer function $\frac{\bar{H}(s)}{\bar{Q}_i(s)}$ (overbar indicates deviation variables) of the process around the steady-state point, $Q_{i,s}$ = 18 cm^3/s and H_s = 36 cm, is

(a) $\frac{1}{100s+1}$ (b) $\frac{2}{200s+1}$
(c) $\frac{3}{300s+1}$ (d) $\frac{4}{400s+1}$

[2009 : 2 M]

2.20 Match the location of the poles/zeros in the s-plane, listed in **Group-I** with the system response characteristics in **Group-II**:

Group-I
P. Pole in the right half plane
Q. Pole at origin
R. Zero in the right half plane

Group-II
1. Stable response
2. Integrating response
3. Unstable response
4. Inverse response

(a) P-1, Q-2, R-3 (b) P-3, Q-4, R-1
(c) P-3, Q-2, R-4 (d) P-1, Q-4, R-3

[2010 : 1 M]

2.21 A thermometer initially at 100°C is dipped at $t = 0$ into an oil bath, maintained at 150°C. If the recorded temperature is 130°C after 1 minute, then the time constant of thermometer (in min) is
(a) 1.98 (b) 1.35
(c) 1.26 (d) 1.09

[2012 : 1 M]

2.22 A unit gain 2nd order underdamped process has a period of oscillation 1 second and decay ratio 0.25. The transfer function of the process is

(a) $\frac{1}{0.024s^2+0.067s+1}$

(b) $\frac{1}{0.067s^2+0.024s+1}$

(c) $\frac{1}{0.021s^2+0.1176s+1}$

(d) $\frac{1}{0.1176s^2+0.021s+1}$

[2013 : 2 M]

2.23 Assume that an ordinary mercury-in-glass thermometer follows first order dynamics with a time constant of 10s. It is at a steady state temperature of 0°C. At time $t = 0$, the thermometer is suddenly immersed in a constant temperature bath at 100°C. The time required (in s) for the thermometer to read 95°C, approximately is
(a) 60 (b) 40
(c) 30 (d) 20

[2014 : 1 M]

2.24 A step change of magnitude 2 is introduced into a system having the following transfer function

$$G(s) = \frac{2}{s^2+2s+4}$$

The percent overshoot is ______.

[2014 : 2 M]

2.25 Which one of the following transfer functions, upon a unit step change in disturbance at t = 0, will show a stable time domain response with a negative initial slope (i.e., slope at t = 0):

(a) $G(s) = \frac{1}{s+1} - \frac{2}{s+4}$

(b) $G(s) = \frac{1}{s+1} + \frac{2}{s+4}$

(c) $G(s) = \frac{1}{s+1} + \frac{2}{s-4}$

(d) $G(s) = \frac{1}{s-1} + \frac{2}{s-4}$

[2015 : 2 M]

2.26 What is the order of response exhibited by a U-tube manometer?

(a) Zero order (b) First order
(c) Second order (d) Third order

[2016 : 1 M]

2.27 A system exhibits inverse response for a unit step change in the input. Which one of the following statement must necessarily be satisfied?

(a) The transfer function of the system has at least one negative pole
(b) The transfer function of the system has at least one positive pole
(c) The transfer function of the system has at least one negative zero
(d) The transfer function of the system has at least one positive zero

[2016 : 1 M]

2.28 For a unit step input, the response of a second order system is

$$y(t) = K_p\left[1 - \frac{1}{\sqrt{1-\zeta^2}} e^{-\frac{\zeta t}{\tau}} \sin\left(\frac{\sqrt{1-\zeta^2}}{\tau} t + \phi\right)\right]$$

where, K_p is the steady state gain, ζ is the damping coefficient, τ is the natural period of oscillation and ϕ is the phase lag. The overshoot of the system is $\exp\left(-\frac{\pi\zeta}{\sqrt{1-\zeta^2}}\right)$. For a unit step input, the response of the system from an initial state condition at $t = 0$ is shown in the figure below.

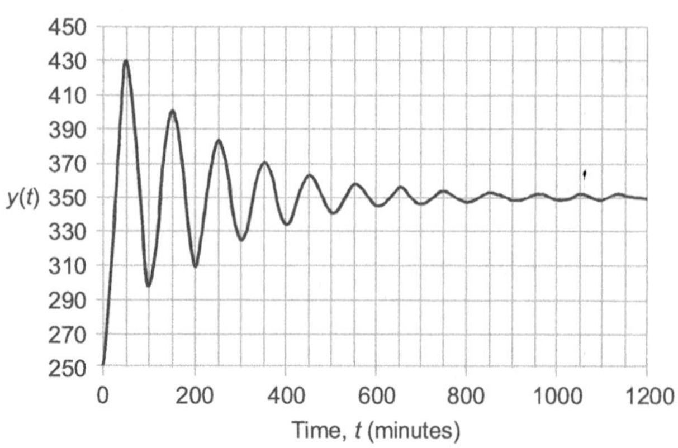

What is the natural period of oscillation (in seconds) of the system?

(a) 15.9 (b) 50
(c) 63.2 (d) 100

[2016 : 2 M]

2.29 The transfer function of a system is

$$\frac{1}{4s^2 + 1.2s + 1}$$

For a unit step increase in the input, the fractional overshoot, rounded to 2 decimal places, is ______.

[2017 : 2 M]

2.30 The decay ratio for a system having complex conjugate poles as $\left(-\frac{1}{10} + j\frac{2}{15}\right)$ and $\left(-\frac{1}{10} - j\frac{2}{15}\right)$ is

(a) 7×10^{-1} (b) 8×10^{-2}
(c) 9×10^{-3} (d) 10×10^{-4}

[2018 : 2 M]

2.31 Match the items in **Column A** with the items in **Column B:**

Column A	Column B
(P) Pure dead-time	(1) $\phi = -90°$
(Q) Pure capacitive	(2) $\frac{K_1}{s} - \frac{K_2}{\tau s + 1}$ with $K_2 > \tau K_1$
(R) Inverse response	(3) $0 < AR < 1$
(S) First-order process with unit gain	(4) $AR = 1$

Here, ϕ denotes the phase shift, K_1 and K_2 the process gains, τ the time constant, and AR the amplitude ratio.

(a) P-2, Q-3, R-4, S-1 (b) P-3, Q-2, R-4, S-1
(c) P-1, Q-4, R-2, S-3 (d) P-4, Q-1, R-2, S-3

[2018 : 2 M]

2.32 Consider two non-interacting tanks-in-series as shown in figure. Water enters Tank 1 at q cm^3/s and drains down to Tank 2 by gravity at a rate $k\sqrt{h_1}$ (cm^3/s). Similarly, water drains from Tank 2 by gravity at a rate of $k\sqrt{h_2}$ (cm^3/s) where h_1 and h_2 represent levels of Tank1 and Tank 2, respectively (see fig.). Drain valve constant $k = 4$

$cm^{2.5}/s$ and cross-sectional areas of the two tanks are $A_1 = A_2 = 28\ cm^2$.

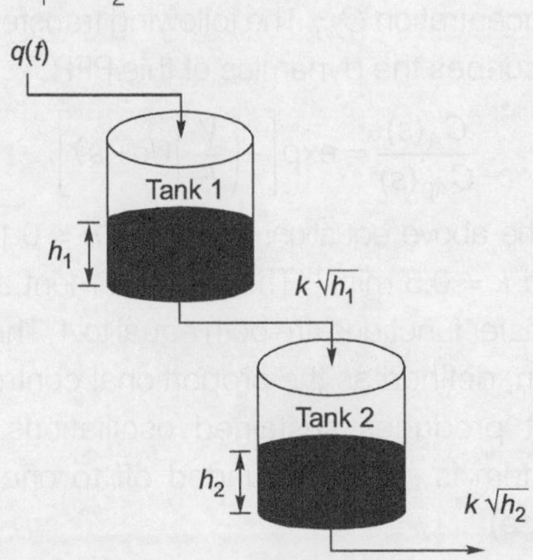

At steady state operation, the water inlet flow rate is $q_{ss} = 16\ cm^3/s$. The transfer function relating the deviation variables $\bar{h}_2$ (cm) to flow rate $\bar{q}$ (cm^3/s) is

(a) $\dfrac{2}{(56s+1)^2}$ (b) $\dfrac{2}{(62s+1)^2}$

(c) $\dfrac{2}{(36s+1)^2}$ (d) $\dfrac{2}{(49s+1)^2}$

[2019 : 2 M]

2.33 Choose the option that correctly matches the step response curves on the left with the appropriate transfer functions on the right. The step input change occurs at time $t = 0$.

Step response

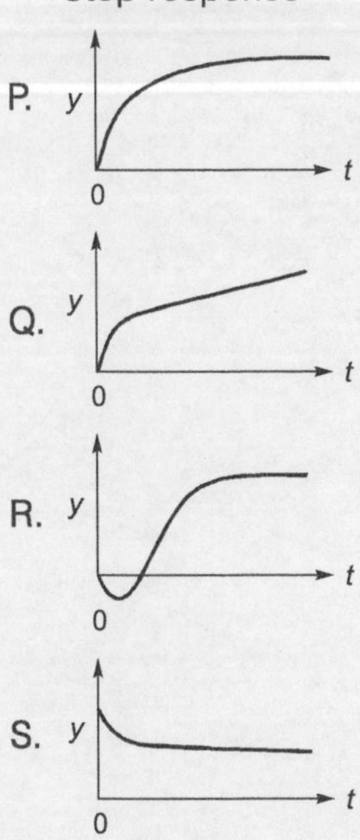

Transfer function

1. $\dfrac{K(\xi s+1)}{s(\tau s+1)}$ $K>0, \xi>0, \tau>0$
2. $\dfrac{K(\xi s+1)}{(\tau s+1)}$ $K>0, \xi>\tau>0$
3. $\dfrac{K}{(\tau s+1)}$ $K>0, \tau>0$
4. $\dfrac{K(\xi s+1)}{(\tau_1 s+1)(\tau_2 s+1)}$ $K>0, \tau<0$, $\tau_1>0, \tau_2>0$

(a) P-3, Q-4, R-2, S-1
(b) P-3, Q-1, R-4, S-2
(c) P-4, Q-3, R-2, S-1
(d) P-3, Q-2, R-4, S-1

[2019 : 2 M]

2.34 A process has a transfer function $G(s) = \dfrac{Y(s)}{X(s)} = \dfrac{20}{90000s^2 + 240s + 1}$. Initially the process is at steady state with $x(t=0) = 0.4$ and $y(t=0) = 100$. If a step change in x is given from 0.4 to 0.5, the maximum value of y that will be observed before it reaches the new steady state is ________ (round off to 1 decimal place).

[2021 : 2 M]

2.35 A liquid surge tank has F_{in} and F_{out} as the inlet and outlet flow rates respectively, as shown in the figure below. F_{out} is proportional to the square root of the liquid level h. The cross-sectional area of the tank is $20\ cm^2$. Density of the liquid is constant everywhere in the system. At steady state, $F_{in} = F_{out} = 10\ cm^3\ s^{-1}$ and $h = 16$ cm. The variation of h with F_{in} is approximated as a first order transfer function. Which one of the following is the CORRECT value of the time constant (in seconds) of this system?

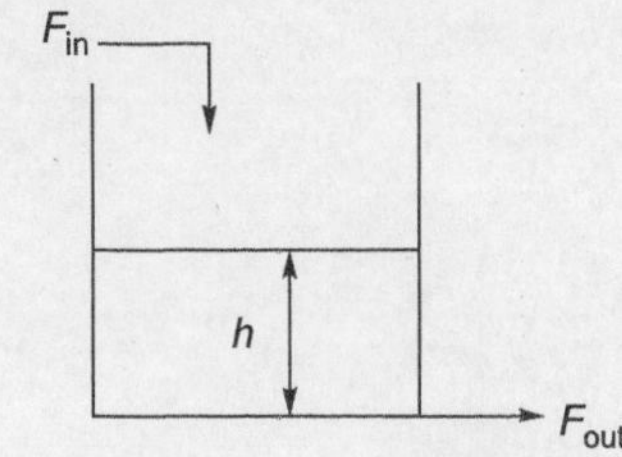

(a) 20 (b) 32
(c) 64 (d) 128

[2023 : 1 M]

2.36 Level (h) in a steam boiler is controlled by manipulating the flow rate (F) of the make-up (fresh) water using a proportional (P) controller. The transfer function between the output and the manipulated input is

$$\frac{h(s)}{F(s)} = \frac{0.25(1-s)}{s(2s+1)}$$

The measurement and valve transfer functions are both equal to 1. A process engineer wants to tune the controller so that the closed-loop response gives decaying oscillations under servo mode. Which one of the following is the CORRECT value of the controller gain to be used by the engineer?

(a) 0.25
(b) 2
(c) 4
(d) 6

[2023 : 2 M]

2.37 The outlet concentration C_A of a plug flow reactor (PFR) is controlled by manipulating the inlet concentration C_{A0}. The following transfer function describes the dynamics of this PFR.

$$\frac{C_A(s)}{C_{A0}(s)} = \exp\left[-\left(\frac{V}{F}\right)(k+s)\right]$$

In the above equation, $V = 1\ m^3$, $F = 0.1\ m^3\ min^{-1}$ and $k = 0.5\ min^{-1}$. The measurement and valve transfer functions are both equal to 1. The ultimate gain, defined as the proportional controller gain that produces sustained oscillations, for this system is ______ (rounded off to one decimal place).

[2023 : 2 M]

2.38 The transfer function of a measuring instrument is

$$G_m(s) = \frac{1.05}{2s+1}\exp(-s)$$

At time $t = 0$, a step change of +1 unit is introduced in the input of this instrument. The time taken by the instrument to show an increase of 1 unit in its output is _____ (rounded off to two decimal places).

[2023 : 2 M]

■■■■

Answers Dynamic Behaviour of First and Second Order System

2.1	(c)	**2.2**	(b)	**2.3**	(64)	**2.4**	(a)	**2.5**	(2.89, 1.93)	**2.6**	(3.125×10^{-4})		
2.7	(b)	**2.8**	(a)	**2.9**	(d)	**2.10**	(a)	**2.11**	(b)	**2.12**	(d)	**2.13**	(c)
2.14	(d)	**2.15**	(a)	**2.16**	(c)	**2.17**	(b)	**2.18**	(d)	**2.19**	(d)	**2.20**	(c)
2.21	(d)	**2.22**	(a)	**2.23**	(c)	**2.24**	(16.3)	**2.25**	(a)	**2.26**	(c)	**2.27**	(d)
2.28	(a)	**2.29**	(0.37)	**2.30**	(c)	**2.31**	(d)	**2.32**	(a)	**2.33**	(b)	**2.34**	(102.5)
2.35	(c)	**2.36**	(b)	**2.37**	(148.4)	**2.38**	(7.1)						

Explanations Dynamic Behaviour of First and Second Order System

2.1 (c)

Given function transfer is

$$\frac{Y(s)}{X(s)} = \frac{(2s-1)}{(3s+1)(4s+1)}$$

Or we can write it as,

$$\frac{Y(s)}{X(s)} = \frac{5}{(3s+1)} - \frac{6}{(4s+1)} \quad ...(1)$$

For unit step input, we have

$$X(s) = \frac{1}{s}$$

Eq. (1) becomes

$$Y(s) = \frac{1}{s}\left[\frac{5}{(3s+1)} - \frac{6}{(4s+1)}\right]$$

$$Y(s) = \frac{5}{s(3s+1)} - \frac{6}{s(4s+1)}$$

$$Y(s) = \frac{-1}{s} - \frac{5}{\left(s+\frac{1}{3}\right)} + \frac{6}{\left(s+\frac{1}{4}\right)}$$

After taking inverse Laplace, we get

$$y(t) = -1 + 6e^{-t/4} - 5e^{-t/3}$$

$$y(\infty) = -1$$

From the above relation, the slope is positive, and response initially increases, and gain is negative.

2.2 (b)

Given transfer function is

$$G(s) = \frac{1}{(s^2+2s+3)}$$

or $$G(s) = \frac{1}{3\left(\frac{1}{3}s^2 + \frac{2}{3}s + 1\right)}$$

or $$G(s) = \frac{1/3}{\left(\frac{1}{3}s^2 + \frac{2}{3}s + 1\right)}$$

The standard equation of transfer function is

$$G(s) = \frac{K}{\tau^2 s^2 + 2\xi\tau s + 1}$$

On comparing with standard equation, we get

$$\tau^2 = \frac{1}{3};\ 2\xi\tau = \frac{2}{3}$$

$$\Rightarrow \quad 2\times\xi\times\frac{1}{\sqrt{3}} = \frac{2}{3}$$

After solving, we get

$$\xi = 0.577$$

Therefore, the value of ξ is in range of $0 < \xi < 1$, this condition is underdamped response.

2.3 (64)

We know that for first order system the step response is given by,

$$Y(t) = A(1 - e^{-t/\tau_p})$$

or $$y(t) - y_s = A(1 - e^{-t/\tau_p}) \quad ...(1)$$

Given : $A = (80°C - 25°C) = 55°C$

$t = 40$ seconds

$y_s = 25°C$ (at steady state)

$y(t) = 0.632 \times 80°C = 50.56°C$

Substituting the values in Eq. (1), we get

$$50.56°C - 25°C = 55°C \times (1 - e^{-40/\tau})$$

$$25.56°C = 55°C(1 - e^{-40/\tau})$$

After solving, we get

$$\tau = 64 \text{ sec}$$

2.5 (2.89, 1.93)

True body temperature of patient is 312 K

Thermometer initial at 300 K

Thermometer is first order with time constant 40 sec

(a) Thermometer reading with time can be found by

$$Y(t) = X(t)\left[1 - e^{t/\tau_p}\right]$$

Here, $Y(t) = y(t) - y_{ss}$

$y(t)$ = Temperature at any time

$X(t) = x(t) - x_{ss}$

(a) $Y(t) = X(t)(1 - e^{-t/\tau})$

$y(t) = T(t) - T(0)$

$X(t) = T_B - T(0)$

$T(t) - T(0) = [T_B - T(0)](1 - e^{-t/\tau}]$

(b) $\frac{T_B - T(t)}{T_B} = 0.0005$

where, $T_B = 312$ K

$T(t) = 311.844$ K

$Y(t) = T(t) - T(0)$

$= 311.844 - 300 = 11.844$ K

$X(t) = T_B - T(0) = 12$ K

$11.844 = 12(1 - e^{-t/40})$

$t = 173.728$ s $= 2.89$ min

(c) Arpit measurement is 0.5 K less than true body temperature

$$\frac{(312-0.5)-T}{312-0.5} = 0.0005$$

$T = 311.344$ K

$Y(t) = 311.344 - 300 = 11.344$ K

$X(t) = 312 - 300 = 12$ K

$11.344 = 12(1 - e^{-t/40})$

$t = 116$ s $= 1.93$ min

2.6 (3.125×10^{-4})

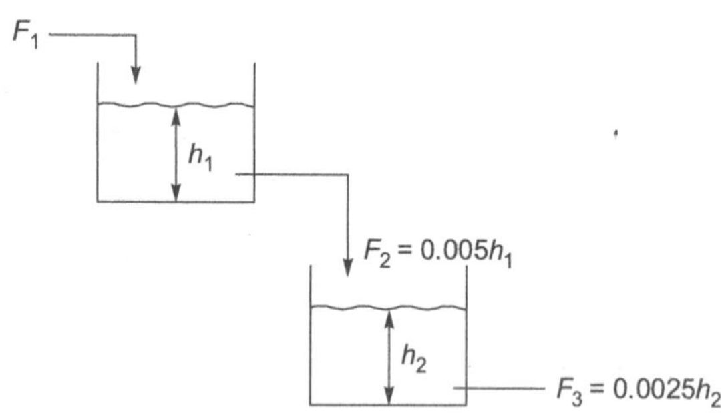

(a) Both tank have area = 1 m^2

$$F = \frac{h}{R}$$

So, $R_1 = 200 = \frac{1}{\left.\frac{dF_2}{dh_1}\right|_{h_2=h_s}} = \frac{1}{0.005}$

Similarly, $R_2 = 400 = \frac{1}{\left.\frac{dF_3}{dh_2}\right|_{h_2=h_s}} = \frac{1}{0.0025}$

Both tanks are in non-interacting manner so transfer function

$$G(s) = \frac{h_2(\bar{s})}{F_1(s)} = \frac{R_2}{(\tau_1 s+1)(\tau_2 s+1)}$$

(b) $\tau_1 = A_1R_1 = 200\times1 = 200$

$\tau_2 = A_2R_2 = 400\times1 = 400$

So, $G(s) = \frac{400}{(200s+1)(400s+1)}$

Since $\tau_1 \neq \tau_2$ so system is overdamped.

(c) Proportional controller with gain K_c is used so characteristic equation of close loop transfer function is given by

$1 + G_cG_p = 0$

$$1+\frac{K_c\times400}{(200s+1)(400s+1)}$$

$80000s^2 + 600s + 1 + 400K_c = 0$

On comparing with $\tau^2 s^2 + 2\xi\tau s + 1 = 0$

$$\tau = 100\sqrt{\frac{8}{1+400K_c}}$$

$$2\xi\tau = \frac{600}{1+400K_c}$$

$$\xi = \frac{300}{100\sqrt{8(1+400K_c)}}$$

For critically damped system $\xi = 1$

So $\frac{3}{\sqrt{8+(1+400K_c)}} = 1$

$K_c = 3.125\times10^{-4}$

2.7 (b)

Transfer function is given as

$$G_c(s) = \frac{10\exp(-0.1s)}{0.5s+1}$$

Possible error is given as ± 20%

So largest value of proportional controller gain K_c can be calculated by using following values

$K_P = 10\times1.2 = 12$

$\tau_D = 0.1\times1.2 = 0.12$

$\tau_P = 0.5\times1.2 = 0.6$

2.8 (a)

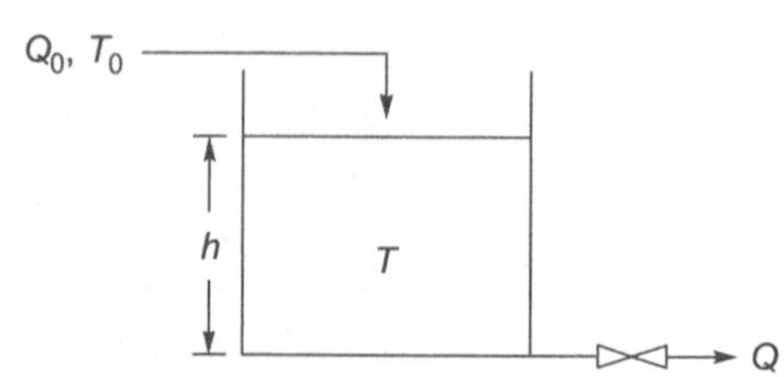

Area of cross-section of tank is A_c

From material balance

$$\rho Q_0 - \rho Q = \frac{d(\rho V)}{dt}$$

$$Q_0 - Q = A_c \frac{dh}{dt} \quad ...(i)$$

From energy balance

$$\rho Q_0 C_P T_0 - \rho Q C_P T = \frac{d}{dt}(\rho V C_P T)$$

$$Q_0 T_0 - QT = A_c d\frac{(hT)}{dt}$$

$$Q_0 T_0 - QT = A_c T\frac{dh}{dt} + A_c h\frac{dT}{dt}$$

From equation (i) by putting value of $A_c \frac{dh}{dt}$

$$Q_0 T_0 - QT = (Q_0 - Q)T + A_c h\frac{dT}{dt}$$

$$Q_0(T_0 - T) = A_c h\frac{dT}{dt}$$

2.9 (d)

Transfer function of thermometer

$$G(s) = \frac{K_p}{\tau_p s + 1} = \frac{1}{\frac{mC_p}{hA}s + 1}$$

Transfer function of mixing tank

$$G(B) = \frac{K_p}{\tau_p s + 1} = \frac{1}{\frac{V}{q}s + 1}$$

2.10 (a)

Step input response of first order system is given by

$$Y(t) = MK_p(1 - e^{-t/\tau_p})$$

For unit step $M = 1$

2.11 (b)

Following system are example of open loop second order under damped system

1. U-tube monometer
2. Externally mounted level indicator
3. Viscous damper
4. Pressure transducer

2.12 (d)

(a) To find out $\frac{Y_1(s)}{U_2(s)}$

By putting $U_1(s) = 0$

$$(\tau_1 s + 1)\, Y_1(s) = k_2\, U_2(s)$$

$$\frac{Y_1(s)}{U_2(s)} = \frac{k_2}{\tau_1 s + 1} \text{ so gain} = k_2$$

(b) $(\tau_2 s + 1)\, Y_2(s) = k_3\, U_2(s) + k_4\, Y_1(s)$

$$(\tau_2 s + 1)\, Y_2(s) = k_3\, U_2(s) + k_4\left(\frac{k_2}{\tau_1 s + 1}\right) U_2(s)$$

$$(\tau_2 s + 1)\, Y_2(s) = U_2(s)\left[k_3 + \frac{k_2 k_4}{(\tau_1 s + 1)}\right]$$

$$\frac{Y_2(s)}{U_2(s)} = \frac{k_3 + \frac{k_2 k_4}{\tau_1 s + 1}}{\tau_2 s + 1}$$

$$\text{Gain} = k_3 + k_2\, k_4$$

2.13 (c)

Output of process is given as

$$\bar{Y}(s) = \frac{KA\omega}{(\tau s + 1)(s^2 + \omega^2)} \text{ if } Y(0) = 0$$

Input signal $U(t) = A \sin \omega t$

$$\bar{U}(s) = \frac{A\omega}{s^2 + \omega^2}$$

$$\bar{Y}(s) = \frac{K\bar{U}(s)}{\tau s + 1}$$

$$\tau s\bar{Y}(s) + \bar{Y}(s) = K\bar{U}(s)$$

$$\tau\frac{dy(t)}{dt} + y(t) = Ku(t)$$

So option (c) is correct option.

2.14 (d)

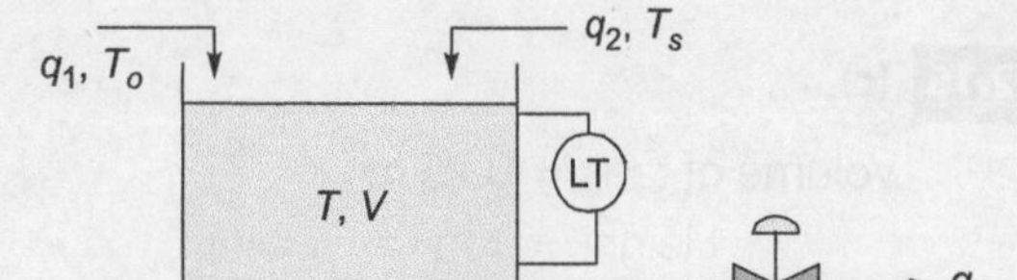

Total mass balance

(Rate of water in) – (Rate of water out) ± (Rate of water consumption or generation) = (Rate of accumulation)

$$\rho q_1 + \rho q_2 - \rho q_3 - \rho q_4 = \frac{d}{dt}(\rho V)$$

$$\frac{dV}{dt} = q_1 + q_2 - q_3 - q_4 \quad ...(A)$$

Since it is mixing tank so temperature inside tank is same as outlet stream temperature.

So from energy balance

$$\rho q_1 C_p T_o + \rho C_p q_2 T_s - \rho C_p q_3 T - \rho C_p q_4 T = \frac{d}{dt}(\rho V C_p T)$$

$$T\frac{dV}{dt} + V\frac{dT}{dt} = q_1 T_o + q_2 T_s - q_3 T - q_4 T$$

By putting the value of $\frac{dV}{dt}$ from equation (A)

$$\left[T(q_1 + q_2 - q_3 - q_4) + V\frac{dT}{dt}\right] = q_1 T_o + q_2 T_s - q_3 T - q_4 T$$

$$V\frac{dT}{dt} = (q_1)(T_o - T) + (q_2)(T_s - T)$$

2.15 (a)

(i) $\frac{-2.5(-4s+1)}{4s^2+4s+1}$

This transfer function has positive zero so it will show a inverse response so curve *e* is the response of it for unit step input.

(v) $\frac{4s+3}{2s+1}$

From initial value theorem for unit step input

$$\text{Initial value} = \lim_{s\to\infty} s\left(\frac{4s+3}{2s+1}\right)\cdot\frac{1}{s} = 2$$

So curve (b) is response of it.

(iv) $\frac{-0.1}{s}$

For unit step input

$$Y(s) = \frac{-0.1}{s}\cdot\frac{1}{s}$$

$$Y(s) = -0.1t \Rightarrow \text{curve (d)}$$

2.16 (c)

Volume of tank = 0.25 m^3

Height = 1 m

Flow rate, $Q = 0.05\ m^3/min$

Output flow rate is given by

$$F_{out} = 0.1h \qquad \text{...(i)}$$

Here, h = Height of water in tank

Inlet flow rate changes suddenly from 0.05 to 0.15 m^3/min

$$F_{out} = \frac{h}{R}$$

From equation (i),

$$R = 10$$

For liquid tank

$$K_p = R = 10$$

$$\tau_p = AR = 0.25 \times 10 = 2.5 \text{ min}$$

$$Q_{in} = 0.05 = F_{out} \text{ (at steady state)}$$

$$0.05 = 0.1 \times h_s$$

$$h_s = 0.5 \text{ (at steady state)}$$

Time required to over flow (means change in height from 0.5 to 1 m) is given by

$$h(t) - h_s = mK_p(1 - e^{-t/\tau_p})$$

(For first order system with step input)

$$0.5 = 0.1 \times 10\,(1 - e^{-t/2.5})$$

$$t = 1.732 \text{ minute}$$

2.17 (b)

To check positive gain we can put $s = 0$ and check the value whether it is positive or negative.

By putting $s = 0$ only option (b) and option (d) shows positive value.

For inverse response at least one zero must be positive:

$$\text{Option (d) is} \Rightarrow \frac{5}{s+1} - \frac{3}{2s+1}$$

$$= \frac{10s+5-3s-3}{(s+1)(2s+1)}$$

$$= \frac{7s+3}{(s+1)(2s+1)}$$

Zero is not positive so it will not show a inverse response

$$\text{Option (b) is} \Rightarrow \frac{2}{s+1} - \frac{5}{s+10}$$

$$= \frac{2s+20-5s-5}{(s+1)(s+10)}$$

$$= \frac{-3s+15}{(s+1)(s+10)}$$

For zero $-3s + 15 = 0$

$s = 5$ positive zero

So option (b) is correct option.

2.18 (d)

For 2nd order system characteristic equation can be written as

$\tau^2 s^2 + 2\xi\tau s + 1 = 0$

$$\text{Roots} = \frac{-2\xi\tau \pm \sqrt{(2\xi\tau)^2 - 4\tau^2}}{2\tau^2}$$

$$\Rightarrow \frac{-\xi}{\tau} \pm \frac{\sqrt{\xi^2 - 1}}{\tau}$$

For underdamped system $\xi < 1$

So we will get complex conjugate roots.

2.19 (d)

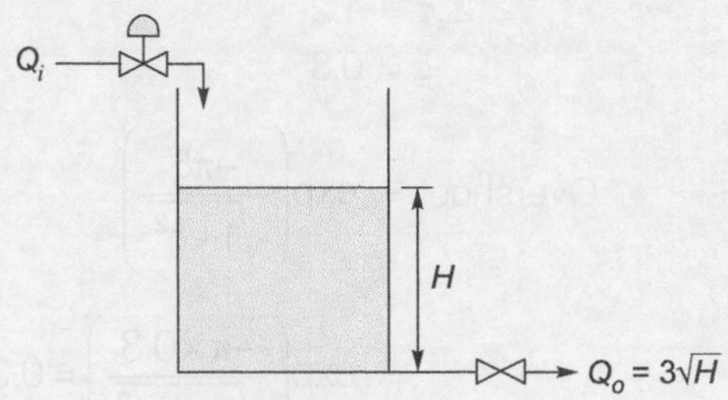

Area = 100 cm^2

At steady state point

$$Q_{i,s} = 18 \text{ cm}^3/\text{s}, \quad H_s = 36 \text{ cm}$$

For liquid tank system transfer function $G(s)$ is given by

$$G(s) = \frac{R}{ARs+1} = \frac{K_p}{\tau_p s+1}$$

Since output flow rate is not varies linearly with liquid height in tank. So gain K_p is given as

$$\frac{1}{K_p} = \left.\frac{dQ_o}{dH}\right|_{H=H_s}$$

$$\frac{1}{K_p} = \frac{3}{2\sqrt{H_s}} = \frac{3}{2\sqrt{36}} = \frac{1}{4}$$

$$K_p = 4$$

$$\tau = A\,K_p = 100 \times 4 = 400$$

So, $$\frac{\bar{H}(s)}{\bar{Q}_i(s)} = \frac{4}{400s+1}$$

2.20 (c)

Poles with positive real part means unstable system and atleast one positive zero means inverse response.

2.21 (d)

Temperature obtained by thermometer in time t can be calculated by

$$Y(t) = X(t)\,(1 - e^{-t/\tau})$$

$Y(t)$ = Change in output = $y(t) - y_{ss}$

$X(t)$ = Change in input = $x(t) - x_{ss}$

τ = Time constant of thermometer

$$(130 - 100) = (150 - 100)\,(1 - e^{-1/\tau})$$

$$\tau = 1.09 \text{ min}$$

2.22 (a)

Transfer function of 2nd order system is given below:

$$G(s) = \frac{K_p}{\tau^2 s^2 + 2\xi\tau s + 1} \quad \text{...(i)}$$

Period of oscillation = 1 second

Decay ratio = 0.25

Now we know that for 2nd order under damped system

$$\text{Decay ratio} = \exp\left(-\frac{2\pi\xi}{\sqrt{1-\xi^2}}\right) = 0.25$$

On solving, $\xi = 0.2154$

Period of oscillation is given by

$$T = \frac{2\pi\tau}{\sqrt{1-\xi^2}} = 1$$

$$\tau = 0.1554$$

On comparing option (a) with equation (i)

$$\tau = 0.155, \quad \xi = 0.216$$

So option (a) is correct option.

2.23 (c)

Response of first order thermometer can be found by

$$Y(t) = mK_p(1 - e^{-t/\tau_p})$$

$Y(t) = y(t) - y_{ss}$

m = Step input $= x(t) - x_{ss}$

$K_p = 1$ In case of thermometer

So, $y(t) - 0 = (100 - 0)\,(1 - e^{-t/10})$

$y(t) = 95°\text{C}$

$t = 29.95 \approx 30$ sec

2.24 (16.3)

$$G(s) = \frac{2}{s^2 + 2s + 4}$$

Comparing with 2nd order system transfer function

$$G(s) = \frac{K_p}{\tau^2 s^2 + 2\xi\tau s + 1}$$

$$K_p = \frac{1}{2}, \quad \tau = \frac{1}{2}$$

$$2\xi\tau = \frac{1}{2}, \quad \xi = 0.5$$

$$\%\text{ Overshoot} = \left[\exp\left(-\frac{\pi\xi}{\sqrt{1-\xi^2}}\right)\right] \times 100$$

$$= \left[\exp\left(-\frac{\pi \times 0.5}{\sqrt{1-0.5^2}}\right)\right] \times 100$$

$$= 16.3\%$$

2.25 (a)

Initial slope of response of any transfer function can be found by following method.

$$\text{Initial slope} = \lim_{s\to\infty} s^2 Y(s)$$

From option (a)

$$G(s) = \frac{1}{s+1} - \frac{2}{s+4}$$

Unit step input

$$\Rightarrow \quad X(s) = \frac{1}{s}$$

So, $\quad Y(s) = G(s)X(s)$

$$Y(s) = \left[\frac{1}{s+1} - \frac{2}{s+4}\right]\frac{1}{s}$$

$$\text{Initial slope} = \lim_{s\to\infty} s^2 Y(s)$$

$$= \lim_{s\to\infty} s^2\left(\frac{1}{s+1} - \frac{2}{s+4}\right)\frac{1}{s}$$

$$= \lim_{s\to\infty} s^2\left[\frac{1}{1+\frac{1}{s}} - \frac{2}{1+\frac{4}{s}}\right] = 1 - 2$$

$= -1$ (Negative initial slope)

2.26 (c)

Following systems shows 2nd order response inherently:

1. U tube manometer
2. Pneumatic valve
3. Externally mounted level indicator
4. Spring-mass damper or viscous damper
5. Pressure transducer

2.27 (d)

The transfer function of the system has at least one positive zero.

2.29 (0.37)

$$G(s) = \frac{1}{4s^2 + 1.2s + 1}$$

On comparing with standard 2nd order system

$$G(s) = \frac{1}{\tau^2 s^2 + 2\xi\tau s + 1}$$

$$\tau = 2$$

$$2\xi\tau = 1.2$$

$$\xi = 0.3$$

$$\text{Overshoot} = \exp\left(\frac{-\pi\xi}{\sqrt{1-\xi^2}}\right)$$

$$= \exp\left(\frac{-\pi\times 0.3}{\sqrt{1-0.3^2}}\right) = 0.37$$

2.30 (c)

Complex conjugate poles is given as $-\frac{1}{10} \pm j\frac{2}{15}$

Transfer function of a 2nd order system is written below:

$$G(s) = \frac{k_p}{\tau^2 s^2 + 2\xi\tau s + 1}$$

$$\text{Poles} = \frac{-2\xi\tau \pm \sqrt{(2\xi\tau)^2 - 4\tau^2}}{2\tau^2}$$

$$= -\frac{\xi}{\tau} \pm \sqrt{\frac{\xi^2 - 1}{\tau}}$$

On comparing with poles we find $\xi = 0.6$

$$\text{Delay ratio} = \exp\left(-\frac{2\pi\xi}{\sqrt{1-\xi^2}}\right)$$

$$= \exp\left(-\frac{2\pi\times 0.6}{\sqrt{1-0.36}}\right) = 9\times 10^{-3}$$

2.31 (d)

Pure dead time always has AR = 1, phase angle (ϕ) = –90° for pure capacitive system. Inverse response has atleast one positive zero and option (2) of column B matches with this criteria. First order system AR is define as $\frac{K_1}{\sqrt{1+\tau^2\omega^2}}$.

For all values of τ and ω with $K_1 = 1$, AR lies between 0 and 1.

So option (d) is correct option.

2.32 (a)

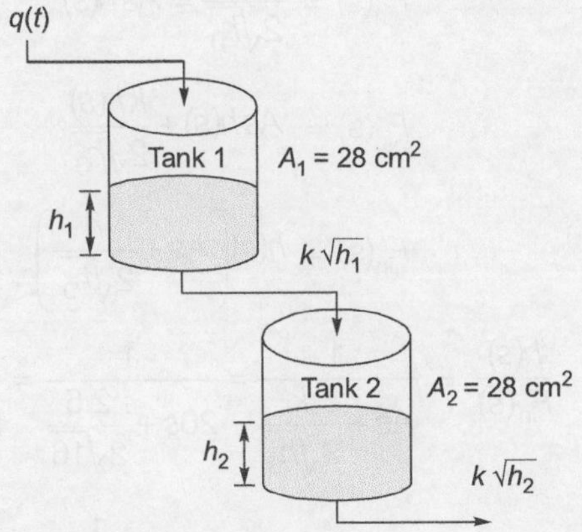

For non-interacting tanks

$$\frac{H_2(s)}{Q(s)} = \frac{1}{(\tau_1 s+1)}\cdot\frac{R_2}{(\tau_2 s+1)}$$

For a tank with

$$q = k\sqrt{h}$$

$$R = \frac{2(h_s)^{1/2}}{k}$$

and $\tau = A \times R$

At steady state

$$q_{ss} = 16 \text{ cm}^3/\text{s (Given)}$$

$$q_{ss} = k\sqrt{h_s}$$

$$16 = 4\sqrt{h_s}$$

$$h_s = 16 \text{ cm}$$

$$R = \frac{2\sqrt{16}}{4} = 2 \text{ s/cm}^2$$

$$R_1 = R_2 = R$$

So, $\tau_1 = A_1R_1,\ \tau_2 = A_2R_2$

$$\tau_1 = \tau_2 = 56$$

$$\frac{H_2(s)}{Q(s)} = \frac{R_2}{(\tau_1 s+1)(\tau_2 s+1)}$$

$$\frac{H_2(s)}{Q(s)} = \frac{2}{(56s+1)(56s+1)}$$

$$\frac{H_2(s)}{Q(s)} = \frac{2}{(56s+1)^2}$$

2.33 (b)

Response (R) shows a inverse response so its transfer function must contain a positive zero. Transfer function has

$$\left\{\frac{K(\xi s+1)}{(\tau_1 s+1)(\tau_2 s+1)}\quad \begin{matrix} K>0,\ \xi<0 \\ \tau_1>0,\ \tau_2>0\end{matrix}\right\} \text{ a positive zero.}$$

So, R - (IV)

Response (S) has a positive initial value, Initial value can be calculated by $\lim_{s\to\infty}[S\cdot Y(s)]$ (initial value theorem) by checking initial value theorem the transfer function only which has positive initial value is $\frac{K(\xi s+1)}{(\tau s+1)}$ $K>0,\ \xi>\tau>0$

S - (II)

So correct option is (b).

Response (P) and response (Q) has same initial value that is equal to zero, both have positive final value and both have positive initial slope of response so better to check response (R) and response (S) first.

2.34 (102.5)

$$G(s) = \frac{Y(s)}{X(s)} = \frac{20}{90000s^2+240s+1}$$

$$x_s = 0.4$$

and $y_s = 100$

$$x(t) = x - x_s = 0.1$$

Ultimate value of $y(t)$

$$= \lim_{s\to 0}[s.y(s)]$$

$$= \frac{0.1}{s}.s\left[\frac{20}{90000s^2+240s+1}\right]$$

Ultimate value $(y(t)) = 2$

$$\text{Overshoot} = e^{-\frac{\pi\xi}{\sqrt{1-\xi^2}}}$$

where $2\tau\xi = 240$

$$\tau = \sqrt{90000} = 300$$

$$\xi = \frac{240}{600} = 0.4$$

$$\text{Overshoot} = e^{\frac{-\pi\times 0.4}{\sqrt{1-0.4^2}}} = 0.2538$$

$$y(t)_{max} - y_s = y(t)|_{ultimate}(1+\text{overshoot})$$

$$y(t)_{max} - y_s = 2[1+0.2538]$$

$$y(t)|_{max} - 100 = 2.5076$$

$$y(t)|_{max} = 102.5076$$

2.35 (c)

Applying law of conservation of mass, we get

$$\rho F_{in} - \rho F_{out} = \rho A\frac{dh}{dt}$$

$$F_{in} - k\sqrt{h} = A\frac{dh}{dt}$$

$$F_{in} - k\left(\frac{\sqrt{h_0}}{2} + \frac{h}{2\sqrt{h_0}}\right) = A\frac{dh}{dt} \quad \text{...(i)}$$

$$F_{in} - \frac{k\sqrt{h_0}}{2} - \frac{kh}{2\sqrt{h_0}} = A\frac{dh}{dt}$$

$$F_{in} - \frac{k}{2\sqrt{h_0}}.h = A\frac{dh}{dt} + \frac{k\sqrt{h_0}}{2}$$

$$f(x) = f(x_0) + (x - x_0)\left.\frac{\partial t}{\partial x}\right|_{x_0}$$

$$\sqrt{h} = \sqrt{h_0} + (h - h_0)\times\frac{1}{2\sqrt{h_0}}$$

$$\sqrt{h} = \sqrt{h_0} + \frac{h}{2\sqrt{h_0}} - \frac{h_0}{2\sqrt{h_0}}$$

$$\sqrt{h} = \sqrt{h_0} + \frac{h}{2\sqrt{h_0}} - \frac{h_0\times\sqrt{h_0}}{2\sqrt{h_0}}$$

$$\sqrt{h} = \sqrt{h_0} - \frac{\sqrt{h_0}}{2} + \frac{h}{2\sqrt{h_0}}$$

$$\sqrt{h} = \frac{\sqrt{h_0}}{2} + \frac{h}{2\sqrt{h_0}}$$

At S.S.

$$(F_{in})_{SS} - k = \left(\frac{\sqrt{h_0}}{2} + \frac{h}{2\sqrt{h_0}}\right) = A\frac{dh_s}{dt} = 0 \quad \text{...(ii)}$$

Equation (i) – (ii)

$$[F_{in} - (F_{in})_{SS}] - \frac{k(h - h_s)}{2\sqrt{h_0}} = A\frac{d(h - h_s)}{dt}$$

$$\bar{F}_{in} = \frac{k\bar{h}}{2\sqrt{h_0}} = A\frac{d\bar{h}}{dt}$$

As S.S, $F_{in} = F_{out} = 10\ \text{cm}^3/\text{sec}$

$h_s = 16\ \text{cm} = h_0 \qquad \sqrt{h_0} = 4$

At S.S, $F_{out} = 10$

$$k\sqrt{h_s} = 10$$

$$k = 2.5$$

$$F_{in}(s) = \frac{kh(s)}{2\sqrt{h_0}} = Ash(s)$$

$$F_{in}(s) = Ash(s) + \frac{kh(s)}{2\sqrt{h_0}}$$

$$F_{in}(s) = h(s)\left(As + \frac{h}{2\sqrt{h_0}}\right)$$

$$\frac{h(s)}{F_{in}(s)} = \frac{1}{As + \frac{k}{2\sqrt{h_0}}} = \frac{1}{20s + \frac{2.5}{2\sqrt{16}}} = \frac{1}{20s + \frac{2.5}{8}}$$

$$\frac{h(s)}{F_{in}(s)} = \frac{1}{20s + 0.3125} = \frac{\frac{1}{0.3125}}{\frac{20}{0.3125s} + 1} = \frac{k_p}{\tau_p s + 1}$$

$$\tau_p = \frac{20}{0.3125} = 64$$

2.38 (7.1)

Given:

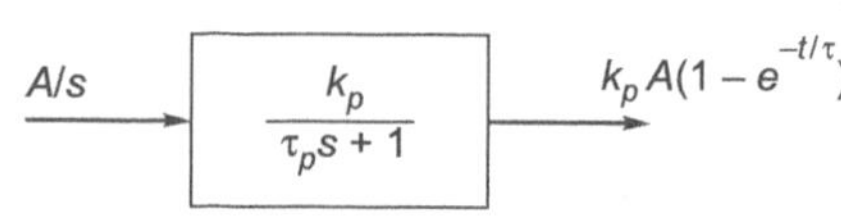

$$G_m(s) = \frac{1.05}{2s + 1}\exp(-s)$$

$$\frac{y(s)}{x(s)} = \frac{1.05}{(2s + 1)}\exp(-s)$$

$$y(s) = \frac{1}{s}.\frac{1.05}{(2s + 1)}\exp(-s)$$

$$y(t) = k_p A[1 - e^{-(t-1)/\tau}]$$

$$1 = 1.05 \times 1 \times [1 - e^{-(t-1)/2}]$$

$$\frac{1}{1.05} = 1 - e^{-(t-1)/2}$$

$$0.952 = 1 - e^{-(t-1)/2}$$

$$e^{-(t-1)/2} = 0.048$$

$$\frac{-(t-1)}{2} = \ln 0.048$$

$$\frac{-(t-1)}{2} = -3.0365$$

$$t - 1 = 6.073$$

$$t = 1 + 6.073 = 7.073$$

■■■■

3 Close Loop Transfer Functions

3.1 The block diagram of an integrating level process is given below. For unit step change in the set point $\Delta h_{set} = 1$ with $\Delta d = 0$, the offset exhibited by the system is

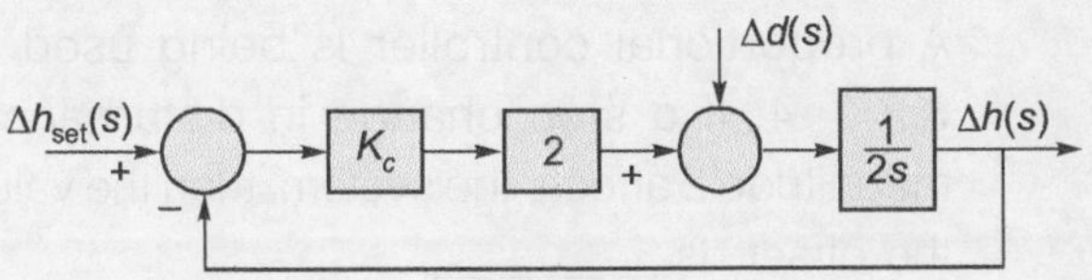

(a) $\dfrac{K_c}{1+K_c}$ (b) $\dfrac{1}{1+K_c}$

(c) 0 (d) $\dfrac{2K_c}{1+2K_c}$

[2001 : 2 M]

3.2 The closed loop poles of a stable second order system could be

(a) both real and positive

(b) complex conjugate with positive real parts

(c) both real and negative

(d) one real positive and the other real negative

[2002 : 1 M]

3.3 For the block diagram shown below, the characteristic equation is

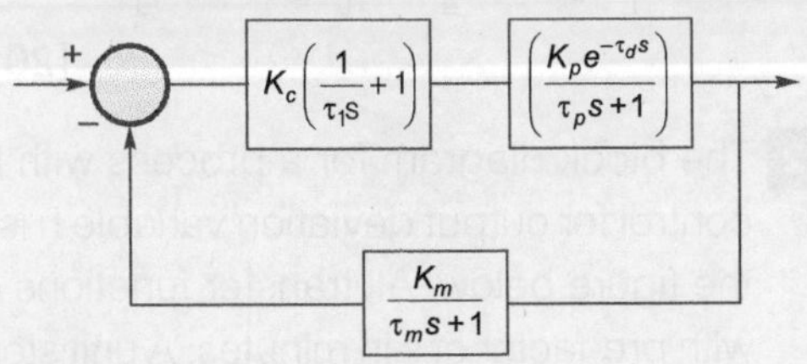

(a) $\tau_I s(\tau_p s + 1) + K_c K_p(\tau_I s + 1)e^{-\tau_d s} = 0$

(b) $(\tau_m s + 1)(\tau_p s + 1) + K_m K_p e^{-\tau_d s} = 0$

(c) $[\tau_I s(\tau_p s + 1)(\tau_m s + 1)] + K_p K_c K_m(\tau_I s + 1)e^{-\tau_d s} = 0$

(d) $(\tau_m s + 1)(\tau_p s + 1) + K_c K_p K_m e^{-\tau_d s} = 0$

[2004 : 2 M]

Common Data for Questions (3.4 & 3.5) :

The block diagram of a closed loop control system is shown in the figure below. Y is the controlled variable, D is disturbance, Y_{sp} is the set point, G_1, G_2, and G_3, are transfer functions, and K_c is the proportional controller.

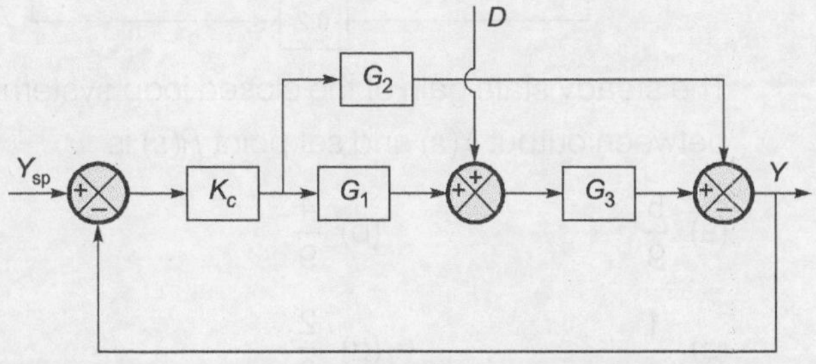

3.4 The closed loop transfer function $\dfrac{Y(s)}{D(s)}$ is given by

(a) $\dfrac{G_3G_1}{1+(G_1G_3+G_2)K_c}$ (b) $\dfrac{G_1}{1+(G_1G_3+G_2)K_c}$

(c) $\dfrac{G_3}{1+(G_1+G_2)G_3K_c}$ (d) $\dfrac{G_3}{1+(G_1G_3+G_2)K_c}$

[2006 : 2 M]

3.5 Let $G_1(s) = 1$ and $G_2(s) = G_3(s) = \dfrac{1}{(s+1)}$. A step change of magnitude M is made in the set point. The steady state offset of the closed loop response Y is

(a) $\dfrac{M}{1+2K_c}$ (b) $\dfrac{M}{1+K_c}$

(c) $\dfrac{M(K_c-1)}{1+2K_c}$ (d) Zero

[2006 : 2 M]

3.6 The block diagram for a control system is shown below :

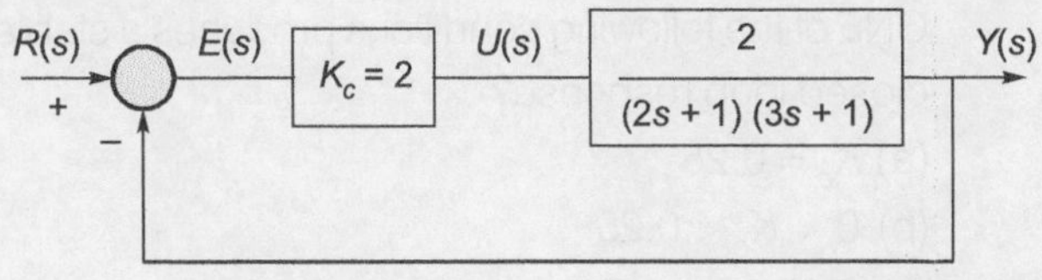

For a unit step change in the set point, $R(s)$, the steady state offset in the output $Y(s)$ is

(a) 0.2 (b) 0.3

(c) 0.4 (d) 0.5

[2009 : 2 M]

3.7 A block diagram for a control system is shown below:

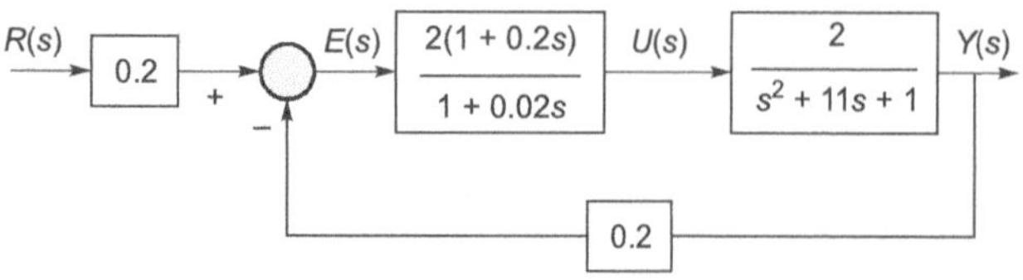

The steady state gain of the closed loop system, between output $Y(s)$ and set point $R(s)$ is

(a) $\frac{5}{9}$ (b) $\frac{4}{9}$

(c) $\frac{1}{3}$ (d) $\frac{2}{9}$

[2010 : 2 M]

3.8 The block diagram of a system with a proportional controller is shown below:

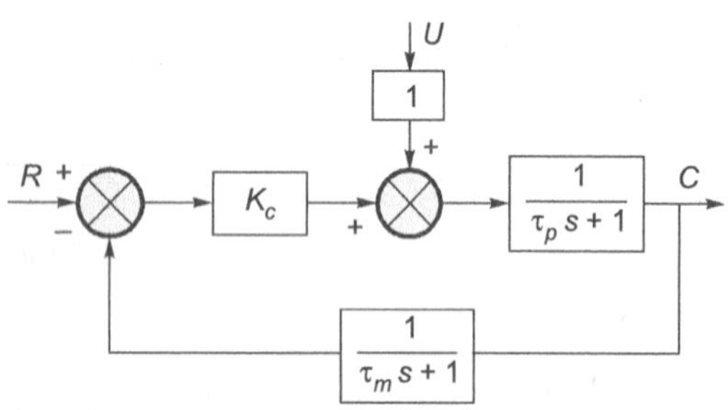

A unit step input is introduced in the set point. The value of K_c to provide a critically damped response for $U = 0$, $\tau_p = 8$ and $\tau_m = 1$ is

(a) 3.34 (b) 2.58

(c) 1.53 (d) 1.12

[2012 : 2 M]

3.9 A process with transfer function, $G_p = \frac{2}{s-1}$ is to be controlled by a feedback proportional controller with a gain K_c. If the transfer functions of all other elements in the control loop are unity, then which ONE of the following conditions produces a stable closed loop response?

(a) $K_c = 0.25$

(b) $0 < K_c < 0.25$

(c) $0.25 < K_c < 0.5$

(d) $K_c > 0.5$

[2014 : 2 M]

3.10 Consider the following block diagram for a closed-loop feedback control system

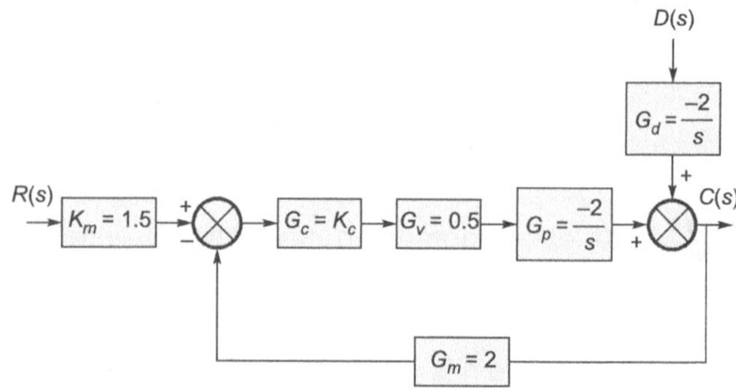

A proportional controller is being used with $K_c = -4$. If a step change in disturbance of magnitude 2 affects the system, then the value of the offset is ________.

[2014 : 2 M]

3.11 The transfer function for the disturbance response in an open-loop process is given by $G_d^{open}(s)$. The corresponding transfer function for the disturbance response in a closed-loop feedback control system with proportional controller is given by $G_d^{closed}(s)$. Select the option that is ALWAYS correct {$O[G(s)]$ represents order of transfer function $G(s)$}:

(a) $O\left[G_d^{open}(s)\right] = O\left[G_d^{closed}(s)\right]$

(b) $O\left[G_d^{open}(s)\right] \neq O\left[G_d^{closed}(s)\right]$

(c) $O\left[G_d^{open}(s)\right] \geq O\left[G_d^{closed}(s)\right]$

(d) $O\left[G_d^{open}(s)\right] \leq O\left[G_d^{closed}(s)\right]$

[2015 : 1 M]

3.12 The block diagram for a process with feedback control for output deviation variable h is shown in the figure below. All transfer functions are given with pre-factor of s in minutes. A unit step change is made in the set-point at $t = 0$. The time required for h to reach 50% of its ultimate value, in minutes (up to two decimal places), is ______ .

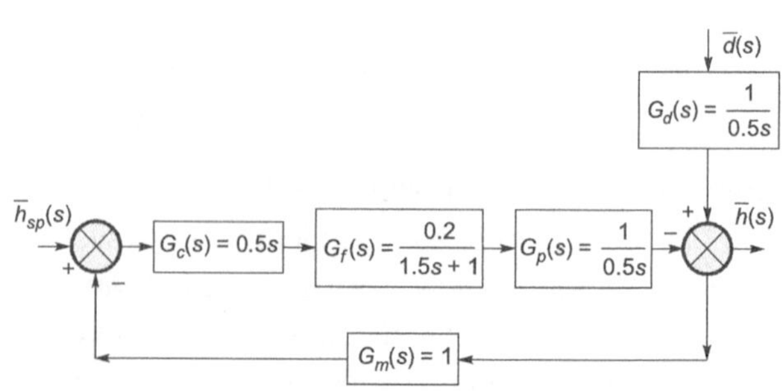

[2015 : 2 M]

3.13 A first-order process having a transfer function,

$$G_p = \frac{2}{7s+1}$$ is controlled by a proportional controller with gain of 3.2. The process time constant is in minutes. Addition of the integral control action with an integral time constant of 5 minutes leads to increase in

(a) offset
(b) speed of response
(c) order of the closed loop system
(d) proportional band

[2018 : 1 M]

3.14 For a closed-loop system, consider the following transfer functions :
process $G_p(s)$, controller $G_c(s)$, measuring device $G_m(s)$, and final control element $G_f(s)$

$$G_p(s) = \frac{2}{7s+1};\ G_c(s) = 2;\ G_m(s) = 1;\ G_f(s) = 1$$

The offset in the closed loop response due to a unit step change introduced in the set point of the output variable is ________.

[2018 : 2 M]

3.15 Consider the following closed loop system with G_p and G_c as the transfer functions of the process and the controller, respectively.

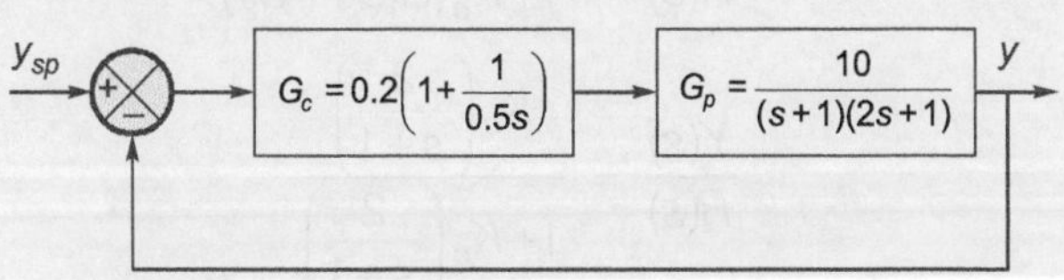

For a unit step change in the set point (y_{sp}), the change in the value of the response (y) at steady state is ________ (round off to 1 decimal place).

[2020 : 1 M]

3.16 It is required to control the volume of the contents in the jacketed reactor shown in the figure.

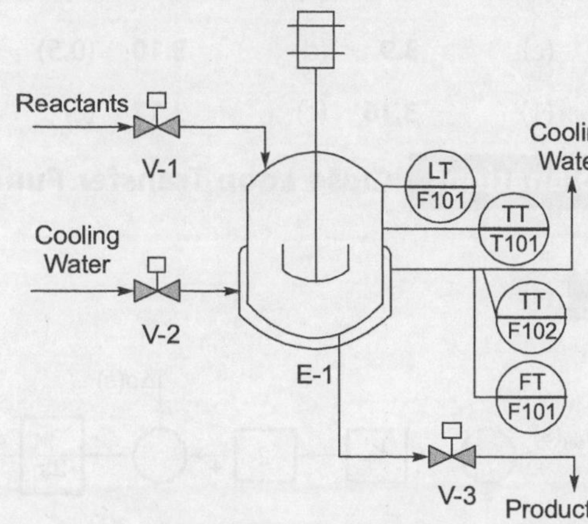

Which one of the following schemes can be used for feedback control?

(a) Measure L101 and manipulate valve V-2
(b) Measure T101 and manipulate valve V-1
(c) Measure L101 and manipulate valve V-3
(d) Measure F101 and manipulate valve V-1

[2021 : 2 M]

3.17 For the block diagram shown in the figure, the correct expression for the transfer function

$$G_d = \frac{y_1(s)}{d(s)}$$ is

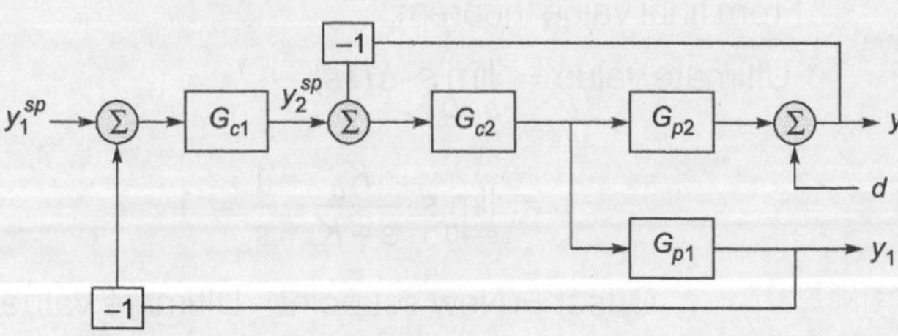

(a) $\dfrac{-G_{p1}G_{c2}}{(1+G_{c1}G_{c2}G_{p1})(1+G_{c2}G_{p2})}$

(b) $\dfrac{-G_{p1}G_{c2}}{1+G_{c2}G_{p2}+G_{c1}G_{c2}G_{p1}G_{p2}}$

(c) $\dfrac{-G_{p1}G_{c2}}{1+G_{c2}G_{p2}+G_{c1}G_{c2}G_{p1}}$

(d) $\dfrac{1}{1+G_{c2}G_{p2}+G_{c1}G_{c2}G_{p1}G_{p2}}$

[2024 : 2 M]

■■■■

Answers Close Loop Transfer Functions

3.1	(c)	**3.2**	(c)	**3.3**	(c)	**3.4**	(d)	**3.5**	(a)	**3.6**	(a)	**3.7**	(b)
3.8	(c)	**3.9**	(d)	**3.10**	(0.5)	**3.11**	(d)	**3.12**	(0.866)	**3.13**	(c)	**3.14**	(0.2)
3.15	(1)	**3.16**	(c)	**3.17**	(c)								

Explanations Close Loop Transfer Functions

3.1 (c)

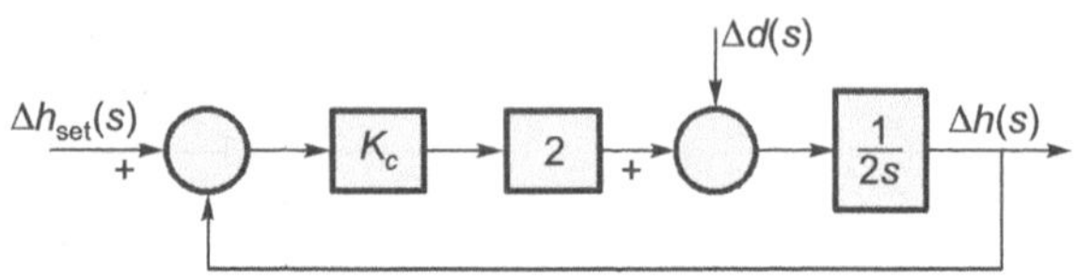

Close loop transfer function is given as

$$\frac{\Delta h(s)}{\Delta h_{set}(s)} = G(s) = \frac{\frac{2K_c}{2s}}{1+\frac{2K_c}{2s}} = \frac{K_c}{s+K_c}$$

Unit step change is given in set point.

So, $\Delta h_{set}(s) = \frac{1}{s}$

$$\Delta h(s) = \frac{K_c}{s+K_c}\cdot\frac{1}{s}$$

From final value theorem:

$$\text{Ultimate value} = \lim_{s\to 0} s\cdot\Delta h(s)$$

$$= \lim_{s\to 0} s\cdot\frac{K_c}{s+K_c}\cdot\frac{1}{s} = 1$$

Offset = New set point – Ultimate value

$\Rightarrow$ 1 – 1 = 0

Note: Pure capacitive process with measuring device transfer function $G_M = 1$ and no change in load (servo problem) gives zero offset with *P*-only controller.

3.3 (c)

Characteristic equation of a close loop is given by

$$1 + G_cG_pG_fG_m = 0$$

$$= 1+K_c\left(\frac{1}{\tau_1 s}+1\right)\left(\frac{K_pe^{-\tau_D s}}{\tau_p s+1}\right)\left(\frac{K_m}{\tau_m s+1}\right) = 0$$

$$(\tau_p s + 1)(\tau_m s + 1)\tau_1 s + K_cK_pK_m(\tau_1 s + 1)e^{-\tau_D s} = 0$$

3.4 (d)

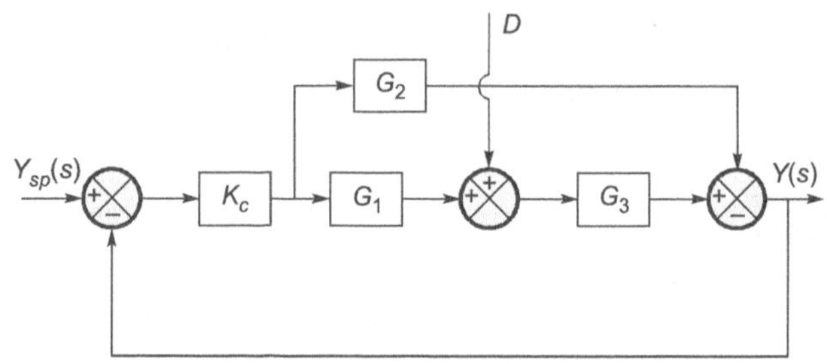

$$\frac{Y(s)}{D(s)} = ?$$

From block diagram,

$$-Y(s)\,[K_c\,(G_1G_3 + G_2)] + D(s)\,G_3 = Y(s)$$

$$\frac{Y(s)}{D(s)} = \frac{G_3}{1+(G_1G_3+G_2)K_c}$$

3.5 (a)

$$\frac{Y(s)}{R(s)} = \frac{K_c(G_1G_3+G_2)}{1+K_c(G_1G_3+G_2)}$$

$$\frac{Y(s)}{R(s)} = \frac{K_c\left(\frac{2}{s+1}\right)}{1+K_c\left(\frac{2}{s+1}\right)}$$

$$\frac{Y(s)}{R(s)} = \frac{2K_c}{s+1+2K_c}$$

$$R(s) = \frac{M}{s}$$

So, $$Y(s) = \frac{2K_c}{s+1+2K_c}\cdot\frac{M}{s}$$

From final value theorem

$$\text{Ultimate value} = \lim_{s\to 0} sY(s)$$

$$= \lim_{s\to 0}\cdot s\frac{2K_c}{s+1+2K_c}\cdot\frac{M}{s} = \frac{2MK_c}{1+2K_c}$$

Offset = New set point – ultimate value

$$= M-\frac{2MK_c}{1+2K_c} = \frac{M}{1+2K_c}$$

3.6 (a)

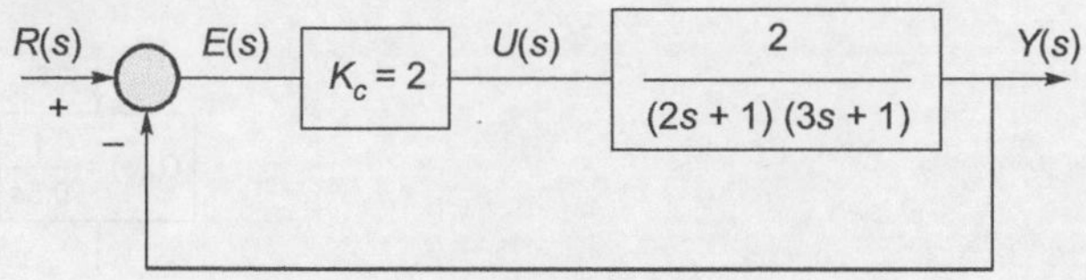

Close loop characteristic equation of above block diagram

$$G(s)_{\text{close loop}} = \frac{\frac{2\times 2}{(2s+1)(3s+1)}}{1+\frac{4}{(2s+1)(3s+1)}}$$

$$G(s)_{\text{close loop}} = \frac{4}{(2s+1)(3s+1)+4}$$

For unit step change in set point

$$R(s) = \frac{1}{s}$$

$$Y(s) = G(s)\,R(s)$$

$$Y(s) = \frac{4}{(2s+1)(3s+1)+4}\cdot\frac{1}{s}$$

From final value theorem

$$\text{Ultimate value} = \lim_{s\to 0} s\cdot Y(s)$$

$$= \lim_{s\to 0} s\frac{4}{(2s+1)(3s+1)+4}\cdot\frac{1}{s}$$

$$= \frac{4}{5} = 0.8$$

$$\text{Offset} = \text{New set point} - \text{Ultimate value}$$

$$= 1 - 0.8 = 0.2$$

3.7 (b)

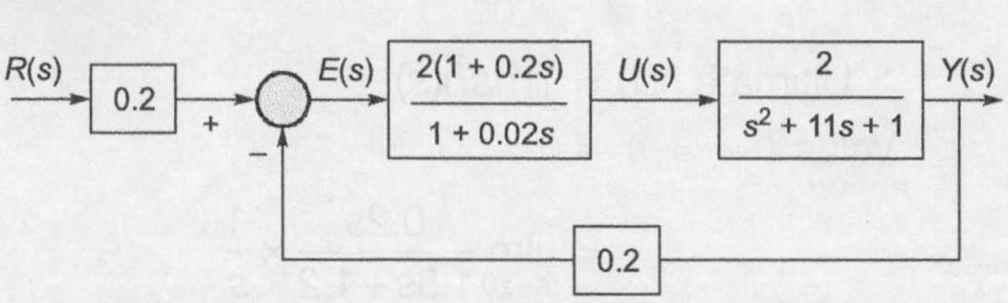

Close loop transfer function of system is given by

$$\frac{Y(s)}{R(s)} = \left[\frac{\frac{2(1+0.2s)}{(1+0.02s)}\left(\frac{2}{s^2+11s+1}\right)}{1+0.2\left[\frac{2(1+0.2s)}{1+0.02s}\right]\left(\frac{2}{s^2+11s+1}\right)}\right]\times 0.2$$

$$G(s) = \frac{0.8(1+0.2s)}{(1+0.02s)(s^2+11s+1)+0.8(1+0.2s)}$$

For steady state gain ⇒ put $s = 0$ in transfer function

$$\text{So steady state gain} = \frac{0.8}{1+0.8} = \frac{4}{9}$$

3.8 (c)

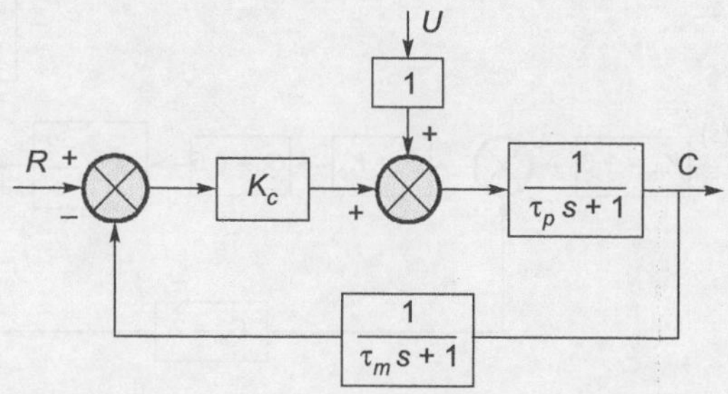

Close loop transfer function for above diagram and data

$$\frac{C(s)}{R(s)} = \frac{\frac{K_c}{\tau_p s+1}}{1+\frac{K_c}{(\tau_p s+1)}\frac{1}{(\tau_m s+1)}}$$

$$\tau_p = 8,\quad \tau_m = 1 \text{ (given)}$$

On putting value,

$$\frac{C(s)}{R(s)} = \frac{K_c(s+1)}{(8s+1)(s+1)+K_c}$$

$$= \frac{K_c(s+1)}{8s^2+9s+(1+K_c)}$$

For critically damped system ξ must be equal to 1. On comparing with standard 2nd order system

$$\tau_I = \sqrt{\frac{8}{1+K_c}}$$

$$2\xi\tau_I = \frac{9}{(1+K_c)}$$

$$\xi = \frac{9}{2\sqrt{8(1+K_c)}} = 1$$

$$K_c = 1.53$$

3.9 (d)

Characteristic equation of feed back controller system is given by

$$1 + G_cG_pG_fG_m = 0$$

$$1+K_c\left(\frac{2}{s-1}\right) = 0$$

$$s - 1 + 2K_c = 0$$

For stable system all poles must have negative real parts so

$$\Rightarrow \quad -1 + 2K_c > 0$$

$$K_c > 0.5$$

3.10 (0.5)

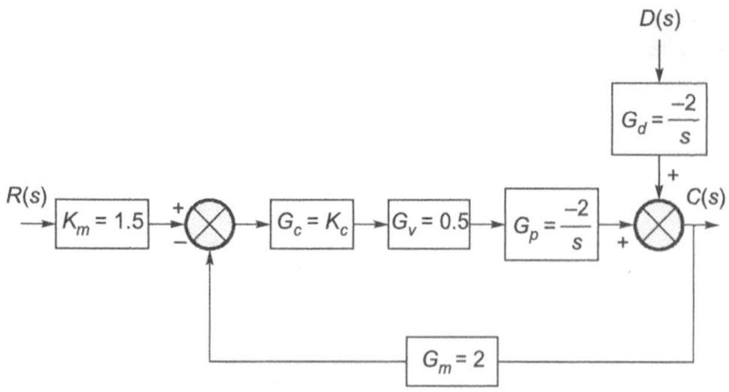

$$K_c = -4 \text{ (given)}$$

Step change of magnitude 2 in disturbance affects the system so

$$R(s) = 0$$

Now corresponding close loop transfer function is written below

$$G(s) = +\frac{G_d(s)}{1+G_cG_vC_pG_m}$$

$$G(s) = +\frac{-\frac{2}{s}}{1+(-4)(0.5)\left(-\frac{2}{s}\right)\times 2}$$

$$G(s) = -\frac{\frac{2}{s}}{1+\frac{8}{s}} = \frac{-2}{s+8}$$

Now, $\quad D(s) = \frac{2}{s}$ (given)

So, $\quad Y(s) = -\frac{2}{s+8}\cdot\frac{2}{s}$

From final value theorem

$$\text{Final value} = \lim_{s\to 0} sY(s)$$

$$= s\cdot\frac{-2}{s+8}\cdot\frac{2}{s} = -0.5$$

Offset = Steady state error

= New set point – Ultimate value

= 0 – (–0.5) = 0.5

3.11 (d)

Order of close loop is always greater than or equal to open loop.

If (G_m = 1) measuring element has unity transfer function then both becomes equal.

3.12 (0.866)

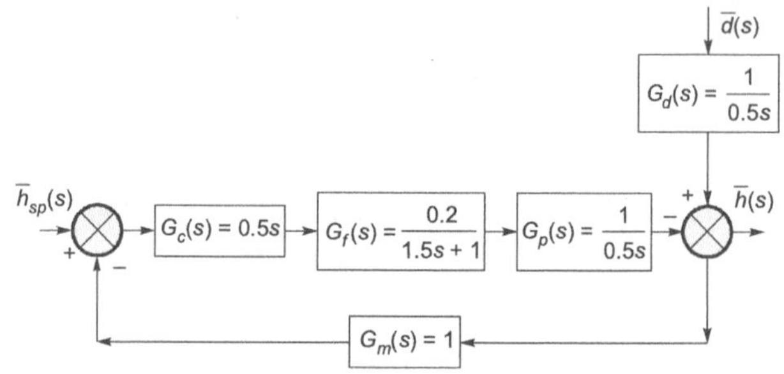

Close loop transfer function

$$\bar{h}(s) = \frac{G_cG_fG_p}{1+G_cG_fG_pG_m}h_{sp}(s) + \frac{G_d(s)\bar{d}(s)}{1+G_cG_pG_fG_m}$$

For unit step change in set point

$$h_{sp}(s) = \frac{1}{s} \text{ and } \bar{d}(s) = 0$$

$$\bar{h}(s) = \frac{0.5s\left(\frac{0.2}{1.5s+1}\right)\times\frac{1}{0.5s}}{1+0.5s\left(\frac{0.2}{1.5s+1}\right)\times\frac{1}{0.5s}}\times\frac{1}{s}$$

$$\bar{h}(s) = \frac{0.2}{1.5s+1.2}\cdot\frac{1}{s}$$

$$\bar{h}(s) = \frac{-\frac{1}{4}}{1.5s+1.2}+\frac{\frac{1}{6}}{s}$$

$$h(t) = \frac{1}{6}-\frac{1}{6}e^{-0.8t}$$

$$h(t) = \frac{1}{6}\left[1-e^{-0.8t}\right] \quad \text{...(A)}$$

or

$$\text{Ultimate } h(t) = \lim s\bar{h}(s)$$

Value $t \to \infty$, $s \to 0$

$$= \lim_{s\to 0}\frac{0.2s}{1.5s+1.2}\times\frac{1}{s}$$

$$= \frac{0.2}{1.2} = \frac{1}{6}$$

Ultimate value is value at steady state at $t = \infty$.

So from equation (A) ultimate value is $\frac{1}{6}$.

Time required to reach 50% of ultimate value is given by

$$\frac{1}{6}\times 0.5 = \frac{1}{6}\left[1-e^{-0.8t}\right]$$

$$t = 0.866 \text{ minute}$$

3.13 (c)

Addition of the integral control action leads the following conclusion:

(i) Integral action eliminates the offset.

(ii) Increases the order of close loop.

(iii) Speed of response becomes slower.

3.14 (0.2)

Close loop transfer function

$$G(s) = \frac{G_c G_p G_f}{1 + G_c G_p G_f G_m}$$

$$G(s) = \frac{2 \times \frac{2}{7s+1}}{1 + \frac{2 \times 2}{7s+1}} = \frac{4}{7s+5}$$

Unit step change is introduced in set point

$$X(s) = \frac{1}{s}$$

$$Y(s) = G(s)\,X(s)$$

$$Y(s) = \frac{4}{7s+5} \times \frac{1}{s}$$

From final value theorem

$$Y(t)\big|_{t\to\infty} = \lim_{s\to 0} sY(s)$$

$$Y(t)\big|_{t\to\infty} = \lim_{s\to 0} s \cdot \frac{4}{7s+5} \cdot \frac{1}{s} = \frac{4}{5} = 0.8$$

Offset = New set point – Steady state value

= 1 – 0.8 = 0.2

3.15 (1)

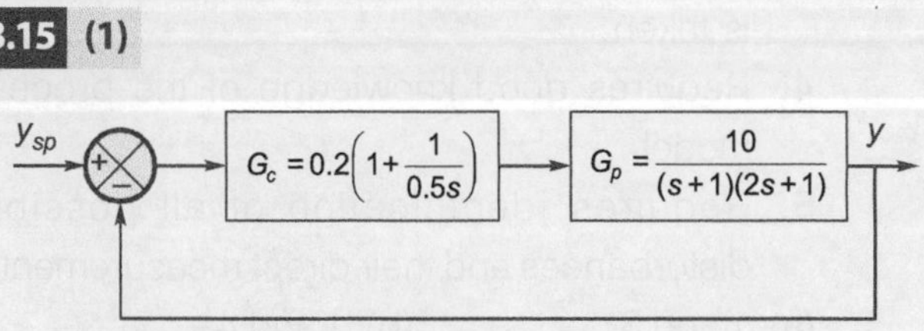

Overall transfer function of above loop is

$$G(s) = \frac{G_c G_p}{1 + G_c G_p}$$

$$= \frac{0.2\left(1+\frac{1}{0.5s}\right)\left(\frac{10}{(s+1)(2s+1)}\right)}{1+0.2\left(1+\frac{1}{0.5s}\right)\left(\frac{10}{(s+1)(2s+1)}\right)}$$

$$= \frac{2\left(\frac{s+2}{s}\right)\left(\frac{1}{(s+1)(2s+1)}\right)}{1+2\left(\frac{s+2}{s}\right)\left(\frac{1}{(s+1)(2s+1)}\right)}$$

$$G(s) = \frac{2(s+2)}{s(s+1)(2s+1)+2(s+2)}$$

For unit step change in set point

$$Y(s) = \frac{2(s+2)}{s(s+1)(2s+1)+2(s+2)}\,Y_{sp}(s)$$

$$\left(\text{Here } Y_{sp}(s) = \frac{1}{s}\right)$$

From final value theorem

$$Y(t)\big|_{t\to\infty} = \lim_{s\to 0} sY(s)$$

$$\therefore\ Y(t)\big|_{t\to\infty}\ \lim s \to 0 = 0\left[s\frac{2(s+2)}{s(s+1)(2s+1)+2(s+2)}\frac{1}{s}\right]$$

$$Y(s) = 1$$

3.17 (c)

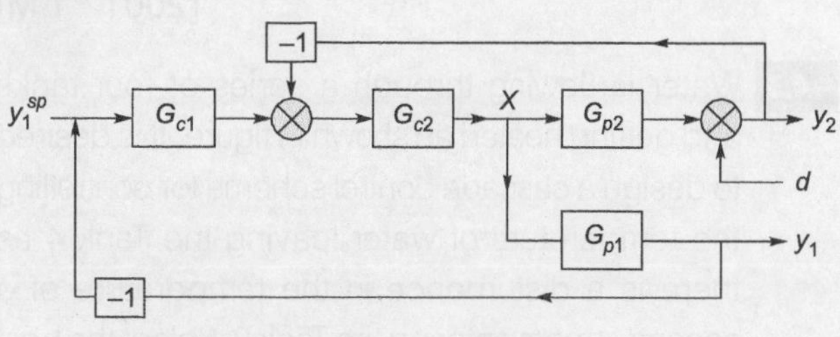

Assuming output from controller G_{c2} is X

From the above given diagram

$$XG_{p1} = Y_1 \quad \ldots\text{(i)}$$

$$(-y_1 G_{c1} - XG_{p2} - d)G_{c2} = X$$

By putting the value of X from equation (i)

$$\left(-y_1 G_{c1} - \frac{y_1}{G_{p1}} G_{p2} - d\right) G_{c2} = \frac{y_1}{G_{p1}}$$

$$(-y_1 G_{c1} \cdot G_{p1} - y_1 G_{p2} - dG_{p1})G_{c2} = y_1$$

$$-G_{c2} \cdot G_{p1} d = y_1(1 + G_{c2}G_{p2} + G_{c1}G_{p1}G_{p2})$$

$$\frac{y_1(s)}{d(s)} = \frac{-G_{c2}G_{p1}}{(1+G_{c2}G_{p2}+G_{c1}G_{c2}G_{p1})}$$

■■■■

4 Controller and Advanced Control Strategies

4.1 A feedback control loop with a proportional controller has an open loop transfer function

$$G_L(s) = \frac{K_c}{s(5s+1)^2}$$

where time is in minutes. Obtain the crossover frequency and the ultimate controller gain.

[2000 : 5 M]

4.2 The inherent characteristics of an equal percentage valve relating flow rate q with valve stem movement x are described by the equation

(a) $\frac{dq}{dx} = k$ (b) $\frac{dq}{dx} = kq$

(c) $\frac{dq}{dx} = \frac{k}{q}$ (d) $\frac{dq}{dx} = kq^2$

[2001 : 1 M]

4.3 Water is flowing through a series of four tanks and getting heated as shown in figure. It is desired to design a cascade control scheme for controlling the temperature of water leaving the Tank 4 as there is a disturbance in the temperature of a second stream entering the Tank 2. Select the best place to take the secondary measurement for the secondary loop

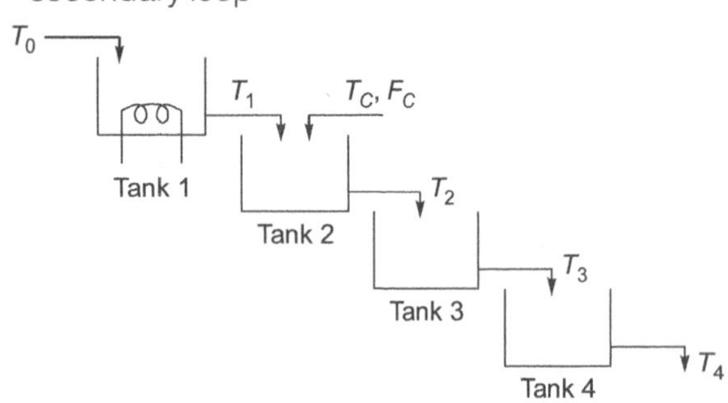

(a) Tank 1 (b) Tank 2

(c) Tank 3 (d) Tank 4 [2003 : 1 M]

4.4 Match the type of controller given in **Group-2** that is most suitable for each application given in **Group-1**:

Group-1

A. Distillation column bottoms level to be controlled with bottoms flow.

B. Distillation column pressure to be controlled by manipulating vapor flow from the top plate.

C. Flow control of a liquid from a pump by positioning the valve in the line.

D. Control of temperature of a CSTR with coolant flow in the jacket.

Group-2

1. P control
2. P-I control
3. P-I-D control

Codes:

	A	B	C	D
(a)	1	1	2	3
(b)	2	2	3	3
(c)	2	2	1	1
(d)	2	3	2	3

[2003 : 2 M]

4.5 In the case of a feed forward control scheme, which of the following is NOT true?

1. It is insensitive to modeling errors.
2. Cannot cope with unmeasured disturbances.
3. It waits until the effect of the disturbance has been felt by the system before control action is taken.
4. Requires good knowledge of the process model.
5. Requires identification of all possible disturbances and their direct measurement.

(a) 1 and 3 (b) 1 and 4

(c) 2 and 5 (d) 3 and 4

[2003 : 2 M]

4.6 Temperature control of an exothermic chemical reaction taking place in a CSTR is done with the help of cooling water flowing in a jacket around the reactor. The type of valve and controller action to be recommended are

(a) air to open valve with the controller direct acting

(b) air to close valve with the controller indirect acting

(c) air to open valve with the controller indirect acting

(d) air to close valve with the controller direct acting

[2003 : 2 M]

4.7 The experimental response of the controlled variable $y(t)$ for a step change of magnitude P in the manipulated variable $x(t)$ is shown below.

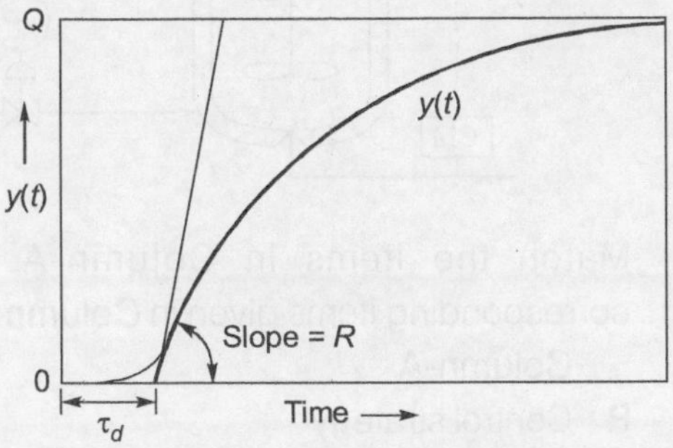

The appropriate transfer function of the process is

(a) $\dfrac{\left(\frac{Q}{P}\right)e^{-(Q/R)s}}{\tau_d s+1}$ (b) $\dfrac{\left(\frac{Q}{R}\right)e^{-\tau_d s}}{\left(\frac{Q}{P}\right)s+1}$

(c) $\dfrac{\left(\frac{Q}{P}\right)e^{-\tau_d s}}{\left(\frac{Q}{R}\right)s+1}$ (d) $\dfrac{\left(\frac{Q}{R}\right)e^{-(P/Q)s}}{\tau_d s+1}$

[2004 : 2 M]

4.8 The process and disturbance transfer functions for a system are given by

$$G_p(s)=\frac{\bar{y}(s)}{\bar{m}(s)}=\frac{2}{(2s+1)(5s+1)}$$

$$G_d(s)=\frac{\bar{y}(s)}{\bar{d}(s)}=\frac{1}{(2s+1)(5s+1)}$$

The feed forward controller transfer function that will keep the process output constant for changes in disturbance is

(a) $\dfrac{2}{(2s+1)^2(5s+1)^2}$ (b) $\dfrac{(2s+1)^2(5s+1)^2}{2}$

(c) $\dfrac{1}{2}$ (d) $(2s+1)(5s+1)$

[2004 : 2 M]

4.9 The control valve characteristic is selected such that the product of process gain and the valve gain

(a) is a linearly increasing function of the manipulated variable

(b) is a linearly decreasing function of the manipulated variable

(c) remains constant as the value of the manipulated variable changes

(d) is an exponentially increasing function of the manipulated variable

[2005 : 1 M]

4.10 Cascade control comes under the control configuration which uses

(a) one measurement and one manipulated variable

(b) more than one measurement and one manipulated variable

(c) one measurement and more than one manipulated variable

(d) more than one measurement and more than one manipulated variable

[2005 : 1 M]

4.11 Given the process transfer function $G_P=\dfrac{4}{(\tau s+1)^2}$ and the disturbance transfer function $G_d=\dfrac{2}{(\tau s+1)}$ select the correct transfer function for the Feed Forward Controller for perfect disturbance rejection.

(a) $-2(\tau s+1)$ (b) -1

(c) $-0.5(\tau s+1)$ (d) $-(\tau s+1)^2$

[2005 : 2 M]

4.12 The value of ultimate period of oscillation P is 3 minutes, and that of the ultimate controller gain K_{cu} is 2. Select the correct set of tuning parameters (controller gain K. The derivative time constant τ_D in minutes, and the integral time constant τ_1 in minutes) for a PID controller using Ziegler-Nichols controller settings.

(a) $K_c = 1.1$; $\tau_1 = 2.1$; $\tau_D = 1.31$

(b) $K_c = 1.5$; $\tau_1 = 1.8$; $\tau_D = 0.51$

(c) $K_c = 15$; $\tau_1 = 1.8$; $\tau_D = 0.51$

(d) $K_c = 1.2$; $\tau_1 = 1.5$; $\tau_D = 0.38$

[2005 : 2 M]

4.13 The control valve characteristics for three types of control valves (P, Q and R) are given in the figure below : Match the control valve with its characteristics.

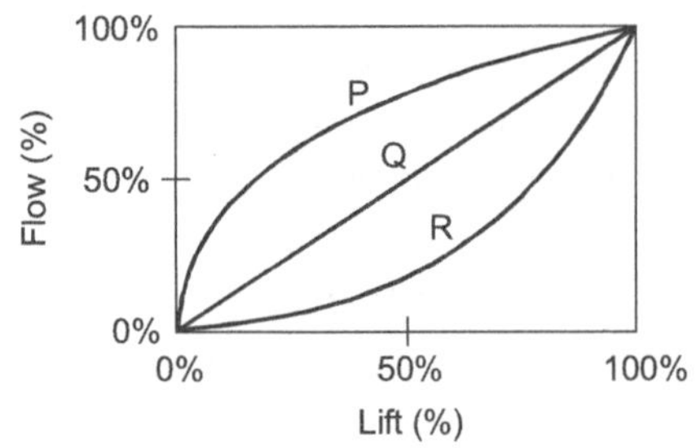

(a) P – Quick opening, Q – Linear, R – Equal percentage
(b) P – Linear, Q – Square root, R – Equal percentage
(c) P – Equal percentage, Q – Linear, R – Quick opening
(d) P – Square root, Q – Quick opening, R – Linear

[2006 : 1 M]

4.14 A liquid level control system is configured as shown in the figure below. If the level transmitter (LT) is direct acting and the pneumatic control valve is air-to-open. What kind of control action should the controller (LC) have and why?

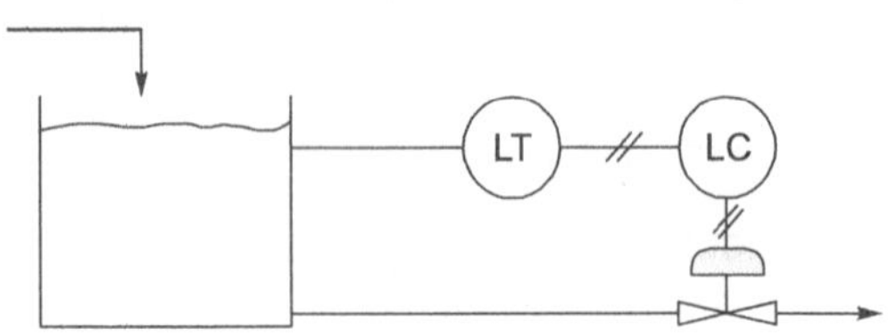

(a) Direct acting since the control valve is direct acting
(b) Reverse acting since the control valve is reverse acting
(c) Direct acting since the control valve is reverse acting
(d) Reverse acting since the control valve is direct acting

[2006 : 2 M]

4.15 An operator was told to control the temperature of a reactor at 60°C. The operator set the set-point of the temperature controller at 60. The scale actually indicated 0 to 100% of a temperature range of 0 to 200°C. This caused a runaway reaction by over-pressurizing the vessel, which resulted in injury to the operator. The actual set-point temperature was

(a) 200°C (b) 60°C
(c) 120°C (d) 100°C

[2007 : 1 M]

4.16 Consider the following instrumentation diagram for a chemical reactor. C_{sp} represents a concentration set-point

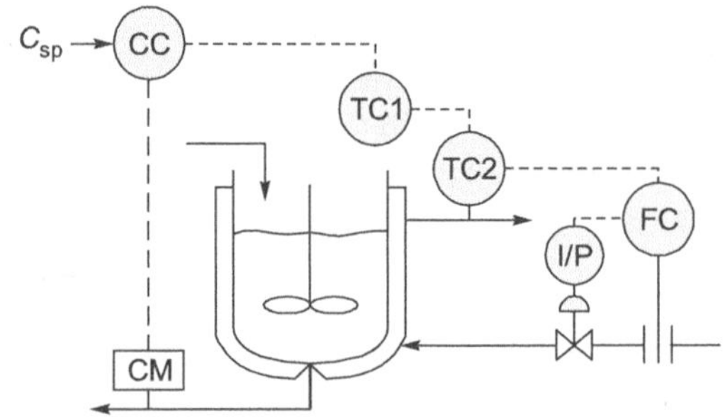

Match the items in **Column-A** with the corresponding items given in **Column-B**:

Column-A

P. Control strategy
Q. Primary control variable
R. Slowest controller
S. Fastest controller

Column-B

1. Feed forward control
2. Cascade control
3. Concentration in the reactor
4. Reactor temperature
5. Jacket temperature
6. Concentration controller
7. Reactor temperature controller
8. Jacket temperature controller
9. Flow controller
10. Selective control

(a) P-2, Q-3, R-6, S-9 (b) P-1, Q-4, R-8, S-7
(c) P-10, Q-7, R-9, S-6 (d) P-1, Q-8, R-5, S-9

[2007 : 2 M]

Common Data for Questions (4.17 - 4.19):

4.17 A cascade control system for pressure control is shown in the figure given below. The pressure transmitter has a range of 0 to 6 bar (g) and the flow transmitter range is 0 to 81 nm^3/hr. The normal flow rate through the valve is 32.4 nm^3/hr corresponding to the value of set point for pressure = 1 bar (g) and to give the flow, the valve must be 40% opened. The control valve has linear characteristics and is fail-open (air to close). Error, set point and control variable are expressed in percentage transmitter output (% TO). Proportional gain is expressed in the units of % controller output (%CO).

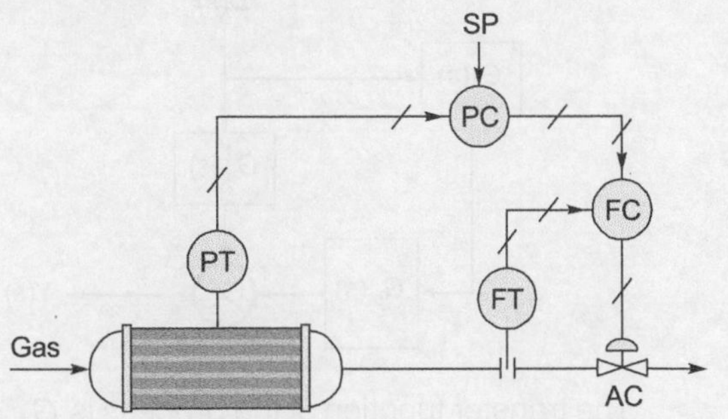

The types of action for the two controllers are
(a) direct acting for the pressure control and direct acting for the flow control
(b) indirect acting for the pressure control and indirect acting for the flow control
(c) direct acting for the pressure control and indirect acting for the flow control
(d) indirect acting for the pressure control and direct acting for the flow control

[2007 : 2 M]

4.18 The bias values for the two controllers, so that no offset occurs in either controller are
(a) Pressure controller : 40%, Flow controller : 60%
(b) Pressure controller : 33%, Flow controller : 67%
(c) Pressure controller : 67%, Flow controller : 33%
(d) Pressure controller : 60%, Flow controller : 40%

[2007 : 2 M]

4.19 Given that the actual tank pressure is 4 bar (g) and a proportional controller is employed for pressure control, the proportional band setting of the pressure controller required to obtain a set point to the flow controller equal to 54 nm^3/hr is
(a) 50% (b) 100%
(c) 150% (d) 187%

[2007 : 2 M]

4.20 Match the following:

Group-1
P. Ziegler Nichols
Q. Under damped response
R. Feed-forward control

Group-2
1. Process reaction curve
2. Decay ratio
3. Frequency response
4. Disturbance measurement

(a) P-3, Q-2, R-4 (b) P-1, Q-2, R-3
(c) P-3, Q-4, R-2 (d) P-1, Q-4, R-2

[2008 : 2 M]

4.21 Consider the process as shown below:

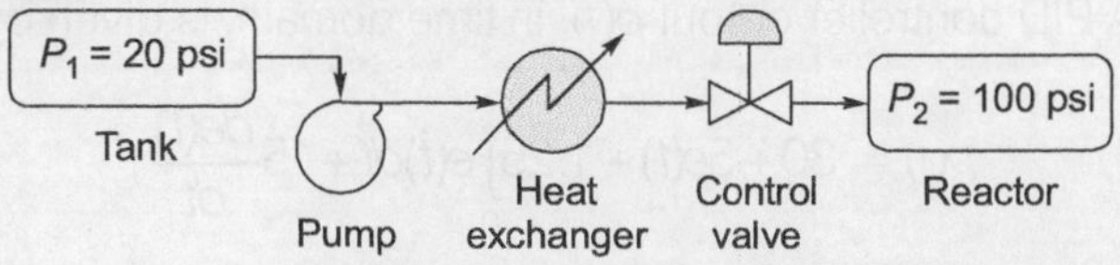

A constant head pump transfers a liquid from a tank maintained at 20 psi to a reactor operating at 100 psi, through a heat exchanger and a control valve. At the design conditions, the liquid flow rate is 1000 litres/min, while the pressure drop across the heat exchanger is 40 psi, and that across the control valve is 20 psi. Assume that the pressure drop across the heat exchanger varies as the square of the flow rate. If the flow is reduced to 500 litres/min, then the pressure drop across the control valve is
(a) 30 psi (b) 50 psi
(c) 80 psi (d) 150 psi

[2010 : 2 M]

4.22 The following diagram shows a CSTR with two control loops. A liquid phase, endothermic reaction is taking place in the CSTR, and the system is initially at steady state. Assume that the changes in physical properties of the system are negligible.

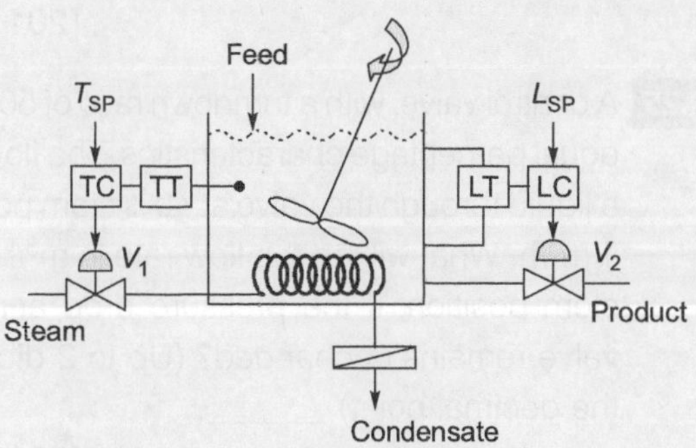

TC : Temperature controller, LC : Level controller, TT : Temperature transmitter, LT : Level transmitter, V_1 and V_2 : Control valves

Which one of the following statements is true?
(a) Changing the level controller set point affects the opening of V_2 only.
(b) Changing the temperature controller set point affects the opening of V_2 only.
(c) Changing the temperature controller set point affects the opening of both V_1 and V_2.
(d) Changing the level controller set point affects the opening of both V_1 and V_2.

[2011 : 2 M]

Statement for Linked Answer Question 4.23 & 4.24:

A PID controller output $p(t)$, in time domain, is given by

$$p(t) = 30 + 5e(t) + 1.25\int_0^t e(t)dt + 15\frac{de(t)}{dt}$$

where $e(t)$ is the error at time t. The transfer function of the process to be controlled is $G_p(s) = \frac{10}{(200s+1)}$. The measurement of the controlled variable is instantaneous and accurate.

4.23 The transfer function of the controller is

(a) $\frac{5(12s^2+4s+1)}{3s}$ (b) $\frac{5(12s^2+3s+1)}{3s}$

(c) $\frac{5(12s^2+4s+1)}{4s}$ (d) $\frac{5(12s^2+3s+1)}{4s}$

[2011 : 2 M]

4.24 The characteristic equation of the closed loop is

(a) $6s^2 + 102s + 1 = 0$

(b) $700s^2 + 102s + 25 = 0$

(c) $100s^2 - 196s - 25 = 0$

(d) $240s^3 + 812s^2 + 204s + 1 = 0$

[2011 : 2 M]

4.25 A control valve, with a turndown ratio of 50, follows equal percentage characteristics. The flow rate of a liquid through the valve at 40% stem position is 1 m^3/h. What will be the flow rate in m^3/h at 50% stem position, if the pressure drop across the valve remains unchanged? (Up to 2 digits after the decimal point) _______.

[2013 : 2 M]

4.26 Integral of the time-weighted absolute error (ITAE) is expressed as

(a) $\int_0^\infty \frac{|\varepsilon(t)|}{t^2}dt$ (b) $\int_0^\infty \frac{|\varepsilon(t)|}{t}dt$

(c) $\int_0^\infty t|\varepsilon(t)|dt$ (d) $\int_0^\infty t^2|\varepsilon(t)|dt$

[2014 : 1 M]

4.27 Given below is a simplified block diagram of a feedforward control system.

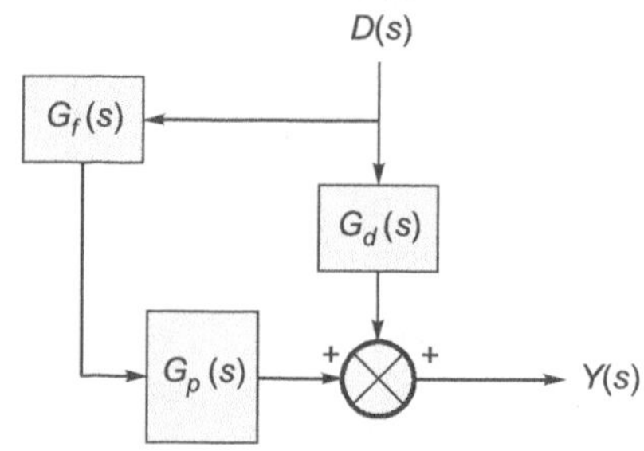

The transfer function of the process is $G_p = \frac{5}{s+1}$ and the disturbance transfer function is $G_d = \frac{1}{s^2+2s+1}$. The transfer function of the PERFECT feedforward controller, $G_f(s)$ is

(a) $\frac{-5}{(s+1)^3}$ (b) $\frac{-5}{(s+1)}$

(c) $\frac{-1}{5(s+1)}$ (d) $-5(s+1)$

[2014 : 2 M]

4.28 A liquid flows through an "equal percentage" valve at a rate of 2 m^3/h when the valve is 10% open. When the valve opens to 20% the flow rate increases to 3 m^3/h. Assume that the pressure drop across the valve and the density of the liquid remain constant. When the valve opens to 50%, the flow rate (in m^3/h, rounded off to the second decimal place) is _______. [2016 : 2 M]

4.29 The liquid flow rate through an equal percentage control valve, when fully open, is 150 gal/min and the corresponding pressure drop is 50 psi. If the specific gravity of the liquid is 0.8, then the valve coefficient, C_V, in gal/(min $psi^{0.5}$) is _______ (rounded off to two decimal places).

[2019 : 1 M]

4.30 Flow of water through an equal percentage valve is 900 liter h^{-1} at 30% opening, and 1080 liter h^{-1} at 35% opening. Assume that the pressure drop across the valve remains constant. The flow rate (in liter h^{-1}) through the valve at 45% opening is _______ (round off to nearest integer).

[2020 : 2 M]

4.31 A feed-forward controller can be used only if

(a) the disturbance variable can be measured.

(b) the disturbance variable can be manipulated.

(c) the disturbance variable can be ignored.

(d) regulatory control is not required.

[2021 : 1 M]

4.32 The inherent characteristics of three control valves *P*, *Q* and *R* are shown in the figure.

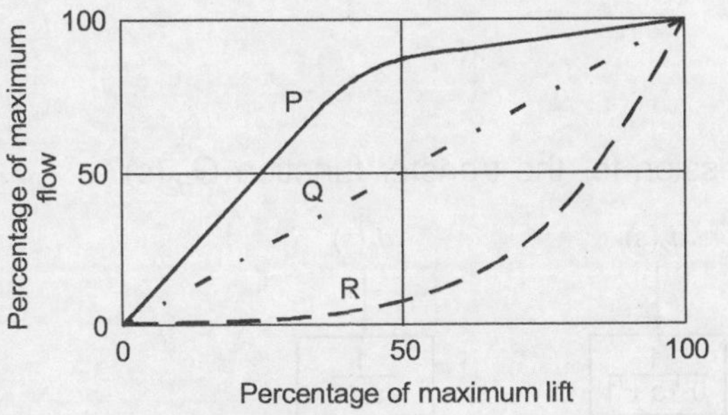

The correct option(s) is/are

(a) *P* is a quick opening valve.

(b) *Q* is a quick opening valve.

(c) *P* is an equal percentage valve.

(d) *R* is an equal percentage valve.

[2021 : 1 M]

4.33 A control system on the jacket side of a reactor is shown in the figure. Pressurized water flows through the jacket to cool the reactor. The heated water flashes in the boiler. The exothermic reaction heat thus generates steam. Fresh boiler feed water (BFW) is added to make-up for the loss of water as steam. Assume that all control valves are air-to-open. The controller action, 'direct' or 'reverse', is defined with respect to the controller. Select the option that correctly specifies the action of the controllers.

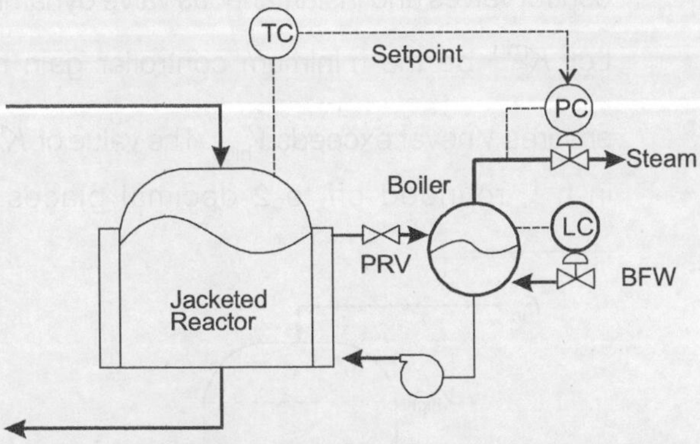

LC : Level Controller, PC : Pressure Controller

TC : Temperature Controller, PRV : Pressure Reducing Valve

(a) PC: Reverse, LC: Direct, TC: Reverse

(b) PC: Direct, LC: Reverse, TC: Direct

(c) PC: Direct, LC: Reverse, TC: Reverse

(d) PC: Reverse, LC: Direct, TC: Direct

[2022 : 1 M]

4.34 The appropriate feedforward compensator, G_{ff}, in the shown block diagram is

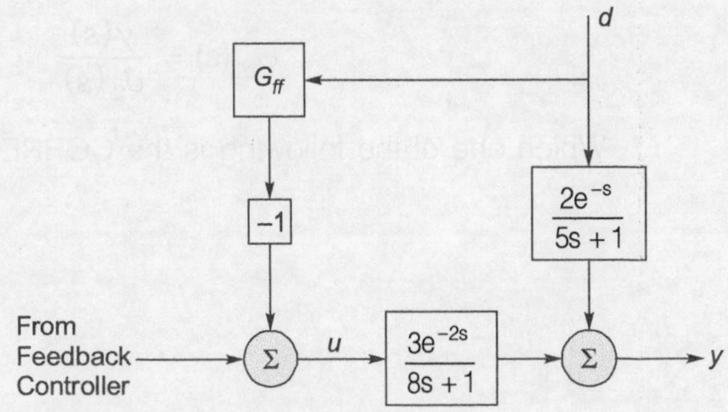

(a) $G_{ff} = \dfrac{2(8s+1)}{3(5s+1)}$

(b) $G_{ff} = -\dfrac{2(8s+1)}{3(5s+1)}$

(c) $G_{ff} = \dfrac{3(5s+1)}{2(8s+1)}e^{-s}$

(d) $G_{ff} = -\dfrac{3(5s+1)}{2(8s+1)}e^{-s}$

[2022 : 1 M]

4.35 Consider the tank level control system shown in the figure, where the cross-section area of the tank is *A*. Assume perfect flow controllers (FC). The level controller (LC) is proportional-integral (PI). For an integral time, τ_I, the level controller gain, K_c, is tuned for critical damping. The value of $\dfrac{K_c \tau_I}{A}$ is ______ (rounded off to the nearest integer).

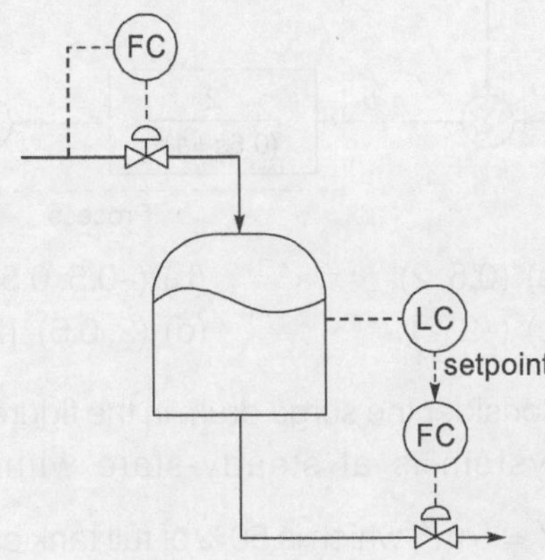

[2022 : 2 M]

4.36 A cascade control strategy is shown in the figure below. The transfer function between the output (y) and the secondary disturbance (y_2) is defined as

$$G_{d2}(s) = \frac{y(s)}{d_2(s)}$$

Which one of the following is the CORRECT expression for the transfer function $G_{d2}(s)$?

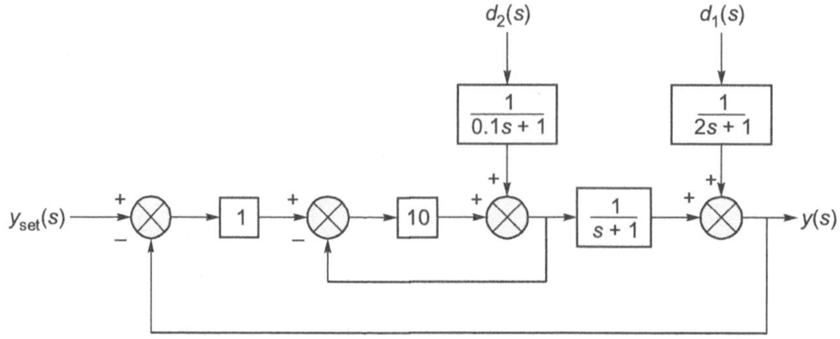

(a) $\frac{1}{(11s+21)(0.1s+1)}$ (b) $\frac{1}{(s+1)(0.1s+1)}$

(c) $\frac{(s+1)}{(s+2)(0.1s+1)}$ (d) $\frac{(s+1)}{(s+1)(0.1s+1)}$ **[2023 : 2 M]**

4.37 Consider the block diagram in the figure with control input u, disturbance d and output y. For the feedforward controller, the ordered pair $\left(K, \frac{\alpha}{\beta}\right)$ is

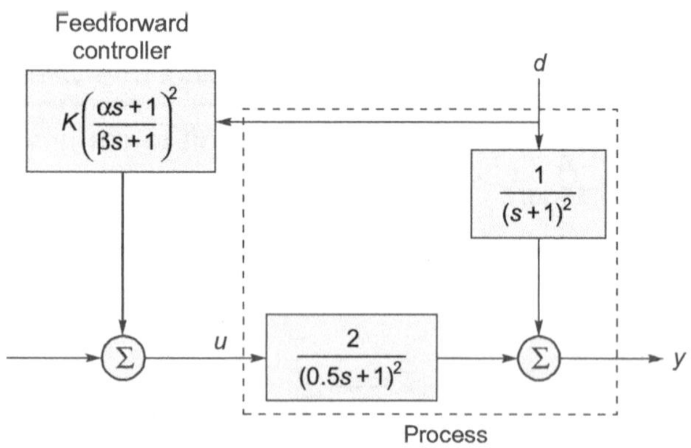

(a) (0.5, 2) (b) (–0.5, 0.5)
(c) (–2, 2) (d) (2, 0.5) **[2024 : 1 M]**

4.38 Consider the surge drum in the figure. Initially the system is at steady-state with a hold-up $\bar{V} = 5$ m^3, which is 50% of full tank capacity, V_{full}, and volumetric flow rates $\bar{F}_{in} = \bar{F}_{out} = 1$ m^3 h^{-1}. The high hold-up alarm limit $V_{high} = 0.8\ V_{full}$ while the low hold-up alarm limit $V_{low} = 0.2\ V_{full}$. A proportional (P-only) controller manipulates the outflow to regulate the hold-up V as $F_{out} = K_c\left(V - \bar{V}\right) + \bar{F}_{out}$. At $t = 0$, F_{in} increases as a step from 1 m^3 h^{-1} to 2 m^3 h^{-1}. Assume linear control valves and instantaneous valve dynamics. Let K_c^{min} be the minimum controller gain that ensures V never exceeds V_{high}. The value of K_c^{min}, in h^{-1}, rounded off to 2 decimal places, is ________.

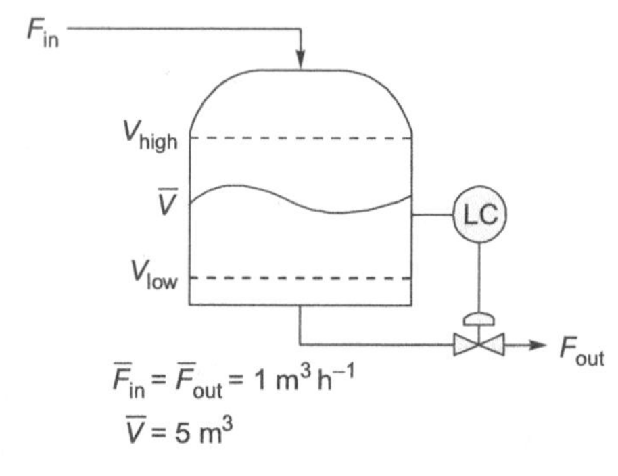

[2024 : 2 M]

Answers Controller and Advanced Control Strategies

4.1	(0.2, 0.4)	4.2	(b)	4.3	(b)	4.4	(d)	4.5	(a)	4.6	(b)	4.7	(c)
4.8	(c)	4.9	(d)	4.10	(b)	4.11	(c)	4.12	(d)	4.13	(a)	4.14	(a)
4.15	(c)	4.16	(a)	4.17	(c)	4.18	(d)	4.19	(b)	4.20	(a)	4.21	(b)
4.22	(d)	4.23	(c)	4.24	(b)	4.25	(1.478)	4.26	(c)	4.27	(c)	4.28	(10.124)
4.29	(18.974)	4.30	(1595.21)	4.31	(a)	4.32	(a, d)	4.33	(c)	4.34	(a)	4.35	(4)
4.36	(a)	4.37	(b)	4.38	(0.33)								

Explanations Controller and Advanced Control Strategies

4.1 (0.2,0.4)

The given open loop transfer function is

$$G_{OL}(s) = \frac{K_C}{s(5s+1)^2}$$

The above equation can be written as

$$G_{OL}(s) = \left(\frac{K_C}{s}\right)\cdot\left(\frac{1}{5s+1}\right)\cdot\left(\frac{1}{5s+1}\right) \quad ...(1)$$

$$\text{Phase lag } \phi = -90° + \tan^{-1}(-5\omega) + \tan^{-1}(-5\omega)$$

$$\phi = -90° + 2\tan^{-1}(-5\omega) \quad ...(2)$$

At crossover frequency,

$$\phi = -180° \text{ and } \omega = \omega_{CO}$$

Substituting the values in Eq. (2), we get

$$-180° = -90° + 2\tan^{-1}(-5\omega_{CO})$$

$$\Rightarrow \quad -90° = 2\tan^{-1}(-5\omega_{CO})$$

$$\Rightarrow \tan^{-1}(-5\omega_{CO}) = -45°$$

$$\Rightarrow \quad -5\omega_{CO} = \tan(-45°)$$

$$\Rightarrow \quad -5\omega_{CO} = -1$$

$$\Rightarrow \quad \omega_{CO} = \frac{1}{5} = 0.2 \text{ rad/s}$$

For ultimate controller gain, where AR = 1

Amplitude ratio is given by,

$$AR = \frac{K_C}{\omega_{CO}}\cdot\frac{1}{\sqrt{(5\omega_{CO})^2+1}}\cdot\frac{1}{\sqrt{(5\omega_{CO})^2+1}}$$

$$\Rightarrow \quad 1 = \frac{K_C}{\omega_{CO}}\cdot\frac{1}{\sqrt{(25\omega_{CO}^2+1)}}$$

$$= \frac{K_C}{0.2}\cdot\frac{1}{[25\times(0.2)^2+1]}$$

After solving, we get

$$K_C = 0.4$$

4.2 (b)

For equal percentage value

$$\frac{dq}{dx} = kq$$

For quick opening

$$\Rightarrow \quad \frac{dq}{dx} = \frac{k}{q}$$

For linear characteristic

$$\Rightarrow \quad \frac{dq}{dx} = k$$

4.3 (b)

Because taking secondary measurement at tank 2 gives more time to controller for taking action instead of tank 3 and tank 4.

4.4 (d)

A : Normally level does not change rapidly and P control or better PI control is recommended.

B : Column pressure is a critical parameter and needs precise and rapid control against any fluctuation. Therefore PID control is recommended.

C : Flow control of a liquid normally remains steady but needs quick control action against fluctuations, PI is recommended.

D : It is very critical parameter and needs quick and precise control at the set point. PID is recommended.

4.5 (a)

Feed forward control scheme is sensitive to modeling error.

It eliminates the effect of disturbance before it has been felt by system.

It requires good knowledge of all process model and identification of all possible disturbances in system. It is good for sluggish system.

4.6 (b)

In exothermic reaction valve should be always open, hence air to close valve should be used.

4.7 (c)

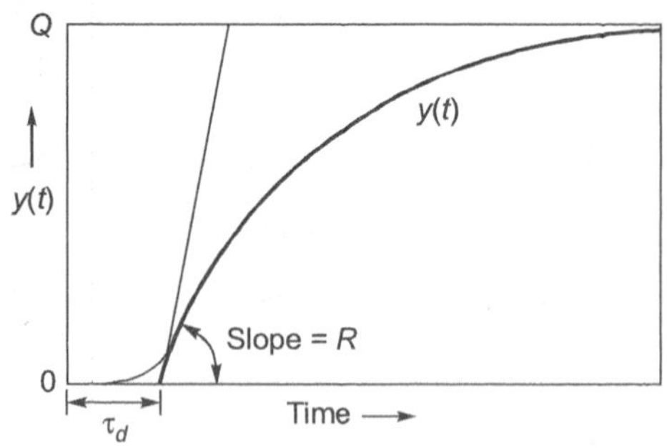

Step change of magnitude P in manipulated variable $X(t)$ is given

Ultimate value = PK_p

From graph ultimate value = Q

So, $$K_p = \left(\frac{Q}{P}\right)$$

Initial slope of response is given by $\frac{PK_p}{\tau_P}$

From graph, $$\frac{PK_p}{\tau_P} = R$$

$$\tau_P = \frac{PK_p}{R}$$

$$\tau_P = \frac{Q}{R}$$

From figure dead time = τ_D

So transfer function is

$$G(s) = \frac{\left(\frac{Q}{P}\right)e^{-\tau_D s}}{\left(\frac{Q}{R}\right)s+1}$$

4.8 (c)

For perfect disturbance rejection transfer function of feed forward control system is given by

$$G(c) = \frac{G_d}{G_p} = \frac{\frac{1}{(2s+1)(5s+1)}}{\frac{2}{(2s+1)(5s+1)}} = \frac{1}{2}$$

4.11 (c)

For perfect disturbance rejection transfer function of feed forward controller

$$G_c = -\frac{G_d}{G_p}$$

$$G(c) = \frac{-\frac{2}{\tau s+1}}{\frac{4}{(\tau s+1)^2}}$$

$$G_c = -0.5\,(\tau s+1)$$

4.12 (d)

Ultimate period of oscillation P_u = 3 minutes

Ultimate controller gain K_{cu} = 2

Control tuning parameter like K_c, τ_I and τ_D can be found by Ziegler-Nichols method by using following table:

Controller type	K_c	τ_I	τ_D
P only	$\frac{K_{cu}}{2}$		
PI	$\frac{K_{cu}}{2.2}$	$\frac{P_u}{1.24}$	
PID	$\frac{K_{cu}}{1.7}$	$\frac{P_u}{2}$	$\frac{P_u}{8}$

By using above table PID controller

$$K_c = 1.17$$
$$\tau_I = 1.5$$
$$\tau_D = 0.38$$

4.15 (c)

Operator set the point of the temperature controller at 60. But the scale was indicating 0 to 100 for temperature range 0° to 200°C. So, the actual set point was 120°C instead of 60°C.

4.16 (a)

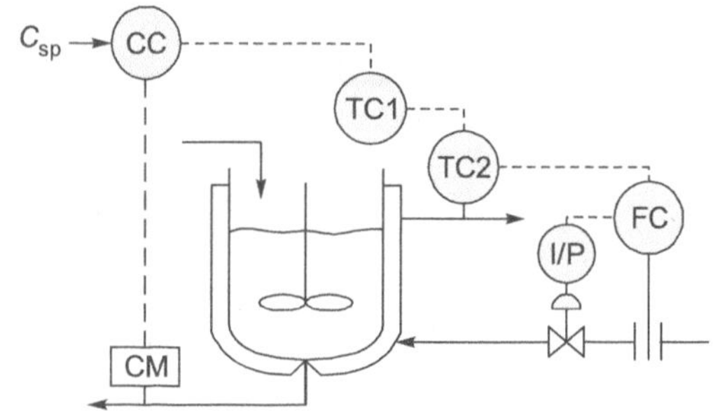

In this control strategy there is 3 measuring variable (temperature, flow and concentration) and only one manipulating variable (flow rate) so it is cascade control strategy.

Primary control variable is concentration in reactor.

Primary controller is slowest controller means concentration controller is slowest.

Fastest controller is secondary controller (flow controller).

4.17 (c)

It is given in the problem that the flow valve is air to close, that is direct acting valve. In order to maintain stability of the overall system, the next valve should be air to open. So that pressure valve will be air to open, that is, indirect acting valve. Therefore, for pressure control-indirect acting valve and for flow control direct acting valve.

4.18 (d)

For linear characteristics

Stem position ∝ Flow rate

Therefore, for 40% opening,

Flow controller = 40%

and Pressure controller = 60%

4.19 (b)

The value of K_c is given by,

$$K_c = \frac{\text{Change in output}}{\text{Change in input}}$$

$$K_c = \frac{81-0}{6-0} = \frac{81}{6}$$

Change in O/P = K_c × Change in I/P

$$\text{Change in O/P} = \frac{81}{6} \times 4$$

$$= 54 \text{ nm}^3/\text{h}$$

Since, the change in calculated O/P value is same as the given O/P value. Therefore, given actual operating range and ideal operating range are same. Hence, the proportional band is 100%.

4.21 (b)

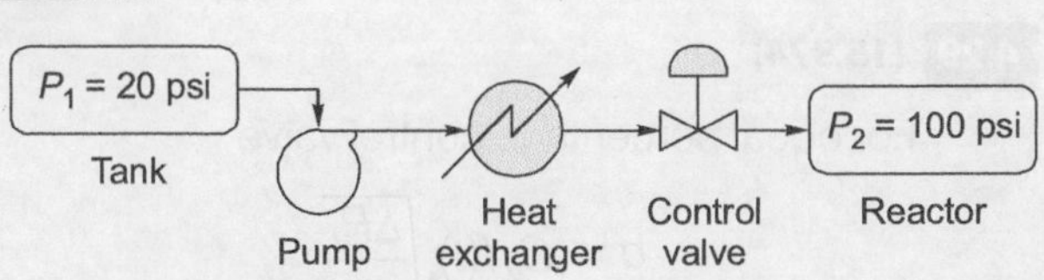

Constant head pump is used so head difference between pump and reactor will always constant. Now pressure difference across heat exchangers varies as square of flow rate.

So new pressure drop across heat exchangers

$$\Delta P_2 = \frac{Q_2^2}{Q_1^2} \Delta P_1$$

$$\Delta P_2 = \left(\frac{500}{1000}\right)^2 \times 40 = 10 \text{ psi}$$

New pressure drop across control valve can be found by equating the new pressure drop across instruments with previous one as constant head pump is used so

$$\Delta P_{valve} + 10 = 20 + 40$$

$$\Delta P_{valve} = 50 \text{ psi}$$

4.22 (d)

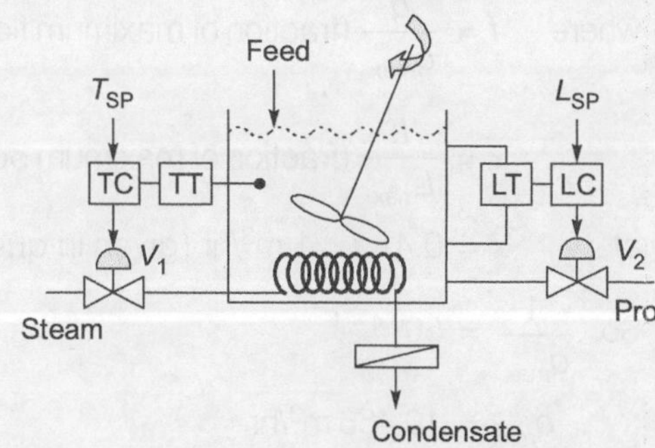

Changing the level controller set point affects the valve V_2 because flow rate through valve V_2 is proportional to liquid height in CSTR. If level of liquid is changed in CSTR, then valve V_1 opening will also change because requirement of steam for heating will change according to liquid level.

4.23 (c)

PID controller output $p(t)$ in time domain is given by

$$p(t) = 30 + 5e(t) + 1.25\int_0^t e(t)dt + 15\frac{de(t)}{dt}$$

$$P(t) = p(t) - 30$$

So,
$$P(t) = 5e(t) + 1.25\int_0^t e(t)dt + 15\frac{de(t)}{dt}$$

Taking Laplace,

$$P(s) = 5e(s) + \frac{1.25}{s}e(s) + 15s\ e(s)$$

$$G(s) = \frac{P(s)}{e(s)} = \frac{5(12s^2 + 4s + 1)}{4s}$$

4.24 (b)

Characteristic equation is given by

$1 + G_cG_pG_fG_m = 0$

$$1+\frac{5(12s^2+4s+1)}{4s}\left(\frac{10}{200s+1}\right) = 0$$

$800s^2 + 600s^2 + 200s + 4s + 50 = 0$

$1400s^2 + 204s + 50 = 0$

$700s^2 + 102s + 25 = 0$

4.25 (1.478)

Control valve turndown ratio (R) = 50

For equal percentage valve

$$f = R^{x-1}$$

where, $f = \frac{q}{q_{max}}$ (fraction of maximum flow rate)

$x = \frac{L}{L_{max}}$ (fraction of maximum opening)

at, $x = 0.4$, $q = 1$ m^3/hr (given in question)

So, $\frac{1}{q_{max}} = 50^{0.4-1}$

$q_{max} = 10.456$ m^3/hr

Now flow rate at 50% opening is

$$\frac{q}{10.456} = 50^{0.5-1}$$

$q = 1.478$ m^3/hr

4.26 (c)

(i) Square of absolute error

$$\int_0^{\infty} |\epsilon(t)|^2\, dt$$

(ii) Integral of time weighted absolute error

$$\int_0^{\infty} t|\epsilon(t)|\, dt$$

When error is vary with time this method of Tuning is used.

(iii) Integral of absolute error

$$\int_0^{\infty} |\epsilon(t)|\, dt$$

This method of Tuning is used when error is small.

4.27 (c)

Transfer function of process $G_p = \frac{5}{s+1}$

Transfer function of disturbance is $G_d = \frac{1}{s^2+2s+1}$

$$G_d = \frac{1}{(s+1)^2}$$

Transfer function of perfect feed forward controller $G_f(s)$ is given by

$$G_f(s) = -\frac{G_d}{G_p} = \frac{\frac{-1}{(s+1)^2}}{\frac{5}{(s+1)}}$$

$$G_f(s) = \frac{-1}{5(s+1)}$$

4.28 (10.124)

Flow through equal percentage valve is given by

$$F = R^{x-1} \quad \text{...(i)}$$

Here, $F = \frac{q}{q_{max}}$ (Fraction of maximum flow)

R = Rangibility = $\frac{q_{max}}{q_{min}}$

$x = \frac{L}{L_{max}}$ (Fraction opening)

At 10%, flow rate given is 2 m^3/hr

Similarly at 20%, flow rate = 3 m^3/hr

From equation (i)

$$\frac{2}{3} = \frac{R^{-0.9}}{R^{-0.8}}$$

$R = 57.66$

$q_{max} = 76.88$ m^3/hr

So flow rate at 50% opening is given by

$$\frac{q}{76.88} = 57.66^{-0.5}$$

$q = 10.124$ m^3/hr

4.29 (18.974)

For equal percentage control valve

$$q = C_v f(l)\sqrt{\frac{\Delta P_v}{s.g}}$$

where s.g. is specific gravity

$$f = R^{(l-1)}$$

When valve is fully open then lift (l) = 1

$$f = R^0 = 1$$

$$q = C_v\sqrt{\frac{\Delta P_v}{s.g.}}$$

Given: $q = 150$ gal/min

$\Delta P_v = 50$ psi

s.g. = 0.8

$$C_v = \frac{q}{\sqrt{\frac{\Delta P_v}{\text{s.g.}}}} = \frac{150}{\sqrt{\frac{50}{0.8}}}$$

$$C_v = 18.974 \text{ gal/(min) psi}^{0.5}$$

4.30 (1595.21)

Flow through equal % valve is given with following conditions:

at 30% opening q = 900 liter/hr

at 35% opening q = 1080 liter/hr

For equal percentage value

$$F = R^{x-1}$$

where, $F = \frac{q}{q_{max}}$

$$x = \text{Faction opening} = \frac{L}{L_{max}}$$

$$R = \text{Rangibility} = \frac{q_{max}}{q_{min}}$$

From given that

$$\frac{900}{q_{max}} = R^{-0.7} \quad \text{...(i)}$$

$$\frac{1080}{q_{max}} = R^{-0.65} \quad \text{...(ii)}$$

By solving equation (i) and (ii)

q_{max} = 11555.266 liter/hr

R = 38.337

So flow rate at 45% opening is given by

$$\frac{q}{q_{max}} = R^{x-1}$$

$$\frac{q}{11599.266} = 38.337^{(0.45-1)}$$

$$q = 1595.21 \text{ liter/hr}$$

4.31 (a)

A feed-forward controller can be used only if the disturbance variable can be measured.

4.33 (c)

All are air to open.

Exothermic reaction

At one place both direct or indirect can't be used.

(a) Level should be increase option (a) is incorrect.

(b) All the situation is correct but system may unstable.

(c) All the situation are correct.

(d) Level should be increased.

4.34 (a)

$$[-G_{ff}\, d(s) + C]\frac{3e^{-2s}}{8s+1} + \frac{2e^{-s}}{5s+1}d(s) = y(s)$$

For feed forward controller $y(s) = 0$

$$-G_{ff}\frac{3e^{-2s}}{8s+1}d(s) + \frac{2e^{-s}}{5s+1}d(s) = 0$$

$$d(s)\left[-G_{ff}\left(\frac{3e^{-2s}}{8s+1}\right) + \frac{2e^{-s}}{5s+1}\right] = 0$$

$$G_{ff} = \frac{2}{3}\left[\frac{8s+1}{5s+1}\right]$$

4.35 (4)

Critical damping, $\xi = 1$

For the 2nd order, controller,

Only for PI

$$\frac{R}{(AR)S+1},\ q_0 = \frac{h}{R} \text{ (Not possible, no resistance)}$$

It is pure capacitive system.

$$q_0 = C$$

$$G_P(s) = \frac{1}{AS} \quad G_C(s) = K_c\left(1 + \frac{1}{\tau_I s}\right)$$

$$G_P(s)_{\text{close loop}} = \frac{\frac{K_c}{AS}\left(\frac{1+\tau_I s}{\tau_I s}\right)}{1 + \frac{1}{AS}\times\left(\frac{1+\tau_I s}{\tau_I s}\right)}$$

$$= \frac{(1+\tau_I s)K_c}{\tau_I AS^2 + K_c\tau_I s + K_c}$$

$$= \frac{1+\tau_I s}{\left(\frac{\tau_I A}{K_c}\right)s^2 + \tau_I s + 1}$$

$$\tau^2 = \frac{\tau_I A}{K_c}, \quad 2\xi\tau = \tau_I$$

$$\tau = \sqrt{\frac{\tau_I A}{K_c}}, \quad \xi = \frac{1}{2}\sqrt{\frac{K_c\tau_I}{A}}$$

Given, $\xi = 1$

So, $$2 = \sqrt{\frac{K_c\tau_I}{A}}, \quad 4 = \frac{K_c\tau_I}{A}$$

4.36 (a)

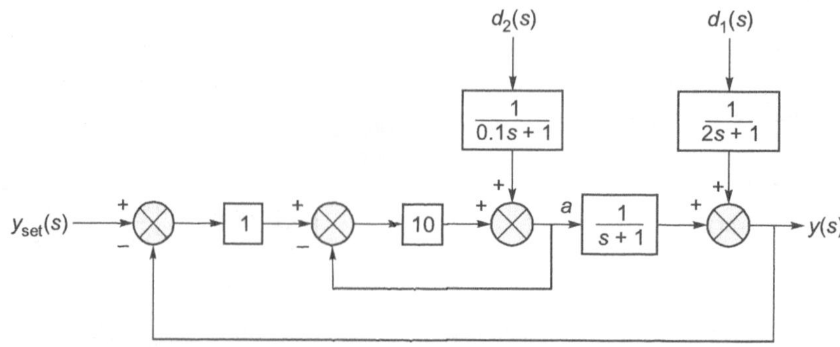

$$\left[(y_{sp}-y)(1)-a\right]10+d_s(s).\frac{1}{0.1s+1} = a \qquad ...(i)$$

$$a\left(\frac{1}{s+1}\right)+d_1(s)\frac{1}{(2s+1)} = y \qquad ...(ii)$$

$$(y_{sp}-y)10-10a+d_2(s).\frac{1}{0.1s+1} = a$$

$$(y_{sp}-y)10+\frac{d_2(s)}{(0.1s+1)} = 11a$$

$$(y_{sp}-y)\frac{10}{11}+\frac{d_2(s)}{11(0.1s+1)} = a \qquad Gd_2(s)=\frac{y(s)}{d_2(s)}$$

$$\left[(y_{sp}-y)\frac{10}{11}+\frac{d_2(s)}{11(0.1s+1)}\right]\frac{1}{(s+1)}+d_1(s)\frac{1}{(2s+1)} = y$$

$$(y_{sp}-y)\frac{10}{11}\times\frac{1}{(s+1)}+\frac{d_2(s).1}{11(0.1s+1)(s+1)}+d_1(s)\frac{1}{(2s+1)} = y$$

$$(0-y)\frac{10}{11}\times\frac{1}{(s+1)}+\frac{d_2(s)}{11(0.1s+1)(s+1)} = y$$

$$\frac{d_2(s)}{11(0.1s+1)(s+1)} = y+\frac{10}{11}y.\frac{1}{(s+1)}$$

$$\frac{d_2(s)}{11(0.1s+1)(s+1)} = y(s)\left(1+\frac{10}{11}.\frac{1}{(s+1)}\right)$$

$$\frac{d_2(s)}{11(0.1s+1)(s+1)} = y(s)\left(\frac{11(s+1)+10}{11(s+1)}\right)$$

$$\frac{d_2(s)}{(0.1s+1)} = y(s)\,[11s+11+10]$$

$$\frac{d_2(s)}{(0.1s+1)} = y(s)\,[11s+21]$$

$$\frac{d_2(s)}{(0.1s+1)} = y(s)\,[11s+21]$$

$$\frac{y(s)}{d_s(s)} = \frac{1}{(0.1s+1)(11s+21)}$$

4.37 (b)

Transfer function of feed forward controller

$$G_c = -\frac{G_d}{G_p} = \frac{-\dfrac{1}{(s+1)^2}}{\dfrac{2}{(0.5s+1)^2}} = \frac{0.5(0.5s+1)^2}{(s+1)^2}$$

On comparing with $K\left(\dfrac{\alpha s+1}{\beta s+1}\right)^2$

$K = -0.5$

$\alpha = 0.5, \quad \beta = 1$

$\dfrac{\alpha}{\beta} = 0.5$

■■■■

5 Stability and Frequency Analysis of Control Systems

5.1 Select the correct statement from the following:
(a) The frequency response of a pure capacity process is unbounded.
(b) The phase lag of a pure time delay system decreases with increasing frequency.
(c) The amplitude ratio of a pure capacity process is inversely proportional to the frequency.
(d) The amplitude ratio of a pure time delay system increases with frequency.

[2000 : 1 M]

5.2 For a feedback control system to be stable, the
(a) roots of the characteristic equation should be real.
(b) poles of the closed loop transfer function should lie in the left half of the complex plane.
(c) Bode plots of the corresponding open loop transfer function should monotonically decrease.
(d) poles of the closed loop transfer function should lie in the right half of the complex plane.

[2000 : 1 M]

5.3 An ideal PID controller has the transfer function $\left[1+\frac{1}{0.5s}+0.2s\right]$. The frequency at which the Magnitude Ratio of the controller is 1, is

(a) $\frac{0.5}{0.2}$ (b) $\frac{0.2}{0.5}$

(c) 0.2×0.5 (d) $\frac{1}{\sqrt{0.2 \times 0.5}}$

[2001 : 2 M]

5.4 A first order system with unity gain and time constant τ is subjected to a sinusoidal input of frequency $\omega = \frac{1}{\tau}$. The amplitude ratio for this system is

(a) 1 (b) 0.5

(c) $\frac{1}{\sqrt{2}}$ (d) 0.25

[2002 : 1 M]

5.5 Find the ultimate gain and frequency for a proportional controller in the case of a process having the following transfer function

$$G_p(s) = \frac{1}{(4s+1)(2s+1)(s+1)}$$

(a) $\omega = \frac{1}{\sqrt{14}}; K_c = \frac{45}{7\sqrt{14}}$

(b) $\omega = \sqrt{\frac{7}{6}}; K_c = \frac{46}{3}$

(c) $\omega = 1;\ K_c = 13$

(d) $\omega = \sqrt{\frac{7}{8}}; K_c = \frac{45}{4}$

[2003 : 2 M]

5.6 Consider a system with open loop transfer function

$$G(s) = \frac{1}{(s+1)(2s+1)(5s+1)}$$

Match the range of ω (frequency) in **Group-I** with the slope of the asymptote of the log AR (amplitude ratio) versus log ω plot in **Group-II**:

Group-I	Group-II
A. $0 < \omega < 0.2$	1. −5
B. $\omega > 1$	2. −3
	3. −2
	4. −1
	5. Zero

Codes:

	A	B
(a)	5	2
(b)	4	2
(c)	5	3
(d)	4	1

[2004 : 2 M]

5.7 Given the characteristic equation below, select the number of roots which will be located to the right of the imaginary axis, $s^4 + 5s^3 - s^2 - 17s + 12 = 0$

(a) One (b) Two
(c) Three (d) zero

[2005 : 2 M]

5.8 Given the process transfer function $G_P = \frac{20}{(s-2)}$,

and controller transfer function $G_c = K_c$, and assuming the transfer functions of all other elements in the control loop are unity, select the range of K_c for which the closed loop response will be stable.

(a) $K_c < \frac{1}{10}$ (b) $K_c < \frac{1}{100}$

(c) $\frac{1}{100} < K_c < \frac{1}{10}$ (d) $K_c > \frac{1}{10}$

[2005 : 2 M]

Statement for Linked Answer Question (5.9 & 5.10):

For the system shown below, $G_1(s) = \frac{1}{\tau_1 s+1}$,

$G_2(s) = \frac{1}{\tau_2 s+1}$ and $\tau_2 = 2\tau_1$

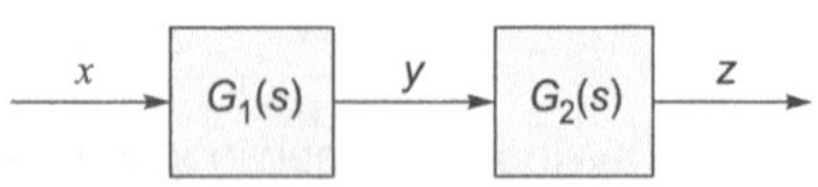

When the system is excited by the sinusoidal input $X(t) = \sin \omega t$, the intermediate response Y is given by

$$Y = A\sin(\omega t + \phi)$$

5.9 If the response of Y lags behind the input X by 45° and $\tau_1 = 1$, then the input frequency ω is

(a) 1 (b) $\frac{\pi}{4}$

(c) Zero (d) –1

[2006 : 2 M]

5.10 For the same input, the amplitude of the output Z will be

(a) 1.00 (b) 0.62

(c) 0.42 (d) 0.32

[2006 : 2 M]

5.11 The first two rows of Routh's tabulation of a third order equation are :

s^3	2	2
s^2	4	4

Select the correct answer from the following choices:

(a) The equation has one root in the right half s-plane.

(b) The equation has two roots on the j axis at $s = j$ and $-j$. The third root is in the left half plane.

(c) The equation has two roots on the j axis at $s = 2j$ and $s = -2j$. The third root is in the left half plane.

(d) The equation has two roots on the j axis at $s = 2j$ and $s = -2j$. The third root is in the right half plane.

[2007 : 2 M]

Statement for Linked Answer Question (5.12 & 5.13) :

The cross-over frequency associated with a feedback loop employing a proportional controller to control the process represented by the transfer function

$$G_p(s) = \frac{2e^{-s}}{(\tau s+1)^2}, \text{ (units of time is minutes)}$$

is found to be 0.6 rad/min. Assume that the measurement and valve transfer functions are unity.

5.12 The time constant, τ (in minutes) is

(a) 1.14 (b) 1.92

(c) 3.23 (d) 5.39

[2008 : 2 M]

5.13 If the control loop is to operate at a gain margin of 2.0, the gain of the proportional controller must equal

(a) 0.85 (b) 2.87

(c) 3.39 (d) 11.50

[2008 : 2 M]

5.14 The characteristic equation of a closed loop system using a proportional controller with gain K_c is

$$12s^3 + 19s^2 + 8s + 1 + K_c = 0$$

At the onset of instability, the value of K_c is

(a) $\frac{35}{3}$ (b) 10

(c) $\frac{25}{3}$ (d) $\frac{20}{3}$

[2009 : 2 M]

5.15 The transfer function, $G(s)$, whose asymptotic Bode diagram is shown below, is

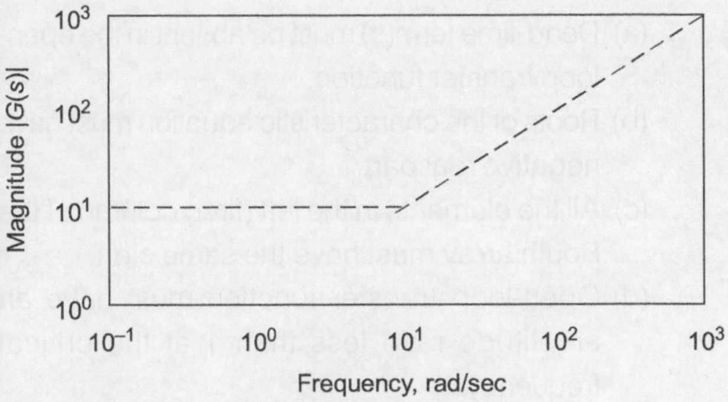

(a) $10s + 1$ (b) $s - 10$
(c) $s + 10$ (d) $10s - 1$

[2010 : 2 M]

5.16 Consider the cascade control configuration shown in the figure below:

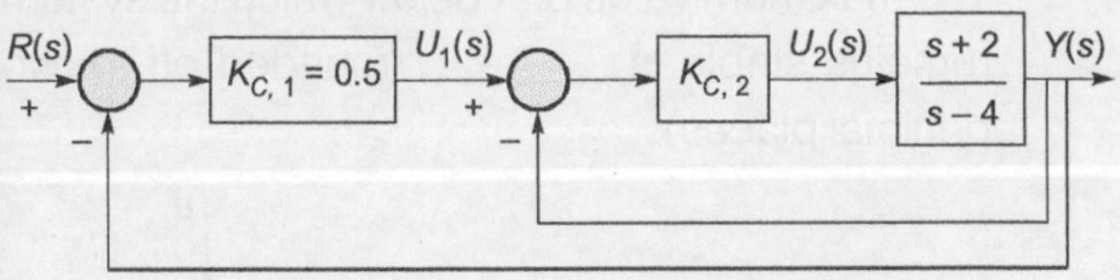

The system is stable when $K_{C,2}$ is

(a) $\frac{3}{4}$ (b) 1
(c) $\frac{5}{4}$ (d) $\frac{3}{2}$

[2010 : 2 M]

5.17 The Bode stability criterion is applicable when
(a) Gain and phase curves decrease continuously with frequency
(b) Gain curve increases and phase curve decreases with frequency
(c) Gain curve and phase curve both increase with frequency
(d) Gain curve decreases and phase curve increases with frequency

[2012 : 1 M]

5.18 The characteristic equation for a system is

$$s^3 + 9s^2 + 26s + 12(2 + K_c) = 0$$

Using the Routh test, the value of K_c that will keep the system on the verge of instability is
(a) 20.9 (b) 18.4
(c) 17.5 (d) 15.3

[2012 : 2 M]

Statement for Linked Answer Questions 5.19 & 5.20:

Consider the following transfer function:

$$G_p(s) = \frac{5}{(2s+1)^4}$$

(**Note:** The unit of the process time constant is in seconds)

5.19 The crossover frequency (in rad/s) of the process is
(a) 20 (b) 0.1
(c) 0.5 (d) 0.05

[2013 : 2 M]

5.20 For the computation of Ziegler-Nichols settings, the ultimate period (in s/cycle) and the ultimate gain are
(a) π and 0.8, respectively
(b) 4π and 0.8, respectively
(c) 4π and 1.25, respectively
(d) π and 1.25, respectively

[2013 : 2 M]

5.21 Consider a control system with the open loop transfer function given by:

$$G_{OL}(s) = \frac{K_c e^{-0.3s}}{1.5s+1}$$

In the above function, pre-factor of s is in minutes and K_c is the gain of proportional controller. The frequency for phase margin of 30° is 4.04 rad/min. The value of K_c for a gain margin of 1.7 (up to one decimal place) is ________.

[2015 : 2 M]

5.22 A PI controller with integral time constant of 0.1 min is to be designed to control a process with transfer function

$$G_p(s) = \frac{10}{s^2 + 2s + 100}$$

Assume the transfer functions of the measuring element and the final control element are both unity ($G_m = 1$, $G_f = 1$). The gain (rounded off to the first decimal place) of the controller that will constitute the critical condition for stability of the PI feedback control system is ________.

[2016 : 2 M]

5.23 The open loop transfer function of a process with a proportional controller (gain K_c) is

$$G_{OL} = K_c \frac{e^{-2s}}{s}$$

Based on the Bode criterion for closed-loop stability, the ultimate gain of the controller, rounded to 2 decimal places, is _______ .

[2017 : 2 M]

5.24 The characteristic equation of a closed-loop system is

$$6s^3 + 11s^2 + 6s + (1 + K) = 0, \text{ where } K > 0$$

The value of K beyond which the system just becomes unstable, rounded to the nearest integer, is _______. [2017 : 2 M]

5.25 Consider the following transfer function:

$$G(s) = \frac{3}{(5s+1)^2}$$

where, the natural period of oscillation is in min. The amplitude ratio at a frequency of 0.5 rad/min is _______ (rounded off to second decimal place).

[2018 : 2 M]

5.26 For the closed loop system shown in figure, the phase margin (in degrees) is _____. (rounded off to one decimal place).

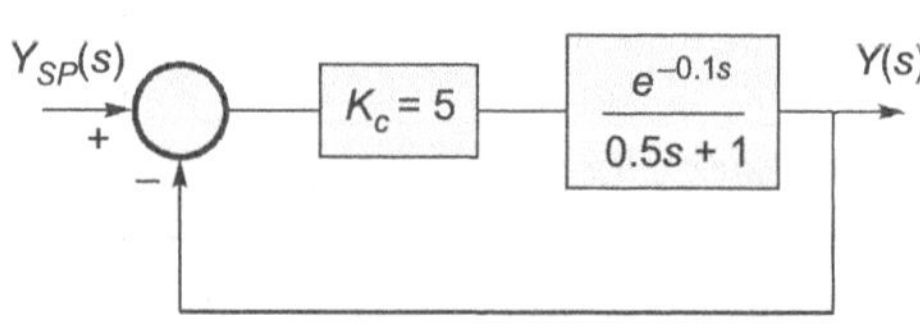

[2019 : 2 M]

5.27 Consider the following closed loop system:

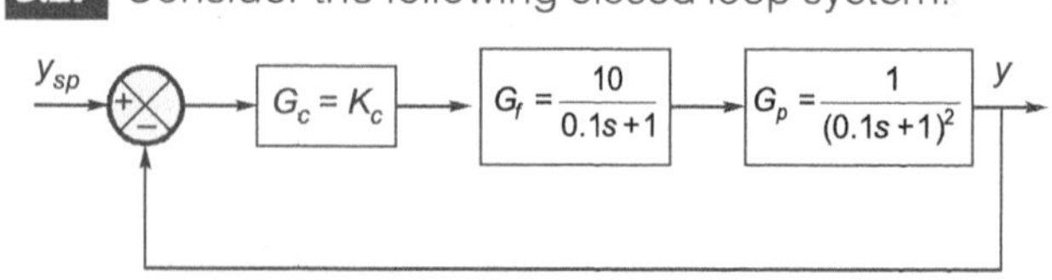

G_c, G_f and G_p are the transfer functions of the controller, the final control element and the process, respectively. y and y_{sp} are the response and its set point, respectively. For a gain margin of 1.6, the design value of K_c is _____ (correct upto one decimal place).

[2020 : 2 M]

5.28 Which of the following is NOT a necessary condition for a process under closed-loop control to be stable?

(a) Dead-time term(s) must be absent in the open-loop transfer function.

(b) Roots of the characteristic equation must have negative real part.

(c) All the elements in the left (first) column of the Routh array must have the same sign.

(d) Open-loop transfer function must have an amplitude ratio less than 1 at the critical frequency.

[2021 : 2 M]

5.29 In the block diagram shown in the figure, the transfer function $G = \frac{K}{(\tau s+1)}$ with $K > 0$ and $\tau > 0$.

The maximum value of K below which the system remains stable is _______ (rounded off to two decimal places).

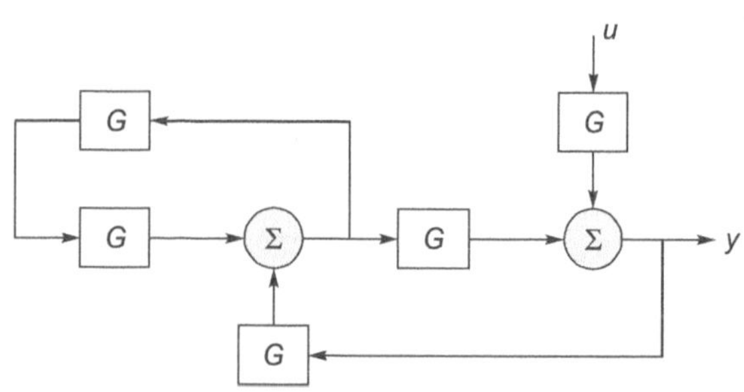

[2022 : 2 M]

5.30 A PD controller with transfer function G_c is used to stabilize an open-loop unstable process with transfer function G_p, where

$$G_c = K_c \frac{\tau_D s + 1}{\left(\frac{\tau_D}{20}\right)s + 1}, \quad G_p = \frac{1}{(s-1)(10s+1)}$$

and time is in minutes. From the necessary conditions for closed-loop stability, the maximum feasible value of τ_D, in minutes, rounded off to 1 decimal place, is _________.

[2024 : 2 M]

■■■■

Answers Stability and Frequency Analysis of Control Systems

5.1	(c)	**5.2**	(b)	**5.3**	(d)	**5.4**	(c)	**5.5**	(d)	**5.6**	(a)	**5.7**	(b)
5.8	(d)	**5.9**	(a)	**5.10**	(d)	**5.11**	(b)	**5.12**	(d)	**5.13**	(b)	**5.14**	(a)
5.15	(c)	**5.16**	(d)	**5.17**	(a)	**5.18**	(c)	**5.19**	(c)	**5.20**	(b)	**5.21**	(5.06)
5.22	(2.50)	**5.23**	(0.785)	**5.24**	(10)	**5.25**	(0.413)	**5.26**	(45.38)	**5.27**	(0.5)	**5.28**	(a)
5.29	(0.707)	**5.30**	(22.2)										

Explanations Stability and Frequency Analysis of Control Systems

5.1 (c)

For a pure capacitive process

$$\text{Amplitude ratio (AR)} = |G(j\omega)| = \left|\frac{k}{j\omega}\right| = \frac{k}{\omega}$$

Therefore, from above equation, we can analyze that the amplitude ratio of a pure capacity process is inversely proportional to frequency.

5.2 (b)

When the poles of the closed loop transfer function lie in the left half of the complex plane, the feedback control system will be stable.

5.3 (d)

Transfer function of PID controller is given as

$$G_c(s) = \left[1+\frac{1}{0.5s}+0.2s\right]$$

Frequency ω at which magnitude ratio = 1 can be calculated by following method

Put, $s = j\omega$

$$G_c(j\omega) = \left[1+\frac{1}{0.5j\omega}+0.2j\omega\right]$$

$$G_c(j\omega) = \left[1-\frac{j}{0.5\omega}+0.2j\omega\right]$$

$$G_c(j\omega) = \left[1+\left(\frac{-1}{0.5\omega}+0.2\omega\right)j\right]$$

$$\text{Real part} = 1$$

$$\text{Imaginary part} = \left(\frac{-1}{0.5\omega}+0.2\omega\right)$$

$$\text{AR} = \sqrt{\text{Real}^2+\text{Imaginary}^2}$$

$$= \sqrt{1^2+\left(\frac{-1}{0.5\omega}+0.2\omega\right)^2} = 1$$

$$\omega = \frac{1}{\sqrt{0.2\times 0.5}}$$

5.4 (c)

Amplitude ratio of first order system is given by

$$\text{AR} = \frac{1}{\sqrt{1+\tau^2\omega^2}} \quad \text{(for unity gain)}$$

$$\omega = \frac{1}{\tau} \text{ (given)}$$

$$\text{AR} = \frac{1}{\sqrt{2}}$$

5.5 (d)

Transfer function is given as below:

$$G_p(s) = \frac{1}{(4s+1)(2s+1)(s+1)}$$

For ultimate frequency ω_{co} phase lag $(-\phi)$ should be 180°

$$\phi = -\tan^{-1}(\tau_1\omega) - \tan^{-1}(\tau_2\omega) - \tan^{-1}(\tau_3\omega)$$

$$-180° = -\tan^{-1}(4\omega) - \tan^{-1}(2\omega) - \tan^{-1}(\omega)$$

$$180 = \tan^{-1}\left(\frac{4\omega+2\omega}{1-8\omega^2}\right)+\tan^{-1}(\omega)$$

$$180 = \tan^{-1}\left[\frac{\frac{6\omega}{1-8\omega^2}+\omega}{1-\frac{6\omega}{1-8\omega^2}\times\omega}\right]$$

$$\frac{6\omega}{1-8\omega^2} = -\omega$$

$$\omega = \sqrt{\frac{7}{8}}$$

For ultimate gain (K_c) AR must be equal to 1 at phase cross over frequency.

$$\frac{K_c}{\sqrt{1+\tau_1^2\omega^2}\times\sqrt{1+\tau_2^2\omega^2}\times\sqrt{1+\tau_3^2\omega^2}} = 1$$

$$K_c = \sqrt{\left(1+16\times\frac{7}{8}\right)\left(1+4\times\frac{7}{8}\right)\left(1+\frac{7}{8}\right)}$$

$$K_c = \frac{45}{4}$$

5.6 (a)

Open loop transfer function is given as

$$G(s) = \frac{1}{(s+1)(2s+1)(5s+1)}$$

Corner frequency (ω)	Slope
$\omega_1 = \frac{1}{5} = 0.2$	−1
$\omega_2 = \frac{1}{2} = 0.5$	−2
$\omega_3 = 1$	−3

So, $0 < \omega < 0.2$ Slope = 0

$\omega > 1$ Slope = −3

5.7 (b)

Characteristic equation is given as

$s^4 + 5s^3 - s^2 - 17s + 12 = 0$

Arranging Routh array

s^4	1	−1	12
s^3	5	−17	0
s^2	2.4	12	
s	−42	0	
s^0	12	0	

In first column sign changes two times so two roots lies right to imaginary axis.

5.8 (d)

$$G_p = \frac{20}{s-2} \text{ (Process transfer function)}$$

$G_c = K_c$ (Controller transfer function)

$G_f = G_m = 1$

Close loop characteristic equation is given by

$$1 + G_cG_pG_fG_m = 0$$

$$1+K_c\times\left(\frac{20}{s-2}\right) = 0$$

$$\frac{s-2+20K_c}{(s-2)} = 0$$

For stable system

$$-2 + 20K_c > 0$$

$$K_c > \frac{1}{10}$$

5.9 (a)

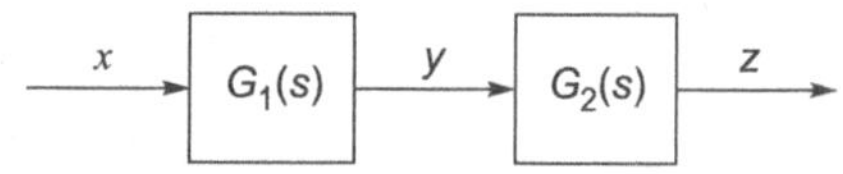

$$G_1(s) = \frac{1}{\tau_1 s+1},\quad G_2(s) = \frac{1}{\tau_2 s+1}$$

$$\tau_2 = 2\tau_1$$

System is excited by the sinusoidal input

$$X(t) = \sin\omega t$$

$$Y(t) = A\sin(\omega t+\phi)$$

$\phi = -45°$ and $\tau_1 = 1$

(Given Y lags behind input X by 45°C)

$$\phi = -\tan^{-1}(\tau_1\omega)$$

$$-45 = -\tan^{-1}(\omega)$$

$$\omega = 1$$

5.10 (d)

Amplitude of output z = ?

$$\text{Amplitude ratio} = \frac{1}{\sqrt{1+\tau_1^2\omega^2}}\times\frac{1}{\sqrt{1+\tau_2^2\omega^2}}$$

$$= \frac{1}{\sqrt{1+1}\times\sqrt{1+4}} = \frac{1}{\sqrt{10}}$$

So amplitude of z = 0.316

5.11 (b)

s^3	2	2
s^2	4	4

Characteristic equation will be

$2s^3 + 4s^2 + 2s + 4 = 0$

$s^3 + 2s^2 + s + 2 = 0$

$s^2(s + 2) + 1(s + 2) = 0$

$(s + 2)(s^2 + 1) = 0$

$s = -2$

$s = \pm j$

So third root on left half plane and two roots are $+j$ and $-j$.

5.12 (d)

$$G_p(s) = \frac{2e^{-s}}{(\tau s+1)^2}$$

At crossover frequency $\phi = -180°$

$$-180 = -2\tan^{-1}(\tau\cdot\omega_{co}) - \frac{180}{\pi}\times\omega_{co}$$

ω_{co} = 0.6 rad/min (given)

So, $$-180 = -2\tan^{-1}(\tau\times 0.6) - \frac{180}{\pi}\times 0.6$$

τ = 5.39 minutes

5.13 (b)

Gain margin (GM) = 2.0

$$GM = \left(\frac{1}{AR}\right)_{\text{at } \omega_{co}}$$

Amplitude ratio of $\frac{2Ke^{-s}}{(\tau s+1)^2}$ is given by

$$= \frac{2K}{\left(\sqrt{1+\tau^2\omega^2}\right)^2}$$

$$= \frac{2K}{\left(\sqrt{1+5.39^2 \times 0.6^2}\right)^2}$$

$$= \frac{2K}{11.458} = \frac{1}{GM}$$

$$K = \frac{11.458}{2\times 2} = 2.87$$

5.14 (a)

Characteristic equation of close loop system is given as

$12s^3 + 19s^2 + 8s + 1 + K_c$

Arranging Routh array

s^3	12	8	0
s^2	19	$1 + K_c$	0
S	$\frac{19\times 8 - 12(1+K_c)}{19}$		
S^0	$1 + K_c$		

At onset of instability

$19 \times 8 - 12(1 + K_c) = 0$

$$K_c = \frac{35}{3}$$

5.15 (c)

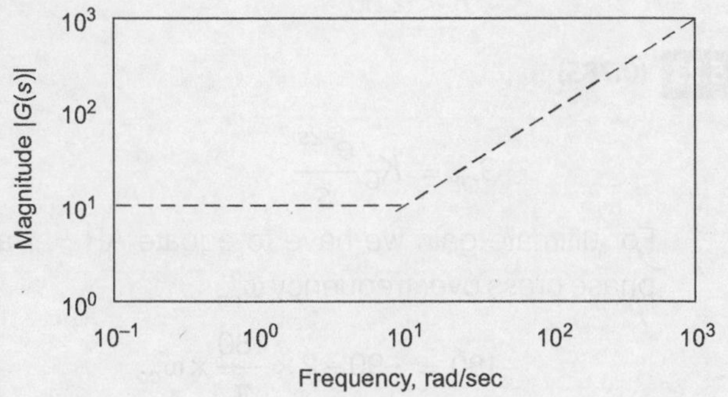

Gain, $k_p = 10$

and slope is positive at frequency $\omega_{co} = 10$ rad/s

So value of time constant

$$(\tau) = \frac{1}{\omega_{co}} = \frac{1}{10} = 0.1 \text{ sec}$$

$$G(s) = k_p(1 + \tau s)$$

$$= 10(1 + 0.1s) = 10 + s$$

5.16 (d)

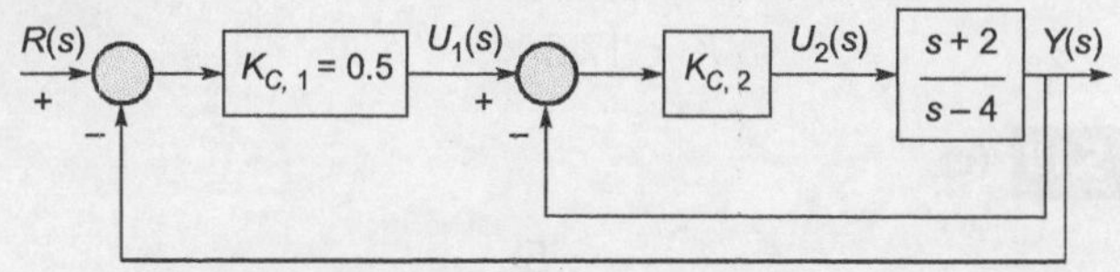

Close loop transfer function of given diagram is given by

$$\frac{Y(s)}{R(s)} = \frac{\left(\dfrac{\dfrac{K_{C,2}(s+2)}{s-4}}{1+K_{C,2}\left(\dfrac{s+2}{s-4}\right)}\right)0.5}{1+\left(\dfrac{K_{C,2}(s+2)}{s-4+K_{C,2}(s+2)}\right)0.5}$$

$$= \frac{\dfrac{0.5K_{C,2}(s+2)}{s-4+K_{C,2}(s+2)}}{\dfrac{s-4+K_{C,2}(s+2)+0.5K_{C,2}(s+2)}{(s-4)+K_{C,2}(s+2)}}$$

$$= \frac{0.5K_{C,2}(s+2)}{s-4+K_{C,2}(s+2)+0.5K_{C,2}(s+2)}$$

So characteristic equation is

$(s-4) + 1.5K_{C,2}(s+2) = 0$

$s(1 + 1.5K_{C,2}) + 3K_{C,2} - 4 = 0$

$$s = \frac{4-3K_{C2}}{1+1.5K_{C,2}}$$

For system to be stable poles must have negative real part.

Only option (d) which has value of $K_{C,2}$ of $\frac{3}{2}$ is satisfies this criteria.

5.17 (a)

Gain and phase curve decrease continuously with frequency.

5.18 (c)

Characteristic equation given in question is

$s^3 + 9s^2 + 26s + 12(2 + K_c) = 0$

Creating Routh array from above equation

s^3	1	26	0
s^2	9	$12(2+K_c)$	0
S	$\dfrac{9\times 26-12(2+K_c)}{9}$	0	
S^0	$12(2+K_c)$		

For system to be stable all elements from first column must be positive

$9 \times 26 - 12(2 + K_c) > 0$

$$K < 17.5$$

5.19 (c)

$$G_p(s) = \frac{5}{(2s+1)^4}$$

Cross over frequency is frequency at which phase angle $\phi = -180°$

So, $-180 = -4\tan^{-1}(2\omega)$

$$\omega = 0.5 \text{ rad/s}$$

5.20 (b)

Ultimate gain and ultimate period can be calculated by following method:

Ultimate period, $P_u = \dfrac{2\pi}{\omega_{co}}$

where, ω_{co} = Crossover frequency

$$P_u = \frac{2\pi}{0.5} = 4\pi$$

Ultimate gain (K_u) can be calculated by putting AR = 1 at crossover frequency

$$\text{AR of}\left(\frac{K_c\times 5}{(2s+1)^4}\right) = 1$$

$$\frac{5K_u}{\left(\sqrt{1+(2\times 0.5)^2}\right)^4} = 1$$

$$K_u = 0.8$$

5.21 (5.06)

[5 to 5.2]

$$G_{OL}(s) = \frac{K_c e^{-0.3s}}{1.5s+1}$$

Gain margin given is 1.7

For G.M. amplitude ratio of open loop transfer function at phase cross frequency is should be equated to 1:

$$\text{Gain margin} = \left(\frac{1}{\text{AR}}\right)_{\text{at } \omega_{co}}$$

For calculation of ω_{co} phase angle of open loop transfer function must be equated to $-180°$.

$$-180° = -\tan^{-1}(1.5\omega_{co}) - 0.3\times\frac{180}{\pi}\omega_{co}$$

From trial and error method

$$\omega_{cs} = 5.7 \text{ rad/min}$$

$$\text{AR} = \frac{K_c}{\sqrt{1+(1.5\times 5.7)^2}} = \frac{1}{1.7}$$

$$K_c = 5.06$$

5.22 (2.50)

PI controller with integral time constant $\tau_1 = 0.1$ used to control following process

$$G_p(s) = \frac{10}{s^2+2s+100}$$

$$G_m = 1, \quad G_f = 1 \quad \text{(given)}$$

Close loop transfer function characteristic equation is given by

$1 + G_c G_p G_f G_m = 0$

$$1+K_C\left(1+\frac{1}{\tau_1 s}\right)\left(\frac{10}{s^2+2s+100}\right) = 0$$

$$1+K\left(1+\frac{10}{s}\right)\left(\frac{10}{s^2+2s+100}\right) = 0$$

$s^3 + 2s^2 + 100s + 10K(s + 10) = 0$

$s^3 + 2s^2 + (100 + 10K)s + 100K = 0$

From Routh array arrangement

s^3	1	$100+10K$
s^2	2	$100K$
S	$\dfrac{200+20K-100K}{2}$	0
S^0	$100K$	

For stability all elements of first column should be positive so

$200 + 20K - 100K > 0$

$$K < 2.50$$

5.23 (0.785)

$$G_{OL} = K_C\frac{e^{-2s}}{s}$$

For ultimate gain we have to equate AR = 1 at phase cross over frequency ω_{co}

$$-180 = -90 - 2\times\frac{180}{\pi}\times\omega_{co}$$

$$\omega_{co} = 0.785$$

$$AR = \frac{K}{\omega_{co}} = \frac{K}{0.785}$$

For K_u, $AR = 1$

$$K_u = 0.785$$

5.24 (10)

$6s^3 + 11s^2 + 6s + (1 + K) = 0$

Routh array for above close loop characteristic equation can be written as

s^3	6	6	0
s^2	11	$1 + K$	
	↓	↓	
S^1	$\frac{66 - 6(1+K)}{11}$	0	
S^0	$1 + K$		

For stable system all the elements of first column should be positive so

$$11 \times 6 - 6(1 + K) \geq 0$$

$$K \leq 10$$

Beyond ($K = 10$) system becomes just unstable.

5.25 (0.413)

$$G(s) = \frac{3}{(5s+1)^2}$$

Amplitude ratio is given as

$$AR = \frac{3}{\left(\sqrt{(1+5^2\omega^2)}\right)^2}$$

Here, $\omega = 0.5$ rad/min

So,
$$AR = \frac{3}{\left(\sqrt{(1+5^2 \times 0.5^2)}\right)^2} = 0.413$$

5.26 (45.38)

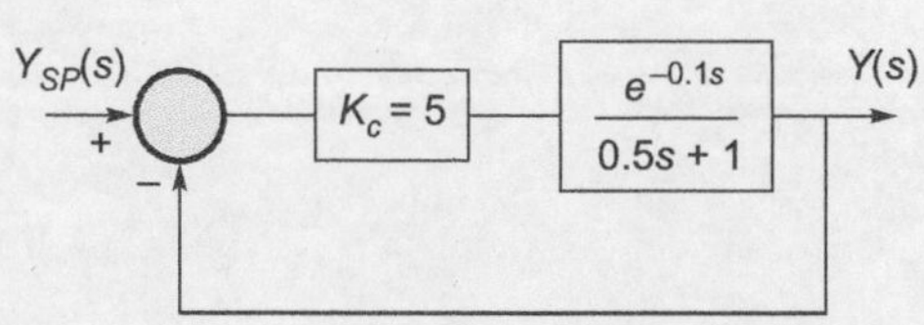

Open loop transfer function of above process loop is given by

$$G_{op} = K_c \frac{e^{-0.1s}}{0.5s+1} = \frac{5e^{-0.1s}}{0.5s+1}$$

For calculating phase margin amplitude ratio of transfer function must be equated to 1.

$$A.R = 1$$

$$\frac{5}{\sqrt{(0.5\omega)^2 + 1}} = 1 \Rightarrow \omega_{pc} = 9.8$$

$$\text{Phase angle } (\phi) = -\tan^{-1}(\tau \cdot \omega_{pc}) - t_d \times \frac{180}{\pi}\omega_{pc}$$

$$\phi = -\tan^{-1}(9.8 \times 0.5) - 0.1 \times \frac{180}{\pi} \times (9.8)$$

$$\phi = -134.615$$

$$\text{Phase margin} = 180 - |\phi|$$

$$= 180 - |134.615| = 45.38$$

5.27 (0.5)

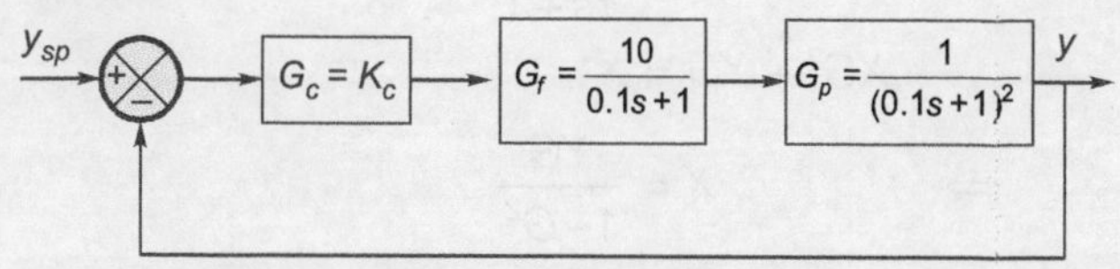

Open loop transfer function of above block diagram is given by

$$G_{op}(s) = K_C\left(\frac{10}{(0.1s+1)}\right)\left(\frac{1}{(0.1s+1)^2}\right)$$

$$= \frac{K_c \times 10}{(0.1s+1)^3}$$

For phase cross over frequency, phase angle should to equated to −180°

$$-180 = -3\tan^{-1}(\tau\omega_{co})$$

$$-180 = -3\tan^{-1}(0.1\,\omega_{co})$$

$$\omega_{co} = 17.32 \text{ rad/sec}$$

Gain margin is 1.6 given in the question

$$\text{Gain margin} = \left(\frac{1}{AR}\right)_{\text{at } \omega_{co}}$$

AR is amplitude ratio of corresponding transfer function

So,
$$A.R = \frac{K_C \times 10}{\sqrt{1+\tau^2\omega^2}} \text{ (for first order system)}$$

$$A.R = \frac{K_C \times 10}{\left(\sqrt{1+0.1^2 \times 17.32^2}\right)^3} = \frac{1}{\text{G.M.}}$$

$$\frac{K_C \times 10}{7.999} = \frac{1}{1.6}$$

$$K_C = 0.5$$

5.28 (a)

Dead-time term(s) must be absent in the open-loop transfer function is not a necessary condition for a process under closed-loop control to be stable.

5.29 (0.707)

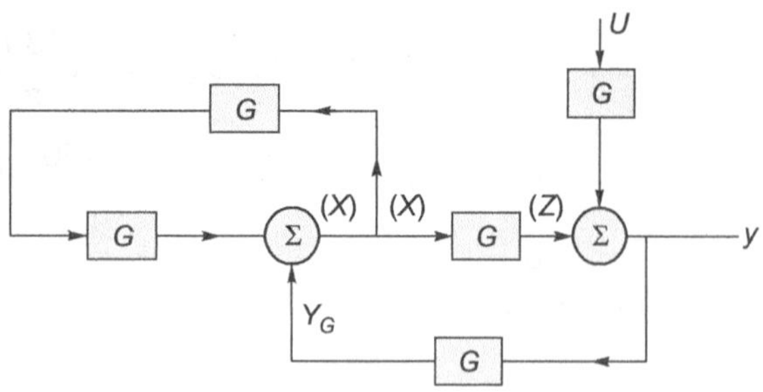

Given: $G = \dfrac{k}{\tau s+1}$

$$XG^2 + YG = X$$

$$\Rightarrow \quad X = \frac{YG}{1-G^2}$$

$$Z = XG$$

$$Y = Z + UG$$

$$Y = XG + UG$$

$$Y = \frac{YG^2}{1-G^2} + UG$$

$$Y\left[1-\frac{G^2}{1-G^2}\right] = UG$$

$$\frac{Y}{U} = \frac{G}{\left[1-\dfrac{G^2}{1-G^2}\right]} = \frac{(1-G^2)G}{1-2G^2}$$

Characteristic equation $= 1 - 2G^2$

$$= 1-\frac{2k^2}{(\tau s+1)^2}$$

$$(\tau s + 1)^2 - 2k^2 = 0$$

$$\tau^2 s^2 + 2\tau s + 1 - 2k^2 = 0$$

Routh array

s^2	τ^2	$1-2k^2$
s^1	2τ	0
s^0	$1-2k^2$	

For system to be stable, given $\tau > 0$ and $k > 0$

$$1 - 2k^2 \geq 0$$

$$2k^2 \leq 1$$

$$k^2 \leq \frac{1}{2}$$

$$k \leq \frac{1}{\sqrt{2}} = 0.707$$

■■■■

6 Sensors, Transducers and P & ID Symbol

6.1 The calibration data of a thermocouple with its cold junction at 0°C are given below:

Hot junction temperature (°C)	0	20	40	60	80	100
Thermo emf (mV)	0.00	0.80	1.61	2.43	3.26	4.10

The hot junction of the thermocouple is placed in a bath at 80°C while its cold junction is at 20°C. What is the emf of the thermocouple?

(a) 3.26 mV (b) 0.80 mV
(c) 2.46 mV (d) 2.43 mV

[2001 : 1 M]

6.2 Match the measured process variables with the list of measuring devices given below:

List-I (Measured process variables)
A. Temperature
B. Pressure
C. Flow
D. Liquid level
E. Composition

List-II (Measuring devices)
1. Bourdon tube element
2. Orifice plates
3. Infrared analyzer
4. Displacer devices
5. Pyrometer

Codes:

	A	B	C	D	E
(a)	5	1	2	4	3
(b)	3	1	4	2	5
(c)	1	3	4	2	5
(d)	3	1	2	4	5

[2003 : 1 M]

6.3 Match the process variables **(Group-I)** given below with the measuring devices **(Group-II)**:

Group-I
A. High temperature
B. Flow
C. Composition

Group-II
1. Orifice meter
2. Chromatograph
3. Radiation pyrometer
4. Bi-metallic thermometer

Codes:

	A	B	C
(a)	1	2	3
(b)	1	3	2
(c)	3	1	2
(d)	4	2	1

[2005 : 2 M]

6.4 A weighing machine is calibrated at 25°C and the output reading R (in mm) is related to the weight W (in kg) by the equation $R = sW$ where the sensitivity s = 20 mm/kg. At a temperature of 30°C, the weighing machine undergoes a zero drift (change in instrument output reading at zero value of weight) of 2 mm and its sensitivity changes to 20.5 mm/kg. The weighing machine when used at 30°C shows a reading of 50 mm. The true weight (in kg) of the object is

(a) 2.34 (b) 2.40
(c) 2.44 (d) 2.50

[2006 : 2 M]

6.5 Match the following:

Group-1
P. Temperature
Q. Pressure
R. Flow

Group-2
1. Hot wire anemometry
2. Strain gauge
3. Chromatographic analyzer
4. Pyrometer

(a) P-1, Q-2, R-3 (b) P-4, Q-1, R-3
(c) P-1, Q-2, R-4 (d) P-4, Q-2, R-1

[2008 : 2 M]

6.6 Which one of the following sensors is used for the measurement of temperature in a combustion process (T > 1800°C)?

(a) Type J thermocouple
(b) Thermistor
(c) Resistance temperature detector
(d) Pyrometer

[2009 : 1 M]

6.7 Flow measuring instruments with different specifications (zero and span) are available for an application that requires flow rate measurements in the range of 300 litres/h to 400 litres/h. The appropriate instrument for this application is the one whose specifications are

(a) zero = 175 litres/h, span = 150 litres/h
(b) zero = 375 litres/h, span = 100 litres/h
(c) zero = 275 litres/h, span = 150 litres/h
(d) zero = 475 litres/h, span = 100 litres/h

[2010 : 2 M]

6.8 Match the process parameters in **Group-I** with the measuring instruments in **Group-II**:

Group-I
P. Flame temperature
Q. Composition of LPG
R. Liquid air temperature

Group-II
1. Thermocouple
2. Radiation pyrometer
3. Gas chromatograph

(a) P-3, Q-1, R-2
(b) P-1, Q-3, R-2
(c) P-2, Q-3, R-1
(d) P-2, Q-1, R-3

[2011 : 1 M]

6.9 The range of standard current signal in process instruments is 4 to 20 mA. Which one of the following is the reason for choosing the minimum signal as 4 mA instead of 0 mA?

(a) To minimise resistive heating in instruments.
(b) To distinguish between signal failure and minimum signal condition.
(c) To ensure a smaller difference between maximum and minimum signal
(d) To ensure compatibility with other instruments

[2011 : 1 M]

6.10 A thermocouple having a linear relationship between 0°C and 350°C shows an emf of zero and 30.5 mV, respectively at these two temperatures. If the cold junction temperature is shifted from 0°C to 30°C, then the emf correction (in mV) is

(a) 3.13 (b) 2.92
(c) 2.61 (d) 2.02

[2012 : 2 M]

6.11 Match the following:

Group-1	Group-2
P. Viscosity	1. Pyrometer
Q. Pressure	2. Hot wire anemometer
R. Velocity	3. Rheometer
S. Temperature	4. Piezoelectric element

(a) P-4, Q-3, R-1, S-2
(b) P-3, Q-4, R-2, S-1
(c) P-3, Q-4, R-1, S-2
(d) P-4, Q-3, R-2, S-1

[2013 : 1 M]

6.12 Match the output signals as obtained from four measuring devices in response to a unit step change in the input signal.

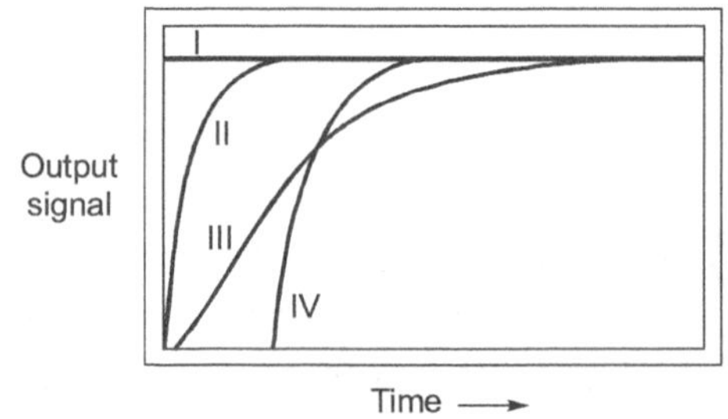

P : Gas chromatograph, with a long capillary tube
Q : Venturi tube
R : Thermocouple with first order dynamics
S : Pressure transducer with second order dynamics

(a) P-IV, Q-III, R-II, S-I
(b) P-III, Q-I, R-II, S-IV
(c) P-IV, Q-I, R-II, S-III
(d) P-II, Q-IV, R-III, S-I

[2015 : 1 M]

6.13 Match the instruments in **Group-1** with process variables in **Group-2**:

Group-I	Group-2
P. Conductivity meter	I. Flow
Q. Turbine meter	II. Pressure
R. Piezoresistivity element	III. Composition

(a) P-II, Q-I, R-III (b) P-II, Q-III, R-I
(c) P-III, Q-II, R-I (d) P-III, Q-I, R-II

[2016 : 1 M]

6.14 Match the variables in **Group-1** with the instruments in **Group-2**:

Group-I	Group-2
P. Temperature	I. Capacitance probe
Q. Liquid level	II. McLeod gauge
R. Vacuum	III. Chromatograph
S. Concentration	IV. Thermistor

Choose the correct set of combinations.
(a) P-IV, Q-III, R-II, S-I
(b) P-I, Q-II, R-IV, S-III
(c) P-IV, Q-I, R-II, S-III
(d) P-IV, Q-I, R-II, S-IV

[2017 : 1 M]

6.15 An LVDT (Linear Variable Differential Transformer) is a transducer used for converting
(a) displacement to voltage
(b) voltage to displacement
(c) resistance to voltage
(d) voltage to current

[2017 : 1 M]

6.16 A thermocouple senses temperature based on the
(a) Nernst effect
(b) Maxwell effect
(c) Seebeck effect
(d) Peltier effect

[2019 : 1 M]

6.17 Hot wire anemometer is used for the measurement of
(a) composition (b) flow
(c) pressure (d) temperature

[2020 : 1 M]

6.18 Which of the following is NOT a standard to transmit measurement and control signals?
(a) 4–20 mA (b) 3–15 psig
(c) 0–100% (d) 1–5 VDC

[2021 : 1 M]

6.19 Which of these symbols can be found in piping and instrumentation?

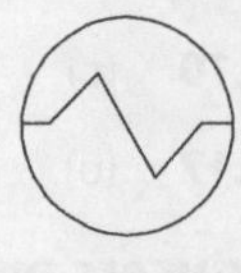
(P)

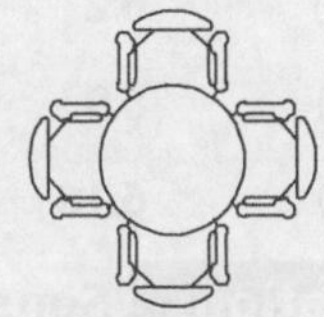
(Q)

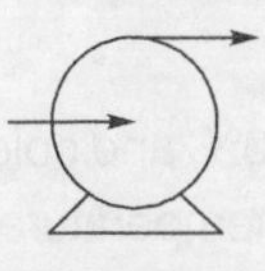
(R)

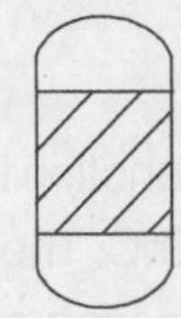
(S)

(a) (Q) and (S) only
(b) (P), (Q) & (R) only
(c) (P), (R) and (S) only
(d) (P), (Q), (R) and (S)

[2021 : 2 M]

6.20 Choose the option that correctly pairs the given measurement devices with the quantities they measure.

S No.	Measurement Device	S No.	Measured Quantity
I	Bourdon gauge	A	Temperature
II	Orifice plate meter	B	Concentration
III	Pyrometer	C	Pressure
IV	Colorimeter	D	Flow rate
V	Pirani gauge	E	Liquid level

(a) I - E, II - C, III - D, IV - B, V - A
(b) I - C, II - D, III - A, IV - B, V - C
(c) I - C, II - D, III - E, IV - A, V - D
(d) I - D, II - C, III - A, IV - E, V - C

[2022 : 1 M]

■■■■

Answers Sensors, Transducers and P & ID Symbol

6.1	(d)	**6.2**	(a)	**6.3**	(c)	**6.4**	(a)	**6.5**	(d)	**6.6**	(d)	**6.7**	(c)
6.8	(c)	**6.9**	(b)	**6.10**	(c)	**6.11**	(b)	**6.12**	(c)	**6.13**	(d)	**6.14**	(c)
6.15	(a)	**6.16**	(c)	**6.17**	(b)	**6.18**	(c)	**6.19**	(c)	**6.20**	(b)		

Explanations Sensors, Transducers and P & ID Symbol

6.1 (d)

If hot junction is in a bath of 80°C and cold junction is at 20°C. Then difference in temperature between hot junction and cold junction is 80 – 20 = 60°C emf of thermocouple only depends upon temperature difference between hot junction and cold junction so from data emf = 2.43 mV.

6.4 (a)

Machine shows a zero drift of + 2 mm means 2 mm reading for no weight.

So for reading $(R-2) = 20.5 \times W$

When, $R = 50$ mm, $W = ?$

$$W = \frac{50-2}{20.5} = 2.341 \text{ kg}$$

6.6 (d)

For high temperature measurement we use radiation pyrometer because radiation pyrometer works on Stefan Boltzmann law and need not to come in contact with object for temperature measurement.

6.7 (c)

Instrument required to measure flow rate in range of 300 liter/hr to 400 liter/hr

For option (c)

Zero = 275 liter/hr

Span = 150 liter/hr

So instrument can measure flow rate in range of 275 to (275 + 150) = 425 liters/hr.

6.9 (b)

To distinguish between signal failure and minimum signal condition.

6.10 (c)

For thermocouple, emf $\propto$ temperature difference between hot and cold junctions

emf $\propto \Delta T$

So emf for shifted cold junctions thermocouple is

$$\Rightarrow \quad \frac{320}{350} \times 30.5 = 27.88 \text{ mV}$$

So emf correction = 30.5 – 27.88 = 2.61 mV

6.12 (c)

Gas chromatograph with a long capillary tube will shows a significance time delay. IV will be response of it.

Venturi tube is instruments which shows a steady state response as soon as it get the flow rate in pipe I will be response of it.

Curve II shows a unit step response of first order system. Similarly curve III shows a 2nd order unit step response.

6.13 (d)

Turbine meter measures flow rate

Conductivity meter measures composition

Pressure is measured by piezoresistivity element

6.15 (a)

An LVDT (linear variable differential transformer) is a transducer used for converting displacement to voltage.

6.16 (c)

Seebeck effect is when any two points across an electrically conducting material is subjected to a temperature difference then an emf will develop across these points. Thermocouple senses temperature based on this phenomenon.

6.17 (b)

Hot-wire anemometer is used for the flow measurement.

6.18 (c)

0–100% is not a standard to transmit measurement and control signals.

■■■■

Plant Design and Economics

UNIT

IX

CONTENTS

Plant Design and Economics

Syllabus

Principles of Process Economics and Cost Estimation including Depreciation and Total Annualized Cost, Cost Indices, Rate of Return, Payback Period, Discounted Cash Flow, Optimization in Process Design and Sizing of Chemical Engineering Equipments such as Heat Exchangers and Multistage Contactors.

Analysis of Previous GATE Papers

Exam Year	1 Mark Ques.	2 Marks Ques.	5 Marks Ques.	Total Marks
2000	1	–	1	**6**
2001	3	3	–	**9**
2002	1	3	–	**7**
2003	2	3	–	**8**
2004	–	2	–	**4**
2005	–	5	–	**10**
2006	–	4	–	**8**
2007	1	1	–	**3**
2008	2	1	–	**4**
2009	1	1	–	**3**
2010	1	3	–	**7**
2011	1	3	–	**7**
2012	–	2	–	**4**
2013	–	2	–	**4**
2014	–	2	–	**4**
2015	–	2	–	**4**
2016	3	1	–	**5**
2017	1	2	–	**5**
2018	1	2	–	**5**
2019	–	2	–	**4**
2020	1	–	–	**1**
2021	3	2	–	**7**
2022	1	1	–	**3**
2023	–	3	–	**6**
2024	–	1	–	**2**

1 Cost Estimation

1.1 Direct costs component of the fixed capital consists of

(a) Contingency
(b) Onsite and offsite costs
(c) Labour costs
(d) Raw material costs

[2003 : 1 M]

1.2 For a solid processing plant, the delivered equipment cost is Rs. 10 lakhs. Using Lang multiplication method, the total capital investment, in lakhs of rupees, is

(a) 46 (b) 57
(c) 100 (d) 200

[2005 : 2 M]

1.3 The cost of a drum dryer is Rs. 10 lakhs. The cost of a drum dryer with double the surface area in lakhs of rupees is

(a) 2×10 (b) $3^{0.6} \times 10$
(c) $5^{0.6} \times 10$ (d) $2^{0.6} \times 10$

[2005 : 2 M]

1.4 The cost of a distillation column in the year 2000 is x rupees. What is the cost of the column in rupees in the year 2010 given the cost indices for the years 2000 and 2010 are 480 and 520 respectively?

(a) $\left(\frac{520}{480}\right)^2 x$ (b) $\left(\frac{480}{520}\right) x$
(c) $\left(\frac{520}{480}\right) x$ (d) $\left(\frac{520}{480}\right)^{0.6} x$

[2005 : 2 M]

1.5 In a desalination plant, an evaporator of area 200 m^2 was purchased in 1996 at a cost of \$3,00,000. In 2002, another evaporator of area 50 m^2 was added. What was the cost of the second evaporator in year 2002? Assume that the cost of evaporator scales as $(\text{capacity})^{0.54}$. The Marshall and Swift index was 1048.5 in 1996 and 1116.9 in 2002.

(a) 1,30,500 (b) 1,39,100
(c) 1,41,900 (d) 1,51,200

[2006 : 2 M]

1.6 The total fixed cost of a chemical plant is Rs. 10.0 lakhs; the internal rate of return is 15%, and the annual operating cost is Rs. 2.0 lakhs. The annualized cost of the plant (in lakhs of Rs.) is

(a) 1.8 (b) 2.6
(c) 3.5 (d) 4.3

[2009 : 1 M]

1.7 A continuous fractionator system is being designed. The following cost figures are estimated for a reflux ratio of 1.4

Fixed cost including all accessories (Rs.) for			Operating cost (Rs./year) for	
Column	Condenser	Reboiler	Condenser cooling water	Reboiler heating steam
6×10^6	2×10^6	4×10^6	8×10^6	1×10^6

The annualised fixed charge is 15% of the fixed cost. The total annualised cost (in Rs.) is

(a) 10.8×10^6 (b) 13.35×10^6
(c) 15.9×10^6 (d) 3.15×10^6

[2011 : 2 M]

1.8 The purchase cost of a heat exchanger of 20 m^2 area was Rs. 500000 in 2006. What will be the estimated cost (in Rs. to the nearest integer) of a similar heat exchanger of 50 m^2 area in the year 2013? Assume the six-tenths factor rule for scaling and the cost index for 2006 as 430.2. The projected cost index for the year 2013 is 512.6. ________.

[2013 : 2 M]

1.9 Terms used in engineering economics have standard definitions and interpretations. Which one of the following statements is INCORRECT?

(a) The profitability measure 'return on investment' does not consider the time value of money
(b) A cost index is an index value for a given time showing the cost at that time relative to a certain base time

(c) The 'six-tenths factor rule' is used to estimate the cost of an equipment from the cost of a similar equipment with a different capacity

(d) Payback period is calculated based on the payback time for the sum of the fixed and the working capital investment [2016 : 1 M]

1.10 In the year 2005, the cost of a shell and tube heat exchanger with 68 m^2 heat transfer area was Rs. 12.6 Lakh. Chemical Engineering Index for cost in 2005 was 509.4 and now the index is 575.4. Based on index of 0.6 for capacity scaling, the present cost (in Lakhs of rupees) of a similar heat exchanger having 100 m^2 heat transfer area is estimated to be

(a) 17.94 (b) 19.94
(c) 20.94 (d) 22.94

[2018 : 2 M]

1.11 A furnace installed at a cost of Rs. 24 Lakh is expected to serve its useful life of 5 years. Salvage value of the furnace is Rs. 8 Lakh. The interest rate compounded annually is 8%. The estimated capitalized cost (in Lakhs of rupees) is

(a) 30 (b) 34.09
(c) 34.9 (d) 58.09

[2018 : 2 M]

1.12 A company invests in a recovery unit to separate valuable metals from effluent streams. The total initial capital investment of this unit is Rs. 10 lakhs. The recovered metals are worth Rs. 4 lakhs per year.

If the annual return on this investment is 15%, the annual operating costs should be _____ lakhs of rupees (correct to 1 decimal place).

[2021 : 1 M]

1.13 The purchase cost of a new distillation column is Rs.10 lakhs with an installation factor of 5.8. The cost of the capital is to be annualized over a period of 6 years at a fixed rate of interest of 5 % per annum, compounded annually. The annual cost (in lakhs of rupees) of the installed capital is _______ (rounded off to one decimal place).

[2023 : 2 M]

Answers Cost Estimation

1.1	(b)	1.2	(a)	1.3	(d)	1.4	(c)	1.5	(d)	1.6	(c)	1.7	(a)
1.8	(1032386.23)	1.9	(d)	1.10	(a)	1.11	(d)	1.12	(2.5)	1.13	(11.4) [11.0 - 11.9]		

Explanations Cost Estimation

1.2 (a)

Total capital investment = Lang multiplication factor × Delivered equipment cost

∴ Total capital investment = 4.6 × 10
= Rs. 46 lakhs

1.3 (d)

Cost of drum dryer with double surface area is given as

$$C_2 = C_1 \times (2)^{0.6} = \text{Rs. } 10 \times 2^{0.6}$$

1.4 (c)

$$\frac{\text{Cost in year } X}{\text{Cost index in year } X} = \frac{\text{Cost in year } Y}{\text{Cost index in year } Y}$$

So, cost in year 2010 = Rs. $\left(\frac{520}{480}\right)x$

1.5 (d)

Cost of new evaporator can be calculated by following formula:

$$C_2 = C_1\left(\frac{P_2}{P_1}\right)^n\left(\frac{\text{Cost index in 2002}}{\text{Cost index in 1996}}\right)$$

where, P = Represent capacity in m^2
n = Capacity index
C = Cost of evaporator

So, $$C_2 = 300000\left(\frac{50}{200}\right)^{0.54}\left(\frac{1116.9}{1048.5}\right)$$

C_2 = Rs. 151166

1.6 (c)

Total fixed cost of plant is Rs. 10 lakhs.
Internal rate of return is 15%.
Annual operating cost is Rs. 2 lakhs.

$$2 + (10 \times 0.15) = 3.5 \text{ lakhs}$$

1.7 (a)

The annualised fixed charge is 15% of fixed cost.
So, annualised fixed charge

$$F_c = (6 + 2 + 4) \times 0.15 \times 10^6$$
$$= \text{Rs. } 1.8 \times 10^6$$

Annualised operating cost

$$= (8 + 1) \times 10^6 = \text{Rs. } 9 \times 10^6$$

Total annual cost

$$= \text{Fixed cost} + \text{Operating cost}$$
$$= (1.8 + 9) \times 10^9 = \text{Rs. } 10.8 \times 10^6$$

1.8 (1032386.23)

Purchase cost of a new similar equipment in present year (C_2) can be calculated based on cost (C_1) of same equipment with different capacity in a different year by following formula :

$$C_2 = C_1\left(\frac{P_2}{P_1}\right)^n \left(\frac{\text{Cost index in present year}}{\text{Cost index in old year}}\right)$$

where P represents capacity and n represents six-tenths factor.

So, cost of heat exchanger of 50 m² in 2013

$$C_2 = 5\times10^5\left(\frac{50}{20}\right)^{0.6}\left(\frac{512.6}{430.2}\right)$$
$$C_2 = \text{Rs. } 1032386.23$$

1.9 (d)

In the calculation of payback period we do not consider working capital investment.

$$\text{Payback period} = \frac{\text{Fixed capital investment}}{\left(\dfrac{\text{Average profit}}{\text{Year}}\right) + \left(\dfrac{\text{Average depreciation}}{\text{Year}}\right)}$$

So, option (d) is incorrect.

1.10 (a)

Present cost of heat exchanger can be calculated by following formula.

Present cost = (Cost in year 2005) ×

$$\left(\frac{\text{Present cost index}}{\text{Cost index in 2005}}\right) \times \left(\frac{C_2}{C_1}\right)^n$$

where, C_2 = New capacity of heat exchanger in m²

C_1 = Old capacity of heat exchanger in m²

n = Cost index

So, present cost of 100 m² heat exchanger

$$= 12.6\left(\frac{575.4}{509.4}\right)\left(\frac{100}{68}\right)^{0.6}$$
$$= \text{Rs. } 17.94 \text{ lakhs}$$

1.11 (d)

Capitalized cost (K) is given as $= C_V + \dfrac{C_R}{(1+i)^n - 1}$

where, C_V = Original cost

C_R = Replacement cost = Original cost – Salvage value

i = Annual interest rate

n = Service life

Here, C_V = Rs. 24 lakhs

n = 5 years

i = 0.08

$C_R = 24 - 8 =$ Rs. 16 lakhs

So,

$$K = 24 + \frac{16}{(1.08)^5 - 1}$$
$$= \text{Rs. } 58.09 \text{ lakhs}$$

1.12 (2.5)

Annual rate of return = 15%

So, $\dfrac{\text{Profit}}{\text{FCI}} \times 100 = 15$

$\dfrac{4 - C}{10} \times 100 = 15$ (C is annual operating cost)

$C = 2.5$ lakhs/year

■■■■

2 Interest and Investment Cost

2.1 An investment of Rs. 1000 is carrying an interest of 10% compounded quarterly. The value of the investment at the end of five years will be

(a) $1000\left(1+\frac{0.1}{4}\right)^{20}$ (b) $1000(1+0.10)^{20}$

(c) $1000\left(1+\frac{0.1}{4}\right)^{5}$ (d) $1000\left(1+\frac{0.1}{2}\right)^{5}$

[2001 : 2 M]

2.2 If an amount R is paid at the end of every year for n years, then the net present value of the annuity at an interest rate of i is

(a) $R\left[\frac{(1+i)^n-1}{i}\right]$ (b) $R\left[\frac{(1+i)^n-1}{i(1+i)^n}\right]$

(c) $R(1+i)^n$ (d) $\frac{R}{(1+i)^n}$

[2002 : 2 M]

2.3 A bond has a maturity value of Rs. 20,000/- at the end of 4 years. The interest is compounded at the rate of 5% per year.

The initial investment to be made, rounded to the nearest integer, is Rs. ______.

[2017 : 2 M]

2.4 A principal amount is charged a nominal annual interest rate of 10%. If the interest rate is compounded continuously, the final amount at the end of one year would be

(a) higher than the amount obtained when the interest rate is compounded monthly.
(b) lower than the amount obtained when the interest rate is compounded annually.
(c) equal to 1.365 times the principal amount.
(d) equal to the amount obtained when using an effective interest rate of 27.18%.

[2021 : 1 M]

2.5 A design engineer needs to purchase a membrane module (M) for a plant. Details about the two available options, M1 and M2, are given in the table below. The overall plant has an expected life of 7 years. If the interest rate is 8% per annum, compounded annually, the difference in the net present value (NPV) of these two options, in lakhs of rupees, is ______ (rounded off to one decimal place).

	M1	M2
Purchase cost (in lakhs of rupees)	10	5
Expected life (years)	5	3

[2023 : 2 M]

2.6 Pumps A and B are being considered for purchase in a chemical plant. Cost details for these two pumps are given in the table below. The interest rate is 10% per annum, compounded annually. For both the pumps to have the same capitalized cost, the salvage value (in Rs.) of pump B should be ______ (rounded off to the nearest integer).

Item	**Pump *A***	**Pump *B***
Installed cost (Rs.)	16000	32000
Uniform end of year maintenance (Rs.)	2400	1600
Salvage value (Rs.)	1000	?
Service life (years)	1	2

[2023 : 2 M]

2.7 For purchasing a batch reactor, three alternatives P, Q and R have emerged, as summarized in the table. For a compound interest rate of 10% per annum, choose the correct option that arranges the alternatives, in order, from the least expensive to the most expensive.

	P	**Q**	**R**
Installed cost (lakh rupees)	15	25	35
Equipment life (years)	3	5	7
Maintenance cost (lakh rupees per year)	4	3	2

(a) P, Q, R (b) R, P, Q
(c) R, Q, P (d) Q, R, P [2024 : 2 M]

■■■■

Answers Interest and Investment Cost

2.1 (a) 2.2 (b) 2.3 (16454) 2.4 (a) 2.5 (4.7) 2.6 (2180) 2.7 (c)

Explanations Interest and Investment Cost

2.1 (a)

The amount after compound is given as

$$S = P\left(1+\frac{i}{m}\right)^{m \times n}$$

where, m = Number of time compounded per year
n = Number of years
P = Present value
i = Interest rate per year

$$S = 1000\left(1+\frac{0.1}{4}\right)^{4\times 5}$$

$$= 1000\left(1+\frac{0.1}{4}\right)^{20}$$

2.2 (b)

The net present value of the annuity (R) at an interest rate of i is given as

$$= \frac{R[(1+i)^n - 1]}{i(1+i)^n}$$

where, R = Annuity

2.3 (16454)

Maturity value of bond at the end of 4 years

= Rs. 20000

Interest rate = 5% per year

Maturity value of a bond can be related with initial investment by following formula:

$$s = P(1+i)^n$$

where, s = Maturity value
P = Initial investment
i = interest rate (nominal)
n = years

So, $20000 = P(1+0.05)^4$

P = Rs. 16454.05

2.4 (a)

Future value of any investment for continuous compounding can be calculated by

$$S = Pe^{rn}$$

Here, $r = 0.1$
n = 1 year
$S = Pe^{0.1} = 1.105P$
$i_{\text{eff}} = e^r - 1 = e^{0.1} - 1 = 10.5\%$

Above calculation indicates that final amount at the end of 1 year is higher than amount obtained if interest rate is compounded monthly.

2.5 (4.7)

For the first module:

$$10+\frac{10}{(1+0.08)^5} = 16.80$$

Similarly for the second module:

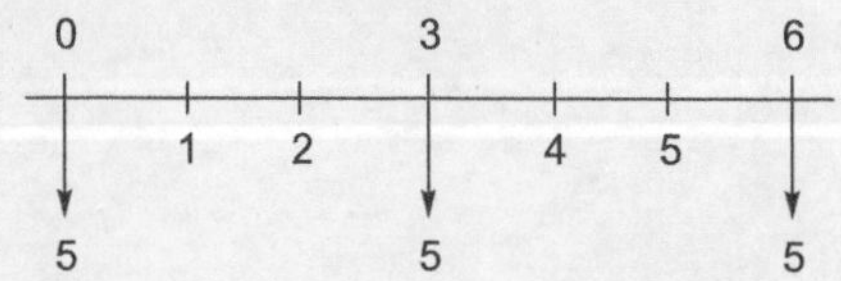

$$5+\frac{5}{(1.08)^3}+\frac{5}{(1.08)^6} = 12.12$$

$$NPV_1 - NPV_2 = 4.67$$

2.6 (2180)

Given:

Item	Pump *A*	Pump *B*
Installed cost (Rs.)	16000	32000
Uniform end of year maintenance (Rs.)	2400	1600
Salvage value (Rs.)	1000	?
Service life (years)	1	2

i = 10% per year

Now, given [both the instrument have equal capitalized cost], therefore

$$\left[V_0+\frac{V_0-V_s}{(1+i)^n-1}+\frac{V_m}{i}\right]_A = \left[V_0+\frac{V_0-V_s}{(1+i)^n-1}+\frac{V_m}{i}\right]_B$$

$$16000+\frac{16000-1000}{(1.1)^1-1}+\frac{2400}{0.1} = 32000+\frac{32000-V_s}{(1.1)^2-1}+\frac{1600}{0.1}$$

$$142000 = \frac{32000-V_s}{(1.1)^2-1}$$

Solving we get, V_s = 2180 Rs.

2.7 (c)

For purchasing a batch reactor three alternatives P, Q and R are summarized in the table:

	P	Q	R
Installed cost (lakh rupees)	15	25	35
Equipment life (years)	3	5	7
Maintenance cost (lakh rupees per year)	4	3	2

$$\text{Capitalized cost} = C_V + \frac{C_R}{(1+i)^n - 1}$$

$$\text{Capitalized maintenance cost} = \frac{C_M}{i}$$

C_V = Original cost

C_R = Replacement cost

C_M = Maintenance cost

i = 0.1 (given)

Now total capitalized cost:

For reactor P $15 + \dfrac{15}{1.1^3 - 1} + \dfrac{4}{0.1} = 100.317$ lakh

For reactor Q $25 + \dfrac{25}{1.1^5 - 1} + \dfrac{3}{0.1} = 95.949$ lakh

For reactor R $35 + \dfrac{35}{1.1^7 - 1} + \dfrac{2}{0.1} = 91.891$ lakh

So reaction R is least expensive and reaction P is most expensive.

■■■■

3 Taxes and Depreciation

3.1 P is the investment made on an equipment, S is its salvage value and n is the life of the equipment in years. The depreciation for the mth year by the Sum-of-Years- Digits method will be

(a) $\frac{P-S}{n}$ (b) $1-\left(\frac{P}{S}\right)^{1/m}$

(c) $\frac{m}{n}(P-S)$ (d) $\frac{2(n-m+1)}{n(n+1)}(P-S)$

[2001 : 2 M]

3.2 The original value of an equipment is Rs. 10000/-. The salvage value is Rs. 500/- at the end of its useful life period of 5 years. What is the asset value in rupees after two years by textbook declining balance method?

(a) 3025/- (b) 4010/-
(c) 5020/- (d) 6050/-

[2005 : 2 M]

3.3 A pump has an installed cost of Rs. 40,000 and a 10 year estimated life. The salvage value of the pump is zero at the end of 10 years. The pump value (in rupees), after depreciation by the double declining balance method, at the end of 6 years is

(a) 4295 (b) 10486
(c) 21257 (d) 37600

[2007 : 2 M]

3.4 A column costs Rs. 5.0 lakhs and has a useful life of 10 years. Using the double declining balance depreciation method, the book value of the unit at the end of five years (in lakhs of Rs.) is

(a) 1.21 (b) 1.31
(c) 1.64 (d) 2.05

[2009 : 2 M]

3.5 A process plant has a life of 7 years and its salvage value is 30%. For what minimum fixed-percentage factor will the depreciation amount for the second year, calculated by declining balance method be equal to that calculated by the straight line depreciation method?

(a) 0.1 (b) 0.113
(c) 0.527 (d) 0.887

[2011 : 2 M]

3.6 The cost of a new pump (including installation) is 24000 Rupees, The pump has a useful life of 10 years. Its salvage value is 4000 Rupees. Assuming straight line depreciation, the book value of the pump at the end of 4th year, rounded to the nearest integer, is _____ Rupees.

[2017 : 1 M]

3.7 Information for a proposed greenfield project is provided in the table. The discounted cash flow for the fourth year is Rs _______ crores (rounded off to one decimal place).

Fixed capital investment (excluding land) : Rs 250 crores

Salvage value : Rs 0

Yearly revenue from product sales : Rs 120 crores

Yearly manufacturing cost (excluding depreciation): Rs 30 crores

Interest rate : 10% compounded annually

Annual taxation rate : 30%

Depreciation method : Double declining balance* over seven years

Plant start-up : 2 years after project initiation

* $d_k = \frac{2}{7}BV_{k-1}$ k : years post start-up

d : Depreciation amount

BV : Book value

[2022 : 2 M]

Answers Taxes and Depreciation

3.1 (d) 3.2 (a) 3.3 (b) 3.4 (c) 3.5 (b) 3.6 (16000) 3.7 (53.4)

Explanations Taxes and Depreciation

3.1 (d)

The depreciation by the sum of year digit method is given as:

$$d = \frac{(n-m+1)}{\frac{n(n+1)}{2}}(P-S)$$

$$d = \frac{2(n-m+1)}{n(n+1)}(P-S)$$

where P is the investment, S is the salvage value, n is the life of the equipments in years.

3.2 (a)

Original value of equipment, V = Rs. 10000

Salvage value, V_s = Rs. 900

Useful life, n = 5 years

By textbook declining balance method

$$V_s = V(1-F)^n$$

$$500 = 10000(1-F)^5$$

$$F = 0.4507$$

Book value after 2 years

$$= V(1-F)^2$$

$$= 10000\,(1-0.4507)^2 = \text{Rs. } 3017$$

$\therefore$ Closet value option (a).

3.3 (b)

Installation cost of pump (V) = Rs. 40000

Estimated life (n) = 10 years

Salvage value (V_s) = 0

For double declining balance method,

Depreciation factor, $F = \frac{2}{n} = \frac{2}{10} = 0.2$

Book value of pump after 6 years

$$= V(1-F)^6$$

$$= 40000(1-0.2)^6$$

$$= \text{Rs. } 10485.6$$

3.4 (c)

Column cost = Rs. 5.0 lakhs

Useful life = 10 years

Assuming salvage value is zero for double declining balance method.

$$F = \frac{2}{n} = \frac{2}{10} = 0.2$$

Book value at the end of five years

$$= 5(1-F)^5$$

$$= 5 \times 0.8^5$$

$$= \text{Rs. } 1.638 \text{ lakhs}$$

3.5 (b)

Life of process plant = 7 years

Assuming initial investment is V.

Salvage value = $0.3V$

Depreciation by straight line method

$$d = \frac{V - 0.3v}{7} = 0.1V$$

Depreciation for second year is same for both straight line method and declining balance method.

So, $0.1v = v(1-f)f$

$$0.1 = f - f^2$$

$$f^2 - f + 0.1 = 0$$

Solving for f = 0.113, 0.887

Minimum value of fixed percent factor is 0.113.

3.6 (16000)

Cost of new pump (including installation)

= Rs. 24000

Useful life of pump = 10 years

Salvage value of pump = Rs. 4000

Book value of pump (assuming straight line depreciation) can be calculated using following formula :

$$V_a = V - a \times d$$

where, V_a = Book value after a years depreciation

d = Depreciation per year

So, book value at the end of four years

$$V_4 = 24000 - 4 \times d \quad ...(1)$$

From salvage value

$$4000 = 24000 - 10 \times d$$

$$d = \text{Rs. } 2000$$

Now, from equation (1)

$$V_4 = 24000 - 2000 \times 4 = \text{Rs. } 16000$$

3.7 (53.4)

From given data:

For double declining method, $f = \frac{2}{7} = 0.286$

It takes 2 years to plant start up, so d_2 will be calculated

$$d_2 = \frac{2}{7}(B.V)_1$$

$$(B.V)_1 = 250\,(1 - 0.286) = 178.5$$

$$d_2 = \frac{2}{7} \times 178.5 = 51$$

$$A = (120 - 30 - 51)\,(1 - 0.30) + 51 = 78.3$$

This is the cash flow after tax paid.

Discounted cash flow = Present value at time $t = 0$

$$\text{Discounted cash flow} = \frac{78.3}{(1.1)^4} = 53.45$$

■■■■

4 Profitability Analysis

4.1 For a typical project, the cumulative cash flow is zero at the
(a) end of the project life
(b) break-even point
(c) start-up
(d) end of the design stage

[2000 : 1 M]

4.2 A plant is designed to produce 1.2×10^8 kg/yr of an agrochemical. The estimated fixed capital investment is Rs. 1.5×10^9. The working capital is Rs. 2×10^8 and the start-up cost (only in the first year of commissioning and to be accounted for in the first year) is Rs. 1.5×10^8. The following cost data are available:
Raw materials : Rs. 0.89/kg product
Labour and utilities, etc. : Rs. 0.27/kg product
Selling price of product : Rs. 10/kg product
Other costs (on per year basis) including maintenance, insurance, etc. @ 10% of fixed capital.
Indirect costs of administration, R&D, marketing, etc. @ 20% of sale proceeds.
The plant will be fully depreciated over a period of 5 years using the straight-line method. The rate of income tax is 40%. Calculate
(a) the net profit at the end of first year
(b) the payout period.

[2000 : 5 M]

4.3 An investment of Rs. 100 lakhs is to be made for construction of a plant which will take two years to start production. The annual profit from operation of the plant is Rs. 20 lakhs. What will be the payback time?
(a) 5 years (b) 7 years
(c) 12 years (d) 10 years

[2001 : 1 M]

4.4 The total investment in a project is Rs. 10 lakhs and the annual profit is Rs. 1.5 lakhs. If the project life is 10 years, then the simple rate of return on investment is
(a) 15% (b) 10%
(c) 1.5% (d) 150%

[2002 : 1 M]

4.5 A company has a depreciable investment of Rs. 36300 which is depreciated in equal installments in two years. Assume that the tax rate is 50% and interest rate is 10%. The net present value of tax that the company would have saved, if it had depreciated 2/3 of the investment in the first year and the rest in the second year, is
(a) 0 (b) 250
(c) 375 (d) 500

[2002 : 2 M]

4.6 A series of equal payments (e.g., deposit or cost) made at equal intervals of time known as
(a) perpetuity (b) capital charge factor
(c) annuity (d) future worth

[2003 : 1 M]

4.7 Two pumps under consideration, for installation at a plant have the following capital investments and salvage values. Pump A: C_I – Rs. 40,000, C_{sal} = Rs. 3,900. Pump B: C_I = Rs. 50,000, C_{sal} = Rs. 20,000. Using capitalized cost, determine what should be the common life of the pumps for both to be competitive (economically equivalent). Interest rate is 10% per annum. Maintenance and operational costs are negligible.
(a) 3 years (b) 5 years
(c) 6 years (d) 8 years

[2003 : 2 M]

Q.4.8 - Q.4.9 are based on the data supplied in the paragraph below:
A process has fixed capital investment of Rs. 150 lakhs, working capital Rs. 30 lakhs and salvage value zero. Annual revenues from sales are Rs. 250 lakhs, manufacturing costs 145 lakhs and other expenses 10% of the revenue. Assume project life span of 11 years, tax life of 5 years and interest rate to be 10 %. Tax rate is 40 % and straight line depreciation, i.e. 20% per year, is applicable

4.8 Discounted value (to present time) of the profit before tax (for the total plant life period) in rupees is
(a) 228 lakhs (b) 400 lakhs
(c) 520 lakhs (d) 660 lakhs
[2003 : 2 M]

4.9 Discounted value of the depreciation benefit over the tax life in rupees is
(a) 12 lakhs (b) 24 laths
(c) 46 lakhs (d) 60 lakhs
[2003 : 2 M]

Common Data for Questions (4.10 & 4.11):
Fixed capital investment for a chemical plant is Rs. 40 million with an estimated useful life of 6 years and a salvage value of Rs. 4 million. The rate of interest is 15%. Tax is 25% of the annual taxable income. In the first year of operation, the income from sales is Rs. 20 million and manufacturing expenses are Rs. 5 million. The plant depreciates on a straight line basis.

4.10 The rate of return on investment is given by
(a) 50% (b) 37.5%
(c) 32% (d) 20%
[2004 : 2 M]

4.11 The net present value (NPV) in million Rs. at the start and at the end of the first year of operation is respectively given by
(a) zero and –28.9 (b) –40 and –28.9
(c) –40 and 12.75 (d) zero and 12.75
[2004 : 2 M]

4.12 The depreciable fixed cost is Rs. 100 lakhs. The average profit per year is Rs. 15 lakhs. The average depreciation cost per year is Rs. 10 lakhs. What is the payout period in years, if there is no interest charge?
(a) 8 (b) 4
(c) 10 (d) 20
[2005 : 2 M]

4.13 Due to a 20% drop in the product selling price, the pay-back period of a new plant increased to 1.5 times that estimated initially, the production cost and the production rate remaining unchanged. If the production cost is C_p, and the new selling price is C_s ,then $\frac{C_p}{C_s}$ is
(a) 0.2 (b) 0.4
(c) 0.5 (d) 0.6
[2006 : 2 M]

4.14 A sale contract signed by a chemical manufacturer is expected to generate a net cash flow of Rs. 2,50,000/- per year at the end of each year for a period of three years. The applicable discount rate (interest rate) is 10%. The net present worth of the total cash flow is Rs.
(a) 7,50,000 (b) 6,83,750
(c) 6,21,500 (d) 3,32,750
[2006 : 2 M]

4.15 For the case of single lump-sum capital expenditure of Rs. 10 crores which generates a constant annual cash flow of Rs. 2 crores in each subsequent year, the payback period (in years), if the scrap value of the capital outlay is zero is
(a) 10 (b) 20
(c) 1 (d) 5
[2008 : 1 M]

4.16 The relation between capital rate of return ratio (CRR), net present value (NPV) and maximum cumulative expenditure (MCE) is
(a) $CRR = \frac{NPV}{MCE}$ (b) $CRR = \frac{MCE}{NPV}$
(c) $CRR = NPV \times MCE$ (d) $CRR = \frac{MCE}{(NPV + MCE)}$
[2008 : 1 M]

4.17 A reactor has been installed at a cost of Rs. 50,000 and is expected to have a working life of 10 years with a scrap value of Rs. 10,000. The capitalized cost (in Rs.) of the reactor based on an annual compound interest rate of 5% is
(a) 1,13,600 (b) 42,000
(c) 52,500 (d) 10,500
[2008 : 2 M]

4.18 A reactor needs to be lined with a corrosion resistant lining. One type of lining costs Rs. 5 lakhs, and is expected to last for 2 years. Another type of lining lasts for 3 years. If both choices have to be equally economical, with the effective interest rate being 18%, compounded annually, the price one should pay for the second type of lining is

(a) Rs. 6.1 lakhs (b) Rs. 6.5 lakhs
(c) Rs. 6.9 lakhs (d) Rs. 7.6 lakhs

[2010 : 2 M]

4.19 Heat integration is planned in a process plant at an investment Rs. 2×10^6. This would result in a net energy savings of 20 GJ per year. If the nominal rate of interest is 15% and the plant life is 3 years, then the break even cost of energy, in Rs. per GJ (adjusted to the nearest hundred), is

(a) 33500 (b) 43800
(c) 54200 (d) 65400

[2012 : 2 M]

4.20 A cash flow of Rs. 12,000 per year is received at the end of each year (uniform periodic payment) for 7 consecutive years. The rate of interest is 9% per year compounded annually. The present worth (in Rs.) of such cash flow at time zero is ________

[2014 : 2 M]

4.21 A polymer plant with a production capacity of 10,000 tons per year has an overall yield of 70%, on mass basis (kg of product per kg of raw material). The raw material costs Rs. 50,000 per ton. A process modification is proposed to increase the overall yield to 75% with an investment of Rs. 12.5 crore. In how many years can the invested amount be recovered with the additional profit?

[2014 : 2 M]

4.22 A proposed chemical plant is estimated to have a fixed capital (FC) of Rs. 24 crores. Assuming other costs to be small, the total investment may be taken to be same as FC. After commissioning (at t = 0 years), the annual profit before tax is Rs. 10 crores /year (at the end of each year) and the expected life of the plant is 10 years. The tax rate is 40% per year and a linear depreciation is allowed at 10% per year. The salvage value is zero. If the annual interest rate is 12%, the NPV (net present value or worth) of the project in crores of rupees (up to one decimal place) is ______.

[2015 : 2 M]

4.23 Two design options for a distillation system are being compared based on the total annual cost Information available is as follows:

	Option P	Option Q
Installed cost of the system (Rs. in lakhs)	150	120
Cost of cooling water for condenser (Rs. in lakhs/year)	6	8
Cost of system for reboiler (Rs. in lakhs/year)	16	20

The annual fixed charge amounts to 12% of the installed cost. Based on the above information, what is the total annual cost (Rs in lakhs /year) of the better option?

(a) 40 (b) 42.4
(c) 92 (d) 128

[2016 : 1 M]

4.24 Consider two competing equipment A and B. For a compound interest rate of 10% per annum, in order for equipment B to be the economically cheaper option, its minimum life (in years) is ______ (rounded off to the next higher integer).

Equipment	Capital Cost (Rs.)	Yearly Operating Cost (Rs.)	Equipment Life (Years)
A	80,000	20,000	4
B	1,60,000	15,000	?

[2019 : 2 M]

4.25 A taxi-car is bought for Rs. 10 lakhs. Its salvage value is zero. The expected yearly income after paying all expenses and applicable taxes is Rs. 3 lakhs. The compound interest rate is 9% per annum. The discounted payback period (in years), is _____ (rounded off to the next higher integer).

[2019 : 2 M]

4.26 Annual capacity of a plant producing phenol is 100 metric tonnes. Phenol sells at INR 200 per kg and its production cost is INR 50 per kg. The sum of annual fixed charges, overhead costs and general expenses is INR 30,00,000. Taxes are payable at 18% on gross profit. Assuming the plant runs at full capacity and that all the phenol produced is sold, the annual net profit of the plant (in INR) is

(a) 1,39,40,000 (b) 1,50,00,000
(c) 98,40,000 (d) 1,20,00,000

[2020 : 1 M]

4.27 Turnover ratio is defined as

(a) $\dfrac{\text{Fixed capital investment}}{\text{Gross annual sales}}$

(b) $\dfrac{\text{Gross annual sales}}{\text{Fixed capital investment}}$

(c) $\dfrac{\text{Fixed capital investment}}{\text{Average selling price of the product}}$

(d) $\dfrac{\text{Gross annual sales}}{\text{Average selling price of the product}}$

[2021 : 1 M]

4.28 A compressor with a life of 10 years costs Rs 10 lakhs. Its yearly operating cost is Rs 0.5 lakh. If the annual compound interest rate is 8%, the amount needed at present to fund perpetual operation of the compressor is Rs. ________ lakhs (rounded to first decimal place).

[2022 : 1 M]

■■■■

Answers Profitability Analysis

4.1	(b)	4.2	(1.52×10^8, 1.76)	4.3	(a)	4.4	(a)	4.5	(b)	4.6	(c)		
4.7	(b)	4.8	(c)	4.9	(c)	4.10	(b)	4.11	(b)	4.12	(b)	4.13	(b)
4.14	(c)	4.15	(d)	4.16	(a)	4.17	(a)	4.18	(c)	4.19	(b)	4.20	(60395)
4.21	(2.62)	4.22	(15.3)	4.23	(a)	4.24	(8)	4.25	(5)	4.26	(c)	4.27	(b)
4.28	(24.9)												

Explanations Profitability Analysis

4.1 (b)

Cumulative cash flow is zero at break even point because at break even point, total production cost and total sales becomes equal.

4.2 (1.52×10^8, 1.76)

Plant is designed to produce 1.2×10^8 kg/year of an agrochemical.

The estimated fixed capital investment is Rs. 1.5×10^9.

Working capital investment is Rs. 2×10^8.

Start up cost in first year = Rs. 1.5×10^8

Raw material cost = Rs. 0.89/kg product

Total raw material cost

$$= 1.2 \times 10^8 \times 0.89$$
$$= \text{Rs. } 1.068 \times 10^8/\text{year}$$

Selling price = Rs. 10/kg product

$$\text{(S) Total sales} = 10 \times 1.2 \times 10^8 = \text{Rs. } 12 \times 10^8/\text{year}$$

$$\text{(C) Total cost} = 1.5 \times 10^8 + 1.068 \times 10^8 + 1.5 \times 10^8 + 2.4 \times 10^8 = \text{Rs. } 6.468 \times 10^8/\text{year}$$

Depreciation per year,

$$d = \frac{1.5\times10^9}{5} = \text{Rs. } 3 \times 10^8/\text{year}$$

(a) Net profit $= (s - c - d)(1 - \phi)$

$$= (12 \times 10^8 - 6.468 \times 10^8 - 3 \times 10^8)(1 - 0.4)$$
$$= \text{Rs. } 1.5192 \times 10^8/\text{year}$$

(b) Payout period

$$= \frac{\text{Fixed capital investment}}{\frac{\text{Avg. profit}}{\text{Year}} + \frac{\text{Avg. dep.}}{\text{Year}}}$$

$$= \frac{1.5\times10^9}{(5.532+3)\times10^8} = 1.76 \text{ years}$$

4.3 (a)

$$\text{Payback time} = \frac{\text{Total investment}}{\text{Average annual cash flow}} = \frac{100}{20} = 5 \text{ years.}$$

4.4 (a)

$$\text{Rate of return} = \frac{\text{Net annual profit}}{\text{Total investment}}\times 100 = \frac{1.5}{10}\times 100 = 15\%.$$

4.5 (b)

Case 1 : Total money invested = Rs. 36300

Depreciation after 1 year

$$= 36300 \times 0.5 = \text{Rs. } 18150$$

Tax saved if depreciation not occur

$$= 18150 \times 0.5 = \text{Rs. } 9075$$

Present value of tax saved after 1 year

$$= \frac{9075}{(1+i)^n} = \frac{9075}{1.1} = \text{Rs. } 8250$$

Depreciation after 2 years = Rs. 18150

Tax saved if depreciation not occur

$$= 18150 \times 0.5 = \text{Rs. } 9075$$

Present value of tax saved after 2 years

$$= \frac{9075}{(1.1)^2} = \text{Rs. } 7500$$

Total net present value of tax saved

$$= 8250 + 7500 = \text{Rs. } 15750$$

Case 2 : Depreciation after 1 year

$$= 36300\times\frac{2}{3} = \text{Rs. } 24200$$

Tax saved if depreciation had not occurred

$$= 24200 \times 0.5 = \text{Rs. } 12100$$

Present value of tax saved

$$= \frac{12100}{1.1} = \text{Rs. } 11000$$

Depreciation after 2 years

$$= 36300 \times \frac{1}{3} = \text{Rs. } 12100$$

Tax saved = 12100×0.5 = Rs. 6050

Present value of tax after 2 years

$$= \frac{6050}{(1.1)^2} = \text{Rs. } 5000$$

Total net present value of tax

$$= 11000 + 5000 = \text{Rs. } 16000$$

Saving on net present value of tax saved

$$= 16000 - 15750 = \text{Rs. } 250$$

4.7 (b)

Given:

For Pump A:

C_I = Rs. 40,000

C_{sal} = Rs. 3,900

i = 10%

For Pump B:

C_I = Rs. 50,000

C_{sal} = Rs. 20,000

$$\text{Capitalized cost } (K) = C_I + \frac{C_I - C_{sal}}{(1+i)^n - 1}$$

For economically equivalent

$$k_A = k_B$$

$$40000 + \frac{40000 - 3900}{(1.1)^n - 1} = 50000 + \frac{50000 - 20000}{(1.1)^n - 1}$$

$$\frac{6100}{(1.1)^n - 1} = 10000$$

$$(1.1)^n - 1 = \frac{6100}{10000} = 0.61$$

$$n \ln(1.1) = \ln(1.61)$$

$$n = \frac{\ln(1.61)}{\ln(1.1)}$$

$$n = 4.996 \simeq 5$$

4.8 (c)

Given:

Fixed capital investment = 150 lakh

Annual revenue = 250 lakh/year

Manufacturing cost = 145 lakh/year

Other expenses = 10% of revenue

$$= 0.1 \times 250 = 25 \text{ lakh/year}$$

Total expenses = (145 + 25) = 170 lakh/year

Profit before tax = 250 − 170 = 80 lakh/year

For present value,

$$P = \frac{80[(1+i)^n - 1]}{i(1+i)^n}$$

$$i = 10\%$$

$$n = 11 \text{ years}$$

$$P = \frac{80[(1+i)^{11} - 1]}{0.1 \times (1.1)^{11}}$$

$$= 519.60 \text{ lakh} \simeq 520 \text{ lakh}$$

4.9 (c)

When there is tax on depreciation amount,

Profit after tax = 80×0.6 = 48 lakh/year

When there is no tax on depreciation amount

Profit before tax = 80 lakh/year

Depreciation amount = 30 lakh/year

Profit after tax = $(80 - 30) \times 0.6 + 30$

$$= 60 \text{ lakh/year}$$

Dep. benefit = 60 − 48 = 12 lakh/year

Discounted value of the depreciation benefit over the tax life

$$= \frac{12[(1+i)^n - 1]}{i(1+i)^n}$$

$$= \frac{12[(1.1)^5 - 1]}{0.1(1.1)^5} = 45.489 \text{ lakh}$$

4.10 (b)

Fixed capital investment = Rs. 40 million

Useful life = 6 years

Salvage value = Rs. 4 million

Interest rate = 15%

Tax rate = 25%

Assuming straight line depreciation

$$d = \frac{40-4}{6} = \text{Rs. } 6 \text{ million}$$

Profit = Sales − Manufacturing cost

$$= 20 - 5 = 15 \text{ millions}$$

$$\text{Rate of return} = \frac{\text{Annual Profit}}{\text{Total capital investment}}$$

Here, Total capital investment

= Fixed capital investment

$$\frac{15}{40} = 37.5\%$$

4.11 (b)

Net present value = Discounted cash flow – Fixed capital investment

$$\text{Cash flow} = (s - c - d)(1 - \phi) + d$$

$$= (20 - 5 - 6)(1 - 0.25) + 6 = 12.75$$

Discounted cash flow

$$= \text{Cash Flow}\left[\frac{(1+i)^n - 1}{(1+i)^n \times i}\right]$$

Discounted cash flow

$$= 12.75\left[\frac{1.15 - 1}{1.15 \times 0.15}\right] = 11.087$$

So, NPV at start of first year

$$= 0 - 40 = \text{Rs. } -40 \text{ million}$$

NPV at end of first year

$$= 11.087 - 40$$

$$= \text{Rs. } -28.91 \text{ millions}$$

4.12 (b)

$$\text{Pay-out period} = \frac{\text{Depreciable Fixed Capital Investment}}{\frac{\text{Avg. Profit}}{\text{Years}} + \frac{\text{Avg. Depreciation}}{\text{Year}}}$$

$$= \frac{100}{15 + 10} = 4 \text{ years}$$

4.13 (b)

Pay-back period

$$= \frac{\text{Fixed capital investment (F)}}{\frac{\text{Average Profit}}{\text{Year}} + \frac{\text{Average Depreciation}}{\text{Year}}}$$

Old selling price is C_s.

New selling price after 20% drop = $0.8C_s$

Now pay-back period increases 1.5 times. So,

$$(1.5)\frac{\text{F.C.I.}}{C_s - C_P} = \frac{\text{F.C.I.}}{0.8C_s - C_P}$$

$$1.5\,(0.8C_s - C_P) = C_s - C_P$$

$$1.2C_s - 1.5C_P = C_s - C_P$$

$$0.2C_s = 0.5C_P$$

$$\frac{C_P}{C_s} = \frac{0.2}{0.5} = 0.4$$

4.14 (c)

Net present value can be calculated by following formula:

$$A = P\left[\frac{(1+i)^n - 1}{i(1+i)^n}\right]$$

where, n = period of payment

p = payment per year

i = interest rate

So, net present value

$$A = 250000\left[\frac{(1.1)^3 - 1}{0.1 \times (1.1)^3}\right]$$

$$= \text{Rs. } 621712$$

4.15 (d)

$$\text{Payback period} = \frac{\text{Fixed capital investment}}{\frac{\text{Avg. Profit}}{\text{Year}} + \frac{\text{Avg. Depreciation}}{\text{Year}}}$$

$$\text{Payback period} = \frac{\text{Fixed capital investment}}{\text{Annual cash flow}}$$

$$= \frac{10}{2} = 5 \text{ years}$$

4.16 (a)

Capital rate of return ratio

$$= \frac{\text{Net Present Value}}{\text{Max. Cumulative Expenditure}}$$

4.17 (a)

Installation cost of reactor (V) = Rs. 50000

Working life (n) = 10 years

Scrap value (V_s) = Rs. 10000

Annual compound interest rate (i) = 0.05

Replacement cost = Original cost – Salvage value

$$V_R = V - V_s$$

$$= 50000 - 10000$$

$$= \text{Rs. } 40000$$

Capitalized cost (K)

$$= V + \frac{V_R}{(1+i)^n - 1}$$

$$= 50000 + \frac{40000}{(1.05)^{10} - 1}$$

$$= \text{Rs. } 113603$$

4.18 (c)

For both choices to be economical equal capitalized cost of both lining should be equal.

So, $$C_1\left(1 + \frac{1}{(1+i)^{n_1} - 1}\right) = C_2\left(1 + \frac{1}{(1+i)^{n_2} - 1}\right)$$

$$5\left(1 + \frac{1}{(1.18)^2 - 1}\right) = C_2\left(1 + \frac{1}{(1.18)^3 - 1}\right)$$

$$C_2 = \text{Rs. } 6.943 \text{ lakhs}$$

4.19 (b)

Heat integration is planned at an investment of Rs. 2×10^6.

Energy saving due to this integration

$= 20$ GJ/year.

Plant life (n) = 3 years

Nominal interest rate = 15%

Assuming break even cost of energy (in Rs. per GJ) is C.

So, $$2\times 10^6 = C\times 20\left[\frac{(1+i)^n-1}{i(1+i)^n}\right]$$

$$2\times 10^6 = C\times 20\left[\frac{(1.15)^3-1}{0.15\times(1.15)^3}\right]$$

$$C = \text{Rs. } 43798 \text{ per GJ}$$

4.20 (60395)

The present worth of any periodic payment (P) for n consecutive years with interest rate (i) can be obtained by following formula :

$$P = A\left[\frac{(1+i)^n-1}{i(1+i)^n}\right]$$

Here, A = Rs. 12000

$n = 7,\quad i = 0.09$

So, present worth

$$P = 12000\left[\frac{(1+0.09)^7-1}{0.09\times(1.09)^7}\right]$$

$=$ Rs. 60395.43

4.21 (2.62)

Production capacity of polymer plant = 10000 tons per year

Yield of polymer plant on mass basis = 70%

Cost of raw material = Rs. 50000 per ton

With an investment of Rs. 12.5 crores, overall yield increases to 75%.

Saving per year due to increment in overall yield

$$= 10000\left[\frac{1}{0.70}-\frac{1}{0.75}\right]\times 50000$$

$=$ Rs. 4.7619×10^7

Assuming n years required to recover the investment of Rs. 12.5 crores. So,

$n \times 4.7619 \times 10^7 = 12.5 \times 10^7$

$n = 2.625$ years

4.22 (15.3)

Fixed capital of plant = Rs. 24 crores

Total cost = Fixed cost (given)

Profit before tax = Rs. 10 crores/year

Expected life of plant (n) = 10 years

Tax rate (ϕ) = 40%

Depreciation (d) = $24 \times 0.1 = 2.4$ crores

Annual interest rate, $i = 0.12$

Profit after tax $= (s-c-d)(1-\phi)$

$= (10-2.4)(1-0.4)$

$=$ Rs. 4.56 crores/year

Cash flow $= (s-c-d)(1-\phi)+d$

$=$ Rs. (4.56 + 2.4) crores/year

$=$ Rs. 6.96 crore/year

Discounted cash flow for 10 years

$$= \text{Cash Flow}\left[\frac{(1+i)^n-1}{(1+i)^n\times i}\right]$$

$$= 6.96\left[\frac{(1.12)^{10}-1}{(1.12)^{10}\times 0.12}\right]$$

$=$ Rs. 39.32 crores

Net present value = Discounted cash flow − Fixed capital

$= 39.32 - 24 =$ Rs. 15.32 crores

4.23 (a)

Annual fixed charge amount is 12% of installed cost of system.

So, total cost for option P

$= 150 \times 0.12 + 6 + 16$

$=$ Rs. 40 lakh/year

Similarly, total cost for option Q

$= 120 \times 0.12 + 8 + 20$

$=$ Rs. 42.4 lakhs/year

So, better option is P having total cost of Rs. 40 lakhs/year.

4.24 (8)

Equipment	Capital Cost (Rs.)	Yearly Operating Cost (Rs.)	Equipment Life (Years)
A	80,000	20,000	4
B	1,60,000	15,000	?

Compound interest rate of 10% per annum is given. By using capital recovery method, where the equivalent uniform annual cost is calculated to compare between the alternatives.

$$\text{EUAC (A)} = \text{EUAC (B)}$$

$$80{,}000\left(\frac{A}{P},10\%,4\right)+20{,}000$$

$$= 1{,}60{,}000\left(\frac{A}{P},10\%,n\right)+15{,}000$$

where, $\frac{A}{P} = \frac{i(1+i)^n}{(1+i)^n-1}$

$$80{,}000\times\frac{0.1(1+0.1)^4}{(1.1)^4-1}+20{,}000$$

$$= 1{,}60{,}000\times\frac{0.1(1.1)^n}{(1.1)^n-1}+15{,}000$$

$$45237.664 = 15000+160000\left[\frac{0.1\times1.1^n}{(1.1^n-1)}\right]$$

$$\frac{0.1\times1.1^n}{(1.1^n-1)} = \frac{45237.664-15000}{160000}$$

$$\frac{0.1\times1.1^n}{(1.1^n-1)} = 0.189$$

$$(1.1)^n = 1.89(1.1)^n - 1.89$$

$$(1.1)^n = \frac{1.89}{0.89} = 2.1236$$

$$n\log(1.1) = \log(2.1236)$$

$$n = \frac{\log(2.1236)}{\log(1.1)} = 7.90 \simeq 8 \text{ years}$$

4.25 (5)

Taxi car is bought for Rs. 10 lakhs and its salvage value is zero.

Expected yearly income after paying all expenses and taxes is Rs. 3 lakhs.

Compound interest rate is 9% per annum.

To calculate discounted pay-back period we should calculate present value of profit and that should be equal to initial investment.

So, $10\times10^5 = 3\times10^5\left[\frac{(1+i)^n-1}{(1+i)^n\times i}\right]$

$$\frac{10}{3} = \frac{1.09^n-1}{1.09^n\times0.09}$$

$$0.3\times1.09^n = 1.03^n-1$$

$$0.7\times1.09^n = 1$$

$$n = 4.14 \text{ years} = 5 \text{ years}$$

(Rounded-off to the next higher integer).

4.26 (c)

Annual capacity of plant is 100 metric tonnes phenol.

Selling price of phenol is INR 200 per kg.

Production cost of phenol is INR 50 per kg.

Sum of annual fixed charges, overhead costs and general expenses is INR 30,00,000.

Annual revenue from phenol

$= 100\times1000\times200 = \text{INR } 20{,}000{,}000$

Annual production cost

$= 100\times1000\times50 = \text{INR } 50{,}00{,}000$

Total annual product cost = Fixed and other charges + product cost

$= 50{,}00{,}000 + 30{,}00{,}000$

$= \text{INR } 80{,}00{,}000$

Gross Profit = Revenue – Total production cost

$= 20{,}000{,}000 - 8{,}000{,}000$

$= \text{INR } 12{,}000{,}000$

Net profit = Gross profit – Taxes

$= 12{,}000{,}000(1-0.18)$

$= \text{INR } 98{,}40{,}000$

4.28 (24.9)

Given: $V_o = 10$ lakh, $n = 10$ year

Yearly operating cost = 0.5 lakh

$i = 8\%$

$$\text{Capitalized cost} = V_o + \frac{V_o - V_s}{(1+i)^n-1} + \frac{M}{i}$$

$$\text{C.C.} = 10 + \frac{10}{(1.8)^{10}-1} + \frac{0.5}{0.08}$$

Amount needed at present = 24.87 lakh

5 Optimization and Storage Vessel

5.1 A perforated plate has holes of diameter d_h arranged in a pitch, p_h. Each hole has a tube of *i.d.*, d_t, passing through it. The ligament efficiency is given by

(a) $\frac{p_h - d_h}{p_h}$ (b) $\frac{p_h - d_h d_t / 2}{p_h}$

(c) $\frac{p_h - d_t}{p_h}$ (d) $\frac{p_h - d_h}{p_h}$

[2001 : 1 M]

5.2 The ends of a cylindrical vessel can be closed by a head, which can be one of the four shapes. For the same thickness, choose the one which can withstand the highest pressure.

(a) Flat plate (b) Hemispherical
(c) Torispherical (d) Ellipsoidal

[2001 : 1 M]

5.3 In a cylindrical vessel subjected to internal pressure, the longitudinal stress σ_L, and the circumferential stress, σ_h, are related by

(a) $\sigma_h = 2\sigma_L$ (b) $\sigma_h = \sigma_L$

(c) $\sigma_h = \frac{\sigma_L}{2}$ (d) No relation exists

[2001 : 2 M]

5.4 A cylindrical pressure vessel of volume 6π m^3 has to be designed to withstand a maximum internal pressure of 10 atm. The allowable design stress of the material is 125 N/mm^2 and corrosion allowance is 2 mm. The thickness of the vessel for a length/diameter ratio of 3 will be close to

(a) 5 mm (b) 6 mm
(c) 8 mm (d) 10 mm

[2002 : 2 M]

5.5 Obtain the optimal diameter of a cylindrical storage vessel of volume V. The curved shell costs C_s (in Rs/m^2), and the flat top and bottom plates cost C_p (in Rs/m^2)

(a) $D = \frac{C_s}{C_p}\left[\frac{4V}{\pi}\right]^{1/3}$ (b) $D = \left[\frac{8VC_s}{\pi C_p}\right]^{1/3}$

(c) $D = \left[\frac{VC_s}{C_p}\right]^{1/3}$ (d) $D = \left[\frac{4VC_s}{\pi C_p}\right]^{1/3}$

[2006 : 2 M]

5.6 A cylindrical storage tank can have a self supported conical roof,

(a) if its diameter is less than 20 m
(b) if its diameter is more than 50 m
(c) if the thickness of the roof is more than that of the cylindrical shell
(d) whatever is the diameter

[2007 : 1 M]

5.7 The term, **knuckle radius**, is associated with

(a) flat heads
(b) torispherical heads
(c) hemispherical heads
(d) conical heads

[2010 : 1 M]

Common Data for Question 5.8 and 5.9:

A plant produces phenol. The variable cost in rupees per tonne of phenol is related to the plant capacity P (in tonnes/day) as 45,000 + 5P. The fixed charges are Rs. 100,000 per day. The selling price of phenol is Rs. 50,000 per tonne.

5.8 The optimal plant capacity (in tonnes per day) for minimum cost per tonne of phenol, is

(a) 101 (b) 141
(c) 283 (d) 422

[2010 : 2 M]

5.9 The break-even capacity in tonnes per day is

(a) 50 (b) 40
(c) 30 (d) 20

[2010 : 2 M]

5.10 Minimum input required to calculate the 'blank diameter' for a torispherical head is

(a) crown radius
(b) crown radius, knuckle radius and length of straight flange
(c) knuckle radius and length of straight flange
(d) crown radius and knuckle radius

[2011 : 1 M]

5.11 In a thin-walled cylindrical vessel of thickness t with inside radius r, the internal gauge pressure is p. The hoop stress and the longitudinal stress in the shell are σ_h and σ_l respectively. Which one of the following statement is true?

(a) $\sigma_h = \frac{pr}{t}, \sigma_l = \frac{pr}{4t}$ (b) $\sigma_h = \frac{pr}{4t}, \sigma_l = \frac{pr}{t}$

(c) $\sigma_h = \frac{pr}{2t}, \sigma_l = \frac{pr}{t}$ (d) $\sigma_h = \frac{pr}{t}, \sigma_l = \frac{pr}{2t}$

[2011 : 2 M]

5.12 A batch reactor produces 1×10^5 kg of a product per year. The total batch time (in hours) of the reactor is $k\sqrt{P_B}$, where P_B is the product per batch in kg and $k = \frac{1.0h}{\sqrt{kg}}$. The operating cost of the reactor is Rs. 200/h. The total annual fixed charges are Rs.$340 \times P_B$ and the annual raw material cost is Rs. 2×10^6. The optimum size (in kg) of each batch (adjusted to the nearest integer) is

(a) 748 (b) 873

(c) 953 (d) 1148

[2012 : 2 M]

5.13 A plant manufactures compressors at the rate of N units/day. The daily fixed charges are Rs. 20000 and the variable cost per compressor is Rs. 500 + 0.2 $N^{1\text{-}3}$. The selling price per compressor is Rs. 1000. The number of compressors to be manufactured, to the nearest integer, in order to maximize the daily profit is ______.

[2013 : 2 M]

5.14 The cost of two independent process variables f_1 and f_2 affects the total cost C_T (in lakhs of rupees) of the process as per the following function:

$$C_T = 100f_1 + \frac{1000}{f_1 f_2} + 20f_2^2 + 50$$

The lowest total cost C_T, in lakhs of rupees (up to one decimal place), is ______.

[2015 : 2 M]

5.15 Standard pipes of different schedule numbers and standard tubes of different BWG numbers are available in the market. For a pipe / tube of a given nominal diameter, which one of the following statements is TRUE?

(a) Wall thickness increases with increase in both the schedule number and the BWG number

(b) Wall thickness increases with increase in the schedule number and decreases with increase in the BWG number

(c) Wall thickness decreases with increase in both the schedule number and the BWG number

(d) Neither the schedule number, nor the BWG number has any relation to wall thickness

[2016 : 1 M]

5.16 A vertical cylindrical tank with a flat roof and bottom is to be constructed for storing 150 m^3 of ethylene glycol. The cost of material and fabrication for the tank wall is Rs 6000 per m^2 and the same for the roof and the tank bottom are Rs 2000 and Rs 4000 per m^2, respectively. The cost of accessories, piping and instruments can be taken as 10% of the cost of the wall. 10% of the volume of the tank needs to be kept free as vapour space above the liquid storage. What is the optimum diameter (in m) for the tank?

(a) 3.5 (b) 3.9

(c) 7.5 (d) 7.8

[2016 : 2 M]

5.17 The total cost (C_T) of an equipment in terms of the operating variables x and y is

$$C_l - 2x + \frac{12000}{xy} + y + 5$$

The optimal value of C_T, rounded to 1 decimal place, is ______.

[2017 : 2 M]

5.18 The Annual Fixed Charges (AFC) and Annual Utilities Costs (AUC) of a distillation column being designed are expressed in terms of the reflux ratio (R) as AFC (Rs. Lakh)

$$= 641 \times R^2 - 1796 \times R + 1287 + \frac{1}{R - 1.16}$$

AUC (Rs. Lakh) = $80 + 62.5 \times R$

The reflux ratio (R_{opt}) for optimizing the total cost of the distillation column may be found by solving

(a) $1282 \times R_{opt} - 1796 - \frac{1}{(R_{opt} - 1.16)^2}$

(b) $62.5 + 1282 \times R_{opt} - 1796 - \frac{1}{(R_{opt} - 1.16)^2} = 0$

(c) $80 + 62.5 \times R_{opt} + 641 \times R_{opt}^2 - 1796 \times R_{opt} +$

$$1287 + \frac{1}{(R_{opt} - 1.16)}$$

[2018 : 1 M]

5.19 Operating labor requirements L in the chemical process industry is described in terms of the plant capacity C (kg day^{-1}) over a wide range ($10^3 - 10^6$) by a power law relationship

$$L = \alpha C^{\beta}$$

where α and β are constants. It is known that
L is 60 when C is 2×10^4
L is 70 when C is 6×10^4
The value of L when C is 10^5 kg day^{-1} is ______ (round off to nearest integer).

[2021 : 2 M]

5.20 A viscous liquid is pumped through a pipe network in a chemical plant. The annual pumping cost per unit length of pipe is given by

$$C_{pump} = \frac{48.13q^2\mu}{D^4}$$

The annual cost of the installed piping system per unit length of pipe is given by

$$C_{piping} = 45.92D$$

Here, D is the inner diameter of the pipe in meter, q is the volumetric flow rate of the liquid in m^3s^{-1} and μ is the viscosity of the liquid in Pa.s.
If the viscosity of the liquid is 20×10^{-3} Pa.s and the volumetric flow rate of the liquid is 10^{-4} m^3 s^{-1}, the economic inner diameter of the pipe is ______ meter (round off to 3 decimal places).

[2021 : 2 M]

Answers Optimization and Storage Vessel

5.1	(c)	**5.2**	(b)	**5.3**	(a)	**5.4**	(d)	**5.5**	(d)	**5.6**	(a)	**5.7**	(b)
5.8	(b)	**5.9**	(d)	**5.10**	(b)	**5.11**	(d)	**5.12**	(c)	**5.13**	(217)	**5.14**	(572.8)
5.15	(b)	**5.16**	(d)	**5.17**	(91.5)	**5.18**	(b)	**5.19**	(75)	**5.20**	(0.015)		

Explanations Optimization and Storage Vessel

5.1 (c)

When a shell drum is drilled for tubes in a line parallel to the axis of shell, the ligament efficiency is $\frac{p_h - d_t}{p_h}$,

where, p_h = Pitch of tube holes
d_t = Diameter of tube

5.2 (b)

End of cylindrical vessel closed by a hemispherical head can withstand maximum pressure because pressure force in the vessel is divided equally across the surface of head.

5.3 (a)

In a cylindrical vessel longitudinal stress (σ_l) is $\frac{Pd}{4t}$ and circumferential stress (σ_h) is $\frac{Pd}{2t}$.

where, P = Design pressure
d = Diameter of vessel
t = Thickness of vessel

So, $\sigma_h = 2\sigma_L$

5.4 (d)

Given: $\frac{L}{D} = 3,\ V = 6\pi$

$$\therefore \quad \frac{\pi}{4}D^2 \times 3D = 6\pi \Rightarrow D = 2 \text{ m}$$

$$\Rightarrow \quad t = \frac{Pd}{2\sigma} + 2 \text{ mm}$$

$$= \frac{10^6 \times 2}{2 \times 125 \times 10^6} \times 10^3 + 2$$

$$= 8 + 2 = 10 \text{ mm}$$

5.5 (d)

Volume of cylindrical storage vessel

$$V = \frac{\pi}{4}d^2h$$

Curved shell costs in C_s (in Rs./m^2)
Flat top and bottom plates costs in C_p (in Rs./m^2)

Total cost, $C_T = \pi dh \times C_S + 2 \times \frac{\pi}{4}d^2 \times C_p$

$$= \pi d \times \frac{4V}{\pi d^2} \times C_S + 2 \times \frac{\pi}{4} d^2 \times C_p$$

$$= \frac{4VC_s}{d} + 2 \times \frac{\pi}{4} d^2 \times C_p$$

For optimal diameter,

$$\frac{\partial C_T}{\partial d} = 0$$

$$\Rightarrow \frac{-4VC_s}{d^2} + \pi d C_p = 0$$

$$d = \left[\frac{4VC_s}{\pi C_p}\right]^{1/3}$$

5.8 (b)

$$(\text{Rs/tonne}) C_T = 45000 + 5P + \frac{100000}{P}$$

For optimal plant capacity

$$\frac{\partial C_T}{\partial P} = \frac{dC_T}{dP} = 0$$

$$\frac{dC_T}{dP} = 5 - \frac{1 \times 10^5}{P^2} = 0$$

$$P^2 = 20000$$

$$P = 141.42 \text{ tonnes/day}$$

5.9 (d)

At break even point

Revenue = Total production cost

$$50000P = 100000 + (45000 + 5P)P$$

$$5P^2 - 5000P + 100000 = 0$$

$$P = 20.42 \text{ tonnes per day}$$

5.12 (c)

Batch reactor produces 10^5 kg of phenol per year.

Total batch time is $k\sqrt{P_B}$, where P_B is the product per batch in kg.

Operating cost = Rs. 200/hour

Total annual fixed charges = Rs. 340 × P_B

Annual raw material cost = Rs. 2 × 10^6

Total productions cost of phenol (in Rs.)

$$C_T = 340 \times P_B + 2 \times 10^6 + 200 \times k\sqrt{P_B} \times \frac{10^5}{P_B}$$

$$C_T = 340 \times P_B + 2 \times 10^6 + \frac{200 \times 10^5}{\sqrt{P_B}}$$

For optimum size

$$\frac{dC_T}{dP_B} = 0$$

$$\Rightarrow 340 + \frac{200 \times 10^5}{P_B^{3/2}}\left(-\frac{1}{2}\right)$$

$$P_B = 952.8 \approx 953 \text{ kg}$$

5.13 (37)

Plant manufactures N compressors/day

Daily fixed charges = Rs. 20000

Variable cost per compressor = Rs. 500 + $0.2N^{1.3}$

Selling price of compressor = Rs. 1000 per unit

Total cost of compressor (in Rs./day)

$$= 20000 + (500 + 0.2N^{1.3})$$

Profit (in Rs./day)

$$= 1000N - 20000 - 500N - 0.2N^{2.3}$$

$$P = 500N - 0.2N^{2.3} - 20000$$

For maximizing the daily profit

$$\frac{dP}{dN} = 0$$

$$\Rightarrow 500 - 0.2 \times 2.3N^{1.3} = 0$$

$$N = 216.54 \approx 217 \text{ (to nearest integer)}$$

5.14 (572.8)

Total cost (in lakhs of rupees) of the process is given as

$$C_T = 100f_1 + \frac{1000}{f_1 f_2} + 20f_2^2 + 50$$

Now, $$\frac{dC_T}{df_1} = 0$$

$$\Rightarrow 100 - \frac{1000}{f_1^2 f_2} = 0$$

$$f_1^2 f_2 = 10 \quad \text{...(1)}$$

Similarly, $$\frac{dC_T}{df_2} = 0$$

$$\Rightarrow -\frac{1000}{f_1 f_2^2} + 40f_2 = 0$$

$$40f_1 f_2^3 = 1000 \quad \text{...(2)}$$

On solving equation (1) and (2)

$$f_1 = 2.08, \quad f_2 = 2.29$$

Now, optimal total cost

$$C_{T,\text{ optimum}} = 100 \times 2.08 + \frac{1000}{2.08 \times 2.29} + 20 \times 2.29^2 + 50$$

$$C_T = \text{Rs. } 572.8 \text{ lakhs}$$

5.15 (b)

For a given nominal wall thickness of pipe increases with increase in schedule number and decreases with increase in BWG number.

5.16 (d)

10% of volume of tank needs to be kept free as vapour space above liquid.

So, $0.9 \times V = 150\ m^3$

$V = 166.667\ m^3$

$$\pi \frac{d^2}{4} l = 166.667$$

$$l = \frac{4 \times 166.667}{\pi d^2} \quad ...(1)$$

Now, total cost of tank is given by

$$C_T = 6000 \times \pi dl + \frac{\pi d^2}{4} \times 2000 + \frac{\pi}{4} d^2 \times 4000 + 6000 \times \pi dl \times 0.1$$

By putting values of l from eqn. (1)

$$C_T = 6600 \times \pi d \times \frac{4 \times 166.67}{\pi d^2} + 6000 \times \frac{\pi d^2}{4}$$

For optimum diameter, $\frac{\partial C_T}{\partial d} = 0$

$$-\frac{6600 \times 4 \times 166.67}{d^2} + 3000\pi d = 0$$

$\Rightarrow \quad d = 7.75$ m (closest value)

5.17 (91.5)

Total cost of an equipment in terms of operating variable x and y is given as

$$C_T = 2x + \frac{12000}{xy} + y + 5$$

For optimal value of C_T,

$$\frac{\partial C_T}{\partial x} = 0 \Rightarrow 2 - \frac{12000}{x^2 y} = 0$$

$$x^2 y = 6000 \quad ...(1)$$

Similarly,

$$\frac{\partial C_T}{\partial y} = 0 \Rightarrow \frac{-12000}{xy^2} + 1 = 0$$

$$xy^2 = 12000 \quad ...(2)$$

On solving eqn. (1) and (2)

$x = 14.42, \quad y = 28.84$

Now, optimal value of C_T

$$= 2 \times 14.42 + \frac{12000}{14.42 \times 28.84} + 28.84 + 5 = 91.53$$

5.18 (b)

Annual fixed charges of distillation column (in Rs. lakh)

$$= 641 \times R^2 - 1796 \times R + 1287 + \frac{1}{(R - 1.16)}$$

Annual utility cost of distillation column (in Rs. lakh) = $80 + 62.5 \times R$

where R is reflux ratio of distillation column.

Total cost of distillation column = Fixed cost + Utilities cost

$$C_T = 641 \times R^2 - 1796 \times R + 1287 + \frac{1}{(R - 1.16)} + 80 + 62.5 \times R$$

For optimum total cost, $\frac{dC_T}{dR} = 0$. So,

$$62.5 + 1282 \times R_{opt} - 1796 - \frac{1}{(R_{opt} - 1.16)^2} = 0$$

5.19 (75)

$$L = \alpha C^{\beta}$$

Putting the values given in question,

$$60 = \alpha(2 \times 10^4)^{\beta} \quad ...(1)$$

$$70 = \alpha(6 \times 10^4)^{\beta} \quad ...(2)$$

Equation (1) / Equation (2)

$$\frac{60}{70} = \frac{1}{3^{\beta}}$$

$$\beta = 0.1403$$

To find the value of L for $C = 10^5$

$$L = \alpha(10^5)^{\beta} = \alpha(5 \times 2 \times 10^4)^{\beta}$$

$$= 5^{\beta} \times [\alpha(2 \times 10^4)^{\beta}] = 5^{\beta} \times 60$$

$$L = 5^{0.1403} \times 60$$

$= 75.2 = 75$ (nearest integer)

5.20 (0.015)

Annual pumping cost per unit length

$$C_{pump} = \frac{48.13 q^2 \mu}{D^4}$$

Annual cost of installed piping cost

$$C_{piping} = 45.92D$$

Total cost per unit length

$$C_T = C_{pump} + C_{piping}$$

$$= \frac{48.13 q^2 \mu}{D^4} + 45.92D$$

For economic diameter, $\frac{dC_T}{dD} = 0$

$$\frac{-4}{D^5}(48.13 q^2 \mu) + 45.92 = 0$$

On solving, $D = 0.0153$ m

■■■■

Chemical Technology

UNIT

X

CONTENTS

Chemical Technology

Syllabus

Inorganic Chemical Industries (Sulfuric Acid, Phosphoric Acid, Chlor-Alkali Industry), Fertilizers (Ammonia, Urea, SSP and TSP); Natural Products Industries (Pulp and Paper, Sugar, Oil and Fats); Petroleum Refining and Petrochemicals; Polymerization Industries (Polyethylene, Polypropylene, PVC and Polyester Synthetic Fibers).

Analysis of Previous GATE Papers

Exam Year	1 Mark Ques.	2 Marks Ques.	5 Marks Ques.	Total Marks
2000	3	1	–	5
2001	3	2	–	7
2002	3	2	–	7
2003	3	6	–	15
2004	2	5	–	12
2005	–	7	–	14
2006	2	5	–	12
2007	2	4	–	10
2008	1	3	–	7
2009	2	3	–	8
2010	3	2	–	7
2011	5	1	–	7
2012	3	2	–	7
2013	1	2	–	5
2014	3	1	–	5
2015	3	1	–	5
2016	3	–	–	3
2017	3	1	–	5
2018	4	–	–	4
2019	5	1	–	7
2020	4	–	–	4
2021	1	1	–	3
2022	3	1	–	5
2023	3	2	–	7
2024	3	–	–	3

1 Inorganic Chemical Industries

1.1 Triple super phosphate is manufactured by reacting
(a) phosphate rock with phosphoric acid.
(b) phosphate rock with sulphuric acid.
(c) phosphate rock with nitric acid.
(d) ammonium phosphate with phosphoric acid.
[2000 : 1 M]

1.2 In the converter of the contact process for the manufacture of H_2SO_4, the equilibrium conversion of SO_2 –(a)– with increase in the temperature and –(b)– with increase in the mole ratio of SO_2 to air.
(a) (a) increases (b) decreases.
(b) (a) decreases (b) increases.
(c) (a) increases (b) increases.
(d) (a) decreases (b) decreases.
[2001 : 2 M]

1.3 In the manufacture of sulphuric acid from elemental sulphur, the following sequence of major operations is followed
(a) furnace → converter → absorber
(b) furnace → evaporator → absorber
(c) furnace → converter → evaporator
(d) converter → furnace → absorber
[2002 : 2 M]

1.4 Match the process in **Group-I** with the product in **Group-II**:

Group-I
A. DCDA process
B. Mercury cell

Group-II
1. Sodium hydroxide
2. Sulfuric acid
3. Sodium carbonate
4. Nitric acid

Codes:

	A	B
(a)	1	4
(b)	1	2
(c)	2	3
(d)	2	1

[2004 : 2 M]

1.5 Match the product in **Group-I** with the raw material in **Group-II**:

Group-I
A. Caustic soda
B. Soda ash

Group-II
1. Ammonia and sulfuric acid
2. Sodium carbonate and slaked lime
3. Salt and limestone
4. Salt and sulfuric acid

Codes:

	A	B
(a)	2	3
(b)	1	2
(c)	3	4
(d)	4	1

[2005 : 2 M]

1.6 In the Sulfite process for paper manufacture, the 'cooking liquor' is
(a) magnesium bisulfite and sulfur dioxide in acid medium
(b) magnesium sulfite and magnesium dicarbonate
(c) sodium sulfite and magnesium sulfite
(d) sodium sulfite, sodium bisulfite and sulfur dioxide
[2007 : 1 M]

1.7 In the manufacture of caustic soda from brine, which one of the following statements is True?
(a) The membrane cell CANNOT produce concentrated NaOH solution and CANNOT tolerate calcium and magnesium ions in the feed brine.
(b) The membrane cell CANNOT produce concentrated NaOH solution, but CAN tolerate calcium and magnesium ions in the feed brine.
(c) The membrane cell CAN produce concentrated NaOH solution, but CANNOT tolerate calcium and magnesium ions in the feed brine.
(d) The membrane cell CAN produce concentrated NaOH solution, and CAN tolerate calcium and magnesium ions in the feed brine.
[2010 : 1 M]

1.8 For making superphosphate by acidulation of phosphate rock, use of nitric acid is desirable, because

(a) nitric acid is less expensive than sulphuric acid.
(b) the availability of nitrogen enhances the value of the superphosphate as a fertilizer.
(c) the process produces non-hygroscopic superphosphate.
(d) the process produces superphosphate having higher phosphorus content than the sulphuric acid process.

[2010 : 1 M]

1.9 The correct sequence of process equipment used in the production of sulphuric acid from sulphur by contact process is

(a) burner, catalytic converter, 98% sulphuric acid absorption tower, oleum absorption column
(b) catalytic converter, oleum absorption column, 98% sulphuric acid absorption tower, burner
(c) burner, catalytic converter, oleum absorption column, 98% sulphuric acid absorption tower
(d) burner, oleum absorption column, catalytic converter, 98% sulphuric acid absorption tower

[2011 : 1 M]

1.10 In the manufacture of sulphuric acid by the contact process, the catalytic oxidation of SO_2 is carried out in multiple stages mainly to

(a) increase the reaction rate by providing inter-stage heating
(b) increase the overall conversion by providing inter-stage heating
(c) increase the overall conversion by providing inter-stage cooling
(d) decrease the overall conversion by removing sulphur trioxide between stages

[2013 : 1 M]

1.11 Decomposition efficiency (η_D) of an electrolytic cell used for producing NaOH is defined as

(a) η_D = (grams of NaOH produced / grams of NaCl decomposed) × 100
(b) η_D = (grams of NaOH produced / grams of NaCl charged) × 100
(c) η_D = (gram equivalents of NaOH produced/ gram equivalents of NaCl charged) × 100
(d) η_D = (theoretical current to produce one gram equivalent / actual current to produce one gram equivalent) × 100

[2014 : 1 M]

1.12 Match the raw materials of **Groups-1** and **2** with the final products of **Group-3**.

Group-1
P_1 : Ethylene
P_2 : Propylene
P_3 : Adipic acid
P_4 : Terephthalic acid

Group-2
Q_1 : Ammonia
Q_2 : 1-Butene
Q_3 : Ethylene glycol
Q_4 : Hexamethylene diamine

Group-3
R_1 : Synthetic fibre
R_2 : Nylon 66
R_3 : LLDPE
R_4 : Acrylonitrile

(a) $P_1 + Q_2 \rightarrow R_3$; $P_2 + Q_1 \rightarrow R_4$; $P_3 + Q_4 \rightarrow R_2$; $P_4 + Q_3 \rightarrow R_1$
(b) $P_1 + Q_1 \rightarrow R_3$; $P_2 + Q_3 \rightarrow R_1$; $P_3 + Q_4 \rightarrow R_4$; $P_4 + Q_2 \rightarrow R_2$
(c) $P_1 + Q_2 \rightarrow R_2$; $P_2 + Q_3 \rightarrow R_1$; $P_3 + Q_4 \rightarrow R_3$; $P_4 + Q_1 \rightarrow R_4$
(d) $P_1 + Q_1 \rightarrow R_4$; $P_2 + Q_2 \rightarrow R_3$; $P_3 + Q_4 \rightarrow R_2$; $P_4 + Q_3 \rightarrow R_1$

[2014 : 2 M]

1.13 Match the chemicals written on the left with the raw materials required to produce them mentioned on the right.

Group-1
I. Single superphosphate (SSP)
II. Triple superphosphate (TSP)
III. Diammonium phosphate (DAP)
IV. Caustic soda

Group-2
P. Rock phosphate + Sulfuric Acid + Ammonia
Q. Brine
R. Rock phosphate + Sulfuric Acid
S. Rock phosphate + Phosphoric Acid

(a) I - Q, II - R, III - S, IV - P
(b) I - S, II - P, III - Q, IV - R
(c) I - R, II - S, III - P, IV - Q
(d) I - S, II - R, III - P, IV - Q

[2015 : 1 M]

1.14 India has no elemental sulphur deposits that can be economically exploited. In India, which one of the following industries produces elemental sulphur as a by-product?
(a) Coal carbonisation plants
(b) Petroleum refineries
(c) Paper and pulp industries
(d) Iron and steel making plants

[2016 : 1 M]

1.15 Two paper pulp plants P and Q use the same quality of bamboo as a raw material. The chemicals used in their digester are as follows:

	Plant P	Plant Q
NaOH	Yes	No
Na_2S	Yes	No
Na_2CO_3	Yes	Yes
$NaHCO_3$	No	Yes
Na_2SO_3	No	Yes

Which one of the following statements is CORRECT?
(a) Plant P and Plant Q both use the Sulfite process
(b) Plant P and Plant Q both use the Kraft process
(c) Plant P uses Sulfite process
(d) Plant P uses Kraft process

[2016 : 1 M]

1.16 The DCDA (Double Contact Double Absorption) process is used for the manufacture of
(a) urea
(b) sulphuric acid
(c) nitric acid
(d) ammonia

[2017 : 1 M]

1.17 Match the equipment in **Group-1** with the process in **Group-2**:

Group-1
P. Fluidized bed
Q. Multistage adiabatic reactor with inter-stage cooling
R. Fourdrinier machine
S. Diaphragm cell

Group-2
I. Paper-making
II. Sodium hydroxide manufacture
III. SO_2 oxidation
IV. Catalytic cracking

Choose the correct set of combinations.
(a) P - IV, Q - III, R - I, S - II
(b) P - IV, Q - III, R - II, S - I
(c) P - III, Q - IV, R - I, S - II
(d) P - III, Q - IV, R - II, S - I

[2017 : 2 M]

1.18 In Kraft process, the essential chemical reagents used in the digester are
(a) caustic soda, mercaptans and ethylene oxide
(b) caustic soda, sodium sulphide and soda ash
(c) quick lime, salt cake and dimethyl sulphide
(d) baking soda, sodium sulphide and mercaptans

[2019 : 1 M]

1.19 The combination that correctly matches the process in **Group-1** with the entries in **Group-2** is

Group-1
P. Wulff process
Q. Sulphite process
R. Solvay process
S. Frasch process

Group-2
1. Sulfur mining
2. Soda ash production
3. Acetylene production
4. Pulp production

(a) P - 2, Q - 4, R - 3, S - 1
(b) P - 3, Q - 4, R - 2, S - 1
(c) P - 4, Q - 1, R - 2, S - 3
(d) P - 2, Q - 1, R - 4, S - 3

[2019 : 2 M]

1.20 Consider the processes in **Group-1** and the reactions in **Group-2**:

Group-1
P. Solvay process
Q. Oxo process
R. Saponification

Group-2
1. $RCOOH + NaOH \rightarrow RCOONa + H_2O$
2. $CH_2 = CH_2 + CO + H_2 \rightarrow CH_3CH_2CHO$
3. $CaCO_3 + 2NaCl \rightarrow Na_2CO_3 + CaCl_2$

The correct combination is
(a) P - 2, Q - 1, R - 3
(b) P - 3, Q - 2, R - 1
(c) P - 3, Q - 1, R - 2
(d) P - 1, Q - 3, R - 2

[2020 : 1 M]

1.21 Match the common name of chemicals in **Group-1** with their chemical formulae in **Group-2**.

Group - 1	Group - 2
P. Gypsum	1. $Ca(H_2PO_4)_2$
Q. Dolomite	2. $CaSO_4.2H_2O$
R. Triple superphosphate	3. $CaCO_3.MgCO_3$

The correct combination is

(a) P - 3, Q - 2, R - 1 (b) P - 3, Q - 1, R - 2
(c) P - 2, Q - 3, R - 1 (d) P - 2, Q - 1, R - 3

[2021 : 1 M]

1.22 The chemical formula of Glauber's salt, used in the Kraft process, is

(a) $Na_2CO_3.10H_2O$ (b) $Na_2S_2O_4.2H_2O$
(c) $Na_2HPO_4.2H_2O$ (d) $Na_2SO_4.10H_2O$

[2022 : 1 M]

1.23 Match the product in **Group-1** with the manufacturing process in **Group-2**. The correct combination is

Group-1	Group-2
P. Nitric acid	I. Trona process
Q. Phosphoric acid	II. Twitchell process
R. Potassium chloride	III. Ostwald's process
S. Stearic acid	IV. Haifa process

(a) P - III, Q - I, R - IV, S - II
(b) P - IV, Q - I, R - II, S - III
(c) P - III, Q - IV, R - I, S - II
(d) P - I, Q - IV, R - II, S - III

[2022 : 2 M]

1.24 Which one of the following options represents the major components of oleum?

(a) Sulfuric acid and nitric acid
(b) Concentrated sulfuric acid and petroleum jelly
(c) Sulfuric acid and hydrochloric acid
(d) Sulfuric acid and sulfur trioxide

[2023 : 1 M]

1.25 Match the products in Group 1 with the manufacturing processes in Group 2 listed in the table below.

Group 1	Group 2
P. Acetaldehyde	I. Sulfate process
Q. Sulfuric acid	II. Electric furnace process
R. Pulp	III. Wacker process
S. Phosphorous	IV. Contact process

Codes:

(a) P - III, Q - IV, R - I, S - II
(b) P - III, Q - I, R - IV, S - II
(c) P - IV, Q - I, R - II, S - III
(d) P - IV, Q - I, R - II, S - III

[2023 : 2 M]

1.26 For the electrolytic cell in a chlor-alkali plant, which of the following statements is/are correct?

(a) A membrane cell operates at a higher brine concentration than a diaphragm cell.
(b) Chlorine gas is produced at the cathode.
(c) Hydrogen gas is produced at the cathode.
(d) The caustic product stream exits the cathode compartment.

[2024 : 1 M]

Answers Inorganic Chemical Industries

1.1	(a)	1.2	(b)	1.3	(a)	1.4	(d)	1.5	(a)	1.6	(a)	1.7	(d)
1.8	(d)	1.9	(c)	1.10	(c)	1.11	(c)	1.12	(a)	1.13	(c)	1.14	(b)
1.15	(d)	1.16	(b)	1.17	(a)	1.18	(b)	1.19	(b)	1.20	(b)	1.21	(c)
1.22	(d)	1.23	(c)	1.24	(d)	1.25	(a)	1.26	(a, c, d)				

Explanations Inorganic Chemical Industries

1.1 (a)

Triple superphosphate is a component of fertilizer that primarily consist of monocalcium phosphate and is obtained by treating phosphate rock with phosphoric acid.

1.4 (d)

- Contact process is the most widely used method for the production of sulphuric acid.
- Mercury cell (Castner Kellner process) is a method of electrolysis of an aqueous alkali chloride solution to produce the corresponding alkali hydroxide.

1.5 (a)

Caustic soda

Sodium carbonate

Sodium hydroxide: $Ca(OH)_2 + Na_2CO_3 \rightarrow CaCO_3\downarrow + 2NaOH$

Slaked line

Soda ash

(Salt) (Limestone)

Solvay process: $2\,NaCl + CaCO_3 \rightarrow Na_2CO_3 + CaCl_2$

1.6 (a)

Sulphite process: It produces wood pulp which is almost pure cellulose.

$S + O_2 \rightarrow SO_2$

$SO_2 + H_2O \rightleftharpoons H_2SO_3$

$MgCO_3 + 2H_2SO_3 \rightarrow Mg\,(HSO_3)_2 + CO_2 + H_2O$

$Mg\,(HSO_3)_2 + MgCO_3 \rightarrow 2MgSO_3 + CO_2 + H_2O$

Cooking liquor composition is 7% by weight SO_2, of which 4.5% is in combined form as acid and 2.5% as calcium or $Mg\,(HSO_3)_2$.

1.7 (d)

Manufacture of caustic soda:

$2NaCl\,(aq) + 2H_2O\,(l) \rightarrow H_2(g) + Cl_2\,(g) + 2\,NaOH\,(aq)$

Anode: Cl^- are oxidized to chlorine gas

Cathode: Na^+ and water are reduced to $H_2(g)$ and NaOH (aq)

Calcium and Mg: Temporary hardness is the water hardness caused due to the presence of $CaCO_3$ and $MgCO_3$.

Ca^{+2} and Mg^{+2} form precipitate on membrane layer. Therefore in the feed brine content of Ca^{+2} and Mg should be low.

1.8 (d)

The main advantage of nitric acid is that no sulphur or less sulphur is required. This is important during a shortage of sulphur. Acidulation of phosphate rock with nitric acid produces a hygroscopic super phosphate having higher phosphorous content than the sulphuric acid.

1.9 (c)

$H_2SO_4 + SO_3 \rightarrow H_2S_2O_7$ (oleum)

Sequence of process equipment;

Burner, catalytic converter, oleum absorption column, 98% sulphuric acid absorption tower.

1.10 (c)

The intermediate cooling keeps the gas temperature in the operating range of the catalyst which is from 400°C – 625°C, otherwise the large amount of heat produced by the exothermic reaction would overheat and chemically degrade the catalyst.

1.11 (c)

Decomposition efficiency (η_D) of an electrolytic cell is defined as:

$$\eta_D = \frac{\text{Gram equivalents of NaOH produced}}{\text{Gram equivalents of NaCl charged}} \times 100$$

1.12 (a)

For LLDPE: Ethylene and 1-butene are raw materials
For Nylon 66: Hexamethylene diamine and adipic acid are raw materials
For Synthetic fibre: Terephthalic acid and ethylene glycol are raw materials
For Acrylonitrile: Propylene and ammonia are raw materials.

1.13 (c)

Single superphosphate (SSP) $\rightarrow$ Rock phosphate + Sulfuric acid
Triple superphosphate (TSP) $\rightarrow$ Rock phosphate + Phosphoric acid
Diammonium phosphate (DAP) $\rightarrow$ Rock phosphate + Sulfuric acid + Ammonia
Caustic soda $\rightarrow$ Brine

1.14 (b)

"Petroleum refineries produces elemental sulphur as a by-product".

1.15 (d)

"In a kraft process the chemicals used in the digester are Na_2S".

1.16 (b)

The DCDA (Double Contact Double Absorption) process is used for the manufacture of sulphuric acid.

1.17 (a)

Fluidized bed $\rightarrow$ Catalytic cracking
Multistage adiabatic reactor with inter-state cooling $\rightarrow SO_2$ oxidation
Fourdrinier machine $\rightarrow$ Paper-making
Diaphragm cell $\rightarrow$ NaOH manufacture

1.18 (b)

The essentials chemicals reagents used in the digester in sulfate kraft process are
- Caustic soda
- Sodium sulphide
- Soda ash

1.19 (b)

Wulff process $\rightarrow$ Acetylene production
Sulfite process $\rightarrow$ Pulp production
Solvay process $\rightarrow$ Soda ash production
Frasch process $\rightarrow$ Sulfur mining

1.20 (b)

Solvay process $\Rightarrow CaCO_3 + 2NaCl \rightarrow Na_2CO_3 + CaCl_2$
Oxo process $\Rightarrow CH_2 = CH_2 + CO + H_2 \rightarrow CH_3CH_2CHO$
Saponification $\Rightarrow$ RCOOH + NaOH $\rightarrow$ RCOONA + H_2O

1.21 (c)

Gypsum: Gypsum is composed of calcium sulphate and water. Its chemical name is calcium sulphate dihydrate [$CaSO_4 . 2H_2O$].
Dolamite is a type of limestone. It is rich in magnesium and calcium carbonate Ca $Mg(CO_3)_2$.
Triple superphosphate Ca $(H_2PO_4)_2$ is highly concentrated phosphorous fertilizer with 48% content of P_2O_5.

1.22 (d)

Glauber's salt is the decahydrate form of sodium sulfate acid its chemical formula can be written as $Na_2SO_4.10H_2O$.

1.24 (d)

Oleum is a term referring to solutions of various compositions of SO_3 in H_2SO_4 having chemical formula of $H_2S_2O_7$.

1.26 (a, c, d)

For the electrolytic cell in a chlor-alkali plant, membrane cell operates at higher concentration than diaphragm cell and produces comparatively more concentrated product along with chlorine at anode and hydrogen at the cathode and caustic product steam exists the cathode compartment.

■■■■

2 Fertilizer Industry

2.1 Synthesis gas is a mixture of
(a) CO and H_2 (b) N_2 and H_2
(c) H_2, CH_4 and CO (d) CO_2 and H_2
[2000 : 2 M]

2.2 Consider the production of ammonia from methane and air as raw materials. The catalysts used are (i) for steam reforming of methane and (ii) for ammonia synthesis
(a) (i) Ni/Al_2O_3; (ii) $Cu\text{-}ZnO/Al_2O_3$
(b) (i) Fe/Al_2O_3; (ii) $Cu\text{-}ZnO/Al_2O_3$
(c) (i) Ni/Al_2O_3; (ii) Fe/Al_2O_3
(d) (i) Fe/Al_2O_3; (ii) Ni/Al_2O_3
[2002 : 2 M]

2.3 The composition of fresh feed to the high temperature high pressure urea autoclave
(a) excess liquid ammonia and liquefied CO_2
(b) excess liquid ammonia and compressed CO_2 gas
(c) liquid ammonia and excess compressed CO_2
(d) compressed NH_3 gas and excess compressed CO_2
[2003 : 1 M]

2.4 Match the groups:

Group 1
A. Calcium ammonium nitrate
B. $CaCl_2$ - NaCl liquor

Group 2
1. Fertilizer industry
2. Paper and Pulp industry
3. Soda ash industry

Codes;

	A	B
(a)	1	3
(b)	1	2
(c)	3	1
(d)	2	3

[2003 : 2 M]

2.5 Prilling tower is found in the flowsheet for the manufacture of
(a) ammonia (b) urea
(c) superphosphate (d) triple superphosphate
[2004 : 1 M]

2.6 Match the product in **Group-I** with the raw materials in **Group-II**:

Group I
A. Urea
B. Polyester

Group II
1. Ammonia and carbon dioxide
2. Dimethyl terephthalate and ethylene glycol
3. Ammonia and carbon monoxide
4. Hexamethylene diamine and adipic
5. acid

Codes:

	A	B
(a)	1	4
(b)	3	2
(c)	3	2
(d)	1	2

[2004 : 2 M]

2.7 Match the product in **Group-I** with the name of the process in **Group-II**.

Group I	Group II
P. Sodium carbonate	1. Haber
Q. Ammonia	2. Solvay
R. Sulphuric acid	3. Fischer-Tropsch
	4. Contact

(a) P-2, Q-1, R-4 (b) P-4, Q-1, R-2
(c) P-3, Q-4, R-2 (d) P-2, Q-1, R-3
[2009 : 2 M]

2.8 Match the technologies in **Group 1** with the entries in **Group 2**:

Group-1
P. Urea manufacture
Q. Coal gasification
R. Controlled release of chemicals
S. Deep hydrodesulphurization

Group-2
I. Microencapsulation
II. Ultra-low sulphur diesel
III. Shale oil
IV. Prilling tower
V. Gas hydrates
VI. Gas-solid non-catalytic reaction

(a) P-I, Q-V, R-II, S-VI
(b) P-IV, Q-VI, R-I, S-II
(c) P-IV, Q-I, R-III, S-II
(d) P-V, Q-VI, R-IV, S-II [2015 : 1 M]

2.9 The purpose of methanation reaction used in ammonia plants is to
(a) remove CO as it is a catalyst poison
(b) increase the amount of hydrogen
(c) remove sulphur as it is a catalyst poison
(d) utilize methane as a catalyst for ammonia synthesis

[2017 : 1 M]

2.10 The operating temperature range for the Haber process is 350-500°C. It is used for the production of ammonia at
(a) 20 MPa using Fe catalyst in an exothermic reaction.
(b) 0.1 MPa using Fe catalyst in an exothermic reaction.
(c) 20 MPa using Fe catalyst in an endothermic reaction.
(d) 20 MPa using zeolite catalyst in an endothermic reaction.

[2020 : 1 M]

2.11 In an ammonia manufacturing facility, the necessary hydrogen is generated from methane. The facility consists of the following process units:
P : Methanator
Q : CO shift convertor
R : CO_2 stripper
S : Reformer
T : Ammonia convertor
The correct order of these units, starting from methane feed is
(a) S, Q, R, P, T
(b) P, Q, R, S, T
(c) S, P, Q, R, T
(d) P, S, T, Q, R [2024 : 1 M]

■■■■

Answers Fertilizer Industry

2.1	(a)	**2.2**	(c)	**2.3**	(b)	**2.4**	(a)	**2.5**	(b)	**2.6**	(d)	**2.7**	(a)
2.8	(b)	**2.9**	(a)	**2.10**	(a)	**2.11**	(a)						

Explanations Fertilizer Industry

2.3 (b)

"Excess liquid ammonia and compressed CO_2 gas".

2.4 (a)

Calcium ammonia nitrate or CAN is widely used inorganic fertilizer. It contains 20 to 30% calcium carbonate and 70 to 80% ammonium nitrate.

$CaCl_2 - NaCl$ liquor $\rightarrow$ Soda ash industry

2.5 (b)

"Urea is manufactured by reacting ammonia and carbon dioxide in an autoclave to form ammonium carbonate. Prilling tower is a dryer unit in this process".

2.6 (d)

"Urea is produced through synthesis of ammonia and carbon dioxide".

Polyster: This artificial fiber is comprised of purified terephthalic acid (PTA) and monoethylene glycol (MEG).

2.7 (a)

Haber process is used for ammonia production.

$N_2 + 3H_2 \rightleftharpoons 2NH_3 \qquad \{\Delta H = -92.22 \text{ kJ/mol}\}$

Solvay process: It is used for the production of soda ash

$2\,NaCl + Ca_2CO_3 \rightarrow Na_2CO_3 + Ca_2Cl_2$

Contact process: Widely used method for the production of H_2SO_4.

$H_2SO_4 + SO_3 \rightarrow H_2S_2O_7(l)$

$H_2S_2O_7(l) + H_2O(l) \rightarrow 2H_2SO_4(l)$

2.8 (b)

Urea manufacture $\rightarrow$ Prilling tower

Coal gasification $\rightarrow$ Gas solid non catalytic reaction

Controlled release of chemicals $\rightarrow$ Microencapsulation

Deep hydrodesulphurization $\rightarrow$ Ultra low sulphur diesel.

2.9 (a)

Methanation is the final stage in the purification of synthesis gas and most modern ammonia plants plus some older hydrogen plants use the simple and convenient methanation reaction to remove traces of carbon oxides from the process gas.

2.10 (a)

The operating temperature for the Haber process is around 20 MPa and it is an exothermic reaction.

2.11 (a)

The correct sequence for ammonia manufacturing is:

Reformer $\rightarrow$ CO shift convertor $\rightarrow$ CO_2 stripper $\rightarrow$ Methanator $\rightarrow$ Ammonia convertor

■■■■

3 Organic Industries

3.1 One of the steps during refining of cane sugar consists of addition of hydrated lime to the sugar syrup followed by carbonation of the resulting solution. The purpose of this step is to
(a) Adjust the pH of the syrup.
(b) Remove the coloring matter from the syrup.
(c) Reduce the viscosity of the syrup.
(d) Improve the rate of crystallization of sugar.

[2001 : 1 M]

3.2 For the hydrogenation of oils, –(a)– is commonly used as catalyst, arid –(b)– is a catalyst poison.
(a) (a) Platinum (b) Sulfur
(b) (a) Palladium (b) Oxygen
(c) (a) Nickel (b) Sulfur
(d) (a) Nickel (b) Oxygen

[2001 : 2 M]

3.3 Match the groups:

Group-1
A. Black liquor
B. Activated silica alumina
C. Press mud

Group-2
1. Petroleum refining
2. Sugar factory

Codes:
(a) A - 1, C - 2 (b) A - 2, B - 1
(c) B - 1, C - 2 (d) B - 2, C - 1

[2003 : 2 M]

3.4 Multiple effect evaporators are commonly used in the manufacture of
A. Paper B. Super phosphate
C. Sugar D. Fats
(a) A and B (b) A and C
(c) A and D (d) C and D

[2004 : 2 M]

3.5 Match the process in **Group-I** with the catalyst used in **Group-II**.

Group-I
A. Sulfuric acid manufacture
B. Vegetable oil hydrogenation

Group-II
1. Platinum
2. Vanadium pentoxide
3. Iron
4. Raney nickel

Codes:

	A	B
(a)	3	1
(b)	2	1
(c)	2	4
(d)	4	2

[2004 : 2 M]

3.6 Which of the following is a detergent?
(a) Benzene hexachloride
(b) Cellulose nitrate
(c) Polyvinyl chloride
(d) Alkyl benzene sulfonate

[2005 : 2 M]

3.7 Pair the following reactors with their products

List-I	List-II
A. Arc Furnace	1. Citric acid
B. Fermenter	2. Calcium carbide
C. Hydrogenator	3. Saturated fats
	4. Alum

Codes:

	A	B	C
(a)	4	1	3
(b)	2	1	3
(c)	4	3	1
(d)	2	3	1

[2006 : 2 M]

3.8 Choose the most appropriate pairs from the following:

List-I	List-II
A. Nitration	1. Detergent
B. Sulphonation	2. Pulp and paper
C. Carbonation	3. Explosives
	4. Sugar

Codes :

	A	B	C
(a)	1	2	4
(b)	3	1	2
(c)	3	1	4
(d)	3	2	4

[2006 : 2 M]

3.9 Match the product in **Group-1** with its application in the industries of **Group-2**:

Group-1

P. Lithium stearate

Q. Magnesium stearate

R. Aluminium sulphate

Group-2

1. Cosmetics
2. Paper
3. Dry cleaning
4. Greases

(a) P-2, Q-3, R-1 (b) P-3, Q-4, R-2

(c) P-4, Q-2, R-3 (d) P-4, Q-1, R-2

[2007 : 2 M]

3.10 Match the product in **Group-1** with its application in the industries of **Group-2**:

Group-1

P. Saponification

Q. Calcination

R. Alkylation

Group-2

1. Petroleum refining
2. Synthetic fibres
3. Cement
4. Soaps and detergents

(a) P-1, Q-3, R-4 (b) P-2, Q-3, R-4

(c) P-4, Q-2, R-1 (d) P-4, Q-3, R-1

[2008 : 2 M]

3.11 Which one of the following process sequences is used in the sugar industry?

(a) Ca_2HPO_4/Lime Treatment → Crystallization → Crushing

(b) Ca_2HPO_4/Lime Treatment → Multiple stage evaporation → Crystallization

(c) Crushing → Crystallization → Ca_2HPO_4/Lime Treatment

(d) Multiple stage evaporation → Crystallization → Ca_2HPO_4/Lime Treatment

[2008 : 2 M]

3.12 Identify which of the following statements are FALSE.

P : Oils with an oleic radical (1 double bond) are more suitable than oils with a linolenic radical (3 double bonds) as film forming vehicles for paints.

Q : Production of synthesis gas from coal and steam is an endothermic process.

R : Use of chlorine for bleaching of wood pulp results in the release of dioxins.

S : In the manufacture of urea from ammonia, the main intermediate product formed is ammonium bicarbonate.

(a) P and Q only (b) R and S only

(c) Q and R only (d) P and S only

[2013 : 2 M]

3.13 Which ONE of the following statements is CORRECT?

(a) The major components of biodiesel are triglycerides

(b) Biodiesel is essentially a mixture of ethyl esters

(c) Biodiesel is highly aromatic

(d) Biodiesel has a very low aniline point

[2014 : 1 M]

3.14 The molecular formula of the predominant chemical compound in commercial sugar is

(a) $C_{12}H_{22}O_{11}$ (b) $C_{12}H_{24}O_{12}$

(c) $C_6H_{10}O_5$ (d) $C_6H_{12}O_6$

[2018 : 1 M]

3.15 In soap manufacturing, the triglycerides present in oils and fats are hydrolyzed to mainly produce

(a) Fatty acids and glycerol

(b) Glycerol only

(c) Fatty acids only

(d) Glycerol and paraffins

[2022 : 1 M]

■■■■

Answers Organic Industries

3.1	(a)	3.2	(c)	3.3	(c)	3.4	(b)	3.5	(c)	3.6	(d)	3.7	(b)
3.8	(b)	3.9	(d)	3.10	(d)	3.11	(b)	3.12	(d)	3.13	(b)	3.14	(a)
3.15	(a)												

Explanations Organic Industries

3.1 (a)

Addition of hydrated lime to the sugar syrup followed by carbonation of the resulting solution is basically done to maintain the pH of the syrup around 7.

3.3 (c)

Press mud: Is the compressed sugar industry waste produced during filtration of cane juice.
Activated silica alumina: For controlling humidity silica and alumina gets are used as adsorbents in petroleum refining.

3.4 (b)

"Multiple effect evaporators commonly used in the manufacturing process of sugar. Black liquor is concentrated in a pulp and paper industry".

3.5 (c)

Sulfuric acid → Vanadium pentoxide
Vegetable oil hydrogenation → Raney nickel

3.6 (d)

Alkyl benzene sulfonates are a class of anionic surfactants and is produced by sulfonation which produces high foam and have excellent detergent properties.

3.7 (b)

- Calcium carbide is produced in electric arc furnace from mixture of lime and coke
  $CaO + 3C \rightarrow CaC_2 + CO$
- Citric acid is produced by fermentation and is mainly used as an acidifier.
- Hydrogenation is the process that can be used to turn liquid oil into solid fat.

3.8 (b)

Nitration: Is a process of addition of a nitro group into an organic chemical compound. These reactions are used for explosive production.

Sulphonation: Sulfonic acid is produced by sulphonation process
$RC_6H_5 + SO_3 \rightarrow RC_6H_4\,SO_3H$
Sulfonic acid exhibit detergent like properties.
Carbonation: Dissolving of CO_2 in liquid. This reaction is used in pulp and paper industries.

3.9 (d)

Lithium stearate → is a white soft solid prepared by the reaction of lithium hydroxide and stearic acid are components of lithium grease.
Magnesium stearate → is a salt that forms when stearate molecules bond with a magnesium ion and it is used in cosmetics.
Aluminium sulphate → paper

3.10 (d)

Saponification: Is a process that involves the conversion of fat, oil or lipid in to soap and alcohol by the action of aqueous alkali.
Calcination: Refers to heating (thermal treatment of) a solid chemical compound (e.g. carbonate ores) to high temperature in absence of or limited supply of air. Decomposition of calcium carbonate to calcium oxide and CO_2 is done to produce cement.
Alkylation: Is the transfer of an alkyl group from one molecule to another. It combines low molecular weight olefins with isobutene in presence of a catalyst (H_2SO_4 or HF) and the product thus obtained is mixture of highly branched chain paraffinic hydrocarbons with a high octane number.

3.11 (b)

"Cane sugar processing consist of the following steps: Sugar cane is crushed, the juice is heated and filtered, then sent to a series of crystallization steps to create crystals of raw sugar followed by centrifugation to remove any remaining juice or syrup".

3.12 (d)

Out of given four statements first and last one are incorrect i.e.

- (P) oils with an oleic radical (1 double bond) are more suitable than oils with a linolenic radical (3 double bonds) as film forming vehicles for points.
- (S) in the manufacture of urea from ammonia the main intermediate product formed is ammonium bicarbonate.

3.13 (b)

"Biodiesel is defined as the mixtures of ethyl esters derived from renewable lipid sources".

3.14 (a)

Chemical formula for commercial sugar is:

$C_{12}H_{22}O_{11}$

3.15 (a)

"Soap is produced by the saponification process of a triglyceride (fat oil). In this process the triglyceride is reacted with a strong base such as sodium or KOH to produce glycerol and fatty acid salts".

■■■■

4 Polymer Industries

4.1 'Nylon 66' is so named because
(a) the average degree of polymerization of the polymer is 1966
(b) the number of carbon atoms between two nitrogen atoms are 6
(c) the number of nitrogen atoms between two carbon atoms are 6
(d) the polymer was first synthesized in 1966.
[2003 : 1 M]

4.2 Identify the group in which all the polymers mentioned can be used to make fibers
(a) Butadiene copolymers, polyamides, Urea aldehydes
(b) Cellulose derivatives, polyisoprene, polyethylene
(c) Cellulose derivatives, polyamides, polyurethanes
(d) Polypropylenes, polyvinylchloride, silicones
[2003 : 2 M]

4.3 Match the product in **Group-I** with the nature of the reaction in **Group-II**:

Group-I
A. Polyethylene
B. Nylon
C. Polystyrene

Group-II
1. Condensation polymerization
2. Addition polymerization

Codes:

	A	B	C
(a)	1	1	2
(b)	2	2	1
(c)	1	2	1
(d)	2	1	2

[2004 : 2 M]

4.4 Which of the following is NOT employed in the commercial production of linear polyvinyl chloride?
(a) Emulsion polymerization
(b) Suspension polymerization
(c) Addition polymerization
(d) Condensation polymerization [2005 : 2 M]

4.5 Match the raw materials in **Group-II** with the products in **Group-I**:

Group-I
A. Nylon-66
B. Terylene

Group-II
1. Chlorodifluromethane
2. Dimethyl terephthalate and ethylene glycol
3. Acetylene and hydrogen cyanide
4. Hexamethylene diamine and adipic acid

Codes:

	A	B
(a)	3	4
(b)	4	3
(c)	4	2
(d)	1	2

[2005 : 2 M]

4.6 Pair the following polymers with their chain characteristics.

List-I
A. HDPE
B. LDPE
C. LLDPE

List-II
1. Very few branches
2. Short and regular branches
3. High branching with both short and long chain branches

Codes:

	A	B	C
(a)	1	2	3
(b)	2	1	3
(c)	1	3	2
(d)	2	3	1

[2006 : 2 M]

4.7 Match the following:

List-I	List-II
A. Carbon disulphide	1. Nylon-6
B. Caprolactum	2. Nylon-66
C. Gypsum	3. Phosphoric acid
	4. Viscose rayon

Codes :

	A	B	C
(a)	4	2	3
(b)	2	1	3
(c)	3	1	4
(d)	4	1	3

[2006 : 2 M]

4.8 Match the synthetic fibres in **Group-1** with their classification in **Group-2**:

Group-1	Group-2
P. Rayon	1. Polyamide
Q. Orlon	2. Polyester
R. Dacron	3. Cellulose
	4. Acrylic

(a) P-2, Q-3, R-1 (b) P-3, Q-4, R-2
(c) P-3, Q-1, R-3 (d) P-3, Q-3, R-4

[2007 : 2 M]

4.9 Match each of the following techniques of polymerization in **Group-I**, with the corresponding process characteristics in **Group-II**.

Group-I

P. Bulk
Q. Solution
R. Suspension
S. Emulsion

Group-II

1. Polymer with very high molecular weight can be obtained.
2. Heat removal is crucial but very difficult.
3. Small amount of undesired low molecular weight polymer is formed.
4. Polymer concentration in the product stream is low.

(a) P-1, Q-2, R-3, S-4
(b) P-2, Q-1, R-3, S-4
(c) P-2, Q-3, R-4, S-1
(d) P-3, Q-2, R-4, S-1

[2010 : 2 M]

4.10 Match each of the polymers in **Group-I**, with the raw material in **Group-II**, from which they are made.

Group-I

P. Polyester
Q. Polyamide
R. Viscose rayon
S. Epoxy resin

Group-II

1. Ethylene Glycol
2. Adipic acid
3. Cellulose
4. Bisphenol

(a) P-1, Q-2, R-3, S-4
(b) P-2, Q-1, R-3, S-4
(c) P-1, Q-2, R-4, S-3
(d) P-3, Q-2, R-4, S-1

[2010 : 2 M]

4.11 Match the polymerisation mechanisms in **Group-I** with the corresponding polymers in **Group-II**.

Group-I

P. Chain growth/addition polymerisation
Q. Step growth/condensation polymerisation

Group-II

1. Polyethylene
2. Polyvinyl chloride
3. Polyethylene terephthalate

(a) P-3; Q-1, 2 (b) P-1, 2; Q-3
(c) P-2, 3; Q-1 (d) P-1; Q-2, 3

[2011 : 1 M]

4.12 Match the polymer in **Group-I** to the polymer characteristic in **Group-II**:

Group-I

P. Polyethylene
Q. Phenol-formaldehyde polymer
R. Polyisoprene
S. Polyester

Group-II

I. Elastomer
II. Fiber
III. Thermoplastic
IV. Thermosetting polymer

(a) P-III, Q-IV, R-I, S-II
(b) P-IV, Q-II, R-III, S-I
(c) P-III, Q-II, R-I, S-IV
(d) P-IV, Q-III, R-I, S-II

[2012 : 2 M]

4.13 Match the polymer mentioned on the left with the catalyst used for its manufacture given on the right.

Group-1

I. Low density Polyethylene
II. High density Polyethylene
III. Polyethylene Terephthalate
IV. Polyvinyl Chloride

Group-2

P. Ziegler-Natta catalyst
Q. Traces of Oxygen
R. Butyl Lithium
S. Antimony

(a) I-Q, II-R, III-S, IV-R
(b) I-S, II-P, III-Q, IV-R
(c) I-Q, II-P, III-S, IV-R
(d) I-S, II-R, III-P, IV-Q

[2015 : 1 M]

4.14 Match the polymerization processes in **Group-1** with the polymers in **Group-2**.

Group-1

P. Free radical polymerization
Q. Ziegler Natta polymerization
R. Condensation polymerization

Group-2

I. Nylon 6.6
II. Polypropylene
III. Poly vinyl chloride

Choose the correct set of combinations.

(a) P-I, Q-II, R-III (b) P-III, Q-II, R-I
(c) P-I, Q-III, R-II (d) P-II, Q-I, R-III

[2017 : 1 M]

4.15 Polyvinyl chloride is produced by

(a) co-polymerization
(b) addition-type kinetics
(c) reacting chlorine with polyethylene
(d) reacting hydrochloric acid with polyethylene

[2018 : 1 M]

4.16 Choose the **correct** statement in viscose rayon manufacturing process,

(a) carbon disulphide used as reactant for xanthate formation is regenerated in a later step.
(b) caustic soda used as reactant for steeping of cellulose is regenerated in a later step.
(c) sulphuric acid is used in steeping process of cellulose.
(d) the spun viscose rayon is hardened in an alkali bath.

[2018 : 1 M]

4.17 The combination that correctly matches the polymer in **Group-1** with the polymerization reaction type in **Group-2** is

Group-1

P. Nylon 6
Q. Polypropylene
R. Polyester

Group-2

1. Condensation polymerization
2. Ring opening polymerization
3. Addition polymerization

(a) P-2, Q-1, R-3
(b) P-1, Q-3, R-2
(c) P-3, Q-2, R-1
(d) P-2, Q-3, R-1

[2019 : 1 M]

4.18 Which one of the following is NOT correct?

(a) nylon-6, 6 is produced by condensation polymerization.
(b) phenol-formaldehyde resin is a thermosetting polymer.
(c) high density polyethylene (HDPE) is produced by condensation polymerization.
(d) poly (ethylene terephthalate) (PET) is a polyester.

[2020 : 1 M]

4.19 Nitrile rubber is manufactured via polymerization process. Which one of the following options is the CORRECT pair of monomers used in this process?

(a) Acrylonitrile and styrene
(b) Acrylonitrile and butadiene
(c) Butadiene and styrene
(d) Butadiene and isoprene

[2023 : 1 M]

■■■■

Answers Polymer Industries

4.1	(b)	**4.2**	(c)	**4.3**	(d)	**4.4**	(c)	**4.5**	(c)	**4.6**	(a)	**4.7**	(d)
4.8	(b)	**4.9**	(b)	**4.10**	(a)	**4.11**	(b)	**4.12**	(a)	**4.13**	(c)	**4.14**	(b)
4.15	(b)	**4.16**	(a)	**4.17**	(d)	**4.18**	(c)	**4.19**	(b)				

Explanations Polymer Industries

4.1 (b)

"Nylon 66 is made from 2 monomers with each one having 6 carbon atoms, hence the nylon 66 name".

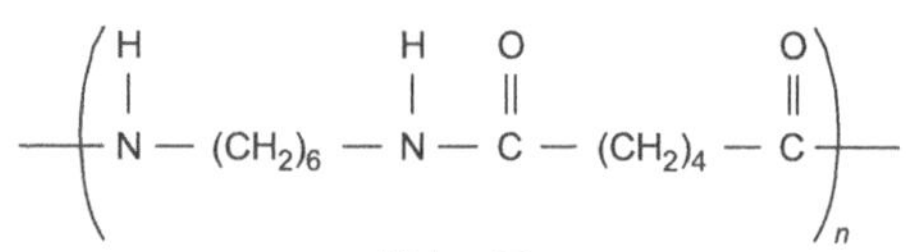

Nylon 66

4.2 (c)

"Fibers can be made by cellulose derivatives, polyamides and polyurethanes".

4.3 (d)

Addition polymerization: A polymer that forms by simple linking of monomers without the co-generation of other products.

Examples: Polythenes, PVC, polystyrene etc.

Condensation polymerization: It is a form of step-growth polymerization in which monomers react with each others to form larger structural units while releasing smaller molecules as a by product.

Examples: PET, Nylon etc.

4.4 (c)

PVC is high strength thermoplastic material produced by the monomer vinyl chloride most step growth polymers are also classified as condensation polymers.

4.5 (c)

- Nylon-66 is synthesized by polycondensation of hexamethylenediamine and adipic acid.
- Ethylene glycol and terephthalic acid undergo condensation reaction to produce terylene.

4.6 (a)

HDPE: It is a thermoplastic polymer made from petroleum. If has very few branches giving it stronger intermolecular forces and tensile strength than LDPE.

LDPE: It is a soft, flexible plastic material and it has more branching than HDPE, less molecular force and lower tensile strength.

LLDPE: It has short branches which is made by copolymerization of ethylene with longer chain olefins.

4.7 (d)

"Viscose rayon is derived from cellulose and it is then treated with carbon disulfide to form a solution of sodium cellulose xanthate".

Caprolactum: Nylon-6 is synthesized by ring opening polymerization of caprolactam.

Gypsum: It is composed of calcium sulphate and water. It is produced from the fabrication of phosphoric acid by reacting phosphate ore with sulphuric acid.

4.8 (b)

Rayon is a synthetic fiber, made from natural sources of regenerated cellulose such as wood, and related agricultural products.

Orlon a synthetic acrylic fibre used for textiles and knitwear or a fabric made from it.

Dacron is a trade name of polyester fiber. It is a condensation polymer which is obtained from ethylene glycol and terephthalic acid.

4.9 (b)

Bulk polymerization: In this process the real challenge concerns the dissipation of the heat produced by polymerization. Heat transfer and mixing become difficult as the viscosity of reaction mass increases.

Solution polymerization: A monomer is dissolved in non-reactive solvent that contains a catalyst polymer with very high molecular weight can be obtained.

Emulsion polymerization: Polymer concentration in the product stream is low.

Suspension: Under this low molecular weight polymer is formed in small amount which is undesired.

4.10 (a)

Polyester → Ethylene glycol
Polyamide → Adipic acid
Viscous rayon → Cellulose
Epoxy resin → Bisphenol

4.11 (b)

Polyethylene, polypropylene and polyvinylchloride (PVC) are chain growth polymers. Chain growth polymers involves the linking together of molecules incorporating double or triple chemicals bonds. These unsaturated monomers break and link up with other monomers to form the repeating chain.

4.12 (a)

Thermoplastic → Polymer that becomes mouldable above specific temperature and returns to a solid state upon cooling e.g. polyethylene, PVC etc.
Thermosetting → These are polymer materials which are malleable at low temperatures but which change irreversible to become hard at high temperatures e.g. Polyamides, phenol-formaldehyde
Fiber → Nylon, polyester etc.
Elastomer → Polyisoprene, styrene butadiene, polybutadiene

4.13 (c)

Low density polyethylene → Traces of oxygen
High density polyethylene → Ziegler-Natta catalyst
Polyethylene terephthalate → Antimony
Polyvinyl chloride → Butyl Lithium

4.14 (b)

Free radical polymerization → Poly vinyl chloride
Ziegler Natta polymerization → Polypropylene
Condensation polymerization → Nylon 6, 6

4.15 (b)

Addition type kinetics is used in the PVC production.

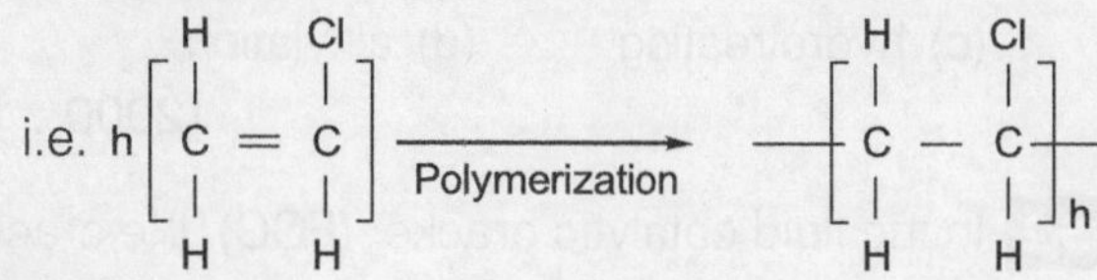

4.16 (a)

In viscose rayon manufacturing process, carbon disulphide used as reactant for xanthate formation is regenerated in a later step.

4.17 (d)

Nylon 6 → Ring opening polymerization in which caprolactum is used
Polypropylene → Addition polymerization
Polyester → Condensation polymerization

4.18 (c)

"High density polyethylene (HDPE) is produced by additional type polymerization".

4.19 (b)

Nitrile rubber is manufactured via polymerization, the correct monomers are acrylonitrile and butadiene.
Nitrile rubber also known as nitrile butadiene rubber i.e. (BUNA-N) which is composed of acrylonitrile and butadiene.

■■■■

5 Petroleum Industries

5.1 In petroleum refining, the process used for conversion of hydrocarbons to aromatics is
(a) catalytic cracking (b) catalytic reforming
(c) hydrotreating (d) alkylation
[2000 : 1 M]

5.2 In the fluid catalytic cracker (FCC), the cracking reaction is ---(a)--- and the regeneration is ---(b)--
(a) (a) Exothermic (b) Endothermic
(b) (a) Exothermic (b) Exothermic
(c) (a) Endothermic (b) Endothermic
(d) (a) Endothermic (b) Exothermic
[2001 : 1 M]

5.3 The commonly used solvent in supercritical extraction is
(a) Methyl-ethyl-ketone
(b) Water
(c) Carbon tetrachloride
(d) Carbon dioxide [2002 : 1 M]

5.4 The average boiling point of aviation turbine fuel is closest to that of
(a) Lubricating oils (b) LPG
(c) Diesel (d) Kerosene
[2002 : 1 M]

5.5 The catalytic converter for conversion of SO_2 to SO_3 by contact process should have feed with SO_2 content between
(a) 2 - 5% (b) 7 - 10%
(c) 12 - 15% (d) 20 - 25%
[2003 : 1 M]

5.6 In a refinery, petroleum crude is fractionated into gas fraction, light ends, intermediate distillates, heavy distillates, residues and by products. The group of products including gas oil, diesel oil and heavy fuel oil belongs to the fraction
(a) heavy distillates (b) intermediate distillates
(c) light ends (d) residues
[2003 : 2 M]

5.7 The order of preference for feedstock to a catalytic reformer is
(a) catalytic naphtha – coking naphtha – virgin naphtha
(b) coking naphtha – virgin naphtha – catalytic naphtha
(c) virgin naphtha – catalytic naphtha – coking naphtha
(d) virgin naphtha – coking naphtha – catalytic naphtha [2003 : 2 M]

5.8 The proper arrangement of the petroleum fractions, in the order of their boiling points is
(a) Lubricating oils > diesel > petrol > LPG
(b) Lubricating oils > petrol > diesel > LPG
(c) Petrol > lubricating oils > diesel > LPG
(d) Petrol > diesel > LPG > lubricating oil
[2004 : 1 M]

5.9 Match the product in **Group-I** with the catalyst used for its production in **Group-II**:

Group-I	Group-II
A. Nitric acid	1. Silver oxide
B. Formaldehyde	2. Raney nickel
	3. Activated carbon
	4. Platinum- Rhodium

Codes:

	A	B
(a)	1	2
(b)	2	3
(c)	3	4
(d)	4	1

[2005 : 2 M]

5.10 Match the feed in **Group-I** with the process in **Group-II**:

Group-I	Group-II
A. Gas oil	1. Acetylene production
B. Residual crude	2. Ethylene production
	3. Coking
	4. Cracking

Codes:

	A	B
(a)	3	4
(b)	4	3
(c)	2	3
(d)	1	4

[2005 : 2 M]

5.11 In a fluid catalytic cracking unit, the nature of the reactions occurring in the reactor and the regenerator is
(a) Reactor-Exothermic, Regenerator-Exothermic
(b) Reactor-Exothermic, Regenerator-Endothermic
(c) Reactor-Endothermic, Regenerator-Exothermic
(d) Reactor-Endothermic, Regenerator-Endothermic
[2006 : 1 M]

5.12 Which of the following is desirable in gasoline but undesirable in kerosene?
(a) Aromatics (b) Mercaptans
(c) Naphthenic acid (d) Paraffins
[2007 : 1 M]

5.13 Which one of the following is NOT a major constituent of crude oil?
(a) Paraffins (b) Olefins
(c) Naphthenes (d) Aromatics
[2008 : 1 M]

5.14 Which one of the following process sequences is used in the production of synthesis gas?
(a) Desulphurization → Steam reforming → Hot K_2CO_3 cycle
(b) Steam reforming → Desulphurization → Hot K_2CO_3 cycle
(c) Hot K_2CO_3 cycle → Steam reforming → Desulphurization
(d) Hot K_2CO_3 cycle → Desulphurization → Steam reforming
[2008 : 2 M]

5.15 In petroleum refining operations, the process used for converting paraffins and naphthenes to aromatics is
(a) catalytic reforming (b) catalytic cracking
(c) hydrocracking (d) alkylation
[2009 : 1 M]

5.16 The active component of catalysts used in steam reforming of methane to produce synthesis gas is
(a) Nickel (b) Iron
(c) Platinum (d) Palladium
[2009 : 1 M]

5.17 Match the product in **Group-I** to the raw material in **Group-II**:

Group-I	Group-II
P. Ethylene	1. Natural gas
Q. Methanol	2. Synthesis gas
R. Phthalic anhydride	3. Naphtha
	4. Naphthalene

(a) P - 1, Q - 2, R - 3 (b) P - 2, Q - 1, R - 4
(c) P - 3, Q - 1, R - 4 (d) P - 3, Q - 2, R - 4
[2009 : 2 M]

5.18 Match the unit process in **Group-I** with the industry in **Group-II**.

Group-I	Group-II
P. Steam cracking	1. Petroleum refining
Q. Hydrocracking	2. Petrochemicals
R. Condensation	3. Polymers
	4. Soaps and detergents

(a) P - 1, Q - 2, R - 3 (b) P - 2, Q - 3, R - 3
(c) P - 1, Q - 2, R - 4 (d) P - 2, Q - 1, R - 3
[2009 : 2 M]

5.19 Match the process in **Group-I** with the products in **Group-II**:

Group-I	Group-II
P. Claus process	1. Syngas
Q. Linde process	2. Oxygen
R. Lurgi process	3. Sulphur

(a) P - 1, Q - 2, R - 3 (b) P - 2, Q - 1, R - 3
(c) P - 3, Q - 1, R - 2 (d) P - 3, Q - 2, R - 1
[2010 : 1 M]

5.20 Which one of the following sequences is arranged according to increasing calorific value?
(a) Producer gas, natural gas, water gas
(b) Natural gas, producer gas, water gas
(c) Producer gas, water gas, natural gas
(d) Water gas, natural gas, producer gas
[2011 : 1 M]

5.21 Hydrotreating is used for
(a) removal of water from crude oil
(b) treatment of crude oil with water
(c) improving octane number of gasoline
(d) removal of sulphur and nitrogen from petroleum fractions
[2011 : 1 M]

5.22 Zeolite ZSM-5 is added to commercial FCC catalyst for
(a) promoting SO_2 reduction
(b) promoting CO oxidation
(c) improving tolerance to metal
(d) enhancing Octane number
[2011 : 1 M]

5.23 In petroleum refining, catalytic reforming is used to convert
(a) Paraffins and naphthenes to aromatics
(b) Paraffins to hydrogen and carbon monoxide
(c) Gas oil to diesel and gasoline
(d) Light olefins to gasoline [2012 : 1 M]

5.24 The final boiling points of gasoline, diesel, atmospheric gas oil (AGO) and lubricating oils vary as
(a) gasoline > diesel > AGO > lubricating oils
(b) lubricating oils > AGO > diesel > gasoline
(c) AGO > lubricating oils > diesel > gasoline
(d) lubricating oils > diesel > AGO > gasoline
[2012 : 1 M]

5.25 The main unit processes used for the production of hydrogen from natural gas are steam reforming (SR), pressure swing adsorption (PSA), low temperature water gas shift reaction (LTWGS) and high temperature water gas shift reaction (HT WGS). The correct sequence of these in the plant is
(a) SR; LT WGS; HT WGS; PSA
(b) PSA; SR; LT WGS; HT WGS
(c) SR; HT WGS; LT WGS; PSA
(d) PSA; HT WGS; LT WGS; SR [2012 : 1 M]

5.26 Catalytic cracking is
(a) a hydrogen addition process
(b) a carbon rejection process
(c) an exothermic process
(d) a coking process [2014 : 1 M]

5.27 Select the WRONG statement regarding water gas shift converters from the list below.
(a) Inter-stage cooling is provided between the two stages of shift converters.
(b) Usually high temperature shift (HTS) reactor has a iron-based catalyst and low temperature shift (LTS) reactor has a copper-based catalyst.
(c) HTS reactor is followed by LTS reactor.
(d) LTS reactor is followed by HTS reactor.
[2015 : 2 M]

5.28 Match the industrial processes in **Group-1**, with the catalyst materials in **Group-2**:

Group-1
P. Ethylene polymerisation
Q. Petroleum feedstock cracking
R. Oxidation of SO_2 to SO_3
S. Hydrogenation of oil

Group-2
I. Nickel
II. Vanadium pentoxide
III. Zeolite
IV. Aluminium triethyl with titanium chloride promoter

(a) P - IV, Q - III, R - II, S -I
(b) P - I, Q - IV, R - III, S - II
(c) P - I, Q - II, R - III, S - IV
(d) P - II, Q - III, R - IV, S - I
[2016 : 1 M]

5.29 In connection with petroleum refining, identify the **incorrect** statement among the following options.
(a) Desalting of crude oil is done before processing it in atmospheric distillation unit.
(b) A stream of hydrogen is produced in catalytic reforming of naphtha.
(c) Asphalt used for paving is a petroleum product.
(d) Cetane number indicates the quality of petrol/motor spirit.
[2018 : 1 M]

5.30 Producer gas is obtained by
(a) passing air through red hot coke
(b) thermal cracking of naphtha
(c) passing steam through red hot coke
(d) passing air and steam through red hot coke
[2019 : 1 M]

5.31 In petroleum refining operations, the process used for converting paraffins and naphthenes to aromatics is
(a) alkylation (b) catalytic reforming
(c) hydrocracking (d) isomerization
[2019 : 1 M]

5.32 Consider the refinery processes in **Group-1** and the catalysts in **Group-2**:

Group-1
P. Hydrodesulfurization
Q. Fluid catalytic cracking (FCC)
R. Naphtha reforming.

Group-2
1. Zeolites
2. Pt/Al_2O_3
3. $Co\text{-}Mo/Al_2O_3$

The correct combination is

(a) P - 2, Q - 1, R - 3 (b) P - 3, Q - 2, R - 1
(c) P - 3, Q - 1, R - 2 (d) P - 1, Q - 3, R - 2

[2020 : 1 M]

5.33 Catalytic reforming is commonly used in the petroleum industry to improve fuel quality. The undesirable reaction in the catalytic reforming of naphtha is

(a) Hydrocracking of paraffins
(b) Dehydrogenation of naphthenes
(c) Isomerization of naphthenes
(d) Cyclization of paraffins [2022 : 1 M]

5.34 Which one of the following statements related to octane number is NOT correct?

(a) Linear alkanes with higher carbon number have higher octane number.
(b) Branching in linear alkanes increases their octane number.
(c) Catalytic reforming of hydrocarbons increases their octane number.
(d) Gasoline quality is measured in terms of octane number. [2023 : 1 M]

5.35 Which of the following statements with reference to the petroleum/petrochemical industry is/are correct?

(a) Catalytic hydrocracking converts heavier hydrocarbons to lighter hydrocarbons.
(b) Catalytic reforming converts straight-chain hydrocarbons to aromatics.
(c) Cumene is manufactured by the catalytic alkylation of benzene with propylene.
(d) Vinyl acetate is manufactured by reacting methane with acetic acid over a palladium catalyst.

[2024 : 1 M]

■■■■

Answers Petroleum Industries

5.1 (b)	5.2 (d)	5.3 (d)	5.4 (d)	5.5 (b)	5.6 (a)	5.7 (d)
5.8 (a)	5.9 (d)	5.10 (b)	5.11 (c)	5.12 (a)	5.13 (c)	5.14 (d)
5.15 (a)	5.16 (a)	5.17 (d)	5.18 (d)	5.19 (d)	5.20 (c)	5.21 (d)
5.22 (d)	5.23 (a)	5.24 (b)	5.25 (c)	5.26 (b)	5.27 (d)	5.28 (a)
5.29 (d)	5.30 (d)	5.31 (b)	5.32 (c)	5.33 (a)	5.34 (a)	5.35 (a, b, c)

Explanations Petroleum Industries

5.1 (b)

Catalytic reforming is used for conversion of hydrocarbons to aromatics i.e. under this rearrangement of molecules takes place without much affecting the average molecular weight.

5.2 (d)

During FCC higher boiling petroleum fractions is converted into gasoline, and olefinic gases.

i.e., Catalytic cracking → Endothermic reaction
Regenerator → Exothermic reaction.

5.3 (d)

CO_2 is the most used solvent in the super critical extraction.

5.4 (d)

Aviation turbine fuel is a mixture of a large number of different hydrocarbons. Average B.P of aviation turbine fuel is close to that of kerosene.

5.5 (b)

The SO_2 content in the catalytic converter for conversion of SO_2 to SO_3 is 7 to 10%

$$2SO_2 + O_2 \rightleftharpoons 2SO_3$$

5.6 (a)

Light ends: fraction (70 - 200°C) gasoline, naphtha, jet fuel, etc.
Intermediate distillate: (200°C to 350°C)
Heavy distillate: > 350°C mineral oil, lubricating oil

5.7 (d)

The correct order would be:

Virgin naphtha → Coking naphtha → Catalytic naphtha

"Main application of naphthas is as a feedstock in the petrochemical production of olefins".

5.8 (a)

The correct sequence for their boiling points are:

Lubricating oils > Diesel > Petrol > LPG

5.9 (d)

- The production of the nitric acid is done in the presence of catalyst platinum with 10% rhodium.
- Formaldehyde is produced by the catalytic oxidation of methanol. Most common catalysts are silver metal or mixture of mixture of iron and molybdenum or vanadium oxides. Silver based catalyst operates at higher temperature at 650°C.

5.10 (b)

Gas oil → cracking

Residual crude → coking

5.11 (c)

In the process of fluid catalytic cracking, the reaction is endothermic while the crude oil regeneration reaction is exothermic.

5.12 (a)

Aromatics are desirable in gasoline as it raise its octane rating. These are hydrocarbons with the highest octane number for the same number of carbon atoms.

Aromatics decreases the smoke point of kerosene and hence undesirable in kerosene.

5.13 (c)

"Crude oil is a mixture of volatile liquid hydrocarbons i.e. mainly consist of paraffins, olefins and aromatics".

5.14 (d)

"Synthesis gas is produced from natural gas via catalytic processes. The CO_2 produced from gasification process must be removed before syngas to avoid the formation of unwanted products. Mixture of (K_2CO_3, Na_2CO_3, Li_2ZrO_3) is used to test the CO_2 absorption rates.

Gasification for power generation requires desulfurization. H_2S and carbonyl sulphide are removed from the synthesis gas by physical absorption. Steam reforming is preferred if high H_2/CO ratio is desired.

5.15 (a)

Catalytic reforming is used to convert paraffin and naphthenes to aromatics which have high octane rating.

5.16 (a)

For steam reforming of methane to produce synthesis gas mainly (Ni) is used as catalyst

$$C_nH_{2n+2} + n\,H_2O \xrightarrow{(Ni)} n\text{Co} + (2n+1)H_2$$

5.17 (d)

Naphtha is a predominant feed for steam crackers for production of ethylene.

- Methanol is produced by syngas
  $CO + 2H_2 \rightarrow CH_3OH$
- Phthalic anhydride is produced by oxidation of naphthalene

5.18 (d)

Steam cracking: This is a petrochemical process in which saturated hydrocarbons are broken into unsaturated hydrocarbons.

Hydrocracking: Saturated hydrocarbons are the products of this process. Products from this unit process are jet fuels, diesel, LPG etc.

Condensation: Widely used process in polymer industries.

5.19 (d)

Claus process: (Gas desulfurization process)

Hydro-desulfurization: Naphtha's and other petroleum oil are converted to sulphur in claus process.

$$2H_2S + O_2 \rightarrow S_2 + 2H_2O$$

Lurgi process: Process used for production of syngas.

Linde process: Process of the preparation of liquid air in two column Hampson-mechanical cycle using Joule Thomson effect.

5.20 (c)

The correct sequence for calorific values:
Producer gas < Water gas < Natural gas

5.21 (d)

Hydrotreating is used to remove sulphur, nitrogen and other contaminants like arsenic from crude oil.

5.22 (d)

ZSM-5 is a 10 membered sieve containing channel openings used for boosting gasoline octane number and for increasing C_3 and C_4 LPG selectivity in FCC.

5.23 (a)

Catalytic reforming is a chemical process used to convert petroleum refinery naphthas distilled from crude oil into high octane liquid products called reformates i.e., it converts paraffins and napthanes to aromatics.

5.24 (b)

The correct sequence of boiling points are lubricating oils > AGO > diesel > gasoline

5.25 (c)

Steam reforming: $CH_4 + H_2O \rightarrow CO + 3H_2$
High temperature : 700°C – 1100°C
Low temperature water gas shift reaction : (350°C)
PSA is used for the removal of CO_2 as a final step in production of hydrogen $CO + H_2O \rightarrow CO_2 + H_2$

5.26 (b)

"Catalytic cracking is a process of breaking of carbon chain i.e. carbon rejection process".

5.27 (d)

"In order to take advantage of both the thermodynamics and kinetics of the reaction, the industrial scale water gas shift reaction is conducted in multiple adiabatic stages consisting of a high temperature shift (HTS) followed by a low temperature shift (LTS) with intersystem cooling".

5.28 (a)

Ethylene polymerization $\rightarrow$ Aluminium triethyl with titanium chloride promoter
Petroleum feedstock cracking $\rightarrow$ Zeolite
Oxidation of SO_2 to $SO_3 \rightarrow$ Vanadium pentaoxide
Hydrogenation of oil $\rightarrow$ Nickel

5.29 (d)

Cetane number is used to define the quality of diesel, while octane number is used for petrol, therefore option (d) is incorrect.

5.30 (d)

Producer gas is produced by passing both air and steam through hot bed of solid coke.

5.31 (b)

"Catalytic reforming are the process in which paraffins and naphthenes are converted to aromatics".

5.32 (c)

The correct combinations of the refinery process with their respective catalysts are
hydrodesulfurization $\rightarrow$ Co-Mo/Al_2O_3
Fluid catalytic cracking (FCC) $\rightarrow$ Zeolites
Naphtha reforming $\rightarrow$ Pt/Al_2O_3

5.33 (a)

The typical reaction in catalytic reforming are dehydrocyclization, isomerization and dehydrogenation. Dealkylation and hydrocracking reactions are undesirable owing to the low value of the resulting light hydrocarbon products.

5.35 (a, b, c)

In petroleum or petrochemical industry:

- Catalytic hydrocracking converts heavier hydrocarbons to lighter hydrocarbon in the presence of tungsten sulphide for vapour phase and iron catalyst in liquid phase.
- Catalytic reforming converts straight-chain hydrocarbon to aromatic in pressure of platinum catalyst.
- Cumene is manufactured by catalytic alkylation of benzene with propylene.
- Vinyl acetate is produced by vapour phase reaction of ethylene and acetic acid over palladium catalyst.

■■■■

6 Petrochemical Industries

6.1 Commercially, ethylene is produced from naphtha by
(a) catalytic cracking
(b) catalytic dehydrogenation
(c) pyrolysis
(d) hydrocracking

[2000 : 1 M]

6.2 Styrene is produced from ethylbenzene by the process of
(a) Dehydrogenation (b) Oxidation
(c) Alkylation (d) Dehydration

[2001 : 1 M]

6.3 From among the following, choose one that is not an exothermic process
(a) Methanol synthesis
(b) Catalytic cracking
(c) Ammonia synthesis
(d) Oxidation of sulphur

[2002 : 1 M]

6.4 The preferred reacting system for oxidation of o-xylene to phthalic anhydride is
(a) jacketed liquid phase CSTR
(b) jacketed steam heated multitubular reactor
(c) multitubular reactor with cooling
(d) multistage multitubular reactor with interstage cooling

[2003 : 2 M]

6.5 The chief raw material for the commercial production of methanol is
(a) Synthesis gas
(b) Formaldehyde
(c) Acetic acid
(d) Ethanol

[2005 : 2 M]

6.6 The oxo reaction is used for converting
(a) alcohol to aldehyde
(b) paraffin to olefin
(c) olefin to aldehyde
(d) aldehyde to alcohol

[2006 : 1 M]

6.7 Pair the following industrial processes with the catalysts used:

List-I
A. Oxidation of o-xylene to phthalic anhydride
B. Oxidation of ethanol to acetaldehyde
C. Oxidation of ammonia to oxides of nitrogen

List-II
1. V_2O_5
2. Pd
3. Ag
4. Pt

Codes:

	A	B	C
(a)	1	3	4
(b)	2	1	4
(c)	4	3	1
(d)	1	3	2

[2006 : 2 M]

6.8 Match the chemicals in **Group-1** with their function in **Group-2**:

Group-1
P. Styrene
Q. Tert-dodecylmercaptan
R. Potassium pyrophosphate

Group-2
1. Buffer
2. Catalyst
3. Modifier
4. Monomer

(a) P-1, Q-4, R-4 (b) P-4, Q-1, R-2
(c) P-4, Q-1, R-3 (d) P-4, Q-3, R-1

[2007 : 2 M]

6.9 Match the petrochemical derivative in **Group-1** with the raw materials in **Group-2**:

Group-1	Group-2
P. Acrylonitrile	1. Methane
Q. Ammonia	2. Ethane
R. Dodecene	3. Ethylene
	4. Propylene

(a) P-1, Q-2, R-1 (b) P-2, Q-1, R-2
(c) P-3, Q-4, R-3 (d) P-4, Q-1, R-4

[2007 : 2 M]

6.10 Match the reactions in **Group-I** with the products in **Group-II**:

Group-I
P. Ammoxidation
Q. Nitration
R. Dehydrogenation
S. Oxidation

Group-II
1. Aniline from benzene
2. Benzoic acid from toluene
3. Acrylonitrile from propylene
4. Styrene from ethylbenzene

(a) P-3, Q-1, R-4, S-2 (b) P-4, Q-1, R-3, S-2
(c) P-1, Q-3, R-4, S-2 (d) P-1, Q-2, R-3, S-4

[2011 : 2 M]

6.11 Match the process in **Group-I** with the catalyst in **Group-II**:

Group-I
P. Fischer-Tropsch synthesis
Q. Formaldehyde from methanol
R. Hydrogenation of vegetable oils
S. Dehydrogenation of ethylbenzene

Group-II
I. Nickel
II. Fe_2O_3
III. Silver
IV. Cobalt

(a) P-III, Q-IV, R-I, S-II (b) P-IV, Q-II, R-I, S-III
(c) P-IV, Q-III, R-I, S-II (d) P-III, Q-IV, R-II, S-I

[2012 : 2 M]

6.12 Match the reactant-product combination in **Group-1** with the unit process in **Group-2**:

Group-1
P. Propylene - butanol
Q. Cumene - phenol
R. Butane - butadiene
S. Ethylene dichloride - vinyl chloride

Group-2
1. Pyrolysis
2. Dehydrogenation
3. Hydroformylation
4. Peroxidation

(a) P-3, Q-2, R-4, S-1
(b) P-2, Q-4, R-3, S-1
(c) P-1, Q-3, R-2, S-4
(d) P-3, Q-4, R-2, S-1

[2013 : 2 M]

6.13 The most common catalyst used for oxidation of o-oxylene to phthalic anhydride is

(a) V_2O_5 (b) Pd
(c) Pt (d) Ag

[2019 : 1 M]

6.14 Match the reaction in **Group-1** with the reaction type in **Group-2**:

Group-1
P. Methylcyclohexane $\rightarrow$ Toluene + $3H_2$
Q. Ethylcyclopentane $\rightarrow$ Methylcyclohexane
R. *n*-Octane $\rightarrow$ Ethylbenzene + $4H_2$
S. *n*-Octane $\rightarrow$ *n*-Pentane + Propylene

Group-2
1. Dehydrocyclization
2. Cracking
3. Dehydrogenation
4. Isomerization

The correct combination is
(a) P-2, Q-3, R-1, S-4
(b) P-3, Q-4, R-1, S-2
(c) P-3, Q-4, R-2, S-1
(d) P-1, Q-4, R-3, S-2

[2021 : 2 M]

6.15 Match the reactions in **Group 1** with the catalysts in **Group 2** listed in the table below:

Group 1
P. $C_2H_6 + Cl_2 \rightarrow$ Chlorobenzene + HCl

Q. $H_2C=CH_2 + \frac{1}{2}O_2 \rightarrow$ Ethylene oxide

R. $CH_3OH + \frac{1}{2}O_2 \rightarrow$ Formaldehyde + H_2O

S. Naphthalene + $\frac{9}{2}O_2 \rightarrow$ Phthalic anhydride + $2H_2O + 2CO_2$

Group 1
I. Mixed oxide of Mo and Fe
II. V_2O_5
III. $FeCl_3$
IV. Ag_2O

(a) P - III, Q - IV, R - II, S - I
(b) P - III, Q - IV, R - I, S - II
(c) P - IV, Q - II, R - I, S - III
(d) P - IV, Q - III, R - I, S - II

[2023 : 2 M]

Answers Petrochemical Industries

6.1	(d)	6.2	(a)	6.3	(b)	6.4	(a)	6.5	(a)	6.6	(c)	6.7	(a)
6.8	(d)	6.9	(d)	6.10	(a)	6.11	(c)	6.12	(d)	6.13	(a)	6.14	(b)
6.15	(b)												

Explanations Petrochemical Industries

6.1 (d)

Ethylene production by the hydro-cracking of naphtha is an energy intensive process and most widely used method.

6.2 (a)

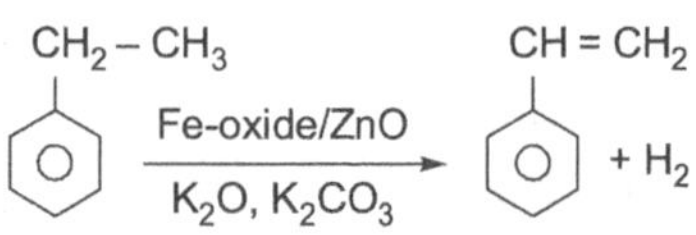

Styrene is produced by catalytic dehydrogenation.

6.3 (b)

Catalytic cracking is an endothermic process.

6.4 (a)

"Jacketed liquid phase CSTR" is most widely reactor system for oxidation of o-xylene to phthalic anhydride.

6.5 (a)

$CO + 2H_2 \rightarrow CH_3OH$ [300°C to 400°C]

CO and H_2 react to produce methanol.

Catalyst: mixture of Cu, Zn, Al

6.6 (c)

In the exo process, a mixture of alkene, CO and hydrogen under pressure and elevated temperature (in presence of catalyst) to forms aldehydes.

6.7 (a)

O-xylene + $O_2 \xrightarrow{V_2O_5}$ Phthalic anhydride [$C_6H_4(CO_2)_0$]

Ethanol + $O_2 \xrightarrow{Ag}$ CH_3CHO (Acetaldehyde)

$NH_3 \xrightarrow{Pt}$ Nitrogen oxide

6.8 (d)

Styrene → Monomer

Tert-dodecyl mercaptan → Modifier

Potassium pyrophosphate → PH buffer

6.9 (d)

- Propylene, ammonia and air is used in the manufacturing process of acrylonitrile.
- In haber process for production of ammonia, nitrogen and hydrogen gas are used. The main source of hydrogen is methane from natural gas.
- Nonene and dodecene are referred to as propylene trimer and propylene tetramer. Nonene and dodecene are obtained by oligomerization of a propylene containing feedstock.

6.10 (a)

Ammoxidation : Acrylonitrile from propylene

$$CH_3-CH=CH_2+\frac{3}{2}O_2+NH_3 \rightarrow NC-CH=CH_2 + \frac{3}{2}H_2O \text{ (Acrylonitrile)}$$

Nitration:

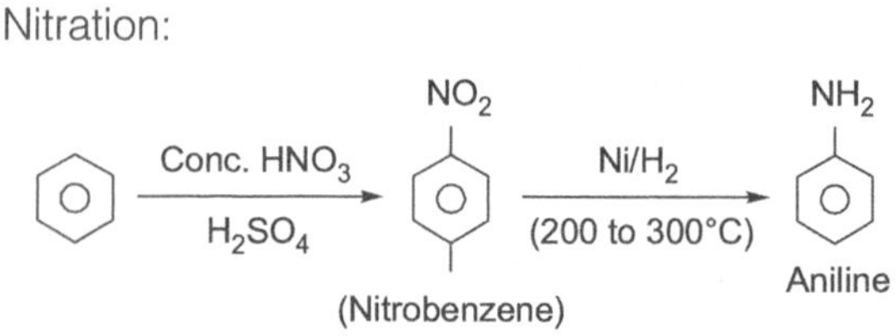

Dehydrogenation:

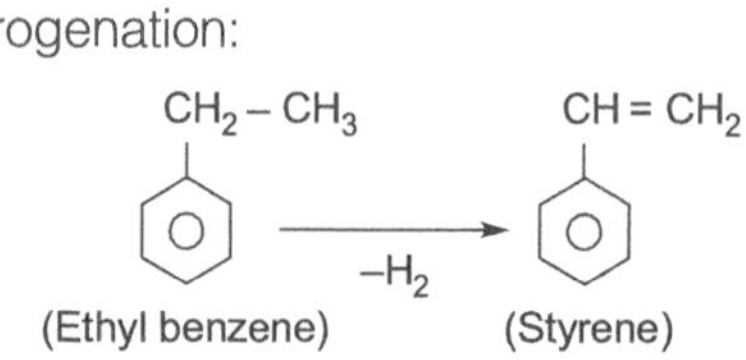

Oxidation:

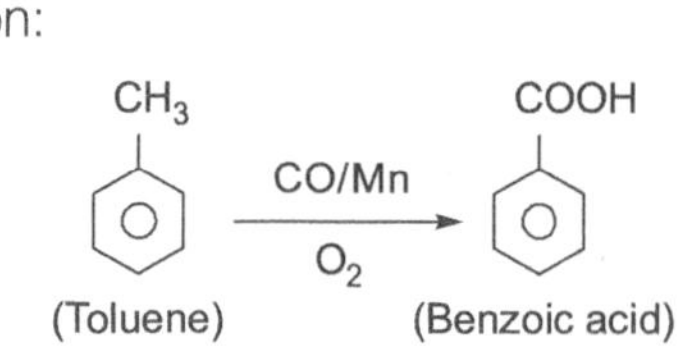

6.11 (c)

The correct matches for the catalyst are

Fischer Tropsch synthesis: Iron, cobalt and ruthenium

Formaldehyde from methanol: The most common catalysts are silver metal or a mixture of an iron and molybdenum or vanadium oxide.

Hydrogenation of vegetable oil: Palladium or Nickel catalyst

Dehydrogenation of ethylbenzene: Fe_2O_3

6.12 (d)

Dehydrogenation: Butadiene is produced by dehydrogenation of butane.

Pyrolysis:

$$CH_2 = CH_2 + 2HCL + \frac{1}{2}O_2 \rightarrow ClCH_2CH_2Cl + H_2O$$

$$ClCH_2 \,.\, CH_2Cl \rightarrow CH_2 = CHCl + HCL$$

n-butanol is produced by propylene by OXo reaction (Hydroformylation). Propylene is hydroformylated to butyraldehyde in the presence of rhodium-based catalyst. Butyraldehyde is then hydrogenated to produced *n*-butanol.

Peroxidation:

$$C_6H_5CH(CH_3) + O_2 \rightarrow C_6H_5^{C}(CH_3)_2^{OOH}$$

$$C_6H_5C(CH_3)_2^{OOH} + H_2SO_4 \rightarrow C_6H_5^{OH} + CH_3COCH_3$$

It involves the liquid phase air oxidation of cumene peroxide, which in turn is decomposed to phenol and acetone by action of acid.

6.13 (a)

$[V_2O_5]$ is used for oxidation of o-xylene to phthalic anhydride.

6.14 (b)

Methylcyclohexane $\rightarrow$ Toluene + $3H_2$ (dehydrogenation)

Ethylcyclopentane $\rightarrow$ Methylcyclohexane (Isomerization)

n-octane $\rightarrow$ Ethylbenzene + $4H_2$ (Dehydrogenation)

n-octane $\rightarrow$ *n*-Pentane + Propylene (Cracking)

■■■■

Engineering Mathematics

UNIT

XI

CONTENTS

Engineering Mathematics

Syllabus

Linear Algebra: Matrix Algebra, Systems of Linear Equations, Eigenvalues, Eigenvectors.

Calculus: Mean Value Theorems, Theorems of Integral Calculus, Evaluation of Definite and Improper Integrals, Partial Derivatives, Maxima and Minima, Multiple Integrals, Fourier Series, Vector Identities, Directional Derivatives, Line Integral, Surface Integral, Volume Integral, Stokes's Theorem, Gauss's Theorem, Divergence Theorem, Green's Theorem.

Differential Equations: First Order Equations (Linear and Non-Linear), Higher Order Linear Differential Equations with Constant Coefficients, Method of Variation of Parameters, Cauchy's Equation, Euler's Equation, Initial and Boundary Value Problems, Partial Differential Equations, Method of Separation of Variables.

Complex Variables: Analytic Functions, Cauchy's Integral Theorem, Cauchy's Integral Formula, Taylor Series, Laurent Series, Residue Theorem, Solution Integrals.

Probability and Statistics: Sampling Theorems, Conditional Probability, Mean, Median, Mode, Standard Deviation, Random Variables, Discrete and Continuous Distributions, Poisson Distribution, Normal Distribution, Binomial Distribution, Correlation Analysis, Regression Analysis.

Analysis of Previous GATE Papers

Exam Year	1 Mark Ques.	2 Marks Ques.	5 Marks Ques.	Total Marks
2000	3	3	2	**19**
2001	2	2	–	**6**
2002	2	1	2	**14**
2003	3	3	–	**9**
2004	4	8	–	**20**
2005	3	11	–	**25**
2006	4	6	–	**16**
2007	4	6	–	**16**
2008	5	10	–	**25**
2009	4	7	–	**18**
2010	4	4	–	**12**
2011	2	3	–	**8**
2012	4	3	–	**10**
2013	4	4	–	**12**
2014	4	5	–	**14**
2015	5	3	–	**11**
2016	3	4	–	**11**
2017	5	4	–	**13**
2018	2	2	–	**6**
2019	3	5	–	**13**
2020	4	4	–	**12**
2021	4	3	–	**10**
2022	5	4	–	**13**
2023	5	4	–	**13**
2024	5	4	–	**13**

1 Linear Algebra

1.1 The inverse of the matrix $\begin{bmatrix} 1 & -1 \\ -1 & -1 \end{bmatrix}$ is

(a) Does not exist (b) $\begin{bmatrix} 1 & 1 \\ 1 & -1 \end{bmatrix}$

(c) $\begin{bmatrix} 0.5 & -0.5 \\ -0.5 & -0.5 \end{bmatrix}$ (d) $\begin{bmatrix} -0.5 & 0.5 \\ 0.5 & 0.5 \end{bmatrix}$

[2000 : 2 M]

1.2 Find whether or not the vectors (1,1,2), (1, 2, 1) and (0, 3, –3) are linearly independent.

[2000 : 5 M]

1.3 The value of the following determinant $\begin{vmatrix} 1 & 0 & 0 & 0 & 0 \\ -2 & 2 & 0 & 0 & 0 \\ 3 & 5 & 3 & 0 & 0 \\ -1 & 4 & 7 & 4 & 0 \\ -5 & -6 & 2 & 1 & 1 \end{vmatrix}$ is

(a) 24 (b) 32

(c) –112 (d) 0

[2001 : 1 M]

1.4 The inverse of the matrix $\begin{bmatrix} 0.2 & 0 & 0 \\ 0 & 1 & 0 \\ 0 & 0 & 0.5 \end{bmatrix}$ is

(a) $\begin{bmatrix} -0.2 & 0 & 0 \\ 0 & -1 & 0 \\ 0 & 0 & -0.5 \end{bmatrix}$ (b) $\begin{bmatrix} -0.2 & 0 & 0 \\ 0 & -1 & 0 \\ 0 & 0 & -0.5 \end{bmatrix}$

(c) $\begin{bmatrix} 5 & 0 & 0 \\ 0 & 1 & 0 \\ 0 & 0 & 2 \end{bmatrix}$ (d) $\begin{bmatrix} 0.5 & 0 & 0 \\ 0 & 1 & 0 \\ 0 & 0 & 0.2 \end{bmatrix}$

[2002 : 1 M]

1.5 Matrix $A = \begin{bmatrix} 0.1 & 0.5 & 0 \\ 0.8 & 0 & 0.4 \\ 0.1 & 0.5 & 0.6 \end{bmatrix}$ has the property that it satisfies $Ax = x$, for any vector x.

(a) Write the characteristic equation to be solved for eigen values of A.

(b) Based on visual observation, find one of the eigen values of A.

(c) Find the other two eigen values of A.

[2002 : 5 M]

1.6 The sum of the eigen values of the matrix $\begin{bmatrix} 3 & 4 \\ x & 1 \end{bmatrix}$ for real and negative values of x is

(a) greater than zero

(b) less than zero

(c) zero

(d) dependent on the value of x [2004 : 2 M]

1.7 The system of equations:

$4x + 6y = 8$

$7x + 8y = 9$

$3x + 2y = 1$

has

(a) no solution

(b) only one solution

(c) two solutions

(d) infinite number of solutions [2004 : 2 M]

1.8 How many solutions does the following system of equations have?

$$4x + 2y + z = 7$$
$$x + 3y + z = 3$$
$$3x + 4y + 2z = 2$$

(a) 0 (b) 1

(c) 2 (d) ∞ [2005 : 2 M]

1.9 The matrix A is given by,

$$A = \begin{pmatrix} 1 & 4 \\ a & 2 \end{pmatrix}$$

The eigen values of the matrix A are real and non-negative for the condition

(a) $\left(-\frac{1}{16}\right) \le a \le \left(\frac{1}{16}\right)$ (b) $\left(-\frac{1}{2}\right) \le a \le \left(\frac{1}{2}\right)$

(c) $\left(-\frac{1}{2}\right) \le a \le \left(\frac{1}{16}\right)$ (d) $\left(-\frac{1}{16}\right) \le a \le \left(\frac{1}{2}\right)$

[2005 : 2 M]

1.10 The value of α for which the following three vectors are coplanar is

$$a = i + 2j + k$$
$$b = 3j + k$$
$$c = 2i + \alpha j$$

(a) 4 (b) zero
(c) −2 (d) −10

[2006 : 1 M]

1.11 If the following represents the equation of a line

$$\begin{vmatrix} x & 2 & 4 \\ y & 8 & 0 \\ 1 & 1 & 1 \end{vmatrix} = 0$$, then the line passes through the point

(a) (0, 0) (b) (3, 4)
(c) (4, 3) (d) (4, 4)

[2006 : 2 M]

1.12 If $A = \begin{bmatrix} 2 & 1 \\ 2 & 3 \end{bmatrix}$, then the eigen values of A^3 are

(a) 27 and 8 (b) 64 and 1
(c) 12 and 3 (d) 4 and 1

[2006 : 2 M]

1.13 The value of "a" for which the following set of equations

$$y + 2z = 0$$
$$2x + y + z = 0$$
$$ax + 2y = 0$$

have non-trivial solution is,

(a) 0 (b) 8
(c) −2 (d) 3

[2007 : 1 M]

1.14 If $A = \begin{bmatrix} 1 & 2 \\ 2 & 1 \end{bmatrix}$, then the eigen values of A^3 are

(a) 5, 4 (b) 3, −1
(c) 9, −1 (d) 27, −1

[2008 : 2 M]

1.15 The system of linear equation $Ax = 0$, where A is an $n \times n$ matrix, has a non-trivial solution only if

(a) rank of $A > n$
(b) rank of $A = n$
(c) rank of $A < n$
(d) A is an identity matrix

[2009 : 1 M]

1.16 The eigen values of matrix $A = \begin{bmatrix} 1 & 2 \\ 4 & 3 \end{bmatrix}$ are 5 and −1. Then the eigen values of $-2A + 3I$ (I is a 2×2 identity matrix) are

(a) −7 and 5 (b) 7 and −5
(c) $-\frac{1}{7}$ and $\frac{1}{5}$ (d) $\frac{1}{7}$ and $-\frac{1}{5}$

[2009 : 2 M]

1.17 The inverse of the matrix $\begin{bmatrix} 1 & 2 \\ 3 & 4 \end{bmatrix}$ is

(a) $\begin{bmatrix} -2 & -1 \\ \frac{3}{2} & -\frac{1}{2} \end{bmatrix}$ (b) $\begin{bmatrix} -2 & \frac{3}{2} \\ 1 & -\frac{1}{2} \end{bmatrix}$

(c) $\begin{bmatrix} -2 & 1 \\ \frac{3}{2} & -\frac{1}{2} \end{bmatrix}$ (d) $\begin{bmatrix} 2 & -\frac{3}{2} \\ -1 & \frac{1}{2} \end{bmatrix}$

[2010 : 1 M]

1.18 Let $\lambda_1 = -1$ and $\lambda_2 = 3$ be the eigen values and $V_1 = \begin{pmatrix} 1 \\ 0 \end{pmatrix}$ and $V_2 = \begin{pmatrix} 1 \\ 1 \end{pmatrix}$ be the corresponding eigen vectors of a real 2×2 matrix R. Given that $P = (V_1 \;\; V_2)$, which one of the following matrices represents $P^{-1}RP$?

(a) $\begin{pmatrix} 0 & -1 \\ 3 & 0 \end{pmatrix}$ (b) $\begin{pmatrix} 0 & 3 \\ -1 & 0 \end{pmatrix}$

(c) $\begin{pmatrix} 3 & 0 \\ 0 & -1 \end{pmatrix}$ (d) $\begin{pmatrix} -1 & 0 \\ 0 & 3 \end{pmatrix}$

[2011 : 1 M]

1.19 Consider the following set of linear algebraic equations

$$x_1 + 2x_2 + 3x_3 = 2$$
$$x_2 + x_3 = -1$$
$$2x_2 + 2x_3 = 0$$

The system has

(a) a unique solution
(b) no solution
(c) an infinite number of solutions
(d) only the trivial solution

[2012 : 1 M]

1.20 Consider the following (2 × 2) matrix $\begin{pmatrix} 4 & 0 \\ 0 & 4 \end{pmatrix}$.

Which one of the following vectors is NOT a valid eigenvector of the above matrix?

(a) $\begin{pmatrix} 1 \\ 0 \end{pmatrix}$ (b) $\begin{pmatrix} -2 \\ 1 \end{pmatrix}$

(c) $\begin{pmatrix} 4 \\ -3 \end{pmatrix}$ (d) $\begin{pmatrix} 0 \\ 0 \end{pmatrix}$

[2012 : 1 M]

1.21 Which of the following statements are TRUE?

P. The eigenvalues of a symmetric matrix are real.

Q. The value of the determinant of an orthogonal matrix can only be +1.

R. The transpose of a square matrix A has the same eigenvalues as those of A.

S. The inverse of an '$n \times n$' matrix exists if and only if the rank is less than 'n'.

(a) P and Q only (b) P and R only

(c) Q and R only (d) P and S only

[2013 : 1 M]

1.22 The following set of three vectors

$\begin{pmatrix} 1 \\ 2 \\ 1 \end{pmatrix}, \begin{pmatrix} x \\ 6 \\ x \end{pmatrix}$ and $\begin{pmatrix} 3 \\ 4 \\ 2 \end{pmatrix}$

is linearly dependent when x is equal to

(a) 0 (b) 1

(c) 2 (d) 3

[2015 : 1 M]

1.23 For the matrix $\begin{pmatrix} 4 & 3 \\ 3 & 4 \end{pmatrix}$, if $\begin{pmatrix} 1 \\ 1 \end{pmatrix}$ is an eigenvector, the corresponding eigen value is ______.

[2015 : 1 M]

1.24 A set of simultaneous linear algebraic equations is represented in a matrix form as shown below.

$$\begin{bmatrix} 0 & 0 & 0 & 4 & 13 \\ 2 & 5 & 5 & 2 & 10 \\ 0 & 0 & 2 & 5 & 3 \\ 0 & 0 & 0 & 4 & 5 \\ 2 & 3 & 2 & 1 & 5 \end{bmatrix} \begin{bmatrix} x_1 \\ x_2 \\ x_3 \\ x_4 \\ x_5 \end{bmatrix} = \begin{bmatrix} 46 \\ 161 \\ 61 \\ 30 \\ 81 \end{bmatrix}$$

The value (rounded off to the nearest integer) of x_3 is ______. [2016 : 2 M]

1.25 For the matrix $A = \begin{pmatrix} \cos\theta & -\sin\theta \\ \sin\theta & \cos\theta \end{pmatrix}$ if **det** stands for the determinant and A^T is the transpose of A then the value of det (A^TA) is ______.

[2018 : 2 M]

1.26 A system of n homogeneous linear equations containing n unknowns will have non-trivial solutions if and only if the determinant of the coefficient matrix is

(a) 1 (b) −1

(c) 0 (d) ∞

[2019 : 1 M]

1.27 The product of the eigen values of the matrix $\begin{pmatrix} 2 & 3 \\ 0 & 7 \end{pmatrix}$ is ______ (rounded off to one decimal place).

[2019 : 1 M]

1.28 Sum of the eigen values of the matrix $\begin{bmatrix} 2 & 4 & 6 \\ 3 & 5 & 9 \\ 12 & 1 & 7 \end{bmatrix}$ is ______ (round off to nearest integer).

[2020 : 1 M]

1.29 A, B, C and D are vectors of length 4.

$A = [a_1 \;\; a_2 \;\; a_3 \;\; a_4]$,

$B = [b_1 \;\; b_2 \;\; b_3 \;\; b_4]$,

$C = [c_1 \;\; c_2 \;\; c_3 \;\; c_4]$,

$D = [d_1 \;\; d_2 \;\; d_3 \;\; d_4]$

It is known that B is not a scalar multiple of A. Also, C is linearly independent of A and B. Further, $D = 3A + 2B + C$.

The rank of the matrix $\begin{bmatrix} a_1 & a_2 & a_3 & a_4 \\ b_1 & b_2 & b_3 & b_4 \\ c_1 & c_2 & c_3 & c_4 \\ d_1 & d_2 & d_3 & d_4 \end{bmatrix}$ is ____.

[2021 : 1 M]

1.30 Given matrix $\begin{bmatrix} x & 1 & 3 \\ y & 2 & 6 \\ 3 & 5 & 7 \end{bmatrix}$ the ordered pair (x, y) for which det$(A) = 0$ is

(a) (1, 1) (b) (1, 2)

(c) (2, 2) (d) (2, 1)

[2022 : 1 M]

1.31 If a matrix M is defined as $M = \begin{bmatrix} 10 & 6 \\ 6 & 10 \end{bmatrix}$, the sum of all the eigenvalues of M^3 is equal to ________ (in integer).

[2023 : 2 M]

1.32 Consider a linear homogeneous system of equations $Ax = 0$, where A is an $n \times n$ matrix, x is an $n \times 1$ vector and 0 is an $n \times 1$ null vector. Let r be the rank of A. For a non-trivial solution to exist, which of the following conditions is/are satisfied?

(a) Determinant of $A = 0$
(b) $r = n$
(c) $r < n$
(d) Determinant of $A \neq 0$

[2024 : 1 M]

1.33 Consider a matrix $A = \begin{bmatrix} -5 & a \\ -2 & -2 \end{bmatrix}$ where a is a constant. If the eigenvalues of A are –1 and –6, then the value of a, rounded off to the nearest integer, is ________.

[2024 : 1 M]

Answers Linear Algebra

1.1	(c)	**1.2**	(Sol.)	**1.3**	(a)	**1.4**	(c)	**1.5**	(Sol.)	**1.6**	(a)	**1.7**	(d)
1.8	(b)	**1.9**	(d)	**1.10**	(c)	**1.11**	(b)	**1.12**	(b)	**1.13**	(b)	**1.14**	(d)
1.15	(c)	**1.16**	(a)	**1.17**	(c)	**1.18**	(d)	**1.19**	(b)	**1.20**	(d)	**1.21**	(b)
1.22	(d)	**1.23**	(7)	**1.24**	(15)	**1.25**	(1)	**1.26**	(c)	**1.27**	(14)	**1.28**	(14)
1.29	(3)	**1.30**	(b)	**1.31**	(4160)	**1.32**	(a, c)	**1.33**	(–2)				

Explanations Linear Algebra

1.1 (c)

Given matrix:

$$I = \begin{bmatrix} 1 & -1 \\ -1 & -1 \end{bmatrix}$$

For matrix $A = \begin{bmatrix} a & b \\ c & d \end{bmatrix}_{2\times2}$

Its adjoint will be $\begin{bmatrix} d & -b \\ -c & a \end{bmatrix}$

Hence, for $I = \begin{bmatrix} 1 & -1 \\ -1 & -1 \end{bmatrix}$, the inverse matrix will be,

$\because \quad |I| = -1 - 1 = -2$

$$\Rightarrow \quad I^{-1} = \frac{\text{adj }(I)}{|I|}$$

$$= \frac{1}{2}\begin{bmatrix} -1 & 1 \\ 1 & 1 \end{bmatrix}$$

$$= \begin{bmatrix} 0.5 & -0.5 \\ -0.5 & -0.5 \end{bmatrix}$$

1.2 Sol.

For given vectors to be independent, determined formed by them should be non zero.

$$\Rightarrow \begin{vmatrix} 1 & 1 & 2 \\ 1 & 2 & 1 \\ 0 & 3 & -3 \end{vmatrix} = 1(-6-3) - 1(-3-0) + 2(3)$$

$$= -9 + 3 + 6 = 0$$

So, the given vectors are linearly dependent.

1.3 (a)

$$\text{Given : } \begin{vmatrix} 1 & 0 & 0 & 0 & 0 \\ -2 & 2 & 0 & 0 & 0 \\ 3 & 5 & 3 & 0 & 0 \\ -1 & 4 & 7 & 4 & 0 \\ -5 & -6 & 2 & 1 & 1 \end{vmatrix}$$

Since corresponding matrix of the determine is triangle matrix, the determinant will be given by product of loading diagonal elements.

$\therefore \quad 2 \times 3 \times 4 = 24$

1.4 (c)

$$\text{Given:} \quad A = \begin{bmatrix} 0.2 & 0 & 0 \\ 0 & 1 & 0 \\ 0 & 0 & 0.5 \end{bmatrix}$$

adj. (A) =

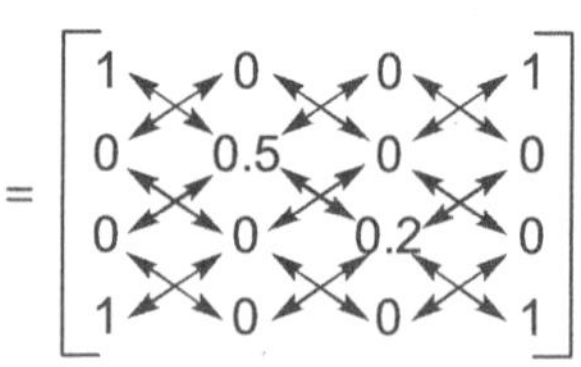

$$\text{adj. }(A) = \begin{bmatrix} 0.5 & 0 & 0 \\ 0 & 0.1 & 0 \\ 0 & 0 & 0.2 \end{bmatrix}$$

$$|A| = 0.2 \times 1 \times 0.5 = 0.1$$

$$\text{So,} \quad A^{-1} = \frac{\text{adj }(A)}{|A|}$$

$$= \frac{1}{0.1}\begin{bmatrix} 0.5 & 0 & 0 \\ 0 & 0.1 & 0 \\ 0 & 0 & 0.2 \end{bmatrix}$$

$$= \begin{bmatrix} 5 & 0 & 0 \\ 0 & 1 & 0 \\ 0 & 0 & 2 \end{bmatrix}$$

1.5 Sol.

$$\text{Given :} \quad A = \begin{bmatrix} 0.1 & 0.5 & 0 \\ 0.8 & 0 & 0.4 \\ 0.1 & 0.5 & 0.6 \end{bmatrix}$$

Characteristic equation $|A - \lambda I| = 0$

$$\Rightarrow \begin{vmatrix} 0.1-\lambda & 0.5 & 0 \\ 0.8 & -\lambda & 0.4 \\ 0.1 & 0.5 & 0.6-\lambda \end{vmatrix} = 0$$

$\Rightarrow (0.1 - \lambda)(-0.6\lambda + \lambda^2 - 0.2) - \frac{1}{2}(0.48 - 0.8\lambda - 0.04) = 0$

$\Rightarrow (0.1 - \lambda)(\lambda^2 - 0.6\lambda - 0.2) - \frac{1}{2}(0.44 - 0.8\lambda) = 0$

$\Rightarrow 0.1\lambda^2 - 0.06\lambda - 0.02 - \lambda^3 + 0.6\lambda^2 + 0.20\lambda - 0.22 + 0.4\lambda = 0$

$\Rightarrow -\lambda^3 + 0.7\lambda^2 + 0.54\lambda - 0.24 = 0$

$\Rightarrow \lambda^3 - 0.7\lambda^2 - 0.54\lambda + 0.24 = 0$ is the characteristic equation of A.

$\Rightarrow \lambda = -0.66, 1, 0.36$ are the eigen values of A.

1.6 (a)

$$\text{Matrix} = \begin{bmatrix} 3 & 4 \\ x & 1 \end{bmatrix}$$

Sum of eigen value = 3 + 1 = 4 > 0

1.7 (d)

$$\begin{bmatrix} 4 & 6 & 0 \\ 7 & 8 & 0 \\ 3 & 2 & 0 \end{bmatrix}\begin{bmatrix} x \\ y \\ z \end{bmatrix} = \begin{bmatrix} 8 \\ 9 \\ 1 \end{bmatrix}$$

Augmented matrix

$$A/B = \begin{bmatrix} 4 & 6 & 0 & : & 8 \\ 7 & 8 & 0 & : & 9 \\ 3 & 2 & 0 & : & 1 \end{bmatrix}$$

$$A/B = \begin{bmatrix} 3 & 2 & 0 & : & 1 \\ 4 & 6 & 0 & : & 8 \\ 7 & 8 & 0 & : & 9 \end{bmatrix}$$

$R_2 \to R_2 - R_1$ and
$R_3 \to R_3 - 2R_1$

$$= \begin{bmatrix} 3 & 2 & 0 & : & 1 \\ 1 & 4 & 0 & : & 7 \\ 1 & 4 & 0 & : & 7 \end{bmatrix}$$

$R_3 \to R_3 - R_2$

$$= \begin{bmatrix} 3 & 2 & 0 & : & 1 \\ 1 & 4 & 0 & : & 7 \\ 0 & 0 & 0 & : & 0 \end{bmatrix}$$

$\rho(A/B) = \rho(A) = 2 < 3$

So, infinite number of solutions exist.

1.8 (b)

$$4x + 2y + z = 7$$
$$x + 3y + z = 3$$
$$3x + 4y + 2z = 2$$

$$\begin{bmatrix} 4 & 2 & 1 \\ 1 & 3 & 1 \\ 3 & 4 & 2 \end{bmatrix}\begin{bmatrix} x \\ y \\ z \end{bmatrix} = \begin{bmatrix} 7 \\ 3 \\ 2 \end{bmatrix}$$

$$A/B = \begin{bmatrix} 4 & 2 & 1 & : & 7 \\ 1 & 3 & 1 & : & 3 \\ 3 & 4 & 2 & : & 2 \end{bmatrix}$$

$R_2 \leftrightarrow R_1$

$$\rho[A/B] = \begin{bmatrix} 1 & 3 & 1 & : & 3 \\ 4 & 2 & 1 & : & 7 \\ 3 & 4 & 2 & : & 2 \end{bmatrix}$$

$R_2 \to R_2 - 4R_1$ and $R_3 \to R_3 - 3R_1$

$$= \begin{bmatrix} 1 & 3 & 1 & : & 3 \\ 0 & -10 & -3 & : & -5 \\ 0 & -5 & -1 & : & -7 \end{bmatrix}$$

$\because \quad \rho[A/B] = \rho[A] = 3$

So, unique solution exists.

1.9 (d)

$$\text{Matrix} = \begin{pmatrix} 1 & 4 \\ a & 2 \end{pmatrix}$$

Characteristic equation $|A - \lambda I| = 0$

$$\begin{vmatrix} 1-\lambda & 4 \\ a & 2-\lambda \end{vmatrix} = 0$$

$$(\lambda - 1)(\lambda - 2) - 4a = 0$$
$$\lambda^2 - 3\lambda + 2 - 4a = 0$$
$$\lambda^2 - 3\lambda + 2 - 4a = 0$$

For real root : $D = b^2 - 4ac > 0$

$$9 - 4(1)(2 - 4a) \geq 0$$
$$9 - 8 + 16a \geq 0$$
$$16a + 1 \geq 0$$
$$a \geq -\frac{1}{16}$$

Non negative root :

$$2 - 4a \geq 0$$
$$4a \leq 2 \Rightarrow a \leq \frac{1}{2}$$

So, $\frac{1}{2} \leq a \leq \frac{1}{16}$

1.10 (c)

$$a = \hat{i} + 2\hat{j} + \hat{k}$$
$$b = 3\hat{j} + \hat{k}$$
$$c = 2\hat{i} + \alpha\hat{j}$$

For coplanar vectors $\begin{vmatrix} 1 & 2 & 1 \\ 0 & 3 & 1 \\ 2 & \alpha & 0 \end{vmatrix} = 0$

$$1[-\alpha] - 2[-2] + 1[-6] = 0$$
$$-\alpha + 4 - 6 = 0$$
$$-\alpha - 2 = 0$$
$$\alpha = -2$$

1.11 (b)

$$\begin{vmatrix} x & 2 & 4 \\ y & 8 & 0 \\ 1 & 1 & 1 \end{vmatrix} = 0$$

$$x(8-0) - 2(y-0) + 4(y-8) = 0$$
$$8x - 2y + 4y - 32 = 0$$
$$8x + 2y - 32 = 0$$
$$4x + y - 16 = 0$$

Only option (b) satisfies the equation.

1.12 (b)

$$A = \begin{bmatrix} 2 & 1 \\ 2 & 3 \end{bmatrix}$$
$$|A - \lambda I| = 0$$
$$\Rightarrow \begin{vmatrix} 2-\lambda & 1 \\ 2 & 3-\lambda \end{vmatrix} = 0$$
$$\Rightarrow (\lambda-2)(\lambda-3) - 2 = 0$$
$$\Rightarrow \lambda^2 - 5\lambda + 6 - 2 = 0$$
$$\Rightarrow \lambda^2 - 5\lambda + 4 = 0$$
$$\Rightarrow (\lambda-4)(\lambda-1) = 0$$
$$\Rightarrow \lambda = 1, 4$$

So, eigen values of A^3 matrix $= (1)^3, (4)^3 = 1, 64$

1.13 (b)

For non-trivial solutions, $|A| = 0$

$$\therefore \quad \begin{vmatrix} 0 & 1 & 2 \\ 2 & 1 & 1 \\ \alpha & 2 & 0 \end{vmatrix} = 0$$

$$-1(-\alpha) + 2(4-\alpha) = 0$$
$$\alpha + 8 - 2\alpha = 0$$
$$8 - \alpha = 0$$
$$\alpha = 8$$

1.14 (d)

$$A = \begin{bmatrix} 1 & 2 \\ 2 & 1 \end{bmatrix}$$
$$|A - \lambda I| = 0 \Rightarrow \begin{vmatrix} 1-\lambda & 2 \\ 2 & 1-\lambda \end{vmatrix} = 0$$
$$\Rightarrow (1-\lambda)^2 - 4 = 0$$
$$\Rightarrow \lambda^2 + 1 - 2\lambda - 4 = 0$$
$$\Rightarrow \lambda^2 - 2\lambda - 3 = 0$$
$$\Rightarrow (\lambda-3)(\lambda+1) = 0$$
$$\Rightarrow \lambda = -1, 3$$

So, eigen value of A^3 matrix are

$$= (-1)^3, (3)^3 = -1, 27$$

1.16 (a)

Eigen value of matrix $A = \begin{bmatrix} 1 & 2 \\ 4 & 3 \end{bmatrix}$ are 5 and –1.

Then eigen values of $-2A + 3I$ are

$\lambda_1 = -2(5) + 3(1) = -10 + 3 = -7$

$\lambda_2 = -2(-1) + 3(1) = 2 + 3 = 5$

1.17 (c)

Given: Matrix $A = \begin{bmatrix} 1 & 2 \\ 3 & 4 \end{bmatrix}$

$$|A| = 4 - 6 = -2$$
$$\text{adj.}(A) = \begin{bmatrix} 4 & -2 \\ -3 & 1 \end{bmatrix}$$
$$A^{-1} = \frac{\text{adj}(A)}{|A|}$$
$$= -\frac{1}{2}\begin{bmatrix} 4 & -2 \\ -3 & 1 \end{bmatrix} = \begin{bmatrix} -2 & 1 \\ \frac{3}{2} & -\frac{1}{2} \end{bmatrix}$$

1.18 (d)

$$P = \begin{bmatrix} 1 & 1 \\ 0 & 1 \end{bmatrix}$$
$$|P| = 1$$
$$P^{-1} = \frac{\text{adj}(P)}{|P|} = \frac{1}{1}\begin{bmatrix} 1 & -1 \\ 0 & 1 \end{bmatrix}$$
$$\Rightarrow \quad P^{-1} = \begin{bmatrix} 1 & -1 \\ 0 & 1 \end{bmatrix}$$

Let, matrix $R = \begin{bmatrix} a & b \\ c & d \end{bmatrix}$

Then, $R V_1 = \lambda_1 V_1$

$$\begin{bmatrix} a & b \\ c & d \end{bmatrix}\begin{bmatrix} 1 \\ 0 \end{bmatrix} = -1\begin{bmatrix} 1 \\ 0 \end{bmatrix}$$

$$\begin{bmatrix} a \\ c \end{bmatrix} = -1\begin{bmatrix} 1 \\ 0 \end{bmatrix}$$

$$a = -1 \text{ and } c = 0$$

Similarly,

$$R V_2 = \lambda_2 V_2$$

$$\begin{bmatrix} a & b \\ c & d \end{bmatrix}\begin{bmatrix} 1 \\ 1 \end{bmatrix} = 3\begin{bmatrix} 1 \\ 1 \end{bmatrix}$$

$$a + b = 3 \Rightarrow b = 4$$

$$c + d = 3 \Rightarrow d = 3$$

So, $R = \begin{bmatrix} -1 & 4 \\ 0 & 3 \end{bmatrix}$

$$\Rightarrow \quad P^{-1}RP = \begin{bmatrix} 1 & -1 \\ 0 & 1 \end{bmatrix}\begin{bmatrix} -1 & 4 \\ 0 & 3 \end{bmatrix}\begin{bmatrix} 1 & 1 \\ 0 & 1 \end{bmatrix}$$

$$= \begin{bmatrix} -1 & 1 \\ 0 & 3 \end{bmatrix}\begin{bmatrix} 1 & 1 \\ 0 & 1 \end{bmatrix} = \begin{bmatrix} -1 & 0 \\ 0 & 3 \end{bmatrix}$$

1.19 (b)

$$A/B = \left[\begin{array}{ccc|c} 1 & 2 & 3 & 2 \\ 0 & 1 & 1 & -1 \\ 0 & 2 & 2 & 0 \end{array}\right]$$

$$R_3 \to R_3 - 2R_2$$

$$= \left[\begin{array}{ccc|c} 1 & 2 & 3 & 2 \\ 0 & 1 & 1 & -1 \\ 0 & 0 & 0 & 2 \end{array}\right]$$

$$\therefore \quad \rho(A) = 2, \quad \rho(A/B) = 3$$

$$\Rightarrow \quad \rho(A) \neq \rho\left(\frac{A}{B}\right)$$

$\therefore$ No solution exists.

1.20 (d)

Given matrix $\begin{bmatrix} 4 & 0 \\ 0 & 4 \end{bmatrix}$ (Since the martix is triangular)

Eigen value are = 4, 4

So, $AX = \lambda X$

$$\begin{bmatrix} 4 & 0 \\ 0 & 4 \end{bmatrix}\begin{bmatrix} x_1 \\ x_2 \end{bmatrix} = 4\begin{bmatrix} x_1 \\ x_2 \end{bmatrix}$$

$$\begin{bmatrix} 4x_1 \\ 4x_2 \end{bmatrix} = \begin{bmatrix} 4x_1 \\ 4x_2 \end{bmatrix}$$

So, $\begin{bmatrix} x_1 \\ x_2 \end{bmatrix}$ can take any value except $\begin{bmatrix} 0 \\ 0 \end{bmatrix}$.

1.22 (d)

For vectors to be dependent

$$\Rightarrow \begin{vmatrix} 1 & x & 3 \\ 2 & 6 & 4 \\ 1 & x & 2 \end{vmatrix} = 0$$

$$\Rightarrow 1(12 - 4x) - x(4 - 4) + 3(2x - 6) = 0$$

$$\Rightarrow 12 - 4x + 6x - 18 = 0$$

$$\Rightarrow 2x - 6 = 0$$

$$\Rightarrow x = 3$$

1.23 (7)

$$AX = \lambda X$$

$$\begin{bmatrix} 4 & 3 \\ 3 & 4 \end{bmatrix}\begin{bmatrix} 1 \\ 1 \end{bmatrix} = \lambda\begin{bmatrix} 1 \\ 1 \end{bmatrix}$$

$$\begin{bmatrix} 7 \\ 7 \end{bmatrix} = \begin{bmatrix} \lambda \\ \lambda \end{bmatrix}$$

Hence $\lambda = 7$

1.24 (15)

$$4x_4 + 13x_5 = 46 \quad \text{...(i)}$$

$$2x_1 + 5x_2 + 5x_3 + 2x_4 + 10x_5 = 161 \quad \text{...(ii)}$$

$$2x_3 + 5x_4 + 3x_5 = 61 \quad \text{...(iii)}$$

$$4x_4 + 5x_5 = 30 \quad \text{...(iv)}$$

$$2x_1 + 3x_2 + 2x_3 + x_4 + 5x_5 = 81 \quad \text{...(v)}$$

From (i) and (iv)

$$x_5 = 2, \; x_4 = 5$$

From (iii)

$$x_3 = 15$$

1.25 (1)

$$A = \begin{bmatrix} \cos\theta & -\sin\theta \\ \sin\theta & \cos\theta \end{bmatrix}$$

$$A^T = \begin{bmatrix} \cos\theta & \sin\theta \\ -\sin\theta & \cos\theta \end{bmatrix}$$

$$A^T \cdot A = \begin{bmatrix} \cos\theta & \sin\theta \\ -\sin\theta & \cos\theta \end{bmatrix}\begin{bmatrix} \cos\theta & -\sin\theta \\ \sin\theta & \cos\theta \end{bmatrix}$$

$$= \begin{bmatrix} \cos^2\theta + \sin^2\theta & 0 \\ 0 & \cos^2\theta + \sin^2\theta \end{bmatrix}$$

$$A^T \cdot A = \begin{bmatrix} 1 & 0 \\ 0 & 1 \end{bmatrix}$$

$$|A^T \cdot A| = 1$$

1.27 (14)

For matrix $\begin{bmatrix} 2 & 3 \\ 0 & 7 \end{bmatrix}$

Eigen value = 2, 7 (triangular matrix)

Product of eigen values = 2 × 7 = 14

1.28 (14)

$$\begin{vmatrix} 2-\lambda & 4 & 6 \\ 3 & 5-\lambda & 9 \\ 12 & 1 & 7-\lambda \end{vmatrix} = 0$$

$\Rightarrow$ $(2-\lambda)[(5-\lambda)(7-\lambda)-9]-4[3(7-\lambda)-108]$
$+ 6[3-12(5-\lambda)] = 0$

$\Rightarrow$ $(2-\lambda)[\lambda^2-12\lambda+35-9]-4[21-3\lambda-108]$
$+ 6[3-60+12\lambda)] = 0$

$\Rightarrow$ $(2-\lambda)[\lambda^2-12\lambda+26]-4[-3\lambda-87]+6$
$[12\lambda-57] = 0$

$\Rightarrow$ $2\lambda^2-24\lambda+52-\lambda^3+12\lambda^2-26\lambda+12\lambda+348$
$+72\lambda-342 = 0$

$\Rightarrow$ $-\lambda^3+14\lambda^2+34\lambda+58 = 0$

$\Rightarrow$ $\lambda^3-14\lambda^2-34\lambda-58 = 0$

$$\lambda_1+\lambda_2+\lambda_3 = \frac{-\text{Coefficient of } \lambda^2}{\text{Coefficient of } \lambda}$$

$$\lambda_1+\lambda_2+\lambda_3 = 14$$

1.29 (3)

$$A = \begin{bmatrix} a_1 & a_2 & a_3 & a_4 \\ b_1 & b_1 & b_3 & b_4 \\ c_1 & c_2 & c_3 & c_4 \\ d_1 & d_2 & d_3 & d_4 \end{bmatrix}$$

$$R_4 \to R_4 - (3R_1 + 2R_2 + R_3)$$

$$A = \begin{bmatrix} a_1 & a_2 & a_3 & a_4 \\ b_1 & b_1 & b_3 & b_4 \\ c_1 & c_2 & c_3 & c_4 \\ 0 & 0 & 0 & 0 \end{bmatrix}$$

$\rho A) = 3$

1.30 (b)

$$A = \begin{bmatrix} x & 1 & 3 \\ y & 2 & 6 \\ 3 & 5 & 7 \end{bmatrix}; |A| = 0$$

For, $|A| = 0$

Row 2 element is 2 times of Row 1

$\Rightarrow$ $y = 2x$

So, option (b) satisfies this equation.

1.31 (4160)

Given: $A = \begin{bmatrix} 10 & 6 \\ 6 & 10 \end{bmatrix}$

For eigen values, we have

$$|A-\lambda I| = 0$$

$$\begin{vmatrix} 10-\lambda & 6 \\ 6 & 10-\lambda \end{vmatrix} = 0$$

$$(10-\lambda)^2 - 36 = 0$$

$$100 + \lambda^2 - 20\lambda - 36 = 0$$

$$\lambda^2 - 20\lambda + 64 = 0$$

$$\lambda = \frac{20 \pm \sqrt{400 - 4 \times 64}}{2}$$

$$\lambda = 16, 4$$

$$\lambda^3 = (16)^3 \text{ and } (4)^3$$

Sum of eigen values = $(16)^3 + (4)^3 = 4160$

1.32 (a, c)

For a homogeneous equation $Ax = 0$ where A is on $n \times n$ matrix.

For a non-trivial solution

$$|A| = 0$$

Rank of A $\quad r < n$

1.33 (–2)

matrix $A = \begin{bmatrix} -5 & a \\ -2 & -2 \end{bmatrix}$

Given eigen values are –1 and –6

Multiplication of eigenvalues = Determinant of matrix

$$(-1) \times (-6) = (-5) \times (-2) + 2a$$

$$a = -2$$

■■■■

2 Calculus

2.1 For an even function $f(x)$,

(a) $\int_{-a}^{a} f(x)dx = 0$ (b) $\int_{-a}^{a} f(-x)dx = 0$

(c) $f(x) = -f(-x)$ (d) $f(x) = f(-x)$

[2000 : 1 M]

2.2 Find the directional derivative of $u = xyz$ at the point (1, 2, 3) in the direction from (1, 2, 3) to (1, –1, –3).

[2000 : 5 M]

2.3 The function $f(x, y) = x^2 + y^2 - xy - x - y + 5$ has the

(a) Maximum at (1, 1)
(b) Saddle point at (1, 1)
(c) Minimum at (1, 1)
(d) None of the above at (1, 1)

[2001 : 2 M]

2.4 The coefficient of x^2 in the Taylor series of $\cos^2 x$ about 0 is

(a) 2 (b) 0
(c) 1 (d) –1

[2002 : 2 M]

2.5 The directional derivative of $f(x, y, z) = x^2 + y^2 + z^2$ at the point (1, 1, 1) in the direction $\hat{i} - \hat{k}$ is

(a) 0 (b) 1
(c) 2 (d) $2\sqrt{2}$

[2003 : 1 M]

2.6 The Taylor series expansion of the function:

$$f(x) = \frac{x}{(1+x)} \text{ around } x = 0 \text{ is}$$

(a) $x + x^2 + x^3 + x^4 .....$
(b) $1 + x + x^2 + x^3 + x^4 .....$
(c) $2x + 4x^2 + 8x^3 + 16x^4 .....$
(d) $x - x^2 + x^3 - x^4 .....$

[2003 : 1 M]

2.7 The function $f(x) = 3x(x - 2)$ has a

(a) minimum at $x = 1$
(b) maximum at $x = 1$
(c) minimum at $x = 2$
(d) maximum at $x = 2$

[2004 : 1 M]

2.8 Value of the integral $\int_{-2}^{2} \frac{dx}{x^2}$ is

(a) 0 (b) 0.25
(c) 1 (d) ∞

[2004 : 2 M]

2.9 The value of $\lim_{x \to 9} \frac{\sqrt{x} - 3}{x^2 - 81}$ is

(a) 0 (b) $\frac{1}{27}$

(c) $\frac{1}{108}$ (d) ∞

[2004 : 2 M]

2.10 The divergence of a vector field A is always equal to zero, if the vector field A can be expressed as

(a) the gradient of any scalar field ϕ
(b) the divergence of any scalar field ϕ
(c) the divergence of any vector field B
(d) the curl of any vector field B

[2005 : 2 M]

2.11 In the limit $x \to 0$, what is the limiting value of the function $F(x)$ given below?

$$F(x) = \frac{1 - \cos(2x)}{x^2}$$

(a) 0 (b) 1
(c) 2 (d) ∞

[2005 : 2 M]

2.12 In the domain $-\infty < x < \infty$, the function $y(x) = x^3 e^{-x}$ has

(a) no maximum and no minimum
(b) one maximum and no minimum
(c) no maximum and one minimum
(d) one maximum and one minimum

[2005 : 2 M]

2.13 Given the characteristic equation below, select the number of roots which will be located to the right of the imaginary axis $s^4 + 5s^3 - s^2 - 17s + 12 = 0$

(a) One (b) Two
(c) Three (d) zero

[2005 : 2 M]

2.14 The derivative of $|x|$ with respect to x when $x \neq 0$ is

(a) $\frac{|x|}{x}$ (b) -1
(c) 1 (d) Undefined

[2006 : 1 M]

2.15 Determine the following integral

$$I = \int_S r.dS$$

where r is the position vector field ($r = ix + jy + kz$) and S is the surface of a sphere of radius R

(a) $4\pi R^2$ (b) $\frac{3}{4}\pi R^2$
(c) πR^2 (d) $4\pi R^3$

[2006 : 2 M]

2.16 The direction derivative of $f = \frac{1}{2}\sqrt{x^2 + y^2}$ at (1, 1) in the direction of $\vec{b} = \vec{i} - \vec{j}$ is

(a) 0 (b) $\frac{1}{\sqrt{2}}$
(c) $\sqrt{2}$ (d) 2 [2007 : 2 M]

2.17 Evaluate the following integral ($n \neq 0$) $\int_c (-xy^n dx + x^n y dy)$ within the area of a triangle with vertices (0, 0), (1, 0) and (1, 1) (counter-clockwise)

(a) 0 (b) $\frac{1}{(n+1)}$
(c) $\frac{1}{2}$ (d) $\frac{n}{2}$ [2007 : 2 M]

2.18 The limit of $\frac{\sin x}{x}$ as $x \to \infty$ is

(a) -1 (b) 0
(c) 1 (d) ∞

[2008 : 1 M]

2.19 The unit normal vector to the surface of the sphere $x^2 + y^2 + z^2 = 1$ at the point $\left(\frac{1}{\sqrt{2}}, 0, \frac{1}{\sqrt{2}}\right)$ is ($\hat{i}, \hat{j}, \hat{k}$ are unit normal vectors in the Cartesian coordinate system)

(a) $\frac{1}{\sqrt{2}}\hat{i} + \frac{1}{\sqrt{2}}\hat{j}$ (b) $\frac{1}{\sqrt{2}}\hat{i} + \frac{1}{\sqrt{2}}\hat{k}$
(c) $\frac{1}{\sqrt{2}}\hat{j} + \frac{1}{\sqrt{2}}\hat{k}$ (d) $\frac{1}{\sqrt{3}}\hat{j} + \frac{1}{\sqrt{3}}\hat{j} + \frac{1}{\sqrt{3}}\hat{k}$

[2008 : 1 M]

2.20 The value of the surface integral $\oiint_S (x\hat{i} + y\hat{j}) \cdot \hat{n} dA$ evaluated over the surface of a cube having sides of length a is ($\hat{n}$ is unit normal vector)

(a) 0 (b) a^3
(c) $2a^3$ (d) $3a^3$

[2008 : 2 M]

2.21 The first four terms of the Taylor series expansion of $\cos x$ about the point $x = 0$ are

(a) $1 + x + \frac{x^2}{2!} + \frac{x^3}{3!}$ (b) $1 - x + \frac{x^2}{2!} - \frac{x^3}{3!}$
(c) $1 - \frac{x^2}{2!} + \frac{x^4}{4!} - \frac{x^6}{6!}$ (d) $x - \frac{x^3}{3!} + \frac{x^5}{5!} - \frac{x^7}{7!}$

[2008 : 2 M]

2.22 The direction of largest increase of the function $xy^3 - x^2$ at the point (1, 1) is

(a) $3\hat{i} + \hat{j}$ (b) $\hat{i} + 3\hat{j}$
(c) $-\hat{i} + 3\hat{j}$ (d) $-\hat{i} - 3\hat{j}$

[2009 : 1 M]

2.23 The value of the limit $\lim_{x \to \frac{\pi}{2}} \frac{\cos x}{\left(x - \frac{\pi}{2}\right)^3}$ is

(a) $-\infty$ (b) 0
(c) 1 (d) ∞ [2009 : 2 M]

2.24 Consider the integral $\iint (2x\hat{i} - 2y\hat{j} + 5z\hat{k}) \cdot \hat{n} dS$ over the surface of a sphere of radius = 3 with center at the origin, and surface unit normal $\hat{n}$ pointing away from the origin. Using the Gauss divergence theorem, the value of this integral is

(a) -180π (b) 0
(c) 90π (d) 180π

[2009 : 2 M]

2.25 If $\bar{u} = y\hat{i} + xy\hat{j}$ and $\bar{v} = x^2\hat{i} + xy^2\hat{j}$, then $\text{curl}(\bar{u} \times \bar{v})$ is

(a) $(2xy^2)\hat{i} - (x + y^2)\hat{j}$
(b) $(xy - x^2)\hat{i} - (y - 3xy)\hat{j}$
(c) $(2x^2y^2 - 3x^3)\hat{i} - (y^3 - 3xy^2)\hat{k}$
(d) $(3xy^2 - x^3)\hat{i} - (y^3 - 3x^2y)\hat{j}$

[2010 : 2 M]

2.26 For a function $g(x)$, if $g(0) = 0$ and $g'(0) = 0$, then $\lim\limits_{t\to 0}\int_0^{g(t)} \frac{2t}{x}dt$ is equal to

(a) ∞ (b) 2
(c) 0 (d) $-\infty$

[2010 : 2 M]

2.27 Which one of the following functions $y(x)$ has the slope of its tangent equal to $\frac{ax}{y}$? (**Note :** a and b are real constants)

(a) $y = \frac{x+b}{a}$ (b) $y = ax + b$
(c) $y = \sqrt{\frac{x^2+b}{a}}$ (d) $y = \sqrt{ax^2 + b}$

[2011 : 1 M]

2.28 Unit vectors in x and z directions are $\vec{i}$ and $\vec{k}$ respectively. Which one of the following is the directional derivative of the function $F(x, z) = \ln(x^2 + z^2)$ at the point $P : (4, 0)$, in the direction of $(\vec{i} - \vec{k})$?

(a) $\frac{\vec{i}}{2\sqrt{2}}$ (b) $\vec{i}$
(c) 1 (d) $\frac{1}{2\sqrt{2}}$

[2011 : 2 M]

2.29 The value of the improper integral $\int_{-\infty}^{\infty} \frac{dx}{(1+x^2)}$ is

(a) -2π (b) 0
(c) π (d) 2π

[2011 : 2 M]

2.30 For the function $f(t) = e^{-t/\tau}$, the Taylor series approximation for $t << \tau$ is

(a) $1+\frac{t}{\tau}$ (b) $1-\frac{t}{\tau}$
(c) $1-\frac{t^2}{2\tau^2}$ (d) $1 + t$

[2012 : 1 M]

2.31 If a is a constant, then the value of the integral $a^2\int_0^{\infty} xe^{-ax}dx$ is

(a) $\frac{1}{a}$ (b) a
(c) 1 (d) 0

[2012 : 2 M]

2.32 Evaluate $\int \frac{dx}{e^x - 1}$

(**Note:** C is a constant of integration)

(a) $\frac{e^x}{e^x-1} + C$ (b) $\frac{\ln(e^x-1)}{e^x} + C$
(c) $\ln\left(\frac{(e^x)}{e^x-1}\right) + C$ (d) $\ln(1 - e^{-x}) + C$

[2013 : 1 M]

2.33 Gradient of a scalar variable is always

(a) a vector (b) a scalar
(c) a dot product (d) zero

[2014 : 1 M]

2.34 If $f^*(x)$ is the complex conjugate of $f(x) = \cos(x) + i\sin(x)$, then for real a and b, $\int_a^b f^*(x)f(x)dx$ is **ALWAYS**

(a) positive (b) negative
(c) real (d) imaginary

[2014 : 1 M]

2.35 If $f(x)$ is a real and continuous function of x, the Taylor series expansion of $f(x)$ about its minima will **NEVER** have a term containing

(a) first derivative
(b) second derivative
(c) third derivative
(d) all higher derivative

[2014 : 1 M]

2.36 A scalar function in the xy-plane is given by $\phi(x, y) = x^2 + y^2$. If $\hat{i}$ and $\hat{j}$ are unit vectors in the x and y directions, the direction of maximum increase in the value of ϕ at (1, 1) is along:

(a) $-2\hat{i} + 2\hat{j}$ (b) $2\hat{i} + 2\hat{j}$

(c) $-2\hat{i} - 2\hat{j}$ (d) $2\hat{i} - 2\hat{j}$

[2015 : 1 M]

2.37 A vector $u = -2y\hat{i} + 2x\hat{j}$, where $\hat{i}$ and $\hat{j}$ are unit vector in x and y-directions, respectively. Evaluate the line integral.

$$I = \oint_C \vec{u} \cdot d\vec{r}$$

where C is a closed loop formed by connecting points (1, 1), (3, 1), (3, 2) and (1, 2) in that order. The value of I is ______. [2015 : 2 M]

2.38 The Lagrange mean-value theorem is satisfied for $f(x) = x^3 + 5$, in the interval (1, 4) at a value (rounded off to the second decimal place) of x equal to ______.

[2016 : 2 M]

2.39 The value of $\lim\limits_{x \to 0} \dfrac{\tan(x)}{x}$ is ______.

[2017 : 1 M]

2.40 The number of positive roots of the function $f(x)$ shown below in the range $0 < x < 6$ is ______.

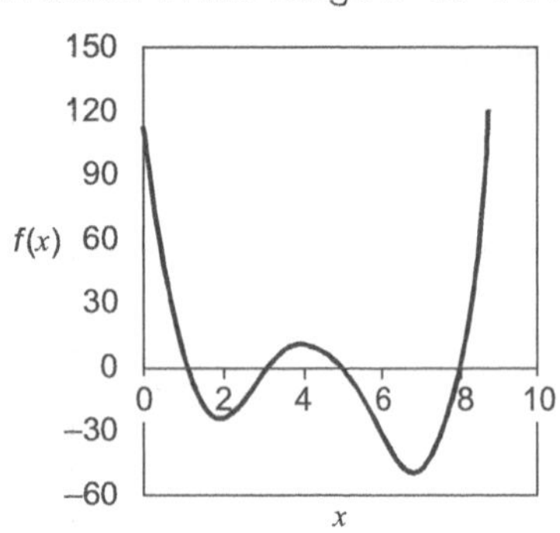

[2017 : 1 M]

2.41 Let i and j be the unit vectors in the x and y directions, respectively. For the function

$F(x, y) = x^3 + y^2$

the gradient of the function, i.e.. ∇F is given by

(a) $3x^2i - 2yj$ (b) $6x^2y$

(c) $3x^2i + 2yj$ (d) $2yi - 3x^2j$

[2017 : 1 M]

2.42 If $y = e^{-x^2}$ then the value of $\lim\limits_{x \to \infty} \dfrac{1}{x} \cdot \dfrac{dy}{dx}$ is ______.

[2018 : 2 M]

2.43 The value of the expression $\lim\limits_{x \to \frac{\pi}{2}} \left| \dfrac{\tan x}{x} \right|$ is

(a) ∞ (b) 0

(c) 1 (d) –1

[2019 : 1 M]

2.44 If x, y and z are directions in a Cartesian coordinate system and $\vec{i}$, $\vec{j}$ and $\vec{k}$ are the respective unit vectors, the directional derivative of the function $u(x, y, z) = x^2 - 3yz$ at the point (2, 0, –4) in the direction $\dfrac{(\vec{i} + \vec{j} - 2\vec{k})}{\sqrt{6}}$ is ______ (rounded off to two decimal places).

[2019 : 2 M]

2.45 Consider the hyperbolic functions in **Group-1** and their definitions in **Group-2**.

Group-1	Group-2
P. $\tan hx$	1. $\dfrac{e^x + e^{-x}}{e^x - e^{-x}}$
Q. $\cot hx$	2. $\dfrac{2}{e^x + e^{-x}}$
R. $\sec hx$	3. $\dfrac{2}{e^x - e^{-x}}$
S. $\text{cosec}\, hx$	4. $\dfrac{e^x - e^{-x}}{e^x + e^{-x}}$

The correct combination is

(a) P-4, Q-1, R-3, S-2

(b) P-2, Q-3, R-1, S-4

(c) P-4, Q-1, R-2, S-3

(d) P-1, Q-2, R-4, S-3 [2020 : 1 M]

2.46 Consider the following continuously differentiable function

$$\hat{v}(x,y,z) = 3x^2y\hat{i} + 8y^2z\hat{j} + 5xyz\hat{k}$$

where $\hat{i}, \hat{j}$ and $\hat{k}$ represent the respective unit vectors along the x, y and z directions in the Cartesian coordinate system. The curl of this function is

(a) $-3x^2\hat{i} - 8y^2\hat{j} + 5z(x+y)\hat{k}$
(b) $6xy\hat{i} - 16yz\hat{j} + 5xy\hat{k}$
(c) $(5xz - 8y^2)\hat{i} - 5yz\hat{j} - 3x^2\hat{k}$
(d) $y(11x + 16z)$

[2020 : 1 M]

2.47 The maximum value of the function

$$f(x) = -\frac{5}{3}x^3 + 10x^2 - 15x + 16$$

in the interval (0.5, 3.5) is
(a) 0 (b) 8
(c) 16 (d) 48

[2020 : 2 M]

2.48 The function cos(x) is approximated using Taylor series around $x = 0$ as $\cos(x) \approx 1 + ax + bx^2 + cx^3 + dx^4$. The value of a, b, c and d are
(a) $a = 1, b = -0.5, c = -1, d = -0.25$
(b) $a = 0, b = -0.5, c = 0, d = 0.042$
(c) $a = 0, b = 0.5, c = 0, d = 0.042$
(d) $a = -0.5, b = 0, c = 0.042, d = 0$

[2021 : 1 M]

2.49 For the function $f(x) = \begin{cases} -x, & x < 0 \\ x^2, & x \geq 0 \end{cases}$, the correct statement(s) is/are
(a) $f(x)$ is continuous at $x = 1$.
(b) $f(x)$ is differentiable at $x = 1$.
(c) $f(x)$ is continuous at $x = 0$.
(d) $f(x)$ is differentiable at $x = 0$.

[2021 : 1 M]

2.50 Let $f(x) = e^{-|x|}$, where x is real. The value of $\frac{df}{dx}$ at $x = -1$ is
(a) $-e$ (b) e
(c) $\frac{1}{e}$ (d) $-\frac{1}{e}$

[2022 : 1 M]

2.51 The value of the real variable $x \geq 0$, which maximizes the function $f(x) = x^e e^{-x}$ is
(a) e (b) 0
(c) $\frac{1}{e}$ (d) 1 [2022 : 1 M]

2.52 The directional derivative of $f(x, y, z) = 4x^2 + 2y^2 + z^2$ at the point (1,1,1) in the direction of the vector $\vec{v} = \hat{i} - \hat{k}$ is ________ (rounded off to two decimal places).

[2022 : 2 M]

2.53 Consider a sphere of radius 4, centered at the origin, with outward unit normal $\hat{n}$ on its surface S. The value of the surface integral

$$\iint_S \left(\frac{2x\hat{i} + 3y\hat{j} + 4z\hat{k}}{4\pi}\right) \cdot \hat{n}\, dA$$ is ________ (rounded off to one decimal place)

[2022 : 2 M]

2.54 Which one of the following is the CORRECT value of y, as defined by the expression given below?

$$y = \lim_{x \to 0} \frac{2x}{e^x - 1}$$

(a) 1 (b) 2
(c) 0 (d) ∞

[2023 : 1 M]

2.55 The vector $\vec{v}$ is defined as

$$\vec{v} = zx\,\hat{i} + 2xy\,\hat{j} + 3yz\,\hat{k}$$

which one of the following is the CORRECT value of divergence of $\vec{v}$, evaluated at the point $(x, y, z) = (3, 2, 1)$?
(a) 0 (b) 3
(c) 14 (d) 13

[2023 : 1 M]

2.56 For a two-dimensional plane, the unit vectors, $(\hat{e}_r, \hat{e}_\theta)$ of the polar coordinate system and $(\hat{i}, \hat{j})$ of the Cartesian coordinate system, are related by the following two equations.

$$\hat{e}_r = \cos\theta\,\hat{i} + \sin\theta\,\hat{j}$$
$$\hat{e}_\theta = -\sin\theta\,\hat{i} + \cos\theta\,\hat{j}$$

Which one of the following is the CORRECT value of $\frac{\partial(\hat{e}_r + \hat{e}_\theta)}{\partial\theta}$?
(a) 1 (b) $\hat{e}_\theta$
(c) $\hat{e}_r + \hat{e}_\theta$ (d) $-\hat{e}_r + \hat{e}_\theta$

[2023 : 1 M]

2.57 The first non-zero term in the Taylor series expansion of $(1 - x) - e^{-x}$ about $x = 0$ is

(a) 1 (b) –1

(c) $\frac{x^2}{2}$ (d) $-\frac{x^2}{2}$

[2024 : 1 M]

2.58 Consider the line integral $\int_C F(r) \cdot dr$, with $F(r) = x\hat{i} + y\hat{j} + z\hat{k}$, where $\hat{i}, \hat{j}$ and $\hat{k}$ are unit vectors in the (x, y, z) Cartesian coordinate system. The path C is given by $r(t) = \cos(t)\,\hat{i} + \sin(t)\,\hat{j} + t\hat{k}$, where $0 \le t \le \pi$. The value of the integral, rounded off to 2 decimal places, is ________.

[2024 : 2 M]

2.59 Consider the function $f(x, y, z) = x^4 + 2y^3 + z^2$. The directional derivative of the function at the point $P(-1, 1, -1)$ along $(\hat{i} + \hat{j})$, where $\hat{i}$ and $\hat{j}$ are unit vectors in the x and y directions, respectively, rounded off to 2 decimal places, is ________.

[2024 : 2 M]

■■■■

Answers Calculus

2.1	(d)	**2.2**	$(-7/\sqrt{5})$	**2.3**	(c)	**2.4**	(d)	**2.5**	(a)	**2.6**	(d)	**2.7**	(a)
2.8	(d)	**2.9**	(c)	**2.10**	(d)	**2.11**	(c)	**2.12**	(b)	**2.13**	(a)	**2.14**	(a)
2.15	(d)	**2.16**	(a)	**2.17**	(c)	**2.18**	(b)	**2.19**	(b)	**2.20**	(c)	**2.21**	(c)
2.22	(c)	**2.23**	(a)	**2.24**	(d)	**2.25**	(d)	**2.26**	(c)	**2.27**	(d)	**2.28**	(d)
2.29	(c)	**2.30**	(b)	**2.31**	(c)	**2.32**	(b)	**2.33**	(a)	**2.34**	(c)	**2.35**	(a)
2.36	(b)	**2.37**	(8)	**2.38**	(2.645)	**2.39**	(1)	**2.40**	(4)	**2.41**	(c)	**2.42**	(0)
2.43	(a)	**2.44**	(6.53)	**2.45**	(c)	**2.46**	(c)	**2.47**	(c)	**2.48**	(b)	**2.49**	(a,b,c)
2.50	(c)	**2.51**	(a)	**2.52**	(4.24)	**2.53**	(192)	**2.54**	(b)	**2.55**	(d)	**2.56**	(d)
2.57	(d)	**2.58**	(4.93)	**2.59**	(1.41)								

Explanations Calculus

2.1 (d)

$f(x) = f(-x)$ for an even function.

2.2 $(-7/\sqrt{5})$

Given : $u = xyz$

$$\text{grad}(u) = \frac{\partial(u)}{\partial x}\hat{i} + \frac{\partial}{\partial y}(u)\hat{j} + \frac{\partial}{\partial z}(u)\hat{k}$$

$$\text{grad}(u) = yz\hat{i} + xz\hat{j} + xy\hat{k}$$

$$\text{grad}(u)|_{(1,2,3)} = 6\hat{i} + 3\hat{j} + 2\hat{k}$$

Direction vector, $\hat{p} = (1-1)\hat{i} + (-1-2)\hat{j} + (-3-3)\hat{k}$

$$\vec{P} = -3\hat{j} - 6\hat{k}$$

$$\hat{P} = \frac{-3\hat{j}-6\hat{k}}{\sqrt{9+36}} = \frac{-3\hat{j}-6\hat{k}}{3\sqrt{5}} = \frac{-\hat{j}-2\hat{k}}{\sqrt{5}}$$

So, directional derivative

$$= (6\hat{i}+3\hat{j}+2\hat{k})\cdot\left(\frac{-\hat{j}-2\hat{k}}{\sqrt{5}}\right)$$

$$= \frac{-3-4}{\sqrt{5}} = \frac{-7}{\sqrt{5}}$$

2.3 (c)

$$f(x, y) = x^2 + y^2 - xy - x - y + 5$$

$$p = \frac{\partial f}{\partial x} = 2x - y - 1 = 0$$

$$q = \frac{\partial f}{\partial y} = 2y - x - 1 = 0$$

So, from $p, q = 0$

$$x = 1, y = 1$$

$$r = \frac{\partial^2 f}{\partial x^2} = 2,\ s = \frac{\partial^2 f}{\partial x \partial y} = -1,\ t = \frac{\partial^2 t}{\partial y^2} = 2$$

$$rt - s^2 = 4 - 1 = 3 > 0 \text{ and } r = 2 > 0$$

So, $f(x, y)$ has minima at (1, 1).

2.4 (d)

Taylor's series is given by,

$$f(x) = f(0) + \frac{(x-0)}{1!}f'(0) + \frac{(x-0)^2}{2!}f''(0)$$

$$f(x) = \cos^2 x$$

$$f(0) = 1$$

$$f'(x) = 2\cos x(-\sin x) = -\sin 2x$$

$$f'(0) = 0$$

$$f''(x) = -2\cos 2x$$

$$f''(0) = -2$$

So, $f(x) = 1 + \frac{(x-0)}{1!}\times 0 + \frac{(x-0)^2}{2!}(-2) + ...$

$$= 1 + \frac{x^2}{2}\times(-2)$$

$$f(x) = 1 - x^2$$

So, coefficient of x^2 is -1.

2.5 (a)

$$f = x^2 + y^2 + z^2$$

$$\text{Grad } f = 2x\hat{i} + 2y\hat{j} + 2z\hat{k}$$

$$\text{Grad } f|_{(1,1,1)} = 2\hat{i} + 2\hat{j} + 2\hat{k}$$

Directional derivative

$$= 2\hat{i} + 2\hat{j} + 2\hat{k}\cdot\frac{(\hat{i}-\hat{k})}{\sqrt{2}} = \frac{2-2}{\sqrt{2}} = 0$$

2.6 (d)

$$f(x) = \frac{x}{1+x}$$

Maclaurin series is

$$f(x) = f(0) + \frac{(x-0)}{1!}f'(0) + \frac{(x-0)^2}{2!}f''(0) + \ldots$$

$\Rightarrow \quad f(0) = 0$

$$f'(x) = \frac{(1+x)-x}{(1+x)^2} = \frac{1}{(1+x)^2} \Rightarrow f'(0) = 1$$

$$f''(x) = \frac{-2}{(1+x)^3} \Rightarrow f''(0) = -2$$

$$f'''(x) = \frac{6}{(1+x)^4} \Rightarrow f'''(0) = 6$$

$$f^{iv}(x) = \frac{-24}{(1+x)^5} \Rightarrow f^{iv}(0) = -24$$

So, $f(x) = 0 + \frac{x}{1!}(1) + \frac{x^2}{2!}(-2) + \frac{x^3}{3!}(6) + \frac{x^4}{4!}(-24) + \ldots$

$$f(x) = x - x^2 + x^3 - x^4 + \ldots$$

2.7 (a)

$$f(x) = 3x(x-2)$$
$$f(x) = 3x^2 - 6x$$
$$f'(x) = 6x - 6 = 0$$
$$x = 1$$
$$f''(1) = 6 > 0$$

So, $x = 1$ is local minima.

2.8 (d)

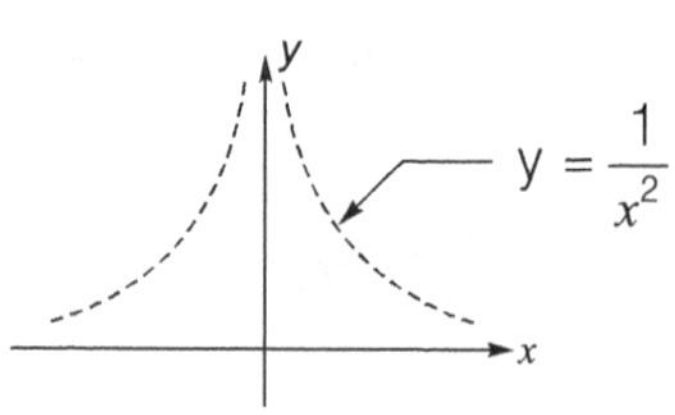

Given : $\int_{-2}^{2} \frac{dx}{x^2}$

$\because f(x) = \frac{1}{x^2}$ is discontinues at $x = 0$.

$$\Rightarrow \quad \int_{-2}^{2} \frac{dx}{x^2} = 2\int_{0}^{2} \frac{dx}{x^2} = 2\left[\frac{-1}{x}\right]_0^2$$

$$= -2\left[\frac{1}{2} - \infty\right] = \infty$$

2.9 (c)

$$\lim_{x\to 9} \frac{\sqrt{x}-3}{x^2-81} \qquad \left(\frac{0}{0} \text{ form}\right)$$

Applying L'Hospital rule

$$\lim_{x\to 9} \frac{\frac{1}{2}x^{-1/2}}{2x} = \lim_{x\to 9} \frac{1}{4x^{3/2}} = \frac{1}{4\times 27} = \frac{1}{108}$$

2.10 (d)

$$\because \quad \vec{A} = \text{curl}\,(\vec{B})$$

$$\therefore \quad \text{div}\,(\vec{A}) = \text{div}\left[\text{curl}\,(\vec{B})\right] = \nabla\cdot(\nabla\times\vec{B}) = 0$$

2.11 (c)

$$\because \quad F(x) = \frac{1-\cos(2x)}{x^2}$$

$$\lim_{x\to 0} \frac{1-\cos(2x)}{x^2} \qquad \left(\frac{0}{0} \text{ form}\right)$$

$$\lim_{x\to 0} \frac{\sin(2x)\cdot 2}{2x} = 2\left[\lim_{x\to 0} \frac{\sin 2x}{2x}\right] = 2 \times 1 = 2$$

2.12 (b)

$$f(x) = x^3 e^{-x}$$
$$f'(x) = 3x^2 e^{-x} + x^3(e^{-x} \times -1)$$
$$f'(x) = e^{-x}(3x^2 - x^3) = 0$$
$$x = 0, 0, 3$$
$$f''(x) = -e^{-x}(3x^2 - x^3) + e^{-x}(6x - 3x^2)$$
$$- e^{-x}(-3x^2 + x^3 + 6x - 3x^2)$$
$$f''(x) = e^{-x}(x^3 - 6x^2 + 6x)$$

At $x = 3$, $f''(x) = e^{-3}(27 - 54 + 18) = -9e^{-3} < 0$

So, $f(x)$ has local minima at $x = 3$.

2.13 (a)

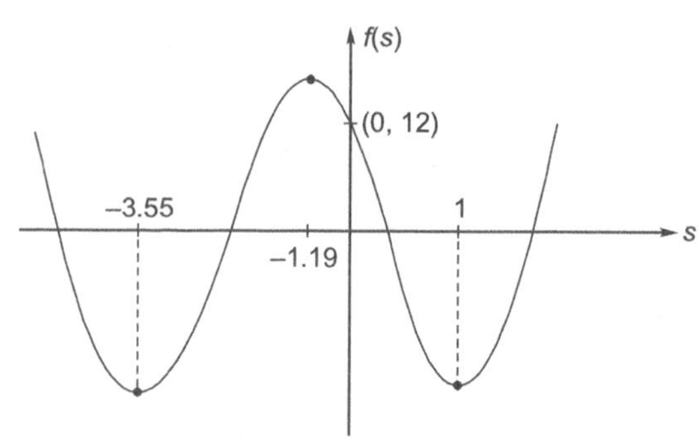

$$f(s) = s^4 + 5s^3 - s^2 - 17s + 12 = 0$$
$$f'(s) = 4s^3 + 15s^2 - 2s - 17 = 0$$
$$s = 1, -3.55, -1.19$$
$$f''(s) = 12s^2 + 30s - 2 = 2(6s^2 + 15s - 1)$$

At $s = 1$, $\quad f''(s) = 40 > 0$ (Local minima)

At $s = -3.55$, $f''(s) = 42.73 > 0$ (Local minima)
At $s = -1.19$, $f''(s) = -20.7 < 0$ (Local maxima)
So, number of roots to the right of imaginary axis = One.

2.14 (a)

$|x|$ is continuous at $x = 0$ but not differentiable at $x = 0$.
When $x \neq 0$, then

$$\frac{d(|x|)}{dx} = \frac{|x|}{x}$$

2.15 (d)

$$\iint F.ds = \iiint \text{div } FdV$$

[Gauss divergence theorem]

$$\vec{F} = \vec{r} = x\hat{i} + y\hat{j} + z\hat{k}$$

$$\text{div } \vec{F} = 3$$

So, $$\iint F.ds = \iiint 3dv$$

$$= 3 \times V$$

$$= 3 \times \frac{4}{3}\pi R^3 = 4\pi R^3$$

2.16 (a)

$$f = \frac{1}{2}\sqrt{x^2 + y^2}$$

$$\text{grad}(f) = \frac{\partial f}{\partial x} + \hat{i} + \frac{\partial f}{\partial y}\hat{j}$$

$$= \frac{1}{2} \times \frac{1}{2\sqrt{x^2 + y}} \times 2x\hat{i} + \frac{1}{2} \times \frac{2y}{2\sqrt{x^2 + y^2}}\hat{j}$$

$$\nabla(f) = \frac{x\hat{i}}{2\sqrt{x^2 + y^2}} + \frac{y\hat{j}}{2\sqrt{x^2 + y^2}} = \frac{x\hat{i} + y\hat{j}}{2\sqrt{x^2 + y^2}}$$

$$\nabla(f)|_{(1,1)} = \frac{\hat{i} + \hat{j}}{2\sqrt{2}}$$

So, direction derivative

$$= \frac{\hat{i} + \hat{j}}{2\sqrt{2}} \cdot \frac{(\hat{i} - \hat{j})}{\sqrt{2}} = \frac{1-1}{4} = 0$$

2.17 (c)

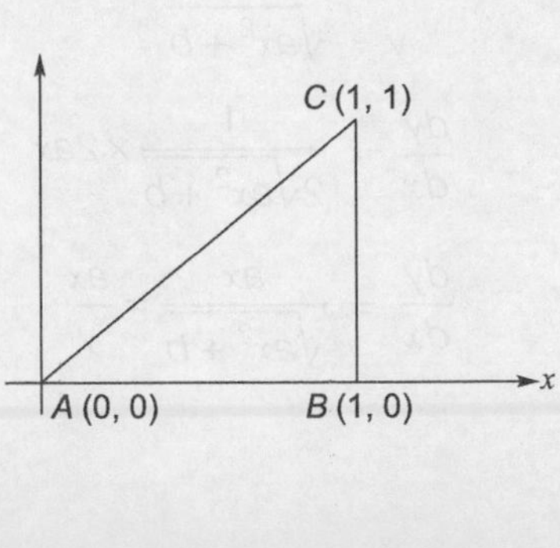

$$I = \int_C (-xy^n dx + x^n y dy)$$

$\because$ From Green's theorem

$$\int_C Mdx + Ndy = \int_R \left(\frac{\partial N}{\partial x} - \frac{\partial M}{\partial y}\right) dxdy$$

So, $$\iint_R \left(\frac{\partial N}{\partial x} - \frac{\partial M}{\partial y}\right) dxdy$$

$$= \iint_R (nx^{n-1}y + x \times ny^{n-1}) dxdy$$

$$\Rightarrow \int_{y=0}^{y=1}\int_{x=y}^{x=1} (nx^{n-1}y + x \times ny^{n-1}) dxdy$$

$$\Rightarrow \int_{y=0}^{y=1} \left[\frac{nx^n}{n} y + \frac{x^2}{2} ny^{n-1}\right]_y^1 dy$$

$$\Rightarrow \int_0^1 \left\{[1 - y^n]y + \frac{(1-y^2)}{2} . ny^{n-1}\right\} dy$$

$$\Rightarrow \int_0^1 \left[y - y^{n+1} + \frac{1}{2}(ny^{n-1} - ny^{n+1})\right] dy$$

$$\Rightarrow \int_0^1 \left[y - y^{n+1} + \frac{1}{2}(ny^{n-1} - ny^{n+1})\right] dy$$

$$\Rightarrow \left\{\frac{y^2}{2} - \frac{y^{n+2}}{n+2} + \frac{1}{2}\left(\frac{ny^n}{n} - \frac{ny^{n+2}}{n+2}\right)\right\}_0^1$$

$$\Rightarrow \frac{1}{2} - \frac{1}{n+2} + \frac{1}{2}\left[1 - \frac{n}{n+2}\right]$$

$$\Rightarrow \frac{n}{2(n+2)} + \frac{1}{n+2} = \frac{n+2}{2(n+2)} = \frac{1}{2}$$

2.18 (b)

$$\lim_{x \to \infty} \frac{\sin x}{x} = \frac{\text{Finite value in } (-1,1)}{\text{Infinite}} = 0$$

2.19 (b)

$S : x^2 + y^2 + z^2 = 1$

$$\text{grad}(S) = 2x\hat{i} + 2y\hat{j} + 2z\hat{k}$$

$$\nabla(S)\Big|_{\left(\frac{1}{\sqrt{2}}, 0, \frac{1}{\sqrt{2}}\right)} = \frac{2}{\sqrt{2}}\hat{i} + \frac{2}{\sqrt{2}}\hat{k}$$

$$\nabla(S) = \sqrt{2}\hat{i} + \sqrt{2}\hat{k}$$

Unit vector, $$\hat{n} = \frac{\sqrt{2}\hat{i} + \sqrt{2}\hat{k}}{\sqrt{2+2}} = \frac{\hat{i}}{\sqrt{2}} + \frac{\hat{k}}{\sqrt{2}}$$

2.20 (c)

$$\iint_S (x\hat{i} + y\hat{j})\hat{n}ds = \iiint_V \text{div } F.dv$$

$$\text{div } F = 2$$

$$= \iiint 2dv$$

$$= 2\iiint dv = 2 \times a^3 = 2a^3$$

2.21 (c)

$$f(x) = f(0) + \frac{(x-0)}{1!}f'(0) + \frac{(x-0)^2}{2!}f''(0) + \frac{(x-0)^3}{3!}f'''(0) + ...$$

$\because$

$$f(x) = \cos x,\ f(0) = 1$$
$$f'(x) = -\sin x,\ f'(0) = 0$$
$$f''(x) = -\cos x,\ f''(0) = -1$$
$$f'''(x) = \sin x,\ f'''(0) = 0$$
$$f^{iv}(x) = \cos x,\ f^{iv}(0) = 1$$
$$f^{v}(x) = -\sin x,\ f^{v}(0) = 0$$
$$f^{vi}(x) = -\cos x,\ f^{vi}(0) = -1 \text{ and so on.}$$

So, $f(x) = 1 + \frac{x}{1!}(0) + \frac{x^2}{2!}(-1) + \frac{x^3}{3!}(0) + \frac{x^4}{4!}(1) + \frac{x^5}{5!}(0) + \frac{x^6}{6!}(-1) + ...$

$$f(x) = 1 - \frac{x^2}{2!} + \frac{x^4}{4!} - \frac{x^6}{6!} + ...$$

2.22 (c)

$$f(x) = xy^3 - x^2$$

$$\nabla(f) = \frac{\partial}{\partial x}(xy^3 - x^2)\hat{i} + \frac{\partial}{\partial y}(xy^3 - x^2)\hat{j} + \frac{\partial}{\partial z}(xy^3 - x^2)\hat{k}$$

$$\nabla(f) = (y^3 - 2x)\hat{i} + (3xy^2)\hat{j} + 0\hat{k}$$

$$\nabla f|_{(1,1)} = (1-2)\hat{i} + 3\hat{j}$$

$$\nabla f|_{(1,1)} = -\hat{i} + 3\hat{j}$$

2.23 (a)

$$\lim_{x\to\frac{\pi}{2}} \frac{\cos x}{\left(x - \frac{\pi}{2}\right)^3} \quad \left(\frac{0}{0} \text{ form}\right)$$

Using L-Hospital rule

$$\lim_{x\to\frac{\pi}{2}} \frac{-\sin x}{3\left(x - \frac{\pi}{2}\right)^2} = -\infty$$

2.24 (d)

For Gauss divergence theorem

$$\iint \vec{F}\cdot\hat{n}ds = \iiint_V \text{div } \vec{F}dV$$

$$\because \quad \vec{F} = (2x\hat{i} - 2y\hat{j} + 5z\hat{k})$$

$$\text{div } F = 2 - 2 + 5 = 5$$

So, $\iiint \text{div } FdV = \iiint 5dV = 5\text{ V}$

$$= 5 \times \frac{4}{3}\pi \times (3)^3 = 5 \times 4\pi \times 9$$

$$= 180\pi$$

2.25 (d)

For curl $(\vec{u} \times \vec{v})$

$$\vec{u} \times \vec{v} = \begin{vmatrix} \hat{i} & \hat{j} & \hat{k} \\ y & xy & 0 \\ x^2 & xy^2 & 0 \end{vmatrix}$$

$$= \hat{i}(0-0) - \hat{j}(0-0) + \hat{k}(xy^3 - x^3y)$$

$$(\vec{u} \times \vec{v}) = (xy^3 - x^3y)\hat{k}$$

$$\nabla \times (\vec{u} \times \vec{v}) = \begin{vmatrix} \hat{i} & \hat{j} & \hat{k} \\ \frac{\partial}{\partial x} & \frac{\partial}{\partial y} & \frac{\partial}{\partial z} \\ 0 & 0 & xy^3 - x^3y \end{vmatrix}$$

$$= \hat{i}(3xy^2 - x^3 - 0) - \hat{j}(y^3 - 3x^2y) + \hat{k}(0-0)$$

$$= (3xy^2 - x^3)\hat{i} - (y^3 - 3x^2y)\hat{j}$$

2.26 (c)

$$\lim_{t\to 0} \int_0^{g(t)} \frac{2t}{x} = \frac{0}{0}$$

$$\lim_{t\to 0} \frac{2g(t)g'(t)}{1} = 2g(0)g'(0)$$

$$= 2 \times 0 \times 0 = 0$$

2.27 (d)

$$y = \sqrt{ax^2 + b}$$

$$\frac{dy}{dx} = \frac{1}{2\sqrt{ax^2 + b}} \times 2ax$$

$$\frac{dy}{dx} = \frac{ax}{\sqrt{ax^2 + b}} = \frac{ax}{y}$$

2.28 (d)

$$F(x, z) = \ln(x^2 + z^2)$$

$$\text{grad}(F) = \frac{\partial}{\partial x}\left[\ln(x^2+z^2)\right]\hat{i} + \frac{\partial}{\partial y}\left[\ln(x^2+z^2)\right]\hat{j} + \frac{\partial}{\partial z}\left[\ln(x^2+z^2)\right]\hat{k}$$

$$= \frac{2x}{x^2+z^2}\hat{i} + \frac{2z\hat{k}}{x^2+z^2} = \frac{8\hat{i}}{16+0} = \frac{\hat{i}}{2}$$

So, directional derivative $= \frac{\hat{i}}{2}\cdot\left(\frac{\hat{i}-\hat{k}}{\sqrt{2}}\right) = \frac{1}{2\sqrt{2}}$

2.29 (d)

$$\int_{-\infty}^{\infty}\frac{dx}{(1+x^2)} = \tan^{-1}(x)\Big|_{-\infty}^{\infty}$$

$$= 2\left[\frac{\pi}{2}-\frac{\pi}{2}\right] = 2\pi$$

2.30 (b)

Taylor series expansion of

$$e^x = 1+\frac{x}{1!}+\frac{x^2}{2!}+\frac{x^3}{3!}$$

$\because \quad t << \tau$

$\therefore \quad \frac{t}{\tau} << 1$

So, by ignoring higher terms

$$e^{-t/\tau} = 1+\frac{\left(-\frac{t}{\tau}\right)}{1!} = 1-\frac{t}{\tau}$$

2.31 (c)

$$a^2\int_0^{\infty} xe^{-ax}dx = a^2\left\{\left[\frac{xe^{-ax}}{-a}\right]_0^{\infty} + \frac{1}{a}\times-\frac{1}{a}\left[e^{-ax}\right]_0^{\infty}\right\}$$

$$= a^2\left\{-\frac{1}{a}[0-0]-\frac{1}{a^2}[0-1]\right\}$$

$$= a^2\frac{1}{a^2} = 1$$

2.32 (b)

$$\int\frac{dx}{e^x-1} = \int\frac{e^x-(e^x-1)}{e^x-1}dx$$

$$= \int\frac{e^x}{e^x-1}dx - \int\frac{(e^x-1)}{(e^x-1)}dx$$

$$= \ln(e^x-1) - x + c$$

$$= \ln(e^x-1) - x\ln e + c$$

$$= \ln\left(\frac{e^x-1}{e^x}\right)+c$$

2.33 (a)

If $f(x, y, z) = c$ is a scalar

Then grad $f = \nabla f = \frac{\partial f}{\partial x}\hat{i}+\frac{\partial f}{\partial y}\hat{j}+\frac{\partial f}{\partial z}\hat{k}$ is a vector.

2.34 (c)

$$f(x) = \cos x + i\sin x$$

$$f^*(x) = \cos x - i\sin x$$

$$= \int_a^b(\cos x - i\sin x)(\cos x + i\sin x)dx$$

$$= \int_a^b(\cos^2 x+\sin^2 x)dx$$

$$= \int_a^b dx = [x]_a^b = b-a$$

$\because \quad (b-a)$ is a real value.

Hence $\int_a^b f^*(x)\,f(x)dx$ is real.

2.35 (a)

Suppose $f(x)$ is having minima at $x = a$

Then, $\quad f'(a) = 0$

$\quad f''(a) > 0$

So, Taylor series expansion about $x = a$

$$f(x) = f(a)+\frac{(x-a)}{1!}f'(a)+\frac{(x-a)^2}{2!}f''(a)+\ldots$$

$$= f(a)+\frac{(x-a)}{1!}\times 0+\frac{(x-a)^2}{2!}f''(a)+\ldots$$

$$= f(a)+\frac{(x-a)^2}{2!}f''(a)+\ldots$$

So, first derivative will be absent in the Taylor series expansion.

2.36 (b)

$$\phi(x, y) = x^2 + y^2$$

$$\text{grad}(\phi) = 2x\hat{i}+2y\hat{j}$$

$$\nabla\phi|_{(1,1)} = 2\hat{i}+\hat{2}j$$

So, direction of maximum increase $= 2\hat{i}+2\hat{j}$

2.37 (8)

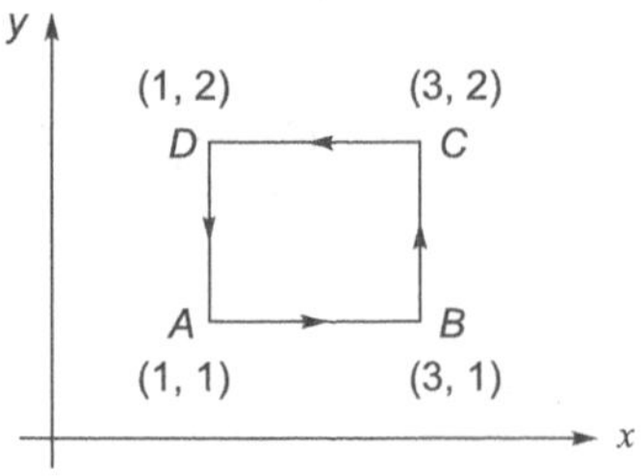

$$u = -2y\hat{i} + 2x\hat{j}$$

$$\oint_C u \cdot dr = \oint_C -\underbrace{2y}_{M}\,dx + \underbrace{2x}_{N}\,dy$$

$$= \iint_R \left(\frac{\partial N}{\partial x} - \frac{\partial M}{\partial y}\right) dx\,dy$$

$$= \iint_R (2+2)\,dx\,dy$$

$$= 4\iint_R dx\,dy = 4 \times 1 \times 2 = 8$$

2.38 (2.645)

According to LMVT:

$$f'(c) = \frac{f(b)-f(a)}{b-a}$$

So, $$f'(c) = \frac{f(4)-f(1)}{4-1}$$

$$3x^2 = \frac{(4^3+5)-(1^3+5)}{3}$$

$$3x^2 = \frac{69-6}{3} = \frac{63}{3} = 21$$

$$x^2 = 7$$

$$x = \sqrt{7} = 2.645$$

2.39 (1)

$$\lim_{x\to 0}\frac{\tan x}{x} = 1$$

2.40 (3)

The value of x where $f(x) = 0$ is a root of the function $f(x)$.

2.41 (c)

$$F(x, y) = x^3 + y^2$$

$$\text{grad } F = \frac{\partial}{\partial x}(x^3+y^2)\hat{i} + \frac{\partial}{\partial y}(x^3+y^2)\hat{j} + \frac{\partial}{\partial z}(x^3+y^2)\hat{k}$$

$$= 3x^2\hat{i} + 2y\hat{j}$$

3.32 (0)

$$y = e^{-x^2}$$

$$\frac{dy}{dx} = e^{-x^2} \times -2x$$

$$\frac{1}{x}\frac{dy}{dx} = -2e^{-x^2}$$

$$\lim_{x\to\infty}\frac{1}{x}\left(\frac{dy}{dx}\right) = \lim_{x\to\infty} -2e^{-x^2} = -2 \times e^{-\infty} = 0$$

2.43 (a)

$$\lim_{x\to\frac{\pi}{2}}\left|\frac{\tan x}{x}\right|$$

LHL $$\lim_{x\to\frac{\pi^-}{2}}\left[\frac{\tan x}{x}\right] = \infty$$

RHL $$\lim_{x\to\frac{\pi^+}{2}}\left[\frac{\tan x}{x}\right] = \infty$$

$\because$ LHL = RHL = ∞

2.44 (6.53)

$$u(x, y, z) = x^2 - 3yz$$

$$\text{grad}(u) = 2x\hat{i} - 3z\hat{j} - 3y\hat{k}$$

$$\text{grad}(u)|_{(2,0,-4)} = 4\hat{i} + 12\hat{j} - 0\hat{k} = 4\hat{i} + 12\hat{j}$$

$$\text{Directional derivative} = (4\hat{i}+12\hat{j}) \cdot \frac{(\hat{i}+\hat{j}-2\hat{k})}{\sqrt{6}}$$

$$= \frac{4+12}{\sqrt{6}} = \frac{16}{\sqrt{6}} = 6.53$$

2.46 (c)

$$v(x, y, z) = 3x^2y\hat{i} + 8y^2z\hat{j} + 5xyz\hat{k}$$

$$\text{curl } v = \nabla \times v$$

$$= \begin{vmatrix} \hat{i} & \hat{j} & \hat{k} \\ \frac{\partial}{\partial x} & \frac{\partial}{\partial y} & \frac{\partial}{\partial z} \\ 3x^2y & 8y^2z & 5xyz \end{vmatrix}$$

$$= \hat{i}(5xz - 8y^2) - \hat{j}(5yz - 0) + \hat{k}(0 - 3x^2)$$

$$= (5xz - 8y^2)\hat{i} - 5yz\hat{j} - 3x^2\hat{k}$$

2.47 (c)

$$f(x) = \frac{-5}{3}x^3 + 10x^2 - 15x + 16$$

$$f'(x) = \frac{-5}{3}\cdot 3x^2 + 20x - 15 = 0$$

$$= -5x^2 + 20x - 15 = 0$$
$$= -5(x^2 - 4x + 3) = 0$$
$$= (x-3)(x-1) = 0$$
$$x = 1, 3$$
$$f''(x) = -5(2x-4)$$

At $x = 1$
$$f''(x) = -5(2-4) = 10 > 0$$
$\therefore$ $x = 3$ is local minima

At $x = 3$
$$f''(x) = -5(2 \times 3 - 4) = -5 \times 2 = -10 < 0$$
$\therefore$ $x = 3$ is local maxima

So, $f(3) = \frac{-5}{3} \times 27 + 10 \times 9 - 15 \times 3 + 16$
$$= -45 + 90 - 45 + 16 = 16$$
$$f(0.5) = \frac{-5}{3} \times \frac{1}{8} + 10 \times \frac{1}{4} - \frac{15}{2} + 16$$
$$= \frac{-5}{24} + \frac{10}{4} - \frac{15}{2} + 16 = 10.79$$
$$f(3.5) = \frac{-5}{3} \times (3.5)^3 + 10 \times 3.5^2 - 15 \times 3.5 + 16$$
$$= 14.54$$
So, global maxima = 16

2.48 (b)

$$\cos x = 1 - \frac{x^2}{2!} + \frac{x^4}{4!} + \ldots$$
$$a = 0,\ b = \frac{-1}{2!} = -0.5$$
$$c = 0,\ d = \frac{1}{4!} = \frac{1}{24} = 0.042$$

2.49 (a, b, c)

$$f(x) = \begin{cases} -x, & x < 0 \\ x^2, & x \geq 0 \end{cases}$$

L.H.L = 0, R.H.L = 0

$\because$ L.H.L = R.H.L = $f(0)$

Hence it is continuous at $x = 0$.

L.H.D. = −1, R.H.D. = 0

$\because$ L.H.D ≠ R.H.D

$\therefore$ $f(x)$ is not differentiable at $x = 0$.

2.50 (c)

$$f(x) = e^{-|x|} = \begin{Bmatrix} e^{x} & x < 0 \\ e^{-x} & x \geq 0 \end{Bmatrix}$$

$$f'(x)\Big| = \frac{df(x)}{dx}\Big| = +\frac{1}{e^{|x|}}$$
$$f'(x)_{x=-1}\Big| = \frac{1}{e}$$

2.51 (a)

Given, $f(x) = x^e e^{-x}$

For the function to be maximum, $f''(x) < 0$

Which is determined by $f'(x) = 0$
$$f'(x) = ex^{e-1} e^{-x} - x^e e^{-x} = 0$$
$(x = e)$

2.52 (4.24)

$$f = 4x^2 + 2y^2 + z^2 \text{ at } (1, 1, 1)$$
$$\vec{V} = (\hat{i} - \hat{k})$$
$$\hat{V} = \frac{\hat{i} - \hat{k}}{\sqrt{1^2 + 1^2}} = \frac{1}{\sqrt{2}}(\hat{i} - \hat{k})$$

Directional derivative = $\text{grad}(f)_{at(1,1,1)} \cdot \vec{V}$
$$= \frac{\left[(8x\hat{i} + 4y\hat{j} + 2z\hat{k}) \cdot (\hat{i} - \hat{k})\right]_{(1,1,1)}}{\sqrt{2}}$$
$$= \frac{(8x - 2z)_{at}(1,1,1)}{\sqrt{2}} = \frac{8-2}{\sqrt{2}} = \frac{6}{\sqrt{2}} = 3\sqrt{2}$$

2.53 (192)

$$\iint_S \left(\frac{2x\hat{i} + 3y\hat{j} + 4z\hat{k}}{4\pi} \right) \cdot \hat{n}\, dS$$

By gauss divergence theorem
$$= \frac{1}{4\pi} \iiint_V \text{div}(\vec{A})\, dV$$
$$= \frac{1}{4\pi} \iiint_V \nabla \cdot \vec{A}\, dV$$
$$= \frac{1}{4\pi} \iiint_V \left(\frac{\partial}{\partial x}(2x) + \frac{\partial}{\partial y}(3y) + \frac{\partial}{\partial z}(4z) \right) dv$$
$$= \frac{1}{4\pi} \int_V (2 + 3 + 4)\, dv$$
$$= \frac{1}{4\pi} \times 9 \times \text{Volume of sphere}$$
$$= \frac{9}{4\pi} \times \frac{4}{3} \times \pi \times (4)^3 = 3 \times 4^3 = 192$$

2.54 (b)

Given: $y = \lim_{x \to 0} \frac{2x}{e^x - 1}$

Applying limit the given form is in $\frac{0}{0}$, thus using L-Hospital rule we get

$$\therefore \lim_{x \to 0}\left(\frac{2x}{e^x - 1}\right) = \lim_{x \to 0}\left(\frac{2}{e^x}\right) = \frac{2}{e^0} = \frac{2}{1} = 2$$

2.55 (d)

Given: $\vec{v} = zx\hat{i} + 2xy\hat{j} + 3yz\hat{k}$

For divergence, we know

$$\nabla \cdot \vec{v} = \frac{\partial}{\partial x}(zx) + \frac{\partial}{\partial y}(zxy) + \frac{\partial}{\partial z}(zyx)$$

$$\nabla \cdot \vec{v} = z + 2x + 3y$$

Putting the values of x, y, z i.e. 3, 2, 1.

$$\nabla \cdot \vec{v} = 1 + 2 \times 3 + 3 \times 2 = 13$$

2.56 (d)

Given: $\hat{e}_r = \cos\theta\hat{i} + \sin\theta\hat{j}$

$$\hat{e}_\theta = -\sin\theta\hat{i} + \cos\theta\hat{j}$$

$$\hat{e}_r + \hat{e}_\theta = (\cos\theta - \sin)\hat{i} + (\sin\theta + \cos\theta)\hat{j} = f(z)$$

$$\frac{\partial f(z)}{\partial 0} = -\sin\theta\hat{i} - \cos\hat{i} + \cos\theta\hat{j} - \sin\theta\hat{j}$$

$$I = -(\cos\theta\hat{i} + \sin\theta\hat{j}) = -\hat{e}_r$$

$$II = -\sin\theta\hat{i} + \cos\theta\hat{j} = \hat{e}_\theta$$

$$\frac{\partial f(z)}{\partial \theta} = -\hat{e}_r + \hat{e}_\theta$$

2.57 (d)

From Taylor series expression about x_0

$$f(x) = f(x_0) + \left.\frac{df}{dx}\right|_{x=0}(x - x_0) + \left.\frac{d^2f}{dx^2}\right|_{x=0}\frac{(x-x_0)^2}{2}$$

Here, $x_0 = 0$ and $f(x)\ (1 - x) - e^{-x}$

$$f(x) = 1 - 0 - e^{-0} + (-1 + e^{-0})(x - 0) + e^{-0}(-1)\frac{(x-0)^2}{2}$$

$$f(x) = 0 + 0 - \frac{x^2}{2}$$

So first non-zero term is $-\frac{x^2}{2}$.

Correct option is (d).

2.59 (1.41)

Given function $f(x) = x^4 + 2y^3 + z^2$

Directional derivative $= \nabla\phi \cdot \hat{n}$

$$\nabla\phi = \frac{\partial\phi}{\partial x}\hat{i} + \frac{\partial\phi}{\partial y}\hat{j} + \frac{\partial\phi}{\partial z}\hat{k}$$

$$= 4x^3\hat{i} + 6y^2\hat{j} + 2z\hat{k}$$

$$\nabla\phi \text{ at } (-1, 1, -1) = -4\hat{i} + 6\hat{j} - 2\hat{k}$$

$$\hat{n} = \frac{\hat{i} + \hat{j}}{\sqrt{2}}$$

Directional derivative of ϕ in direction of $\hat{n}$

$$= \nabla\phi \cdot \hat{n}$$

$$= \frac{(-4\hat{i} + 6\hat{j} - 2\hat{k})}{\sqrt{2}} \cdot (\hat{i} + \hat{j})$$

$$\frac{-4+6}{\sqrt{2}} = \sqrt{2} = 1.414$$

■■■■

3 Differential Equations

3.1 The integrating factor for the differential equation

$$\left(\cos^2 x\right)\frac{dy}{dx} + y = \tan x \text{ is}$$

(a) $e^{\tan x}$ (b) $\cos 2x$
(c) $e^{-\tan x}$ (d) $\sin 2x$

[2000 : 1 M]

3.2 The general solution of $\frac{d^4y}{dx^4} + 2\frac{d^2y}{dx^2} + y = 0$ is

(a) $(C_1 x + C_2)e^x + (C_3 + C_4 x)e^{-x}$
(b) $C_1 \cos x + C_2 \sin x + C_3 e^x + C_4 e^{-x}$
(c) $C_1 e^{ix} + C_2 e^{-ix}$
(d) $(C_1 + C_2 x)\cos x + (C_3 + C_4 x)\sin x$

where C_1, C_2, C_3 and C_4 are constants.

[2000 : 2 M]

3.3 Reduce the following differential equation to linear form

$$\frac{dz}{dt} + zt = z^{-1}t$$

(a) Find a general solution to the linearised equation.
(b) Determine the integration constants if $z(0) = 0$.

[2002 : 5 M]

3.4 The value of y as $t \to \infty$ for the following differential equation for an initial value of $y(1) = 0$ is

$$(4t^2 + 1)\frac{dy}{dt} + 8yt - t = 0$$

(a) 1 (b) $\frac{1}{2}$
(c) $\frac{1}{4}$ (d) $\frac{1}{8}$

[2003 : 2 M]

3.5 The differential equation $\frac{d^2x}{dt^2} + 10\frac{dx}{dt} + 25x = 0$ will have a solution of the form (where C_1 and C_2 are constants)

(a) $(C_1 + C_2 t)e^{-5t}$ (b) $C_1 e^{-2t}$
(c) $C_1 e^{-5t} C_2 e^{5t}$ (d) $C_1 e^{-5t} + C_2 e^{2t}$

[2003 : 2 M]

3.6 The differential equation

$$\frac{d^2y}{dx^2} + \sin x\frac{dy}{dx} + ye^x = \sinh x \text{ is}$$

(a) first order and linear
(b) first order and non-linear
(c) second order and linear
(d) second order and non-linear

[2004 : 1 M]

3.7 The differential equation for the variation of the amount of salt x in a tank with time t is given by

$\frac{dx}{dt} + \frac{x}{20} = 10$. x is in kg and t is in minutes.

Assuming that there is no salt in the tank initially, the time (in min) at which the amount of salt increases to 100 kg is

(a) 10 ln 2 (b) 20 ln 2
(c) 50 ln 2 (d) 100 ln 2

[2004 : 2 M]

3.8 The differential equation can be reduced to

$$\left(\frac{dy}{dx}\right)^2 + y\frac{d^2y}{dx^2} = 0 \text{ (where } \alpha \text{ is a constant)}$$

(a) $\left(\frac{dy}{dx}\right)^2 = \alpha - \frac{3y^2}{2}$

(b) $\left(\frac{dy}{dx}\right)^2 = \alpha - 2y$

(c) $\frac{dy}{dx} = \frac{\alpha}{y^2}$

(d) $\frac{dy}{dx} = \frac{\alpha}{y}$

[2004 : 2 M]

3.9 What condition is to be satisfied so that the solution of the differential equation is of the form $y = (C_1 + C_2 x)e^{mx}$, where C_1 and C_2 are constants of integration?

(a) $a^2 = b$ (b) $b^2 = a$
(c) $a^2 = 4b$ (d) $b^2 = 4a$

[2005 : 2 M]

3.10 If $f(x)$ is the solution of the equation $\frac{dy}{dx}+2xy+2x = 0$ and $g(x)$ is the solution of the equation $\frac{dy}{dx}+2xy-2x=0$, and the constant of integration in $f(x)$ is equal to that in $g(x)$, then which of the following is true?

(a) $g(x) = f(x) + 2$
(b) $g(x) = f(x) + 1$
(c) $g(x) = f(x) - 1$
(d) $g(x) = f(x) - 2$

[2005 : 2 M]

3.11 With $y = e^{ax}$, if the sum $S=\frac{dy}{dx}+\frac{d^2y}{dx^2}+.....+\frac{d^ny}{dx^n}$ approaches $2y$ as $n \to \infty$, then the value of a is

(a) $\frac{1}{3}$ (b) $\frac{1}{2}$
(c) $\frac{2}{3}$ (d) 2

[2006 : 2 M]

3.12 The initial condition for which the following equation

$$(x^2+2x)\frac{dy}{dx} = 2(x+1)y;\ y(x_0) = y_0$$

has infinitely many solutions, is

(a) $y(x = 0) = 5$
(b) $y(x = 0) = 1$
(c) $y(x = 2) = 1$
(d) $y(x = -2) = 0$

[2007 : 1 M]

3.13 The solution of the following differential equation $x\frac{dy}{dx}+y(x^2-1) = 2x^3$ is

(a) 0 (b) $2+ce^{-\frac{x^2}{2}}$
(c) $c_1x + c_2x^2$ (d) $2x+cxe^{-\frac{x^2}{2}}$

[2007 : 2 M]

3.14 The family of curves that is orthogonal to $xy = c$ is

(a) $y = c_1x$ (b) $y = \frac{c_1}{x}$
(c) $y^2 + x^2 = c_1$ (d) $y^2 - x^2 = c_1$

[2007 : 2 M]

3.15 Which one of the following is NOT an integrating factor for the differential equation $xdy - ydx = 0$?

(a) $\frac{1}{x^2}$ (b) $\frac{1}{y^2}$
(c) $\frac{1}{xy}$ (d) $\frac{1}{(x+y)}$

[2008 : 1 M]

3.16 Which one of the following is NOT a solution of the differential equation $\frac{d^2y}{dx^2}+y = 1$?

(a) $y = 1$ (b) $y = 1 + \cos x$
(c) $y = 1 + \sin x$ (d) $y = 2 + \sin x + \cos x$

[2008 : 1 M]

3.17 Which one of the following transformation $\{u = f(y)\}$ reduces $\frac{dy}{dx}+Ay^3+By = 0$ to a linear differential equation? (A and B are positive constants)

(a) $u = y^{-3}$ (b) $u = y^{-2}$
(c) $u = y^{-1}$ (d) $u = y^2$

[2008 : 2 M]

3.18 The general solution of the differential equation

$$\frac{d^2y}{dx^2}-\frac{dy}{dx}-6y = 0,$$

with C_1 and C_2 as constants of integration, is

(a) $C_1e^{-3x} + C_2e^{-2x}$ (b) $C_1e^{3x} + C_2e^{-2x}$
(c) $C_1e^{3x} + C_2e^{2x}$ (d) $C_1e^{-3x} + C_2e^{2x}$

[2009 : 2 M]

3.19 The solution of the differential equation

$$\frac{d^2y}{dt^2}+2\frac{dy}{dt}+2y = 0$$

with the initial condition $y(0) = \left.\frac{dy}{dt}\right|_{t=0} = -1$ is

(a) $-t \sin t$ (b) $-e^{-t}(1 - \cos t)$
(c) $-\frac{(t+\sin t)}{2}$ (d) $-e^{-t} \sin t$

[2010 : 2 M]

3.20 Which one of the following choices is a solution of the differential equation given below? (**Note:** c is a real constant)

$$\frac{dy}{dx}=\frac{y^2}{x}+\frac{y}{x}-\frac{2}{x}$$

(a) $y = \dfrac{c - x^2}{c + 2x^2}$ (b) $y = \dfrac{c + 2x^2}{c - x^2}$

(c) $y = \dfrac{c - x^3}{c + 2x^2}$ (d) $y = \dfrac{c + 2x^3}{c - x^3}$

[2011 : 2 M]

3.21 If a and b are arbitrary constants, then the solution to the ordinary differential equation $\dfrac{d^2y}{dx^2} - 4y = 0$ is

(a) $y = ax + b$
(b) $y = ae^{-x}$
(c) $y = a \sin 2x + b \cos 2x$
(d) $y = a \cosh 2x + b \sinh 2x$

[2012 : 1 M]

3.22 The solution of the differential equation $\dfrac{dy}{dx} - y^2 = 0$, given $y = 1$ at $x = 0$

(a) $\dfrac{1}{1+x}$ (b) $\dfrac{1}{1-x}$

(c) $\dfrac{1}{(1-x)^2}$ (d) $\dfrac{x^3}{3} + 1$

[2013 : 2 M]

3.23 The solution of the differential equation

$\dfrac{d^2y}{dx^2} - \dfrac{dy}{dx} + 0.25y = 0$, given $y =$ and $\dfrac{dy}{dx} = 1$ at $x = 0$ is

(a) $xe^{0.5x} - x e^{-0.5x}$ (b) $0.5\, xe^{x} - 0.5\, x e^{-x}$
(c) $x e^{0.5x}$ (d) $-x e^{0.5x}$

[2013 : 2 M]

3.24 The integrating factor for the differential equation $\dfrac{dy}{dx} - \dfrac{y}{1+x} = (1+x)$ is

(a) $\dfrac{1}{1+x}$ (b) $(1 + x)$

(c) $x(1 + x)$ (d) $\dfrac{x}{1+x}$

[2014 : 2 M]

3.25 The differential equation $\dfrac{d^2y}{dx^2} + x^2\dfrac{dy}{dx} + x^3 y = e^x$ is

(a) non-linear differential equation of first order
(b) linear differential equation of first order
(c) linear differential equation of second order
(d) non-linear differential equation of second degree

[2014 : 2 M]

3.26 Consider a linear ordinary differential equation: $\dfrac{dy}{dx} + p(x)y = r(x)$. Functions $p(x)$ and $r(x)$ are defined and have a continuous first derivative. The integrating factor of this equation is non-zero. Multiplying this equation by its integrating factor converts this into a:

(a) Homogeneous differential equation
(b) Non-linear differential equation
(c) Second order differential equation
(d) Exact differential equation

[2015 : 1 M]

3.27 What is the solution for the second order differential $\dfrac{d^2y}{dx^2} + y = 0$, with the initial conditions $y|_{x=0} = 5$ and $\left.\dfrac{dy}{dx}\right|_{x=0} = 10$?

(a) $y = 5 + 10 \sin x$
(b) $y = 5 \cos x - \sin x$
(c) $y = 5 \cos x + 10\, x$
(d) $y = 5 \cos x + 10 \sin x$

[2016 : 2 M]

3.28 For the initial value problem,

$$\frac{dx}{dt} = \sin(t), \quad x(0) = 0$$

the value of x at $t = \dfrac{\pi}{3}$, is ______.

[2017 : 2 M]

3.29 Consider the following two equations:

$$\frac{dx}{dt} + x + y = 0$$

$$\frac{dy}{dt} - x = 0$$

The above set of equations is represented by

(a) $\dfrac{d^2y}{dt^2} - \dfrac{dy}{dt} - y = 0$ (b) $\dfrac{d^2x}{dt^2} - \dfrac{dx}{dt} - y = 0$

(c) $\dfrac{d^2y}{dt^2} + \dfrac{dy}{dt} + y = 0$ (d) $\dfrac{d^2x}{dt^2} + \dfrac{dx}{dt} + y = 0$

[2018 : 1 M]

3.30 The solution of the ordinary differential equation $\frac{dy}{dx} + 3y = 1$, subject to the initial condition $y = 1$ at $x = 0$, is

(a) $\frac{1}{3}(1+2e^{-x/3})$ (b) $\frac{1}{3}(5-2e^{-x/3})$

(c) $\frac{1}{3}(5-2e^{-3x})$ (d) $\frac{1}{3}(1+2e^{-3x})$

[2019 : 2 M]

3.31 Given $\frac{dy}{dx} = y - 20$ and $y|_{x=0} = 40$, the value of y at $x = 2$ is _____ (round off to nearest integer).

[2020 : 2 M]

3.32 For the ordinary differential equation

$$\frac{d^3y}{dt^3} + 6\frac{d^2y}{dt^2} + 11\frac{dy}{dt} + 6y = 1$$

with initial conditions $y(0) = y'(0) = y''(0) = y'''(0) = 0$, the value of $\lim_{t\to\infty} y(t)$ = ________. (Round off to 3 decimal places).

[2021 : 2 M]

3.33 The partial differential equation:

$$\frac{\partial u}{\partial t} = \frac{1}{\pi^2}\frac{\partial^2 u}{\partial x^2}$$

where, $t \geq 0$ and $x \in [0, 1]$, is subjected to the following initial and boundary conditions.

$u(x, 0) = \sin(\pi x)$

$u(0, t) = 0$

$u(1, t) = 0$

The value of t at which $\frac{u(0.5,t)}{u(0.5,0)} = \frac{1}{e}$ is

(a) 1 (b) e

(c) π (d) $\frac{1}{e}$

[2022 : 2 M]

3.34 The position $x(t)$ of a particle, at constant ω, is described by the equation

$$\frac{d^2x}{dt^2} = -w^2x$$

The initial conditions are $x(t = 0) = 1$ and $\left.\frac{dx}{dt}\right|_{t=0} = 0$ then position of particle at $t = \left(\frac{3\pi}{\omega}\right)$ is _______ (in integer).

[2023 : 1 M]

3.35 The first derivative of the function

$$U(r) = 4\left[\left(\frac{1}{r}\right)^{12} - \left(\frac{1}{r}\right)^{6}\right]$$

evaluated at $r = 1$ is _____ (in integer).

[2023 : 2 M]

3.36 Consider the ordinary differential equation $x^2\frac{d^2y}{dx^2} - x\frac{dy}{dx} - 3y = 0$, with the boundary conditions $y(x = 1) = 2$ and $y(x = 2) = \frac{17}{2}$. The solution $y(x)$ at $x = \frac{3}{2}$, rounded off to 2 decimal places, is _________.

[2024 : 2 M]

■■■■

Answers Differential Equations

3.1	(a)	**3.2**	(d)	**3.3**	(Sol.)	**3.4**	(d)	**3.5**	(a)	**3.6**	(c)	**3.7**	(b)
3.8	(d)	**3.9**	(d)	**3.10**	(a)	**3.11**	(c)	**3.12**	(d)	**3.13**	(d)	**3.14**	(d)
3.15	(d)	**3.16**	(d)	**3.17**	(b)	**3.18**	(b)	**3.19**	(d)	**3.20**	(d)	**3.21**	(d)
3.22	(b)	**3.23**	(c)	**3.24**	(a)	**3.25**	(c)	**3.26**	(d)	**3.27**	(d)	**3.28**	(0.5)
3.29	(c)	**3.30**	(d)	**3.31**	(167.8)	**3.32**	(0.167)	**3.33**	(a)	**3.34**	(–1)	**3.35**	(–24)
3.36	(4.04)												

Explanations Differential Equations

3.1 (a)

Given : $\left(\cos^2 x\right)\dfrac{dy}{dx} + y = \tan x$

$$\frac{dy}{dx} + (\sec^2 x)y = \tan x \sec^2 x$$

$$\text{I.F.} = e^{\int P dx} = e^{\int \sec^2 dx} = e^{\tan x}$$

3.2 (d)

Given : $\dfrac{d^4 y}{dx^4} + 2\dfrac{d^2 y}{dx^2} + y = 0$

$$m^4 + 2m^2 + 1 = 0$$

$$(m^2 + 1)^2 = 0$$

$$m^2 = -1, -1$$

$$m = \pm i, \pm i$$

So, $y = (C_1 + C_2 x)\cos x + (C_3 + C_4 x)\sin x$

3.3 Sol.

(a) $\dfrac{dz}{dt} + zt = z^{-1}t$

$$\frac{dz}{dt} + zt = \frac{t}{z}$$

$$z\frac{dz}{dt} + z^2 t = t$$

Put, $z^2 = u$

$$2z\frac{dz}{dt} = \frac{du}{dt}$$

$$\frac{1}{2}\frac{du}{dt} + ut = t$$

$$\frac{du}{dt} + u(2t) = 2t$$

$$\text{I.F.} = e^{\int 2t dt} = e^{t^2}$$

So, solution is

$$u(e^{t^2}) = \int e^{t^2} 2t dt + C$$

Let, $t^2 = P$

$$2t dt = dP$$

So, $u(e^{t^2}) = \int e^P dP + C$

$$u(e^{t^2}) = e^P + C$$

Put, $P = t^2$

$$u(e^{t^2}) = e^{t^2} + C$$

$$u = 1 + Ce^{-t^2}$$

Put, $u = z^2$

$z^2 = 1 + Ce^{-t^2}$ is the general solution.

(b) For $z(0) = 0$

$$0 = 1 + C$$

$$C = -1$$

3.4 (d)

$$(4t^2 + 1)\frac{dy}{dt} + 8yt - t = 0$$

$$(4t^2 + 1)\frac{dy}{dt} + 8yt = t$$

$$\frac{dy}{dt} + \frac{8t}{4t^2 + 1}y = \frac{t}{4t^2 + 1}$$

$$\text{I.F.} = e^{\int \frac{8t}{4t^2+1}dt = e^{\ln(4t^2+1)}} = (4t^2 + 1)$$

$$y(4t^2 + 1) = \int \frac{t}{(4t^2 + 1)} \cdot (4t^2 + 1)dt + C$$

$$y(4t^2 + 1) = \frac{t^2}{2} + C$$

$\because$ $y(1) = 0$

$$0 \times (5) = \frac{1}{2} + C$$

$$C = -\frac{1}{2}$$

So, $y = \dfrac{t^2 - 1}{2(4t^2 + 1)}$

$$\lim_{t\to\infty} y = \lim_{t\to\infty} \frac{t^2 - 1}{2(4t^2 + 1)} = \frac{1}{8}$$

3.5 (a)

$$\frac{d^2x}{dt^2}+10\frac{dx}{dt}+25x = 0$$

It's A.E. is

$m^2 + 10m + 25 = 0$

$(m + 5)^2 = 0$

$m = -5, -5$

So, solution is

$$x = (C_1 + C_2 t)e^{-5t}$$

3.7 (b)

$$\frac{dx}{dt}+\frac{x}{20} = 10$$

$$\text{I.F.} = e^{\int 1/20 dt} e^{t/20}$$

So, $x(e^{t/20}) = \int 10\left(e^{t/20}\right)dt + C$

$$= 10\times e^{t/20}\times\frac{1}{\frac{1}{20}}+C$$

$x(e^{t/20}) = 200e^{t/20} + C$

$0 = 200 + C$ [Since no salt initially]

$C = -200$

$\Rightarrow x(e^{t/20}) = 200(e^{t/20} - 1)$

For $x = 100$

$100e^{t/20} = 200(e^{t/20} - 1)$

$$\frac{1}{2} = 1 - e^{-t/20}$$

$$e^{-t/20} = \frac{1}{2}$$

$$\frac{t}{20} = \ln(2)$$

$t = 20\ln(2)$

3.8 (d)

$$\frac{dy}{dx} = \frac{\alpha}{y}$$

By differentiating w.r.t. x

$$\Rightarrow \frac{d^2y}{dx^2} = \frac{-\alpha}{y^2}\left(\frac{dy}{dx}\right)$$

$$\Rightarrow \frac{d^2y}{dx^2}+\frac{\alpha}{y^2}\left(\frac{dy}{dx}\right) = 0$$

Put, $\dfrac{\alpha}{y} = \dfrac{dy}{dx}$

$$\Rightarrow \frac{d^2y}{dx^2}+\frac{1}{y}\left(\frac{dy}{dx}\right)^2 = 0$$

$$\Rightarrow y\frac{d^2y}{dx^2}+\left(\frac{dy}{dx}\right)^2 = 0$$

3.9 (d)

For the solution $y = (C_1 + C_2x)e^{mx}$, both the roots should be equal and for that $b^2 = 4a$.

3.10 (a)

$$\frac{dy}{dx}+2xy+2x = 0$$

$$\frac{dy}{dx}+2xy = -2x$$

$$\text{I.F.} = e^{\int 2xdx} = e^{x^2}$$

$$y(e^{x^2}) = \int -2xe^{x^2}dx + C$$

$y(e^{x^2}) = -e^{x^2} + C$

$y = -1 + Ce^{-x^2}$

$\because f(x)$ is the solution of this differential equation.

Hence, $f(x) = -1 + Ce^{-x^2}$

and $\dfrac{dy}{dx}+2xy = 2x$

$$\text{I.F.} = e^{\int 2xdx} = e^{x^2}$$

$$y(e^{x^2}) = \int 2xe^{x^2}dx + C$$

$y(e^{x^2}) = e^{x^2} + C$

$y = 1 + Ce^{-x^2}$

$g(x) = 1 + Ce^{-x^2}$ ...(2)

$\because$ C has same value in both the equation.

$\therefore$ From (1) and (2)

$g(x) = 1 + 1 + f(x) = f(x) + 2$

3.11 (c)

$$S = \frac{dy}{dx}+\frac{d^2y}{dx^2}+\frac{d^3y}{dx^3}\cdots\cdots\cdots\frac{d^ny}{dx^n}$$

$S = e^{ax}(a + a^2 \ldots\ldots a^n) = 2e^{ax}$

$a + a^2 \ldots\ldots a^n = 2$

$$\Rightarrow \frac{a}{1-a} = 2$$

$\Rightarrow a = 2 - 2a$

$\Rightarrow 3a = 2$

$$\Rightarrow a = \frac{2}{3}$$

3.12 (d)

$$\frac{dy}{dx} = \frac{2(x+1)}{(x^2+2x)}y$$

$$\int \frac{dy}{y} = \int \frac{2(x+1)}{(x^2+2x)}dx$$

$$x^2 + 2x = t$$

$$(2x+2)dx = dt$$

$$\int \frac{dy}{y} = \int \frac{dt}{t}$$

$$\ln y = \ln t + \ln C$$

$$y = Ct = C(x^2+2x) = Cx(x+2)$$

at $x = -2$, $y = 0 \Rightarrow C = 0$

So, $y = 0$ is having infinitely many solution.

3.13 (d)

$$x\frac{dy}{dx} + y(x^2-1) = 2x^3, \quad \frac{dy}{dx} + y\left(x - \frac{1}{x}\right) = 2x^2$$

It is linear D.E.

$$\text{I.F.} = e^{\int\left(x-\frac{1}{x}\right)dx} = e^{\left(\frac{x^2}{2}-\ln x\right)}$$

$$= e^{\frac{x^2}{2}} \cdot e^{-\ln x} = \frac{1}{x}e^{\frac{x^2}{2}}$$

$$\Rightarrow \quad y\left(\frac{1}{x}e^{\frac{x^2}{2}}\right) = \int 2x^2 \times \frac{1}{x}e^{\frac{x^2}{2}}dx + C$$

$$y = 2x + Cxe^{-\frac{x^2}{2}}$$

3.14 (d)

$xy = C$ is a given function.

$$y + \frac{xdy}{dx} = 0$$

$$\frac{dy}{dx} = \frac{-y}{x}$$

For orthogonal curve

$$\frac{dy}{dx} = -\frac{1}{(-y/x)} = \frac{x}{y}$$

So, $$\frac{dy}{dx} = \frac{x}{y}$$

$$\int ydy = \int xdx$$

$$\Rightarrow \quad \frac{y^2}{2} = \frac{x^2}{2} + k$$

$$\Rightarrow \quad y^2 - x^2 = 2k = c'$$

$$\Rightarrow \quad y^2 - x^2 = c'$$

3.15 (d)

$$xdy - ydx = 0$$

$$\frac{\partial M}{\partial y} = -1, \quad \frac{\partial N}{\partial x} = 1$$

$$\because \quad \frac{\partial M}{\partial y} \neq \frac{\partial N}{\partial x}$$

So, the D.E. is not exact.

I.F. are :

1. $$\frac{\frac{\partial M}{\partial y} - \frac{\partial N}{\partial x}}{N} = \frac{-1-1}{x} = \frac{-2}{x}$$

$$\Rightarrow \quad \text{I.F.} = e^{\int -\frac{2}{x}dx} = e^{-\ln x^2} = \frac{1}{x^2}$$

2. $$\frac{\frac{\partial N}{\partial x} - \frac{\partial M}{\partial y}}{M} = \frac{1+1}{-y} = \frac{-2}{y}$$

$$\Rightarrow \quad \text{I.F.} = e^{\int -\frac{2}{y}dy} = e^{-2\ln y} = \frac{1}{y^2}$$

3. Multiplying by $\frac{1}{xy}$

$$\frac{dy}{y} - \frac{dx}{x} = 0 \quad \text{(Variable separable)}$$

$$\frac{\frac{\partial N}{\partial x} - \frac{\partial M}{\partial y}}{M} = \frac{1+1}{-y} = \frac{-2}{y}$$

3.16 (d)

$$\frac{d^2y}{dx^2} + y = 1$$

$$m^2 + 1 = 0$$

$$m = \pm i$$

$$CF = C_1 \cos x + C_2 \sin x$$

$$\text{P.I.} = \frac{e^{0.x}}{D^2+1}$$

Put $D = 0$,

$$\text{P.I.} = \frac{e^{0.x}}{0+1} = \frac{e^0}{1} = 1$$

So, complete solution

$$y = C_1 \cos x + C_2 \sin x + 1$$

So, (a, b, c) can possibly be the solution of differential equation.

3.17 (b)

$$\frac{dy}{dx} + Ay^3 + By = 0, \quad \frac{dy}{du} + By = -Ay^3$$

$$y^{-3}\frac{dy}{dx}+By^{-2} = -A$$

(Bernoulli's form to reduce to linear form)

Put $u = y^{-2}$

3.18 (b)

$$\frac{d^2y}{dx^2}-\frac{dy}{dx}-6y = 0$$

It's A.E. is

$$m^2 - m - 6 = 0$$
$$(m-3)(m+2) = 0$$
$$m = -2, 3$$
$$y = C_1e^{3x} + C_2e^{-2x}$$

3.19 (d)

$$\frac{d^2y}{dt^2}+2\frac{dy}{dt}+2y = 0$$

Its A.E. is

$$m^2 + 2m + 2 = 0$$
$$m = \frac{-2\pm\sqrt{4-8}}{2} = \frac{-2\pm 2i}{2} = -1 \pm i$$

So, $\quad y = e^{-t}(C_1 \cos t + C_2 \sin t)$

$\because \quad y(0) = 0$

$$0 = e^{-0}[C_1 \cos(0) + C_2 \sin(0)]$$
$$0 = C_1$$

So, $\quad y(t) = e^{-t}(C_2 \sin t)$

$$y'(t) = -e^{-t}(C_2 \sin t) + e^{-t}(C_2 \cos t)$$
$$-1 = -e^{-0}(C_2 \times 0) + e^{-0}(C_2 \times 1)$$
$$-1 = C_2$$

So, $\quad y(t) = e^{-t}(-\sin t) = -e^{-t} \sin t$

3.20 (d)

Given,

$$\Rightarrow \quad \frac{dy}{dx} = \frac{y^2}{x}+\frac{y}{x}-\frac{2}{x}$$
$$\Rightarrow \quad \frac{dy}{dx} = \frac{y^2+y-2}{x}$$
$$\Rightarrow \quad \frac{dx}{x} = \frac{dy}{y^2+y-2}$$
$$\frac{dx}{x} = \frac{1}{3}\left[\frac{1}{y-1}-\frac{1}{y+2}\right]dy$$
$$\Rightarrow \quad 3\int\frac{dx}{x} = \int\frac{dy}{y-1}-\int\frac{dy}{y+2}$$
$$\Rightarrow \quad 3 \ln x = \ln(y-1) - \ln(y+2) + \ln c$$
$$\Rightarrow \quad \ln x^3 = \ln\left(\frac{y-1}{y+2}\right)+\ln c$$
$$\Rightarrow \quad x^3 = c\left(\frac{y-1}{y+2}\right)$$
$$\Rightarrow \quad yx^3 + 2x^3 = cy - c$$
$$\Rightarrow \quad 2x^3 + c = y(c - x^3)$$
$$\Rightarrow \quad y = \frac{2x^3+c}{c-x^3}$$

3.21 (d)

$$\frac{d^2y}{dx^2}-4y = 0$$

It's A.E. is $m^2 - 4 = 0$

$$m = \pm 2$$
$$y = c_1 e^{2x} + c_2 e^{-2x}$$

Since, $\sinh(2x) = \dfrac{e^{2x}-e^{-2x}}{2}$

Since, $\cosh(2x) = \dfrac{e^{2x}+e^{-2x}}{2}$

$$y = c_1(\sin h\, 2x + \cos h\, 2x) - c_2(\sin h\, 2x - \cos h\, 2x)$$
$$y = (c_1 - c_2) \sin h\, 2x + (c_1 + c_2) \cos h\, 2x$$

Let, $\quad c_1 - c_2 = a$ and $c_1 + c_2 = b$

where a, b are constants.

Then, $\quad y = a \sin h\, 2x + b \cos h\, 2x$

3.22 (b)

$$\frac{dy}{dx}-y^2 = 0 \Rightarrow \int\frac{dy}{y^2} = \int dx$$
$$-\frac{1}{y} = x + c$$

At $\quad x = 0,\ y = 1$

$$-1 = c$$
$$\Rightarrow \quad c = -1$$
$$-\frac{1}{y} = x - 1 \Rightarrow \frac{1}{y} = -x + 1,\ y = \frac{1}{1-x}$$

3.23 (c)

Its A.E. is

$$m^2 - m + \frac{1}{4} = 0$$
$$4m^2 - 4m + 1 = 0$$
$$(2m-1)^2 = 0$$
$$m = \frac{1}{2}, \frac{1}{2}$$

So, $y = (c_1 + c_2 x)e^{x/2}$

$0 = c_1 \times e^0 = c_1 = 0$

$y = c_2 \times x e^{x/2}$

$$\frac{dy}{dx} = c_2 e^{x/2} + \frac{c_2 x}{2} e^{x/2}$$

$$\left.\frac{dy}{dx}\right|_{x=0} = 1$$

$$1 = c_2 + \frac{c_2 \times 0}{2}$$

$c_2 = 1$

So, $y = x\,e^{0.5x}$

3.24 (a)

$$\frac{dy}{dx} - \frac{y}{(1+x)} = (1+x)$$

$$\text{I.F.} = 1 - e^{\int \frac{-1}{1+x}dx} = e^{-\ln(1+x)} = \frac{1}{x+1}$$

3.26 (d)

$$\frac{dy}{dx} + P(x)y = r(x)$$

$$\text{I.F.} = \int\limits_e P\,dx$$

So, $$e^{\int P dx}\frac{dy}{dx} + e^{\int P dx} Py = e^{\int P dx} r(x)$$

$$e^{\int P dx}\frac{dy}{dx} + (Py - r)\,e^{\int P dx} = 0$$

$$e^{\int P(x)dx}\,dy + e^{\int P(x)dx}\,[P(x)y - r(x)]\,d(x) = 0$$

$$M = [P(x)y - r(x)]\,e^{\int P(x)dx}$$

$$\frac{\partial M}{\partial y} = P(x)\cdot e^{\int P(x)dx}$$

$$N = e^{\int P(x)dx}$$

$$\frac{\partial N}{\partial x} = e^{\int P(x)dx}\cdot P(x)$$

$\because$ $$\frac{\partial M}{\partial y} = \frac{\partial N}{\partial x}$$

Hence, multiplying by I.F. converts it into exact differential equation.

3.27 (d)

$$\frac{d^2y}{dx^2} + y = 0,\ y(0) = 5,\ y'(0) = 10$$

A.E. is $m^2 + 1 = 0$

$m = \pm i,\ y = c_1 \cos x + c_2 \sin x$

$5 = c_1$

$$\frac{dy}{dx} = -c_1 \sin x + c_2 \cos x$$

$10 = -c_1 \times 0 + c_2 \times 1$

$c_2 = 10$

So, $y = 5\cos x + 10 \sin x$

3.28 (0.5)

$$\frac{dx}{dt} = \sin t,\quad x(0) = 0$$

$$\int dx = \int \sin t\,dt$$

$x = -\cos t + c,\ 0 = -1 + c,\ c = 1$

$$x = -\cos t + 1 = -\cos\frac{\pi}{3} + 1 = -\frac{1}{2} + 1 = \frac{1}{2}$$

3.29 (c)

$$\frac{dx}{dt} + x + y = 0 \quad ...(i)$$

$$\frac{dy}{dt} - x = 0 \quad ...(ii)$$

From (ii), $$\frac{d^2y}{dt^2} - \frac{dx}{dt} = 0$$

$$\frac{d^2y}{dt^2} + (x+y) = 0$$

$$\Rightarrow \frac{d^2y}{dt^2} + x + y = 0$$

$$\Rightarrow \frac{d^2y}{dt^2} + \frac{dy}{dt} + y = 0$$

3.30 (d)

$$\frac{dy}{dx} + 3y = 1,\quad y(0) = 1$$

$$\text{I.F.} = \int e^{3dx} = e^{3x}$$

$$y(e^{3x}) = \int 1 \times e^{3x}\,dx + c$$

$$ye^{3x} = \frac{e^{3x}}{3} + c$$

$$1 \times e^0 = \frac{e^0}{3} + c \text{ (Substituting } y(0) = 1)$$

$$c = 1 - \frac{1}{3} = \frac{2}{3}$$

$$y(e^{3x}) = \frac{e^{3x}}{3} + \frac{2}{3}$$

$$y = \frac{1}{3} + \frac{2}{3}e^{-3x} = \frac{1}{3}\left(1 + 2e^{-3x}\right)$$

3.31 (168)

$$\frac{dy}{dx} = y - 20,\ y(0) = 40,\ y(2) = ?$$

$$\frac{dy}{dx} - y = -20$$

$$\text{I.F.} = e^{-\int dx} = e^{-x}$$

$$y(e^{-x}) = \int -20e^{-x}dx + c = \frac{-20e^{-x}}{-1} + c$$

$$\Rightarrow \quad y(e^{-x}) = 20e^{-x} + c$$

$$\Rightarrow \quad y = 20 + c\,e^{x}$$

$$\Rightarrow \quad 40 = 20 + c, \quad c = 20, \quad y = 20(1 + e^{x})$$

$$y(2) = 20(1 + e^2) = 167.78 \simeq 168$$

3.32 (0.167)

$$\frac{d^3y}{dt^3} + 6\frac{d^2y}{dt^2} + 11\frac{dy}{dt} + 6y = 1$$

$$m^3 + 6m^2 + 11m + 6 = 0$$

$$(m+1)(m+2)(m+3) = 0$$

$$m = -1, -2, -3$$

$$CF = C_1 e^{-t} + C_2 e^{-2t} + C_3 e^{-3t}$$

$$\text{P.I.} = \frac{1}{D^3 + 6D^2 + 11D + 6} = \frac{e^{0t}}{D^3 + 6D^2 + 11D + 6}$$

Put, $D = 0$

$$\text{P.I.} = \frac{1}{0^3 + 6\times 0^2 + 11\times 0 + 6} = \frac{1}{6}$$

So, complete solution is

$$y = C_1 e^{-t} + C_2 e^{-2t} + C_3 e^{-3t} + \frac{1}{6}$$

$$\lim_{t\to\infty} y(t) = \frac{1}{6} = 0.167$$

3.33 (a)

$$u(x, t) = X(x), T(t)$$

$$u(x, 0) = \sin(\pi, x)$$

$$\Rightarrow \quad X(x) = \sin(\pi, x)$$

$$u = \sin \pi X\, T(t)$$

$$u_T = \frac{1}{\pi^2} u_{xx}$$

$$X(x)\, T'(t) = \frac{1}{\pi^2}\left[-\sin(\pi x)\right]\pi^2 . T(t)$$

$$\frac{T'(t)}{T(t)} = -1$$

$$\ln T(t) = -t + c$$

$$T(t) = e^{-t+c}$$

$$u(x, t) = \sin(\pi, x) . e^{-t+c}$$

$$u(x, 0) = \sin \pi x . e^{-c} = \sin(\pi x)$$

$$e^{-c} = 1, \ c = 0$$

$$\therefore \quad u = \sin(\pi x) . e^{-t}$$

$$\because \quad \frac{u(0.5, t)}{u(0.5, 0)} = \frac{1}{e}$$

$$\frac{\sin\left(\frac{\pi}{2}\right)e^{-t}}{\sin\left(\frac{\pi}{2}\right)1} = \frac{1}{e}$$

$$t = 1$$

3.34 (–1)

Given: $$\frac{d^2x}{dt^2} + w^2 x = 0$$

$$(D^2 + w^2) = 0$$

$$D = \pm wi$$

$$i = C_1 \cos wt + C_2 \cos wt$$

$$\frac{dx}{dt} = -C_1 w \cos wt + C_2 w \cos wt$$

Using Boundary conditions, we get

$$x = 1 \text{ at } t = 0$$

$$[1 = C_1]$$

and $$\frac{dx}{dt} = 0 \text{ at } t = 0$$

$$[0 = C_2]$$

Therefore, $x = \cos wt$

Now, at $$t = \left(\frac{3\pi}{w}\right)$$

$$x = \cos w \cdot \frac{3\pi}{w} = -1$$

3.35 (–24)

Given: $$U(r) = 4\left[\left(\frac{1}{r}\right)^{12} - \left(\frac{1}{r}\right)^{6}\right]$$

The derivative is given as

$$\frac{du}{dr} = 4[-12r^{-13} + 6r^{-7}]$$

At, $r = 1$

$$\frac{du}{dr} = 4[-12 + 6] = -24$$

3.36 (4.04)

The given ordinary differential equation is

$$x^2\frac{d^2y}{dx^2}-x\frac{dy}{dx}-3y = 0$$

Boundary conditions:

$$x = 1,\quad y = 2$$

$$x = 2,\quad y = \frac{17}{2}$$

$$x^2\frac{d^2y}{dx^2}-x\frac{dy}{dx}-3y = 0$$

Assume $x = e^z$ or $z = \log x$

$$D(D-1)y - Dy - 3y = 0$$
$$(D^2 - 2D - 3)y = 0$$
$$(D-3)(D+1)y = 0$$
$$(D-3)(D+1) = 0$$
$$D = 3, -1$$

C.F. $\Rightarrow$ $y = C_1e^{3z} + C_2e^{-z}$

Solution $y = C_1e^{3z} + C_2e^{-z}$

By putting $2 = \log x$

$$y = C_1x^3 + \frac{C_2}{x}$$

From first boundary condition

$$2 = C_1 + C_2 \quad \text{...(i)}$$

From second boundary condition

$$\frac{17}{2} = 8C_1 + \frac{C_2}{2} \quad \text{...(ii)}$$

From equation (i) and (ii)

$$C_1 = 1 \text{ and } C_2 = 1$$

So, $y = x^3 + \frac{1}{x}$

At $x = \frac{3}{2}$

$$y = \left(\frac{3}{2}\right)^3 + \frac{2}{3} = 4.04$$

■■■■

4 Complex Variables

4.1 The complex conjugate of $\frac{1}{1+i}$ is

(a) $\frac{1}{1-i}$

(b) $(1 + i)$

(c) $0.5(1 - i)$

(d) in the first quadrant of the complex plane

[2000 : 2 M]

4.2 The value of $(1 + i)^8$, where $i = \sqrt{-1}$ is

(a) $8 + 4i$ (b) $8 - 4i$

(c) 16 (d) 8

[2001 : 1 M]

4.3 In the complex plane, the angle between lines $1 + i$ and $-1 + i$ (where $i = i = \sqrt{-1}$) is

(a) $\frac{\pi}{4}$ (b) $\frac{\pi}{2}$

(c) $\frac{3\pi}{4}$ (d) π [2002 : 1 M]

4.4 The most general complex analytical function $f(z) = u(x, y) + iv(x, y)$ for $u = x^2 - y^2$ is

(a) z (b) z^2

(c) z^3 (d) $\frac{1}{z^2}$

[2003 : 2 M]

4.5 The complex number $2(1 + i)$ can be represented in polar form as

(a) $2\sqrt{2}e^{i\pi/4}$ (b) $\frac{\pi}{4}e^{i2\sqrt{2}}$

(c) $\sqrt{2}e^{i\pi/4}$ (d) $\frac{\pi}{4}e^{i/\sqrt{2}}$

[2004 : 1 M]

4.6 Given $i = \sqrt{-1}$, the ratio $\frac{(1+2i)}{(i-2)}$ is given by

(a) 1 (b) −1

(c) i (d) $-i$

[2005 : 1 M]

4.7 If $z = x + iy$ is a complex number, where $i = \sqrt{-1}$ then which of the following is an analytic function of z?

(a) $x^2 + y^2$

(b) $2ixy$

(c) $x^2 + y^2 - 2ixy$

(d) $x^2 - y^2 + 2ixy$ [2005 : 2 M]

4.8 Given $i = \sqrt{-1}$, the ratio $\frac{(i+3)}{(i+1)}$ is given by

(a) i (b) −2

(c) $-i + 2$ (d) $i + 1$

[2007 : 1 M]

4.9 An analytic function $w(z)$ is defined as $w = u + iv$, where $i = \sqrt{-1}$ and $z = x + iy$. If the real part is given by $u = \frac{y}{x^2 + y^2}$, $w(z)$ is

(a) $\frac{1}{z}$ (b) $\frac{1}{z^2}$

(c) $\frac{i}{z}$ (d) $\frac{1}{iz}$

[2008 : 2 M]

4.10 The modulus of the complex number $\frac{1+i}{\sqrt{2}}$ is

(a) $\frac{1}{2}$ (b) $\frac{1}{\sqrt{2}}$

(c) 1 (d) $\sqrt{2}$

[2009 : 1 M]

4.11 Using the residue theorem, the value of the integral (counter-clockwise) $\oint \frac{8-7z}{z-4}dz$ around a circle with center at $z = 0$ and radius = 8 (where z is a complex number and $i = \sqrt{-1}$), is

(a) $-20\pi i$ (b) -40π

(c) $-40\pi i$ (d) $40\pi i$

[2009 : 2 M]

4.12 Given that $i = \sqrt{-1}$. i^i is equal to

(a) $\frac{\pi}{2}$ (b) -1

(c) $i \ln i$ (d) $e^{-\pi/2}$ [2010 : 1 M]

4.13 If $i = \sqrt{-1}$, the value of the integral is

$$\oint_C \frac{7z+i}{z(z^2+1)}dz, \qquad |z|<2,$$

using the Cauchy residue theorem is

(a) $2\pi i$ (b) 0

(c) -6π (d) 6π [2012 : 2 M]

4.14 For the function $f(z) = \frac{1}{(2-z)(z+2)}$ the residue at $z = 2$ is ________. [2013 : 2 M]

4.15 A complex-valued function, $f(z)$, given below is analytic in domain D:

$f(z) = u(x, y) + iv(x, y) \qquad z = x + iy$

Which of the following is NOT correct?

(a) $\frac{df}{dz} = \frac{\partial v}{\partial y} + i\frac{\partial u}{\partial y}$ (b) $\frac{df}{dz} = \frac{\partial u}{\partial x} + i\frac{\partial v}{\partial x}$

(c) $\frac{df}{dz} = \frac{\partial v}{\partial y} - i\frac{\partial u}{\partial y}$ (d) $\frac{df}{dz} = \frac{\partial v}{\partial y} + i\frac{\partial v}{\partial y}$

[2015 : 1 M]

4.16 For complex variable z, the value of the contour integral $\frac{1}{2\pi i}\int_{C^z} \frac{3^{-2z}}{(z-3)}dz$ along the clockwise contour $C: |z| = 2$ (up to two decimal places) is ________. [2015 : 2 M]

4.17 What are the modulus (r) and argument (θ) of the complex number $3 + 4i$?

(a) $r = \sqrt{7}, \theta = \tan^{-1}\left(\frac{4}{3}\right)$

(b) $r = \sqrt{7}, \theta = \tan^{-1}\left(\frac{3}{4}\right)$

(c) $r = 5, \theta = \tan^{-1}\left(\frac{3}{4}\right)$

(d) $r = 5, \theta = \tan^{-1}\left(\frac{4}{3}\right)$ [2016 : 1 M]

4.18 The real part of $6\, e^{i\pi/3}$ is ________.

[2017 : 1 M]

4.19 The value of the complex number $i^{-1/2}$ (where $i = \sqrt{-1}$) is

(a) $\frac{1}{\sqrt{2}}(1-i)$ (b) $-\frac{1}{\sqrt{2}}i$

(c) $\frac{1}{\sqrt{2}}i$ (d) $\frac{1}{\sqrt{2}}(1+i)$

[2019 : 2 M]

4.20 The value of $(1 + i)^{12}$, where $i = \sqrt{-1}$, is

(a) $-4i$ (b) $64i$

(c) 64 (d) -64

[2022 : 1 M]

4.21 Given that

$$F = \frac{|z_1 + z_2|}{|z_1| + |z_2|}$$

where $z_1 = 2 + 3i$ and $z_2 = -2 + 3i$ with $i = \sqrt{-1}$ which one of the following options is CORRECT?

(a) $F < 0$ (b) $F < 1$

(c) $F > 0$ (d) $F = 1$

[2023 : 1 M]

4.22 If $z_1 = -1 + i$ and $z_2 = 2i$, where $i = \sqrt{-1}$ then $\text{Arg}\left(\frac{z_1}{z_2}\right)$ is

(a) $\frac{3\pi}{4}$ (b) $\frac{\pi}{4}$

(c) $\frac{\pi}{2}$ (d) $\frac{\pi}{3}$ [2024 : 1 M]

■■■■

Answers Complex Variables

4.1	(d)	**4.2**	(c)	**4.3**	(b)	**4.4**	(b)	**4.5**	(a)	**4.6**	(d)	**4.7**	(d)
4.8	(c)	**4.9**	(c)	**4.10**	(c)	**4.11**	(c)	**4.12**	(d)	**4.13**	(b)	**4.14**	(–0.25)
4.15	(a)	**4.16**	(0.33)	**4.17**	(d)	**4.18**	(3)	**4.19**	(a)	**4.20**	(d)	**4.21**	(b)
4.22	(b)												

Explanations Complex Variables

4.1 (d)

$$z = \frac{1}{1+i} = \frac{1-i}{(1-i)(1+i)}$$

$$= \frac{1-i}{2} = 0.5(1-i)$$

which lies in first quadrant of the complex plane.

4.2 (c)

$$(1+i)^8 = \left(\sqrt{2}\right)^8 \frac{(1+i)^8}{\left(\sqrt{2}\right)^8}$$

$$= \left(\sqrt{2}\right)^8 \left(\frac{1}{\sqrt{2}} + \frac{i}{\sqrt{2}}\right)^8$$

$$= \left(\sqrt{2}\right)^8 \left(\cos\frac{\pi}{4} + i\sin\frac{\pi}{4}\right)^8$$

$$= (2)^4 \times e^{i\times\frac{\pi}{4}\times 8}$$

$$= (2)^4 \times e^{2\pi i}$$

$$= (2)^4(\cos 2\pi + i\sin 2\pi)$$

$$= 2^4 \times 1 = 16$$

4.3 (b)

$$z_1 = 1 + i$$

$$\arg(z_1) = \tan\theta_1 = \frac{1}{1} = 1$$

$$\theta_1 = \tan^{-1}(1)$$

So, $\arg(z_1) = \dfrac{\pi}{4}$

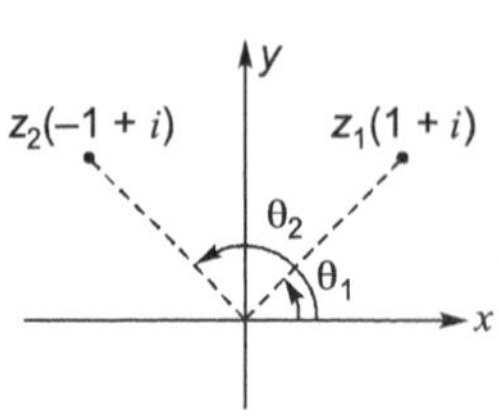

$$z_2 = -1 + i$$

$$\tan\theta_2 = \frac{1}{1} = 1$$

$$\theta_2 = \tan^{-1}(1) = \frac{\pi}{4}$$

$$\arg(z_2) = \pi - \theta = \frac{3\pi}{4}$$

So, $\arg(z_2) - \arg(z_1) = \dfrac{3\pi}{4} - \dfrac{\pi}{4} = \dfrac{\pi}{2}$

4.4 (b)

$\because$ $f(z) = u(x, y) + iv(x, y)$

$$u(x, y) = x^2 - y^2$$

So, $v(x, y) = \dfrac{\partial v}{\partial x}dx + \dfrac{\partial v}{\partial y}dy$

($\because$ for analytic $\dfrac{\partial u}{\partial x} = \dfrac{\partial v}{\partial y}$ and $\dfrac{\partial u}{\partial y} = -\dfrac{\partial v}{\partial x}$)

So, $dv = -\dfrac{\partial u}{\partial y}dx + \dfrac{\partial u}{\partial x}dy$

$\because$ $u(x, y) = x^2 - y^2$

$$u_x = 2x,\ u_y = -2y$$

$$dv = 2ydx + 2xdy$$

$$v = \int 2ydx + \int_{\substack{\text{free}\\ \text{from } x}} 2xdy$$

$$v = 2xy + 0 + C$$

$$v = 2xy + C$$

So, $f(z) = x^2 - y^2 + 2xy + C$

$$= x^2 + (iy)^2 + 2xyi + C$$

$$f(z) = (x + iy)^2 + C$$

$$f(z) = z^2 + C$$

Alternate solution:

$\because$ $f(z) = u(x, y) + iv(x, y)$

$$f'(z) = \frac{\partial u}{\partial x} + i\frac{\partial v}{\partial x}$$

$$f'(z) = \frac{\partial u}{\partial x} - i\frac{\partial u}{\partial y}$$

$\left\{\because \text{ For analytic } \dfrac{\partial u}{\partial x} = \dfrac{\partial v}{\partial y} \text{ and } \dfrac{\partial u}{\partial y} = \dfrac{-\partial v}{\partial x}\right\}$

$$f'(z) = \frac{\partial u}{\partial x} - i\frac{\partial u}{\partial y}$$

$$f'(z) = 2x - i(-2y)$$

$$f'(z) = 2x + i2y$$

Put $x = z$, $y = 0$

$$f'(z) = 2z$$

$$f(z) = z^2 + C$$

4.5 (a)

$$2(1+i) = 2+2i = re^{i\theta}$$

$$r = |z| = \sqrt{2^2+2^2} = 2\sqrt{2}$$

$$\theta = \tan^{-1}\left(\frac{y}{x}\right) = \tan^{-1}\left(\frac{2}{2}\right) = \frac{\pi}{4}$$

So, $z = re^{i\theta} = 2\sqrt{2}e^{i\times\frac{\pi}{4}}$

4.6 (d)

$$\frac{(1+2i)}{(i-2)} = \frac{(1+2i)}{(i-2)} \times \frac{(i+2)}{(i+2)}$$

$$\Rightarrow \frac{i+2+2i^2+4i}{i^2-4} = \frac{i+2-2+4i}{-5} = \frac{5i}{-5} = -i$$

4.7 (d)

$$f(x) = x^2 - y^2 + 2xyi$$

$$u = x^2 - y^2,\ v = 2xy$$

$$u_x = 2x,\ v_y = 2x$$

$$u_y = -2y,\ v_x = 2y$$

So, $u_x = v_y = 2x$

$$u_y = -v_x = -2y$$

Hence, $f(x) = x^2 - y^2 + 2xyi$ is an analytic function.

4.8 (c)

$$\frac{(i+3)}{(i+1)} \times \frac{(1-i)}{(1-i)} = \frac{i-i^2+3-3i}{1-i^2}$$

$$= \frac{i+1+3-3i}{2}$$

$$= \frac{4-2i}{2} = 2-i$$

4.9 (c)

$$w = u + iv$$

where, $u = \frac{y}{x^2+y^2}$

$$w'(z) = u_x + iv_x$$

(For analytic $u_x = v_y,\ u_y = -v_x$)

$$w'(z) = u_x - iu_y$$

$$u_x = \frac{(x^2+y^2)\cdot(0) - y(2x)}{(x^2+y^2)^2}$$

$$u_x = -\frac{2xy}{(x^2+y^2)^2}$$

$$u_y = \frac{(x^2+y^2)\times 1 - y(2y)}{(x^2+y^2)^2}$$

$$= \frac{x^2+y^2-2y^2}{(x^2+y^2)^2} = \frac{x^2-y^2}{(x^2+y^2)^2}$$

$$w'(z) = \frac{-2xy}{(x^2+y^2)^2} - i\frac{(x^2-y^2)}{(x^2+y^2)^2}$$

Put, $x = z,\quad y = 0$

$$w'(z) = \frac{-iz^2}{z^4} = \frac{-i}{z^2}$$

$$w(z) = \int \frac{-i}{z^2}dz + C$$

$$w(z) = \frac{-i \times z^{-1}}{-1} + C$$

$$w(z) = \frac{i}{z} + C$$

4.10 (c)

$$|z| = \sqrt{\left(\frac{1}{\sqrt{2}}\right)^2 + \left(\frac{1}{\sqrt{2}}\right)^2} = 1$$

4.11 (c)

$$\oint \frac{8-7z}{z-4}dz\ ;\ C: |z| = 8$$

$z = 4$ is a pole which lies inside C.

$$\underset{z=4}{\text{Res}} = \lim_{z\to 4}(z-4)\frac{8-7z}{z-4}$$

$$= 8 - 7\times 4 = -20$$

So, $\oint \frac{8-7z}{z-4}dz = 2\pi i(-20) = -40\pi i$

4.12 (d)

$$i^i = \left(\cos\frac{\pi}{2} + i\sin\frac{\pi}{2}\right)^i$$

$$= \left(e^{i\frac{\pi}{2}}\right)^i = e^{i^2\times\frac{\pi}{2}} = e^{-\frac{\pi}{2}}$$

4.13 (b)

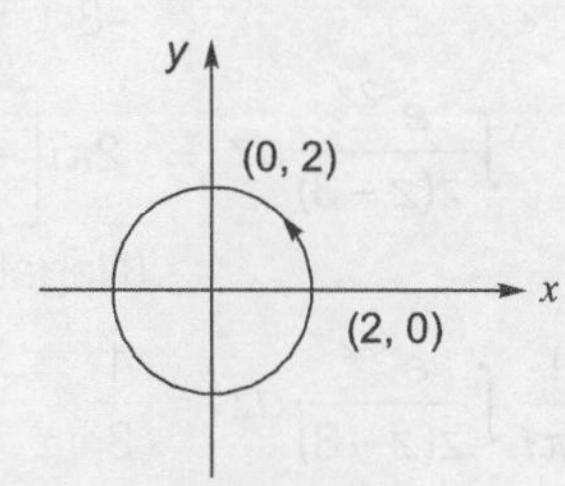

$$\oint \frac{7z+i}{z(z^2+1)}dz = \oint \frac{7z+i}{z(z+i)(z-i)}dz$$

Poles, $z = 0, i, -i$

$$\underset{z=0}{\text{Res}} = \lim_{z\to 0}(z-0)\frac{7z+i}{z(z+i)(z-i)} = \frac{i}{1} = i$$

$$\underset{z=i}{\text{Res}} = \lim_{z\to i}(z-i)\frac{7z+i}{z(z-i)(z+i)} = \frac{7i+i}{i(2i)} = -4i$$

$$\underset{z=-i}{\text{Res}} = \lim_{z\to -i}(z+i)\frac{7z+i}{z(z-i)(z+i)} = \frac{-7i+i}{-i(-2i)} = \frac{-6i}{-2} = 3i$$

So,

$$\oint\frac{7z+i}{z(z^2+1)}dz = 2\pi i\,(\text{sum of residue inside } c)$$

$$= 2\pi i\,(i-4i+3i) = 4\pi i\times 0 = 0$$

4.14 (–0.25)

Given, $$f(z) = \frac{1}{(2-z)(z+2)} = \frac{-1}{(z-2)(z+2)}$$

$$\underset{z=2}{\text{Res}} = \lim_{z\to 2}(z-2)\times\left[\frac{-1}{(z-2)(z+2)}\right]$$

$$= -\frac{1}{4} = -0.25$$

4.15 (a)

$$f(z) = u+iv$$

$$\frac{\partial f}{\partial z} = \frac{\partial u}{\partial x}+i\frac{\partial v}{\partial x} = \frac{\partial v}{\partial y}-i\frac{\partial u}{\partial y} = \frac{\partial v}{\partial y}+i\frac{\partial v}{\partial x}$$

4.16 (0.33)

$$\frac{1}{2\pi i}\int_C\frac{e^{-2z}}{z(z-3)}dz;\ |z|=2$$

Poles: $z(z-3) = 0$

$z = 0, z = 3$

$\because$ Pole $z = 0$ lies inside C

$$\therefore\quad \underset{z=0}{\text{Res}} = \lim_{z\to 0}(z-0)\times\frac{e^{-2z}}{z(z-3)}$$

$$= \frac{1}{-3} = \frac{-1}{3}$$

So, $$\int\frac{e^{-2z}}{z(z-3)}dz = 2\pi i\left[-\left(-\frac{1}{3}\right)\right]$$

(Negative for clockwise)

$$\Rightarrow \frac{1}{2\pi i}\int\frac{e^{-2z}}{z(z-3)}dz = \frac{1}{3}$$

4.17 (d)

$$z = 3+4i$$

$$r = \sqrt{3^2+4^2} = 5$$

$$\theta = \tan^{-1}\left(\frac{4}{3}\right)$$

4.18 (3)

$$6e^{i\pi/3} = 6\left(\cos\frac{\pi}{3}+i\sin\frac{\pi}{3}\right)$$

$$= 6\left(\frac{1}{2}+i\times\frac{\sqrt{3}}{2}\right) = 3+i3\sqrt{3}$$

Real part = 3

4.19 (a)

$$i^{-1/2} = \left(\cos\frac{\pi}{2}+i\sin\frac{\pi}{2}\right)^{-1/2}$$

$$= e^{\left(i\times\frac{\pi}{2}\right)^{-1/2}} = e^{-i\times\frac{\pi}{4}}$$

$$= \cos\left(-\frac{\pi}{4}\right)+i\sin\left(-\frac{\pi}{4}\right)$$

$$= \frac{1}{\sqrt{2}}-\frac{i}{\sqrt{2}} = \frac{(1-i)}{\sqrt{2}}$$

4.20 (d)

$$(1+i)^{12} = [(1+i)^2]^6$$
$$= [1^2+i^2+2i]^6$$
$$= [1-1+2i]^6$$
$$= 2^6\times i^6$$
$$= 64\times i^2\times i^4$$
$$= 64\times(-1)\times 1 = -64$$

4.21 (b)

Given: $z_1 = 2+3i$

$$|z_1| = \sqrt{4+9} = \sqrt{13}$$

$$z_2 = -2+3i$$

$$|z_2| = \sqrt{(-2)^2+9} = \sqrt{13}$$

$$z_1+z_2 = 0+6\hat{i}$$

$$\Rightarrow\quad |z_1+z_2| < |z_1|+|z_2|$$

Hence, $F < 1$

4.22 (b)

$z_1 = -1+i$ and $z_2 = 2i$

$$\frac{z_1}{z_2} = \frac{(-1+i)}{2i} = \frac{1}{2}\left[\frac{(-1+i)i}{i^2}\right]$$

$$= \frac{-1}{2}[-1-i] = \frac{1}{2}+\frac{i}{2}$$

$$\text{Arg}\left(\frac{z_1}{z_2}\right) = \tan^{-1}\left(\frac{\frac{1}{2}}{\frac{1}{2}}\right) = \tan^{-1}1 = \frac{\pi}{4}$$

■■■■

5 Probability and Statistics

5.1 A pair of fair dice is rolled simultaneously. The probability that the sum of the numbers from the dice equals six is

(a) $\frac{1}{6}$ (b) $\frac{7}{36}$

(c) $\frac{5}{36}$ (d) $\frac{1}{12}$ [2000 : 1 M]

5.2 A fair die is rolled four times. Find the probability that six shows up twice.

(a) $\frac{1}{2}$ (b) $\frac{16}{325}$

(c) $\frac{1}{36}$ (d) $\frac{25}{216}$ [2001 : 2 M]

5.3 A box contains 6 red balls and 4 green balls, one ball is randomly picked and then a second ball is picked without replacement of the first ball. The probability that both the balls are green is

(a) $\frac{1}{15}$ (b) $\frac{2}{25}$

(c) $\frac{2}{15}$ (d) $\frac{4}{25}$ [2003 : 1 M]

5.4 A box contains three blue balls and four red balls. Another identical box contains two blue balls and five red balls. One ball is picked at random from one of the two boxes and it is red. The probability that it came from the first box is

(a) $\frac{2}{3}$ (b) $\frac{4}{9}$

(c) $\frac{4}{7}$ (d) $\frac{2}{7}$ [2004 : 2 M]

5.5 Two bags contain ten coins each, and the coins in each bag are numbered from 1 to 10. One coin is drawn at random from each bag. The probability that one of the coins has value 1, 2, 3 or 4, while the other has value 7, 8, 9 or 10, is

(a) $\frac{2}{5}$ (b) $\frac{4}{25}$

(c) $\frac{8}{25}$ (d) $\frac{16}{25}$ [2005 : 1 M]

5.6 A pair of fair dice is rolled three times. What is the probability that 10 (sum of the numbers on the two faces) will show up exactly once?

(a) $\frac{121}{1728}$ (b) $\frac{363}{1728}$

(c) $\frac{121}{576}$ (d) $\frac{363}{576}$

[2006 : 2 M]

5.7 The thickness of a conductive coating in micrometers has a probability density function of $600x^{-2}$ for $100\ \mu m < x < 120\ \mu m$. The mean and the variance of the coating thickness is

(a) 1 μm, 108.39 μm² (b) 33.83 μm, 1 μm²

(c) 105 μm, 11 μm² (d) 109.39 μm, 33.83 μm²

[2007 : 2 M]

5.8 The normal distribution is given by

$$f(x) = \frac{1}{\sqrt{2\pi}\sigma}\exp\left(-\frac{(x-\mu)^2}{2\sigma^2}\right), \quad -\infty < x < \infty$$

The points of inflexion to the normal curve are

(a) $x = -\sigma, \sigma$ (b) $x = \mu + \sigma, \mu - \sigma$

(c) $x = \mu + 2\sigma, \mu - 2\sigma$ (d) $x = \mu + 3\sigma, \mu - 3\sigma$

[2008 : 2 M]

5.9 The Poisson distribution is given by $P(r) = \frac{m^r}{r!}\exp(-m)$. The first moment about the origin for this distribution is

(a) 0 (b) m

(c) $\frac{1}{m}$ (d) m^2 [2008 : 2 M]

5.10 A fair die is rolled. Let R denote the event of obtaining a number less than or equal to 5 and S denote the event of obtaining an odd number. Then which one of the following about the probability (P) is true?

(a) $P\left(\frac{R}{S}\right) = 1$ (b) $P\left(\frac{R}{S}\right) = 0$

(c) $P\left(\frac{S}{R}\right) = 1$ (d) $P\left(\frac{S}{R}\right) = 0$

[2009 : 2 M]

5.11 A box contains three red and two black balls. Four balls are removed from the box one by one, without replacement. The probability of the ball remaining in the box being red, is

(a) $\frac{609}{225}$ (b) $\frac{3}{5}$
(c) $\frac{2}{5}$ (d) $\frac{81}{625}$ [2010 : 2 M]

5.12 The number of emails received on six consecutive days is 11, 9, 18, 18, 4 and 15, respectively. What are the median and the mode for these data?
(a) 18 and 11, respectively
(b) 13 and 18, respectively
(c) 13 and 12.5, respectively
(d) 12.5 and 18, respectively [2013 : 1 M]

5.13 For two rolls of a fair die, the probability of getting a 4 in the first roll and a number less than 4 in the second roll, up to 3 digits after the decimal point, is _______. [2013 : 1 M]

5.14 Consider the following two normal distributions:

$$f_1(x) = \exp(-\pi x^2)$$

$$f_2(x) = \frac{1}{2\pi}\exp\left(-\frac{1}{4\pi}(x^2 + 2x + 1)\right)$$

If μ and σ denote the mean and standard deviation respectively then
(a) $\mu_1 < \mu_2$ and $\sigma_1^2 < \sigma_2^2$ (b) $\mu_1 < \mu_2$ and $\sigma_1^2 > \sigma_2^2$
(c) $\mu_1 > \mu_2$ and $\sigma_1^2 < \sigma_2^2$ (d) $\mu_1 > \mu_2$ and $\sigma_1^2 > \sigma_2^2$
[2014 : 2 M]

5.15 In rolling of two fair dice, the outcome of an experiment is considered to be the sum of the numbers appearing on the dice. The probability is highest for the outcome of _______.
[2014 : 2 M]

5.16 The marks obtained by a set of students are:
38, 84, 45, 70, 75, 60, 48
The mean and median marks, respectively, are
(a) 45 and 75 (b) 55 and 48
(c) 60 and 60 (d) 60 and 70
[2017 : 1 M]

5.17 A box has 6 red balls and 4 white balls. A ball is picked at random and replaced in the box, after which a second ball is picked.
The probability of both the balls being red, rounded to 2 decimal places, is _______. [2017 : 2 M]

5.18 Two unbiased dice are thrown. Each dice can show any number between 1 and 6. The probability that the sum of the outcomes of the two dice is divisible by 4 is _____. (Rounded off to two decimal places).
[2019 : 2 M]

5.19 In a box, there are 5 green balls and 10 blue balls. A person picks 6 balls randomly. The probability that the person has picked 4 green balls and 2 blue balls is
(a) $\frac{42}{1001}$ (b) $\frac{45}{1001}$
(c) $\frac{240}{1001}$ (d) $\frac{420}{1001}$ [2020 : 2 M]

5.20 The probability distribution function of a random variable X is shown in the following figure.

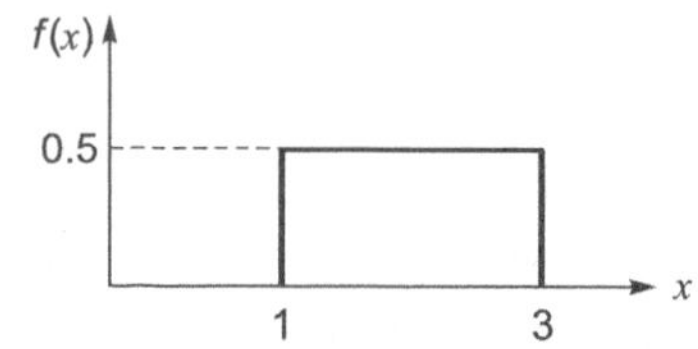

From this distribution, random samples with sample size $n = 68$ are taken. If $\bar{X}$ is the sample mean, the standard deviation of the probability distribution of $\bar{X}$, i.e., $\sigma_{\bar{X}}$ is _____ (round off to 3 decimal places). [2021 : 2 M]

5.21 An exhibition was held in a hall on 15 August 2022 between 3 PM and 4 PM during which any person was allowed to enter only once. Visitors who entered before 3:40 PM exited the hall exactly after 20 minutes from their time of entry. Visitors who entered at or after 3:40 PM, exited exactly at 4 PM. The probability distribution of the arrival time of any visitor is uniform between 3 PM and 4 PM. Two persons X and Y entered the exhibition hall independent of each other. Which one of the following values is the probability that their visits to the exhibition overlapped with each other?
(a) $\frac{5}{9}$ (b) $\frac{4}{9}$
(c) $\frac{2}{9}$ (d) $\frac{7}{9}$ [2023 : 2 M]

5.22 Consider the normal probability distribution function

$$f(x) = \frac{4}{\sqrt{2\pi}}e^{-8(x+3)^2}$$

If μ and σ are the mean and standard deviation of $f(x)$ respectively, then the ordered pair (μ, σ) is
(a) $\left(3, \frac{1}{4}\right)$ (b) $\left(-3, \frac{1}{4}\right)$
(c) (3, 4) (d) (–3, 4) [2024 : 1 M]

■■■■

Answers Probability and Statistics

5.1 (c)	5.2 (d)	5.3 (c)	5.4 (b)	5.5 (b)	5.6 (c)	5.7 (d)
5.8 (b)	5.9 (b)	5.10 (a)	5.11 (b)	5.12 (b)	5.13 (0.083)	5.14 (c)
5.15 (7)	5.16 (c)	5.17 (0.36)	5.18 (0.25)	5.19 (b)	5.20 (0.5773)	5.21 (a)
5.22 (b)						

Explanations Probability and Statistics

5.1 (c)

Possible outcomes = {(1, 5), (2, 4), (3, 3), (4, 2), (5, 1)}

$$P = \frac{5}{36}$$

5.2 (d)

$$p(6) = \frac{1}{6},\ q = 1-\frac{1}{6} = \frac{5}{6}$$

So, probability of getting six twice

$$= {}^4C_2 p^2 q^2$$

$$= \frac{4!}{2!2!}\times\left(\frac{1}{6}\right)^2\left(\frac{5}{6}\right)^2$$

$$= 6\times\frac{1}{36}\times\frac{25}{36} = \frac{25}{216}$$

5.3 (c)

∵ This is hypergeometric distribution.

So, required probability is

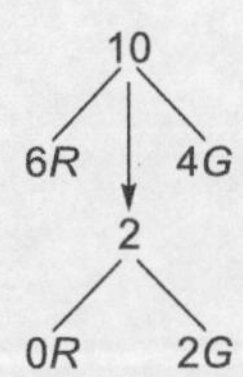

$$= \frac{{}^4C_2\times{}^6C_0}{{}^{10}C_2} = \frac{6\times 1}{45} = \frac{2}{15}$$

5.4 (b)

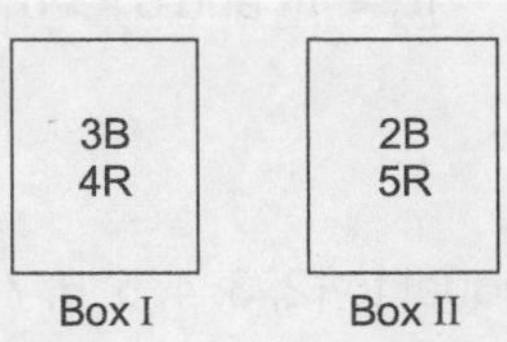

$$P\left(\frac{I}{R}\right) = \frac{P\left(\frac{R}{I}\right)\cdot P(I)}{P\left(\frac{R}{I}\right)\cdot P(I)+P\left(\frac{R}{II}\right)\cdot P(II)}$$

$$= \frac{\frac{1}{2}\times\frac{4}{7}}{\frac{1}{2}\times\frac{4}{7}+\frac{1}{2}\times\frac{5}{7}}$$

$$= \frac{\frac{4}{14}}{\frac{4}{14}+\frac{5}{14}} = \frac{4}{14}\times\frac{14}{9} = \frac{4}{9}$$

5.5 (b)

Required probability

$$= \frac{4}{10}\times\frac{4}{10} = \frac{16}{100} = \frac{4}{25}$$

5.6 (c)

$$P(\text{sum} = 10) = \frac{13-\text{sum}}{36}$$

$$= \frac{13-10}{36} = \frac{3}{36} = \frac{1}{12}$$

$$q = 1-\frac{1}{12} = \frac{11}{12}$$

So, getting once in three time rolling

$$= {}^3C_1(p)^1(q)^2$$

$$= {}^3C_1\left(\frac{1}{12}\right)^1\left(\frac{11}{12}\right)^2$$

$$= 3\times\frac{1}{12}\times\frac{121}{144} = \frac{121}{576}$$

5.7 (d)

$$f(x) = \frac{600}{x^2};\quad 100 < x < 120$$

$$f(x) = \int_{100}^{120} x\cdot\frac{600}{x^2}dx$$

$$= \int_{100}^{120}\frac{600}{x}dx = (600\ln(x))_{100}^{120}$$

$$= 600\ln\left(\frac{120}{100}\right) = 109.39\ \mu\text{m}$$

$$f(x^2) = \int_{100}^{120} x^2\cdot\frac{600}{x^2}dx = \int_{100}^{120}600dx$$

$$= 600\times[x]_{100}^{120}$$

$$= 600\times 20 = 12{,}000\ \mu\text{m}$$

$$\text{Var}(x) = E(x^2)-[E(x)]^2$$

$$= 12000-(109.39)^2 = 33.185\ \mu\text{m}^2$$

5.8 (b)

Given: $$f(x) = \frac{1}{\sqrt{2\pi}\sigma} e^{-\frac{1}{2}\left\{\left(\frac{x-\mu}{\sigma}\right)^2\right\}}$$

Point of inflexion on the curve will be where

$$f''(x) = 0$$

So, $$f'(x) = \frac{1}{\sqrt{2\pi}\sigma} e^{-\frac{1}{2}\left\{\left(\frac{x-\mu}{\sigma}\right)^2\right\}} \cdot \frac{-1}{2} \times 2\left\{\frac{x-\mu}{\sigma^2}\right\}$$

Consider, $$\frac{1}{\sqrt{2\pi}\sigma} = k$$

$$f'(x) = ke^{-\frac{1}{2}\left(\frac{x-\mu}{\sigma}\right)^2} - \left(\frac{x-\mu}{\sigma^2}\right)$$

$$f''(x) = k\left(-\frac{1}{\sigma^2}\right)e^{-\frac{1}{2}\left(\frac{x-\mu}{\sigma}\right)^2} + ke^{-\frac{1}{2}\left(\frac{x-\mu}{\sigma}\right)^2} \times \left(\frac{x-\mu}{\sigma^2}\right)^2 = 0$$

$$\Rightarrow ke^{-\frac{1}{2}\left(\frac{x-\mu}{\sigma}\right)^2}\left[\frac{-1}{\sigma^2} + \left(\frac{x-\mu}{\sigma^2}\right)^2\right] = 0$$

$$\frac{1}{\sigma^2} = \frac{(x-\mu)^2}{\sigma^4}$$

$$(x-\mu)^2 = \sigma^2$$

$$x - \mu = \pm\sigma$$

$$x = \mu \pm \sigma$$

5.9 (b)

First moment = Mean of Poisson distribution = m

5.10 (a)

$$R = \{1, 2, 3, 4, 5\};\ S = \{1, 3, 5\}$$

$$\text{(a) } P(R/S) = \frac{P(R \cap S)}{P(S)} = \frac{P(1,3,5)}{\frac{3}{6}} = \frac{\frac{3}{6}}{\frac{3}{6}} = 1$$

$$\text{(b) } P(S/R) = \frac{P(S \cap R)}{P(R)} = \frac{\frac{3}{6}}{\frac{5}{6}} = \frac{3}{5}$$

5.11 (b)

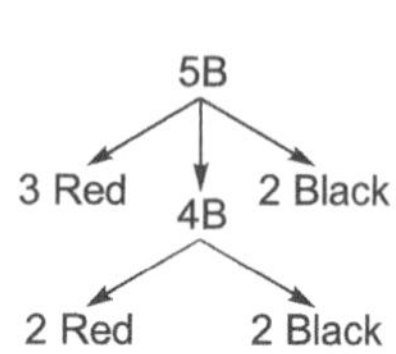

This is the case of hypegeometric distribution.

So, required probability $= \frac{{}^3C_2 \times {}^2C_2}{{}^5C_4} = \frac{3 \times 1}{5} = \frac{3}{5}$

5.12 (b)

Arrange in ascending order: 4, 9, 11, 15, 18, 18

$$\text{Mode} = 18$$

$$\text{Median} = \frac{11+15}{2} = \frac{26}{2} = 13$$

5.13 (0.083)

Required probability $= \frac{1}{6} \times \frac{3}{6} = \frac{3}{36} = \frac{1}{12} = 0.083$

5.14 (c)

$$f_1(x) = e^{-\pi x^2}$$

Normal distribution

$$f(x) = \frac{1}{\sqrt{2\pi}\,\sigma} e^{-\frac{1}{2}\left(\frac{x-\mu}{\sigma}\right)^2}$$

$$\frac{1}{\sqrt{2\pi}\,\sigma} = 1,\ (x-\mu)^2 = x^2$$

$$\sigma_1 = \sqrt{2\pi},\ \mu_1 = 0$$

$$f_2(x) = \frac{1}{2\pi} e^{-\frac{(x+1)^2}{4\pi}}$$

$$\frac{1}{\sqrt{2\pi}\,\sigma} = \frac{1}{2\pi}$$

$$\sigma_2 = \sqrt{2\pi}$$

$$(x-\mu)^2 = (x+1)^2$$

$$\mu_2 = -1$$

$$\because \quad \mu_2 < \mu_1 \text{ and } \sigma_2^2 > \sigma_1^2$$

5.15 (7)

Case-1:

For sum equal to {2, 3, 4, 5, 6, 7}

Required probability $= \frac{\text{Sum} - 1}{36}$

Case-2:

For sum equal to {8, 9, 10, 11, 12}

Required probability $= \frac{13 - \text{Sum}}{36}$

For the above two case we can observe that for sum equal to 7 the probability

$$= \frac{\text{Sum} - 1}{36} = \frac{7-1}{36} = \frac{1}{6}$$

Which will be highest from the rest other cases.

5.16 (c)

Arranging in ascending order 38, 45, 48, 60, 70, 75, 84

$$\text{Mean} = \frac{38+45+48+60+70+75+84}{7} = 60$$

Median = 60

5.17 (0.36)

$$P(R_1) = \frac{^6C_1}{^{10}C_1} = \frac{6}{10}$$

$$P(R_2) = \frac{^6C_1}{^{10}C_1} = \frac{6}{10}$$

$$\text{So, required probability} = \frac{6}{10}\times\frac{6}{10} = \frac{36}{100} = 0.36$$

5.18 (0.25)

The possible outcomes are such that the sum can be 4, 8, 12.

$$P(\text{sum} = 4) = \frac{\text{Sum}-1}{36} = \frac{4-1}{36} = \frac{3}{36}$$

$$P(\text{sum} = 8) = \frac{13-\text{Sum}}{36} = \frac{13-8}{36} = \frac{5}{36}$$

$$P(\text{sum} = 12) = \frac{13-12}{36} = \frac{1}{36}$$

$$\text{So, required probability} = \frac{3}{36}+\frac{5}{36}+\frac{1}{36}$$

$$= \frac{9}{36} = \frac{1}{4} = 0.25$$

5.19 (b)

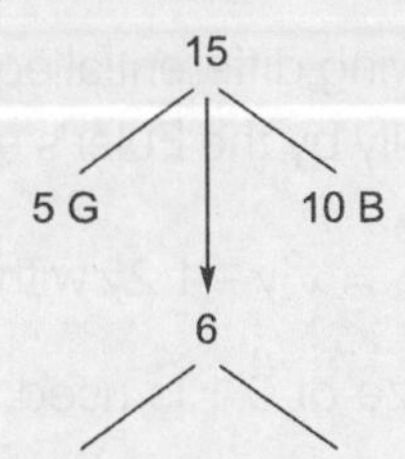

$$\text{Required probability} = \frac{^{10}C_2 \times {}^5C_4}{^{15}C_6} = \frac{45\times5}{5005} = \frac{45}{1001}$$

5.20 (0.5773)

For uniform distribution

$$f(x) = \begin{cases} \dfrac{1}{\beta-\alpha} & \alpha \le x \le \beta \\ 0 & \text{Otherwise} \end{cases}$$

$$f(x) = \begin{cases} 0.5 & 1 \le x \le 3 \\ 0 & \text{Otherwise} \end{cases}$$

$$\text{Mean} = \frac{\alpha+\beta}{2} = \frac{1+3}{2} = 2$$

$$\text{Variance} = \frac{(\beta-\alpha)^2}{12} = \frac{(3-1)^2}{12} = \frac{4}{12} = \frac{1}{3}$$

$$\text{Standard deviation} = \sqrt{\frac{1}{3}} = 0.5773$$

5.21 (a)

Given case can be expressed as:

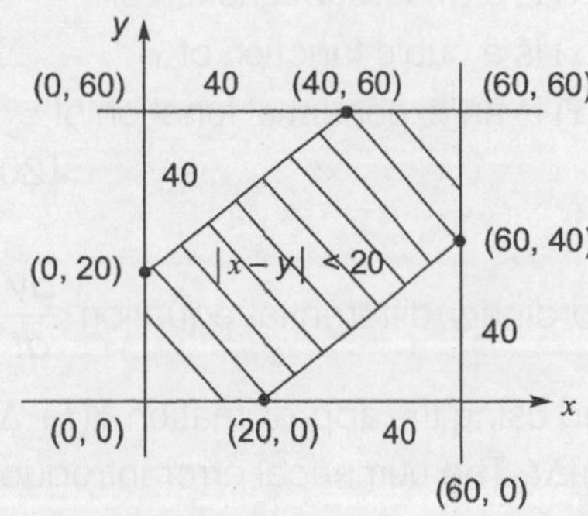

Let arrival time of two persons be x and y then according to question $|x-y| < 20$

or $-20 < x - y < 20$

Shaded area is favourable area

Total area = 60 × 60 = 3600

$$\text{Unfavourable area} = 2\times\frac{1}{2}\text{ Base}\times\text{Height}$$

$$= 40 \times 40 = 1600$$

So, Favourable area = 3600 – 1600 = 2000

$$\text{Hence } P(|x-y| < 20) = \frac{\text{Favourable area}}{\text{Total area}}$$

$$= \frac{2000}{3600} = \frac{5}{9}$$

5.22 (b)

Normal probability distribution function is given as $f(x) = \dfrac{4}{\sqrt{2\pi}}e^{-8(x+3)^2}$

Normal distribution function is $\dfrac{1}{\sqrt{2\pi\sigma^2}}e^{\frac{-(x-\mu)^2}{2\sigma^2}}$

and comparing the above function.

$$\frac{1}{\sqrt{2\pi\sigma^2}} = \frac{4}{\sqrt{2\pi}} \Rightarrow \sigma = \frac{1}{4}$$

and $\mu = -3$

6 Numerical Methods

6.1 The function $f(x)$ satisfies the equation $f(x) = 0$ at $x = x_c$. The Newton-Raphson iterative method converges to the solution in one step, regardless of the initial guess, if

(a) $f(x)$ is a linear function of x

(b) $f(x)$ is a quadratic function of x

(c) $f(x)$ is a cubic function of x

(d) $f(x)$ is an exponential function of x

[2005 : 2 M]

6.2 The ordinary differential equation $\frac{dy}{dt} = f(Y)$ is solved using the approximation $Y(t + \Delta t) = Y(t) + f[Y(t)]\Delta t$. The numerical error introduced by the approximation at each step is

(a) proportional to Δt

(b) proportional to $(\Delta t)^2$

(c) independent of Δt

(d) proportional to $\left(\frac{1}{\Delta t}\right)$

[2006 : 1 M]

6.3 The trapezoidal rule of integration when applied to $\int_a^b f(x)dx$ will give the exact value of the integral

(a) if $f(x)$ is a linear function of x

(b) if $f(x)$ is a quadratic function of x

(c) for any $f(x)$

(d) for no $f(x)$

[2006 : 1 M]

6.4 The Newton-Raphson method is used to solve the equation, $(x - 1)^2 + x - 3 = 0$. The method will fail in the very first iteration if the initial guess is

(a) Zero (b) 0.5

(c) 1 (d) 3

[2006 : 2 M]

6.5 A non-linear function $f(x)$ is defined in the interval $-1.2 < x < 4$ as illustrated in the figure below. The equation $f(x) = 0$ is solved for x within this interval by using the Newton-Raphson iterative scheme. Among the initial guesses (I_1, I_2, I_3 and I_4), the guess that is likely to lead to the root most rapidly is

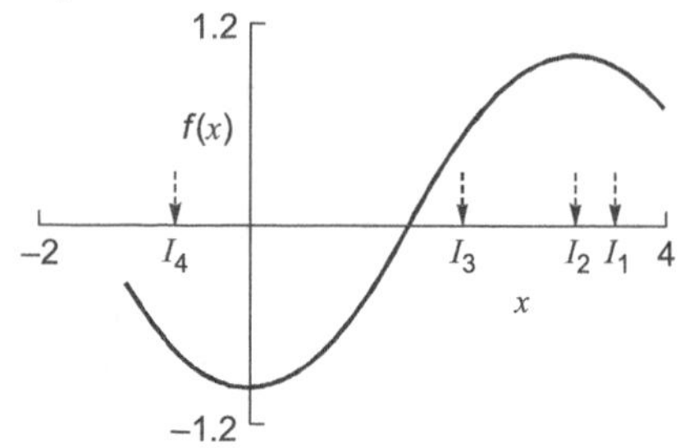

(a) I_1 (b) I_2

(c) I_3 (d) I_4

[2008 : 1 M]

6.6 Using Simpson's $\frac{1}{3}$ rule and four equally spaced intervals ($n = 4$), estimate the value of the integral

$$\int_0^{\pi/4} \frac{\sin x}{\cos^3 x} dx$$

(a) 0.3887 (b) 0.4384

(c) 0.5016 (d) 0.5527

[2008 : 2 M]

6.7 The following differential equation is to be solved numerically by the Euler's explicit method

$$\frac{dy}{dx} = x^2 y - 1.2y \text{ with } y(0) = 1$$

A step size of 0.1 is used. The solution for y at $x = 0.1$ is

(a) 0.880 (b) 0.905

(c) 1.000 (d) 1.100

[2008 : 2 M]

6.8 Using the trapezoidal rule and 4 equal intervals ($n = 4$), the calculated value of the integral (rounded to the first place of decimal) $\int_0^{\pi} \sin\theta d\theta$ is

(a) 1.7 (b) 1.9

(c) 2.0 (d) 2.1

[2009 : 2 M]

6.9 A root of the equation $x^4 - 3x + 1 = 0$ needs to be found using the Newton-Raphson method. If the initial guess, x_0, is taken as 0, then the new estimate, x_1 after the first iteration is

(a) $\frac{1}{3}$ (b) $-\frac{1}{3}$

(c) 3 (d) –3

[2010 : 1 M]

6.10 The Newton Raphson method is used to find the roots of the equation:

$$f(x) = x - \cos \pi x; \quad 0 \le x \le 1$$

If the initial guess for the root is 0.5, then the value of x after the first iteration is

(a) 1.02 (b) 0.62

(c) 0.55 (d) 0.38

[2012 : 2 M]

6.11 The value of the integral $\int_{0.1}^{0.5} e^{-x^3} dx$ evaluated by Simpson's rule using 4 subintervals (up to 3 digits after the decimal point) is ______.

[2013 : 2 M]

6.12 Consider the following differential equation:

$$\frac{dy}{dx} = x + \ln(y); \; y = 2 \text{ at } x = 0$$

The solution of this equation at $x = 0.4$ using Euler method with a step size of $h = 0.2$ is ______.

[2014 : 2 M]

6.13 The solution of the non-linear equation $x^3 - x = 0$ is to be obtained using Newton-Raphson method. If the initial guess is $x = 0.5$, the method converges to which one of the following values:

(a) –1 (b) 0

(c) 1 (d) 2

[2015 : 2 M]

6.14 Which one of the following is an iterative technique for solving a system of simultaneous linear algebraic equations?

(a) Gauss elimination (b) Gauss-Jordan

(c) Gauss-Seidel (d) LU decomposition

[2016 : 1 M]

6.15 Values of $f(x)$ in the interval [0, 4] are given below.

x	0	1	2	3	4
$f(x)$	3	10	21	36	55

Using Simpson's $\frac{1}{3}$ rule with a step size of 1, the numerical approximation (rounded off to the second decimal place) of $\int_0^4 f(x)dx$ is ______.

[2016 : 2 M]

6.16 Match the problem type in **Group-1** with the numerical method in **Group-2**.

Group-1

P. System of linear algebraic equations

Q. Non-linear algebraic equations

R. Ordinary differential equations

S. Numerical integration

Group-2

I. Newton-Raphson

II. Gauss-Seidel

III. Simpson's rule

IV. Runge-Kutta

Choose the correct set of combinations.

(a) P-II, Q-I, R-III, S-IV

(b) P-I, Q-II, R-IV, S-III

(c) P-IV, Q-III, R-II, S-I

(d) P-II, Q-I, R-IV, S-III

[2017 : 2 M]

6.17 The fourth order Runge-Kutta (RK4) method to solve an ordinary differential equation $\frac{dy}{dx} = f(x, y)$ is given as

$$y(x+h) = y(x) + \frac{1}{6}(k_1 + 2k_2 + 2k_3 + k_4)$$

$$k_1 = hf(x, y)$$

$$k_2 = hf\left(x + \frac{h}{2}, y + \frac{k_1}{2}\right)$$

$$k_3 = hf\left(x + \frac{h}{2}, y + \frac{k_2}{2}\right)$$

$$k_4 = hf(x + h, y + k_3)$$

For a special case when the function f depends solely on x, the above RK4 method reduces to

(a) Euler's explicit method

(b) Trapezoidal rule

(c) Euler's implicit method

(d) Simpson's $\frac{1}{3}$ rule

[2018 : 1 M]

6.18 The Newton-Raphson method is used to determine the root of the equation $f(x) = e^{-x} - x$. If the initial guess for the root is 0, the estimate of the root after two iterations is _______ (rounded off to three decimal places).

[2019 : 2 M]

6.19 Consider the following data set:

x	1	3	5	15	25
$f(x)$	6	8	10	12	5

The value of the integral $\int_1^{25} f(x)\,dx$ using Simpson's 1/3rd rule is ____ (round off to 1 decimal place.)

[2020 : 2 M]

6.20 An ordinary differential equation (ODE), $\frac{dy}{dx} = 2y$, with an initial condition $y(0) = 1$, has the analytical solution $y = e^{2x}$.

Using Runge-Kutta second order method, numerically integrate the ODE to calculate y at $x = 0.5$ using a step size of $h = 0.5$.

If the relative percentage error is defined as,

$$\varepsilon = \left| \frac{y_{\text{analytical}} - y_{\text{numerical}}}{y_{\text{analytical}}} \right| \times 100$$

then the value of ε at $x = 0.5$ is _______.

(a) 0.06
(b) 0.8
(c) 4.0
(d) 8.0

[2021 : 1 M]

6.21 To solve an algebraic equation $f(x) = 0$, an iterative scheme of the type $x_{n+1} = g(x_n)$ is proposed, where $g(x) = x - \frac{f(x)}{f'(x)}$.

At the solution, $x = s$, $g'(s) = 0$ and $g''(s) \neq 0$. The order of convergence for this iterative scheme near the solution is _______.

[2021 : 2 M]

6.22 The area of a circular field is 25 m^2. The radius, r, is to be determined using the Newton-Raphson iterative method. For an initial guess of $r = 2.500$ m, the revised estimate of r after one iteration is _______ m (rounded off to three decimal places).

[2022 : 1 M]

6.23 The equation $\frac{dy}{dx} = xy^2 + 2y + x - 4.5$ with the initial condition $y(x = 0) = 1$ is to be solved using a predictor-corrector approach. Use a predictor based on the implicit Euler's method and a corrector based on the trapezoidal rule of integration, each with a full-step size of 0.5. Considering only positive values of y, the value of y at $x = 0.5$ is _______ (rounded off to three decimal places).

[2022 : 2 M]

6.24 Simpson's one-third rule is used to estimate the definite integral

$$I = \int_{-1}^{1} \sqrt{(1 - x^2)}\,dx$$

with an interval length of 0.5. Which one of the following is the CORRECT estimate of I obtained using this rule?

(a) $\frac{1}{3} - \frac{1}{\sqrt{3}}$
(b) $\frac{1}{3} + \frac{2}{\sqrt{3}}$
(c) $\frac{1}{3} + \frac{1}{\sqrt{3}}$
(d) $\frac{1}{3} - \frac{2}{\sqrt{3}}$

[2023 : 2 M]

6.25 The Newton-Raphson method is used to solve $f(x) = 0$, where $f(x) = e^x - 5x$.

If the initial guess $x^{(0)} = 1.0$, the value of the next iterate, $x^{(1)}$, rounded off to 2 decimal places, is _______.

[2024 : 2 M]

■■■■

Answers Numerical Methods

6.1	(a)	**6.2**	(b)	**6.3**	(a)	**6.4**	(b)	**6.5**	(c)	**6.6**	(c)	**6.7**	(a)
6.8	(b)	**6.9**	(a)	**6.10**	(d)	**6.11**	(0.385)	**6.12**	(2.33)	**6.13**	(c)	**6.14**	(c)
6.15	(94.67)	**6.16**	(d)	**6.17**	(a)	**6.18**	(0.566)	**6.19**	(60.7)	**6.20**	(d)	**6.21**	(1)
6.22	(2.841)	**6.23**	(0.875)	**6.24**	(b)	**6.25**	(0)						

Explanations Numerical Methods

6.4 (b)

$$f(x) = (x-1)^2 + (x-3)$$

In Newton Raphson method, if $f'(x) = 0$ then iteration fails.

So, $f'(x) = 2(x-1) + 1 = 0$

$$x = 0.5$$

6.6 (c)

From Simpson's rule,

$$\int_a^0 f(x)dx = \frac{h}{3}\left[y_0 + y_n + 2(y_2 + y_4 \ldots) + 4(y_1 + y_3 \ldots)\right]$$

Here, $a = 0$ and $b = \frac{\pi}{4}$ and $n = 4$

$$h = \frac{\frac{\pi}{4} - 0}{4} = \frac{\pi}{16}$$

x	$x_0 = 0$	$x_1 = \frac{\pi}{16}$	$x_2 = \frac{\pi}{8}$	$x_3 = \frac{3\pi}{16}$	$x_4 = \frac{\pi}{4}$
y	0	0.2068	0.4853	0.9665	2

$$\therefore \int_0^{\pi/4} \frac{\sin x}{\cos^3 x} = \frac{\pi}{16 \times 3}\left[\begin{matrix} 0 + 2 + 2(0.4853) \\ +4(0.2068 + 0.9665) \end{matrix}\right]$$

$$= 0.5016$$

6.7 (a)

$$\frac{dy}{dx} = x^2 y - 1.2y;\ y(0) = 1,\ h = 0.1$$

$$y_{n+1} = y_n + hf(x_n, y_n)$$

$$y(0.1) = y(0) + hf(x_0, y_0)$$

$$= 1 + 0.1 f(0, 1)$$

$$= 1 + 0.1 \times (0 - 1.2 \times 1)$$

$$= 1 + 0.1 \times (-1.2)$$

$$y(0.1) = 1 - 0.12 = 0.88$$

6.8 (b)

x	0	$\frac{\pi}{4}$	$\frac{\pi}{2}$	$\frac{3\pi}{4}$	π
$f(x)$	0	$\frac{1}{\sqrt{2}}$	1	$\frac{1}{\sqrt{2}}$	0
	y_0	y_1	y_2	y_3	y_4

$$\int_0^{\pi} \sin\theta\, d\theta = \frac{h}{2}[(y_0 + y_4) + 2(y_1 + y_2 + y_3)]$$

$$= \frac{\pi}{8}\left[(0+0) + 2\left(\frac{1}{\sqrt{2}} + 1 + \frac{1}{\sqrt{2}}\right)\right]$$

$$= \frac{\pi}{8}\left[2\left(\sqrt{2} + 1\right)\right] = 1.896 \simeq 1.9$$

6.9 (a)

$$f(x) = x^4 - 3x + 4,\ f'(x) = 4x^3 - 3$$

$$x_1 = x_0 - \frac{f(x_0)}{f'(x_0)}$$

$$f(0) = 1,\ f'(0) = -3$$

$$x_1 = 0 - \frac{1}{-3} = \frac{1}{3}$$

6.10 (d)

$$f(x) = x - \cos \pi x;\ f(0.5) = 0.5 - \cos\frac{\pi}{2} = 0.5$$

$$f'(x) = 1 + \sin \pi x \times (\pi);$$

$$f'(0.5) = 1 + \pi \times \sin\frac{\pi}{2} = 1 + \pi$$

$$x_{n+1} = x_n - \frac{f(x_n)}{f'(x)}$$

$$x_1 = x_0 - \frac{f(0.5)}{f'(0.5)} = 0.5 - \frac{0.5}{(1+\pi)}$$

$$= 0.379 \simeq 0.38$$

6.11 (0.385)

$$h = \frac{b-a}{n} = \frac{0.5-0.1}{4} = 0.1$$

0.1	0.2	0.3	0.4	0.5
$\frac{1}{e^{0.1^3}}$	$\frac{1}{e^{0.2^3}}$	$\frac{1}{e^{0.3^3}}$	$\frac{1}{e^{0.4^3}}$	$\frac{1}{e^{0.5^3}}$
y_0	y_1	y_2	y_3	y_4

$$\int_{0.1}^{0.5} e^{-x^3}dx = \frac{h}{3}[(y_0+y_4)+4\times(y_1+y_3)+2\times y_2]$$

$$= \frac{0.1}{3}\left[\left(\frac{1}{e^{0.001}}+\frac{1}{e^{0.125}}\right)+4\times\left(\frac{1}{e^{0.008}}+\frac{1}{e^{0.064}}\right)+2\times\frac{1}{e^{0.027}}\right]$$

$$= \frac{0.1}{3}[1.8814+7.7201+1.9467]$$

$$= 0.3849 \simeq 0.385$$

6.12 (2.33)

$$\frac{dy}{dx} = x + \ln(y)$$

$$y_0 = y(0) = 2,\ x_0 = 0$$

$$y_1 = y_0 + h\,f(x_0, y_0)$$
$$= 2 + 0.2\,f(0, 2)$$
$$= 2 + 0.2\,[0 + \ln 2]$$
$$= 2 + 0.2 \ln 2 = 2.138$$

$$x_1 = x_0 + h = 0.2$$

$$y_2 = y(0.4)$$
$$= y_1 + h\,f(x_1, y_1)$$
$$= 2.138 + 0.2\,(0.2, 2.138)$$
$$= 2.138 + 0.2\,[0.2 + \ln(2.138)]$$
$$= 2.329 \simeq 2.33$$

6.13 (c)

Slope of function at $x = 0.5$

$f'(0.5) = 3(0.5)^2 - 1 = -0.25$

Thus, solution will converge to 1.

6.15 (94.67)

x	0	1	2	3	4
$f(x)$	3	10	21	36	55
	y_0	y_1	y_2	y_3	y_4

$$\int_0^4 f(x)dx = \frac{h}{3}[(y_0+y_4)+4\times(y_1+y_3)+2\times y_2]$$

$$= \frac{1}{3}[(3+55)+4\times(10+36)+2\times 21]$$

$$= 94.67$$

6.18 (0.566)

$$f(x) = e^{-x} - x,\ x_0 = 0,\ f'(x) = -e^{-x} - 1$$

$$x_1 = x_0 - \frac{f(x_0)}{f'(x_0)} = 0 - \frac{f(0)}{f'(0)}$$

$$x_1 = 0 - \frac{(e^{-0}-0)}{(-e^{-0}-1)} = 0 - \frac{(1-0)}{(-1-1)} = \frac{-1}{-2} = \frac{1}{2}$$

$$x_2 = x_1 - \frac{f(x_1)}{f'(x_1)} = \frac{1}{2} - \frac{f(0.5)}{f'(0.5)}$$

$$= \frac{1}{2} - \frac{(e^{-0.5}-0.5)}{(-e^{-0.5}-1)}$$

$$x_2 = 0.5 + \frac{(e^{-0.5}-0.5)}{(e^{-0.5}-1)} = 0.566$$

6.19 (60.7)

$$\int_1^{25} f(x)dx = \frac{h}{3}\left[\begin{matrix}(y_0+y_n)+2(y_2+y_4\ldots)\\ +4(y_1+y_3\ldots.)\end{matrix}\right]$$

$$= \frac{2}{3}[(6+5)+2(10)+4(8+12)]$$

$$= \frac{2}{3}[11+20+60] = 60.7$$

6.20 (d)

$$y(0.5) = y_0 + k$$
$$y_0 = y(0) = 1$$
$$k = \frac{k_1+k_2}{2}$$
$$k_1 = hf(x_0, y_0)$$
$$= 0.5f(0, 1) = 0.5 \times [2] = 1$$
$$k_2 = h_f(x_0 + h, y_0 + k_1)$$
$$= 0.5f(0.5, 2) = 0.5 \times 4 = 2$$
$$k = \frac{1+2}{2} = 1.5$$

So, $y(0.5) = 1 + 1.5 = 2.5$

$$\varepsilon = \left[\frac{e^1 - 2.5}{e}\right]\times 100 = 8.0$$

6.22 (2.841)

$$\text{Area of circle} = 25\text{ m}^2$$
$$\pi R^2 = 25$$
$$f(R) = \pi R^2 - 25 = 0$$
$$\Rightarrow\quad f'(R) = 2\pi R$$

Initial guess, $R = 2.5$ m

So, by Newton Raphson's method,

$$x_{i+1} = x_i - \frac{f(x_i)}{f'(x_i)}$$

$$R_1 = 2.5 - \frac{f(R)}{f'(R)} = 2.5 - \frac{\pi \times 2.5^2 - 25}{2\pi \times 2.5}$$

$$= 2.5 + 0.341 = 2.841$$

6.23 (0.875)

$$\frac{dy}{dx} = xy^2 + 2y + x - 4.5$$

$$y(0) = 1$$

Given $h = 0.5$

$\Rightarrow$ $x_1 = x_0 + h = 0 + 0.5 = 0.5$

By implicit Euler formula

$$y_1 = y_0 + h\, f(x_1, y_1)$$

$$dy = (xy^2 + 2y + x - 4.5)\, dx$$

$$\int_0^{0.5} dy = \int_0^{0.5} \left(x^2 + 2y + x - 4.5\right) dx$$

By trapezoidal

$$\Rightarrow y(0.5) - y(0) = \int_0^{0.5} \left(xy^2 + 2y + x - 4.5\right) dx$$

x	0	0.5
$f(x, y)$	−2.5	2

By trapezoidal rule,

$$y(0.5) = y(0) + \frac{h}{2}\left[f(x_0, y_0) + f(x_1, y_1)\right]$$

$$= 1 + \frac{0.5}{2}[-2.5 + 2]$$

$$= 1 + \frac{0.5}{2}(-0.5)$$

$$= 1 - \frac{0.25}{2} = \frac{7}{8} = 0.875$$

6.24 (b)

The given function is

$$f(x) = \int_{-1}^{1} \sqrt{1 - x^2}\, dx$$

$$h = 0.5 \text{ \{Interval length\}}$$

From Simpson's $\frac{1}{3}$rd rule, we have

$$\int f(x)\, dx = \frac{h}{3}\begin{bmatrix} y_0 + y_n + 2[\text{even terms}] \\ + 4\,(\text{odd terms}) \end{bmatrix}$$

x	−1	−0.5	0	0.5	1
$f(x)$	0	0.87	1	0.87	0
	y_0	y_1	y_2	y_3	y_4

$$\int_{-1}^{1} f(x)\, dx = \frac{0.5}{3}[2x_1 + 4[0.87 + 0.87]]$$

$$= 1.49 \cong 1.5$$

6.25 (0)

Given equation is

$$f(x) = e^x - 5x$$

Initial guess $x^{(0)} = 1$

From Newton-Raphson method

$$x_1 = x_0 - \frac{f(x_0)}{f'(x_0)}$$

$$= 1.0 - \frac{e^x - 5x}{e^x - 5}$$

$$= 1.0 - \frac{e^1 - 5}{e^1 - 5} = 1 - 1 = 0$$

■■■■

7 Transform Theory

7.1 The inverse Laplace transform of the function $f(s) = \frac{1}{s(1+s)}$ is

(a) $1 + e^t$ (b) $1 - e^t$
(c) $1 + e^{-t}$ (d) $1 - e^{-t}$ [2004 : 1 M]

7.2 The series $\sum_{n=1}^{\infty} \frac{(z+2)^n}{n!}$ converges for

(a) all z (b) $z > 2$
(c) $|z| > 2$ (d) $|z| < 2$ [2004 : 2 M]

7.3 Given that the Laplace transform of the function below over a single period $0 < t < 2$ is $\frac{1}{s^2}(1-e^{-s})^2$, the Laplace transform of the period function over $0 < t < \infty$ is

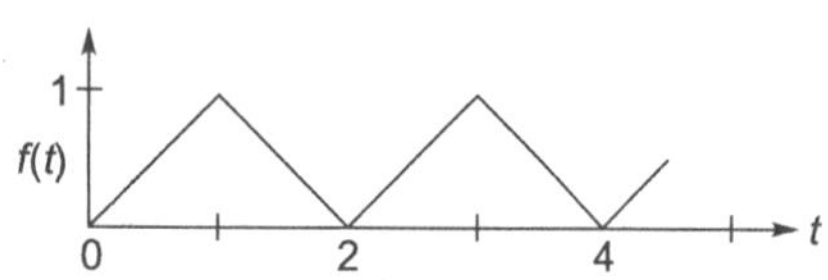

(a) $\frac{1}{s}(1-e^{-s})^2$ (b) $\frac{1}{s}(1-e^{-s})^2$

(c) $\frac{1}{s^2}\frac{(1-e^{-s})}{(1+e^{-s})}$ (d) $\frac{1}{s^2}\tanh\frac{s}{2}$

[2007 : 1 M]

7.4 The Laplace transform of $f(t) = \frac{1}{\sqrt{t}}$ is

(a) $\sqrt{\frac{\pi}{s}}$ (b) $\frac{1}{\sqrt{s}}$

(c) $\frac{1}{s^{3/2}}$ (d) Does not exist

[2007 : 2 M]

7.5 The Laplace transform of the function $f(t) = t\sin t$ is

(a) $\frac{2s}{(s^2+1)^2}$ (b) $\frac{1}{s^2(s^2+1)}$

(c) $\frac{1}{s^2} + \frac{1}{(s^2+1)}$ (d) $\frac{1}{(s-1)^2+1}$

[2008 : 2 M]

7.6 The inverse Laplace transform of $\frac{1}{2s^2+3s+1}$ is

(a) $e^{-t/2} - e^{-t}$ (b) $2e^{-t/2} - e^{-t}$
(c) $e^{-t} - 2e^{-t/2}$ (d) $e^{-t} - e^{-t/2}$

[2009 : 2 M]

7.7 The Laplace transform of the function shown in the figure below is

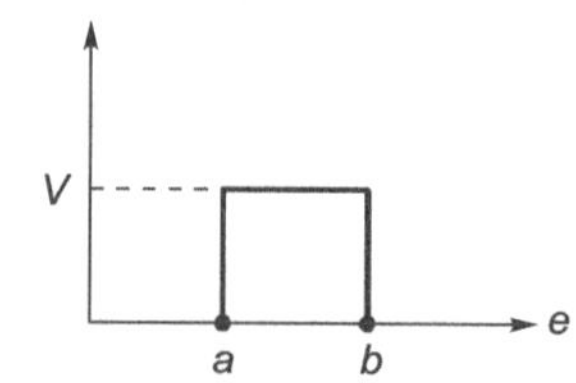

(a) $\frac{Ve^{(ab)}}{s}$ (b) $\frac{V}{s}(e^{-a} - e^{-b})$

(c) $\frac{V}{s}(e^{-a} - e^{-b})$ (d) $\frac{V}{s^2}(e^{a} - e^{b})$

[2010 : 1 M]

7.8 For the time domain function $f(t) = t^2$, which ONE of the following is the Laplace transform of $\int_0^t f(t)dt$?

(a) $\frac{3}{s^4}$ (b) $\frac{1}{4s^4}$

(c) $\frac{2}{s^3}$ (d) $\frac{2}{s^4}$

[2014 : 1 M]

7.9 The Laplace transform of $e^{at}\sin(bt)$ is

(a) $\frac{b}{(s-a)^2+b^2}$ (b) $\frac{(s-a)}{(s-a)^2+b^2}$

(c) $\frac{(s-a)}{(s-a)^2-b^2}$ (d) $\frac{b}{(s-a)^2-b^2}$

[2016 : 1 M]

7.10 The Laplace transform of a function is $\frac{s+1}{s(s+2)}$.

The initial and final values, respectively, of the function are

(a) 0 and 1 (b) 1 and $\frac{1}{2}$

(c) $\frac{1}{2}$ and 1 (d) $\frac{1}{2}$ and 0

[2017 : 2 M]

7.11 Consider the following unit step function:

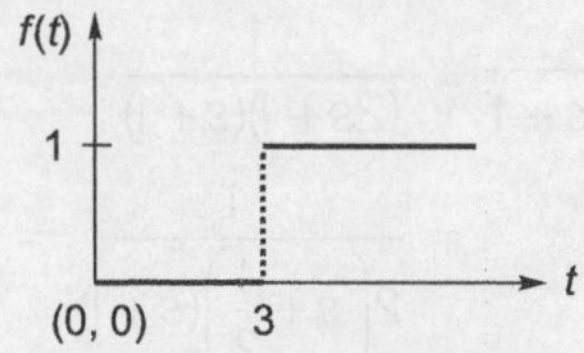

The Laplace transform of this function is

(a) $\frac{e^{-3s}}{s}$ (b) $\frac{e^{-3s}}{s^2}$

(c) $\frac{e^{-3s}}{3s}$ (d) $\frac{e^{-6s}}{s}$

[2020 : 1 Mark]

Answers Transform Theory

7.1	(d)	**7.2**	(d)	**7.3**	(c)	**7.4**	(a)	**7.5**	(a)	**7.6**	(a)	**7.7**	(c)
7.8	(d)	**7.9**	(a)	**7.10**	(b)	**7.11**	(a)						

Explanations Transform Theory

7.1 (d)

$$F(s) = \frac{1}{s(s+1)}$$

$$F(s) = \frac{1}{s} - \frac{1}{s+1}$$

Taking inverse Laplace

$$F(t) = 1 - e^{-t}$$

7.2 (d)

$$\sum_{n=1}^{\infty} \frac{(z+2)^n}{n!} = \sum_{n=1}^{\infty} \frac{2^n}{n!}\left(1+\frac{z}{2}\right)^n$$

For convergence,

$$\left|\frac{z}{2}\right| < 1; \ |z| < 2$$

7.3 (c)

For a periodic function having time period T the Laplace transform

$$F(s) = \frac{1}{1-e^{-Ts}} \times \int_0^T e^{-st} f(t)dt$$

$$\because \quad \int_0^T e^{-st} f(t)dt = \frac{1}{s^2}(1-e^{-s})^2$$

So, $$F(s) = \frac{1}{s^2} \times \frac{(1-e^{-s})^2}{(1-e^{-s})(1+e^{-s})}$$

$$= \frac{1}{s^2} \times \frac{(1-e^{-s})}{(1+e^{-s})}$$

7.4 (a)

$$\because \quad L(t^n) = \frac{\Gamma(n+1)}{S^{n+1}}$$

$$\therefore \quad L(t^{-1/2}) = \frac{\Gamma\left(-\frac{1}{2}+1\right)}{S^{-\frac{1}{2}+1}} = \frac{\Gamma\left(\frac{1}{2}\right)}{S^{\frac{1}{2}}} = \frac{\sqrt{\pi}}{\sqrt{S}} = \sqrt{\frac{\pi}{S}}$$

7.5 (a)

Laplace transform of $\sin t$ is given by

$$L[\sin t] = \frac{1}{s^2+1}$$

If Laplace transform of $f(t)$ is $F(s)$

Then Laplace transform of $t^n f(t)$ is given by

$$L[t^n f(t)] = (-1)^n \frac{d^n F(s)}{ds^n}$$

So, $$L[t \sin t] = (-1)\frac{d}{ds}\left(\frac{1}{s^2+1}\right) = \frac{2s}{(s^2+1)^2}$$

7.6 (a)

$$\frac{1}{2s^2+3s+1} = \frac{1}{(2s+1)(s+1)}$$

$$= \frac{1}{2\left(s+\frac{1}{2}\right)(s+1)} = \frac{1}{s+\frac{1}{2}} - \frac{1}{s+1}$$

$$f(t) = L^{-1}\left(\frac{1}{s+\frac{1}{2}} - \frac{1}{s+1}\right)$$

$$= L^{-1}\left(\frac{1}{s+\frac{1}{2}}\right) - L^{-1}\left(\frac{1}{s+1}\right)$$

Taking inverse Laplace

$\Rightarrow$ $f(t) = e^{(-1/2)t} - e^{-t}$

7.7 (c)

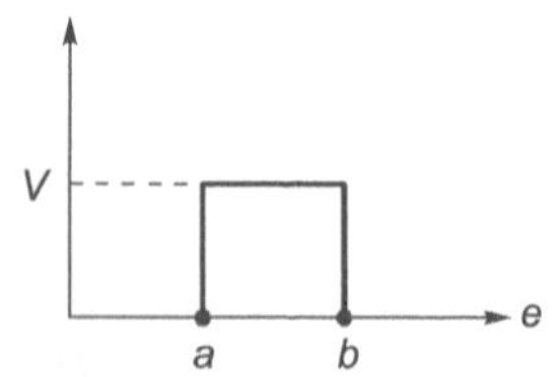

Laplace of above diagram can be found by using 2nd shifting theorem

$$L[(t-a)\,\mu(t-a)] = \frac{e^{-as}}{s}$$

So Laplace of above diagram is given by

$$\frac{V}{s}e^{-as} - \frac{V}{s}e^{-bs} = \frac{V}{s}\left[e^{-as} - e^{-bs}\right]$$

7.8 (d)

$$\int_0^t f(t)\,dt = \int_0^t t^2 dt = \left[\frac{t^3}{3}\right]_0^t = \frac{t^3}{3}$$

$$L\left(\frac{t^3}{3}\right) = \frac{1}{3}L(t^3) = \frac{1}{3}\times\frac{3!}{s^4} = \frac{1}{3}\times\frac{6}{s^4} = \frac{2}{s^4}$$

7.9 (a)

$L[e^{at}\sin bt]$

$$\because \quad L[\sin bt] = \frac{b}{s^2+b^2}$$

$$\therefore \quad L[e^{at}\sin bt] = \frac{b}{(s-a)^2+b^2}$$

7.10 (b)

$$\frac{s+1}{s(s+2)} = \frac{1}{s} - \frac{1}{s(s+2)} = \frac{1}{s} - \frac{1}{2}\left[\frac{1}{s} - \frac{1}{s+2}\right]$$

$$\Rightarrow \frac{2}{2s} - \frac{1}{2s} + \frac{1}{2(s+2)} = \frac{1}{2s} + \frac{1}{2(s+2)}$$

$$f(t) = L^{1}\left[\frac{1}{2s} + \frac{1}{2(s+2)}\right]$$

$$= \frac{1}{2}\left[1+e^{-2t}\right] = \frac{1+e^{-2t}}{2}$$

$$\lim_{t\to\infty}[f(t)] = \frac{1+0}{2} = \frac{1}{2} \text{ (Final value)}$$

$$\lim_{t\to 0} = \frac{1+1}{2} = 1 \text{ (Initial value)}$$

7.11 (a)

$$L(f(t)) = \int_0^\infty e^{-st} f(t)\,dt$$

$$= \int_0^3 e^{-st}\times 0\,dt + \int_3^\infty e^{-st}\times 1\,dt$$

$$= \left.\frac{e^{-st}}{-s}\right|_3^\infty$$

$$= -\frac{1}{s}\left[e^{-\infty} - e^{-3s}\right] = \frac{e^{-3s}}{s}$$

8 Second Order Linear Partial Differential Equations

8.1 Match the following, where x is the spatial coordinate and t is time,

Group-I

A. Wave equation

B. Heat equation

Group-II

1. $\dfrac{\partial c}{\partial t} = \alpha \dfrac{\partial c}{\partial x}$ 2. $\dfrac{\partial c}{\partial t} = \alpha^2 \dfrac{\partial^2 c}{\partial x^2}$

3. $\dfrac{\partial^2 c}{\partial t^2} = \alpha^2 \dfrac{\partial c}{\partial x}$ 4. $\dfrac{\partial^2 c}{\partial t^2} = \alpha^2 \dfrac{\partial^2 c}{\partial x^2}$

Codes:

	A	B
(a)	4	2
(b)	2	4
(c)	3	1
(d)	1	3

[2005 : 1 M]

8.2 If the constants A_n are suitably chosen so as to satisfy the initial conditions, and n is an integer, which of the following is a valid solution for the unsteady one-dimensional diffusion equation, $\dfrac{\partial c}{\partial t} = D\dfrac{\partial^2 c}{\partial x^2}$ in the domain $0 \le x \le L$ with boundary conditions $c = 0$ at $x = 0$ and $c = 0$ at $x = L$?

(a) $\sum_n A_n \sin\left(\dfrac{n\pi x}{L}\right) e^{\left(\frac{Dn^2\pi^2 t}{L^2}\right)}$

(b) $\sum_n A_n \cos\left(\dfrac{n\pi x}{L}\right) e^{\left(\frac{Dn^2\pi^2 t}{L^2}\right)}$

(c) $\sum_n A_n \sin\left(\dfrac{n\pi x}{L}\right) e^{-\left(\frac{Dn^2\pi^2 t}{L^2}\right)}$

(d) $\sum_n A_n \cos\left(\dfrac{n\pi x}{L}\right) e^{-\left(\frac{Dn^2\pi^2 t}{L^2}\right)}$

[2005 : 2 M]

■■■■

Answers Second Order Linear Partial Differential Equations

8.1 (a) **8.2** (c)

Explanations Second Order Linear Partial Differential Equations

8.1 (a)

Wave equation

$$\frac{\partial^2 c}{\partial t^2} = \alpha^2 \frac{\partial^2 c}{\partial x^2}$$

Heat equation

$$\frac{\partial c}{\partial t} = \alpha^2 \frac{\partial^2 c}{\partial x^2}$$

■■■■

General Aptitude

UNIT

XII

SYLLABUS

Numerical Ability : Numerical Computation, Numerical Estimation, Numerical Reasoning and Data Interpretation.

Verbal Ability : English Grammar, Sentence Completion, Verbal Analogies, Word Groups, Instructions, Critical Reasoning and Verbal Deduction.

CONTENTS

Analysis of Previous GATE Papers

Exam Year	1 Mark Ques.	2 Marks Ques.	Total Marks
2010	5	5	15
2011	5	5	15
2012	5	5	15
2013	5	5	15
2014	5	5	15
2015	5	5	15
2016	5	5	15
2017	5	5	15
2018	5	5	15
2019	5	5	15
2020	5	5	15
2021	5	5	15
2022	5	5	15
2023	5	5	15
2024	5	5	15

1 Numerical Ability

1.1 If 137 + 276 = 435 how much is 731 + 672?

(a) 534 (b) 1403
(c) 1623 (d) 1531

[2010 : 2 M]

1.2 Hari (H), Gita (G), Irfan (I) and Saira (S) are siblings (i.e. brothers and sisters). All were born on I[st] January. The age difference between any two successive siblings (that is born one after another) is less than 3 years. Given the following facts:

1. Hari's age + Gita's age > Irfan's age + Saira's age.
2. The age difference between Gita and Saira is 1 year. However, Gita is not the oldest and Saira is not the youngest.
3. There are no twins.

In what order were they born (oldest first)?

(a) HSIG (b) SGHI
(c) IGSH (d) IHSG

[2010 : 2 M]

1.3 5 skilled workers can build a wall in 20 days; 8 semi-skilled workers can build a wall in 25 days; 10 unskilled workers can build a wall in 30 days. If a team has 2 skilled, 6 semi-skilled and 5 unskilled workers, how long will it take to build the wall?

(a) 20 days (b) 18 days
(c) 16 days (d) 15 days

[2010 : 2 M]

1.4 Given digits 2, 2, 3, 3, 3, 4, 4, 4, 4 how many distinct 4 digit numbers greater than 3000 can be formed?

(a) 50 (b) 51
(c) 52 (d) 54

[2010 : 2 M]

1.5 If log(*P*) = (1/2)log(*Q*) = (1/3) log(*R*), then which of the following options is TRUE?

(a) $P^2 = Q^3R^2$ (b) $Q^2 = PR$
(c) $Q^2 = R^3P$ (d) $R = P^2 Q^2$

[2011 : 1 M]

1.6 The variable cost (*V*) of manufacturing a product varies according to the equation $V = 4q$, where q is the quantity produced. The fixed cost (*F*) of production of same product reduces with q according to the equation $F = 100/q$. How many units should be produced to minimize the total cost $(V + F)$?

(a) 5 (b) 4
(c) 7 (d) 6

[2011 : 2 M]

1.7 *P*, *Q*, *R* and *S* are four types of dangerous microbes recently found in a human habitat. The area of each circle with its diameter printed in brackets represents the growth of a single microbe surviving human immunity system within 24 hours of entering the body. The danger to human beings varies proportionately with the toxicity, potency and growth attributed to a microbe shown in the figure below:

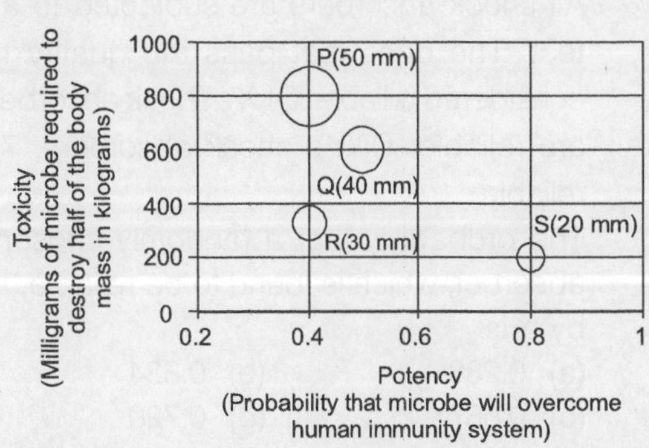

A pharmaceutical company is contemplating the development of a vaccine against the most dangerous microbe. Which microbe should the company target in its first attempt?

(a) *P* (b) *Q*
(c) *R* (d) *S*

[2011 : 2 M]

1.8 A container originally contains 10 litres of pure spirit. From this container 1 litre of spirit is replaced with 1 litre of water. Subsequently, 1 litre of the mixture is again replaced with 1 litre of water and this process is repeated one more time. How much spirit is now left in the container?

(a) 7.58 litres (b) 7.84 litres
(c) 7 litres (d) 7.29 litres

[2011 : 2 M]

1.9 A transporter receives the same number of orders each day. Currently, he has some pending orders (backlog) to be shipped. If he uses 7 trucks, then at the end of the 4th day he can clear all the orders. Alternatively, if he uses only 3 tucks, then all the orders are cleared at the end of the 10th day. What is the minimum number of trucks required so that there will be no pending order at the end of the 5th day?

(a) 4 (b) 5
(c) 6 (d) 7 [2011 : 2 M]

1.10 The cost function for a product in a firm is given by $5q^2$, where q is the amount of production. The firm can sell the product at a market price of Rs. 50 per unit. The number of units to be produced by the firm such that the profit is maximized is

(a) 5 (b) 10
(c) 15 (d) 25 [2012 : 1 M]

1.11 An automobile plant contracted to buy shock absorbers from two suppliers X and Y. X supplies 60% and Y supplies 40% of the shock absorbers. All shock absorbers are subjected to a quality test. The ones that pass the quality test are considered reliable. Of X's shock absorbers, 96% are reliable. Of Y's shock absorbers, 72% are reliable.

The probability that a randomly chosen shock absorber, which is found to be reliable, is made by Y is

(a) 0.288 (b) 0.334
(c) 0.667 (d) 0.720

[2012 : 2 M]

1.12 A political party orders an arch for the entrance to the ground in which the annual conventions is being held. The profile of the arch follows the equation $y = 2x - 0.1x^2$ where y is the height of the arch in meters. The maximum possible height of the arch is

(a) 8 meters (b) 10 meters
(c) 12 meters (d) 14 meters

[2012 : 2 M]

1.13 Which of the following assertions are CORRECT?

P : Adding 7 to each entry in a list adds 7 to the mean of the list

Q : Adding 7 to each entry in a list adds 7 to the standard deviation of the list

R : Doubling each entry n a list doubles the mean of the list

S : Doubling each entry in a list leaves the standard deviation of the list unchanged

(a) P, Q (b) Q, R
(c) P, R (d) R, S

[2012 : 2 M]

1.14 If $3 \le X \le 5$ and $8 \le Y \le 11$ then which of the following options is TRUE?

(a) $\frac{3}{5} \le \frac{X}{Y} \le \frac{8}{5}$ (b) $\frac{3}{11} \le \frac{X}{Y} \le \frac{5}{8}$

(c) $\frac{3}{11} \le \frac{X}{Y} \le \frac{8}{5}$ (d) $\frac{3}{5} \le \frac{X}{Y} \le \frac{8}{11}$

[2013 : 1 M]

1.15 Velocity of an object fired directly in upward direction is given by $V = 80 - 32t$, where t (time) is in seconds. When will the velocity be between 32 m/sec and 64 m/sec?

(a) $\left(1, \frac{3}{2}\right)$ (b) $\left(\frac{1}{2}, 1\right)$

(c) $\left(\frac{1}{2}, \frac{3}{2}\right)$ (d) (1, 3)

[2013 : 2 M]

1.16 In a factory, two machines M1 and M2 manufacture 60% and 40% of the autocomponents respectively. Out of the total production, 2% of M1 and 3% of M2 are found to be defective. If a randomly drawn autocomponent from the combined lot is found defective, what is the probability that it was manufactured by M2?

(a) 0.35 (b) 0.45
(c) 0.5 (d) 0.4

[2013 : 2 M]

1.17 Following table gives data on tourists from different countries visiting India in the year 2011.

Country	Number of Tourists
USA	2000
England	3500
Germany	1200
Italy	1100
Japan	2400
Australia	2300
France	1000

Which two countries contributed to the one third of the total number of tourists who visited India in 2011?

(a) USA and Japan
(b) USA and Australia
(c) England and France
(d) Japan and Australia

[2013 : 2 M]

1.18 If $|-2X + 9| = 3$ then the possible value of $|-X| - X^2$ would be

(a) 30 (b) –30
(c) –42 (d) 42

[2013 : 2 M]

1.19 If $y = 5x^2 + 3$, then the tangent at $x = 0, y = 3$

(a) passes through $x = 0, y = 0$
(b) has a slope of +1
(c) is parallel to the x-axis
(d) has a slope of –1

[2014 (Set-1) : 1 M]

1.20 A foundry has a fixed daily cost of Rs. 50,000 whenever it operates and a variable cost of Rs. 800Q, where Q is the daily production in tonnes. What is the cost of production in Rs. per tonne for a daily production of 100 tonnes?

[2014 (Set-1) : 1 M]

1.21 The smallest angle of a triangle is equal to two thirds of the smallest angle of a quadrilateral. The ratio between the angles of the quadrilateral is 3 : 4 : 5 : 6. The largest angle of the triangle is twice its smallest angle. What is the sum, in degrees, of the second largest angle of the triangle and the largest angle of the quadrilateral?

[2014 (Set-1) : 2 M]

1.22 One percent of the people of country X are taller than 6 ft. Two percent of the people of country Y are taller than 6 ft. There are thrice as many people in country X as in country Y. Taking both countries together, what is the percentage of people taller than 6 ft?

(a) 3.0 (b) 2.5
(c) 1.5 (d) 1.25

[2014 (Set-1) : 2 M]

1.23 The monthly rainfall chart based on 50 years of rainfall in Agra is shown in the following figure. Which of the following are true? (k percentile is the value such that k percent of the data fall below that value)

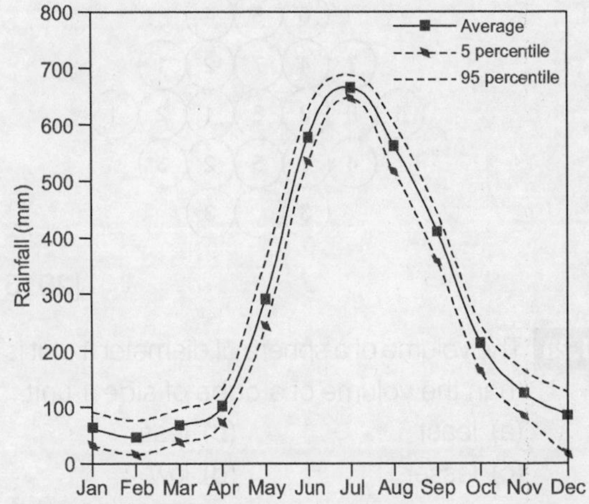

(i) On average, it rains more in July than in December.
(ii) Every year, the amount of rainfall in August is more than that in January.
(iii) July rainfall can be estimated with better confidence than February rainfall.
(iv) In August, there is at least 500 mm of rainfall.

(a) (i) and (ii) (b) (i) and (iii)
(c) (ii) and (iii) (d) (iii) and (iv)

[2014 (Set-1) : 2 M]

1.24 If $\log_x\left(\frac{5}{7}\right) = -\frac{1}{3}$, then the value of x is

(a) $\frac{343}{125}$ (b) $\frac{12}{343}$
(c) $-\frac{25}{49}$ (d) $-\frac{49}{25}$

[2015 : 1 M]

1.25 Operations □, ◇ and → are defined by $a \,\square\, b = \frac{a-b}{a+b}$; $a \diamond b = \frac{a+b}{a-b}$; $a \rightarrow b = ab$.

Find the value (66 □ 6) → (66 ◇ 6).

(a) –2 (b) –1
(c) 1 (d) 2

[2015 : 1 M]

1.26 A cube of side 3 units is formed using a set of smaller cubes of side 1 unit. Find the proportion of the number of faces of the smaller cubes visible to those which are NOT visible.

(a) 1 : 4 (b) 1 : 3
(c) 1 : 2 (d) 2 : 3

[2015 : 2 M]

1.27 Fill in the missing value:

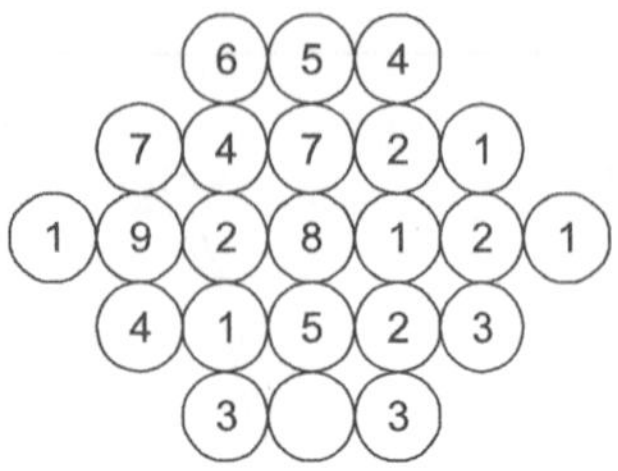

[2015 : 2 M]

1.28 The volume of a sphere of diameter 1 unit is ______ than the volume of a cube of side 1 unit.

(a) least (b) less
(c) lesser (d) low

[2016 (Set-2) : 1 M]

1.29 A window is made up of a square portion and an equilateral triangle portion above it. The base of the triangular portion coincides with the upper side of the square. If the perimeter of the window is 6 m, the area of the window in m^2 is ________.

(a) 1.43 (b) 2.06
(c) 2.68 (d) 2.88

[2016 (Set-2) : 1 M]

1.30 Students taking an exam are divided into two groups, *P* and *Q* such that each group has the same number of students. The performance of each of the students in a test was evaluated out of 200 marks. It was observed that the mean of group *P* was 105, while that of group *Q* was 85. The standard deviation of group *P* was 25, while that of group *Q* was 5. Assuming that the marks were distributed on a normal distribution, which of the following statements will have the highest probability of being TRUE?

(a) No student in group *Q* scored less marks than any student in group *P*.
(b) No student in group *P* scored less marks than any student in group *Q*.
(c) Most students of group *Q* scored marks in a narrower range than students in group *P*.
(d) The median of the marks of group *P* is 100.

[2016 (Set-2) : 2 M]

1.31 Which of the following curves represents the function $y = \ln\left(\left|e^{[\sin C(|x|)]}\right|\right)$ for $|x| < 2p$? Here, x represents the abscissa and y represents the ordinate.

(a)
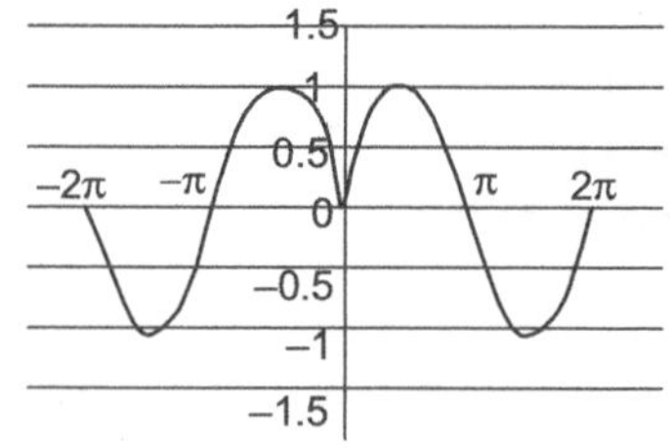

(b)
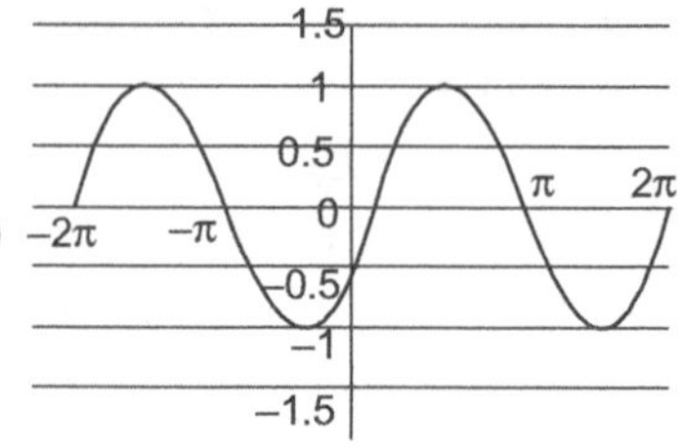

(c)
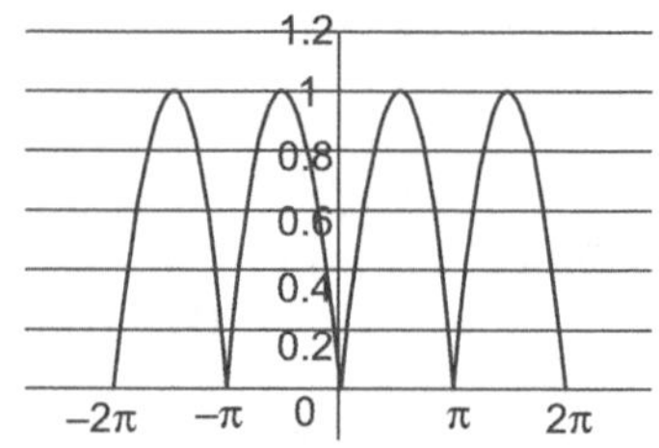

(d)
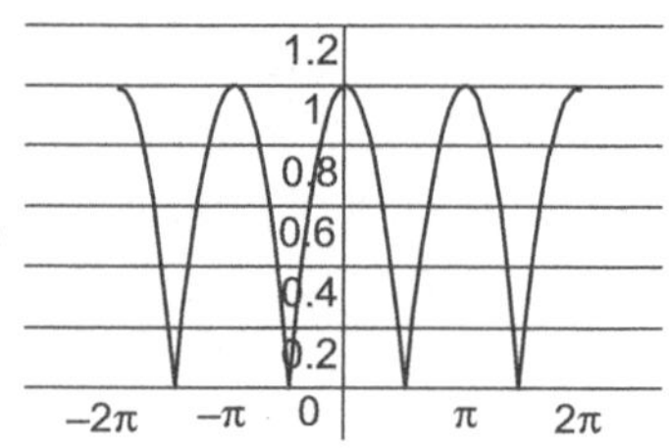

[2016 (Set-2) : 2 M]

1.32 If the radius of a right circular cone is increased by 50%, its volume increases by

(a) 75% (b) 100%
(c) 125% (d) 237.5%

[2017 (Set-1) : 1 M]

1.33 The following sequence of numbers is arranged in increasing order: 1, x, x, x, y, y, 9, 16, 18. Given that the mean and median are equal, and are also equal to twice the mode, the value of y is

(a) 5 (b) 6
(c) 7 (d) 8

[2017 (Set-1) : 1 M]

1.34 The last digit of $(2171)^7 + (2172)^9 + (2173)^{11} + (2174)^{13}$ is

(a) 2 (b) 4
(c) 6 (d) 8

[2017 (Set-1) : 2 M]

1.35 Two machines $M1$ and $M2$ are able to execute any of four jobs, P, Q, R and S. The machines can perform one job on one object at a time, Jobs P, Q, R and S take 30 minutes, 20 minutes, 60 minutes and 15 minutes each respectively. There are 10 objects each requiring exactly 1 job. Job P is to be performed on 2 objects, Job Q on 3 objects, Job R on 1 object and Job S on 4 objects. What is the minimum time needed to complete all the jobs?

(a) 2 hours (b) 2.5 hours
(c) 3 hours (d) 3.5 hours

[2017 (Set-1) : 2 M]

1.36 The bar graph below shows the output of five carpenters over one month, each of whom made different items of furniture; chairs, tables and beds.

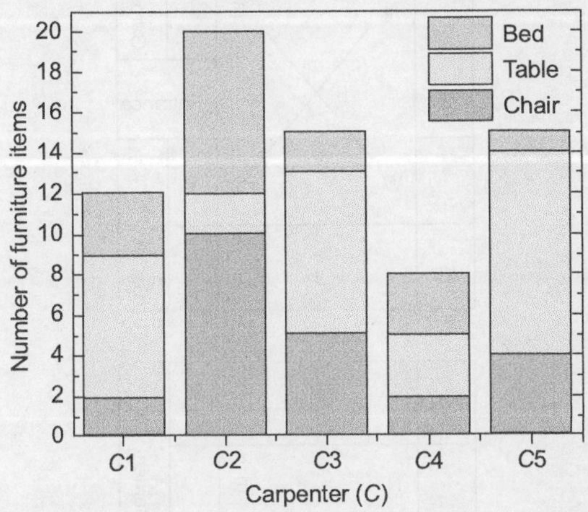

Consider the following statements:

(i) The number of beds made by carpenter $C2$ is exactly the same as the number of tables made by carpenter $C3$.

(ii) The total number of chairs made by all carpenters is less than the total number of tables.

Which one of the following is true?

(a) Only (i) (b) Only (ii)
(c) Both (i) and (ii) (d) Neither (i) nor (ii)

[2017 (Set-1) : 2 M]

1.37 For $0 \le x \le 2\pi$, $\sin x$ and $\cos x$ are both decreasing functions in the interval ________.

(a) $\left(0, \frac{\pi}{2}\right)$ (b) $\left(\frac{\pi}{2}, \pi\right)$
(c) $\left(\pi, \frac{\pi}{2}\right)$ (d) $\left(\frac{3\pi}{2}, 2\pi\right)$

[2018 : 1 M]

1.38 The area of an equilateral triangle is $\sqrt{3}$. What is the perimeter of the triangle?

(a) 2 (b) 4
(c) 6 (d) 8

[2018 : 1 M]

1.39 Arrange the following three-dimensional objects in the descending order of their volumes:

(i) A cuboid with dimensions 10 cm, 8 cm and 6 cm.
(ii) A cube of side 8 cm.
(iii) A cylinder with base radius 7 cm and height 7 cm.
(iv) A sphere of radius 7 cm.

(a) (i), (ii), (iii), (iv) (b) (ii), (i), (iv), (iii)
(c) (iii), (ii), (i), (iv) (d) (iv), (iii), (ii), (i)

[2018 : 1 M]

1.40 An automobile travels from city A to city B and returns to city A by the same route. The speed of the vehicle during the onward and return journeys were constant at 60 km/h and 90 km/h, respectively. What is the average speed in km/h for the entire journey?

(a) 72 (b) 73
(c) 74 (d) 75

[2018 : 2 M]

1.41 A set of 4 parallel lines intersect with another set of 5 parallel lines. How many parallelograms are formed?

(a) 20 (b) 48
(c) 60 (d) 72

[2018 : 2 M]

1.42 If $x^2 + x - 1 = 0$ what is the value of $x^4 + \frac{1}{x^4}$?

(a) 1 (b) 5
(c) 7 (d) 9

[2018 : 2 M]

1.43 In a detailed study of annual crow births in India, it was found that there was relatively no growth during the period 2002 to 2004 and a sudden spike from 2004 to 2005. In another unrelated study, it was found that the revenue from cracker sales in India which remained fairly flat from 2002 to 2004, saw a sudden spike in 2005 before declining again in 2006. The solid line in the graph below refers to annual sale of crackers and the dashed line refers to the annual crow births in India. Choose the most appropriate inference from the above data.

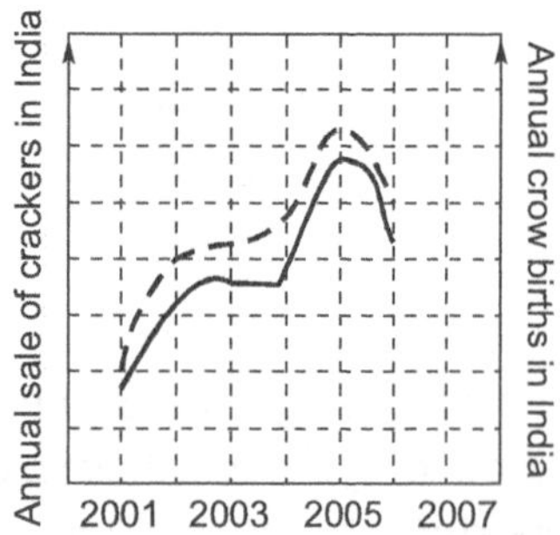

(a) There is a strong correlation between crow birth and cracker sales.
(b) Cracker usage increases crow birth rate.
(c) If cracker sale declines, crow birth will decline.
(d) Increased birth rate of crows will cause an increase in the sale of crackers.

[2018 : 2 M]

1.44 Two cars start at the same time from the same location and go in the same direction. The speed of the first car is 50 km/h and the speed of the second car is 60 km/h. The number of hours it takes for the distance between the two cars to be 20 km is _______.

(a) 1 (b) 2
(c) 3 (d) 6

[2019 : 1 M]

1.45 Ten friends planned to share equally the cost of buying a gift for their teacher. When two of them decided not to contribute, each of the other friends had to pay Rs. 150 more. The cost of the gift was Rs. _______.

(a) 12000 (b) 6000
(c) 3000 (d) 666

[2019 : 1 M]

1.46 In a college, there are three student clubs. Sixty students are only in the Drama club. 80 students are only in the Dance club, 30 students are only in the Maths club. 40 students are in both Drama and Dance clubs. 12 students are in both Dance and Maths clubs, 7 students are in both Drama and Maths clubs, and 2 students are in all the clubs. If 75% of the students in the college are not in any of these clubs, then the total number of students in the college is _______.

(a) 225 (b) 1000
(c) 975 (d) 900

[2019 : 2 M]

1.47 Three of the five students allocated to a hostel put in special requests to the warden. Given the floor plan of the vacant rooms, select the allocation plan that will accommodate all their requests.

Request by X : Due to pollen allergy. I want to avoid a wing next to the garden.

Request by Y : 1 want to live as far from the washrooms as possible, since I am very sensitive to smell.

Request by Z : I believe in Vaastu and so want to stay in the South-West wing.

The shaded rooms are already occupied. WR is washroom.

(a)

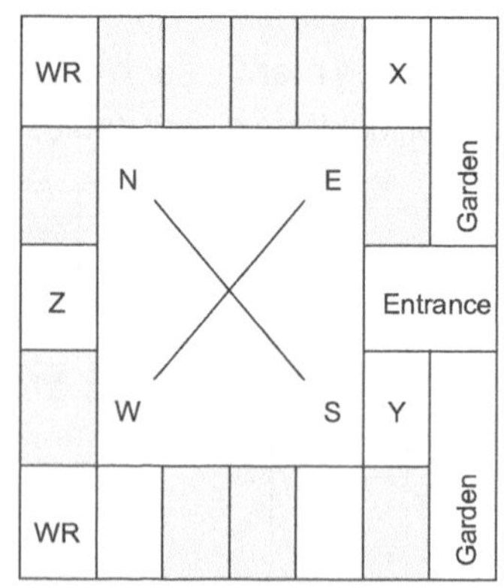

(b)

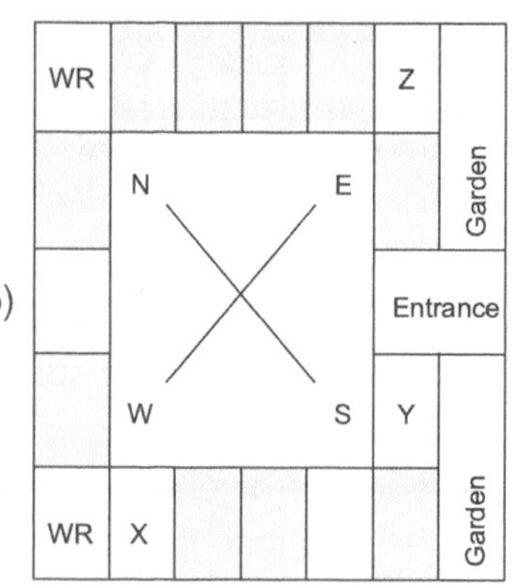

(c)

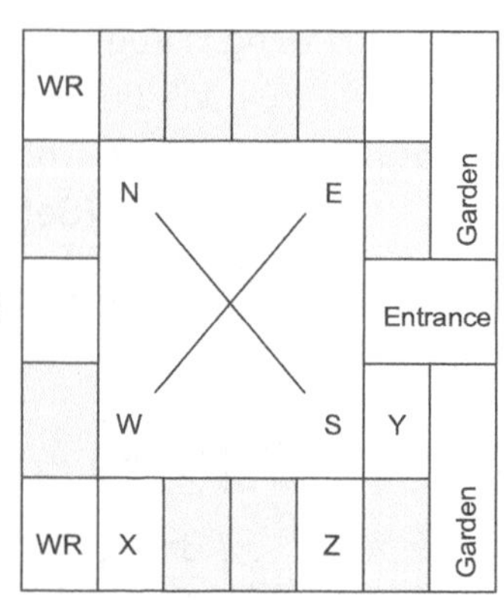

(d) 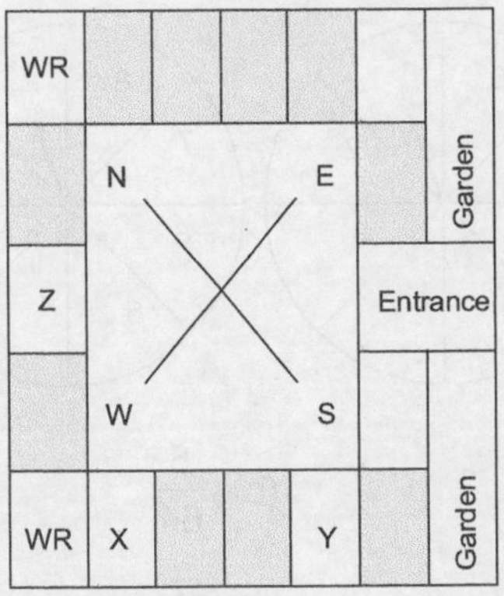

[2019 : 2 M]

1.48 The difference between the sum of the first $2n$ natural numbers and the sum of the first n odd natural numbers is ________.

(a) $n^2 - n$
(b) $n^2 + n$
(c) $2n^2 - n$
(d) $2n^2 + n$

[2020 : 1 M]

1.49 The distance between Delhi and Agra is 233 km. A car *P* started travelling from Delhi to Agra and another car *Q* started from Agra to Delhi along the same road 1 hour after the car *P* started. The two cars crossed each other 75 minutes after the car *Q* started. Both cars were travelling at constant speed. The speed of car *P* was 10 km/hr more than the speed of car *Q*. How many kilometers the car *Q* had travelled when the cars crossed each other?

(a) 66.6 (b) 75.2
(c) 88.2 (d) 116.5

[2020 : 2 M]

1.50 For a matrix $M = [m_{ij}]$; $i, j = 1, 2, 3, 4$, the diagonal elements are all zero and $m_{ij} = -m_{ji}$. The minimum number of elements required to fully specify the matrix is ________.

(a) 0 (b) 6
(c) 12 (d) 16

[2020 : 2 M]

1.51 The profit shares of two companies *P* and *Q* are shown in the figure. If the two companies have invested a fixed and equal amount every year, then the ratio of the total revenue of company *P* to the total revenue of company *Q*, during 2013-2018 is ________.

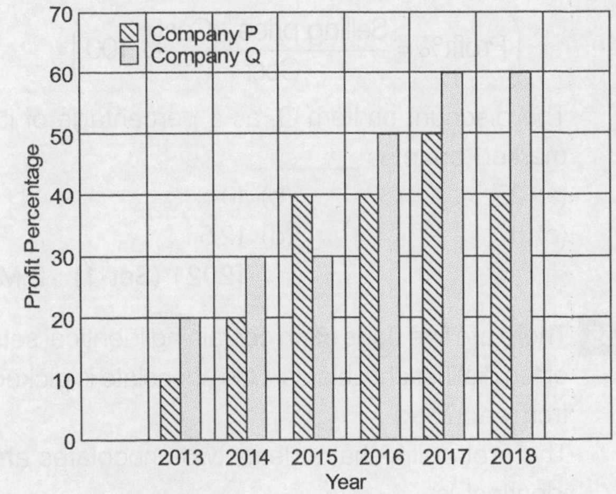

(a) 15 : 17 (b) 16 : 17
(c) 17 : 15 (d) 17 : 16

[2020 : 2 M]

1.52 The ratio of boys to girls in a class is 7 to 3. Among the options below, an acceptable value for the total number of students in the class is:

(a) 37 (b) 50
(c) 21 (d) 73

[2021 (Set-1) : 1 M]

1.53 We have 2 rectangular sheets of paper, M and N, of dimensions 6 cm × 1 cm each. Sheet M is rolled to form an open cylinder by bringing the short edges of the sheet together. Sheet N is cut into equal square patches and assembled to form the largest possible closed cube. Assuming the ends of the cylinder are closed, the ratio of the volume of the cylinder to that of the cube is ________.

(a) $\frac{9}{\pi}$ (b) 3π
(c) $\frac{\pi}{2}$ (d) $\frac{3}{\pi}$

[2021 (Set-1) : 2 M]

1.54

Items	Cost (₹)	Profit%	Marked price (₹)
P	5400	–	5860
Q	–	25	10000

Details of prices of two items P and Q are presented in the above table. The ratio of cost of item P to cost of item Q is 3 : 4. Discount is calculated as the difference between the marked price and the selling price. The profit percentage is calculated as the ratio of the difference between selling price and cost, to the cost

$$\left(\text{Profit\%} = \frac{\text{Selling price} - \text{Cost}}{\text{Cost}} \times 100\right)$$

The discount on item Q, as a percentage of its marked price, is ________.

(a) 25 (b) 10
(c) 5 (d) 12.5

[2021 (Set-1) : 2 M]

1.55 There are five bags each containing identical sets of ten distinct chocolates. One chocolate is picked from each bag.

The probability that at least two chocolates are identical is ________.

(a) 0.6976 (b) 0.3024
(c) 0.4235 (d) 0.8125

[2021 (Set-1) : 2 M]

1.56 A sphere of radius r cm is packed in a box of cubical shape.

What should be the minimum volume (in cm^3) of the box that can enclose the sphere?

(a) $\frac{r^3}{8}$ (b) r^3
(c) $2r^3$ (d) $8r^3$

[2022 : 1 M]

1.57 Pipes P and Q can fill a storage tank in full with water in 10 and 6 minutes, respectively. Pipe R draws the water out from the storage tank at a rate of 34 litres per minute. P, Q and R operate at a constant rate.

If it takes one hour to completely empty a full storage tank with all the pipes operating simultaneously, what is the capacity of the storage tank (in litres)?

(a) 26.8 (b) 60.0
(c) 120.0 (d) 127.5

[2022 : 1 M]

1.58 A game consists of spinning an arrow around a stationary disk as shown below. When the arrow comes to rest, there are eight equally likely outcomes. It could come to rest in any one of the sectors numbered 1, 2, 3, 4, 5, 6, 7 or 8 as shown. Two such disks are used in a game where their arrows are independently spun.

What is the probability that the sum of the numbers on the resulting sectors upon spinning the two disks is equal to 8 after the arrows come to rest?

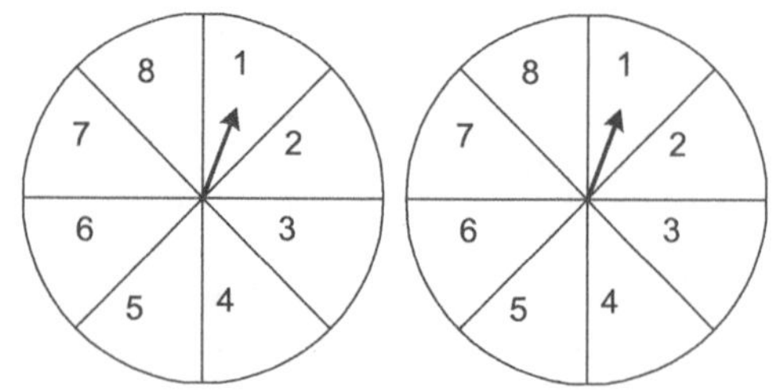

(a) $\frac{1}{16}$ (b) $\frac{5}{64}$
(c) $\frac{3}{32}$ (d) $\frac{7}{64}$

[2022 : 2 M]

1.59 Consider the following inequalities:

(i) $3p - q < 4$

(ii) $3q - p < 12$

Which one of the following expressions below satisfies the above two inequalities?

(a) $p + q < 8$ (b) $p + q = 8$
(c) $8 \le p + q < 16$ (d) $p + q \ge 16$

[2022 : 2 M]

1.60 In the given figure, $PQRS$ is a parallelogram with $PS = 7$ cm, $PT = 4$ cm and $PV = 5$ cm. What is the length of RS in cm? (The diagram is representative.)

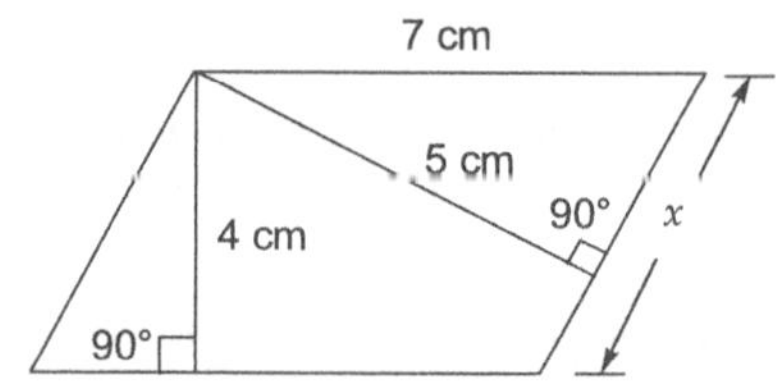

(a) $\frac{20}{7}$ (b) $\frac{28}{5}$
(c) $\frac{9}{2}$ (d) $\frac{35}{4}$

[2023 : 1 M]

1.61 A line of symmetry is defined as a line that divides a figure into two parts in a way such that each part is a mirror image of the other part about that line.

The given figure consists of 16 unit squares arranged as shown. In addition to the three black squares, what is the minimum number of squares that must be coloured black, such that both PQ and MN form lines of symmetry? (The figure is representative)

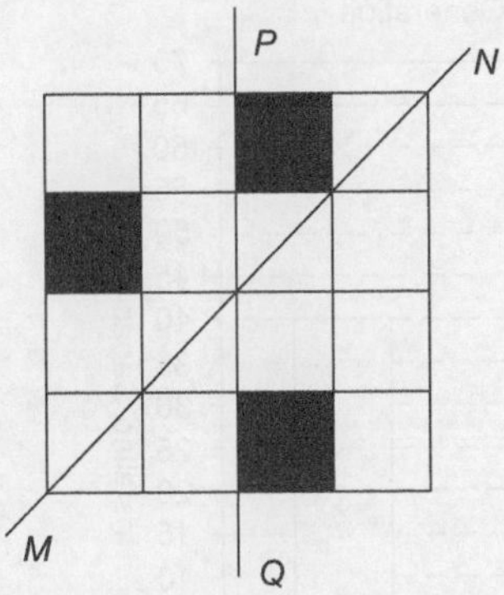

(a) 3 (b) 4
(c) 5 (d) 6

[2023 : 1 M]

1.62 To construct a wall, sand and cement are mixed in the ratio of 3 : 1. The cost of sand and that of cement are in the ratio of 1 : 2.
If the total cost of sand and cement to construct the wall is 1000 rupees, then what is the cost (in rupees) of cement used?
(a) 400 (b) 600
(c) 800 (d) 200

[2023 : 2 M]

1.63 The coefficient of x^4 in the polynomial $(x-1)^3(x-2)^3$ is equal to ______.
(a) 33 (b) –3
(c) 30 (d) 21

[2023 : 2 M]

1.64 Which one of the following shapes can be used to tile (completely cover by repeating) a flat plane, extending to infinity in all directions, without leaving any empty spaces in between them? The copies of the shape used to tile are identical and are not allowed to overlap.
(a) circle
(b) regular octagon
(c) regular pentagon
(d) rhombus

[2023 : 2 M]

1.65 In a locality, the houses are numbered in the following way:
The house-numbers on one side of a road are consecutive odd integers starting from 301, while the house-numbers on the other side of the road are consecutive even numbers starting from 302. The total number of houses is the same on both sides of the road.
If the difference of the sum of the house-numbers between the two sides of the road is 27, then the number of houses on each side of the road is
(a) 27 (b) 52
(c) 54 (d) 26

[2024 : 1 M]

1.66 For positive integers p and q, with

$$\frac{p}{q} \neq 1, \left(\frac{p}{q}\right)^{\frac{p}{q}} = p^{\left(\frac{p}{q}-1\right)}.$$

Then,
(a) $q^p = p^q$
(b) $q^p = p^{2q}$
(c) $\sqrt{q} = \sqrt{p}$
(d) $\sqrt[p]{q} = \sqrt[q]{p}$

[2024 : 1 M]

1.67 Which one of the given options is a possible value of x in the following sequence?

3, 7, 15, x, 63, 127, 255

(a) 35 (b) 40
(c) 45 (d) 31

[2024 : 1 M]

1.68 On a given day, how many times will the second-hand and the minute-hand of a clock cross each other during the clock time 12:05:00 hours to 12:55:00 hours?
(a) 51 (b) 49
(c) 50 (d) 55

[2024 : 1 M]

1.69 Three distinct sets of indistinguishable twins are to be seated at a circular table that has 8 identical chairs. Unique seating arrangements are defined by the relative positions of the people.
How many unique seating arrangements are possible such that each person is sitting next to their twin?
(a) 12 (b) 14
(c) 10 (d) 28

[2024 : 2 M]

1.70 The chart given below compares the Installed Capacity (MW) of four power generation technologies, T1, T2, T3, and T4, and their Electricity Generation (MWh) in a time of 1000 hours (h).

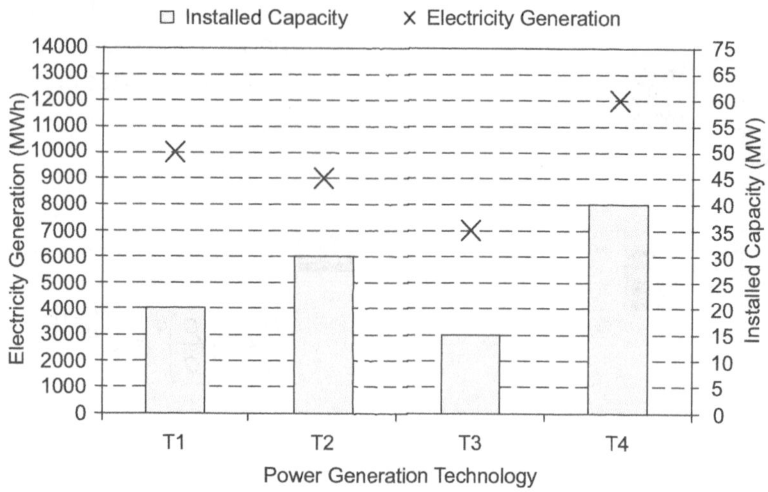

The Capacity Factor of a power generation technology is:

$$\text{Capacity Factor} = \frac{\text{Electricity Generation (MWh)}}{\text{Installed Capacity (MW)} \times 1000\text{ (h)}}$$

Which one of the given technologies has the highest Capacity Factor?

(a) T1
(b) T2
(c) T3
(d) T4

[2024 : 2 M]

1.71 During a half-moon phase, the Earth-Moon-Sun form a right triangle. If the Moon-Earth-Sun angle at this half-moon phase is measured to be 89.85°, the ratio of the Earth-Sun and Earth-Moon distances is closest to

(a) 328
(b) 382
(c) 238
(d) 283

[2024 : 2 M]

Answers Numerical Ability

1.1	(c)	**1.2**	(b)	**1.3**	(d)	**1.4**	(b)	**1.5**	(b)	**1.6**	(a)	**1.7**	(d)
1.8	(d)	**1.9**	(c)	**1.10**	(a)	**1.11**	(b)	**1.12**	(b)	**1.13**	(c)	**1.14**	(b)
1.15	(c)	**1.16**	(c)	**1.17**	(c)	**1.18**	(b)	**1.19**	(c)	**1.20**	(1300)	**1.21**	(180°)
1.22	(d)	**1.23**	(b)	**1.24**	(a)	**1.25**	(c)	**1.26**	(c)	**1.27**	(3)	**1.28**	(b)
1.29	(b)	**1.30**	(c)	**1.31**	(c)	**1.32**	(c)	**1.33**	(d)	**1.34**	(b)	**1.35**	(a)
1.36	(c)	**1.37**	(b)	**1.38**	(c)	**1.39**	(d)	**1.40**	(a)	**1.41**	(c)	**1.42**	(c)
1.43	(a)	**1.44**	(b)	**1.45**	(b)	**1.46**	(d)	**1.47**	(c)	**1.48**	(b)	**1.49**	(b)
1.50	(b)	**1.51**	(b)	**1.52**	(b)	**1.53**	(a)	**1.54**	(b)	**1.55**	(a)	**1.56**	(d)
1.57	(c)	**1.58**	(d)	**1.59**	(a)	**1.60**	(b)	**1.61**	(c)	**1.62**	(a)	**1.63**	(a)
1.64	(d)	**1.65**	(a)	**1.66**	(a)	**1.67**	(d)	**1.68**	(c)	**1.69**	(a)	**1.70**	(a)
1.71	(b)												

Explanations Numerical Ability

1.1 (c)

$$137 + 276 = 435$$

This is a octahedral addition thus,

$$731 + 672(8) = 1623$$

Alternative Method:

7 and 6 added is becoming five means the given two numbers are added on base 8.

$$\begin{array}{r} (137)_8 \\ +(276)_8 \\ \hline (435)_8 \end{array}$$

Hence we have to add the another two given set of numbers also on base 8.

$$\begin{array}{r} (731)_8 \\ +(672)_8 \\ \hline (1623)_8 \end{array}$$

Hence the overall problem was based on identifying base which was 8 and adding number on base 8.

1.2 (b)

Suppose : Hari's age : H, Gita's age : G, Saira's age : S, Irfan's age : I

- $H + G > I + S$
- Using Statement (2) both $G - S = 1$ or $S - G = 1$; G can't be oldest and S can't be youngest.
- There are no twins thus using statement (2) either GS or SG possible.

(a) **HSIG:** not possible as there is I between S and G which is not possible using statement (3).

(b) **SGHI:** SG order is possible, $S > G > H > I$ and $G + H > S + I$ (possible).
Because if $\{S = G + 1$; and $G = H + 1$ and $H = I + 2$ then $G + (I + 2) > (G + 1) + I\}$

(c) **IGSH:** according to this $I > G$ and $S > H$ thus adding these both inequalities we get $I + S > G + H$ which is opposite of statement (2) thus not possible.

(d) **IHSG:** according to this $I > H$ and $S > G$ thus adding both inequalities $I + S > H + G$ which is opposite of statement (2) thus not possible.

1.3 (d)

Per day work or rate of 5 skilled workers $= \dfrac{1}{20}$

$\Rightarrow$ Per day work or rate of one skill worker

$$= \frac{1}{5 \times 20} = \frac{1}{100}$$

Similarly Per day work or rate of 8 semiskilled workers $= \dfrac{1}{25}$

$\Rightarrow$ Per day work or rate of one semi-skill worker

$$= \frac{1}{8 \times 25} = \frac{1}{200}$$

And per day work or rate of 10 unskilled workers

$$= \frac{1}{30}$$

$\Rightarrow$ Per day work or rate of one semi-skill worker

$$= \frac{1}{10 \times 30} = \frac{1}{300}$$

Thus total per day work of 2 skilled, 6 semiskilled and 5 unskilled workers

$$= \frac{2}{100} + \frac{6}{200} + \frac{5}{300} = \frac{12 + 18 + 10}{600}$$

$$= \frac{40}{600} = \frac{1}{15}$$

Thus time to complete the work is 15 days.

Alternative method:

Let one day work of skilled semi-skilled and unskilled worker be a, b, c units respectively.

$5a \times 20 = 8b + 25 = 10c \times 30 =$ Total unit of work,

$$100a = 200b = 300c$$

$$a = 2b = 3c$$

$$\Rightarrow \quad b = \frac{a}{2} \text{ and } c = \frac{a}{3}$$

Given that 2 skilled, 6 semi-skilled and 5 unskilled workers are working. Let they finish the work in 'x' days.

$$(2a + 6b + 5c)x = 5a + 20$$

$$= \text{Total units of work}$$

$$\left(2a + 3a + \frac{5}{3}a\right)x = 5a \times 20$$

$$\frac{20a}{3}x = 5a \times 20$$

$$x = 15 \text{ days}$$

1.4 (b)

We have to make 4 digit numbers, so the number should be start with 3 or 4, two cases possible;

Case (1) thousands digit is 3

Now other three digits may be any of 2, 2, 3, 3, 4, 4, 4, 4.

(a) Using 2, 2, 3
$\Rightarrow$ 223, 232, 322 -------
$\left(\frac{3!}{2!} = 3 \text{ numbers are possible}\right)$

(b) Using 2, 2, 4 $\Rightarrow$ 224, 242, 422
$\left(\frac{3!}{2!} = 3 \text{ numbers are possible}\right)$

(c) Using 2, 3, 3 $\Rightarrow$ 233, 323, 332
$\left(\frac{3!}{2!} = 3 \text{ numbers are possible}\right)$

(d) Using 2, 3, 4 $\Rightarrow$ 234, 243, 324, 342, 423, 432
(3! = 6 numbers are possible)

(e) Using 2, 4, 4 $\Rightarrow$ 244, 424, 442
$\left(\frac{3!}{2!} = 3 \text{ numbers are possible}\right)$

(f) Using 3, 3, 4 $\Rightarrow$ 334, 343, 433
$\left(\frac{3!}{2!} = 3 \text{ numbers are possible}\right)$

(g) Using 3, 4, 4 $\Rightarrow$ 344, 434, 443
$\left(\frac{3!}{2!} = 3 \text{ numbers are possible}\right)$

(h) Using 4, 4, 4 $\Rightarrow$ 444
$\left(\frac{3!}{3!} = 1 \text{ numbers are possible}\right)$

Total 4 digit numbers in case 1 = 3 + 3 + 3 + 6 + 3 + 3 + 3 + 1 = 25

Case (2) thousands digit is 4;

Now other three digits may be any of 2, 2, 3, 3, 3, 4, 4, 4.

(a) Using 2, 2, 3 $\Rightarrow$ 223, 232, 322
$\left(\frac{3!}{2!} = 3 \text{ numbers are possible}\right)$

(b) Using 2, 2, 4 $\Rightarrow$ 224, 242, 422
$\left(\frac{3!}{2!} = 3 \text{ numbers are possible}\right)$

(c) Using 2, 3, 3 $\Rightarrow$ 233, 323, 332
$\left(\frac{3!}{2!} = 3 \text{ numbers are possible}\right)$

(d) Using 2, 3, 4 $\Rightarrow$ 234, 243, 324, 342, 423, 432
$\left(\frac{3!}{2!} = 3 \text{ numbers are possible}\right)$

(e) Using 2, 4, 4 $\Rightarrow$ 244, 424, 442
$\left(\frac{3!}{2!} = 3 \text{ numbers are possible}\right)$

(f) Using 3, 3, 3 $\Rightarrow$ 333
$\left(\frac{3!}{3!} = 1 \text{ numbers are possible}\right)$

(g) Using 3, 3, 4 $\Rightarrow$ 334, 343, 433
$\left(\frac{3!}{2!} = 3 \text{ numbers are possible}\right)$

(h) Using 3, 4, 4 $\Rightarrow$ 344, 434, 443
$\left(\frac{3!}{2!} = 3 \text{ numbers are possible}\right)$

(i) Using 4, 4, 4 $\Rightarrow$ 444
$\left(\frac{3!}{3!} = 1 \text{ numbers are possible}\right)$

Total 4 digit numbers in case 2 = 3 + 3 + 3 + 6 + 3 + 3 = 1 + 3 + 1 = 26

Thus total 4 digits numbers using case (1) and case (2) = 25 + 26 = 51

***Alternative Method / Shortcut method:**

As the number is greater than 3000. So thousand's place can be their 3 or 4.

Let's consider the following two cases

Case (I) When thousand's place is 3.

3 a b c

If there is no restriction on number of two's, three's and four's. Then each of *a*, *b*, *c* can be filled with 2 or 3 or 4 each in 3 ways.

So 3 × 3 × 3 = 27 numbers are there. Out of which 3222, 3333 are invalid as 2 can be used twice and three thrice only so number of such valid numbers beginning with 3 are 27 – 2 = 25.

Case (II) When thousand's place is 4.

4 a b c

Without restriction on number of 2's, 3's and 4's *a*, *b*, *c* (as explained in case I) can be filled in 27 ways.

Out of these 27 numbers, 4 2 2 2 is only invalid as two have to be used twice only.

So valid numbers are 27 – 1 = 26.

Total numbers from Case (I) and Case (II) 25 + 26 = 51.

1.5 (b)

$$\log(P) = \frac{1}{2}\log(Q) = \frac{1}{3}\log(R)$$

$$\Rightarrow \log(P) = \log(Q)^{1/2}$$

$$= \log(R)^{1/3} = K$$

$$\Rightarrow P = (Q)^{1/2} = (R)^{1/3} = K$$

$$\Rightarrow P = K, Q = K^2, R = K^3 \quad \text{...(i)}$$

Now, only option (*B*) $Q^2 = PR$ satisfies

$$Q^2 = (K^2)^2 = K^4 \quad \text{From ...(i)}$$

$$P.R = K * K^3 = K^4 \quad \text{From...(i)}$$

Here, $Q^2 = PR$ holds true

1.6 (a)

(T.C.) Total cost = *V* + *f*

$$\boxed{T.C = 4q + \frac{100}{q}}$$

As we have to minimize total cost.

Using options,

(a) $q = 5$, T.C. $= 4 \times 5 + \frac{100}{5} = 40$

(b) $q = 4$, T.C. $= 4 \times 4 + \frac{100}{4} = 41$

(c) $q = 7$, T.C. $= 4 \times 7 + \frac{100}{7} = 42.285$

(d) $q = 6$, T.C. $= 4 \times 6 + \frac{100}{6} = 40.\overline{66}$

Hence, T.C. is minimum at $q = 5$.

∗ always put options to get answer fast.

Alternate solution:

or $$\text{T.C.} = 4q + \frac{100}{q}$$

For minimum $$\frac{d}{dq}\left(4q + \frac{100}{q}\right) = 0$$

$$\Rightarrow q^2 = 25 \text{ or } q = 5$$

$$\frac{d}{dq}\left(4q + \frac{100}{q}\right)_{q=5} > 0$$

Hence, TC (Total cost) is minimum at $q = 5$.

1.7 (d)

As we can understand that the danger of a microbe to human being will be directly proportional to potency and growth. At the same time, it will be inversely proportional to toxicity defined (more dangerous will a microbe be if lesser of its milligram is required). So level of dangerous

$\alpha \frac{P\uparrow \times G\uparrow}{T\downarrow}$ where *P*, *G* and *T* are the potency, growth and toxicity as defined in question.

So $$D_i = \frac{KPG}{T} \quad \text{...(i)}$$

Where *K* is constant of proportionality. So level of dangerous of *S* will be maximum given by

$$D_S = \frac{0.8 \times \pi(10\text{mm})^2}{200}.$$

Similar calculation for D_P, D_Q, D_R can be done based on (i) to find out that D_S is maximum and so most dangerous among them.

1.8 (d)

Shortcut method:

Every time if we take 1 litre of mixture out and replace with water, content of pure spirit will keep on reducing by 10%.

So, final quantity of spirit after 3 such operations is 10 × 0.9 × 0.9 × 0.9 = 7.29 litres.

Alternate solution:

$$\frac{\text{Quantity of spirit left after } n^{\text{th}} \text{ operation}}{\text{Intial quantity of spirit}}$$

$$= \left(\frac{a-b}{a}\right)^n = \left(1 - \frac{b}{a}\right)^n$$

where 'a' is initial quantity of pure spirit and 'b' is quantity taken out and replaced every time.

Hence, quantity of spirit left after 3rd operation

$$= \text{Initial quantity} \times \left(1 - \frac{1}{10}\right)^3$$

$$= 10 \times 0.9 \times 0.9 \times 0.9 = 7.29 \text{ litres}$$

1.9 (c)

Let 'y' be the backlog with transporter and 'x' be the number of orders each day. So, as per conditions given in question,

$$4x + y = 28 \quad \text{...(i)}$$

$$10x + y = 30 \quad \text{...(ii)}$$

Solving (i) and (ii),

$$x = \frac{1}{3} \text{ and } y = \frac{80}{3}$$

Since, we need to find out number of trucks so that no pending order will be there at the end of 5th day.

$$5x + y = n \times 5$$

So we need to find 'n', where 'n' is the number of trucks required.

$$n = \frac{5x + y}{5} = \frac{5 \times \frac{1}{3} + \frac{80}{3}}{5} = \frac{\frac{85}{3}}{5}$$

Hence, 5.66 truck will be required. As number of trucks have to be natural number. Hence, 6 trucks will be required.

1.10 (a)

Total Cost $C = 5q^2$

Total Sales Revenue,

$$S = 50q$$

$$\text{Profit } (P) = (S - C) = 50q - 5q^2$$

As the profit have to be maximized,

$$P = 50q - 5q^2$$

Using options,

(a) $q = 5$

$$P = 50 \times 5 - 5 \times 5^2 = 125$$

(b) $q = 10$

$$P = 50 \times 10 - 5 \times 10^2 = 0$$

(c) $q = 15$

$$P = 50 \times 15 - 5 \times 15^2 = \text{–ive}$$

(d) $q = 25$

$$P = 50 \times 25 - 5 \times 25^2 = \text{–ive}$$

Hence the maximum profit will happen at $q = 5$

Alternative solution:

$$P = 50q - 5q^2$$

For max/min $\frac{dp}{dq} = 0$

$\Rightarrow \quad 50 - 10q = 0$

or, $\quad q = 10$

$$\left.\frac{d^2p}{dq^2}\right]_{q=5} = -10 \text{ which is negative}$$

Hence maximum profit will happen at,

$$q = 5$$

1.11 (b)

Let total 100 shock absorber are supplied

So X supplies = 60 (40% of 100)

Y supplies = 40 (60% of 100)

Reliable supply by X = 96% of 60 = 57.6

Reliable supply by Y = 72% of 40 = 28.8

So the required probability that random chosen reliable shock absorber is made by Y is

Required probability

$$= \frac{28.8}{57.6 + 28.8} = 0.333 \approx 0.334$$

1.12 (b)

$$y = 2x - 0.1x^2$$

For y (height) to be maximum $\frac{dy}{dx} = 0$

$$\frac{dy}{dx} = 2x - 2x = 0$$

$\Rightarrow \quad x = 10$

$$\left.\frac{d^2y}{dx^2}\right]_{x=10} = -0.2 \text{ (–ive)}, \quad \frac{d^2y}{dx^2} < 0$$

Hence maximum height will be at $x = 10$

$$y = 2 \times 10 - 0.1 \times (10)^2 = 10$$

Maximum height = 10 meters.

1.13 (c)

Adding each and every observation by constant increases the arithmetic mean by same constant. Also multiplying each and every observation by constant multiplies the arithmetic mean by same constant. While the standard deviation will not show similar effect.

Alternative solution:

$$\frac{x_1 + x_2 + x_3 \cdots x_N}{N} = \bar{x}$$

$$\frac{(x_1+7)+(x_2+7)+\cdots+(x_n+7)}{N} = \bar{x}+7$$

and $$\frac{2(x_1+x_2\cdots x_N)}{N} = 2\bar{x}$$

Hence only P, R are correct.

1.14 (b)

Given : $3 \le X \le 5$ and $8 \le Y \le 11$

So, $$\frac{1}{11} \le \frac{1}{Y} \le \frac{1}{8}$$

$$\frac{3}{11} \le \frac{X}{Y} \le \frac{5}{8}$$

1.15 (c)

$$V = 80 - 32t$$

when $V = 32$ m/s; when $V = 64$ m/s

$32 = 80 - 32t$; $64 = 80 - 32t$

$32t = 48$; $32t = 16$

$t = \frac{3}{2}$; $t = \frac{1}{2}$

So, the correct answer is $\left(\frac{1}{2}, \frac{3}{2}\right)$.

1.16 (c)

$P(M1) = 0.6$, $P(M2) = 0.4$

$P(D/M1) = 0.02$, $P(D/M2) = 0.03$

$$P(M2/D) = \frac{P(D/M2)\cdot P(M2)}{P(D/M1)\cdot P(M1) + P(D/M2)\cdot P(M2)}$$

$$= \frac{0.03\times 0.4}{0.02\times 0.6 + 0.03\times 0.4}$$

$$= \frac{0.012}{0.012+0.012} = \frac{0.012}{0.024} = 0.5$$

1.17 (c)

Total number of tourists = 13500

(England + France)'s tourists = 4500

$\because$ (England + France)'s tourist = $\frac{1}{3}$ (total tourists)

1.18 (b)

$$|-2X+9| = 3$$

$-2X + 9 = -3$; $-2X + 9 = 3$

$-2X = -12$; $-2X = -6$

$X = 6$; $X = 3$

So, $|-X| - X^2$ at $X = 6$

$\Rightarrow$ $6 - 36 = -30$

$|-X| - X^2$ at $X = 3$

$\Rightarrow$ $3 - 9 = -6$

1.19 (c)

$$y = 5x^2 + 3 \Rightarrow \frac{dy}{dx} = 10$$

Slope of tangent = $\left(\frac{dy}{dx}\right)_{x=0,y=3} = 10 \times 0 = 0$

Slope = 0 $\Rightarrow$ tangent is parallel to x-axis.

1.20 (1300)

Total product cost = 50000 + 800Q

= 50000 + 800(100) = 13000

This is the cost for 10 tonnes.

So, per ton product cost = $\frac{130000}{100}$ = Rs. 1300

1.21 (180°)

Sum of the angles of quadrilateral = 360°

$\therefore$ $3x + 4x + 5x + 6x = 360°$

$18x = 360$

$$x = \frac{360}{18} = 20$$

$\therefore$ Smallest angle of quadrilateral = 3 × 20 = 60

Largest angle of quadrilateral = 6 × 20 = 120

$\therefore$ Smallest angle of triangle = $60 \times \frac{2}{3} = 40$

Largest angle of triangle = 2 × 40 = 80

$\therefore$ Third angle of triangle = 180 – 40 – 80 = 60

Required sum = 60 + 120 = 180°

1.22 (d)

Let the population of county Y is P

$\therefore$ Population of country X is $3P$

% of people taken than 6 ft

$$= \frac{\frac{3P\times 1}{100} + \frac{P\times 2}{100}}{3P+P} = 1.25$$

1.24 (a)

$$\log_x\left(\frac{5}{7}\right) = -\frac{1}{3}$$

$$\frac{5}{7} = x^{-1/3}$$

$$\Rightarrow \quad x = \left(\frac{7}{5}\right)^3 = \frac{343}{125}$$

1.25 (c)

$$66 \,\square\, 6 = \frac{66-6}{66+6} = \frac{5}{6}$$

$$66 \,\diamond\, 6 = \frac{66+6}{66-6} = \frac{6}{5}$$

$$(66 \,\square\, 6) \rightarrow (66 \,\diamond\, 6) = \frac{5}{6} \times \frac{6}{5} = 1$$

1.26 (c)

Number of faces per cube of 1 unit = 6
Total number of cubes of 1 unit = 9 × 3 = 27
Total number of faces of cubes of 1 unit = 27 × 6 = 162
Total number of faces of cubes of 1 unit visible = 9 × 6 = 54
Total number of faces of cubes of 1 unit non-visible = 162 – 54 = 108

$$\therefore \frac{\text{Number of faces of cubes of 1 unit visible}}{\text{Number of faces of cubes of 1 unit non-visible}} = \frac{54}{108} = \frac{1}{2}$$

1.27 (3)

In all five rows the middle number is equal to the sum of the remaining numbers divided by 2.

First row:

$$5 = \frac{6+4}{2}$$

Second row:

$$7 = \frac{7+4+2+1}{2}$$

Third row:

$$8 = \frac{1+9+2+1+2+1}{2}$$

Fourth row:

$$5 = \frac{4+1+2+3}{2}$$

$$\therefore \text{ Missing number} = \frac{3+3}{2} = 3$$

1.29 (b)

Following figure can be drawn:

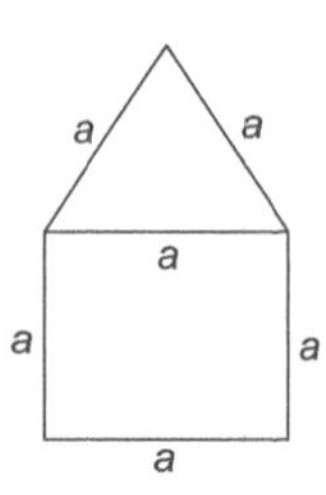

Perimeter of window,

$$5a = 6 \text{ m}$$

$$a = 1.2 \text{ m}$$

Complete area for window

= Area of square + Area of equi Δ

$$= (1.2)^2 + \frac{\sqrt{3}}{4}(1.2)^2 \approx 2.06 \text{ m}^2$$

1.30 (c)

Lesser standard deviation for group Q will mean that the observation sin group Q are closer and hence narrower is the range of marks scored in group Q as compared to group P. Other options (a), (b), (d) remains unconfined.

1.32 (c)

$$V = \frac{1}{3}\pi(R)^2 H$$

$$V' = \frac{1}{3}\pi(1.5R)^2 H$$

$$= \frac{1}{3}\pi\, 2.25\, R^2\, H = 2.25\text{ V}$$

Hence increase 125% as,

$$\left[\frac{2.25V - V}{V}\right] \times 100 = 125\%$$

1.33 (d)

1, x, x, x, y, y, 9, 16, 18

$$\text{Mean} = \frac{1+3x+2y+9+16+18}{9}$$

$$= \frac{3x+2y+44}{9} \quad \text{...(i)}$$

Median ⇒ As number of terms are odd i.e. 9, so median will be middle i.e. y.
Mode ⇒ Number showing maximum frequency of repetition. So mode is x.
Given ⇒ Mean = Median = 2 Mode

$$\Rightarrow \quad y = 2x \quad \text{...(ii)}$$

From equation (i) and (ii), we get,

$$x = 4,\ y = 8$$

Value of y is 8

1.34 (b)

Unit digit:

$$(2171)^7 + (2172)^9 + (2173)^{11} + (2174)^{13}$$

$$(1)^{any} + (2)^{4n+1} + (3)^{4n+3} + (4)^{2n+1}$$

$$1 + 2 + 7 + 4 = 14$$

1.35 (a)

P	30 min ×	2 = 60 min
Q	20 min ×	3 = 60 min
R	60 min ×	1 = 60 min
S	15 min ×	4 = 60 min

M_1 [P] [Q] = 2 hrs

M_2 [R] [S] = 2 hrs

1.36 (c)

Clearly we can see from graph both (i) and (ii) are correct.

1.37 (b)

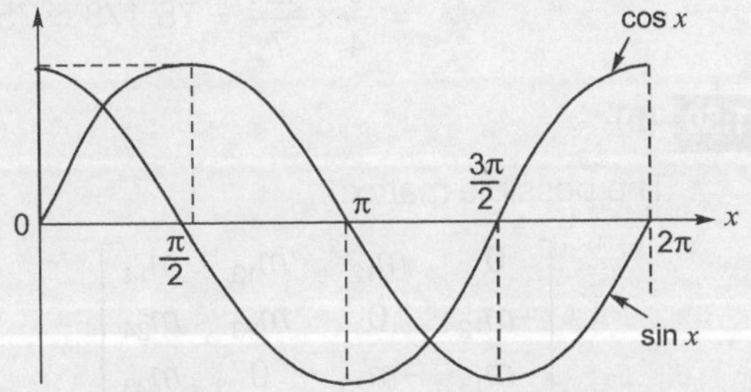

From the curve it is clear that sin$x$ and cos$x$ both are decreasing in the interval $\left(\frac{\pi}{2}, \pi\right)$.

1.38 (c)

Area of equilateral triangle = $\sqrt{3}$

$$\frac{\sqrt{3}}{4}a^2 = \sqrt{3}$$

$$a^2 = 4 \Rightarrow a = 2$$

$$\text{perimeter} = 3a = 3 \times 2 = 6$$

1.39 (d)

(i) Cuboid volume = $8 \times 10 \times 6 = 480$ cm^3

(ii) Cube volume = $8 \times 8 \times 8 = 512$ cm^3

(iii) Cylinder volume = $\pi r^2 h$

$$= \frac{22}{7} \times 7 \times 7 \times 7 = 1078 \text{ cm}^3$$

(iv) Sphere volume = $\frac{4}{3}\pi r^3 = \frac{4}{3}\pi \times (7)^3$

$$= 1436.75 \text{ cm}^3$$

Hence, the descending order of their volumes is (iv), (iii), (ii), (i).

1.40 (a)

Total distance = x

(Onward journey) S_1 = 60 km/h

(Return journey) S_2 = 90 km/h

$$\text{Average speed} = \frac{\text{Total distance}}{\text{Total time}}$$

$$= \frac{x + x}{\frac{x}{60} + \frac{x}{90}} = \frac{2x}{x\left[\frac{3+2}{90 \times 2}\right]}$$

$$= \frac{2 \times 90 \times 2}{5} = 72 \text{ km/h}$$

1.41 (c)

Number of parallelogram

$$= {}^4C_2 \times {}^5C_2 = 6 \times 10 = 60$$

1.42 (c)

$$x^2 + x = 1$$

$$x(x+1) = 1$$

$$x + 1 = \frac{1}{x}$$

$$\left(x - \frac{1}{x}\right) = (-1)$$

Squaring above equation,

$$x^2 + \frac{1}{x^2} - 2 = (+1)$$

$$x^2 + \frac{1}{x^2} = 3$$

Squaring above equation,

$$x^4 + \frac{1}{x^4} + 2 = 9$$

$$x^4 + \frac{1}{x^4} = 7$$

1.44 (b)

Speed of car A = 50 km/hr

Speed of car B = 60 km/hr

Since, both cars A and B are moving in same direction, the relative speed = 60 – 50 = 10 km/hr

Distance required between them = 20 km

$$\therefore \quad \text{Time} = \frac{\text{Distance}}{\text{Speed}} = \frac{20}{10} = 2 \text{ hrs}$$

1.45 (b)

Let share of each student = x

Total cost of gift= $10 \times x$

$$x = 8(x + 150) = 600$$

$$\text{Total cost} = 10 \times 600 = 6000$$

1.46 (d)

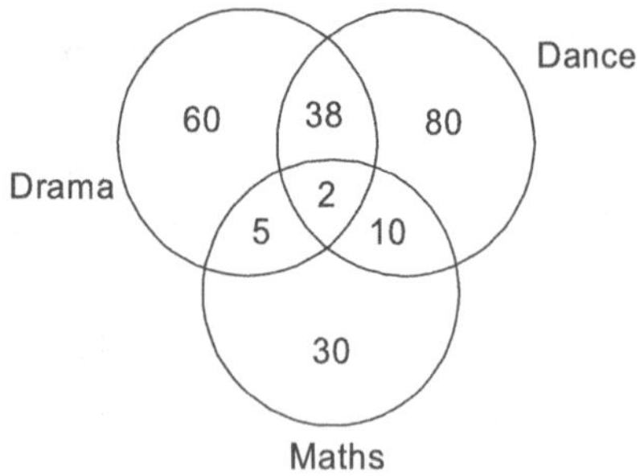

Total number of students = 60 + 80 + 30 + 38 + 5 + 10 + 2 = 225

$$25\% = 225$$

$$\Rightarrow \quad 100\% = \frac{225}{25} \times 100 = 900$$

1.48 (b)

Sum of first $2n$ natural number :
1, 2, 3, 4, ..., $2n$

So, $\frac{2n}{2}[1+2n] = n(2n+1)$

Sum of first n odd natural number
1, 3, 5, ..., n^{th} terms.
n^{th} term of this series

$$T_n = a + (n-1)d$$
$$= 1 + (n-1) \times 2$$
$$= 1 + 2n - 2 = 2n - 1$$

So, series is, 1, 3, 5, 7, ... $(2n - 1)$

$$S_n = \frac{n}{2}[a + l]$$
$$= \frac{n}{2}[1 + 2n - 1] = \frac{n}{2} \times 2n = n^2$$

So, the difference is

$$= n(2n+1) - n^2$$
$$= 2n^2 + n - n^2$$
$$= n^2 + n$$

1.49 (b)

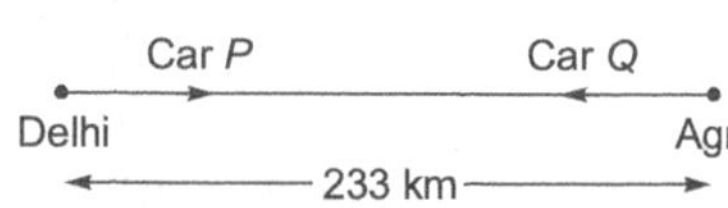

Let the speed of car Q is V km/hr, then the speed of car P is $(V + 10)$ km/hr.

Travelling time is $\frac{5}{4}$ hours for Q and $\frac{9}{4}$ hours for P before they meet each other.

Hence,

$$\frac{9}{4}(V+10) + \frac{5}{4} \times V = 233$$
$$\frac{9V}{4} + \frac{90}{4} + \frac{5V}{4} = 233$$
$$\frac{14V}{4} = 233 - \frac{45}{2}$$
$$\frac{14V}{4} = \frac{44}{2}$$
$$V = \frac{421}{7} \text{ km/hr}$$

So, distance travelled by Q before they crossed each other

$$= \frac{5}{4} \times \frac{421}{7} = 75.178 \simeq 75.2 \text{ km}$$

1.50 (b)

The possible matrix:

$$\begin{bmatrix} 0 & m_{12} & m_{13} & m_{14} \\ -m_{12} & 0 & m_{23} & m_{24} \\ -m_{13} & -m_{23} & 0 & m_{23} \\ -m_{14} & -m_{24} & -m_{34} & 0 \end{bmatrix}$$

So, the required element to fully specify the matrix are : $m_{12}, m_{13}, m_{14}, m_{23}, m_{24}, m_{34}$.
Hence, 6 elements are required to fully specify the matrix.

1.51 (b)

Let x is the amount invested by both each year.
Then, revenue of company P

$= x \times 1.1 + x \times 1.2 + x \times 1.4 + x \times 1.4 + x \times 1.5 + x \times 1.4$

$= x(1.1 + 1.2 + 1.4 + 1.4 + 1.5 + 1.4) = 8x$

Revenue of company Q

$= x(1.2 + 1.3 + 1.3 + 1.5 + 1.6 + 1.6) = 8.5x$

$$\text{So, } \frac{\text{Revenue of } P}{\text{Revenue of } Q} = \frac{8x}{8.5x} = \frac{16}{17} \text{ m}$$

1.52 (b)

Boys = $7x$, Girls = $3x$
Total number of students = $10x$
Now, $7x$ and $3x$ should be integers.
Hence, x should be integer.
So $10x$ should be an integer.
Only 50 satisfies the above constraints.

1.53 (a)

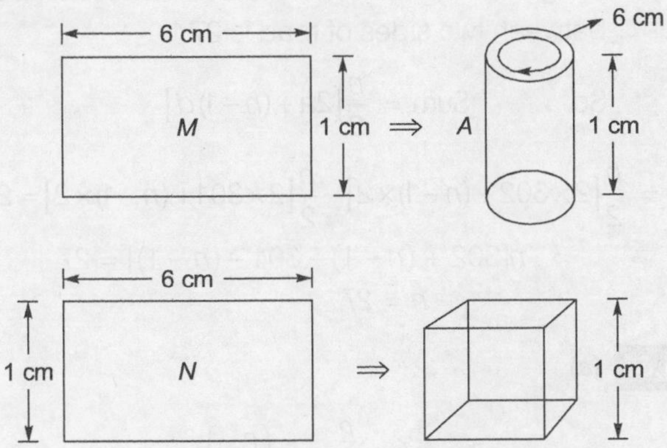

$$\text{Volume of cylinder} = \pi r^2 h$$

Now, $2\pi r = 6$ (figure A)

$$r = \frac{3}{\pi}$$

$$\text{Volume of cylinder} = \pi \times \frac{3}{\pi} \times \frac{3}{\pi} \times 1 = \frac{9}{\pi}$$

$$\text{Volume of cube} = (1)^3$$

$$\text{Ratio} = \frac{\frac{9}{\pi}}{1} = \frac{9}{\pi}$$

1.54 (b)

$$CP_P = 5400,\ MP_P = 5860$$

$$\frac{5400 \times 4}{5} = CP_Q$$

$$CP_Q = 7200,\ SP_Q = ?$$

$$\text{Profit \%} = \frac{\text{Profit}}{\text{CP}} \times 100 \text{ [for Q]}$$

$$25 = \left(\frac{SP - 7200}{7200}\right) 100$$

$$SP_Q = 9000$$

$$\text{Discount Q} = MP_Q - SP_Q = 1000$$

$$\text{Discount \%} = \frac{1000}{10000} \times 100 = 10\%$$

1.55 (a)

10 distinct chocolates, 5 bags

$$P(\text{at least 2}) = 1 - P(\text{exactly 1})$$

$$P(\text{exactly 1}) = P(\text{all different})$$

$$= \frac{10 \times 9 \times 8 \times 7 \times 6}{10^5}$$

$$P(\text{atleast 2}) = 1 - \frac{10 \times 9 \times 8 \times 7 \times 6}{10^5}$$

$$= 1 - 0.3024 = 0.6976$$

1.56 (d)

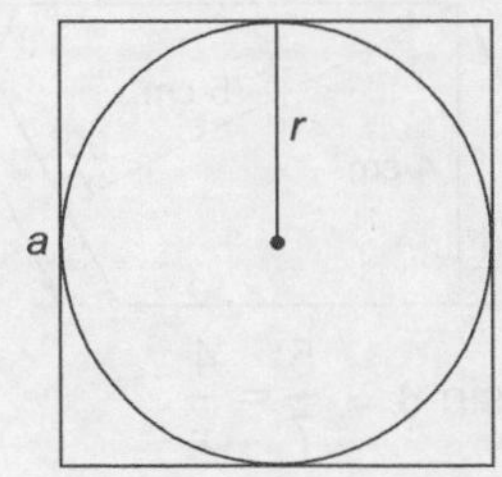

$$\text{Diameter of sphere} = \text{Side of cube}$$

$$a = 2r$$

$$\text{Volume of cube} = (2r)^3 = 8r^3$$

1.57 (c)

Let the capacity of storage tank is x liters. Amount of water drawn only by pipe R is in one minute.

$$\frac{x}{60} + \left(\frac{x}{6} + \frac{x}{10}\right) = +\frac{17x}{60}$$

Total time taken R to draw a storage tank

$$= \frac{x}{\frac{17x}{60}} = \frac{60}{17} \text{min}$$

$$\text{Capacity of tank} = \frac{60}{17} \times 34 = 120 \text{ liters}$$

1.58 (d)

$$\text{Probability of getting only one number} = \frac{1}{8}$$

As they are independent spins probability of getting one number for each disc is

$$P = \frac{1}{8} \times \frac{1}{8}$$

Possible outcomes to the sum of 8

= (1, 7), (2, 6), (3, 5), (4, 4), (5, 3), (6, 2), (7, 1)

$$\text{Required probability} = \frac{7}{64}$$

1.59 (a)

Adding both equation

$$2p + 2q < 16$$

$$p + q < 8$$

1.60 (b)

Area of parallelogram = Base × Height

So, $4 \times 7 = 5 \times x$

$$x = \frac{28}{5} \text{ cm}$$

Alternate Method:

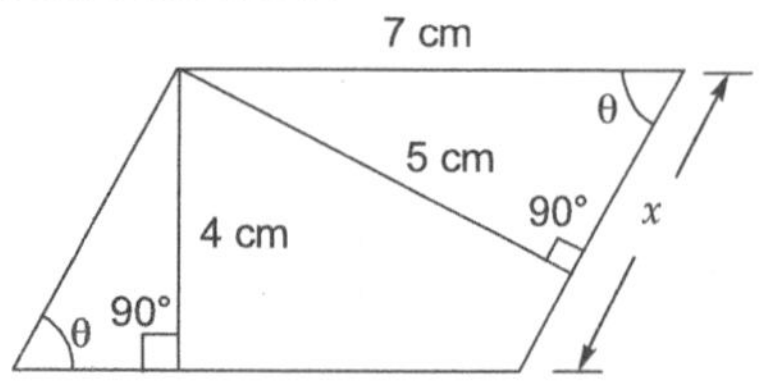

$$\sin\theta = \frac{5}{7} = \frac{4}{x}$$

$$\Rightarrow \qquad x = \frac{28}{5} \text{ cm}$$

1.61 (c)

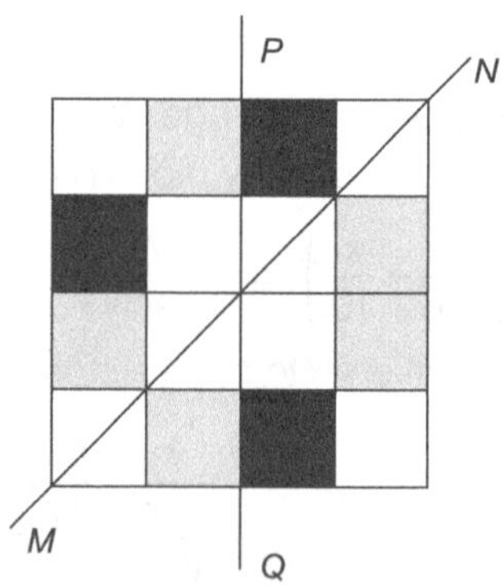

1.62 (a)

	Sand	Cement
Quantity	$3x$	x
Cost	y	$2y$
Total cost	$3xy$	$2xy = 1000$

$$\frac{S}{C} = \frac{3xy}{2xy} = \frac{3}{2} = \frac{600}{400}$$

$$C = \frac{4}{10} \times 1000$$

Cost of cement = 400

1.63 (a)

Given: $(x-1)^3(x-2)^3$

$[x^3 - 1 - 3x^2 + 3x]\,[x^3 - 8 - 6x^2 + 12x]$ here we consider the coefficient of x^4 only.

$$12 + 18 + 3 = 33$$

1.65 (a)

House numbers on one side of road are 301, 303, 305 ________

Similarly on other side of road house number are 302, 304, 306 ________

Assuming total number of house on each side of road is n.

It is clear that both the above are A.P. series.

It is given that difference of sum of house numbers between two sides of road is 27.

$$\text{So,} \qquad \text{Sum} = \frac{n}{2}[2a + (n-1)d]$$

$$= \frac{n}{2}[2\times 302 + (n-1)\times 2] - \frac{n}{2}[2\times 301 + (n-1)\times 2] = 27$$

$$\Rightarrow n[302 + (n-1) - 301 - (n-1)] = 27$$

$$n = 27$$

1.66 (a)

$$\frac{p}{q} \neq 1, \qquad \left(\frac{p}{q}\right)^{\frac{p}{q}} = p^{\left(\frac{p}{q}-1\right)}$$

$$\Rightarrow \qquad \frac{(p)^{p/q}}{(q)^{p/q}} = \frac{(p)^{p/q}}{p}$$

$$q^{p/q} = p$$

$$\text{So,} \qquad p^q = q^p$$

1.67 (d)

The given sequence is as following 3, 7, 15, x, 63, 127, 255

It is clear by observation that next term in given sequence is obtained by multiplying by two and adding one.

3 → 7 → 15 → 31 → 63 → 127 → 255

3 × 2 + 1   7 × 2 + 1   15 × 2 + 1   31 × 2 + 1   63 × 2 + 1   127 × 2 + 1

So the value of x is 31.

1.68 (c)

In every minute, second-hand of a clock crosses minute-hand ones.

The difference of minute from 12:05:00 to 12:55:00 is 55 – 5 = 50 minute.

So second-hand and minute hand will cross 50 times to each other.

1.69 (a)

There are three sets of indistinguishable twins and 8 identical chairs.

We can assume identical twins as a single person and similarly two identical chairs can be assumed as a single unit.

So by the above assumptions there are two cases to be taken.

Case-1: Two vacant chairs are always together:

This case is similar to three person to be sitted in

four chair in a circular manner so total possible cases can be calculated as below:

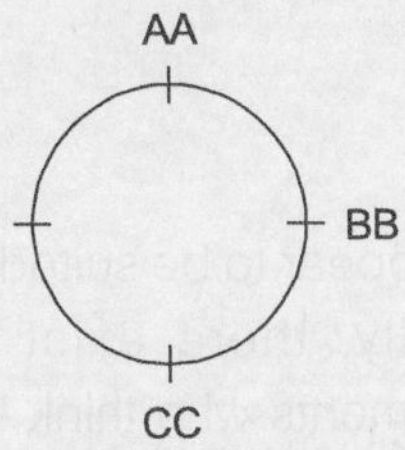

First we have to fix position of a particular twins after that there are three ways for 2nd pair of twins and two ways for last pair of twins.

So total possible ways are $3 \times 2 = 6$

Case-2: Two vacant identical chairs are not together. In this case there will be five vacant space and we have fix a particular twins first.

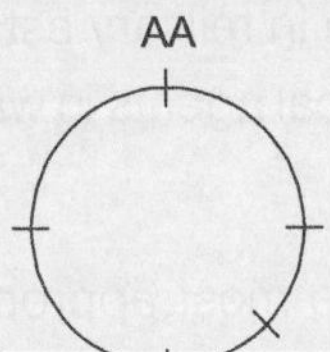

The two chairs which are vacant can be placed between three places between identical set of twins.

So the possibilities are $3_{P_2} = \frac{3!}{(3-2)!} = 6$

Total ways = 6 + 6 = 12

1.70 (a)

Capacity factor of plant is given by

$$\text{Capacity Factor} = \frac{\text{Electricity Generation (MWh)}}{\text{Installed Capacity (MW)} \times 1000\text{ (h)}}$$

For technology T1 capacity factor

$$= \frac{10000}{20 \times 1000} = 0.5$$

For technology T2 capacity factor

$$= \frac{9000}{30 \times 1000} = 0.3$$

For technology T3 capacity factor

$$= \frac{7000}{15 \times 1000} = 0.467$$

For technology T4 capacity factor

$$= \frac{12000}{40 \times 1000} = 0.3$$

So technology T1 has highest capacity factor.

1.71 (b)

Earth-Moon-Sun form a right angle triangle. Moon-Earth-Sun angle at this half-moon phase form an angle of 89.85°

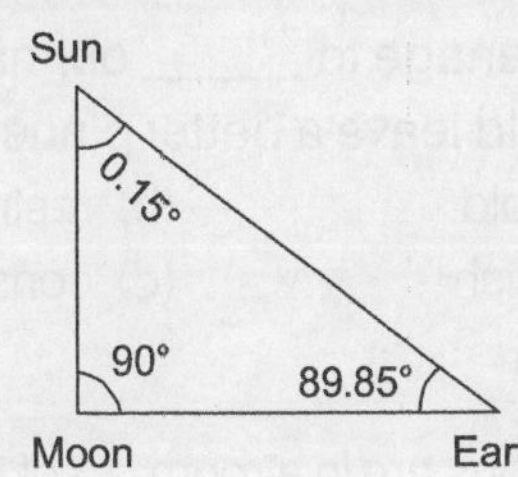

From figure,

$$\frac{\text{Earth sun distance}}{\text{Earth moon distance}} = \frac{1}{\cos 89.85°} = 382$$

■■■■

2 Verbal Ability

2.1 Choose the most appropriate word from the options given below to complete the following sentence:

His rather casual remarks on politics ______ his lack of seriousness about the subject.

(a) masked (b) belied
(c) betrayed (d) suppressed

[2010 : 1 M]

2.2 Which of the following options is the closest in meaning to the word below?

Circuitous

(a) Cyclic (b) Indirect
(c) Confusing (d) Crooked

[2010 : 1 M]

2.3 Choose the most appropriate word from the options given below to complete the following sentence:

If we manage to ______ our natural resources, we would leave a better planet for our children.

(a) uphold (b) restrain
(c) cherish (d) conserve

[2010 : 1 M]

2.4 25 persons are in a room. 15 of them play hockey, 17 of them play football and 10 of them play both hockey and football. Then the number of persons playing neither hockey nor football is

(a) 2 (b) 17
(c) 13 (d) 3

[2010 : 1 M]

2.5 The question below consists of a pair of related words followed by four pairs of words. Select the pair that best expresses the relation in the original pair.

Unemployed: Worker

(a) Fallow : Land (b) Unaware : Sleeper
(c) Wit : Jester (d) Renovated : House

[2010 : 1 M]

2.6 Modern warfare has changed from large scale clashes of armies to suppression of civilian populations. Chemical agents that do their work silently appear to be suited to such warfare; and regretfully, there exist people in military establishments who think that chemical agents are useful fools for their cause.

Which of the following statements best sums up the meaning of the above passage?

(a) Modern warfare has resulted in civil strife.
(b) Chemical agents are useful in modern warfare.
(c) Use of chemical agents in warfare would be undesirable.
(d) People in military establishments like to use chemical agents in war.

[2010 : 2 M]

2.7 Choose the most appropriate word(s) from the options given below to complete the following sentence.

I contemplated ________ Singapore for my vacation but decided against it.

(a) to visit (b) having to visit
(c) visiting (d) for a visit

[2011 : 1 M]

2.8 Which of the following options is the closest in the meaning to the word below:

Inexplicable

(a) Incomprehensible (b) Indelible
(c) Inextricable (d) Infallible

[2011 : 1 M]

2.9 Choose the word from the options given below that is most nearly opposite in meaning to the given word:

Amalgamate

(a) merge (b) split
(c) collect (d) separate

[2011 : 1 M]

2.10 Chose the most appropriate word from the options given below to complete the following sentence.

If you are trying to make a strong impression on your audience, you cannot do so by being understated, tentative or ________.

(a) hyperbolic (b) restrained
(c) argumentative (d) indifferent

[2011 : 1 M]

2.11 **Few school curricula include a unit on how to deal with bereavement and grief, and yet all students at some point in their lives suffer from losses through death and parting.**
Based on the above passage which topic would not be included in a unit on bereavement?
(a) how to write a letter of condolence
(b) what emotional stages are passed through in the healing process
(c) what the leading causes of death are
(d) how to give support to a grieving friend

[2011 : 2 M]

2.12 Which one of the following options is the closest in meaning to the word given below?
Mitigate
(a) Diminish (b) Divulge
(c) Dedicate (d) Denote

[2012 : 1 M]

2.13 Choose the most appropriate alternative from the options given below to complete the following sentence:
Despite several ______ the mission succeeded in its attempt to resolve the conflict.
(a) attempts (b) setbacks
(c) meetings (d) delegations

[2012 : 1 M]

2.14 Choose the most appropriate alternative from the options given below to complete the following sentence.
Suresh's dog is the one ______ was hurt in the stampede.
(a) that (b) which
(c) who (d) whom

[2012 : 1 M]

2.15 Choose the grammatically INCORRECT sentence:
(a) They gave us the money back less the service charges to Three Hundred rupees.
(b) This country's expenditure is not less than that of Bangladesh.
(c) The committee initially asked for a funding of Fifty Lakh rupees, but later settled for a lesser sum.
(d) This country's expenditure on educational reforms is very less.

[2012 : 1 M]

2.16 Wanted Temporary, Part-time persons for the post of Field Interviewer to conduct personal interviews to collect and collate economic data. Requirements: High School-pass, must be available for Day, Evening and Saturday work. Transportation paid, expenses reimbursed.
Which one of the following is the best inference from the above advertisement?
(a) Gender-discriminatory
(b) Xenophobic
(c) Not designed to make the post attractive
(d) Not gender-discriminatory

[2012 : 2 M]

2.17 Given the sequence of terms, AD, CG, FK, JP, the next term is
(a) OV (b) OW
(c) PV (d) PW

[2012 : 2 M]

2.18 The Headmaster ______ to speak to you.
Which of the following option is incorrect to complete the above sentence?
(a) is wanting (b) wants
(c) want (d) was wanting

[2013 : 1 M]

2.19 Mahatama Gandhi was known for his humility as
(a) he played an important role in humiliating exit of British from India.
(b) he worked for humanitarian causes.
(c) he displayed modesty in his interactions.
(d) he was a fine human being.

[2013 : 1 M]

2.20 All engineering students (I) should learn mechanics, (II) mathematics and (III) how to do computation (IV).
Which of the above underlined parts of the sentence is not appropriate?
(a) I (b) II
(c) III (d) IV

[2013 : 1 M]

2.21 Select the pair that best expresses a relationship similar to that expressed in the pair:
water : pipe ::

(a) cart : road
(b) electricity : wire
(c) sea : beach
(d) music : instrument

[2013 : 1 M]

2.22 All professors are researchers.
Some scientists are professors.
Which of the given conclusions is logically valid and inferred from the above arguments:
(a) All scientists are researchers
(b) All professors are scientists
(c) Some researchers are scientists
(d) No conclusion follows

[2013 : 2 M]

2.23 A student is required to demonstrate a high level of comprehension of the subject, especially in the social sciences.
The word closest in meaning to comprehension is
(a) understanding
(b) meaning
(c) concentration
(d) stability

[2014 (Set-1) : 1 M]

2.24 Choose the most appropriate word from the options given below to complete the following sentence.
One of his biggest ______ was his ability to forgive.
(a) vice
(b) virtues
(c) choices
(d) strength

[2014 (Set-1) : 1 M]

2.25 Rajan was not happy that Sajan decided to do the project on his own. On observing his unhappiness, Sajan explained to Rajan that he preferred to work independently.
Which one of the statements below is logically valid and can be inferred from the above sentences?
(a) Rajan has decided to work only in a group.
(b) Rajan and Sajan were formed into a group against their wishes.
(c) Sajan had decided to give in to Rajan's request to work with him.
(d) Rajan had believed that Sajan and he would be working together.

[2014 (Set-1) : 1 M]

2.26 Find the odd one in the following group: ALRVX, EPVZB, ITZDF, OYEIK
(a) ALRVX
(b) EPVZB
(c) ITZDF
(d) OYEIK

[2014 (Set-1) : 2 M]

2.27 Anuj, Bhola, Chandan, Dilip, Eswar and Faisal live on different floors in a six-storeyed building (the ground floor is numbered 1, the floor above it 2, and so on). Anuj lives on an even-numbered floor. Bhola does not live on an odd numbered floor. Chandan does not live on any of the floors below Faisal's floor. Dilip does not live on floor number 2. Eswar does not live on a floor immediately above or immediately below Bhola. Faisal lives three floors above Dilip. Which of the following floor-person combinations is correct?

	Anuj	Bhola	Chandan	Dilip	Eswar	Faisal
(a)	6	2	5	1	3	4
(b)	2	6	5	1	3	4
(c)	4	2	6	3	1	5
(d)	2	4	6	1	3	5

[2014 (Set-1) : 2 M]

2.28 Choose the word most similar in meaning to the given word : Educe
(a) Exert
(b) Educate
(c) Extract
(d) Extend

[2015 : 1 M]

2.29 Choose the most appropriate word from the options given below to complete the following sentence.
The principal presented the chief quest with a ________, as token of appreciation.
(a) momento
(b) memento
(c) momentum
(d) moment

[2015 : 1 M]

2.30 Choose the appropriate word/phase, out of the four options given below, to complete the following sentence:
Frogs ________
(a) Croak
(b) roar
(c) hiss
(d) patter

[2015 : 1 M]

2.31 Humpty Dumpty sits on a wall every day while having lunch. The wall sometimes breaks. A person sitting on the wall falls if the wall breaks.
Which one of the statements below is logically valid and can be inferred from the above sentence?
(a) Humpty Dumpty always falls having lunch.
(b) Humpty Dumpty does not fall sometimes while having lunch.

(c) Humpty Dumpty never falls during dinner.
(d) When Humpty Dumpty does not sit on the wall, the wall does not break.

[2015 : 2 M]

2.32 The following question presents a sentence, part of which is underlined. Beneath the sentence you find four ways of phrasing the underlined part. Following the requirements of the standard written English, select the answer that produces the most effective sentence.

Tuberculosis together with its effects <u>ranks one of the leading causes of death</u> in India.

(a) Ranks as one of the leading causes of death
(b) Rank as one of the leading causes of death
(c) Has the rank of one of the leading causes of death
(d) Are one of the leading causes of death

[2015 : 2 M]

2.33 Read the following paragraph and choose the correct statements:

Climate change has reduced human security and threatened human well being. An ignored reality of human progress is that human security largely depends upon environmental security. But on the contrary, human progress seems contradictory to environmental security. To keep up both at the required level is a challenge to be addressed by one and all. One of the ways to curb the climate change may be suitable scientific innovations, while the other may be the Gandhian perspective on small scale progress with focus on sustainability.

(a) Human progress and security are positively associated with environmental security.
(b) Human progress is contradictory to environmental security.
(c) Human security is contradictory of environmental security.
(d) Human progress depends on environmental security.

[2015 : 2 M]

2.34 The unruly crowd demanded that the accused be __________ without trial.

(a) hanged (b) hanging
(c) hankering (d) hung

[2016 (Set-2) : 1 M]

2.35 Choose the statement(s) where the underlined word is used correctly:

(i) A <u>prone</u> is a dried plum.
(ii) He was lying <u>prone</u> on the floor.
(iii) People who eat a lot of fat are <u>prone</u> to heart disease.

(a) (i) and (iii) only
(b) (iii) only
(c) (i) and (ii) only
(d) (ii) and (iii) only

[2016 (Set-2) : 1 M]

2.36 **Fact:** If it rains, then the field is wet.

Read the following statements:

(i) It rains
(ii) The field is not wet
(iii) The field is wet
(iv) It did not rain

Which one of the options given below is **NOT** logically possible, based on the given fact?

(a) If (iii), then (iv) (b) If (i), then (iii)
(c) If (i), then (ii) (d) If (ii), then (iv)

[2016 (Set-2) : 1 M]

2.37 A smart city integrates all modes of transport, uses clean energy and promotes sustainable use of resources. It also uses technology to ensure safety and security of the city, something which critics argue, will lead to a surveillance state. Which of the following can be logically inferred from the above paragraph?

(i) All smart cities encourage the formation of surveillance states.
(ii) Surveillance is an integral part of a smart city.
(iii) Sustainability and surveillance go hand in hand in a smart city.
(iv) There is a perception that smart cities promote surveillance.

(a) (i) and (iv) only
(b) (ii) and (iii) only
(c) (iv) only
(d) (i) only

[2016 (Set-2) : 2 M]

2.38 Find the missing sequence in the letter series.

B, FH, LNP, _ _ _ _.

(a) SUWY (b) TUVW
(c) TVXZ (d) TWXZ

[2016 (Set-2) : 2 M]

2.39 The binary operation □ is defined as a □ b = ab + (a + b), where a and b are any two real numbers. The value of the identity element of this operation, defined as the number x such that a □ x = a, for any a, is ________.
(a) 0 (b) 1
(c) 2 (d) 10
[2016 (Set-2) : 2 M]

2.40 The bacteria in milk are destroyed when it ________ heated to 80 degree Celsius.
(a) would be (b) will be
(c) is (d) was
[2017 (Set-1) : 1 M]

2.41 ________ with someone else's email account is now a very serious offense.
(a) Involving (b) Assisting
(c) Tampering (d) Incubating
[2017 (Set-1) : 1 M]

2.42 Consider the following sentences:
All benches are beds. No bed is a bulb. Some bulbs are lamps.
Which of the following can be inferred?
(i) Some beds are lamps.
(ii) Some lamps are beds.
(a) Only (i) (b) Only (ii)
(c) Both (i) and (ii) (d) Neither (i) nor (ii)
[2017 (Set-1) : 1 M]

2.43 The old concert hall was demolished because of fears that the foundation would be affected by the construction of the new metro line in the area. Modern technology for underground metro construction tried to mitigate the impact of pressurized air pockets created by the excavation of large amounts of soil. But even with these safeguards, it was feared that the soil below the concert hall would not be stable.
From this, one can infer that
(a) the foundations of old buildings create pressurized air pockets underground, which are different to handle during metro construction.
(b) metro construction has to be done carefully considering its impact on the foundations of existing buildings.
(c) old buildings in an area form an impossible hurdle to metro construction in that area.
(d) pressurized air can be used to excavate large amount of soil from underground areas.
[2017 (Set-1) : 2 M]

2.44 Students applying for hostel rooms are allotted rooms in order of seniority. Students already staying in a room will move if they get a room in their preferred list. Preferences of lower ranked applicants are ignored during allocation.
Given the data below, which room will Ajit stay in?

Names	Student Seniority	Current room	Room preference list
Amar	1	P	R, S, Q
Akbar	2	None	R, S
Anthony	3	Q	P
Ajit	4	S	Q, P, R

(a) P (b) Q
(c) R (d) S
[2017 (Set-1) : 2 M]

2.45 "When she fell down the ________. She received many ________ but little help." The words that best fill the blanks in the above sentence are
(a) stairs, stares (b) stairs, stairs
(c) stares, stairs (d) stares, stares
[2018 : 1 M]

2.46 "In spite of being warned repeatedly, he failed to correct his ________ behaviour." The word that best fills that blank in the above sentence is
(a) rational (b) reasonable
(c) errant (d) good
[2018 : 1 M]

2.47 To pass a test, a candidate needs to answer at least 2 out of 3 questions correctly. A total of 6,30,000 candidates appeared for the test. Question A was correctly answered by 3,30,000 candidates. Question B was answered correctly by 2,50,000 candidates. Question C was answered correctly by 2,60,000 candidates. Both questions A and B were answered correctly by 1,00,000 candidates. Both questions B and C were answered correctly by 90,000 candidates. Both questions A and C were answered correctly by 80,000 candidates. If the number of students answering all questions correctly is the same as the number answering none, how many candidates failed to clear the test?

(a) 30,000 (b) 2,70,000
(c) 3,90,000 (d) 4,20,000

[2018 : 2 M]

2.48 The expenditure on the project ________ as follows; equipments Rs. 20 lakhs, salaries Rs. 12 lakhs, and contingency Rs.3 lakhs.

(a) break (b) breaks
(c) breaks down (d) break down

[2019 : 1 M]

2.49 The search engine's business model ________ around the fulcrum of trust.

(a) plays (b) bursts
(c) revolves (d) sinks

[2019 : 1 M]

2.50 A court is to a judge as ________ is to a teacher.

(a) a syllabus (b) a school
(c) a student (d) a punishment

[2019 : 1 M]

2.51 The police arrested four criminals – P, Q, R and S. The criminals knew each other. They made the following statements:

P says "Q committed the crime."
Q says "S committed the crime."
R says "I did not do it."
S says "What Q said about me is false."

Assume only one of the arrested four committed the crime and only one of the statements made above is true. Who committed the crime?

(a) R (b) P
(c) S (d) Q

[2019 : 2 M]

2.52 In the given diagram, teachers are represented in the triangle, researchers in the circle and administrators in the rectangle. Out of the total number of the people, the percentage of administrators shall be in the range of ________.

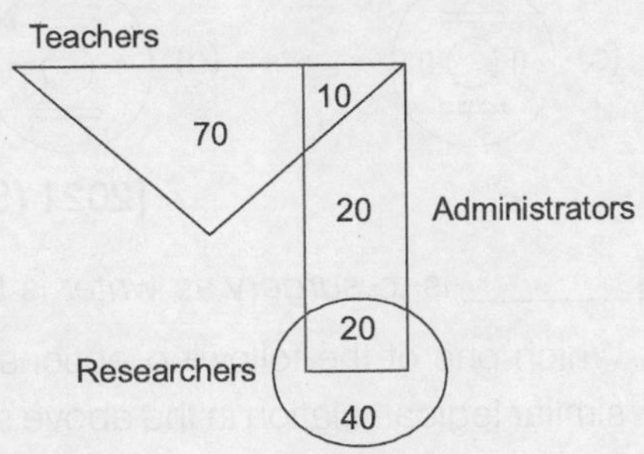

(a) 46 to 60 (b) 0 to 15
(c) 31 to 45 (d) 16 to 30

[2019 : 2 M]

2.53 "A recent High Court judgement has sought to dispel the idea of begging as a disease — which leads to its stigmatization and criminalization — and to regard it as a symptom. The underlying disease is the failure of the state to protect citizens who fall through the social security net".

Which one of the following statements can be inferred from the given passage?

(a) Begging is an offence that has to be dealt with firmly.
(b) Beggars are created because of the lack of social welfare schemes.
(c) Begging has to be banned because it adversely affects the welfare of the state.
(d) Beggars are lazy people who beg because they are unwilling to work. [2019 : 2 M]

2.54 Rajiv Gandhi Khel Ratna Award was conferred ________ Mary Kom, a six-time world champion in boxing, recently in a ceremony _____ the Rashtrapati Bhawan (the President's official residence) in New Delhi.

(a) with, at (b) on, in
(c) on, at (d) to, at

[2020 : 1 M]

2.55 Despite a string of poor performances, the chances of K.L. Rahul's selection the team are ________

(a) slim (b) bright
(c) obvious (d) uncertain

[2020 : 1 M]

2.56 Select the word that fits the analogy:

Cover : Uncover :: Associate : ________

(a) Unassociate (b) Inassociate
(c) Misassociate (d) Dissociate

[2020 : 1 M]

2.57 Hit by floods, the kharif (summer sown) crops in various parts of the country have been affected. Officials believe that the loss in production of the kharif crops can be recovered in the output of the rabi (winter sown) crops so that the country can achieve its food-grain production target of 291 million tons in the crop year 2019-20 (July-June). They are hopeful that good rains in July-August will help the soil retain moisture for a longer period, helping winter sown crops such as wheat and pulses during the November-February period.

Which of the following statements can be inferred from the given passage?

(a) Officials declared that the food-grain production target will be met due to good rains.
(b) Officials want the food-grain production target to be met by the November-February period.
(c) Officials feels that the food-grain production target cannot be met due to floods.
(d) Officials hope that the food-grain production target will be met due to a good rabi produce.

[2020 : 1 M]

2.58 Repo rate is the rate at which Reserve Bank of India (RBI) lends commercial banks, and reverse repo rate is the rate at which RBI borrows money from commercial banks.

Which of the following statements can be inferred from the above passage?

(a) Decrease in repo rate will increase cost of borrowing and decrease lending by commercial banks.
(b) Increase in repo rate will decrease cost of borrowing and increase lending by commercial banks.
(c) Increase in repo rate will decrease cost of borrowing and decrease lendign by commercial banks.
(d) Decrease in repo rate will decrease cost of borrowing and increase lending by commercial banks.

[2020 : 2 M]

2.59 P, Q, R, S, T, U, V and W are seated around a circular table.

I. S is seated opposite to W.
II. U is seated at the second place to the right of R.
III. T is seated at the third place to the left of R.
IV. V is a neighbour of S.

Which of the following must be true?

(a) P is a neighbour of R.
(b) Q is a neighbour of R.
(c) P is not seated opposite to Q.
(d) R is the left neighbour of S.

[2020 : 2 M]

2.60 A polygon is convex if, for every pair of points. P and Q belonging to the polygon, the line segment PQ lies completely inside or on the polygon.

Which one of the following is NOT a convex polygon?

(a) 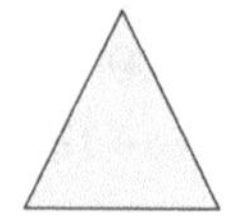

(b) 

(c) 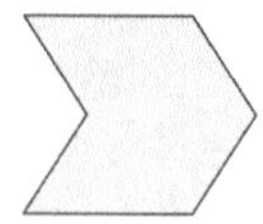

(d) 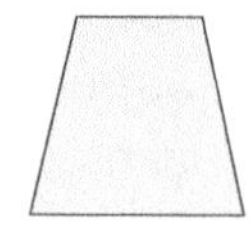

[2021 (Set-1) : 1 M]

2.61 Consider the following sentences:

(i) Everybody in the class is prepared for the exam.
(ii) Babu invited Danish to his home because he enjoys playing chess.

Which of the following is the CORRECT observation about the above two sentences?

(a) (i) is grammatically incorrect and (ii) is unambiguous
(b) (i) is grammatically correct and (ii) is ambiguous
(c) (i) is grammatically incorrect and (ii) is ambiguous
(d) (i) is grammatically correct and (ii) is unambiguous.

[2021 (Set-1) : 1 M]

2.62 A circular sheet of paper is folded along the lines in the directions shown. The paper, after being punched in the final folded state as shown and unfolded in the reverse order of folding, will look like ________.

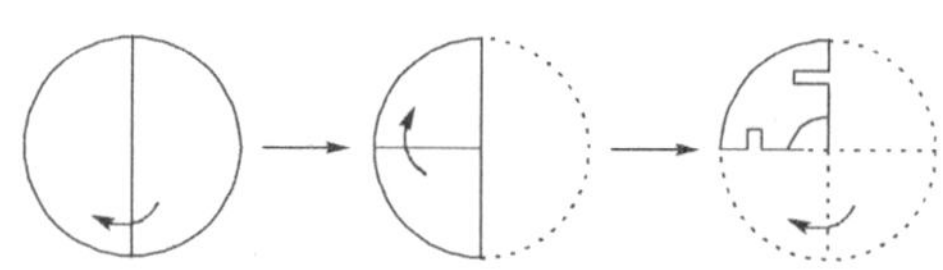

(a) 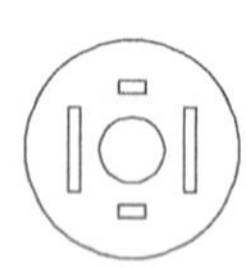

(b) 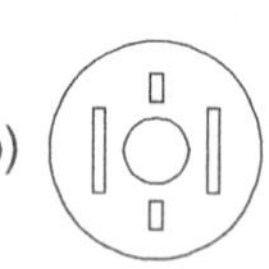

(c) 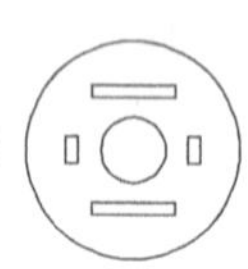

(d) 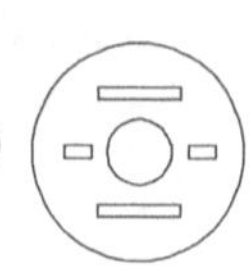

[2021 (Set-1) : 1 M]

2.63 ________ is to *surgery* as *writer* is to ________

Which one of the following options maintains a similar logical relation in the above sentence?

(a) Plan, outline
(b) Hospital, library
(c) Doctor, book
(d) Medicine, grammar

[2021 (Set-1) : 1 M]

2.64 Given below are two statements I and II and two conclusions I and II:

Statement:

I. All bacteria are microorganisms.

II. All pathogens are microorganisms.

Conclusions:

I. Some pathogens are bacteria.

II. All pathogens are not bacteria.

Based on the above statements and conclusions, which one of the following options is logically CORRECT?

(a) Only conclusion I is correct

(b) Either conclusion I or II is correct.

(c) Only conclusion II is correct

(d) Neither conclusion I nor II is correct.

[2021 (Set-1) : 2 M]

2.65 Some people suggest anti-obesity measures (AOM) such as displaying calorie information in restaurant menus. Such measures sidestep addressing the core problems that cause obesity: poverty and income inequality.

Which one of the following statements summarizes the passage?

(a) The proposed AOM addresses the care problems that cause obesity.

(b) If obesity reduces, poverty will naturally reduce, since obesity causes poverty.

(c) AOM are addressing the problem superficially.

(d) AOM are addressing the core problems and are likely to succeed.

[2021 (Set-1) : 2 M]

2.66 Inhaling the smoke from a burning ________ could ________ you quickly.

(a) tire / tier (b) tire / tyre

(c) tyre / tire (d) tyre / tier

[2022 : 1 M]

2.67 Six persons P, Q, R, S, T and U are sitting around a circular table facing the center not necessarily in the same order. Consider the following statements:

- P sits next to S and T.
- Q sits diametrically opposite to P.
- The shortest distance between S and R is equal to the shortest distance between T and U.

Based on the above statements, Q is a neighbor of

(a) U and S (b) R and T

(c) R and U (d) P and S

[2022 : 1 M]

2.68 A building has several rooms and doors as shown in the top view of the building given below. The doors are closed initially.

What is the minimum number of doors that need to be opened in order to go from the point P to the point Q?

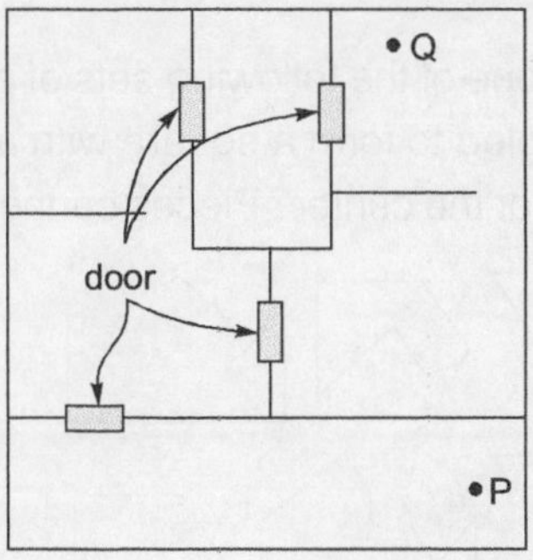

(a) 4 (b) 3

(c) 2 (d) 1

[2022 : 1 M]

2.69 Rice, a versatile and inexpensive source of carbohydrate, is a critical component of diet worldwide. Climate change, causing extreme weather, poses a threat to sustained availability of rice. Scientists are working on developing Green Super Rice (GSR), which is resilient under extreme weather conditions yet gives higher yields sustainably.

Which one of the following is the CORRECT logical inference based on the information given in the above passage?

(a) GSR is an alternative to regular rice, but it grows only in an extreme weather.

(b) GSR may be used in future in response to adverse effects of climate change.

(c) GSR grows in an extreme weather, but the quantity of produce is lesser than regular rice.

(d) Regular rice will continue to provide good yields even in extreme weather.

[2022 : 2 M]

2.70 Given below are three statements and four conclusions drawn based on the statements.

Statement 1: Some engineers are writers.

Statement 2: No writer is an actor.

Statement 3: All actors are engineers.

Conclusion I: Some writers are engineers.

Conclusion II: All engineers are actors.

Conclusion III: No actor is a writer.

Conclusion IV: Some actors are writers.

Which one of the following options can be logically inferred?

(a) Only conclusion I is correct
(b) Only conclusion II and conclusion III are correct
(c) Only conclusion I and conclusion III are correct
(d) Either conclusion III or conclusion IV is correct

[2022 : 2 M]

2.71 Which one of the following sets of pieces can be assembled to form a square with a single round hole near the center? Pieces cannot overlap.

(a) 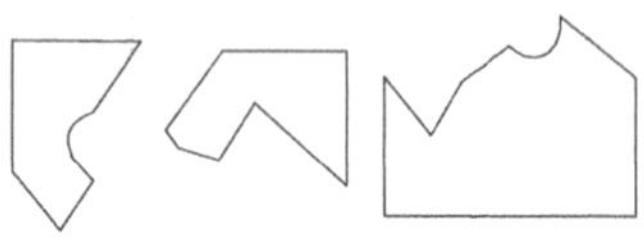

(b) 

(c) 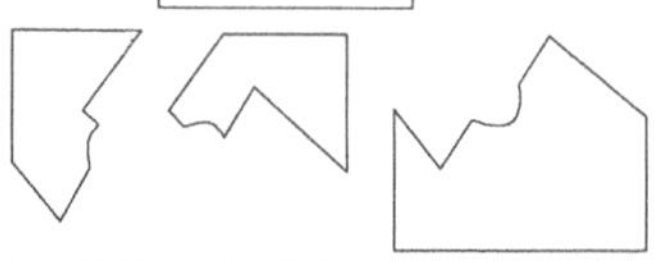

(d) 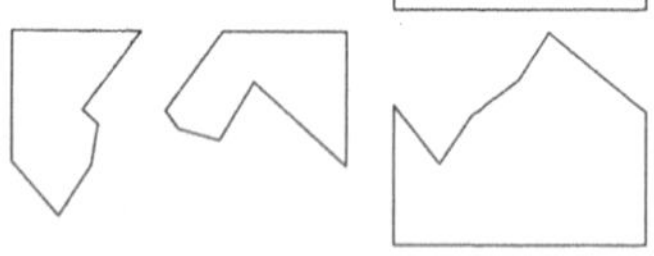

[2022 : 2 M]

2.72 "You are delaying the completion of the task. Send _______ contributions at the earliest".

(a) you are (b) your
(c) you're (d) yore

[2023 : 1 M]

2.73 References : ______ : : Guidelines : Implement (By word meaning)

(a) Sight (b) Site
(c) Cite (d) Plagiarise

[2023 : 1 M]

2.74 In 2022, June Huh was awarded the Fields medal, which is the highest prize in Mathematics.
When he was younger, he was also a poet. He did not win any medals in the International Mathematics Olympiads. He dropped out of college.
Based only on the above information, which one of the following statements can be logically inferred with *certainty*?

(a) Every Fields medalist has won a medal in an International Mathematics Olympiad.
(b) Everyone who has dropped out of college has won the Fields medal.
(c) All Fields medalists are part-time poets.
(d) Some Fields medalists have dropped out of college.

[2023 : 1 M]

2.75 Human beings are one among many creatures that inhabit an imagined world. In this imagined world, some creatures are cruel. If in this imagined world, it is given that the statement "Some human beings are not cruel creatures" is FALSE, then which of the following set of statement(s) can be logically inferred with certainty?

(i) All human beings are cruel creatures.
(ii) Some human beings are cruel creatures.
(iii) Some creatures that are cruel are human beings.
(iv) No human beings are cruel creatures.

(a) only (i) (b) only (iii) and (iv)
(c) only (i) and (ii) (d) (i), (ii) and (iii)

[2023 : 2 M]

2.76 The World Bank has declared that it does not plan to offer new financing to Sri Lanka, which is battling its worst economic crisis in decades, until the country has an adequate macroeconomic policy framework in place. In a statement, the World Bank said Sri Lanka needed to adopt structural reforms that focus on economic stabilization and tackle the root causes of its crisis. The latter has starved it of foreign exchange and led to shortages of food, fuel, and medicines. The bank is repurposing resources under existing loans to help alleviate shortages of essential items such as medicine, cooking gas, fertilizer, meals for children, and cash for vulnerable households.
Based only on the above passage, which one of the following statements can be inferred with *certainty*?

(a) According to the World Bank, the root cause of Sri Lanka's economic crisis is that it does not have enough foreign exchange.
(b) The World Bank has stated that it will advise the Sri Lankan government about how to tackle the root causes of its economic crisis.
(c) According to the World Bank, Sri Lanka does not yet have an adequate macroeconomic policy framework.
(d) The World Bank has stated that it will provide Sri Lanka with additional funds for essentials such as food, fuel, and medicines.

[2023 : 2 M]

2.77 If '→' denotes increasing order of intensity, then the meaning of the words [simmer → seethe → smolder] is analogous to [break → raze → ______]. Which one of the given options is appropriate to fill the blank?

(a) obfuscate (b) obliterate
(c) fracture (d) fissure [2024 : 1 M]

2.78 In the given text, the blanks are numbered (i) – (iv). Select the best match for all the blanks.

From the ancient Athenian arena to the modern Olympic stadiums, athletics __(i)__ the potential for a spectacle. The crowd __(ii)__ with bated breath as the Olympian artist twists his body, stretching the javelin behind him. Twelve strides in, he begins to cross-step. Six cross-steps __(iii)__ in an abrupt stop on his left foot. As his body __(iv)__ like a door turning on a hinge, the javelin is launched skyward at a precise angle.

(a) (i) hold (ii) waits (iii) culminates (iv) pivot
(b) (i) holds (ii) wait (iii) culminates (iv) pivot
(c) (i) hold (ii) wait (iii) culminate (iv) pivots
(d) (i) holds (ii) waits (iii) culminate (iv) pivots

[2024 : 2 M]

2.79 In the 4 × 4 array shown below, each cell of the first three columns has either a cross (X) or a number, as per the given rule.

1	1	2	
2	X	3	
2	X	4	
1	2	X	

Rule: The number in a cell represents the count of crosses around its immediate neighboring cells (left, right, top, bottom, diagonals).

As per this rule, the **maximum** number of crosses possible in the empty column is

(a) 0 (b) 1
(c) 2 (d) 3 [2024 : 2 M]

■■■■

Answers Verbal Ability

2.1	(c)	2.2	(b)	2.3	(d)	2.4	(d)	2.5	(a)	2.6	(d)	2.7	(c)
2.8	(a)	2.9	(d)	2.10	(b)	2.11	(c)	2.12	(a)	2.13	(b)	2.14	(a)
2.15	(b)	2.16	(d)	2.17	(a)	2.18	(c)	2.19	(c)	2.20	(d)	2.21	(b)
2.22	(c)	2.23	(a)	2.24	(b)	2.25	(d)	2.26	(d)	2.27	(b)	2.28	(c)
2.29	(b)	2.30	(a)	2.31	(b)	2.32	(a)	2.33	(d)	2.34	(a)	2.35	(d)
2.36	(c)	2.37	(c)	2.38	(c)	2.39	(a)	2.40	(c)	2.41	(c)	2.42	(d)
2.43	(b)	2.44	(b)	2.45	(a)	2.46	(c)	2.47	(d)	2.48	(c)	2.49	(c)
2.50	(b)	2.51	(a)	2.52	(c)	2.53	(b)	2.54	(c)	2.55	(b)	2.56	(d)
2.57	(d)	2.58	(d)	2.59	(c)	2.60	(c)	2.61	(b)	2.62	(c)	2.63	(c)
2.64	(b, d)	2.65	(c)	2.66	(c)	2.67	(c)	2.68	(c)	2.69	(b)	2.70	(c)
2.71	(c)	2.72	(b)	2.73	(c)	2.74	(d)	2.75	(d)	2.76	(c)	2.77	(b)
2.78	(d)	2.79	(c)										

Explanations Verbal Ability

2.1 (c)

(a) Masked: Hide under a false appearance ⇒ opposite.

(b) Belied: Be in contradiction with ⇒ not appropriate.

(c) Betrayed: Reveal unintentionally ⇒ most appropriate.

(d) Suppressed: To put down by force or authority ⇒ irrelevant.

2.2 (b)

Circuitous: Deviating from a straight course ⇒ Indirect.

(a) **Cyclic:** Recurring in cycle

(b) **Indirect:** Not leading by straight line

(c) **Confusing:** Lacking clarity

(d) **Crooked:** For shapes (irregular in shape)

2.3 (d)

(a) Uphold: cause to remain ⇒ not appropriate

(b) Restrain: keep under control ⇒ not appropriate

(c) Cherish: be fond of ⇒ not related

(d) Conserve: Keep in safety and protect from harm, decay, loss, or destruction ⇒ most appropriate.

2.4 (d)

Using the set theory formula,

$n(A)$: Number of people who play hockey = 15

$n(B)$: Number of people who play football = 17

$n(A \cap B)$: Persons who play both hockey and football = 10

$n(A \cup B)$: Persons who play either hockey or football or both

Using the formula,

$$n(A \cup B) = n(A) + n(B) - n(A \cap B)$$

$$n(A \cup B) = 15 + 17 - 10 = 22$$

Thus people who play neither hockey nor football = 25 − 22 = 3

Alternative method:

Refer to Venn diagram given below:

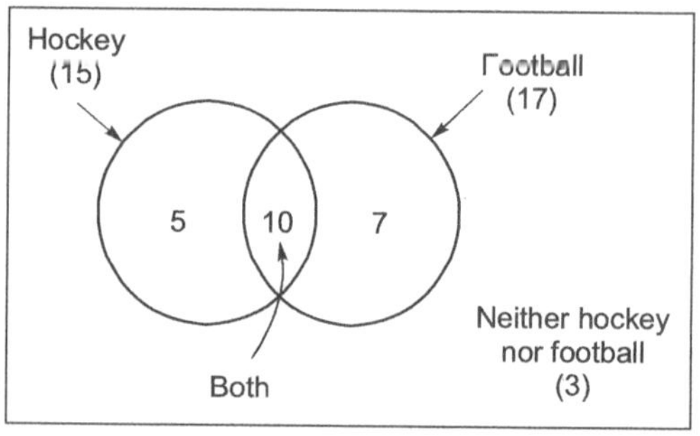

2.5 (a)

Unemployed: Worker ⇒ Here one is opposite to other.

(a) Fallow: Land ⇒ Fallow means undeveloped land.

(b) Unaware: sleeper ⇒ Both are same unaware or asleep.

(c) Wit: Jester ⇒ Wit means ability to make jokes and jester is a joker.

(d) Renovated : House ⇒ Renovate means to make better and house can be renovated.

2.6 (d)

(a) Modern warfare has resulted in civil strife: There is no direct consequence of warfare given, so it is not appropriate.
(b) Chemical agents are useful in modern warfare: Passage does not say whether chemical agents are useful or not, so not appropriate.
(c) Use of chemical agents in warfare would be undesirable: Given that people in military think these are useful, undesirable is wrong
(d) People in military establishments like to use chemical agents in war; Correct choice as last statement tells that military people think that chemical agents are useful tools for their cause (work silently in warfare).

2.12 (a)

Mitigate means to reduce, to lessen etc. So only the word Diminish is close. Rest all choices have no link with the given word.
Divulge means to disclose or reveal which has no link with the given word.

2.13 (b)

Despite several setbacks the mission succeeded in its attempt to resolve the conflict.
The word 'Despite' indicates that there has to be a contrast in the sentence, use of the word 'Setbacks' in the blank indicates that despite many problems the mission was successful.

2.14 (a)

Suresh's dog is the one that was hurt in the stampede.
That is used with restrictive clauses.

2.15 (b)

The country's expenditure is not less than that of Bangladesh.
The correct statement should be: The country's expenditure is not less than that of Bangladesh's. Because the country's expenditure cannot be compared with Bangladesh. It should be compared with Bangladesh's expenditure.

2.16 (d)

Not gender-discriminatory.
Choice (*A*) cannot be considered since there is no gender discrimination mentioned in the argument.
Choice (*B*) Xenophobic is one who has fear of foreigners, no link with the given argument.
Choice (*C*) It is wrong to say that the profile has not been designed to make the post attractive, since there are certain features which have been added to make the profile lucrative (which are given towards the end of the advertisement, like Transportation paid, expenses reimbursed).

2.17 (a)

AD, CG, FK, JP are given terms. The first letter in each term A, C, F, J are respectively 1, 3, 6, 10 letter of alphabetical sequence.
A = 1; 1st letter of alphabetical sequence
C = 1 + 2 = 3 ; 3rd letter of alphabetical sequence
F = 1 + 2 + 3 = 6 ; 6th letter of alphabetical sequence
G = 1 + 2 + 3 + 4 = 10; 10th letter of alphabetical sequence
The next term should begin with
1 + 2 + 3 + 4 + 5 = 15th letter of alphabetical sequence. So next term should begin with O.
Also in the series AD, CG, FK, JP, two letters in each term are separated by 2, 3, 4, 5 letters in between them respectively. Hence next term should be OV as there should be a gap of 6 alphabets between O and V.
Hence answer is OV.

2.19 (c)

Mahatama Gandhi was known for his humility as he displayed modesty in his interaction.

2.20 (d)

All engineering students should learn mechanics, mathematics and computation.

2.21 (b)

Water flow through pipes. Similarly, electricity flow through wires.

2.22 (c)

The possible cases are:

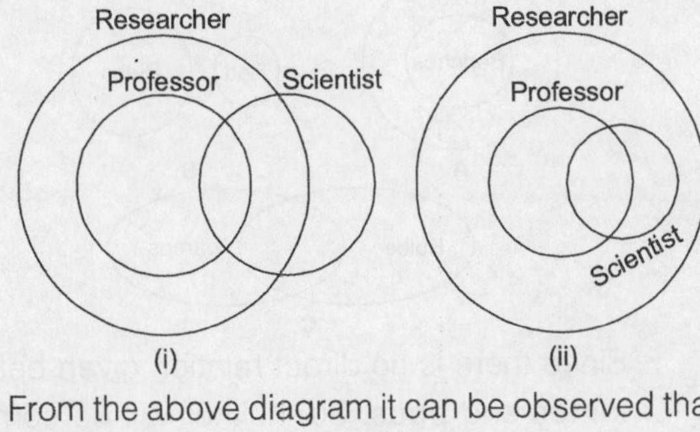

From the above diagram it can be observed that it follows only conclusion III.

2.27 (b)

Option (a) : Bhola's floor number – 2 and Eswar floor number – 3.
Option (b) : Bhola's floor number – 6 and Eswar floor number – 3.
Option (c) : Bhola's floor number – 2 and Eswar floor number – 1.
Option (d) : Bhola's floor number – 4 and Eswar floor number – 3.
Here, we can clearly see that in option (a), (c) and (d), the statement is not followed. Therefore, the answer is option (b).

2.38 (c)

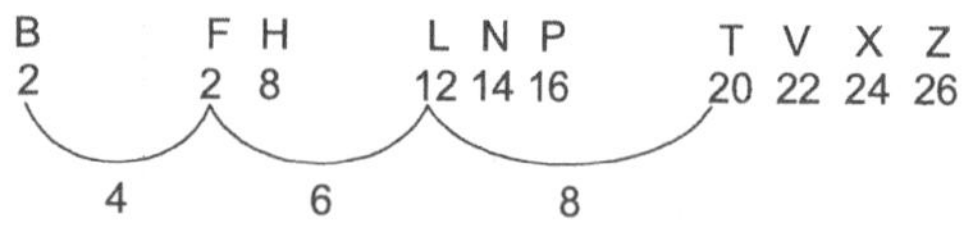

So the required alphabets have to start with 20 is *T* and there after common difference in term will be 2 each hence answer is TVXZ.

2.39 (a)

Use options only **(a) → 0**, satisfies

$a \square b = ab - (a + b)$

$a \square b = a \times 0 + (a + 0) = \mathbf{a}$

Given $a \square x = \text{a}$ which satisfied for $(x = 0)$

2.40 (c)

The bacteria is in milk are destroyed when it is heated to 80° celcius. Use of simple present tense is apt for indication factual statements/ scientifically proven statements / universal truth.

2.41 (c)

Tampering with someone else's email account is a very serious offence. Tampering refers to interfering in a harmful or disruptive manner.

2.42 (d)

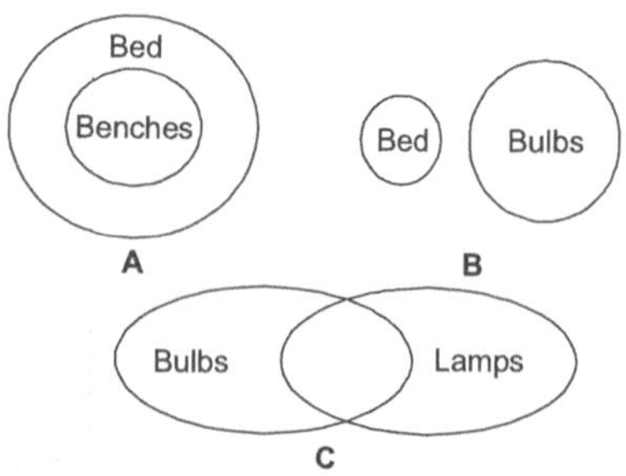

Since there is no direct relation given between **lamps** and **beds**. So, neither will be correct.

2.44 (b)

Amar ⟶ R
Akbar ⟶ S
Antony ⟶ P
Ajit ⟶ Q

As per their preferences given.

2.45 (a)

Stairs - A construction designed to bridge a large vertical distance by dividing it into smaller vertical distances, called steps.
Stares - To look at someone for a long time.

2.46 (c)

Errant means misbehaving, exhibiting inappropriate behavior / offending conduct.

2.47 (d)

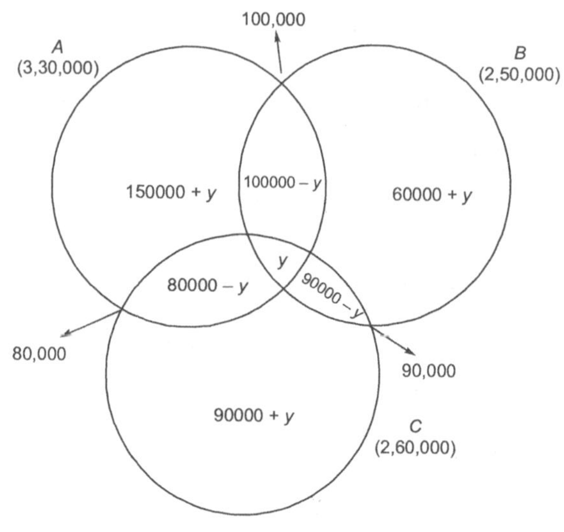

$6,30,000 = 2y + 1,50,000 + 100000 + 80000 + 60000 + 90000 + 90000$

$6,30,000 - 5,70,000 = 2y$

$y = 30000$

Students who failed to clear the test

$= 150000 + 60000 + 90000 + 4y$

$= 300000 + 4 \times 30000$

Students who failed to clear the test = 420000

2.48 (c)

'Breaks down' is a transitive phrasal verb which means to divide something such as a total amount into separate parts. So, option (c) is most appropriate.

2.49 (c)

The search engine business model <u>resolves</u> around the fulcrum of trust.
Fulcrum is any thing that plays a central or essential role in an activity, event, or situation. Thus, out of the given options, 'revolves' is the most appropriate.

2.50 (b)

A 'court' is for a 'judge' as a 'school' is for a 'teacher'.
Court is a place where a judge works.
Similarly, school is a place where a teacher works.

2.51 (a)

Suffix C as criminal, NC as not criminal.

Case I:

Criminals	P	Q	R	S
Assumption	T	F	F	F
Result	Q_C	S_{NC}	R_C	S_C

Q and R are criminal in the result is impossible because only one person committed the crime.

Case II:

Criminals	P	Q	R	S
Assumption	F	T	F	F
Result	Q_{NC}	S_C	R_C	S_C

S and R are criminal in the result is impossible because only one person committed the crime.

Case III:

Criminals	P	Q	R	S
Assumption	F	F	T	F
Result	Q_{NC}	S_{NC}	R_{NC}	S_C

S_{NC} and S_C in the result which is contradiction. [S committed crime and same time not committed crime which is contradiction]

Case IV:

Criminals	P	Q	R	S
Assumption	F	F	F	T
Result	Q_{NC}	S_{NC}	R_C	S_{NC}

R is criminal in the result.
Hence this case satisfies only one person committed the crime.
So option (a) is correct.

2.52 (c)

Percentage of Administrators

$$= \frac{\text{Administrators}}{\text{Total}} \times 100$$

$$= \frac{50}{160} \times 100 = 31.25$$

2.53 (b)

The passage states that the underlying disease behind begging is the failure of the state to protect citizens who fall through the social security net. Only option (b) can be inferred from this.

2.56 (d)

Uncover is the opposite of cover. Similarly, dissociate is the opposite of associate.

2.57 (d)

Official did not declare anything. So, option (a) is wrong. Official feel that production target can be recovered. So, option (c) is wrong. November-February is the period when the moisture is required in the soil to grow the rabi crop.
So, option (b) is wrong.

2.58 (d)

Decrease in repo rate means cost of borrowing money from RBI will be less, which will encourage commercial banks to borrow more money.

2.59 (c)

The possible seating arrangement is

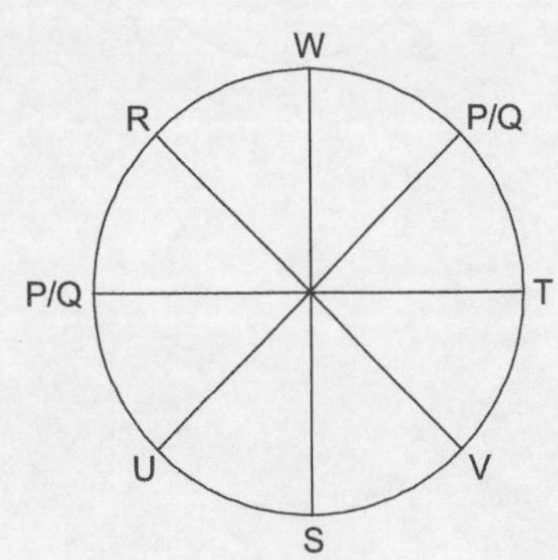

2.61 (b)

Everybody is singular and takes singular verb is, first statement is correct and the purpose of invitation is unclear so second statement is vague and that sounds ambiguous.

2.63 (c)

Doctor performs surgery as writer writes book.

2.64 (b, d)

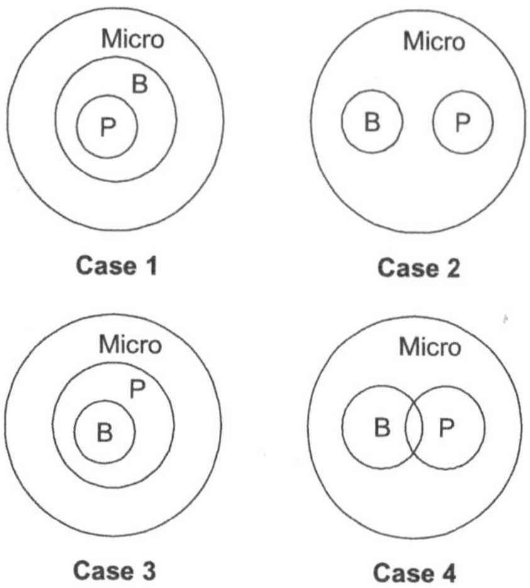

None of the two conclusions will satisfy all the 4 cases.

2.65 (c)

Superficially is the deciding key word which means apparently/seemingly.

2.67 (c)

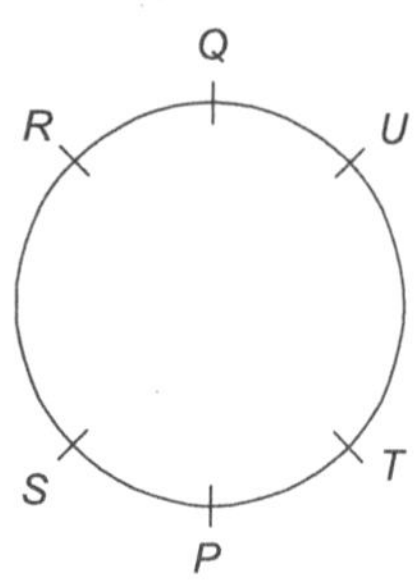

2.69 (b)

GSR may be used in future in response to adverse effects of climate change.

2.70 (c)

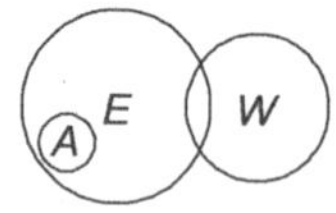

2.77 (b)

break → raze → ________

- break means separating something into two or more pieces.
- rage means to destroy something that nothing is left.
- obfuscate means creating illusion.
- fracture means breaking of bones and fissure means a narrow opening or crack.
- obliterate meaning to eliminate something.

So correction option is (b).

2.78 (d)

In present indefinite tense with singular subject verb's main form comes with 's' or 'es' and with plural subject verb's first form comes without 's' or 'es'.

Only option (d) satisfies this criteria.

2.79 (c)

As per the rule given in the question if we fill first row last column with a cross then only one more cross will be possible and that will be in third row last column. Similarly we can check for all the options possible.

Maximum number of cross for all possibilities will be 2.

■■■■